Biology of the Cell

Biology of the Cell

Stephen L. Wolfe
University of California, Davis

Wadsworth Publishing Company, Inc.
Belmont, California

Biological drawings: Darwen & Vally Hennings
Graphs & diagrams: Carleton Brown

BIOLOGY OF THE CELL BY STEPHEN L. WOLFE

ISBN–0–534–00106–8

L. C. Cat. Card No. 70–171896

Printed in the United States of America

2 3 4 5 6 7 8 9 10–76 75 74 73 72

Preface

Cell biology has become established as a standard college course through a synthesis of the subject matter of biochemistry and molecular biology with the materials covered in more traditional courses in plant and animal cytology. In many ways, the development of this new discipline has simply paralleled the emphasis and results of scientific investigation, which has of necessity turned from separate activity in morphology and function to an integrated and unified approach. This book is designed to reflect the contemporary integration of approaches and thus attempts to provide as even a balance as possible between structure and function.

Within this general aim, an effort has been made to analyze the most important concepts and hypotheses of cell biology and to present them in simple and direct language. The discussion flows from a survey of membranes to the role of membranes in cellular energetics, the use of energy in synthesis, and the main result of synthesis, cellular reproduction. Descriptions range from molecular to morphological and include both classical and contemporary information. Evidence from plants, animals, and prokaryotic microorganisms is integrated into the text. In the presentation of this evidence, copious literature references have been avoided. Instead, evidence is presented by describing key experiments, and only representative literature citations are given in areas of lesser interest. But in all cases, the reader is directed to papers and reviews in which extensive references can be found if needed. The book contains a complete selection of micrographs and diagrams (almost 500) illustrating the topics covered in the text. The inclusion of this illustrative material should make the book useful as an atlas of cell structure as well as a textbook of cell biology. While simplicity of language is retained, sophisticated topics are included that are not usually a part of books at the undergraduate level, such as redundancy and strandedness of DNA and chromosomes, recombination, and cytoplasmic inheritance. In the discussion of these and other topics, an effort has been made to emphasize problems and areas of uncertainty rather than simply cataloging facts.

The book is divided into four sections. The introductory unit, after a survey of cell structure and a brief history of cell biology, describes the molecular subunits of membranes, discusses the techniques used for studying membranes and other cellular structures, and evaluates the current status of research into membrane structure. The survey of techniques in the introductory unit emphasizes light and electron microscopy; other techniques of importance in cell biology, such as X-ray diffraction, autoradiography, and analytical centrifugation are described in later chapters where they form a part of key experiments. The introductory unit prepares the student for the consideration of the role and structural configurations of membranes in cellular energetics taken up in the second unit, which includes chapters on introductory thermodynamics, the chloroplast, and the mitochondrion. The book then turns to the activities of the cell in synthesis (Unit 3) and reproduction (Unit 4). Unit 3 concentrates on synthesis in interphase and includes chapters on the structure of the interphase nucleus, RNA metabolism, and protein synthesis. In Unit 4, the dividing cell is discussed. This unit includes chapters on mitosis, meiosis, gametogenesis and fertilization, cytoplasmic inheritance, and the evolutionary origins of cellular life.

The preparation of this book in its present form would not have been possible without the help of friends and colleagues who reviewed various chapters and offered many suggestions, including Max Alfert, Peter B. Armstrong, Dennis Barrett, Harris Bernstein, John H. Crowe, David W. Deamer, Robert D. Grey, Jerry L. Hedrick, William A. Jensen, Bernard John, B. E. Juniper, Park S. Nobel, Lester Packer, Irwin Segal, L. Keith Shumway, G. Ledyard Stebbins, Hewson Swift, and T. Elliot Weier. Special thanks are due to my postdoctoral student, Barry R. Zirkin, who reviewed the entire text and made many helpful comments and suggestions, and to Ronald J. Baskin, who provided a first draft of parts of the chapter on thermodynamics. I am indebted to my production editor, Mary Arbogast, who greatly improved the readability of the text and who was most helpful throughout the preparation of the manuscript for publication. I am also indebted to numerous colleagues throughout the world who have generously supplied micrographs, tables, and diagrams for the book, and to Darwen Hennings, who provided the original artwork and, for the sake of uniformity, redrew many of the diagrams sent by other authors.

Stephen L. Wolfe

Contents

Unit One Membrane Systems and the Generalized Cell

1
Introduction 3

Historical Development of the Cell Concept 3
- Development of the "Cell Theory" 5
- Discovery of the Role of the Nucleus in Heredity 6
- Final Establishment of the Nucleus and Chromosomes as the Bearers of Heredity 6

The Beginnings of Biochemistry 7
- The Integration of Biochemistry and Cytology 8

An Introduction to Cell Structure 9
- Units of Measurement Used in Cell Biology 9
- Cellular Organization 9
- The Generalized Prokaryotic Cell 11
- The Generalized Eukaryotic Cell 13
- Specializations of the Cell Surface 19
- Specializations of the Cells of Plants 19

Suggestions for Further Reading 24

2
The Structure of Lipids and Proteins 25

Lipids 26

The Neutral Lipids 26

Phospholipids 28

Sterols 29

Proteins 31

The Amino Acid Subunits of Proteins 31

The Peptide Bond 32

The Three-Dimensional Structure of Proteins 34

Suggestions for Further Reading 39

3
The Light and Electron Microscopes 41

The Light Microscope 41

Elementary Geometrical Optics 41

Construction of the Light Microscope 44

The Phase Contrast Microscope 45

The Electron Microscope 49

Resolution in the Electron Microscope 50

The Effect of Spherical Aberration on Resolution in the Electron Microscope 50

Construction of the Electron Microscope 52

Contribution of Diffraction and Phase Contrast to the Electron Image 56

Speciman Preparation for Electron Microscopy 57

Fixation, Embedding, and Thin Sectioning 58

Isolation of Cellular Components for Electron Microscopy 61

Suggestions for Further Reading 65

4
The Structure of Membranes 67

The Lamellar and Particulate Models for Membrane Structure 68

The Lipid and Protein of Membranes 68

Lipid Subunits of Cell Membranes 68

Properties of Phospholipid Suspensions in Water 70

Electron Microscopy of Lipid Bilayers 71

The Protein Component of Membranes 72

Artificial Lipid–Protein Membranes 73

The Lamellar and Particulate Models Revisited 74

The Danielli–Davson Model 74

Robertson's "Unit Membrane" Concept 76

Resolution of the Lamellar-Particulate Controversy from Freeze-Etch Studies of Natural Membranes 79

The Dynamic State of Membranes 81

Suggestions for Further Reading 82

Unit Two The Role of Membrane Systems in the Provision of Energy for Synthesis

5
The Flow of Energy in Biological Systems 87

An Introduction to Elementary Thermodynamics 88

The First Law of Thermodynamics 88

The Second Law of Thermodynamics 89

Free Energy and the Chemical Potential of Biological Reactions 91

Free Energy and Chemical Reactions 92

Free Energy of Hydrolysis of ATP 93

The Energetics of Biological Catalysis 95
The Mechanism of Enzymatic Catalysis 98
Suggestions for Further Reading 99

6
The Mitochondrion and the Provision of Energy for Cellular Activities 101

Some Historical Perspectives 101
Structure and Occurrence of the Mitochondrion 102
Cellular Respiration 105
- Glycolysis 106
- The Krebs Citric Acid Cycle 110
- The Electron Transport System 112
- Summary of the Energetics of Glucose Metabolism 118
- Cellular Oxidation of Other Metabolites: Fats and Proteins 118
- The Pentose "Shunt" 119

Localization of Oxidative Metabolism in Mitochondrial Fine Structure 121
Integration of Mitochondrial Activity with the Cell Environment 123
Peroxisomes and Glyoxisomes (Microbodies) 124
Suggestions for Further Reading 126

7
Photosynthesis and the Chloroplast 127

Chloroplast Structure 127
The Structure of the Photosynthetic Apparatus in Bacteria and Blue-Green Algae 133
Molecular Components of Chloroplast Membranes 134
Chloroplast Pigments 137
Ultrastructure of the Thylakoid Membrane 139
Photosynthesis 142
- The Overall Reactions of Photosynthesis 143
- The "Light" Reactions in Photosynthesis 144
- The Fixation of Carbon Dioxide in Photosynthesis: The "Dark" Reactions 149
- The Energetics and Efficiency of Photosynthesis 150
- An Important Alternative Pathway for CO_2 Fixation 151
- Other Synthetic Mechanisms in the Chloroplast 152

Localization of Photosynthesis in Chloroplast Fine Structure 153
The Development of Chloroplasts in Higher Plants 154
Suggestions for Further Reading 159

Unit Three The Interphase Cell

8
The Structure of Chromosomes in the Interphase Nucleus 163

DNA as the Molecular Basis of Heredity 164
- The Establishment of DNA as the Hereditary Molecule 164
- Discovery of the Molecular Structure of DNA 165
- Proteins Associated with DNA in the Cell Nucleus 169

The Molecular Structure of Nucleoprotamine and Nucleohistone 171
Nucleoproteins and Chromosome Structure 173
- The Light Microscope Image of the Interphase Nucleus 173
- The Electron Microscope Image of the Interphase Nucleus 174
- The Ultrastructure of Chromosome Fibers in Thin Sections 175
- The Ultrastructure of Chromatin as Seen in Isolated Chromosome Fibers 176
- Isolated Chromosomes 177

Prokaryote Nuclei 180

Kellenberger Fixation of Bacterial DNA 182

Some Evolutionary Implications of Nuclear Structure in Bacteria, Blue-Green Algae, and Dinoflagellates 184

Suggestions for Further Reading 188

9

The Metabolism of RNA in the Nucleus 191

A Brief Overview of Protein Synthesis 192

The Molecular Structure of RNA 194

The Synthesis of RNA on DNA 196

Origin of Cellular RNA 196

RNA Synthesis 197

The Major RNA Classes Transcribed on the Chromosome 200

Messenger RNA (mRNA) 201

Transfer RNA (tRNA) 203

Ribosomal RNA (rRNA) 206

The Role of the Nucleolus in the Synthesis of Ribosomal RNA 208

Suggestions for Further Reading 211

10

Morphological Aspects of RNA Metabolism and Transport in the Nucleus 213

Nucleolar Structure 213

Cytochemical Identification of the Four Nucleolar Components 216

The Relationship of Nucleolar Morphology to the Maturation of Ribosomal RNA 217

Other Ribonucleoprotein Structures in the Nucleus 220

A Review of the Fine Structure of the Interphase Nucleus 223

The Transport of Ribonucleoprotein to the Cytoplasm 224

The Nuclear Envelope 224

Permeability of the Nuclear Envelope: Are the Pores Open or Closed? 229

Polytene Chromosomes 238

Suggestions for Further Reading 249

11

Protein Synthesis 251

Preliminary Reactions of Protein Synthesis 251

The Assembly and Structure of Ribosomes 251

The Activation and Attachment of Amino Acids to Transfer RNA 255

The Mechanism of Protein Synthesis at the Ribosome 256

Initiation 256

Termination of Synthesis 258

Morphological Demonstration of the mRNA-Ribosome Complex 260

The Genetic Code 261

The Regulation of Protein Synthesis 264

Transcriptional Control of Protein Synthesis 264

Morphological Evidence of Transcriptional Control in Eukaryotes 268

Heterochromatin and Euchromatin 269

Post-Transcriptional Control in Eukaryotes 269

Translational Control of Protein Synthesis 269

Morphological Aspects of Protein Synthesis in the Cytoplasm 270

The Endoplasmic Reticulum 270

Smooth Endoplasmic Reticulum 274

The Golgi Complex 276

Suggestions for Further Reading 285

Unit Four Cellular Reproduction

12
Mitosis I: Replication of the Chromosomes 289

An Overview of Replication, Mitosis, and Cytokinesis 290

The Cell Cycle 290

Mitosis 291

Prophase 291

Metaphase 295

Anaphase 297

Telophase 299

The Time Course of Mitosis 299

Synthesis and Respiration during Mitosis 299

The Significance of Mitosis 302

Cytokinesis 303

Replication of the Chromosomes 303

Experimental Demonstration of the Semiconservative Pathway 305

The Synthesis of DNA 310

DNA Synthesis and Semiconservative Replication 311

Mutations 316

Some Notes on Replication in the Eukaryote Chromosome 316

Redundant or Repetitive DNA 317

Suggestions for Further Reading 327

13
Mitosis II: Division of the Nucleus and Cytoplasm 329

The Structure of Chromosomes at Prophase and Metaphase 329

Coiling of Chromosomes to the Metaphase State 329

The Structure of the Primary Constriction: The Centromeres 331

Secondary Constrictions 336

A Summary of Metaphase Chromosome Structure 336

Independent Movements of the Chromosomes during Prophase and Early Metaphase 336

The Spindle 339

Formation of the Spindle 339

Structure of the Spindle 340

Function of the Spindle: The Anaphase Movement 342

Hypotheses of the Anaphase Movement 343

Some Comparative Aspects 347

Asters and Centrioles 350

Ultrastructure of the Centriole 351

The Role of the Centrioles and Aster during Mitosis 353

The Nucleolar Cycle during Mitosis 354

The Nuclear Envelope during Mitosis 359

Cytokinesis 360

Furrowing 362

Formation of the Cell Plate in Plants 365

Suggestions for Further Reading 372

14
Meiosis 375

An Overview of Meiosis 376

Premeiotic Interphase 376

Prophase 377

Metaphase I of Meiosis 382

Anaphase and Telophase I 383

The Second Meiotic Division 384

The Time and Place of Meiosis 384

Synthesis of Nucleic Acids and Protein during Meiosis 385

The Centrioles in Meiosis 387

The Sex Chromosomes during Meiosis 387

Recombination 388

Early Hypotheses of Recombination 389

Experimental Support for the Breakage and Exchange Hypothesis 390

The Copy-Choice Model For Recombination 394

The Development of Molecular Models for Recombination 396

Special Characteristics of Intragenic Recombination 396

Molecular Models for Recombination 399

The Morphology of Recombination in Eukaryotes 403

The Synaptinemal Complex 403

The Relationship of the Synaptinemal Complex to Pairing and Recombination 407

The Problem of Polycomplexes 408

The Synaptinemal Complex and Recombination: Some Possibilities 412

Lampbrush Chromosomes: A Configuration of the Diplotene Nucleus 415

The Structure of Lampbrush Chromosomes 415

The Functional Significance of Lampbrush Chromosomes 418

Nucleolar Synthesis during the Lampbrush Stage 419

Suggestions for Further Reading 422

15
Gametogenesis and Fertilization 425

Spermiogenesis 426

Morphological Changes in Development of Spermatozoa in Animals 426

Some Hypotheses Concerning the Molecular Basis of Flagellar Motion 444

Spermiogenesis in Plants 446

A Summary of Spermiogenesis 450

Oogenesis 452

Oogenesis in Animals 453

The Secretion and Deposition of Yolk during Oogenesis 459

Fertilization 462

Fertilization in Animals 462

Gametogenesis and Fertilization in Higher Plants 474

Suggestions for Further Reading 480

16
Cytoplasmic Inheritance 483

The Cytoplasmic Control of Inheritance in the Mitochondrion 484

The Genetic Evidence 484

The Search for DNA in the Mitochondrion 485

The Physical Properties of Mitochondrial DNA 487

Replication of Mitochondrial DNA 493

RNA in the Mitochondrion 494

Possible Functions of the Nucleic Acids of Mitochondria 496

The Origin and Continuity of Mitochondria 497

The Cytoplasmic Control of Inheritance in the Chloroplast 499

Genetic Evidence 499

DNA in the Chloroplast 500

RNA in the Chloroplast 506

Possible Functions of the Nucleic Acids of the Chloroplast 508

The Continuity of Chloroplasts in Dividing Cells 509

Cytoplasmic Inheritance of the Centriole 511

DNA in the Centriole 513

The Possible Evolutionary Origins of Mitochondria and Chloroplasts 515

Suggestions for Further Reading 518

17
The Origin of Cellular Life 521

The Primordial Atmosphere 523

The Spontaneous Appearance of Organic Molecules on the Primordial Earth 524

The Concentration of the Organic Molecules of the Primative Environment in Functional Aggregates 526

Aggregation by the Spontaneous Formation of Primitive Membranes 527

The Origin of Life in the Primitive Aggregates 527

The Origins of Photosynthesis and the Change from a Reducing to Oxidizing Atmosphere 530

The Appearance of Eukaryotes 531

The Time Course of the Evolution of Life 531

Suggestions for Further Reading 531

Index 533

To my wife,
Betty

and to my brother,
Milt

Unit One *Membrane Systems and the Generalized Cell*

1
Introduction

The concept of the cell as the working unit of living organisms arose in the last century, during the years from about 1830 to 1880. The problem of the fundamental organization of living matter had always been of central interest, but investigations before the nineteenth century were seriously hampered by the imperfections of the early light microscopes. Greatly improved microscopes became available in the 1830's, and rapid developments in the burgeoning science of cell biology quickly followed. Progress in optics and in cytology kept pace during the nineteenth century; by the 1890's both the light microscope and the cell concept had essentially reached their modern form. Some of the more important stages in the progression of ideas leading to full development of the cell concept shortly after 1900 will be reviewed in the section that follows. In broad outline, these developments include the generalizations that the structural unit of living organisms is the cell, that cells arise only by division of preexisting cells, and that the control of cell heredity lies primarily in the nucleus of the cell.

Historical Development of the Cell Concept

The first important event in cell biology occurred when Robert Hooke, an Englishman and curator of the Royal Society, first described cells in his *Micrographia*, published in 1665. Hooke, using an early version of the compound microscope, was able to trace out small compartment-like units in the woody tissues of plants; he called these compartments "pores" or "cells." In some of the plant tissue observed, such as cork, the cells were simply empty spaces delineated by residual cell walls. But in other tissue, such as the "inner pulp or pith of an Elder, or almost any other tree," Hooke "plainly enough discover'd these cells or Pores fill'd with juices." These first descriptions of the organization of living matter into cells were the beginning of the sciences of cytology and cell biology.

Soon after the publication of *Micrographia*, other accounts of investigations into the finer structure of plants and animals were written. The most important investigations were reported by another Englishman, Nehemiah Grew, and a Dutchman, Anton van Leeuwenhoek. Grew, a contemporary of Hooke, and his successor as curator of the Royal Society, also studied the microscopic structure of plants. He reported the existence of cells, or "bladders," as he called them, in a variety of plant tissues. His first publication, *The Anatomy of Vegetables Begun*, appeared in 1672. In the following year, the Royal Society received the first in a long series of letters from Leeuwenhoek, a linen draper and amateur microscopist, that described his observations of the microscopic structure of protozoa, spermatozoa, blood cells, and many other "animalcules." His communications to the Royal Society, some 200 of them, arrived at intervals over a period of about 50 years.

After the last of Leeuwenhoek's letters, in the early part of the 1700's, cell biology entered a period of quiescence that continued for a hundred years, well into the nineteenth century. This delay in the evolution of the cell concept resulted in large part from the imperfections of the microscopes available in the seventeenth and eighteenth centuries. Hooke and Grew both used compound microscopes, which consisted of two lenses mounted at either end of a tube. These early microscopes gave a total magnification of approximately 30 diameters. Leeuwenhoek used a simple microscope, constructed from a single lens mounted between two metal plates. Although Leeuwenhoek's microscopes, made by hand in his own shop, were of considerably greater magnifying power than those used by Hooke and Grew, the conditions of observation were much worse. It is not understood today quite how Leeuwenhoek was able to see the details so accurately drawn in his letters to the Society, unless, as has been suggested, he possessed exceptional eyesight in addition to his skill at grinding lenses.

The most critical lens defects in the early microscopes were chromatic and spherical aberration. In chromatic aberration, light of different colors is spread into different points of focus, producing concentric halos of color around an image point. Spherical aberration causes light rays passing through the lens axis to be focused at a different point from rays passing through the lens closer to the edges (see Fig. 3–12). This lens defect blurs the image of a point into a disc. The combined effect of the two aberrations obscured the finer details of cell structure, including even organelles as large as the nucleus.

A practical method for the correction of chromatic aberration was developed late in the eighteenth century and became commercially available in about 1825. At about the same time, in 1830, the first solution to the problem of spherical aberration was made. With improved microscopes, in which the corrections were made by using combinations of lenses with compensating errors, objects as small as one micron or slightly less could be resolved. Other improvements were made in the optical system and in specimen mounting until, by about 1870 to 1880, the light microscope had almost reached its maximum development, allowing visualization of objects about 0.2 microns in diameter. These developments in optics took place in continental Europe; by the 1800's interest in the use of the improved microscopes to study living objects had also shifted from England to the continent.

The early observations of Hooke and Grew, made in plant tissues, had given the impression that cells were units of living matter delineated by conspicuous walls. At this time, the essential feature of the cell was considered to be the wall itself rather than the enclosed contents. Because the cells of animal tissues lack distinct walls, units equivalent to plant cell walls could not be resolved in animals, and the basic similarities in microscopic structure between plants and animals at first escaped notice.

The parallels in structure between animal and plant tissue were not drawn until 1839, when Theodor Schwann observed that cartilaginous tissues of animals contained a microscopic structure that "exactly resembles [the] parenchymatous cellular tissue of plants." Schwann's work was aided by the fact that the extracellular material in cartilage occupies a position analogous to the cell wall in plants. Thus, while not possessing walls as an integral structural component, the cells of cartilage are separated by partitions that are more easily resolved in the light microscope than the boundaries of cells in most other animal tissues. This enabled Schwann to recognize the cellular nature of the animal tissue. Schwann also remarked on the presence of nuclei in the cartilage cells, but was at first not as impressed by the nucleus as by the walls, which at this time seemed to provide the main basis for the parallel between the microscopic structure of plants and animals.

The improvements in light optics in the 1830's soon allowed recognition of the much thinner cell boundaries in other types of animal tissues. As the details of structure in the cell interior became discernible, emphasis gradually shifted from the walls to the contents, and the term "cell" began to take on modern connotations. By the 1830's, it was recognized that the fluid contents of the cell are the primary substance of living organisms; J. E. Purkinje adapted the term *protoplasm* for this substance in a publication appearing in 1840.

Development of the "Cell Theory"

In 1833, a few years before the definition of protoplasm by Purkinje, Robert Brown published a paper describing the microscopic structure of the reproductive organs of plants. In this paper, Brown drew attention to the nucleus as a constant feature of all of the plant cells investigated in his studies. This work, which subsequently received wide notice, established that the nucleated cell was the unit of living tissue in plants and laid the foundations for later development of the cell theory by Schleiden and Schwann.

Soon after publication of his work with cartilage in 1839, Schwann came under the influence of an eminent personage of the day, the botanist M. J. Schleiden. Schleiden had developed a series of hypotheses about the growth of cells in plants. As visualized by Schleiden, the cell nucleus took on a central role in the reproduction of cells. Although the details of Schleiden's hypotheses eventually turned out to be completely erroneous, he did pass on to Schwann the concept that the nucleus is of central importance in the structure and development of cells. After analyzing the morphological evidence from his own studies and the existing reports from other investigators, Schwann realized the universality of the nucleated cell as the structural unit of living matter. From his studies, Schwann developed two postulates that were to provide the basis for the cell theory: that all living organisms are composed of one or more nucleated cells, and that the cell is the unit of function of living organisms. Historians usually attribute origin of the theory jointly to Schwann and Schleiden.

Unfortunately, Schwann had also incorporated Schleiden's erroneous views on cell growth into his hypotheses. Schleiden maintained that the nucleus was formed, as a first step in cell reproduction, by the aggregation of small protoplasmic granules. The nucleus then grew in volume, eventually developing into the daughter cell while still within the boundaries of the parent cell. This idea, that nuclei are derived from protoplasmic granules and later mature into daughter cells, was called "free-cell formation." The hypothesis held sway for many years, partly, at least, because of Schleiden's influence over other scientists and his vigor in defense of his ideas.

Not many investigators had the temerity to challenge Schleiden's views. However, a few, including Hugo von Mohl, were unimpressed. In one criticism of Schleiden's hypotheses, von Mohl remarked that "The entire representation proves that Schleiden has never once observed the division of a cell." Soon, experimental evidence began to appear that directly contradicted Schleiden. Observations of tissue growth in plants indicated that cells multiplied by division rather than free cell formation. In 1846, these observations were combined and formalized by Karl von Nägeli, who from this work and his own extensive investigations in plant development concluded that all plant cells are derived by the direct division of one cell into two daughter cells. However, with many cytologists of this period, Nägeli felt that the cell wall was the significant structure in the division mechanism, and he did not take exception to Schleiden's hypothesis that nuclei were formed anew in cells by aggregation of protoplasmic granules. Nägeli thus subscribed to free formation of nuclei but not of cells.

Equivalent developments were also taking place for animal cells. Robert Remak, in 1841, observed the division of embryonic blood cells into two equal daughter cells, and the cytologists K. E. von Baer and R. A. von Kölliker correctly interpreted the furrowing of the surface of fertilized eggs as a process of cell division. Finally, in 1855, Rudolf Virchow, after an exacting study of the growth of cells in human tissues, stated clearly that all cells arise only by the division of preexisting cells. Virchow's expression of this concept, "*omnis cellula e cellula*," has become a celebrated aphorism in the history of biology. This concept led to the appreciation, stated by Virchow and others, that an unbroken sequence of cell divisions links the earliest cellular organisms with the animals and plants of our day. Thus, with the elimination of Schleiden's theory of free cell formation, the significance of the cell in the transmission of hereditary traits became obvious.

Discovery of the Role of the Nucleus in Heredity

The importance of the nucleus in the transmission of heredity was not realized until the 1870's, when fertilization and the early development of animal eggs became a matter of central interest in cytology. From this work it was established that, just after fertilization, there are two nuclei in the egg; these subsequently fuse into a single nucleus before further development is initiated. In 1875, Oskar Hertwig demonstrated that one of these two nuclei is a part of the egg, and the second is introduced by the sperm during fertilization. Equivalent observations were made in plant tissue by Eduard Strasburger in 1877. These findings were of fundamental importance to the developing appreciation of the role of the nucleus in heredity.

If the nucleus was to be seriously considered as the carrier of hereditary traits, it was necessary to establish that the nuclear substance was physically continuous during the division cycle of the cell. Many cytologists of the time believed otherwise, claiming that the nucleus simply dissolved during division and was formed anew after organization of the daughter cells. The physical continuity of the internal matter of the nucleus was finally confirmed in 1870 by Strasburger and Walter Flemming, who were able to follow the nuclear stages in cell division in plant and animal tissues with great accuracy. The essential feature in this process, termed *mitosis* by Flemming, is that the substance of the nucleus is transformed into elongated threads during division; when fully formed the threads (chromosomes) are clearly seen to be double. As the stages of division progress, the two halves of each chromosome are split apart and pass into separate daughter cells, where they are transformed into the daughter nuclei. Later work by Flemming, Carl Rabl, and others showed that the numbers of chromosomes appearing during division are constant for all members of a given species, and the significance of mitosis in the preservation of the chromosome number became obvious. The continuity of the nuclear substance during division of the cell was now established; it was left for Rabl to demonstrate that the chromosomes, although invisible in the nucleus of a nondividing cell, probably remain physically intact throughout the cell cycle. In 1885, Rabl accomplished this by showing that the chromosomes reappear at the onset of mitosis in approximately the same arrangement in which they were last seen at the close of the previous division.

The final link in the concept of the physical continuity of the nucleus was provided by studies of chromosome behavior during formation of gametes and fertilization. In 1883, the Belgian cytologist Edouard van Beneden followed this process in *Ascaris,* in which only a single pair of chromosomes is present in the zygote nucleus after fertilization. With this favorable material, van Beneden demonstrated that the egg and sperm each donate one chromosome to the pair found in the zygote. He generalized from this that in fertilization the egg and sperm donate equal numbers of chromosomes to the zygote nucleus.

This work made it obvious that at some time before the union of gametes the chromosome number must be reduced to one-half the somatic value. Otherwise, the chromosomes would double in number with each successive fertilization. In the late 1880's, the anticipated reduction divisions, predicted on theoretical grounds by August Weismann, were discovered by Oskar Hertwig and T. Boveri, who found that in animals the number of chromosomes is halved during the last series of divisions giving rise to sperm and eggs. Equivalent observations in plant tissues were made a few years later by Strasburger and E. Overton. The details of the reduction mechanism were filled in by J. B. Farmer and J. E. Moore, who termed this special form of nuclear division maiosis (from the Greek *meioun*–to diminish). This was later changed to the present-day form, *meiosis.* Extensive research in plants and animals soon led to the recognition that meiosis takes place in all sexually reproducing organisms.

Final Establishment of the Nucleus and Chromosomes as the Bearers of Heredity

The discovery and description of the cellular mechanisms that preserve the physical continuity of the nucleus and the constancy of the chromosome number resulted in a general acceptance of the hypothesis that the factors controlling the heredity of the cell were probably located in the nucleus. In 1884, both Hertwig and Strasburger had already declared that the physical basis of heredity resides in the nucleus; in 1885, in his famous publication *The Germ Plasma, a Theory of Heredity,* August Weismann identified the nucleus with the hereditary substance and concluded further that "the complex mechanism for cell division exists practically for the sole purpose of dividing the chromatin, and . . . thus the [chromatin] is without doubt the most important part of the nucleus."

A quantitative assessment of the behavior of the hereditary factors had been made and published some years before these cytological developments took place. In 1865, an Austrian monk, Gregor Mendel, announced the results of his work with inherited traits in garden peas. Mendel formulated and successfully tested a series of hypotheses concerning heredity. Briefly, these hypotheses stated: hereditary factors occur in pairs; one member of the pair is derived from each parent of an individual; the pairs are segregated independently in the formation of gametes; and the members of the pair, if different in effect, are alternate forms of the same entity. The significance of his work was not appreciated at the time of publication, because descriptive cytology had not progressed far enough to allow Mendel's mathematical abstraction of the heredity factors to be related to physical units in the cell. Mendel's results were unnoticed and lay forgotten until just after the turn of the century, when the same relationships were rediscovered independently by Hugo De Vries, C. Correns, and E. von Tschermak. By this time, the continuity of the nucleus and the behavior of the chromosomes in division and reproduction were so well established that the correlation between the gene and the chromosome was very quickly made. In 1902 and 1903 the American cytologists W. A. Cannon, E. B. Wilson, and Wilson's student Walter S. Sutton pointed out the precise equivalence between the behavior of genes and chromosomes. This correlation was extended to a specific gene and chromosome in 1916, when C. B. Bridges demonstrated the precise equivalence in inheritance between the X chromosome and a gene for white eye color in *Drosophila.* These findings joined together genetics and cytology and completed the foundations for the contemporary science of cell biology.

The Beginnings of Biochemistry

The application of physics and chemistry to the study of living organisms also has its roots in the eighteenth and nineteenth centuries. The real beginnings in this field were made in 1772 when Joseph Priestley discovered that oxygen is released by plants; at almost the same time, Antoine Lavoisier recognized that "respiration is a . . . combustion, slow, it is true, but otherwise perfectly similar to that of charcoal." The philosophical climate of the time, however, was such that the fundamental importance of these discoveries was not appreciated. During the latter part of the eighteenth century it was still held that the substances and processes in living organisms were basically different from those of the inorganic world and that the techniques used for studying inanimate objects could not be applied to life. As a part of this point of view, it was thought that organic substances could not be artificially synthesized.

The first movement away from these attitudes came in 1828 when Friedrich Wöhler synthesized urea, at about the same time that cytologists were beginning to formulate the cell theory. Wöhler was able to convert the inorganic chemical ammonium cyanate to urea, an organic substance commonly excreted by animals. This synthesis of a chemical characteristic of living organisms and the many other examples that followed gradually established that the same substances occur in both living and inanimate objects and are governed by the same chemical and physical laws. By the end of the nineteenth century, investigators had synthesized many of the chemicals found in plants and animals; this work reached its highest development in the researches of Emil Fischer, who extracted, degraded, and resynthesized many substances from living organisms and laid the foundation for the chemical description of proteins, fats, and sugars. In 1902, Fischer and F. Hofmeister, in independent work, discovered the peptide bond, which links amino acids together to form the primary structure of proteins.

This descriptive work was complemented in the early nineteenth century by the beginnings of functional biochemical studies. Before this time, the substances and the reactions occurring in living systems were thought to be moved by a "vital force." The discovery of catalysts at the beginning of the nineteenth century provided an insight into real nature of the chemical force moving the interactions in living organisms. In 1836, Jons Jacob Berzelius, a Swedish chemist, wrote that it is "justifiable . . . to suppose that, in living plants and animals, thousands of catalytic processes take place . . . and result in the formation of the great number of dissimilar chemical compounds, for whose formation out of the common raw material . . . no probable cause could be assigned."

Berzelius' intuitions were later proved correct as a result of investigations into the nature of alcoholic fermentation, which has been of central interest to mankind since ancient times. In the 1850's Louis Pasteur began his efforts to determine the cause of fermentation. Pasteur found that fermentation occurred only if particular microorganisms were present; if the microorganisms were eliminated or killed, sugar was not converted to alcohol. The first insights into the catalytic nature of the reactions occurring in fermentation came toward the end of the century. In 1897, Hans and Eduard Buchner were working with the problem of isolating and preserving extracts from yeast cells for medicinal purposes. In order to preserve their extract, made by grinding and pressing yeast cells, the Buchners added sugar, commonly used then and today as a method for preserving foods. To their surprise, the sugar was rapidly fermented by the cell-free yeast extract. Intensive studies of the yeast extracts followed. This work led directly to the systematic characterization and description of the proteinaceous catalysts of living systems, the *enzymes*. In fact, the word "enzyme" was coined from a Greek expression that, translated literally, means "in yeast." The study of fermentation and other biochemical reactions in cell-free extracts received great attention in the years following the Buchners' discovery and continues to the present day. Among the important early discoveries arising from this approach was the elucidation of oxidative respiration, which had its beginnings in the work of Wieland, Warburg, and Keilin in the early part of this century.

The Integration of Biochemistry and Cytology

The cytologists and biochemists of the nineteenth and early twentieth centuries took little notice of each other, and few attempts were made to integrate the findings of the two disciplines. Significantly enough, the first isolated efforts to integrate the approaches of biochemistry and cytology involved the cell nucleus and stemmed directly from the growing realization in the 1860's that this organelle is of central importance in cell function and heredity.

Friedrich Miescher, a physician and physiological chemist, became interested in the chemical composition of the nucleus and devised a method of isolating leucocyte nuclei in quantity for chemical analysis. By using dilute acid, pepsin, and ether, Miescher was able to obtain a fairly pure extract of whole nuclei. The purity of the extract was determined by examination of the nuclei under the light microscope. From this preparation, he isolated a previously unknown substance distinguished by properties then considered unusual for organic matter including high acidity and a large phosphorus content. Miescher, in announcing his discovery in 1871, called the new substance "nuclein." Later, he discovered a closely similar material in the nuclei of salmon sperm. Here he found the nuclein, or nucleic acid, as it was later called, in association with a nitrogenous substance which he termed "protamine." Soon after the announcement of Miescher's findings, Flemming concluded that if the chromatin of the nucleus and nuclein were not one and the same substance, "one carries the other."

A method for purifying nucleic acids was worked out by R. Altmann in 1889, and the major constituents, the purine and pyrimidine bases and the pentose sugars, were soon identified. Two nucleic acids were distinguished. One, of the type first discovered by Miescher, was thought to occur only in animal nuclei. The second was for a time thought to be characteristic of the plant nucleus. However, in 1914, a biochemist, R. Feulgen, showed that the "animal" type of nucleic acid interacts specifically with the dye acid fuchsin after hydrolysis with dilute acid. In a later application of this method to different tissues, Feulgen was surprised to find that both animal and plant nuclei show a positive reaction for the nucleic acid formerly thought to be confined to animals. It was subsequently established that this type of nucleic acid, deoxyribonucleic acid (DNA), is characteristic of the nucleus of all cells. The second type, ribonucleic acid (RNA), was also found to be present in all cells, distributed in both the nucleus and cytoplasm. Later, in the 1920's and 1930's, extensive use of Feulgen's reaction confirmed that DNA is localized not only in the nucleus but specifically in the chromosomes, and many cytologists began to suspect direct involvement of this substance in heredity. Finally, in the 1940's and 1950's, this was indeed shown to be the case (the more recent discoveries linking DNA with heredity are reviewed in detail in Chapter 8).

Similar explorations in the integration of cytology and biochemistry were made with other organelles, beginning soon after the turn of the century with

Table 1–1 *A Comparison of the Units of Measurement Used in Cell Biology*

Millimeter	*Micron*	*Millimicron*	*Angstrom*
1	1,000	1,000,000	10,000,000
0.001 (1×10^{-3})	1	1,000	10,000
0.000001 (1×10^{-6})	0.001 (1×10^{-3})	1	10
0.0000001 (1×10^{-7})	0.0001 (1×10^{-4})	0.1	1

investigation into the function of chloroplasts and mitochondria. But the synthesis of the two sciences was never rapid until the 1950's, when the development of powerful new methods such as electron microscopy, the use of radioactive tracers in biochemistry, analytical centrifugation, and other sophisticated techniques led to the present understanding of cell biology, in which structure and biochemistry are inseparable and must be considered together in all investigations. These recent developments in cell biology and the unity of structure and function in cells revealed by these methods make up the subject matter of this book.

An Introduction to Cell Structure

Units of Measurement Used in Cell Biology

Before giving a brief survey of cell structure, it is first necessary to review the units of measurement used in cell biology and cytology. The units of measurement most frequently used are the *micron* (μ), the *millimicron* (mμ), and the *Angstrom* (Å or A: see Table 1–1 for a comparison of these units). The micron, equivalent to 1/1000 millimeter, is probably already familiar to the reader. The lower limit of human visual acuity is about 200 microns; objects below this size are not discernible by the naked eye. In terms of a rigorous definition of resolution (see Chapter 3) it is more accurate to state that small objects closer together than 200 microns cannot be distinguished as separate points by the human eye. Most cells range between 10 and 100 microns in size, although a few protozoa and the eggs of some animals are considerably larger.

The micron is used primarily to describe whole cells or multicellular aggregates. This unit is less useful in dealing with the subparts of cells because most organelles, with the exception of the nucleus, are smaller than one micron in dimensions. To avoid continual use of parts of a micron, smaller units have become widely employed in cell biology. The most convenient of these is the Ångstrom, named for the Swedish physicist A. J. Ångstrom. The Ångstrom is equal to 1/10,000 microns. At first encounter it is difficult to relate the Ångstrom to more familiar units of measurement, but after some experience a relative evaluation can be made. The Ångstrom is a molecular-level dimension. For example, lipid molecules are about 20 Å and amino acids approximately 10 Å in length, and proteins may be 100 Å or so in diameter. On the level of cell organelles, two structures have fairly regular dimensions that are repeated from cell to cell: the membrane, with a diameter from 75–100 Å, and the ribosome, with a diameter of 200–250 Å. Once these structures have become familiar they can be used as points of reference for the estimation of the magnitude dimensions given for other structures.

Another unit of measurement used by cell biologists is the *millimicron* (mμ), equivalent to 10 Å. This unit, also called the *nanometer* in recent years, has not been as widely employed as the Ångstrom and is encountered less frequently in the biological literature. For this reason, the Ångstrom rather than the millimicron will be used throughout this book. The Å has the added advantage that all cellular dimensions, no matter how minute, can be given in whole Ångstroms; no significant measurements can be made with the electron microscope in parts of an Ångstrom. Further, as noted, the Å is also useful as a unit of measurement at the molecular level.

Cellular Organization

The cytological investigations of the nineteenth century established that cells are the structural and functional units of all living organisms. In essence, the cell is a highly organized collection of molecules capable of all of the activities associated with the quality

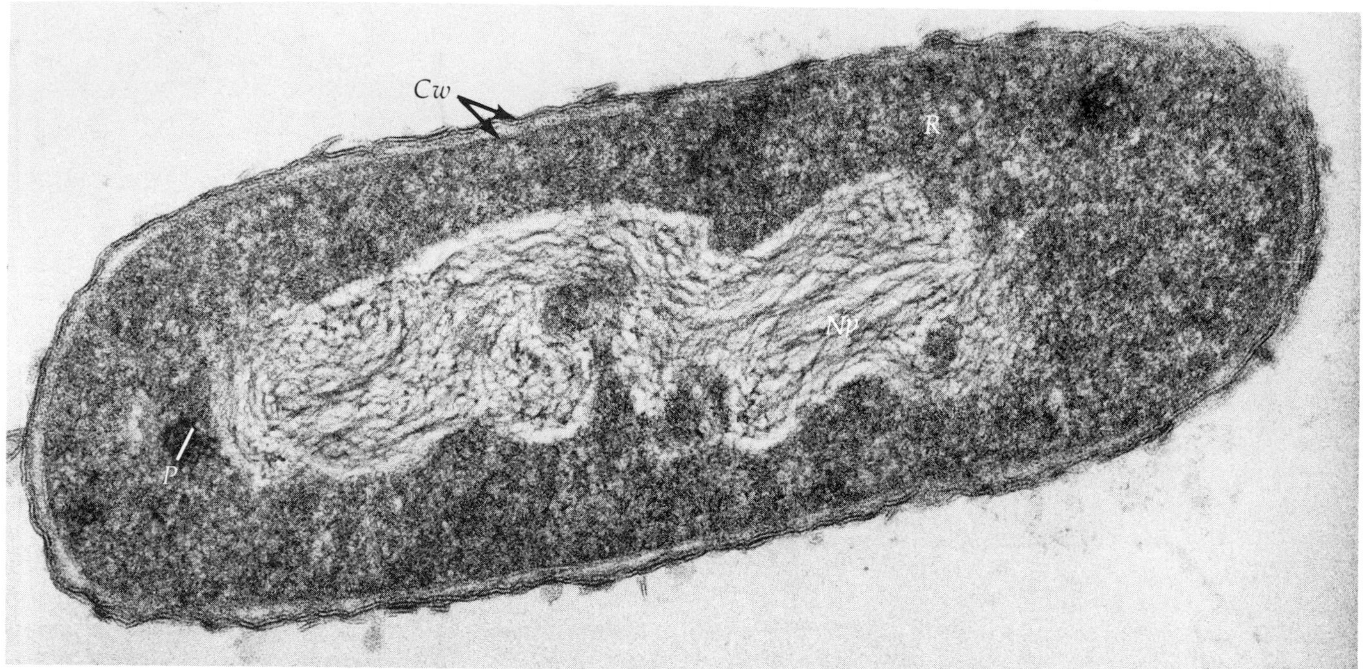

1–1. *A typical prokaryotic cell, the bacterium* Escherichia coli. *A conspicuous nucleoid* (Np) *occupies the center of the cell. The dense cytoplasm surrounding the nucleoid is packed with ribosomes* (R), *which are not individually visible in this micrograph. The cell is surrounded by the* cell wall (Cw); *just beneath this exterior structure lies the plasma membrane. A polyphosphate body* (P) *is located in the cytoplasm at one end of the nucleus.* × *28,500. Courtesy of G. Cohen-Bazire and Wadsworth Publishing Company, Inc.*

of life: it can grow, respond to stimuli, move, and reproduce. In accomplishing these tasks, the cell expends energy derived from light or the metabolism of fuel substances. Each cell is maintained as a distinct environment and collection of matter by a surface membrane, a layer of lipid and protein only a few molecules thick. If this membrane, called the *plasma membrane,* is ruptured, death of the cell quickly follows.

Cells can be classified into one of two major categories, depending on the complexity of internal organization. The most simple cells are the bacteria and blue-green algae. In these forms, there is only one major system of membranes, the plasma membrane and its derivatives; any additional internal membranes are either directly connected to or are derived from the plasma membrane. The chromosomal material of the bacteria and the blue-green algae is not separated from the rest of the cell by a continuous system of membranes as in all other organisms. Because this level of organization is believed to be much like the primitive cells that first arose in evolution, Hans Ris of the University of Wisconsin has suggested that these organisms be called the *prokaryotes* (*pro* = before, *karyon* = nucleus).

The cells of all other organisms contain internal membrane systems that divide the cell interior into separate compartments, each with distinctive concentrations and organization of internal substances. The most conspicuous division separates the cell interior into *nucleus* and *cytoplasm;* other continuous membranes surround internal organelles such as mitochondria, chloroplasts, and the vesicular components of the cytoplasm. Other differences from the prokaryotes are found in the degree of complexity of organization of molecular systems, particularly in the nucleus. This group, which includes the vast majority of living organisms, has been termed by Ris the *eukaryotes* (*eu* = true, *karyon* = nucleus). The typical morphology of these two cell types, the prokaryotes and eukaryotes, is described in some detail in the following sections.

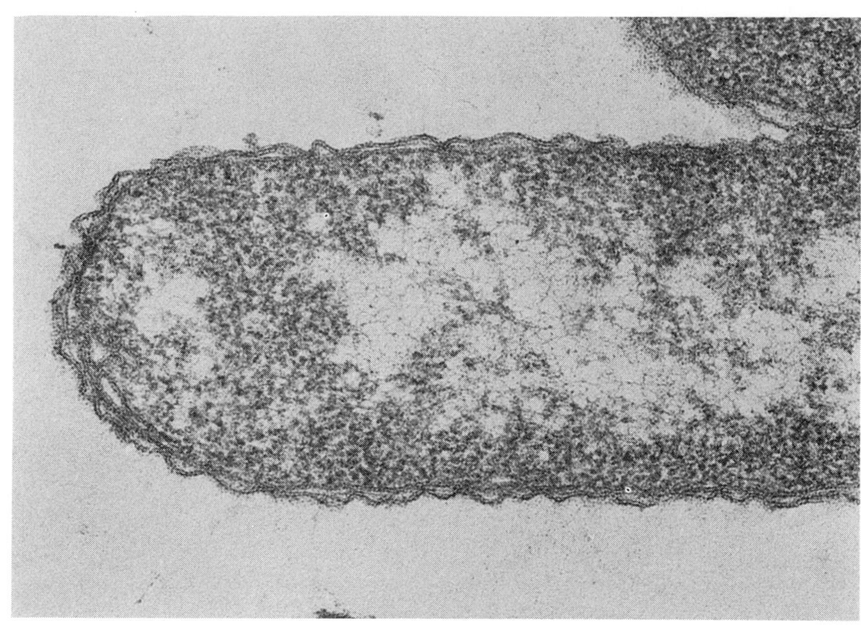

1–2. Another micrograph of the bacterium E. coli, *in which individual ribosomes can be discerned in large numbers in the cytoplasm. Note also the infoldings and additional layers of the plasma membrane at the tip of the cell.* × *100,000. Courtesy of W. Van Iterson, from* Bact. Rev. *29:299 (1965).*

The Generalized Prokaryotic Cell

Prokaryotes are relatively small cells, usually no more than a few microns in length and a micron or slightly less in diameter. Figs. 1–1 and 1–2 show typical bacterial cells in longitudinal section, as viewed under the electron microscope. The limits of the cell, as mentioned, are marked by the plasma membrane. This membrane, which is 75–100 Å thick, is most frequently designated as the "plasma" rather than "cell" membrane in order to avoid confusion with other cellular membranes.

Just outside the plasma membrane is a rigid exterior surface coat, the *cell wall,* which may range from 150 to 1000 Å or more in thickness. While the cell wall is outside the limits of the cell proper, it is still an indispensable part of the cell and occurs in all prokaryotes. The chemical substances that make up the cell wall vary in different prokaryotes; in bacteria the cell wall is a complex structure containing lipids, polysaccharides, proteins, and complexes between sugars and other substances such as amino acids and peptides. Different groups, such as the gram-negative and gram-positive bacteria, show regular variations in the structure and chemical composition of the cell wall. Outside the wall may be a jelly-like layer of variable thickness, the *capsule.* This sticky, mucilaginous coating usually contains polysaccharides and traces of other substances.

Blue-green algae (see Fig. 1–3) have exterior walls that are intermediate in chemical structure between the bacterial type and the walls of eukaryotic plant cells. Although cellulose, the most characteristic component of eukaryotic cell walls (see p. 370) is not found in the cell wall of blue-green algae, it occurs as a component of the capsule (called the *sheath* in blue-green algae).

The protoplasm of the prokaryotic cell is differentiated into two regions of distinct morphology and function. One of these contains the chromosomal material and under the electron microscope appears as one or more centrally located, fibrous deposits of irregular outline. Surrounding these fibrous masses is the cytoplasm, which under the electron microscope appears almost homogeneous in prokaryotes.

The chromosomal material is not set off from the surrounding cytoplasm by a system of membranes and is thus not strictly analogous to the nucleus of eukaryotes. For this reason, other terms are frequently employed to describe this structure, including *nucleoid, nuclear equivalent, chromatin body,* or *genophore.* The term "nucleoid," or in the case of bacteria, "bacterial nucleus" or "bacterial chromosome" will be used here.

The nucleoid appears in electron micrographs as a structure of indefinite outline, less dense than the surrounding cytoplasm, containing tangled masses of fibers 30–50 Å in thickness. By various means, the fibrous material of the bacterial nucleus can be shown

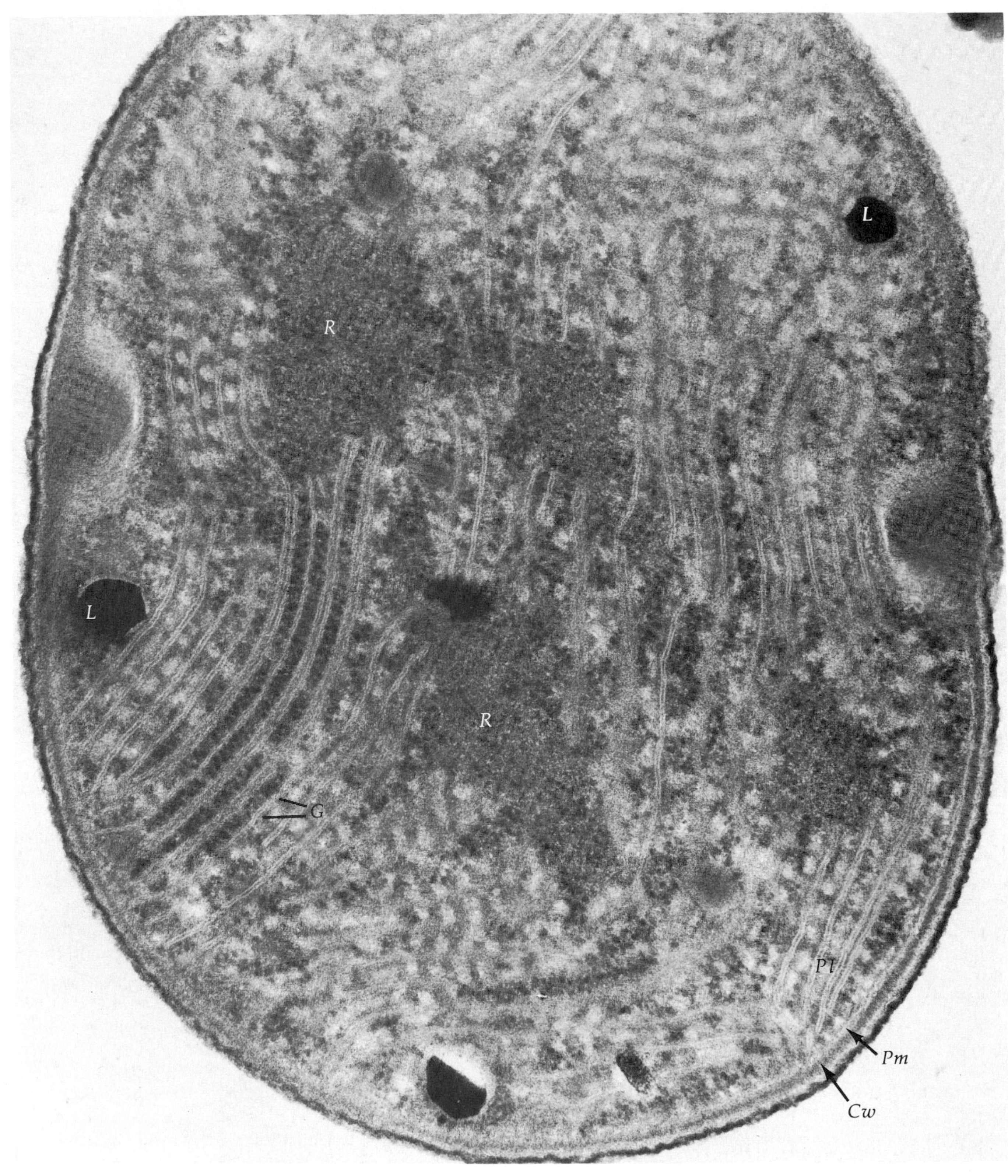
L
R
L
R
G
Pl
Pm
Cw

to consist of a single molecule of DNA, irregularly folded into a compact mass. When isolated, this DNA proves to form a single closed circle. Apparently, the DNA of prokaryotes is not complexed with any detectable quantities of histone or the other proteins that are regular constituents of the chromosomes of eukaryotic cells. No structures other than the 30–50 Å fibers are regularly seen in the prokaryote nucleoid.

The cytoplasm surrounding the nucleoid usually appears to be highly electron-dense. At higher magnification, it can be seen that this density is caused by the presence of large numbers of spherical particles 150–200 Å in diameter (Fig. 1–2). These particles, the *ribosomes*, are the sites of protein synthesis in the cell. In spite of their small size, ribosomes are exceedingly complex structures that contain at least three types of RNA and as many as 40 or more different proteins.

The prokaryote cytoplasm sometimes also contains complex infoldings of the plasma membrane. These may be single or arranged into multiple parallel sheets, vesicles, or concentric layers (Figs. 1–2 and 1–3). The precise function of these internal membranes in bacteria is uncertain, but they undoubtedly serve as a framework for molecular systems having specialized functions in the cell. If suspected of a respiratory activity similar to the mitochondria of higher organisms, the membranes are sometimes called *chondroids* (from *chondriosome*, an archaic name for mitochondria); if a photosynthetic function is suspected, the membranes are termed *chromatophores* or *photosynthetic lamellae*. The pattern of photosynthetic membranes in the cytoplasm of blue-green algae may be highly complex (Fig. 1–3). Another type of internal membrane system appears as a spherical array of closely packed, concentric membranes (see Fig. 13–19). Although the function of this system of membranes, called the *mesosome* (*meso* = within, *some* = body) is unknown, it may be involved in division of the nucleoid or possibly in respiration.

1–3. Thin section of the blue-green alga Gleocapsa alpicola, *an example of a prokaryotic cell exhibiting a complex array of internal lamellae and other highly organized internal structures. The cell was fixed in glutaraldehyde-osmium; membranes appear in negative instead of the usual positive image.* Pl, *photosynthetic lamellae;* R, *ribosomes;* L, *polyphosphate granule;* G, *glycogen particles;* Pm, *plasma membrane;* Cw, *cell wall. The nucleus is not visible in this plane of section. × 112,000. Courtesy of M. Allen.*

No other structures are regularly seen in the cytoplasm of prokaryotes, except in a few specialized cells that may contain small vacuoles or deposits of lipid or glycogen (see Fig. 1–3). Certain motile bacteria exhibit an additional specialization at the surface of the cell, the *flagellum*. In living bacteria, these are massed together into bundles forming a sort of tail or oar that extends from the cell surface. When prepared for electron microscopy (Fig. 1–4), the flagellar bundles split into individual hair-like units about 100 Å wide. These individual threads consist almost entirely of multiple subunits of a single protein called *flagellin*. The protein subunits are linked together in rows that are twisted into a helix to form the shaft of the individual flagellum. Just how this simple structure accomplishes movement, or indeed whether it does, is a matter of some discussion at present among bacteriologists. The bacterial flagellum is fundamentally different from the much larger and more complex flagellum of a eukaryotic cell (see Chapter 15).

When compared to the eukaryotes, the cells of bacteria and blue-green algae are relatively simple in morphology. This apparent simplicity is deceptive, because these organisms are exceedingly complex in biochemistry. Some, like the photosynthetic bacteria and blue-green algae, are able to exist in an environment containing only inorganic salts and water. With these simple precursors and an input of energy in the form of light, the organisms are able to synthesize all of the nucleic acids, proteins, lipids, and carbohydrates required for existence. A chemical laboratory capable of manufacturing the same variety of compounds, if this were possible, would occupy many floors of a large building.

The prokaryotes are not given attention in separate chapters in this book. However, various sections deal with photosynthesis in prokaryotes (Chapter 7), the prokaryote nucleoid (Chapter 8), various aspects of protein synthesis in prokaryotes (Chapter 11), the possible origin of the mitochondria and chloroplasts of higher organisms from prokaryotes (Chapter 16), and the evolutionary origin of prokaryotic cells (Chapter 17).

The Generalized Eukaryotic Cell

The eukaryotic cell is surrounded, as is the prokaryotic cell, by a plasma membrane. In addition, the cell interior is divided into distinct regions by systems of

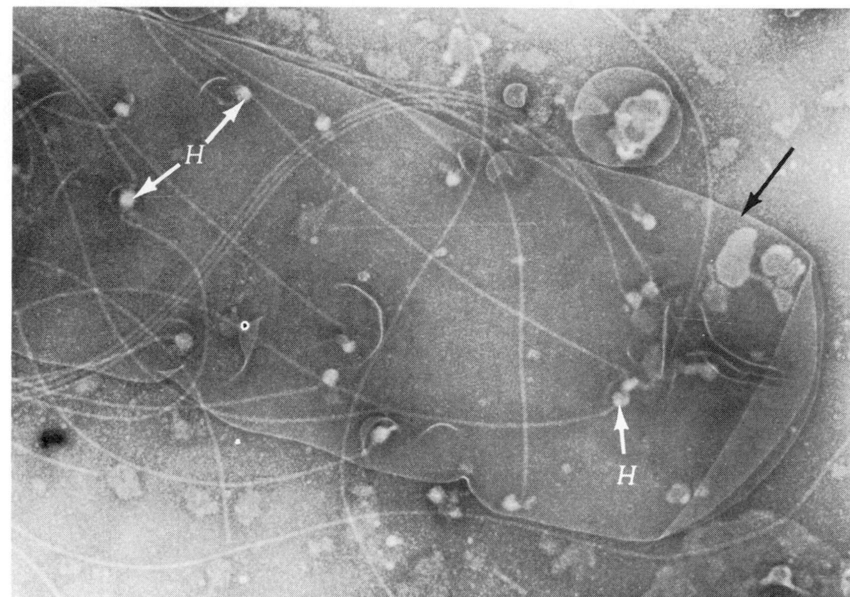

1–4. A portion of the cell wall (arrow) with attached flagellar fibers, isolated from the bacterium Proteus mirabilis. *Each fiber is about 100 Å wide. The spherical particles* (H) *in the figure are basal organelles. Each flagellum appears to arise from one of these structures.* × *60,500. Courtesy of W. Van Iterson and North-Holland Publishing Company, Amsterdam.*

internal membranes (see Figs. 1–5, 1–6, and 1–7). The most conspicuous of these regions, the nucleus, is separated from the surrounding cytoplasm by the continuous membranes of the nuclear envelope. Within the cytoplasm, additional membranous systems set off and define a number of discrete organelles: *mitochondria,* the *endoplasmic reticulum,* the *Golgi complex, vacuoles,* and smaller vesicular components of various sizes. These membranous organelles are present in all animal eukaryotes; in addition, the cells of plants contain *chloroplasts* and sometimes vacuoles of exceedingly large size.

The nucleus. Most of the interior of the nucleus (Figs. 1–5 and 1–6) is occupied by masses of thin fibrils. These fibrils, which are approximately 100 Å in diameter, make up the *chromatin* of the nucleus. The 100 Å fibers may either be sparsely distributed or clumped into dense aggregates. The more densely clumped aggregations are most frequently found just inside the nuclear envelope, often forming a shell of chromatin around the nucleus. In the more loosely aggregated areas, collections of granules or fibrils that are distinct in structure from the chromatin can sometimes be seen; these are identified simply as *extrachromosomal material* or *interchromatin granules* and *fibrils.* By various cytochemical tests, the 100 Å chromatin fibers have been shown to consist of DNA in combination with histone and possibly other proteins. At least some of the extrachromosomal material has been identified as an RNA-protein complex.

Within the nucleus, either suspended in the interior or lying close to the nuclear envelope, is a body of irregular outline and variable size, the *nucleolus.* Although no membranes separate the nucleolus from the rest of the nucleus, this structure is morphologically distinct and can be readily identified as a concentrated mass of dense granules and fibrils (Fig. 1–5) in the nucleoplasm. The granules of the nucleolus, often numbering in the thousands, are about 150 Å in diameter and somewhat indistinct or "fuzzy" in outline. These granular zones alternate with the masses of fibrils and spaces containing parts of the chromatin and nucleoplasm that extend into the substance of the nucleolus. The nucleolus contains large quantities of RNA and protein; by various indirect means this structure has been shown to be the site of the synthesis and assembly of the subunits of cytoplasmic ribosomes. The nucleolus is directly connected to a segment of the chromosomes called the *nucleolar organizer.* While this connection is not apparent in interphase cells, such as those shown in Figs. 1–5 and 1–6, the attachment becomes visible as the chromatin condenses into distinct chromosomes during division of the nucleus. In most cell types, a nucleolar organizer site is present on both members of at least one set of homologous chromosomes. These either remain separate at interphase to form multiple nucleoli or fuse to form a single large nucleolus.

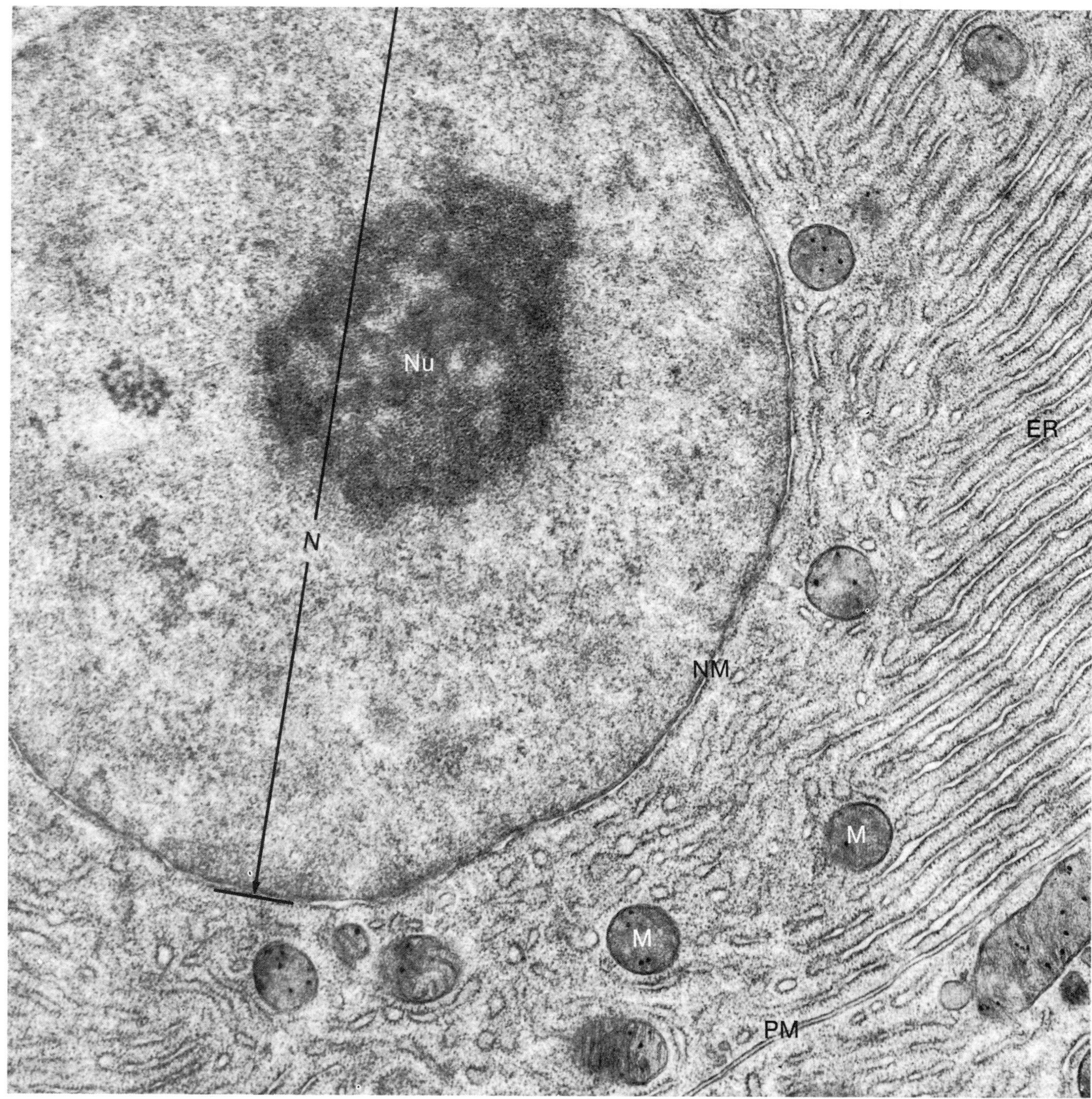

1–5. A cell from the pancreas of a bat. Many structures of the typical eukaryotic cell are visible in this micrograph. The nucleus (N) *contains more or less evenly distributed chromatin, visible in this thin section as very short lengths of 100 Å fibrils. The dark, granular mass suspended in the center of the nucleus is the nucleolus* (Nu). *The nucleus is separated from the surrounding cytoplasm by the nuclear envelope* (NM); *just outside the envelope are mitochondria* (M), *and elements of the rough endoplasmic reticulum* (ER). *A portion of the plasma membranes* (PM) *from this and the adjoining cell is visible at the lower right.* × *29,000. Courtesy of D. W. Fawcett and Wadsworth Publishing Company, Inc.*

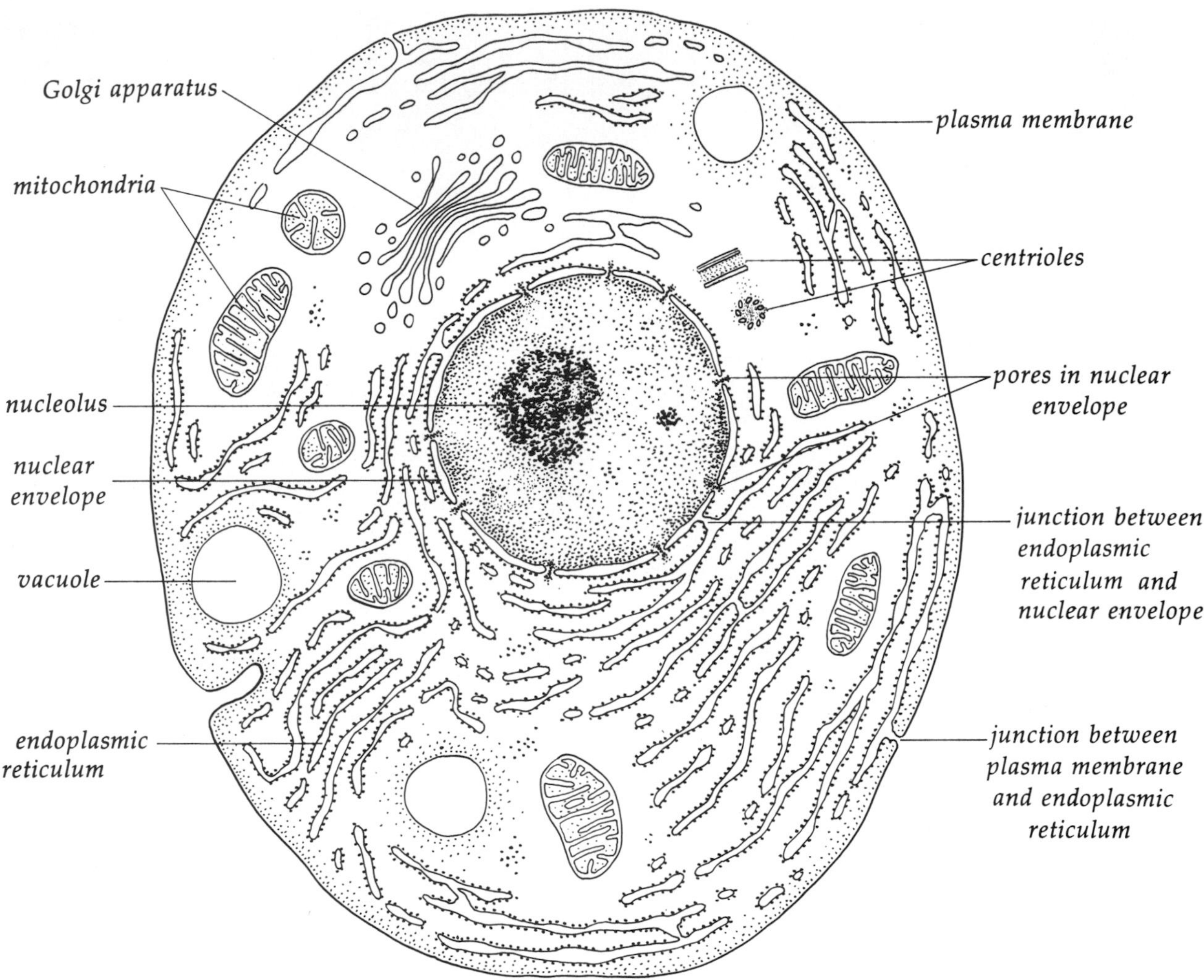

1–6. A diagram of a typical eukaryotic cell as seen in thin sections prepared for electron microscopy.

The membranous envelope separating the nucleus from the cytoplasm is a double system consisting of two concentric membranes (Figs. 1–5 and 1–6). The two membranes of the envelope are separated from each other by a narrow enclosed space 200–300 Å in diameter, the *perinuclear space.* At intervals, the envelope is interrupted by *pore complexes,* circular "holes" approximately 650–700 Å in diameter (see Figs. 10–17 and 10–23), filled in and surrounded by dense *annular material.* The annular material, an amorphous substance of unknown composition, evidently forms an effective barrier to diffusion of ions and larger molecules through the pores.

The nucleus and its functions are the subject of a large part of this text. The nucleus of the interphase cell is considered in detail in Chapter 8; the nucleus in division is the subject of Chapters 12, 13, and 14. The role of the nucleus and nucleolus in synthesis is considered in detail in Chapters 9, 10, and 11. A survey of the morphology and function of the nucleolus, the extrachromatin material, and the nuclear envelope is included in Chapter 10.

The cytoplasm. The cytoplasm of the eukaryotic cell is crowded with various organelles, most of which are defined by systems of enclosing membranes (Figs.

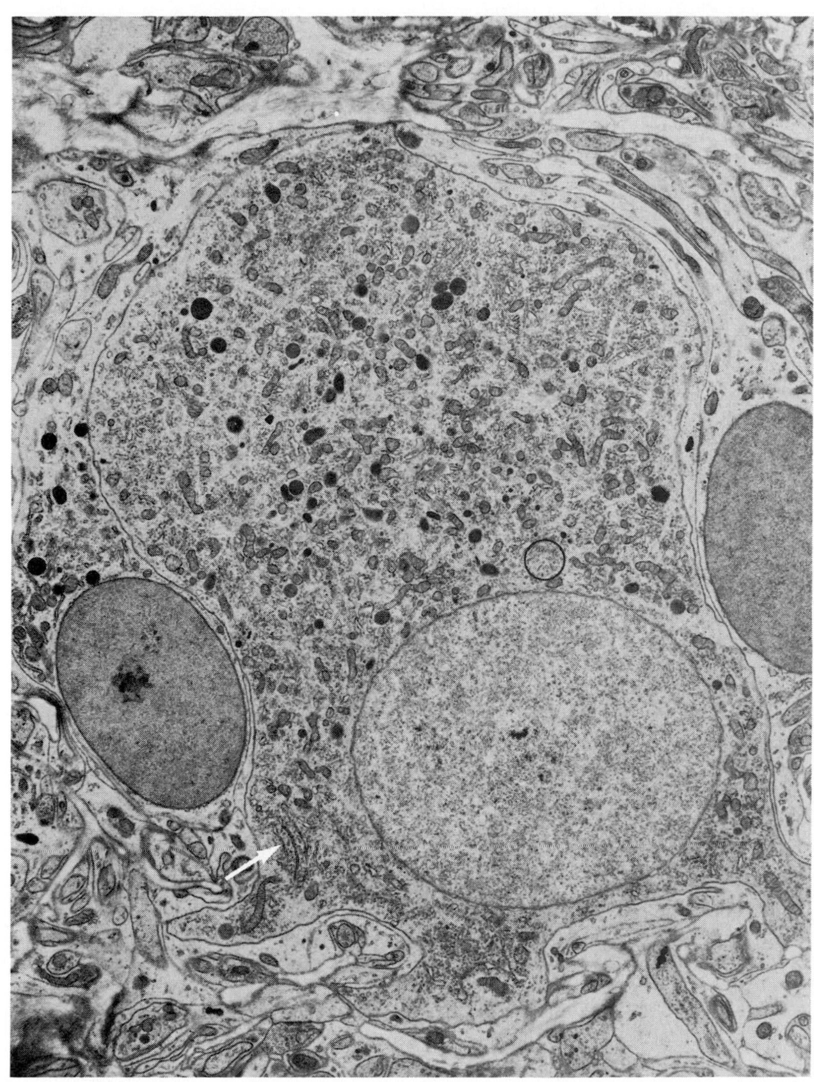

1–7. An example of a eukaryotic cell without a highly organized endoplasmic reticulum. Almost all of the ribosomes are freely suspended in the cytoplasm except at a few restricted locations. A few elements of rough endoplasmic reticulum (arrow) are visible toward the left of the cell, just outside the nucleus. The circled area contains the vesicular form of the Golgi complex. × 4,400. From An Atlas of Ultrastructure *by J. A. G. Rhodin. Courtesy of J. A. G. Rhodin and W. B. Saunders Company.*

1–5 and 1–6). The most prominent of these are the *mitochondria* (*mitos* = thread, *chondros* = grain), small spherical or elongate organelles about the size of a bacterium. Each mitochondrion consists of two separate membrane systems. The outer membrane, which surrounds and completely encloses the mitochondrion, is smooth and continuous. The inner membrane is thrown into folds or *cristae* that extend into the inner cavity of the mitochondrion, called the *matrix*. The mitochrondrial matrix is completely enclosed by the inner membrane. Between the inner and the outer mitochondrial membranes is another enclosed region called simply the "intermembrane space." Within the mitochondrial matrix are scattered ribosomes, one or two deposits of DNA closely similar in structure to the DNA fibrils of prokaryotes, and dense particles considerably larger than ribosomes called "intramitochondrial granules."

Mitochondria are large enough to be visible in the light microscope and were first described in the late 1800's by R. Altmann. At least several hundred mitochondria are found in most eukaryotic cells. However, some, such as a few protozoa, contain only one or two; others, such as the oocytes of amphibians, contain 100,000 or more. In the early years of the present century, good indications were obtained that mitochondria were intimately involved in cellular metabolism; finally,

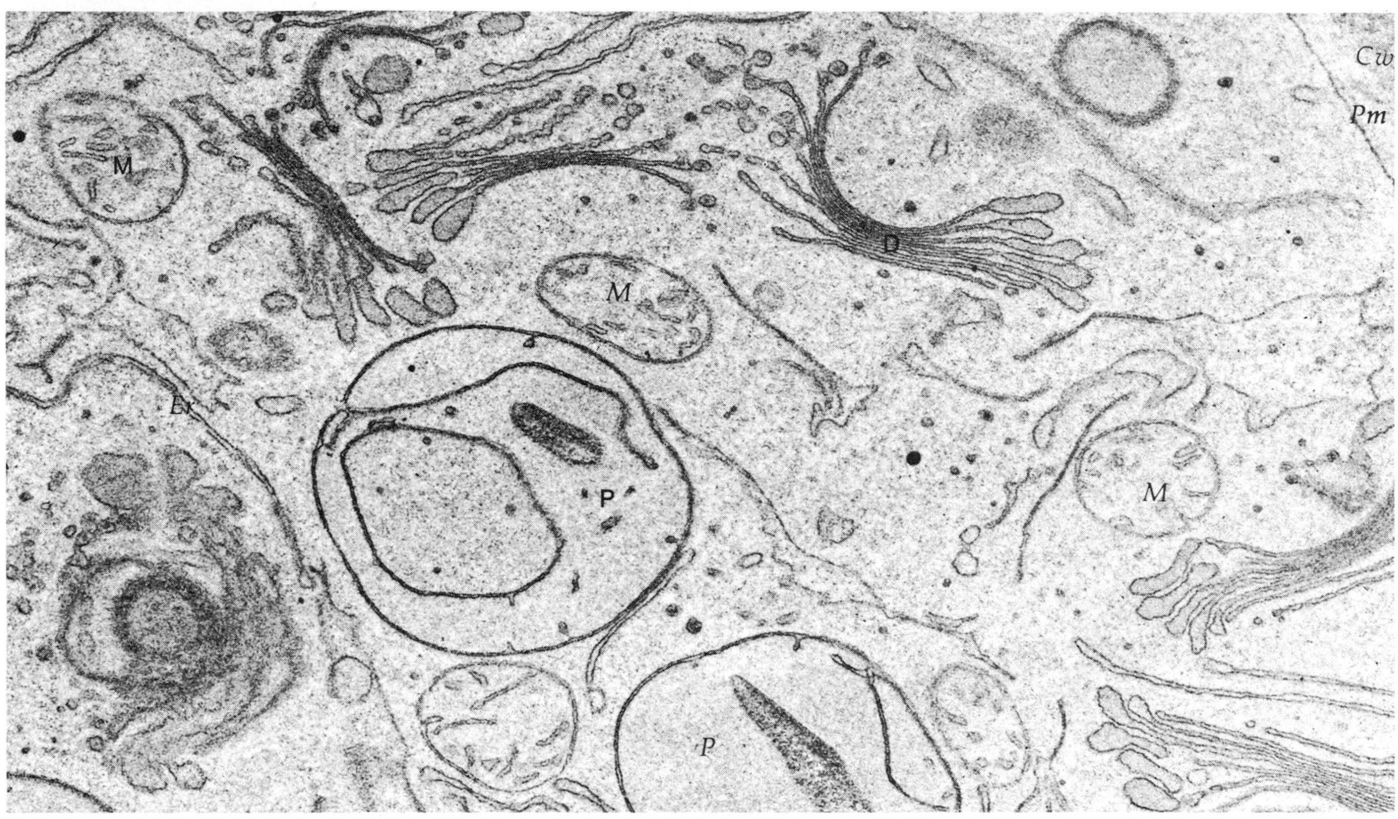

1–8. *Elements of the Golgi complex* (D) *in a cell from the root cap of corn. This configuration illustrates the more highly organized form of the Golgi complex. Note that the edges of the flattened sacs forming the central mass of each element of the complex are swollen into vesicular structures; at many points these can be seen to be separate from the sac membranes. Also visible are elements of the endoplasmic reticulum* (Er), *plastids* (P), *mitochondria* (M), *plasma membrane* (Pm), *and a portion of the cell wall* (Cw). *The curiously "empty" appearance of the cytoplasmic background substance is a characteristic feature of cells fixed in potassium permanganate, which does not preserve ribosomes or chromosomal structures.* × *30,000. Courtesy of W. Gordon Whaley and H. Mollenhauer and Wadsworth Publishing Company, Inc.*

in 1948, G. H. Hogeboom and his collaborators, working with isolated mitochondria, provided definitive evidence that these organelles are the site of oxidative respiration in the cell.

In recent years, the discovery that mitochondria contain DNA and ribosomes and are able to synthesize proteins has led to the hypothesis that they are at least partly independent from control by the nucleus. This hypothesis is supported by evidence from genetic crosses, which show that some mutations affecting the structure and function of mitochondria follow cytoplasmic rather than nuclear patterns of inheritance. That this genetic independence is only partial is indicated by the fact that most traits affecting the mitochondria follow the usual Mendelian patterns of inheritance and are linked to genes on the chromosomes. The ultrastructure and biochemistry of the mitochondrion are discussed in Chapter 6. In Chapter 16, attention is given to the nucleic acids, synthetic machinery, inheritance, and possible evolutionary origins of the mitochondria.

The cytoplasm also contains large numbers of ribosomes. These particles, with a diameter of 200–250 Å, are somewhat larger than the ribosomes of prokaryotes. Eukaryote ribosomes are either attached to systems of membranes (Figs. 1–5 and 1–6) or are freely suspended in the cytoplasm (Fig. 1–7). Most cells exhibit both patterns of distribution, with all possible variations between the extremes that are shown in Figs. 1–5 and 1–7. Where the ribosomes are attached, the supporting membranes form large, flattened vesicles called *cisternae.* The cisternae, with attached ribosomes, collectively make up the *rough endoplasmic reticulum.* The adjective "rough" in this term refers to the appearance given to the membranes by the ribosomes, which are attached to the outer, cytoplasmic surfaces of the cisternae. Other more tubular segments of a similar membrane system lack ribosomes and are called *smooth endoplasmic reticulum.* The cisternae of both types of endoplasmic reticulum enclose a space that is distinctly separate from the surrounding cytoplasm. This space can sometimes be seen to be continuous with the perinuclear space of the nuclear envelope and at other points to communicate with the cell exterior (diagramed in Fig. 1–6). These connections to the cell exterior are not often observed in animals and are very rarely seen in plants.

The ribosomes and membranes of the endoplasmic reticulum form a mechanism for the synthesis and transport of proteins. The protein, synthesized on the ribosomes, penetrates into the cisternal space, where it is concentrated and packed into vesicles that eventually become separate from the endoplasmic reticulum. No doubt, large quantities of protein are also released directly into the cytoplasm from both free and attached ribosomes without entering the cisternae of the reticulum.

The proteins synthesized in the endoplasmic reticulum have been shown in some cases to be transported to the Golgi complex, a system of flattened sacs and vesicles named for its original discoverer, C. Golgi, who first described the system in 1898. The complex may take the form of either a collection of small vesicles (visible in the cell in Fig. 1–7) or a more highly ordered array consisting of a stack of flattened membranous sacs (Fig. 1–8).[1] In addition to the secretion of vesicles containing proteins, the Golgi complex has also been implicated in the synthesis and secretion of polysaccharides and complexes containing both proteins and carbohydrates. The synthesis and transport of protein and other substances in the endoplasmic reticulum and Golgi complex are discussed in detail in Chapter 11.

The cytoplasm also contains other membranous sacs of indefinite function termed *vesicles* if small and *vacuoles* if larger. These characteristically enclose material of very low electron density. Other sacs enclosed by a single membrane are packed with material and appear very dense in the electron microscope. These are often classified under the rather nondescriptive heading of *microbodies.* Two types of microbodies, the *glyoxisomes* and *peroxisomes,* are known to be involved in oxidative metabolism in the cytoplasm. Another group of membrane-bound vesicles of this type, grouped under the generic term *lysosome,* contains enzymes capable of breaking down most substances of the cell. Lysosomes play a role in the elimination of cell organelles or in the "digestion" of materials penetrating from outside the cell. In other instances, lysosomes are secreted to the cell exterior; in cell death, they burst to release their contents into the surrounding cytoplasm. Lysosomes are apparently limited almost entirely to animal cells, although one report of a lysosome-like organelle in plant tissue exists (see p. 282).

Specializations of the cell surface. The plasma membrane of animal cells may be smooth, or it may be thrown into coarse folds or more regular finger-like projections called *microvilli* (Fig. 1–9). In some tissues, the regions of contact between cells may show localized differentiations of the adjacent plasma membranes called *desmosomes* (*desmos* = binding). In the region of the desmosome (visible in Fig. 1–9), the adjacent membranes appear slightly thickened and are separated by a very regular intercellular space 200–500 Å in diameter. On the cytoplasmic side of the thickened regions, dense fibers called *tonofibrils* or *tonofilaments* radiate from the membrane into the underlying cytoplasm. As the derivation of the name suggests, desmosomes are believed to serve as binding points between cells.

Specializations of the Cells of Plants

The eukaryotic plant cell (Figs. 1–10 and 1–11) contains all of the internal structures described in the

---

[1] Frequently, each individual collection of flattened sacs and associated vesicles is called a *dictyosome* (*dictyon* = net, *some* = body), and the terms "Golgi complex" or "Golgi apparatus" are reserved for the entire complement of dictyosomes within a single cell. However, in contemporary usage the term "Golgi complex" is applied both to single elements and to the whole collection of a cell, and for the sake of simplicity it will be used in this sense here.

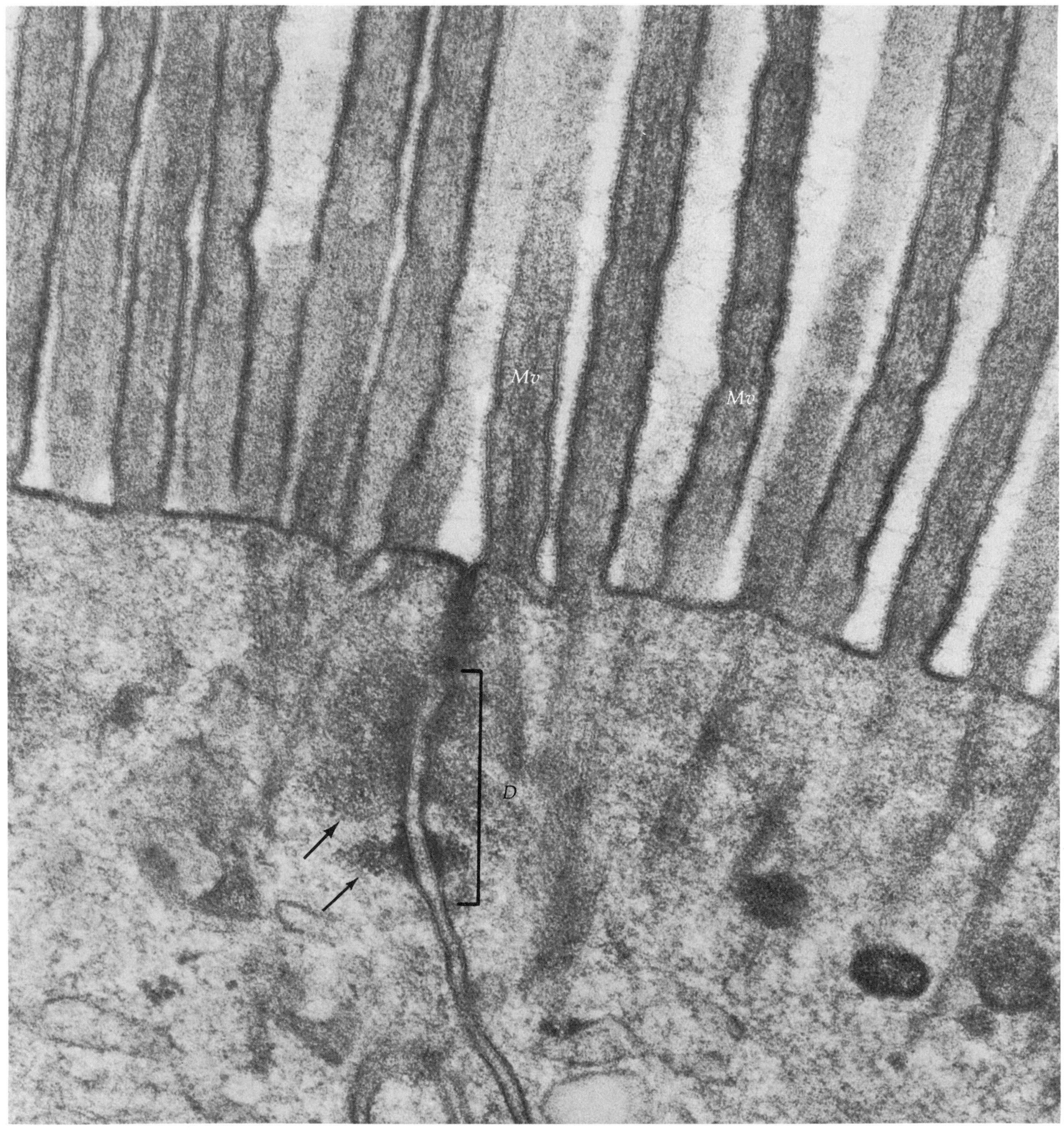
Mv
Mv
D

preceding section and in addition possesses a number of organelles and specialized structures not found in animals. The most conspicuous of these are the *cell wall,* the *plastids,* and large *vacuoles.*

The cell wall. With very few exceptions, the cells of all plants are surrounded by a cell wall (Fig. 1–10). The wall takes on a different form and composition in different groups of plants. In the vascular plants, the wall may either be semirigid and capable of extension (the *primary* wall) or very hard and nonextensible (the *secondary* wall). As the name suggests, the primary wall is the first complete wall laid down by a developing plant cell. The solid material of the primary wall, secreted through the plasma membrane, consists of long fibers of cellulose, which make up about 20–30 percent of the mass of the wall, variable quantities of pectin, and small amounts of other material. Surprisingly, the major constituent of the primary cell wall is water, which in some cases may make up more than 80 percent of the total mass.

Secondary walls, laid down as the cell ceases growth, are much harder than primary walls and give woody tissues their characteristic hardness, tensile strength, and flexibility. The principal differences in secondary walls are a greatly reduced water content, increased amounts of cellulose, and the presence of additional hardening substances such as lignin. Frequent interruptions, through which the *plasmodesmata* (protoplasmic interconnections between adjacent cells) are maintained, occur in both the primary and the secondary wall. Further details of the structure and synthesis of cell walls are included in Chapter 13.

Plastids. Various kinds of plastids are found in plant cells. In green, photosynthesizing tissue the characteristic plastid is the *chloroplast,* a membrane-bound organelle somewhat like the mitochondrion in ultrastructure (Fig. 1–11). The chloroplast is separated from the surrounding cytoplasm by two concentric membranes, the outer and inner membranes. The outer membrane is smooth and continuous. The inner membrane is closely applied to the outer membrane; the two together enclose a central region called the *stroma,* analogous in location to the matrix of the mitochondrion. Suspended within the stroma are complex systems of membranes that are not necessarily connected to the inner boundary membrane of the chloroplast. These systems, which in higher plants occur in interconnected, highly ordered stacks called *grana,* have been shown by isolation and biochemical analysis to contain chlorophyll and accessory pigments and to be the site of conversion of light to chemical energy. The stroma contains the enzymes and other substances involved in the use of this chemical energy to "fix" or reduce carbon dioxide and synthesize a variety of organic compounds. The stroma also contains large, densely staining granules called *osmiophilic granules,* clearly defined ribosomes (similar to bacterial ribosomes in structure and function), and scattered deposits of DNA, with structure and properties similar to prokaryote DNA. Like the mitochondrion, the chloroplast is at least partly independent of genetic control by the nucleus.

Other plastid types occur in both green and nongreen tissues of plants. In seedlings grown in the dark, plastids are colorless and contain a paracrystalline array of internal membranes. These structures, which are capable of differentiating into functional chloroplasts, are called *etioplasts.* Other plastids, packed with stored materials rather than photosynthetic membranes, are called *leucoplasts* (*leukos* = colorless or white). If the stored material is primarily starch, these plastids are sometimes called *amyloplasts.* At times, such as in ripening fruit or the development of fall colors, plastids are formed that contain large quantities of red and yellow pigments, primarily the carotenoids. These plastids, the *chromoplasts,* may either develop from chloroplasts originally containing chlorophyll or may appear as separate plastid types never containing chlorophyll at any stage. In all cases, the various kinds of plastids probably represent different developmental responses of the same organelle to differing environmental conditions in the plant cell. The structure and function of the chloroplast and the development of the various plastid types is included in Chapter 7. Other features of the plastid, including the nucleic acid component and a consideration of the possible origins and the control of inheritance in plastids, are discussed in detail in Chapter 16.

1–9. *Specializations of the cell surface at the boundary region between two intestinal cells. Microvilli* (Mv) *are visible at the top of the picture. A desmosome* (D) *with tonofilaments (arrows) can be seen along the line of juncture of the two cells.* × *111,000. Courtesy of T. S. LeCount.*

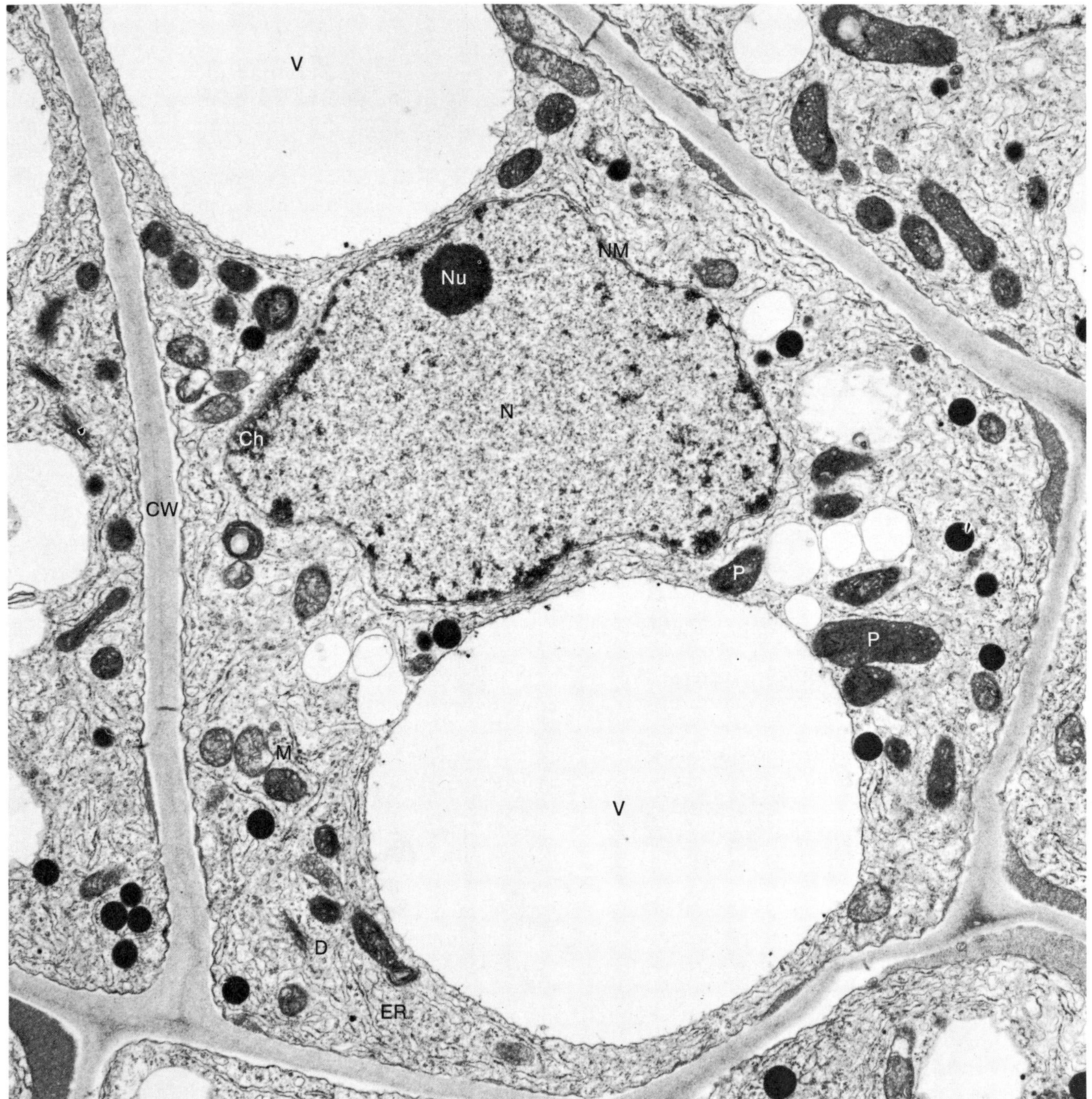

1–10. The ultrastructure of a cell from a higher plant, a parenchyma cell from the developing fruit of cotton. The nucleus (N) *shows some localized aggregation of chromatin fibers* (Ch) *just inside the nuclear envelope* (NM). *A portion of the nucleolus* (Nu) *is present inside the nucleus. Two large vacuoles* (V) *take up much of the volume of the cell. Also visible within the cell are plastids* (P), *mitochondria* (M), *elements of the endoplasmic reticulum* (ER) *and the Golgi complex* (D), *and numerous dense, rounded bodies of unknown identity. Outside the plasma membrane is a conspicuous cell wall* (CW), *which completely surrounds the cell. Glutaraldehyde-osmium fixation.* × *12,800. Courtesy of W. A. Jensen and Paula Stetler and Wadsworth Publishing Company, Inc.*

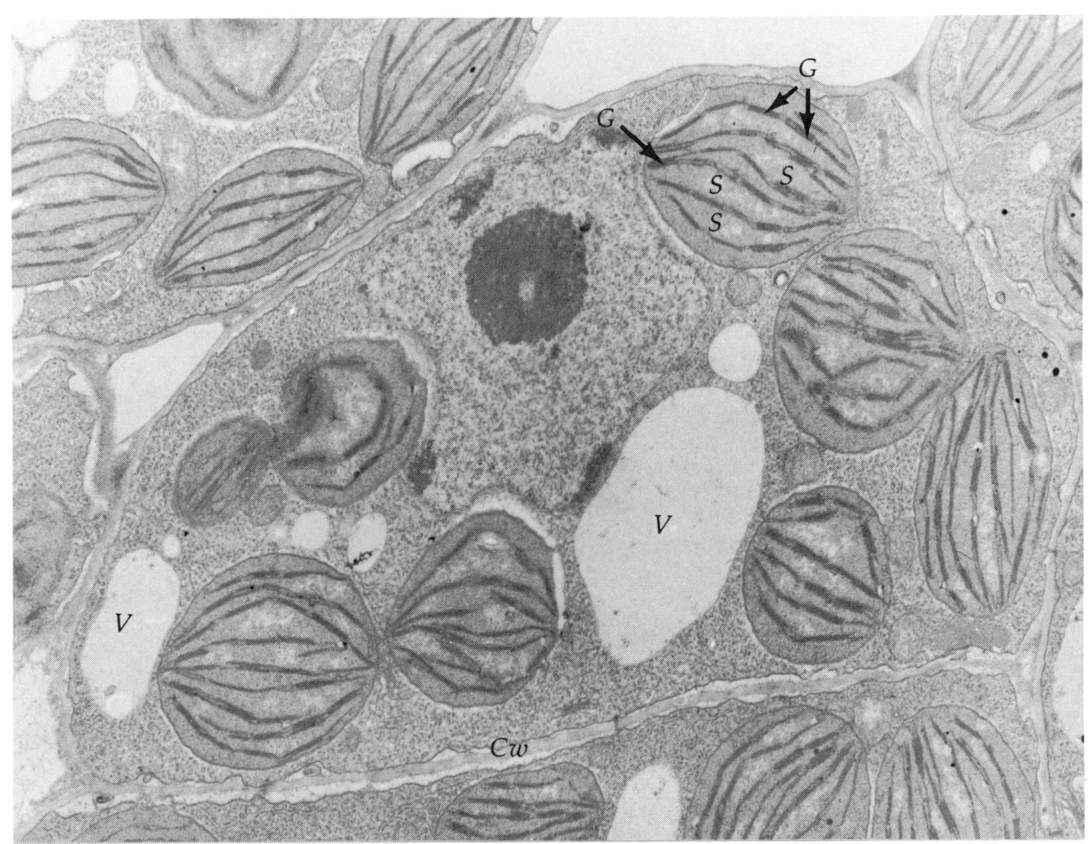

1–11. A cell from a bean seedling (Phaseolus vulgaris) *containing a number of large chloroplasts clearly showing internal differentiation into stroma* (S) *and grana* (G). *Note also the prominent vacuoles* (V) *and the cell wall* (Cw). *× 6,800. Courtesy of T. E. Weier and John Wiley & Sons, Inc.*

Vacuoles. The vacuoles of plant cells are similar to the membrane-bound vesicles of animal cells except for the extreme size attained by these structures in some plant cells (Fig. 1–10). Often, in mature plant tissues, most of the cell's volume is occupied by one or more large vacuoles; in these cells the cytoplasm and nucleus are relegated to a thin layer coating the inside of the cell wall. In many instances, the enlargement of individual cells during the growth of the plant is accomplished primarily by increases in the size of the vacuole. Typically, the vacuole exhibits only a slight electron density in thin sections, although dense inclusions occur occasionally. On analysis, vacuoles prove to contain dilute solutions of molecules such as inorganic salts, organic acids, sugars, various pigments, and small quantities of soluble proteins.

A Summary of Prokaryote and Eukaryote Structure

Prokaryotes differ from eukaryotes both in size and in the complexity of internal morphology. The prokaryotic cell has only one major system of membranes, the plasma membrane. While inner membranes may be present, these are considered to be derived from the plasma membrane. The prokaryote nucleus is relatively simple in structure and is not separated from the surrounding cytoplasm by membranes. Although ribosomes are present in prokaryote cytoplasm, none of the other organelles characteristic of the eukaryote are present. Eukaryotes, in contrast, show a proliferation of internal membrane systems. These primarily cytoplasmic systems delimit the mitochondria, chloroplasts, endoplasmic reticulum, Golgi complex, and diverse membrane-bound vesicles, vacuoles, and microbodies. The nucleus is also separated from the cytoplasm by a highly differentiated membranous envelope. Within the nucleus is a conspicuous nucleolus, chromatin fibers of a greater diameter than prokaryote nuclear fibers, and interchromosomal granules and fibrils. The more prominent differences between prokaryotes and eukaryotes are summarized in Table 1–2.

Table 1–2 A Comparison of Prokaryote and Eukaryote Cells

	Prokaryotes	*Eukaryotes*	
		Animals	*Plants*
Plasma membrane	+	+	+
Ribosomes	+	+	+
Nucleus	+*	+	+
Nuclear membranes	—	+	+
Nucleolus	—	+	+
Interchromosomal material	—	+	+
Mitochondria	—	+	+
Endoplasmic reticulum	—	+	+
Golgi complex	—	+	+
Plastids	—	—	+

*Not set off from the cytoplasm by membranes.

Plan of this Book

It is obvious from the foregoing sections that membranes form the primary framework of eukaryotic cells. The first unit of the book, including Chapters 2–4, describes the molecules from which membranes are structured (lipids and proteins), the primary techniques for studying membranes, and the current status of membrane biology. The next unit of the book takes up the role of the mitochondrion and chloroplast membrane systems in the provision of energy for synthesis in the cell. This unit includes chapters on bioenergetics (5), the mitochondrion (6), and the chloroplast (7). This is followed by a unit on synthesis in the cell, in which the interphase nucleus and its role in synthesis are described, followed by a survey of protein synthesis in the cytoplasm (Chapters 8–11). The final section of the book considers the division and origin of cells, including mitosis (12 and 13), meiosis (14), gametogenesis and fertilization (15), cytoplasmic inheritance (16), and the evolutionary origins of the cell (17).

Suggestions for Further Reading

Hughes, A., 1959. *A History of Cytology*. Abelard-Schuman, Ltd., New York.

Lima-de-Faria, A., ed., 1970. *Handbook of Molecular Cytology*. North-Holland Publishing Company, Amsterdam.

2
The Structure of Lipids and Proteins

The special environment of the cell interior is maintained by the plasma membrane; all substances passing in and out of the cell are regulated by this surface layer of lipid and protein. If the plasma membrane is ruptured so extensively that continuity is not reestablished almost instantaneously, the cell quickly dies. In eukaryotic cells, and to a lesser extent in prokaryotes, internal functions are also regulated and defined by membranes. In eukaryotes, the nucleus, mitochondria, chloroplasts, endoplasmic reticulum, Golgi complex, and numerous vesicular components are all membranous structures. In each of these organelles, the system of continuous membranes maintains an internal environment separate from the surrounding cytoplasm and serves as a structural framework supporting molecular systems that carry out a great diversity of biological functions.

Direct measurement shows that the membranes defining the borders of the cell and the internal organelles cannot be more than a few molecules thick. Observations of living cells reveal that these membranes are highly flexible and can undergo very rapid changes in shape and surface area. This is true not only of the plasma membrane, which undergoes swift alterations in contour and area in cells such as the amoeba, but also of internal organelles such as the mitochondria, which have been observed in living cells to change shape, divide, and fuse with surprising rapidity. Other observations show that in spite of their thinness and flexibility, membranes possess considerable strength and are able to resist mechanical disruption. For example, when membranous organelles such as mitochondria and chloroplasts are isolated, they remain intact through disruptive procedures such as grinding and centrifugation at high speeds.

These remarkable properties are functions of a complex between two major classes of biochemical molecules, the lipids and the proteins. To understand something of the nature and function of membranes—to which the first two units of this book are in large part devoted—it is first necessary to survey the molecular structure of these lipid and protein building blocks.

Lipids

Lipids are a heterogeneous group of substances that have never been successfully categorized under a single, all-inclusive description. Classically, lipids were defined as biological substances soluble in organic solvents such as acetone, alcohol, ether, chloroform, and benzene and only slightly soluble or insoluble in water. More recently, the lipids have been described as esters[1] of alcohols and fatty acids and closely related derivatives of these esters. Even this description is somewhat vague, and authorities still do not agree on the substances to be included as lipids. The problem is simplified if the discussion is limited to the major types of lipids extractable from membranes—*neutral lipids* (fats and oils), *phospholipids*, and *sterols*.

The Neutral Lipids

Almost all of the neutral lipids (so called because they bear no charged groups at physiological *p*H) consist of esters of the alcohol *glycerol* with various long-chain fatty acids. Glycerol (Fig. 2–1) is a molecule with three sites available for the formation of ester linkages

$$\begin{array}{c} \mathrm{H} \\ | \\ \mathrm{H{-}C{-}OH} \\ | \\ \mathrm{H{-}C{-}OH} \\ | \\ \mathrm{H{-}C{-}OH} \\ | \\ \mathrm{H} \end{array}$$

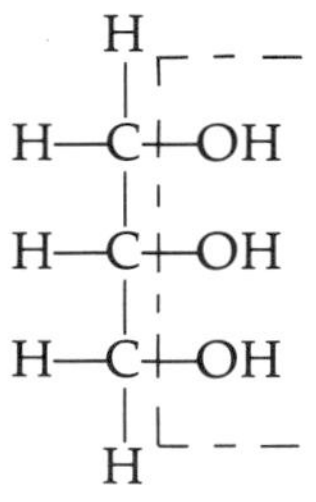

2–1. The alcohol glycerol. Three sites, enclosed by the dotted lines, are available for the formation of ester linkages.

[1] An ester linkage, formed by the interaction between an acid and an alcohol, is characterized by the carbonyl ($\mathrm{R{-}\overset{\displaystyle O}{\overset{\|}{C}}{-}O{-}R}$) group.

with fatty acids (indicated by the dotted lines in Fig. 2–1). Depending upon whether one, two, or three of these sites are bound to fatty acids, the lipid is termed a *mono-*, *di-*, or *triglyceride*.

The fatty acids (Fig. 2–2 and Table 2–1) consist of long, straight chains of carbon atoms. A carboxyl

Table 2–1 Some Fatty Acids of Natural Triglycerides

Saturated fatty acids	
Butyric acid	$CH_3(CH_2)_2CO_2H$
Caproic acid	$CH_3(CH_2)_4CO_2H$
Caprylic acid	$CH_3(CH_2)_6CO_2H$
Capric acid	$CH_3(CH_2)_8CO_2H$
Lauric acid	$CH_3(CH_2)_{10}CO_2H$
Myristic acid	$CH_3(CH_2)_{12}CO_2H$
Palmitic acid	$CH_3(CH_2)_{14}CO_2H$
Stearic acid	$CH_3(CH_2)_{16}CO_2H$
Arachidic acid	$CH_3(CH_2)_{18}CO_2H$
Lignoceric acid	$CH_3(CH_2)_{22}CO_2H$
Unsaturated fatty acids	
Crotonic acid	$CH_3CH{=}CHCO_2H$
Palmitoleic acid	$CH_3(CH_2)_5CH{=}CH(CH_2)_7CO_2H$
Oleic acid	$CH_3(CH_2)_7CH{=}CH(CH_2)_7CO_2H$
Linoleic acid	$CH_3(CH_2)_3(CH_2CH{=}CH)_2(CH_2)_7CO_2H$
Linolenic acid	$CH_3(CH_2CH{=}CH)_3(CH_2)_7CO_2H$
Arachidonic acid	$CH_3(CH_2)_3(CH_2CH{=}CH)_4(CH_2)_3CO_2H$
Nervonic acid	$CH_3(CH_2)_7CH{=}CH(CH_2)_{13}CO_2H$

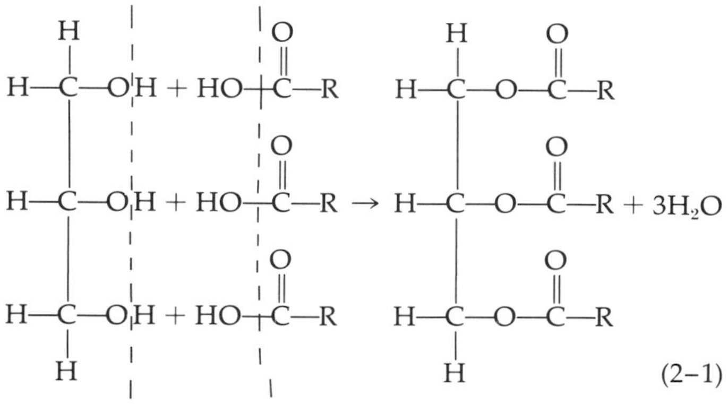

(—COOH) group at one end gives the molecule its acidic properties. In the combination between fatty acids and glycerol producing a triglyceride, an ester linkage is formed between the alcoholic groups of the glycerol and the carboxyl groups of the fatty acids:

$$\begin{array}{l} \mathrm{H} \\ | \\ \mathrm{H{-}C{-}OH + HO{-}\overset{O}{\overset{\|}{C}}{-}R} \\ | \\ \mathrm{H{-}C{-}OH + HO{-}\overset{O}{\overset{\|}{C}}{-}R} \\ | \\ \mathrm{H{-}C{-}OH + HO{-}\overset{O}{\overset{\|}{C}}{-}R} \\ | \\ \mathrm{H} \end{array} \rightarrow \begin{array}{l} \mathrm{H} \\ | \\ \mathrm{H{-}C{-}O{-}\overset{O}{\overset{\|}{C}}{-}R} \\ | \\ \mathrm{H{-}C{-}O{-}\overset{O}{\overset{\|}{C}}{-}R} \\ | \\ \mathrm{H{-}C{-}O{-}\overset{O}{\overset{\|}{C}}{-}R} \\ | \\ \mathrm{H} \end{array} + 3H_2O \qquad (2\text{–}1)$$

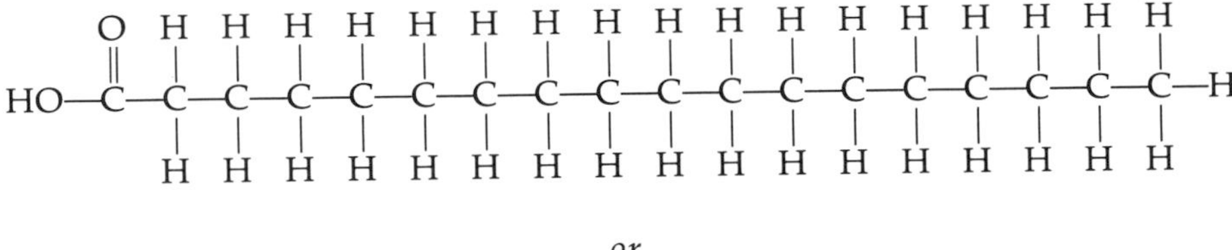

or

$CH_3(CH_2)_{16}COOH$

2–2. *Stearic acid, an example of a fatty acid. The carboxyl (—COOH) group gives the molecule its acidic properties. Because the maximum possible number of hydrogen atoms are bound into the chain, this fatty acid is said to be saturated.*

One molecule of water is eliminated for each ester linkage formed. The "R" in equation 2–1 represents the long terminal chain of carbon atoms on each fatty acid.

Triglycerides (Fig. 2–3) are much more commonly found in nature than mono- or diglycerides. The three fatty acid residues linked to glycerol may all be the same, as in the example given in Fig. 2–3, or each may be different. Because a large number of different fatty acids combine with glycerol to form triglycerides, many permutations of the basic triglyceride format are known. In the triglycerides of fish, for example, at least 20 different fatty acids are found in various combinations with glycerol, giving more than 1500 possible permutations. Individual species usually have distinctive mixtures of fatty acids in their triglycerides; these are somewhat labile and change to a degree depending on the diet of the organism.

The various naturally occurring fatty acids are almost exclusively unbranched chains containing an even number of carbon atoms. Although a few with odd numbers are found in all organisms, these make up only a minor fraction of the total. The chains vary in length from as few as four carbons (in butyric acid) to much longer structures containing 24 or more carbons. Within the chains, hydrogen atoms are bound to the carbon atoms that make up the backbone of the molecule. If the maximum possible number of hydrogen atoms are bound into the chain, the fat is said to be *saturated*. The molecule of stearic acid shown in Fig. 2–2 represents a fully saturated fatty acid. In other

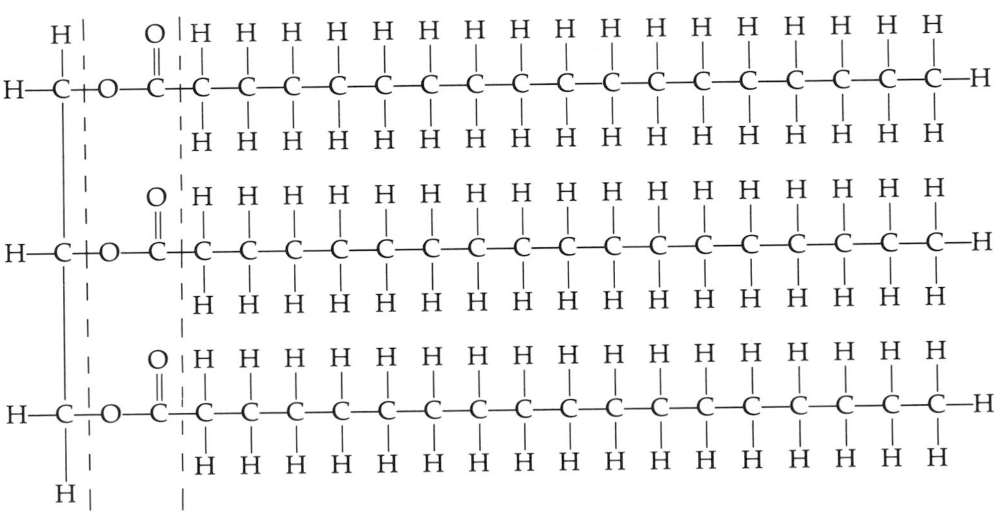

2–3. *Tristearin, a triglyceride consisting of a glycerol residue bound to three chains derived from stearic acid. The dotted lines enclose the ester linkages between the glycerol and the three stearic acid residues.*

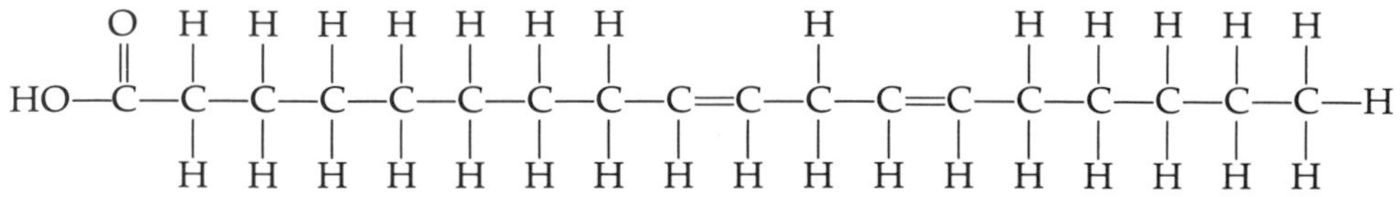

```
     O  H  H  H  H  H  H  H        H        H  H  H  H  H
     ‖  |  |  |  |  |  |  |        |        |  |  |  |  |
HO—C—C—C—C—C—C—C—C—C=C—C—C=C—C—C—C—C—C—H
        |  |  |  |  |  |  |  |  |  |  |  |  |  |  |  |  |
        H  H  H  H  H  H  H  H  H  H  H  H  H  H  H  H  H
```

2–4. *Linoleic acid, a doubly unsaturated fatty acid.*

fatty acids, hydrogen atoms may be absent from adjacent carbons in the interior of the chain. These carbons share a double bond instead of a single bond (Fig. 2–4), and the molecule is said to be *unsaturated.* This arrangement may occur at multiple points in the interior of the chain, producing the so-called *polyunsaturated* fatty acids. Examples of common saturated and unsaturated fatty acids are given in Table 2–1.

Phospholipids

The triglycerides, primarily "depot" or storage fats, are found in membranes in only very small quantities. Of much greater importance in membrane structure are the *phospholipids.* These lipids are similar to the triglycerides in form, except that a phosphate-containing group is substituted for one of the three fatty acids. All of the naturally occurring phospholipids appear to be derived from *glycerophosphoric acid* (Fig. 2–5), in which a phosphate group is bound to one of the "end" carbons (designated the α-carbon) of glycerol. Although an alternate glycerophosphoric acid, with the phosphate bound to the "middle" or β-carbon of glycerol, appears spontaneously in preparations of glycerophosphoric acid, this form, called β-glycerophosphoric acid, apparently does not occur in nature. Phospholipids are derived from α-glycerophosphate by esterification of the two remaining alcohol groups of the glycerol with fatty acids (Fig. 2–5).

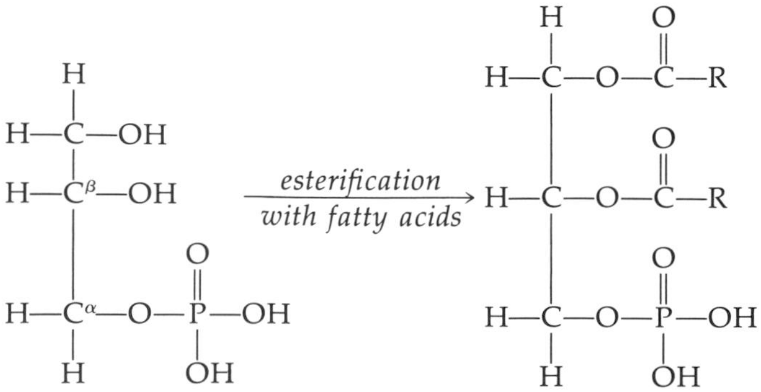

```
                                            H      O
     H                                      |      ‖
     |                                   H—C—O—C—R
H—C—OH                                      |
     |                                      |      O
H—Cβ—OH      esterification                |      ‖
     |       ———————————————→            H—C—O—C—R
     |       with fatty acids               |
     |      O                               |      O
     |      ‖                               |      ‖
H—Cα—O—P—OH                              H—C—O—P—OH
     |      |                               |      |
     H      OH                              H      OH
α-glycerophosphoric acid               a phospholipid
                                    (a phosphatidic acid)
```

2–5. *The formation of phosphatidic acid from α-glycerophosphoric acid.*

This most simple form of the phospholipid is rarely found in living tissues; a number of authorities consider that even the limited quantities detected are artificial breakdown products of a slightly more complicated class of phospholipids, the *phosphatidyl esters.* In these molecules, the phosphate group, linked to the α-carbon of the glycerol, is in turn linked to a group of other substances containing nitrogen (Fig. 2–6). In Fig. 2–6a, the phosphate is linked to the nitrogen-containing alcohol choline; this forms the phospholipid *lecithin,* a major component of cellular membranes. The closely related substances ethanolamine and serine are also commonly found linked to the phosphate group in the same manner, producing membrane phospholipids similar in properties to lecithin (see Fig. 2–6b). The latter two phospholipids are so close in properties that for many years they were considered together under the name "cephalin," a lipid mixture extractable from nerve tissue. In other important phospholipids, inositol (Fig. 2–6b) or glycerol may be added to the phosphate group, forming *phosphatidyl inositol* and *phosphatidyl glycerol.*

These phospholipids each represent a *family* of substances, as different fatty acids may be attached to the other two binding groups of the glycerol. For example, the phospholipid lecithin represents a family of closely related lipids, all with phosphate and choline attached to the α-carbon, but with different permutations of a group of fatty acids bound to the other carbons. In humans, the lecithin of membranes probably consists of at least 20 different molecules.

Phospholipids have affinities for both polar and nonpolar solvents. If a nonpolar solvent, such as benzene, is layered on top of water, a strongly polar solvent, the phospholipid molecules will become aligned in the

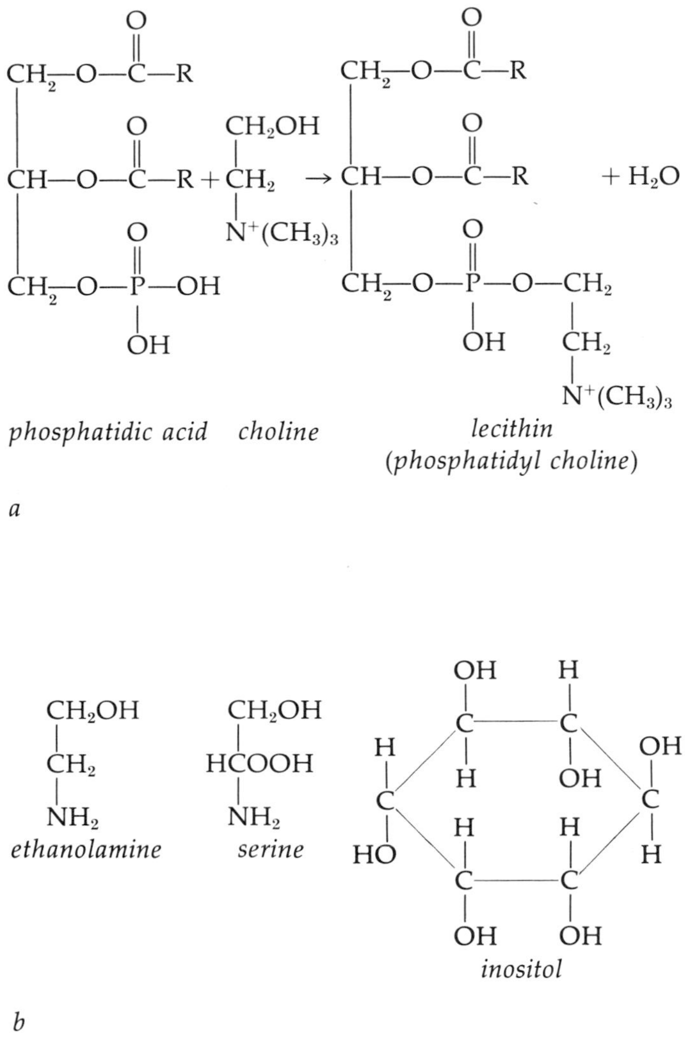

$$\begin{array}{l} CH_2{-}O{-}C({=}O){-}R \\ CH{-}O{-}C({=}O){-}R \\ CH_2{-}O{-}P({=}O)(OH){-}OH \end{array} + \begin{array}{l} CH_2OH \\ CH_2 \\ N^+(CH_3)_3 \end{array} \rightarrow \begin{array}{l} CH_2{-}O{-}C({=}O){-}R \\ CH{-}O{-}C({=}O){-}R \\ CH_2{-}O{-}P({=}O)(OH){-}O{-}CH_2{-}CH_2{-}N^+(CH_3)_3 \end{array} + H_2O$$

phosphatidic acid *choline* *lecithin (phosphatidyl choline)*

a

$CH_2OH{-}CH_2{-}NH_2$ *ethanolamine*

$CH_2OH{-}HCOOH{-}NH_2$ *serine*

Inositol ring: six C atoms, substituents OH, H; H, OH; OH, H; H, OH; H, OH; HO, H. *inositol*

b

2–6. a, *The formation of lecithin, a phospholipid, from a combination between phosphatidic acid and choline.* b, *Examples of other substances that may bind to the phosphate group of phosphatidic acid to form a phospholipid.*

interface between the two solvents, with the long hydrocarbon chains of the phospholipid extended into the nonpolar solvent and the phosphate groups associated with the water (see Fig. 2–7). Thus, one end of the molecule is nonpolar or *hydrophobic* in activity, and the other end is polar or *hydrophilic*. These affinities are diagramed by a standard format, using a line for the hydrophobic, nonpolar portion of the molecule and a circle for the hydrophilic, polar end (Fig. 2–7b). This dual property of the phospholipid molecule forms the basis for all current hypotheses of membrane structure, which consider that the interior of the membrane is nonpolar; in this region, the long hydrocarbon chains of the fatty acid residues of membrane phospholipids are associated together. The phosphate groups are believed to be directed toward the outside surfaces of the membrane, which face the polar, aqueous environment of the cytoplasm or cell exterior.

Phospholipids added to water may spontaneously form into layers two molecules thick called *bilayers* (Fig. 2–7d). In bilayers, the hydrophobic ends of the phospholipid molecules are associated in the interior; the polar ends face the aqueous medium. These films of phospholipid exhibit many of the properties of cellular membranes. (The formation of bilayers, evidence for their structure, and bilayer properties are discussed in more detail in Chapter 4.)

Sterols

The *sterols* are part of a larger group of biological molecules, the *steroids*, which all have a common derivation from a system of four interconnected carbon rings. In the sterols (Fig. 2–8a), one end of this system contains a hydroxyl group; to the opposite end of the molecule is bound a side chain that takes different forms in the various sterols. The most important of these sterols is *cholesterol* (Fig. 2–8b), a constituent of the membranes of all animals. Varying quantities of other sterols are found in mixtures in the membranes of higher plants, algae, and fungi. As is obvious, the sterols have no structural relationship to the neutral lipids or phospholipids; these substances are classified as lipids mainly because they behave similarly in organic solvents. In membranes, cholesterol is believed to form a complex with membrane phospholipids. In this complex, the cholesterol lies in the nonpolar interior of the membrane in association with the long hydrocarbon chains of the phospholipid molecules (see Fig. 4–2).

These components—the phospholipids, sterols, and very limited quantities of triglycerides—make up the lipid component of the membrane. Further details of the role of these substances in the structure and function of membranes will be taken up in Chapter 4.

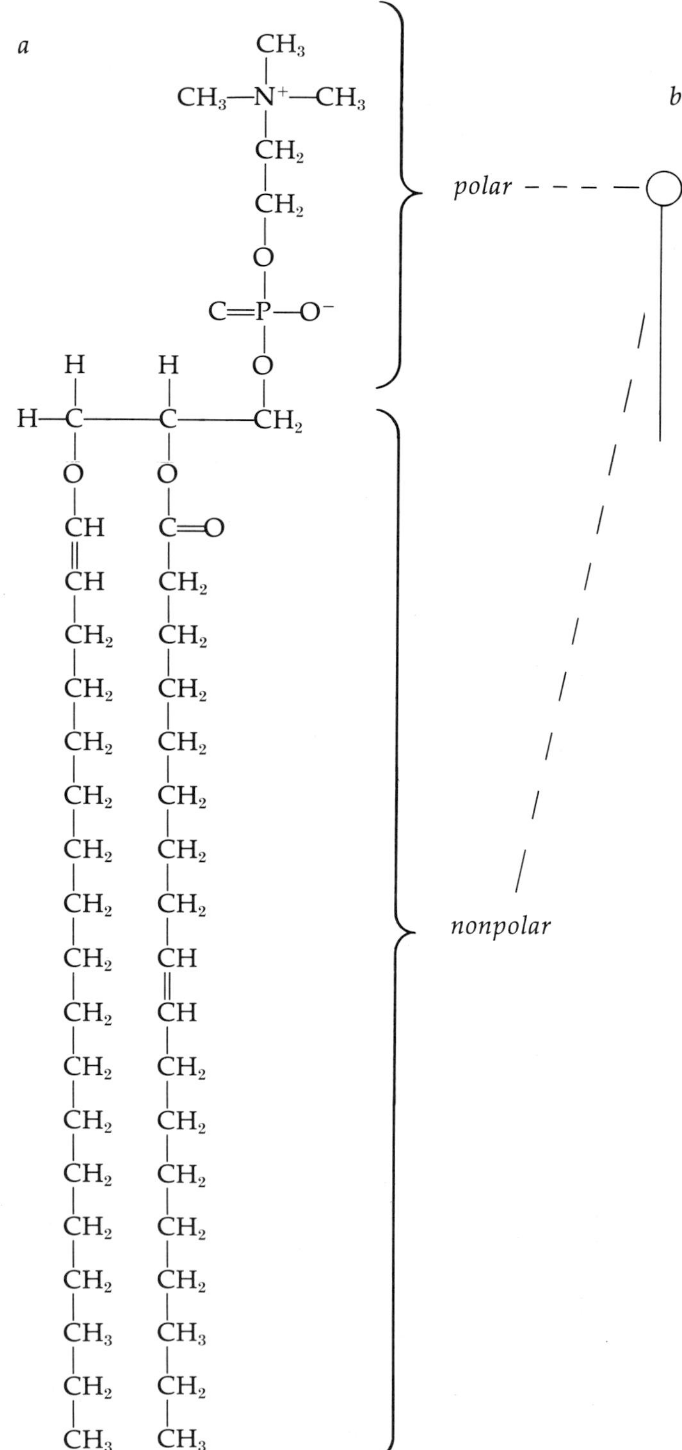

c

benzene

water

d

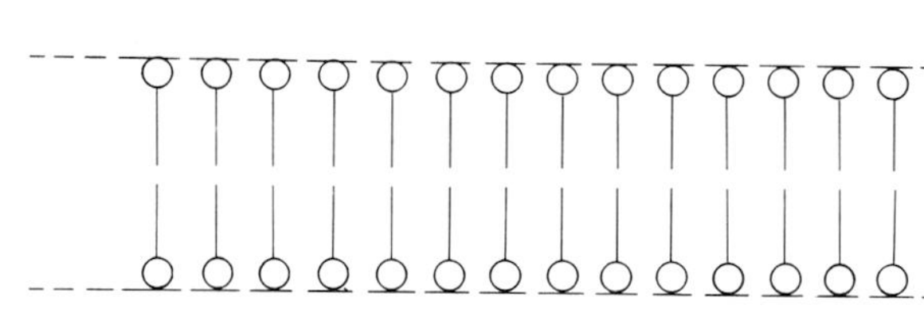

2–7. a, *A phospholipid, showing the polar (hydrophilic) and nonpolar (hydrophobic) portions of the molecule.* b, *The standard diagram of a phospholipid molecule, in which the small sphere designates the polar end and the line the hydrophobic, nonpolar chains.* c, *The alignment of phospholipid molecules at a benzene-water interface.* d, *A bilayer of phospholipids in aqueous medium. The polar regions of the molecules are directed toward the medium and the nonpolar chains associate in the interior of the bilayer.*

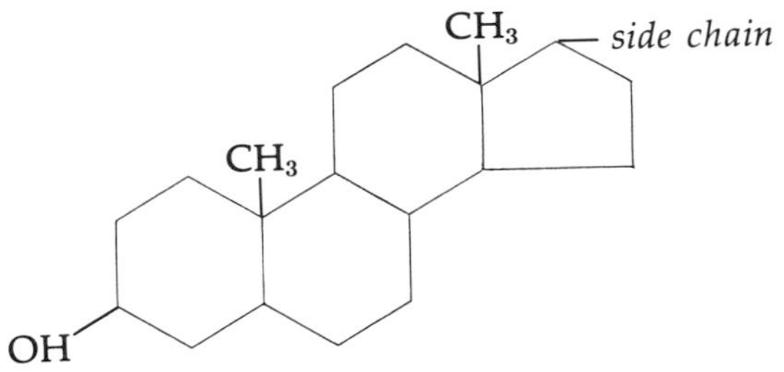

a 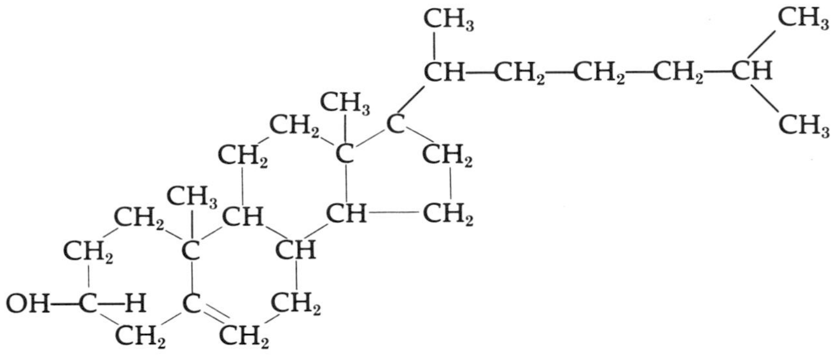 *b*

2–8. a, *A general plan of the sterols.* b, *Cholesterol, an important constituent of the membranes of animal cells.*

Proteins

Proteins occur in an astonishing variety of forms, with an equally astonishing diversity of functions in the cell. In this unit, proteins are considered from the standpoint of their structural role in membranes; however, other functions, such as enzymatic catalysis (see Chapter 5), antigen-antibody interactions, electron transport in respiration and photosynthesis (Chapters 6 and 7) are also carried out by proteins; the structural and functional basis of contractile systems such as muscle also reside in systems of protein molecules. In other structures, such as the chromosomes and ribosomes, proteins, in combination with DNA and RNA, also provide a supporting framework and fill an enzymatic role.

In spite of this diversity in function, all proteins are basically similar in primary structure and consist of one or more long, unbranched chains of subunits. The individual subunits, or links in the chain, are the amino acids. Only 20 different amino acids occur in various sequences in different proteins; the three-dimensional configuration and biochemical properties of each protein are determined by this sequence. Because, in different proteins, any amino acid is an equally probable link at any point in the interior of the chain, the fact that only 20 amino acids occur places no significant restriction on the total possible number of distinct proteins in nature. R. E. Dickerson (Dickerson and Geis, 1969) has estimated that for chains of amino acids 61 subunits in length, equivalent to some of the smallest proteins, enough permutations exist to allow about six different proteins for each atom of the entire universe. Because many other chain lengths are possible, from 50 up to hundreds of amino acids long, the potential number becomes inconceivably large. Each protein may differ significantly in configuration and functional capability.

The Amino Acid Subunits of Proteins

The 20 amino acids found in natural proteins are basically similar in molecular structure. One part of the molecule, repeated in each amino acid, consists of an amino (—NH_2) group and a carboxyl (—COOH) group, both joined to a single linking carbon atom. The carbon atom linking together the amino and carboxyl groups also has a side group; this is indicated in Fig. 2–9. Twenty different side groups may be attached to the basic amino acid structure at this point, producing the 20 different amino acids (Fig. 2–10). The side groups vary in complexity from a single hydrogen atom (in glycine) to complex chains or rings, some containing nitrogen and sulfur atoms in addition to carbon and hydrogen. The configuration of these atoms in the side chains gives each amino acid special properties. Some are nonpolar, some polar but uncharged, and some polar and charged (see Table 2–2). These side chains are retained by the amino acids after linkage into the chains of protein; the locations

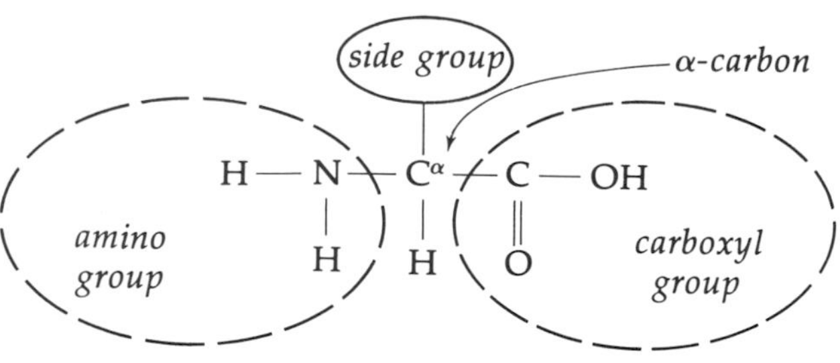

2–9. The basic structure of an amino acid. The side group may be one of twenty different chains of various lengths.

and types of side chains within the protein give the molecule overall solubility properties and determine whether the net reaction of the molecule is acidic or basic. Acidic groups are carried by aspartic and glutamic acid; basic groups by histidine, lysine, and arginine. If arginine, for example, is present in large proportion in the total amino acid component of a protein, the net reaction will be basic. A number of amino acids carry other reactive groups, such as the hydroxyls (—OH) of serine and threonine, the amides ($—NH_2$) of asparagine and glutamine, and the sulfhydryl (—SH) group of cysteine. All of these side groups may either enter interactions within the interior of a protein molecule or complex the protein with other proteins or non-proteinaceous substances such as metallic ions, lipids, nucleic acids, or saccharides.

The Peptide Bond

Amino acids can be readily linked together into chains of two or more units. The bond linking the units together, the *peptide bond* or *peptide linkage,* is produced by a reaction between the amino and carboxyl groups of adjacent amino acids (Fig. 2–11). For each peptide bond formed, the equivalent of one molecule of water is eliminated. An input of energy is required to form the bond; in natural systems, each peptide linkage is formed through an exceedingly complex series of enzymatically catalyzed reactions taking place in the cytoplasm, primarily on the ribosome (outlined in Chapter 11).

The linkage of two amino acids into a chain with two units produces a *dipeptide.* Once the amino acids are linked together, they are usually called "amino acid *residues*" rather than amino acids. Chains of more than two units are called *polypeptides.* Note, in the dipeptide product shown in Fig. 2–11, that the R_1 and R_2 groups, which designate the side chains of the amino acid residues, extend outward on opposite sides of the dipeptide. This arrangement, called the *trans* configuration, is found, with rare exceptions, in all polypeptides. If additional amino acids are added to the chain, the side groups of the subunits continue to alternate in this way. The carbon atoms carrying the side chains (the carbons to which the R_1 and R_2 are attached in Fig. 2–11) are called the α-carbons; these carbons are important in the folding of completed chains into three-dimensional configurations (see below).

The dipeptide shown in Fig. 2–11 retains an amino group at one end and a carboxyl group at the other. Thus, the dipeptide is basically similar to an amino acid in form and can enter into the formation of additional peptide linkages at either the carboxyl (called the C-terminal) end or the amino (N-terminal) end of the chain. In nature, because of the peculiarities of the systems that synthesize proteins on the ribosome, additional amino acids are added only to the C-terminal end of the chain.

Table 2–2 Some Characteristics of the 20 Amino Acids

Nonpolar	Polar, uncharged	Polar, charged	
Glycine	serine	aspartic acid	acidic
Alanine	threonine	glutamic acid	
Valine	asparagine	histidine	basic
Leucine	glutamine	lysine	
Isoleucine	tyrosine	arginine	
Proline	tryptophan		
Phenylalanine	cysteine		
Methionine			

Restrictions on folding imposed by the peptide bond. The carbon and nitrogen atoms immediately forming the peptide bond and the carbons, hydrogen, and oxygen attached to them (see Fig. 2–12) tend to occupy a flat plane in space. This distribution, which is fairly rigid, is assumed primarily because the electrons associated with the double bond between the oxygen and carbon (the lower left-hand corner of the diagram in Fig. 2–12) become distributed in orbitals extending from the oxygen, over the carbon, to the nitrogen of

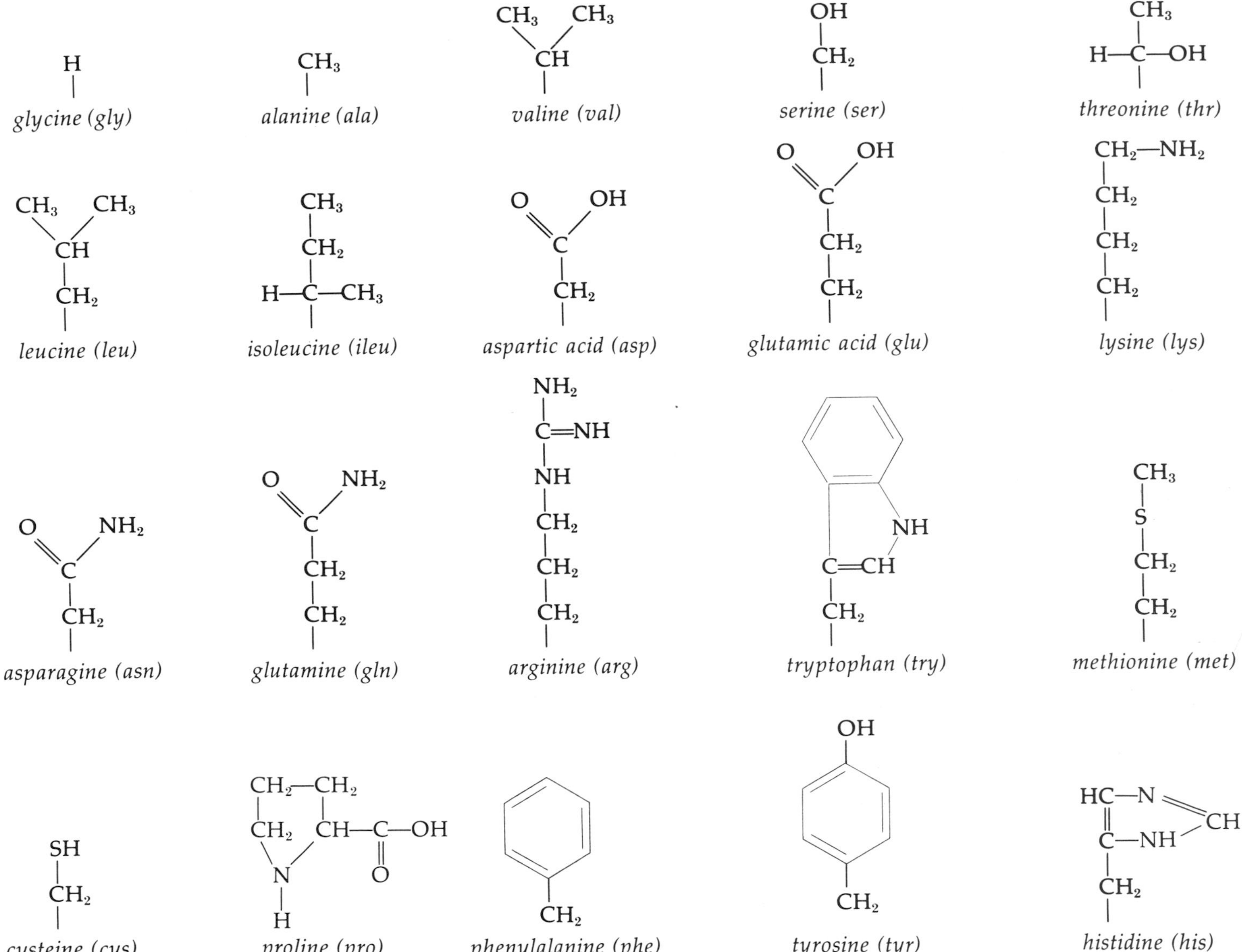

2–10. The twenty amino acid side groups. Proline is shown in its entirety.

the O=C—N group. This can be considered as a resonance between the two states:

$$\begin{matrix} C & & H \\ & C-N & \\ O & & C \end{matrix} \rightleftharpoons \begin{matrix} C & & H \\ & C=\overset{+}{N} & \\ O^- & & C \end{matrix} \qquad (2\text{–}2)$$

This distribution, designated by the dotted lines in Fig. 2–12, has the effect of introducing a partial double bond between the C—N part of the peptide linkage. This, in turn, increases the energy required to rotate the constituent atoms around the C—N bond, and the total structure, including all the atoms in Fig. 2–12, tends to remain in a flat plane in space.

This resistance to twisting of the C—N bond has great significance for the folding configurations taken

carboxyl *amino*

amino acid 1 + *amino acid 2* →

peptide linkage

$OH + H_2O$

a dipeptide

2–11. *Formation of the peptide linkage.*

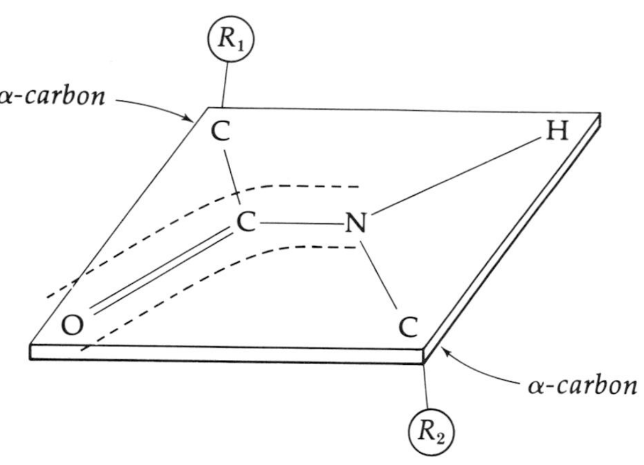

2–12. *Spatial arrangement of the atoms in the peptide linkage. The dotted lines represent the average distribution of the electrons shared between the O⋯C⋯N atoms (see text). Redrawn by permission from R. E. Dickerson and I. Geis,* The Structure and Action of Proteins, *Harper & Row, 1969. All Rights Reserved.*

by a polypeptide chain within a protein molecule. Because of this resistance, all twists of the "backbone" of the polypeptide chain are restricted to the bonds on either side of the α-carbons, the carbons carrying the side groups of the amino acid residues (see Fig. 2–13). Even the degree of rotation around these bonds is limited, because in some positions the various groups attached to these carbons approach too closely. By analyzing the relative probability of different degrees of rotation around these bonds, protein chemists have been able to determine that only a few folding arrangements of the polypeptide chain are possible. Of these configurations, two—the *α-helix* and the *random coil*—are particularly important in the structure of the protein of membranes and other cellular components, such as the enzymes. Different combinations of these two configurations appear to underlie the three-dimensional framework of all spherical, elliptical, or globular proteins.

The Three-Dimensional Structure of Proteins

The α-helix and the random coil configurations. In Fig. 2–13 the degree of rotation around the bonds on either side of the α-carbons is designated by the symbols ϕ and ψ. If ϕ and ψ take on regularly repeated values along a polypeptide, the chain will assume a helical configuration in space. The diameter, pitch, and number of amino acid residues per complete turn of the helix will all be precisely determined by the angles ϕ and ψ repeated along the chain. Not all values for ϕ and ψ are possible, because, as mentioned, some degrees of rotation around these bonds would bring various side atoms carried by the "backbone" too closely together. While several different helices are possible and are found in various natural proteins, a particularly stable configuration is assumed by the polypeptide chain if the bonds on either side of the α-carbons take on the regularly repeated values $\phi = 132°$ and $\psi = 123°$. The helix produced, called the α-helix (Fig. 2–14), is one of the fundamental configurations of the amino acid chain in natural proteins.

The α-helix has a pitch of 5.4 Å, with 3.6 amino acid residues included in each complete turn (dashed lines in Fig. 2–14). The backbone of the α-helix is formed by the repeating $(—N—\alpha C—C—)_n$ groups along the polypeptide chain, indicated by the heavy black bond lines in Fig. 2–14. The hydrogen and oxygen atoms carried by the nitrogen and carbon of the C—N linkage extend outward from the backbone; in the α-helix, these hydrogen and oxygen atoms become arranged in space in such a way that the hydrogens are directed toward oxygens carried by carbons three residues farther along the chain. These opposed groups are held closely enough for hydrogen bonds to form at these points (indicated by the dotted lines in Fig. 2–14). The regularly repeated hydrogen bonds give the α-helix stability and make the whole structure fairly rigid. The α-helix and the concept that a polypeptide chain can assume a helical configuration in space were originally proposed in 1951 by Linus Pauling and R. B. Corey.

Not all of the polypeptide chains of a globular protein can exist in the form of an α-helix, because in this configuration the chain is essentially linear and is too rigid to be bent or coiled upon itself to produce a molecule that is spherical or elliptical in total three-dimensional configuration. Thus major segments of the component chain must be in an alternate, more flexible form. In these flexible regions, the polypeptide probably assumes a configuration known as the *random coil* (Fig. 2–15). In the random coil, the angles of rotation ϕ and ψ do not take on any regularly repeated value, and the chain as a result is irregularly folded or coiled. Although stabilizing hydrogen bonds may be formed at scattered points along the random coil, these are not as closely spaced and do not regularly repeat as in the α-helix. This produces a chain with greater degrees of freedom in spatial arrangement than the α-helix.

A part of the perturbation of the α-helix to produce a random coil may be introduced by proline or cysteine residues if these are present in the polypeptide chain. Because the backbone nitrogen of the proline residue does not possess an attached hydrogen atom (refer to Fig. 2–10), a stabilizing hydrogen bond cannot be formed at this point, and the regular α-helix is disturbed. Cysteine, although possessing a hydrogen at the regular position, may break the regular α-helix through the formation of cross-links with other cysteine residues in neighboring chains (Fig. 2–16). These cross-links, formed between the sulfhydryls (—SH) carried by the side groups of cysteine residues, anchor the chain in a folded configuration. Other factors, such as the reaction of side groups of the amino acid residues to bind metals and other substances (such as lipids and saccharides), will also disturb the α-helical configuration and introduce randomly coiled polypeptide segments.

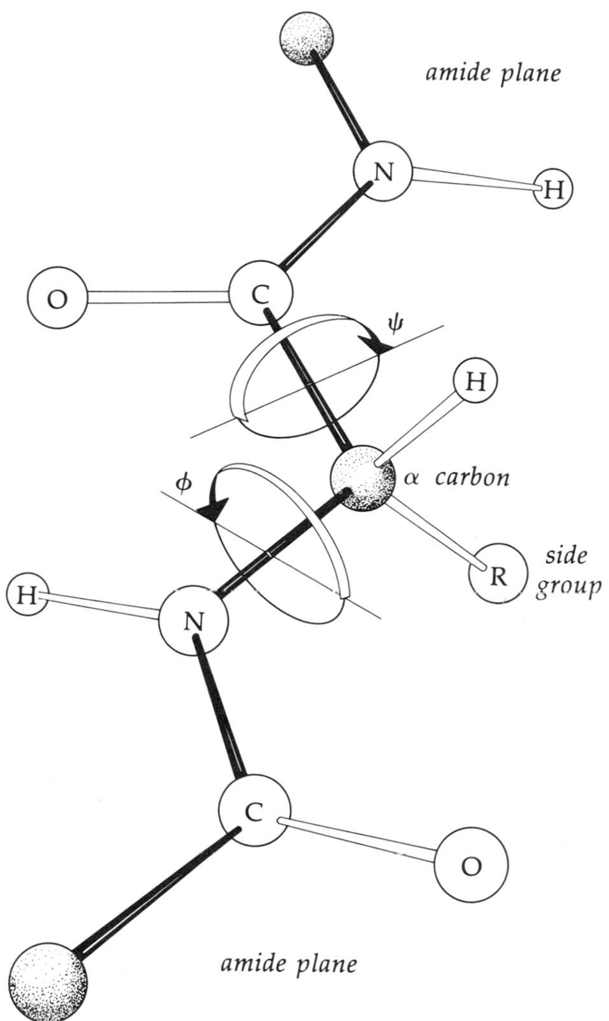

2–13. *The spatial arrangement of two adjacent amino acids in a peptide chain. Because the rotation of the atoms in the peptide linkage is restricted, the amino acid residues of the chain rotate only at the bonds on either side of the α carbons. Rotation in degrees around the C—α C bond is designated ψ; rotation around the N—α C bond is designated ϕ. Not all values of ψ and ϕ are equally possible, because in some positions other atoms of the amino acid residues approach too closely. Redrawn by permission from R. E. Dickerson and I. Geis,* The Structure and Action of Proteins, *Harper & Row, 1969. All Rights Reserved.*

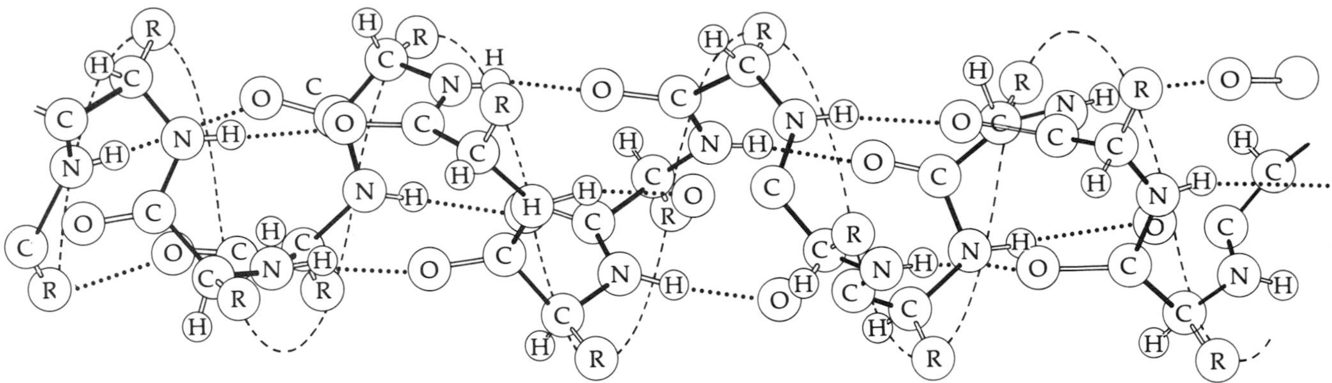

2–14. The helix. The spiraled, dashed line traces out the helical path taken by the backbone of the polypeptide chain in the α-helix. The small spheres designated R *are the side chains of the amino acid residues; other spheres identify atoms of carbon, oxygen, hydrogen, and nitrogen. Hydrogen bonds formed between adjacent hydrogen and oxygen atoms (dotted lines) stabilize the α-helix and make the total configuration relatively rigid. Redrawn from "Proteins" by Paul Doty,* Scientific American *197:173 (1957). Copyright © 1957 by Scientific American, Inc. All rights reserved.*

The existence of both α-helical and randomly coiled segments in the polypeptide chains of globular proteins was demonstrated in the late 1950's by Paul Doty and his collaborators. Working first with synthetic protein chains, about 100 to 1,000 amino acid residues in length, Doty was able to show that the same polypeptide may undergo reversible transformations from random coil to α-helical configurations, depending on the acidity of the medium. This reversible change was detected by characteristic alterations in the scattering of light by the protein solution as the protein coiled from completely α-helical (long rodlets) to random chain configuration (globular molecules).

A different method was used to detect the presence of mixed segments of α-helical and randomly coiled chains in natural proteins. This method took advantage of the rotation of polarized light by amino acids. Polarized light consists of rays vibrating in only one plane. A solution of asymmetrical molecules, such as the amino acids, will cause the plane of vibration of the polarized light to change in angle as the light is transmitted through the solution. This property in amino acids is a result of the asymmetrical attachment of the side chains; in natural amino acids, this asymmetry causes a rotation of polarized light to the left. The amino acids retain this rotational effect when they are linked into a polypeptide. In a solution of proteins in the complete random coil configuration, the rotation of light is about 100° to the left. Using the synthetic polypeptides, Doty found that complete transformation to the α-helical coil caused a change of the rotation of light, completely canceling out the effect of the amino acid side chains, so that the net rotation was zero. Testing natural proteins, Doty and his colleagues discovered a rotation of between 30° and 60° to the left. This was more rotation than expected for proteins consisting entirely of α-helical segments, but less than if all the polypeptides were in the random chain configuration. Doty was able to conclude from this that natural proteins contain mixed segments of the two coiling configurations.

The three-dimensional structure of proteins. At a given pH, the distribution of randomly coiled and α-helical segments in a protein is determined by the amino acid sequence of the constituent polypeptide chains. This sequence will also determine the number and position of disulfide linkages between cysteine residues, the location of proline residues, and the posi-

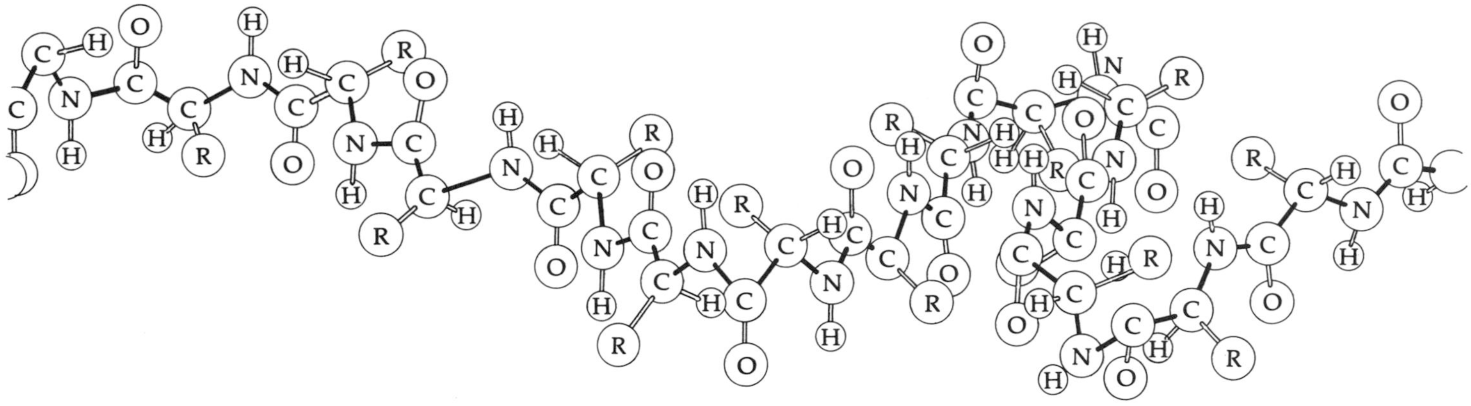

2–15. *The random coil. Values for* ψ *and* ϕ *in the random coil vary from residue to residue along the polypeptide. This configuration is relatively flexible; few hydrogen bonds are formed in the chain. Redrawn from "Proteins" by Paul Doty,* Scientific American *197:173 (1957). Copyright © 1957 by Scientific American, Inc. All rights reserved.*

tion at which other substances, such as metallic ions, lipids, and saccharides, may bind. Also important, and equally dependent on the amino acid sequence, are the position of hydrogen bonds, attraction between positively and negatively charged side groups, and the mutual attraction of the nonpolar side groups—which will tend to become associated in the interior of the protein—directed away from the strongly polar aqueous medium. The net effect of this will be to fix, for each protein of unique sequence, a unique three-dimensional configuration and shape. Because, even for polypeptides of fixed sequence, the folding may change with alterations in *p*H and the binding of other substances such as metallic ions, the number of possible three-dimensional shapes and configurations of proteins is incalculable. This diversity in form is reflected in the diversity of function of proteins in the cell.

The three-dimensional structure of a relatively simple protein, the enzyme *ribonuclease,* is shown in Fig. 2–17a. This protein contains 124 amino acids (Fig. 2–17b) and has a molecular weight of 13,700. The molecule is held in configuration at physiological *p*H by a combination of hydrogen bonds, hydrophobic interactions, and four disulfide bridges, indicated by the bent double lines in Fig. 2–17a. The active site of the enzyme (see Chapter 5) is located in the crevice at the midpoint of the molecule.

amino acid chain
disulfide linkage
amino acid chain

2–16. *The disulfide linkage, an inter-chain bond that partly determines the pattern of folding and three-dimensional configuration of a polypeptide.*

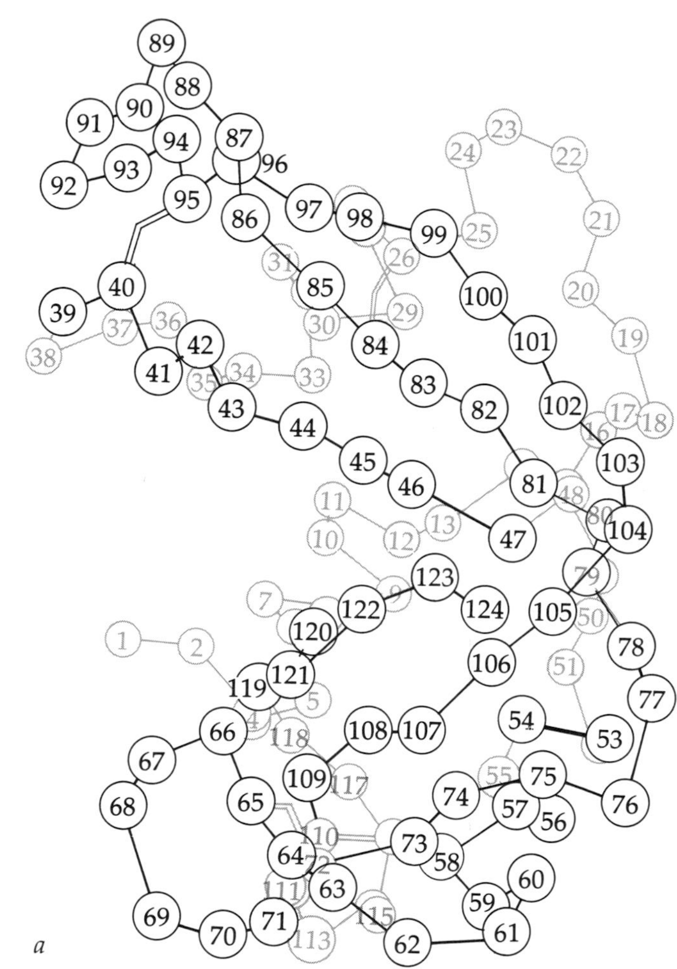

2–17. a, *Three dimensional configuration of* ribonuclease A, *an enzymatic protein. The active site of the enzyme is in the cleft at the left center of the molecule. The positions of individual amino acid residues are marked by the numbered circles; these correspond to the list in 2–17b. Disulfide bridges are indicated by a bent double line.* b, *Sequence of amino acids in ribonuclease. The numbers correspond to the circles in 2–17a; cysteine residues joined by disulfide linkages are enclosed in a box. The number indicated by the arrow at the bottom of each box shows the amino acid residue to which the disulfide linkage is joined. Reproduced by permission from R. E. Dickerson and I. Geis,* The Structure and Action of Proteins, *Harper & Row, 1969. All Rights Reserved.*

b

1				**5**					**10**					**15**					**20**	
lys	glu	thr	ala	ala	ala	lys	phe	glu	arg	gln	his	met	asp	ser	ser	thr	ser	ala	ala	ser

			25					**30**					**35**					**40**	
ser	ser	asn	tyr	cys	asn	gln	met	met	lys	ser	arg	asn	leu	thr	lys	asp	arg	cys	lys
				↓ **84**														↓ **95**	

			45					**50**					**55**					**60**		
pro	val	asn	thr	phe	val	his	glu	ser	leu	ala	asp	val	gln	ala	val	cys	ser	gln	lys	asn
																↓ **110**				

		65					**70**					**75**					**80**			
val	ala	cys	lys	asn	gly	gln	thr	asn	cys	tyr	gln	ser	tyr	ser	thr	met	ser	ilu	thr	asp
		↓ **72**							↓ **65**											

	85					**90**					**95**					**100**				
cys	arg	glu	thr	gly	ser	ser	lys	tyr	pro	asn	cys	ala	tyr	lys	thr	thr	gln	ala	asn	lys
↓ **26**											↓ **40**									

105					**110**					**115**					**120**				**124**
his	ilu	ilu	val	ala	cys	glu	gly	asn	pro	tyr	val	pro	val	his	phe	asp	ala	ser	val
					↓ **58**														

Combination of proteins with other substances. Proteins, by virtue of the different reactive groups carried on the amino acid residues, are able to enter into a variety of intermolecular interactions. Several types of reactive groups are important in this interaction. One of these is the charged groups, such as the acidic or basic amino acid residues (see Table 2–2), which are capable of binding metallic ions or other charged substances through electrostatic attraction. A second type is formed by the various uncharged polar groups carried by the amino acids, which are capable of forming hydrogen bonds. In proteins, these form between OH, NH_2, NH, and SH groups as proton (H+) donors, and C=O and N groups as proton acceptors. A third important type of binding is the attraction set up between nonpolar groups, primarily the nonpolar hydrocarbon side chains of some amino acid residues (see Table 2–2). In an aqueous medium, these groups tend to associate and exclude the strongly polar water molecules. This produces a binding force between nonpolar substances that may be reflected as an association between the nonpolar amino acid residues, or as an association between these residues and other nonpolar substances such as the hydrocarbon chains of lipids.

All these forces—the electrostatic, uncharged polar, and nonpolar—act separately or in combination to bind proteins to other substances. The inorganic ions Na^+, K^+, Cl^-, Ca^{+2}, Mg^{+2}, PO_4^{-2}, CO_3^{-2}, and the heavy metals copper and iron, among others, bind to various proteins to form important biological complexes. For example, CO_2 dissolved in the blood, forming the charged ion CO_3^{-2}, combines reversibly with the protein of the hemoglobin molecule; this interaction is important in regulating the acidity of the bloodstream. Nonpolar residues are important in the conjugation of proteins and lipids to form *lipoproteins;* these complexes are the basic structural units of membranes. Still another important group is formed by the combination of proteins and carbohydrates to form *glycoproteins* and *mucoproteins.* In these conjugates, the carbohydrate present may either be a monosaccharide, disaccharide, or polysaccharide. As suggested by the terminology, these protein-lipid complexes are frequently highly viscous in solution and are the primary components of such substances as mucus and the gelatinous or sticky extracellular coats of some cells. Because a protein-associated polysaccharide is frequently observed to be associated with the exterior surface of the plasma membrane, this substance may also be important in membrane structure. Last but not least must be mentioned the primarily electrostatic binding of proteins to nucleic acids to form the *nucleoproteins,* the major constituent of the chromosomes of higher organisms.

Suggestions for Further Reading

Deuel, H. J., 1951. *The Lipids.* Interscience, New York.

Dickerson, R. E., and I. Geis, 1969. *The Structure and Action of Proteins.* Harper & Row, New York.

Doty, P., 1957. Proteins. *Scientific American,* vol. *197,* pp. 173–184.

Hanahan, D. J., F. R. N. Gurd, and I. Zabin, 1960. *Lipide Chemistry.* John Wiley & Sons, New York.

Haurowitz, F., 1963. *The Chemistry and Function of Proteins.* Academic Press, New York.

Lovern, J. A., 1957. *The Chemistry of Lipids of Biological Significance.* John Wiley & Sons, New York.

3
The Light and Electron Microscopes

The electron microscope has supplanted other methods for the investigation of cell structure to such an extent that most biologists, from biochemists to morphologists, visualize the cell primarily in terms of the electron image. For this reason a discussion of electron microscopy, with emphasis on the formation and interpretation of the electron image, is particularly important for an understanding of the present-day concepts of cell structure and function. But the light microscope cannot be ignored in this discussion, because it still plays a great part in contemporary experimentation and is much more likely to be part of the immediate experience of students in cell biology and cytology. The phase contrast light microscope remains the only method for observing the subcellular details of living tissue because materials must be fixed and dried for electron microscopy. Further, the sophistication of analytical cytochemistry in electron microscopy has not yet begun to approach the techniques developed for the light microscope. Because the principles on which both the light and electron microscopes are based are much more easily understood in the light microscope, this instrument will be considered first.

The Light Microscope

Elementary Geometrical Optics

The refraction of light. Light waves transmitted through a vacuum travel at a fixed velocity. If transmitted through a medium such as air, water, or glass, the velocity of the light waves is slowed. Each medium transmits light at a characteristic velocity; the ratio between the velocity of light in a vacuum and in a transmitting medium is termed the *refractive index:*

$$\frac{\text{velocity of light in a vacuum}}{\text{velocity in transmitting medium}} = \text{refractive index}$$

The velocity of light in air is almost equivalent to its velocity in a vacuum, and for most calculations a refractive index of 1.00 is used (the correct value for the refractive index of air is 1.00029). Glass, with a refractive index of 1.5, transmits light more slowly.

As light rays pass through the boundary between a transmitting medium of one refractive index to a second medium of different refractive index, the light rays are bent, or refracted, from their path. The degree of bending depends on two factors—the refractive index and the angle at which the light strikes the surface of the second medium. If the light strikes the surface at an angle exactly perpendicular to the surface, no refraction occurs and the light is transmitted through without a change in path (Fig. 3–1a). At any other angle, some degree of bending occurs; as the angle of incidence decreases from the perpendicular, the degree of bending increases (Fig. 3–1b). For a given medium, light passing through the boundary into the interior at a fixed angle is always refracted to the same degree. This constant angle of bending is directly related to the refractive index. Therefore, if parallel rays strike a flat glass plate at an angle to the surface, as shown in Fig. 3–1b, the refracted rays transmitted through the glass will remain parallel and will not converge or diverge.

This is not the case if the glass surface transmitting the light is curved. As a beam of light containing parallel rays strikes a convexly curved surface, each point on the surface is presented to the approaching light at a different angle, and the angle of refraction at all points will be different. If the glass surface is a part of a sphere, the light rays will be bent more strongly at the periphery, and the refracted rays will converge on a point (Fig. 3–2). The light passing through the center, or axis, of the piece of curved glass at an angle of 0° will not be bent. All other rays, passing through at greater distances from the axis, will be bent to greater degrees. The point at which these rays converge on the axis is called the focal point; the distance between this point and the center of the lens is the focal length. For lenses of a smaller radius of curvature, this point will be located closer to the center of the lens.

One significant property of the system is that light rays traveling either direction will follow the same paths. Therefore, if a source of light is placed at the focal point, the rays passing through the glass lens will be refracted into a parallel beam as they leave the other side of the lens. Similarly, a glass lens with symmetrically curved surfaces will possess a focal point on either side of the lens at the same distance from the lens center.

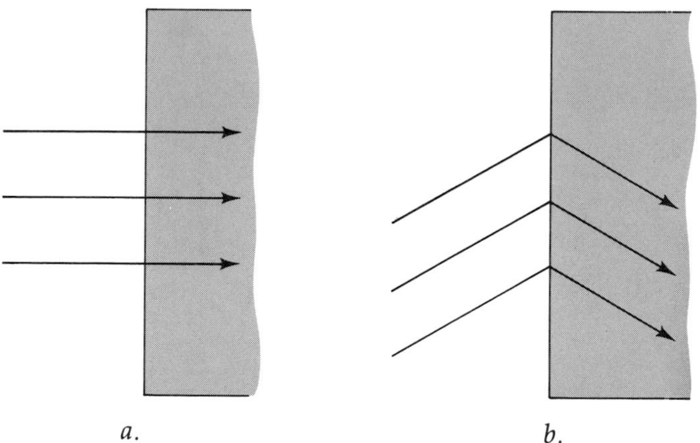

3–1. Refraction of light at the boundary between two media of different refractive indices, as for example between air (unshaded) and glass (shaded). a, *Light striking the boundary at an angle perpendicular to the surface is unrefracted.* b, *Light striking at any other angle is refracted as it passes through the boundary.*

Image formation and magnification by lenses. Light rays reflected or issuing from any object placed outside the focal point of a converging lens will be focused into an *image* of the object beyond the other side of the lens. The way the image is formed can be understood in simple graphical terms (Fig. 3–3). Consider an object placed at the point *A* outside the focal point of the lens. Whether this object is a light source, or merely reflects light, rays will radiate outward in all directions from each point on the object surface. Rays originating from a specific point (*B* in Fig. 3–3) will pass through the lens in the following way. Rays from point *B* traveling parallel to the lens axis (ray *BC*) will be refracted by the lens through the focal point (F_2) on the opposite side of the lens; this follows from the properties of converging lenses diagramed in Fig. 3–2. By the same relationship, rays from point *B* passing through the focal point (F_1) of the lens on the same side as the object (ray *BD*) will be refracted to a line parallel to the lens axis beyond the far side of the lens. These two rays, *BC* and *BD*, con-

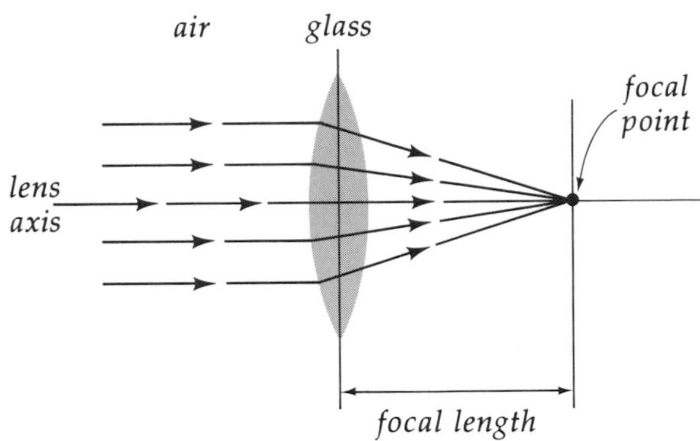

3–2. *Refraction of rays of light passing through a convex lens (see text).*

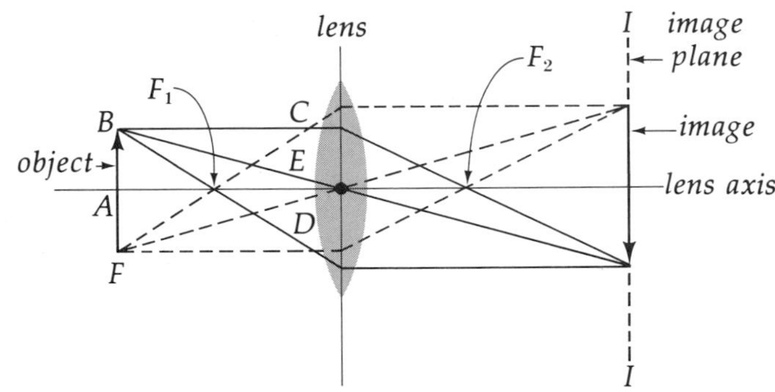

3–3. *Formation of a magnified, real image by a converging lens (see text).*

verge at a fixed location beyond the opposite focal point of the lens. This point lies in a plane defined as the *image plane* (*I* in Fig. 3–3). Rays from the object point passing through the center of the lens (ray *BE*) will not be refracted, and will leave the lens at the same angle to pass through the image point. All other rays from the object point *B* will also converge and be brought to focus at the same point in the image plane. By drawing an equivalent diagram for any other point in the object, it can be shown that rays originating from all points in the object will be brought to focus at different locations in the image plane (set of dotted lines from point *F* in Fig. 3–3). Because all points of the object are brought to focus at plane *I*, an image of the object will be constructed in this plane. If a screen is placed in this plane, the image will be directly visible to an observer.

This type of image, which can be focused on a screen placed in the image plane, is defined as a *real image.* Real images are formed only from objects placed outside the focal point of the lens. The real image exhibits two important characteristics: it is inverted, and it differs in size from the object. The size of the image depends on the focal length of the lens and the placement of the object with respect to the focal point. As the object is moved closer to the focal point, the image plane will be formed farther away from the opposite side of the lens and the image will be increased in size; that is, it will be *magnified.* It can be seen from this that the greatest magnification will be obtained from lenses having a very short focal length, with the object as close as possible to the focal point.

If an object is placed *inside* the focal point of a converging lens, a real image is not formed. However, an

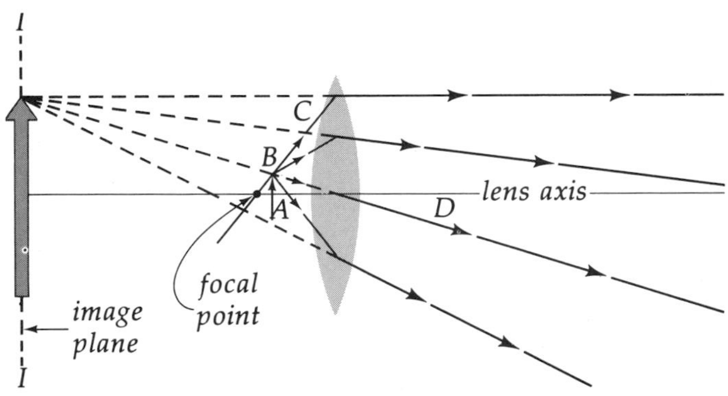

3–4. *Formation of a magnified, virtual image of an object placed inside the focal point of a converging lens (see text).*

image can still be seen by looking directly into the opposite side of the lens. This image, called the *virtual image*, cannot be focused on a screen, and no points exist at any plane in space at which rays radiating from the object are brought to focus. The basis for the formation of a virtual image is diagramed in Fig. 3–4. The object, at point *A*, is placed between the focal point and the lens. Note the ray *BC*, representing a ray passing from the object point to the lens. This line, if extended backward, passes through the focal point of the lens. Therefore, rays leaving the object point *B* at this angle will follow a path equivalent to rays originating from the focal point, and will be refracted to a line parallel to the lens axis on the opposite side of the lens. Other rays, leaving the object at point *B* and passing through the center of the lens will not be bent from their path and will leave the opposite side of the lens at the same angle (ray *BD*). These rays, leaving the opposite side of the lens from the object, diverge and are not brought to a point of focus in space. All other rays leaving the object point will diverge in this way. Because of this divergence, the rays, to an observer looking into the lens, will appear to radiate from a point in space behind the real object. By extending the diverging lines back through the lens, this point in space can be located (dotted lines). Construction of a similar diagram for all object points shows that all of these apparent points of origin lie in the same plane (*I* in Fig. 3–4). This is the image plane; the image formed is a virtual image. This image can be seen by the eye because the rays apparently radiate outward from all points in the image plane; this is precisely equivalent to the rays that radiate outward from all points on a physical object. These rays are focused into a real image on the retina by the lens of the eye, and the virtual image is perceived as is a real object. Note that the virtual image is magnified, as is the real image, but not inverted. Both real and virtual images are used in the compound microscope.

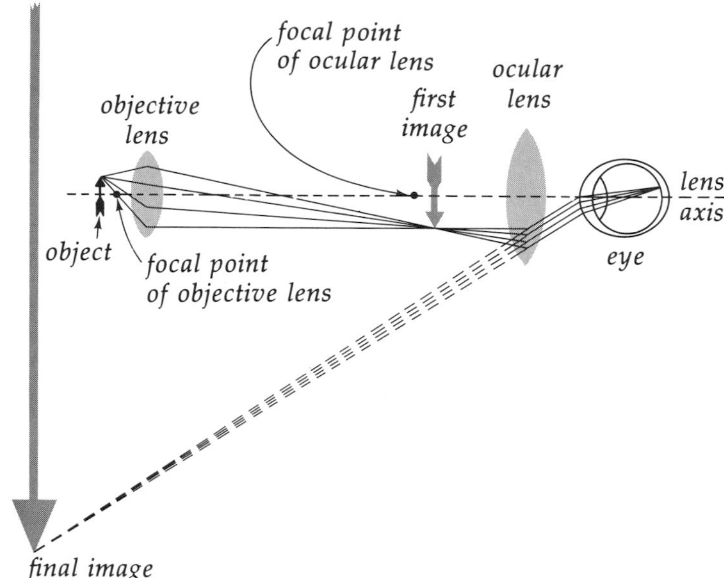

3–5. The placement of lenses and image formation in a compound microscope. The objective and ocular lenses are fixed at either end of a movable tube; focusing is accomplished by moving the tube with relation to the object. The image formed by the objective lens is focused just inside the focal point of the ocular lens. Thus, the final image viewed by the observer looking into the ocular lens is seen as a virtual image (see text).

Construction of the Light Microscope

Because a converging lens acts as a magnifier, a simple microscope can be constructed from a single lens of very short focal length. Anton van Leeuwenhoek's microscopes (see p. 4) were constructed in this way. These microscopes, capable of magnifying between 200 and 300 times, consisted of a single, small sphere of glass mounted between two brass plates.

Greater magnifying power can be achieved by using two lenses, placed as shown in Fig. 3–5. This type of instrument is called a compound microscope. The lens closest to the object, called the *objective lens,* has a very short focal length. The object, the specimen, is placed just outside the focal point of this lens. Therefore, a real image of the object is formed by the objective lens, indicated as the first image in Fig. 3–5. This intermediate image acts as the "object" for the second or the *ocular lens.* The ocular lens is positioned so that the image formed by the objective lens falls inside the focal point of this lens. Thus a virtual image, rather than a real image, is formed by the second lens. This image (the "final image" in Fig. 3–5) is observed by looking into the ocular lens.

In the construction of the light microscope (Fig. 3–6), the objective and ocular lenses are fixed at either end of a tube. Both the objective and ocular systems are made up of a series of lenses, usually as many as eight to ten for the objective and two or three for the ocular. These multiple lenses are combined primarily to correct various aberrations that are encountered if single lenses are used. As long as the systems are constructed so that the multiple lenses are reasonably close together, the net effect is closely similar to the focusing diagramed in Figs. 3–3 and 3–4 for single lenses. The object to be viewed is placed on the specimen stage, just below the objective lens. Provision is made for the precise adjustment of the specimen so that the intermediate real image, focused by the objective lens, falls in the correct plane inside the focal point of the ocular lens. This is accomplished by a focusing adjustment that controls the position of either the lens mounting tube or the specimen stage. Illumination of the specimen is provided by the condenser lens, a third system mounted just below the specimen state. This system can be adjusted so that the cone of light leaving the specimen enters and just fills the objective lens.

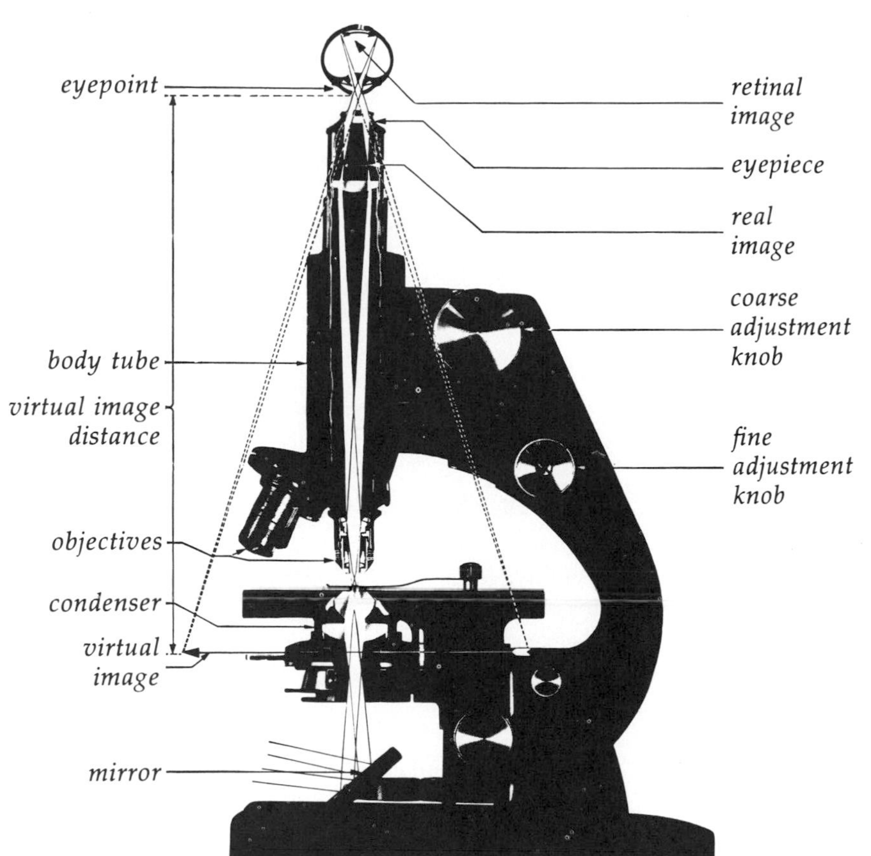

3–6. Construction of a typical compound light microscope. Objective lenses of different magnifying power can be selected by rotating the nosepiece. Light is focused on the specimen by the condenser lens, located just below the stage. Courtesy of the American Optical Company.

The Phase Contrast Microscope

Most of the structures within the cell are highly transparent and do not have sufficient contrast to be seen with any clarity in living, unstained cells. From the time light optics was perfected in the last century until the present day, this difficulty has been routinely surmounted by fixing and staining tissues to improve the definition and contrast of details within the cell. However, the desirability of observing living specimens prompted the development of an optical system to improve the contrast of unfixed and unstained cells. In the 1930's, these efforts culminated in the design of the phase contrast microscope, which allows many small details in living, unstained cells to be readily seen. As will be demonstrated in later sections of this chapter, phase contrast effects are also an important factor in image formation in the electron microscope.

The optical system used in phase contrast microscopes takes advantage of the fact that the velocity of light is slowed as the component rays pass through material with a refractive index higher than 1.00. Many components of the cell, although practically transparent to light, have relatively high refractive indices and significantly affect the velocity of transmitted light rays. These structures remain poorly contrasted in the ordinary light microscope because even though the light transmitted through them is momentarily retarded in velocity, it is not diminished in amplitude and appears just as bright to the eye. But because the light transmitted through these structures is slowed briefly, it is out of phase with other parts of the beam when it emerges from the organelle. The phase contrast microscope converts these differences in phase to differences in brightness. In order to understand the basis for converting differences in refractive index to differences in brightness or intensity it is first necessary to consider the behavior of light rays as *waves.*

Diffraction of light waves. Light leaving a point source advances in spherical wavefronts that travel outward in all directions (Fig. 3–7). This effect is duplicated in two dimensions by dropping a stone into a quiet pool of water. If an opaque edge is placed in a position that interrupts the spherical wavefronts at some point (see Fig. 3–7), the original wavefront is completely absent from all regions directly below the edge, producing a shadow of the edge. However, the shadow region will not be completely free of advancing wavefronts, for, as each original wavefront strikes the edge, a new wave is created, with its source the edge of the object. This new wave, called the diffracted wave, travels into both the shadowed and unshadowed regions below the edge with the same wavelength as the original undiffracted wave.

In the unshadowed region below the edge, an interaction called *interference* occurs between the original uninterrupted waves and the diffracted waves. As the two waves travel together into the unshadowed region below the edge, the peaks cross at regular points (indicated by the lines in Fig. 3–7). At these points, the crests of two waves come together and reinforce each other to produce maximum brightness. At other points halfway between the crossing points, the crest of one wave coincides with the trough of the other. Here the effect is cancellation, with the production of minimum brightness at this location. If a screen is placed just below the edge, these alternating maxima and minima can be seen as an alternating series of light and dark lines running parallel to the shadow of the edge. The effect is strongest close to the shadow of the edge. If the object is a small opaque disc instead of an edge, the alternating light and dark lines will appear as concentric circles surrounding the shadow of the disc. These alternating light and dark lines surrounding an opaque object are also present and arise from the same source in images focused by a lens (Fig. 3–8).

Diffraction in the phase contrast microscope. The phenomena of diffraction and interference are central to the operating principle of the phase contrast microscope. As light rays pass through components of the cell with differing refractive indices, they are diffracted and leave the path followed by the undiffracted waves passing through other portions of the specimen. The relative differences between the refractive indices of various parts of the specimen are relatively small, and the diffracted and undiffracted waves, focused back into the image plane by the ordinary microscope, do not interfere enough to cause detectable changes in intensity in the image. In the phase contrast microscope the diffracted waves, bent from the path followed by

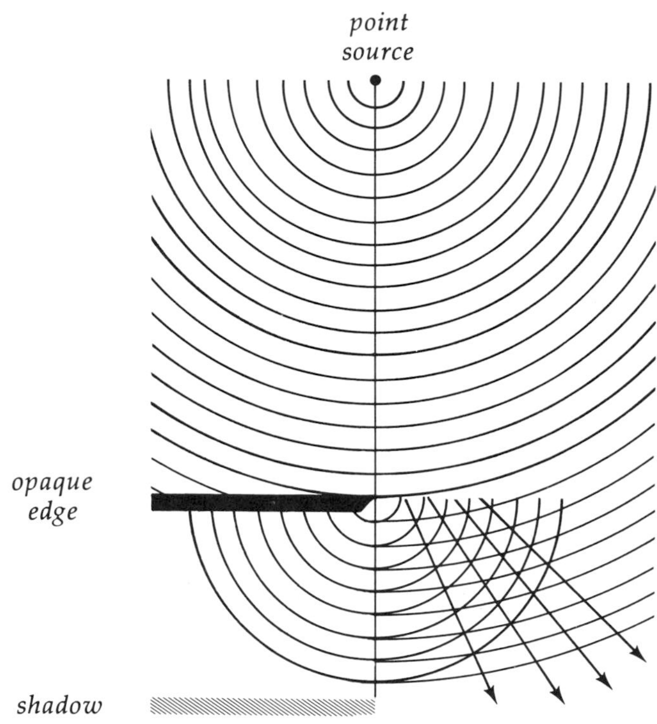

3–7. Diffraction and interference at an opaque edge. The curved lines radiating from the point source represent spherical wavefronts. As these strike the opaque edge, the original wavefront is interrupted and does not fall into the shadowed region below the edge. New wavefronts of reduced intensity, but at the same wavelength as the original wavefront, will arise from the edge. These fall in both the shadowed and unshadowed regions; in the unshadowed region, interference will occur between the original and diffracted wavefronts. Where crests intersect, points of maximum brilliance will be created (arrows). At intermediate points, where crests and troughs are superimposed, the waves cancel out, and points of minimum brightness are created. These are seen in an image as alternating light and dark lines that parallel the outline of the edge.

the unrefracted light, are channeled through a phase plate, a piece of glass coated with a layer of refracting material with just the right thickness to delay the waves by one-fourth wavelength. The undiffracted light rays, transmitted through the specimen without disturbance, pass through the central, uncoated region of the phase plate. These two rays are focused back together into the image plane, where interference occurs between the diffracted light, delayed by one-fourth wavelength, and

3–8. The diffraction pattern around an image point imaged by a lens (see text). From A. C. Hardy and F. H. Perrin, Principles of Optics. *Courtesy of McGraw-Hill Book Company.*

the transmitted light. This produces maxima and minima of light intensity related to the refractive index differences in the specimen, and regions of the specimen with differing refractive indices stand out clearly as objects having varying degrees of brightness (Fig. 3–9).

Resolution in the light microscope. The resolution of a microscope refers to the power of the instrument to distinguish the separateness of (to resolve) small details in the image. If two small image points lie very close together, the concentric light and dark lines (see Fig. 3–8) produced by diffraction will overlap. If these overlap to the extent that the images merge, the separate points will be seen as a single point and, therefore, are not resolved.

Improving the resolution of two small points in the image, as the points are separated by a fixed distance, can be accomplished only by reducing the diameter of the diffraction lines around the points (Fig. 3–8). The diameter of the diffraction patterns depends on the

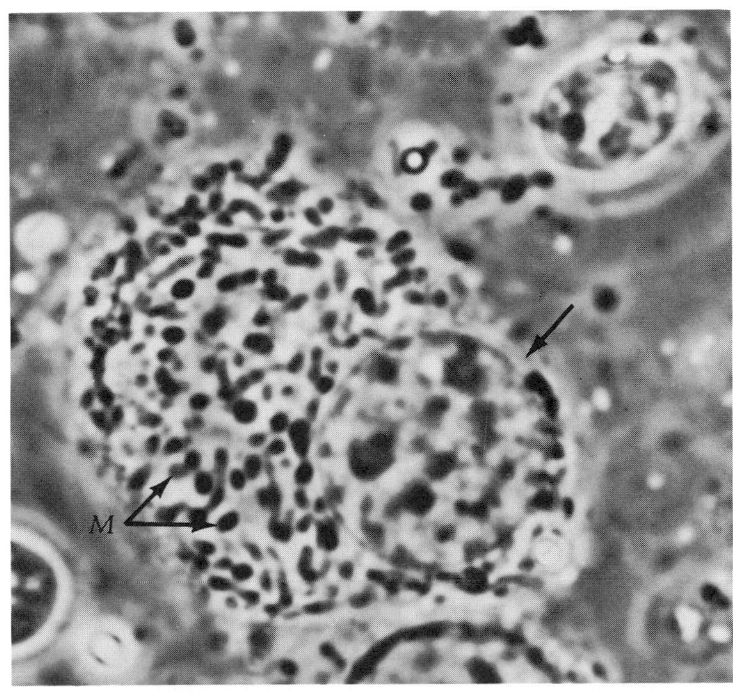

a

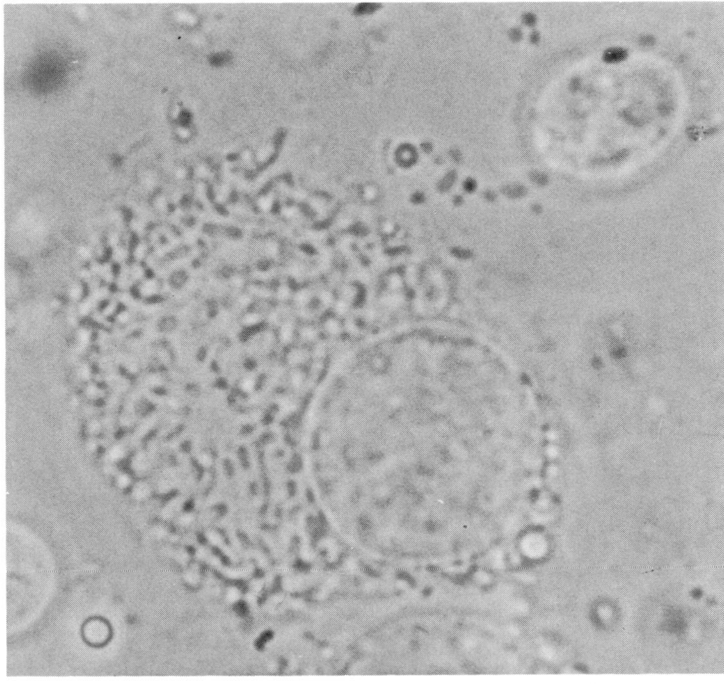

b

3–9. a, *A phase contrast image of a living cell from the liver of the frog* Rana pipiens. *Compare the details visible in this micrograph with the image of the same cell as seen in* b *by ordinary bright field microscopy. The nucleus is marked by an arrow; the cytoplasm contains numerous mitochondria* (M) *and other inclusions.* × *2,300. Courtesy of Barry R. Zirkin.*

ratio of the wavelength of light to the size of the points being resolved. With shorter wavelengths of light, the diffraction patterns have a smaller diameter, and the resolution is improved. Obviously there is a lower limit to improvement from this source, because the wavelength of visible light does not extend below 3800 Å.

The resolution-wavelength relationship is expressed in the equation:

$$\text{Resolution} = d = \frac{0.61\lambda}{n \sin \alpha} \tag{3–1}$$

In this equation, λ = the wavelength of light, and 0.61 is a constant related to the arbitrarily defined overlap allowed for two points to be recognized as separate by a human observer. According to this equation, improving the resolution (obtaining smaller values for d) requires that 0.61λ approach the smallest value and $n \sin \alpha$ the largest value possible. For contemporary microscopes of the best quality, $n \sin \alpha$ can be considered to have a constant magnitude. In this quantity, n is the refractive index of the medium surrounding the object and filling the space between the object and the objective lens. For air, $n = 1$; a maximum value for n can be obtained by the use of immersion oils, which have a refractive index of 1.4. The second part of this quantity, $\sin \alpha$, is related to the angle of the cone of light entering the objective lens. Under the best possible conditions, approached by good lenses, this cone has a maximum angle of 180°. In the expression used, α = one-half this angle, and therefore $\sin \alpha$ approaches a maximum value of $\sin 90° = 1$, and the quantity $n \sin \alpha$, called the *numerical aperture*, approaches a maximum value of 1.4.

This leaves as a variable in the right hand side of equation 3–1 only the wavelength of light, λ. For the smallest values of d, λ takes on a minimum value. For visible light at the shorter wavelengths, with $\lambda = 5000$ Å (green light) the smallest details resolved are about 0.2 to 0.25 microns, or 2000 to 2500 Å in diameter. While this allows resolution of the larger organelles, such as the nucleus, nucleolus, mitochondria, and chloroplasts, much of the internal structure of these components and the smaller parts of the cell, such as the ribosomes, lie below the limits of visibility.

Resolution and the electron microscope. The limitation to the resolving power of the light microscope, imposed by the minimum wavelengths of visible light, prompted efforts to use the shorter wavelengths of

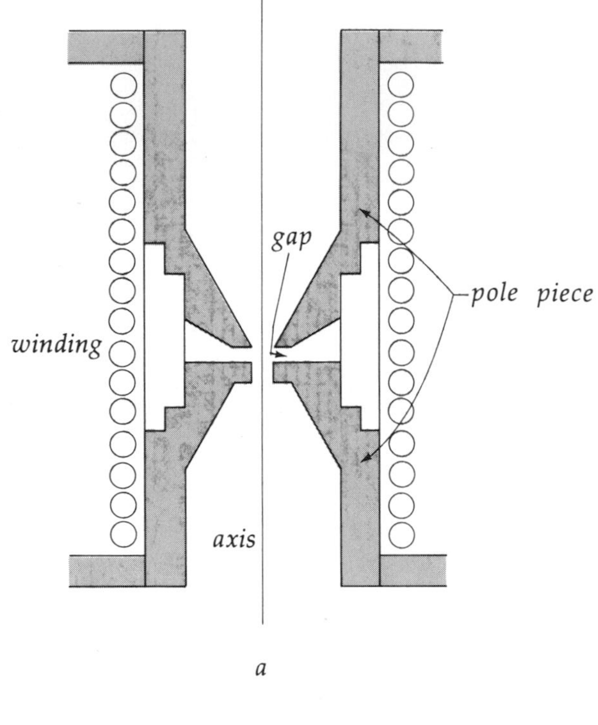

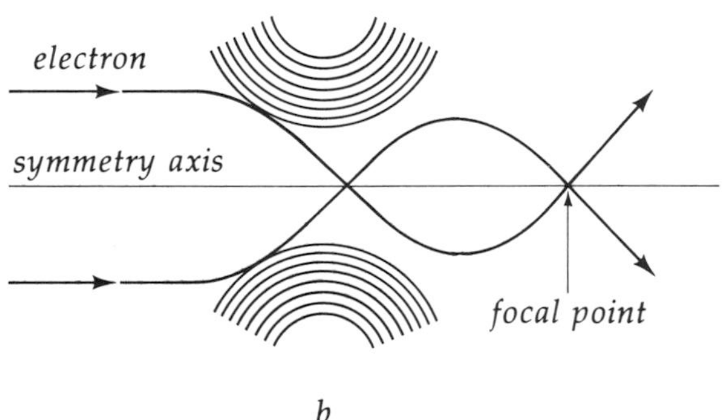

3–10. a, The pole piece of an electron microscope. The pole piece is a cylindrical iron insert (shown in cross section in this view) that shapes and concentrates in the region of the gap the magnetic field generated by current passed through the surrounding lens coils. The cylindrical bore of the pole piece is machined to extremely close tolerances. b, The focusing effect of the magnetic field on electrons passing through the bore of an electron lens. The path described by an electron passing through the lens is helical and is more complex than shown in this simplified diagram. Modified from R. B. Setlow and E. C. Pollard, Molecular Biophysics. *Courtesy of R. B. Setlow.*

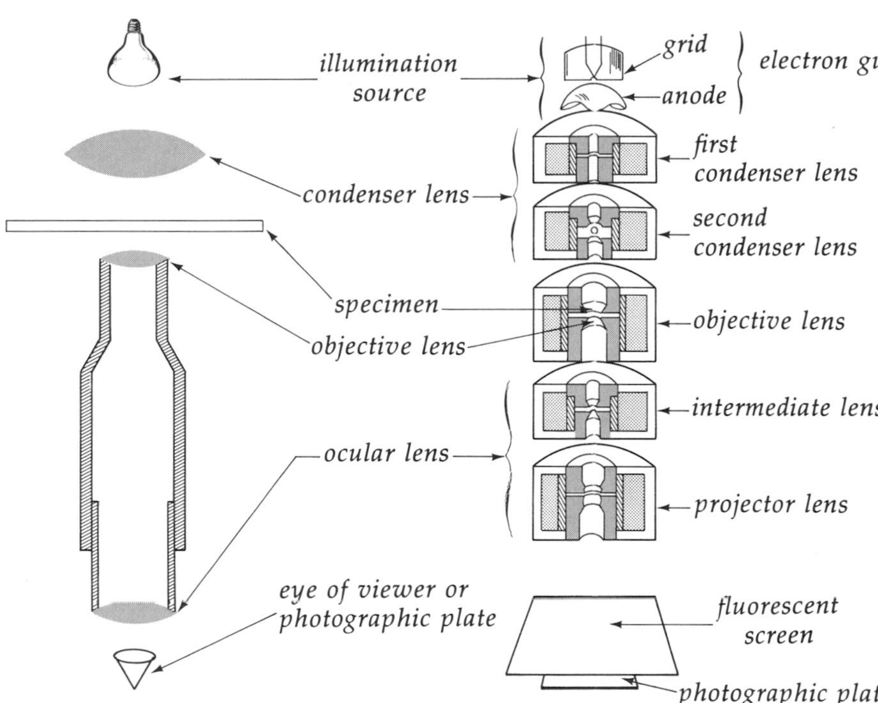

3–11. Diagrammatic comparison of the light and electron microscopes. The lenses of the light microscope, shown on the left, are inverted for sake of comparison with the electron microscope system shown on the right.

electromagnetic radiation. Some limited success was obtained with ultraviolet light, pushing the resolution down to about 1100 to 1200 Å ($\lambda = 2500$ Å). However, the greatest possibilities were offered by X-ray and electron beams, with wavelengths much shorter than visible light. For electrons, wavelengths of 0.05 Å are easily obtainable; the trick was to design a system of lenses capable of focusing electrons into an image. Intensive research into solution of this problem began in Germany in the 1930's, leading finally to the design of a practical electron microscope by E. Ruska and B. von Borries in 1938, and commercial production of this microscope began in Germany in 1939. These developments were paralleled in America, where in 1941 J. Hillier and W. W. Vance designed a commercially successful electron microscope.

The Electron Microscope

In overall design the electron microscope closely resembles the light microscope. The primary difference is in the type of lens used. Instead of glass lenses, the electron microscope uses magnetic or electrostatic fields to focus electrons. In the most commonly employed type of magnetic lens, the fields are generated by massive coils of wire, through which precisely controlled electric currents are passed. The magnetic field is shaped into the correct three-dimensional configuration for focusing by *pole pieces* (Fig. 3–10a), iron inserts placed into the bore of the electromagnets. Although the paths followed by electrons through the magnetic fields are helical and quite complex, the net effect is the same as a converging lens in light microscopy, and the electrons approaching the lens in a parallel beam are focused into a focal point just beyond the lens (Fig. 3–10b). By adjusting the current applied to the electron lenses, the position of the focal point and thus the focal length and magnification of the lenses can be changed at will.

A second major difference is that the electron microscope column must be under a vacuum. Otherwise the electrons of the beam, which have relatively poor penetrating power, would be completely scattered by molecules of gas in the microscope. This places special requirements on the specimen, which must be dried before being introduced into the microscope. For this reason, living specimens cannot be viewed in the electron microscope, at least not without isolation from the vacuum of the column. Attempts to produce an isolating specimen holder thin enough to allow transmission of electrons through living specimens have not yet been successful.

For the sake of comparison, diagrams of the light and electron microscopes are shown side by side in Fig. 3–11. Specific details of the construction of the electron microscope will be deferred until consideration is given to the resolution obtainable by electron optics.

Resolution in the Electron Microscope

Although electrons are particulate and have mass, they can also be described as a beam of radiant energy with wavelength and velocity. The velocity and wavelength of an electron are related according to the following equation:

$$\lambda = \frac{h}{mv} \qquad (3\text{–}2)$$

in which λ is the wavelength, h is a constant (Planck's constant, 6.6252×10^{-27} erg sec), m is the mass of the electron (9.1085×10^{-28}g), and v is its velocity. For this discussion, the mass of the electron can be regarded as having a constant value, leaving as variables the wavelength and velocity. It can be seen from this equation that the wavelength decreases as the velocity of the electron increases. The velocity, in turn, is directly proportional to the voltage difference between two points in space traversed by the electron. This voltage difference, called the accelerating voltage, is easily measured and can be substituted in equation 3–2:

$$\lambda = \frac{12.3}{V^{1/2}} \qquad (3\text{–}3)$$

In this expression V is the voltage difference, and 12.3 represents a value derived from Planck's constant and the mass and charge of the electron. The wavelength is given in Ångstrom units. At an accelerating voltage of 50,000 volts, routinely achieved by commercially available electron microscopes, the wavelength of the electron is 0.05 Å. This is shorter than the wavelength of blue light by a factor of about 100,000, and allows a considerable improvement to be made in resolving power. The resolution equation for light microscopes (equation 3–1) is not strictly applicable to electron microscopy because of special operating conditions. For electron microscopes, the resolution, d, is more simply related to the wavelength:

$$d = \frac{0.61\lambda}{\alpha} \qquad (3\text{–}4)$$

Note that α, rather than $n \sin \alpha$, is used in the expression for resolving power in electron microscopy. Because the microscope is operated under a vacuum, the refractive index, n, of the medium separating the image and the objective lens does not have significance and is omitted. Sin α is replaced by α because the angle of the cone of light entering the objective lens must be kept so small that α and $\sin \alpha$ have equivalent magnitude. The value used for α, approximately 4×10^{-3} radians, is fixed by the operating limitations imposed by spherical aberration (see below). Using this value and a wavelength of 0.05 Å in equation 3–4 gives a theoretical resolving power of 0.01 Å for electron microscopes operated at 50,000 volts.

The effect of spherical aberration on resolution in the electron microscope. In practice, it has not been possible to resolve points separated by distances as small as expected from equation 3–4. While some differences of opinion exist, the resolution achieved under the best operating conditions actually lies somewhere between 5 and 10 Å. The difference between the value and the theoretical minimum of 0.01 Å is almost entirely due to *spherical aberration* in the electron microscope. All magnetic lenses used in the electron microscope show spherical aberration. While this lens error is correctable in light optics, no practical systems have been devised to allow correction in the electron microscope.

In spherical aberration, illustrated in Fig. 3–12, the rays passing through the axis of a lens are focused at a different point from rays passing through the periphery of the lens. In Fig. 3–12, the rays a and a_1 originating from the object point P pass through the lens near the axis and are focused at point c. Other rays (b and b_1) travel through the lens at the periphery and are more strongly focused at point d; d is closer to the lens than c. Other rays will be focused between points c and d, depending on how far from the axis they pass through the lens. Spherical aberration thus effectively decreases the focal length of the lens from the axis to the periphery. The distance between c and d is a measure of the magnitude of this aberration. Because the image of the focused point P is spread out between c and d on the axis, P will be broadened into a disc rather than a point. The diameter of this disc, (called a disc of confusion in op-

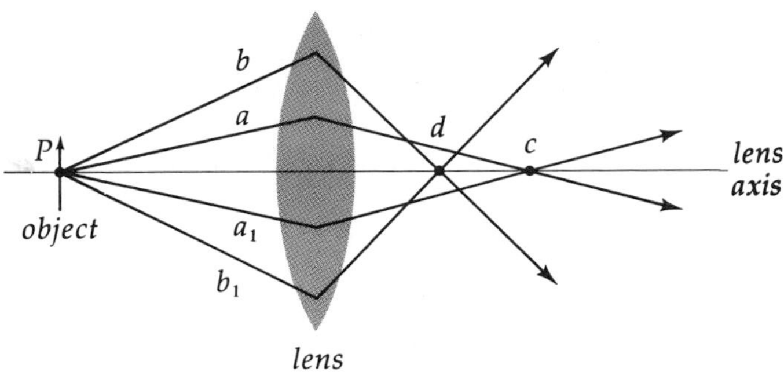

3–12. *Spherical aberration. Rays originating from an object point* P *are brought to different points of focus depending on whether they pass through the lens near the axis or the periphery (see text).*

tics) has a direct effect on the resolving power of the lens. If two specimen points are closer together in space than one-half the diameter of the discs of confusion for these points, they will not be separately resolved. The larger the discs of confusion, the farther apart points in the specimen must be to be resolved.

The presence of spherical aberration limits the resolving power of the electron microscope to values much larger than expected from calculations based on the wavelength of the electron. Where the angle of the cone of light entering the objective lens is small, the effect of spherical aberration on resolving power is expressed by the equation:

$$d = Cf\alpha^3 \qquad (3\text{–}5)$$

in which d = resolution, C is a constant, f is the focal length of the objective lens, and α is the angle of the cone of light entering the lens. As can be seen from this expression, d will be kept at a small value only at short focal lengths and very small angles of illumination, because d increases as the third power of α. In a given lens at a fixed focal length, resolution is thus related directly to the angle of illumination, which must be kept as small as possible to minimize spherical aberration. This poses a problem, because the angle of illumination α also enters into the equation for resolution based on the wavelength:

$$d = \frac{0.61\lambda}{\alpha} \qquad (3\text{–}4)$$

To improve resolution (smaller values for d) according to this equation α must be as *large* as possible.

For best possible resolution, the final value used for α in electron lenses is a compromise between these opposing requirements. Fig. 3–13 shows the angle of illumination of the objective lens (the critical lens as far as resolution is concerned) providing the best compromise. As α decreases to values of 3×10^{-3} radians or less, the resolving power decreases (larger values for d) according to the relationship $d = 0.61\lambda/\alpha$. As values for α increase past 4×10^{-3} radians, resolving power again decreases, because of an increase in spherical

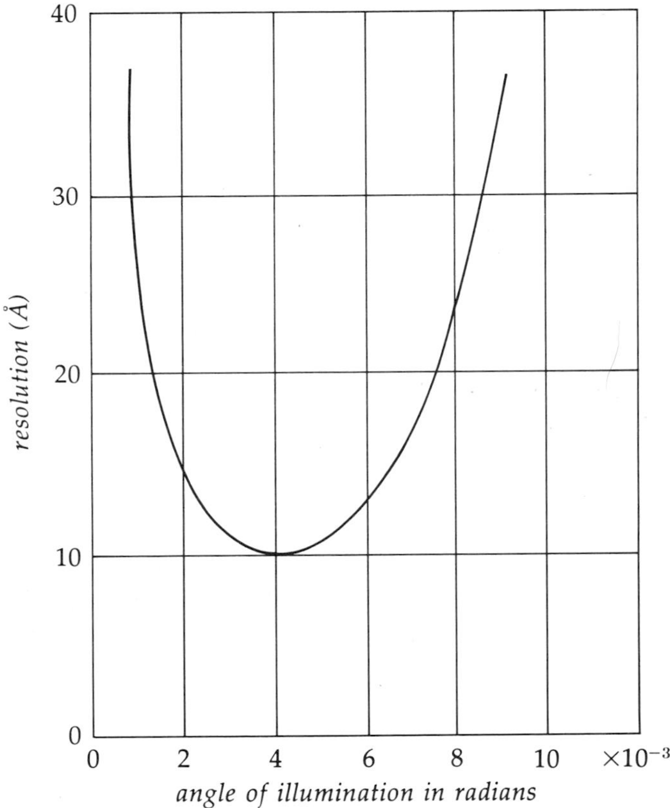

3–13. *Limit of resolution as determined by the combined effects of the illuminating aperture and spherical aberration in the objective lens of an electron microscope operated at 50 kV. At a value of 4×10^{-3} radians, objects with a diameter of 10 Å can be resolved. Resolution is worsened to the left of this point on the graph by the effects of small apertures, according to the resolution equation (d = $0.61\lambda/\alpha$). To the right of the minimum value, resolution is worsened by the effects of spherical aberration, according to the expression d = $Cf\alpha^3$ (see text).*

aberration according to the expression $d_s = Cf\alpha^3$. At approximately 4×10^{-3} radians the smallest value for d is obtained, giving the best resolution obtainable with electron lenses at the present time, approximately 5–10 Å.

At this level of resolution, the electron microscope appears to have reached its maximum development, unless an unexpected breakthrough occurs in the correction of spherical aberration. At the 5–10 Å level, resolution is at a molecular or near-molecular level. If resolution can be improved by correction of spherical aberration, the electron microscope will resolve detail at the atomic level. During the early years of development of the electron microscope, most physicists were convinced that spherical aberration was fundamentally uncorrectable and that resolution could not be improved significantly past the 5–10 Å level. However, some recent developments give hope that at least partial correction can be made. It remains to be seen whether a commercially practical solution will be possible.

Some gains in resolution can be obtained by increasing the accelerating voltage, which further shortens the wavelength of the electron. Efforts to improve resolution by this method have resulted in the production of microscopes with accelerating voltages of 1,000,000 V or more. Some improvement of theoretical resolving power is obtained by this method, as Table 3–1 indicates.

Table 3–1 *Improvement of Theoretical Resolving Power by Increasing Accelerating Voltage*

Accelerating voltage (kV)	*Wavelength (Å)*	*Resolution (Å)*
10	0.122	3.7
20	0.086	2.9
50	0.054	2.1
100	0.037	1.7
200	0.025	1.3
500	0.014	0.9
1000	0.009	0.7

From *Electron Microscopy of Cells and Tissues* by F. S. Sjöstrand, 1967. Courtesy of F. S. Sjöstrand and Academic Press, Inc.

Magnification and resolution. Magnification has not yet been mentioned as a parameter in our discussion of resolution. Magnification is important primarily as a method of increasing the size of the image to a level permitting resolution by the human eye, which can discern as separate details objects approximately 0.2 millimeters apart. To see the smallest objects resolved by the light microscope requires a magnification of approximately 2000 to 4000 times. Increases past this value will result merely in making an indistinct image larger but not clearer. In electron microscopy, human resolution of 10 Å detail requires a final magnification of approximately 200,000 to 300,000 times. This is usually accomplished by some combination of instrumental magnification and photographic enlargement.

Construction of the Electron Microscope

There is very little difference between the performance capabilities of different commercially available electron microscopes. Fig. 3–14 shows a high performance instrument capable of resolution on the order of 5–10 Å. The massive central column houses the magnetic lenses and lens coils and provides shielding from X-rays that arise when metal surfaces are struck by the electron beam. At the top of the column (Fig. 3–15) is the electron gun, consisting of a filament, grid, and anode. The filament, a thin tungsten wire, is heated by an electric current, causing electrons to leave the surface. The filament and grid, well insulated from the rest of the column, are maintained at −50,000 to −100,000 volts. The anode is grounded and thus is strongly positive with respect to the filament and grid (cathode). Electrons are attracted through the grid to the anode and, as they traverse this space, are accelerated to high velocity, with wavelengths in the region 0.054–0.037 Å. The voltage applied must be closely controlled to insure that all electrons crossing the space between the cathode and anode are accelerated to the same wavelength.

Electrons passing through the anode are focused on the specimen by the condenser lens. In modern instruments, two condenser lenses are placed in tandem to allow focusing of a very small, intense spot of electrons. The specimen is movable and in high performance instruments can be traversed and controlled within limits defined by a few Ångstroms. The electrons passing through the specimen are focused into an image by a system of lenses—the objective, intermediate, and projector lenses (Fig. 3–15). Of these, the objective lens is the most critical, because it determines the resolving power of the microscope. Spherical aberration is of significance only in this lens. The image is

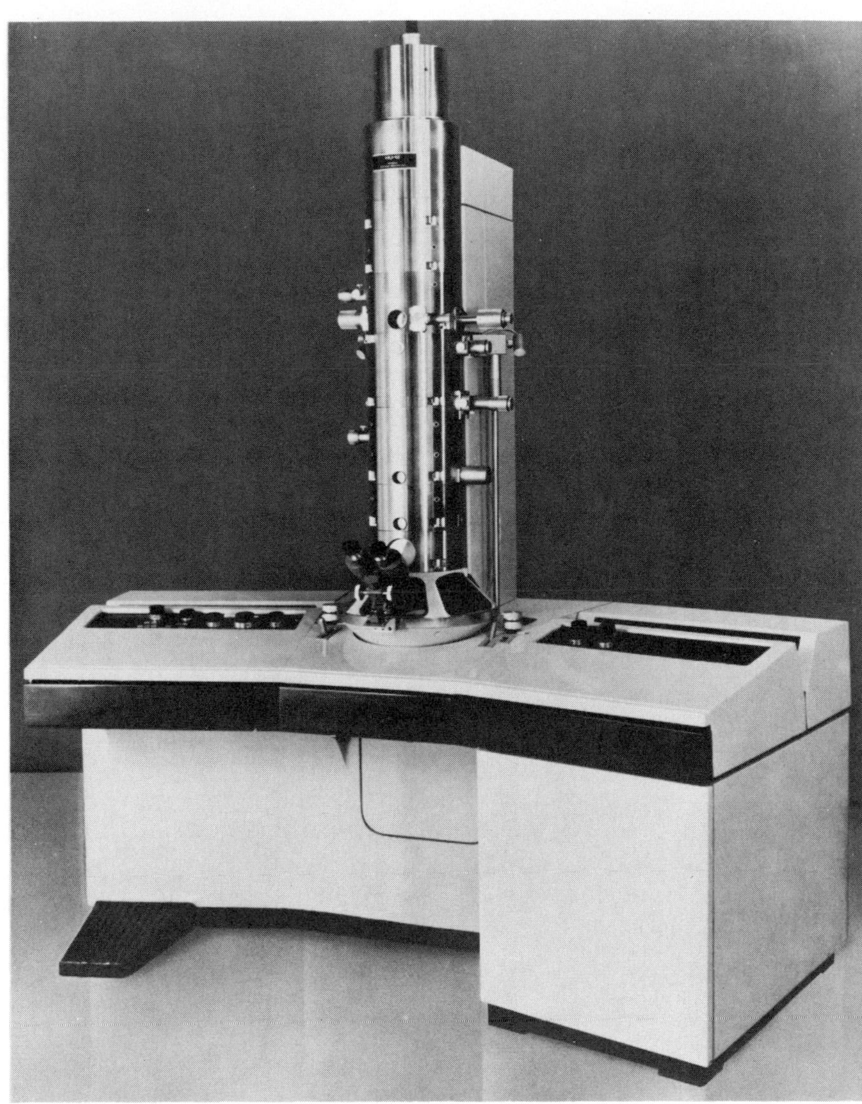

3–14. A high-performance electron microscope. The electron lenses are housed in the central column, which is maintained under a high vacuum. To the right and left of the column are banks of controls by which lens current (focus) can be regulated. The image can be directly viewed on a fluorescent screen through the binocular microscope; for permanent high resolution records, photographs are made. Specimens and photographic plates can be introduced without breaking the vacuum in the main column. Courtesy of the Perkin-Elmer Corporation.

magnified by these three lenses between 300 and 300,000 times. For most specimens, values between 5,000 and 50,000 times are routinely used.

The magnified image is projected onto a fluorescent screen. This screen is coated with a layer of crystals that respond to bombardment by electrons by emitting visible light. In this way, the electron image is converted to a visual image. The resolution obtained on the fluorescent screen, viewed through a binocular microscope, is limited by the comparatively large crystals present on the screen surface. Permanent, high resolution records of the image are made by exposing a photographic plate to the electron beam. The resolution of the recorded image is significantly improved by this method, because the crystal size in a photographic emulsion is considerably smaller than in the fluorescent screen. Ordinary films and plates are used, because the response of the photographic emulsion to electrons and light is essentially the same. As the entire microscope column, including the specimen, viewing screen, and camera, are kept under a high vacuum, provision is made for the exchange of specimens and photographic plates without disturbing the microscope vacuum.

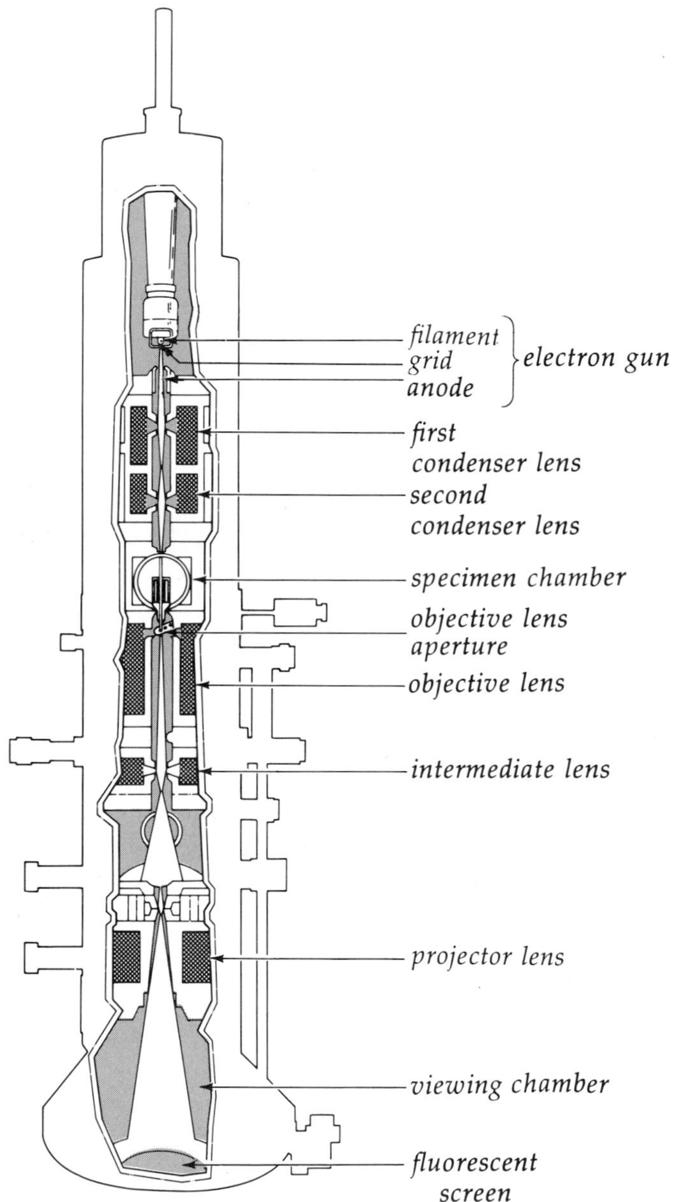

3–15. Cutaway view of the lens column of the instrument shown in Fig. 3–14. Modified from the original diagrams supplied by the Perkin-Elmer Corporation.

Factors Affecting Image Formation in the Electron Microscope

Scattering. The formation of the image in the electron microscope differs to a considerable extent from that in light microscopy. Three elements contribute to the electron microscope image: *scattering, diffraction,* and *phase contrast.* Of these, only scattering is well understood. Scattering refers to a disturbance in path and velocity experienced by electrons as they pass through the specimen. If the change induced by the specimen results only in an alteration in an electron's path with no change in velocity and wavelength, the scattering is *elastic.* A change in path that includes absorption of energy from the electron, resulting in a change in velocity and wavelength, is termed *inelastic* scattering. Only elastic scattering is significant in image formation, because electrons that have changed in wavelength are focused by the lenses on different image planes from the main body of elastically scattered electrons and fall on the plate or screen as background fog. This background does not interfere too greatly with the focused image.

Both elastic and inelastic scattering result from an interaction between beam electrons and specimen atoms. Deflections in path are caused when beam electrons are attracted by positively charged atomic nuclei in the specimen or repelled by the negatively charged electron cloud surrounding the nuclei. This effect increases as electrons pass closer to specimen atoms. The effect of *deflection* (attraction) caused by the positively charged atomic nucleus depends on the atomic number of the specimen atom, in the following manner:

$$\theta = \frac{Ze}{Vr} \tag{3–6}$$

where θ is the angle of deflection of the passing electron, Z the atomic number, e the unit charge of the atomic electrons, V the beam voltage, and r the distance separating the deflected electron from the nucleus. Because the unit charge and voltage are constant values, the deflection in this case increases both as beam electrons pass more closely to an atomic nucleus and as the atomic number increases. For this reason the atoms of heavy metals, with high atomic numbers, have proved of great value as electron "stains."

The electron cloud surrounding the nucleus, which is negatively charged, deflects (repels) beam electrons according to a similar relationship:

$$\theta = \frac{e}{Vr} \tag{3–7}$$

where θ is the angle of deflection, e the unit charge, V

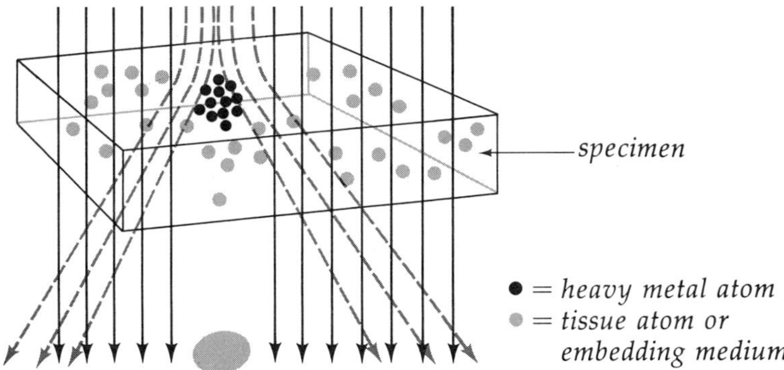

3–16. Scattering of electrons by an object point, represented by an aggregation of atoms of high density or atomic number in the specimen. As a result of scattering, fewer electrons fall on the corresponding region of the focused image, producing a "shadow" of the object point at this level.

the beam voltage, and *r* the distance separating the electron from the electron cloud. Note that this equation is the same as equation 3–6, except that the factor Z, for the atomic number, is not included. Thus the nucleus is Z times more effective in deflecting the beam than the electron cloud. Where Z is a large number, such as in uranium (Z = 238), the nucleus may be considerably more effective in deflecting the electron beam.

A related factor that contributes to scattering is the total number of atoms per unit of volume in the specimen: the greater the number of atoms, the greater the probability of a collision or near miss by a beam electron. This factor, the number of atoms per unit volume, is the *density*. It is important to distinguish between the effects of atomic number and density in scattering. The optimum condition to produce scattering by a given image point, according to these relationships, would be to pack as many atoms as possible (density) of the heaviest atom (atomic number) into the image point.

With this information, we can now examine the interaction of the electron beam with the specimen (Fig. 3–16). Suppose that within the specimen is a concentrated aggregation of atoms with high atomic number. This can be considered as an "image point." Electrons encountering the image point will be scattered according to the relationships discussed in the foregoing paragraphs. If the angle of scattering is wide enough, the stray electrons will be eliminated from the

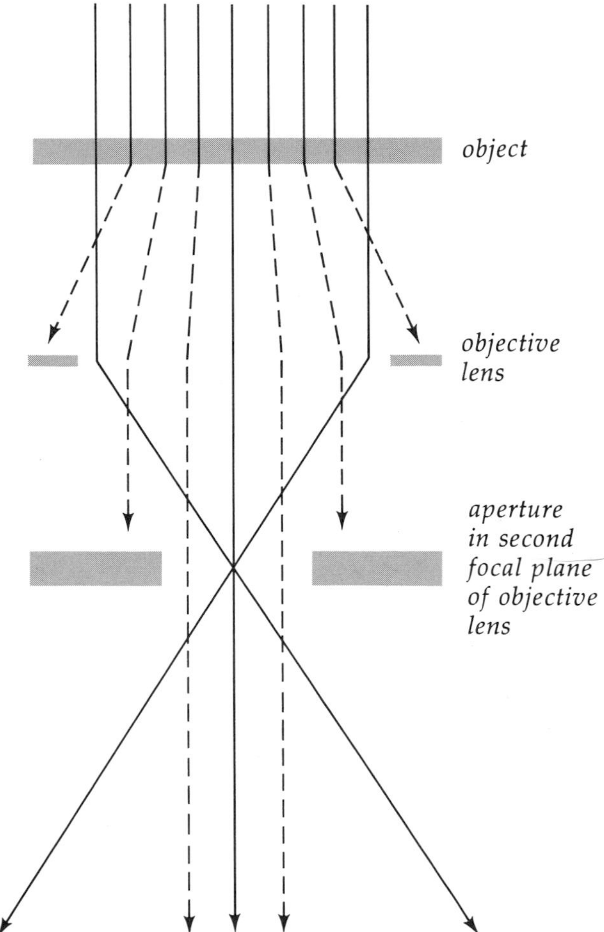

3–17. Elimination of widely scattered electrons by a physical aperture placed at the focal point of the objective lens. This aperture is so small that a large proportion of the scattered electrons, which otherwise would fall on the image plane as a general background fog, are eliminated. The result is an improvement in the contrast and sharpness of the image. Redrawn from Electron Microscopy of Cells and Tissues *by F. S. Sjöstrand, 1967. Courtesy of F. S. Sjöstrand and Academic Press, Inc.*

beam, some by collisions with the sides of the lenses and some by an aperture (Fig. 3–17), which is placed below the objective lens at the focal point. This aperture is so small, on the order of 50 microns or less, that only the unscattered electrons and a relative few of the scattered electrons pass through. In eliminating the scattered electrons a "hole," representing the image point, is produced in the focused image.

After magnification by the objective, intermediate, and projector lenses, the image point, focused on the fluorescent screen, appears darker than surrounding regions, because fewer electrons per unit time strike the surface of the screen in this area. The images of all other parts of the specimen containing scattering points will also appear dark on the screen. These points collectively trace out the electron image. Because density in the specimen is represented by darker areas on the fluorescent screen, the electron image on the screen is considered to be "positive." The opposite effect is noted in photographic emulsions, because points on which fewer electrons fall will appear *lighter* after development. This "negative" photographic image is usually reversed to produce a positive image similar to the density pattern seen on the fluorescent screen (except for "shadowed" specimens—see below).

Contribution of Diffraction and Phase Contrast to the Electron Image

The effects of scattering have been treated by viewing electrons as charged particles traveling through the specimen. Electrons, as waves, also interact to produce diffraction and phase contrast as in the light microscope. The image patterns produced by phase contrast and diffraction are superimposed on the image made from scattering. While these effects are not completely understood, each has an important effect on the electron image, and all must be taken into account in the interpretation of electron micrographs.

Diffraction. The diffraction lines around an image point, it will be recalled, arise from interference between undisturbed rays and the rays diffracted by an image point. Diffraction in the electron microscope differs from the diffraction of light in two significant ways. The source in the electron microscope is relatively large with respect to the object and thus does not represent a point source. This causes confusion of the spherical wavefronts leaving the source and reduces the precision with which the crests and troughs of interfering waves intersect. The resultant confusion in the image usually reduces to invisibility all except the first set of light and dark lines around an image point. A second difference, very significant for electron microscopy, is an unexplained diffraction that causes a bright fringe to appear either just inside or outside the shadow of an edge or point (Fig. 3–18). In images that are slightly *underfocused* (objective lens current not strong enough), this bright fringe falls just outside a shadow, increasing the contrast and giving the object greater apparent sharpness. If the image is slightly overfocused, the bright fringe falls just inside the shadow edge, causing the edge to be apparently outlined with an extra dark line. These unexplained maxima, or lines of greater intensity, are superimposed on the usual diffraction image. Because of the unexplained maxima, the alternating light and dark lines surrounding an edge or point in the image are strongest in images that are just out of focus and are at an absolute minimum in either in-focus or greatly out-of-focus images. Electron micrographs are usually slightly underfocused to take advantage of the greater contrast and apparent image sharpness produced by the underfocus diffraction pattern (see Fig. 3–19).

Phase contrast. It has not been possible to improve the visibility of image points in the electron microscope by the introduction of large phase differences in diffracted rays, as is done with the phase plate in the light microscope. However, an analogous effect does take place. The path taken by electrons through the magnetic lenses, as mentioned, is highly complex. This path is helical, and electrons passing through the lens closer to the periphery travel along a helical path of significantly greater diameter than electrons nearer the axis. The greater distance traveled is sufficient to place the peripheral electrons out of phase with axial electrons, and interference between these electrons takes place in the image plane. Because electrons scattered by object points, if not deflected so widely as to be eliminated from the beam, are more likely to traverse the lenses near the periphery, these are more likely to be delayed than unscattered electrons. The effect in the image, where interference between the scattered and unscattered electrons takes place, is to increase the contrast of scattering points in the image.

According to some authorities in electron optics, these accidental phase effects are the most significant factor in increasing the contrast of fine detail in the image. As in diffraction, the phase contrast effect is at a minimum at focus, and is enhanced by slight defocusing of the image.

Fig. 3–19 shows the combined effects of scattering, diffraction, and phase contrast as the image goes from underfocus to overfocus. Note that significant graininess is present in the underfocus and overfocus image (*a* and *d*). This background grain results from diffraction and phase contrast effects and is frequently mis-

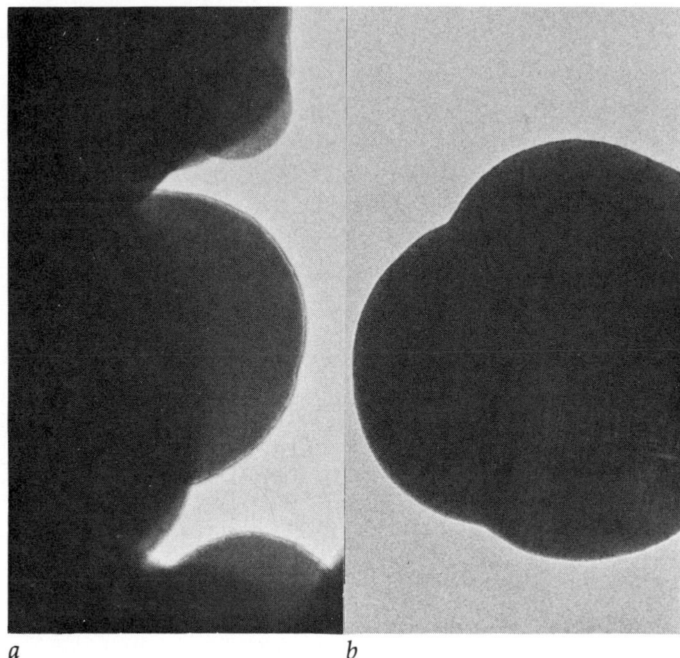

3–18. Unexplained bright fringes in the electron image. Both overfocus (a) *and underfocus* (b) *fringes are visible in this micrograph. Carbon particles.* × *90,000. Courtesy of Barry R. Zirkin.*

interpreted as a structural configuration. Thus, it is always necessary to be skeptical about apparent subunit detail in electron micrographs, especially when graininess of the same periodicity as the subunit is present over the entire micrograph. Graininess arising from diffraction and phase contrast has a periodicity in the 30–40 Å range near focus, but almost disappears at focus. Subunit detail in this range is much more likely to be "real" if it persists at focus. A through-focal series, of the type shown in Fig. 3–19, is therefore most useful in determining the authenticity of fine detail in the image.

Specimen Preparation for Electron Microscopy

The special operating requirements of the electron microscope closely limit the conditions under which the specimen can be examined in the instrument. Two requirements are critical: the specimen must be dry and nonvolatile, and it must be very thin. The second condition depends on the fact that electrons, even at high velocities, have very low penetrating power. Specimens more than about 1500 to 2000 Å thick are essentially opaque to electrons. Specimen thickness also has an effect on resolution. As thickness of the specimen increases, resolution is worsened because of the superimposition of details and the general scattering of the electron beam.

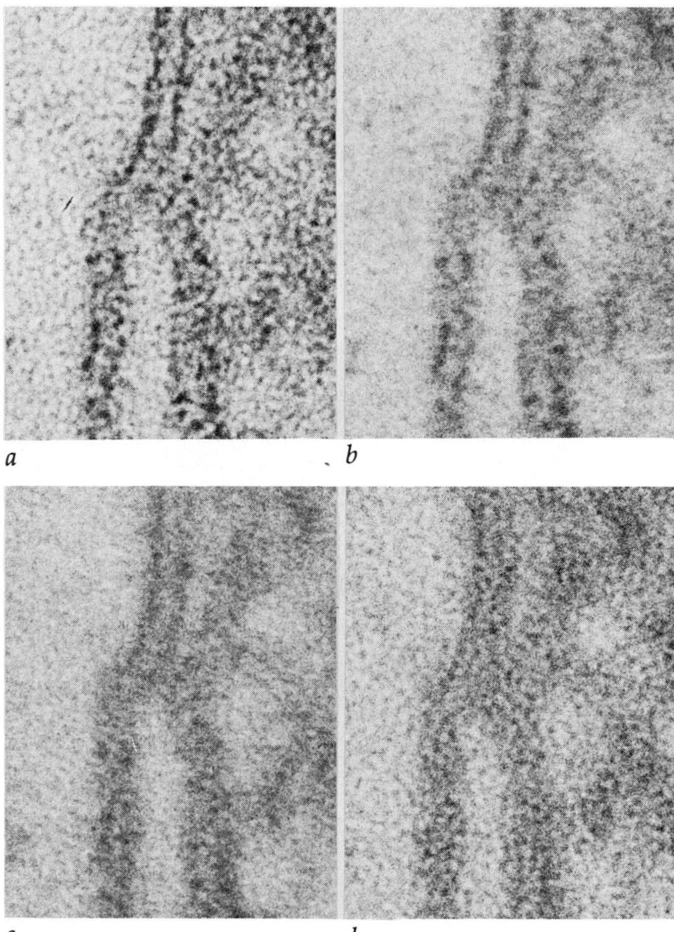

3–19. Combined effects of diffraction and phase contrast in a highly magnified image of the outer membrane of a mitochondrion. a, *Underfocused;* b, *slightly underfocused;* c, *slightly overfocused;* d, *overfocused. The increased periodicity or graininess at underfocus and overfocus is caused by diffraction and phase contrast in the image plane; these effects are at a minimum at focus.* × *1,100,000. From* Electron Microscopy of Cells and Tissues *by F. S. Sjöstrand, 1967. Courtesy of F. S. Sjöstrand and Academic Press, Inc.*

In spite of the difficulties presented by these special operating conditions, a number of preparative techniques have been successfully developed for electron microscopy. Most of these involve either the fixation, embedding, and thin sectioning of specimens, or the isolation and drying of particulate subfractions with dimensions small enough to allow viewing in the microscope.

Fixation, Embedding, and Thin Sectioning

Fixation. Adequate chemical fixation and the allied cytochemical techniques, such as the development of stains specific for particular macromolecular complexes, are incompletely solved problems in electron microscopy. Fixation has depended either on the use of buffered solutions of osmium tetroxide alone, or on combinations of fixatives in which aldehydes such as formalin and glutaraldehyde are followed by osmium postfixation. The quality of fixation achieved by solutions of osmium tetroxide came as something of a surprise because this particular fixative was not highly regarded by light microscopists. Cells fixed by osmium tetroxide appear almost featureless, or "empty," under the light microscope. We now know that this apparent emptiness results from the fact that osmium fixes cellular details at levels far below the resolving power of the light microscope, giving good preservation at levels approaching 10–20 Å.

In addition to the good quality of fixation produced, osmium tetroxide carries with it the advantage that the osmium atom has a high atomic number. After fixation with osmium tetroxide, ordered deposits of metal or bound ion are left in the tissue in density sufficient for visibility in the electron image. In cell structures that react readily with osmium tetroxide, such as the membranes of the mitochondrion or chloroplast, a pattern of deposition is created that very clearly traces the outlines of the organelle.

The aldehyde fixatives, which do not contain atoms of high atomic number, impart no density to fixed preparations, even though the quality of fixation is good. The value of these substances as fixatives for electron microscopy was missed until recently for this reason. However, when combined with osmium as a postfixative, aldehydes have proved to yield a slightly better fixation than osmium alone. At least some of the improvement rests on the fact that the aldehydes penetrate and fix tissue blocks much more rapidly than

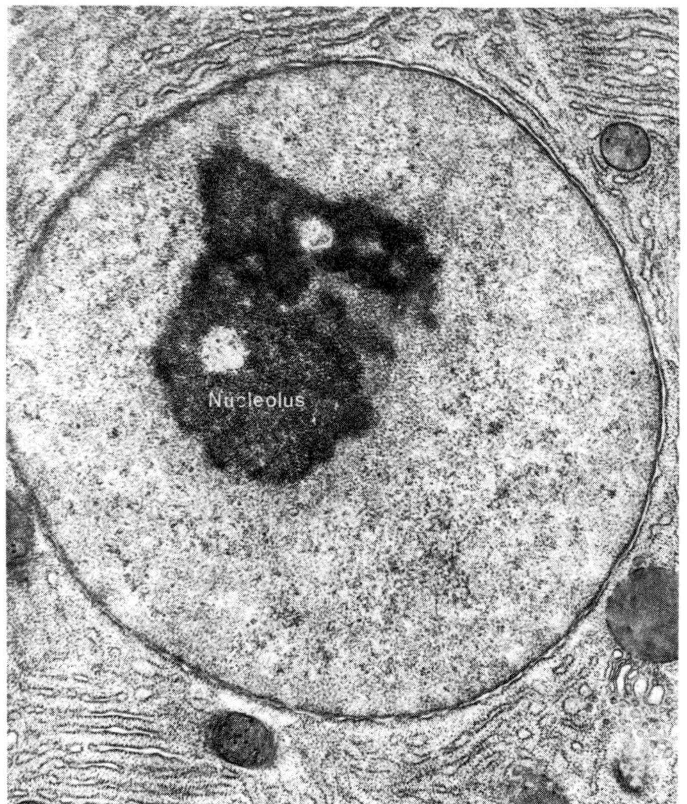

3–20. Cell from bat pancreas fixed in buffered osmium tetroxide. Compare with Fig. 3–21, showing a cell from the same tissue fixed in aldehyde and postfixed with osmium. Typically, the nucleus in tissue fixed with osmium shows a more homogeneous distribution of the nuclear material outside the nucleolus. Where visible, the individual fibers of the chromatin material are frequently seen to be of slightly smaller diameter than in aldehyde-osmium fixed material. × 14,000. From The Cell *by D. W. Fawcett, 1966. Courtesy of D. W. Fawcett and W. B. Saunders Company.*

osmium, which diffuses into tissue very slowly. Although basically the same in appearance, tissues fixed by osmium tetroxide alone and tissues fixed by aldehydes followed with osmium postfixation show some differences (Figs. 3–20 and 3–21). Nuclear fibers are somewhat thicker after aldehyde-osmium fixation, and often the distinction between aggregated and dispersed chromatin in the nucleus is enhanced (the light and dark areas in the nucleus in Fig. 3–21). In addition,

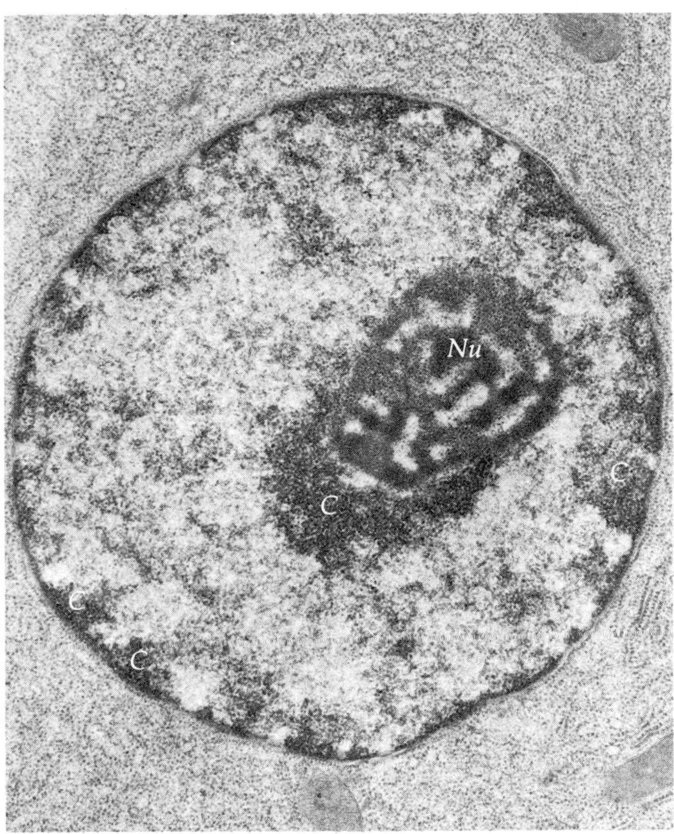

3–21. *Cell from bat pancreas fixed in glutaraldehyde and postfixed in osmium. The nuclear material outside the nucleolus* (Nu) *is coarser, and is differentiated into areas of greater and lesser aggregation. This differentiation of the chromatin* (C) *is typical of aldehyde-osmium fixation. Other conspicuous differences are seen in membranes, which usually appear more regular after aldehyde-osmium fixation as compared to fixation with osmium alone.* × *15,000. From* The Cell *by D. W. Fawcett, 1966. Courtesy of D. W. Fawcett and W. B. Saunders Company.*

the spacing between the two membranes of the nuclear envelope is more regular after aldehyde-osmium fixation. Other differences are more subtle but distinctive enough so that an experienced observer can usually recognize whether aldehyde-osmium or standard osmium fixation has been used.

Another, less valuable, fixative is potassium permanganate. This fixative was formerly much used for plant tissue, because osmium frequently yields poor results in plants for reasons that are not clearly understood. Permanganate fixes and adds density almost exclusively to membranes (see Fig. 3–22). Even membranes do not appear to be well fixed, because mitochondria are often rounded, and many gaps or separations are visible in other membrane systems after permanganate fixation. The nucleus is characteristically almost featureless after permanganate fixation. Because of the general appearance of permanganate-fixed tissue, this fixative is not considered as effective as osmium tetroxide. In recent years, aldehyde-osmium mixtures, which give much improved fixation in plant tissues, have almost entirely supplanted permanganate as a fixative for botanical specimens.

The battery of fixatives used in electron microscopy has been developed largely by cut-and-try adaptation of established light microscope techniques. Almost nothing is understood about the chemical interaction of fixatives with biological molecules. To date, attempts to interpolate the known chemical reactions of fixatives with the image produced in the electron microscope have met with little success. For example, many of the reactions of the osmium ion with various chemical substances have been well studied. In the test tube, osmium is known to interact with lipids, being reduced and precipitated by double bonds in the hydrocarbon chains of the fatty acid residues. However, it has not been possible to satisfactorily interpret the electron microscope image of osmium-fixed membranes in the light of these chemical findings. Most hypotheses of membrane structure consider that the nonpolar, hydrophobic part of the lipid molecule, containing the double bonds, is inside the membrane. The density pattern, however, usually indicates binding or precipitation of osmium at the surface of the membrane. The reaction of nonmembranous structures with osmium is similarly uninterpretable. Evidence that such interactions do occur is provided by the obvious density of such structures as chromosomes, the nucleolus, and ribosomes after osmium fixation. Some of this density may result from the known reaction of osmium with the disulfide linkages of proteins; nucleic acids, at least in the test tube, appear to be completely inert to osmium tetroxide. Therefore, the protein rather than the nucleic acid of the chromosome may interact with osmium to produce the density seen in the image. This interpretation is merely speculative; the same can be said for almost all current hypotheses about the interaction of fixatives with the parts of the cell. Along with

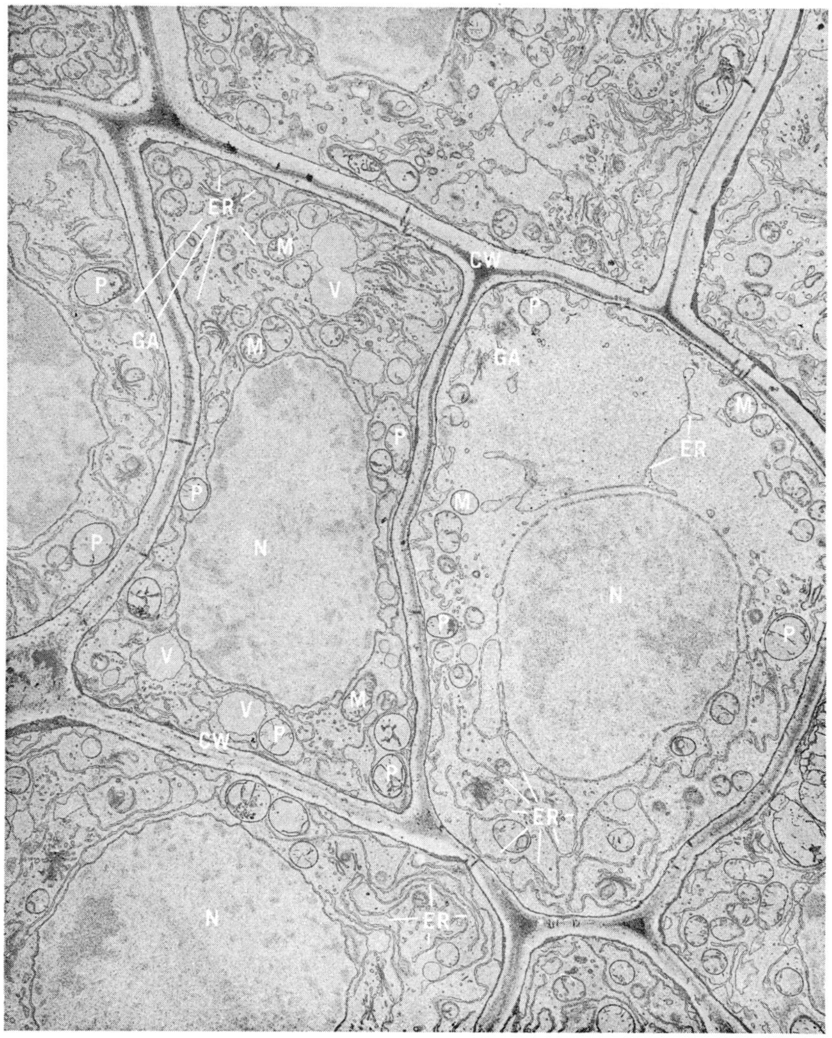

3–22. Plant tissue fixed in potassium permanganate. The cell appears empty of organized structures except for membranes. Even these are distorted; separations appear at various points, and mitochondria (M) are rounded and swollen in appearance. The nucleus (N) shows little internal differentiation or density after permanganate fixation. ER, endoplasmic reticulum; CW, cell wall; P, plastid; GA, Golgi apparatus; V, vacuole. Courtesy of W. A. Jensen and Wadsworth Publishing Company, Inc.

the stains (see below) used in electron microscopy, fixation remains a highly underdeveloped area of investigation in electron microscopy.

Embedding and sectioning. The main problem presented by embedding and sectioning for electron microscopy is the requirement that the sections must be very thin, from about 300 to 1000 Å, but still able to resist both mechanical disruption during handling and the heat generated by the electron beam. Surprisingly enough, these requirements have been met very successfully. As with fixatives, embedding and sectioning techniques were developed by adaptation of methods previously used in light microscopy. By the late 1950's it was discovered that diamond knives or knives made by fracturing plate glass are satisfactory for cutting very thin sections, on the order of 500 Å or less. To support such thin material, it has been necessary to turn from the paraffin or agar media used for light microscopy to plastics such as methacrylates and epoxy resins. The methacrylates, widely used in the late 1950's and early 1960's, were found to melt and evaporate in the electron beam, however, and have since been supplanted almost completely by the epoxy resins. These less volatile plastics have generally excellent characteristics for electron microscopy.

Tissue stains for electron microscopy. A number of substances, such as uranyl acetate and lead citrate, are used as stains to increase the density of structures within sectioned tissue. Because the substances used, all salts of heavy metals, are similar in reactivity to osmium, no new structures are revealed in tissues after exposure to these stains. In general, organelles or substructures made dense by precipitation or binding of the osmium in the fixative are simply made more dense by the stain. Some limited selectivity is noted; for example, uranyl ion is more specific than osmium in binding to structures containing nucleic acids, and the chromosomes, nucleolus, and ribosomes are enhanced in contrast after use of this stain. Lead staining, although closely similar to osmium, appears to be slightly more selective for structures containing protein. Phosphotungstic acid is also frequently used as a general protein stain.

Efforts to develop stains with greater specificity have met with little success. The highly specific staining reactions used in light microscopy have been of no use, because they typically employ organic dyes with density and atomic number not significantly different from tissue. Thus these dyes, although recognizable as color under the light microscope, do not scatter electrons sufficiently to have a visible effect on the electron image.

Some indication of the possible trend of future developments in this area may be taken from the work of M. L. Watson and W. G. Aldridge. Watson and Aldridge developed a technique with high specificity for nucleic acids, using as a stain indium trichloride. Normally, this substance is relatively nonselective and reacts with amino, hydroxyl, carboxyl, carbonyl, aldehyde, sulfate, and phosphate groups. By a combination of aldehyde fixatives, reduction by lithium borohydroxide, acetylation by acetic anhydride, and washing to remove membrane phospholipids, the reactivity of tissue for indium was practically limited to the phosphate groups of nucleic acids. This staining procedure yields density with a high degree of selectivity in structures containing nucleic acids (Fig. 3–23). The part of the technique significant for future developments in cytochemistry is the concept of blocking nonspecific reactivity. This approach will surely become central in future work on stains for electron microscopy. Another approach of value may be the linking of heavy metals to organic molecules that bind specifically to biological macromolecules.

Few additional analytical staining methods are available to the electron microscopist (see also the Bernhard technique for RNA staining in Chapter 10). However, this situation may improve. The sectioning and staining techniques for electron microscopy have existed for only twenty-five years; at an equivalent period in light microscopy, in the mid-1800's, the prospects for developing specific cytochemical methods must have appeared just as doubtful.

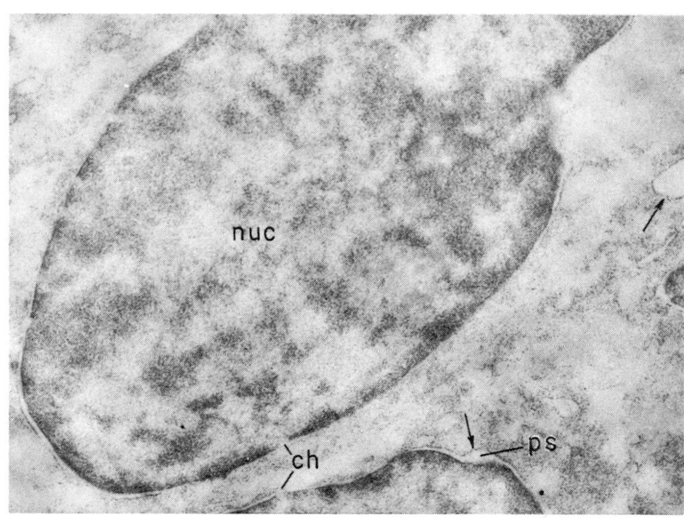

3–23. *Rat mammary tissue fixed with acrolein (an aldehyde) and stained by the indium technique of Watson and Aldridge. Only chromatin and ribosomes are stained. Rows of ribosomes outline elements of the endoplasmic reticulum and the outer membrane of the nuclear envelope (arrows). The perinuclear space* (ps) *is outlined on the nuclear side by chromatin* (ch) *and on the cytoplasmic side by a layer of ribosomes.* nuc, *nucleus.* × *16,500. Courtesy of M. L. Watson and W. G. Aldridge and The Rockefeller University Press.* [J. Cell Biol. *11:257 (1961)*]

Isolation of Cellular Components for Electron Microscopy

The techniques used to isolate cell organelles in electron microscopy have a history that predates thin sectioning and actually extends back to the first application of the instrument to biological specimens. Because apparatus for thin sectioning was not available to the early electron microscopists, isolation was the principal approach used in pioneer studies. Isolation methods, as a rule, have two major sources of difficulty: damage to the organelle during isolation and drying for electron microscopy, and identification once removed from the cell. As a part of the latter difficulty,

3–24. Positively stained microtubules isolated from the salamander Taricha granulosa. *A central lumen of lower electron density is just visible within the microtubules. × 130,000. Photograph by the author.*

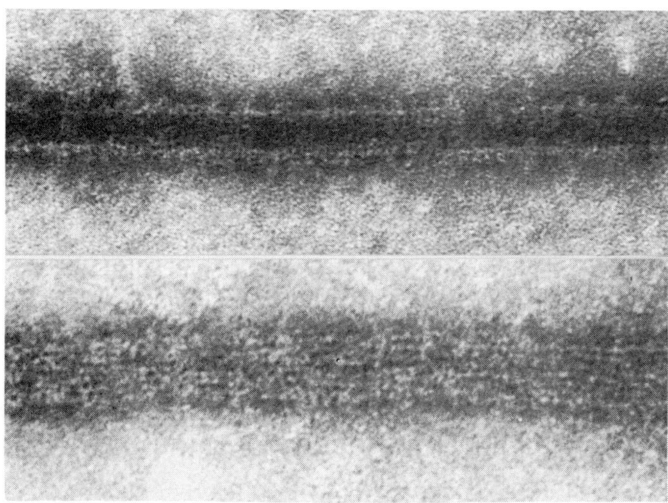

3–25. Negatively stained microtubules isolated from the newt Triturus viridescens. *Stain has not penetrated into the lumen of the microtubules, but the subunits of the tubule walls are clearly outlined. These wall subunits cannot be discerned in positively stained microtubules and are only indistinctly seen in thin-sectioned preparations. × 260,000. Courtesy of J. G. Gall and The Rockefeller University Press.* [J. Cell Biol. *31:639 (1966)*]

it is often impossible to determine what is debris and what is organelle. Before thin sectioning gave some insight into the ultrastructure of the cell, the identification of objects in isolated preparations was particularly difficult. Consequently, although isolation methods predated sectioning by a number of years, these techniques produced very few concrete results until the development of successful thin-sectioning techniques in the late 1950's.

Positive and negative staining. Isolated preparations are usually stained or shadowed before examination in the microscope. Staining may be "positive" or "negative." Positive staining simply increases the density of a preparation by reacting it with the stain. Uranyl acetate is most often employed for this purpose. Negative staining is an interesting technique, in which the stain is allowed to dry around the surface of the isolated specimen. The stain is deposited into surface crevices, revealing external details down to a level of 10 Å or slightly less. An amorphous rather than crystalline deposit is required for negative staining to be successful. Uranyl acetate and phosphotungstic acid, when used as negative stains reveal surface details otherwise invisible in isolated preparations. Figs. 3–24 and 3–25 show positively and negatively stained isolated microtubules. The surface detail in the negatively stained microtubules in Fig. 3–25 is in the vicinity of 30 Å or less.

Shadowing. The contrast and surface detail of specimens is frequently increased by a technique known as *shadowing*. In this technique, the specimen is placed in a vacuum chamber, at a specified distance and angle from a filament consisting of a heavy metal such as platinum (see Fig. 3–26). As the filament is heated to high temperatures by an electric current, the heavy metal evaporates and deposits on the specimen. Because the filament is located to one side, the evaporated metal deposits only on the surfaces facing the source. This produces a shadowed effect in the specimen. Under the electron microscope, shadowed preparations appear as if a bright, sharply focused light were directed at the specimen from one side (Fig. 3–27). In this case, a "negative" image is always used in the finished electron micrograph. The light areas in Fig. 3–27 represent a layer of the evaporated metal. These in reality are "dark" to the electron beam, which passes through the specimen only in the "shadow" areas, darkening the corresponding regions on the photographic plate. This produces the usual negative image, in the sense that electron-dense regions in the specimen are light on the photographic plate. However, the negative image is most useful in this case, because it gives the illusion of a bright light striking the specimen from the side.

Resolution in shadowed preparations is limited by the size of the crystallites, or small aggregates of metal, that deposit on the surface of the specimen. Because these crystallites average about 20 Å in diameter, specimen details smaller than this value are obscured. Although this places a lower limit on the resolution obtainable in shadowed preparations, this technique has been used to great advantage in studies of both isolated organelles and macromolecules. Along with negative staining, shadowing forms the basis for interpretation of the three-dimensional surface structure of isolated specimens and produces some of the most esthetically pleasing electron micrographs. Improving the resolution of shadowed preparations will depend on development of methods to reduce crystallite size.

Freeze etching. Freeze etching, a recently developed technique, combines some characteristics of both sectioned and isolated preparations. In this method, a block of living tissue is rapidly frozen by immersion in liquid nitrogen. The freezing of small tissue blocks at the low temperatures obtainable with liquid nitrogen is almost instantaneous and causes little apparent disturbance to the specimen. The unfixed, frozen tissue block is then transferred to a vacuum chamber and frac-

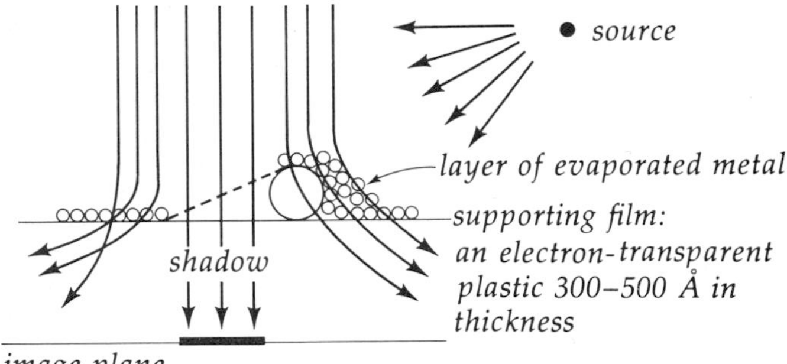

3–26. Diagram of the shadowing technique. Metal is evaporated from a source placed to one side of the specimen. A layer of metal is deposited along areas of the specimen facing the source; other areas, in the "shadow" cast from the source, are not coated. The effect in the electron image is the creation of surface relief, as if a bright light were directed at the specimen from one side.

3–27. Thick filaments (myosin) isolated from barnacle muscle and prepared for electron microscopy by the shadowing technique. Each filament measures between 175 and 200 Å in diameter. × 30,000. Courtesy of C. R. Zobel.

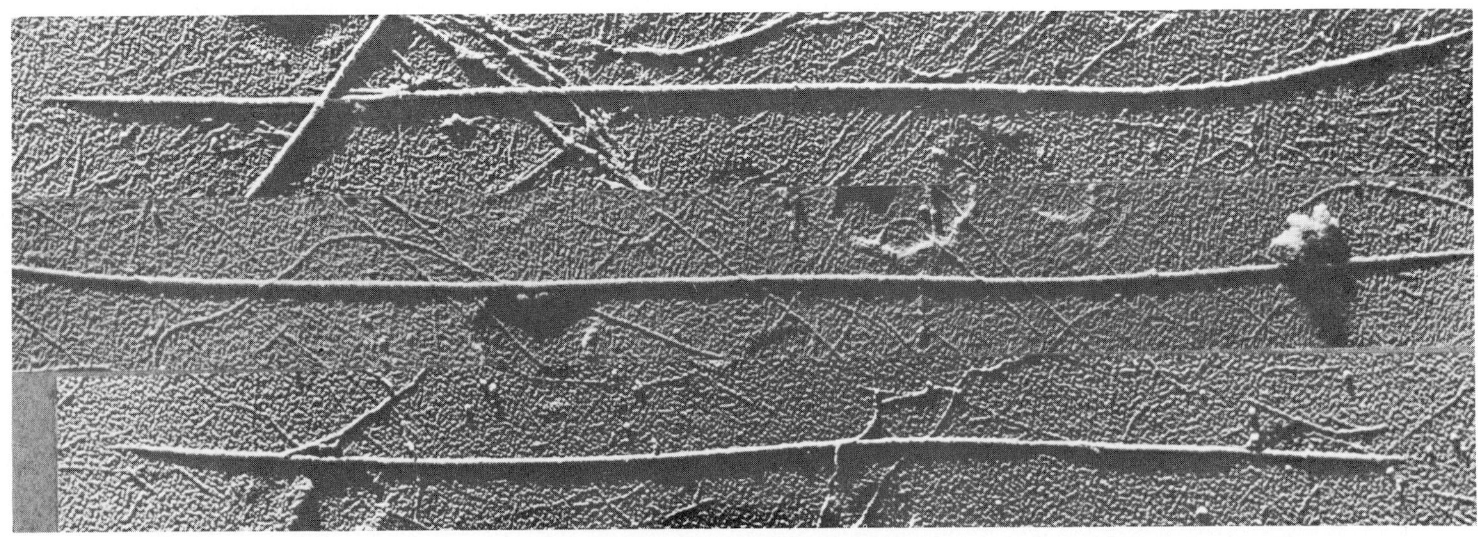

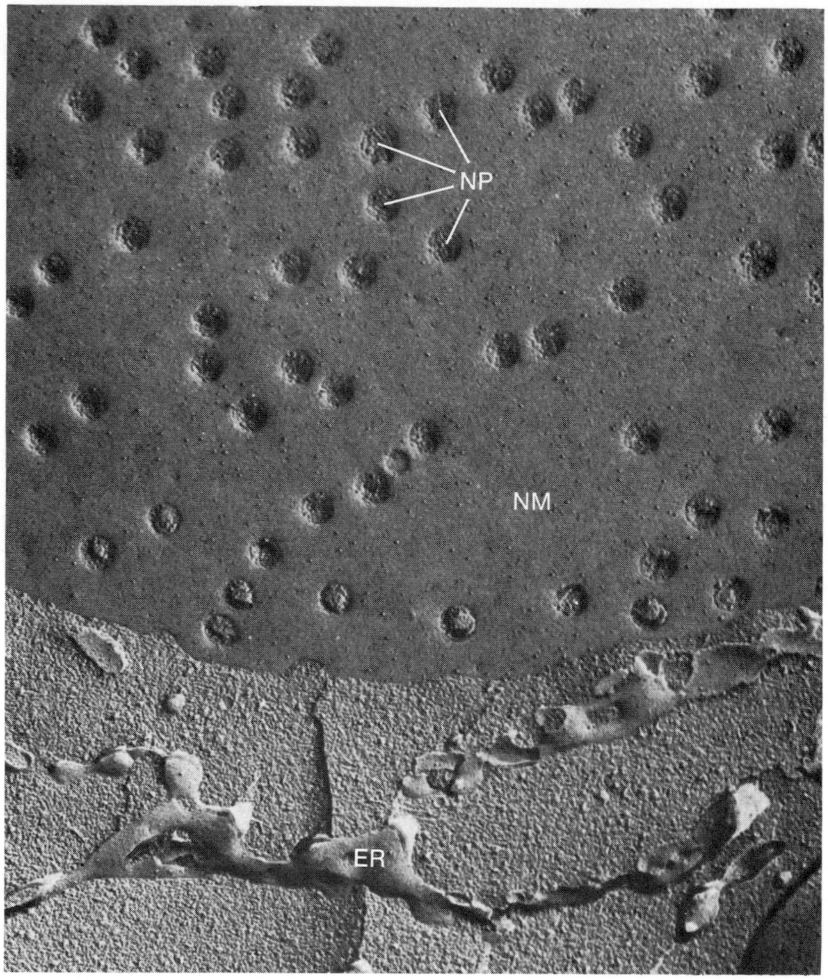

3–28. A replica of a fractured cell prepared by the freeze-etch technique. The lines of fracture in these preparations often follow the surfaces of membranes. In this preparation, an onion root tip cell, the nuclear envelope (NM) *is exposed, showing numerous pores* (NP). *Fragments of the endoplasmic reticulum* (ER) *are visible in the cytoplasm outside the nucleus.* $\times$ *40,000. Courtesy of D. Branton.*

tured under vacuum by a knife edge. The fractured surface contains depressions and ridges that follow ultrastructural features in the specimen. The fractured surface is then immediately shadowed with metal and coated with a heavy layer of evaporated carbon, or it is first exposed to the vacuum for a brief period, about one minute, before shadowing and coating. In either case, sublimation of water from the fractured surface takes place, increasing the relief of surface details and "etching" the specimen. The preparation is then removed from the vacuum and the tissue dissolved away, leaving only the shadowing metal backed by carbon. This *replica,* viewed under the electron microscope, shows shadowed patterns that follow the ridges and depressions in the original fractured tissue surface (Fig. 3–28). The resolution obtainable is limited both by the fidelity with which the fractured surface is followed by the replica and by the usual limitations of the shadowing technique. Resolution is nevertheless surprisingly good, depending on the specimen, and in some cases has approached 20 Å.

The lines of fracture in the freeze-etch preparations reveal or follow membranes in the specimen. No other cellular structures are seen with any clarity, and no new organelles, previously undescribed with the more conventional embedding techniques, have been made visible by this technique. However, freeze etching has nevertheless been of inestimable value, because the replicas are prepared from tissue frozen directly from the living state without the use of fixatives or stains. Even though this technique differs basically from the methods used in preparation of osmium-fixed embedded and sectioned tissues, the appearance of the various organelles is closely similar. This confirmation of structure from two widely different sources greatly reduces the chance that the structures revealed by electron microscopy are merely artifacts.

Summary

Light and electron microscopy has been considered in some detail in this chapter, because these techniques are the primary sources of the morphological and structural information stressed in this book. It should be obvious from the foregoing material that the present limitations in electron microscopy stem from the preparative techniques rather than the microscope. Few investigations use the ultimate resolving power of the instrument. Rarely are micrographs published that carry the necessary total magnification required to make a resolution of 10 Å visible to the naked eye. As mentioned, this must exceed 200,000 times. Nevertheless, much useful information lies in the 50–2000 Å range. In this range, both the instrument and the preparative techniques excel. The frontier of electron microscopy lies at the 5–10 Å level, with research being carried out actively in both the physics of the instrument and in new methods of fixation, staining, and isolation. The biological aim of the research is to extend electron microscopy into the molecular and atomic levels and to develop methods answering functional as well as structural questions.

The techniques outlined in this chapter are of general importance in light and electron microscopy and form the basis for the micrographs presented in this book. Rather than cataloging additional, more specialized techniques at this time, other methods—such as autoradiography, selective digestion of tissue by enzymes, analytical methods for fixation and staining, and cell fractionation in combination with electron microscopy—are described in later chapters as part of the supporting evidence for the structure and function of cell organelles.

Suggestions for Further Reading

Pease, Daniel C., 1960. *Histological Techniques for Electron Microscopy*. Academic Press, New York, 1960.

Sjöstrand, Fritjof S., 1967. *Electron Microscopy of Cells and Tissues,* Volume 1, *Instrumentation and Techniques.* Academic Press, New York.

Watson, M. L., and W. G. Aldridge, 1961. Methods for the use of indium as an electron stain for nucleic acids. *Journal of Biophysical and Biochemical Cytology,* vol. 11, pp. 257–272.

Wischnitzer, Saul, 1970. *Introduction to Electron Microscopy* (2nd ed.). Pergamon Press, New York.

Zobel, C. R., and M. Beer, 1965. The use of heavy metal salts as electron stains. *International Review of Cytology,* vol. 18, pp. 363–400.

4
The Structure of Membranes

The brief survey of lipid and protein structure presented in Chapter 2, and the description of microscopical techniques in the last chapter place us in a position to take up the ultrastructure of membranes and some models currently advanced for the possible arrangement of lipid and protein molecules into the framework of the membrane. Even though much effort has been spent in investigation, extensive disagreement exists on the possible molecular structure of membranes, and no single model presently accounts for all observations of membrane behavior.

As mentioned in the introductory chapter, the eukaryotic cell is now regarded as built on a framework of membranes. But as recently as the 1940's and 1950's, the cell was regarded simply as an unorganized "bag of enzymes." This concept did not seem unreasonable at the time, because most of the known cellular reactions could be duplicated with only the enzymes and substrates present. Hints of the importance of structure began to emerge with continuing investigations in biochemistry, particularly in mitochondria and chloroplasts, in which particular reaction sequences could be studied only if intact organelles, or at least subfractions of the organelle, were retained. These subfractions were identified in many cases as systems of membranes; the organization of enzymes on or within membranes was found to greatly modify their properties. Often, dissolution of membrane structure completely inhibited enzymatic activity. At about this time, electron microscopy revealed the abundance of membranes in the cell, and a new concept began to take precedence in cell biology. This concept was that the cell is highly ordered, rather than unorganized, and that much of the order is based on a framework of membranes.

Realizing the importance of membranes spurred a tremendous burst of research into membrane ultrastructure. This effort has not yet led to a complete solution of the problem of membrane structure, but many insights have been gained from this work. In recent years several tentative models that extend speculation to the molecular level have been advanced for membrane structure.

The Lamellar and Particulate Models for Membrane Structure

A number of hypotheses of membrane structure are currently favored. These rely heavily on evidence from electron microscopy and other physical and chemical techniques. While the models differ in detail, all assume that the membrane is either lamellar or particulate in substructure. The lamellar model states that the membrane is constructed from laminations of sheet-like layers of protein and lipid. In the most widely favored lamellar model, the membrane is considered to consist of a double layer of lipids between two layers of protein, in a sort of sandwich in which the lipid forms the meat and the protein the bread. The particulate hypothesis, in contrast, considers that the membrane is made up of a single layer of globular subunits, much like a layer of marbles. This model holds, in common with the lamellar model, that the globular subunits are a complex of lipid and protein. In the particulate model, the lipid is usually considered to form the core of the subunit particle, with the protein bound to the surface of the lipid, forming an exterior coat.

Neither model can be eliminated on the basis of the evidence at hand. Even though the models are quite different, both are based on the known properties of membrane lipids and protein. It is even possible that both models may be correct and that membranes may occur in either the lamellar or particulate form, depending on their location and function in the cell. This hypothesis, which combines features of both the lamellar and particulate models, actually has good evidence in its favor, as will be demonstrated later in this chapter. But before considering these models further, it is first necessary to discuss additional details of membrane lipids and proteins that have a direct bearing on membrane structure.

The Lipid and Protein of Membranes

Lipid Subunits of Cell Membranes

Analysis of isolated cell fractions has established that all membranes contain at least two major classes of macromolecules; lipids and protein. Although a polysaccharide component is also sometimes found to be associated with membranes, the possible structural role of this substance is less certain. The lipid part of the membrane is highly variable; membranes isolated from different cell organelles or from different cells often show surprising differences in the species present (Table 4–1).

The most common lipids present in membranes are phospholipids consisting of a phosphatidyl group complexed with choline (lecithin), ethanolamine, serine, or inositol, and the phospholipids diphosphatidyl glycerol (cardiolipin) and sphingomyelin (for details of the structure of these phospholipids, see p. 28 and Fig. 2–6). The fatty acid chains of these membrane phospholipids are variable and respond to changes in the diet of an organism by changes in length or degree of saturation.

Often a sterol, almost exclusively cholesterol (see Fig. 2–8), occurs in association with the phospholipid. Plasma membranes contain a significant quantity of cholesterol, up to 30 percent; phospholipid forms most of the remainder. In mitochondria, almost all of the lipid, up to 90 percent, is phospholipid. In this case, cholesterol forms only 5–10 percent of the total lipid, most of which is concentrated in the outer membrane. Chloroplast membranes apparently contain no cholesterol.

Another lipid class important in membrane structure is the *glycolipids* (Fig. 4–1). The glycolipids contain a

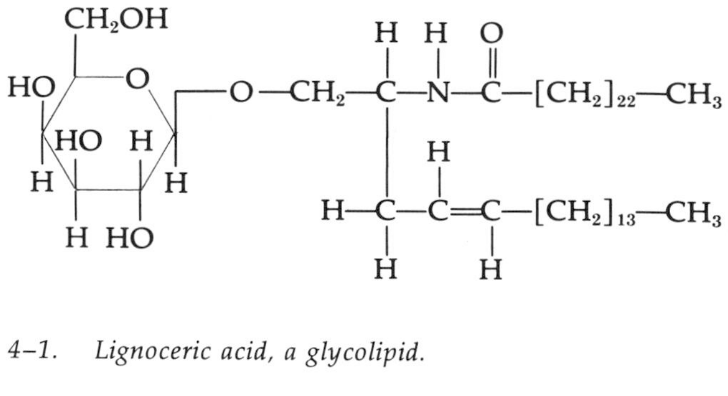

```
  CH2OH                    H  H  O
   |                       |  |  ||
HO/    \O      ——O—CH2—C—N—C—[CH2]22—CH3
 / HO  H \                 |
H \      / H               |     H
   |    |                  |     |
  H HO                 H—C—C=C—[CH2]13—CH3
                           |     |
                           H     H
```

4–1. Lignoceric acid, a glycolipid.

Table 4–1 Lipid Composition of Animal and Bacterial Membranes (in percent)

	Myelin	*Erythro-cyte*	*Mito-chondria*	*Microsomal Fractions* 1	2	Escherichia coli	Bacillus megaterium	*Chloro-plasts*
Cholesterol	25	25	5	6	*	0	0	*
Phosphatidyl ethanolamine	14	20	28	17	18	100	45	0.6
Phosphatidyl serine	7	11	0	0	9	0	0	*
Phosphatidyl choline	11	23	48	64	48	0	0	4.1
Phosphatidyl inositol	0	2	8	11	6	0	0	1.4
Phosphatidyl glycerol	0	0	1	2	0	0	45	5.3
Cardiolipin	0	0	11	0	2	0	0	0.6
Sphingomyelin	6	18	0	0	9	0	0	*
Cerebroside	21	0	0	0	0	0	0	*
Cerebroside sulfate	4	0	0	0	0	0	0	*
Ceramide	1	0	0	0	0	0	0	*
Lysyl phosphatidyl glycerol	0	0	0	0	0	0	10	*
Galactosyl diglyceride								14.7
Digalactosyl diglyceride								35.3
Sulfoquinovosyl diglyceride								4.9
Chlorophyll								23.5
Carotenoids								4.9
Unknown or other	12	2	0	0	0	0	0	*

*Not analyzed.

Adapted from E. D. Korn, Structure of biological membranes. *Science* 153:1491 (1966). Chloroplast data from J. S. O'Brien, *J. Theoret. Biol.* 15:307 (1967).

polar carbohydrate group, usually galactose, but no phosphate. This lipid class, which includes the cerebrosides, gangliosides, and sulfatides, are found in significant quantities in plasma membranes and occur in high concentrations, approaching 50 percent, in myelin. Glycolipids are also a major constituent of chloroplast membranes.

Table 4–2 summarizes the lipid composition of a frequently studied membrane, the erythrocyte plasma membrane. From this table it can be seen that the phospholipid-cholesterol ratio is fairly uniform between the different species, in each roughly 26 percent cholesterol and 62–63 percent phospholipid. These percentages are based on the weight of the lipids in the membranes; in terms of molar ratios, the cholesterol and phospholipid occur in a 1:1 proportion. The phospholipid and cholesterol are believed to form a stable complex, which is undoubtedly responsible for many of the properties of the plasma membrane. There are several possibilities for the structure of this complex, and as yet no agreement exists concerning the details of this association. Because the cholesterol molecule is nonpolar, it is considered to be associated with the nonpolar hydrocarbon chains of the phospholipid in the interior of the membrane. The other end of the phospholipid is polar and electrostatically charged at physiological *p*H, primarily because of the phosphate group, and presumably lies at the surface of the membrane, facing the aqueous medium. A possible configuration for the phospholipid-cholesterol complex, advanced by F. A. Vandenheuvel, is shown in Fig. 4–2. Note that the total length of the proposed complex, which is roughly cylindrical, is between 30 and 35 Å. In the model shown in Fig. 4–2, the part of the complex toward the left, bearing the phosphate group, would face the surface of the membrane. The hydrocarbon

Table 4–2 *Lipid Distribution in Erythrocytes from Various Mammalian Species*

	Total lipid mg/ml packed	Total lipid gm/cell	Cholesterol (%)	Total ganglioside (%)	Other glycolipids (%)	Phospholipid (%)
Cat	6.04	3.45×10^{-13}	26.8	8.8	3.1	61.3
Cow	4.44	2.58×10^{-13}	27.5	5.5	2.2	64.8
Dog	5.76	4.84×10^{-13}	24.7	11.8	10.9	52.6
Goat	6.14	1.23×10^{-13}	26.2	5.7	17.9	50.2
Guinea pig	5.72	4.41×10^{-13}	27.0	2.2	15.2	55.6
Horse	5.37	2.58×10^{-13}	24.5	15.5	8.0	52.0
Pig	4.33	2.52×10^{-13}	26.8	3.3	10.1	59.8
Rabbit	4.57	4.15×10^{-13}	28.9	4.5	0.8	65.8
Rat	5.08	3.15×10^{-13}	24.7	6.3	2.0	67.0
Sheep	4.91	1.62×10^{-13}	26.5	7.8	2.5	63.2

Adapted from G. Rouser et al., Lipid composition of animal cell membranes, organelles, and organs. D. Chapman, ed. *Biological Membranes.* New York: Academic Press, 1968, p. 5. See this paper for original references.

"tails," extending toward the right in the figure, would lie in the interior of the membrane. The complex, as shown, would form one-half of the thickness of the lipid part of the membrane. How fundamental this complex is to membrane structure in general is problematical, as cholesterol exists in mitochondrial membranes in only very small quantities and is absent from chloroplast membranes. The role of cholesterol, when present in membranes, remains uncertain. Its function may be related to stabilization of the lipid interior of the membrane, which some evidence indicates is in a fluid state (see below). Formation of the cholesterol-phospholipid complex also leads to tighter packing, or compaction, of the hydrocarbon chains in the membrane interior.

Properties of Phospholipid Suspensions in Water

Much has been learned about the possible structural configuration of phospholipid in membranes by X-ray diffraction of suspensions of phospholipid in water. In this technique, a beam of X-rays is passed through the specimen. If sufficient order is present in the arrangement of atoms and molecules in the specimen, X-rays will be reflected from regularly repeating groups of atoms. These reflections are refracted at various angles, depending on the spatial arrangement of the constituent atoms or molecules. If the atoms of the specimen are highly ordered, as in a crystal, a pattern of discrete spots is produced. Lesser degrees of order cause a blurring and spreading of the spots into concentric rings. In either case, measurement of the separations between the spots or rings in the diffraction pattern can be related to the separation and spatial arrangement of atoms and molecules in the specimen.

X-ray diffraction of phospholipid-water systems produces a pattern of spots partially blurred into arcs of a circle. The pattern obtained has been interpreted to mean that the phospholipid molecules in the suspensions are aligned in flat sheets, separated by alternating layers of water. The thickness of the water layer is variable, but the phospholipid layers are of approximately constant dimensions, between 40 and 50 Å in thickness (see, for example, Bear et al., 1941). The X-ray evidence, combined with evidence from other sources, also indicates that the phospholipid molecules are stacked vertically within the sheets, with the long axis of the molecules perpendicular to the plane of the sheet. Because the artificial membranes are thick enough to accommodate two layers of vertically stacked molecules, the constituent lipids are considered to be present in a layer two molecules in thickness, called a *bilayer* (Fig. 4–3).

The most probable configuration of phospholipid molecules in a bilayer is an orientation such that the phosphate groups, which are negatively charged at physiological *p*H, face the aqueous medium, and the

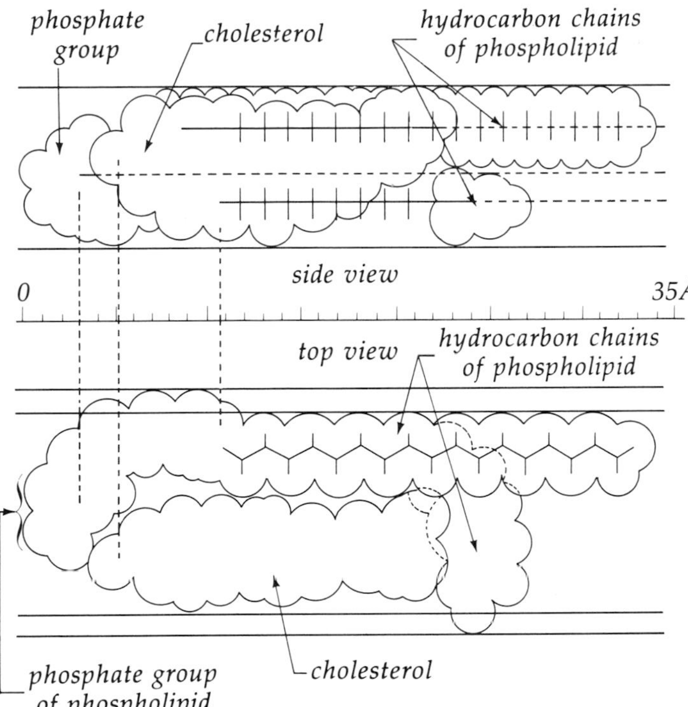

4–2. Dimensions and arrangement of the phospholipid-cholesterol complex, as visualized by Vandenheuvel. Redrawn from F. A. Vandenheuvel, J. Amer. Chem. Soc. *40:455 (1963).*

nonpolar lipid "tails" associate end-to-end in the interior of the bilayer. Within the nonpolar interior region of the bilayer, the hydrocarbon chains are considered to be free to rotate and flex around particular bonds. For this reason, the hydrocarbon interior of the bilayer is often described as being in a "liquid" state.

Electron Microscopy of Lipid Bilayers

Lipid bilayers can be prepared for electron microscopy by fixation in osmium tetroxide and other standard fixatives. When embedded and transversely sectioned, fixed bilayers reveal a substructure of alternating light and dark lines (Fig. 4–4), very similar in spacing to the dimensions indicated by X-ray diffraction. It is uncertain whether the dark lines represent binding of osmium to the hydrophobic or hydrophilic parts of the bilayers. On the basis of the chemistry of osmium, binding would be expected in the nonpolar, hydrophobic interior of the bilayer, because osmium tetroxide is known to be reduced by double bonds. This is supported by the observation of many investigators that fully saturated phospholipids (lipids in which no double bonds are present; see Chapter 2) do not interact readily with osmium tetroxide at room temperature (see for example Chapman and Fluck, 1966).

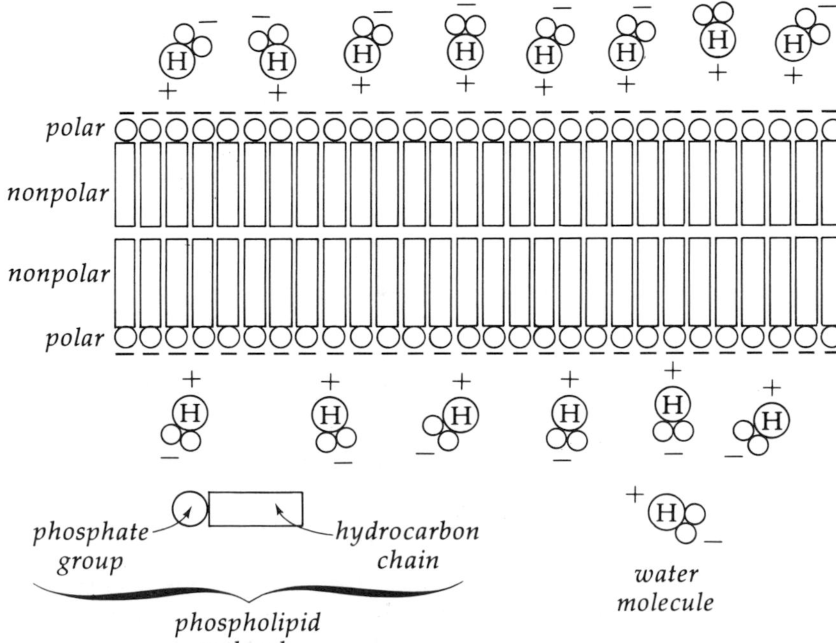

4–3. The presumed arrangement of phospholipid molecules in a bilayer.

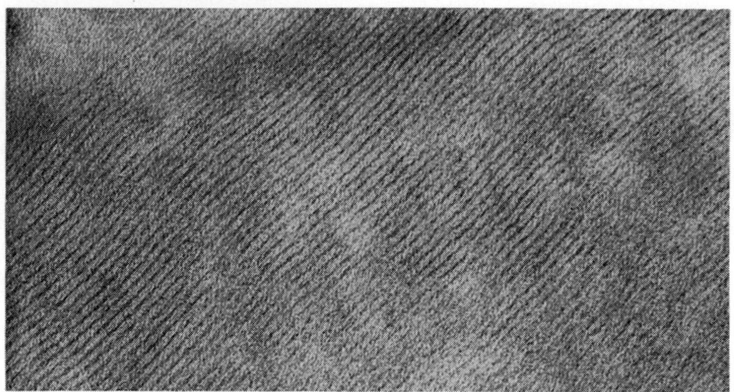

4–4. Closely opposed phospholipid bilayers, fixed with osmium and thin sectioned. × *280,000. Courtesy of W. Stoeckenius and The Rockefeller University Press, from* J. Cell Biol. *12:221 (1962).*

This would suggest that the dark lines represent the nonpolar interior of the phospholipid bilayers. However, other experiments indicate instead that the dark lines result from the binding or precipitation of osmium at the hydrophilic surfaces of the bilayer. For example, Walther Stoeckenius reacted phospholipid films with uranyl ions rather than osmium tetroxide. When embedded and thin sectioned, these preparations showed a pattern of lines and spaces with the same dimensions as the osmium-fixed films. This is most significant, for the uranyl ion is known to bind very strongly to phosphate groups, which, as noted, are thought to be located at the surface of the bilayer and to represent the hydrophilic portion of the system. Stoeckenius also added osmium tetroxide to the uranyl-treated films. When this was done, the lines already present were merely slightly darkened. If osmium deposited in the hydrophobic layer, the lines would be expected to double in number, reflecting deposition in the previously unstained alternating clear spaces (see Stoeckenius, 1960).

If the osmium binds at the surface of the membrane, as these results suggest, the chemical basis for this reaction remains a mystery. One proposal is that the osmium reacts as expected with the nonpolar hydrocarbon chains in the interior of the bilayer and then somehow migrates to the hydrophilic surfaces of the bilayer. Needless to say, this explanation is not too widely accepted, and the general discrepancies in the expected and observed reactions of osmium in bilayers

Table 4–3 Comparison of Lipid Bilayers and Cell Membranes

	Cell membranes (at 20–25°C)	*Lipid bilayers (at 36°C)*
Refractive index	1.6	1.66
Electron microscope image	trilaminar	trilaminar
Thickness, Å	60–100	60–75
Capacitance, μmf/cm^2	0.5–1.3	0.38–1.0
Resistance, Ω cm^2	10^2–10^5	10^6–10^9
Dielectric breakdown, mV	100	150–200
Surface tension, dynes/cm	0.03–1	0.5–2
Water permeability, μ/sec	0.37–400	31.7 (25°C)
Activation energy for water permeation, kcal/mole	9.6	12.7 (25°C)
Urea permeability, μ/sec $\times 10^2$	0.015–280	4.2 (20°C)
Glycerol permeability, μ/sec $\times 10^2$	0.003–27	4.6 (20°C)
Erythritol permeability, μ/sec $\times 10^2$	0.007–5	0.75 (20°C)

Adapted from T. E. Thompson and F. A. Henn, Experimental phospholipid model membranes. In E. Racker, ed., *Membranes of Mitochondria and Chloroplasts.* American Chemical Society Monograph 165. See this paper for original references.

remain an area of intensive discussion and controversy.

Phospholipid bilayers and living membranes. In 1962, P. Meuller and his colleagues succeeded in producing a phospholipid film one bilayer in thickness, stretched across an aperture separating two aqueous solutions. These films, which form spontaneously after application of a solution of phospholipid to the aperture with a brush, exhibit a number of physical properties similar to the living membrane (Table 4–3).

The similarities extend to a close correspondence between the dimensions and morphology of transversely sectioned films and membranes (Fig. 4–5). This correspondence in properties and the morphology of the single phospholipid films have provided strong evidence for the lamellar model of membrane organization.

The Protein Component of Membranes

It is interesting that phospholipid films in the pure state, without attached protein, exhibit so many properties closely similar to living membranes. This invites

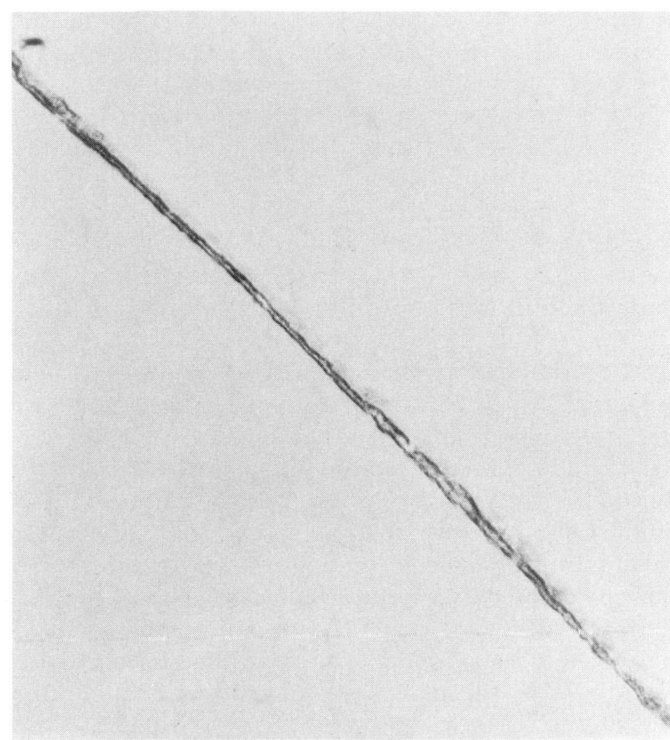

4–5. A single bilayer formed from the phospholipid phosphatidyl ethanolamine, fixed in lanthanum nitrate and potassium permanganate. × 64,500. Courtesy of G. Decker and J. Greenawalt and Van Nostrand Reinhold Company.

the question of whether proteins are important in a structural role in membranes or are found only as an attached functional, enzymatic component. Although a presumed "structural protein" has been isolated from various membranes by D. E. Green and his co-workers, much controversy surrounds the reality of the so-called structural protein; many investigators are convinced that this proteinaceous material merely represents enzymes denatured by the isolation techniques (see p. 121).

At any rate, whether structural or enzymatic or both, all membranes contain a protein component. Many of the enzymes bound to membranes have been isolated and purified and are well characterized as to identity and properties. But no discrete and well-characterized proteins have been identified as a structural component of membranes. This results partly from the fact that much of the membrane protein is highly insoluble, except under conditions of *p*H and salt concentration so extreme that denaturation is likely. As a result, in contrast to the lipids of the membrane, very little is known about specific proteins that may occur as part of the structural framework of the membrane.

Early studies concluded that the ratio of protein to lipid in membranes is almost constant. More recent data have established that wide variations occur in the protein-lipid ratio between types of membranes and between species. Myelin, common in nerve tissue, contains enough protein to cover only about 43 percent of the total lipid surface of the membrane. Erythrocyte ghosts contain enough for a monolayer of protein with 2–5 times the area of the lipid; bacterial membranes apparently contain even greater proportions of protein. It is possible that contamination from other cellular proteins is responsible for some of the variation in these figures. In any event, from available information it is difficult to arrive at any fixed lipid-protein ratio as fundamental to membrane structure.

Artificial Lipid-Protein Membranes

If a soluble protein, such as cytochrome *c*, is added to a system of artificial lipid bilayers in aqueous medium, the protein binds readily to the phospholipid. The initial attraction is probably electrostatic between the negatively charged phosphate groups of the lipid and positively charged groups on the surface of the protein. Many possibilities exist, however, for the formation of nonpolar bonds between the hydrocarbon chains of phospholipid and nonpolar amino acid residues of the protein. Although extensive unfolding of the protein component would be required for the 4–5 Å side chains of the amino acid residues to penetrate into the hydrophobic bilayer interior, it is nevertheless likely that such nonpolar interactions, as well as electrostatic interactions, contribute to formation of the lipid-protein complex in membranes.

David E. Green and Sidney Fleischer have studied both types of interaction in phospholipid suspensions. Cytochrome *c*, an enzymatic protein easily removed from mitochondrial and chloroplast membranes, was found to completely neutralize the phosphate groups of the phospholipid bilayers. This attraction is electrostatic, and binding is easily reversed by adjustment of salt concentration. But a presumed "structural" protein isolated from membranes, when reacted with phospholipids, left many phosphate groups unoccupied. After interaction with this protein a large percentage of the phosphate groups were still capable of binding cytochrome *c*. Green and Fleischer speculated that a significant part of the presumed structural protein was bound to the nonpolar, hydrophobic interior of the phospholipid bilayer, leaving most of the phosphate groups free to interact with added cytochrome *c*.

The degree to which nonpolar interactions occur between lipid and protein in living membranes is difficult to assess. The work of Green and Fleischer suggests that this type of interaction does occur. As mentioned, it is likely that this would require extensive unfolding of the protein. However, some enzymes, if reacted with phospholipid films, retain activity even when tightly bound to the bilayer. This would not be possible if extensive unfolding of the protein took place and indicates instead that the primary binding may be electrostatic rather than nonpolar. Other methods confirm that very little membrane protein exists in the β, or completely unfolded, state in natural membranes. Optical methods, for example, indicate that the protein of membranes is distributed approximately equally between the α-helical and random chain configurations (for details, see Chapman and Wallach, 1968).

Electron microscopy of artificial lipoprotein membranes. Electron microscopy of the protein-phospholipid bilayer complex adds evidence supporting the view that osmium binds at the polar surface of the bilayer. Walther Stoeckenius fixed protein-phospholipid bilayers with osmium tetroxide and compared them with pure phospholipid films. In comparing the two types, Stoeckenius noted that, in the lipid-protein films, the dark line between alternating layers was thicker than in pure lipid films. Where single bilayers could be seen in either preparation, the dense line at the surface of the bilayer was thicker in the protein-phospholipid films. This indicates that the protein binds at the surface of the bilayer and also provides confirmation of the probable binding of osmium in the same region (see Stoeckenius, 1959).

The Lamellar and Particulate Models Revisited

The Danielli-Davson Model

With the information presented in the foregoing sections in mind, we can now turn to a more complete evaluation of the lamellar and particulate models for membrane structure. The evidence obtained from the study of phospholipid and lipoprotein bilayers, as noted, appears to provide strong support for the lamellar model, and for a time this hypothesis was almost universally accepted. But more recent morphological investigations, particularly in chloroplast and mitochondrial membranes, have forced a reassessment of the lamellar model. These developments are most easily understood if traced out in historical sequence.

Historical developments leading to the formulation of the lamellar model. The lipoprotein bilayer model for the structure of natural membranes took form before the introduction of electron microscopy, during the period from 1920 to 1940. Studies on the permeability of living cells, carried out even earlier, had shown that small nonpolar molecules penetrate cells more readily than polar substances, and led to the proposition that cells are externally coated by a lipid layer. In 1925, E. Gorter and F. Grendel extracted lipids from isolated erythrocyte membranes and showed that enough lipid was present in the membranes to provide a coating two molecules in thickness over the entire surface of the cell. The lipid extraction method used by Gorter and Grendel was incomplete, and as a result their conclusions are often criticized. However, these workers also underestimated the surface area of the erythrocytes by a factor that compensated for the extraction error, so that their results, by coincidence, have been shown on reinvestigation to be approximately correct.

Not long after Gorter and Grendel's work, J. F. Danielli and Hugh Davson, after a consideration of earlier findings and their own observations on the surface tension of various lipids, advanced a new model for membrane ultrastructure. They proposed that the cell is covered by a bimolecular layer of lipids, with the polar parts of the lipid molecules extending into the surrounding aqueous phase and the nonpolar hydrocarbon chains associated together in the center of the bilayer. Noting that the surface tension of living cells is much lower than the values obtained for oil droplets, Danielli and Davson proposed that the lipid bilayer is coated on both the internal and external surfaces by a layer of protein, which acts to reduce the surface tension. This hypothesis was later modified to include polar pores of small dimensions extending through the lipid bilayer, in order to account for the high penetration rate of water through membranes. By 1954 this model had reached the form shown in Fig. 4–6.

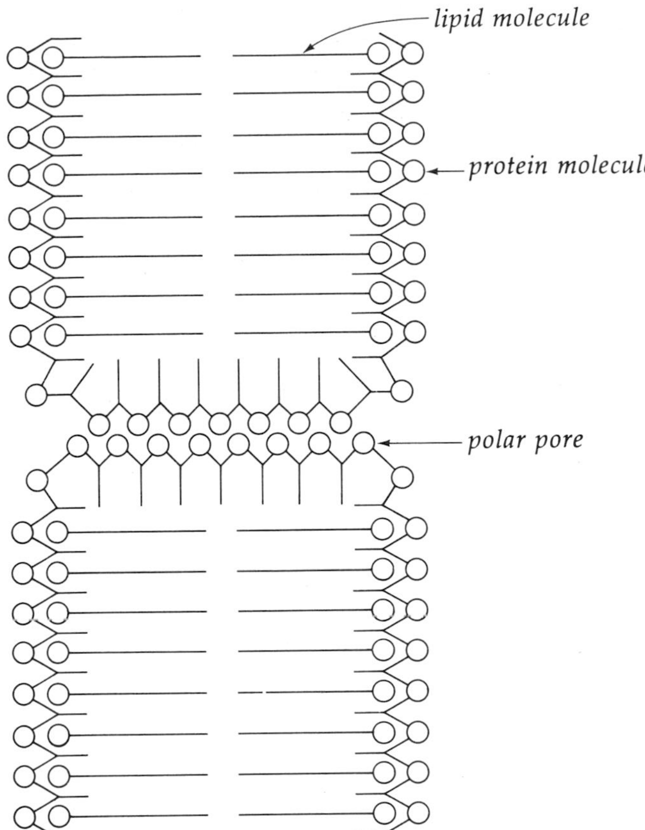

4–6. *The Danielli-Davson model for membrane structure. A lipid bilayer is stabilized by layers of protein on both outer surfaces. At intervals, pores, presumably coated with protein, extend through the membrane. These "pores" are extremely minute, and give the membrane in these regions polar properties, with the result that water, but not larger polar molecules, penetrates the membrane readily. Redrawn, original courtesy of J. F. Danielli.*

As mentioned, the proposed distribution of a layer of protein over the surfaces of the membrane complex was based on the observation that living membranes have a lower surface tension than oil droplets. But the supporting observations were based on triglycerides and other lipids that are not readily absorbed by or occur in cell membranes. These substances gave a surface tension 20 to 30 times higher than living membranes. Phospholipids, however (see Table 4–3), have a surface tension in water more closely approaching the cell membrane. Thus, it is not necessary to assume a coating layer of protein on the basis of surface tension. But other lines

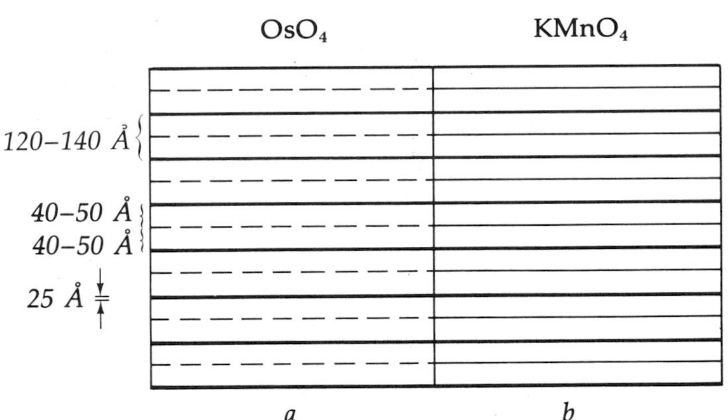

4–7. *Comparison of the pattern of density seen in cross sections of osmium- and permanganate-fixed myelin.* a, *In osmium-fixed preparations, 25 Å lines of heavy density are visible, separated by a 120–140 Å interspace. An intermediate line of very light density is placed between the dense lines.* b, *With permanganate fixation, the intermediate line takes up equivalent density, giving the appearance of repeating lines of approximately equal density spaced 40–50 Å apart.*

of evidence from X-ray diffraction and electron microscopy indicate that this model may nevertheless be correct.

Myelin membranes and the Danielli-Davson model. Studies of natural membranes in the myelin sheath of nerve axons gave strong support to the Danielli-Davson model. The existence of a highly ordered system in the myelin sheath was first revealed by the polarizing light microscope. Isolated myelin under polarized light (light vibrating in only one plane) reflects a strong beam in a new plane. This condition, called birefringence, occurs when polarized light encounters an ordered array of microscopic or submicroscopic dimensions. X-ray diffraction of myelin also showed that a basic repeating unit was present in the sheath. From data obtained by these methods, F. O. Schmitt proposed in 1941 that the sheath is composed of multiple lipid layers wrapped concentrically around the axon, with successive lipid layers separated by layers of protein. A 170 Å repeating unit, revealed by X-ray diffraction, was proposed to represent two bimolecular layers of lipid, 140 Å in total thickness, separated by alternating 25 Å layers of protein. Although Schmitt did not relate this hypothetical structure to the plasma membrane, it was shown later by Betty Geren that the

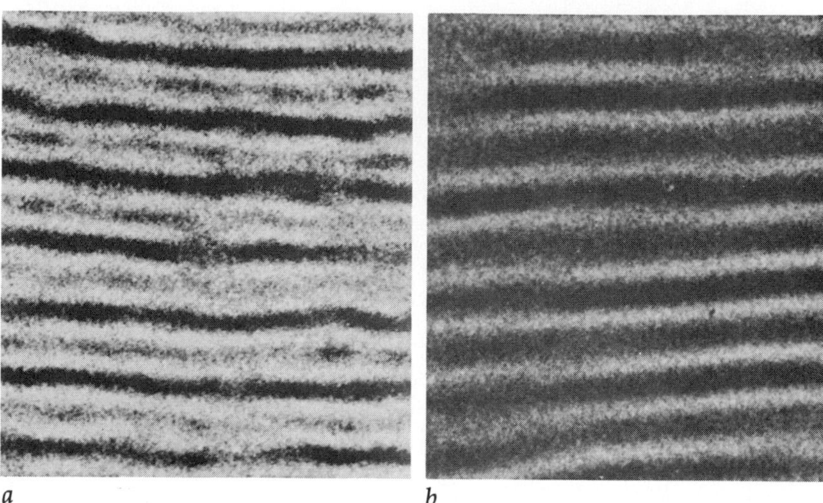

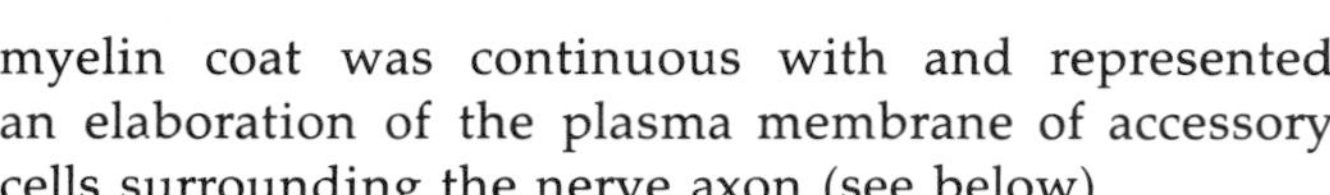

4–8. a, *Myelin from the sciatic nerve of the rat, fixed in osmium tetroxide. Compare with Figs. 4–4 and 4–7a. The major period lines are spaced about 110 Å apart in this preparation.* × *800,000.* b, *Myelin from the sciatic nerve of a toad, fixed in potassium permanganate. The intermediate lines have density equivalent to the major period lines in* a, *producing the image of dense lines spaced about 50 Å apart.* × *1,000,000. Photographs courtesy of H. Fernández-Morán.*

myelin coat was continuous with and represented an elaboration of the plasma membrane of accessory cells surrounding the nerve axon (see below).

The structure deduced for myelin from X-ray diffraction was confirmed by electron microscopy in the early 1950's, when H. Fernández-Morán and Fritjof Sjöstrand detected repeating dense lines about 25Å in thickness in sections of osmium-fixed myelin. The dense lines were separated by a 120–140 Å interspace. A faint line lying in the center of the interspace was just visible in the osmium-fixed myelin (Figs. 4–7a and 4–8a). After prolonged fixation with potassium permanganate, the faint line was darkened, producing a regularly alternating pattern of dense lines separated by lighter areas approximately 50 Å in thickness (Figs. 4–7b and 4–8b).

The electron microscope image revealed in these studies is consistent with Schmitt's hypothesis that protein and lipid form alternating layers in myelin, if the dark lines are assumed to represent protein and the 40–50 Å layers of low electron density an alternating lipid bilayer. The dark lines would then be formed by the protein of two facing membranes pressed closely together. According to this interpretation, the thickness of individual membranes of the stack is expected to be about 75 Å (40–50 Å lipid, with two surface layers of protein, each $\frac{1}{2} \times 25$ Å thick).

Robertson's "Unit Membrane" Concept

The emerging picture of membrane structure was formalized into the "unit membrane" concept by J. D. Robertson. After pointing out that electron microscopists have consistently reported that the plasma membrane appears in cross section as two dark lines separated by a less dense interzone, Robertson reviewed the relationship of this configuration to the fine structure of the myelin sheath. In unmyelinated nerve, axons are partially or wholly surrounded by single layers of Schwann cell membranes (Fig. 4–9). Myelinated nerve has a similar appearance early in development (Figs. 4–10 and 4–11). During subsequent growth of the myelin sheath, the plasma membrane of the Schwann cell elongates and progressively extends in a spiral around the axon, forming the myelin sheath. The multiple layers of membrane formed by this spiraling become closely stacked by elimination of both the intercellular space and the Schwann cell cytoplasm (Fig. 4–12). The dark period lines, which appear lighter after osmium fixation, represent fused "outside" surfaces of the Schwann cell plasma membrane; these surfaces were originally separated by the intercellular space. The dark line alternating between lighter lines is formed after the elimination of Schwann cell cytoplasm separating the "inside" surfaces of the plasma membrane. Each of the repeating dense lines in myelin therefore represents either the fused inside or outside surfaces of two opposed cell membranes.

After reviewing the X-ray data, Robertson took the view that osmium binds to protein rather than to lipid in membranes. Thus the light, interperiod zones would represent the nonpolar hydrocarbon chains of the lipid bilayers. The period lines, both the denser and the lighter lines, represent layers of protein. Because the "inside" and "outside" dark lines stain differently

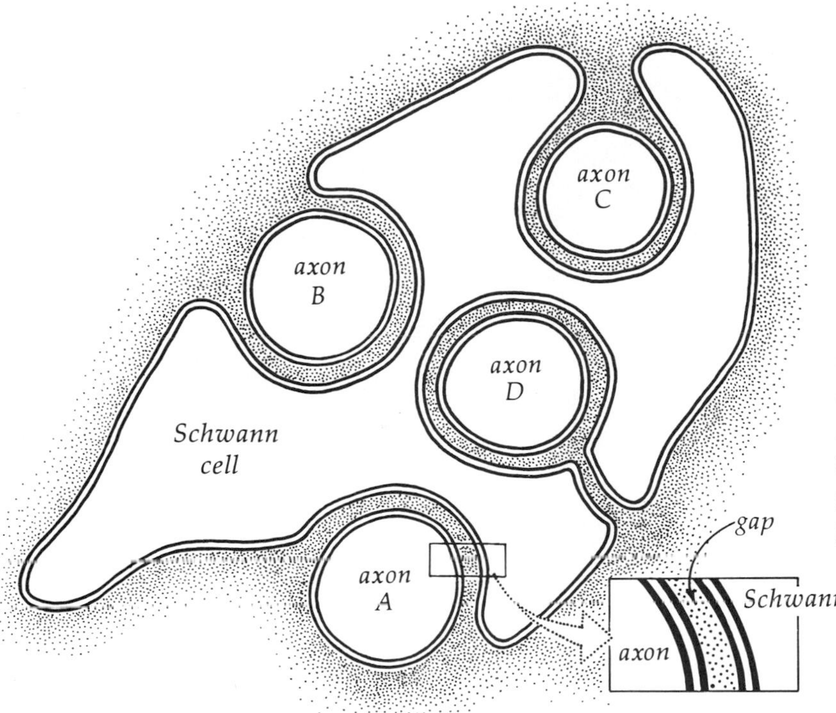

4–9. Diagram of unmyelinated nerve fibers. Four nerve axons are shown embedded in processes of a Schwann cell. Intercellular material extends into the space (gap) *between opposed membranes of the axons and the Schwann cell. Redrawn, original courtesy of J. D. Robertson.*

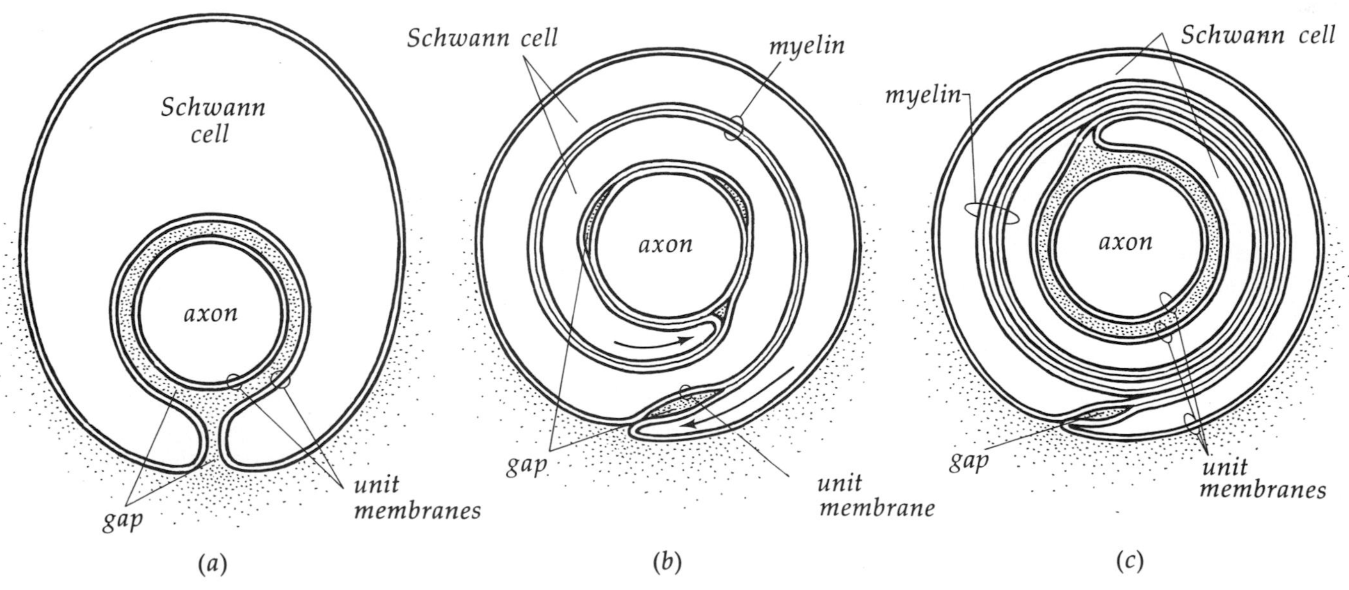

4–10. Development of the myelin sheath. In a, *the earliest stage of development, a single axon is shown embedded in a Schwann cell. In* b, *the Schwann cell membrane has begun to overlap in a spiral around the axon. Note that the intercellular gap between opposed membranes is eliminated. At a later stage,* c, *the spiral is more extensive. The membranes of the Schwann cell become closely packed, in some regions eliminating both the intercellular and cytoplasmic spaces. Redrawn, original courtesy of J. D. Robertson.*

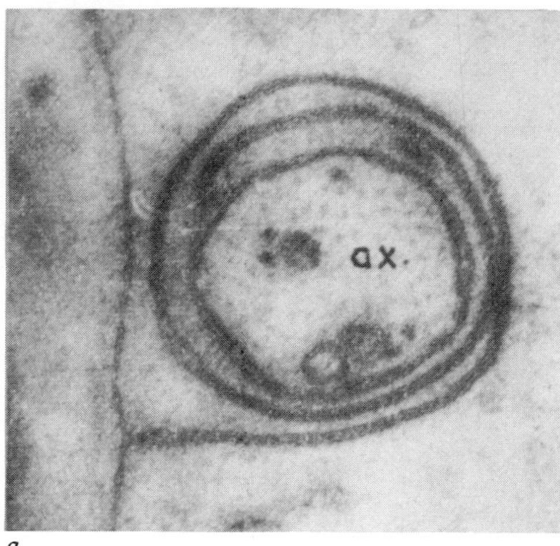

a

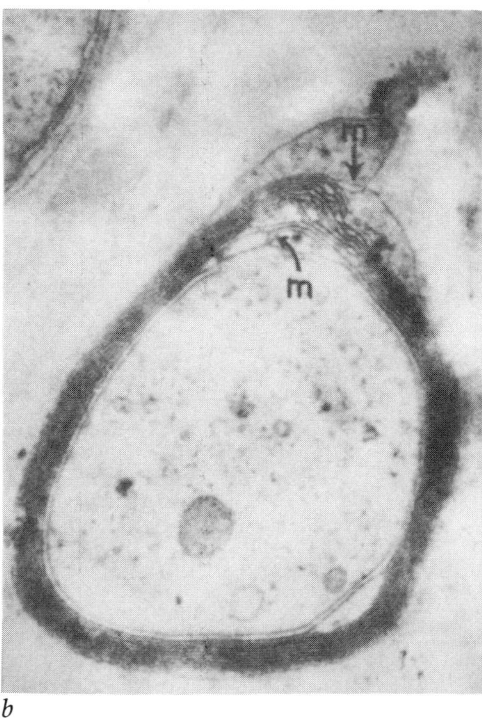

b

4–11. a, *Section of the developing myelin sheath around a sciatic nerve fiber of the mouse. At this stage, development of the sheath is closely equivalent to the diagram shown in Fig. 4–10b.* ax, *axon.* × *93,500.* b, *Section of a myelinated nerve at a later stage of development. In most regions, the membranes are so closely opposed that the intercellular and cytoplasmic spaces have been eliminated.* m, *single membrane.* × *46,000. Photographs courtesy of J. D. Robertson.*

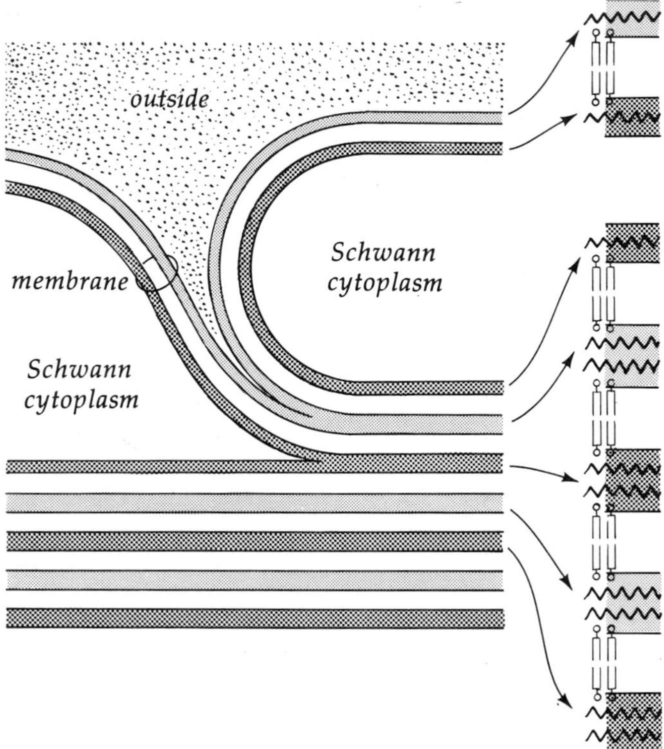

4–12. *Diagram showing Robertson's interpretation of the period line in myelin. According to Robertson, the slightly lighter period lines represent fused "outside" surfaces of closely opposed cell membranes; the intercellular space ("outside") originally separating these surfaces has been eliminated. The darker lines are fused "inside" membrane surfaces. These surfaces, which originally faced the cytoplasm of the Schwann cell, have become so closely aligned that the intervening cytoplasm has been eliminated. Redrawn, original courtesy of J. D. Robertson.*

with osmium, Robertson proposed that these might represent differences in the protein layers of the inside and outside surfaces of the plasma membrane. Alternatively, the difference might be caused by the presence of a polysaccharide in the exterior coat. The Robertson model therefore closely follows and extends the Danielli-Davson model, except for the hypothesis that the inner and outer membrane surfaces are different.

Robertson also compared the structure of other cell organelles in which tripartite membrane structure is visible and proposed that all membranes have a basically similar structure. This he termed the "unit membrane." Robertson concluded further that the unit membrane has a total thickness of 75 Å, and consists of two 25 Å dark lines separated by a 25 Å interspace.

By the early 1960's, this lamellar model had been extended and strengthened by Robertson and others into a far-reaching statement of membrane structure that included all cellular membranes. Dissenting voices were heard during the mid 1960's, however, and by the end of this decade serious objections were raised to the simplicity and inclusiveness of this model.

Recent objections to the Robertson hypothesis. The Robertson unit membrane hypothesis is based on the assumption that all membranes have as a fundamental structure a lipoprotein bilayer. One serious objection to this hypothesis is the fact that the model was based primarily on the results of electron microscopy and X-ray diffraction of myelin, a relatively inert proliferation of the plasma membrane. Myelin has none of the complex enzyme systems associated with other membranes, such as the internal lamellae of mitochondria and chloroplasts, and appears to function primarily as an electrical insulator. Other membranes with greater specialization in enzymatic roles, such as the photosynthetic lamellae, are likely to be quite different in structure from myelin. Also significant for this line of argument are the striking differences in lipid-protein ratios observed in the membranes of these organelles (Table 4–4).

Table 4–4 Lipid-Protein Ratios of Various Cellular Membranes

Membranes	*Lipid (%)*	*Protein (%)*
Myelin	80	20
Chloroplast lamellae	50	50
Erythrocyte	40	60
Mitochondrial inner membranes	25	75

From *Papers on Biological Membrane Structure* by Daniel Branton and Roderic B. Park. Copyright © 1968 by Little, Brown and Company (Inc.). Reprinted by permission.

Other difficulties arise from the dimensions proposed by Robertson as common to all membranes. Fritjof Sjöstrand discovered that significant differences in total thickness are revealed by high-resolution microscopy of membranes from different cell organelles. These actually vary from a minimum of 50 Å in smooth endoplasmic reticulum to 85 to 95 Å for the plasma membrane.

But the most serious objections to the lamellar model arise from the frequent observation of an apparent particulate or globular subunit in high-resolution electron micrographs of some cellular membranes. Sjöstrand, for example, detected extensive particulate organization in mitochondrial membranes (Fig. 4–13) and to a lesser degree in other cytoplasmic membranes. Similar globular units have been detected in chloroplast membranes (see Fig. 7–15), and in recent years, other observations of particulate organization have appeared with increasing frequency. As a result, an opposing school of thought has appeared, supporting the interpretation that the membrane consists not of sandwiched layers of lipid and protein, but of rows of globular or particulate lipoprotein subunits.

Resolution of the Lamellar-Particulate Controversy from Freeze-Etch Studies of Natural Membranes

Freeze etching of cellular membranes has provided new information that permits a resolution of the opposing viewpoints concerning membrane structure. The work of Daniel Branton illustrates the results obtained by this recently developed method. In the freeze-etch technique (see also the description in Chapter 3), unfixed, living tissue is quick-frozen and fractured under vacuum. After etching of the tissue by sublimation of water from the frozen specimen, a shadowed replica is made of the fractured surface. In his freeze-etch preparations, Branton determined that the line of fracture often follows the interior of membranes, causing the membranes to split into separated inner and outer surfaces (Fig. 4–14). Unsplit membranes showed a typical trilaminar structure. Because it is possible to see this configuration in a replica of unfixed tissue, it seems likely that the trilaminar structure is "real" and does not represent a rearrangement of lipid resulting from fixation, dehydration, or embedding.

Branton also found that some membranes in his preparations contained a particulate subunit. This subunit appeared to be embedded between the interior and exterior surfaces of the membranes. The internal membrane systems of chloroplasts, for example, show a great abundance of these particles (Fig. 4–15). On the other hand, the relatively inert membranes of myelin appear to have no subunits (Fig. 4–16).

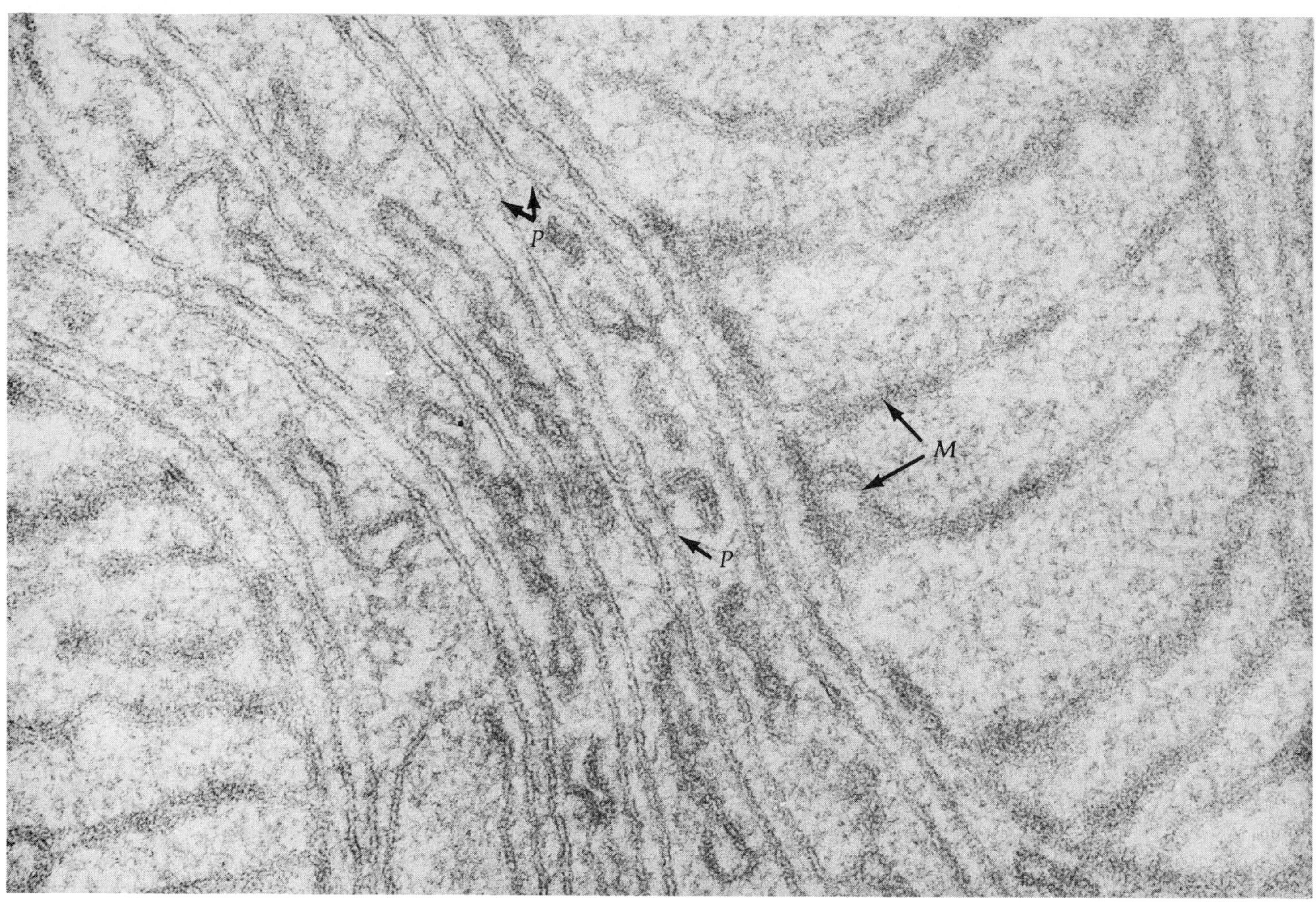

4–13. *Apparent subunit organization in mitochondrial membranes* (M) *from mouse kidney, fixed with potassium permanganate. The globular or particulate substructure of the mitochondrial membranes, sectioned in oblique plane in this preparation, is contrasted with the clearly trilaminar image of adjacent elements of the plasma membrane* (P). × *176,000. Courtesy of F. S. Sjöstrand and Academic Press, Inc.* [J. Ultrastr. Res. *9:561 (1963)*]

If these membranes are considered to lie on two ends of a continuum, it is possible to reconcile the two major opposing viewpoints in membrane substructure. The inert myelin sheath, with little or no enzymatic activity, exists primarily as an unmodified protein-lipid bilayer that produces a distinct trilaminar image in the electron microscope. More active membranes contain additional enzymatic components bound within the trilaminar structure, giving the appearance of particulate structure within the membrane. Organelles, such as the chloroplast, containing highly complex and tightly packed systems of enzymes bound to their membranes have such an extensive granular substructure that the underlying trilaminar framework is almost completely obscured. This hypothesis carries with it the supposition that the lipid-protein bilayer underlies all membranes as a structural framework into which a greater or lesser number of enzymes may be bound, depending on the function of the membrane.

The key to an evolutionary trend is discernible within this framework. Because phospholipids form bilayers spontaneously in an aqueous medium, it is possible

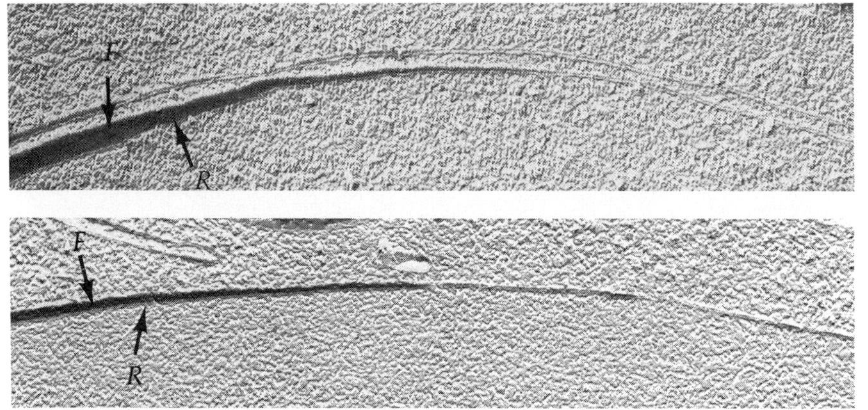

4–14. Trilaminar structure visible in freeze-etched tissue from an onion root-tip cell. Vacuolar membranes were exposed by etching of the fractured surfaces. In these replicas of the surface, typical trilaminar structure is visible to the right of both micrographs. Toward the left, the fracture appears to have traveled along the interior of the membrane, splitting the inner and outer surfaces (F *and* R *in the micrographs). upper,* × *72,000. lower,* × *65,000. Courtesy of D. Branton.* [Proc. Nat'l Acad. Sci., *55:1048 (1966)*]

4–15. Fractured chloroplast membranes. Compare the abundance of particles on these membranes, which are known to contain complex enzyme systems, with the smooth aspect of the membranes of the fractured myelin sheath in Fig. 4–16. The myelin sheath is a relatively inert system and probably contains few enzymes. × *92,500. Courtesy of D. Branton and Academic Press, Inc.* [J. Ultrastr. Res. *19:238 (1967)*]

that the primitive membrane represented a spontaneous lipid film of this type. The addition of protein to this primitive diffusion barrier increased its stability and introduced specificity of function, including localized areas of permeability to polar reagents. As complexity increased, greater numbers of protein molecules, both structural and enzymatic, were added to the primitive lipid film (see also Chapter 17).

The Dynamic State of Membranes

The electron microscope image is static, and the temptation is strong to regard the membrane as a rigidly defined and permanent framework within the cell. But observations of living cells reveal that membranes are continually being assembled and disassembled, sometimes with great rapidity. How are membranes synthesized with the cell? One possibility is that a building block of lipid and protein that can be readily assembled into or removed from the membrane exists in the cytoplasm. If such a subunit exists, however, it has not been detected by X-ray diffraction or electron microscopy. Another possibility is that phospholipid bilayers are built up from pools of lipid within the cytoplasm, followed by binding of protein to the surface of the bilayer. While this is entirely possible, it goes against the grain of current philosophy, which holds that specificity of function in the cell resides within protein. According to this line of reasoning, the protein component of membranes would be synthesized first, because proteins are known to be the initial product of the synthetic machinery of the cell. This specific membrane protein would then assemble into a framework, followed by the addition of lipid to complete

4–16. *Membrane surfaces of a myelin sheath from a five-day-old rat. The exposed surfaces of the membranes are smooth; no suggestion of particulate substructure is present. Compare with Fig. 4–15. × 59,000. Courtesy of D. Branton.* [Exptl. Cell Res. *45:703 (1967)*]

the membrane. Other mechanisms are possible. For example, the lipid and protein may be combined together from separate pools at the same instant in growing membranes. Which of these possibilities actually occurs is a completely open question.

Suggestions for Further Reading

Books

Branton, D., and R. B. Park, 1968. *Papers on Biological Membrane Structure.* Little, Brown & Co., Boston. A collection of papers of historical importance in the development of theories of membrane structure.

Chapman, D., 1968. *Biological Membranes.* Academic Press, New York.

Racker, E., 1970. *Membranes of Mitochondria and Chloroplasts.* American Chemical Society Monograph 165.

Papers and Review Articles

Bear, R. S., K. J. Palmer, and F. O. Schmitt, 1941. X-ray diffraction studies of nerve lipids. *Journal of Cellular and Comparative Physiology,* vol. 17, pp. 355–367.

Chapman, D., and D. J. Fluck, 1966. Physical studies of phospholipids III. Electron microscope studies of some pure, fully saturated 2,3-diacyl-DL-phosphatidyl-ethanolamines and phosphatidyl-cholines. *Journal of Cell Biology,* vol. 30, pp. 1–11.

Chapman, D., and D. F. H. Wallach, 1968. Recent physical studies of phospholipids and natural membranes. In D. Chapman, ed., *Biological Membranes.* Academic Press, New York, pp. 125–202.

Finean, J. B., 1966. The molecular organization of cell membranes. *Progress in Biophysical and Molecular Biology,* vol. 16, p. 145.

Green, D. E., and S. Fleischer, 1963. The role of lipids in mitochondrial electron transfer and oxidative phosphorylation. *Biochemica et Biophysica Acta,* vol. 70, pp. 554–582.

Korn, E. D., 1966. Structure of biological membranes. *Science,* vol. 153, p. 1491.

Korn, E. D., 1969. Cell membranes: structure and synthesis. *Annual Review of Biochemistry,* vol. 38, pp. 263–288.

Meuller, P., D. O. Rudin, H. Ti Tien, and W. C. Wescott, 1962. Reconstitution of cell membrane structure *in vitro* and its transformation into an excitable system. *Nature,* vol. 194, pp. 979–980.

Rouser, G., G. J. Nelson, S. Fleischer, and G. Simon, 1968. Lipid composition of animal cell membranes, organelles, and organs. In D. Chapman, ed., *Biological Membranes.* Academic Press, New York, pp. 5–69.

Sjöstrand, F. S., 1963. A comparison of plasma membranes, cytomembranes, and mitochondrial membrane elements with respect to ultrastructural features. *Journal of Ultrastructure Research,* vol. 9, pp. 561–580.

Stoeckenius, W., 1959. An electron microscope study of myelin figures. *Journal of Biophysical and Biochemical Cytology,* vol. 5, pp. 491–500.

Stoeckenius, W., 1960. Osmium tetroxide fixation of lipids. *Proceedings of the European Regional Conference on Electron Microscopy,* Delft, pp. 716–725.

Stoeckenius, W., and D. M. Engelman, 1969. Current models for the structure of biological membranes. *Journal of Cell Biology,* vol. 42, pp. 613–646.

Thompson, T. E., and F. A. Henn, 1970. Experimental phospholipid model membranes. In E. Racker, ed., *Membranes of Mitochondria and Chloroplasts.* American Chemical Society Monograph 165, pp. 1–52.

Vandenheuvel, F. A., 1963. Study of biological structure at the molecular level with steromodel projections I. The lipids in the myelin sheath of nerve. *Journal of the American Oil Chemists' Society,* vol. 40, pp. 455–471.

Unit Two

The Role of Membrane Systems in the Provision of Energy for Synthesis

5
The Flow of Energy in Biological Systems

All activities of the cell, including growth, movement, and response to changes in the environment require an expenditure of energy. These activities are the *work* of the cell; the energy used in accomplishing these tasks can be defined as the cell's capacity to do work. The energy used by individual cells ultimately originates in the sunlight that falls continuously on the surface of the earth. Living organisms capture part of this energy through *photosynthesis* in green plants. The "capture" takes the form of converting the radiant energy of light to chemical energy. This chemical energy is used by the cells of photosynthesizing organisms to carry out all the activities of life, including the synthesis of complex organic molecules.

These complex molecules represent chemical energy in potential form; the energy captured in this form by photosynthesizers is passed on to other organisms that survive by eating green plants. In these organisms, the complex substances are broken down, and the chemical energy released is used for cellular work. Herbivores eating green plants for survival serve in turn as food for other animals, and so on down the line until final degradation by organisms such as fungi and bacteria. Thus there is a flow of energy from the sun through the photosynthesizers, down through several levels of consumers, until the energy is dissipated as heat or is converted to a form in which it can no longer be transformed into chemical energy. Thus, in order to survive, living organisms on earth require a continuous input of energy.

In this flow of energy through the biological world there are two stages of greatest interest to cell biologists. One is the initial conversion of light to chemical energy, which takes place almost exclusively in the chloroplasts of green plants. The second is the series of reactions in which the complex molecules built up during photosynthesis are degraded with the release of chemical energy. In both plants and animals, the latter processes take place in mitochondria and the extramitochondrial cytoplasm.

These processes, in which energy is captured and converted into a form that can be used by the cell, are the subject of this unit. Before taking up the details of these mechanisms, it is first necessary to describe energy transformations in biological systems in more general terms. The most fruitful approach to the study of these transformations has been through the principles of *thermodynamics.* While it is impossible in a text at this level to give more than an elementary and limited description of the elegant abstractions of thermodynamics, it is useful to develop familiarity with the net energy changes taking place in chemical reactions. This information is of value in understanding the direction taken by biochemical interactions, how energy released by one reaction system is used in biological systems to "drive" another energy-requiring system, and the basis of the catalysis of biological reaction systems by enzymes.

An Introduction to Elementary Thermodynamics

Most simply stated, thermodynamics deals with energy and seeks to establish the principles and laws that govern the transformation of energy in reacting systems. Thermodynamics is based on a few basic laws or premises that summarize experimental observations and experience with physical objects. From these relatively simple statements, all of the remaining relationships are derived by logical reasoning. For a discussion of thermodynamics in relation to biological systems, two basic principles, known as the first and second laws of thermodynamics, are of central importance.

Before stating these principles it is first necessary to define a *system* as viewed by thermodynamics. A system is simply a collection of matter we wish to study with regard to changes in energy. The system is arbitrary but has definite limits; all parts of the universe not specified as belonging to the system constitute the *surroundings.* The system may be very simple, and may contain a single type of matter, such as an enclosed volume of a gas, or it may be exceedingly complex and consist of many species of molecules, as in a series of sequential biochemical reactions between complex substances.

To be of thermodynamic interest, the system must undergo a change in energy. Only two points in this change are considered and compared by thermodynamics: the instant the change begins, called the *initial state,* and the instant the change ceases, called the *final state.* Depending on whether the energy level increases or decreases in the transition from initial to final states, a system either absorbs energy from its surroundings, or delivers energy to its surroundings. Any change in energy level of the system is precisely counterbalanced by an equivalent and opposite change in the surroundings. All energy lost by the system is absorbed completely by the surroundings; systems may increase their energy content only by absorbing energy from the surroundings. If, during the transition, matter is transferred to or from the system, the system is said to be *open.* If the amount of matter in the system remains unchanged, the system is *closed.*

The First Law of Thermodynamics

With these definitions of the system and surroundings we can now consider the basic laws of thermodynamics. The First Law of Thermodynamics, often called the Principle of Conservation of Energy, was developed from a study of the many unsuccessful efforts to construct perpetual motion machines, that is, machines that will do work without an input of energy. The impossibility of such machines is expressed by the first law, which states that the total energy of a system cannot be increased or decreased by any reactions *within* the system. If the system is defined as the universe, this law means that the total energy of the universe remains constant.

The first law may be related to measurable quantities by a consideration of heat (Q) and work (W). Heat is a form of energy, and the transfer of heat can be measured by thermometers or other devices that react in regular increments according to the quantities of heat delivered. Work is also a familiar quantity, and can be defined as the displacement of a mass from one point to another by application of a force. Other forms of work are also significant in biological systems, including the work of moving substances in solution against a concentration gradient (osmotic work); the biosynthesis of complex substances from simple inorganic precursors (chemical work); and the displacement of charged particles through an electric field (electrical work). These various forms of work are interrelated, because the electrical work done by biological systems, as in the movement of an impulse along the axon of a nerve cell, is accomplished by the movement of ions (charged particles) against a concentration gradient (osmotic work).

According to the first law, a flow of heat to or from the system, or any work accomplished as the system goes from the initial to the final states, can only take place through a change in energy (ΔE) of the system. Therefore, in terms of the quantities Q and W, the first law states that

$$\Delta E = Q - W \tag{5–1}$$

There are two important points that must be emphasized about the relationship given in equation 5–1. Because the change in energy described by the equation refers only to a comparison of the initial and final states, it does not specify in any way the specific paths taken during the change. Thus, Q and W may vary considerably for a given ΔE. Another significant fact is that the equation, derived from the first law, does not evaluate the absolute energy content in a system, but only describes the energy change.

These points can be most easily understood in mechanical systems. Imagine a heavy weight at the top of a building (the initial state), being lowered to the ground by a crane. In order to control the rate of descent, a brake is applied constantly until the weight comes to rest on the ground (the final state). In this system, ΔE represents the difference in potential energy between the two positions of the weight. No work is done by the weight as it travels to the ground, and all the energy appears as heat in the braking system of the crane. Other paths are possible as this system goes from the initial to the final states. Imagine that, instead of a crane, the weight is connected to a rope over a pulley at the top of the building. The other end of the rope, at the ground, is attached to a slightly smaller weight. If the first weight is pushed from the building, the smaller weight will be raised to the top as the first travels the distance to the ground. In this example, only a fraction of ΔE will appear as heat; this will be limited primarily to the effects of friction in the pulley system. The remainder of ΔE will be expended in the work accomplished by moving the second weight to the top of the building. These two alternative pathways illustrate that for a given ΔE, Q and W may take on widely different values. The second point, that the equation does not describe absolute energy content, can be illustrated by the same example. In terms of gravity, the absolute potential energy of the weight is determined by its distance from the center of the earth. But the equation does not touch on this absolute content; it is concerned only with the *change* in energy that occurs between the defined initial and final states.

As mentioned, a major reason for introducing thermodynamics into the study of biology is to be able to predict whether a given biochemical or biophysical change is possible under specified conditions. It is not possible from the first law alone to predict whether processes resulting in a decrease in the total energy of a system are feasible. Although this type of prediction would coincide with our everyday experience in which physical objects fall "down" (that is, to a state of lower potential energy) and not "up," it would, in general, be valid only for mechanical processes that do not involve a heat change. Once a process involves a change in the heat content of a system, the first law alone cannot be used to predict the feasibility of a given transformation.

For example, consider a system represented by the collection of water molecules in a block of ice. As long as the temperature of the surroundings is above 0°C, the ice will melt. This occurs *spontaneously*, even though the energy content of the system at the final state is higher than at the initial state, and ΔE therefore increases in value. Thus ΔE as defined by the first law cannot be used to predict the feasibility of a given transformation or to predict whether a system will go to a higher energy state by absorbing energy from its surroundings, or to a lower state by loss of energy.

The Second Law of Thermodynamics

The key to the prediction of feasibility is provided by the second law of thermodynamics. This law supplies the desired "criterion of spontaneity," that is, the ability to predict whether a change of any type, including biochemical reactions, will run spontaneously to completion. In the example given, the melting of ice to water, the final energy state is higher than the initial state; from the first law we know that to accomplish this the system must absorb energy from its surroundings. But how can the feasibility of such a change be determined? Obviously, something must be added to the information provided by the first law.

Two physicists of the last century, Kelvin and Clausius, turned to the increment of heat, Q, as a criterion. The heat absorbed or generated does not allow prediction by itself, because for various reactions, Q, just as ΔE, may be positive, negative, or zero. However, a function of Q proved useful in predicting spontaneity. If various kinds of work are analyzed, each is characterized by two factors, *intensity* and *capacity* (see Table 5–1). For example, in estimating how much work can be done by gravitation, the height of the object alone does not provide enough information. To this must be added the mass, the weight that will drop to provide energy. Thus the mass involved provides a measure of the *capacity* of the system to do work. Similarly, in the case of a gas expanding against a piston, such as in an automobile engine, pressure alone does not provide a measure of the total work possible; to this must be added the volume of gas under consideration. Thus an engine having large displacement and using large volumes of gas develops more horsepower and can do more work than a smaller engine. In a consideration of the work possible by a flow of heat, the temperature is analogous to height and pressure in these examples. What is the equivalent capacity factor for the work to be done by heat?

Table 5–1 Intensity and Capacity Factors for Various Kinds of Work

Work	Intensity factor	Capacity factor
Gravitation	height	mass
Electricity	voltage	charge
Expansion	pressure	volume
Heat	temperature	Q/T

Adapted from I. M. Klotz, *Energy Changes in Biological Systems*. New York: Academic Press, 1967.

Much investigation has shown that the required capacity factor can be represented by Q/T, the ratio of the heat in calories to the absolute temperature. This can be understood in terms of the *specific heat* of a substance. The specific heat of water is 1; this means that one calorie is required to raise the temperature of one gram of water 1°C. The specific heat of iron is about 0.10. Only about 0.1 calorie is required to raise the temperature of 1 gram of iron 1°C. Thus, at any temperature above absolute zero, the heat content per degree of temperature (Q/T) of water is greater than iron. From this it is obvious that if 100 grams of water and an equal weight of iron are cooled in temperature from 100° to 50°C, a greater quantity of heat will flow from the water than from the iron. If this heat is used, for example, to expand a gas against a piston, the water will accomplish more work than the iron as it cools. Thus the Q/T function is a measure of the capacity to do work.

As it turns out, the factor for the work capacity of heat is of fundamental value in predicting whether a system will react spontaneously. The basis for this relationship is most easily understood in *reversible* systems at equilibrium. A reversible system is one that, as the name suggests, can run in either direction. In a reversible system at equilibrium, the tendency to run in either direction is exactly counterbalanced; only an infinitesimal change is required to displace the balance in either direction. This can be approximated by bringing two blocks of the same metal and of equal mass into contact. If one block is at a higher temperature, heat will flow from the warmer to the cooler block until the temperature of the two blocks is the same. At this point the system represented by the two blocks is at equilibrium. This system is fully reversible, because heat can be made to flow equally well in either direction simply by slightly increasing the temperature of one block (at the expense of energy derived from the surroundings). Conversely, as long as no heat flows between the surroundings and the system, no further change in the temperature of the two blocks is expected. Therefore, at equilibrium,

$$\frac{\Delta Q}{T}_{\text{system}} + \frac{\Delta Q}{T}_{\text{surroundings}} = 0 \qquad (5\text{–}2)$$

This equation simply repeats the statement that no total change is expected in the heat capacity of either the system or the surroundings at equilibrium.

If this statement is accepted as fact, then it follows for any reversible system not at equilibrium,

$$\frac{\Delta Q}{T}_{\text{system}} + \frac{\Delta Q}{T}_{\text{surroundings}} \neq 0 \qquad (5\text{–}3)$$

Equation 5–3 therefore states that, in a reversible system not in equilibrium, the sum of the $\Delta Q/T$, or $\Sigma(\Delta Q/T)$, for system and surroundings will be greater or less than zero, that is, a positive or negative number. By experience, we know that a system at some point other than equilibrium will probably run spontaneously to the equilibrium point. Therefore, a value for $\Sigma(\Delta Q/T)$ of any magnitude indicates that the reaction in question will proceed spontaneously.

Whether $\Sigma(\Delta Q/T) < 0$ or $\Sigma(\Delta Q/T) > 0$ is related to the *direction*, but not the spontaneity, of an interaction. Assume that for any reversible system, the criterion of spontaneity is defined as $\Sigma(\Delta Q/T) > 0$ (as is the case). Therefore, if $\Sigma(\Delta Q/T)$ is measured in a transformation running from state a to state b, and a positive number is obtained, the prediction can be made that the transformation will proceed spontaneously. Suppose instead that $\Sigma(\Delta Q/T') < 0$ for the transformation $a \rightarrow b$. This simply means that for the reverse reaction $b \rightarrow a$, $\Sigma(\Delta Q/T) > 0$. The transformation will therefore still run spontaneously, but in the opposite direction. For various reasons (see below), the criterion for spontaneity in reversible transformations is defined as

$$\Sigma\left(\frac{\Delta Q}{T}\right) > 0 \tag{5–4}$$

and for a transformation at equilibrium,

$$\Sigma\left(\frac{\Delta Q}{T}\right) = 0 \tag{5–5}$$

The Q/T function is of such importance that it has been specially designated as the *entropy* function, and is usually designated with the special symbol ΔS (read "change in entropy"). For a spontaneous change, therefore (from equation 5–4)

$$\Sigma \Delta S > 0 \tag{5–6}$$

Because ΔS is a positive number greater than 0 for all spontaneous reactions, the universe, in which spontaneous reactions are continuously proceeding, exists under conditions of constantly increasing entropy. This is really a way of expressing the Second Law of Thermodynamics, which simply states that the entropy of the universe always increases. Taken together, the first law states that the total energy of the universe, considered as a system, remains constant, while the second law states that as long as any reactions proceed spontaneously anywhere in the universe, the entropy increases.

Entropy is primarily a mathematical abstraction and is difficult to visualize in intuitive terms. One difficulty is that, in a spontaneous reaction, entropy *increases* as the system goes from the initial to final states. Thus we must visualize a quantity that becomes greater in magnitude as a reaction proceeds. This violates our usual criteria, because in ordinary experience we expect things to run down or *decrease* in magnitude during a spontaneous reaction. A rock will travel spontaneously downhill; gas will flow from a region of higher to lower pressure, and so on. Perhaps the best verbal representation of entropy is to regard it as the *randomness* or *disorder* of a system. Consider two closed vessels connected by a stopcock. In one is a quantity of gas under pressure. The second vessel is under a vacuum. We know by experience that if the stopcock is opened, the gas will flow spontaneously from the first vessel to the second. Once the pressure is equal in the two vessels, the net flow of gas will cease, and we know that it is almost impossible for the original conditions to occur again unless energy is added to the system from outside. That is, chances are that the gas will never spontaneously redistribute itself so that one vessel is under a vacuum. In the change from the initial to the final states, the randomness or disorder of the total system increases. Gas arranged into two vessels so that all the gas is confined to one vessel is a more ordered state than equal distribution between the two vessels. The decrease of order, or increase of disorder, in going to an even distribution is a measure of the increase of entropy of the reacting system. Irving Klotz (1967) points out that, for intuitive understanding, it would be somewhat more useful if entropy were defined as a negative rather than a positive function. Entropy could then be regarded as order rather than disorder, and we could easily see how systems would tend to run from an ordered to disordered state. This is a legitimate way of viewing entropy, and would make verbal representation of the entropy function easier to understand. However, for computation of the *free energy* (see below) of reactions, regarding entropy as an increase in disorder (Q/T) instead of a loss of order ($-Q/T$) turns out to be more convenient.

Free Energy and the Chemical Potential of Biological Reactions

The introduction of entropy into the analysis of reactions allows the prediction of spontaneity of a reaction. However, to apply the entropy function to a given situation, the ΔS for both the system and surroundings must somehow be measured. Measuring the ΔS for the system does not present too many problems, but the requirement that the surroundings must also be included in the measurements is inconvenient and sometimes impossible. What is needed is a method whereby the criterion of spontaneity—that is, the entropy change—can be evaluated by considering the system alone.

This problem was solved by Willard Gibbs, who combined the first and second laws into a single mathematical relationship. From the law of conservation of energy (the first law), any change in energy of the surroundings must be compensated for by a change in the system. A change in the entropy, $\Delta Q/T$, of the surroundings, must therefore be balanced by a change in the system. The compensatory energy change in the system can be expressed in terms of energy, pressure, volume, and temperature as follows:

$$\left(\frac{\Delta Q}{T}\right)_{\text{surroundings}} = \Delta S_{\text{surroundings}} = \frac{-\Delta(E + PV)_{\text{system}}}{T} \quad (5\text{–}7)$$

where P and V are the pressure and volume. In biological systems, the pressure and temperature can be assumed to remain constant during a transformation. Under these conditions, substitution of $-\Delta(E + PV)/T$ for $(\Delta Q/T)$ surroundings, and ΔS for $(\Delta Q/T)$ system in equations 5–3 and 5–4 gives:

$$\Delta S + \frac{-\Delta(E + PV)}{T} > 0 \quad (5\text{–}8)$$

then, by algebraic computation

$$\Delta S - \frac{\Delta E - P\Delta V}{T} > 0$$

$$T\Delta S - \Delta E - P\Delta V > 0$$

and

$$\Delta E + P\Delta V - T\Delta S < 0 \quad (5\text{–}9)$$

for spontaneous interactions.

The significance of equation 5–9 is that all quantities are now expressed in terms of the system.[1] This function, in honor of Gibbs, is termed the Gibbs free energy function, represented by ΔG:

$$\Delta G = \Delta E + P\Delta V - T\Delta S \quad (5\text{–}10)$$

For spontaneous reactions, as noted, $\Delta G < 0$. Often the terms $\Delta E + P\Delta V$ are combined into a single term, ΔH, called the *enthalpy change*, and equation 5–10 is written

$$\Delta G = \Delta H - T\Delta S \quad (5\text{–}11)$$

In biological systems, where the change in volume, ΔV, is small, the term $P\Delta V$ is negligible and ΔH is equivalent to ΔE. Under these conditions, the enthalpy change can be considered to be a change in energy, the net flow of heat between the system and the surroundings. In this situation the change in free energy, ΔG, will depend on a change in heat and entropy in the system. Because, for a spontaneous reaction, $\Delta H - T\Delta S$ is less than zero (from equations 5–8 to 5–10), the ΔG for a spontaneous reaction will have a negative value. Conversely, if a reaction is described as having a negative value for ΔG, we know that it will proceed spontaneously. Note, however, that the value for ΔG does not provide information on the rate of a reaction. Even if a transformation has a negative ΔG of great magnitude, its spontaneous rate of reaction may be so slow as to be unmeasurable. This will be considered further in the section of this chapter dealing with enzymatic catalysis.

[1]These computations apply only under conditions of constant temperature and pressure. However, the assumption of constant temperature and pressure in biological systems is often valid.

Free Energy and Chemical Reactions

In any chemical reaction in which one substance, A, is undergoing conversion into a second substance, B, a point is reached at which the rate of conversion of A to B is exactly counterbalanced by the rate of conversion of B to A. At this point, the reaction is said to be at equilibrium. For any reaction, under a specific set of conditions of temperature and pressure, the concentration of A and B at equilibrium is fixed and is thus a constant. This constant is called the *equilibrium constant*, and is equal to the ratio of concentrations of substances A and B at equilibrium:

$$K_{eq} = \frac{[B]}{[A]} \quad (5\text{–}12)$$

If more than single substances are present as reactants or products, such as in the reaction

$$A + B \rightarrow C + D \quad (5\text{–}13)$$

the equilibrium constant is equal to the product of the concentrations[2] of the end products ($[C][D]$) divided by the product of the concentrations of the reactants ($[A][B]$):

$$K_{eq} = \frac{[C][D]}{[A][B]} \quad (5\text{–}14)$$

[2]For substances at high concentrations, the *activity* rather than concentration of reactants and product should be used for accuracy in these calculations. The activity is the product of the concentration and an experimentally determined factor called the *activity coefficient* that corrects a solution for the departures from ideal behavior encountered at higher concentrations. At dilute concentrations, the activity coefficient approaches unity, and the concentration is approximately equal to the activity. In biological systems the solutes are assumed to be present in sufficiently dilute solution to allow the concentrations rather than the activities to be used in computation of the equilibrium constant.

From these relationships, for any system containing reactants and products at given concentrations, the spontaneity and direction of a reacting system can be predicted if the equilibrium constant is known. It is obvious from this that there must be a relationship between the equilibrium constant, K_{eq}, and the free energy, ΔG, which has the same predictive capacity. These quantities prove to be related by a relatively simple expression:

$$\Delta G^\circ = -RT \ln K_{eq} \qquad (5\text{–}15)$$

In this expression ΔG°, the *standard free energy change*, rather than ΔG is used. This means that the reaction is carried out under standard conditions, in which reactants at concentrations of one mole are converted to products at one mole, at a temperature of 25°C and a pressure of one atmosphere. In equation 5–15, R is the gas constant, T is the absolute temperature, and $\ln K_{eq}$ is the natural logarithm of the equilibrium constant. If K assumes values greater than 1, ΔG° is negative, and the reaction proceeds spontaneously with the release of free energy. For values of K less than one, ΔG° is positive, indicating that energy must be added to the system in order for the reaction to proceed. Representative values for ΔG° in relation to K are given in Table 5–2.

Table 5–2 *$\Delta G'$ and the Equilibrium Constant**

K_{eq}	$\Delta G'$ *calories/mole*
0.001	+4,089
0.01	+2,726
0.1	+1,363
1.0	0
10.0	−1.363
100.0	−2,726
1000.0	−4,089

*$\Delta G' = \Delta G^\circ$ at *p*H 7

From A. L. Lehninger, *The Mitochondrian*, Copyright © 1964, W. A. Benjamin, Inc., Menlo Park, California, by permission.

A specific example will be given to illustrate the calculation of ΔG° from the equilibrium constant. By measuring concentrations at equilibrium, the equilibrium constant for the reaction glucose 1-phosphate ⇌ glucose 6-phosphate, a reaction that takes place in living organisms in the oxidation of glucose, has been determined to have a value of 17 at *p*H 7.0. By substitution in equation 5–15:

$$\Delta G^\circ = -RT \ln K_{eq}$$

$$\Delta G^\circ = -(1.987)(298)(\ln 17)$$

$$\Delta G^\circ = -(1.987)(298)(2.303 \times \log_{10} 17)$$

$$\Delta G^\circ = -1700 \text{ cal/mole}$$

Because ΔG° is negative, this reaction will proceed spontaneously under standard conditions. The calculated standard free energy change for a number of reactions of biological interest is given in Table 5–3.

Table 5–3 *Standard Free Energy Changes at pH 7 and 25°C for Some Chemical Reactions of Biological Interest*

Reaction	$\Delta G'$ *cal/mole*
Oxidation	
glucose + $6O_2 \rightarrow 6CO_2 + 6H_2O$	−686,000
lactic acid + $3O_2 \rightarrow 3CO_2 + 3H_2O$	−326,000
palmitic acid + $23O_2 \rightarrow 16CO_2 + 16H_2O$	−2,338,000
Hydrolysis	
sucrose + H_2O → glucose + fructose	−5,500
glucose 6-phosphate + H_2O → glucose + H_3PO_4	−3,300
glycylglycine + H_2O → 2 glycine	−4,600
Rearrangement	
glucose 1-phosphate → glucose 6-phosphate	−1,745
fructose 6-phosphate → glucose 6-phosphate	−400
Ionization	
$CH_3COOH + H_2O \rightarrow H_3O^+ + CH_3COO^-$	+6,310
Elimination	
malate → fumarate + H_2O	+750

From A. L. Lehninger, *The Mitochondrian*, Copyright © 1964, W. A. Benjamin, Inc., Menlo Park, California, by permission.

Free Energy of Hydrolysis of ATP

The energetics of one particular molecular system is so important and central in the accomplishment of work in cells that it is worth special attention. This reaction involves interconversions between adenosine triphosphate (ATP) and adenosine diphosphate (ADP) by addition or removal of a phosphate group:

$$ADP + HPO_4^{=} + H^+ \rightleftharpoons ATP + H_2O \qquad (5\text{–}16)$$

adenine

ribose

ADP *(adenosine diphosphate)*

ATP *(adenosine triphosphate)*

5–1. *The ATP (adenosine triphosphate) molecule. The molecule consists of a purine base,* adenine, *linked to a 5-carbon sugar,* ribose. *Depending on whether one, two, or three phosphates are linked to the adenine-ribose complex, the molecule is termed adenosine monophosphate (AMP), diphosphate (ADP), or triphosphate (ATP).*

This system is reversible; in the right-hand direction, in which ATP is synthesized from ADP and phosphate, the reaction requires an input of about 7000 calories under standard conditions ($\Delta G^\circ = +7000$ cal/mole). Reaction in the opposite direction, in which hydrolysis with removal of phosphate from ATP to produce ADP + $HPO_4^=$ occurs, yields an increment of free energy of the same magnitude ($\Delta G^\circ = -7000$ cal/mole). Note that these values apply only to standard conditions. Under other conditions of temperature, pressure, and concentration of reactants and products encountered in living systems, departures from the values given may be large.

The ATP and ADP molecules are closely similar in structure (Fig. 5–1). Each consists of a nitrogenous base, adenine, attached to a 5-carbon sugar, ribose. A chain of phosphate groups is bound to the 5-carbon of the sugar. In ATP, three phosphates are bound at this point; in ADP, the chain consists of two phosphate groups. In the conversion from ADP to ATP, one phosphate is removed or added to the end of this chain without changes in structure of the remainder of the molecule.

The ATP-ADP system has been aptly described as the currency, or the "dollar" of energy in the cell. In photosynthesis, the energy of light, after conversion to chemical energy, is used to drive the synthesis of ATP from ADP. Thus, one of the first chemical products of photosynthesis is the ATP molecule. Similarly, in respiration, the energy derived from oxidation of complex substances is immediately used to drive the synthesis of ATP from ADP. The energy "captured" as potential energy in ATP is used in almost all cellular activities requiring energy, including growth, movement, and responsiveness. At the site of these activities, ATP is hydrolyzed to ADP, and the energy released is used to do work. Thus the ATP-ADP system cycles between energy-yielding and energy-requiring activities in the cell (Fig. 5–2).

Because of this role, the bond linking the terminal phosphate to ATP is often described as a "high-energy" bond, and the ATP molecule is regarded much like a quantity of fuel that may be used to supply energy

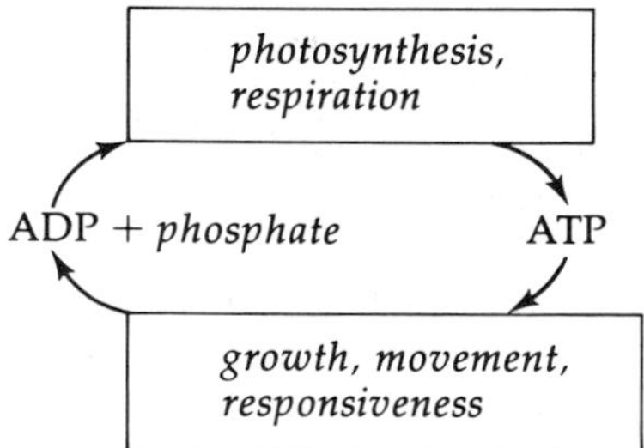

5–2. *The ATP-ADP cycle in living systems.*

directly for cellular synthesis. The actual nature of the energy transformation taking place in this system can be more accurately described by the thermodynamic principles outlined in this chapter. As mentioned, the synthesis of ATP from ADP requires an input of energy of 7000 cal/mole at standard conditions. In the oxidation of glucose, a primary food substance in both plants and animals, combustion in air to CO_2 and H_2O under standard conditions yields a large increment of free energy:

$$C_6H_{12}O_6 + 6O_2 \rightarrow 6CO_2 + 6H_2O \qquad (5\text{–}17)$$

$$\Delta G^\circ = -686{,}000 \text{ cal/mole}$$

The metabolism of glucose in the cell proceeds in many steps in an exceedingly complex manner, and with a different end result in terms of energetics. Although many reactants and products serve as intermediates in this system in the cell, the significant ones for this discussion are the same as given in equation 5–17, to which are added the ATP-ADP system:

$$C_6H_{12}O_6 + 6O_2 + 38ADP + 38HPO_4^{=} \rightarrow 6CO_2 + 44H_2O + 38ATP$$

$$\Delta G^\circ = -420{,}000 \text{ cal/mole} \qquad (5\text{–}18)$$

Note that the ΔG°, representing the change in free energy between the initial and final states, is of smaller magnitude than in the system shown in equation 5–17. While a large negative value is still obtained for the free energy change, indicating that the reaction will proceed spontaneously, not as much energy is yielded from the system. The energy conserved is bound as potential energy in the 38ATP molecules synthesized from 38ADP. By comparing the relative values for ΔG° in the two systems, it can be seen that about $(266{,}000/686{,}000) \times 100 = 39\%$ of the energy is captured in ATP. Because the mechanism acts in this way, the synthesis of ATP is sometimes said to be *coupled* to the metabolic breakdown of glucose.

Precisely the same mechanism, operating in the opposite sense, takes place in the work activities of the cell, such as growth and movement. In one such reaction, a part of growth, individual nucleotides are linked together into chains (Fig. 5–3) to produce the DNA molecule. The linkage of two nucleotides into a dinucleotide requires an input of energy and thus will not proceed spontaneously. Formation of the bond linking the nucleotides together, a *phosphodiester* bond, requires an input of 5000 calories for each mole of dinucleotide synthesized. In the cell, this reaction proceeds by a complex pathway that ultimately includes the hydrolysis of ATP:

$$\text{nucleotide}_1 + \text{nucleotide}_2 + ATP + H_2O \rightarrow \text{dinucleotide} + ADP + HPO_3^{=} + H^+ \qquad (5\text{–}19)$$

$$\Delta G^\circ = -2000 \text{ cal/mole}$$

Since this reacting system proceeds from an initial to final state with a negative value for the free energy change, it will proceed spontaneously to the right. In this system, the synthesis of a phosphodiester bond linking two nucleotides, an "uphill" reaction requiring energy, is coupled to the hydrolysis of ATP to produce an *overall* reaction that may proceed "downhill," and the system becomes thermodynamically feasible. Almost all of the chemical, osmotic, electrical, and mechanical work of the cell is directly or indirectly coupled to the ATP–ADP system and made energetically favorable in this way.

The Energetics of Biological Catalysis

The principles of thermodynamics also provide an insight into the mechanism of catalysis of biological reaction by enzymes. As we have seen, the first and second laws of thermodynamics allow us to predict whether a reaction will proceed spontaneously but do not give an indication of the rate at which a reaction will proceed. For example, glucose can be oxidized by combustion in air to the products CO_2 and H_2O, with $\Delta G^\circ = -686{,}000$ cal/mole. Because the value for ΔG° is negative, the reaction will proceed spontaneously. However, if left in air at physiological temperatures, the conversion to H_2O and CO_2 proceeds so slowly that the rate is unmeasurable and is essentially zero. This illustrates the fact that the rate of a reaction is independent of the thermodynamic feasibility.

The rate at which a reaction proceeds spontaneously at a given temperature can be considered as a function of the *activation energy* required for an interaction to occur. This concept states that even though a reaction may proceed with an overall release of free energy, the molecules of the system must first be raised to a critical energy level before the interaction can proceed (Fig. 5–4a). This may be visualized as an energy barrier over which the molecules must be raised before the overall reaction occurs. The condition is analogous to a rock resting at the top of a hill (Fig. 5–4b). As long as

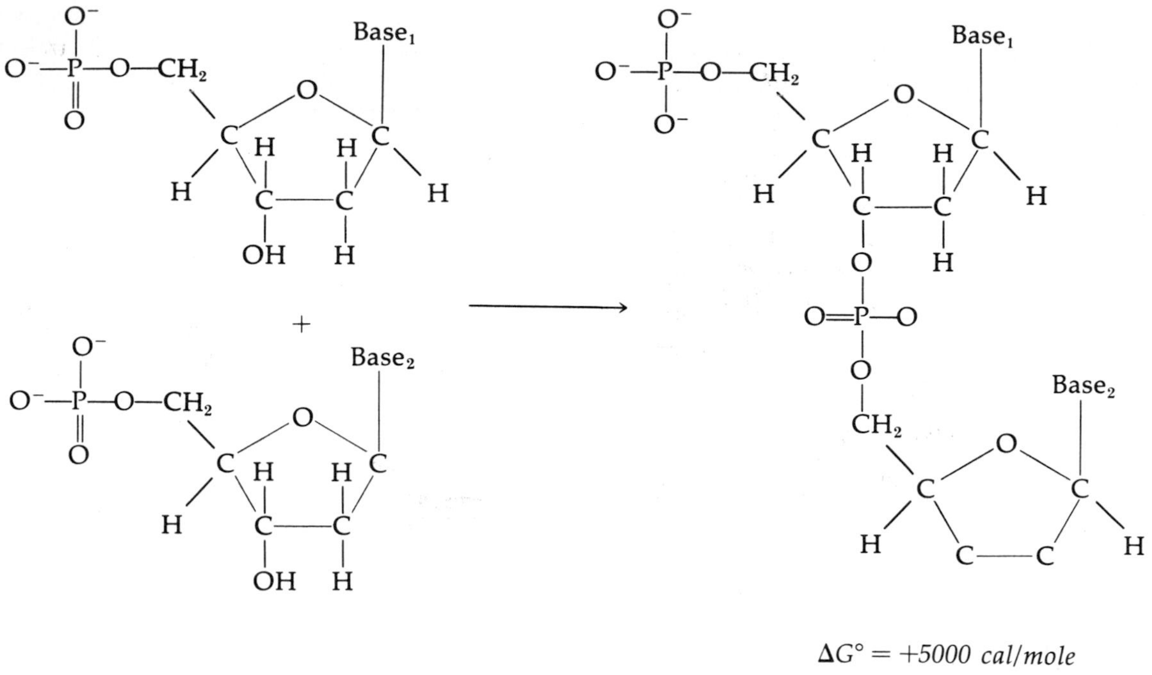

$\Delta G^\circ = +5000$ *cal/mole*

5–3. *The linkage of two nucleotides into a short dinucleotide. Forming the phosphodiester bond requires an input of 5000 cal/mole under standard conditions.*

the rock is undisturbed, it will not spontaneously begin its travel downhill, even though the total "reaction," the progression of the rock downward, is energetically favorable. In this example the activation energy may be regarded as the effort required to overcome the friction holding the rock in position. If the rock is pictured as resting in a small depression at the top of the hill as shown in Fig. 5–4b, the analogy is closer to the situation in a collection of interacting molecules.

The requirement for energy of activation raises the question of why reactions proceed spontaneously at all. In other words, how are some molecules in the system spontaneously raised over the energy barrier? This happens because, instead of being at rest as in the analogy of the rock at the top of a hill, the molecules of any system above a temperature of absolute zero are in constant motion. The amount of motion, or kinetic energy, is not the same in all molecules of the population, although the statistical *average* is below the energy required for activation. Some molecules in the population are below this average, and some are above. The distribution occurs partly as a result of random collisions in which energy is gained and lost by individual molecules depending on factors such as the angle and number of collisions per unit time. Depending on the height of the energy barrier, a number of molecules may be raised to the energy level required for the reaction to proceed. If the activation energy is high, very few molecules will pass over the barrier in unit time, and the collection of molecules will be stable. From this explanation, the effect of heating such a reacting system becomes immediately obvious. Heating the system increases the average kinetic energy of the population of molecules and increases the probability that a significant number will be raised over the energy barrier.

This brings us finally to the role of enzymes in biological systems. Enzymes lower the activation energy required for a reaction to proceed (Fig. 5–5). By combining with the reactants, the enzyme forms a new molecular complex that requires less energy to overcome the activation barrier and is in a sense less stable. This greatly increases the probability that molecules of the system will spontaneously pass over the barrier to complete the reaction. Thus interactions that might otherwise require elevation to high temperatures in order to proceed can take place at physiological levels.

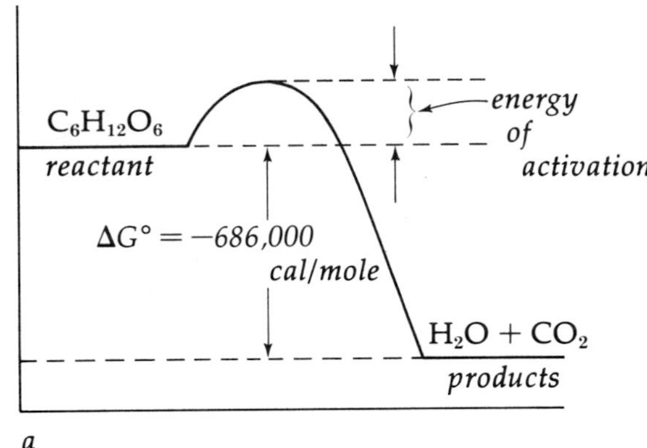

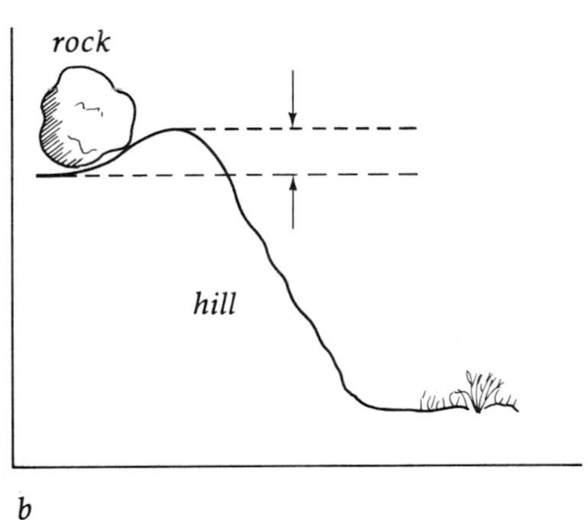

5–4. a, Diagram illustrating the effect on a spontaneous reaction of the requirement for energy of activation. Even though the conversion of the reactant, glucose, to the products is a "downhill" reaction and proceeds with the release of a large increment of free energy, spontaneous oxidation of glucose at room temperature proceeds at an immeasurably slow rate. This is because an increment of energy must first be added to raise the molecule to a state at which spontaneous reaction may occur. This increment is shown as the "energy of activation" in the diagram. b, An analogous mechanical situation in which a rock is poised in a slight depression at the top of a hill. The rock will not spontaneously move downward unless enough "activating energy" is first added to raise it from the depression and push it over the brink of the hill.

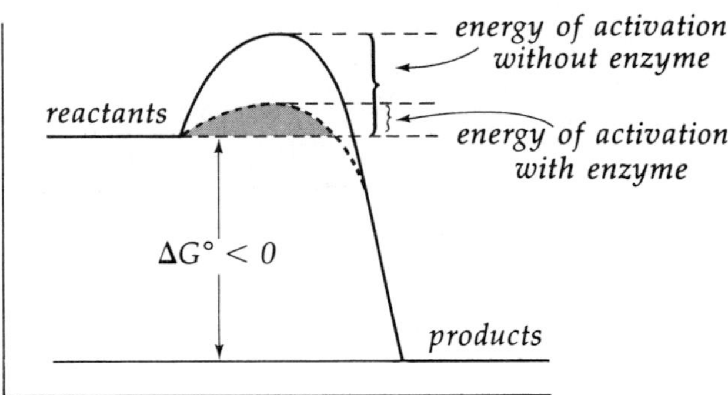

5–5. Diagram illustrating the effect of an enzyme. An enzyme increases the rate of a spontaneous reaction by reducing the energy of activation.

This point deserves emphasis, because in the absence of enzymes many of the reactions occurring in cells would require temperatures high enough to kill the organism.

An enzyme affects only the rate of a biological reaction. The direction of the reaction, the thermodynamic feasibility, and the concentrations of reactants and products at equilibrium remain the same whether or not the catalyzing enzyme is present. Thus an enzyme will not catalyze a reaction that would not take place spontaneously without the enzyme.

Enzymes have other important characteristics. They are not changed by their participation in reactions. Therefore only small amounts of the enzyme are needed to cause a great increase of rate, and each enzyme molecule may catalyze the conversion of more than one molecule of reactant to products. The number of reactant molecules (called the *substrate*) catalyzed by a single enzyme molecule per unit time may be exceedingly high; this quantity, called the *turnover number*, may vary from 100 to more than 3,000,000 molecules of substrate catalyzed by each enzyme molecule per minute. Finally, enzymes are specific in activity, and usually catalyze only a single biochemical reaction.

All known enzymes are proteins. Nearly a thousand different enzymes have been described; a few of these, about 130, have been purified and crystallized. These vary from relatively small molecules having about 100 amino acid residues (see Fig. 2–17a) to large complexes of several amino acid chains with several thousand residues. In many enzymes, the protein is complexed with other nonproteinaceous substances such as metallic ions and nucleotides.

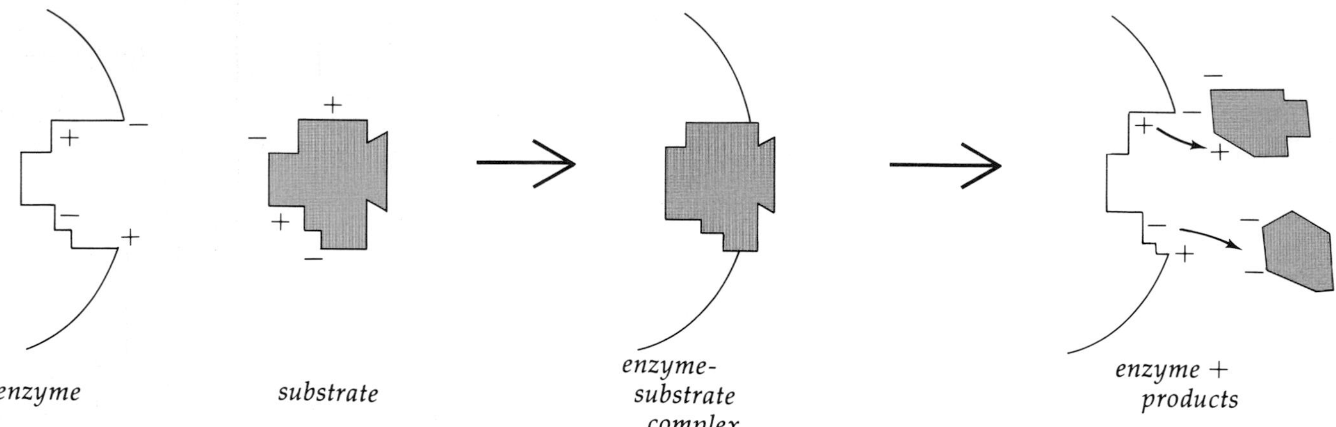

5–6. The lock-and-key model of the active site, as hypothesized by Emil Fischer. The active site is considered to be a rigid arrangement of charged groups that precisely matches the substrate.

The Mechanism of Enzymatic Catalysis

The precise molecular mechanism by which enzymes lower the energy of activation remains obscure. It is known that the enzyme combines briefly with the substrate molecule during catalysis of the reaction. This interaction may involve several separate steps, as indicated by the work of Britton Chance and his colleagues. Some enzymes absorb light at certain wavelengths, producing characteristic colors in solution. Chance followed changes in these colors during the course of enzymatically catalyzed reactions. One enzyme studied, *peroxidase,* catalyzes the reduction of hydrogen peroxide to water and oxygen. Solutions of the enzyme without substrate are brown. On addition of the substrate, the solution turns green, indicating that reaction has taken place between the enzyme and the added hydrogen peroxide. A second color change, to red, takes place immediately after the appearance of green color, indicating that a rearrangement to a second enzyme-substrate complex occurs immediately after the initial interaction. Following conversion to hydrogen peroxide to products, the solution returns to the brown color. Measurement of the time duration of the various colors indicates that the successive complexes last only for milliseconds.

Other more direct evidence of the enzyme-substrate complex has been obtained by David C. Phillips and his co-workers. These investigators studied the conformation of *lysozyme,* an enzyme that lyses, or dissolves, bacterial cells. In Phillips' experiments, X-ray diffraction techniques were used to map out the three-dimensional structure of the enzyme. One interesting feature of the structure of lysozyme is a cleft that runs down the center of the molecule. By X-ray diffraction of the enzyme in the presence of compounds closely resembling the normal substrate molecules, it was shown that a complex between the enzyme and substrate does occur, and that the site of this binding is located in the cleft.

Significantly, these studies also showed that on binding of the substrate to the enzyme the cleft narrows slightly. This observation lends support to some recent proposals concerning the manner of binding of enzyme to substrate. According to the classical hypothesis of enzyme-substrate interaction, advanced by Emil Fischer in 1884, the site on the enzyme binding the substrate, called the *active site,* contains an arrangement of oppositely charged groups and a surface conformation that precisely matches, in lock-and-key fashion, the charge and shape of the substrate molecule (Fig. 5–6). The specificity of enzymatic activity is explained under this hypothesis by the assumption that only the substrate molecule possesses the correct conformation to fit into the active site of the enzyme.

Recently, several important modifications to Fischer's proposed mechanism were advanced by D. E. Koshland. Under the classical concept, it was assumed

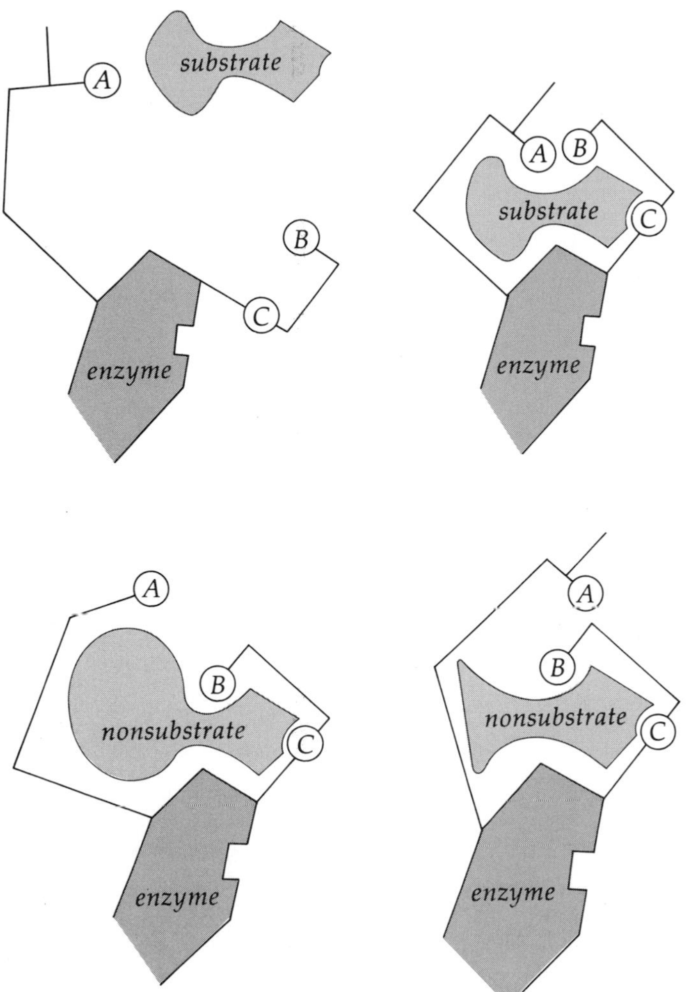

5–7. *A modified conception of the active site, as advanced by D. E. Koshland. In the model shown, the catalytic groups* A *and* B *must be aligned for catalysis to occur. When nonsubstrates are bound, the alignment of the catalytic groups is incorrect and no catalysis occurs. This model implies that the active site is not rigid and changes in conformation with binding of the substrate. Redrawn, original diagram courtesy of D. E. Koshland.*

that the structure of the enzyme is rigid and that conformational changes do not take place on binding of the substrate. Koshland has suggested instead that at the time of binding, the enzyme is induced by the correct substrate to undergo changes in conformation that complete the requirements for catalysis (Fig. 5–7). Under Koshland's hypothesis, therefore, the final pattern of charges and the three-dimensional conformation of the active site is not completed until the substrate is bound. This hypothesis, as mentioned, is supported by the observation from X-ray diffraction that the binding site of lysozyme changes in dimensions during binding to its substrate.

The molecular basis for the reduction of activation energy, which undoubtedly involves some interaction between the enzyme and substrate at the active site, remains unknown. One likely possibility is that by binding at the active site, reactants are brought into close proximity on the surface of the enzyme, with the result that the probability of interaction is greatly increased. Another possibility, in the case of substrates that are broken down, as in the hydrolysis of protein chains into individual amino acids, is that the substrate is "warped" during binding to the enzyme. This warping may place strain on particular bonds within the substrate, lowering the activation energy required for rupture of the bond. However, these ideas are merely speculative, and it must be admitted that we have very little insight into the mechanism of action of enzymes at the present time.

Suggestions for Further Reading

Klotz, Irving M., 1967. *Energy Changes in Biochemical Reactions.* Academic Press, New York.

Koshland, D. E., Jr., and M. E. Kirtley, 1966. Protein structure in relation to cell dynamics and differentiation. In M. Locke, ed., *Major Problems of Developmental Biology.* Academic Press, New York, pp. 217–249.

Lehninger, Albert L., 1965. *Bioenergetics.* W. A. Benjamin, Menlo Park, Calif.

Morowitz, Harold J., 1970. *Entropy for Biologists.* Academic Press, New York.

6
The Mitochondrion and the Provision of Energy for Cellular Activities

The energy required for the cell's chemical work is provided by the oxidation of fuel substances, primarily carbohydrates, fatty acids, and amino acids. As far as is known, the reactions carrying out the oxidation of these substances occur almost entirely within the cytoplasm. Initial breakdown, with some release and capture of energy, occurs in the cytoplasm external to the mitochondrion. Products of these initial reactions enter the mitochondrion, where the major conversion of released energy into a chemically useful form takes place.

The role of the mitochondrion as the "powerhouse" of the cell was not clearly understood until the 1940's, although descriptions of mitochondria, and use of the term, date back to the late 1800's. Until recently, cytological investigation of the mitochondrion and biochemical study of respiration went separate ways, and as late as 1935 the functional role of the mitochondrion in cellular respiration was unknown. These separate efforts combined in the 1940's, leading to our present-day appreciation of the integration of respiratory function with mitochondrial structure. Before the contemporary understanding of mitochondrial structure and function is surveyed, a few of the important milestones in the research will be reviewed.

Some Historical Perspectives

Although early references to structures identifiable as mitochondria appear in the literature published from 1850 to 1890, positive identification and description was not made until the last decade of the nineteenth century. At this time R. Altmann made extensive observations of the occurrence and distribution of these bodies, aided by a new stain that was highly selective for mitochondria. Altmann named the cytoplasmic particles *bioblasts,* and speculated that they were the elementary units of life of the cell. Interestingly enough, Altmann also proposed that "bioblasts" were akin to bacteria, living a potentially independent existence in the cytoplasm. These speculations, as it turns out, have proved to be correct in part, for recent work (see Chapter 16) has demonstrated that mitochondria possess at least partial independence from genetic control by the nucleus. A short time later, at the turn of the century, C. Benda coined the term *mitochondria* for these organelles (from the Greek *mitos* = thread and *chondros* = granule). During these early years, the mitochondrion was thought to be primarily involved in cell heredity. The cytologist F. Meves was particularly vigorous in expounding this point of view, with the result that the real function of mitochondria was missed until much later.

As these cytological results were taking form, purely biochemical investigations began to trace out the oxidation pathways of foodstuffs in living systems, without reference to specific cellular organelles. Particularly important in these developments was the work of O. Warburg and H. Wieland. Warburg showed in the 1920's that oxygen enters into cellular metabolism through the action of a catalyst containing iron, which he named the *atmungsferment* (now called the *cytochrome* system). At about the same period, Wieland developed the principle that biological oxidations proceed by the removal of hydrogen from substrates. Other work, by Warburg and Hugo Theorell led to the discovery in the 1930's of the *flavoproteins,* substances that link the enzymes removing hydrogen from substrates to the cytochromes. This work led to the appreciation that hydrogen, removed from substrates by dehydrogenases, is conducted to the cytochromes by the flavoproteins. The cytochromes, in turn, catalyze the reduction of oxygen to water, giving the sequence

$$\text{Substrates} \rightarrow \text{dehydrogenases} \rightarrow \text{flavoproteins} \rightarrow \text{cytochromes} \rightarrow O_2$$

for the flow of hydrogen (and electrons) in cellular respiration (for details of these early developments see Lehninger, 1965b).

Many details about specific reactions in which oxidation of food substances is accomplished by dehydrogenation were learned in the years immediately following these discoveries. A number of these reactions and the enzymes catalyzing them were identified by the 1930's. From these discoveries, part of the series of cellular interactions that break down carbon chains in respiration was reconstructed. Finally, Hans Krebs postulated that the oxidation of carbohydrate derivatives occurs in a cycle of reactions in which short carbon chains introduced into the cycle are continuously broken down to yield CO_2.

The first indications of a linkage between these oxidative reactions and the formation of ATP came with the Warburg's discovery in 1938 that a specific reaction, the oxidation of glyceraldehyde phosphate (see p. 108) was coupled in some manner to the synthesis of ATP. By 1940, Herman Kalckar had demonstrated that additional ATP formation is closely linked to the respiratory chain of reactions that reduce oxygen to water. This system coupling ATP synthesis to respiration, called *oxidative phosphorylation,* is the main source of ATP in the cell.

Until the late 1940's, little attention was paid to the possible location of respiration and oxidative phosphorylation in a particular cell organelle. In the first half of the present century, the general attitude among biochemists was that the presence of cellular particles in subfractions under study was merely a hindrance to the purification of respiratory enzymes. This attitude began changing in the 1940's when a mitochondrial fraction was isolated by A. Claude, K. R. Porter, and G. E. Palade. A highly purified preparation, with morphology closely similar to the mitochondria of intact cells, was subsequently isolated and partly characterized biochemically by G. H. Hogeboom and W. C. Schneider, who showed that most of the cytochrome activity of the cell was localized in this fraction. Later work by these investigators and by A. L. Lehninger showed that the Krebs cycle and oxidative phosphorylation are also located within the mitochondrion. This confluence of biochemical and cytological investigation of the mitochondrion finally led to the acceptance of this organelle, by the 1950's, as the "powerhouse" of the cell. At about this time, in 1952 and 1953, the first high resolution electron micrographs of mitochondria were obtained by Palade and Fritjof S. Sjöstrand.

Structure and Occurrence of Mitochondria

Under the light microscope, mitochondria are seen as spherical, granular or fibrillar bodies roughly 1 to

2 microns in length and 0.5 microns in diameter. The number of mitochondria per cell is highly variable. They occur in great numbers in some cells, such as amphibian eggs and the giant amoeba *Pelomyxa*, where more than 100,000 mitochondria may be present. In other cells, such as the algae *Micromonas* and *Chromulina*, only a single mitochondrion is found. In most cells, at least several hundred mitochondria are present; vertebrates, for example, contain on the average 500–1000 mitochondria per cell.

Phase contrast microscopy of living cells shows that mitochondria grow in length, branch, divide, and coalesce. These changes, occurring over periods of less than one minute, indicate that mitochondria are probably almost as dynamic in structural configuration as the plasma membrane. This flexibility of structure is worth keeping in mind in the morphological descriptions that follow, because the evidence from electron microscopy often gives the impression that mitochondria are static, rigidly organized structures.

The work of Palade and Sjöstrand on mitochondrial ultrastructure and subsequent investigations with the electron microscope have shown the mitochondrion to contain two separate membrane systems. One membrane forms a single, continuous outer layer around the mitochondrion, completely separating the interior from the rest of the cytoplasm. The second, inner system, is variously folded or convoluted. In the most frequently observed type, the inner membrane is folded into flattened sacs called *cristae*, which extend across the interior of the mitochondrion at angles to the outer membrane (Fig. 6–1).

Two other types of cristae occur less frequently. In one, found in spermatocytes, the inner membrane forms a number of concentric layers within the outer membrane, enclosing a roughly spherical central space (Fig. 6–2). This type may represent a degenerating form, because it is also found in necrotic cells. In a third important type, found in protozoa and some cells of higher organisms, the inner membrane is formed into interdigitating tubules (Fig. 6–3a and b).

The folding of the inner membrane increases its surface area and separates the interior of the mitochondrion into separate compartments. The most common pattern of folding forms two separate regions (see Fig. 6–1b): the *intermembrane space*, between the inner and outer membranes, and the *matrix*, occupying the central cavity of the mitochondrion and enclosed by both the inner and outer membranes. It is not clear whether the spaces enclosed within the folds of the cristae always make connections with the intermembrane space. At some points, such interconnections are clearly seen; other cristae membranes in the plane of section appear to be completely separate from the inner membrane, even where the two systems come into contact. While it cannot be definitely proved from the electron microscopy of thin sections, most investigators agree that the cristae membranes probably make contact and are continuous with the inner mitochondrial membrane at some point around their periphery. This connection may be so narrow that only a few are caught in the plane of section. That this is actually the case is indicated by careful morphological investigations, which demonstrate that the cristae of mouse liver mitochondria make connection with the inner membrane at very limited regions marked by narrow, tubular extensions only 300 Å in diameter (see Daems and Wisse, 1966).

After fixation with permanganate or glutaraldehyde (without osmium postfixation), the two opposed cristae membranes are frequently so closely pressed together that the cristal space is entirely absent. The same arrangement is also seen in cells prepared for electron microscopy by freeze-substitution (quick-freezing, followed by substitution of ice by organic substances that remain liquid at low temperatures). Whether this configuration, reported primarily by Sjöstrand, is found in the living cell remains uncertain.

The intermembrane and cristal spaces appear structureless and show very little density in the electron microscope. However, the matrix, occupying the innermost cavity of the mitochondrion, is considerably more dense than the surrounding extramitochondrial cytoplasmic background and contains a variety of structures. The main substance of the matrix appears faintly granular, with a periodicity of about 40–50 Å. Embedded in this material may be various inclusions, including large granules, fibrils, microtubules, and crystals. The most common large granule is the intramitochondrial or osmiophilic granule, a dense, spherical particle about 250–350 Å in diameter, or about 2–3 times the size of a ribosome (see Fig. 6–1). The function of these large granules is unknown. Other smaller granules identifiable as mitochondrial ribosomes occur in the matrix in large numbers (see Fig. 16–12). Embedded within "clear" regions of the matrix are scattered deposits of thin fibrils 30–40 Å in diameter (see Fig. 16–4). By various techniques, these fibrils have been identified as

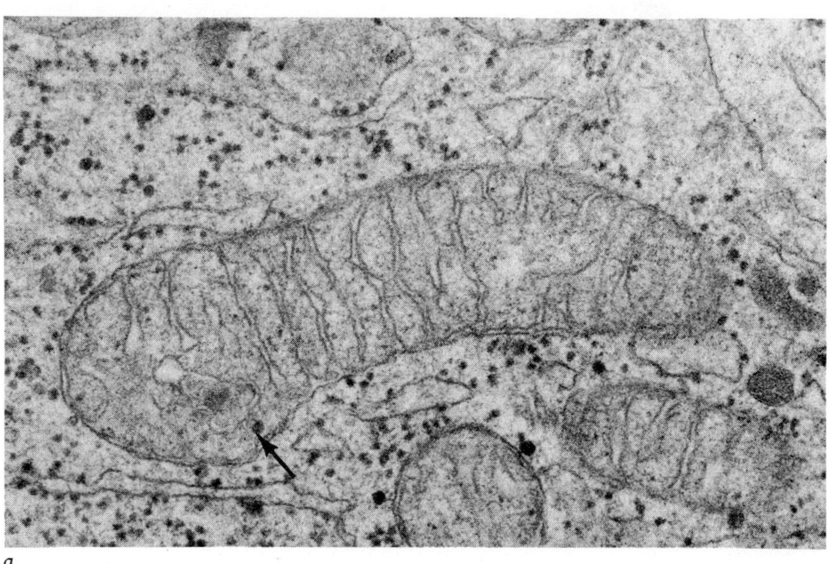

a

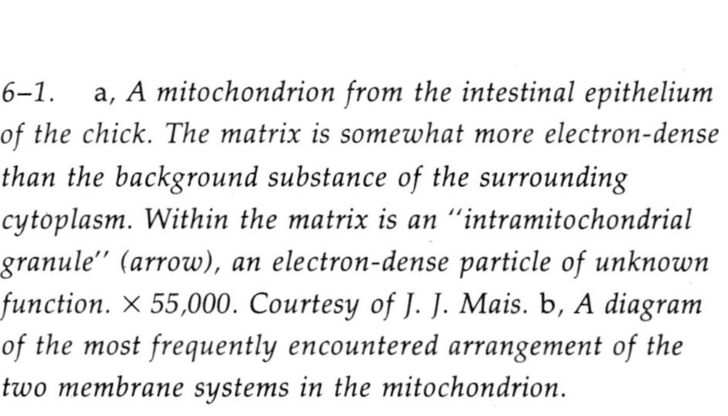

6–1. a, *A mitochondrion from the intestinal epithelium of the chick. The matrix is somewhat more electron-dense than the background substance of the surrounding cytoplasm. Within the matrix is an "intramitochondrial granule" (arrow), an electron-dense particle of unknown function.* × *55,000. Courtesy of J. J. Mais.* b, *A diagram of the most frequently encountered arrangement of the two membrane systems in the mitochondrion.*

b

DNA, closely similar in structure and properties to bacterial DNA (the evidence for the presence of DNA and ribosomes and their function in the mitochondrion is considered in detail in Chapter 16). More rarely, large crystals of proteins, yolk granules (see Fig. 15–47), or deposits of glycogen are found in the matrix, indicating in these cases a possible storage function for this region of the mitochondrion.

These mitochondrial membranes and spaces house the systems of respiration and oxidative phosphorylation. Some subparts of the overall mechanism, such as the Krebs cycle, hydrogen (proton) and electron trans-

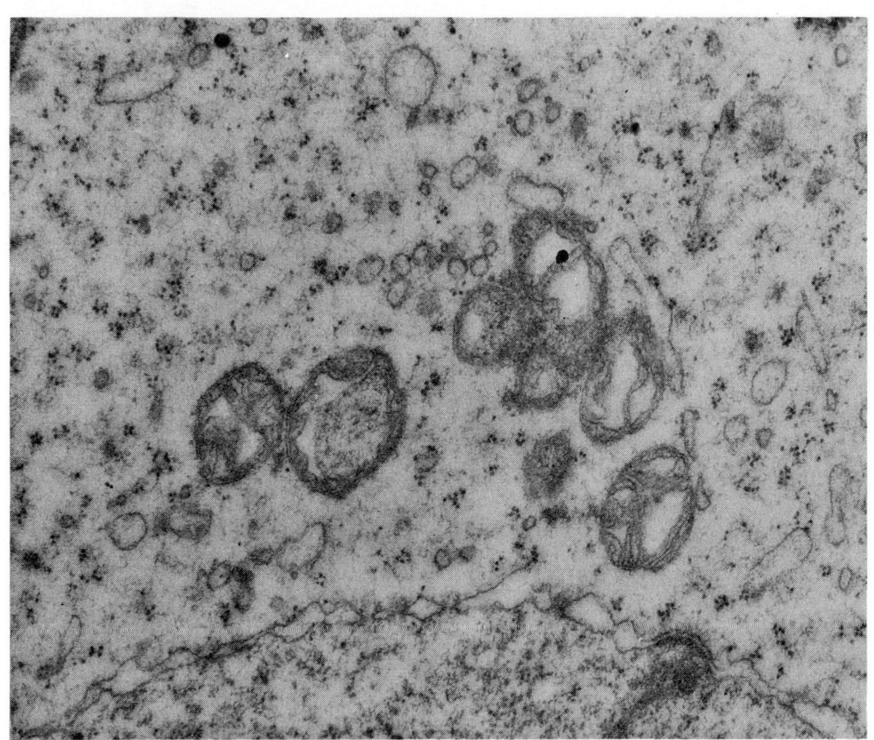

6–2. Mitochondria from rat testis, showing an alternate form of the inner membranes of the mitochondrion. Here the cristae form several concentric layers inside the outer boundary membrane, and the matrix often forms a central cavity. × 21,500. Photograph by the author.

port, and phosphorylation have been localized within specific mitochondrial compartments. Before going into the details of this localization and continuing with some recent insights into the structure and function of the inner mitochondrial membranes, these biochemical interactions will be described in some detail. This discussion will include not only mitochondrial systems, but a series of "make-ready" reactions that take place in the extramitochondrial cytoplasm. Because the reactions of glycolysis, fatty acid oxidation, and the Krebs citric acid cycle are essentially classical information and are likely to be already familiar to the reader, these will be given in descriptive fashion, with little reference to supporting experimental evidence.

Cellular Respiration

The carbon chains oxidized in cellular respiration originate from carbohydrates, fats, and proteins. Each class is converted in the extramitochondrial cytoplasm into molecules of much shorter length before undergoing oxidation within the mitochondrion. The most important reaction sequence of this type, *glycolysis*, converts glucose into two-carbon segments that then enter the mitochondrion for further oxidation. Another series of reactions breaks fatty acids into two-carbon segments as fuel for mitochondrial oxidation.

It is important to remember that oxidation does not require combination of molecular oxygen with a metabolite. Oxidation describes any reaction in which electrons, usually in pairs, are removed from a substrate. Usually, in cellular oxidations, hydrogen ions are removed at the same time. For every oxidation, an electron acceptor is reduced by combining with the removed electrons. Therefore, each oxidation is accompanied by a reduction. Frequently, the reduced electron acceptors also combine with one or both hydrogen ions (protons) removed with a pair of electrons. Depending on the arrangement of the atoms and bonds in the oxidized molecule, more or less energy is associated with the removed electrons; this energy is used to do chemical work in the cell.

The energy of the electron can be expressed as relative *potential* or *voltage* by comparison with an arbitrary standard. The standard used is the characteristic energy of electrons removed from hydrogen in the reaction $H_2 \rightarrow 2H^+ + 2e^-$, catalyzed by platinum. The potential, or voltage, associated with these electrons is arbitrarily set as 0.00 volts, and all other potentials

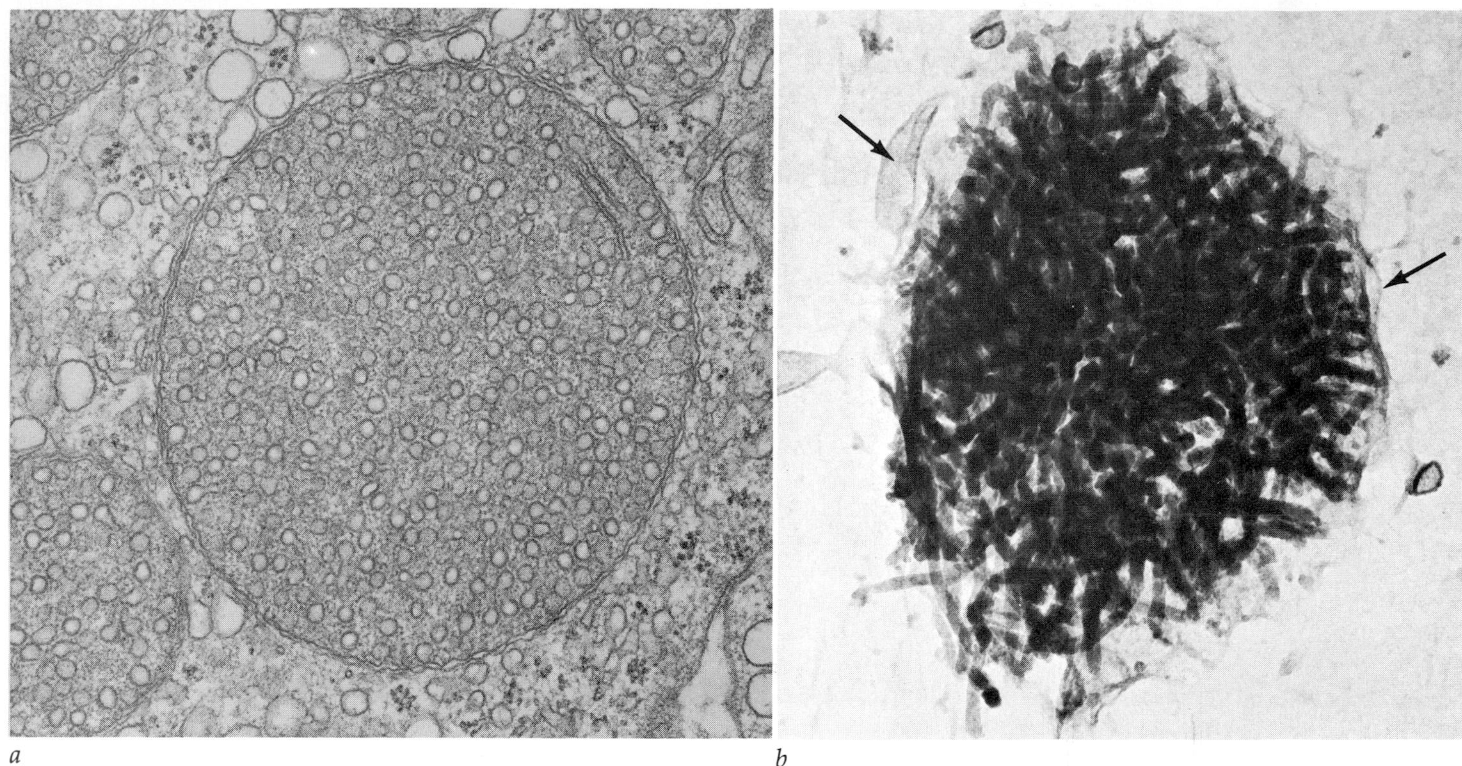

6–3. a, *A large, spherical mitochondrion from the adrenal cortex of the rat with cristae in the form of tubes or vesicles. The matrix is conspicuously granular.* × *43,500. Courtesy of D. S. Friend and G. E. Brassil and The Rockefeller University Press.* [J. Cell Biol. *46:252 (1970)*]. b, *An isolated, partially spread mitochondrion from the protozoan* Euplotes. *This mitochondrion, which was released from a cell disrupted on a water-air interface, contains tubular cristae typical of many ciliate protozoans. The outer membrane (arrows) has been lost except at a few locations. This mitochondrion, which has been positively stained, does not show the presence of "elementary particles" (compare Fig. 6–15a).* × *41,500. Photograph by the author.*

(often called "redox" or reduction-oxidation potentials) are measured and assigned a value with respect to this standard.

Glycolysis

The reactions taking place in glycolysis illustrate the preliminary reaction sequences that split complex substrates into shorter carbon segments for oxidation within the mitochondrion. A few primitive organisms, such as some disease-causing bacteria, depend on a closely similar reaction sequence for all cellular oxidation. Because this type of respiration can be carried out in the absence of molecular oxygen, cells or organisms respiring by this method are called *anaerobes.* Other primitive organisms, including most bacteria and yeasts, and some cells of higher plants and animals can respire either anaerobically or aerobically depending on the supply of oxygen.

In the first two reactions in the glycolytic sequence (Fig. 6–4), glucose is converted into a more reactive derivative containing two phosphate groups. The energy required for the phosphorylation of glucose is provided by the breakdown of two molecules of ATP to ADP. The first phosphorylation will be described in detail, because it illustrates a number of important principles. The reaction is believed to take place on the surface of the enzyme *hexokinase.* Glucose and ATP combine with the enzyme at the active site of the protein molecule. This combination brings glucose and

glucose — *glucose 6-phosphate* — *fructose 6-phosphate* — *fructose diphosphate* — *3-phosphoglyceraldehyde* ⇅ *dihydroxyacetone phosphate* — *1,3-diphosphoglycerate* — *3-phosphoglycerate* — *2-phosphoglycerate* — *phosphoenolpyruvate* — *pyruvic acid*

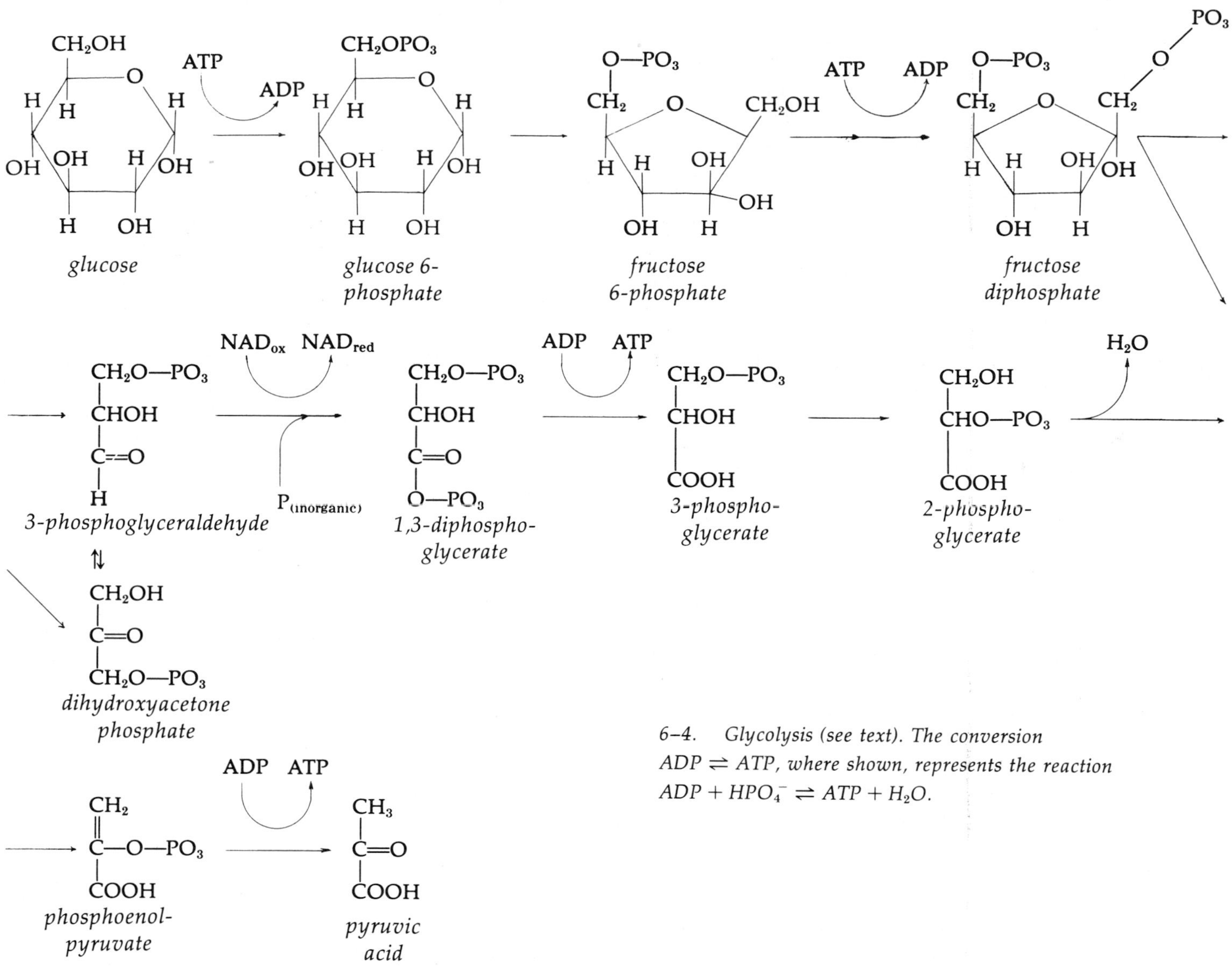

6–4. Glycolysis (see text). The conversion ADP ⇌ ATP, where shown, represents the reaction ADP + HPO_4^- ⇌ ATP + H_2O.

ATP into close proximity on the enzyme surface, increasing the probability of interaction. In some manner, the enzyme regulates this interaction, catalyzing the transfer of the terminal phosphate of ATP to the 6-carbon of glucose. The enzyme hexokinase is specific for this interaction; presumably, some configuration of charged groups at the active site of the enzyme catalyzes the transfer of phosphate from ATP to glucose. The products of the reaction, glucose 6-phosphate and ADP, are then released from the active site. Although the phosphorylation of glucose is an "uphill" reaction, the overall reaction (6–1), including the breakdown of ATP to ADP, proceeds in a "downhill" direction with the release of free energy:

$$\text{Glucose} + \text{ATP} + H_2O \xrightarrow[\text{hexokinase}]{} \text{glucose 6-phosphate} + \text{ADP} \quad (6\text{–}1)$$

$$\Delta G^\circ < 0$$

The product of this reaction, glucose 6-phosphate, becomes the substrate for the next reaction in the sequence, a rearrangement of glucose 6-phosphate to fructose 6-phosphate, which takes place on the surface of the enzyme *phosphoglucomutase*. At the cost of a second ATP molecule, fructose 6-phosphate is phosphorylated to fructose 1, 6-diphosphate. This step is mediated by the enzyme *phosphofructokinase*. The remaining reactions, all of which are catalyzed by specific enzymes, do not require an energy input at the expense of ATP. Fructose 1, 6-diphosphate is broken into two three-carbon segments on the enzyme *aldolase*. These segments are readily converted by another enzyme, *triosephosphate isomerase*, to form a pool of the two three-carbon sugars. One of these phosphorylated three-carbon sugars, 3-phosphoglyceraldehyde, enters into the remaining steps in glycolysis. As 3-phosphoglyceraldehyde is depleted from the pool, it is replaced by conversion of the second 3-carbon product, dihydroxyacetone phosphate.

At the next step in the sequence, two electrons and two hydrogens are removed from 3-phosphoglyceraldehyde. Some of the energy released by this oxidation is "trapped" by the attachment of a second phosphate to the three carbon sugar to form 1, 3-diphosphyglycerate. Note that this phosphate is derived from free inorganic phosphate in the medium, and not from ATP. The electrons removed at this step have a relatively high potential or voltage, and are accepted by a molecule similar to ATP, *nicotinamide adenine dinucleotide* (Fig. 6–5b), abbreviated NAD. The NAD molecule is a carrier of electrons, just as ATP is a carrier of phosphate groups. The reduced NAD produced at this step is one of the important products of the glycolytic sequence.

The 1, 3-diphosphoglycerate can be regarded as a "high energy" molecule. Removal of one of the two phosphates is accompanied by a large release of free energy. The enzyme *diphosphoglycerate kinase* catalyzes the transfer of the phosphate at the 1-carbon to ADP. Much of the energy released by removal of the phosphate is "captured" in this conversion of ADP to ATP.

This reaction pattern is worth discussing in some detail, because it is the method by which the energy of oxidation is converted into a usable chemical form in the cell. If one mole of 1, 3-diphosphoglycerate is hydrolyzed to 3-phosphoglycerate and $HPO_4^=$, the reaction releases about 10,000 to 15,000 calories under standard conditions of temperature and pressure. In glycolysis, the breakdown of the diphosphorylated sugar is coupled to the synthesis of ATP from ADP. The combined reaction releases about 4500 cal/mole, which is presumably lost as heat.

$$\text{1, 3-diphosphoglycerate} + \text{ADP} \rightarrow \text{3-phosphoglycerate} + \text{ATP} \qquad \Delta G° = -4500 \text{ cal/mole} \tag{6–2}$$

As mentioned in the introduction to this chapter, the link between ATP formation and the oxidation of phosphoglycerate was originally discovered in 1938 by Warburg. Direct transfer of the phosphate to ADP does not actually take place on the surface of the enzyme diphosphoglycerate kinase. Since this transfer is catalyzed by the enzyme *ATPase*, at least two enzymes are involved in this step, and at least one other phosphorylated intermediate is formed. In fact, each step of the glycolytic sequence as given must be regarded as a skeleton reaction, because enzymes and intermediate reactions in addition to those shown almost certainly take part in many of the transformations.

The 3-phosphoglycerate produced by reaction 6–2 is rearranged in two enzymatically catalyzed steps into phosphoenolpyruvate, which is subsequently dephosphorylated by the enzyme *pyruvate phosphokinase*. This dephosphorylation is coupled with the synthesis of a second ATP:

$$\text{Phosphoenolpyruvate} + \text{ADP} \rightarrow \text{pyruvic acid} + \text{ATP} \qquad \Delta G° = -5{,}800 \text{ cal/mole} \tag{6–3}$$

Since each glucose molecule entering glycolysis produces two molecules of 3-phosphoglyceraldehyde, a total of four molecules of ATP are produced in the conversion of 3-phosphoglyceraldehyde to puyruvic acid. The reactions attaching two phosphates to the glucose molecule in the first steps in glycolysis require two molecules of ATP. Therefore there is a net gain of 2ATP for every glucose entering glycolysis. Another significant product is reduced NAD,[1] which carries two electrons capable of doing chemical work. The overall

---

[1] Several conventions are in use for representing the alternate oxidized and reduced forms of NAD. The designations "NAD_{ox}" and "NAD_{red}" will be used here. Other designations for the reduced form likely to be encountered by the reader in other sources are "NAD^{+}" or "$NADH + H^{+}$."

adenine

pantothenic acid

ribose

a

adenine

D-ribose

adenine

c

D-ribose

nicotinamide

D-ribose

b

6–5. a, *Coenzyme A, a carrier of acetate groups.* b, *NAD, a carrier of hydrogen and electrons.* c, *ATP, a carrier of phosphate groups. Each molecule is structurally similar and is based on a group consisting of the purine base adenine and the 5-carbon sugar ribose.*

products of glycolysis are summarized in equation 6–4:

$$\text{Glucose} + 2\text{ADP} + 2\text{HPO}_4^{=} + 2\text{NAD}_{ox} \rightarrow 2\text{ pyruvate} + 2\text{NAD}_{red} + 2\text{ATP} \quad (6\text{–}4)$$

Each reduced NAD molecule carries, in addition to a pair of electrons, one of the two hydrogens removed at the oxidation of 3-phosphoglyceraldehyde. The other hydrogen enters the pool of H^+ ions in the medium surrounding the reaction sequence.

The precise location of the glycolytic sequence in the cytoplasm is not known. The traditional biochemical viewpoint has regarded the various glycolytic substrates and enzymes as in solution in the cytoplasm, interacting simply by collisions in the medium. Recent investigations indicate that many biochemical reactions are spatially ordered by structures within the cell. It is possible that the same thing is true of glycolysis, and that the glycolytic enzymes are loosely bound to a structural framework within the cytoplasm. There is much fibrous material faintly visible in the ground substance of the cytoplasm; this material, probably a protein, could provide the necessary framework.

Production of acetyl-coenzyme A. The electrons carried by NAD_{red} in the cytoplasm surrounding the mitochondrion are transferred through the mitochondrial membrane by an unknown mechanism to NAD molecules inside the mitochondrion (see Fig. 6–6). Pyruvic acid enters the mitochondrion directly. It is then oxidized, with NAD as an electron acceptor, and converted into a two-carbon segment. The carbon removed is released as CO_2. The two-carbon (acetate) group produced in this reaction is accepted by *coenzyme A,* an accessory molecule attached to the enzyme that catalyzes the oxidation of pyruvic acid.

Coenzyme A, with the attached acetate group derived from pyruvic acid, is another "high energy" substance and forms the immediate fuel for the oxidative reactions in the mitochondrion. This molecule, which acts as a carrier of acetates, closely resembles ATP and NAD (see Fig. 6–5). Each of these molecules consists of *adenine* (a purine base), the 5-carbon sugar *ribose,* and a short chain of two or three phosphate groups. Coenzyme A and NAD contain, in addition, groups derived from the vitamins *pantothenic acid* and *nicotinamide.* Coenzyme A, NAD, and ATP, respectively, act as carriers of acetates, electrons, and phosphates.

Coenzyme A is central to the metabolism of fuel substances in the cell. The longer carbon chains of carbohydrates, fats, and amino acids are broken into two-carbon acetate segments. Before entering the oxidative sequences in the mitochondrion, these acetate groups are combined with coenzyme A to form acetyl-coenzyme A. In the case of pyruvic acid, the reaction proceeds as shown in equation 6–5:

$$\text{Pyruvic acid} + \text{NAD}_{ox} + \text{coenzyme A} \rightarrow \text{acetyl-CoA} + \text{NAD}_{red} \quad (6\text{–}5)$$

For each molecule of glucose entering the overall sequence:

$$\text{Glucose} + 2\text{ADP} + 2\text{HPO}_4^{=} + 4\text{NAD}_{ox} + 2\text{ coenzyme A} \rightarrow 2\text{ acetyl-CoA} + 2\text{ATP} + 4\text{NAD}_{red} + 2\text{CO}_2 + 2\text{H}_2\text{O} \quad (6\text{–}6)$$

The Krebs Citric Acid Cycle

The acetate group carried by coenzyme A is oxidized inside the mitochondrion in a cyclic series of reactions first described by Hans A. Krebs in the late 1930's. This cycle is usually named either for Krebs or for citric acid, the first product of the interaction of acetate with the cycle. Within the cycle, the final oxidation of the two-carbon segments originally derived from glucose takes place. Very little trapped energy in the form of ATP is directly generated in the citric acid cycle. Instead, the energy of the electrons removed in the cycle, carried by NAD, is used later to drive the synthesis of ATP. The synthesis of ATP is accomplished by a series of enzymes and cofactors closely associated with the citric acid cycle, called the electron transport system.

The overall cycle is shown in Fig. 6–7. As with glycolysis and acetyl-CoA formation, the cycle as depicted must be regarded as a skeletonized system; many of the reactions involve intermediates that are not shown. In the first step in the sequence, the acetate group linked to acetyl-CoA (bound to the enzyme *pyruvic acid dehydrogenase*) is transferred to oxaloacetic acid (4 carbons), forming citric acid (6 carbons). This reaction, catalyzed by *citrate synthetase,* releases coenzyme A, which is then free to recycle through another oxidation of pyruvic acid. Citric acid forms part of an equilibrium pool of three acids, all mutually intraconvertible by the enzyme *aconitase.* At equilibrium, the pool contains about 90 percent citrate, 3 percent *cis*-aconitate, and 7 percent isocitrate. Isocitric acid is oxidized at the

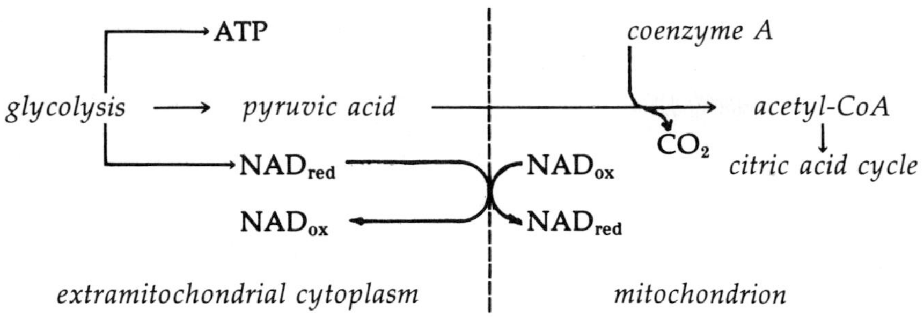

6–6. *Coordination of glycolysis with the mitochondrion (see text).*

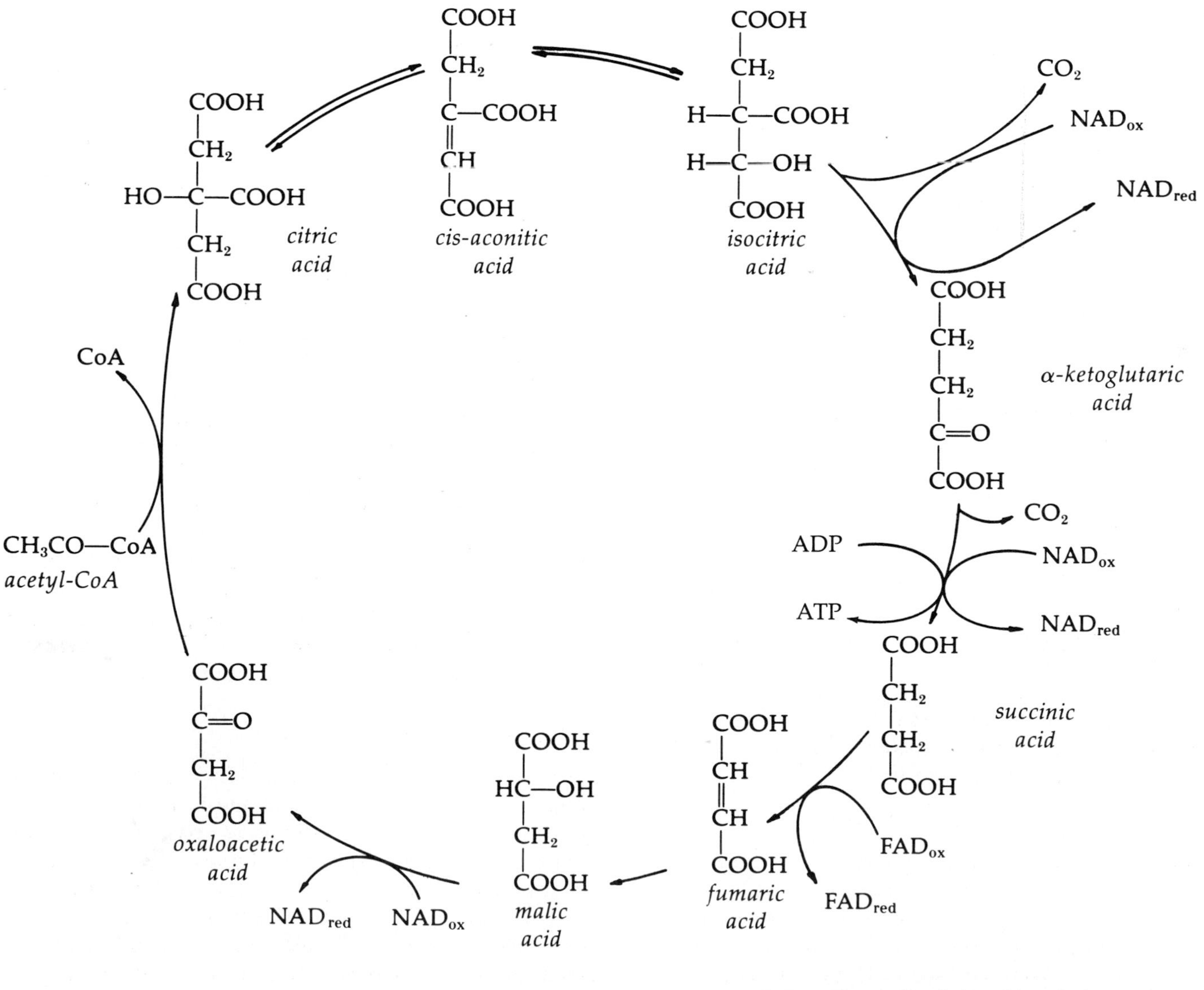

6–7. *The Krebs citric acid cycle (see text).*

next step in the cycle, with removal of two hydrogens. One carbon is removed in the form of carbon dioxide. The electrons removed at this oxidation are accepted by either NAD or the closely similar *NADP* (nicotinamide adenine dinucleotide phosphate). NADP, which contains one additional phosphate group (see Fig. 6–13) is the primary electron acceptor for this oxidation. However, the electrons carried by NADP are transferred to NAD before entering the electron transport system:

$$NADP_{red} + NAD_{ox} \rightarrow NADP_{ox} + NAD_{red} \quad (6\text{–}7)$$

The α-ketoglutarate produced at this step is oxidized to succinic acid in a complex series of reactions which includes coenzyme A, and NAD as an electron acceptor. Two molecules of ADP are converted to ATP as a part of this reaction sequence. One molecule of carbon dioxide is given off, shortening the chain to four carbons. The next step is unusual in that the enzyme catalyzing the oxidation of succinate to fumarate carries with it the electron acceptor *flavine adenine dinucleotide* (FAD). The combination of this acceptor with the enzyme is termed a *flavoprotein.* The electrons removed from succinate do not have sufficient energy, or potential, to reduce NAD and are accepted by the flavoprotein instead. In the next step in the cycle, after a rearrangement of fumarate to malic acid (catalyzed by *fumerase*), malic acid is oxidized by *malate dehydrogenase* to oxaloacetate, with NAD as an electron acceptor. The oxaloacetate regenerated at this step is available for interaction with another entering unit of acetyl-CoA, which starts the cycle again. It is interesting that oxaloacetate is a competitive inhibitor of the enzyme catalyzing the oxidation of succinic acid. If no acetate molecules enter the cycle, creating an excess of oxaloacetic acid, the enzyme catalyzing the oxidation of succinic acid is inhibited by combination with oxaloacetate. This mechanism, one of several internal feedback controls in the system, stops or slows the cycle until acetate is again available.

Overall products of the citric acid cycle. For each complete turn of the cycle, one molecule of acetic acid is oxidized to two molecules of carbon dioxide. At each of four separate steps, two electrons are removed from intermediates in the cycle, producing three molecules of reduced NAD and one of reduced FAD. At the end of the cycle, oxaloacetate, which can enter into the cycle with another acetate group, is regenerated. As an overall equation

$$\text{Acetic acid} + \text{oxaloacetate} + 3NAD_{ox} + FAD_{ox} \rightarrow \text{oxaloacetate} + 2CO_2 + 3NAD_{red} + FAD_{red} \quad (6\text{–}8)$$

Oxaloacetate can be crossed out from both sides of this equation.

With this information, we can add to the results the complete oxidation of one molecule of glucose to carbon dioxide and water via glycolysis, acetyl-CoA formation, and the citric acid cycle. For each molecule of glucose entering the sequence, the citric acid cycle will turn twice. From glycolysis and acetyl-CoA formation, 2ATP, and $4NAD_{red}$ are produced, in addition to two molecules of CO_2. Adding to this the results of the citric acid cycle, we have:

$$\text{Glucose} + 4ADP + 4HPO_4^{=} + 10NAD_{ox} + 2FAD \rightarrow 4ATP + 10NAD_{red} + 2FAD_{red} + 6CO_2 + 4H_2O \quad (6\text{–}9)$$

The 4ATP shown represent two ATP from the glycolytic sequence and two additional ATP produced by the coenzyme A cycle accompanying α-ketoglutaric acid oxidation. Up to this point, little ATP, the real currency of energy transfer in synthesis, has been produced. Of greatest significance are the electrons carried by reduced NAD and FAD, which contain most of the energy resulting from the oxidation of glucose to carbon dioxide.

This energy is released in the cell by the eventual transfer to oxygen of the electrons carried by NAD. Instead of occurring in one step, the transfer is accomplished in a series of smaller steps in the electron transport system. This mechanism releases the energy in small increments, some of which are sufficient to drive the synthesis of ATP from ADP + $HPO_4^{=}$.

The Electron Transport System

Investigators working with isolated mitochondria found that the system of enzymes transporting electrons from oxidized substrates to oxygen was tightly bound to mitochondrial membranes. The carriers can be removed only with difficulty, with one exception (cytochrome *c*). The system has been found to function only when intact and still bound to mitochondrial membranes. As a result, not all of the steps are completely understood, and it is possible that a number of carriers remain to be identified.

The known electron carriers are enzymes, all of which contain an additional nonprotein subunit called a *prosthetic group.* The prosthetic group is the actual electron carrier, and it is alternately oxidized and reduced as electrons flow along the system. The flavoproteins have a prosthetic group similar to NAD, except that a subunit derived from riboflavin, a vitamin of the B group, is also present. The cytochromes, including cytochrome a, a_3, b, c, and c_1 have as a prosthetic group a combination of a porphyrin ring and iron (see Fig. 6–8) very similar to hemoglobin. The various cytochromes differ from one another in the form of the heme prosthetic group. The iron in the center of the porphyrin ring can exist in either the ferrous (Fe^{+2}) or ferric (Fe^{+3}) form, depending on the oxidation or reduction of the carrier. The cytochromes differ from other carriers in that only one electron is accepted, in contrast to the pairs carried by the dehydrogenases and flavoproteins. Another electron carrier, called *ubiquinone* or *coenzyme Q* (see Fig. 7–11), has been implicated in electron transport as a prosthetic group. This carrier, which contains a quinone group that can be alternately oxidized and reduced, has not been conclusively identified as a part of the electron transport system.

Although the exact sequence of the carriers acting to transport electrons from substrate to oxygen is not known, it has been possible to reconstruct much of the probable sequence. Comparisons of the redox potentials (see below) have been useful in sequencing the electron carriers. Carriers releasing electrons at lower voltage are expected to follow carriers of higher (more negative) voltage in a series. Because the redox potential of an electron carrier can be determined experimentally, a hypothetical sequence can be assembled by comparing the potentials and aligning the carriers from most electronegative to most positive. Sequences determined in this way suffer from the fact that the potentials are measured with the electron carriers purified and in solution. The properties of carriers undoubtedly change when they are removed from mitochondrial membranes and purified to allow measurement of redox potentials.

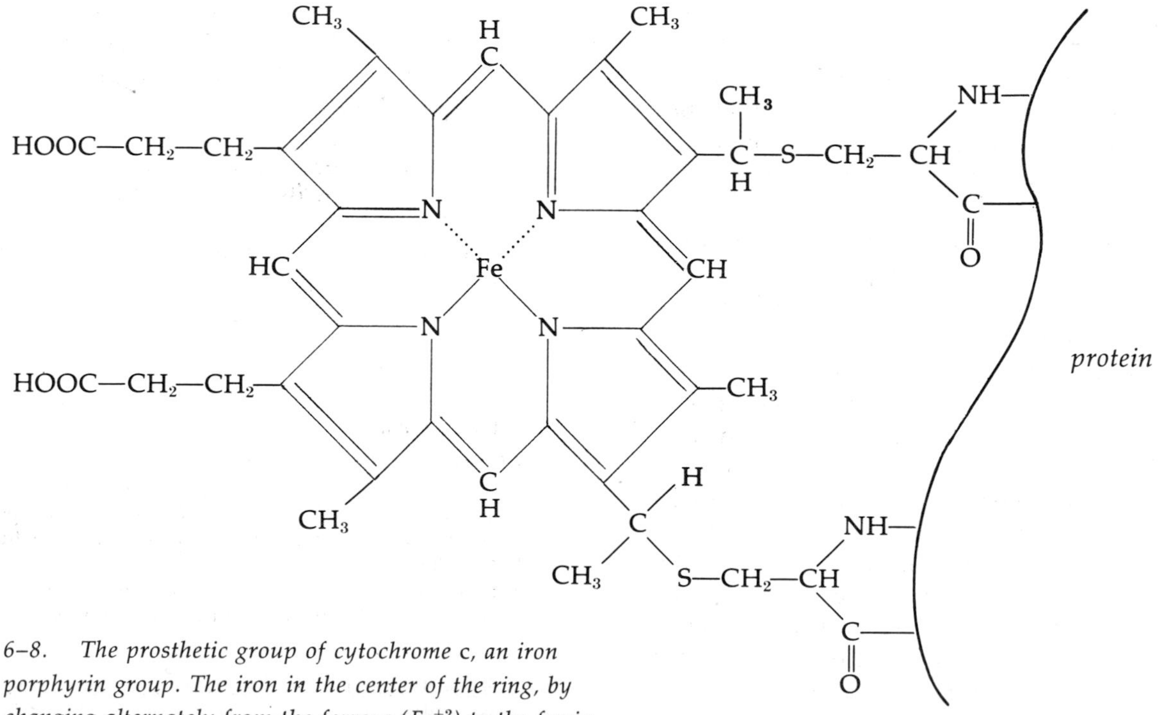

6–8. The prosthetic group of cytochrome c, *an iron porphyrin group. The iron in the center of the ring, by changing alternately from the ferrous (Fe^{+2}) to the ferric (Fe^{+3}) form, acts as an electron carrier. Closely similar porphyrin rings occur as central structures in hemoglobin, chlorophyll, and other molecules of biological importance (compare with Fig. 7–12).*

Other techniques have enabled investigators to study the sequence of electron carriers in intact mitochondria. The prosthetic groups of the carriers absorb visible light differentially and exhibit colors when concentrated and purified. Distinct color changes result when the carrier changes from the reduced to the oxidized state. These characteristic colors are also detectable in suspensions of isolated mitochondria.

Britton Chance used these characteristic color changes to study electron transport, combining his approach with the use of inhibitors specific for particular carriers in the system. Carriers falling after the inhibited carrier in the sequence will gradually become oxidized by removal of electrons and will show the characteristic colors for the oxidized forms. Carriers falling before the "block" will remain in the reduced form and will show colors characteristic for this state. By using different inhibitors and comparing the color changes, Chance was able to reconstruct much of the sequence. Chance's findings, in combination with the redox potentials, reveal the probable sequence shown in Fig. 6–9. Because the cytochromes carry only one electron instead of two as in NAD and FAD, these are regarded either as represented twice in the sequence or as "turning" twice for every cycle of NAD and FAD.

The position of coenzyme Q in the sequence is uncertain. This possible carrier, isolated and described by D. E. Green and his colleagues, carries an oxidation-reduction potential that places it at approximately the same position as cytochrome *b*. It is not clear whether coenzyme Q should be placed before or after cytochrome *b* in the system. The flavoproteins have not been shown to interact directly with cytochrome *b*, which suggests that an additional carrier molecule, possibly coenzyme Q, must form a part of the carrier chain between the flavoprotein and cytochrome *b*.

Electrons flow spontaneously along the chain from NAD or FAD to molecular oxygen, in the direction of more positive electrical potentials. At a given instant, the substrate end of the chain is relatively electronegative and the oxygen end relatively positive. Hydrogen ions (protons) are also considered to flow along the chain. Hydrogen combines directly only with NAD, FAD, the quinones, and oxygen. At the level of the carriers transporting only electrons, hydrogen is considered to be released to the medium as H^+ ions, or in the case of the Mitchell hypothesis (see below) to be expelled across a membrane. At the level of the final acceptor, hydrogen ions reenter the sequence in the reduction of oxygen to water.

The energetics of electron transport from NAD to oxygen. As electrons flow from more electronegative to less electronegative carriers, there is a decrease in their associated energy. The energy lost may be released as heat or may be used to do electrical or chemical work. The greater the difference in redox potential between reductant and oxidant, the greater will be the increment of free energy released as electrons are transferred.

Table 6–1 shows oxidation-reduction potentials for the known electron carriers of the electron transport system. Between NAD and flavoprotein, electrons fall from a potential of −0.32 volts to −0.05 volts. The difference of 0.27 volts is shown as a positive value on the table, even though both these substances are electronegative. "Positive" or "negative" in this table is significant only in a relative sense. As noted on p. 105, the standard from which these voltages are measured as positive or negative is the voltage carried by electrons removed from hydrogen in the reaction $H_2 \rightarrow 2H^+ + 2e$. This voltage is arbitrarily given a relative value of zero. Thus flavoprotein is *relatively* positive (less negative) with respect to NAD, and will oxidize (remove electrons from) NAD.

Table 6–1 Free Energy Changes Associated with the Electron Transport System

Carrier	E_0' *volts*	$\Delta E_0'$ *volts*	$\Delta G_1'$*
NAD	−0.32	0.27	12,200
Flavoprotein	−0.05	0.09	4,050
Cytochrome *b*	+0.04	0.22	9,900
Cytochrome *c*	+0.26	0.03	1,350
Cytochrome *a*	+0.29	0.53	23,800
Oxygen	+0.82		

*The $\Delta G'$ value is the change in free energy for a pair of electrons transferred from one mole of the indicated carrier to the carrier below it in the table.

From A. L. Lehninger, *The Mitochondrian*, Copyright © 1964, W. A. Benjamin, Inc., Menlo Park, California, by permission.

The absolute difference between the oxidation-reduction potentials for a given pair of carriers ($\Delta E_0'$ in the table) gives a measure of the free energy released when electrons pass from a more electronegative to

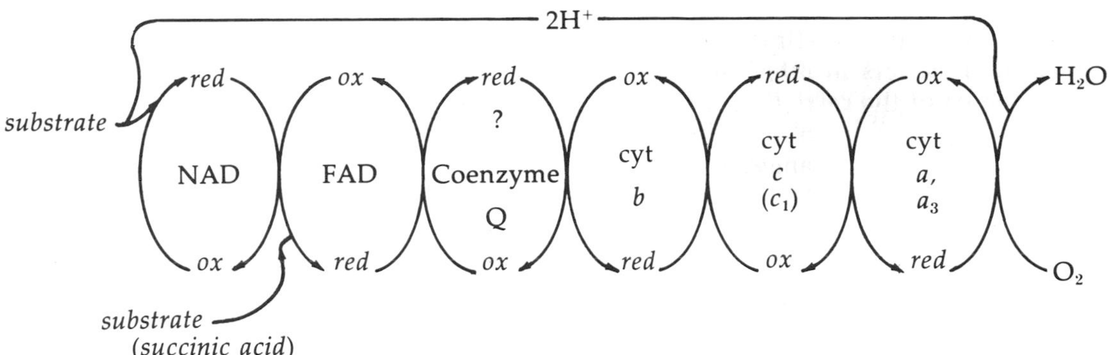

6–9. A proposed sequence of carriers in the electron transport system of the mitochondrion, as deduced from redox potentials and the use of various inhibitors in the system (see text). Hydrogen ions, which are shown without direct attachment to specific carriers, bind directly to some carriers but not others.

a less electronegative carrier, according to the relationship

$$\Delta G' = -n\text{F}\Delta\xi \qquad (6\text{–}10)$$

where $\Delta G'$ is the free energy released, n the number of electrons transferred from reductant to oxidant, F the Faraday (23,040 cal/volt), and $\Delta\xi$ the difference in oxidation-reduction potential between the oxidized and reduced carriers. Table 6–1 gives the calculated change in free energy ($\Delta G'$) for each oxidation-reduction pair listed.

Oxidation of glucose to water and carbon dioxide under standard conditions releases 686,000 cal/mole. Substitution in equation 6–10 of the difference in potential between NAD and oxygen reveals that a free energy change of more than 50,000 cal/mole occurs when a pair of electrons travels from a mole of NAD to a mole of oxygen. Thus, most of the energy released in the oxidation of glucose is associated with the electrons carried by reduced NAD and FAD, for, using 52,000 cal/mole as an average value, the twelve pairs of electrons carried by NAD and FAD (equation 6–9) release $12 \times 52{,}000 = 624{,}000$ cal/mole when transferred along the electron transport system to oxygen. This figure indicates that about 90 percent of the energy of oxidation of glucose is "trapped" in reduced NAD and FAD and released later in the electron transport system.

ATP synthesis in the mitochondrion. Inspection of Table 6–1 reveals that, as electrons travel from NAD to oxygen via the carriers of the electron transport system, energy is released in increments at each step in the chain. Considering that approximately 7,000–10,000 cal/mole (depending on the conditions of concentration and *p*H) are required to drive the conversion of ADP to ATP, it is obvious that a free energy change of sufficient magnitude to convert ADP to ATP is produced at three of the steps.

This is consistent with results obtained from isolated mitochondria, which have been most useful in investigations into the coupling of ATP synthesis to respiration. Lehninger showed that isolated mitochondria could synthesize ATP from ADP if reduced NAD was added to a medium containing mitochondria, an oxidizable substrate such as pyruvate, ADP, phosphate, and Mg^{+2} ions. For each molecule of reduced NAD added, the isolated mitochondria were able to synthesize three molecules of ATP. When ascorbic acid, an electron donor incapable of reducing carriers higher in the sequence than cytochrome *c*, was added to the isolated mitochondria, only one phosphorylation of ADP to ATP occurred for each molecule of ascorbate. This shows that one site for the synthesis of ATP is located somewhere in the sequence from cytochrome *c* to oxygen.

Other possible sites of ATP synthesis have been located by an extension of the light absorbency studies of carriers in reduced and oxidized states. Chance and G. R. Williams used a mitochondrial preparation having a limited supply of ADP and an *excess* of oxidizable substrate. The limited supply of ADP caused a bottleneck at three sites in the electron transport system. These bottlenecks were detectable by color changes indicating an excess of reduced NAD, cytochrome *b*, and cytochrome *c*. This means, for example, that the step in electron transfer from NAD to flavoprotein is

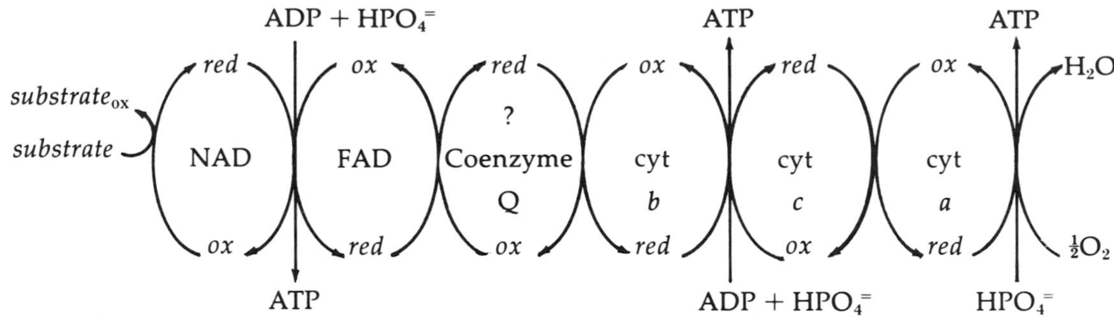

6–10. Possible sites of phosphorylation in the electron transport system (see text).

closely coupled to the synthesis of one ATP from ADP + $HPO_4^=$; a limited supply of ADP causes the accumulation of reduced NAD. According to the combined results of Lehninger, Chance, and Williams, electron transport is coupled to ATP synthesis according to the probable scheme shown in Fig. 6–10.

The coupling mechanism. Although ATP synthesis is known to be closely coupled to electron transport, the actual mechanism of coupling remains unknown. All that is known with certainty is that this mechanism is dependent on the integrity of mitochondrial structure. If mitochondrial membranes are disrupted, or if solvents are used to dissolve lipid away from mitochondrial membranes, phosphorylation ceases.

Two major hypotheses have been advanced to account for the coupling mechanism. The oldest of these, and the most favored until very recently, is the "chemical" hypothesis, developed in 1953 by E. C. Slater. This hypothesis proposes that ATP synthesis is coupled to electron transports by *common intermediates.* A common intermediate is a linking substance that is a *product* of a first reaction and a *reactant* in a second, coupled reaction (equations 6–11):

$$A + B \rightarrow C \quad (1)$$
$$C + D \rightarrow E \quad (2) \qquad (6\text{–}11)$$

In this hypothetical pair of reactions, the common intermediate C is required for the second reaction to take place and *couples* reaction (2) to reaction (1). Common intermediates are proposed to operate in the phosphorylation of ADP as follows. In electron transport, an electron carrier, designated A in equation 6–12, becomes linked to a coupling intermediate C at the instant of oxidation (when electrons are passed to the next carrier in the sequence, designated B in equation 6–12):

$$A_{red} + B_{ox} + C \rightarrow A \sim C + B_{red} \qquad (6\text{–}12)$$

The curved bond between A and C indicates a "high energy" bond. The A ~ C complex then acts as a common intermediate in a subsequent reaction in which C is phosphorylated:

$$A \sim C + HPO_4^= \rightarrow C \sim P + A_{ox} \qquad (6\text{–}13)$$

The complex C ~ P is a common intermediate coupling this reaction to the conversion of ADP to ATP:

$$C \sim P + ADP \rightarrow ATP + C \qquad (6\text{–}14)$$

The chemical hypothesis satisfies many of the facts known about electron transport and phosphorylation. Critical evidence in support of the hypothesis would be provided by the isolation of a phosphorylated common intermediate (C ~ P). This has not yet been accomplished.

Peter Mitchell has challenged the chemical hypothesis with a brilliant proposal known as the chemiosmotic hypothesis. Noting that a phosphorylated intermediate has not been identified, Mitchell proposed that the phosphorylation of ADP is driven by an ionic gradient set up by electron transport rather than by direct chemical coupling with a common intermediate.

In a very simplified form, the Mitchell hypothesis first points out that the breakdown of ATP to ADP proceeds with the addition of one molecule of water:

$$H_2O + ATP \rightarrow ADP + HPO_4^= \qquad (6\text{–}15)$$

This reversible reaction is catalyzed by an enzyme, ATPase, which is known to be bound to mitochondrial

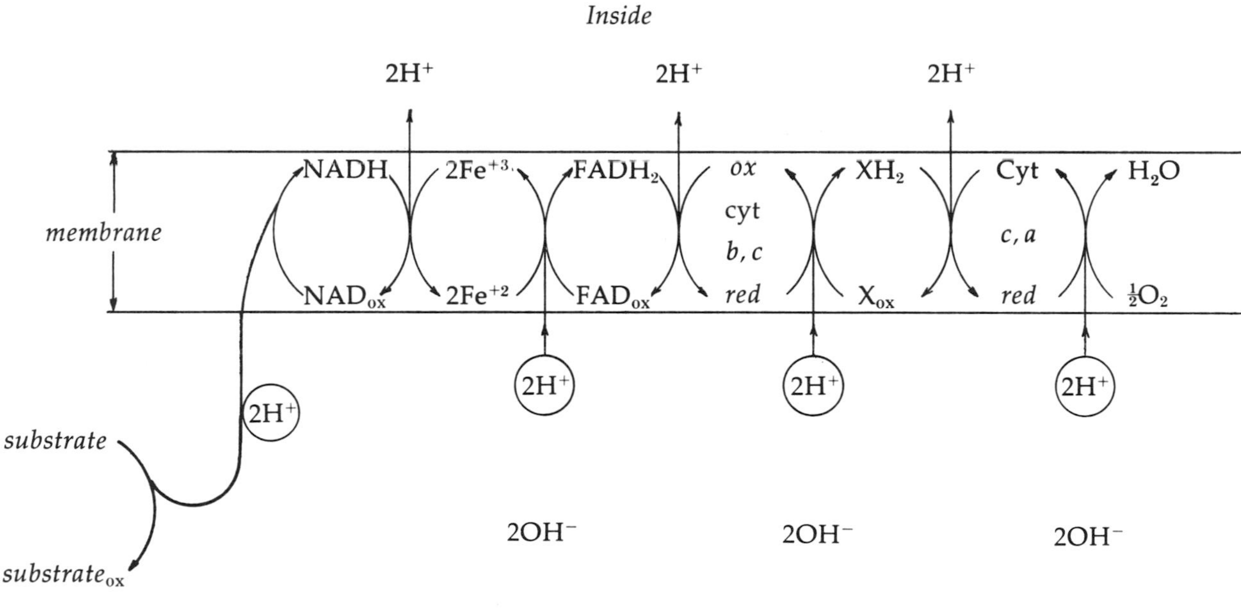

6–11. Establishment of a gradient of H^+ ions on two sides of a membrane during electron transport, according to the Mitchell hypothesis (see text).

membranes. Mitchell points out that, with an excess of ADP and a reduction in concentration of water molecules, the equation can be made to run in reverse:

$$ADP + HPO_4^{=} \rightarrow ATP + [H_2O] \rightarrow \text{removed} \quad (6\text{–}16)$$

Mitchell proposes that the mitochondrial membrane containing the ATPase molecule is selectively impermeable to H^+ and OH^- ions. The concentration of water at the active site of phosphorylation is reduced by the splitting of H_2O, with separation of H^+ and OH^- on opposite sides of the membrane containing ATPase. To achieve this separation, H^+ ions are "driven" across the membrane by the electron transport system. Remember that the electron carriers NAD and FAD carry hydrogens in addition to electrons. Mitchell proposes that the hydrogens carried by these coenzymes are transferred across the membrane as shown in Fig. 6–11. The transfer of H^+ ions requires that carriers of both electrons and hydrogens must alternate with "pure" electron carriers in the transport system. The hydrogens carried by NAD, for example, are discharged across the membrane when the pure electron carrier, designated Fe^{+3}, accepts only electrons from NAD. At the next step, in which the pure electron carrier is followed by an electron-hydrogen carrier, the required hydrogens are derived from water molecules "outside" the membrane. This occurs at three successive steps and results in the splitting of six molecules of water. This achieves the reduction in water concentration required to drive ATP synthesis.

A number of assumptions are introduced in addition to the permeability properties required for the membrane. One is that a pure electron carrier, Fe^{+3}, exists in the transport system between NAD and FAD. There is good evidence that this additional carrier, designated "nonheme iron," may actually occur in mitochondria. Another assumption is that an additional dual electron-hydrogen carrier, designated "X," occurs between cytochrome *b* and oxygen.

The Mitchell hypothesis has received experimental support from several lines of investigation since its inception. The hypothesis infers that *hydrolysis* of ATP to ADP should expel H^+ ions from the inside to the outside of the mitochondrion, in reverse of the pattern shown in Fig. 6–11. Mitchell was able to show that this occurs during ATP breakdown and that approximately two H^+ ions (protons) were ejected from the mitochondrion for each ATP molecule hydrolyzed. A second verification was found in Mitchell's discovery that mitochondria are not freely permeable to protons. Finally, the hypothesis predicts that an artificial H^+ ion gradient produced by addition of acid to the medium will result in a burst of ATP synthesis. This was found to be the case by A. T. Jagendorf and his co-workers. Jagendorf worked with chloroplast ATP synthesis, which proceeds by an electron transport mechanism similar to that in mitochondria (see Chapter 7).

This evidence does not rule out the chemical hypothesis, and it must be emphasized that the actual mechanism of ATP synthesis is an open question. Explanations have been devised for these observations that are compatible with the concept of a coupling common intermediate in phosphorylation. The chemiosmotic hypothesis continues to gain ground, however, and at this writing is given serious consideration as an important alternative to the chemical hypothesis.

Summary of the Energetics of Glucose Metabolism

The mechanism linking phosphorylation to electron transport synthesizes three molecules of ATP for each pair of electrons carried from NAD to oxygen. With this information, the total number of ATP molecules synthesized for each molecule of glucose oxidized to carbon dioxide and water can be calculated. Complete oxidation of one molecule of glucose (through glycolysis, acetyl-CoA formation, and the citric acid cycle) results in the accumulation of a total of 4ATP, $10NAD_{red}$, $2FAD_{red}$, and $6CO_2$ (equation 6–9). The ten electron pairs carried by reduced NAD, traversing the electron transport system, will result in the synthesis of $10 \times 3 = 30$ molecules of ATP. Reduced FAD, carrying electrons that enter the transport system at a point farther along the system, will cause the synthesis of $2 \times 2 = 4$ molecules of ATP. Added to the total from equation 6–9, the overall oxidation of glucose accumulates 38 molecules of ATP:

$$\text{Glucose} + 38ADP + 38HPO_4^{=} + 6O_2 \rightarrow 6CO_2 + 44H_2O + 38ATP \qquad (6\text{–}17)$$

Under standard conditions (reactants and products at 25°C and at a 1 molar concentration), the hydrolysis of ATP to ADP yields 7000 cal/mole. Using this value as the energy required to synthesize ATP from ADP + $HPO_4^{=}$, the total energy trapped during the oxidation of glucose amounts to $38 \times 7000 = 266{,}000$ cal/mole. Combustion of glucose in oxygen yields 686,000 cal/mole. On this basis, the efficiency of glucose metabolism in cells falls at about $266/686 \times 100 = 39\%$.

The energy captured under the conditions of temperature and concentration in the living cell may be considerably higher. The free energy change for the hydrolysis of ATP increases in magnitude as the concentration of the reactants and products is reduced. The concentration of ADP and ATP at the site of phosphorylation is not known with certainty. Estimates of the actual free energy change vary from 9,000 to 15,000 cal/mole depending on the concentrations assumed to occur in the cell. Using these figures gives a higher efficiency for the total oxidation of glucose, from 50 percent at 9,000 cal/mole to 80 percent at 15,000 cal/mole. Precisely where the actual figure falls in this range is unknown, but the efficiency is probably higher than the 39 percent calculated from standard conditions.

Cellular Oxidation of Other Metabolites: Fats and Proteins

Coenzyme A, acting as a carrier of acetate groups to the mitochondrion, forms an important intermediate in the breakdown of lipids and proteins. Fats are first hydrolyzed into glycerol and the component fatty acids. Glycerol, converted into dihydroxyacetone phosphate at the expense of ATP, enters the glycolytic pathway directly. The long fatty acid residues remaining are successively broken into two-carbon acetate segments linked to coenzyme A, either inside the mitochondrion or in the extramitochondrial cytoplasm. These residues are extremely long and supply a large number of acetate groups. Fats, therefore, are an excellent source of energy, because oxidation of each two-carbon segment cleaved from the fatty acid chain is eventually coupled to the synthesis of 17ATP molecules. By weight, fats have twice the energy content of carbohydrates. The complete oxidation of palmitate ($C_{15}H_{31}COOH$) for example, proceeds with the eventual synthesis of 130 molecules of ATP.

Proteins are a source of fuel for cellular oxidation, after hydrolysis into free amino acids by proteolytic enzymes. The oxidation of many amino acids also uses acetyl-CoA as an intermediate, although some amino acids are converted instead into keto acids that enter the citric acid cycle directly. There are many pathways for the different amino acids; it is not necessary to include them all here. One type will illustrate these reactions and is of interest because an intermediate of the citric acid cycle is produced.

$$\begin{array}{c} COOH \\ | \\ C{=}O \\ | \\ CH_2 \\ | \\ COOH \\ \text{oxaloacetic} \\ \text{acid} \end{array} + \begin{array}{c} COOH \\ | \\ H{-}CNH_2 \\ | \\ CH_2 \\ | \\ CN_2 \\ | \\ COOH \\ \text{glutamic} \\ \text{acid} \end{array} \xrightarrow[\text{transaminase}]{} \begin{array}{c} COOH \\ | \\ HC{-}NH_2 \\ | \\ CH_2 \\ | \\ COOH \\ \text{aspartic} \\ \text{acid} \end{array} + \begin{array}{c} COOH \\ | \\ H{-}C{=}O \\ | \\ CH_2 \\ | \\ CH_2 \\ | \\ COOH \\ \alpha\text{-ketoglutaric} \\ \text{acid} \end{array} \qquad (6\text{–}18)$$

The pathways of the various fuel substances are summarized in Fig. 6–12. This figure illustrates the position of coenzyme A as a common link between the oxidative metabolism of carbohydrates, fats, and proteins.

The Pentose "Shunt"

Not all of the glucose metabolized in plants and animals is broken down in the glycolytic pathway. Certain inhibitors, such as fluoride or iodoacetate, block the glycolytic pathway but do not completely halt the oxidation of glucose in the cell. This finding led to the discovery of several alternate pathways for glucose metabolism. The most important of these is one in which glucose is converted during breakdown into a 5-carbon sugar (a pentose). Through this pathway, all reactions of glycolysis and the citric acid cycle are bypassed, or "shunted." Discovery of this pathway also led to the identification, by Warburg, of an alternate electron acceptor, NADP. This electron carrier is almost identical to NAD, except that it contains one additional phosphate group (compare Figs. 6–5 and 6–13).

The metabolism of glucose via the pentose shunt begins with the conversion of glucose 6-phosphate to phosphogluconic acid. This conversion (equation 6–19) takes place in two steps, including an oxidation followed by addition of a molecule of water. Only the overall reaction is illustrated in equation 6–19:

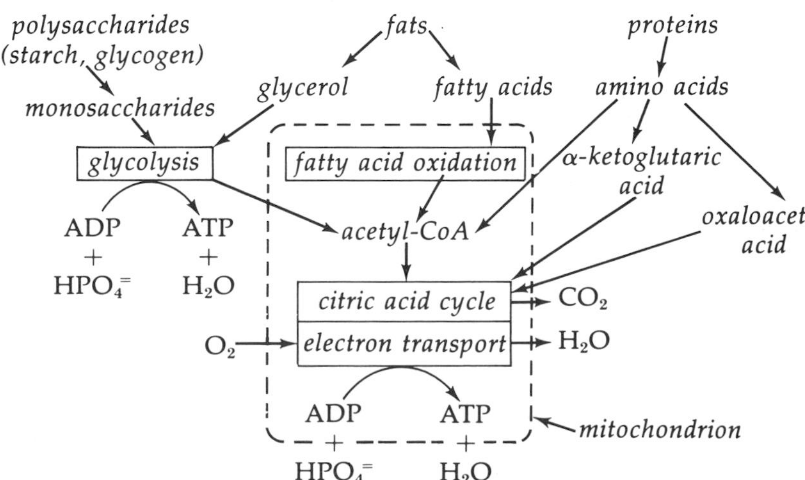

6–12. Diagram showing the integration of biochemical mechanisms that oxidize polysaccharides, lipids, and proteins. Central to the integrated system is the carrier of acetates, coenzyme A. Formation of acetyl-CoA may take place within the mitochondrion or in the extramitochondrial cytoplasm.

(6–19)

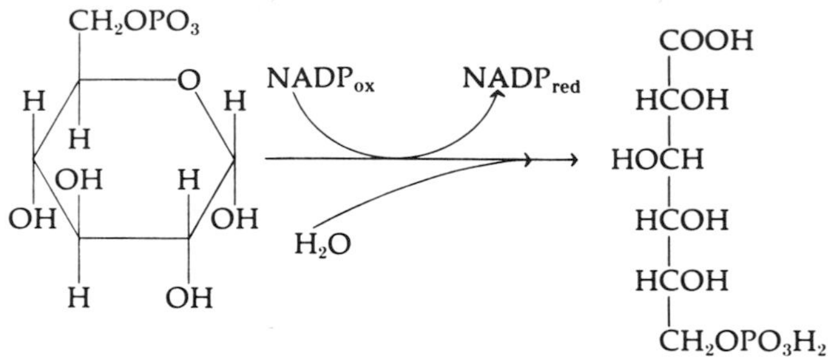

$$\text{glucose 6-phosphate} + H_2O \xrightarrow{NADP_{ox} \rightarrow NADP_{red}} \text{6-phosphogluconic acid } (COOH{-}HCOH{-}HOCH{-}HCOH{-}HCOH{-}CH_2OPO_3H_2)$$

glucose 6-phosphate 6-phosphogluconic acid

Note that the electron acceptor for this oxidation is NADP. Phosphogluconic acid is subsequently oxidized to a 5-carbon sugar, ribulose 5-phosphate, with the removal of one carbon from the chain (equation 6–20):

(6–20)

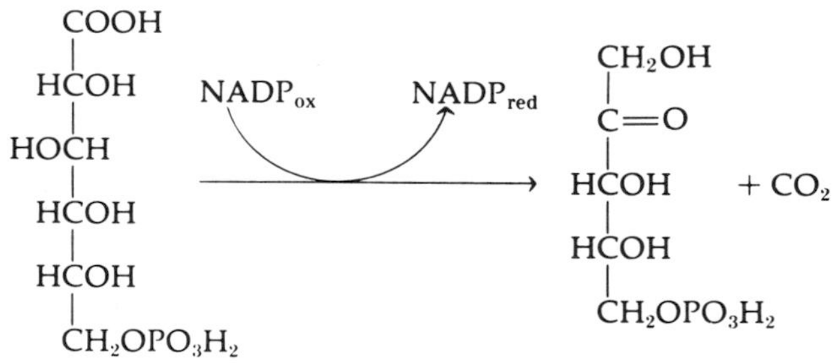

$$COOH{-}HCOH{-}HOCH{-}HCOH{-}HCOH{-}CH_2OPO_3H_2 \xrightarrow{NADP_{ox} \rightarrow NADP_{red}} CH_2OH{-}C{=}O{-}HCOH{-}HCOH{-}CH_2OPO_3H_2 + CO_2$$

6-phosphogluconic acid ribulose 5-phosphate

The two products of reactions 6–19 and 6–20, the ribulose sugar and reduced NADP, are of great significance in cellular metabolism. The pentose sugar, from which the pathway takes its name, is a major source of the various riboses important in the cell. Two of these 5-carbon sugars are synthesized into the backbone structure of DNA and RNA; these molecules are the molecular basis of storage and transfer of hereditary information in the cell (the subject of Unit 3).

Reduced NADP may transfer its electrons to NAD, which is in turn oxidized by the electron transport system of the mitochondrion with the synthesis of ATP or, more importantly, may enter into synthetic reactions in the cell that require a reduction. With a few exceptions, whenever a reduction (*addition* of electrons and hydrogen) takes place in a synthetic step, the electron donor is reduced NADP. NADP thus donates electrons with relatively high negative potential or voltage. Examples of this type of reaction are the synthesis of fats and steroids and the fixation of carbon dioxide in photosynthesis. The source of reduced NADP for these reactions in animal cells is primarily in the pentose pathway, although NADP also acts as an electron acceptor in the citric acid cycle in the oxidation of isocitric acid (see p. 112). In photosynthesizing plants and bacteria, reduced NADP is one of the first chemical products resulting from the absorption of light in the chloroplast. The pentose shunt and the citric acid cycle also act to produce reduced NADP in plants.

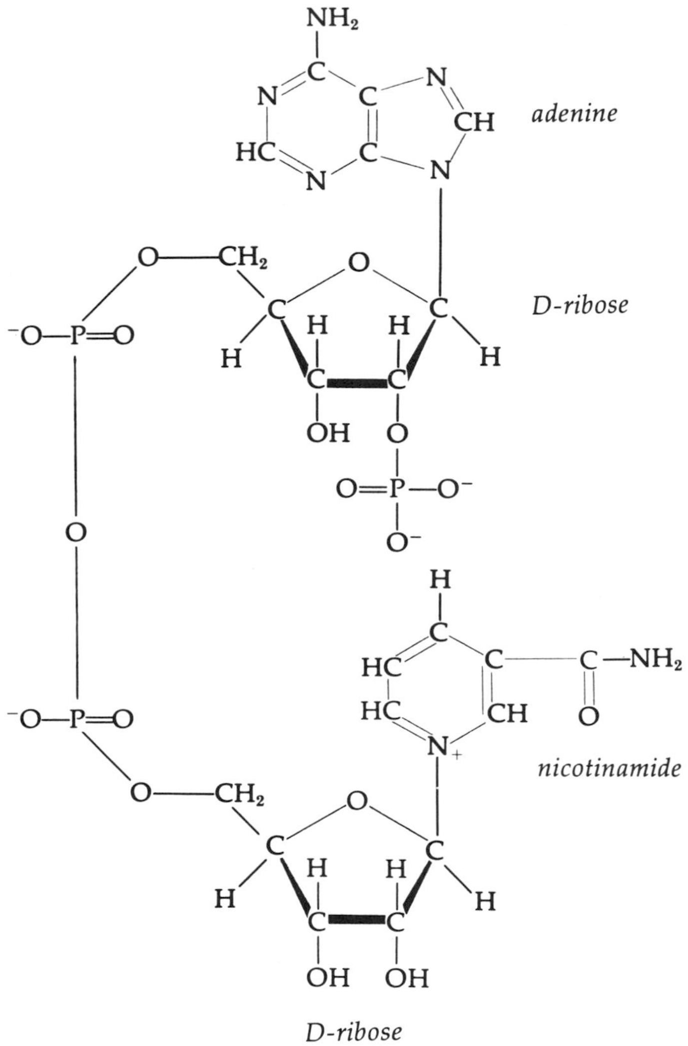

adenine

D-ribose

nicotinamide

D-ribose

6–13. Nicotine adenine dinucleotide phosphate (NADP), a carrier of electrons and hydrogen closely similar in structure and function to NAD.

The reactions synthesizing carbohydrates in photosynthesis use both reduced NADP and ribulose. In both plants and animals, ribulose is the starting point for the synthesis of other sugars with 3-, 4-, 5-, 6-, and 7-carbon chains. In photosynthesis, a phosphorylated derivative of ribulose is combined with CO_2 from the atmosphere to form two 3-carbon sugars as products. A reduction occurs in this reaction; the required electrons are donated by reduced NADP. The two 3-carbon sugars are then synthesized into glucose by a series of reactions that are the reverse of glycolysis, from 3-phosphoglyceraldehyde back to glucose. The energy required to drive these reactions in reverse is provided by light, absorbed and converted into chemical energy (ATP) in the chloroplast. All the intermediates and enzymes of the synthetic reactions of photosynthesis are common to both plants and animals. The major difference in photosynthesis is that the primary source of ATP and reduced NADP is from the light-absorbing molecular systems of the chloroplast (discussed in detail in Chapter 7).

Localization of Oxidative Metabolism in Mitochondrial Fine Structure

Methods developed in recent years have allowed the isolation of reasonably pure subfractions of mitochondria, including outer and inner membrane preparations. Using these methods, inner (cristae) membranes have been shown to contain the enzymes and prosthetic groups of the electron transport system, along with ATPase activity. These enzymes, tightly bound to the cristae membranes, remain active in electron transport and phosphorylation only as long as the inner membranes are intact. Removal of the lipid portion of the membrane stops electron transport and phosphorylation, even though structural lipids of the membranes are not known to enter directly into electron transport.

The enzymes and substrates of the citric acid cycle are released from the interior of the mitochondrion immediately on rupture of the inner and outer mitochondrial membranes. Removal of the outer membrane alone, however, does not release these factors. Therefore, these enzymes and substrates are considered to be "soluble" and to be contained within the mitochondrial matrix.

The outer mitochondrial membrane is relatively devoid of enzymatic activity. Although some controversy exists concerning the enzymatic constituents of this membrane, it is generally agreed that none of the enzymes of the citric acid cycle or electron transport are contained within the outer membrane. One enzyme, associated with the oxidation of amino acids, monamine oxidase, has been identified as a constituent of the outer membrane. Other enzymes, associated with acetyl-CoA formation and oxidation of cytoplasmic NADH, may also be present. It is not known whether the relative permeability of the mitochondrion to various metabolites is a property of the outer or inner membrane.

Analysis of the lipid and protein fractions of isolated mitochondrial membranes reveals that the outer and inner membrane systems differ in their lipid-protein ratio. The outer membranes are similar to the plasma membrane and contain about half protein and half lipid. Membranes of the cristae contain a higher proportion of protein, approximately 70–80 percent. Of the lipid fraction, about 90 percent of the lipids in the cristae are phospholipids. The proteins of the cristae are claimed by Green and his co-workers to be about 50 percent "structural protein." The remainder of the protein is made up of ATPase and the enzymes and prosthetic groups of the respiratory chain. Some recent work by A. E. Senior and David H. MacLennan indicates that a large part of the so-called "structural" protein isolated from mitochondria by Green and R. S. Criddle may consist of denatured enzymes. For example, ATPase was found by Senior and MacLennon to form about 70 percent of "structural protein" isolated by the techniques of Green and Criddle. Therefore the relative proportion of structural and enzymatic protein in mitochondrial membranes is not yet settled.

Ultrastructure of mitochondrial membranes. For a time, mitochondrial membranes were thought to be identical in dimensions and ultrastructure to the plasma membrane. However, careful measurements made by F. S. Sjöstrand on high-resolution photographs of cellular membranes reveal that mitochondrial membranes are somewhat thinner than the plasma membrane, about 50–60 Å in thickness. Furthermore, Sjöstrand's micrographs show an apparent granularity in mitochondrial membranes (Fig. 6–14). Sjöstrand has interpreted this granularity as an indication that mitochondrial membranes are made up of globular subunits and has pictured the mitochondrial membrane as a single layer of 50–60 Å granules. He considers these

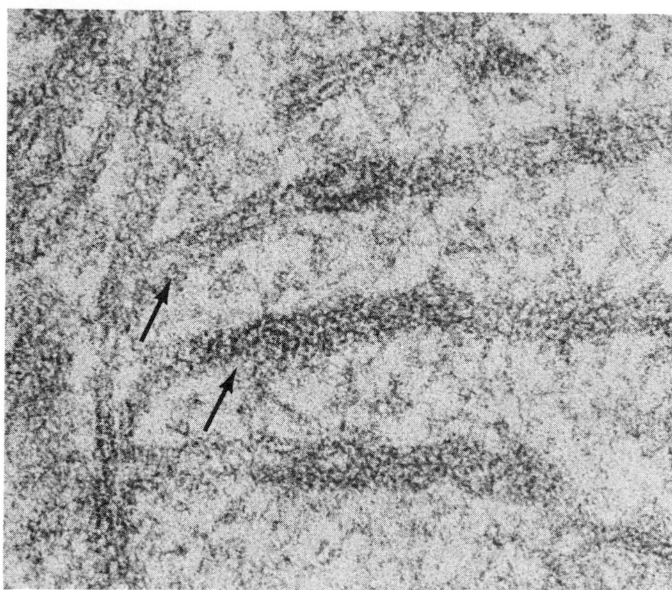

6–14. Highly magnified inner membranes of a mitochondrion, in which granular or particulate substructure is visible (arrows). × 310,000. Courtesy of F. S. Sjöstrand and Academic Press, Inc. [J. Ultrastr. Res. *9:340 (1963)*]

granules to be lipid globules, separated by electron dense "septa" that consist of the proteins of the electron transport system.

Other investigators have agreed that the cristae membranes are fundamentally different in structure from a simple lipoprotein bilayer. After negative staining with phosphotungstic acid, the inner membranes appear to be covered on the side facing the mitochondrial matrix by a layer of spherical particles 80–100 Å in diameter (Fig. 6–15). Less clearly seen are a stalk, connecting the particle to the membrane surface, and a "basepiece," located within the membrane. These particles, first discovered by H. Fernández-Morán, have touched off one of the most heated controversies in cell biology. The stalked particles can be seen only in negatively stained preparations. Granular substructure in sectioned membranes, when visible, is contained *within* the margins of the membrane. In contrast, the particles seen after negative staining appear to be on the membrane surface, connected to the membrane by a thin stalk. Furthermore, both the inner and outer mitochondrial membranes appear to be granular in thin sections. But only the inner membranes show surface particles after negative staining. For these reasons, many investigators

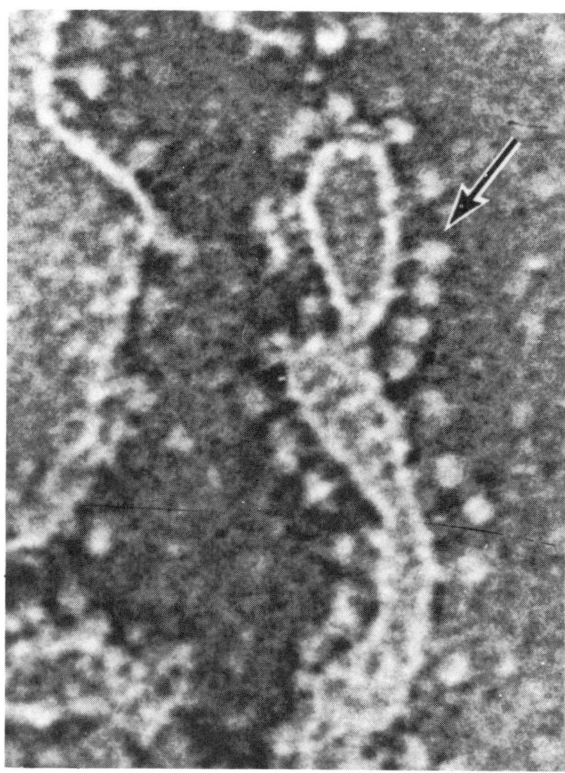

a

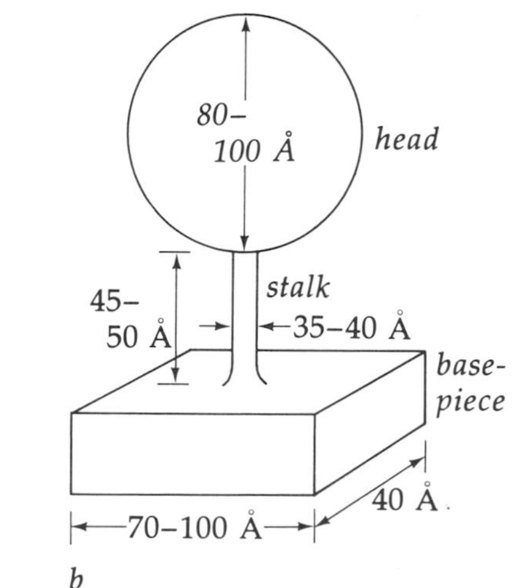

b

6–15. a, *Isolated mitochondrial cristae, negatively stained with phosphotungstic acid (PTA). The stalk and headpiece of the "elementary particles" are clearly visible (arrow). Isolated from beef heart. × 420,000. Courtesy of H. Fernandez-Morán and The Rockefeller University Press.* [J. Cell Biol. *22:63 (1964)*]
b, *Diagram showing the average dimensions of the head, stalk, and basepiece of elementary particles, from various sources.*

have considered the surface particles to be a "regular artifact," a structure extracted or extruded in some way from the membrane by phosphotungstic acid during negative staining. This conclusion is by no means generally agreed on at the present time.

These particles were interpreted by Green and his co-workers as the enzyme site for the electron transport system. When it was pointed out that the 80–100 Å particle was too small to house the components of the electron transport system, based on the estimated total molecular weight of the system, Green modified his hypothesis, stating that the electron transport system was distributed throughout the head, basepiece, and stalk of the structure, now termed the "elementary particle." This hypothesis, elaborated to the extent that particular segments of the electron chain were assigned to basepiece, stalk, or head, stimulated much interest. For a while it gained wide acceptance, although many workers remained highly skeptical of Green's views.

Major revision of this hypothesis was required when Efraim Racker discovered that removal of the head and stalks from inner membranes had no effect on electron transport. Racker was able to show that only ATPase activity paralleled removal or addition of elementary particles on inner membranes. Purified ATPase caused reappearance of the "elementary particles" when added to inner membrane fragments from which particles had been removed. Finally, Racker showed that preparations of mitochondrial ATPase, when negatively stained, appeared as spherical granules 80–90 Å in diameter under the electron microscope. With the announcement of these results, establishing the identity of the membrane particles with ATPase, Green modified his hypothesis accordingly, assuming that the enzymes of the electron transport system are contained in four adjacent basepieces. As pointed out, however, the existence of the basepiece as such within the membrane has not been firmly demonstrated, even in negatively stained preparations. At the moment it may be stated that, although the electron transport system is known to be bound firmly to the inner mitochondrial membranes, the precise location, ultrastructure, and manner of binding remains to be discovered.

Integration of Mitochondrial Activity with the Cell Environment

The rate of mitochondrial oxidation is regulated by a system of controls within the cell. Several hypotheses have been advanced as the basis for this control system. Of these, the most widely accepted is that the rate of oxidation depends on the concentration of ADP in the medium surrounding the mitochondrion. As the concentration of ADP increases, phosphorylation of ADP to ATP in the mitochondrion begins. Phosphorylation is closely coupled to electron transport; as phosphorylation takes place, electrons flow without restriction from NAD to oxygen, and the NAD end of the transport chain becomes relatively oxidized. NAD, once oxidized, is free to recycle as an electron acceptor in reactions oxidizing the substrates of the citric acid cycle. This system closely links the rate of oxidation in the citric acid cycle to the concentration of ADP in the cell.

The linking of glycolysis to mitochondrial oxidation probably occurs through the oxidation-reduction cycle of extramitochondrial NAD. Continuance of glycolysis depends on reoxidation of the NAD reduced during dehydrogenation of 3-phosphoglyceraldehyde. In some manner, incompletely understood, the extramitochondrial NADH transfers its electrons to NAD inside the mitochondrion. In this way, glycolysis proceeds, depending indirectly on the supply of oxidized NAD within the mitochondrion. This, in turn, may be regulated by the concentration of ADP.

Synthesis and transport in the cell require an input of energy, which is supplied in the form of ATP. Synthesis and transport are interpreted here in the broadest sense to include all the activities of the cell, including growth, movement, and irritability. As these processes are carried out, ATP is hydrolyzed to ADP + $HPO_4^{=}$. The increase in ADP concentration causes an increase in the rate of oxidation in the mitochondrion, and the concentration of ATP is restored. All the various activities, from glycolysis through synthesis, thus exist as a delicately balanced open system, ready to run if a demand is placed in the form of hydrolysis of ATP. These relationships are summarized in Fig. 6–16.

As we have seen, many of the systems, such as electron transport, taking part in oxidation within the mitochondrion are intimately bound into membrane structure. Our understanding of the role of structure in the activity of these systems is only beginning to take shape. The arrangement of enzymes and substrates in an ordered fashion on a basic membrane framework causes a significant alteration of properties, particularly with respect to the energy and concentrations of reactants required for a given reaction to occur. Future work will certainly be concerned with an integration of bio-

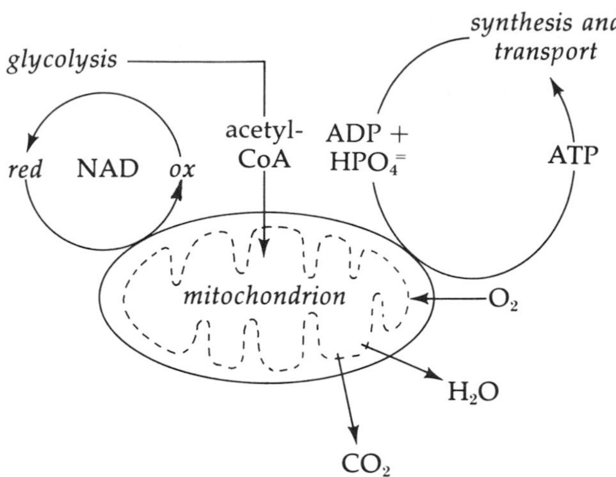

6–16. *The coupling of the oxidative reactions of the mitochondrion to the cell environment through the ATP-ADP and NAD_{redox} cycles.*

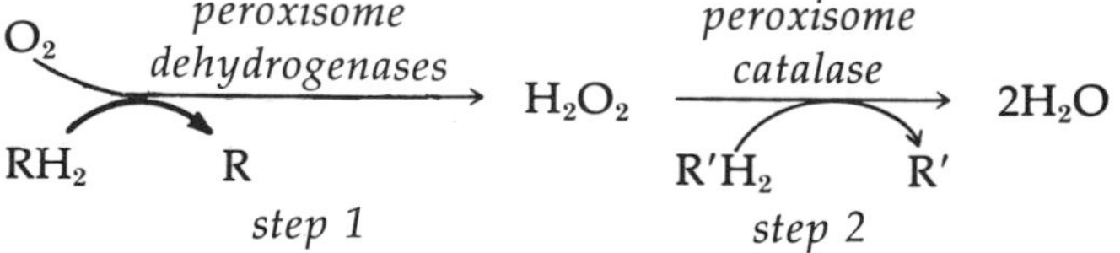

6–17. *The oxidative pathway in peroxisomes, as proposed by DeDuve and Baudhuin (see text).*

chemistry and the molecular structure of membranes. Control mechanisms, understood today in only the barest terms, will also form an area of concentrated research.

Peroxisomes and Glyoxisomes (Microbodies)

Within the last few years it has been established that a number of cellular oxidations are carried out within a previously undescribed group of membrane-bound, cytoplasmic particles collectively termed *peroxisomes, glyoxisomes,* or *microbodies.* These oxidative particles, intensively studied by Christian de Duve and Pierre Baudhuin and their colleagues, typically contain large amounts of *catalase,* an enzyme catalyzing the breakdown of hydrogen peroxide to water, and varying amounts of other enzymes capable of removing hydrogen from a large number of substrates.

De Duve and Baudhuin have proposed that peroxisomes carry out the transport of hydrogen from a number of substrates to oxygen in a two-step reaction (Fig. 6–17). In the first step, dehydrogenases of the peroxisome remove hydrogens from (oxidize) different substrates, designated RH_2 in Fig. 6–17. These may include uric acid, D-amino acids, L-amino acids, lactic acid, and other substances. The hydrogens removed are combined with oxygen to form hydrogen peroxide. In the second step, catalyzed by the enzyme catalase, protons and electrons from a second donor ($R'H_2$) are added to hydrogen peroxide to produce water. The second donor may be a molecule of H_2O_2, which is oxidized back to oxygen, or other substances oxidized by catalase, such as ethanol or formic acid. The end result of the two steps is oxidation of various substrates, with the eventual reduction of oxygen to water.

Peroxisomes have been identified in several organisms, including the liver and kidney cells of rats and in yeasts and protozoa. These particles were first identified by biochemical characterization of fractions purified by centrifugation. The particles (Fig. 6–18) are enclosed by a single membrane and in electron micrographs contain a moderately dense, granular matrix. Within the matrix is a dense, highly ordered array that often appears to be paracrystalline. This structure, the "core" or "nucleoid," is constructed from closely packed tubules or fibers 90–100 Å in diameter. A particle of this type, *in situ* in the cytoplasm of a cell from the castor bean, is shown in Fig. 6–19.

Purified extracts of the peroxisome enzyme *uric acid dehydrogenase* (or "urate oxidase") contain crystallites identical in morphology to the "cores" of intact peroxisomes. Peroxisomes lacking this enzyme are apparently without cores. Therefore, it is assumed that the cores either represent ordered deposits of uric acid dehydrogenase or are structures containing this enzyme.

Closely related particles, found in germinating bean seedlings, have been termed *glyoxisomes* by R. W. Breidenbach and H. Beevers. In addition to catalase and the dehydrogenases of peroxisomes, these particles contain a system of enzymes capable of carrying out the glyoxylate cycle, a series of reactions representing a bypass or "shunt" of the Krebs cycle (Fig. 6–20). Because of the presence of the usual dehydrogenases and

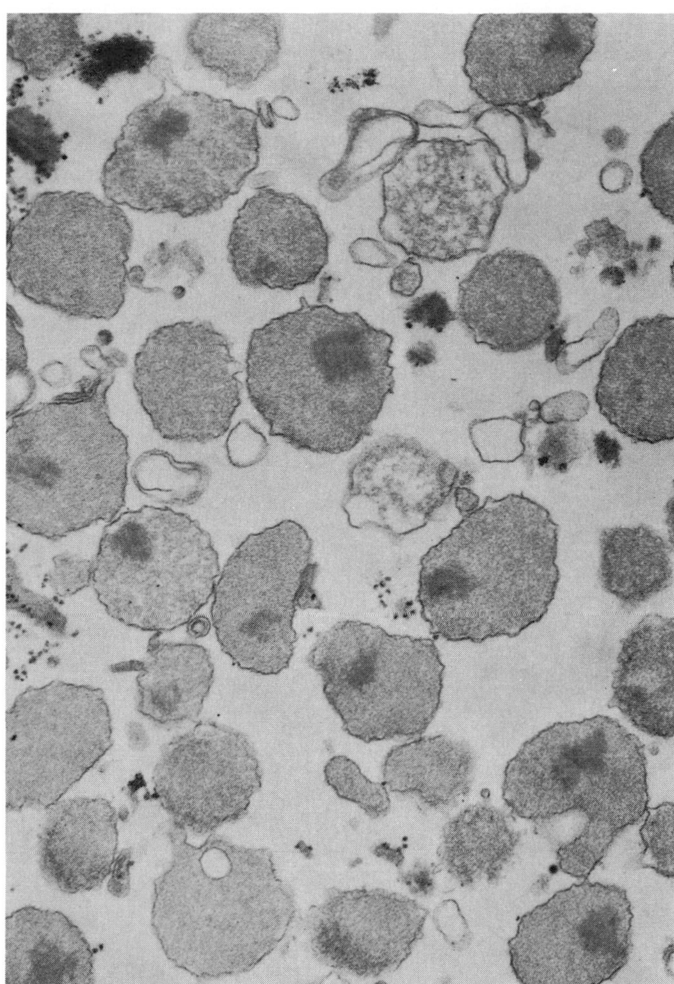

6–18. Peroxisomes isolated from rat liver. × 40,000. Courtesy of P. Baudhuin and North-Holland Publishing Company, Amsterdam.

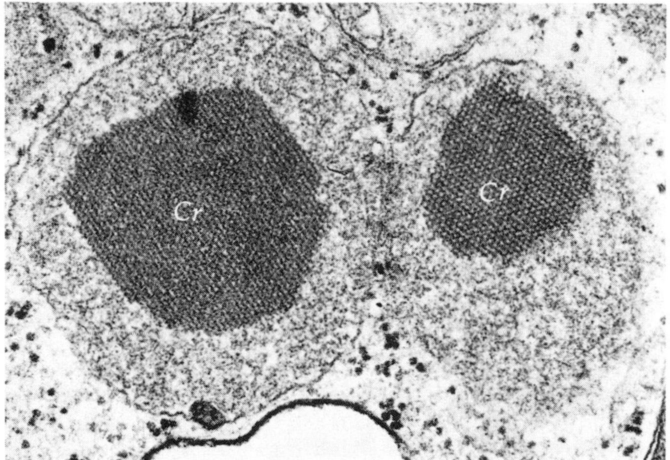

6–19. "Microbodies" in a thin section of castor bean endosperm tissue. Crystallites (Cr) *occupy much of the volume of the matrix. × 58,000. Courtesy of E. L. Vigil and The Rockefeller University Press.* [J. Cell Biol. *46:435 (1970)*]

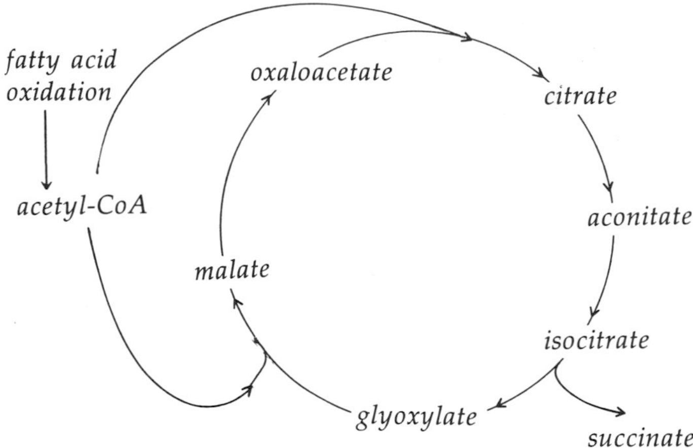

6–20. The glyoxylate cycle (see text).

catalase of peroxisomes, glyoxisomes are considered to be a modification of the basic peroxisome particle.[2]

The functional significance of peroxisomes and glyoxisomes is not yet clear. Because acetyl-CoA derived from fatty acids is converted into intermediates of the glyoxylate pathway in glyoxisomes, these particles may serve as sites for the conversion of fat derivatives to carbohydrates. However, because the glyoxylate cycle has not been identified with the peroxisomes of kidney and liver, this functional role does not appear likely in these tissues. Other possibilities include the removal of H_2O_2, known to have destructive effects in the cell, from the cytoplasm. In addition, the peroxisome

[2] Some authors have called both peroxisomes and glyoxisomes *microbodies.* Because this term has been rather indiscriminately applied in the past to a multitude of small cytoplasmic vesicles, the more descriptive and precise terms peroxisome and glyoxisome are considered preferable.

has been considered as an organelle primarily concerned with the oxidation of NAD reduced elsewhere in the cell. Oxidative reactions coupled to the reduction of NAD, such as the glycolytic sequence, could thus be stimulated by the peroxisomes. One difficulty in relating the peroxisomes to cellular oxidation is presented by the fact that ATP synthesis is apparently not linked in any way to peroxisomal mechanisms. In spite of this, it has been estimated that in liver, 20 percent of the oxygen consumed can be accounted for by peroxisomes.

Suggestions for Further Reading

Books

Annals of the New York Academy of Sciences, 1969. Vol. 168, pp. 209–381. A collection of recent articles on peroxisomes and microbodies.

Lehninger, A. L., 1965a. *Bioenergetics*. W. A. Benjamin, Menlo Park, Calif.

Lehninger, A. L., 1965b. *The Mitochondrion*. W. A. Benjamin, Menlo Park, Calif.

Papers and Review Articles

Baudhuin, Pierre, 1969. Peroxisomes (microbodies, glyoxisomes). In A. Lima-de-Faria, ed., *Handbook of Molecular Cytology*. North-Holland Publishing Company, Amsterdam, pp. 1179–1195.

Borst, P., 1969. Biochemistry and function of mitochondria. In A. Lima-de-Faria, ed., *Handbook of Molecular Cytology*. North-Holland Publishing Company, Amsterdam, pp. 914–942.

Breidenbach, R. W., and H. Beevers, 1968. Association of the glyoxylate enzymes in a novel subcellular particle from castor bean endosperm. *Biochemical and Biophysical Research Communications*, vol. 27, pp. 462–469.

Daems, W. T., and E. Wisse, 1966. Shape and attachment of the cristae mitochondriales in mouse hepatic cell mitochondrion. *Journal of Ultrastructure Research*, vol. 16, pp. 123–140.

DeDuve, C., and P. Baudhuin, 1966. Peroxisomes (microbodies and related particles). *Physiological Review*, vol. 46, p. 323.

Munn, E. A., 1969. Ultrastructure of mitochondria. In A. Lima-de-Faria, ed., *Handbook of Molecular Cytology*. North-Holland Publishing Company, Amsterdam, pp. 875–913.

Novikoff, A. B., 1961. Mitochondria. In J. Brachet and A. E. Mirsky, eds., *The Cell*, vol. II. Academic Press, New York, pp. 299–421.

Pullman, M. E., and G. Schatz, 1967. Mitochondrial oxidations and energy coupling. *Annual Reviews of Biochemistry*, vol. 36, pp. 539–610.

Racker, E., 1968. The membrane of the mitochondrion. *Scientific American*, vol. 218, pp. 32–39.

Senior, A. E., and D. H. MacLennan, 1970. Mitochondrial "structural protein," a reassessment. *Journal of Biological Chemistry*, vol. 245, pp. 5086–5095.

Sjöstrand, F. S., 1963. A comparison of plasma membrane, cytomembranes and mitochondrial membrane elements with respect to ultrastructural features. *Journal of Ultrastructure Research*, vol. 9, pp. 561–580.

Sjöstrand, F. S., 1963. A new ultrastructural element of the membranes in mitochondria and some cytoplasmic membranes. *Journal of Ultrastructure Research*, vol. 9, pp. 340–361.

7
Photosynthesis and the Chloroplast

The reactions of the cell that maintain the level of ATP required for activities such as growth, motility, and response to changes in the environment require a continuous input of oxidizable substrates in order to run. Unless constantly renewed, the existing levels of these substrates would be very quickly exhausted by the great mass of living organisms on earth. The supply is replenished by photosynthesis, in which the energy of sunlight drives the reduction of carbon dioxide to carbohydrate (CH_2O) units. In algae and all higher green plants the electrons and hydrogen required for this reduction are ultimately derived from water, producing as a byproduct molecular oxygen, also required for the survival of life on earth.

From these simple precursors, carbon dioxide and water, and a supply of other elements in the form of mineral salts, green plants are able to synthesize all the complex substances required for life. The ultimate source of energy for this synthesis is sunlight, absorbed and converted into chemical energy in the chloroplast.

The biochemical pathways of photosynthesis are not identical in all plants and bacteria. In this chapter, photosynthesis will be reviewed primarily as it is thought to occur in the chloroplasts of vascular plants. Even with this restriction, there are several possible formulations of the reaction pathways and much conflicting evidence concerning these pathways awaits resolution. An attempt has been made to follow what appears to be the mainstream of opinion in chloroplast structure and biochemistry at the present time.

Chloroplast Structure

The chloroplasts of algae and higher plants are easily visible under the light microscope as discoid or ellipsoid bodies with overall dimensions in the 5–10 micron range. Usually, chloroplasts are considerably larger than the mitochondria in the same plant cell and occur in smaller numbers. This may vary from a single chloroplast, as found in the alga *Micromonas*, to 100–150 in the cells of higher plants. From observations with phase contrast microscopy of living cells it is apparent that chloroplasts are more rigidly defined in structure than mitochondria and are not seen to undergo rapid changes in shape or to fuse. Division of chloroplasts, however, is observed to occur (see Chapter 16). While evidently resistant to mechanical disturbance and not readily labile in form, as mitochondria are, the chloroplast is sensitive to changes in osmotic balance of the medium and responds to changes in osmotic pressure by rapid swelling or shrinking.

The first recorded observations of chloroplasts were made toward the end of the seventeenth century by the English botanist and microscopist Nehemiah Grew. Leeuwenhoek made drawings of the large chloroplasts of *Spirogyra* and other algae. Reports on the general occurrence of chloroplasts in green plants accumulated during the eighteenth and nineteenth centuries. Starch grains were first detected within chloroplasts by H. von Mohl in 1837; by 1887 J. von Sachs demonstrated that carbon dioxide absorbed by a plant is converted into the material of these starch grains through the action of light on the chloroplasts. Chloroplasts were first isolated for independent investigation by W. Menke and S. Granick in 1938.

The green pigments in undamaged chloroplasts appear under the light microscope to be distributed homogeneously throughout the organelle. If the chloroplasts of vascular plants are subjected to slight damage from heat, osmotic shock, or other factors, the arrangement of pigment into disc-like substructures becomes just discernible under the light microscope. These disc-like aggregations of green pigment, called *grana*, are characteristic of the chloroplasts of mosses, ferns, spermatophytes, and a few green algae. In these chloroplasts, the grana appear to be suspended in a colorless matrix substance called the *stroma*.

Organization of the pigment of the chloroplast into grana was suspected by early light microscopists, but the invisibility of these structures in undamaged chloroplasts led most observers to conclude that the granum was an artifact of preparation methods. In the 1930's, E. Heitz described grana in the chloroplasts of flowering plants, but the reality of this organization of pigment was still debated until 1947, when Granick and Keith Porter detected membranous structures equivalent to grana in chloroplasts under the electron microscope.

Grana are readily distinguished in electron micrographs (see Fig. 7–1) of the chloroplasts of land plants and the green algae. In addition to separation of the chloroplast interior into grana and stroma, the electron microscope reveals that the grana and accessory structures are built up from a complex system of folded membranes. The intricacies of this system are reflected in the complexity of the terminology used by investigators working in the morphology and function of chloroplasts. Not all authorities agree on the terms to be used, and often the same term is used differently by various investigators. What appears to be the most commonly used terminology will be given in this chapter; however, the reader is very likely to encounter differences in terminology in other sources.

The membranous subunits of the granum, the individual "coins" from which the stack is assembled, are flattened sacs or vesicles consisting of a single, continuous membrane enclosing an interior space. In 1962 Menke suggested the term *thylakoid* (*thylakos* = sack or pouch) for this membranous subunit. This is a useful term, because a flattened membranous sac enclosing an interior space appears to be the subunit of organization of all chloroplasts, whether or not the sacs are stacked into grana. The space enclosed within the thylakoid membrane is termed the *thylakoid space*. Often the word *lamella* is used interchangeably with thylakoid.

As mentioned, the grana of the chloroplasts of green algae and higher plants, including the mosses, ferns, and spermatophytes, consist of a vertical stack of greatly flattened thylakoids, much like a stack of coins (Figs. 7–2 and 7–3). Where adjacent thylakoids lie in contact in the granum, the membranes are fused into a plate-like, double structure. T. E. Weier has suggested the term *partition* to specifically identify this fused structure. The partition separates adjacent thylakoid spaces in the granum, and, as will be shown later, much of the green pigment of the chloroplast is likely to be bound to its surfaces.

Frequent membranous connections can be seen between the thylakoids of adjacent grana (circle, Fig. 7–1). These connections, called *frets* or *stromal lamellae* enclose a channel that is continuous with the thylakoid spaces of adjacent grana. It is not clear whether all thylakoid spaces make such connections with the thylakoids of other grana. The grana and frets are surrounded on all sides by the stroma. These structures,

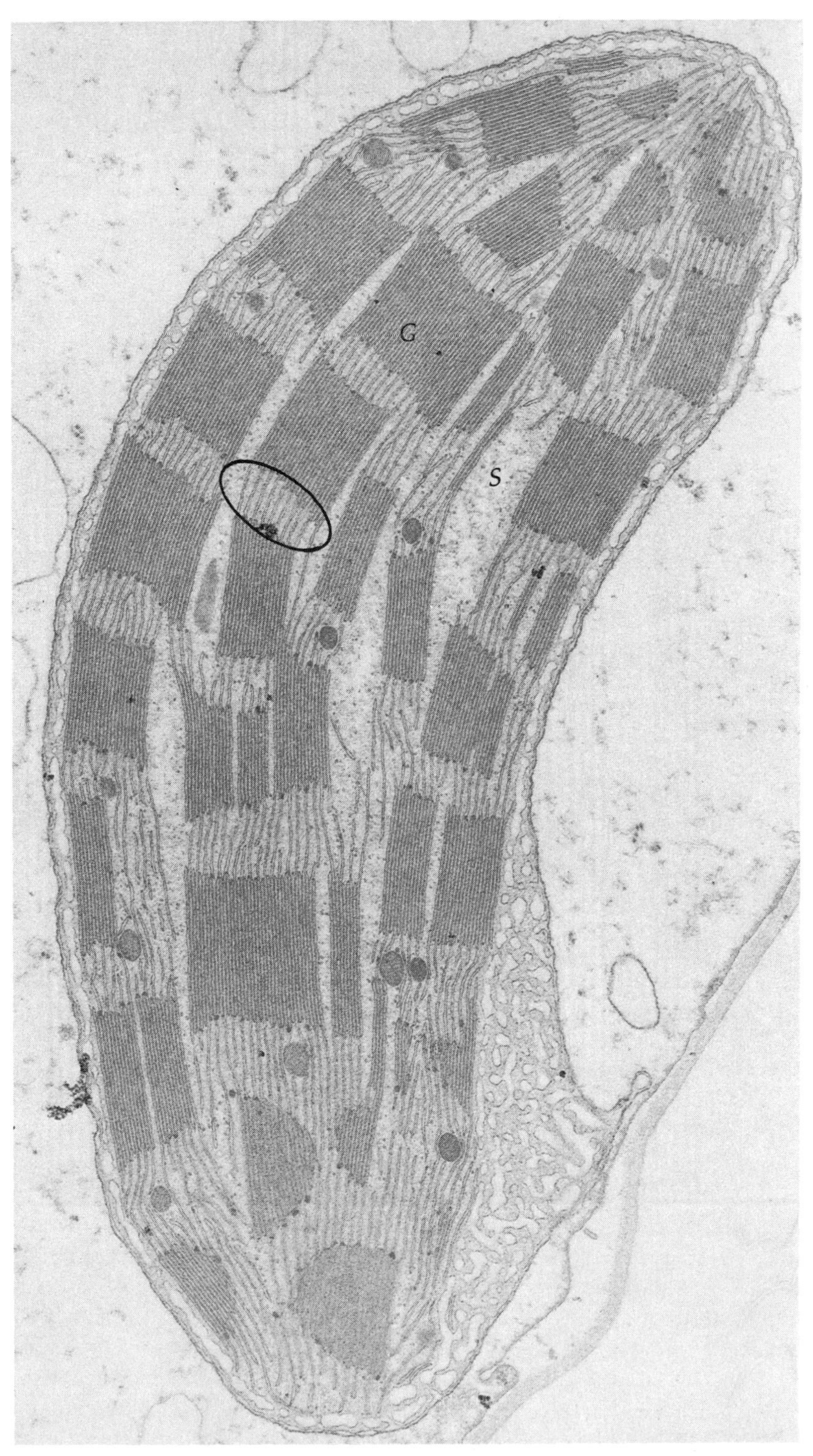

7–1. *Chloroplast from a glutaraldehyde-osmium fixed corn leaf. The grana (G) can be readily distinguished as stacks of membranes of very regular dimensions. The grana stacks are interconnected at frequent intervals by membranes called fret membranes or stromal lamellae (circled). These interconnecting membranes enclose a channel that is continuous with the thylakoid spaces in the granum. The membranes are suspended in a matrix called the stroma (S); also visible within the stroma are electron-dense spherical particles, the osmiophilic granules.* $\times$ *65,000. From* Cell Ultrastructure *by W. A. Jensen and R. B. Park, 1967. Courtesy of L. K. Shumway, Washington State University, and Wadsworth Publishing Company, Inc.*

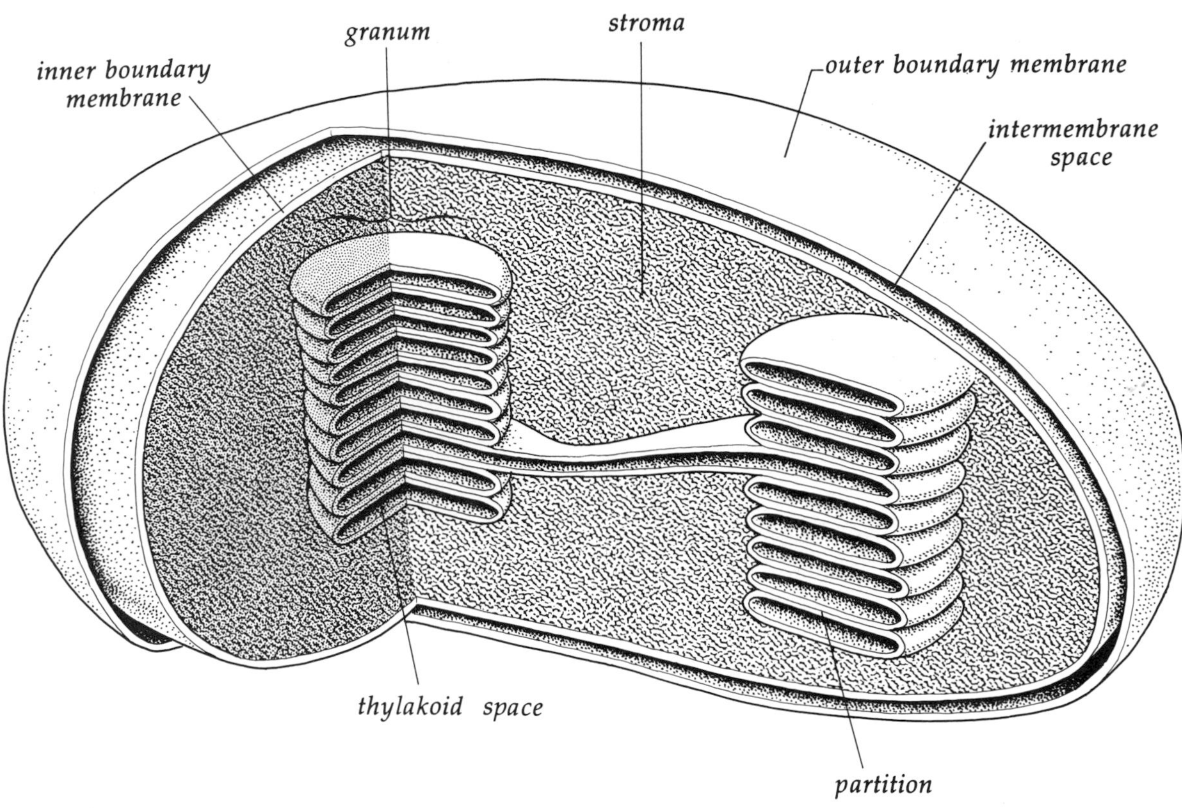

7–2. Diagram of the membranous structures and discrete internal regions of the chloroplast. Compare with Fig. 7–1.

shown diagrammatically in Fig. 7–2, are drawn in three dimensions in Fig. 7–3, and can be located in the sectioned chloroplast in Fig. 7–1.

Grana are about 0.3–0.6 microns in diameter, and number from 40 to 60 per chloroplast. Each granum may contain from two or three to more than a hundred stacked thylakoids. A few chloroplasts of higher plants, such as in the sheath cells surrounding the vascular bundles in corn, do not show organization of thylakoids into grana. Instead, the thylakoids in the chloroplasts of these cells are single and greatly elongated. Grana are apparently also absent from the guard cells of the stomata in grasses.

The stroma surrounding the grana and frets is analogous to the *matrix* of mitochondria, and like the mitochondrial matrix is more electron-dense than the ground substance of the cytoplasm outside the chloroplast. A number of structures are visible in the stroma, suspended in spaces between the grana. *Osmiophilic granules* or *droplets* are often visible (Fig. 7–1). These are electron-dense, roughly spherical bodies, 500–3000 Å in diameter, of unknown function. These may number from a few to hundreds. In addition, starch grains of various sizes are found in numbers of one to four or five per chloroplast. These are absent from the chloroplasts of plants kept in the dark but appear in the stroma after a period of photosynthesis in the light. In algae and some higher plants, *pyrenoid bodies*, of unknown function but traditionally believed to be related to starch synthesis, also appear in the stroma (see Fig. 7–5a). DNA also occurs in the chloroplast and can be seen as faint masses of fibers about 25–30 Å in diameter, dispersed in small aggregations throughout the stroma (Fig. 7–1). Small granules 150–200 Å in diameter,

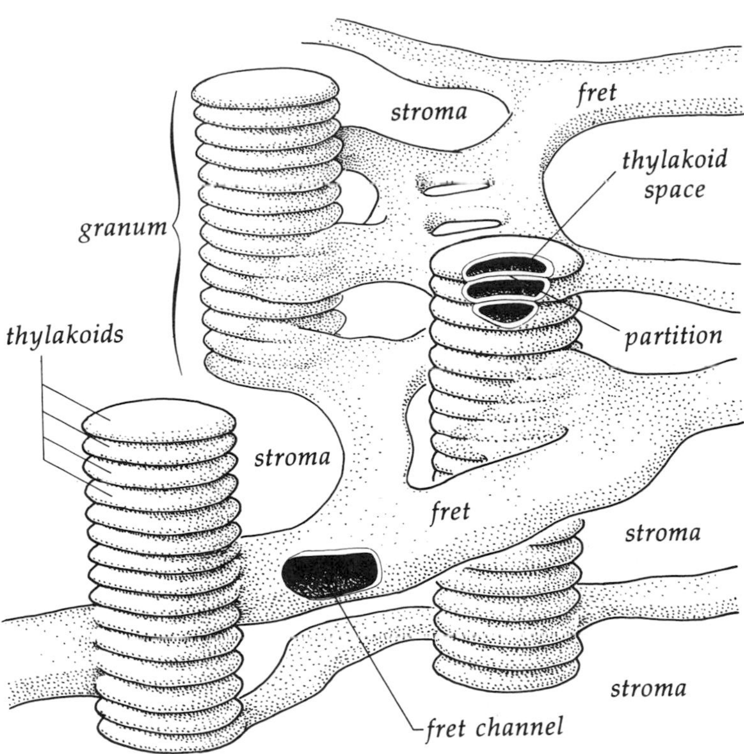

7–3. A diagram of the three-dimensional arrangement of the internal membranes of the chloroplast, as visualized by T. E. Weier. Redrawn, original courtesy of Dr. Weier.

identifiable as ribosomes of the bacterial type, are also scattered throughout the stroma. (The occurrence and possible function of the DNA and ribosomes of the chloroplast are discussed in detail in Chapter 16.)

The chloroplast interior is separated from the surrounding cytoplasm by double membranes, as diagramed in Fig. 7–2. The space between these membranes, designated here as the *intermembrane space,* is not always clearly defined in electron micrographs because of extensive vesiculation or folding of the inner membrane.

The structure of chloroplasts as outlined is typical of the green algae, mosses, ferns, and spermatophytes. The main structural feature of these chloroplasts is the stacking of thylakoids into the granum. The concept of the thylakoid can also be extended to chloroplasts without clearly defined grana. In these forms, which occur in the algae, thylakoids exist singly or in multiple layers, but are not found closely fused as in the granum. In the red algae single thylakoids are visible in the chloroplasts (Fig. 7–4). Each greatly elongated and flattened thylakoid in this chloroplast (arrows, Fig. 7–4) consists of a double membrane with an enclosed, narrow thylakoid space. In other algae thylakoids occur in the chloroplast in multiple layers but still remain separated by a narrow space. Fig. 7–5 shows a chloroplast of this type from a brown alga. Here the thylakoids are typically layered in groups of three (brackets, Fig. 7–5b). In these multiple layers, the space separating the three thylakoids is very regular and approximately equal to the thylakoid space, giving the whole structure the appearance of a group of six dense lines alternating with five spaces. In the regions where the edge of a thylakoid is visible the membrane can be seen to be continuous, completely enclosing the thylakoid space.

The chloroplasts of different algal groups show a diversity of structure. Often the morphology is quite distinctive and is useful in the taxonomy of the various groups of algae. Many, such as the previously mentioned *Micromonas,* contain only a single chloroplast per cell; among others are *Chlamydomonas* and *Spirogyra.* One hundred or more are found in others, including *Euglena.* These chloroplasts may resemble the discoid chloroplasts of higher plants, or may be ribbon-like, as in *Spirogyra,* lobed, as in *Zygnema,* or highly irregular as

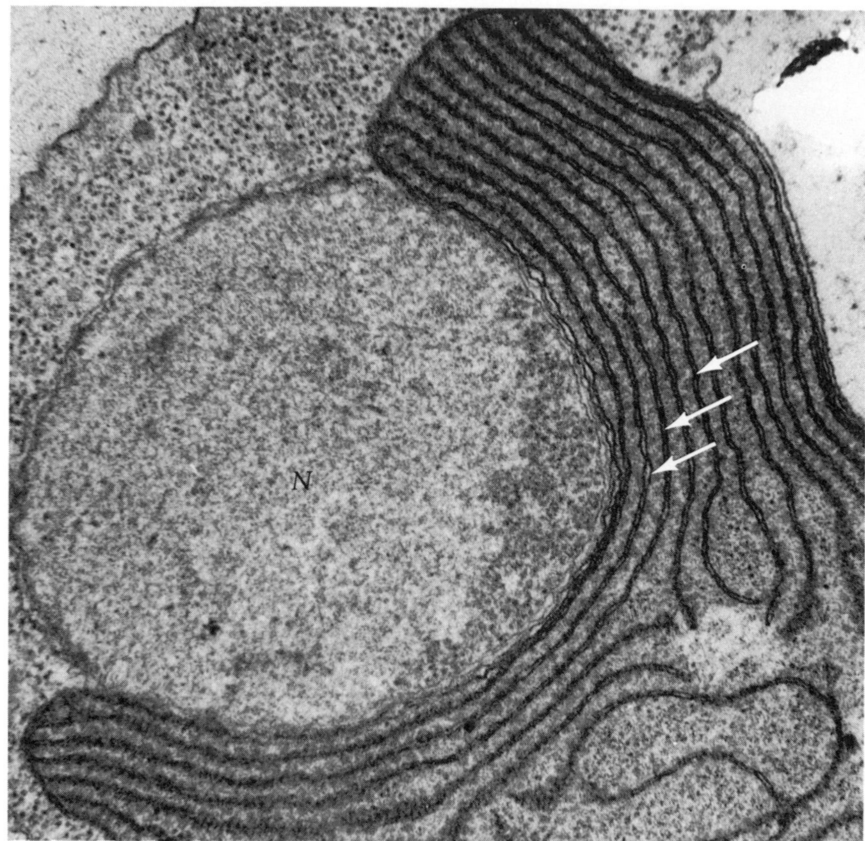

7–4. Chloroplast of a red alga, Rhodymenia palmata, *illustrating the extension of thylakoids into single, elongated sheets (arrows) in this group. Each of the extended sheets consists of two membranes, forming a greatly flattened closed sac.* N, *nucleus. × 47,000. Courtesy of R. L. Chapman.*

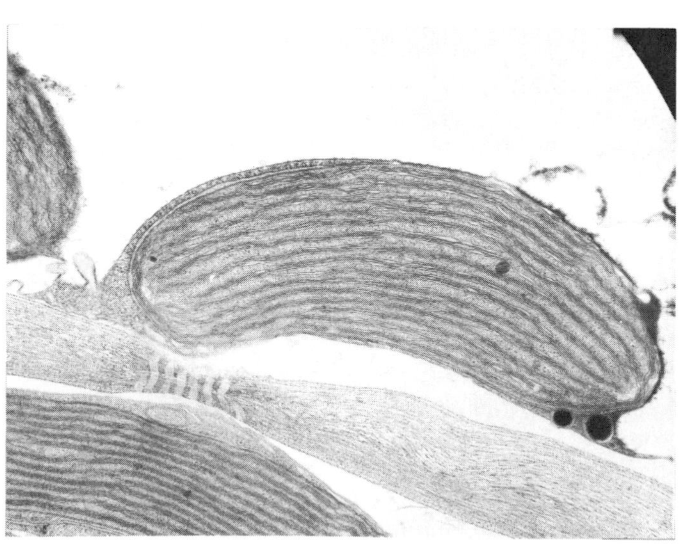

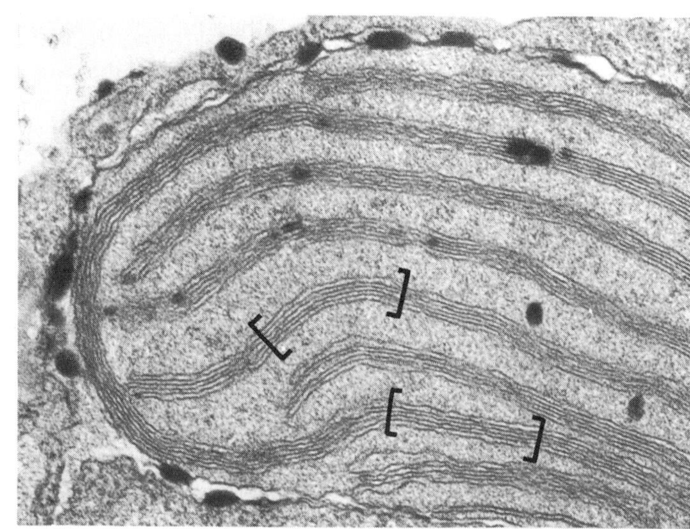

7–5. a, *A chloroplast of a brown alga. × 1,500. At higher magnification* (b), *the thylakoids of the brown algae can be seen to occur in stacks of three (brackets). × 47,000. Courtesy of M. Neushul and E. Y. Chi.*

Table 7–1 Chloroplast Components in Algae and Higher Plants

	Arrangement of thylakoids	*Pyrenoids*	*Starch grains inside chloroplasts*
Blue-green algae	no chloroplasts		
Red algae	single	few species	−
Cryptophytes	in pairs running the length of chloroplast	+	−
Brown algae	in threes	+	−
Green algae	long, may be locally stacked into grana-like aggregations	+	+
Bryophytes	grana-like stacks	few species	+
Vascular plants	grana	−	+

From data presented in J. T. O. Kirk and R. A. E. Tilney-Bassett, *The Plastids*. San Francisco: W. H. Freeman, 1967.

in *Oodegonium*. The red "eye-spot" of some algae, including *Chlamydomonas*, is a specialization of the chloroplast that in ultrastructure appears as a concentrated mass of osmiophilic granules. Some relationships of chloroplast structure between the algae and higher plants are summarized in Table 7–1.

The Structure of the Photosynthetic Apparatus in Bacteria and Blue-Green Algae

The photosynthetic apparatus is also bound to membranes in bacteria and blue-green algae. In these prokaryotic organisms, the thylakoid also takes the form of a flattened vesicle, although in some primitive photosynthetic bacteria (Fig. 7–6), the thylakoids are sperical bodies. In most photosynthetic bacteria, thylakoid membranes occur as flattened vesicles, either singly or in multiple layers (Fig. 7–7). The photosynthetic membranes in blue-green algae are often folded extensively but do not fuse into layered structures (Fig. 7–8). In the prokaryotes, the assemblage of thylakoids is not separated from the cytoplasm by a double membrane system as in the chloroplasts of higher plants.

When bacterial cells are disrupted, a particulate fraction capable of photosynthesis can be centrifuged down and purified. This fraction, which is fairly uniform in density, proves to consist of small membranous vesicles and fragments of membranes when viewed under

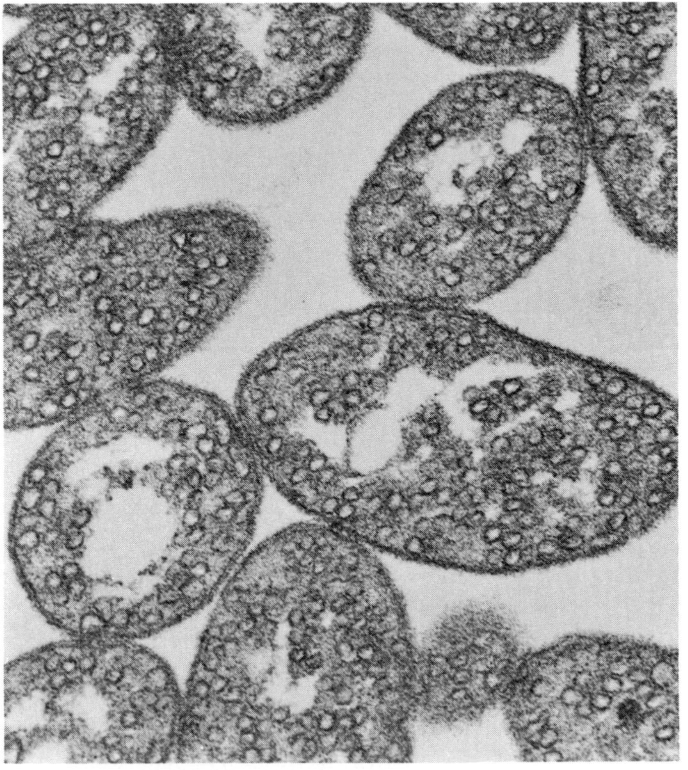

7–6. Spherical thylakoids in the cytoplasm of the purple photosynthetic bacterium Rhodopseudomonas. *× 51,000. From* Biochemistry of Chloroplasts, *Vol. 1, ed. T. W. Goodwin, 1966. Courtesy of W. Menke and Academic Press, Inc.*

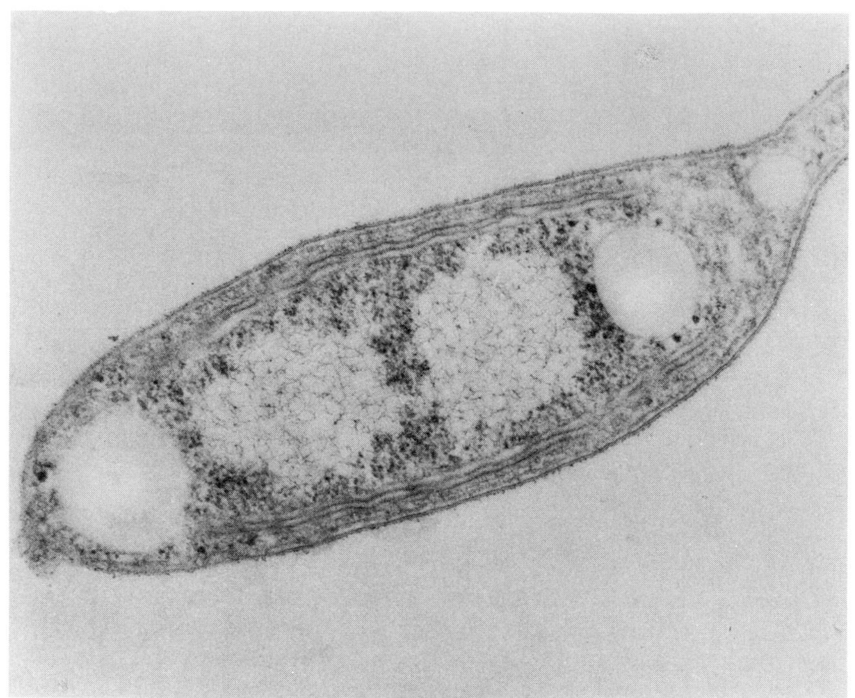

7–7. Granum-like stacks of thylakoid membranes in dividing cells of a purple photosynthetic bacterium. × 45,000. Courtesy of W. C. Trentini, Mt. Allison University, Sackville, N. B., Canada.

the electron microscope. These small fragments have been termed *chromatophores.* The term is something of an abstraction, because it refers only to a centrifuged and purified subfraction of bacterial cells (although the spherical thylakoids of green bacteria are sometimes called chromatophores). Chromatophores contain all the pigments and enzymes necessary for photosynthesis. These particles can be further broken down into membrane fragments only 200–400 Å in diameter. The smallest of these fragments still capable of carrying out photosynthesis are considered by some investigators to represent the ultimate *photosynthetic unit* of bacteria and blue-green algae.

Individual photosynthetic units are not resolved in thin sections of bacteria and blue green algae. Only the internal lamellae, thought to contain the photosynthetic units, are visible. However, the photosynthetic units are fairly well characterized in terms of weight and composition and are widely accepted as representing an actual physical entity within bacteria and blue-green algae. The term is also applied to the unit of photosynthesis in higher plants. As we shall see later in this chapter, the photosynthetic unit of higher plants contains more molecules and is more complex in function than the unit in bacteria and blue-green algae.

Molecular Components of Chloroplast Membranes

Before considering the ultrastructure of the thylakoid, attention must be given to the molecular components of chloroplast membranes. The situation here is similar to the present level of development of molecular morphology in mitochondria. The lipid component is comparatively well characterized, with several classes of lipid known and identified even to the level of the structural formula. The protein component is known primarily in terms of the activity of the various enzymes associated with the membranes in photosynthesis. An additional structural protein may exist in combination with the lipid to form the framework of the membrane. However, this protein component is poorly characterized.

A protein thought to be a "structural protein" has been isolated from chloroplasts by R. S. Criddle. A high proportion of nonpolar amino acids is found to be present in this protein, as in the structural protein of mitochondria. This work suggests that both hydrophobic and electrostatic bonds play an important part in binding structural proteins into membranes. Thylakoid membranes contain about 40 percent "structural

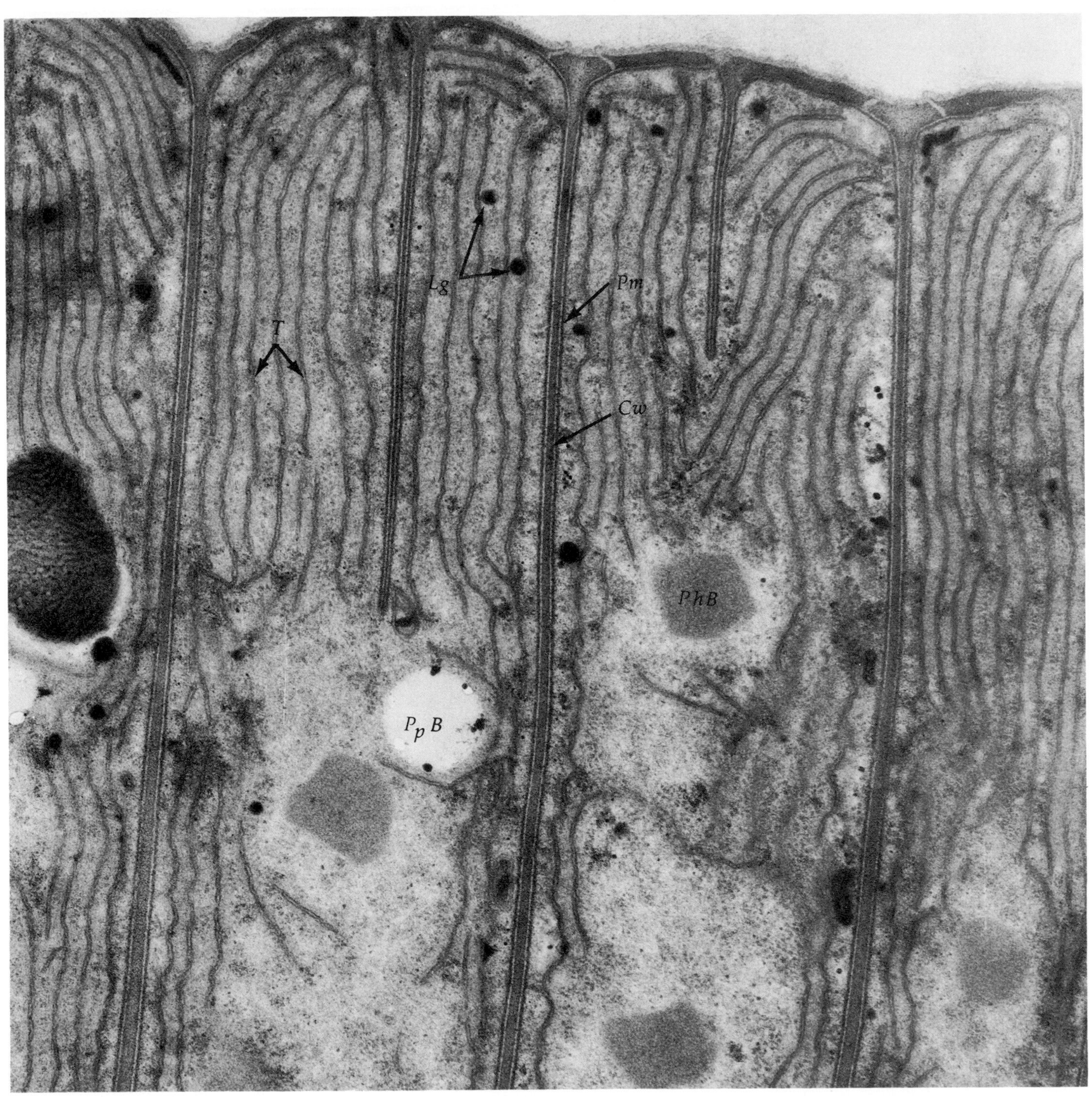

7–8. *Single thylakoids in the cytoplasm of a blue-green alga. This alga, recently isolated from a soil sample, has not yet been assigned a species name.* T, *thylakoids;* Lg, *lipid granules;* Pm, *plasma membrane;* Cw, *cell wall;* PhB, *polyhedral body;* PpB, *polyphosphate body.* × *43,000. Courtesy of Norma J. Lang.*

protein" as isolated by Criddle's techniques. As mentioned in the previous chapter, at least some proportion of the "structural protein" of mitochondria appears to be contaminated by the presence of denatured ATPase (see p. 121, Chapter 6). Therefore, although the chloroplast membranes almost certainly contain structural protein, the relative proportions of structural and enzymatic protein remain in doubt.

About 80 percent of the lipid fraction of chloroplasts is *glycolipid* (Fig. 7–9; see also Fig. 4–1), a neutral lipid with a carbohydrate side chain. Phospholipids are present in thylakoid membranes in a lower proportion than in mitochondrial cristae. An important class of phospholipids, the phosphatidyl glycerols (Fig. 7–10), forms the major phospholipid fraction. Sterols are present only in small quantities, representing about 0.5 percent of the total lipid fraction. Even the presence of this small proportion is somewhat doubtful, because it is difficult to eliminate possible contamination of the chloroplast lipids with cholesterol originating from other cellular membranes.

Quinones, called *plastoquinones* in chloroplasts, form a larger proportion of the total lipid than in mitochondria, being roughly equal to chlorophyll in quantity (about 8 percent). A quinone, *coenzyme Q* or "ubiquinone," is a possible member of the electron transport chain in mitochondria. The same role is suspected for the plastoquinones (including vitamin K) in chloroplasts. Curiously enough, the osmiophilic granules of chloroplasts contain plastoquinones on isolation. As pointed out, the function of these granular inclusions is unknown. It has been suggested that they are a storage place for plastoquinones, with the major function of the quinone fraction remaining in electron transport in the lamellae. In any event, the total quinone fraction is too large to be accounted for by electron transport alone. Six quinones, differing primarily in the length of an isoprenoid side chain (Fig. 7–11) occur commonly

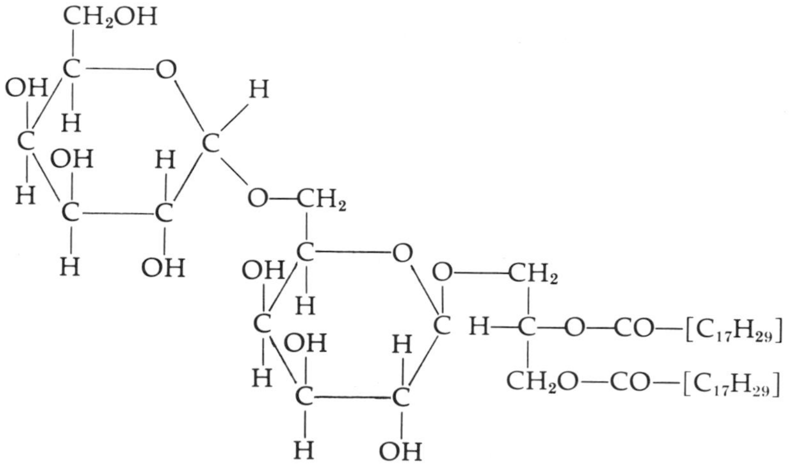

7–9. A glycolipid common in chloroplast membranes. Glycolipids may be important in the binding of enzymatic proteins to thylakoid membranes; removing this lipid causes a drastic reduction in the electron transport of photosynthesis.

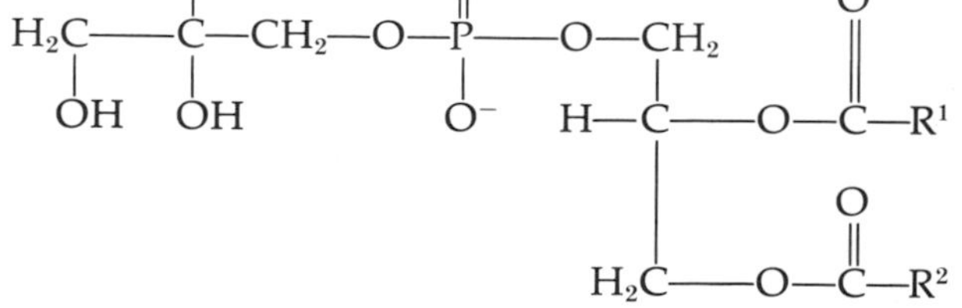

7–10. A phosphatidyl glycerol. R^1 *and* R^2 *are fatty acid residues.*

in chloroplasts. These long, nonpolar side chains may be active in forming bonds to structural proteins. As in mitochondria, the precise details of the binding of lipid to protein in chloroplast membranes are unknown.

Chloroplast Pigments

Associated with the lipid and protein components of internal chloroplast membranes is a group of pigments that absorb light at different wavelengths. These, acting together, absorb almost the entire visible spectrum, with one pigment or another absorbing wavelengths over the region between 4000 (blue) and 7000 Å (red).[1] The predominant pigment is *chlorophyll,* which absorbs red wavelengths most strongly and transmits green light. All photosynthetic organisms use chlorophyll in one of its several forms; the green color of the habitable regions of the earth is due to the abundance of this pigment in plants.

(n = 6–10)

ubiquinones

plastoquinones

(n = 4–9)

K vitamins

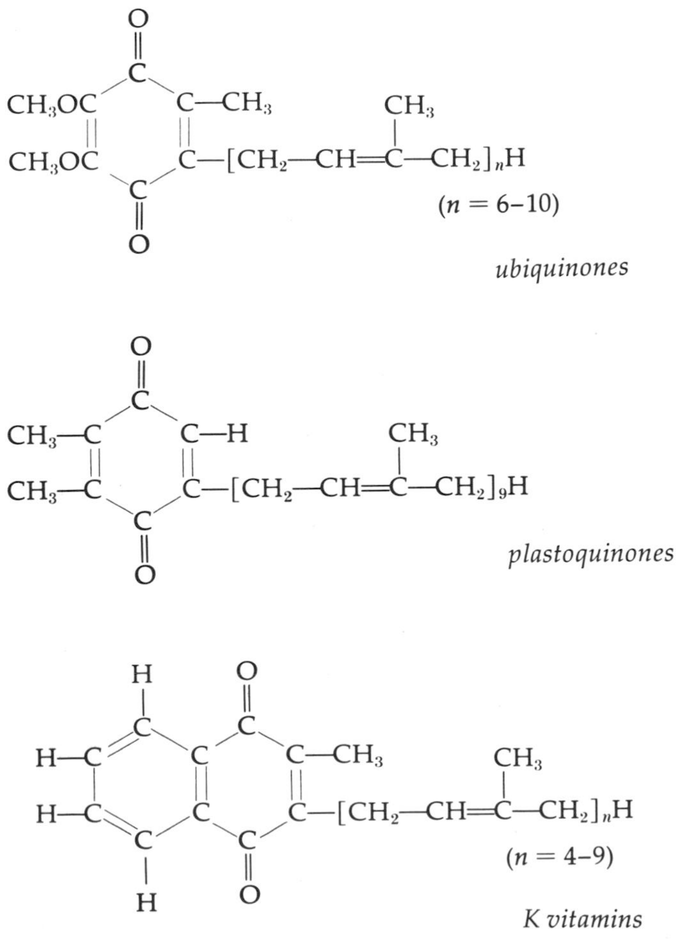

7-11. *Quinones of the photosynthetic apparatus.*

The various chlorophylls consist of a porphyrin ring to which a long phytol side chain is attached (Fig. 7-12). The porphyrin ring of chlorophyll is similar to the prosthetic groups of the cytochromes and hemoglobin. At least four types of chlorophyll occur commonly in different organisms, differing chiefly in the groups bound to the porphyrin ring. Many algae and all higher plants contain chlorophyll *a* and *b*. Chlorophylls *a* and *c* occur in brown algae and diatoms. Green and purple photosynthesizing bacteria contain a different species known as *bacteriochlorophyll.*

The maximum light absorption of the chlorophylls varies, depending on the state and type of binding of the molecule to membrane components. Bacteriochlorophyll, for example, has absorption bands at 8000, 8500, and 8700–8900 Å in the living organism. When bacteriochlorophyll is extracted and purified, these absorption bands disappear and are replaced by a single band at 7700 Å. In the same way, the absorption maxima for purified chlorophyll *a* and *b*, 6750 and 6500 Å, vary depending on the state of binding to chloroplast membranes.

In addition to the chlorophylls, another group of lipid-soluble pigments, the *carotenoids,* is found in all photosynthetic organisms. These yellow pigments, which absorb in the region from 4000 to 5000 Å, are all closely similar in structure and are based on a long chain of 40 carbons. Various substitutions in this chain give rise to the various pigments of the group. Chief in abundance among these in plants is β-carotene (Fig. 7-13).

The function of these accessory pigments appears to be the absorption of light in regions of the visible spectra not absorbed by chlorophyll. This increases the efficiency of photosynthesis in use of light energy. The energy absorbed by these accessory pigments is apparently transmitted to chlorophyll *a*, the pigment directly involved in transforming light into chemical energy. A secondary role of the carotenes may be in the prevention of light damage to the main photoactive center formed by the chlorophylls. It is known that, if an excessive amount of chlorophyll reaches an active

[1] Ordinarily, the millimicron, or more recently the nanometer, both equivalent to 10 Å, are used as the unit of measurement for the wavelength of light. For uniformity the Ångstrom is used here.

7–12. Chlorophyll a. *Chlorophyll* b *has a* —CHO *group substituted for the* —CH_3 *at carbon 3; chlorophyll* d *has a* —CHO *substituted for* —CH=CH_2 *at carbon 2; protochlorophyll has double bonds instead of* —H *at carbons 7 and 8; bacteriochlorophyll has* —H *substituted the double bonds at carbons 3 and 4 and a* —CO—CH_3 *instead of* —CH=CH_2 *at carbon 2.*

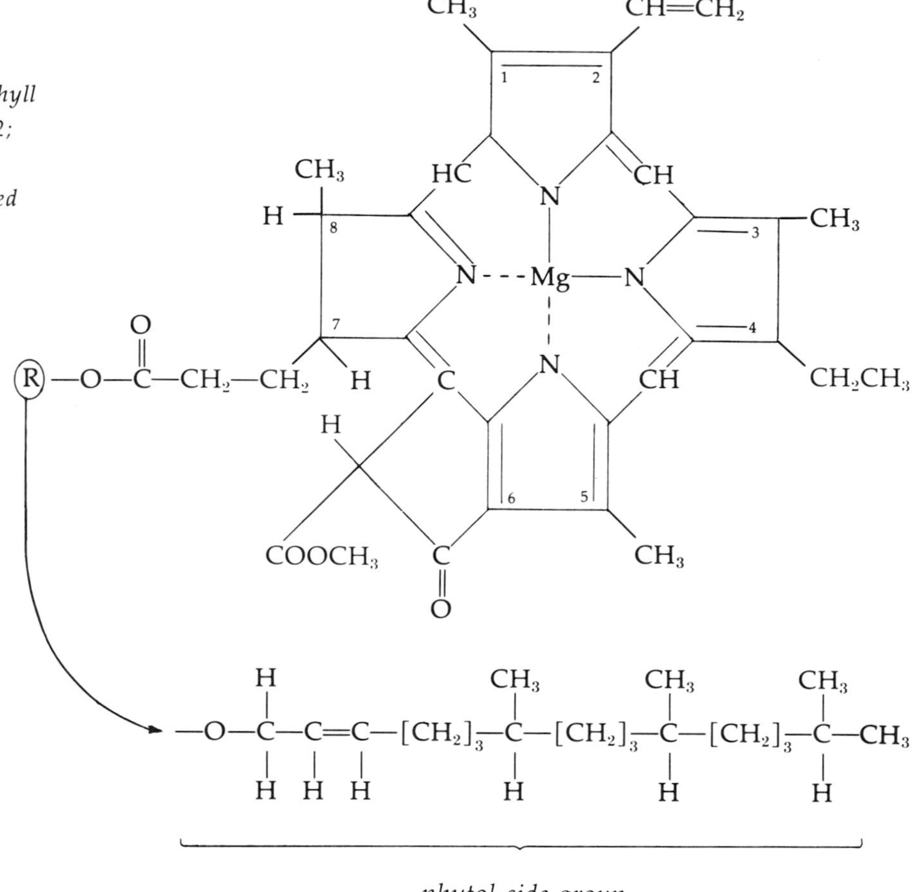

7–13. β-carotene. This molecule, in addition to functioning as an accessory pigment in photosynthesis, absorbs light in the retina of animals.

Table 7–2 Major Photosynthetic Pigments

Pigment	*Organism*	*Absorption maxima**
Chlorophyll *a*	higher plants, algae	6750 Å
Chlorophyll *b*		
Chlorophyll *c*	brown algae, diatoms, dinoflagellates	6400 Å
Bacteriochlorophyll	bacteria	7700 Å
Carotenoids: Carotenes, Xanthophylls	all photosynthetic organisms	4000–5000 Å
Biliproteins: Phycocyanin, Phycoerithin	blue-green and red algae	5000–6000 Å

*The absorption maxima of these pigments vary according to the state of binding to membranes and the metabolism of the chloroplast.

chemical state through the absorption of light, destruction of the chlorophyll may occur. The carotenes evidently "quench" the excess excitation of chlorophyll. How vital this role is to photosynthesis is problematical, because mutant photosynthetic bacteria lacking carotenes are able to photosynthesize normally.

A final group of pigments, absorbing in the range 5000–6000 Å, is of interest because of a curious structural relationship to hemoglobin. These pigments, called *biliproteins,* resemble the bile pigments of mammals and contain a pigmented group resembling protoporphyrin IX, the porphyrin ring of hemoglobin. Because of difficulty in splitting off the firmly bound porphyrin group in the biliproteins, it has not been possible to work out the structure of this chromophoric group in detail. Biliprotein pigments are confined entirely to the blue-green algae, red algae, and the cryptomonads, an algal form sometimes classified as protozoan.

In summary, the primary light-absorbing pigment in all organisms is a form of chlorophyll. The accessory pigments vary in terms of concentration or occurrence; their primary function is to increase the range of visible electromagnetic radiation used as a source of energy for photosynthesis. In all cases, energy trapped by the accessory pigments is transferred to chlorophyll, the primary photoactive molecule in the conversion of light energy to chemical energy. The end result is that, in various organisms, any part of the spectrum from at least 4000 to 7500 Å may be used as an energy source in photosynthesis. A summary of the various pigments and their absorption maxima is given in Table 7–2.

Ultrastructure of the Thylakoid Membrane

Of all the cellular membranes, the internal membranes of chloroplasts and mitochondria show particulate substructure most clearly. But in neither case is the substructure clear enough to allow definite statements to be made about the size, shape, or distribution of the particulate subunit. The picture is further confused by the fact that different preparative methods give widely different appearances, in both dimensions and location, to the particles in these membranes.

Some of the earliest ultrastructural work with internal chloroplast membranes was carried out by shadowing isolated, air-dried grana. In slightly disrupted preparations of grana, Roderick B. Park and N. G. Pon found a large repeating subunit. This subunit was reported as a slightly oblong, flattened sphere, 155 × 180 Å in width and 100 Å in thickness (Fig. 7–14). Four spherical subunits are visible within each particle, tightly packed into a square with a slight depression or hole marking the center. Park and Pon called the large, four-part particle a *quantasome* and equated it to the photosynthetic unit of higher plants. The quantasome particle is large enough to accommodate 250–300 chlorophyll molecules; this agrees well with the number expected for the photosynthetic unit on the basis of biochemical data (see below).

The quantasome particles observed by Park and Pon appeared to be located on the inner surface of the thylakoid membrane, apparently exposed when the membrane surface was torn away. These particles are only rarely seen in a crystalline array, as shown in Fig. 7–14. More often the particles are randomly distributed or at times lined up into scattered rows.

Particles with the dimensions and substructure of quantasomes cannot be seen in thin sections. The globular subunits faintly visible in cross sections of thylakoid membranes are much smaller (Fig. 7–15). Elliot Weier has made extensive measurements of these small subunits, and has determined that each particle is 70 Å in diameter and consists of a lightly stained core, approximately 40 Å in diameter, surrounded by an electron-dense shell 30 Å in thickness. It is possible that these particles represent the spherical subunits of the quantasome. In thin sectioned membranes, however, no regular arrangement into quartets with the

7–14. Crystalline packing of quantasomes in a thylakoid isolated from spinach chloroplasts and prepared for electron microscopy by shadowing. The surface membrane of the thylakoid has been torn off in the region in which the quantasomes are exposed. Note that each quantasome particle contains four subunits of equal size arranged in a square pattern. × 96,500. From Plant Biochemistry, *ed. J. Bonner and J. E. Warner, 1965. Courtesy of R. B. Park and Academic Press, Inc.*

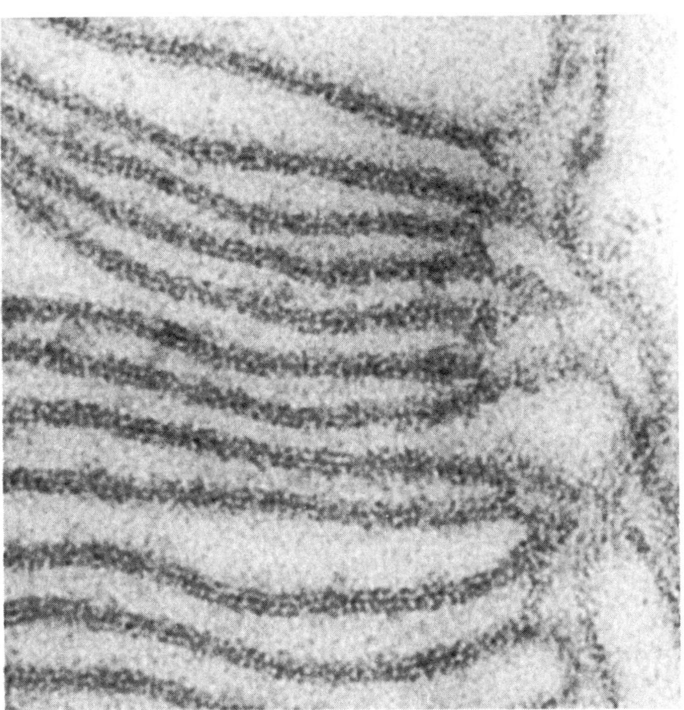

7–15. Subunit structure in the thylakoid membranes of a chloroplast from Aspidistra. *The thylakoid spaces of this granum have been swollen by treatment of isolated chloroplasts with distilled water. × 240,000. From* Biochemistry of Chloroplasts, *Vol. 1, ed. T. W. Goodwin, 1966. Courtesy of T. E. Weier and Academic Press, Inc.*

total dimensions and form of the quantasome can be discerned. Further, the small 60–90 Å particles are very regularly distributed through the membranes, in much greater numbers than would be expected for subunits of the quantasome, which usually occurs in a less densely packed arrangement.

The differences in thylakoid structure as observed in isolated and thin sectioned preparations have not yet been resolved. At times, it has been maintained that either the 70 Å membrane subunit or the quantasome is an artifact. Other alternatives are that the fixed and embedded quantasome is not sufficiently electron dense to be seen in or on the membrane or that it is destroyed during the fixation and embedding procedure. In this case, the layer of 70 Å particles making up the membranes seen in thin sections might represent a supporting framework to which the quantasome particle is bound. This would fit well with Park and Pon's early observation that the quantasome appeared to be *inside* the thylakoid space. This observation would place the quantasome on the inner surface of the thylakoid membrane.

These observations can be reconciled if the membrane forming the walls of the thylakoid is considered to consist of a single layer of 60–70 Å particles, with the quantasome bound to the inner surface of this layer and contained within the thylakoid space. The degree of flattening of the thylakoid is such that only a single layer of quantasomes could be accommodated within the enclosed space. The absence of quantasomes in thin sectioned chloroplasts could be explained by assuming that only the 60–70 Å particles forming the walls of the thylakoid are visible or are preserved by the preparative techniques.

Studies of thylakoid structure using the freeze-etch technique are compatible with this line of reasoning. Two particles are revealed as subunits of the thylakoid in these preparations. One, of smaller dimensions, is

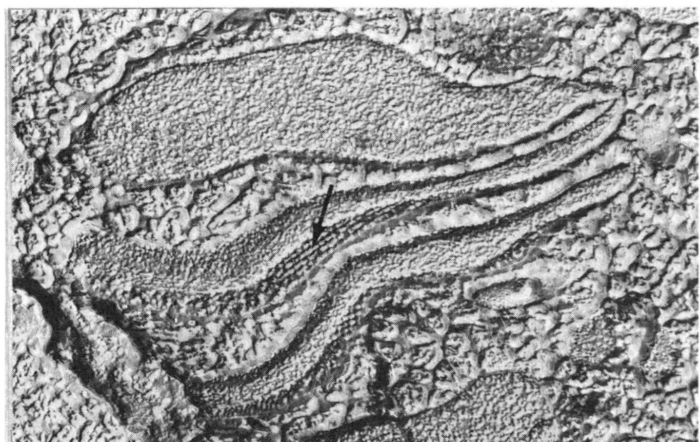

7–16. *A replica of thylakoid membranes from a granum of spinach chloroplasts prepared by the freeze-etch method. Particles of two different sizes are clearly visible; the larger globules, lined up in crystalline array, are the quantasomes (arrow). × 47,000. From* Biochemistry of Chloroplasts, *Vol. 1, ed. T. W. Goodwin, 1966. Courtesy of K. Mühlethaler and Academic Press, Inc.*

approximately 110 Å in diameter. The second, larger subunit is 175 Å in diameter. In spinach chloroplasts, the larger particles are lined up in crystalline arrays and are thus identifiable as the quantasome (Fig. 7–16, arrow). The smaller particles are densely packed into layers that appear to alternate with the layers of the larger quantasome particle.

The two layers are clearly visible in Fig. 7–17, prepared from a chloroplast that has been fractured along the flattened surfaces of the thylakoid membranes. In this micrograph, exterior and interior surfaces are exposed at different levels by the line of fracture. Daniel Branton and Park have proposed that both the smaller and larger particles form the walls of the thylakoid. According to this interpretation, both particles are contained within the structure of the thylakoid membrane, forming a single layer that bounds the thylakoid space. However, it is also possible to conclude from the freeze-etch results that the larger particles are attached to the inner surface of a continuous layer of smaller particles. This is better suited to the evidence from thin sections, which does not reveal a larger particle interspersed with the 60–70 Å particles in thylakoid membranes as the Branton–Park hypothesis would require. The two alternatives are diagramed in Fig. 7–18.

Thylakoid membranes do not show a particle analogous to the "elementary particle" of mitochondrial

7–17. *A replica of the various membrane faces exposed by freeze-etch of a granum from spinach chloroplasts. The letters correspond to the diagram shown in Fig. 7–18a. × 83,000. From* Photophysiology, *Vol. III, ed. A. C. Giese, 1968. Courtesy of D. Branton and Academic Press, Inc.*

cristae when negatively stained. A particle seen in negatively stained thylakoid membranes was at first thought to be attached to the surfaces of thylakoid membranes, but this particle was found to be randomly distributed in the preparations, not necessarily associated with the membranes. Rinsing the isolated thylakoid membranes in dilute buffer before negative staining eliminated the particle from the preparations; the particle was later identified as an enzyme, *carboxydismutase* (see Wettstein, 1966, and Branton and Park, 1967; carboxydismutase is discussed further on p. 149). Carboxydismutase is normally a "soluble" enzyme located in the stroma. Evidently this enzyme appeared in the negatively stained thylakoids as a contaminant, released from the outer membranes during preparation of the chloroplast. If a subunit analogous to the "elementary particle" does indeed exist in the thylakoid membrane, it is too deeply embedded in the membrane structure to be revealed by negative staining.

X-ray diffraction and polarized light studies of chloroplast membranes. X-ray diffraction of chloroplasts does not produce a clear pattern except for one reflection at 35 Å. This periodicity has been interpreted as arising from alternating layers of lipid and protein in chloroplast membranes. Because this periodicity does not change when chloroplasts swell in hypotonic media, Elliot Weier and Andrew A. Benson have suggested that it represents the partition, the only structure in grana not penetrated by water during swelling of the chloroplast in hypotonic medium.

Polarized light studies indicate that most of the chlorophyll in the granum is randomly oriented. Only very weak polarization of light by chlorophyll can be detected. The weakly polarized light from a small fraction of the chlorophyll, about 5 percent, is believed to arise from a small number of ordered porphyrin rings that lie in the plane of the flattened thylakoid membranes. The rest of the chlorophyll, except for this 5 percent, is thought to be randomly oriented. (For a more detailed summary and literature references in the area of X-ray diffraction and polarized light studies of chloroplast structure, see Branton, 1968.)

In summary, studies in the fine structure of thylakoid membranes have indicated that at least two major subunits exist. One forms the framework of the membrane and lies in the size range 70–110 Å. A second particle, the quantasome, is probably either contained within the membrane or attached to inner surfaces of the thylakoid. Chlorophyll is also bound to the membrane, probably in association with membrane particles. Only a small proportion of the chlorophyll is bound in a highly ordered array.

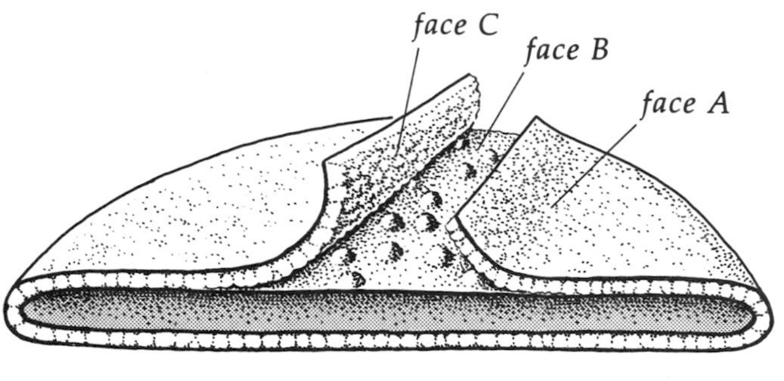

7–18. Two interpretations of the localization of the particulate entities making up the thylakoid membrane. In a, *according to a model advanced by Branton and Park, a smaller subunit forming a part of the membrane substructure and a larger particle believed to represent the quantasome are both located within the membrane, between the inner and outer faces. Redrawn, original courtesy of D. Branton and Academic Press, Inc.* [J. Ultrastr. Res. *19:283.*] *In* b, *the smaller particles form the structural subunit of the membrane as in* a; *the quantasome-like particles are attached to the interior membrane surface of the thylakoid (see text).*

Photosynthesis

Photosynthesis can be defined as the process in green plants, algae, and some bacteria in which the energy of sunlight is used to reduce or "fix" CO_2 into organic compounds. In algae and the higher plants, oxygen is evolved during photosynthesis as an important by-product. The overall reaction in higher plants has been known in skeletal form since the early nineteenth century, when the general equation for photosynthesis of carbohydrates in higher plants was written:

$$6CO_2 + 6H_2O \xrightarrow[\text{light}]{} C_6H_{12}O_6 + 6O_2 \qquad (7\text{–}1)$$

Even before this time, Joseph Priestley had observed that green plants, in the presence of light, were able to "purify" air. Priestley also noted that only the green portions of the plant were able to carry out this activity. Some insight into the types of reactions occurring was gained after the turn of the century by F. F. Blackmann, who studied the effects of light, temperature, and CO_2 concentration on photosynthesis. In the presence of excess CO_2 and a limited amount of light, the rate of photosynthesis was found to be independent of temperature. The independence of reaction rate on temperature is a characteristic of photochemical reactions. Blackmann also noted that, in the presence of excess light and a rate-limiting concentration of CO_2, the rate of photosynthesis was dependent on the temperature in the usual way: an increase in temperature caused an increase in reaction rate.

In the 1920's, Otto Warburg, who has already been mentioned as an important figure in the study of respiration in the mitochondrion, drew an important conclusion from Blackmann's results. Warburg reasoned that there are two classes of reactions in photosynthesis, a temperature-independent "light" reaction, and a temperature-dependent "dark" reaction. Warburg's inferences were supported by subsequent work. The two reactions were shown to be separable by Robert Emerson only a few years later, during the 1930's. Emerson studied photosynthesis in plants exposed to flashes of light lasting only a few milliseconds. Emerson found that after a brief flash of intense light, a period of darkness many times longer than the duration of the light flash could be inserted without any diminution in photosynthetic rate. In fact, a period of darkness was *required* after the bright flash to allow the most efficient use of a given quantity of light in photosynthesis. Emerson interpreted this to mean that two separate reactions occurred in which the products of an initial light reaction were subsequently used to fix CO_2 in a dark reaction. The products of the light reaction accumulated and reached saturation very rapidly; the subsequent dark reactions used these accumulated products much more slowly, probably as an energy source in the fixation of CO_2 into carbohydrates. A cyclic pathway for at least some of the intermediates was indicated by the inability of the "light" system to operate unless the immediate products of the light system were used in the dark system.

The reactions were subsequently localized in the chloroplast. R. Hill found in the 1930's that isolated chloroplasts could evolve molecular oxygen if an artificial electron acceptor was added to the system. (CO_2, if added to isolated chloroplasts, is reduced very slowly.) Daniel I. Arnon later showed that the synthetic machinery for CO_2 fixation and for the phosphorylation of ADP to ATP was also in the chloroplast. This proved that the light and dark reactions were both contained in the chloroplast. Closer localization to the grana and stroma was provided in the late 1950's by Park and Pon, who purified grana from sonicated (disrupted by high frequency sound) chloroplasts by centrifugation. An insoluble precipitate, made up of the grana, was able to carry out the Hill reaction and to phosphorylate ADP to ATP. The chlorophyll was identified with this fraction. The stroma remained in suspension in the form of a colorless solution of proteins. This fraction was found to contain almost all of the enzyme associated with the carbon-fixation reaction, carboxydismutase (also called ribulose 1, 5-diphosphate carboxylase). Although this fraction was able to carry out CO_2 fixation only very slowly, the rate was increased many times if the granal fraction was added. Park and Pon concluded that the rate increase was due to the addition of ATP and a reducing agent, both supplied to the stromal system by the grana.

The Overall Reactions of Photosynthesis

While the separation of photosynthesis into light and dark reactions was being worked out, it was discovered that some bacteria carried out photosynthesis using raw materials other than H_2O and CO_2. Many hydrogen-electron donor substrates, such as H_2S, thiosulfate, secondary alcohols, and hydrogen itself, could be used. This led to a more widely applicable expression for the overall reactions of photosynthesis

$$CO_2 + 2H_2A \rightarrow (CH_2O) + H_2O + 2A \qquad (7\text{–}2)$$

in which the substance A is a hydrogen donor. In algae and higher plants, for example, H_2A is H_2O, and 2A is O_2. Confirmation of this generalized scheme, first proposed by C. B. Van Niel, was provided by the use of labeled oxygen. In this method, CO_2 or H_2O containing O^{13}, an isotope of oxygen, was added to photosynthesizing chloroplasts. The oxygen evolved was found to contain the isotope only if the oxygen of H_2O was labeled. This showed that the oxygen evolved in photosynthesis arises from water, not CO_2. In equation 7–2, water (H_2A) is split, leaving O_2 (2A), the oxidized byproduct of the reaction.

The generalized equation given in 7–2 can be subdivided into "light" and "dark" systems, if the ADP-ATP couple and a hypothetical hydrogen acceptor (designated "X") are introduced:

$$2H_2A + 2X + n(ADP + HPO_4^=) \xrightarrow[\text{light}]{} 2XH_2 + 2A + nATP \quad (7\text{–}3)$$

$$2XH_2 + nATP + CO_2 \rightarrow (CH_2O) + 2X + n(ADP + HPO_4^=) \quad (7\text{–}4)$$

If like terms are canceled on both sides of the equations, equation 7–2 can be reproduced. There is strong evidence that the overall reaction proceeds in this way. Only the light reactions are unique to photosynthesizing organisms; most of the dark reactions, which can also produce amino acids or fatty acids and other intermediates in addition to carbohydrate, are common to all living organisms.

The "Light" Reactions in Photosynthesis

Some background information. Visible light is a form of electromagnetic radiation with wavelengths between almost 4000 and 7500 Å. The treatment of optics presented in Chapter 3 depends on the properties of light as waves of energy. Diffraction and phase contrast, for example, depend on the wave path followed by beams of light in space. In some cases, such as the absorption of light by macromolecules, light behaves as a flow of discrete units of energy rather than a continuous wave. At any given instant of time, these packets of energy are likely to be found in a position in space corresponding to the path described by a beam of light at a given wavelength. This is analogous to the behavior of electrons, which, although particulate, follow a wave path in space. The particulate units of light are called *quanta.* The amount of energy contained in a quantum of light depends on the wavelength according to the relationship:

$$E = hc/\lambda \quad (7\text{–}5)$$

in which E is the energy of a quantum of light, h is Planck's constant, c is the velocity of light, and λ is the wavelength. For light transmitted through a medium of constant refractive index, such as air, all parameters on the right-hand side of the equation except the wavelength of light are constants. Thus, as the wavelength decreases, the energy of a quantum increases. For a given wavelength, the energy of a quantum has a constant value. The energy content of light quanta at various wavelengths is given in Table 7–3, in calories per Einstein. The Einstein relates light energy to a gram molecular weight, and is equivalent to a "mole" of light, 6.023×10^{23} quanta. Thus one mole of chlorophyll, containing 6.023×10^{23} molecules and absorbing one Einstein of red light at 6,500 Å, will absorb energy equivalent to 43,480 calories.

Table 7–3 Energy Associated with Light at Various Wavelengths

Wavelength (Å)	*Color of light*	*cal/Einstein*
3,950	violet	71,800
4,900	blue	57,880
5,900	yellow	48,060
6,500	red	43,480
7,500	far red	37,800

Some molecules, such as chlorophyll, absorb light at characteristic wavelengths. Other wavelengths are transmitted. The absorption of the energy of light is a property of particular electrons in a molecule. These electrons either lie in orbitals surrounding an atomic nucleus or are in molecular orbitals associated with more than one atom of the molecule. Electrons in these orbitals exist at a characteristic energy level. If one of these electrons absorbs the energy of a quantum of light, it undergoes a transition to a new orbital with a higher energy level. In the new orbital, the electron is said to be in an *excited state.* The difference in energy levels between the unexcited state (called the *ground state*) and the excited state has a fixed value for an electron in a given position in a molecule; the energy in the quantum of light absorbed must contain exactly this amount of energy. Another way of stating this is that if a molecule contains an electron capable of absorbing light energy, the light must be at some fixed wavelength. All the energy in the quantum must be absorbed.

The electron in the excited state will eventually return to the ground state. Usually, the excited state lasts only a fraction of a second. On returning to the ground state, the energy absorbed by the electron is released, in an amount exactly equal to the quantum absorbed. The release can take several forms. If it is released as light energy, the light emitted is called *fluorescence.* The fluorescence usually occurs at a greater wavelength than the light absorbed. The difference in energy between the quantum absorbed and the quantum radiated as fluorescence appears as heat. Alternatively, the energy associated with the excited electron may be transformed into chemical energy. One way this may take place is that an electron may absorb a sufficient quantity of energy to leave its orbital entirely. Electrons ejected in this way have velocity and wavelength that depend on the amount of energy absorbed. If these electrons are transferred to a suitable acceptor, the reduced acceptor will have "trapped" the energy of the electron as chemical energy. It is not known whether the initial conversion of excitation energy to chemical energy in chloroplasts occurs in this manner or whether the excitation energy is in some way transferred to an electron arising from a different source. It is known, however, that the energy in a single light quantum absorbed by one chlorophyll molecule is sufficient to raise an electron to an energy level high enough to reduce NADP or acceptors of even greater redox potential.

The photosynthetic unit and the reaction center. In Emerson's research into the response of chloroplasts to brief flashes of light, he and a collaborator, A. Arnold, were able to estimate the total number of chlorophyll molecules required to evolve one molecule of O_2 when just enough light was given in a brief flash to saturate the mechanism. Emerson and Arnold discovered that one O_2 molecule was released for each 2500 chlorophyll molecules per flash. The subsequent dark reactions also reduced one molecule of CO_2 to (CH_2O) for each 2500 chlorophyll molecules. Obviously, 2500 quanta of light are not required to drive the mechanism evolving one molecule of O_2 and fixing one molecule of CO_2. Although estimates vary, the quanta required for these reactions probably approach a total of 8 ± 1 (see p. 151). Because the quanta required are much lower than the total number of chlorophyll molecules involved, Emerson and Arnold proposed that, instead of absorbing individually, the chlorophyll molecules in the chloroplast are organized into light-harvesting assemblies. A quantum of light absorbed by a molecule anywhere in the assembly, consisting of up to 2500 chlorophyll molecules, is conducted to a *reaction center* where the conversion of light energy to chemical energy occurs. This hypothetical assembly of chlorophyll molecules, along with the enzymes required for the reduction of carbon in photosynthesis, is the *photosynthetic unit*. The number of reaction centers in the photosynthetic unit[2] is considered by many to depend on the number of quanta needed for the evolution of O_2 and reduction of CO_2. Because quanta are totally and singly absorbed, a reaction center is considered to cycle once for each quantum absorbed somewhere in the photosynthetic unit. If the total number of quanta required for the evolution of one O_2 and reduction of one CO_2 is eight, then eight reaction centers are combined to drive the reaction sequence. Using all 2500 chlorophyll molecules required to reduce one molecule of CO_2 gives a rough estimate of 300 chlorophyll molecules for each reaction center.

Although the existence of these units is highly conjectural, some experimental evidence indicates that such units do exist. For example, an inhibitor molecule, DCMU[3-(3, 4-dichlorophenyl) 1, 1-dimethylurea], interferes maximally with the primary light reaction in photosynthesis at a concentration of about 1 DCMU to each 200 chlorophyll molecules (see Duysens and Sweers, 1963; Malkin and Kok, 1966). This suggests that the site of inhibition is at a reaction center associated with approximately 200 chlorophyll molecules. A species of chlorophyll *a*, called "P700," which absorbs maximally at 7000 Å, has been detected in chloroplasts. This chlorophyll gives some indication of an ordered arrangement when viewed by polarizing microscopy. At the small concentration noted, it would be present in approximately the right amounts to represent the reaction center.

The precise method of transfer of energy from an absorbing chlorophyll molecule to the reaction center is unknown. Two hypotheses are favored. In one, the absorbing chlorophyll molecule is considered to set up an electromagnetic field because of the vibration of the excited electron. An adjacent chlorophyll molecule, lying within the field created by the excited chlorophyll, will be induced to vibrate (resonate) at the same frequency. This phenomenon has a physical parallel in the transfer of energy from a broadcasting to a receiving antenna in radio or television. The transfer of energy from molecule to molecule in this manner, called *inductive resonance*, can be highly efficient. In the light harvesting units, the excitation is considered to "walk" by inductive resonance from one chlorophyll molecule to another until the reactive center is reached,

---

[2]Some authorities use the term *photosynthetic unit* to describe only the light absorbing unit represented by 300 chlorophyll molecules and the associated reaction center. Because the term is more commonly used to denote all the pigments and enzymes required to fix one molecule of CO_2 to a unit of carbohydrate (CH_2O), it will be used in this sense here.

where it is "trapped." Presumably, many units of energy, arising from different chlorophyll molecules of the photosynthetic unit, could reach the reaction center in a short time interval.

In the second model for energy transfer, the photosynthetic unit is considered a *semiconductor.* In a *conductor,* atoms are packed so closely that the outer electron orbitals overlap. The overlapping orbitals in a conductor are incomplete (valance orbitals), and electrons can easily move through the conductor by occupying adjacent valance orbitals in sequence. A *nonconductor* contains completed outer orbitals and cannot conduct electrons in this way. In a semiconductor, the outer orbitals are also filled, but an electron can leave an outer orbital by absorbing energy, creating a "hole" or incompletely occupied orbital. The electron can move from atom to atom as long as the "hole" moves with it. This process has been visualized by R. K. Clayton as follows. Imagine two billiard tables, suspended one above the other. The lower one is completely packed with balls, so that none can move (the semiconductor). The upper table is empty. If one ball (the excited electron) is moved to the upper table, that ball can move. Underneath it, the vacancy created by removal of the ball can also move, by shifting the balls. The assembly of chlorophyll molecules is considered to operate in this manner. Somewhere in the unit, a chlorophyll absorbs a quantum of light, and an electron is elevated to an excited orbital, leaving a "hole" in the vacated orbital. The electron and hole, together called an *exciton,* migrate together from molecule to molecule until the reaction center is reached.

Of these two models for energy transfer, inductive resonance seems most likely, because inductive resonance can occur in an unordered collection of molecules, as chlorophyll is believed to be in the chloroplast. However, *dried* chloroplasts have been found to behave as semiconductors. On illumination, dried films from chloroplasts can be polarized by an electric current, indicating that some conductance of electrons occurs. Whether this condition occurs in hydrated, living chloroplasts remains to be demonstrated.

The "trapping" of energy at the reaction center of the photosynthetic unit presents special problems. Supposedly, the reaction center is a type of chlorophyll at which an energy "sink" forms. In this chlorophyll molecule, the difference in the energy of orbitals occupied by electrons in the excited and ground state is slightly less than in the rest of the chlorophyll. Although sufficient energy is supplied from the transferring molecule to raise an electron in the trapping molecule to an excited state, the reverse flow is not possible, because energy released by the trapping molecule will not be of sufficient magnitude to excite surrounding "transfer" molecules.

As mentioned, the light absorption maximum is related to the quantum of energy required to raise an electron to an excited orbital. Chlorophyll *a* absorbs light maximally at a wavelength of 6750 Å. The special form of chlorophyll *a,* called P700, detected in small quantities in chloroplasts, absorbs light at 7000 Å. At 7000 Å, a slightly smaller amount of energy is contained in a quantum of light, indicating that a smaller amount of energy is required to raise an electron to the excited state in P700. This form of chlorophyll, therefore, has properties suitable as the energy sink required for the reaction center. There is also evidence that P700 can be alternately oxidized and reduced, allowing it to cycle with enzymes carrying out electron transport.

In summary, current hypotheses hold that light is absorbed somewhere in a unit of about 300 chlorophyll molecules, raising an electron to an excited state. The energy of the absorbing molecule is transferred to the reaction center of the unit either by the migration of excitons or by inductive resonance. After reaching the reaction center, represented by a modified form of chlorophyll *a* (P700), the "trapped" energy is converted to chemical energy in the form of electrons with potential high enough to reduce an electron acceptor such as NADP. Once in this form, the electrons can either enter an electron transport system coupled to phosphorylation of ADP to ATP or enter the series of reactions culminating in the reduction of CO_2.

The primary chemical products of photosynthesis. It was originally believed that the photochemical reactions leading to formation of ATP, reduced NADP, and O_2 were parts of a single pathway linked to the absorption of light energy by chlorophyll. In the late 1950's, however, it became obvious that the pathway was not this direct. Robert Emerson and his co-workers discovered a curious effect that led to the inference that the light reactions of photosynthesis involve *two* separate photosystems that are coordinated to produce ATP, reduced NADP, and O_2. In 1943, Emerson had noted that at long wavelengths the rate of photosynthesis dropped off. This was surprising, because these wavelengths were still within the red region strongly absorbed by chlorophyll. More than ten years later, Emerson found that the reduced rate at long wavelengths was increased if supplemental light was added at a somewhat shorter red wavelength. Most significant was the finding that the increased efficiency was retained if the two wavelengths of light were flashed *alternately*, at intervals of time up to several seconds. This showed that the two systems could act separately and that the products linking the two systems were almost certainly chemical in nature. The energy from an excited state would be expected to return to the ground state, with fluorescence and heat, within fractions of a second unless the energy was converted into a chemical form, such as the attachment of electrons to a carrier.

The two systems, called simply system I and system II, both absorb light in the red region typical of chlorophyll. In green algae and higher plants, the slightly shorter wavelengths absorbed by system II indicate the presence of chlorophyll *b* in combination with chlorophyll *a*. In brown, red, and blue-green algae, the phycobilins are substituted for chlorophyll *b* in system II. Accessory pigments, such as the carotenes, transfer light energy to either system.

Publication of Emerson's results caused a burst of intensive research in the late 1950's and early 1960's, leading to the discovery that the chemical system linking system I and system II includes cytochromes similar to those found in mitochondria. Most significantly, these cytochromes were found to be reduced by system II and reoxidized by system I. This established that electrons flow from system II to system I and that the two systems are linked by an electron transport system. Separating the two systems by the use of selective inhibitors revealed that the splitting of water, with O_2 evolution, occurred in system II and that system I forms a strong reductant that forms reduced NADP. Logically, the numbers given to the systems should be reversed, because the reactions taking place in system II precede those in system I, but the systems were named before the sequence of events was determined, and the original designation has been retained.

Electron transport in chloroplasts. It has not yet been possible to work out the identity and sequence of the carriers in electron transport in chloroplasts or indeed to determine whether there are one or two transport systems associated with the two photoactive systems. Three cytochromes have been identified in chloroplasts, including cytochromes *b* and b_6, closely similar to the cytochrome *b* of mitochondria, and cytochrome *f*, a *c*-type cytochrome unique to chloroplasts. Another carrier found in chloroplasts is *ferredoxin*, a carrier with a redox potential high enough to reduce NADP. Other possible electron carriers have been identified, including a flavoprotein and a highly varied group of quinones, including vitamin K, coenzyme Q, and at least seven plastoquinones. As in mitochondria, the position of the quinones in electron transport is not certain. Cytochrome oxidase, the enzyme catalyzing the transfer of electrons from cytochrome *a* to oxygen in mitochondria, is not found in chloroplasts.

This information has been used to construct a general scheme of electron transport in photosynthesis. The oxidation of water in system II, involving removal of electrons and hydrogen, and the reduction of NADP in system I, involving addition of electrons and hydrogen, suggests that electrons and hydrogens flow from water in system II to NADP in system I. The energy required to raise these electrons to a potential high enough to reduce NADP is produced by the combined systems, linked through the carriers. Fig. 7–19 outlines this hypothesis. The two heavy arrows represent two photosynthetic units, each with an active center consisting of a chlorophyll molecule absorbing at slightly longer wavelengths. In system I, this specialized chlorophyll, as noted, is probably P700.

In the initial reaction in system II, electrons removed from water are raised from a potential of +800 mv to 0 mv by transfer of energy from an active center to these electrons. The exact manner in which this transfer occurs is unknown. No one has been able to discover how water is oxidized into protons, electrons, and molecular oxygen. Undoubtedly, a series of enzymatically catalyzed reactions occurs. The electrons removed are either accepted by an oxidized form of chlorophyll, which is subsequently raised to an excited state (designated chlorophyll* in Fig. 7–19) by absorbed light energy, or energy is transferred from chlorophyll* to the electrons removed from water (carried one at a time), which are thereby raised from +800 mv to the relatively negative value of 0 mv. The electrons are then transferred to a sequence of electron carriers. The most likely sequence of carriers is shown in the figure. At some points in this sequence, at least two molecules of ADP are phosphorylated to ATP (see below). One likely site of phosphorylation is between cytochromes *b* and *f*.

The electrons lose energy as they travel along the chain of carriers from chlorophyll* to cytochrome *f*. Some of this energy, as noted, is captured in the form of ATP. After traveling through the chain, the electrons, at approximately +400 mv, are accepted by chlorophyll P700 of system I, which, by trapping absorbed light energy, becomes raised to an excited state (P700*). Electrons at this point enter two possible pathways.

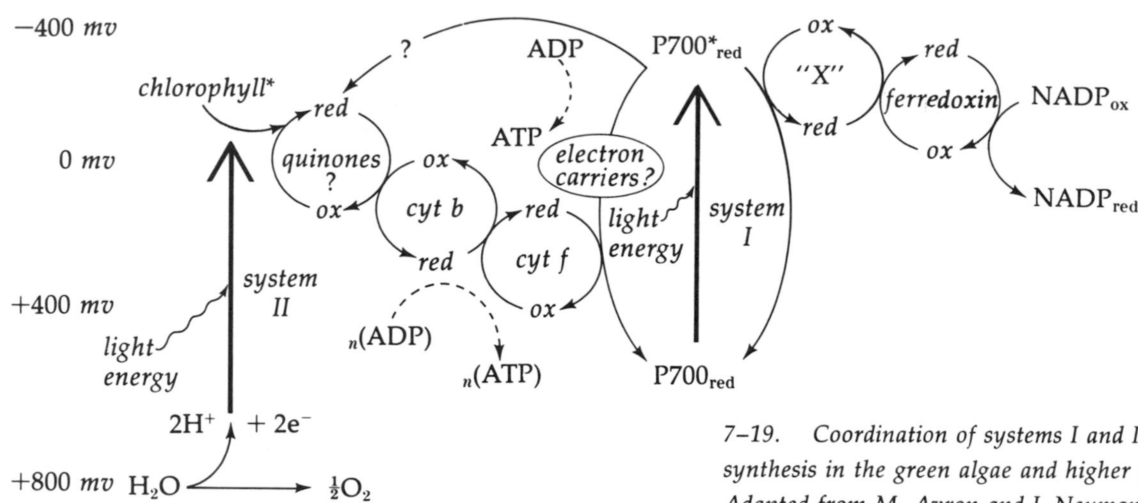

7–19. Coordination of systems I and II in photosynthesis in the green algae and higher plants (see text). Adapted from M. Avron and J. Neumann, Ann. Rev. Plant. Physiol. *19:137(1968), by permission of the authors.*

One chain of carriers goes from an unknown first carrier, designated "X" in the figure, to NADP by several steps that include the carrier ferredoxin. This pathway finally produces reduced NADP, a raw material for the "dark" reactions of photosynthesis that fix CO_2. An alternate pathway is through a sequence of carriers, with eventual return to the oxidized P700 of system I. This series may include either the same set of carriers linking systems II and I, or it may involve a completely different series. In either case, only ATP synthesis occurs in this second pathway.

Note that, with respect to the flow of electrons, system I can operate "cyclically" or "noncyclically." In cyclic flow, electrons travel a route from P700, through a series of carriers, and back to P700 again. The net result of cyclic flow is the conversion of light energy to chemical energy in the form of ATP. In noncyclic flow, electrons removed from water travel through system II and electron transport to produce ATP and then through system I to produce reduced NADP. Thus noncyclic photosynthesis produces ATP, reduced NADP, and molecular oxygen. Hydrogen ions (protons) are also carried by some members of both sequences, either in pairs or singly.

The enhancement of the efficiency of photosynthesis by mixing light of two wavelengths, on which the demonstration of two photoactive systems depends, is not observed in bacteria. It is possible that the development of two coordinated light absorbing systems, observed in eukaryotes, represents an evolutionary advance of the algae and higher plants over a primitive, single system in bacteria. Whether the proposed cyclic flow of electrons, involving only ATP synthesis, actually occurs in system I of higher plants is uncertain, because it has been clearly demonstrated only in bacteria.

Photophosphorylation. Both isolated chloroplasts and extracts of photosynthetic bacteria are able to drive the phosphorylation of ADP to ATP when illuminated. According to the hypothetical scheme presented in Fig. 7–19, ADP is phosphorylated to ATP in both cyclic and noncyclic electron flow in systems I and II. As indicated in the discussion of this proposed mechanism, the sites and the number of phosphorylations that occur in photosynthesis are controversial. Estimates vary from one to three per pair of electrons traversing the electron transport chains. ATP synthesis does not form a control mechanism in chloroplasts as in mitochondria; evidently electron transport can take place whether ATP is synthesized or not. Some indication that cyclic and noncyclic phosphorylation may follow different transport chains has been suggested by the fact that agents uncoupling phosphorylation from electron transport have differing effects on the two pathways. It is generally agreed that no sites of phosphorylation occur in the pathway from P700* to NADP in system I.

Some of the strongest evidence for the Mitchell chemiosmotic hypothesis for phosphorylation comes from chloroplasts. It might be useful at this point to reread the section in Chapter 6 describing the Mitchell hypothesis. According to this hypothesis, the coupling of phosphorylation to electron transport is driven by the splitting of water and the accumulation of H^+ and

OH^- ions on different sides of a membrane. If H^+ ions are actually expelled outside the chloroplast as this hypothesis proposes, electron transport should be accompanied by a change in *p*H around photosynthesizing chloroplasts. On illumination, unbuffered suspensions of chloroplasts do show the expected *p*H change. Further, Andre T. Jagendorf was able to show that the reverse situation was true: an artificial change in *p*H of the suspending medium induced ATP synthesis in isolated chloroplasts; the amount of synthesis was directly related to the difference in magnitude between the initial and final *p*H. These findings confirm a relationship between hydrogen ion concentration and phosphorylation as predicted by the Mitchell hypothesis.

The Fixation of Carbon Dioxide in Photosynthesis: The "Dark" Reactions

The reactions outlined in the foregoing sections, from the initial absorption of light to the production of ATP and reduced NADP, act to provide the raw materials for the production of carbohydrate and other metabolic intermediates in the "dark" reactions of the chloroplast. These products, ATP and reduced NADP, accumulate after illumination of the chloroplast; unless used within a short time by the dark reactions with the regeneration of $NADP_{ox}$ and ADP, the efficiency of photosynthesis in using light energy falls. If a suitable electron acceptor is introduced (the Hill reaction) in a suspension of chloroplasts photosynthesizing at low efficiency because of lagging dark reactions, efficiency is returned to high levels, with increased ATP synthesis and NADP reduction. This demonstrates that the bottleneck imposed by lagging dark reactions is the absence of an acceptor for the electrons carried by reduced NADP. Once oxidized by a suitable acceptor, NADP is free to recycle in the light reaction, and the rate of photosynthesis is increased. Obviously, the artificial acceptor introduced in the Hill reaction takes the place of CO_2 or an immediate product of a carboxylation reaction involving absorbed CO_2.

The dark reactions became the subject of intensive research in the 1940's. A breakthrough came in this work with the introduction of compounds containing radioactive isotopes of carbon. Using a radioactive form of CO_2, Melvin Calvin and his colleagues were able to trace out the first chemical products of photosynthesis. After allowing photosynthesis to run for only a few seconds, extracts were made of photosynthesizing chloroplasts. The bulk of the radioactivity was found to be concentrated in a 3-carbon sugar, 3-phosphoglycerate (also an intermediate in glycolysis: see p. 108). Lowering the CO_2 concentration to rate-limiting levels resulted in the accumulation of a nonradioactive 5-carbon sugar, ribulose 1, 5-diphosphate. This suggested to Calvin that this substance is the first acceptor of absorbed CO_2, forming as products two molecules of the 3-carbon sugar 3-phosphoglycerate. Other intermediates were subsequently identified in chloroplast extracts. In the 1950's, Calvin and his colleagues James A. Bassham and Andrew A. Benson constructed a model for the synthesis of carbohydrate in the dark reaction of photosynthesis. This model, for which Calvin was awarded the Nobel Prize in 1961, is known as the Calvin or pentose cycle.

In this cycle of reactions (Fig. 7–20) CO_2 is incorporated by combination with ribulose 1, 5-diphosphate (reaction 1), a carboxylation catalyzed by the enzyme carboxydismutase. This reaction requires no input of energy (ribulose 1, 5-diphosphate can be regarded as a "high-energy" compound). Reaction 1 produces two molecules of 3-phosphoglycerate. One of the two molecules contains the newly incorporated CO_2 in the 1-position (marked by an asterisk in Fig. 7–20). The reduction of this triose takes place after reaction 2, in which 3-phosphoglycerate is phosphorylated, at the expense of ATP, to form the more reactive phosphoryl 1, 3-phosphoglyceric acid. This activated molecule accepts hydrogens and electrons from NADP in the next reaction (3), resulting in a reduction to 3-phosphoglyceraldehyde. For each molecule of CO_2 attached to ribulose diphosphate, two molecules of phosphoglyceraldehyde are ultimately produced. Thus reactions 2 and 3 in Fig. 7–20 are represented twice, requiring 2ATP and $2NADP_{red}$.

The phosphoglyceraldehyde produced in step 3 can either enter a complex series of reactions (series 4 in Fig. 7–20) that regenerates the ribulose needed for the next turn of the cycle or can enter into the synthesis of glucose or other complex substances. The phosphoglyceraldehyde is supposed to be used in the following manner. In three turns of the Calvin cycle through reaction 3, six molecules of 3-phosphoglyceraldehyde are formed (a total of 18 carbons). Five of these (15 carbons) enter the complex series regenerating three molecules of ribulose phosphate. The remaining 3-phosphoglyceraldehyde is surplus to the cycle and can enter synthetic pathways forming glucose or other metabolites. In a diagrammatic sense the Calvin cycle can be regarded as forming one *subunit* of carbohydrate (CH_2O) for each turn of the cycle. Three subunits, or

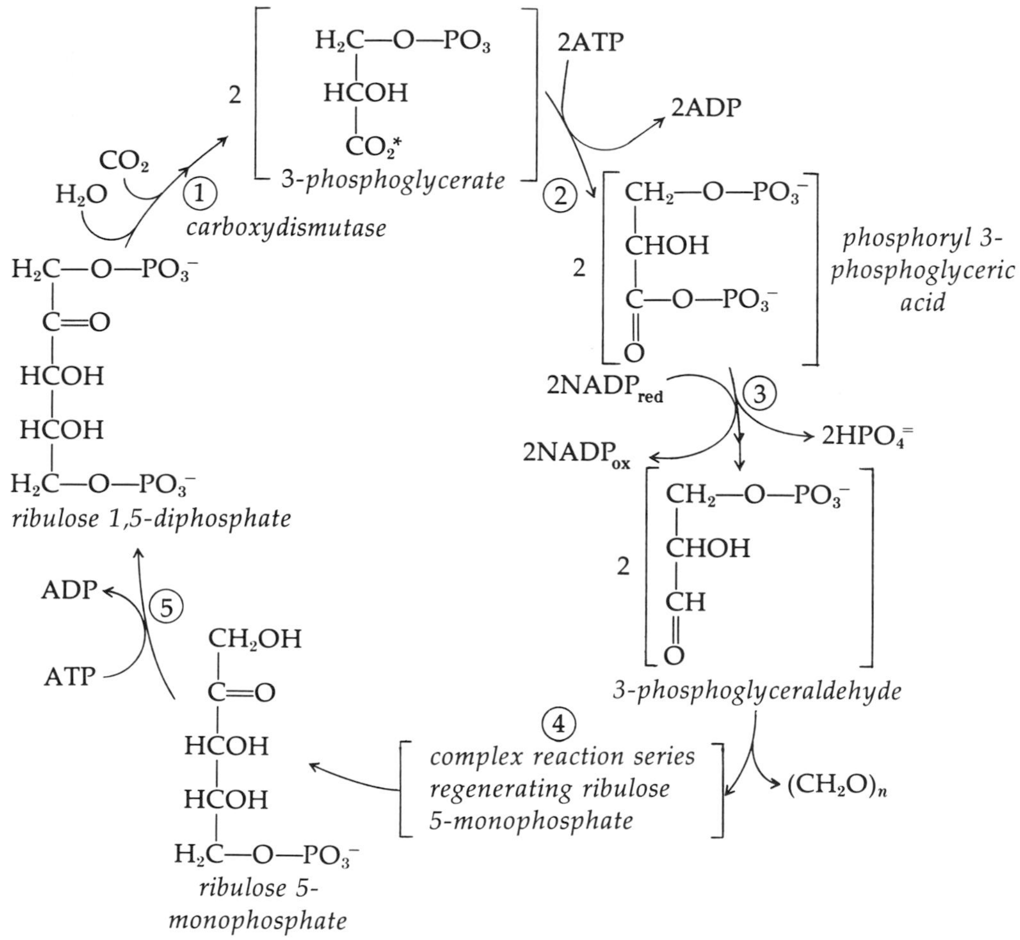

2 [$H_2C—O—PO_3$ / HCOH / CO_2^*] 3-*phosphoglycerate*
CO_2
H_2O
① *carboxydismutase*
2ATP
2ADP
② 2 [$CH_2—O—PO_3^-$ / CHOH / $C—O—PO_3^-$ / O] *phosphoryl 3-phosphoglyceric acid*
$2NADP_{red}$
③
$2HPO_4^=$
$2NADP_{ox}$
2 [$CH_2—O—PO_3^-$ / CHOH / CH / O] 3-*phosphoglyceraldehyde*
④ *complex reaction series regenerating ribulose 5-monophosphate*
$(CH_2O)_n$
$H_2C—O—PO_3^-$ / C=O / HCOH / HCOH / $H_2C—O—PO_3^-$ *ribulose 1,5-diphosphate*
ADP
⑤
ATP
CH_2OH / C=O / HCOH / HCOH / $H_2C—O—PO_3^-$ *ribulose 5-monophosphate*

7–20. *The Calvin cycle (see text).*

three turns, are required to synthesize one molecule of 3-phosphoglyceraldehyde as a surplus to the cycle.

One additional ATP enters the cycle in reaction 5, during the synthesis of ribulose 1, 5-diphosphate, which is formed from the immediate product of reaction series 4, ribulose monophosphate. For each CO_2 fixed, the entire cycle thus requires a total of 3ATP and $2NADP_{red}$. The ribulose 1, 5-diphosphate regenerated by reaction series 4 and 5 combines with another CO_2 at the next step, and the cycle repeats.

Surplus phosphoglyceraldehyde formed in the cycle probably enters into the synthesis of glucose by a reversal of the glycolytic pathway (see p. 107, Chapter 6). If photosynthesis is allowed to run for a slightly longer time before the extraction of products, intermediates of the glycolytic pathway between phosphoglyceraldehyde and glucose show radioactivity. This confirms the likelihood of this pathway for the synthesis of glucose.

The Energetics and Efficiency of Photosynthesis

A rough estimate of the overall energetics of photosynthesis can be made by returning to the most simple expression for photosynthesis given in equation 7–1. In this case an organic compound, $C_6H_{12}O_6$, is synthesized from the inorganic oxides CO_2 and H_2O. This reaction will proceed spontaneously in the reverse direction, from $C_6H_{12}O_6$ to the oxides, with a free energy release of 686,000 cal/mole. To drive it in the reverse direction, as indicated by equation 7–1, requires an input of this much energy. This is supplied in photosynthesis by the absorbed energy of light. In this form, equation 7–1 can be written as:

$$6CO_2 + CH_2O \rightarrow C_6H_{12}O_6 + 6O_2 \qquad \Delta G^\circ = +686 \text{ kcal/mole of glucose} \quad (7\text{–}6)$$

Thus the synthesis of one mole of glucose requires an input of 686,000 calories.

An indication of the efficiency of the photosynthetic mechanisms in synthesis of carbohydrate can be obtained from the total ATP and $NADP_{red}$ entering the Calvin cycle. For each molecule of glucose formed, the cycle turns 6 times, requiring a total of $6 \times 2NADP_{red} + 6 \times 3ATP = 12NADP_{red} + 18ATP$. Using a value of 7,000 cal/mole for the free energy change when ATP is hydrolyzed to ADP + $HPO_4^{=}$, and 62,600 cal/mole for the oxidation of $NADP_{red}$, use of $12NADP_{red}$ and 18ATP represents a free energy increment of 760,000 cal/mole. An input of 760,000 calories for the synthesis of a mole of glucose represents an overall efficiency of $686/760 \times 100 = 90\%$. This figure varies depending on the value used for the hydrolysis of ATP, which is somewhat controversial and depends on the concentrations assumed to exist at the site of reaction. In any event, even the most conservative estimates for these interactions place the dark reactions of photosynthesis among the most efficient systems known.

The efficiency of use of light energy in the overall process of photosynthesis has been debated for many years. The difficulties that arise here are connected with the possible number of quanta required to drive the synthesis to 18ATP and the reduction of $12NADP_{ox}$ in the light reactions. Lengthy and not too productive arguments have been held on this topic since the 1930's. If the reduction of NADP proceeds as outlined in this chapter by the coordination of systems I and II, four quanta are required for each NADP reduced. Two electrons pass from water and are excited by system II, requiring two quanta of red light. After flowing through the electron transport system linking system II and system I, two additional quanta are absorbed in the excitation of this pair for the reduction of NADP (a two-electron reduction). Because reduction of six molecules of CO_2 (or six turns of the Calvin cycle) requires 12 molecules of $NADP_{red}$, the total input of light energy, if only electron flow is considered, will be $12 \times 4 = 48$ quanta for each molecule of glucose synthesized. At 43.4 kcal/Einstein per quanta of red light absorbed, the total energy input is about 2100 kcal for each mole of glucose, giving an overall efficiency of $686/2100 \times 100 = 33\%$. While this figure is not as impressive as the 80 percent obtained for the dark reactions, it is still very high compared to the efficiency of man's machines.

A major uncertainty in this speculative assessment of the quantum efficiency of photosynthesis is the amount of ATP synthesized during electron transport in systems I and II. Estimates for the ATP synthesized by the transport system linking systems II and I vary from 1 to 3 molecules for each pair of electrons traversing the carriers. If it is assumed that two molecules of ATP are synthesized, then ATP and $NADP_{red}$ will be synthesized by the photosystems in the ratio of $2ATP{:}1NADP_{red}$, which is more than sufficient for the requirements of the Calvin cycle at $3ATP{:}2NADP_{red}$. Thus, at the level $2ATP{:}1NADP_{red}$, the overall efficiency of photosynthesis will be slightly increased, reflecting the energy conserved in ATP surplus to the needs of the Calvin cycle. If only one ATP is synthesized for each NADP reduced, additional light quanta will be required to drive the synthesis of ATP by cyclic photosynthesis of system I. This would reduce the overall efficiency slightly.

The latter statement gives a key to the flexibility inherent in the hypothetical organization of systems I and II. The synthesis of ATP in cyclic electron flow is independent of NADP reduction; this cycle can supply ATP required for the synthesis of any metabolite in the chloroplast, even, as some have suggested, to drive the reduction of $NADP_{ox}$ in a dark reaction if required.

An Important Alternative Pathway for CO_2 Fixation

Recent discoveries by M. D. Hatch and C. R. Slack have revealed that in some plants, an important alternative exists to CO_2 fixation by the Calvin cycle. This alternate pathway, observed in corn, sugar-cane, and other tropical grasses, was discovered when Hatch and Slack looked for the earliest radioactive intermediates in these grasses after allowing photosynthesis to proceed for brief periods in radioactive CO_2. Surprisingly, the earliest label appeared in an equilibrium mixture of oxaloacetic, malic, and aspartic acid instead of phosphoglyceric acid as in the Calvin experiments. Phosphoglyceric acid was also labeled in these plants, but only in very small quantities if extracts were made in 2 seconds after the onset of photosynthesis. Hatch and Slack proposed the following scheme to account for the early appearance of the label in these 4-carbon acids (Fig. 7–21). The Calvin cycle is considered to operate in the usual manner, except that ribulose 1, 5-diphosphate does not combine directly with absorbed CO_2. Instead, the absorbed CO_2 combines with an activated form of pyruvic acid, phosphopyruvic acid, to produce either oxaloacetic or malic acid. Aspartic acid, an amino acid, is considered to result from a side reaction in-

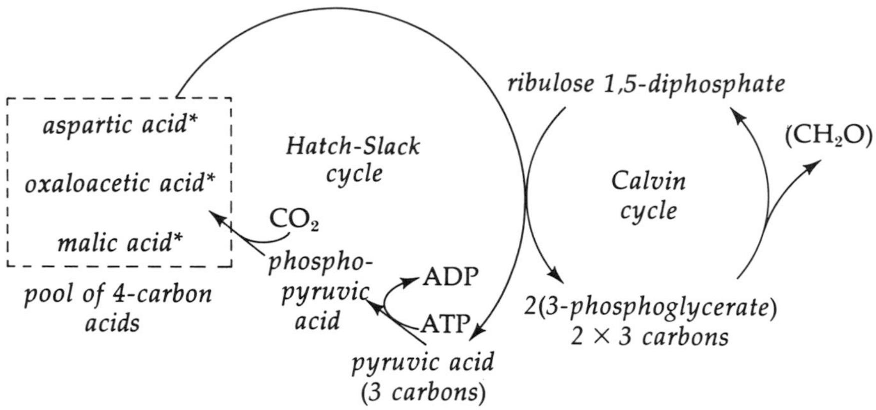

7–21. The Hatch-Slack cycle, an alternative pathway to the Calvin cycle, found in some plants, including the grasses (see text).

volving oxaloacetic acid. One of the members of this pool of 4-carbon acids, probably oxaloacetate, interacts with ribulose 1, 5-diphosphate. The products of this reaction are two molecules of 3-phosphoglycerate and pyruvic acid. The pyruvic acid regenerated at this step, after phosphorylation, interacts with another CO_2 and the Hatch-Slack cycle repeats. The essential difference between this scheme and the Calvin cycle is that CO_2 fixation occurs in an accessory cycle rather than directly in the Calvin cycle.

Other Synthetic Mechanisms in the Chloroplast

The appearance of a labeled amino acid, aspartic acid, as an early product of photosynthesis is most interesting, because it demonstrates that the chloroplast is capable of turning out precursors of proteins as well as carbohydrates. Other amino acids, such as alanine, serine, glutamic acid, and glycine, also become labeled very rapidly in illuminated chloroplasts exposed to radioactive CO_2. In fact, the label appears in these amino acids before it appears in fully synthesized carbohydrates, such as sucrose, suggesting that amino acids are probably synthesized from phosphorylated intermediates in the Calvin cycle (see Stepka et al., 1948). The exact pathways of amino acid synthesis in the chloroplast are only partly understood. These reactions will not be described in detail, except that one sequence will be given as an example (equation 7–7). In this reaction, α-ketoglutaric acid is reduced and aminated to form glutamic acid, an amino acid of primary importance as an intermediate in the synthesis of other amino acids. Presumably, the $NADP_{red}$ entering this sequence is a product of the photosynthetic light reactions.

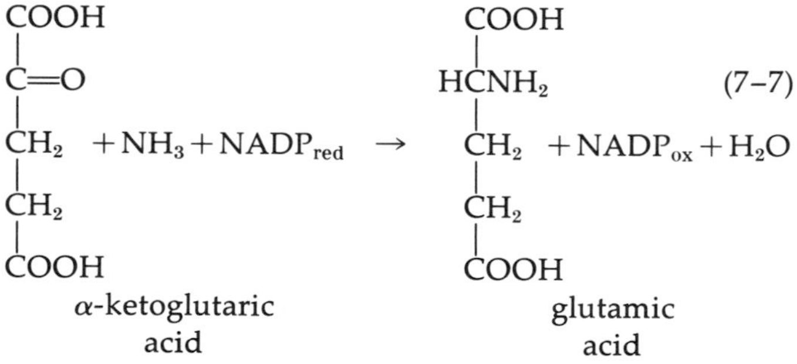

$$\begin{array}{c} COOH \\ | \\ C{=}O \\ | \\ CH_2 \\ | \\ CH_2 \\ | \\ COOH \\ \alpha\text{-ketoglutaric acid} \end{array} + NH_3 + NADP_{red} \rightarrow \begin{array}{c} COOH \\ | \\ HCNH_2 \\ | \\ CH_2 \\ | \\ CH_2 \\ | \\ COOH \\ \text{glutamic acid} \end{array} + NADP_{ox} + H_2O \qquad (7\text{–}7)$$

Glutamic acid can be converted into other amino acids by transaminations, often with the regeneration of α-ketoglutaric acid (equation 7–8):

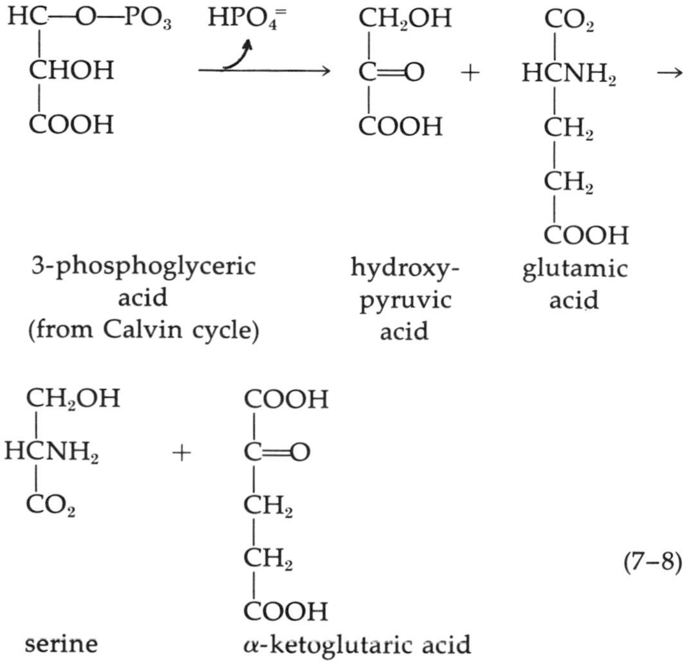

$$\begin{array}{c} HC{-}O{-}PO_3 \\ | \\ CHOH \\ | \\ COOH \\ \text{3-phosphoglyceric acid} \\ \text{(from Calvin cycle)} \end{array} \xrightarrow{HPO_4^{=}} \begin{array}{c} CH_2OH \\ | \\ C{=}O \\ | \\ COOH \\ \text{hydroxy-pyruvic acid} \end{array} + \begin{array}{c} CO_2 \\ | \\ HCNH_2 \\ | \\ CH_2 \\ | \\ CH_2 \\ | \\ COOH \\ \text{glutamic acid} \end{array} \rightarrow$$

$$\begin{array}{c} CH_2OH \\ | \\ HCNH_2 \\ | \\ CO_2 \\ \text{serine} \end{array} + \begin{array}{c} COOH \\ | \\ C{=}O \\ | \\ CH_2 \\ | \\ CH_2 \\ | \\ COOH \\ \alpha\text{-ketoglutaric acid} \end{array} \qquad (7\text{–}8)$$

In most plants, all of the amino acids required for protein synthesis can be synthesized by reactions of this type or by other pathways. While most proteins are synthesized on ribosomes outside the chloroplast, protein synthesis is also known to occur on ribosomes located within the chloroplast (for details, see Chapter 16).

After a few minutes of photosynthesis in labeled CO_2, lipids within the chloroplast become labeled. In some cases, synthesis of labeled lipids, including fatty acids, fats, and galactolipids, may represent as much as 30 percent of the total labeled material after one to two minutes of photosynthesis. Undoubtedly, the synthesis proceeds from a cycle involving linking of acetate units carried by acetyl-coenzyme A (see p. 110). The acetate units may be formed from pyruvic acid, readily derived from intermediates of the Calvin cycle, or from other known pathways. Once formed, fatty acids are combined with glycerol to form fats. Glycerol, in turn, can be synthesized directly from the phosphorylated triose sugars of the Calvin cycle.

The balance among photosynthesis of carbohydrates, fats, and amino acids varies widely among different species of plants. In some plants, nearly all of the CO_2 absorbed is incorporated into carbohydrates such as sucrose, formed from phosphorylated glucose and fructose. In an alga, *Chlorella*, the synthesis rate of fats and amino acids greatly exceeds sucrose synthesis, which may account for 5 percent or less of the CO_2 absorbed. In any event, the chloroplast contains the enzymes and intermediates necessary to synthesize an astounding variety of substances in addition to carbohydrates.

Localization of Photosynthesis in Chloroplast Fine Structure

Analysis of isolated chloroplast subfractions has allowed provisional localization of the photosynthetic reactions in chloroplast structure. The stroma, easily separated from the membranes of the chloroplast, contains the enzymes needed for the dark reactions that fix CO_2 and all of the remaining machinery for synthesis of carbohydrates, fats, and amino acids. These enzymes are in solution or are only loosely bound in aggregates.

Lamellar preparations, originating from grana and frets, contain the photosynthetic units associated with systems I and II, reaction centers (including chlorophyll P700), accessory pigments, and the enzymes and cofactors for electron transport, ATP synthesis, and reduction of NADP. Almost all molecules included in these reaction sequences are tightly bound to the membranes of the grana. Removing membrane lipid arrests absorption and transfer of light energy, phosphorylation of ATP, and NADP reduction, even though few molecules classified as lipids enter directly into these reactions. This confirms that the integrity of membrane structure is necessary to maintain these functions and suggests that the molecules carrying out these reactions are bound to the membranes in highly ordered arrays. The pattern modifies the properties of the various membrane-bound enzymes and cofactors and facilitates the light reactions in terms of both the concentration of reactants and the energy required. Because membrane structure and molecular order have a significant effect on the biochemistry of the chloroplast, it is not possible to duplicate the light reaction systems by merely studying the interaction of the various molecules in solution.

The smaller globular subunits of granal membranes visible in thin sections, with dimensions ranging around 70 Å, probably represent a lipoprotein "building block" of the membrane. Chlorophyll and the various enzymes of the light reactions may form part of the structure of this subunit. Alternatively, the reactive molecules may be localized in the slightly larger particles (the quantasomes) visible in freeze-etch and isolated preparations. Quantasomes are about the right size to accommodate a single light-absorbing unit, consisting of chlorophyll and a reaction center. Quantasomes are tightly bound to or even embedded within the membrane, so the end result, even if the reactive molecules are localized in the quantasome as distinct from the membrane subunits, is close binding between membrane lipoprotein and reactive molecules.

Some comparative aspects. Photosynthesis in bacteria and blue-green algae has been included in this chapter in piecemeal fashion. A brief review of this material will be included at this point, and some new material will be added in order to clarify a few points. About 30–60 molecules of bacteriochlorophyll make up the light-absorbing unit of bacteria, forming a much smaller particle than in the chloroplast. The reaction center of these units is represented by a species of bacteriochlorophyll absorbing maximally at 8900 Å, termed "P890." This is analogous to the P700 of higher plants. Carotenoids are also present in bacteria and blue-green algae as accessory pigments. These systems are bound to membranes in the bacteria as in higher organisms.

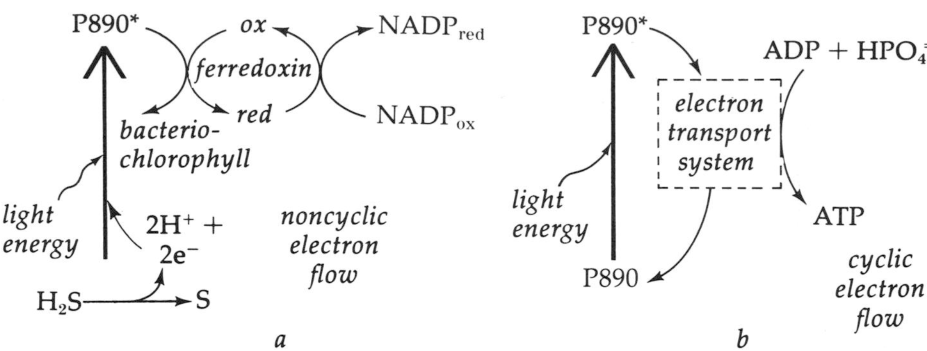

7–22. *A possible mechanism for photosynthesis in bacterial systems. Electrons raised to the excited state by the absorption of light energy can either be accepted by NADP, as in* a, *or may pass through an electron transport system, with phosphorylation of ADP to ATP, as in* b.

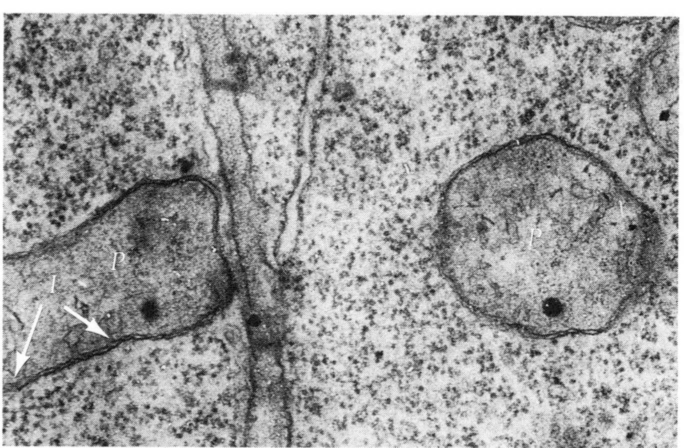

7–23. *Proplastids* (P) *in meristem tissue from a leaf of* Nicotinia tabacum. *A single osmiophilic granule is present in both proplastids; at the positions marked by arrows* (I) *faintly visible interior membranes can be seen to arise from the inner boundary membrane.* × *35,000. Courtesy of D. Von Wettstein.*

Bacterial photosynthesis seems to resemble only system I of higher plants; system II does not appear to be present. Related to this is the inability of bacteria to use water as an electron donor. Only reductants of less positive potential, in terms of electron volts, are usable: hydrogen, H_2S, and organic compounds. The evolution of system II in higher plants may thus represent an adaptation to the use of water as a source of electrons and hydrogen. The bacterial photosystem, as shown in Fig. 7–22, may operate to produce either reduced NADP or to phosphorylate ADP to ATP. Of the electron carriers present, only one, *cytochromoid c*, is not found in higher plants. The carriers include coenzyme Q, flavoproteins, and ferredoxin.

Although morphologically similar to bacteria, blue-green algae carry out photosynthesis that more closely resembles higher plants, in that chlorophyll *a* is present and water is used as a source of electrons and hydrogen. This suggests that both systems I and II are present in these organisms.

The Development of Chloroplasts in Higher Plants

All types of plastids, including the chloroplast, leucoplast, and etioplast, develop from *proplastids*, small bodies 1–1.5 microns in length and about 0.5–1 micron in diameter (Fig. 7–23). These are bounded by a double membrane and have a relatively undifferentiated matrix. A few tubular invaginations of the inner boundary membrane may extend into the matrix; DNA fibrils, starch grains, and osmiophilic granules may also be present. Occasionally, a limited number of thylakoids may be scattered in the matrix. About 10–20 proplastids occur in meristematic cells of the growing stem; somewhat larger numbers, up to 40, are found in root meristem. In stems and leaves grown in the light, these bodies are rapidly transformed into chloroplasts.

Transformation of proplastids to chloroplasts occurs first by tube-like invaginations of the inner boundary membrane that gradually extend into the stroma. These

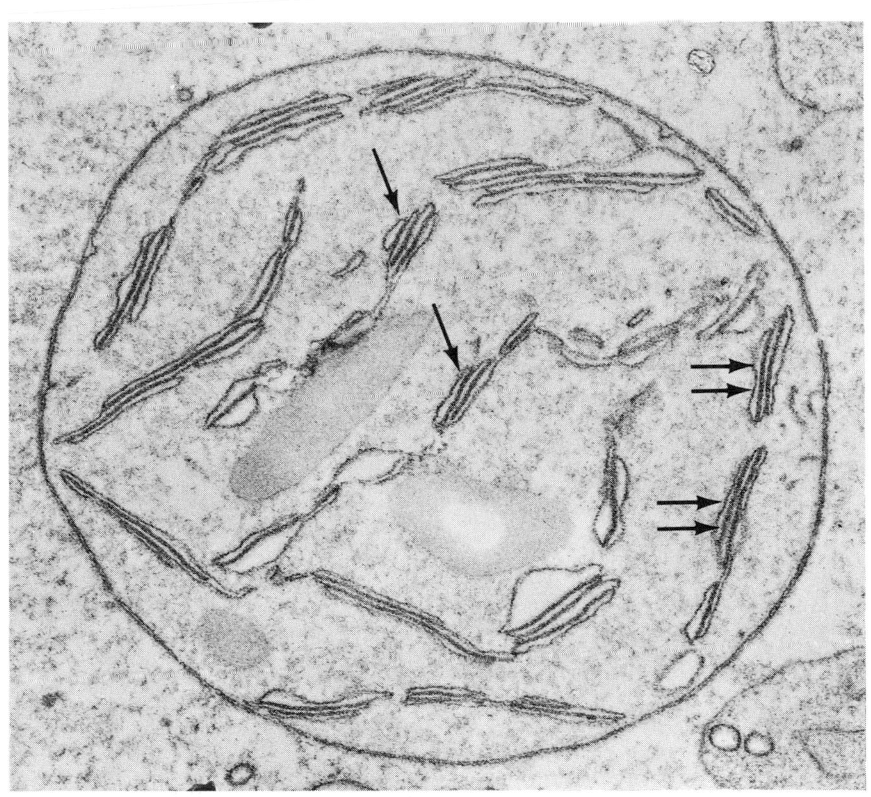

7–24. An early stage in the development of a chloroplast in the leaf meristem of Oenothera hookeri. *At the positions marked by single arrows, initial folding of the membranes in the region of a future grana stack has occurred. Folding has progressed to a slightly more advanced level at the points marked by double arrows. × 46,000. Courtesy of W. Menke.*

then spread into flat sheets consisting of two membranes enclosing an internal space. At points, the flat, thylakoid-like sheets become doubly and triply folded into multiple layers (Fig. 7–24); this process repeats until the grana are formed. During proliferation of the inner membranes of the developing chloroplast, synthesis of chlorophyll and accessory pigments takes place at the same time.

If plants are grown in the dark, proplastids in leaves and stems are transformed into *etioplasts* instead of chloroplasts. These bodies, although larger and more complex internally than proplastids, do not contain typical grana and are a yellow color. On transfer of the plant to light, etioplasts are converted very rapidly into typical chloroplasts. Etioplasts have a characteristic internal structure (Fig. 7–25). The matrix, bounded by double membranes, contains one or more highly regular arrays of membranous tubules. These arrays, called *prolamellar bodies,* consist of a regular lattice of interconnected tubules. The tubule walls are formed from a single membrane; at the narrowest point the outside diameter of the tubule is about 200 Å. B. E. S. Gunning, in a careful analysis of the prolamellar body, demonstrated that the unit structure of the lattice is a cube. Each side of the unit cube is formed by a membranous tubule; at the corners of the unit the tubules are fused together (Fig. 7–26). A single ribosome-like particle is found in the center of the space enclosed by each unit cube. At points within the prolamellar body, a number of irregularly shaped osmiophilic granules may fill the spaces between the tubules; at these points the regularity of the tubules is somewhat disturbed.

At the edges of the prolamellar body, the membranes of the tubules are continuous with thylakoid-like membranes that extend into the matrix of the etioplast. If several prolamellar bodies are present, these may be interconnected by the membranous extensions. Also present in the matrix between the prolamellar bodies are ribosomes and deposits of DNA.

Under the light microscope, etioplasts appear yellow. The prolamellar bodies give off a strong red fluorescence associated with a precursor of chlorophyll, a pigment known as *protochlorophyll.* Protochlorophyll is similar to chlorophyll *a* in structure, except for the absence of two hydrogens in the porphyrin ring (see Fig. 7–12). When etiolated plants are placed in light, the leaves rapidly turn green. This change in color is accompanied by the transformation of etioplasts into

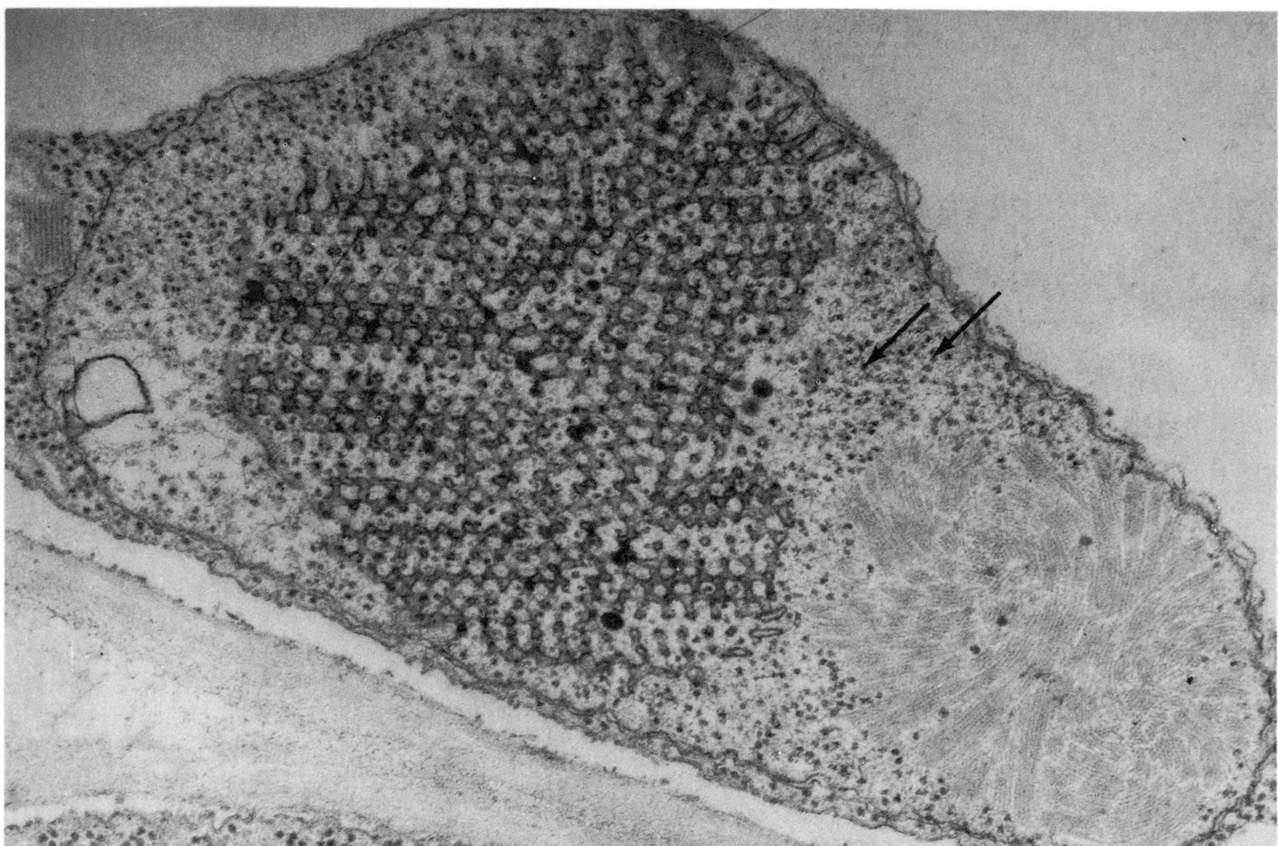

7–25. An etioplast from a leaf of Avena. *A highly ordered* prolamellar body *is present in the matrix; also visible are numerous ribosomes (arrows).* × *51,500. Courtesy of B. E. S. Gunning, from* Protoplasma *60:111(1965).*

chloroplasts. As chlorophyll *a* is formed, protochlorophyll disappears; the wavelengths of light effective in inducing the transformation of etioplasts are those absorbed by protochlorophyll. The conversion of protochlorophyll to chlorophyll *a* may be completed within one minute after exposure to light. Appearance of chlorophyll *b* is much slower; this pigment is not detected until several hours after transfer of plants to lighted conditions.

As the conversion of protochlorophyll to chlorophyll *a* takes place, the regular structure of the prolamellar body becomes disturbed, and the component tubules gradually undergo transformation to thylakoids. The details of this process have been studied by Dieter von Wettstein and his co-workers. If the intensity of light is high, within a few minutes after exposure the tubules extrude into flat sheets which extend into the stroma of the developing chloroplast (Fig. 7–27). The sheets consist of two membranes enclosing an internal space and are thus equivalent in structure to thylakoids. At scattered points, the thylakoids become doubly thick (Fig. 7–28). This process repeats at these points until grana consisting of a large number of stacked thylakoids are formed. With completion of this process, the etioplast is transformed into a typical chloroplast.

It is important to note that formation of the etioplast occurs only in plants grown in the dark. In plants grown in light, proplastids are converted directly into chloro-

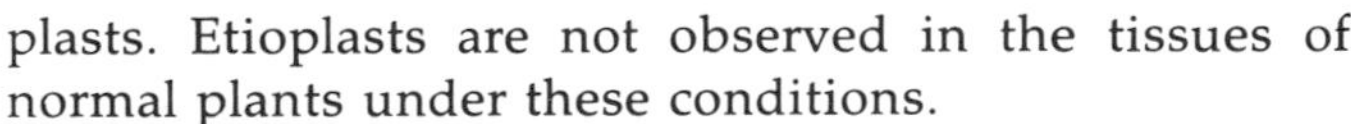

plasts. Etioplasts are not observed in the tissues of normal plants under these conditions.

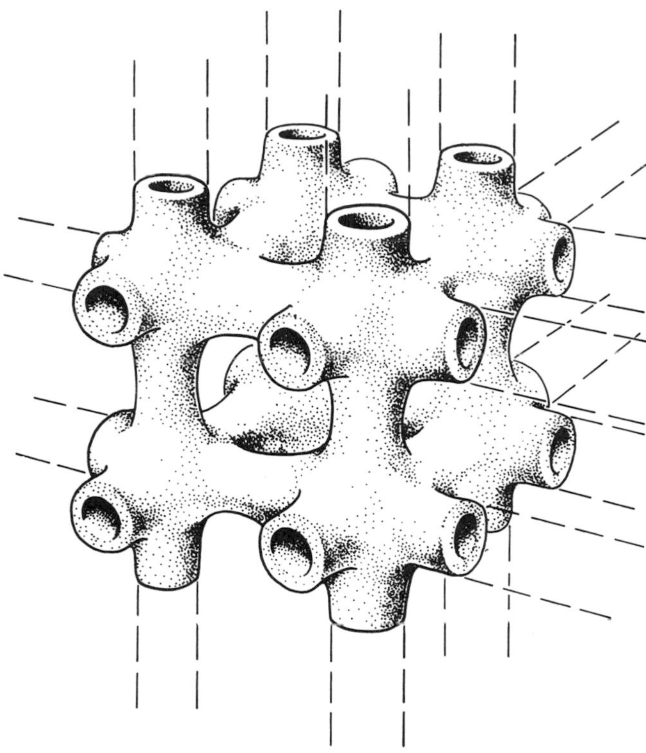

7–26. *Three-dimensional representation of the unit cube of the prolamellar body of* Avena. *A single ribosome-like particle, not visible in the diagram, occurs in the center of each unit cube. Modified from the original by B. E. S. Gunning.*

Summary

In Chapter 6, the mitochondrion was depicted as a generator of ATP for use by the cell as a source of chemical energy. In animal cells and in nonphotosynthesizing plant cells, mitochondrial oxidations are the primary source of ATP for synthesis and transport. Without a source of reduced carbon in the form of carbohydrates, fats, and amino acids, these mitochondrial oxidative reactions would cease. Chloroplasts supply the required reduced carbon, using the energy of sunlight to synthesize a large variety of substances from CO_2, H_2O, and nitrate. As a by-product of photosynthesis, chloroplasts replenish the environment with oxygen, which is used in turn as a final electron acceptor in mitochondrial oxidation. The chloroplast probably also functions as an ATP generator for syn-

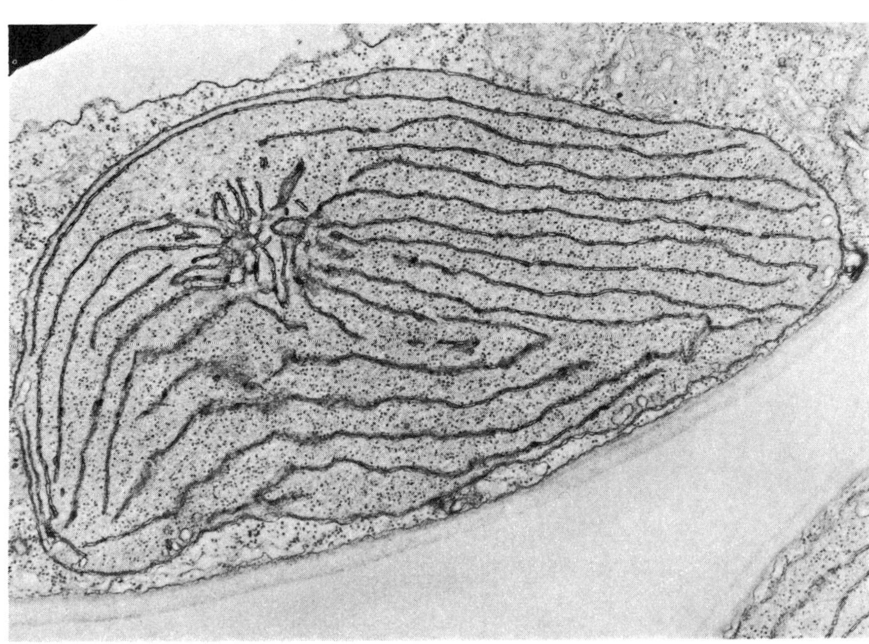

7–27. *Plastid from an etiolated oat leaf after exposure to light for two hours. A residual portion of the prolamellar body is still visible within the developing chloroplast. The sheet-like double membranes extending throughout the stroma arise from the prolamellar body during greening of the chloroplast.* × *19,500. Courtesy of B. E. S. Gunning.*

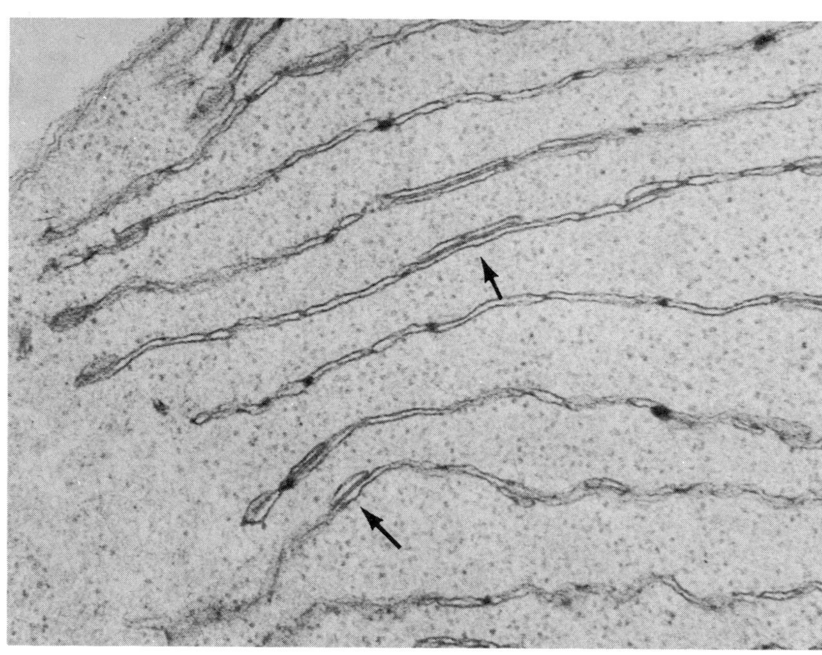

7–28. Interior of a plastid from etiolated oat leaf after exposure to light for four hours. Initial stages in granum formation are visible at the positions marked by arrows. × 28,000. Courtesy of B. E. S. Gunning.

7–29. Diagram illustrating the biochemical relationships between the chloroplast and the mitochondrion. The cells of photosynthesizing plants contain the entire mechanism as shown; animal cells contain only the functions shown to the right of the dotted line.

thesis and transport in plant cells. These interrelationships are summarized in Fig. 7–29.

Many plant tissues contain both chloroplasts and mitochondria. Most of the enzymes for the dark reactions are found both in the chloroplast and in the extra-mitochondrial cytoplasm of plants and animals; evidently, the majority of the dark reactions are not unique to the chloroplast.

A great many problems in chloroplast research have only partial solutions. As in the case of the mitochondrion, the total functions of the chloroplast are known only in skeleton form. Best understood of these are the physics of the initial absorption of light and the biochemistry of the initial dark reactions, which culminate in the synthesis of glucose. Other areas are surrounded with varying degrees of uncertainty. The correlation of ultrastructure with function is only beginning; the detailed molecular structure of chloroplasts is still in the speculation stage. Very little is known about the regulation of the light and dark reactions, that is, what keeps one system from outrunning another. With the exception of the Calvin and Hatch-Slack cycles, only limited details are known about other synthetic pathways leading to synthesis of lipids, carbohydrates, amino acids, and proteins in the chloroplast. Complete mystery surrounds the reaction series leading to the evolution of oxygen in photosynthesis. Efforts to answer these questions have probably required a greater integration of disciplines, including mathematics, physics, biochemistry, and morphology, than any other area of cell biology.

Suggestions for Further Reading

Books

Clayton, R. K., 1965. *Molecular Physics in Photosynthesis.* Blaisdell, Waltham, Mass.

Kamen, Martin D., 1963. *Primary processes in photosynthesis.* Academic Press, New York.

Kirk, J. T. O., and R. A. E. Tilney-Bassett, 1967. *The plastids.* W. H. Freeman, San Francisco.

Lehninger, A. L., 1965. *Bioenergetics.* W. A. Benjamin, Menlo Park, Calif.

Papers and Review Articles

Avron, M., and J. Neumann, 1968. Photophosphorylation in chloroplasts. *Annual Review of Plant Physiology,* vol. 19, pp. 137–166.

Bassham, J. A., 1968. Photosynthesis: the path of carbon. In J. Bonner and J. E. Varner, eds., *Plant Biochemistry.* Academic Press, New York, pp. 875–902.

Branton, D., 1968. Structure of the photosynthetic apparatus. In A. C. Giese, ed., *Photophysiology,* vol. III. Academic Press, New York, pp. 197–224.

Branton, D., and R. B. Park, 1967. Subunits in chloroplast lamellae. *Journal of Ultrastructure Research,* vol. 19, pp. 283–303.

Criddle, R. S., and L. Park, 1964. Isolation and properties of a structural protein from chloroplasts. *Biochemical and Biophysical Research Communications,* vol. 17, pp. 74–79.

Duysens, L. N. M., and H. E. Sweers, 1963. Mechanism of two photochemical reactions in algae as studied by means of fluorescence. In *Studies on Microalgae and Photosynthetic Bacteria: Special Issue of Plant and Cell Physiology.* University of Tokyo Press, pp. 353–372.

Gunning, B. E. S., 1965. The greening process in plastids I: the structure of the prolamellar body. *Protoplasma,* vol. 60, pp. 111–130.

Hatch, M. D., and C. R. Slack, 1966. Photosynthesis by sugarcane leaves: a new carboxylation reaction and the pathway of sugar formation. *Biochemical Journal,* vol. 101, pp. 103–111.

Jagendorf, A. T., 1967. Acid-base transitions and phosphorylation by chloroplasts. *Federation Proceedings,* vol. 26, pp. 1361–1369.

Kok, Bessell, 1965. Photosynthesis: the path of energy. In J. Bonner and J. E. Varner, eds., *Plant Biochemistry.* Academic Press, New York, pp. 903–960.

Malkin, S., and B. Kok, 1966. Fluorescence induction studies in isolated chloroplasts I: number of components involved in the reaction and quantum yields. *Biochimica et Biophysica Acta,* vol. 126, pp. 413–432.

Park, R. B., 1968. The chloroplast. In J. Bonner and J. E. Varner, eds., *Plant Biochemistry.* Academic Press, New York, pp. 124–150.

Park, R. B., and N. G. Pon, 1963. Chemical composition and the substructure of lamellae isolated from *Spinacea oleraceae* chloroplast. *Journal of Molecular Biology,* vol. 6, pp. 105–114.

Robinson, G. W., 1964. Quantum processes in photosynthesis. *Annual Review of Physical Chemistry,* vol. 15, pp. 311–347.

Stepka, W., A. A. Benson, and M. Calvin, 1948. The path of carbon in photosynthesis II: amino acids. *Science,* vol. 108, p. 304.

Weier, T. E., and A. A. Benson, 1967. The molecular organization of chloroplast membranes. *American Journal of Botany,* vol. 54, pp. 389–402.

Wettstein, D. von, 1958. The formation of plastid structures. *Brookhaven Symposia,* vol. 11, pp. 138–157.

Wettstein, D. von, 1966. On the physiology of chloroplast structures. In T. W. Goodwin, ed., *Biochemistry of Chloroplasts,* vol. 1. Academic Press, New York, pp. 19–22.

Unit Three *The Interphase Cell*

8
The Structure of Chromosomes in the Interphase Nucleus

The basic cell theory, as laid down in the 1830's by Schleiden and Schwann, established that the nucleus has a central role in the growth and development of cells. Subsequent investigations in the latter half of the nineteenth century revealed that some components of the nucleus persist during division, and are passed intact to daughter cells. Near the turn of the century, realization of the constancy of chromosome number and discovery of the behavior of the nucleus and chromosomes during meiosis and fertilization led to the growing appreciation that not only cell growth and development but also cell heredity was controlled by the nucleus. This was clearly established by the early 1900's, when Mendel's laws of genetics were rediscovered and correlated with the behavior of chromosomes during meiosis. In 1916, Bridges provided experimental proof of this correlation by showing that the gene controlling "white" eye in *Drosophila* was located on a specific chromosome (for details of these events see Chapter 1).

Recent biochemical and molecular investigation has confirmed that the nucleus is the storage site for most of the information required to direct cell activities. Information is also transcribed at this site into a set of directions to be followed by the synthetic machinery of the cytoplasm and replicated into additional copies for distribution to the daughter cells during cell division.

This chapter will describe the molecular and morphological bases of information storage in the nucleus. Later chapters in this unit will describe the transcription of this information into directions for synthesis and growth and the mechanisms that carry out this synthesis in the cytoplasm. The replication of the information system in the nucleus, which takes place just before cell division, is a part of Unit 4.

DNA as the Molecular Basis of Heredity

The molecular basis of information storage, replication, and transcription in the cell nucleus resides in the structure and chemistry of the nucleic acids *DNA* (*deoxyribonucleic acid*) and *RNA* (*ribonucleic acid*). Development of this level of understanding has spanned nearly a hundred years, from the first discovery of nucleic acids by Friedrich Miescher in 1868 to the establishment of experimental proof that DNA is the molecular basis of heredity during the 1940's and 1950's.

Friedrich Miescher discovered the nucleic acids while serving as a physician during the Franco-Prussian War. He isolated nuclei from leucocytes and found that a previously unknown macromolecular component made up a great part of the isolated nuclei. Miescher established that the new substance was nonproteinaceous and named it *nuclein*. He also found, in sperm cells from salmon, that nuclein occurred in a complex with a protein, which he named *protamine*. Nuclein was subsequently characterized chemically and renamed deoxyribonucleic acid. A second protein, *histone*, was discovered to be much more common than protamine in the nuclei of plant and animal cells. In fact, the protamine-DNA complex was found to occur only in the spermatozoa of some animals.

The Establishment of DNA as the Hereditary Molecule

The discovery that DNA and histone are regular components of nuclei initiated a controversy over which of the two macromolecules was likely to be the molecular basis of heredity. The protein component was slightly favored in this argument until the 1940's because the nucleic acids did not seem complex enough to serve as coding molecules for the storage of information. Proteins were known to contain twenty different amino acids and to exhibit great variety in the arrangement and folding of their component chains. This seemed to offer many possibilities for a coding system. In contrast, DNA is uniform in structure, except for differences in the nitrogenous bases (purines and pyrimidines) that form a part of the molecule. Because there are only four of these, which occur in an approximate 1:1:1:1 ratio, investigators could see no way in which the many thousands of genes could be coded into the DNA molecule.

Gradually, evidence began to accumulate linking DNA with the hereditary mechanism. The first important observations depended on a new staining technique, the Feulgen reaction, which was used in the 1920's to establish that DNA is located in the chromosomes (see p. 8). The Feulgen stain was later found to allow a quantitative estimation of the relative amounts of DNA in the nuclei of different cells because the quantity of stain bound in the reaction is directly proportional to the amount or concentration of DNA present. Although this method does not measure the absolute quantity of DNA present, it allows comparisons to be made among different cells. Using this technique, two independent research teams, A. E. Mirsky and Hans Ris, and Hewson Swift and Max Alfert found in 1950 that the distribution of DNA in cells exactly parallels the distribution of chromosomes. For example, diploid somatic cells, with the 2X number of chromosomes, have twice as much DNA as gametes from the same organism, with the haploid (X) number of chromosomes. In prophase of meiosis (see Chapter 14), in which the chromosome complement is present at the 4X level, there is four times as much DNA present as in haploid cells.[1] While histone follows the same pattern of distribution, it is replaced by protamine during the development of spermatozoa in some animals. Therefore, histone does not have the physical continuity that is expected for the hereditary material.

These early findings suggested but did not prove that DNA is the hereditary molecule. Somewhat better evidence on the side of DNA was provided by the observation that agents such as X-rays, ultraviolet light, and nitrogen mustard that are known to cause inheritable mutations in organisms can also cause chemical alterations in DNA. But this evidence, most of which was established by 1950, was still indirect, and the controversy continued. Final settlement of the dispute came with publication of the results of two major experiments with microorganisms. The first of these was announced in 1944 by Oswald Avery and his co-workers.

Avery and his colleagues were able to purify and identify as DNA the "transforming agent" of pneumonia bacteria. Two forms of this bacterium were

---

[1] An "X" is used here as a descriptive unit for the relative quantity of DNA instead of the more commonly employed "C-value." See the explanatory footnote on p. 290.

known: a virulent, infective form, and a nonvirulent form. In the infective form, each bacterial cell is surrounded by a polysaccharide capsule; noninfective cells lack this capsule. The two forms represent distinct genetic strains of the pneumococcus bacterium. Mutations occur very rarely from one form to the other. Otherwise, the separate strains breed true and retain their differences. In the 1920's, F. Griffith had found that if a mixture of heat-killed capsulated pneumococcus and live, noncapsulated pneumococcus were injected into a mouse, the noncapsulated type could be transformed into the virulent type. Later, it was discovered that a chemical could be extracted from dead cells of the virulent, capsulated line that was capable of inducing the same transformation of nonvirulent bacteria into the infective, capsulated form. This transformation, once accomplished, became a permanent, heritable characteristic of the cells. Avery and his colleagues purified this agent and treated it with hydrolytic enzymes specific for proteins and the two nucleic acids. Only deoxyribonuclease, the enzyme specific for the digestion of DNA, destroyed the transforming agent. On this basis, Avery and his colleagues concluded that the transforming agent was DNA and hypothesized that if DNA is capable of carrying genetic information from one cell to another (in this case the information required to synthesize the polysaccharide coat), it is likely to be the coding molecule for the entire genome.

These conclusions were directly supported by a second series of experiments carried out by Alfred D. Hershey and Martha Chase, who found that the DNA and not the protein of bacteriophage is the agent transmitting the information required for phage reproduction. Bacteriophages, a group of viruses infecting bacteria, consist only of a DNA molecule surrounded by a coat of protein. Hershey and Chase labeled either the protein or DNA of phage particles, using the isotope ^{32}P as a marker for DNA and ^{35}S as a marker for protein. (DNA contains phosphorus but no sulfur; viral proteins contain sulfur but no phosphorus.) During the first stages of infection, the bacteriophage particle attaches to the surface of a bacterial cell. After a brief time interval, virus particles can be detected forming within the bacterial cytoplasm. In some way, the infecting bacteriophage particle converts the bacterial synthetic machinery into a system for replicating viral particles. Because the bacteriophage particle consists only of a protein coat surrounding a DNA core, one or the other of these two parts must enter the bacterial cell at the time of infection to direct the synthesis of new phage particles. Using the ^{35}S and ^{32}P labels, Hershey and Chase were able to show that only the DNA of the viral particles entered the host cells. When only the DNA of the phage particle was labeled, the interior of the bacteria was found to be radioactive after infection. If only the protein of the phage was labeled, radioactivity remained outside the cell and could be easily removed by shaking or sonication. Only the phage DNA entered the bacterial cell and transmitted the information required for the infected bacterium to synthesize the viral particle.

Discovery of the Molecular Structure of DNA

The discoveries of Hershey and Chase, which conclusively established that the chemical basis of heredity resides in the DNA molecule, prompted intensive investigation into the structure and properties of DNA. It was known only that DNA was a long polymer consisting of one or more chains of subunits, the *nucleotides.* The chemical structure of the four individual nucleotides had been known since the turn of the century (Figs. 8–1 and 8–2). Each nucleotide consists of a phosphate group, a 5-carbon sugar (deoxyribose), and a purine or pyrimidine base. The four nucleotides differ only in the particular purine or pyrimidine base bound to the sugar residue (see Fig. 8–2). Within the DNA polymer, the individual nucleotides are linked into chains through phosphodiester bonds between the 5′ and 3′ carbons of adjacent sugar residues (Fig. 8–3).

This was the level of knowledge about DNA structure when Hershey and Chase published their results. At this time, several laboratories began investigating DNA structure further, to determine the number and precise spatial arrangement of the nucleotide chains forming the backbone of the molecule. Success in this investigation came to James D. Watson, an American, and Francis H. C. Crick, an Englishman, working in the Cavendish Laboratory of Cambridge University. For background material in their solution of DNA structure, Watson and Crick relied heavily on X-ray diffraction evidence from the laboratory of Maurice H. F. Wilkins at the University of London.

Watson has set down an entertaining account of the events leading up to the deduction of DNA structure (see Watson, 1968). After inspecting the X-ray evidence

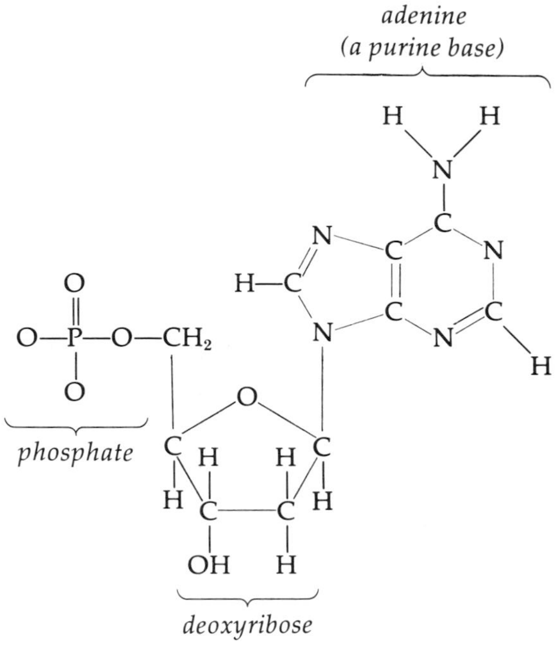

adenine
(a purine base)

phosphate

deoxyribose

8–1. *Deoxyadenosine phosphate, a nucleotide occurring in both DNA and RNA.*

gathered by Wilkins, Crick determined that the nucleotide chains forming the backbone of the DNA molecule were twisted into a helix of regular diameter and pitch. Uncertain as to the number of separate nucleotide chains occurring in the helix, Watson and Crick began to build molecular models, concentrating on forms in which either two or three nucleotide chains were twisted around each other to make up the backbone of the molecule. The general impression at the time was that three nucleotide chains made up the helix. Watson, a geneticist by training, held out for two-chain models, because he felt intuitively that a molecular system related to sexual reproduction was most likely to contain *pairs* of subparts, not threes. After several unsuccessful attempts, Watson and Crick were able to design a workable model in which two nucleotide chains were twisted around each other to form the helical backbone of the molecule. Information from the Wilkins laboratory was consistent with a structure in which the backbone sugar-phosphate chains ran along the outside of the molecule, with the bases facing toward the inside.

Watson and Crick were temporarily baffled by the possible arrangement of the bases inside the molecule, given the restriction from X-ray diffraction evidence that the diameter of the helix was uniform at 20 Å. Some clues to the arrangement of the bases were available. Previous work had shown that a 3.4 Å spacing in X-ray photographs of DNA was most likely to be a result of the separation between bases in the molecule. This was possible if the bases, which are flat molecules, were arranged in a stack, lying perpendicular to the long axis of the molecule. Other work, by Erwin Chargaff, had shown that the bases adenine and thymine were present in equal amounts in DNA from a given organism. Guanine and cytosine were similarly present in equal amounts. Some species showed an excess of guanine and cytosine; others showed an excess of thymine and adenine. This suggested to Watson and Crick that adenine-thymine and guanine-cytosine pairs were formed in some manner inside the molecule.

Watson hit on the solution after considering the possible formation of hydrogen bonds between the base pairs, using models of the bases cut from stiff cardboard:

> When I got to our still empty office the following morning, I quickly cleared away the papers from my desk top so that I would have a large, flat surface on which to form pairs of bases held together by hydrogen bonds. Though I initially went back to my like-with-like prejudices, I saw all too well that they led nowhere. . . . Suddenly I became aware that an adenine-thymine pair held together by two hydrogen bonds was identical in shape to a guanine-cytosine pair held together by at least two hydrogen bonds. All the hydrogen bonds seemed to form naturally; no fudging was required to make the two types of base pairs identical in shape.[2]

Watson had discovered that an adenine-thymine pair had the same total width as a guanine-cytosine pair (Fig. 8–4). Either pair could fit between the backbones of the two nucleotides of the double helix without stretching or reducing the observed 20 Å diameter.

This was the key required. Watson and Crick were able to assemble a model completely fitting the X-ray evidence and the chemical data obtained by Chargaff (Fig. 8–5). In the model, the two phosphate-sugar backbone chains are twisted into a helix with a total diameter of 20 Å. The backbone chains are separated

[2] James D. Watson, *The Double Helix*. New York: Atheneum, 1968.

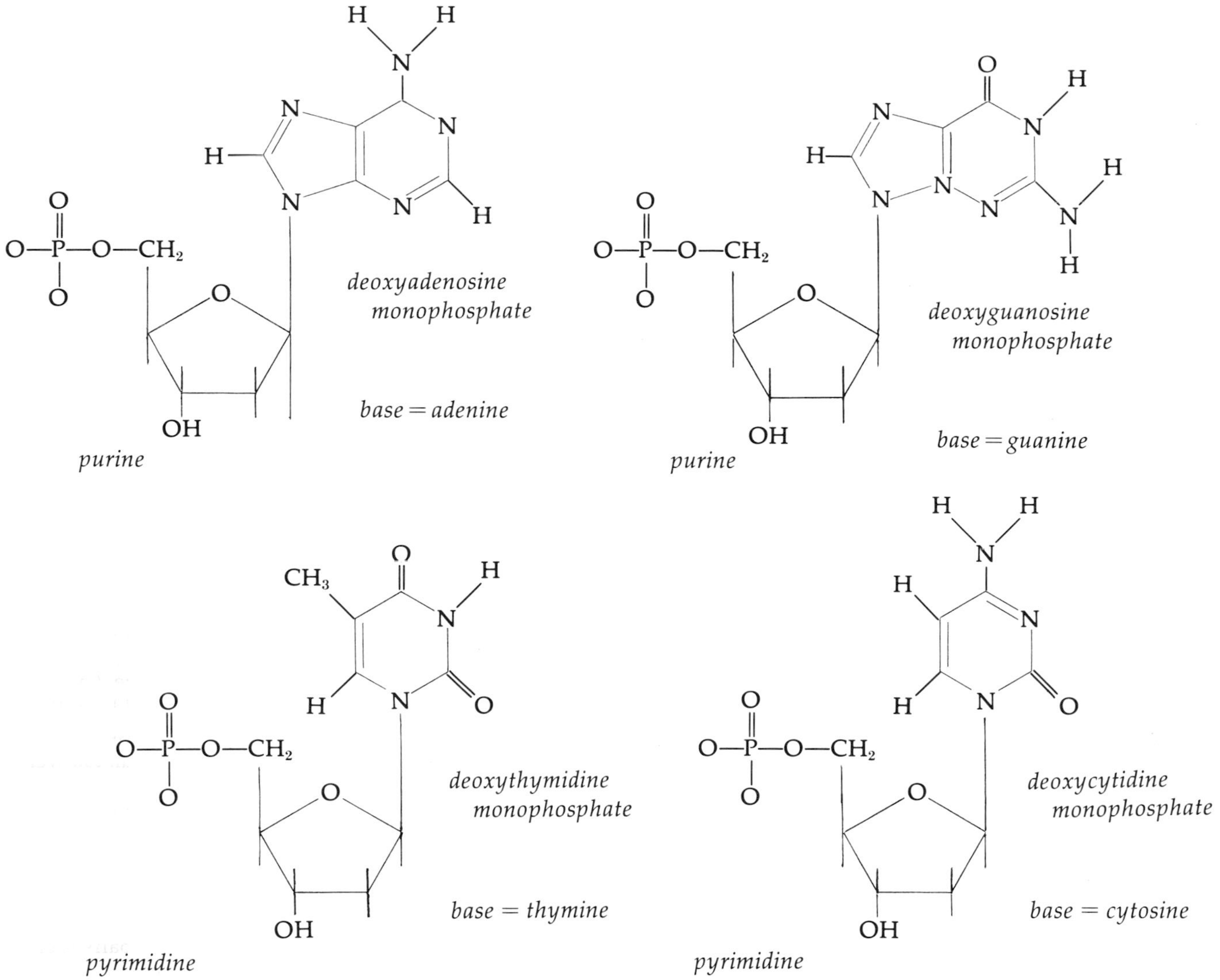

deoxyadenosine monophosphate

base = adenine

purine

deoxyguanosine monophosphate

base = guanine

purine

deoxythymidine monophosphate

base = thymine

pyrimidine

deoxycytidine monophosphate

base = cytosine

pyrimidine

8–2. *The four nucleotides found "as building blocks" in the DNA molecule.*

from each other by a constant distance, approximately 11 Å, across the center of the helix. Note from Fig. 8–5a that the two chains are not evenly spaced in the double helix. A slightly wider spacing alternates with a closer spacing, producing large and small "grooves" running along the outside of the molecule (Fig. 8–5b). These surface grooves are considered by some investigators to be the site of binding of histone or protamine to DNA in the nucleoprotein complex (see below).

The 11 Å spacing in the center of the molecule is filled in by the paired bases. Each pair, consisting of one pyrimidine and one purine base, lies in a flat plane perpendicular to the long axis of the molecule. The planes occupied by successive base pairs along the double helix are spaced at 3.4 Å intervals. These planes are seen from the side in Fig. 8–5a (the horizontal lines); at any level represented by a horizontal line an individual base pair, if viewed from an end of the molecule, would appear as shown in Fig. 8–4. Each complete

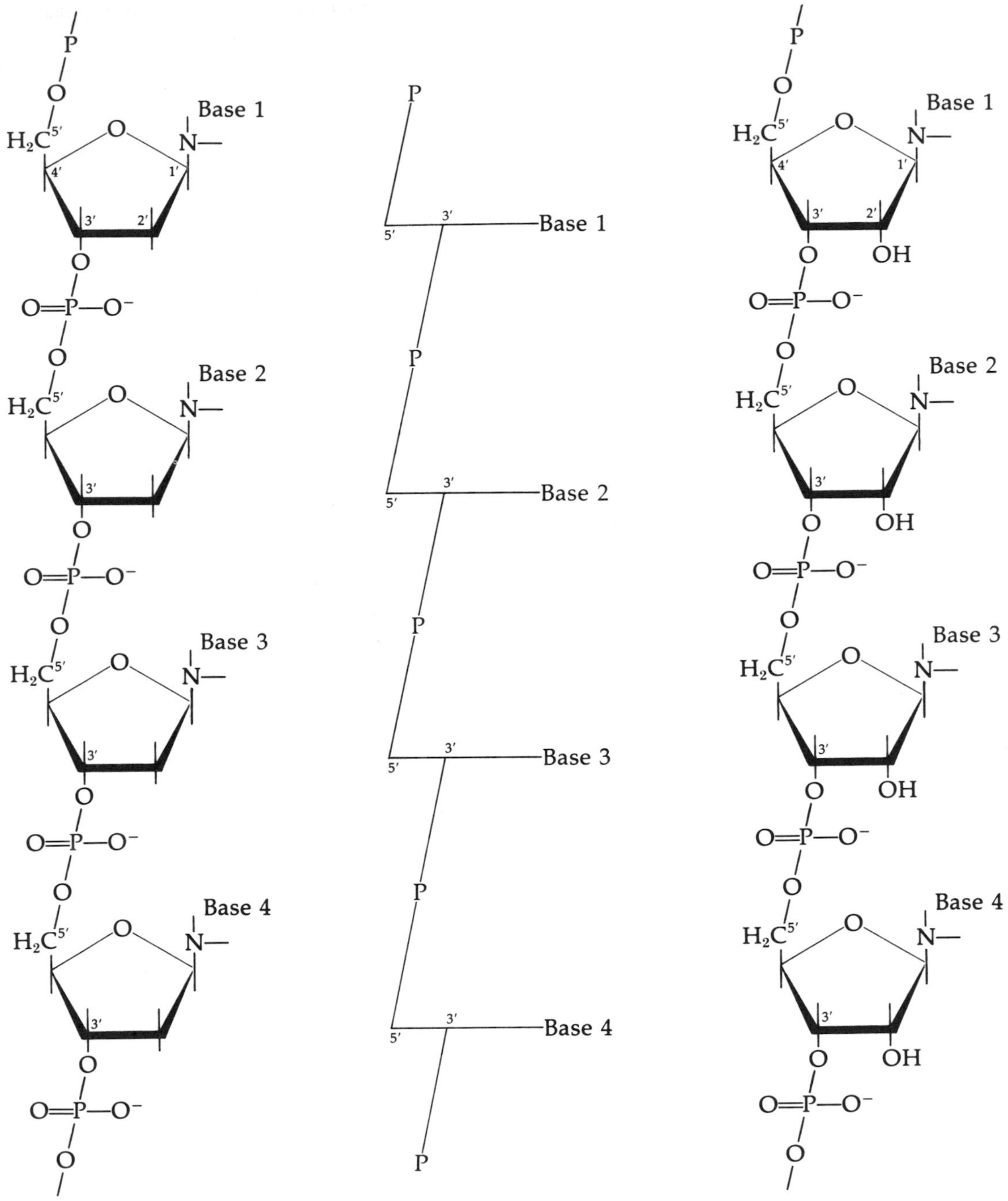

DNA polynucleotide

schematic form of DNA and RNA polynucleotides

RNA polynucleotide

8–3. Linking of nucleotides into the nucleic acid polymers of DNA and RNA. In DNA, the bases adenine (A), guanine (G), cytosine, (C) and thymine (T) may be found with equal probability at the sites marked "base 1, base 2," and so forth. In RNA, the bases A, G, C, and uracil (U) occur at these sites. The 5-carbon sugars differ in the absence of an oxygen at the 2′ carbon in DNA.

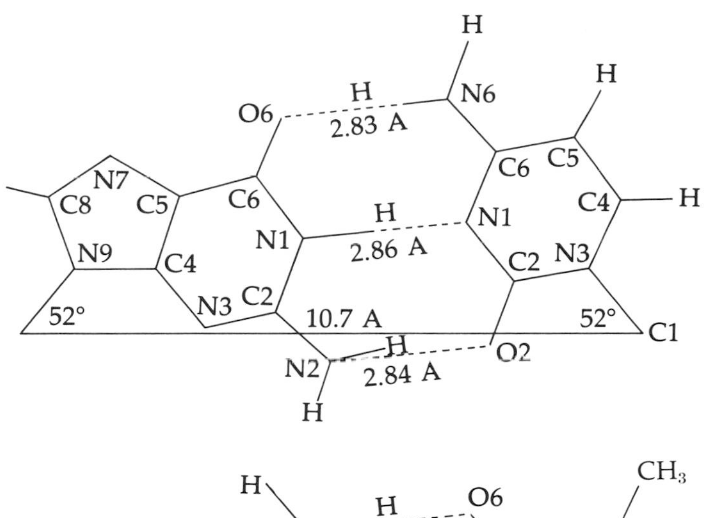

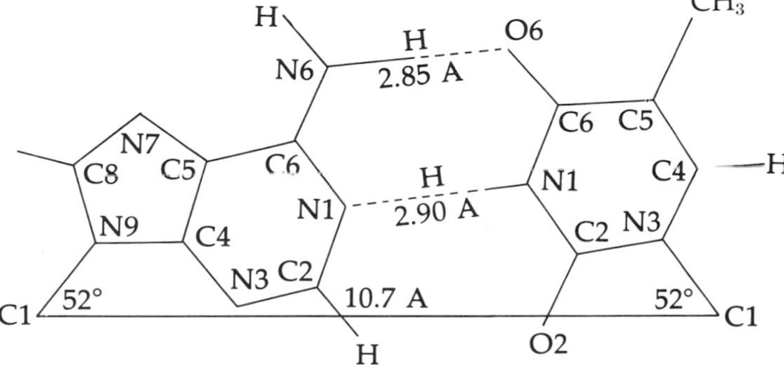

8–4. Hydrogen bonding between guanine-cytosine pairs (top) and adenine-thymine pairs (bottom). Note that the total width of the pairing complexes is approximately the same.

turn of the double helix includes ten base pairs, and occupies 34 Å of the long axis of the molecule.

The two chains of the double helix are held together by the hydrogen bonds formed between base pairs. Although hydrogen bonds are relatively weak intermolecular forces, the great numbers formed between the backbone chains form a highly stable configuration. It has been estimated that a DNA duplex more than 10 nucleotide pairs in length is stable and will remain intact.

Pairs are formed between purines and pyrimidines. Purine-purine or pyrimidine-pyrimidine pairs do not normally occur. This is determined partly by restrictions on the total diameter of the double helix. If two purines or two pyrimidines pair, the total diameter is stretched or reduced from 20 Å at that point. Further restrictions on pairing are imposed by the pattern of hydrogen-bonding between base pairs. Note from Fig. 8–4 that three hydrogen bonds are formed between guanine and cytosine, and two between adenine and thymine. This pattern of hydrogen bonds effectively restricts the pairing to guanine-cytosine and adenine-thymine pairs.

The combined effects of restrictions on helix diameter and on the pattern of hydrogen-bonding between base pairs forms a *complementary base sequence* in the two nucleotide chains of a double helix. This means that a given sequence in one chain is compatible with only one sequence in the opposite chain. For example, if the sequence in one chain is A—T—C—T—G—G—C (A = adenine, T = thymine, G = guanine, C = cytosine) the opposite chain is effectively restricted to the complementary sequence T—A—G—A—C—C—G (see Fig. 8–6a).

The complementarity of the two nucleotide chains of the DNA duplex immediately suggested to Watson and Crick a possible mechanism for the replication of DNA (Fig. 8–6b). If the chains separate, and each separated chain makes a complementary copy of itself by forming the appropriate base pairs, the end result will be two new double helices with a sequence of base pairs identical to the original double helix. Each new molecule will contain one "new" and one "old" chain (Fig. 8–6b). Much evidence that replication actually occurs in this way has accumulated since Watson and Crick first proposed this model (to be discussed in detail in Chapter 12).

Thus, the Watson-Crick model was able to account for both the structure and probable manner of replication of DNA. Also inherent in the structure of the molecule was the key to the coding of genetic information into DNA, for it was obvious that the sequence of nucleotides preserved in replication provides a basis for information storage. Watson, Crick, and Wilkins received the Nobel Prize in 1962 for their model, which has proved to be one of the most fundamental and far-reaching discoveries in the history of biology.

Proteins Associated with DNA in the Cell Nucleus

The DNA studied by Watson, Crick, and Wilkins was isolated in a purified form complexed with a metal ion, usually sodium, to neutralize the phosphate groups of the double helix. These phosphate groups give the

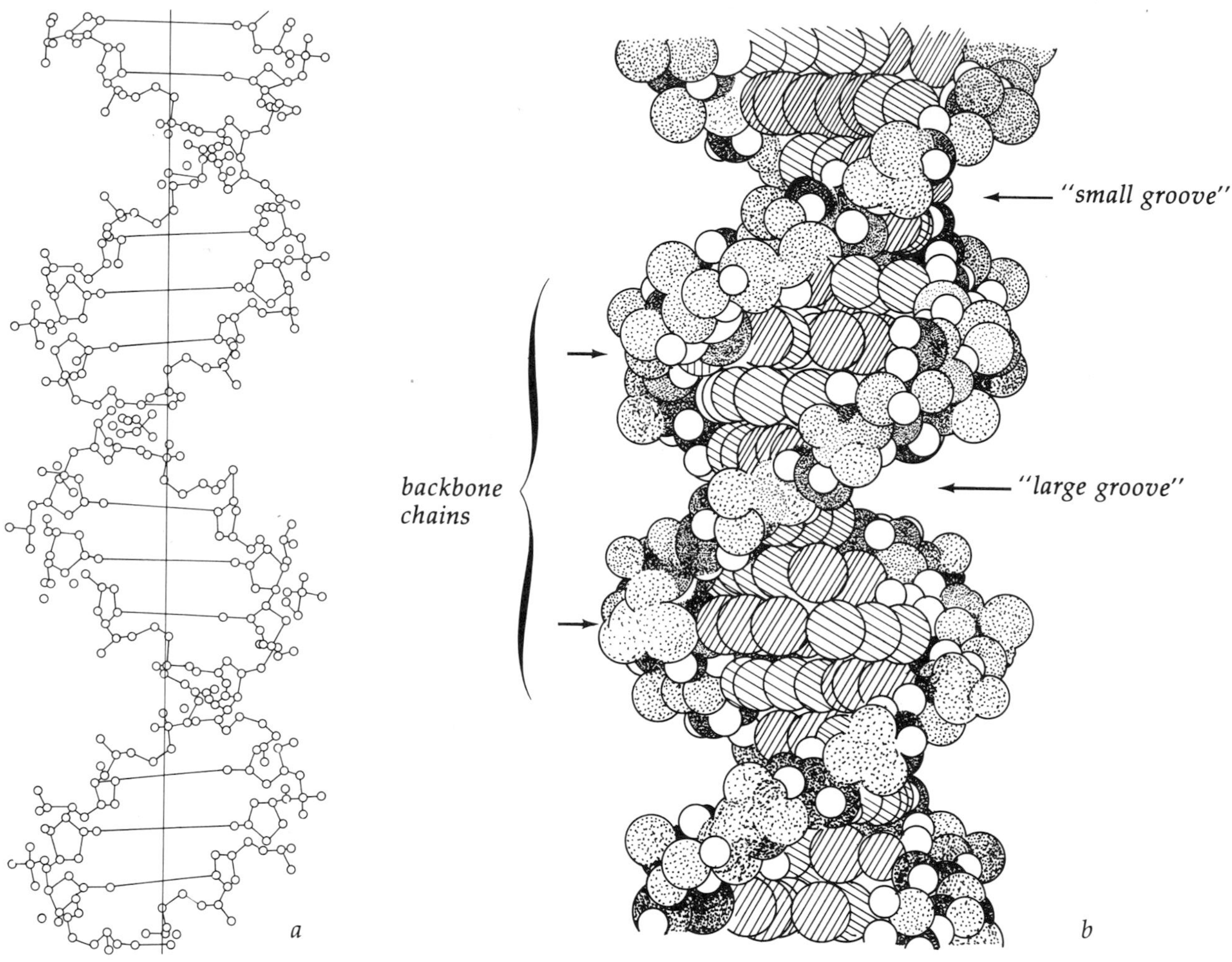

8–5. a, *The double helical model for the DNA duplex, as deduced by Watson, Crick, and Wilkins from X-ray diffraction. The "backbones" of the two chains of the helix consist of alternating deoxyribose sugar residues and phosphate groups. The bases, which lie in a flat plane, are seen on edge from this viewpoint and thus are depicted as straight lines extending between the sugar-phosphate backbones. Courtesy of M. H. F. Wilkins.* b, *The same structure, using atomic radii to show the degree of packing of individual atoms in the helix. Redrawn, original courtesy of V. M. Ingram.*

molecule strongly acidic properties. However, metallic ions are believed to play only a minor role in the neutralization of DNA in the nucleus of higher organisms. DNA is complexed instead with one or more of several proteins to form *nucleoprotein.* Miescher, as mentioned, found the DNA in salmon sperm to be complexed with the strongly basic protein protamine. In the years since Miescher's first discovery of neucleoprotamine, the DNA of the spermatozoa of many animals has been shown to be complexed with protamine or closely similar basic proteins. But in the somatic cells of almost all plants and animals, DNA is complexed instead with another distinct class of more complex basic proteins, the histones. Histones also occur in the sperm nuclei of some species. (For a summary of the nucleoprotein types characteristic of various gametes, see Chapter 15.)

These proteins occur in large quantities in the nuclei of higher organisms; the ratio of DNA to protein by weight in both the protamine and histone types of nucleoprotein is about 1:1. Histone or protamine have not been identified in bacteria or blue-green algae.

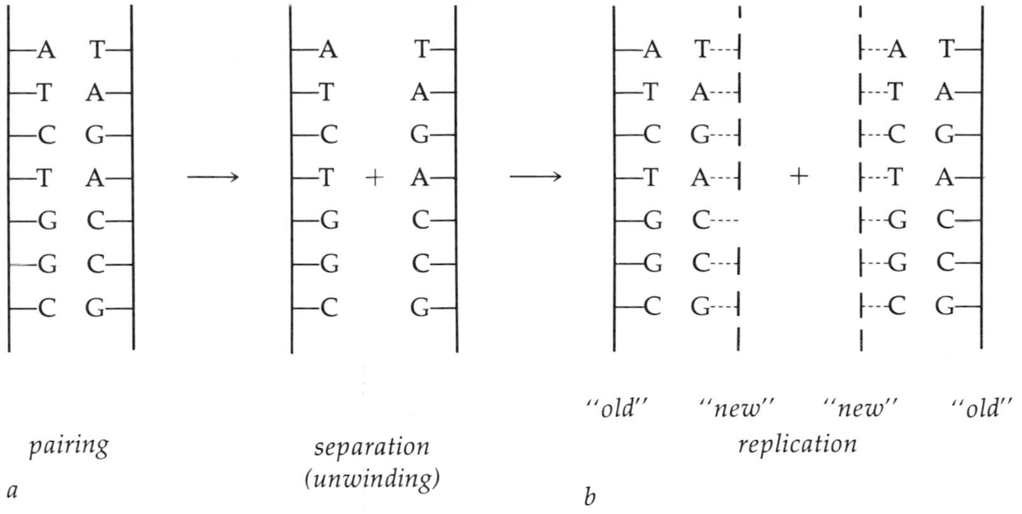

A T
T A
C G
T A
G C
G C
C G
→
A T
T A
C G
T + A
G C
G C
C G
→
A T · A T
T A · T A
C G · C G
T A + T A
G C · G C
G C · G C
C G · C G

"old" "new" "new" "old"

pairing *a*

separation (unwinding)

replication *b*

8–6. Diagram of base pairing, unwinding, and replication of DNA as visualized by Watson and Crick. Replicated DNA duplexes consist of one "old" chain, derived from the parent helix, and one "new" chain, synthesized during replication, as shown in b *(see text).*

In these prokaryotic organisms, DNA is "naked," that is, uncomplexed with a protein, apparently neutralized by a metallic ion such as sodium or calcium.

Other classes of proteins have also been implicated as structural components of the chromosomes of somatic or vegetative nuclei in higher organisms. Variously termed "non-histone," "acidic," "neutral," or "residual," these proteins vary widely in concentration in nuclear extracts, depending on the method of isolation and purification used. It seems probable, for this reason, that at least a part of this material is a contaminant from the cytoplasm or nuclear sap. In any event, the manner of binding, the function, and the precise identity of nonhistone proteins remains obscure.

While histones and protamines are well established as regular components of nucleoproteins, the functions of these substances are unknown. Histone has often been proposed to be involved in regulating the activity of the DNA of the chromosomes, possibly by direct combination with parts of the DNA to turn specific genes "off." There are problems with the specificity required for this function, because histones are relatively simple molecules; only a few different types have been identified. While there are differences of opinion, the total number of different histones is probably less than 20 and possibly less than 10. Nevertheless, some limited evidence shows that the histones actually carry out a regulative function, either as a passive "plug" in the chromosome, controlled by some other molecule such as RNA, or perhaps as the regulating molecule itself. These possibilities will be discussed more fully in Chapter 11, which deals directly with the possible activities of histone and RNA in the regulation of protein synthesis.

Replacement of histone by protamine or other more basic proteins takes place during the maturation of spermatozoa in many animals, particularly in vertebrates and in some invertebrates, including many insects and molluscs. The conversion from histone to protamine in sperm is accompanied by a change in structure in the nucleus, from the more loosely coiled arrangement typical of somatic nuclei to a tightly packed, almost crystalline form in which all of the nuclear constituents except DNA and protamine are eliminated (for details, see Chapter 15).

The Molecular Structure of Nucleoprotamine and Nucleohistone

The semicrystalline packing of the nucleoprotamine in spermatozoa reveals only a very limited periodicity when examined by X-ray diffraction. Within the patterns obtained from spermatozoa, reflections characteristic of the DNA double helix are clearly visible. This indicates that the basic structure of DNA is not altered in the nucleoprotamine complex. Beyond this point, the conclusions from X-ray diffraction of native nucleoproteins are less certain.

More definitive results have been obtained from the study of complexes between DNA and artificial, protein-like substances. One such set of experiments was carried out by M. Feughelman and his co-workers. Knowing that arginine forms two-thirds of the total amino acid content of protamine, Feughelman synthesized an artificial poly-L-arginine, a synthetic protein consisting entirely of arginine residues linked together into a long polypeptide chain. This strongly basic synthetic protein combines readily with DNA through electrostatic bonds formed between the positively charged arginine residues and the negatively charged phosphate groups of the DNA duplex. The more ordered X-ray diffraction patterns produced by this synthetic complex were interpreted by Feughelman to mean that polyarginine winds helically around the surface of DNA, packed into the shallow groove (see Fig. 8–5b) of the molecule. The polyarginine chain is not secondarily folded into an α-helix; the amino acids are thought to be wound around the shallow groove in an extended chain.

X-ray diffraction patterns of nucleoprotamine isolated from sperm heads, although less clear, are comparable to the patterns obtained from polyarginine-DNA complexes. Sequencing studies have shown that arginine occurs in pairs in protamine, evenly spaced by groups of nonarginine residues. This regular arrangement would place the arginine residues in close register with the repeating phosphate groups of the DNA molecule and would permit neutralization and binding closely similar to the pattern suggested by the studies of synthetic polyarginine. From the X-ray data and the pattern of binding of protamine suggested by the spacing of arginine residues, Feughelman constructed the model shown in Fig. 8–7.

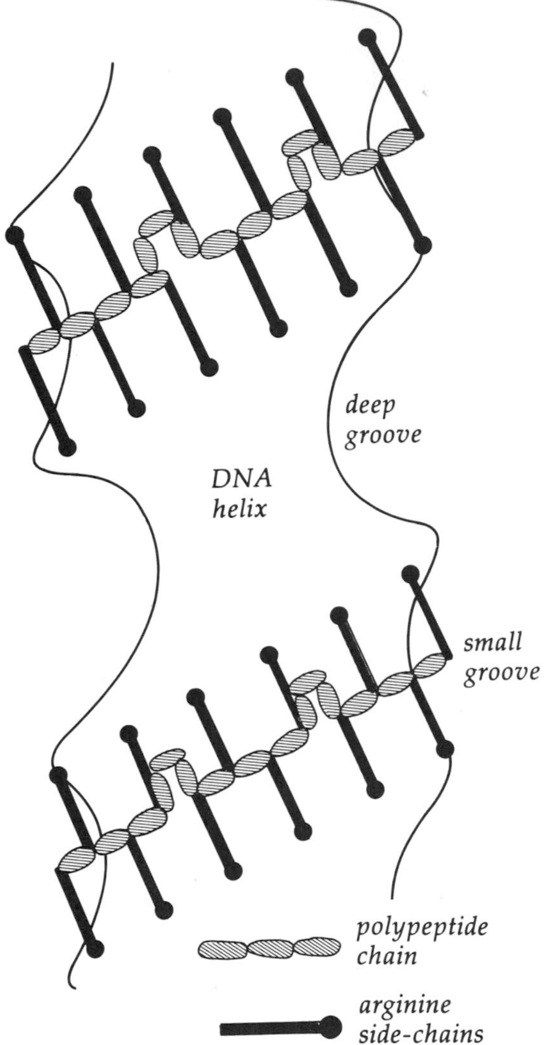

8–7. Binding of protamine into the "shallow groove" of DNA, as deduced by Feughelman from X-ray diffraction evidence (see text). Redrawn; original courtesy of M. H. F. Wilkins.

Unfortunately, results with polyarginine cannot be extended with any confidence to nucleohistone. X-ray diffraction of nucleohistone reveals patterns even less clear than nucleoprotamine. Other difficulties arise from the fact that basic amino acids, primarily arginine or lysine, form only 30 percent of the total amino acid complement of histone instead of 60 percent as in protamine. Although there are still enough basic residues at this ratio to neutralize the phosphate groups of DNA, the nonbasic amino acids are distributed between the arginine and lysine residues in unequal groups of from 1 to 7. This distribution causes an irregular spacing of arginine or lysine along the histone chain that does not match the phosphate groups of the DNA double helix. Still other problems arise from the fact that about 70 percent of the polypeptide chain in histones is secondarily coiled into an α-helix rather than extended, as in protamine. This has an important effect on the possible binding of histone to DNA, because the α-helix is too rigid to wind around either the shallow or wide groove of the DNA molecule.

Some limited evidence indicates that at least a part of the histone is wound around the shallow groove of

DNA. Histone can be separated into a number of subfractions. These differ in strength of binding to the DNA molecule, and can be extracted from nucleohistone in fairly pure form by adjusting the salt concentration of the medium in which the nucleohistone complex is dissolved (Table 8–1). If only the lysine-rich fractions are mixed with DNA, the resulting complex gives X-ray

Table 8–1 Histone Fractions Removed from Nucleohistone by Increasing Salt Concentrations

Fractions Ia, Ib	lysine-rich
Fractions IIa, IIb	slightly lysine-rich
Fraction III, Fraction IV	arginine-rich

diffraction patterns similar to nucleoprotamine. This suggests that the lysine-rich fraction may be wound around the shallow groove of DNA, in a way similar to the binding of protamine in nucleoprotamine (see Wilkins, 1956). Coincidentally, the lysine-rich fractions Ia and Ib make up about 30 percent of the total histone. The percentage of histone bound to DNA is found by other methods to be in the extended, rather than α-helical form. It would be convenient if the extended fraction and the lysine-rich fractions are identical, because this configuration is necessary for binding to the shallow groove of the DNA molecule without the restrictions imposed by the α-helix.

Several investigators have suggested that the remaining 70 percent is wound around the wide groove of the DNA. The difficulty in bending the α-helix to this extent is circumvented by the suggestion that this histone fraction may consist of short lengths of α-helix separated by short lengths of polypeptide chain in the extended form. This would produce a more flexible chain and allow the displacement required for binding the protein into the wide groove of the DNA duplex.

Whatever the precise pattern of binding between DNA and protein, the total diameter of the duplex with associated protein is not expected to be much greater than 30 Å. Additional "supercoiling" of the nucleoprotein complex may result in a configuration of somewhat larger dimensions, possibly 100 Å or more. There is limited evidence that a supercoil of this type may exist. If fragments of "naked" DNA are mixed with histone, association into a nucleohistone complex is accompanied by an average decrease in length of the fragments by 40-60 percent (see Ohba, 1966). This suggests that the DNA-histone complex may be coiled in a 2:1 ratio over extended, "naked" DNA. Recent X-ray data from Wilkin's laboratory support the existence of a nucleohistone "supercoil" and indicate that the coil has a total diameter of 100 Å and a pitch of 120 Å (Pardon et al., 1967). All of this, as will be seen, is compatible with the evidence from morphological studies with the electron microscope, which indicate that the ultrastructural unit of the chromosomes of higher organisms is a fiber about 100 Å in diameter.

The structure of nucleoprotein remains an unsolved problem of fundamental importance in biology. The eventual solution of nucleoprotein structure may prove to be fully as important as the discovery of the DNA double helix. Just as the solution of DNA structure revealed the mechanisms of replication and coding, the solution of the structure of nucleoproteins may reveal the molecular basis for the regulation of gene activity. Intimately connected with this problem is the phenomenon of *differentiation*, a progressive regulation of gene activity occurring during the entire life cycle of plants and animals, including embryonic development, maturation, and senescence. Even a hint as to the molecular interactions underlying these processes would be of central importance to the science of biology.

Nucleoproteins and Chromosome Structure

From the foregoing material it is obvious that much less is known about the structure of nucleoprotein than of the DNA molecule. The number of unanswered questions increases as we progress to the next higher structural level, the organization of nucleohistone into the chromosomes. "Chromosome" in this case refers to the structures, often called *chromatin*, seen in the eukaryotic interphase nucleus under the light and electron microscopes.

The Light Microscope Image of the Interphase Nucleus

Interphase nuclei can be viewed with the light microscope either in the living state or after fixation and staining with a variety of agents. The nucleoplasm of the living nucleus appears faintly fibrous under the phase contrast microscope (see Fig. 3–9a). The fibrous material may either be evenly distributed throughout the nucleus or aggregated into small clumps. Where aggregations occur, these are often distributed around the periphery of the nucleus just inside the nuclear envelope. Suspended in this fibrous material are one

8–8. Thin-sectioned nucleus, fixed with osmium tetroxide. Although some differences of opinion exist, most authors place the diameter of the individual chromatin fibers at 100 Å or slightly less. × 48,500. Courtesy of B. R. Zirkin.

or more nucleoli. The image obtained after fixation and staining is much the same, except that the fibrous material is more clearly delineated. All regions of the nucleoplasm outside of the nucleolus, including both clumped and diffusely extended regions, are stained by the Feulgen reaction. Only differences in staining intensity are noted between these regions, indicating that all parts of the chromatin contain DNA. A similar pattern is noted if histones are selectively stained with *fast green*, a dye used to specifically identify basic proteins. Work with the two staining reactions has shown that the distribution of histone and DNA is always parallel in interphase nuclei.

The Electron Microscope Image of the Interphase Nucleus

The nucleus is ordinarily viewed under the electron microscope after fixation, embedding, and thin sectioning. After fixation with osmium tetroxide, the nucleoplasm outside of the nucleolus reveals a closely packed array of short lengths of dense fibers approximately 100 Å in diameter (Fig. 8–8). The fibers are extensively folded and indistinct in outline; at many points the included length is so short that the fiber appears as a dot instead of a line. This is a reflection of the thinness of the sections, which at 300–600 Å are not much thicker than the fibers.

Changes in the fixative or stain do not greatly alter this image. Aldehyde fixatives used in combination with osmium tetroxide result in a slight coarsening of the unit fibers and reveal greater degrees of clumping or aggregation of the chromatin (compare Figs. 3–20 and 3–21). After permanganate fixation the nucleus appears almost empty, although the component fibers of the nucleoplasm are still faintly visible (see Fig. 3–22). Electron stains enhance the density of fibrils in the nucleus but do not change their appearance or reveal additional structure.

The density observed in the chromatin represents an interaction between macromolecular components of the 100 Å fibers and metallic ions in the fixative or stain. It has never been clearly established whether the DNA or the protein of the chromatin fibers interacts with the fixatives or stains. If DNA is mixed with osmium tetroxide in a test tube, there is no evidence of any interaction (see Bahr, 1954). This was discovered quite early in the development of electron microscope techniques, and for a time it was argued that the thin fibers of the interphase nucleus could not represent the nucleoprotein complex. The weight of evidence since these early studies, although largely indirect, has led to general acceptance of the 100 Å fibers as actually representing some interaction of osmium with either the DNA or more likely the protein of the nucleohistone complex.

The best evidence for the identity of the 100 Å fibrils in thin sections comes from comparisons of light and electron microscope images of adjacent sections of the same nucleus. In some nuclei, as noted, conspicuous aggregations of chromatin occur in clumps just inside the nuclear envelope or randomly distributed throughout the nucleoplasm. If adjacent thick and thin sections of these regions are compared by light and electron microscopy, it can be demonstrated that the clumps, which are intensely stained by the Feulgen reaction for light microscopy, contain only 100 Å fibers in adjacent sections viewed under the electron microscope (see Moses, 1956). Thin sections of metaphase or anaphase chromosomes (see Fig. 12–10), which also stain very intensely with the Feulgen reaction, show no internal structure except dense masses of 100 Å fibers. Finally, digestion of preparations with deoxyribonuclease results in loss of Feulgen staining in thick sections, and disappearance of 100 Å fibers from thin sections. This evidence, although indirect, clearly indicates that the 100 Å fibers in thin sections represent the nucleoprotein complex.

More direct evidence for the identity of the 100 Å fibrils with nucleoprotein has been obtained using autoradiography. Pioneer studies with this technique were carried out by M. J. Moses, and E. D. Hay and J. P. Revel. In this technique, a radioactive precursor of DNA, *tritiated thymidine*,[3] is added to cells undergoing DNA synthesis. As a result of exposure to the label, any DNA synthesized will contain radioactive thymine. The method is selective for DNA, because almost no other macromolecule of the cell will incorporate thymine in significant quantities. The radioactivity is detected by spreading a thin layer of photographic emulsion over sections of the cells. After storage in the dark, crystals of the emulsion are exposed over structures in the cells containing radioactive DNA. After development of the emulsion, silver grains, visible in both the electron and light microscopes, mark the location of radioactive organelles. With these methods, densely clumped areas of the nucleus that contain only 100 Å fibers prove to be radioactive, indicating that DNA is present in these fibers.

These experiments identify the 100 Å fibrils in interphase nuclei as the nucleoprotein complex. Direct evidence could be obtained for the identity of the thin fibers if a stain equivalent to the Feulgen reaction were available for electron microscopy. Unfortunately, there

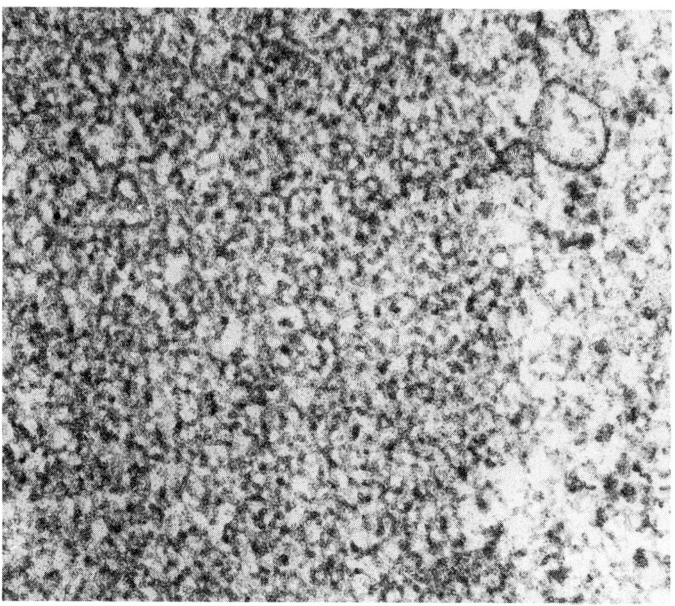

8–9. A thin section of a chromosome from the house cricket. The individual chromatin fibers, which are clearly visible in this micrograph, average between 80 and 90 Å in diameter. × 98,000. Courtesy of J. S. Kaye, R. McMaster-Kaye, and The Rockefeller University Press. [J. Cell. Biol. *31:159 (1966)*].

are no specific stains for DNA or nucleoprotein in electron microscope preparations. A number of methods, best described as selective but not specific for nucleic acids, increase the density of the 100 Å fibers in the nucleus, adding to the weight of evidence in their favor (see the description of indium staining in Chapter 3).

The Ultrastructure of Chromosome Fibers in Thin Sections

The fibers seen in chromosomes or chromatin in thin sections are indefinite in outline and have been reported as varying in diameter from about 20 Å to as much as 500 Å. The great majority of observations places the average diameter of these fibers at approximately 100 Å. From the few statistical analyses that have been made (see Kaye and McMaster-Kaye, 1966) it would appear that slightly less than 100 Å, about 80–90 Å, is a more correct description of chromatin fiber diameter in thin sections (see Fig. 8–9).

At these dimensions, the chromatin fibrils are much larger than the 20–30 Å expected for the nucleoprotein complex, indicating that the nucleohistone is further organized into a superstructure approximating 100 Å. As mentioned in previous sections of this chapter,

---

[3]Tritiated thymidine is a radioactive form of thymidine, in which some hydrogen atoms are replaced by the radioactive isotope *tritium* (^{3}H).

recent evidence from the Wilkins laboratory indicates that the nucleoprotein may be arranged into a 100 Å "supercoil." Other evidence (reviewed on p. 173) also supports this idea. Other possibilities are that two or more nucleohistone complexes are joined side by side in either extended or twisted form to produce the 100 Å diameter or that the 30 Å nucleohistone complex is surrounded by a matrix of nonhistone protein that increases its diameter to nearly 100 Å.

It has not been possible to distinguish between these alternatives with the electron microscope, even though the dimensions involved, from the 30 Å expected for nucleohistone up to the 100 Å fiber seen in thin sections, lie well within the resolving power of the instrument. In even the best micrographs no hint of consistent substructure is visible in the chromatin fibers. The fault in this case lies in the specimen preparation methods, which do not reveal substructural detail within the 100 Å fiber.

Occasionally fibers thinner than 100 Å are reported as consistently appearing in some nuclei. One interesting example occurs in *Euplotes,* a ciliate protozoan. Here the thinner diameter is possibly related to replication of the nucleohistone complex. In this organism, replication occurs in a distinct band that moves progressively along the macronucleus. These replicating regions can be readily identified in thin sections. In the replicating regions, the chromatin fibers of the macronucleus disperse into fine fibrils far smaller than 100 Å in diameter, at times practically invisible in the electron microscope. A reduction in diameter of this type at the time of replication might be anticipated, because it is probable that DNA must dissociate from tightly bound proteins and unwind from the double helix for replication to occur (see p. 311). It is possible that actively synthesizing or replicating DNA-histone regularly occurs in thin sections of all interphase nuclei as a fiber much thinner than 100 Å. Because these activities do not ordinarily occur in a coordinated, localized region, as in *Euplotes,* the smaller diameter fibers would be randomly distributed in the nucleus. These thinner fibers would be unrecognizable as replicating regions and would merely add to the scatter of dimensions observed.

In summary, it seems fairly certain that the 100 Å fibers seen in thin sectioned chromatin represent a configuration of the nucleohistone complex. Because DNA may not interact directly with osmium tetroxide, it is probable that the density seen represents an interaction with the protein component of nucleohistone. The unexpectedly large diameter of the fibers seen in thin sections indicates that a superstructure, probably a "supercoiling" of the nucleohistone complex, occurs. If such a supercoil does occur, no hint of it is visible in the fibers in thin sections, which show no regularly observable substructure.

The Ultrastructure of Chromatin as Seen in Isolated Chromosome Fibers

Almost all highly productive work with X-ray diffraction of DNA and nucleohistone has been carried out in isolated preparations. Such preparations have also been examined under the electron microscope, along with isolated metaphase chromosomes and interphase nuclei. This approach has been much employed in the study of nuclear ultrastructure because the essentially two-dimensional nature of thin sections yields a confusing image of the highly convoluted and folded chromatin fibers.

Isolated nucleohistone. Direct electron microscopic examination of nucleohistone preparations similar to those used for X-ray diffraction reveals little more than a fibrous substructure with a highly variable fiber diameter (Fig. 8–10). Reports of fiber dimensions in these preparations range from 100–500 Å. Aggregation of fibers in isolated nucleohistone probably accounts for some of this variability. Other variations in fiber diameter arise from the extreme sensitivity of the nucleohistone complex to the salt concentration of the isolating medium. Although results are still variable, even when salt concentration is carefully controlled, it appears that isolation in distilled water or media of very low ionic strength (10^{-3} M or less) yields relatively thick fibers averaging between 200 and 300 Å in diameter. This is much larger than expected for the nucleohistone complex and is even larger than the fibers observed in thin sections. Nucleohistone is insoluble in 0.15 molar salt; raising the salt concentration to 1 molar or higher results in almost complete dissociation of the complex into DNA and histone. Much thinner fibers, from 100 Å down to 20–30 Å are visible in these preparations. The very thinnest fibers seen after dissociation in 1 molar salt probably represent "naked" DNA. These effects of changes in salt concentration reflect the electrostatic nature of the bonds between DNA and histone in the complex.

The lack of meaningful results with the electron microscopy of isolated nucleohistone, as compared to X-ray diffraction of the same preparations, is due mainly to what is "seen" by either method. X-ray

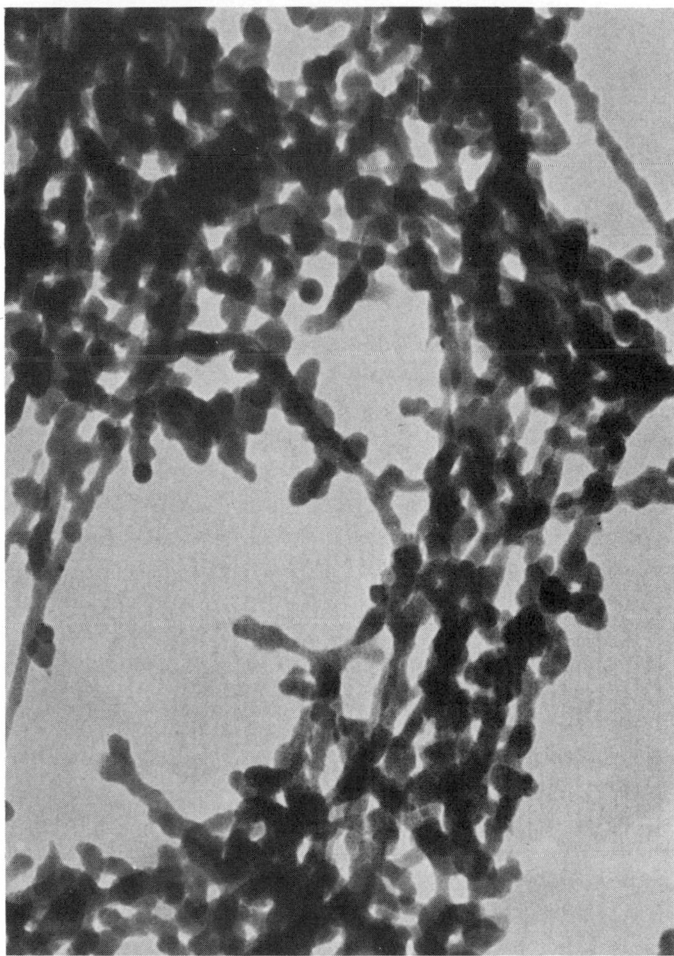

8–10. Fibers of nucleohistone in a preparation similar to those used for X-ray diffraction. Fiber dimensions in these preparations are highly variable. × 115,000. Photograph by the author.

diffraction detects only the ordered parts of a preparation. Disordered or amorphous regions do not produce an X-ray pattern and in a sense are ignored by this method of viewing molecular structure. As an example, a crystal embedded in an amorphous matrix would still produce a fairly clear X-ray diffraction pattern, even though the same preparation viewed under an electron microscope might not reveal the presence of the crystal at all. At present levels of development, extraction procedures for purifying nucleohistone evidently reduce much of the order present to amorphous material. Although sufficient order is retained to give circular reflections in X-ray diffraction patterns, the preparations are too disordered to be useful for electron microscopy. Good preservation for electron microscopy will no doubt require both good fixation and careful control of salt concentration throughout the isolation procedure, including not only the quantity but types of salt present. Although many variations have been tried, there has been little success in these efforts to date.

Isolated Chromosomes

A more successful approach for electron microscopy has been the direct isolation of nuclei or chromosomes from intact cells with as little disturbance as possible to the nuclei during the isolation procedure. This is a fundamentally different approach from the isolation of nucleohistone, in which an attempt is made to extract a quantity of the molecular complex in chemically pure form from a large mass of cells. In the work with isolated nuclei and chromosomes, two areas have been of primary interest: the ultrastructure of the unit fiber and the folding of this fiber into the gross structure of interphase nuclei or metaphase chromosomes.

In the period before the early 1960's, many structures were claimed as chromosomes in isolated preparations, usually on the basis of spiral structure in some part of the specimen. Even contaminating bacterial cells were on occasion confused with chromosomes. While many of these preparations probably contained chromosomes or nuclei, enough cellular debris was carried along to make identification difficult.

Although a few of the earlier attempts provided limited information, the first successful isolation of eukaryotic chromosomes for electron microscopy was not accomplished until the early 1960's, when Joseph G. Gall obtained excellent results by adapting a method previously used for the isolation of bacterial DNA to the cells of higher organisms. In Gall's technique, living, unfixed cells are spread on a water surface. The forces at the surface evidently separate the ground substance of the cell from chromatin without seriously disturbing the arrangement of the component chromosome fibers. The chromosomes are then dried for electron microscopy by the *critical point* method, a technique that accomplishes specimen dehydration without extensive disruption.[4]

---

[4]In this technique, isolated material is transferred in a series of steps from water to liquid carbon dioxide. The specimen, enclosed in a pressure vessel, is then raised in temperature from 20°C to about 45°C. As the temperature rises past 31°C, the *critical temperature* of CO_2, the liquid surrounding the chromosome fibers is almost instantaneously converted to a gas occupying the same space. The critical temperature

8–11. A nucleus isolated from the quail by the Gall technique. At the edge of the nucleus individual chromatin fibers can be distinguished. When isolated from unfixed tissue, the component fibers measure 250 Å in diameter. × 18,000. Photograph by the author.

Chromosomes isolated by the Gall technique contain masses of fibers 200–300 Å in diameter (Figs. 8–11 and 8–12). The fibers are loosely coiled in interphase nuclei and folded and tightly packed in metaphase chromosomes. In both cases, the coiling or folding pattern is apparently irregular; that is, no regular gyres, helices, or spirals are seen. Thus the arrangement of fibers is essentially the same as in thin sectioned preparations, in which no regular coils or gyres are seen either in interphase or metaphase chromosomes. The lack of regular gyres does not necessarily imply a lack of order because the folding pattern, even though irregular, may nevertheless be highly ordered.

The Gall technique has become widely employed because the method allows faithful preservation of the morphology of metaphase chromosomes. In many organisms the chromosomes at metaphase are individually recognizable and have a distinctive pattern of constrictions and swellings and characteristic width and length. These features are retained in the Gall preparations; at the same time the interior ultrastructure is also made visible.

In the isolated chromosomes, individual fibers are relatively smooth and fairly regular in dimensions (Fig. 8–13). Small variations in diameter give the fibers a slightly "lumpy" outline. In well-dried preparations, these variations are not pronounced, and the diameter varies within close limits between 240 and 260 Å. At isolated points, the extreme diameter may vary over wider limits, from 20–30 Å to 500 Å or more. This is dependent to a large extent on the quality of drying.

High resolution microscopy of these fibers has not yet revealed any hint of their interior ultrastructure. The surface of the fibers is smooth, and nothing can be seen internally except a faint, regular graininess with a

---

is the point above which a substance cannot exist as a liquid, no matter how high the pressure. Drying by this method causes little disruption of the specimen because a slowly receding air-liquid interface, with the attendant high forces, is avoided.

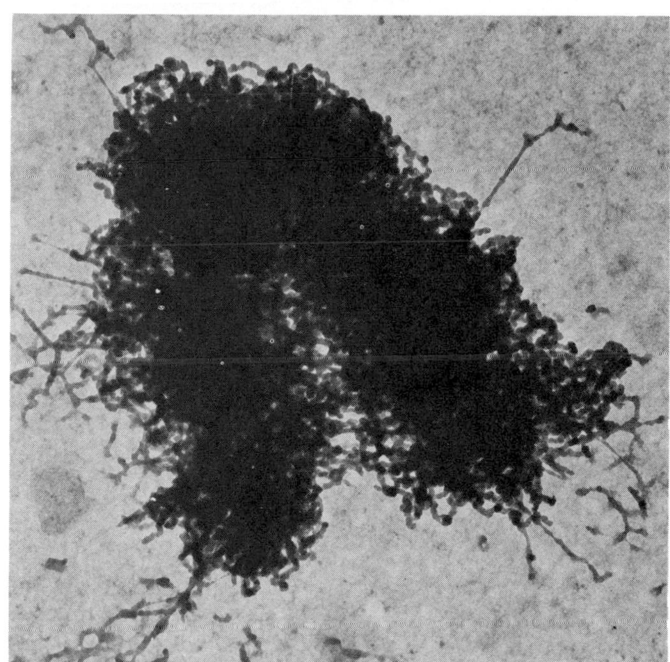

8–12. A metaphase chromosome isolated by the Gall technique from a bovine kidney cell. The component fibers are of the same average dimensions as the unit fibers of interphase nuclei isolated by the same technique. × 28,000. Photograph by the author.

periodicity of about 30 Å. This almost certainly means that denaturation of the fibers occurs during the isolation process. Various investigators claim to have seen multiple subunits within limited regions of fibers isolated by the Gall technique, or have suggested that a coil of nucleohistone exists within the fiber. However, whether these configurations are present or not, they have not yet been clearly demonstrated in isolated fibers.

The fibers in these isolated chromosomes are unexpectedly large, almost ten times the diameter of a DNA duplex and more than twice the size of the 100 Å unit fibers seen in thin-sectioned nuclei. The hypotheses mentioned earlier have also been advanced to account for these discrepancies in dimensions: that the fibers in the isolated preparations result from (1) coiling of nucleohistone into a "supercoil," (2) aggregation of a multiple number of nucleohistone subunits, or (3) deposition of a matrix of material around a single nucleohistone complex. Each hypothesis has its proponents, although at the present time no definitive evidence has been produced in support of any of the proposals.

Some limited indication of the true structure has come from the examination of fibers partly disrupted by either enzymatic digestion or mechanical disturbance during the isolation procedure. Such fibers show lengths of reduced diameter having fairly regular dimensions (Fig. 8–14) from 50–100 Å. These areas have been interpreted in two different ways. One hypothesis, advanced by E. J. DuPraw, maintains that a single coiled subunit forming the 250 Å fiber uncoils or "spins out" to form the region of lessened diameter. An alternate hypothesis, stated by the author, proposes that a matrix has been lost from the thinner regions, exposing a single core fiber. These hypotheses are not mutually exclusive, as it is conceivable that a mechanism involving both processes may produce the thinner areas. In any event, the aspect of these regions suggests that the fiber contains a single subunit. Thus the remaining hypothesis, that the 250 Å fiber contains multiple subunits arranged in parallel fashion (formulated by Hans Ris) does not appear to be compatible with the evidence obtained from slightly disrupted fibers.

Some more recent work indicates that part of the total diameter of the isolated complex may indeed represent a matrix material. The author and his colleagues have studied the effect of prefixation on chromosomes isolated by the Gall technique. In the original technique, chromosomes are isolated from unfixed cells. If cells are prefixed in formalin, the component fibers are of smaller diameter, approaching the 100 Å seen in thin-sectioned nuclei. Loss of a matrix seems the most likely explanation for the reduced diameter in this case, because dissociation of multiple subunits or the uncoiling of a subfiber would be expected to produce serious disarray of nuclei or metaphase chromosomes in the preparations. Implicit in this line of argument is the assumption that much of the "matrix" forming the difference in thickness between the 100 Å and 250 Å diameter is extraneous material that interacts with the nucleohistone complex when the nucleus is disturbed during isolation. This interaction would add material to the surface of the fiber, increasing the diameter from 100 Å to 250 Å.

All of this is highly speculative. Working out the substructure of the component fiber as it appears in both thin-sectioned and isolated nuclei remains one of the most challenging problems in cell biology. The picture that emerges so far is that the nucleohistone complex, placed at 30 Å by X-ray diffraction, is proba-

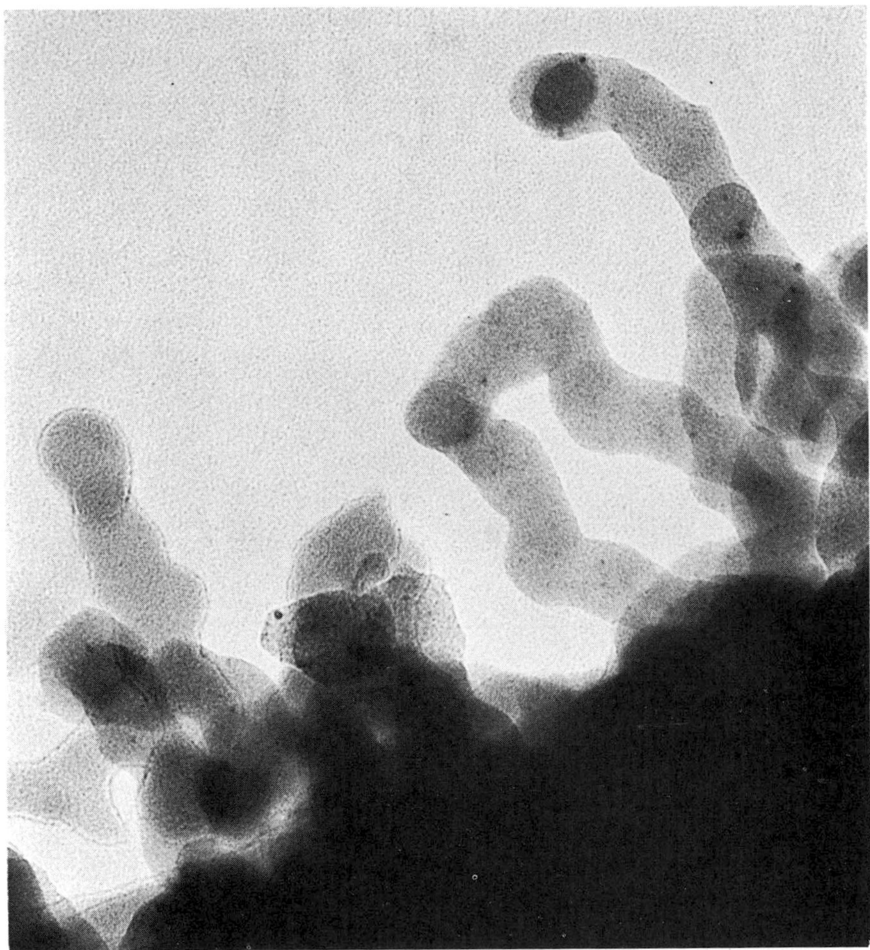

8–13. Chromosome fibers isolated from the newt, Triturus viridescens. *Typically, no substructure can be seen within the fibers. The diameter of the fibers shown averages approximately 250 Å. × 410,000. Photograph by the author.*

bly arranged into a thicker complex in living nuclei. The diameter of the larger complex is not known with certainty, but work with X-ray diffraction and thin sectioning indicates that the diameter is about 100 Å or slightly less. Fibers isolated by the Gall method approach the same diameter if cells are prefixed before isolation. While the precise superstructure of nucleohistone in the 100 Å fiber is unknown, information from X-ray diffraction indicates a supercoil of nucleohistone 100 Å in diameter and 120° in pitch. This degree of coiling would involve a shortening by about 50 percent of coiled over extended, "naked" DNA.

Prokaryote Nuclei

The structure of chromosome fibers described to this point is typical of the nuclei of almost all higher organisms, with very few exceptions. The exceptions that do occur among higher organisms are limited to highly specialized cells, such as erythrocytes and spermatozoa, in which fibers thicker than 100 Å occur in some species. In bacteria and blue-green algae, however, the nucleoid is quite different in appearance under both the light and electron microscopes. The term "prokaryote" applied to these organisms reflects this difference by implying that the nuclear configuration found in bacteria and blue-green algae represents a primitive form that preceded the eukaryote nucleus in evolution.[5]

With phase contrast optics, the nucleoids of living bacteria appear as almost empty, structureless regions within bacterial cells. These regions are variable in

[5] Many terms, including *nucleoid, nuclear equivalent, chromatin body,* and *genophore,* have been applied to the bacterial nucleus to emphasize the structural differences between this form and the eukaryote nucleus. The term "nucleoid" or simply "bacterial nucleus" or "bacterial chromosome" will be used here.

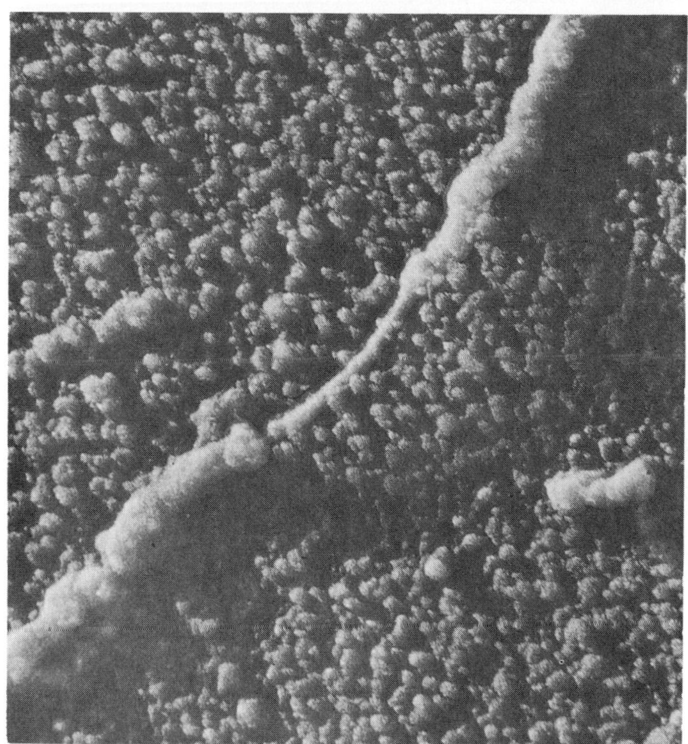

8–14. *A chromosome fiber, isolated from the newt* Triturus viridescens, *in which an apparent single core has been exposed, possibly by loss of a matrix.* × *215,000. Photograph by the author.*

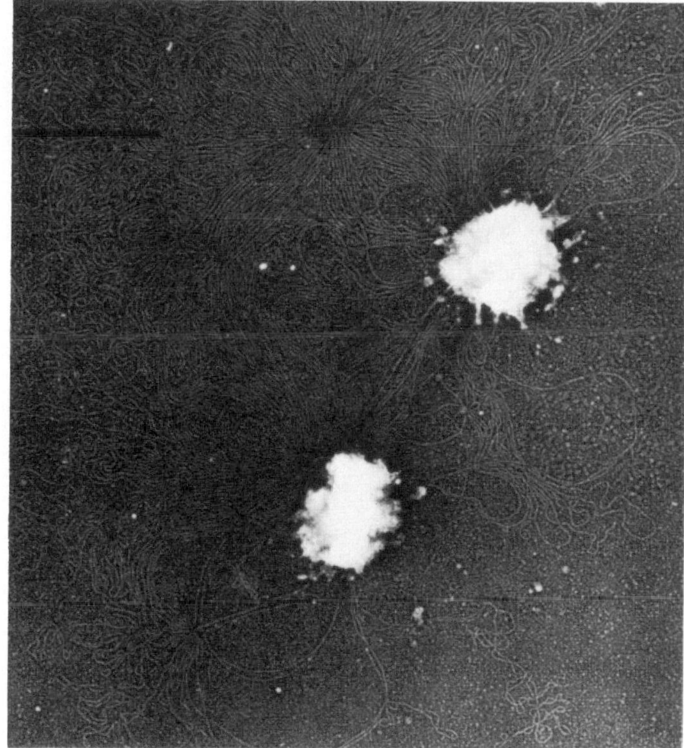

8–15. *Clusters of DNA released by spreading bacterial cells on a water surface. Before spreading, the cell walls were removed from the bacteria by treatment with lysozyme. The DNA molecules were made visible by shadowing, using a modification of the basic technique in which the specimen is rotated 360° during evaporation of the shadowing metal (refer to Fig. 3–26). By this process a layer of metal is built up on all sides of the specimen, greatly increasing the contrast and ease of resolution.* × *20,500. Courtesy of A. K. Kleinschmidt.*

shape and in living cells are seen to undergo transitions in a matter of minutes between compact, more or less ellipsoidal forms and more elongated configurations. These areas, in fixed preparations, are stained by the Feulgen reaction for DNA. In contrast to the eukaryotic nucleus, stains for histone and nonhistone protein produce negative results. Protein either is absent from the bacterial nucleus or exists in quantities too small to be detected by these methods. Some investigators have suggested that the bacterial chromosome may be complexed with small quantities of short polypeptide chains rather than large protein molecules. All studies indicate, however, that most of the DNA in the bacterial nucleus is "naked," that is, not complexed with a protein. Supposedly, the negatively charged phosphates along the molecule are neutralized by combination with a divalent cation.

The Gall technique for isolation of eukaryotic chromosomes was adapted from a technique developed earlier for bacterial cells by A. K. Kleinschmidt. Bacterial chromosomes isolated on a water surface by the Kleinschmidt technique appear as masses of very thin fibers. The component fibers approach the dimensions of a single DNA molecule in thickness (Fig. 8–15). Very few free ends are visible in these masses, suggesting that each bacterial cell contains one long DNA molecule. Other lines of evidence, including "mapping" by genetic crosses, indicate that the bacterial chromosome is circular. This was confirmed in experiments by John Cairns, who made light microscope autoradiographs of isolated bacterial chromosomes. Although the individual fibers of the bacterial chromosomes are not resolved in Cairns' autoradiographs, a pattern of exposed grains over the molecule is visible under the light

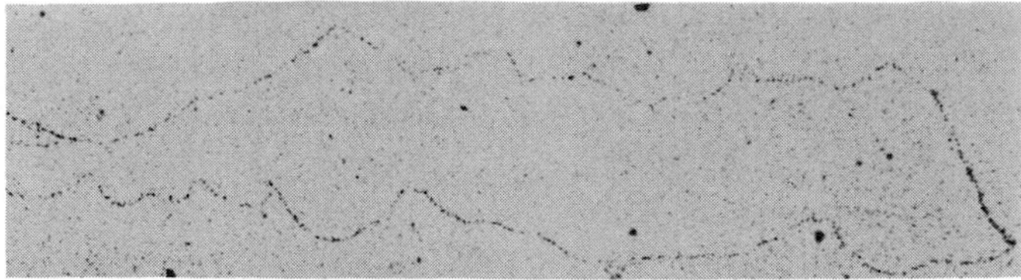

8–16. Pattern of grains exposed in a photographic emulsion by radioactive DNA isolated from the bacterium E. coli. *The DNA was made radioactive by growing cells in tritiated thymidine. The exposed grains clearly indicate that the "chromosome" of this bacterium exists in the form of a closed circle. × 175. Courtesy of J. Cairns.*

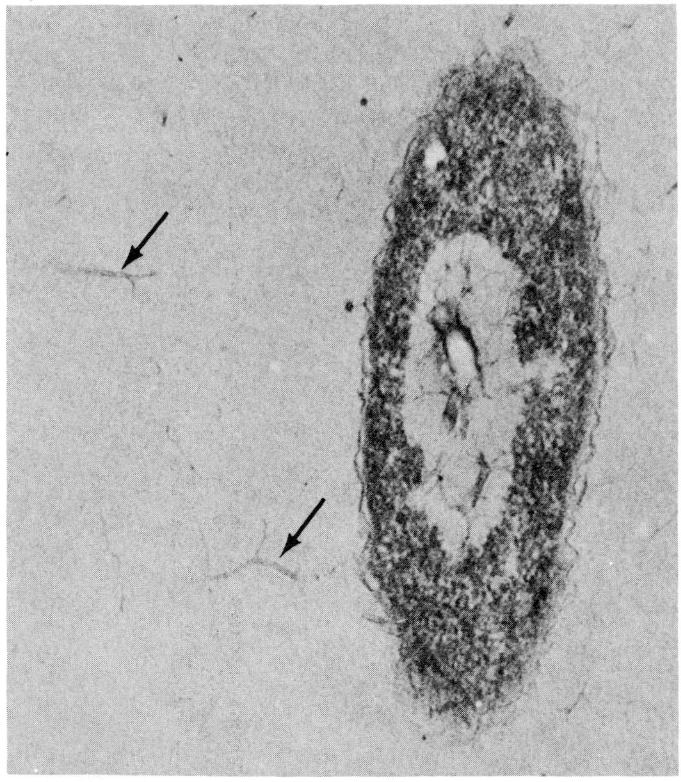

8–17. Results of fixation of a bacterial cell (E. coli) *in osmium tetroxide without addition of calcium ion or amino acids. The nucleoid is coagulated into dense, irregular masses. Surrounding the cell is DNA (arrows), added to the suspension medium, which has also been coagulated by the fixative. × 72,000. Courtesy of W. H. Schreil and The Rockefeller University Press, from* J. Cell Biol. *22:1 (1964).*

microscope (Fig. 8–16). These grains trace a circle, confirming that the bacterial chromosome is a closed circle. Measurement of the circles reveals an included length of 1,000 to 1,400 microns of DNA.

Kellenberger Fixation of Bacterial DNA

The bacterial nucleus has proved to be much more sensitive than eukaryote nuclei to the conditions of fixation for electron microscopy. The first attempts at fixation employed the standard veronal acetate-buffered osmium tetroxide solution. This resulted in coagulation of the nuclear material into dense clumps (Fig. 8–17). The coagulated nuclear material was separated from the cytoplasm by an almost structureless empty space, giving the appearance of a compact "chromosome" suspended in a void inside the cell. Although some investigators were convinced that the dense clump was an accurate representation of the bacterial chromosome, others reasoned that the coagulated appearance indicated destruction and shrinkage of the nuclear material during fixation.

Edward Kellenberger studied carefully the effect of different conditions of fixation on the bacterial nucleus. Kellenberger was able to confirm that osmium does not interact with or fix bacterial DNA. He also observed that the bacterial nucleus was extremely sensitive to the concentration of divalent cations in the fixative, particularly calcium (Ca^{+2}). Working with isolated DNA, Kellenberger also found that a mixture of amino acids, if added to the fixative, caused an interaction between osmium tetroxide and DNA. Finally, Kellenberger noted that a heavy ion, uranyl (UO_2^{+2}), if added to the solutions for rinsing after fixation, "stabilized" the DNA.

All of these findings were combined into a fixation procedure that has appropriately become known as "Kellenberger fixation." The fixative, osmium tetroxide, is used in a buffered solution to which amino acids and calcium ions are added. After fixation, specimens are rinsed in a solution containing uranyl ions. Bacterial nucleoids fixed by this method contain an evenly distributed array of very fine 20–60 Å fibrils instead of a coagulated mass in the nucleus (Fig. 8–18). Often some order is discernible in the arrangement of the fibers (Fig. 8–19).

The Kellenberger fixative is believed to interact with DNA of the bacterial type according to the following

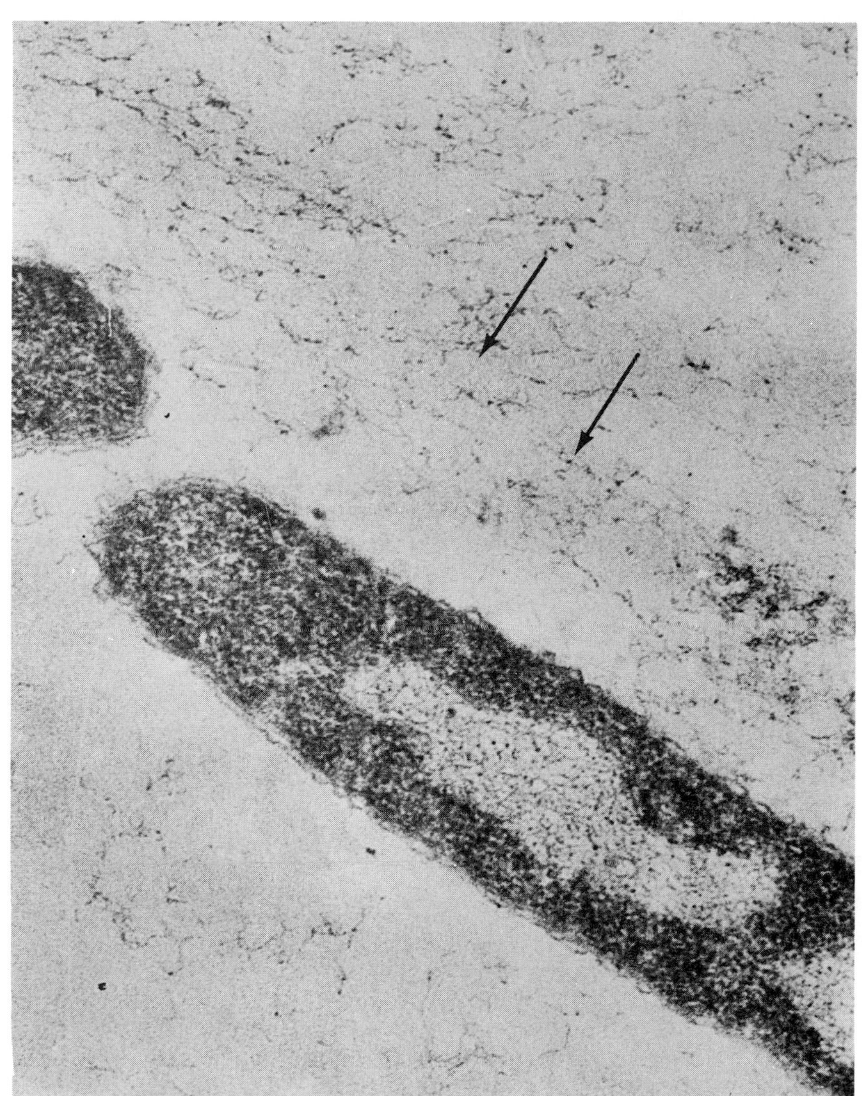

8–18. A bacterial cell (E. coli) *fixed by the Kellenberger method (see text). The nucleoid contains fine 20–60 Å fibrils after this treatment; DNA added to the suspension medium is also fixed in the form of fine fibrils (arrows). × 77,000. Courtesy of W. H. Schreil and The Rockefeller University Press, from* J. Cell Biol. *22:1 (1964).*

pattern. Supposedly, the bacterial DNA, uncomplexed with protein, is neutralized in the undisturbed, living nucleus by interaction with a positive ion, possibly calcium. Unless a divalent cation is added in quantity to the fixative, death of the cell at the time of fixation causes a disturbance in the ionic environment of the DNA, coagulating the nucleus. The presence of calcium in sufficient concentration in the fixative prevents this coagulation by complexing with the phosphate groups on the surface of the DNA molecule. Added calcium has little effect on the quality of fixation in the chromatin of higher organisms, probably because the protein complement is strongly bound to the phosphate groups of DNA and maintains the nucleus in an uncoagulated condition during the transition from the living to the fully fixed condition.

The amino acids added to the bacterial fixative evidently interact with both osmium tetroxide and DNA and serve as a vehicle for attaching the heavy metal to the DNA molecule. Most likely to be active in this process are the basic amino acids arginine and lysine. The stabilization of bacterial DNA by uranyl ions in the final rinse after fixation probably reflects an attachment to any remaining phosphate groups unoccupied by amino acids.

All of these effects are believed to reflect the absence of a complexing protein in the bacterial nucleus. The Kellenberger reaction, highly specific for uncomplexed

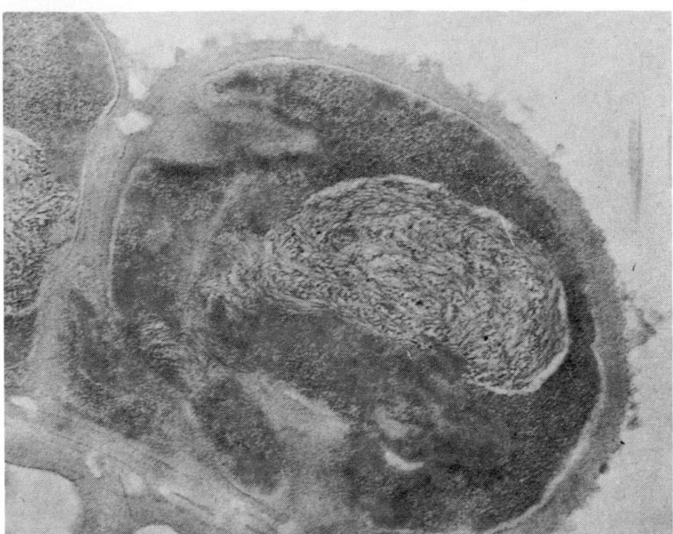

8–19. Bacterial nucleoid of Staphylococcus aureus, *showing an arrangement of the component fibrils in an apparently ordered folding pattern.* × *54,500. Courtesy of A. K. Kleinschmidt.*

or naked DNA, has revealed this type of DNA in some unexpected places. One place in which bacterial DNA was found to occur, this time with little surprise, was in blue-green algae (Fig. 8–20). These organisms show typical prokaryote morphology, including the absence of a nuclear envelope and discrete cytoplasmic organelles (see Chapter 1). The nucleus of blue-green algae is coagulated by standard osmium fixation and contains fine fibrils after Kellenberger fixation. The nuclear fibers in these algae seem to be slightly larger than in bacteria, approximately 50–70 Å as compared to the 20–60 Å reported for bacteria. The closely related "bacterial fungi" (schizomycetes) also give a typical Kellenberger reaction.

A somewhat less expected place for bacterial DNA to appear was in the *dinoflagellates,* a group usually classified with the protozoa. In these protozoa, in contrast to the prokaryotes, the nucleus is separated from the cytoplasm by a nuclear envelope complete with pore complexes in most species. The pores are reduced in number, but otherwise typical in appearance (see also p. 234). The dinoflagellates also possess discrete cytoplasmic organelles, including mitochondria, endoplasmic reticulum, Golgi complex, and typical eukaryotic flagella. Within the dinoflagellate nucleus, chromosome-like bodies are visible. These small, sausage-like structures are intensely stained by the Feulgen reaction. The dinoflagellate chromosomes are peculiar, in that they apparently remain condensed in a coiled condition throughout all stages of the cell cycle. Surprisingly, Kellenberger fixation reveals that these condensed bodies contain 30–80 Å DNA fibers of the bacterial type, folded in the body of the chromosome in a highly ordered pattern (Fig. 8–21).

The Kellenberger technique also reveals that chloroplasts and mitochondria contain deposits of DNA that react in a pattern similar to bacterial DNA. In the early 1960's, when this was first demonstrated by Hewson Swift, Hans Ris, and others, the chloroplasts and mitochondria were most unexpected places to find DNA of any type, even though it had been established for some time that some characteristics of these organelles followed a cytoplasmic rather than nuclear pattern of inheritance. Even with this foreknowledge, it came as a surprise that DNA was a regular component of these organelles. The fact that the DNA present is of the bacterial type is most intriguing and has added support to the hypothesis that mitochondria and chloroplasts originally evolved from parasitic or symbiotic prokaryotes (see Chapter 16).

Some Evolutionary Implications of Nuclear Structure in Bacteria, Blue-Green Algae and Dinoflagellates

The structural relationships between the nuclei of bacteria, blue-green algae, and dinoflagellates revealed by the Kellenberger reaction give a number of clues to the evolution of these groups. The possible evolutionary link between bacteria and blue-green algae is obvious, and it is not difficult to conceive of an ancestral form of the present-day blue-green algae, probably a colonial form similar to the present-day schizomycetes, arising from early bacteria. Dinoflagellates are more difficult to deal with, because these organisms are often classified as protozoa; the protozoa are not considered to be closely related to bacteria. Many evolutionists consider the protozoa to represent a separate line of evolution, branching off from the mainstream leading to the metazoa at about the time of evolution of the first eukaryotic, flagellated cells. It is likely that at a very early time, during the development of the eukaryotic

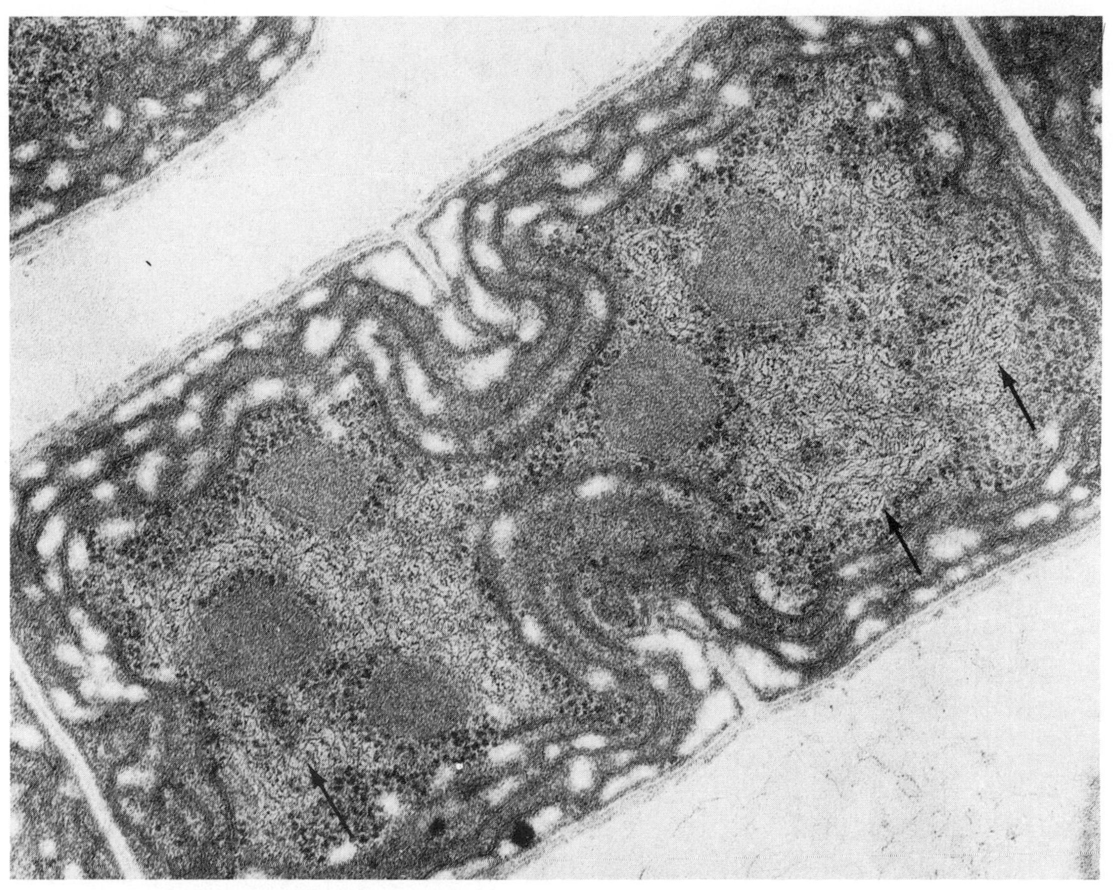

8–20. *A cell of the blue-green alga* Phormidium luridum *after Kellenberger fixation. After such fixation, much of the cell interior contains masses of fine fibrils (arrows).* × *60,000. Courtesy of Mercedes R. Edwards.*

nuclear system in flagellated cells, an ancestral form not yet completely eukaryotic gave rise to the dinoflagellates and possibly other protozoa.

Most interesting from the standpoint of cell biology are some indications in bacteria, blue-green algae, and dinoflagellates of a possible evolutionary sequence in the development of structural complexity in the chromosome. A gradual increase in fiber diameter, from the 20–60 Å noted for bacteria to the 30–80 Å reported in the dinoflagellates, is apparent if these groups are compared (see Table 8–2). A corollary to this is the observation that the blue-green algae are somewhat less sensitive to the conditions of fixation than bacteria. Fine fibers can be seen in the nuclei of these algae after ordinary osmium fixation if the *p*H is carefully controlled. The nuclei of the bacterial fungi also react more readily with osmium; the nuclear areas in both of these forms are frequently described as more dense than the bacterial nucleus. These observations suggest that a protein component is present in increased amounts in the nuclei of the bacterial fungi and blue-green algae.

Another possible evolutionary trend is suggested by the presence of extrachromosomal material in the nucleoplasm of the dinoflagellates. It is obvious that the volume of the nucleus is not completely occupied by the chromatin in these organisms, because the chromosomes are separate, discrete masses even at interphase. Between these chromatin bodies are fine fibrils and granules of various sizes that almost certainly represent extrachromosomal material. Distinction between chromosomal and extrachromosomal material can also be made in more typical eukaryotes, although dispersion of the chromatin makes identification more difficult (Fig. 8–22; see also Fig. 10–11). Development of a distinctive extrachromosomal nucleoplasm is probably a consequence of the appearance of the nuclear envelope in evolution; the fine fibrils and granules no doubt are a product of the interphase chromosome. Other work (Chapter 10) supports this conclusion and indicates that this extrachromosomal material is ribonucleoprotein.

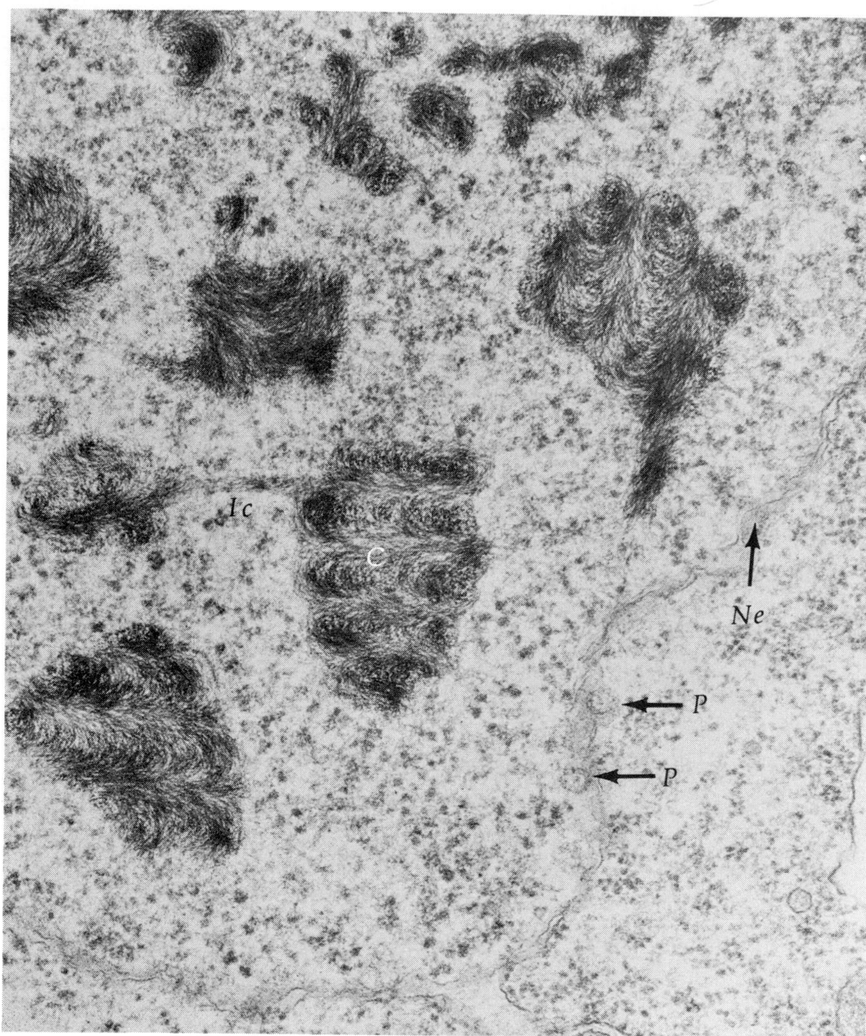

8–21. *An interphase nucleus of the dinoflagellate* Gyrodinium cohnii. *The chromosome-like bodies (C) contain aggregations of DNA closely similar to prokaryotic nucleoids in dimension, structure, and properties. Other characteristics typical of eukaryotes are present, including a nuclear envelope* (Ne) *with pore complexes* (P), *and interchromosomal nucleoplasm* (Ic). × *38,500. Courtesy of D. F. Kubai, H. Ris, and The Rockefeller University Press.* [J. Cell Biol. *40:508 (1969)*]

Table 8–2 Some Interesting Structural Relationships between Prokaryote and Eukaryote Nuclei

	Bacteria	*Blue-green algae*	*Dinoflagellates*	*Eukaryotes*
Bacterial DNA	+	+	+	−
Fiber size in nucleus	20–60 Å	50–70 Å	30–80 Å	80–100 Å
Positive stain for basic proteins in nucleus	−	−	−	+
Nucleolus	−	−	+	+
Extrachromosomal nucleoplasm	−	+?	+	+
Nuclear envelope	−	−	+	+

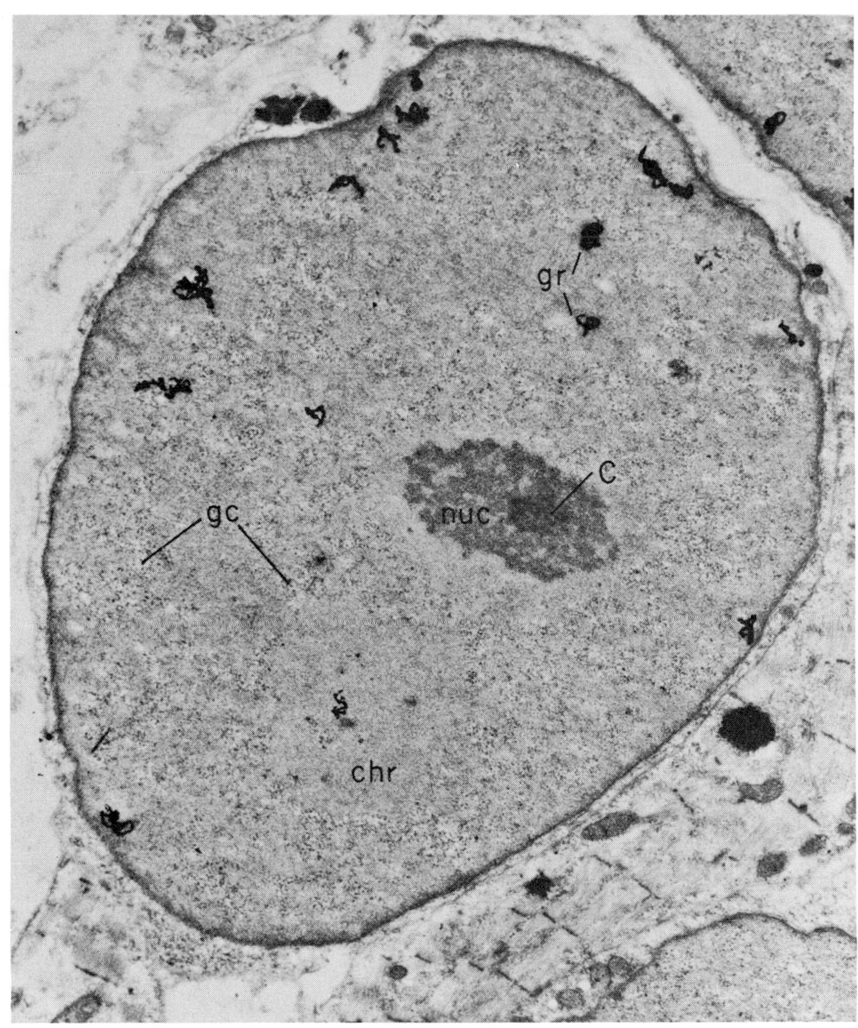

8–22. *A thin-sectioned nucleus from the salamander* Amblystoma. *Extrachromosomal nucleoplasm (gc) can be distinguished as finely textured material containing small granules. chr, chromatin; nuc, nucleolus; C, densely coiled region of nucleolus;* gr, *grains of photographic emulsion, exposed by radioactive DNA labeled by tritiated thymidine. × 8,700. Courtesy of Elizabeth D. Hay, J. P. Revel, and The Rockefeller University Press.* [J. Cell Biol. *16:29 (1963)*]

Summary

The interphase chromosome is a complex between DNA and protein. Several lines of evidence have established that of these constituents, the DNA is the hereditary material. By means of the Feulgen stain, DNA was shown to be a regular component of the chromosomes and to persist during all stages in the life cycle of the cell. In contrast, the major protein, histone, is replaced by protamine during the development of spermatozoa in some animals. Feulgen staining, when applied in quantitative studies, also established that with rare exceptions the amount of DNA per nucleus is constant in the cells of a given species. In 1944, Avery and his co-workers identified the transforming agent of pneumococcus as DNA; finally, in 1952, Hershey and Chase established that the infectivity and reproduction of bacteriophage depends on penetration of the host cell by DNA. These experiments, taken together, provide the fundamental proof of DNA as the molecular basis of heredity.

During the 1950's, the molecular structure of DNA was deduced from X-ray diffraction data by Watson, Crick, and Wilkins. These workers established that the molecule consists of two backbone chains of nucleotides wound around each other into a double helix. At any level, the two chains are held together by hydrogen bonds formed between the nitrogenous bases of opposite nucleotides. The pattern of hydrogen bonding and the restrictions on the total diameter of the double helix specify that at any level only adenine-thymine or guanine-cytosine pairs are formed between the two chains. The base pairs lie in a plane perpendicular to the long axis of the molecule, separated by a space of 3.4 Å. A complete turn of the helix, which occupies 34 Å, includes ten base pairs.

In the interphase nucleus of eukaryotes, the DNA is bound with histone in a 1:1 ratio by weight. Other proteins, identified simply as "nonhistone" proteins, are probably also a regular constituent of the eukaryotic chromosome. The precise manner of binding and function of any of these proteins is unknown. In thin sections, the nucleoprotein complex of the interphase chromosome appears as a dense fiber of indistinct outline approximately 100 Å in diameter. This dimension is larger by several orders of magnitude than the 30 Å diameter expected for the nucleoprotein complex. The relatively large diameter seen in thin sections probably results from "supercoiling" of the basic nucleoprotein complex. In the nucleoids of prokaryotes, smaller fibers, between 20 and 70 Å, are observed. This smaller diameter and the peculiar reaction of prokaryote chromatin to fixatives for electron microscopy probably reflect the absence of detectable quantities of protein in the chromosomes of these organisms. "Naked" DNA of this type also occurs in the dinoflagellates and in small quantities in the chloroplasts and mitochondria of higher organisms.

Suggestions for Further Reading

Books

Bonner, J., and P. Ts'O, eds., 1964. *The Nucleohistones.* Holden-Day, San Francisco. A collection of articles dealing with structure and function of the nucleohistones.

Watson, James D., 1968. *The Double Helix.* Atheneum, New York.

Papers and Review Articles

Avery, O. T., C. M. MacLeod, and M. McCarty, 1944. Studies on the chemical nature of the substance inducing transformation of pneumococcal types. *Journal of Experimental Medicine,* vol. 79, pp. 137–158.

Bahr, G. F., 1954. Osmium tetroxide and ruthenium tetroxide and their reactions with biologically important substances. *Experimental Cell Research,* vol. 7, pp. 457–479.

Feughelman, M., R. Langridge, W. E. Seeds, A. R. Stokes, H. R. Wilson, C. W. Hooper, M. H. F. Wilkins, R. K. Barclay, and L. D. Hamilton, 1955. Molecular structure of deoxyribose nucleic acid and nucleoprotein. *Nature,* vol. 175, pp. 834–838.

Fuhs, G. W., 1965. Fine structure and replication of bacterial nucleoids. *Bacterial Reviews,* vol. 29, pp. 277–293.

Hershey, A. D., and M. C. Chase, 1952. Independent functions of viral protein and nucleic acid in growth of bacteriophage. *Journal of General Physiology,* vol. 36, pp. 39–56.

Kaye, J. S., and R. McMaster-Kaye, 1966. The fine structure and chemical composition of nuclei during spermiogenesis in the house cricket. *Journal of Cell Biology,* vol. 31, pp. 159–179.

Kellenberger, E., A. Ryter, and J. Sechaud, 1958. Electron microscope studies of DNA-containing plasms III: vegetative and mature phage DNA as compared with normal bacterial nucleoids in different physiological states. *Journal of Biophysical and Biochemical Cytology,* vol. 4, pp. 671–678.

Kubai, D. F., and H. Ris, 1969. Division in the dinoflagellate *Cyrodinium cohnii* (Schiller). *Journal of Cell Biology,* vol. 40, pp. 508–528.

Moses, M. J., 1956. Studies on nuclei using correlated cytochemical, light, and electron microscope techniques. *Journal of Biophysical and Biochemical Cytology,* vol. 2 (supplement), pp. 397–406.

Ohba, Y., 1966. Structure of nucleohistone I: hydrodynamic behavior. *Biochimica et Biophysica Acta,* vol. 123, p. 76.

Pardon, J. F., M. H. F. Wilkins, and B. M. Richards, 1967. Superhelical model for nucleohistone. *Nature,* vol. 215, p. 508.

Phillips, D. M. P., 1962. The histones. *Progress in Biophysics and Molecular Biology,* vol. 12, pp. 211–281.

Ris, H., 1969. The molecular organization of chromosomes. In A. Lima-de-Faria, ed., *Handbook of Molecular Cytology.*

North-Holland Publishing Company, Amsterdam, pp. 221–250.

Swift, H., 1950. The constancy of deoxyribose nucleic acid in plant nuclei. *Proceedings of the National Academy of Sciences,* vol. 36, pp. 643–654.

Watson, J. D., and F. H. C. Crick, 1953. Molecular structure of nucleic acids: a structure for deoxyribose nucleic acid. *Nature,* vol. 171, pp. 737–738.

Wilkins, M. H. F., 1956. Molecular structure of deoxyribose nucleic acid and nucleoprotein and possible implications in protein synthesis. *Biochemical Society Symposia,* vol. 14, pp. 13–26.

Wolfe, S. L., 1969. Molecular organization of chromosomes. In E. E. Bittar, ed., *The Biological Basis of Medicine,* vol. IV. Academic Press, New York, pp. 1–42.

9
The Metabolism of RNA in the Nucleus

The DNA of the nucleus enters two major processes in the life cycle of the cell. In one, the information contained in the DNA molecule is *replicated* in preparation for division of the cell. This process, which insures that the directions for synthesis are passed intact to daughter cells, is a part of cellular reproduction and is considered in Unit 4 of this book. The second major functional role of DNA is in the *transcription* of coded information into directions for the synthesis of proteins. The essential feature of transcription is the direct synthesis of RNA on the DNA template. Through this means, the information contained in the DNA is transferred to RNA.

Several types of RNA are synthesized on chromosomal DNA. Each plays a role in the synthesis of protein in the cell. Some types of RNA produced by transcription are well characterized in terms of both molecular structure and function. Others are less well known, and the structure and function of several species remains to be discovered. Most RNA molecules become associated with protein to form *ribonucleoprotein* during or soon after transcription on chromosomal DNA. Except for ribosomal protein, very little is known about the identity, structure, or function of the associated proteins.

The total scheme of protein synthesis will be discussed in this and the following chapters of this unit. This chapter considers the synthesis and metabolism of RNA in the cell nucleus. The next chapter describes the morphological aspects of RNA metabolism and the transport of the synthesized ribonucleoprotein through the nuclear envelope to the cytoplasm. The final chapter in this unit discusses the involvement of RNA in the synthesis of protein in the cytoplasm, the regulation of protein synthesis, and the morphology of the several cytoplasmic organelles engaged in protein synthesis.

A Brief Overview of Protein Synthesis

Protein synthesis takes place throughout interphase and continues at a reduced rate during cell division. The first step in this process is the transfer of the information required for synthesis from DNA, the storage molecule, to the various types of RNA. In the transfer of information from DNA to RNA, the sequence of bases in DNA is copied precisely (transcribed) into a complementary sequence in RNA. Transcription is catalyzed by the enzyme *RNA polymerase*. This enzyme initiates the synthesis of an RNA messenger at the beginning of a length or unit of DNA representing a transcription unit, which may contain one or more genes or *cistrons*. When the end of the unit is reached, the polymerase molecule terminates synthesis, and the newly manufactured *messenger RNA* (mRNA) is released from the chromosome. RNA synthesis on a gene or cistron can be initiated repeatedly by RNA polymerase. Thus each transcription unit can serve as template for hundreds or even thousands of identical RNA molecules.

The specific directions for the synthesis of protein are transcribed into mRNA. Each molecule of mRNA contains the information necessary to assemble a specific sequence of amino acids into a complete polypeptide chain. This chain may be either a complete protein molecule or a polypeptide subunit of a protein molecule.

Very soon after transcription, mRNA probably combines with a protein and passes through the nuclear envelope into the cytoplasm. In the cytoplasm, the messenger associates with one or more ribosomes, small particles approximately 150–200 Å in diameter. Each ribosome is a complex structure consisting of a combination of *ribosomal RNA* (rRNA) with a group of as many as 60 or more different proteins. The ribosome is the site of protein synthesis and may be regarded as a machine that synthesizes protein by assembling individual amino acids into a specific polypeptide sequence. In this assembly, the ribosome follows the directions carried by mRNA. According to current hypotheses, the ribosome moves along the information tape represented by the mRNA, assembling amino acids into a polypeptide chain that gradually grows in size as the ribosome traverses the length of the messenger.

The ribosome uses as raw materials individual amino acids carried by a third class of relatively small RNA molecules called *transfer RNA* (tRNA). Each of the 20 amino acids (see Chapter 2) is carried to the ribosome by specific tRNA molecules. Transfer RNA acts as the adapter molecule, the agent directly linking the amino acid code to particular amino acids.

In the following paragraphs a hypothetical example of protein synthesis will be given to illustrate the total process, without including supporting evidence at this time. Consider the synthesis of a hypothetical protein containing a sequence of only two different amino acids, tryptophan and phenylalanine, which alternate in the polypeptide chain. The information coded in DNA is stored in the sequence of the four nitrogenous bases, adenine (A), thymine (T), guanine (G), and ctyosine (C). These are taken by threes to form the DNA codewords or *codons*. Using the four bases in all different 3-base combinations, 64 different codewords can be formed. This is more than sufficient for the 20 amino acids that make up protein molecules. One DNA codeword for tryptophan is known to be the sequence adenine-cytosine-cytosine (ACC); one codon for phenylalanine is adenine-adenine-guanine (AAG); additional codewords, or "synonyms," exist for these and the remaining amino acids. In the first step of the process, this DNA code is transcribed into the RNA messenger. In the case of our hypothetical model, the DNA sequence[1] consisting of the code ACC/AAG/ACC/AAG . . . will be transcribed into the complementary messenger RNA sequence UGG/UUC/UGG/UUC . . . by the formation of appropriate base pairs. RNA contains uracil (U) instead of thymine as a base pair for adenine. Fig. 9–1 diagrams the formation of the complementary messenger RNA sequence.

After release from DNA, the newly synthesized messenger RNA enters the cytoplasm and attaches to a ribosome. Surrounding the ribosome is a pool of the various tRNAs with attached amino acids. In this pool are tRNAs carrying tryptophan and phenylalanine. Each tRNA contains a recognition site (called an *anticodon*) for the code designating these amino acids in the mRNA (Fig. 9–2). Because the code for tryptophan in the messenger RNA is UGG, the tRNA carrying tryptophan would be expected to contain the anticodon site ACC. The corresponding anticodon for phenylalanine is AAG.

---

[1] This coding sequence is carried in one of the two strands of the DNA duplex, called the "sense" strand. The opposite, complementary strand, called the "missense" strand, carries the complementary sequence, TGG/TTC/TGG/TTC . . . in the example given. The missense strand probably has no direct role in RNA transcription but is required for DNA replication (see Chapter 12).

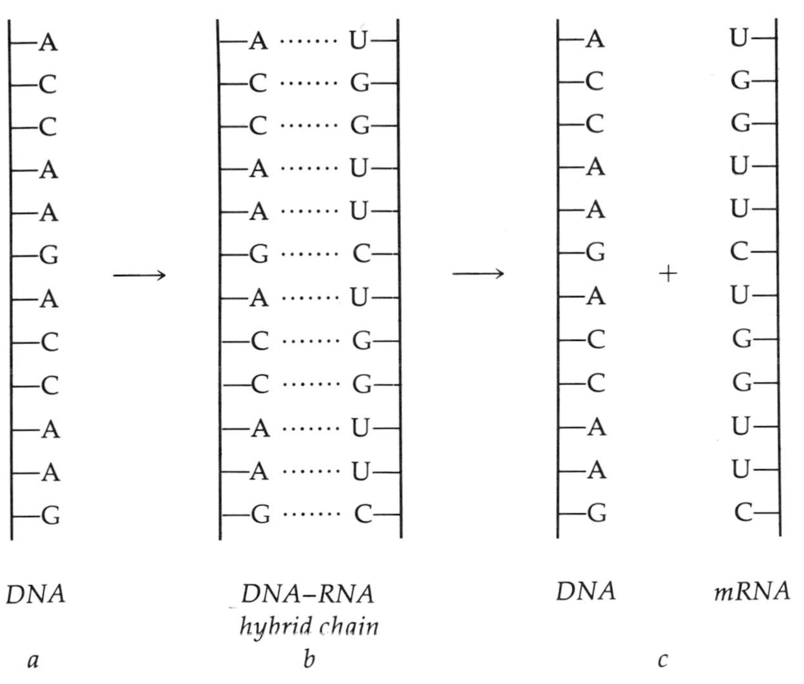

9–1. Synthesis (transcription) of mRNA on the DNA template, shown in diagrammatic form. Only the "sense" half of the DNA molecule is shown in a. b, *Formation of complementary base pairs during synthesis of mRNA.* c, *Release of mRNA.*

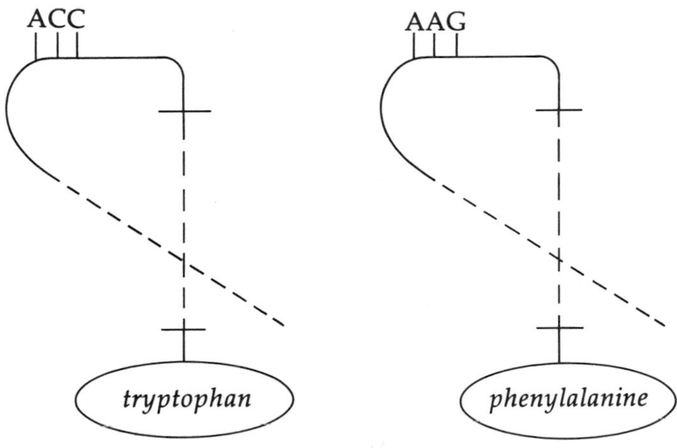

9–2. Diagram of the tRNA molecules for tryptophan and phenylalanine. The recognition codons are shown exposed at one end of the complex. The entire tRNA molecule is a relatively short chain of 75 to 80 nucleotides. The anticodon triplet is located somewhere in the interior of the sequence.

At the ribosome, the anticodons on tRNAs carrying either phenylalanine or tryptophan form complementary base pairs with the appropriate coding triplets of the mRNA. At any instant, only a short region of the mRNA is in contact with the ribosome. In this region of contact, the formation of base pairs between mRNA and tRNA takes place (Fig. 9–3). If the sequence UUC is present at the ribosomal pairing site, a tRNA carrying the anticodon AAG will attach to the mRNA, carrying with it the amino acid phenylalanine. As the ribosome moves to the next triplet on the messenger, it encounters the sequence UGG; a tRNA carrying tryptophan will attach at this point. The amino acids phenylalanine and tryptophan are brought into close proximity by this pairing interaction at the ribosome. Through the expenditure of energy originally brought to the system by ATP (see Chapter 11) a peptide bond is formed between the two amino acids. This process repeats, forming in this example a long polypeptide chain that contains a sequence of alternating tryptophan and phenylalanine residues.

The synthesis of protein as outlined requires the interaction of at least three classes of RNA: messenger, ribosomal, and transfer RNA. A great part of the DNA of the nucleus codes for different mRNA molecules, which in turn code for the synthesis of all structural and enzymatic proteins in the cell. A smaller part of the DNA, probably less than 1–2 percent, codes for the

9–3. *Synthesis of a protein on the ribosome (see text).*

synthesis of rRNA and tRNA. Other RNAs of unknown function are also coded by chromosomal DNA.

A considerable body of experimental data has accumulated to confirm the role of RNA in protein synthesis as outlined in this section. RNA can be synthesized on DNA in cell-free systems if a supply of the four nucleotides, RNA polymerase, and other factors (see Chapter 11) are present. The synthesized RNA has a sequence complementary to the DNA used as a "template" in the cell-free system. Synthetic RNA molecules having a known sequence are able to code for the synthesis of protein in a cell-free system that consists only of isolated ribosomes, amino acids, tRNA, and an energy source. If, for example, a synthetic messenger with a base sequence composed entirely of uracil subunits (UUUUU...) is introduced into a cell-free system, an artificial polypeptide chain consisting entirely of phenylalanine residues will be synthesized. This demonstrates that RNA can act as a "messenger" molecule in protein synthesis, and, incidentally, establishes that one RNA triplet code for the amino acid phenylalanine is UUU, complementary to the sequence AAA (adenine-adenine-adenine) in DNA.

The various parts of the total protein synthesis scheme will be covered in detail in this and the following chapters. While the overall process is well established as occurring according to this outline, many aspects of the mechanism are not yet completely understood.

The Molecular Structure of RNA

RNA is similar in molecular structure to DNA, with some important differences. Both molecules contain a backbone of alternating phosphate and 5-carbon (pentose) sugars. In RNA, the 5-carbon sugar in the backbone of the molecule is *ribose* instead of *deoxyribose* as in DNA; ribose has an additional hydroxyl group at the 2′ carbon position. (The carbon atoms in these sugars are by convention numbered from 1′ to 5′: see Fig. 9–4).

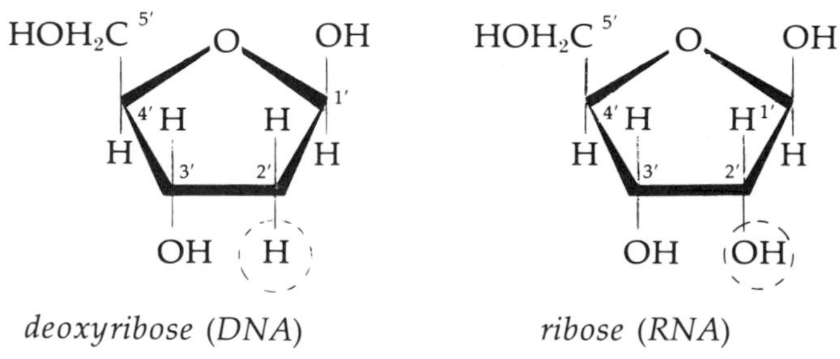

9–4. *The two 5-carbon sugars forming the "backbone" chain of DNA and RNA. Deoxyribose is the component sugar of DNA; ribose occurs in RNA. The two molecules are identical except that an oxygen is absent from the 2′ carbon of deoxyribose.*

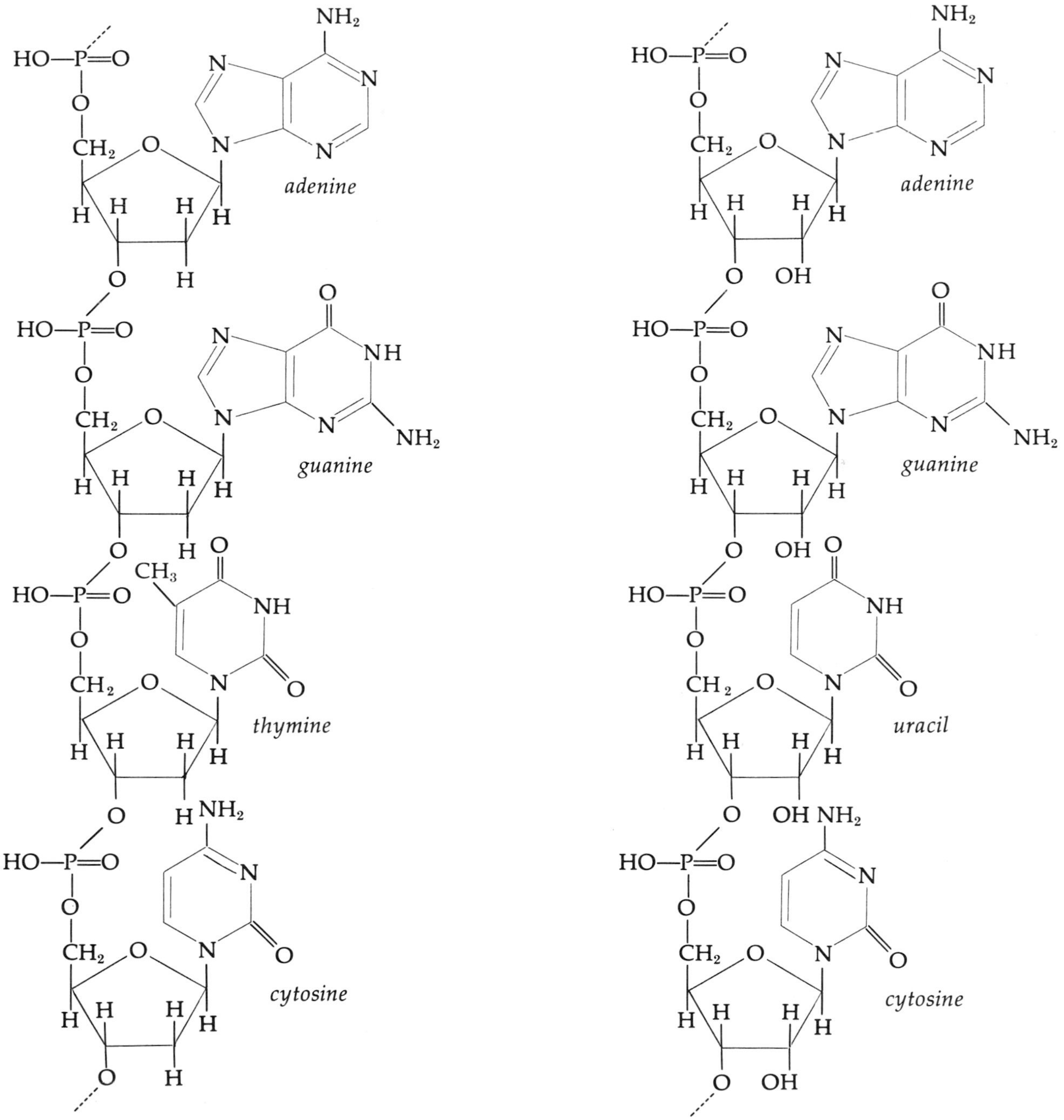

9–5. A single nucleotide chain of DNA (left) and RNA (right).

The pentose sugars of DNA and RNA are joined into the backbone structure by phosphodiester linkages formed between the 5′ and 3′ carbons of adjacent sugars (Fig. 9–5). The 1′ carbon of each sugar residue in the backbone in DNA is bound to one of the four nitrogenous bases adenine, guanine, cytosine, or thymine. In RNA the pyrimidine base uracil is substituted for thymine. Thus there are two differences in the primary structures of DNA and RNA. One is the substitution in RNA of the pentose sugar ribose for deoxy-

double-helical region

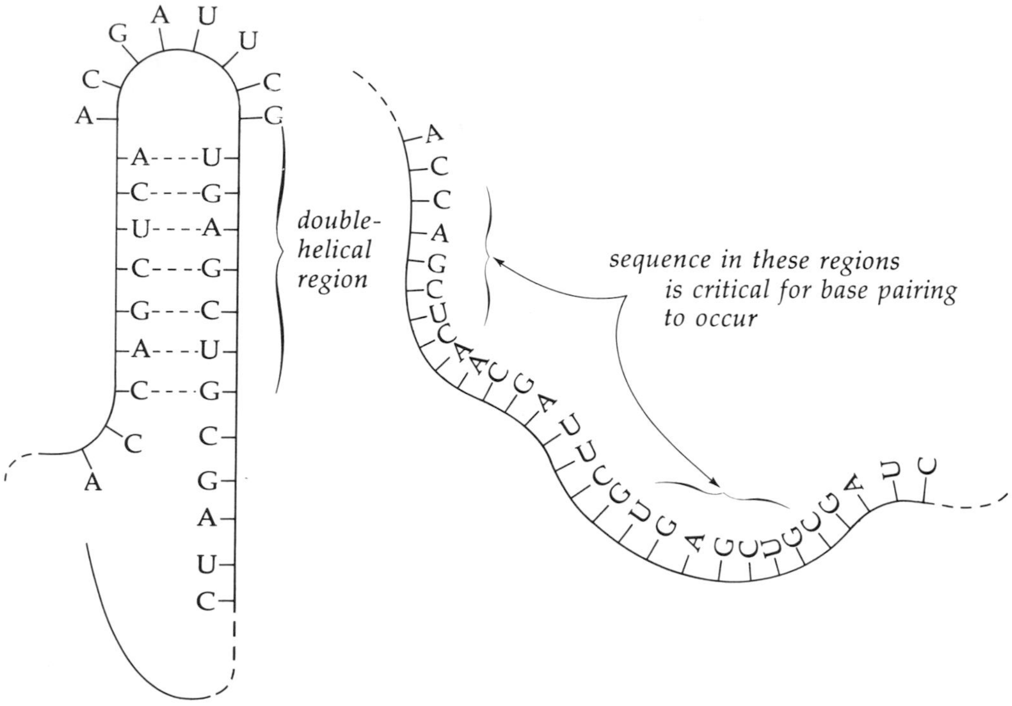

sequence in these regions is critical for base pairing to occur

9–6. RNA in paired configuration (left) and in extended form (right). Transitions from one form to the other are reversible and can be induced by changes in pH, salt concentration, and temperature.

ribose; the second is the substitution in RNA of the nitrogenous base uracil for thymine.

Significant differences are also found in the secondary structure of the two molecules. As noted in Chapter 8, analysis of DNA isolated and purified by X-ray diffraction led to the deduction of a double helical structure for DNA by Watson, Crick, and Wilkins. Most types of RNA, however, do not produce diffraction patterns clear enough to allow estimation of the folding or arrangement of the molecule in space. Both DNA and RNA are birefringent under polarized light (see p. 75); in DNA this birefringence is thought to arise from the parallel arrangement of the purine and pyrimidine bases in planes perpendicular to the long axis of the molecule. Therefore, both molecules are thought to share this arrangement of the bases.

Extreme differences in behavior are noted when the two nucleic acids are subjected to changes in *p*H, concentration of salts, and temperature. RNA is much more flexible in spatial arrangement under these conditions and readily undergoes changes from various degrees of helical coiling to more or less extended forms. These differences in X-ray diffraction analysis and behavior in solution have been found to result from the fact that RNA exists as a single-stranded molecule rather than a double helix. The large numbers of regularly repeating hydrogen bonds in the double helix of DNA are believed to account for the greater "stiffness" of this molecule in solution as compared to RNA.

In limited regions, the single RNA strand may be folded back on itself to form short lengths of double helical structure. This takes place only in regions in which the appropriate base pairs can be formed between adenine-uracil and guanine-cytosine residues. Because the requirement for the correct sequence of bases is rigorous for a folded-back double helix of this type (see Fig. 9–6), it does not at first seem likely that RNA double helices would frequently occur in nature. Surprisingly enough, it appears that as much as 80–90 percent of tRNA, although single-stranded, occurs in the form of a folded-back double helix.

The Synthesis of RNA on DNA

Origin of Cellular RNA

Evidence from a variety of sources indicates that most of the RNA of the cell originates in the nucleus.

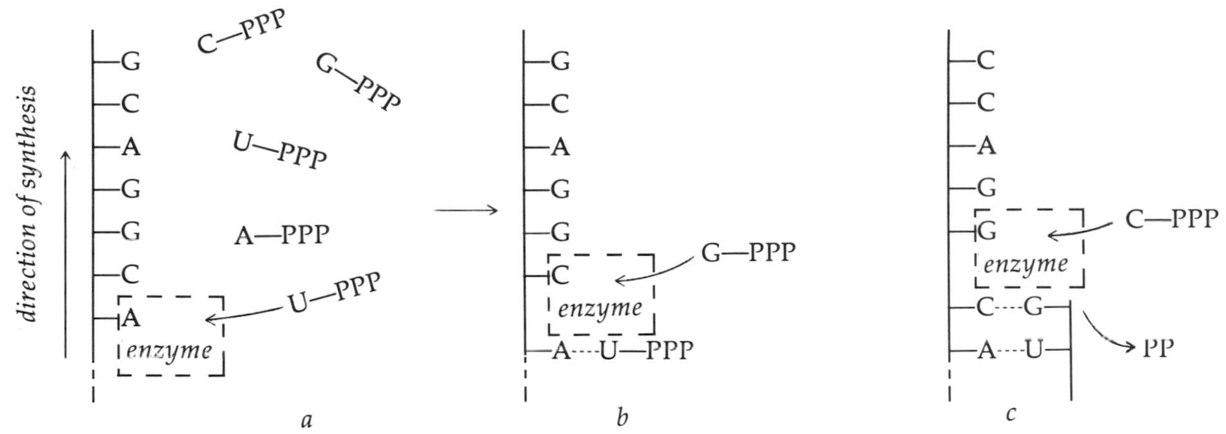

9–7. *Diagram of RNA transcription (see text).*

Autoradiography (see Chapter 8) has been particularly useful in this work, because RNA can be specifically labeled by exposing cells to tritiated uridine (uridine is incorporated in quantity only into RNA). After exposure of cells to tritiated uridine, the first incorporation of label is detected in the nucleus, indicating that this is the site of RNA synthesis. Longer periods of exposure reveal distribution of label in both nucleus and cytoplasm, suggesting that the newly synthesized RNA is transported through the nuclear envelope and enters the cytoplasm shortly after transcription. (For literature references to the many papers published in this area, see Prescott, 1964.) Other work, to be reviewed later in this chapter, indicates that DNA is the nuclear component responsible for RNA synthesis.

In addition to the bulk of cellular RNA originating in the nucleus, smaller amounts are also transcribed on the cytoplasmic deposits of DNA known to be localized in the interior of chloroplasts and mitochondria. Contribution to the total RNA of the cell from this source is relatively limited and is often undetectable by commonly employed experimental techniques such as autoradiography.

RNA Synthesis

RNA synthesis has been extensively studied *in vitro* in systems employing enzymes and other components isolated from cells. Many investigators have contributed to the elucidation of the mechanism of RNA synthesis through the study of these cell-free systems, including Severo Ochoa (who received the Nobel prize for his studies on the *in vitro* synthesis of RNA), Jerard Hurwitz, Samuel B. Weiss, and Audrey Stevens. From these studies, the synthesis of RNA on the DNA template has been determined to proceed as outlined in the following sections.

RNA synthesis requires the presence of a DNA template, the enzyme RNA polymerase, a divalent cation (Mg^{+2} or Mn^{+2}), and the four nitrogenous bases adenine, guanine, cytosine, and uracil. The bases must be present as nucleotide triphosphates. The familiar molecule ATP, which forms the medium of energy capture and transfer in the cell, is a nucleotide triphosphate. The structural formula for ATP is shown on p. 109; the structural formulas for the remaining nucleotide triphosphates, CTP, GTP, and UTP, can be derived by substituting the appropriate base. Under some *in vitro* conditions either single-stranded or double-stranded DNA can act as a template for RNA synthesis. The mechanism for this synthesis is shown diagrammatically in Fig. 9–7. Only one strand of the DNA is shown in this figure (see footnote, p. 192). RNA polymerase attaches to the DNA molecule and catalyzes the addition of the first nucleotide triphosphate. All four triphosphates are equally available to the enzyme in the medium; the particular triphosphate attracted to the DNA-polymerase complex is determined by the base pairing rules for the nucleotides of DNA. This is shown in simplified form in Fig. 9–7. In Fig. 9–7a, the first DNA base encountered by the enzyme is adenine; the base pairing rules (see Chapter 8) will favor the addition of UTP at this site. The enzyme then moves to the next nucleotide on the DNA molecule (cytosine

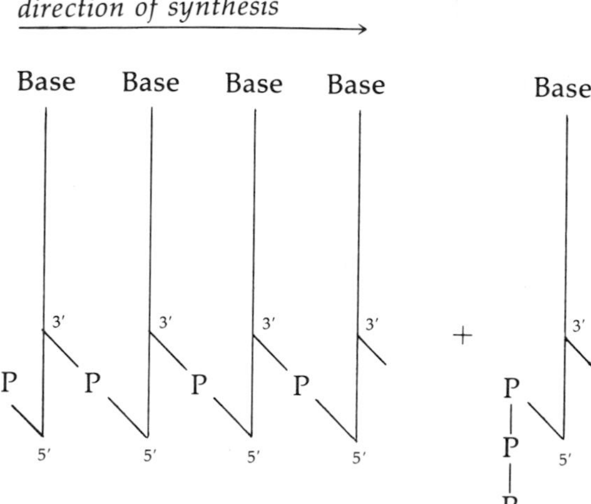

9–8. *Direction of chain growth in RNA synthesis. Because a 3′ —OH group of a ribose sugar is exposed at the growing end of the chain, synthesis is said to proceed in the 5′ to 3′ direction. In the diagram, P represents the phosphate group or phosphodiester linkage and 3′ and 5′ the respective carbons of the ribose residue.*

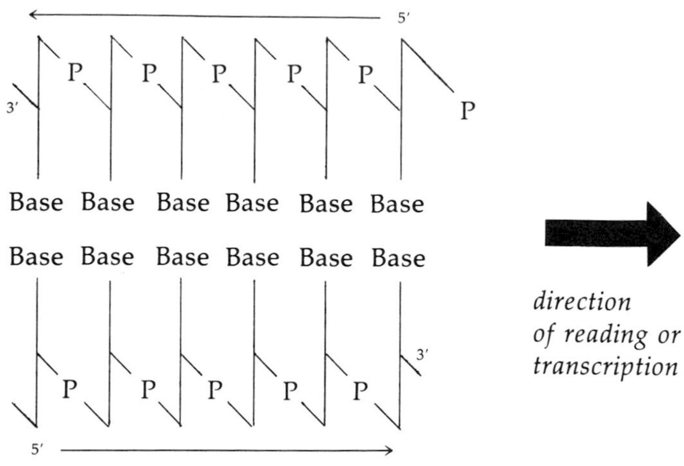

9–9. *Polarity of the two strands of DNA (see text).*

in Fig. 9–7b). Pairing rules will result in the attachment of GTP at this level. At this point, a complex exists between the DNA, the nucleotide triphosphates UTP and GTP, and the enzyme. The enzyme then catalyzes the formation of a phosphate (phosphodiester) linkage between the nucleotides. The energy for the formation of this linkage is provided by the hydrolysis of the two terminal phosphates of UTP, releasing pyrophosphate to the medium. This reaction yields a short dinucleotide chain consisting of the RNA sequence UG. The enzyme then moves to the next nucleotide on the DNA chain, encountering the purine base guanine; at this level the nucleotide triphosphate CTP is added to the growing RNA chain. A second phosphodiester linkage is then formed, with the release of pyrophosphate to the medium, yielding the RNA chain UGC. This process repeats sequentially as the enzyme moves along the DNA chain until synthesis of the RNA is complete.

The phosphodiester linkages formed as each nucleotide is added to the RNA chain extend between the 3′ carbon of a ribose sugar at the growing end of the chain to the 5′ carbon of the ribose sugar of the nucleotide triphosphate being added (Fig. 9–8). The growth of the chain is therefore said to proceed in a 5′ to 3′ direction, because a 5′ carbon is always located at the "old" end of the chain and a 3′ carbon at the "new" end.

RNA polymerase catalyzes transcription at a rate between 3 and 30 nucleotides per second. The slowest rate of 3 nucleotides per second corresponds to movement of the enzyme along the DNA template at about 500 Å per minute. The mechanism by which the enzyme transcribes one of the two strands of an intact DNA duplex is unknown. One possibility is that the template DNA unwinds locally in short regions for synthesis to proceed. This would allow space for the RNA bases to form hydrogen bonds with one of the two DNA strands according to the usual Watson-Crick pairing rules. A second possibility is that pairing occurs in some manner between the growing RNA chain and the unwound double helix. The mechanism for pairing in this case is not immediately obvious, as the bases of the unwound DNA helix face the interior of the molecule and are not sufficiently exposed for the formation of hydrogen bonds. Most evidence favors the unwinding hypothesis. DNA replication is known to proceed with unwinding and separation of the strands of the double helix; it seems most likely that RNA transcription on DNA would follow the same pattern. Only local unwinding would be required, because the growing RNA chain can separate readily from the DNA template, allowing the DNA helix to rewind and close behind the actual point of synthesis.

The asymmetry of RNA transcription. If RNA synthesis is allowed to proceed in the presence of undenatured DNA, only one of the two strands of the DNA double helix is copied. The basis for this characteristic of transcription, called *asymmetry*, is not understood. The best evidence that transcription is asymmetric comes from experiments in which RNA is *hybridized* with its DNA template. Most of these experiments have been carried out using bacteriophage DNA. In these experiments, bacteriophage DNA is used as a template for RNA synthesis, either in the living, *in vivo* state, by infecting bacteria with the bacteriophage, or in cell-free *in vitro* systems. After RNA transcription has taken place the template DNA is isolated and denatured by heating. Heat denaturation breaks the hydrogen bonds between base pairs in the double helix, and the two DNA strands separate. If the separated template DNA is mixed with the newly synthesized RNA, the RNA will form hybrid double helices with no more than 50 percent of its template DNA after the solution is cooled. Thus the RNA is complementary to only one of the two DNA strands, and only one of the two strands serves as a template for RNA transcription (see Hayashi et al., 1964). As far as is known, asymmetry is a characteristic of RNA transcription in all living organisms.

Several hypotheses have been advanced to account for asymmetric transcription. One obvious possibility is that the enzyme might be specific for only one of the two strands of the DNA duplex. However, if DNA is denatured by heating before transcription or broken into short lengths, RNA polymerase will copy both strands. Further, the enzyme will use single stranded DNA as a template, or even, in the case of RNA polymerases from some sources, use RNA as a template for RNA synthesis. Because the enzyme is evidently relatively unspecific as far as template is concerned, the key to the phenomenon of asymmetry must lie in the structure of the double helix itself rather than in the specificity of the enzyme.

Some investigators have attempted to explain asymmetry on the basis of the "polarity" of the two strands of DNA. The two DNA strands are *antipolar*. This means that if the two strands are "read" from a given starting point, one strand reads 5′ to 3′ and the other 3′ to 5′ (Fig. 9–9). If the direction of transcription is in the direction of the large arrow to the right in Fig. 9–9, only the *top* DNA strand in this figure could be copied, because RNA is synthesized in the 5′ to 3′ direction and must be antipolar to the template. Thus the "rule" for transcription might be that RNA synthesis can proceed only in one direction with respect to the whole double helix or chromosome. This would explain asymmetry, provided that the template is always *on the same strand of the DNA helix*. Another way of saying this is that the "sense" or template strand cannot alternate between the two strands of the double helix. Unfortunately for this hypothesis, it is known that in some bacteria the template does alternate between the two DNA strands. The enzyme always copies the correct one in intact DNA from this source (see Hartman et al., 1965).

Initiation and termination of RNA synthesis. Transcription of RNA does not simply begin at one end of the chromosome and proceed through all of the genes or cistrons until the opposite end is reached. Instead, in most organisms only some genes are transcribed at any given time. These may lie at any place along the chromosome, at the ends or in the middle. If only one segment is transcribed along a given length of DNA, the enzyme in some way must "recognize" initiation and termination sites that indicate where to start and stop RNA synthesis at the limits of a transcription unit. Many hypotheses have been advanced to account for this phenomenon. One is that the DNA contains short segments of non-DNA material that separate transcription units. The polymerase molecule would recognize these segments as "initiator" and "terminator" sites. Although this is a plausible mechanism, no such entities have been identified or isolated from the chromosome. It is unlikely that the short segments are intercalated bits of protein, because proteases do not break the chromosome. Another possibility is that the initiator and terminator segments consist of short sequences containing the rare purines and pyrimidines found in small quantities in nucleic acids in addition to the usual bases (see p. 204). The enzyme might then contain a recognition site for a rare "initiator" base or sequence of bases.

Still another possibility is that a particular sequence of the usual four bases identifies the initiator and terminator regions of the gene. Some support for this can be taken from the fact that newly transcribed RNAs frequently begin with a purine nucleotide at the 5′ end. Further, a number of enzymes (the *RNA methylases*) are known to be capable of "recognizing" a specific nucleotide sequence. One difficulty with this hypothetical mechanism arises from the fact that initiation of RNA transcription can take place in cell-free systems if a *homopolymer* (an artificial polynucleotide consisting of a single repeating base) is used as template. As homopolymers are essentially "sequenceless," a particular nucleotide sequence or the presence of rare bases cannot be required to begin transcription in this case. This does not rule out the possibility in natural systems, however.

The latter hypotheses have the advantage of residing in the sequence of DNA itself; it is easy to see how the initiation and termination sites could be replicated with the DNA and passed along to daughter cells. A fourth hypothesis does not have this advantage. This hypothesis proposes that transcription is initiated by a localized unwinding of DNA at the site at which transcription is to begin. Thus, when the unattached enzyme in the medium surrounding the DNA "recognizes" an unwound segment it begins synthesis at that point. While this hypothesis is plausible and has been seriously advanced, it begs the question, for some other control method, outside the DNA, must be imposed to initiate and terminate unwinding at a specific place along the double helix.

Of these several hypotheses, the "initiator sequence" and the "unwinding" hypotheses are most widely favored at the present time. Termination of synthesis can be regarded simply as a reverse of these mechanisms. As mentioned, the "initiator sequence" hypothesis has the advantage of residing in the structure of DNA itself; no additional control systems need be invoked. Unwinding at a specific initiator site, on the other hand, must depend on some control system outside the sequence of the unit being transcribed. The questions surrounding the possible mechanisms operating at the initiator and terminator sites are a part of the larger problem of the *regulation* of genetic activity. Regulation is taken up in detail in Chapter 11, as a part of the discussion of protein synthesis in the cytoplasm.

The Major RNA Classes Transcribed on the Chromosome

Each of the three major types of RNA synthesized on the chromosome, messenger, transfer, and ribosomal RNA, actually represents a class or group of RNAs. Within each class are a number of molecular species with similar physical properties and functions. A number of additional RNA types synthesized on the chromosomes, although classifiable in one of the three major groups, remain of unknown function in the cell.

Transfer RNAs are of relatively small size, with molecular weights ranging from 20,000 to 25,000. In contrast to the other RNA types, tRNA remains soluble in 1 molar sodium chloride. For this reason, this RNA class was once known as "soluble" RNA. The descriptive term "transfer" of this class has gradually gained precedence because of the function of these RNAs in carrying amino acids to the site of protein synthesis at the ribosome. Transfer RNA, forming as much as 10–15 percent of the total cellular RNA, typically contains a high proportion of rare purine and pyrimidine bases in addition to the usual four. Many of these rare bases contain additional methyl groups not found in the usual RNA bases.

The remaining RNA of the cell is almost all of larger molecular size, with molecular weights ranging from approximately 500,000 to 1.5 million. Almost 80 percent of the high molecular weight RNA in the cell is rRNA. Guanine and cytosine occur with a higher frequency in rRNA; the ratio (G + C)/(A + U) for rRNA is approximately 1.6. Other RNAs contain smaller quantities of guanine and cytosine, yielding a (G + C)/(A + U) ratio less than 1, approximately 0.7 to 0.8. The high (G + C)/(A + U) ratio characteristic of rRNA reflects a correspondingly high (G + C)/(A + T) ratio in the DNA template coding for rRNA. This is atypical for DNA if the whole chromosomal complement is taken into consideration and suggests that the DNA coding for rRNA represents only a small proportion of the total. Other, more direct methods (see below) bear this out; the DNA coding for rRNA represents less than 1 percent of the total DNA in most organisms, even though most of the RNA transcribed in the nucleus is rRNA. Ribosomal RNA occurs in all parts of the cell, on the chromosomes, in the nuclear ground substance and nucleolus, and in the cytoplasm. Ribosomes, however, probably occur in functionally complete form only in the cytoplasm.

Three distinct types of RNA, separable by molecular weight, can be extracted from intact ribosomes. Two are of relatively large size, with molecular weights of about 1.2 million and 500,000. One smaller type is found, with a molecular weight of about 40,000. This low molecular weight rRNA fraction can be distinguished from tRNA by a number of criteria, primarily on the basis of a high (G + C)/(A + U) ratio typical for rRNA, slightly higher molecular weight, and the absence of large quantities of the rare purine and pyrimidine bases.

The remaining high molecular weight RNA, approximately 5–10 percent of the total cellular RNA, is DNA-like in (G + C)/(A + U) ratio. This class contains the mRNA of the cell and probably a large proportion of other RNAs of unknown function as well. Messenger RNA has proved difficult to separate from the total DNA-like RNA group and to date has been isolated from only a very few highly specialized cells in which the synthetic mechanisms are limited to single proteins. To be unequivocally identified as a messenger, an RNA species should code for the synthesis of a single identifiable protein in a cell-free system containing ribosomes, tRNA, and amino acids. Provisional identification according to this criterion has been carried out in phage-infected bacteria and in a few highly specialized cells from higher organisms, such as in muscle tissue or in cells destined to become erythrocytes. In more typical cells the wide variety of mRNAs and other DNA-like RNA transcribed makes isolation of a single species impossible.

Classification of RNA by centrifugational analysis. Researchers have found that many RNAs can be conveniently characterized by analytical centrifugation techniques. Each of these RNAs will descend or sediment with a velocity that depends on the mass of the molecule and the speed of centrifugation. For a given RNA, the velocity of sedimentation is related to the speed of centrifugation by a constant value, termed the sedimentation constant:

$$s = \frac{\bar{V}}{a} \qquad (9\text{–}1)$$

where s is the sedimentation constant, $\bar{V}$ is the average velocity of sedimentation, or rate of travel of a given molecule toward the bottom of the tube, and a is the acceleration (rate of centrifugation). In practical terms, this equation means that if a mixed RNA sample is spun in a centrifuge, each RNA will be found after a time interval at some position relative to the top and bottom of the tube; this position is determined by the weight and shape of the molecule and is described by the sedimentation constant.

The sedimentation constants for RNA range from about 4×10^{-13} for the smallest molecules (tRNA) to about 45×10^{-13} for the heaviest (a precursor of rRNA). For convenience, the sedimentation constant is usually expressed as a *Svedberg unit* or S unit, in honor of a pioneer of centrifugation techniques. Each S unit is equivalent to a sedimentation constant of 1×10^{-13}. Accordingly, transfer RNA, with a constant of 4×10^{-13}, is called 4S RNA. Svedberg units are used to describe cellular RNAs, and are closely related to the molecular weight of each species of RNA. In practice, the S units obtained for the various RNA classes are variable, because the configuration of RNA in solution is highly sensitive to *p*H, salt concentration, and temperature. As a result, the S values commonly given for the various RNA types are somewhat arbitrary. In addition to the 4S value accepted for tRNA, rRNA is usually described as consisting of 18S, 28S, and 5S RNA (with variations between animals, plants, and bacteria; see below). Messenger RNA is highly heterogeneous, having S values from 5S to 30S, with a possible median value at 18S.

The S values given above for the various RNAs represent the sedimentation velocity of the purified molecule with little or no associated protein. It is unlikely, however, that much RNA exists in the cell in this "naked" form. It seems probable that most RNA is combined with a protein component to form a ribonucleoprotein particle. In a few RNA types, such as rRNA and its precursors, the ribonucleoprotein equivalent is as well characterized in terms of sedimentation velocity as the corresponding purified RNA fraction. For example, 18S rRNA corresponds to an RNA-protein complex sedimenting at 40S, and 28S rRNA to a ribonucleoprotein complex of 60S. Only a few other RNA types are as clearly related to ribonucleoprotein particles of known S value; in all cases, almost nothing is known about the identity or structure of the associated protein.

Messenger RNA (mRNA)

By definition, messenger RNA is the product of a gene or cistron coding for a structural or enzymatic protein and carries coded information to the cytoplasm, where it is translated into an amino acid sequence. But in chemical terms, mRNA is identifiable only as a part of a larger RNA group similar to DNA in (G+C)/(A+U) ratio. Therefore it has not been possible to identify mRNA except by indirect means. The criteria usually used to identify mRNA do not completely exclude other types of RNA:

1. base ratio similar to the total DNA complement,
2. ability to hybridize *in vitro* with DNA from the same organism,
3. ability to stimulate protein synthesis in a cell-free system, and
4. rapid synthesis in actively growing cells.

The first of these criteria was once considered a definitive characteristic of mRNA. As noted, however, this criterion is not exclusive because the nucleus evidently contains large quantities of DNA-like RNA that do not serve as messengers. This has been established primarily by means of labeling experiments, which have shown that a large portion of the DNA-like RNA remains in the nucleus and does not enter the cytoplasm. It seems unlikely that the fraction remaining in the nucleus could represent a messenger, because most evidence points to the conclusion that protein synthesis is restricted to the cytoplasm (see below).

The remaining criteria are similarly nonexclusive. Almost all forms of RNA, including even "artificial" RNAs, will stimulate protein synthesis in cell-free systems. For a time, it was believed that mRNA was the only rapidly synthesized RNA type, and that it was unstable and broken down within hours or even minutes of synthesis. Recent experiments, however, have shown that rRNA is also rapidly synthesized in actively growing or synthesizing cells and becomes labeled as quickly as mRNA if a radioactive precursor is added. Instability of the messenger, derived as a concept from the study of microorganisms, does not hold in higher organisms; evidently, the messenger may persist from periods of minutes to days in the cytoplasm.

Provisional identification of an mRNA has been carried out in experiments with bacteria infected with bacteriophage. These experiments take advantage of the fact that soon after infection by bacteriophage, synthesis of the normal complement of DNA-like RNA in the bacterial cell ceases, and transcription of viral-type RNA is initiated. If bacterial cells are placed in medium containing a radioactive RNA precursor immediately after infection, only the newly transcribed viral-type RNA is labeled. Thus, it can be separated from the background of unlabeled bacterial RNAs. The experiments by various workers with this system, when taken together, indicate that the first viral-type RNA transcribed after infection is limited to a single messenger coding for viral coat protein or viral lysozyme (see, for example, Spiegelman, 1961, and W. Salser et al., 1967).

The main difficulty encountered in attempting to extend these methods to higher organisms or even to uninfected bacteria is in the purification of a single messenger. In any of these cell types a large number of messengers are simultaneously synthesized along with quantities of DNA-like RNA probably not associated with the messenger function. Addition of this total RNA class to a cell-free system causes synthesis of a wide range of polypeptides that cannot be identified or separated from any specific proteins that might be made.

Greater success has been obtained with cells that are known to synthesize only a limited number of proteins. The examples of maturing erythrocytes and muscle cells have already been given. In one experiment of this type, carried out by S. M. Heywood and M. Nwagwu, ribosomes attached to messenger RNA (*polysomes*, see Chapter 11) were first isolated from muscle cells. These mRNA-ribosome complexes could be separated into distinct size classes depending on the length of the attached messenger. Each size class, if supplied with the factors required for cell-free synthesis of protein, could be identified with a single, specific muscle protein. The mRNA extracted from these distinct classes, in turn, proved to consist of a single species. This was demonstrated by reintroducing the messenger to a cell-free system. The mRNA extracted from myosin-synthesizing polysomes, for example, induced the synthesis of myosin if reintroduced into a cell-free system.

Size of the messenger. The heterogeneity of DNA-like RNA has also resulted in considerable uncertainty over the size of mRNA. Many values, ranging from 5S to 30S, with a few as high as 80S, have been reported after centrifugation of RNA believed to represent a messenger. Part of this heterogeneity is probably due to degradation of RNA into smaller pieces, aggregation, and incomplete separation from protein. Another contributing factor is undoubtedly the extreme lability of RNA in solution. Configurational changes in the molecule, which are readily brought about by changes in ionic environment or *p*H, are reflected in variability of the sedimentation constant for a given sample. In addition to variability arising from these sources, the heterogeneity of mRNA preparations is expected on purely intuitive grounds, because the polypeptide chains of proteins and protein subunits coded for by the messenger are known to vary in length. Proportional differences in length would be expected for the corresponding messengers.

By careful control of isolation methods to avoid degradation and by purification to eliminate salts present as impurities, presumed mRNA can be isolated that sediments with a fairly well defined peak at 18S, corresponding to a molecular weight of 300,000 to 500,000.

But at this value, this RNA fraction corresponds precisely in size to the smaller 18S ribosomal RNA subunit (see below). As a result, many workers in the field have been reluctant to accept the 18S species of RNA as distinct from rRNA. The 18S RNA of the cell exists in amounts equal to 28S rRNA, as far as can be determined. This is to be expected if all of the 18S RNA represents rRNA, because each ribosome consists of one 18S and one 28S subunit. This does not necessarily rule out a part of the 18S RNA as messenger, because the relatively small quantities of mRNA occurring in the cell might not be detectable against the larger quantity of 18S rRNA and might not affect the 18S/28S ratio significantly enough to be detected. At the present time, it is impossible to choose between these possibilities; the 18S peak may represent either the messenger or contaminating rRNA.

Possible roles of the additional DNA-like RNA of the nucleus. Much of the DNA-like RNA transcribed in the nucleus in addition to the messenger is degraded and does not penetrate through the nuclear envelope to the cytoplasm. The role of this additional RNA in the nucleus is unknown. It has been suggested that the extra DNA-like RNA may be a messenger for *nuclear* protein synthesis, being used and degraded entirely within the nucleus. The possibility of protein synthesis within the nucleus has been debated for many years. Isolated nuclei are able to incorporate amino acids into protein, but if these nuclei are treated with detergents to remove the nuclear envelope, protein synthesis ceases. Ribosomes are known to be attached to the cytoplasmic side of the outer membrane of the nuclear envelope; it is now thought that these observations of protein synthesis by isolated nuclei are the result of synthesis by these ribosomes or by cytoplasmic contaminants adhering to the nuclear envelope. The existence of completed, functional ribosomes within the nucleus has also been a debated point for many years. Early claims were based on morphological observation of ribosome-like particles within the nucleus. However, with the introduction of the accurate and sensitive methods used in centrifugational analysis of nuclear RNA extracts, it has become apparent that although the nucleus contains the precursors of 18S and 28S ribosomal RNA, no complete 80S ribosomes are detectable (see below). Thus, if the nucleus or the nucleolus, as has recently been claimed, does synthesize proteins, this synthesis must proceed by an as yet undiscovered mechanism that does not require complete ribosomes.

Another possibility is that the total DNA-like RNA fraction of the nucleus is all precursor of mRNA, and that the maturation of this fraction into a functional messenger involves splitting off of an extra, nonfunctional subunit. This is not entirely unlikely, because the maturation of rRNA proceeds by a similar mechanism (see p. 209). However, the proportion of "unused" RNA is very large and represents more RNA than the final amount reaching the cytoplasm as messenger. Why would so much ultimately nonfunctional RNA be synthesized by the chromosome? The hypothesis remains as a plausible but unlikely possibility.

Another interesting hypothesis, advanced recently by James Bonner and his colleagues, is that some of the extra DNA-like RNA of the nucleus represents regulatory RNA engaged in the control of the gene activity. This hypothesis, which will be taken up in detail in Chapter 11 as a part of the general topic of regulation of protein synthesis, maintains that the specificity required for the recognition of genes and the control of gene expression resides in a hypothetical DNA-like RNA fraction attached to histone proteins. This RNA fraction would provide the capacity for sequence recognition to the histones, which are relatively homogeneous and apparently nonspecific in binding to DNA. The required variability and capacity for recognition and hybridization with specific DNA sequences is inherent in RNA structure. Although this proposed role for the extra DNA-like RNA is plausible, it is hardly more than speculation at the present time, because no direct evidence exists in its support.

Transfer RNA (tRNA)

Transfer RNA has been successfully isolated and purified and is well characterized in terms of structure and properties. A number of tRNAs are known even to the specific sequence of the component nucleotides. Transfer RNAs can easily be separated from the rest of the RNA complement by precipitation of large molecular weight RNA with 1 molar salt. Under these conditions, tRNA remains soluble. The soluble fraction can be further purified by other techniques into the separate tRNAs for the different amino acids. Final identification depends on the ability of the individual tRNAs to bind specifically to an amino acid in a re-

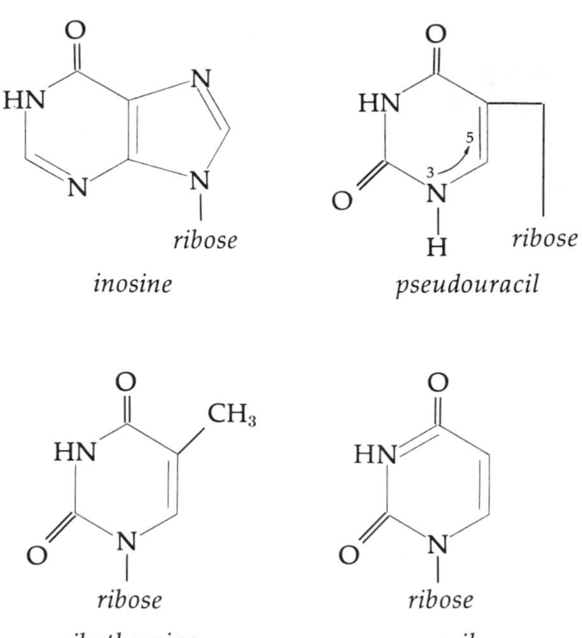

9–10. The rare bases inosine, pseudouracil, and ribothymine, which occur in tRNA. Uracil is also shown for purposes of comparison.

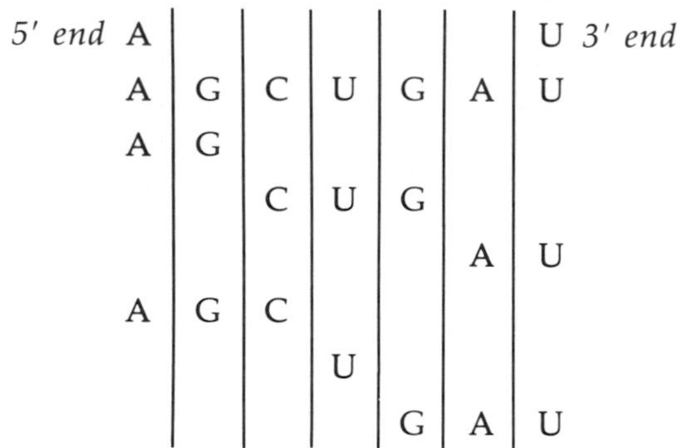

9–11. Sequencing a short polynucleotide from known fragments (see text).

action catalyzed by the enzyme *aminoacyl-tRNA synthetase* (see p. 255).

The total number of different tRNAs in the cell probably corresponds to the number of codons specifying amino acids in the nucleic acid code. Of the 64 codons, 61 have been identified as assigned to amino acids. The remaining three are terminator codons that promote the release of finished polypeptides from the ribosome. Evidently, 61 different tRNAs exist with anticodons specific for these amino acid codons. No tRNAs have been found that bind to the terminator codons. Most of the amino acids are coded by more than one codon; each codon is evidently provided with a separate tRNA. For example, six different tRNAs have been identified as specific in binding to leucine; six different codons are assigned to this amino acid in the nucleic acid code (for details of the nucleic acid code, see Chapter 11). In addition, there are indications that cells contain a distinct group of tRNAs associated with protein synthesis in mitochondria and chloroplasts (see Chapter 16).

All tRNAs consist of a chain 75–80 nucleotides in length with a molecular weight of approximately 25,000. The sequence of nucleotides making up the chain has some features in common in all transfer RNA molecules. In all cases, the sequence at one end terminates with a guanine nucleotide and at the opposite end with the sequence . . . C—C—A (. . .cytosine-cytosine-adenine). The adenine of the CCA terminal sequence acts as the binding site for the amino acids carried to the ribosome by tRNA. Because all tRNAs end identically with the CCA sequence, the specificity of the tRNA molecules for individual amino acids must reside in the sequence of nucleotides in the chain between guanine and the terminal CCA. Likewise, the anticodon site is also to be found somewhere in the intervening sequence.

The high proportion of unusual purine and pyrimidine bases present in tRNAs is an interesting characteristic of this RNA class. Most common of these are the bases *inosine, pseudouracil,* and *ribothymine* (Fig. 9–10). One of two mechanisms is thought to be responsible for the inclusion of these bases in tRNA. Either the rare bases are synthesized as such into the growing tRNA chain by RNA polymerase, or the existing common bases are converted into the rarer forms after transcription of tRNA on DNA. The first possibility seems unlikely, even though RNA polymerase can accept the ribose triphosphates of the rare bases for RNA synthesis, because the rare bases cannot be detected in cells except in fully synthesized tRNA. The more likely possibility is that these bases are formed from the usual four nucleotides after synthesis of a tRNA precursor.

A number of enzymes with this capacity are known. The rare base ribothymine can be synthesized from uracil by the addition of a methyl group (—CH_3) to the

5-carbon in the ring. Methylations of this type are well understood; reactions and enzymes are known for the methylation of all common bases to the rarer forms. These reactions can be accomplished while the bases are in place in the polynucleotide chain. The conversion of uracil to pseudouracil is, in contrast, not well understood. In some manner, the enzyme responsible for this change changes the binding site of the base to the ribose sugar from the nitrogen at the 3-position in the ring to the carbon at the 5-position. Except for the place of attachment, the structures of uracil and pseudouracil are identical.

The functional significance of these alterations in the bases of tRNA is not known. This phenomenon is interesting, because the final sequence produced, after conversion of some (but not all) of the usual bases into the rarer forms, is probably not determined by the DNA segment coding for tRNA. The changes are highly specific as to location within the tRNA nucleotide sequence; for a particular tRNA the location of the rare bases is rigidly determined. Another part of the transfer RNA sequence, the terminal CCA segment, is added to tRNA in the cytoplasm in a reaction catalyzed by a system of enzymes clearly distinct from RNA polymerase.

These changes, including both the specific conversion of parts of the sequence to rarer bases and the addition of the terminal CCA sequence, require that the enzymes involved recognize and catalyze changes at specific points in the nucleotide sequence. The mechanism for this recognition capacity by a protein is unknown but has interesting implications for the possible regulation of transcription of RNA on DNA by proteins such as histone.

Sequencing of transfer RNA. The sequencing of tRNA has been accomplished by biochemical methods taking advantage of the fact that the relatively short chains in this RNA type contain few, if any, repeating sequences. The primary sequencing methods make use of a number of enzymes that attack an RNA chain in specific ways, releasing short fragments with characteristic terminal nucleotides. By combining the results of various treatments and comparing the short fragments obtained, it is often possible to deduce the original sequence. Fig. 9–11 illustrates how this is accomplished with a very short RNA chain containing no repeated sequences. One enzyme, a *phosphomonoesterase*, removes the terminal phosphate from the 3′ end of a chain. After attack by this enzyme and subsequent separation of the chain into separate nucleotides, only the nucelotide fragment originally at the 3′ end will lack a phosphate group. In the example given, this fragment contains the base uracil. Digestion with mild alkali has an analogous effect except that the nucleotide fragment at the 5′ end of the chain is left without a phosphate group; this treatment reveals that the 5′ end terminates with the base adenine. Therefore, the short chain begins with A and ends with U.

Complete digestion of the chain into separate nucleotides (by the enzyme phosphodiesterase) and evaluation of the ratios of the individual bases obtained reveals that the interior of the chain contains one cytosine, one adenine, and one uracil, and two guanine nucleotides. Finally, digestion of other samples of the chain with *pancreatic ribonuclease*, which specifically breaks an RNA chain at the 3′ linkage of uracil and cytosine nucleotides, and *T_1 ribonuclease*, which breaks the 3′ linkage of guanine nucleotides, releases the fragments A—G, C—U—G, A—U, A—G—C, U, and G—A—U. These fragments, and the ratios of bases obtained by complete digestion are compatible only with the internal sequence G—C—U—G—A (see Fig. 9–11).

These ingenious techniques have been used in different combinations to sequence a number of tRNAs. This work was pioneered by R. W. Holley, who received the Nobel Prize in 1968 for his research in the sequencing and structure of tRNA. The complete sequences of tyrosine and alanine RNA from yeast are shown in Fig. 9–12. These tRNAs, sequenced by J. T. Madison, G. A. Everett, and H. K. Kung, are depicted in the form that allows the greatest pairing of nucleotides within the chain, producing three folded-back helices in the form of a cloverleaf. In all tRNAs sequenced to date, the cloverleaf pattern, originally proposed by Holley, allows the largest number of base pairs to be formed and, although hypothetical, may well represent the secondary configuration of tRNA in solution.

The site containing the triplet code recognition site (the anticodon) is expected to be located in one of the unpaired loops at the tips of the cloverleaf arms, because the bases in the paired, double-helical regions face inward and are not believed to be sufficiently exposed to form base pairs with the codon sites of messenger RNA. Many unsolved problems remain in connection with other specific sites of activity probably located within the tRNA sequence. For example, there is probably a ribosomal binding site on tRNA in addi-

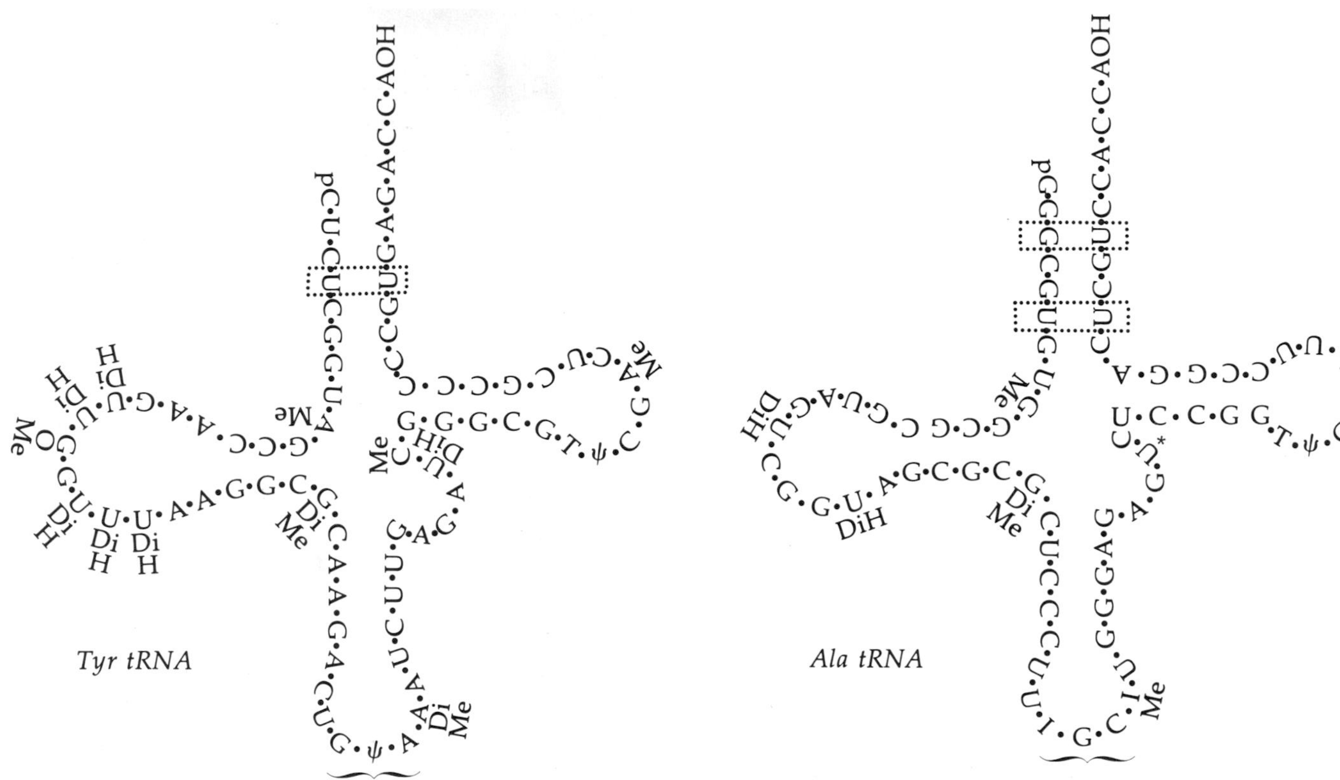

9–12. Probable sequence and secondary structures of yeast tyrosine and alanine tRNAs. The braces mark the presumed location of anticodons. Redrawn, original Courtesy of J. T. Madison.

tion to the anticodon site which binds to the messenger, because the hydrogen bonds formed between the messenger RNA codon and the tRNA anticodon are probably not sufficient in themselves to stabilize the complex. There may also be a site which allows "recognition" of the specific tRNAs by the enzymes that catalyze binding of specific amino acids to the CCA end of the tRNA chain.

Ribosomal RNA (rRNA)

Ribosomal RNA is defined as RNA extracted from ribosomes or RNA from other locations in the cell, including the nucleus, with the same base ratio as the RNA isolated from ribosomes. As mentioned, the base ratio $(G + C)/(A + U) = 1.6$ is a distinguishing characteristic of rRNA; all other cellular RNAs have a $(G + C)/(A + U)$ ratio of unity or less. Although there is some degree of methylation in rRNAs from animals, plants, and bacteria, the rare bases are not encountered with nearly as high a frequency as in tRNA.

Ribosomal RNAs from all eukaryotic cells show some degree of similarity in base composition and internal sequence. This has been shown by experiments revealing that at least partial hybrids can be formed between rRNA and DNA from highly diverse sources. For example, rRNA from *Drosophila* will hybridize to some degree with chick DNA or vice versa, even though there are wide differences in base composition of the total DNA complement from the two species. In plants, cross hybridization of this type has also been noted between such diverse species as cabbage, tobacco, and bean. Cross hybridization has even been found between rRNA and DNA from barley and *Xenopus*, an

Table 9–1 Ribosomal RNA and Ribonucleoprotein Subunits of Ribosomes

Source	rRNA subunits	Ribonucleo-protein subunits	Intact ribosome
Bacteria and	15S	30S	70S
other prokaryotes	20S	50S	
Plants	16S	40S	80S
	25S	60S	
Animals	18S	40S	80S
	28S	60S	

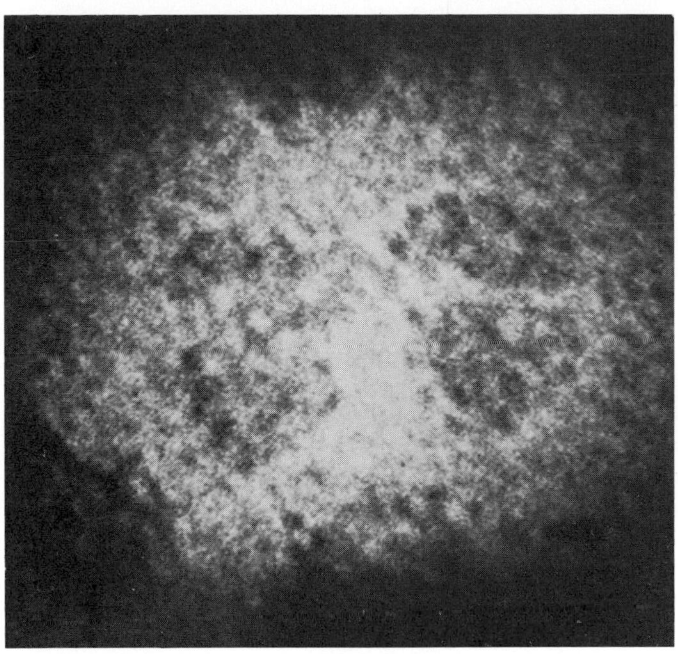

9–13. *Two 80S bean ribosomes, negatively stained with phosphotungstic acid. The cleft marking the line of jointure of the two subunits is clearly visible in the ribosome to the right. × 1,200,000. From* The Ribosome *by A. S. Spirin and L. P. Gavrilova, 1969. Courtesy of A. S. Spirin and Springer-Verlag.*

amphibian. Bacterial rRNA, in contrast, does not appear to be closely related to eukaryotic rRNA. These experiments, which have shown that there are basic similarities in sequence in ribosomal RNA from all plants and animals and between different species of bacteria point to an event establishing rRNA as a functional subclass very early in the evolution of cellular life.

The RNA extracted from ribosomes forms three sharp and well-defined peaks in the ultracentrifuge. The exact S values assigned to these peaks are somewhat arbitrary, as the sedimentation behavior of RNA varies depending upon the salt concentration, *p*H, and temperature. In addition, different investigators have used different media to carry out centrifugation of rRNA. As a result, although three well-defined peaks are always obtained with rRNA, not everyone agrees on the precise S values to be assigned to them.

There are also some differences in sedimentation behavior between ribosomal RNA from prokaryotes, plants, and animals (Table 9–1). Bacterial rRNA generally shows three peaks at 5S, 15S, and 20S; plants at 5S, 16S, and 25S, and animals at 5S, 18S, and 28S. Departures from these average values are noted in all three major groups of organisms. Each RNA is combined in the intact ribosome with a complex of proteins to make a heavier particle of higher S value. In all cases, 5S rRNA is found in association with the larger of the two particles.

The intact ribosome is a combination of the two ribosomal subunits. In animals, for example, each ribosome is made up of one 40S and one 60S subparticle. Note from the table that the S values for the subparticles are not completely additive when combined into the intact ribosome. This probably reflects changes in conformation that occur when the subunits combine to form the complete ribosome. Electron micrographs of negatively stained ribosomes often show a cleft or line of cleavage marking the jointure between the large and small subunits. The two subunits and the line of cleavage are clearly visible in the plant ribosome at the right in Fig. 9–13. The 5S component of rRNA is similar in weight and size to tRNA, which sediments at 4S. These S values reflect relative molecular weights of 25,000–27,000 for tRNA and about 40,000 for 5S rRNA.

The similarity in S values has led to some confusion between 5S rRNA and tRNA. However, a number of biochemical characteristics clearly set the two RNA species apart. 5S rRNA has no methylated bases, and no terminal end group with the CCA sequence found in tRNA. In addition, 5S rRNA does not interact with amino acids and has no sequences in common with tRNA. Further, the two small RNA types do not compete for the same site on DNA in DNA-RNA hybridization experiments. Therefore, it does not seem possible that the two represent the same RNA type nor does it seem likely that 5S RNA is an unmethylated precursor of 4S RNA. Finally, 5S RNA has not been found in the free form in cells as has 4S RNA but is always a component of the large ribosomal subunit.

The Role of the Nucleolus in the Synthesis of Ribosomal RNA

The nucleolus was implicated in RNA metabolism long before any detail was known about the function of RNA in the cell. In 1940, Jean Brachet found that the nucleolus contained large concentrations of RNA. The same investigator later showed that the nucleus is directly involved in RNA metabolism. Much later, in the early 1960's, two independent studies identified the RNA in the nucleolus as ribosomal RNA. In one of these, Jan-Erik Edstrom discovered that the base ratios of nucleolar and cytoplasmic (primarily ribosomal) RNA are closely similar. A second discovery, by Robert Perry, showed that the appearance of rRNA in the cytoplasm was dependent on a functionally intact nucleolus. Perry destroyed the nucleolus in living cells by a microscopic beam of ultraviolet light. After destruction of the nucleolus by this method, the flow of newly synthesized rRNA to the cytoplasm ceased.

This work established that the nucleolus was implicated in the synthetic pathway for rRNA. It was not possible to determine from these experiments whether the nucleolus was the actual site of RNA transcription or served chiefly as a place to modify or store rRNA synthesized elsewhere. Work by Donald D. Brown and J. B. Gurdon with *Xenopus* provided strong evidence that the nucleolus is the site of rRNA synthesis. In most organisms, the nucleolus breaks down and disappears during cell division and is reorganized after division at one or more sites on the chromosomes. This site, called the *nucleolar organizer,* is present in normal *Xenopus* larvae but is absent from a mutant form. The mutant larvae, which survive only to an early swimming stage, were also found to lack cytoplasmic rRNA.

Further work with *Xenopus* established a more direct connection between RNA and the nucleolar site on the chromosome. Fragmented DNA isolated from normal *Xenopus* larvae showed a peculiar component that centrifuged down at a slightly faster rate than the major part of the DNA. This minor component, which comprises about 0.2 percent of the total DNA of *Xenopus,* was found to be capable of forming DNA-RNA hybrids with rRNA much more readily than the rest of the DNA component. DNA prepared from anucleolate mutants of *Xenopus* did not possess this minor component. These experiments linked the minor DNA component with the nucleolar organizer site and established that rRNA was probably synthesized at this site.

Equivalent results were obtained from experiments with *Drosophila,* carried out by F. M. Ritossa and S. Spiegelman. Mutant flies that had the nucleolar organizer site duplicated from zero to five times were obtained. A strict proportionality was found in these flies between the number of nucleolar organizer sites and the amount of DNA coding for rRNA. Other techniques, including examination of nucleoli in stained preparations under the light microscope, have shown that parts of the chromosome can be seen to extend through the nucleolus. These experiments, although not completely definitive, have led to general acceptance of the conclusion that DNA located in the nucleolus is the site of rRNA synthesis.

Although the DNA coding for rRNA is present only in small amounts, representing less than 1 percent of the total DNA complement, more than enough is present to code for the 18S and 28S subunits of the ribosome. In many organisms, the amount of DNA required to synthesize RNA is present in many copies. While some bacteria contain only 2–3 copies of the rRNA cistrons, higher organisms may have 1000–2000 times as much as necessary. By hybridization experiments, this extra DNA has been shown to represent a multiple number of cistrons or genes coding for 18S and 28S rRNA. Wild-type *Drosophila,* with one nucleolar organizer site, have about 130 cistrons for 18S and 28S RNA; human cells have somewhere between 500 and 2000 cistrons for rRNA.

Hybridization experiments have shown that there are approximately equal numbers of cistrons coding for 18S and 28S RNA within the stretches of DNA coding for rRNA. This was demonstrated by D. D. Brown and C. S. Weber, who followed the synthesis of 18S and 28S RNA on fragments of *Xenopus* nucleolar organizer in cell-free systems. The capacity for transcription of both 18S and 28S RNA was found to persist in smaller and smaller fragments of the nucleolar organizer until pieces too short to contain enough DNA for coding both subunits were encountered. These results indicate that the cistrons for the two ribosomal subunits probably alternate within the nucleolar organizer segment of the chromosome, as both sites are usually included in very small segments of the nucleolar organizer DNA.

The small 5S component of rRNA is evidently not synthesized at the nucleolar organizer region, because the amount of DNA coding for 5S RNA is not increased in *Drosophila* with multiple nucleolar organizer sites. Similarly, the DNA from anucleolate mutants in *Xenopus* retains the capacity to hybridize with 5S rRNA. But the level of 5S RNA does not increase in the cytoplasm of anucleolate *Xenopus* mutants with growth of the larva, suggesting that the nucleolus may be involved in further processing of 5S rRNA after synthesis.

At some stage in the maturation of ribosomal subunits, 5S rRNA evidently enters the nucleolus, eventually to become associated with the 28S rRNA subunit. Observation of this low molecular weight RNA in nucleolar fractions led to the early claim that 4S tRNA was synthesized in the nucleus. However, neither the presence or absence of the nucleolus nor the number of nucleolar organizers has any effect on the levels of 4S tRNA in the cytoplasm. Genetic mapping of the chromosomes of *Drosophila* and the bacterium *Bacillus subtilis* has revealed that the chromosomal sites for tRNA synthesis in these organisms are close to, but separate from, the cistrons coding for rRNA. If this is true of most organisms, it is easy to see that the site might inadvertently be included in preparations of isolated nucleoli or nucleolar organizer fractions.

The maturation of ribosomal RNA in the nucleolus. Perry and others have followed the sequence of events in maturation of rRNA in the nucleolus by means of radioactive precursors combined with the effects of the antibiotic *actinomycin D.* In these experiments, cells were briefly exposed to a "pulse" of labeled RNA precursor. Further RNA synthesis was then blocked by actinomycin D, which in sufficient concentrations completely inhibits transcription of RNA on DNA. If cell extracts were made a very short time after exposure to the label and subsequent blockage of further RNA synthesis, radioactivity was found to be associated with a 45S RNA component. A slightly longer interval between exposure to the radioactive precursor and preparation of extracts revealed labeling of a 32-35S component in addition to 45S RNA. At the time the label appeared in the 32-35S component, 18S RNA also became labeled. After longer periods of time, the label appeared in the 28S rRNA fraction in addition to 18S rRNA. If the time period between exposure to the label, blockage, and preparation of extracts was extended to several hours, radioactivity appeared in cytoplasmic ribosomes.

The sequence in distribution of the label between the various RNA components suggested that the first fraction observed to be labeled, the 45S RNA, is a heavy precursor of rRNA. Other lines of evidence support this conclusion. The base composition of the 45S RNA component is closely similar to 18S and 28S rRNA. If mixtures of 45S, 18S, and 28S RNA are hybridized with DNA coding for rRNA, 45S RNA competes with 18S and 28S for the same sites on the DNA. This indicates that the sites coding for 45S and ribosomal subunit RNA are the same. Finally, in anucleolate mutants, synthesis of 45S RNA is greatly diminished or absent.

These findings, and the sequence of labeling noted by Perry and others, notably Joseph Gall and S. Penman, have led to the formulation of a probable sequence of events in the maturation of rRNA. The heavy 45S RNA component, the first detectable rRNA precursor, splits into smaller segments, including a 32–35S subunit and an 18S subunit apparently identical to 18S ribosomal RNA, soon after transcription. The 32S subunit is subsequently split, producing the 28S ribosomal subunit and releasing a small amount of "surplus" RNA. The 18S and 28S subunits, with associated protein, enter the cytoplasm through the nuclear envelope, where they eventually form complete ribosomes.

The sites of these transitions were investigated independently by Gall, Penman, and Harris Busch, who followed the maturation sequence in isolated nuclear and cytoplasmic subfractions. These investigators found that 45S, 32S, and 28S RNA are associated with nucleolar fractions. The intermediate 32–53S subunit could be found only in nucleoli; 28S RNA was found in nucleoli, in the nucleoplasm, and in the cytoplasm. The smaller 18S ribosomal subunit could be detected in the nucleoplasm but not in the nucleolus. Very soon after the appearance of 18S label in nuclear fractions, 18S label also appeared in the cytoplasm; the 28S label took a slightly longer time to appear in cytoplasmic fractions. These findings suggested localization of the various events in the maturation of rRNA in the follow-

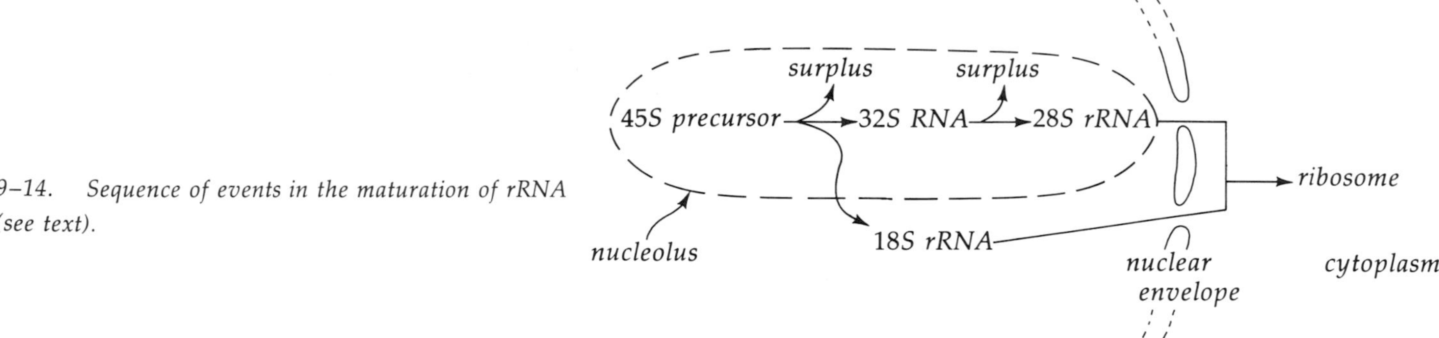

9–14. Sequence of events in the maturation of rRNA (see text).

ing structures of the cell (Fig. 9–14). The 45S rRNA precursor is synthesized within the nucleolus on nucleolar organizer DNA. Splitting the 45S precursor into 32S and 18S segments occurs in the nucleolus. The 18S segment is released from the nucleolus, traverses the nucleoplasm, and enters the cytoplasm. The 32S fraction is subsequently transformed in the nucleolus into 28S RNA, releasing as surplus a small RNA segment that evidently does not contribute to the structure of the mature ribosome. The 28S segment is then released from the nucleus and enters the cytoplasm. The small 5S rRNA subunit, found with 28S RNA in the larger subunit of the mature ribosome, is first detected in the nucleolus in association with 32S RNA.

The precise molecular nature of the transitions from 45S to 18S and 28S RNA are not known. The molecular weight of 45S RNA is estimated to be about 4.5 million; this is just about the right size to contain two 18S and two 28S subunits. However, this leaves almost no RNA as surplus. Some surplus is evidently included in the system, because the 32S to 28S conversion involves the loss of a relatively large quantity of RNA, much larger than any possible surplus if 45S RNA contains two each of the smaller subunits. Therefore, it seems likely that the 45S precursor contains one each of the 28S and 18S ribosomal subunits, with a significant quantity of 45S RNA lost during the maturation process in the nucleolus.

Some methylation of the 45S precursor occurs at the time of synthesis. It is interesting that the maturation of the precursor occurs without a detectable loss in the amount of methylated bases. This shows that the segments lost during maturation are unmethylated and suggests that methylated bases may somehow be important in the splitting mechanism. Perry and others have found that incorporation of an artificial nucleotide containing the base *8-azaguanine* interferes with separation of the 45S precursor into the smaller subunits. This indicates that the sequence of bases in the RNA chain is probably important in the splitting mechanism. However, an inhibitor of protein synthesis, *puromycin,* also interferes with the splitting process, suggesting that a protein component (see below) of the precursor may be involved in addition in the splitting mechanism. Beyond this fragmentary information, nothing is known about the mechanisms determining the precise subdivision of 45S precursor into 18S and 28S RNA (for references, see Perry, 1967, 1970).

Ribosomal nucleoproteins. At each stage in the maturation process, the various rRNA subunits, from precursor to finished ribosome, are found in association with protein in a ribonucleoprotein complex. Sedimentation constants obtained by centrifugation of these RNA-protein particles are highly variable and depend on the media used for centrifugation. Some of the extreme variability observed may result from mixtures of subunits in various stages of transition from 45S to 18S and 28S RNA. The 45S RNA precursor can be extracted from ribonucleoprotein particles sedimenting from 80S to 110S, with a possible peak at 80S. The 32–35S RNA subunit is found in an RNA-protein particle sedimenting from 65S to 80S.

The large and small ribosomal subunits, at the time of their first appearance in the cytoplasm, centrifuge at a slightly different rate from "older" 40S and 60S ribosomal subunits. There are some indications that the transition from "new" to "old" subunits involves removal of a small quantity of protein, because incubation in trypsin for a short time is found to mimic the transition.

The combination of the various RNA precursors and subunits with protein is thought to occur in the nucleus. However, there are no clues that reveal whether or not the complexed protein is actually synthesized in the nucleus or nucleolus or is combined with RNA in the nucleolus after synthesis in the cytoplasm. As noted elsewhere, it does not seem likely that protein synthesis occurs in the nucleus because completed ribosomes are not detected in the cell except in the cytoplasm. Therefore, the nucleolus is a probable assembly site of RNA and protein into ribonucleoprotein complexes but not of synthesis of the protein component from amino acids.

Suggestions for Further Reading

Books

Harris, Henry, 1968. *Nucleus and Cytoplasm*. Clarendon Press, Oxford.

Ingram, Vernon M., 1965. *The Biosynthesis of Macromolecules*. W. A. Benjamin, Menlo Park, Calif.

Spirin, A. S., and L. P. Gavrilova, 1969. *The Ribosome*. Springer-Verlag, New York.

Papers and Review Articles

Birnstiel, Max, 1967. The nucleolus in cell metabolism. *Annual Review of Plant Physiology*, vol. 18, pp. 25–28.

Brown, D. D., and J. B. Gurdon, 1964. Absence of ribosomal RNA synthesis in the anucleolate mutant of *Xenopus laevis*. *Proceedings of the National Academy of Sciences*, vol. 51, pp. 138–146.

Brown, D. D., and C. S. Weber, 1968. Gene linkage by DNA-RNA hybridization II: arrangement of redundant gene sequences for 28S and 18S ribosomal RNA. *Journal of Molecular Biology*, vol. 34, pp. 681–697.

Georgiev, G. P., 1967. The nature and biosynthesis of nuclear ribonucleic acids. *Progress in Nucleic Acid Research and Molecular Biology*, vol. 6, pp. 259–351.

Hartman, P. E., C. Rusgis, and R. C. Stahl, 1965. Orientation of the histidine operon in the *Salmonella typhinurium* linkage map. *Proceedings of the National Academy of Sciences*, vol. 53, pp. 1332–1335.

Hayashi, M., M. N. Hayashi, and S. Spiegelman, 1964. DNA circularity and the mechanism of strand selection in the generation of genetic messages. *Proceedings of the National Academy of Sciences*, vol. 51, pp. 351–361.

Heywood, S. M., and M. Nwagwu, 1968. *De novo* synthesis of myosin in a cell-free system. *Proceedings of the National Academy of Sciences*, vol. 60, pp. 229–234.

Loening, Ulrich E., 1968. RNA structure and metabolism. *Annual Review of Plant Physiology*, vol. 19, pp. 37–70.

Madison, J. T., G. A. Everett, and H. K. Kung, 1966. On the nucleotide sequence of yeast tyrosine transfer RNA. *Cold Spring Harbor Symposia on Quantitative Biology*, vol. 31, pp. 409–416.

Madjiolov, A. A., 1967. Ribonucleic acids and information transfer in animal cells. *Progress in Nucleic Acid Research and Molecular Biology*, vol. 7, pp. 195–242.

Perry, Robert P., 1967. The nucleolus and the synthesis of ribosomes. *Progress in Nucleic Acid Research and Molecular Biology*, vol. 6, pp. 219–257.

Perry, Robert P., 1970. Nucleoli: the cellular sites of ribosome production. In A. Lima-de-Faria, ed., *Handbook of Molecular Cytology*. North-Holland Publishing Company, Amsterdam, pp. 620–636.

Prescott, D. M., 1964. Cellular sites of RNA synthesis. *Progress in Nucleic Acid Research and Molecular Biology*, vol. 3, pp. 33–57.

Richardson, John P., 1969. RNA polymerase and the control of RNA synthesis. *Progress in Nucleic Acid Research and Molecular Biology*, vol. 9, pp. 75–116.

Ritossa, F. M., and S. Spiegelman, 1965. Localization of DNA complementary to ribosomal RNA in the nucleolus organizer region of *Drosophila melanogaster*. *Proceedings of the National Academy of Sciences*, vol. 53, pp. 737–745.

Salser, W., R. F. Gestelland, and A. Bolle, 1967. *In vitro* synthesis of bacteriophage lysozyme. *Nature*, vol. 215, pp. 588–591.

Singer, Maxine F., and Leder, Philip, 1966. Messenger RNA: an evaluation. *Annual Review of Biochemistry*, vol. 35, pp. 195–229.

Spiegelman, S., 1961. The relation of informational RNA to DNA. *Cold Spring Harbor Symposia on Quantitative Biology*, vol. 26, pp. 75–90.

10
Morphological Aspects of RNA Metabolism and Transport in the Nucleus

A number of structures in the nucleus have been identified with RNA metabolism. Most prominent of these is the nucleolus, which is easily visible under the light microscope as a spheroidal body usually one-fifth to one-sixth the total diameter of the nucleus. Until the mid-1960's the function of the nucleolus was obscure. Somewhat earlier, in 1940, Jean Brachet had demonstrated that the nucleolus contains high concentrations of RNA. From this work it was thought likely that the nucleolus is involved in the metabolism of RNA or protein. Many observers believed that it was primarily a storage place for RNA synthesized elsewhere until the work of Perry, Gall, Penman, and others (see Chapter 9) established that the nucleolus is the site of transcription and maturation of ribosomal precursors. Although absent from prokaryotes, the nucleolus is present in the nucleus of almost all higher organisms and takes a basically similar form in both plants and animals.

This chapter describes the morphology of the nucleolus and other nuclear structures associated with RNA metabolism, including the various RNA-containing extrachromosomal granules. Polytene (salivary gland) chromosomes, which have been particularly useful in the accumulation of morphological evidence of RNA metabolism, are also discussed in this chapter. Attention is also given to the transfer of RNA from the nucleus to the cytoplasm, including a discussion of the morphology and function of the nuclear envelope and its role in RNA transport.

Nucleolar Structure

The morphological description of the nucleolus has produced perhaps the most confusing terminology in

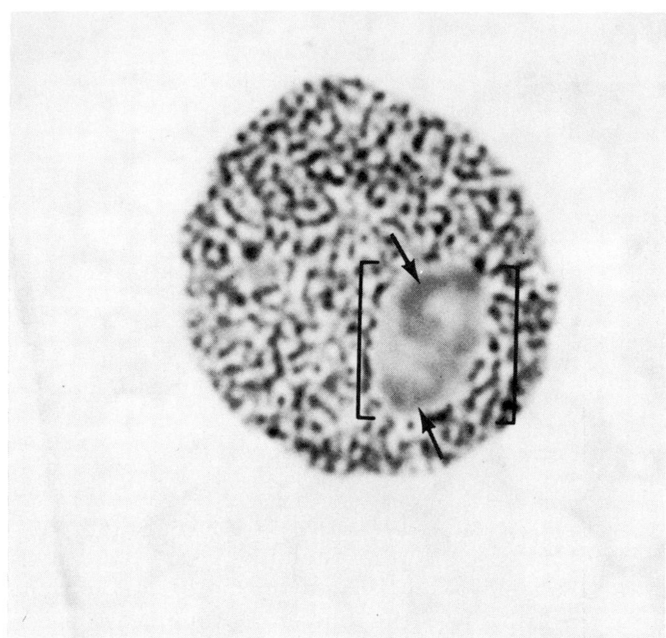

10–1. Light micrographs of a nucleus from meristematic tissue of the plant Allium porum, *containing a single large nucleolus (brackets). The nucleolus is differentiated internally into two regions: a coiled network (the nucleolonema; arrows) and an apparently amorphous matrix, the pars amorpha. The chromatin outside the nucleolus has been Feulgen stained. × 3,200. Courtesy of J.-G. Lafontaine.*

cell biology. A part of this confusion has arisen from attempts to extend the nomenclature used in light microscopy to the electron microscope image of thin-sectioned nucleoli. Under the light microscope (Fig. 10–1), the nucleolus appears to be differentiated into two morphologically distinct regions. One of these, usually forming the central core of the nucleolus, is apparently structureless and is termed the *pars amorpha* (amorphous part). Surrounding the pars amorpha is a coarsely filamentous zone that forms a network or skein around the periphery of the nucleolus, termed by light microscopists the *nucleolonema* (nucleolar threads). The coarse threads making up the network of the nucleolonema are frequently just barely resolvable in the light microscope. Variations in the relative location and distribution of pars amorpha and nucleolonema are common, depending on the organism, tissue, or age of the cell. Stains and specific enzymatic digestion of tissue sections have shown that both of these nucleolar zones contain RNA.

The electron microscope has revealed additional complexity in the nucleolus, including structures not clearly separable as pars amorpha and nucleolonema. Four distinct ultrastructural zones can be recognized (Figs. 10–2 and 10–3). One of these is granular and consists of small electron-dense, spherical particles somewhat smaller than ribosomes, about 150–200 Å in diameter. These granules are indistinct in outline and, as a result, are often described as being "fuzzier" than

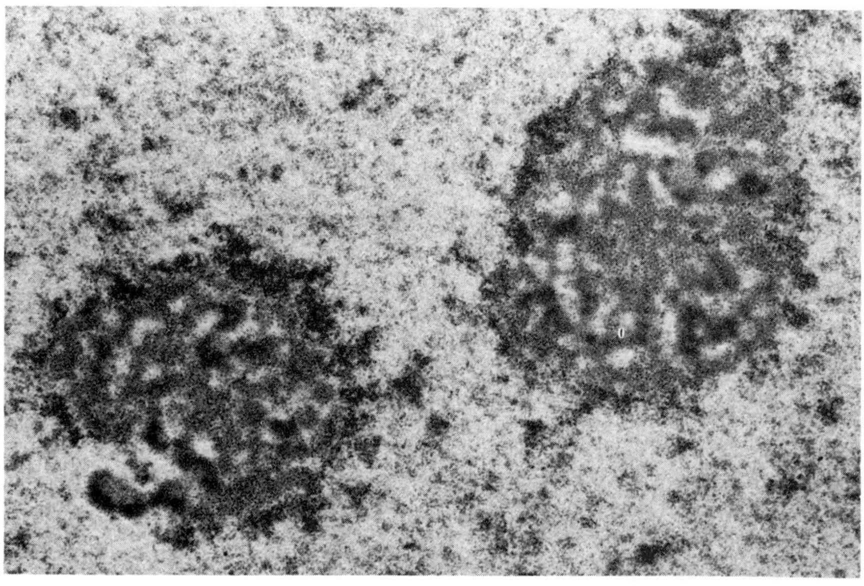

10–2. Nucleoli in a rat liver nucleus. Differentiation of the dense material into fibrillar and granular zones is clearly visible. Glutaraldehyde-osmium fixation; stained with uranyl acetate and lead citrate. × 24,000. From Ultrastructure in Biological Systems, *Vol. 3, ed. A. J. Dalton and F. Haguenau, 1968. Courtesy of W. Bernhard and Academic Press, Inc.*

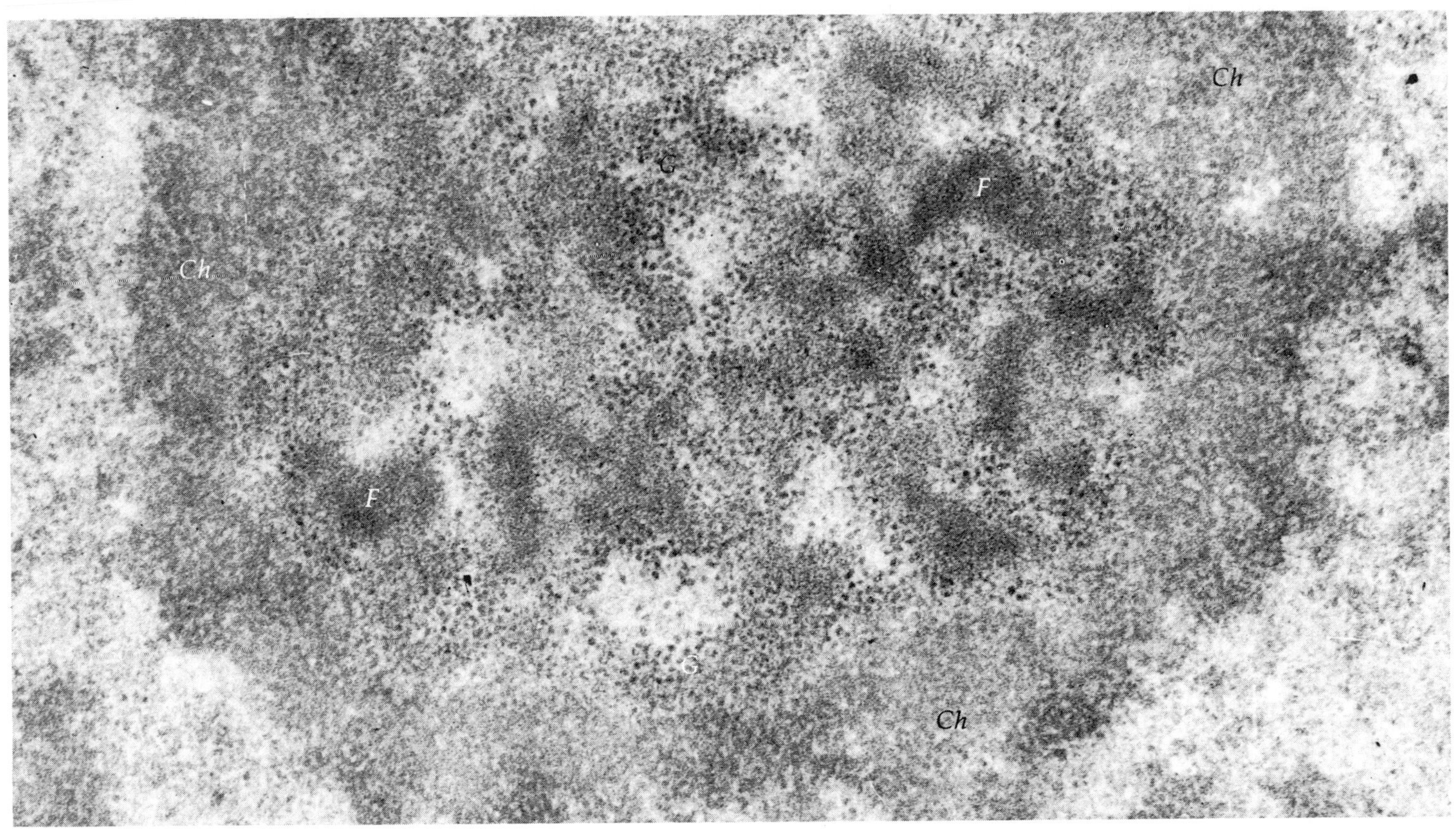

10–3. *Nucleolus of a pancreatic cell of the rat. Fibrillar* (F) *and granular* (G) *zones are visible within the nucleolus; closely associated with the nucleolar periphery are blocks of chromatin* (Ch). × *38,000. Courtesy of V. Marinozzi.*

ribosomes. Several terms have been coined to describe this nucleolar region; the simple and self-explanatory term *granular part* or *zone* will be used here. Often these granules are seen to be interconnected, like beads on a string, by a fine fibril about 50 Å in diameter.

A second nucleolar component appears as a collection of indistinct fibers 50–100 Å in diameter. Connections can be traced from this nucleolar zone, the *fibrillar part* or *zone*, to the granular part of the nucleolus. Both the granular and fibrillar parts of the nucleolus are suspended in a third component, the *matrix*, an apparently structureless background substance that is somewhat more electron dense than the surrounding ground substance of the nucleoplasm.

A fourth nucleolar component is also made up from aggregations of fibers approximately 100 Å in diameter and noticeably less electron-dense than the fibrillar zone of the nucleolus. These 100 Å fibers have been identified as a part of the chromatin that extends into the nucleolus. These fibers may either be localized into pockets within the nucleolus or diffusely spread throughout the nucleolar structure. This chromatin, forming a regular part of nucleolar structure, is termed *nucleolar chromatin*. Other chromatin surrounds the exterior of the nucleolus on all sides, except where the nucleolus touches the nuclear envelope.

These four nucleolar components, the granular zone, fibrillar zone, matrix, and nucleolar chromatin, can be seen in the nucleolus in Fig. 10–3. A great deal of the confusion in terminology has resulted from attempts to use the terms *pars amorpha* and *nucleolonema* to describe these structures. The two nucleoli in Fig. 10–2 contain a network of electron-dense material separated by less dense spaces. This network, which would be identifiable under the light microscope as the nucleolonema, can be seen to contain aggregations of material from both the granular and fibrillar zones. Regions identified as pars amorpha under the light

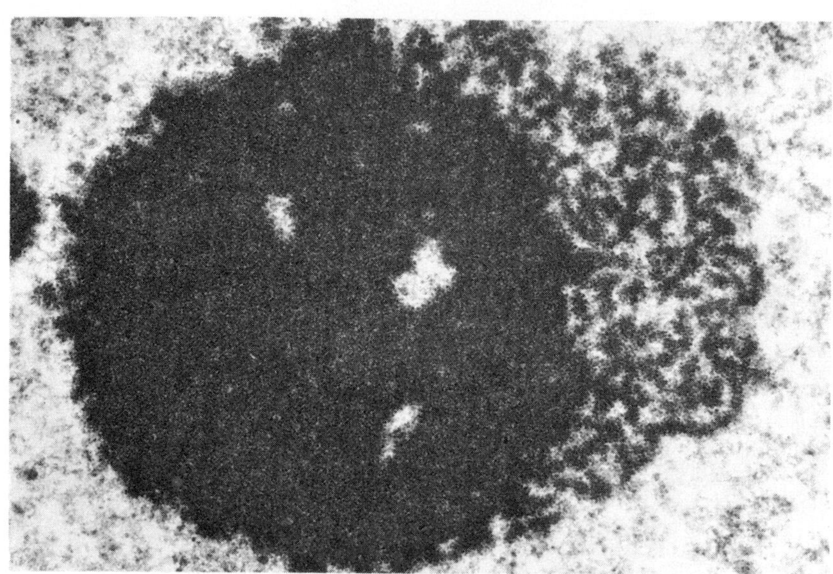

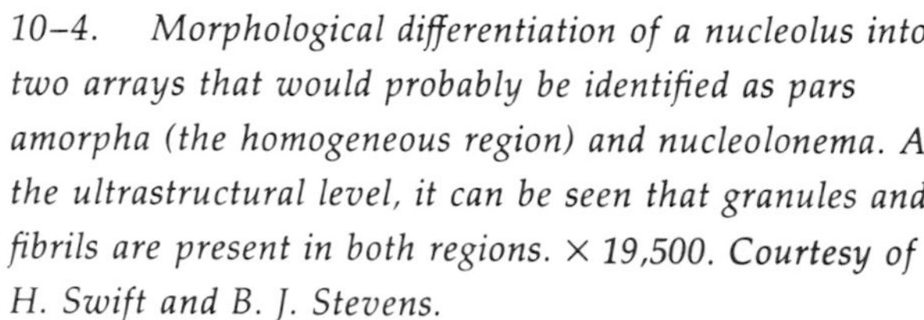

10–4. Morphological differentiation of a nucleolus into two arrays that would probably be identified as pars amorpha (the homogeneous region) and nucleolonema. At the ultrastructural level, it can be seen that granules and fibrils are present in both regions. × 19,500. Courtesy of H. Swift and B. J. Stevens.

microscope have also been shown to contain both granules and fibrils (see Fig. 10–4). These terms therefore denote states of aggregation or gross architecture of the nucleolus and do not differentiate between the granular and fibrillar zones. Unfortunately, many electron microscopists have used pars amorpha to describe the granular zone and nucleolonema to describe the fibrillar zone, or even the reverse. In recent years, the more simple terms fibrillar part, granular part, matrix, and nucleolar chromatin, or slight variations of these terms, have gained precedence.

The relative proportions of the four nucleolar components and their configurations within the nucleolus undergo constant change during the life of the cell. Rapidly synthesizing cells have large nucleoli, which may take up as much as one-fourth of the volume of the nucleus. In these nucleoli, the granular part is frequently increased in proportion to the other components; in quiescent or degenerating cells the relative volume of the granular part is much reduced.

All indications are that the nucleolus is a highly dynamic structure and undergoes alterations in morphology in response to changes in cell activity. Usually, the granular zone is found around the periphery of the nucleolus, partially or completely surrounding the fibrillar zone, but many exceptions can be found to this pattern of distribution, even within the same species. For example, in oocytes the granular zone may be either in the interior or around the nucleolar periphery; in some oocytes the pattern of distribution changes as development proceeds.

Other structural configurations are observed in nucleoli. Often spaces, sometimes called *nucleolar vacuoles,* can be seen inside the nucleolus (Fig. 10–4). These spaces, which occur more frequently in plants, may contain highly dispersed nuclear chromatin or very diffuse and indistinct fibrils or granules. Other inclusions, such as granules or crystals of protein or carbohydrate, may be present. Even lipid inclusions are seen in some cases; in the nuclei of human placental tissue a system of membrane-bound channels extends through the nucleolus (Fig. 10–5). These channels, which are arranged in a regular pattern in the nucleolus, communicate with the perinuclear space (see p. 224) of the nuclear envelope. The function of these nucleolar channels is unknown.

Cytochemical Identification of the Four Nucleolar Components

The localization of DNA, RNA, and protein in the four nucleolar components has been made possible by enzymatic digestion techniques developed for electron microscopy by W. Bernhard and his colleagues. Bernhard developed new embedding media that permit the selective digestion of macromolecular components from thin sections of fixed and embedded tissue. These techniques have been applied with great success to the study of nucleolar cytochemistry.

Digestion of thin sections with ribonuclease removes both the fibrillar and granular parts of the nucleolus (Fig. 10–6), showing that RNA is a major constituent

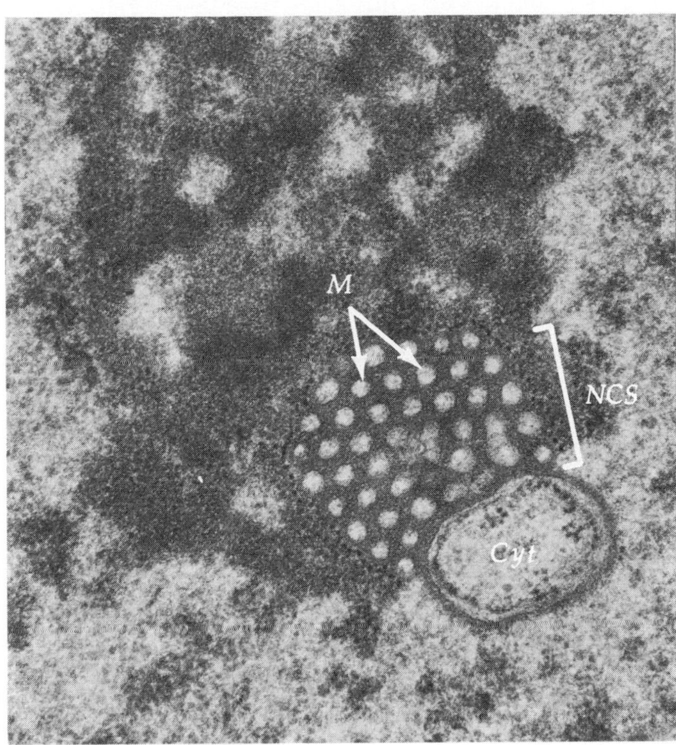

10–5. Nucleolus from human endometrium. A nucleolar channel system (NCS) *is present at one side of the nucleolus; individual channels are 500–600 Å in diameter. The channels, each lined by a membrane, are embedded in amorphous, dense material* (M). *Apparently, the lumen of the channels is in communication with the perinuclear space. Just below the system, an invagination of the cytoplasm* (Cyt) *has been cut in cross section. × 49,500. Courtesy of J. A. Terzakis and The Rockefeller University Press, from* J. Cell Biol. *27:293 (1965).*

of both of these components. A protease, such as pepsin, removes the matrix without apparent disturbance of the granular and fibrillar zones or nucleolar chromatin (Fig. 10–7). Sequential digestion of the same section with both ribonuclease and pepsin results in almost complete removal of density from the interior of the nucleolus (Fig. 10–8). Taken together, these findings indicate that the matrix consists of protein and the fibrillar and granular zones of an RNA-protein complex.

The scattered dense regions that persist after combined digestion with ribonuclease and protease probably represent the chromatin of the nucleus. This conclusion is directly supported by the results of digestion with deoxyribonuclease (Fig. 10–9), which removes density from scattered pockets inside the nucleolus.

The existence of DNA in the nucleolus was a matter of some debate until quite recently. With the Feulgen reaction, only a light pink stain is noted in the nucleolus, which by itself is not enough to prove the presence of DNA. This is complicated by the large amounts of extranucleolar chromatin that are always closely associated with the nucleolus. This material, which lies in contact with the nucleolar surface, undoubtedly contributes to the results obtained with Feulgen staining and makes it difficult to determine what contribution, if any, is made to the binding of stain by nucleolar DNA.

These difficulties were finally resolved in the 1960's by the more sensitive techniques of autoradiography and enzymatic digestion. Another method, using fluorescent dyes that bind specifically to DNA, has also provided direct evidence that DNA is present in the nucleolus. With this technique, the configuration of chromatin inside the nucleolus is made especially distinct (see Barr and Plaut, 1966); from this work it is clear that the chromatin of the nucleolus frequently represents an arm or loop of one or more chromosomes that extends into the nucleolus.[1]

The Relationship of Nucleolar Morphology to the Maturation of Ribosomal RNA

Research by Bernhard, by Elizabeth D. Hay and her colleagues, and by others has made it possible to relate aspects of nucleolar ultrastructure to the sequence of events in the maturation of ribosomal RNA. This work has been done by electron microscope autoradiography. With a pulse label of tritiated uridine, radioactivity in thin sections of the nucleolus is first detected in the fibrillar zones. Biochemical extracts of nuclei from another sample of the same group of cells show, after centrifugation, that the radioactivity is associated with 45S rRNA. After slightly longer periods of time, a label appears in the granular zones of the nucleolus. Extracts centrifuged at this time show the label in 35S and 28S rRNA. These findings indicate that the fibrillar zones of the nucleolus contain 45S rRNA; splitting of 45S precursor into 35S and 28S rRNA can be traced morphologically as a transition from fibrillar to granular

[1] In some species, the DNA of the nucleolus is released and becomes separate from the chromosome. This takes place in the developing oocytes of amphibians and a few insects (see Chapter 15).

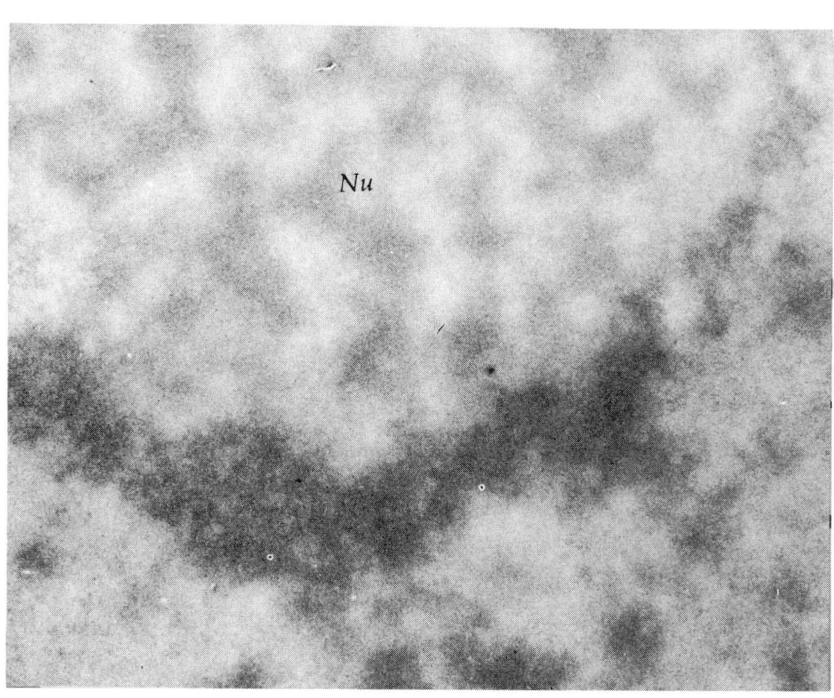

10–6. Nucleolus (Nu) *of rat pancreas, fixed in a mixture of acrolein and formalin and digested for one hour in ribonuclease. The tissue was subsequently stained in lead ion. Neither granules nor fibrils are visible in the nucleolus.* × *45,000. Courtesy of V. Marinozzi and Academic Press, Inc., from* J. Ultrastr. Res. *10:433 (1964).*

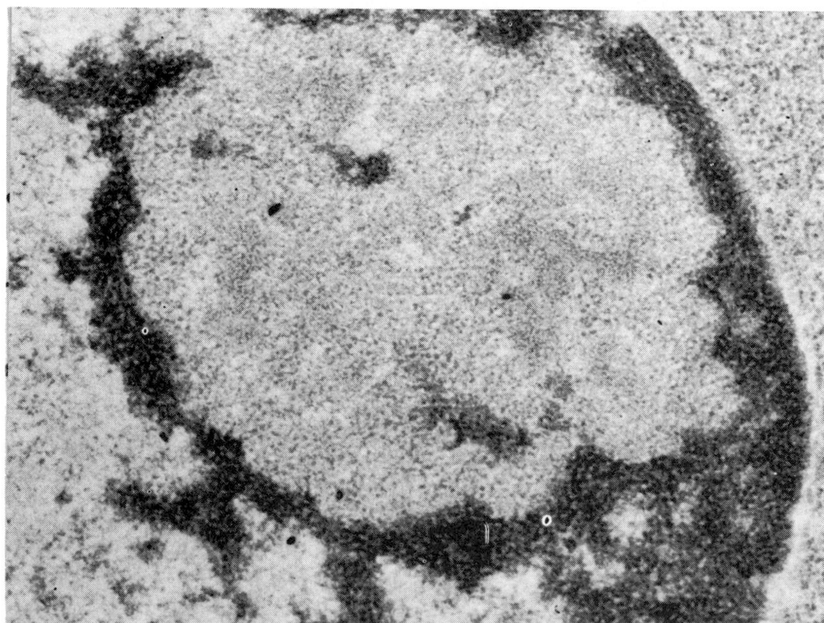

10–7. Nucleolus of rat pancreas, fixed by same procedure as in Fig. 10–6, but digested with pepsin and subsequently stained with uranyl acetate. The matrix material has been removed from between the granules, fibrils, and chromatin of the nucleolus. × *24,500. Courtesy of V. Marinozzi and Academic Press, Inc., from* J. Ultrastr. Res. *10:433 (1964).*

configuration. The granules at the periphery of the nucleolus probably represent 28S rRNA in combination with protein to form the larger 60S ribosomal subunit. Presumably, these particles are subsequently released from the nucleolus, traverse the nuclear envelope, and enter the cytoplasm to combine with 18S rRNA (in 40S rRNA-protein subunits) to make complete 80S ribosomes.

A number of investigators have claimed that the nucleolus contains intact ribosomes, mostly on the basis of the morphology of the granular parts in the nucleolus and the similarity in reactions to stains and enzymatic digestion noted between this nucleolar component and cytoplasmic ribosomes. Although these nucleolar particles do indeed resemble cytoplasmic ribosomes in morphology, centrifugation of nuclear extracts has

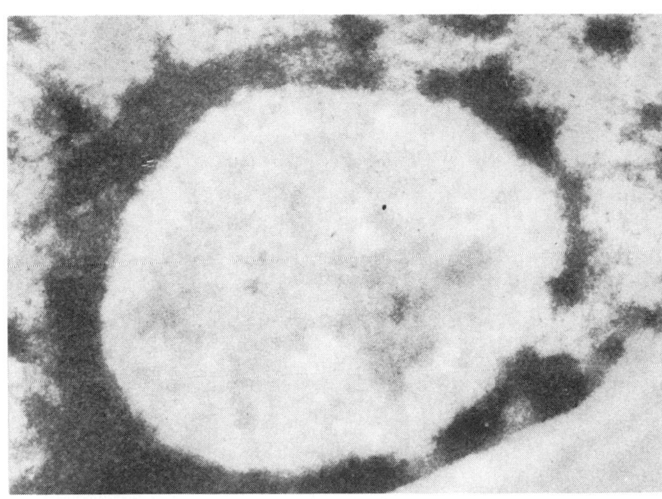

10–8. Nucleolus of rat pancreas, fixed as in Fig. 10–6 and digested with both ribonuclease and pepsin. Almost all density has disappeared from the nucleolus except regions provisionally identified as nucleolar chromatin. × 33,500. Courtesy of V. Marinozzi and Academic Press, Inc., from J. Ultrastr. Res. *10:433 (1964).*

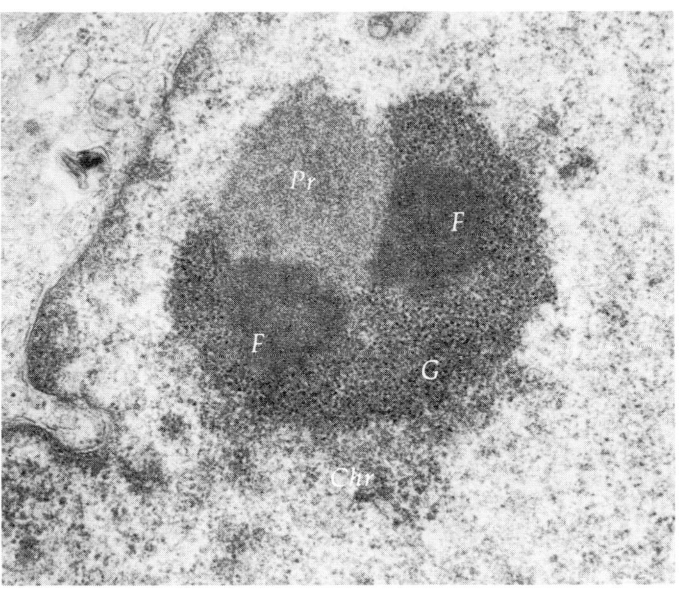

10–10. Nucleolus of a monkey kidney cell exposed to Actinomycin D for one hour. The nucleolar components have segregated into separate fibrillar (F), *granular* (G), *and chromosomal zones* (Chr); *the segment labeled* Pr *is presumably matrix protein. × 18,500. Courtesy of G. I. Schoefl and Academic Press, Inc.* [J. Ultrastr. Res. *10:224 (1964)*]

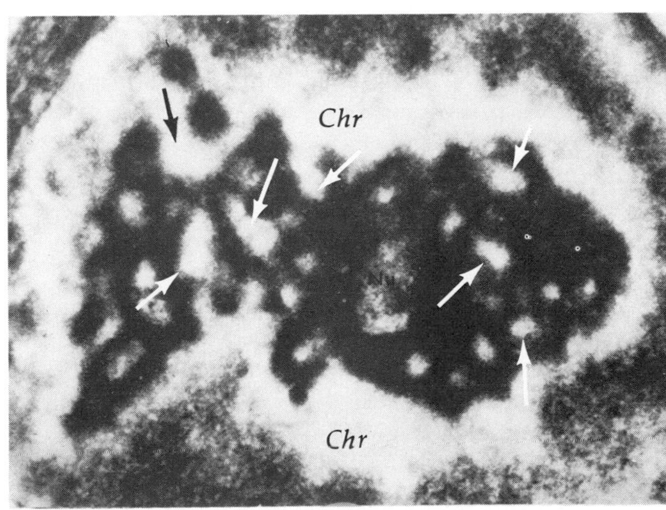

10–9. Nucleolus of a lymphoid leukemia cell from a mouse. Fixed with formalin and digested with deoxyribonuclease. Density has been removed from the nucleoplasm surrounding the nucleolus (Chr) *and from channels within the nucleolus (arrows). × 22,500. Courtesy of Dr. Bernhard, from N. Granboulan and P. Granboulan,* Exptl. Cell Res. *34:71 (1964).*

failed to reveal a ribonucleoprotein particle in the nucleus or nucleolus with the size and weight of completed ribosomes (see Chapter 9).

Experiments in which RNA synthesis is blocked with actinomycin D (see Chapter 9) reveal that this drug has a curious effect on nucleolar ultrastructure. This effect has been investigated in detail by G. I. Schoefl. After exposure to actinomycin D, the four components of the nucleolus sort out or segregate into separate regions (Fig. 10–10). After longer periods of exposure to the drug, the nucleolus degenerates. There is some disagreement among investigators over which component persists after prolonged exposure to actinomycin D. In some cases, the fibrillar zone has been reported to persist and in others the granular zone. Persistence of either part is unexpected, as actinomycin D is not known to affect either the conversion of fibrils (presumably 45S rRNA) to granules (35S + 28S rRNA) or the movement of the completed granules (presumably 60S ribosomal subunits) to the cytoplasm. No doubt the disturbance of synthesis in the cell by actinomycin D causes side effects that eventually disrupt the activity of the nucleolus, even though the drug may not have a direct effect on the maturation of RNA.

During cell division the nucleolus disintegrates, and the component parts disperse and become indistinguishable against the ground substance of the cell. When division is complete, the nucleolus reappears at well-defined regions of the chromosomes called *nucleolar organizers* (described in detail in Chapter 13, which deals with the events of cell division).

Other Ribonucleoprotein Structures in the Nucleus

Other structures, believed to contain RNA in combination with protein, have recently been identified in the nucleus. A number of morphologically distinct entities have been described, all linked in some way with RNA metabolism. While these ribonucleoprotein structures are probably products of the chromosomes and are undoubtedly involved in some manner in the total scheme of protein synthesis, their precise function remains unknown.

A few particulate bodies in the ground substance of the nucleoplasm, quite separate from the chromatin and the nucleolus, were described early in the 1960's. However, the extensive distribution of these interchromatin bodies was not suspected until recently. With the development of a new staining technique by W. Bernhard and his co-workers, it has been possible to distinguish between organelles containing DNA and ribonucleoprotein in the nucleus. Although the specificity of the staining technique is not yet completely proved, it appears to be highly selective for RNA and ribonucleoprotein. The technique depends on the use of EDTA (ethylene-diamine-tetraacetic acid) to produce differential attachment of the uranyl ion (UO_2^{+2}) to DNA and RNA. UO_2^{+2} has strong affinity for the negatively charged phosphate groups of RNA; UO_2^{+2} also binds, but less strongly, to the phosphate groups of DNA. In Bernhard's technique, sections are fixed in an aldehyde but not postfixed in osmium. Thus, the fixed tissues are not electron-dense and are essentially transparent to the electron beam. The tissues are then stained in a solution containing uranyl acetate. After uranyl staining, sections are exposed to EDTA, which evidently removes uranyl ions attached to DNA but not those attached to RNA. Structures containing DNA appear "bleached" after this treatment. Further enhancement of the density of RNA-protein is obtained by post-staining with lead acetate after EDTA treatment. This apparently increases the density of the ribonucleoprotein–UO_2^{+2} complex without adding density to DNA or deoxy ribonucleoprotein.

With this new technique, sections show an extensive and complex array of structures in the nuclear ground substance (Fig. 10–11). Four different types of RNA-containing structures can be discerned. One of these is a scattered collection of very dense granules, approximately 400–450 Å in diameter (see Figs. 10–11 and 10–12). These large granules usually occur within aggregated masses of the chromatin, separated from the chromatin by an electron-transparent shell or halo about 250 Å in thickness, which does not take up stain. Ribonuclease and deoxyribonuclease have no effect on the particles, but if ribonuclease is used in conjunction with a protease such as pronase, the electron density of the granules is greatly reduced. This effect, along with the staining reaction, suggests that these granules contain ribonucleoprotein. Because of their location within and at the edges of the chromatin masses, these particles have been termed *perichromatin granules.* The relative numbers of these granules vary in response to changes in the physiological state of the cell. Small doses of actinomycin D, which interfere with the synthesis of ribosomal RNA but not other types of RNA, cause a sharp increase in the number of perichromatin granules. These granules are morphologically similar to the granules produced in large numbers at the site of large "puffs" or Balbiani rings in salivary gland chromosomes (see p. 241). Because these puffed chromosome sites are believed to manufacture mRNA in large quantities, the Balbiani granules are thought to represent mRNA. The similarities in morphology suggest that the perichromatin granules may also represent a type of mRNA.

A second class of granules, termed *interchromatin granules,* is extensively distributed throughout the nucleus. These granules are aggregated in regions of the nuclear ground substance that are distinctly separate from the chromatin (Figs. 10–11 and 10–13). Interchromatin granules are more variable in size, and are consistently smaller than perichromatin granules, approximately 200–250 Å in diameter. These clusters of particles are similar to perichromatin granules in their reaction to enzymes; deoxyribonuclease is without effect, but a combination of pronase and ribonuclease causes a significant reduction in the electron density of the particles. The numbers of interchromatin granules do not appear to change in response to alterations in the physiological state of the cell.

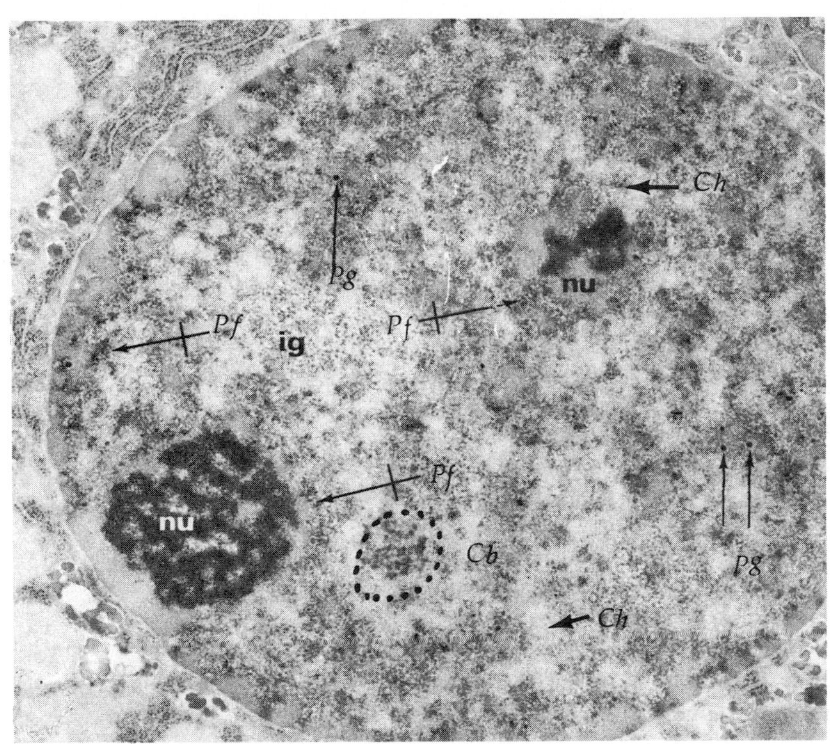

10–11. Nucleus from rat liver, fixed in glutaraldehyde and stained for RNA by the EDTA method (see text). Four types of extrachromosomal structures are visible in this nucleus: perichromatin granules (Pg), *interchromatin granules* (ig), *perichromatin fibrils* (Pf) *and a coiled body* (Cb). *The nucleolus* (nu) *and bleached areas corresponding to aggregates of chromatin* (Ch) *are also present.* × *18,000. Courtesy of A. Monneron, W. Bernhard, and Academic Press, Inc., from* J. Ultrastr. Res. *27:266 (1969).*

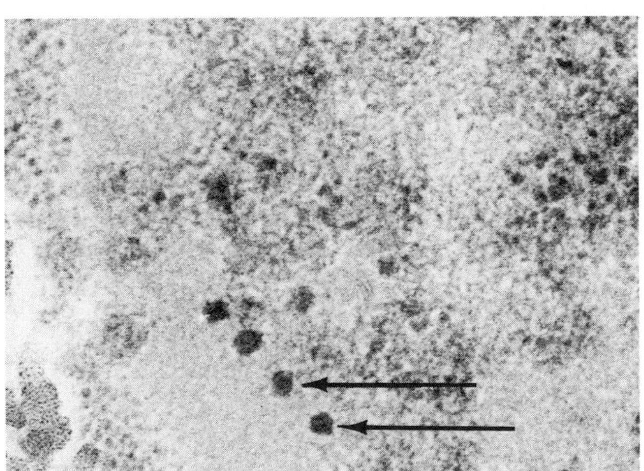

10–12. Perichromatin granules (arrows) in the nucleus of a rat liver cell. Each granule is surrounded by a halo of material of low electron density. × *59,000. Courtesy of A. Monneron, W. Bernhard, and Academic Press, Inc., from* J. Ultrastr. Res. *27:266 (1969).*

Other ribonucleoprotein structures, revealed for the first time in the nucleoplasm by the new staining technique, have been termed by Bernhard *perichromatin fibers* and *coiled bodies.* Perichromatin fibers (Figs. 10–11 and 10–14) are found at the edges of masses of chromatin. These fibers, which appear to be directly connected to 100 Å chromatin fibers, are highly sensitive to digestion with ribonuclease. They may represent a fibrous RNA product of the chromosome or perhaps a coating of RNA on chromosomal fibers at the surface of aggregated clumps of chromatin. Perichromatin fibers, while distinct in structure from perichromatin and interchromatin granules, may possibly represent a precursor of one of these types. The higher sensitivity of perichromatin fibers to ribonuclease suggests that this RNA species is not as tightly bound to protein as other nuclear ribonucleoprotein structures.

Coiled bodies (Figs. 10–11 and 10–15) are masses of relatively coarse fibers 400–600 Å in diameter, which are apparently sparsely distributed in nuclei. Coiled bodies also resist digestion with deoxyribonuclease or ribonuclease if these enzymes are used separately. If pronase treatment is combined with ribonuclease digestion, electron density is greatly reduced.

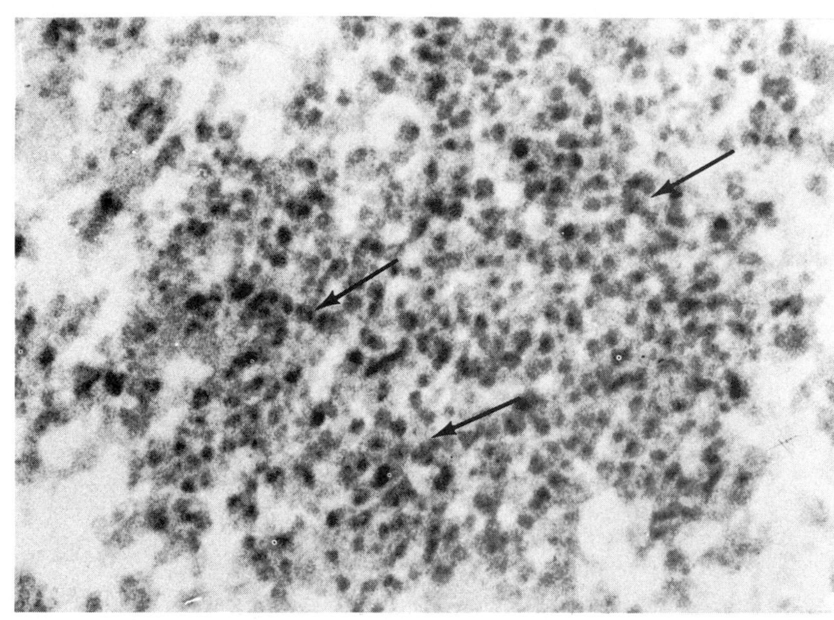

10–13. *Interchromatin granules (arrows) in a nucleus from the adrenal cortex of the rat. Stained by the EDTA method.* × *115,000. Courtesy of A. Monneron, W. Bernhard, and Academic Press, Inc., from* J. Ultrastr. Res. *27:266 (1969).*

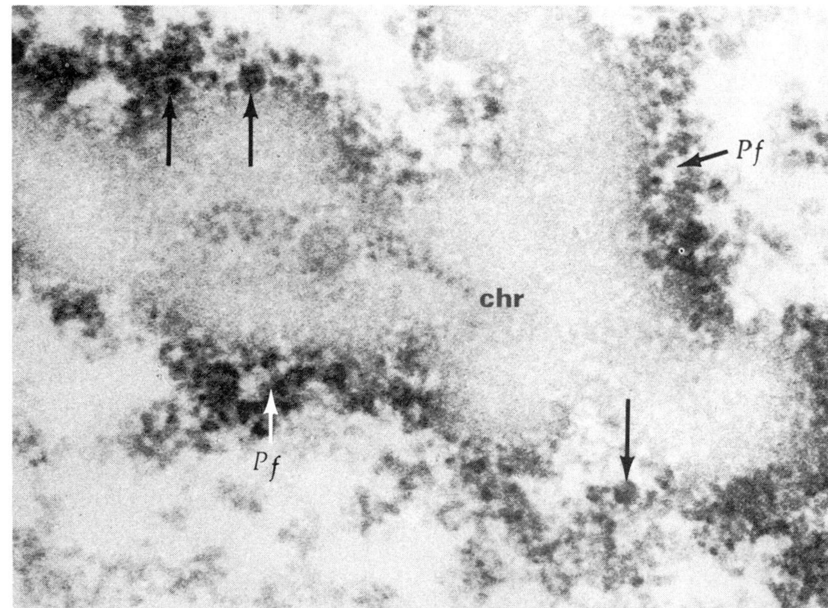

10–14. *An aggregate of chromatin* (chr) *in a nucleus from rat hypothalamus surrounded by perichromatin fibrils* (Pf) *and a few perichromatin granules* (*arrows*). × *72,000. Courtesy of A. Monneron, W. Bernhard, and Academic Press, Inc., from* J. Ultrastr. Res. *27:266 (1969).*

While some very limited information points to a possible relationship between perichromatin granules and mRNA, the precise function of these extrachromosomal and extranucleolar ribonucleoprotein bodies remains unknown. Perichromatin and interchromatin granules were described by others before the development of Bernhard's stain. Much controversy has surrounded the identity and role of these particles. While Bernhard's technique makes it appear likely that these granules contain RNA, their function remains obscure. Not even a hint is available as to the function of perichromatin fibers or coiled bodies, except that all of these structures are almost certainly morphological representations of the transcription of RNA on chromosomal DNA. The resistance of various particles to ribonuclease digestion unless a protease is also used indi-

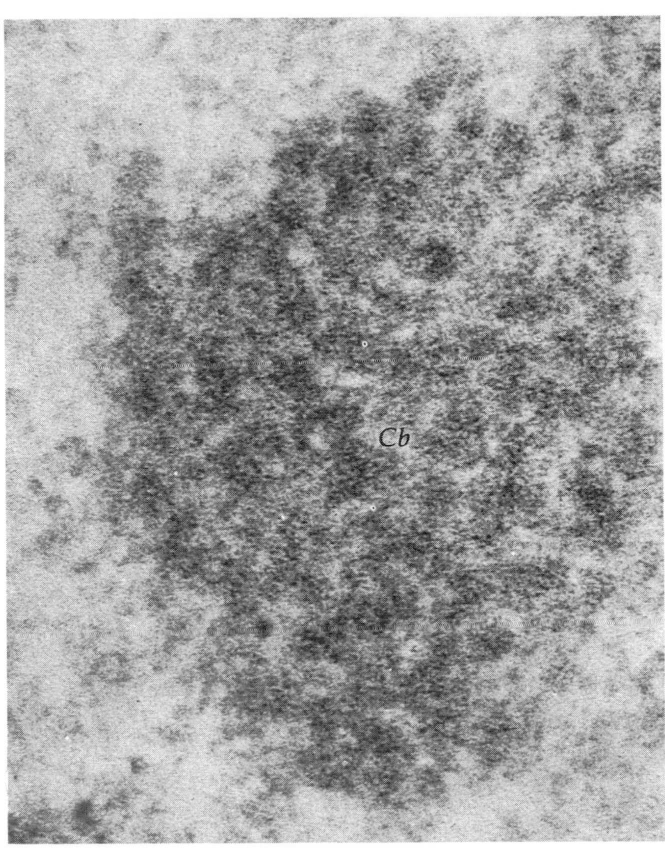

10–15. A coiled body (Cb) *as revealed by the EDTA technique for RNA. The coarse threads are in turn composed of a fine fibrillar substance. × 115,000. Courtesy of A. Monneron, W. Bernhard, and Academic Press, Inc., from* J. Ultrastr. Res. *27:266 (1969).*

cates that each of the particles contains RNA in close association with a protein or perhaps a number of proteins. These particles are not found in close association with the nucleolus; they may therefore represent RNA species, including transfer, messenger, or other DNA-like RNAs, that pass through the nucleoplasm directly to the cytoplasm.

The various extrachromosomal and extranucleolar bodies discussed in the foregoing paragraphs have been described mainly in mammals and in a few insect species. It is not known how widely these structures, or similar RNA-containing bodies, are distributed. Very little work has been done in this area in plant cells. The few existing reports indicate that plants also possess a complex system of extrachromosomal particles in the nuclear ground substance and that structures similar to the four classes described in animals are probably also found in plant nuclei.

A Review of the Fine Structure of the Interphase Nucleus

With the information presented above on the ribonucleoprotein particles of the nuclear ground substance, it is possible to present a fairly complete summary of the ultrastructure of the interphase nucleus (see Fig. 10–16). In Chapter 8, it was shown that the chromatin of the interphase nucleus consists of masses of irregularly folded fibers, indistinct in outline and approximately 80–100 Å in diameter. The chromosome fibers may either be evenly dispersed throughout the nucleus or aggregated into clumps. Where aggregation occurs, the clumped chromatin is often located at the periphery of the nucleus, sometimes forming a complete shell of tightly coiled fibers just inside the nuclear envelope. Embedded within the chromatin are one or more nucleoli that exhibit variations in size, shape, and relative amounts of the four nucleolar components: granular part, fibrillar part, matrix, and nucleolar chromatin.

The granular part of the nucleolus consists of a collection of dense granules, approximately 150 Å in diameter and less distinctly outlined than cytoplasmic ribosomes. Sometimes the granules of the nucleolus are seen to be interconnected by a thin 50 Å fibril. The fibrillar zone consists of indistinct fibers approximately 50–100 Å in diameter. Often connections can be traced running from these fibrils to the granular zone. Both the granular and the fibrillar parts of the nucleolus are embedded within the matrix, a background substance slightly more electron-dense than the nuclear ground substance. Within the nucleolus are fibers identifiable as chromatin fibers; these represent arms of the chromosome that extend into and form a part of the mass of the nucleolus.

In the nuclear ground substance surrounding the chromatin and nucleolus, sometimes called the nuclear sap, variable amounts of other ribonucleoprotein particles are to be found: the 400–450 Å perichromatin granules; the smaller 200–250 Å interchromatin granules; perichromatin fibrils, from 30–50 Å to as much as 200 Å in diameter; and the relatively coarse 400–600 Å fibers of the coiled bodies. The functional significance of these bodies is unknown.

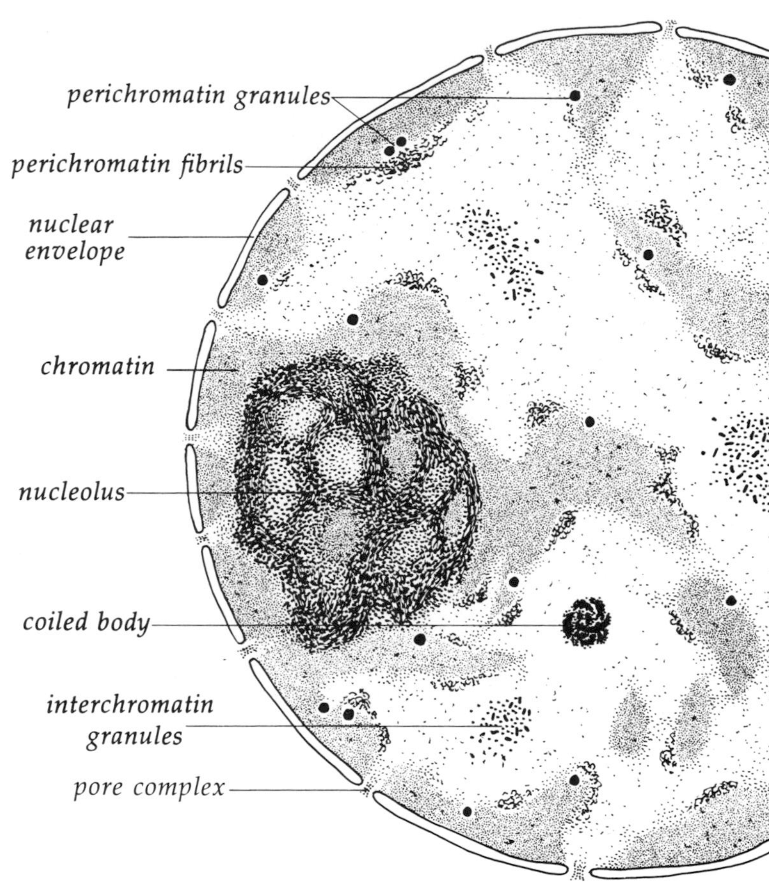

10–16. Diagram of the various structures of the mammalian nucleus, as visualized by Bernhard and his co-workers. All of the extrachromosomal structures are believed to consist of RNA in combination with a protein. Redrawn, original courtesy of A. Monneron, W. Bernhard, and Academic Press, Inc., from J. Ultrastr. Res. *27:266 (1969).*

The Transport of Ribonucleoprotein to the Cytoplasm

The Nuclear Envelope

All material passing from the nucleus to the cytoplasm must traverse the nuclear envelope. Metabolites and precursors of RNA and DNA and probably macromolecules also pass in the opposite direction from cytoplasm to nucleus. The nuclear envelope is highly complex and remarkably uniform in structure from species to species in eukaryotes. This membrane system is absent from prokaryotes; in bacteria and blue-green algae the nuclear material is suspended directly in the cytoplasm.

The nuclear envelope is constructed from two concentric membranes, which completely encircle the nucleus (Fig. 10–17). Each of these membranes, in osmium-fixed thin sections, appears closely similar to the plasma membrane in structure and consists in cross section of two dark lines separated by a less dense central space. The membranes of the nuclear envelope appear close to the plasma membrane in degree of complexity, with little evidence of additional globular substructure superimposed on the basic trilaminar form (see Chapter 4). The cytoplasmic face of the outer membrane of the nuclear envelope is often coated with a layer of ribosomes.

The two nuclear membranes are separated by an included space of variable dimensions, ranging from about 100 to 300 Å or more. A part of the variability in dimension of this separating space, called the *perinuclear space* (Fig. 10–17), is dependent on the quality and type of fixation. With osmium tetroxide fixation the space is more variable in dimensions. Careful control of the conditions of fixation, including buffering, salt concentration, and temperature has the effect of reducing the variability of the perinuclear space. After aldehyde fixation (Fig. 10–17) this space is much more regular.

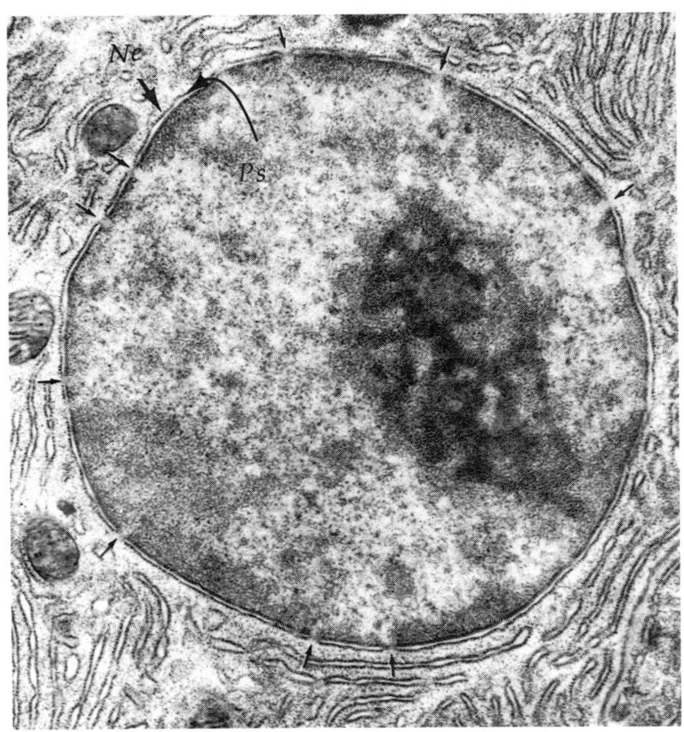

10–17. A nucleus from bat pancreas, clearly showing the nuclear envelope (Ne) *and a perinuclear space* (Ps) *of very regular dimensions. Pores are visible at the arrows. × 12,500. From* The Cell *by D. W. Fawcett, 1966. Courtesy of D. W. Fawcett and W. B. Saunders Company.*

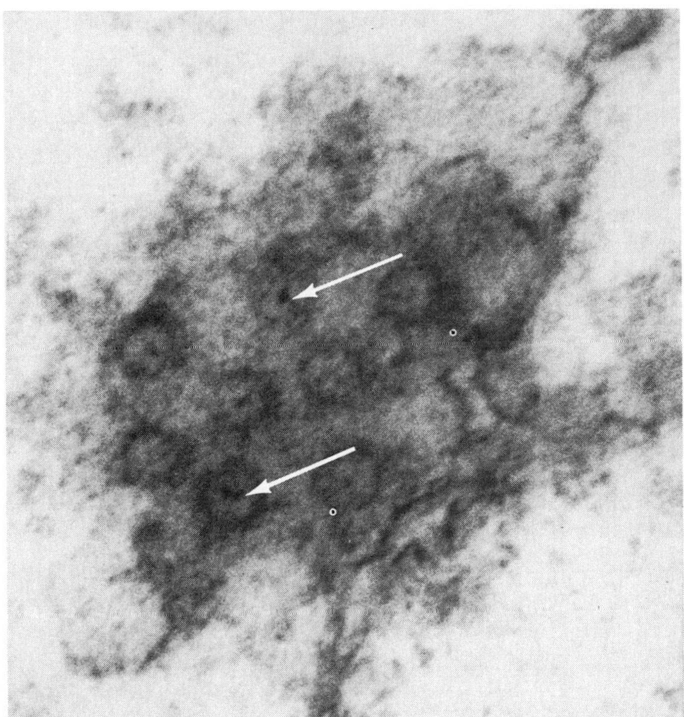

10–18. Nuclear pore complexes in tangential section. From an oocyte of Ciona intestinalis. *Most of the pore complexes shown contain a central dense granule (arrows). × 92,500. Courtesy of R. G. Kessel and Academic Press, Inc.* [J. Ultrastr. Res., *Supplement 10 (1968)*]

At frequent intervals the membranes of the nuclear envelope are interrupted by complex substructures called *pores* (Figs. 10–17 and 10–18). Around the margins of a pore the outer membrane makes a connection with the inner membrane, completely separating the perinuclear space from the channel formed by the pore. When viewed from the surface of the nuclear membrane, in sections on a plane parallel to the surface of the envelope, the pores appear to be circular or octagonal with very regular dimensions.

Although the term "pore" suggests a channel of open communication from nucleus to cytoplasm, the pores are apparently closed to free diffusion of ions and larger molecules (see below). This is reflected structurally by an electron-dense "plug" of faintly fibrous or granular material that fills most of the pore in cross sections. This material, called the *annular material* or *annulus,* is most readily seen in osmium and aldehyde-fixed preparations; permanganate fixation apparently fails to preserve the annular material, giving the impression that the pore is "open."

For a time, the pore was thought to be closed by a membrane-like "pore diaphragm" (Fig. 10–19). This apparent diaphragm has never been shown to contain a trilaminar structure as expected for a cross-sectioned membrane. The image interpreted as a pore diaphragm probably results from inclusion of a part of the margin of the pore in a cross section. This is explained as follows. Pores are about 700 Å in diameter (see below). With this dimension, the average thin section, at 500 to 800 Å, is not likely to cut cleanly through the precise center of a pore. Sections falling on one side or another of the center line of a pore will show a dense line apparently closing the pore (see Fig. 10–20). This line actually represents an included pore margin. Careful work with very thin sections, in which the chance of including the pore margin is greatly lessened, shows a concomitant reduction in the frequency of pores apparently closed by a diaphragm. For these reasons, the "pore diaphragm" is considered to be an illusion created by sectioning techniques.

The annular material occupying the opening formed by the pore (see Fig. 10–23) is not sharply defined in the electron image. In surface views of the membrane, the annulus looks like a doughnut-shaped ring of electron dense, amorphous material superimposed over the pore margin. The outer diameter of the annulus is larger than the pore diameter. As a result, the margin of the pore is obscured by the annulus when the pore complex is viewed from the surface of the nuclear envelope. In cross sections, the annulus can be seen to extend a short distance beyond the pore on both the nuclear and cytoplasmic sides of the nuclear envelope. Because the annular material is less dense toward the center of the pore, the annulus is often described as a hollow cylinder of material that extends through the pore. The bore of this cylinder is indefinite in outline and may vary from 100 to 500 Å.

The diameter of the pore and the configuration of the annulus within the pore complex have been intensively studied by Joseph G. Gall. Gall studied the pore complex in nuclear envelopes isolated from amphibian oocytes. The nuclei in these oocytes are often large enough to be just visible with the unaided eye. If these nuclei are ruptured by micromanipulation over electron microscope grids, a portion of the nuclear envelope may become attached to the grid surface.

Gall fixed and stained nuclear envelopes from oocytes isolated in this manner, using both positive and negative stains. After positive staining by osmium tetroxide, only the annulus is visible (Fig. 10–21). Measurements of these preparations reveal that the outside diameter of the annulus is fairly regular, about 1200 Å. The inside diameter is more variable, ranging from 300–500 Å. After negative staining, the annulus is not clearly visible. Instead, the position of the pore is marked by a light line about 60 Å in thickness, describing a circle or regular octagon (Fig. 10–22). Inside this circle, a uniformly dense area forms the center of the pore. The region just outside the light line is also more or less uniformly dense, although in lightly stained preparations, a circle of eight slightly denser aggregations of stain are visible just outside the circumference of the light line marking the pore margin. Where the sides of the pore clearly form an octagon, the eight dense aggregations are located along each of the sides. Gall interprets the light line as the membrane forming the pore margin. When viewing the nuclear membrane from the surface, as is done in the Gall preparations, the only cross section expected is at the pore margin, where the outer membrane folds inward to connect at the margin with the inner nuclear membrane. Gall hypothesizes that the stain penetrates the perinuclear space, the pore, and the annulus, but not the membranes of the envelope, producing a negative image of the membrane (a less dense, lighter line) where it is seen in cross section at the pore margin.

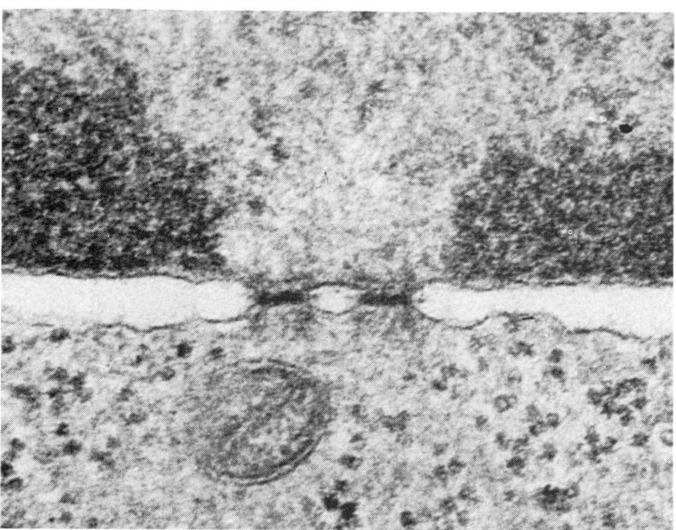

10–19. Apparent "pore diaphragms" closing the pores of a nuclear envelope from guinea pig bone marrow. × 65,000. From The Cell *by D. W. Fawcett, 1966. Courtesy of D. W. Fawcett and W. B. Saunders Company.*

Measurement of the pore diameter in these preparations reveals very regular dimensions, not only in the same organism, but in widely varied species. Gall found the diameter in a salamander, frog, and starfish to be 658 ± 5 Å, 700 ± 9 Å, and 632 ± 8 Å respectively. At this diameter, the margin of the pore is smaller than the outside diameter but larger than the inside diameter of the annulus. This suggests that the pore extends into the sides of the cylinder formed by the annulus (see Fig. 10–23).

In order to establish whether the pore margin describes a regular octagon, Gall used an interesting technique known as Markham rotation. In this method, used to detect radial symmetry in an image, a piece of photographic paper is exposed under an enlarger to a greatly enlarged image of a pore (Fig. 10–22a). The paper is then consecutively rotated and reexposed, with the center of rotation coinciding with the center of the pore in the print. The number of degrees the paper is rotated before the second exposure depends

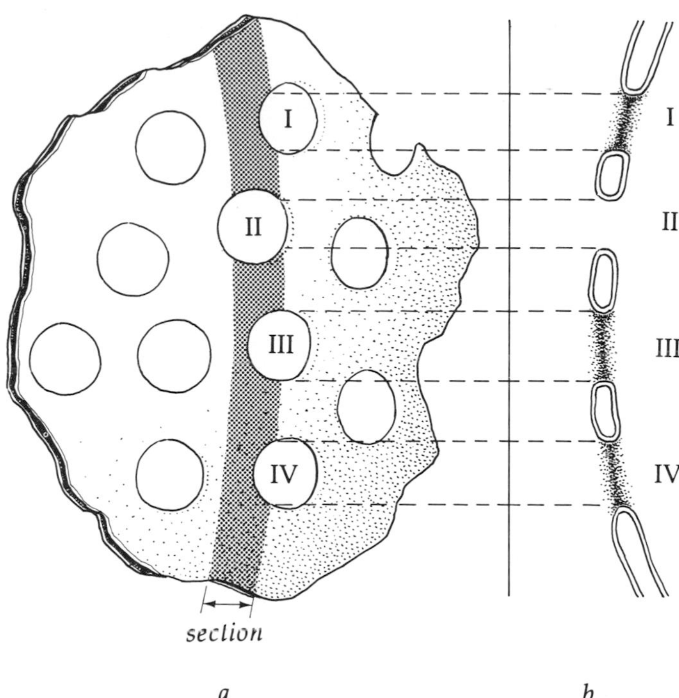

10–20. Origin of the pore diaphragm as an illusion resulting from the depth or thickness of a thin section. A surface view of a nuclear envelope is shown in a. *A section of this membrane made at 90° to the plane of the page, at the position and thickness marked by the shaded area, will produce the electron image shown in* b. *All pores will appear to be closed by a "diaphragm" except pore II.*

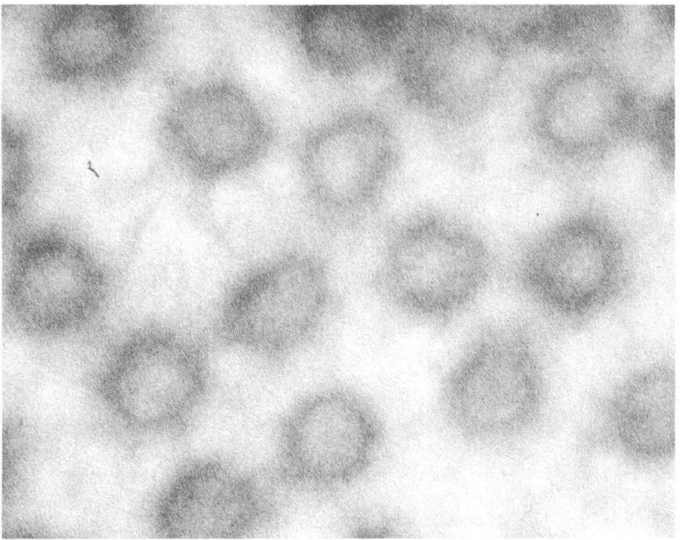

10–21. A nuclear envelope, isolated from the newt Triturus viridescens *and spread on the supporting film of an electron microscope grid. Before drying, the nuclear envelope was fixed in* OsO_4. *Only the annuli are visible.* × *140,000. Courtesy of J. G. Gall and The Rockefeller University Press, from* J. Cell Biol. *32:391 (1967).*

on the symmetry to be detected. If seven-fold symmetry is suspected, for example, the paper is rotated seven times, and seven successive exposures are made on the same piece of paper. Each rotation, in this case, involves a turn of 360/7 degrees. If seven-fold symmetry is present, the symmetry will be reinforced in the composite print; if symmetry is nine-fold, for example, it will not be visible. Fig. 10–22 shows the results of Markham rotation of a pore image. Fig. 10–22a shows the original pore image. When rotated to detect seven-fold symmetry (Fig. 10–22b) or nine-fold symmetry (Fig. 10–22d) no reinforcement is noted. However, rotation for eight-fold symmetry (Fig. 10–22) shows a definite reinforcement, indicating that the outline of the pore and the dense material just outside the pore (probably the outer margin of the annulus) is octagonal.

These results, summarized in Fig. 10–23, show that the pore is a complex structure, consisting of at least two parts: the pore itself, formed by the membranes of the nuclear envelope; and the annulus, which fills in and partly surrounds the pore. The margin of the pore describes a regular octagon approximately 650–700 Å in diameter. The annulus, roughly a cylinder of amorphous material 1200 Å in outside diameter, and with a 300–500 Å central bore, extends through the pore a short distance on either side of the nuclear envelope. The annular material also shows octagonal symmetry when viewed face on, giving the impression that this substance is more densely aggregated at eight equally spaced points corresponding to the sides of the pore margin. The annular material is often described as faintly granular or fibrillar in ultrastructure. It is absent from sections after trypsin digestion, suggesting that protein is a primary component of this structure. Ribonuclease and deoxyribonuclease are without apparent effect on the annulus.

The structure of nuclear pores is remarkably uniform throughout the plant and animal kingdoms. Few other structures show such consistency in different

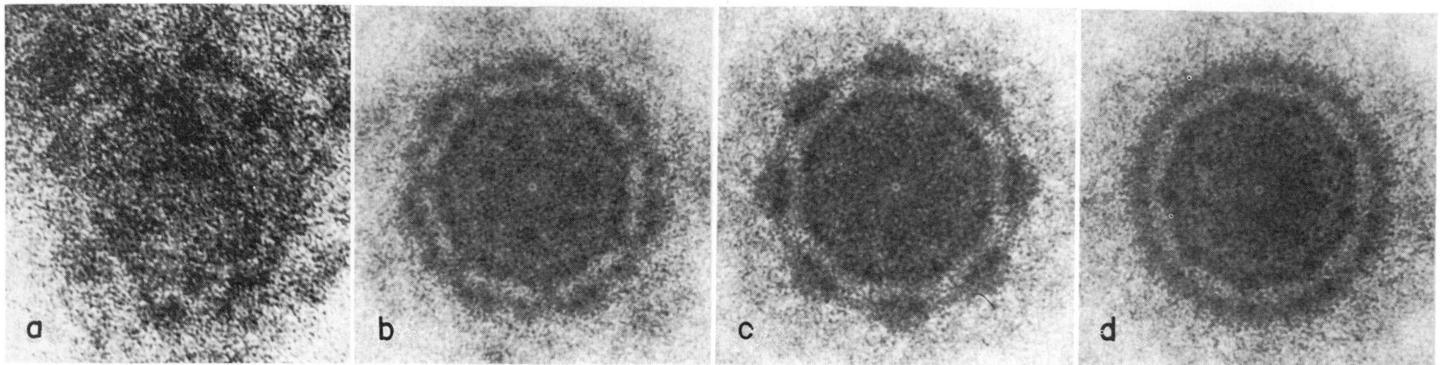

10–22. Markham rotation (see text) of a pore image to detect radial symmetry. a, *the image of a pore;* b, c, *and* d, *the same pore tested for 7-fold, 8-fold, and 9-fold symmetry. Reinforcement of the image is obtained in* c, *indicating the presence of 8-fold symmetry in the pore complex.* × *350,000. Courtesy of J. G. Gall and The Rockefeller University Press, from* J. Cell Biol. *32:391 (1967).*

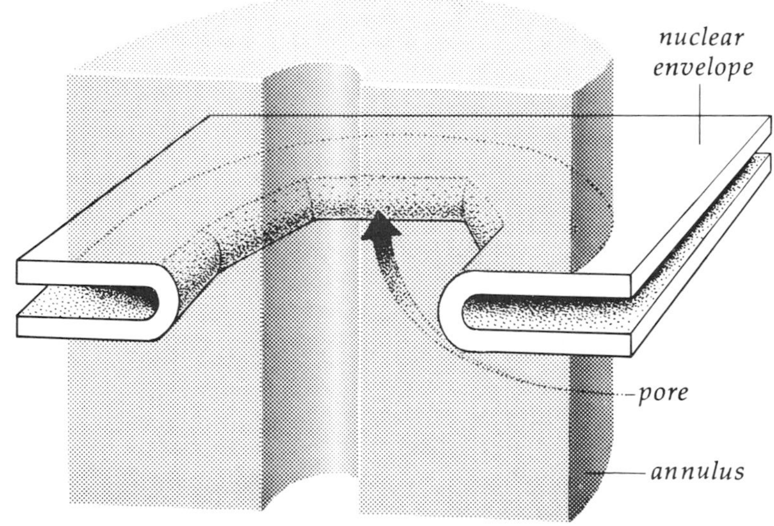

10–23. Diagram of the three-dimensional structure of the pore complex, showing the relationship of annular material to the pore. Redrawn, original courtesy of J. G. Gall and The Rockefeller University Press, from J. Cell Biol. *32:391 (1967).*

cells. Pores do not seem to undergo detectable alterations in structure with either changes in permeability (see below) or mechanical stretching of the nuclear envelope. The latter fact has been demonstrated in the nuclei of muscle cells by Sherman Bloom. As muscle cells contract, the nucleus is compressed to the extent that the nuclear envelope folds. In spite of this, no changes in the dimensions or structure of the pore complex were noted in nuclei from either contracted or relaxed muscle. This suggests that the pore complex in the intact cell is a fairly rigid structure.

Although pore structure in all plants and animals is closely similar, the number of pores per unit area of nuclear envelope shows wide variation from species to species. In most cases, this number varies from approximately 75 to 145 pores per square micron of nuclear envelope surface. At the upper limit, pores take up about one-third of the total area of the nuclear envelope. When most concentrated, the pores are packed in a hexagonal array with approximately 1500 Å separating pore centers. Changes in number of pores per unit area of membrane are observable in cells at different stages of activity or maturation. For example, the number of pores is reduced during formation of the spermatid; pores are rarely detected in the nuclear envelope of mature spermatozoa.

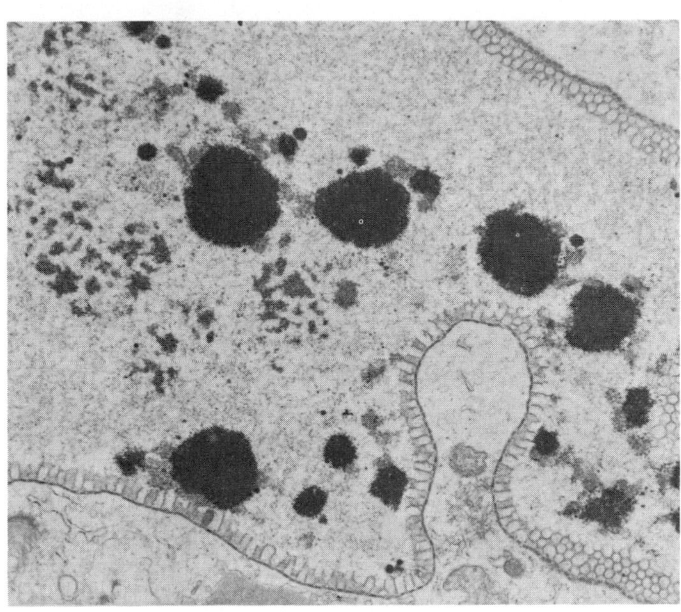

10–24. *A nucleus of* Amoeba proteus *showing dense material arranged into a honeycomb structure just inside the nuclear envelope. Each segment of the honeycomb is centered over a pore complex of the nuclear envelope.* × *6,700. Courtesy of C. J. Flickinger and Academic Press, Inc.* [J. Ultrastr. Res. *23:260 (1968)*]

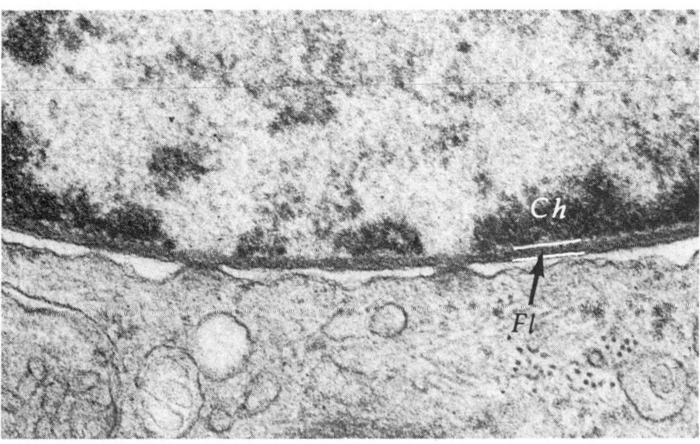

10–25. *The fibrous lamina* (Fl), *a thin layer of intermediate density lying between the chromatin* (Ch) *and the nuclear envelope in some mammalian cells. Courtesy of D. W. Fawcett.*

The fibrous lamina. In some cases, primarily in highly specialized cells, an additional component forms a part of the nuclear envelope. This component is most highly developed in a few ameboid protozoans, including *Amoeba proteus.* Thin sections of the nuclear envelope in *Amoeba* reveal a moderately electron-dense substance occurring in a layer just inside the nuclear membrane (Fig. 10–24). In *Amoeba,* this material forms a honeycomb structure, with each segment of the honeycomb located perpendicularly over a pore. One end of each honeycomb segment faces the nucleoplasm; the other end is attached to the nuclear membrane, pierced at the bottom by a pore. Each segment, with a central space 2000–2200 Å in width and walls approximately 200 Å in thickness, extends 2000–3000 Å into the nucleoplasm. No clue is available as to the function or identity of the material forming the honeycomb structure. At higher resolution, the substance making up the honeycomb is finely fibrillar or granular. A similar, but not quite as highly ordered structure is present in an annelid, the leech *Hirudo medicinalis.*

A similar specialization at the inner surface of the nuclear envelope has been described in many vertebrate and invertebrate cells by Don W. Fawcett. Instead of a honeycomb, the material observed by Fawcett forms a uniform layer 100–300 Å in thickness just inside the inner membrane of the nuclear envelope (Fig. 10–25). At higher magnification, this inner layer appears fibrous; Fawcett has termed the structure the *fibrous lamina.* In the region of a pore, the fibrous lamina may either be continuous, extending over the pore, or perforated. Fawcett has noted that the lamina often appears to form connections with the annular material lining the pores of the nuclear envelope. Because of a basic similarity in structure, electron density, and location, Fawcett has proposed that the honeycomb in *Amoeba* and *Hirudo* and the fibrous lamina are different modifications of the same structure. The function of the fibrous lamina is unknown. Reinforcement of the nuclear envelope comes readily to mind as a possible function for this structure, because many of the nuclei in which it occurs are large. However, other large nuclei, such as the oocyte nucleus of amphibians, do not have a fibrous lamina.

Permeability of the Nuclear Envelope: Are the Pores Open or Closed?

One of the most interesting and most frequently argued questions in cell biology concerns the "openness" of the pore complexes of the nuclear envelope. The nuclear envelope surface between pores, including 80 percent or more of the total envelope in the average cell, is structured from membranes similar to the plasma membrane. Diffusion and transport of ions and metabolites across these membranes would be expected to follow the pattern typical for lipoprotein membranes elsewhere in the cell. Thus in these regions, the membranes would be expected to present a barrier to free diffusion. But what is the effect of the pores? If the pores are open to free diffusion of ions and metabolites, maintenance of a concentration gradient between nucleus and cytoplasm would be impossible because of "leakage" through these openings. At a diameter of 650–700 Å, the pore is large enough to admit not only ions but the largest known protein and nucleoprotein complexes. If, on the other hand, the pores form a barrier to free diffusion, the barrier must depend on an as yet unknown mechanism, for the pores are evidently not closed by anything even remotely similar in morphology to a lipoprotein membrane. This can be clearly seen in permanganate-fixed tissue; permanganate, a good fixative for membranes, does not preserve the annular material of the pore complex. Thin sections of this material clearly show that the membranes of the nuclear envelope are interrupted and do not extend across the pores.

A number of experiments have shown that some ions, such as sodium, phosphate, and potassium, if injected into the cytoplasm of living cells, penetrate the nuclear envelope and reach higher concentrations in the nucleus than in the cytoplasm. Similar results are obtained if isolated nuclei are suspended in a medium containing solutions of these ions. Although these ions become more concentrated inside the nucleus than outside, it is not clear whether they actually become more concentrated as free ions inside the nucleus or whether they are absorbed or bound to macromolecules, such as nucleoprotein, within the nucleus. In the latter case, the ions could apparently become more "concentrated" in the nucleus without establishing an actual concentration gradient in free ions between the inside and outside of the nuclear envelope.

More definitive results have been obtained by measuring voltage differences between nucleus and cytoplasm. If charged particles, such as sodium or potassium ions, are concentrated differentially on two sides of a membrane, a difference in voltage or potential may be established across the membrane. Werner E. Loewenstein and his colleagues measured the potential differences between nucleus and cytoplasm by inserting microelectrodes 0.5 microns in diameter into the nucleus and cytoplasm. In living cells from the salivary glands of *Drosophila* larvae and another dipteran, *Chironomus,* Loewenstein was able to detect potential differences on the order of 15 millivolts between nucleus and cytoplasm. If the nuclear membrane was ruptured, the potential difference fell to zero. The resistance across intact nuclear envelopes in these species was 1000 to 10,000 times greater than expected for a system allowing free diffusion of ions. It is difficult to imagine how potential differences and resistance of this magnitude could be present unless the pores are closed to free diffusion.

In cells from other sources, such as oocytes from the amphibians *Triturus* and *Xenopus,* no potential differences could be detected between nucleus and cytoplasm. This may indicate that two broad classifications of nuclear envelopes exist, one highly permeable and one relatively impermeable to the flow of ions. More probably, it is an indication that the permeability of the nuclear membrane is variable and changes according to the activity or developmental stage of the cell. Support for this has been obtained from studies of *Drosophila,* in which changes in the potential difference between nucleus and cytoplasm can be detected as development of the larva proceeds. These changes, however, are not accompanied by detectable alterations in the morphology of the pore complex. Similarly, no differences in pore morphology can be detected between the highly permeable amphibian oocyte membranes and the less permeable membranes of *Drosophila* larvae.

Loewenstein's experiments make it seem very likely that at times the pores are closed to free diffusion of ions. Other bits and pieces of evidence fit into this interpretation. For example, differences in *p*H can be noted between the nucleus and cytoplasm in some species. Differences in salt concentration can also be detected; measurement of the ratio of sodium to potassium in the nucleus and cytoplasm gives slightly different values for the two locations. Other experiments have shown that amino acids are concentrated in the nucleus against a gradient. In this case, the natural L-isomers of the amino acids are transported across the nuclear envelope and concentrated inside the nucleus, whereas the D-isomers are not. This suggests that an enzymatic apparatus exists on the membranes or in the pore complex, which is specific for the naturally occurring L-isomers (for an excellent review of this information, see Feldherr and Harding, 1964, and Loewenstein, 1964).

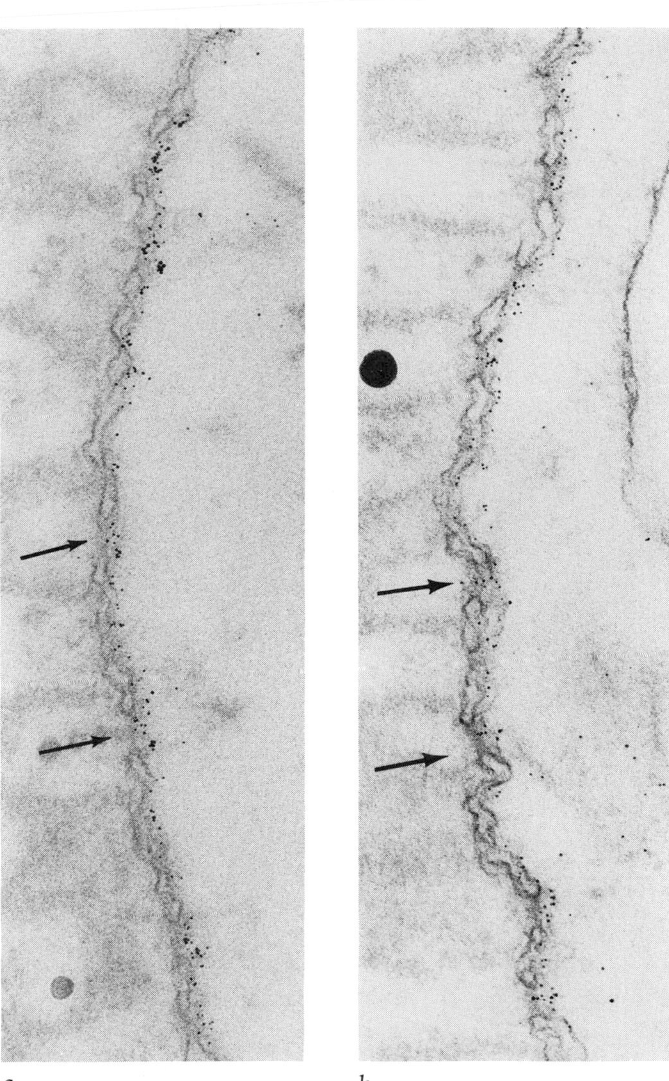

10–26. *Sections of the nuclear envelope of* Amoeba *after injection of the cytoplasm with 25–55 Å gold particles.* a, *7 minutes after injection,* b, *10 minutes after injection. In* a, *gold particles are concentrated over the cytoplasmic surface of the nuclear envelope and the pores (arrows); in* b, *a number of particles have penetrated the pore complexes (arrows). The "honeycomb" structure lying just inside the nuclear envelope is faintly visible.* × *91,000. Courtesy of C. M. Feldherr and The Rockefeller University Press.* [J. Cell Biol. *25:43 (1965)*]

Unless the available fixation methods have failed to reveal an as yet undiscovered structure closing the pore, the control of diffusion must reside in the annular material occupying the pore center. Little is known about the nature of the annulus, except that it is probably proteinaceous. Morphological studies show the annulus to be an amorphous, rather open network of fibers or granules and offer no hint as to a possible mechanism of control exerted by the annulus over diffusion of ions and metabolites of small size.

The transfer of ribonucleoprotein and other large complexes across the nuclear envelope. The barrier to diffusion of ions and small metabolites presented by the nuclear envelope takes on special significance when the relatively large ribonucleoprotein complexes passing from nucleus to cytoplasm are considered. That the transfer of ribonucleoprotein across the nuclear envelope takes place seems obvious on purely intuitive grounds, because the RNA-protein product of the chromosomes must enter the cytoplasm for protein synthesis to occur.

Transfer of large molecules across membranes presents a problem, because classical studies with penetration of various substances through membranes indicate that molecules beyond a particular size, with a molecular weight far below that of the various ribonucleoprotein complexes, do not pass through phospholipid membranes readily, if at all. The pore complexes may therefore represent a specialized adaptation of the nuclear envelope related to the transfer of large macromolecules from nucleus to cytoplasm.

Carl M. Feldherr established very clearly that large particles can cross the nuclear envelope through the pores. Feldherr used colloidal gold particles coated with a very thin plastic film. When injected into the cytoplasm of amoebae, these particles become attached to the cytoplasmic surface of the nuclear envelope and the pore complex (Fig. 10–26a). In preparation made a few minutes after injection, gold particles could be observed in the pores (Fig. 10–26b); tangential sections of the membrane revealed that particles within the pores were located in the axis of the pore complex, suggesting that a channel exists in the center of the annulus. After a number of hours, many gold particles could be found within the nucleus. Feldherr also obtained closely similar results with *ferritin*, a large iron-protein complex that is dense enough to be visible in the electron microscope. Ferritin particles were found within the nucleus as soon as seven minutes after injection into the cytoplasm. Within one hour, large quantities of particles were found to have crossed the nuclear envelope.

Ferritin is a heavy complex with a molecular weight of 550,000 and a diameter of approximately 95 Å. Obviously, particles of at least this size and weight can penetrate through the pore. Feldherr found that the rate of penetration of colloidal gold particles depends on the size of the particle. Particles up to 85 Å in diameter penetrate fairly rapidly; slightly larger particles, 89–106 Å in diameter, were found to pass through the nuclear envelope only in limited numbers. Particles larger than 106 Å were found within the nucleus in very small numbers. From his work, Feldherr has estimated that particles larger than 125–150 Å would not pass through the nuclear envelope at all.

Morphological evidence for transport of macromolecules across the nuclear envelope. Much of the morphological evidence for the passage of ribonucleoprotein and other macromolecular complexes across the nuclear envelope is largely circumstantial. In nuclei with densely aggregated chromatin, channels remain open in the nucleoplasm in the region of the pores (Fig. 10–27). This implies that these channels are open for the transport of materials through the pores of the nuclear envelope. In some cells that synthesize large quantities of ribonucleoprotein, such as oocytes, electron-dense material can be seen in the nucleoplasm near the pores, within the pores, and on the cytoplasmic side of the pores (Fig. 10–28). Although it is not possible to determine conclusively from electron micrographs which way this material is moving, it seems likely that these micrographs represent the transfer of ribonucleoprotein from nucleus to cytoplasm. In a few cases (see the description of work with salivary glands in this chapter), the material traversing the pores has been identified as ribonucleoprotein by enzymatic digestion.

W. Bernhard and his colleagues, in their work with extrachromosomal particles in the nucleus (see p. 220) have also been able to show an association of ribonucleoprotein particles with the pores. The large perichromatin granules, perhaps representing mRNA, are often found in close proximity to and apparently entering the nuclear pores (Fig. 10–29). Other material identified by Bernhard as ribonucleoprotein by the EDTA staining technique can also be seen extending through the nuclear pores (Fig. 10–30). In Fig. 10–30,

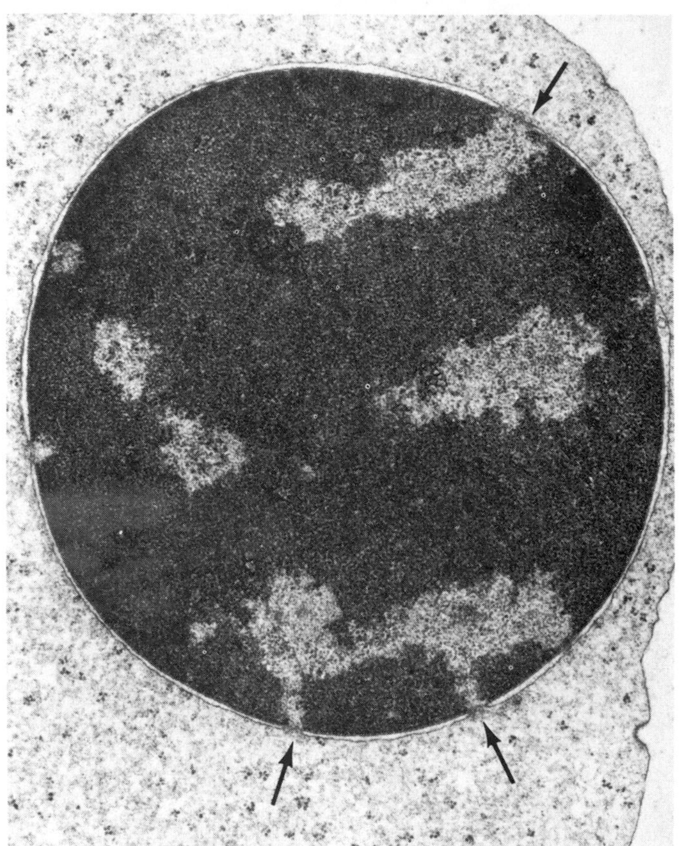

10–27. Developing erythroblast from guinea pig bone marrow. The chromatin has condensed into dense aggregates except for a few more diffusely coiled areas; these remain in communication with the pore complexes (arrows). × 17,000. From The Cell *by D. W. Fawcett, 1966. Courtesy of D. W. Fawcett and W. B. Saunders Company.*

a transition in structure from granules to fine fibrils is detectable as the dense material from the nucleus approaches the pore. This transition in structure, from granules to fine fibrils 30–50 Å in diameter, has been reported by many authors, and seems to be a fundamental structural change in material traversing the pore. However, large granules 150 Å in diameter are frequently seen in the center of a pore (see Fig. 10–18). These resemble the granules of the nucleolus in size and may represent the large subunit of the ribosome, caught in transit through the pore. Treatment of cells with actinomycin D causes a marked decrease in the proportion of nuclear pores containing granules, suggesting that these granules are indeed an RNA-protein product of the chromosomal apparatus. These large granules, however, do not stain as intensely as the

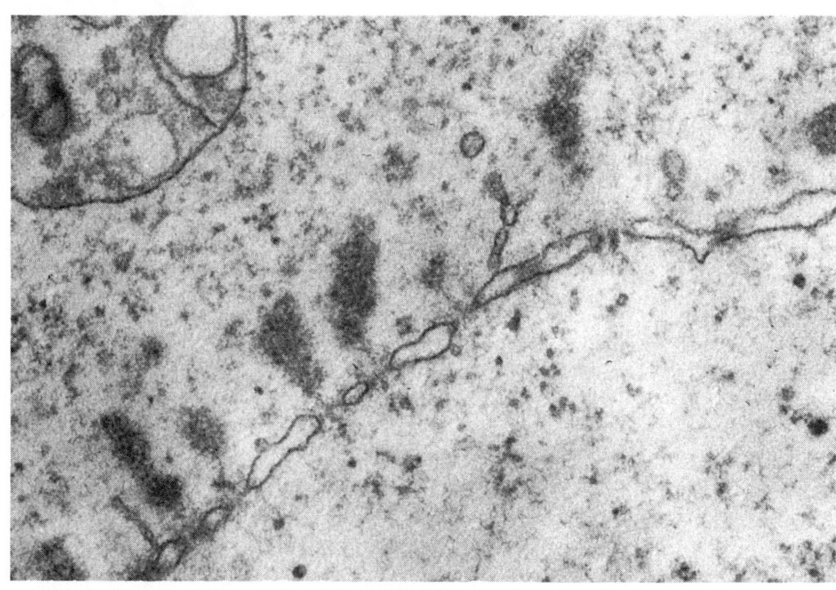

10–28. Apparent movement of electron-dense material from the nucleus into the cytoplasm through the pores in an oocyte of the frog Rana catesbiana. *The collections of dense material in the cytoplasm seem to arise from thin, filamentous material passing through the nuclear pores. × 54,000. Courtesy of H. Swift and W. Massover.*

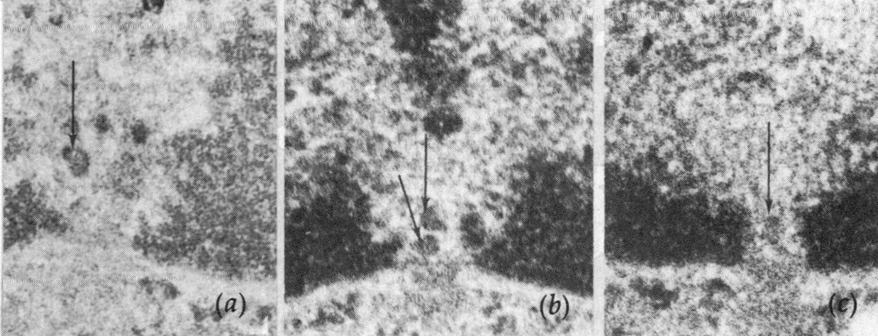

10–29. Apparent passage of perichromatin granules through nuclear pores of rat liver cells. As the granules (arrows) penetrate more deeply into the pore (b *and* c), *a conversion from spherical to extended, fibrous form takes place. × 96,000. Courtesy of A. Monneron, W. Bernhard, and Academic Press, Inc., from* J. Ultrastr. Res. *27:266 (1969).*

granules of the nucleolus and are not sensitive to digestion with ribonuclease. It is possible that, while in the pore interior, the granules are inaccessible to the enzyme; this would explain the insensitivity to ribonuclease. However, at the moment the identity of the pore granules remains unknown.

The fibrils extending through the nuclear pore can be seen to extend into the cytoplasm for distances of 300–400 Å. In a few cases, the material traversing the pore reforms into dense, granular deposits (see Fig. 10–28) on the cytoplasmic side of the nuclear envelope. Material of this type has been observed to move from nucleus to cytoplasm in living cells under the phase contrast microscope.

The precise transfer mechanism of ribonucleoprotein complexes and other large macromolecules through the nuclear envelope is obscure. Many hypotheses have been advanced to account for this transfer, but little evidence exists to support any of them. One proposal suggests that the inner nuclear membrane flows through the pores, becoming the outer membrane, and in the process carries attached macromolecules with it. Another proposal suggests that outpocketings of the outer nuclear membrane occur, forming vesicles that become detached and carry included macromolecules into the cytoplasm. As the perinuclear space rather than the nucleoplasm lies immediately under the outer nuclear membrane, this hypothesis would require that another unknown mechanism transport macromolecules from the nucleoplasm through the inner nuclear membrane into the perinuclear space. Thus, while such vesicular outpocketings of the outer nuclear membrane have been observed, although not regularly, it is difficult to relate them to transfer of substances from nucleoplasm to cytoplasm.

While no definitive evidence exists, the best chances are that large molecules are transferred through the pore complexes and that the pore in some manner supplies the energy required to move large complexes through this space. Work with enzyme histochemistry has shown that an enzyme required for the hydrolysis of ATP to ADP is present in high concentration in the pores (see Scheer and Franke, 1969); this may reflect the presence of an energy-requiring mechanism for the movement of macromolecules in the pore complex. This movement probably occurs through a narrow central channel in the pore, approaching an upper limit in diameter of approximately 150 Å; the limits of the central channel may represent the inner margin of the annular material or may be defined by another material without detectable electron density that lies within the hollow axis of the annulus.

The nuclear envelope is not present in prokaryotes. In dinoflagellates, which combine some of the features of both prokaryotes and eukaryotes (see Chapter 8), a nuclear envelope is present. Pores with the usual configuration are present in some, but not all dinoflagellates (see Fig. 8–21). In at least one case, investigated by Bjorn A. Afzelius, a variant exists that may represent a primitive form of the nuclear envelope. In this organism, *Noctiluca scintillans*, the nuclear envelope consists of two membranes encircling the nucleus as usual, including a perinuclear space as in higher organisms (Fig. 10–31). Few pores are present in the nuclear envelope. Just inside the envelope in the nucleoplasm are vesicles surrounded by double membranes, which resemble a miniature nuclear envelope in structure. Pores are present in the double membrane system surrounding these nuclear vesicles. At many points, the vesicles make contact and fuse with the nuclear envelope. In the region of fusion, the membranes of the vesicles become continuous with the membranes of the nuclear envelope (Fig. 10–32). As these continuities are established, the contents of the vesicles are exposed to and merge with the cytoplasm. During this process, the nucleoplasm and cytoplasm are always separated by intact membranes, because the vesicles, during fusion, become a temporary part of the nuclear envelope (see Fig. 10–32b). It is easy to see how adaptations could convert this system to one in which the included pores remain a functional part of the nuclear envelope. This possible evolutionary mechanism is feasible only if the system in *Noctiluca* represents a primitive rather than degenerate form. In either case, this organism exhibits an interesting and extremely rare example of a significant variation in the structure of the nuclear envelope.

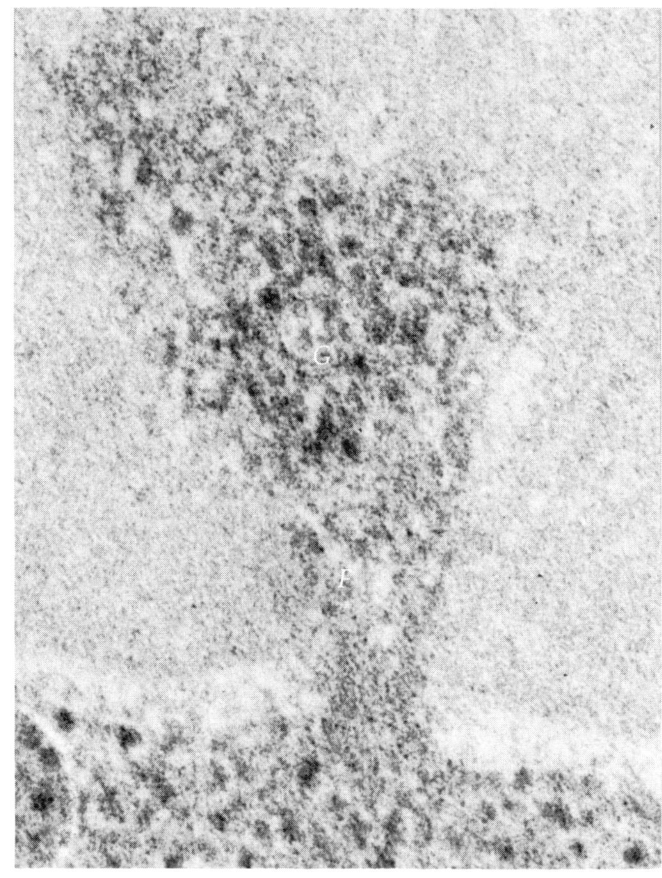

10–30. Apparent conversion of nuclear material in rat liver from coarse granules to fine fibrils during passage through a nuclear pore. Near the pore the material is granular (G); *within the pore only finely fibrillar material* (F) *is seen.* × *79,000. Courtesy of A. Monneron, W. Bernhard, and Academic Press, Inc., from* J. Ultrastr. Res. *27:266 (1969).*

Annulate lamellae. A structure containing multiple layers of membranes curiously similar to the nuclear envelope has been observed in the cytoplasm of developing oocytes and spermatocytes of many animals. These structures, termed *annulate lamellae,* have pore complexes that are apparently identical in morphology to the pores of the nuclear envelope (Figs. 10–33 and 10–34). Annulate lamellae also occur in a few embryonic cells and in the cells of tumors.

Each individual unit of a stack forming the annulate lamellae consists of two membranes that form a flattened sac. The distance separating the two membranes of an individual sac is fairly uniform, from 100 to 300

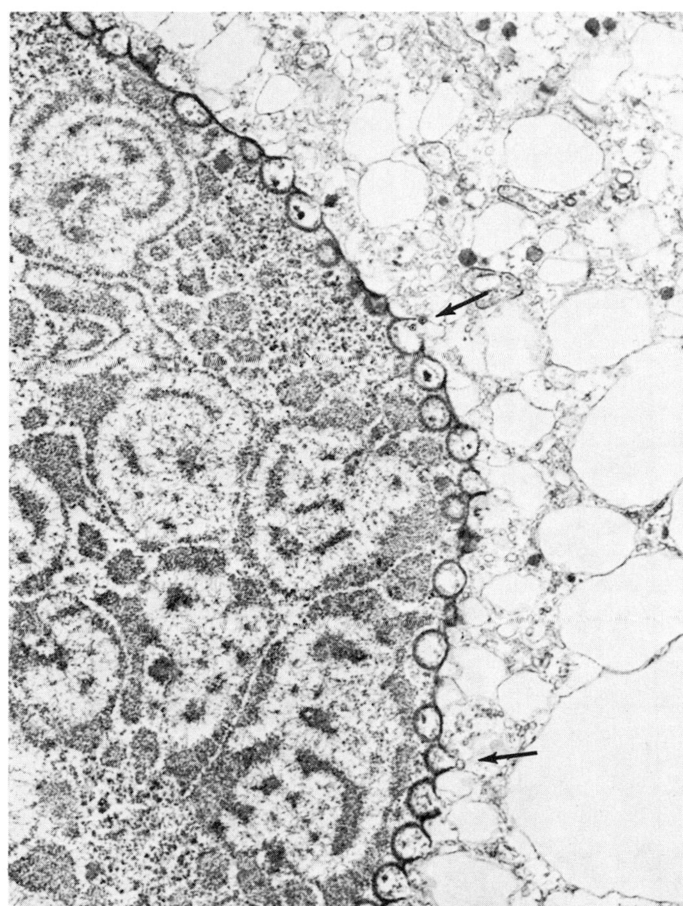

10–31. *Nucleus of the dinoflagellate* Noctiluca scintillans. *Note the numerous vesicles just inside the nuclear envelope. At a few points these vesicles are open to the cytoplasm (arrows). Typical pore complexes are present in the membranes surrounding the vesicles. × 9,000. Courtesy of B. A. Afzelius and The Rockefeller University Press, from* J. Cell Biol. *19:229 (1963).*

Å in width. The space enclosed within the sac is thus similar in dimensions and analogous in location to the perinuclear space of the nuclear envelope. Perforating the sac at regular intervals are pores, at the margin of which the two membranes of the sac are continuous. These pores are octagonal in outline when viewed from the surface in negatively stained preparations of isolated annulate lamellae and are identical in dimensions to the pores of the nuclear envelope. The dense annular material contained within the pores of annulate lamellae is also apparently identical in morphology to the annuli of the nuclear envelope.

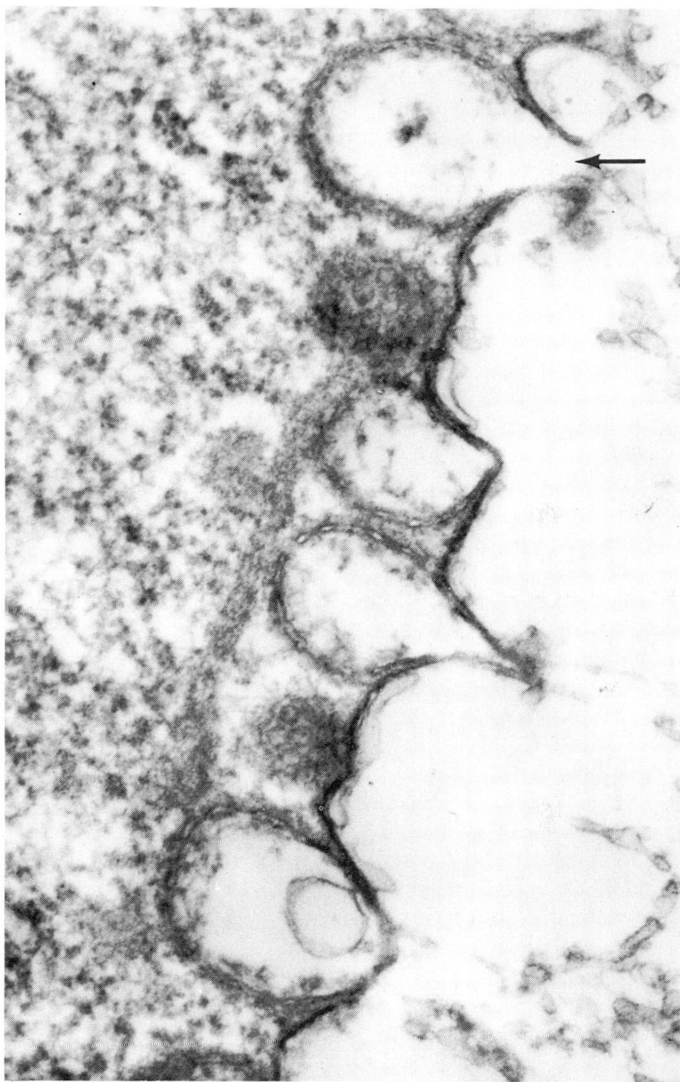

a

10–32. a, *The nuclear vesicles of* Noctiluca scintillans *at higher magnification. One of these (arrow) is open to the cytoplasm. × 36,000. Courtesy of B. A. Afzelius and The Rockefeller University Press, from* J. Cell Biol. *19:229 (1963).*

The sacs making up annulate lamellae are stacked into layers varying from one to as many as 30–40 units; average numbers run from 10 to 15. Each unit is separated from its neighbor by a fairly regular space several hundred angstroms in width. Frequently, the pores of annulate lamellae are more closely spaced than in the nuclear envelope, sometimes in numbers great enough to take up 50 percent of the membrane in surface area.

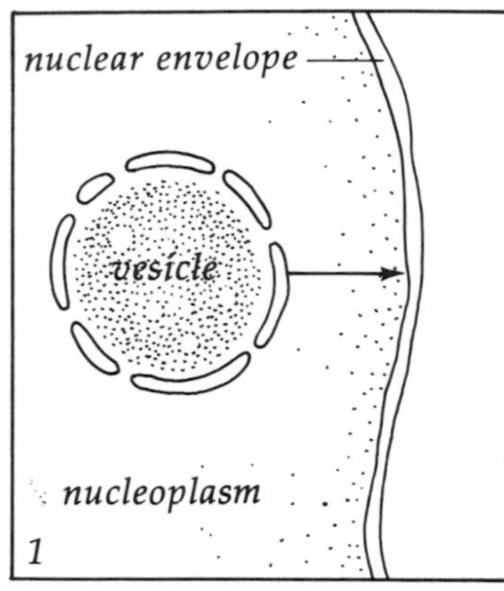

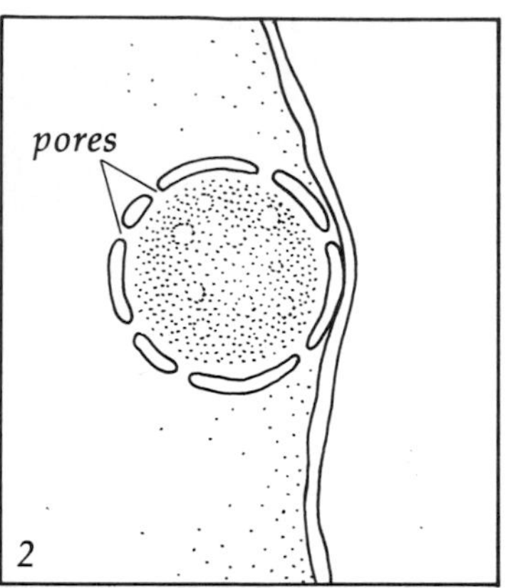

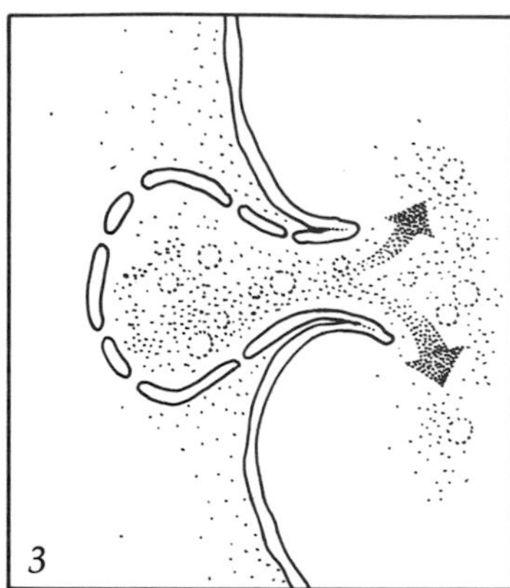

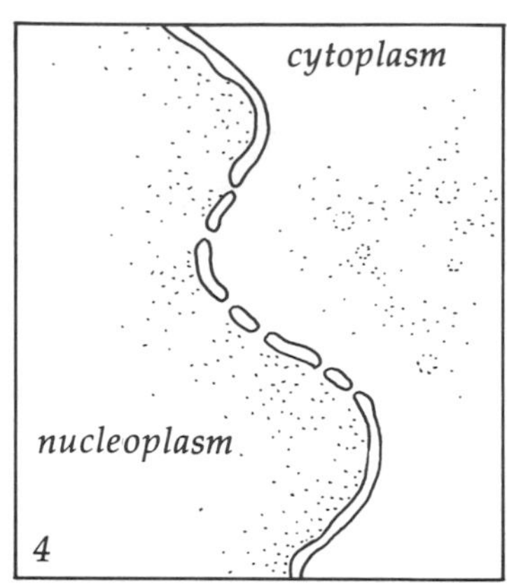

b

10–32. b, *The presumed sequence of events in the expulsion of the vesicular contents to the cytoplasm. 1, movement of a vesicle toward the nuclear envelope; 2, 3, contact and fusion of vesicular membranes with the nuclear envelope and expulsion of vesicle contents to the cytoplasm. 4, temporary inclusion of vesicular pores in the nuclear envelope.*

When present in high frequencies, the pores are separated by distances of 1000–2000 Å. When the pore complexes are closely spaced in this way, the pores in adjacent sacs are often lined up in vertical register, with the annular material of adjacent pores in contact or apparently continuous (see Fig. 10–33).

Annulate lamellae sometimes lie adjacent to the nucleus, with the nuclear envelope forming the innermost layer of the stack of units. Most frequently, however, the lamellar stacks are found in the cytoplasm at a distance from the nucleus, without visible connection with the nuclear envelope. The close proximity found in some species and the morphological similarities have led most investigators to propose that annulate lamellae arise by fragmentation or budding of large segments of the nuclear envelope complete with pores. Careful work by R. G. Kessel indicates that annulate lamellae do arise from the nuclear envelope, but in a somewhat unexpected manner.

Kessel discovered that instead of budding off from the nuclear membrane in finished form, with intact pores, annulate lamellae first appear as proliferations of the outer membrane of the nuclear envelope that extend for long distances into the cytoplasm (Fig. 10–35). These arise in multiple numbers from the outer nuclear membrane and contain an enclosed space that is continuous, for a time, with the perinuclear space. These lamellar extensions, which are at first without pores, become stacked in layers in the cytoplasm. Once this stage is reached, separations, which subsequently differentiate into pore complexes (Fig. 10–36), appear

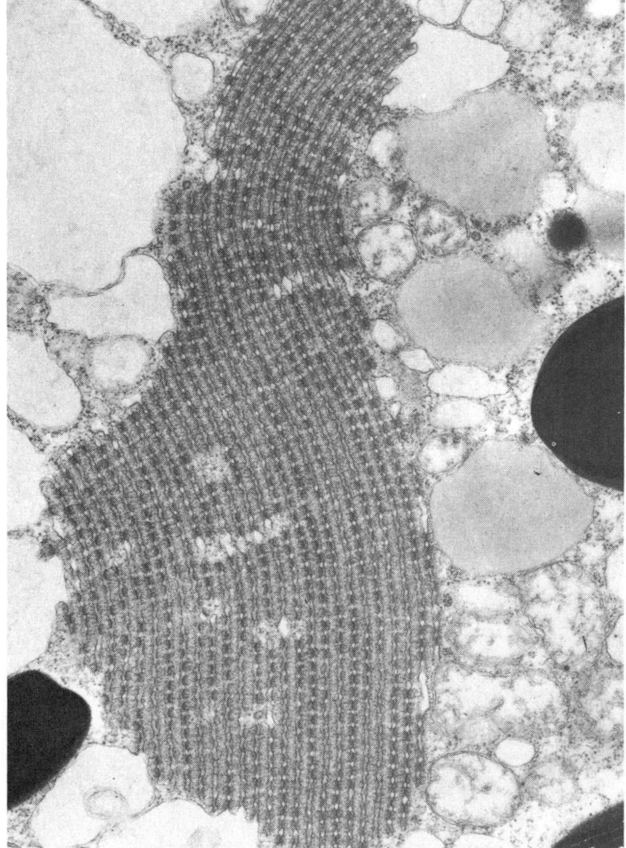

10–33. *An extensive collection of stacked annulate lamellae from an oocyte of* Rana pipiens. *× 14,500. Courtesy of R. G. Kessel and Academic Press, Inc.* [J. Ultrastr. Res., *Supplement 10 (1968)*]

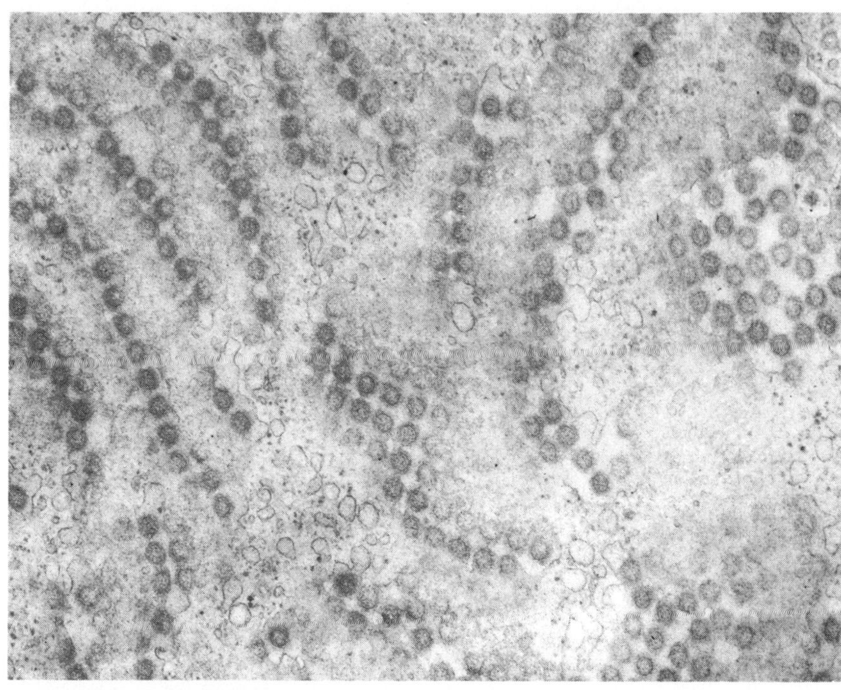

10–34. Stacks of annulate lamellae in tangential section from Rana pipiens *oocyte. × 25,500. Courtesy of R. G. Kessel and Academic Press, Inc.* [J. Ultrastr. Res., *Supplement 10 (1968)*]

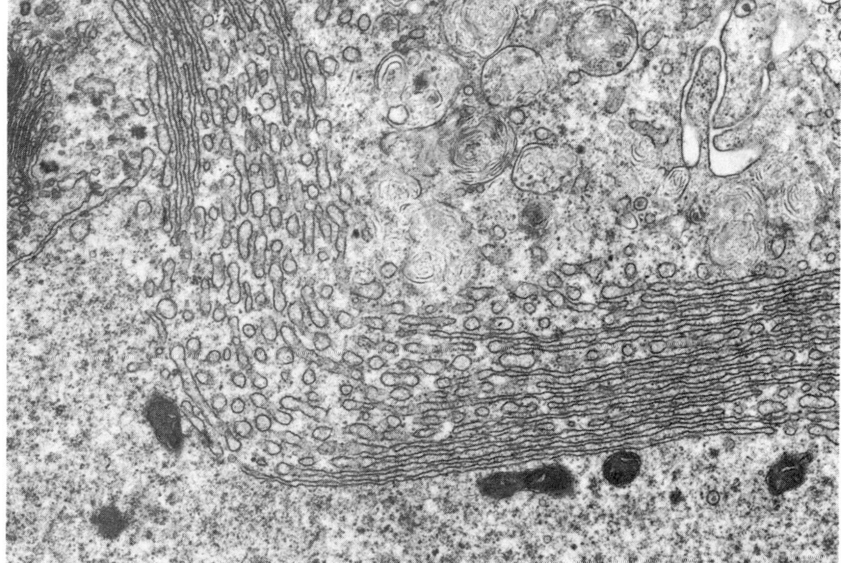

10–35. An early stage in the formation of annulate lamellae in an oocyte of Ophioderma *in which membranous vesicles and flattened sacs extend from the nuclear envelope into the cytoplasm. × 19,000. Courtesy of R. G. Kessel and Academic Press, Inc.* [J. Ultrastr. Res., *Supplement 10 (1968)*]

in the lamellae. Very rarely, stacks of annulate lamellae occur in the nucleoplasm; these stacks arise from a similar process involving proliferation of the inner membrane of the nuclear envelope.

The function of annulate lamellae is unknown. The aggregations of membranes making up the lamellae are large enough to be recognizable under the light microscope as dense granules within the cytoplasm. Cytochemical tests reveal that these bodies contain RNA in high concentrations. Other techniques, using tests of enzymatic activity in tissue prepared for electron microscopy, reveal that high ATPase activity is associated with the pore complexes of annulate lamellae (see Scheer and Franke, 1969) just as in the pores of the nuclear envelope. Although a central 150 Å granule is often present in the pores of annulate lamellae, as it is in the pores of the nuclear envelope, the presence of RNA is noted whether or not granular inclusions are present, either within the pores or in the layers of cytoplasm that extend between the sacs. Ribosomes

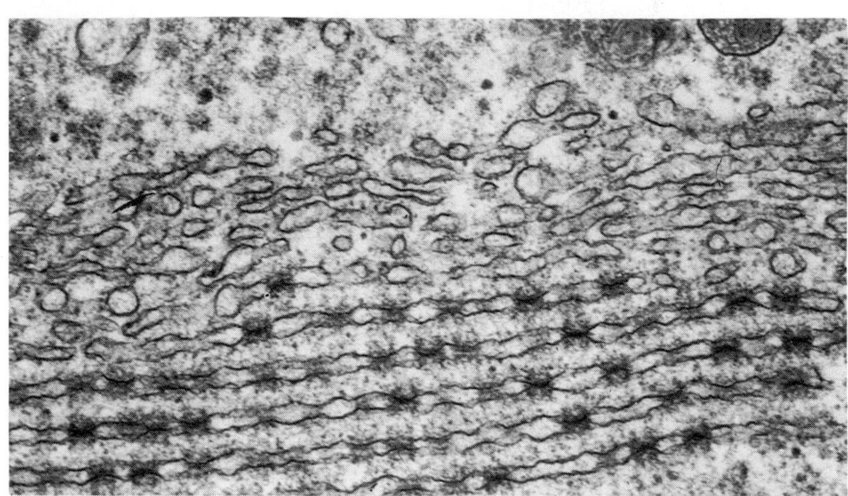

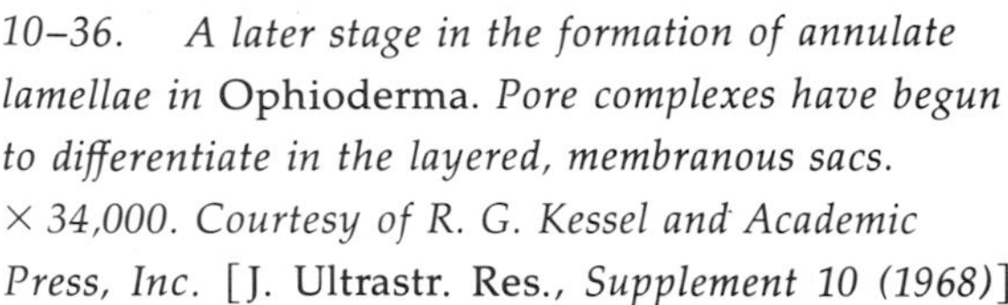

10–36. A later stage in the formation of annulate lamellae in Ophioderma. *Pore complexes have begun to differentiate in the layered, membranous sacs. × 34,000. Courtesy of R. G. Kessel and Academic Press, Inc.* [J. Ultrastr. Res., *Supplement 10 (1968)*]

are frequently present in these cytoplasmic extensions into the annulate lamellae; the presence of RNA in annulate lamellae is also independent of the presence of these ribosomes. The RNA component is apparently localized in the dense material within the pores.

If the pores of the nuclear envelope are considered to be a modification permitting the transfer of large molecular complexes between nucleus and cytoplasm, it is difficult to relate this possible function to the pores of annulate lamellae. Annulate lamellae are too localized to form an effective diffusion barrier between regions of the cytoplasm or between nucleus and cytoplasm. Nor do these lamellae appear to be a vehicle for transferring ribonucleoprotein from nucleus to cytoplasm, because intensive staining for RNA does not appear until after pore complexes differentiate. As we have seen, this differentiation takes place in the cytoplasm some time after separation of the developing lamellae from the nuclear envelope.

Other questions concerning the morphology and function of annulate lamellae remain unanswered. Are these lamellae capable of synthesizing RNA, protein, or something else in the cytoplasm? It is often noted that direct connections occur between the membranes of annulate lamellae and the endoplasmic reticulum (see Chapter 11). This suggests a role in the synthesis of protein. However, this is merely speculation, because no evidence exists to support this possible function. Other questions surround the exact degree of morphological and functional similarity between the pore complexes in the two membrane systems. Are these precisely alike, or are there differences, as yet undiscovered, that reflect differences in function between the two? Finally, the eventual fate of annulate lamellae in the cytoplasm is unknown. Some evidence, although limited, suggests that annulate lamellae break down into scattered vesicles, which eventually disappear or disperse in the cytoplasm late in oogenesis.

It is possible that annulate lamellae form a storage place for ribonucleoprotein in the cytoplasm. The stored RNA could be bound tightly to the pore complexes of annulate lamellae in a form unavailable for breakdown by cytoplasmic ribonucleases. At the time of breakdown or dispersal of the annulate lamellae, the bound RNA, possibly mRNA, could be released to the cytoplasm in an active form. All of this stands as conjecture; at the present time, the function of annulate lamellae remains unknown.

Polytene Chromosomes

A System Illustrating the Morphology of RNA Metabolism in the Nucleus

In the typical cell at interphase, chromatin is diffusely spread throughout the nucleus. Although the chromosome fibers and their products, the ribonucleoprotein granules and fibers, can be seen at interphase under the electron microscope, the RNA synthesized cannot ordinarily be identified with a specific location in the chromosomes. Morphological identification of this type is possible in one highly specialized form of interphase nucleus, in the giant cells in dipteran larvae. The chromosomes in these nuclei are greatly enlarged and contain localized regions that are identifiable as sites intensely active in RNA synthesis.

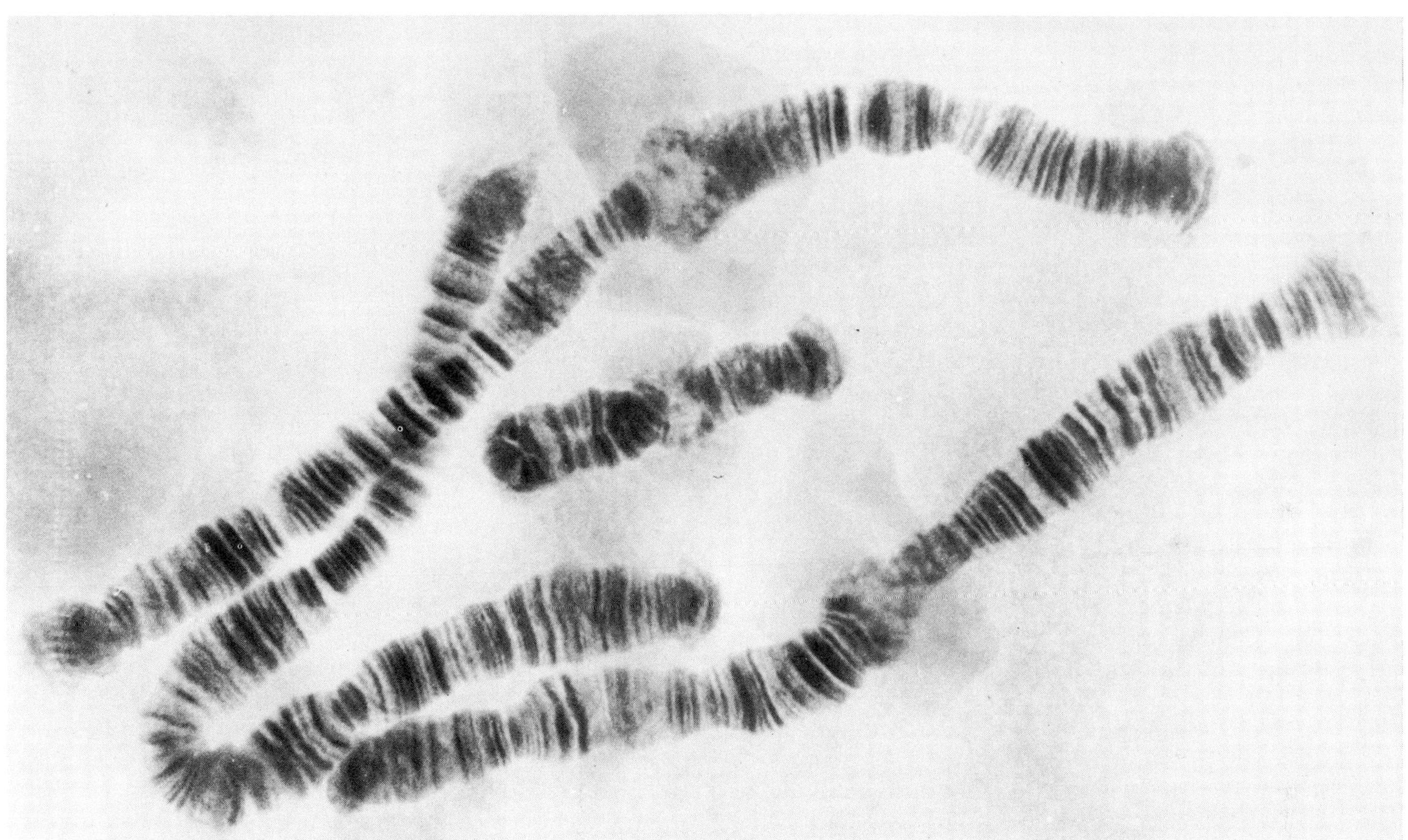

10–37. The polytene chromosomes of the salivary gland of Chironomus tentans. *The longer chromosomes in this micrograph are between 200 and 250 microns in length.* × *750. From* Protoplasmatologia 6D *by W. Beermann, 1966. Courtesy of W. Beermann and Springer-Verlag.*

These giant cells were first described by E. G. Balbiani in the 1880's. However, the giant nuclear structures contained within them were not identified as chromosomes until fifty years later with the work of E. Heitz and H. Bauer in Germany and T. S. Painter in the United States. These large chromosomes usually take the form of roughly cylindrical structures from 5 to 15 microns in diameter and 200 microns or more in length (Fig. 10–37). At these dimensions, the giant chromosomes are more than a hundred times larger than metaphase chromosomes from the same organism.

Intensive work with genetic crosses before the 1930's had allowed an abstract "map" to be made of the chromosomes in *Drosophila.* The genetic experiments indicated that the genes were in linear order on the chromosomes. Many genetic loci had been assigned relative positions with respect to each other on the "map." These maps were strictly an abstraction, however, no morphological location of genes on specific sites of the *Drosophila* chromosomes could be made. With the discovery of the giant chromosomes in *Drosophila* larvae, it became possible to translate the abstract map into physical points on the chromosome. Combined genetic and cytological studies showed that many genes could be identified with specific transverse bands in the giant chromosomes. Feulgen staining showed that these dense, transverse bands contain high concentrations of DNA.

The giant chromosomes are often termed "salivary gland" chromosomes because they are commonly studied in the salivary glands of dipteran larvae. But chromosomes of this type also occur in other organs and tissues of the larva, typically in sites of intense metabolic activity, such as some cells of the intestine and excretory system. Some adult flies also have giant chromosomes in single large cells that occur in the footpads. At some time in their life cycle, probably all except a few diptera have a number of cells containing giant chromosomes. Chromosomes with a very similar appearance also occur as an intermediate stage in the development of the macronucleus in at least one ciliate protozoan (see Alonso and Perez-Silva, 1965). Reports of giant chromosomes of this type have also been made in plants, although these do not closely resemble the chromosomes of the diptera. Because of the occurrence of giant chromosomes in different species and organ systems, the term *polytene* (many-threaded) is more useful than "salivary gland" chromosomes.

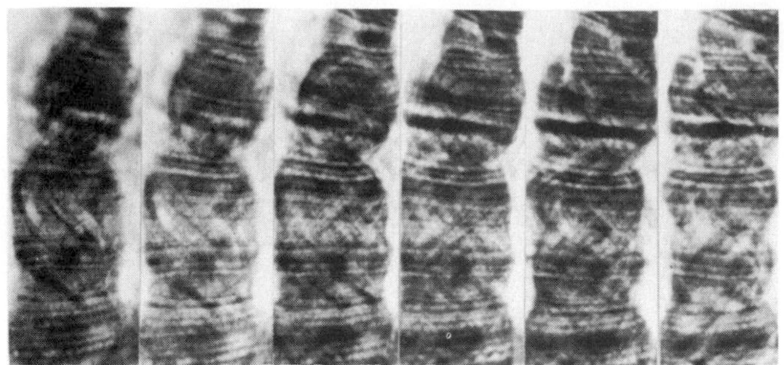

10–38. A series of micrographs taken at different levels of focus through the same chromosome of Chironomus tentans. *Multiple longitudinal subunits within the chromosome are twisted into a cable-like array. × 800. From* Protoplasmatologia 6D *by W. Beermann, 1966. Courtesy of W. Beermann and Springer-Verlag.*

Structure of polytene chromosomes. Much of the basis for contemporary understanding of polytene chromosome structure and function has been developed through experimentation carried out in the laboratory of Wolfgang Beermann. Beermann was able to gain insight into the structure of polytene chromosomes by following their development during differentiation of the larva. In the salivary glands, the cells become greatly enlarged early in the growth of the larva. Within the nuclei of these cells, the quantity of DNA is increased by repeated cycles of replication. This replication takes place without division of the chromosomes or nuclei. As the replication proceeds, the chromosomes become visible in the nuclei as elongated structures with the characteristic transverse bands. The number of chromosomes is one-half the usual somatic (diploid) number because homologous chromosome pairs *synapse,* or join together, during development of polyteny.

Hewson Swift and Ellen Rasch were able to estimate the extent of DNA replication and the probable number of multiple strands in polytene chromosomes of *Drosophila.* This was done by quantitative measurement of the amount of Feulgen stain bound within the giant nuclei as development of the polytene condition proceeds. Swift and Rasch found that the DNA in these nuclei doubled successively in ten well-defined, synchronized steps. This indicated that fully mature polytene nuclei contain more than a thousand times ($2^{10} = 1024$) as much DNA as ordinary somatic nuclei and implies that each chromosome is made up of as many multiple strands of DNA.[2] In other dipterans, this number may be higher. In the larva of the midge *Chironomus,* polytene nuclei may undergo replications producing 16,000 strands in each chromosome.

Direct examination of polytene chromosomes under the light microscope, carried out by Beermann and his co-workers, has also supported the inference that these chromosomes contain multiple numbers of DNA strands. In favorable preparations (Fig. 10–38), polytene chromosomes appear to consist of a thick bundle of fibers much like a cable. Longitudinal fibers can be seen to run through both band and interband regions. Polytene chromosomes sectioned for electron microscopy also show masses of more or less parallel fibers running longitudinally through the chromosome (Fig. 10–39), although reconstruction of the image is made difficult by the fact that a single polytene chromosome may be 100 times thicker than the sections used for electron microscopy. From the results of light and electron microscopy, the individual fibers in polytene chromosomes are thought to extend from one end of the

---

[2] More recent work (G. T. Rudkin, *Genetics* 52:470, 1965; H. D. Berendes and H. G. Keyl, *Genetics* 57:1, 1967) has shown that not all of the chromosomal complement in the larval nuclei of the flies is serially replicated during polytenization. A part of the chromosomes (the *chromocenter*) does not replicate and evidently remains at the diploid level. Thus replication of the total nuclear DNA during polytenization does not proceed in exact geometric multiples as suggested by the work of Swift and Rasch. However, their approximation that ten successive replications occur in *Drosophila* is evidently correct.

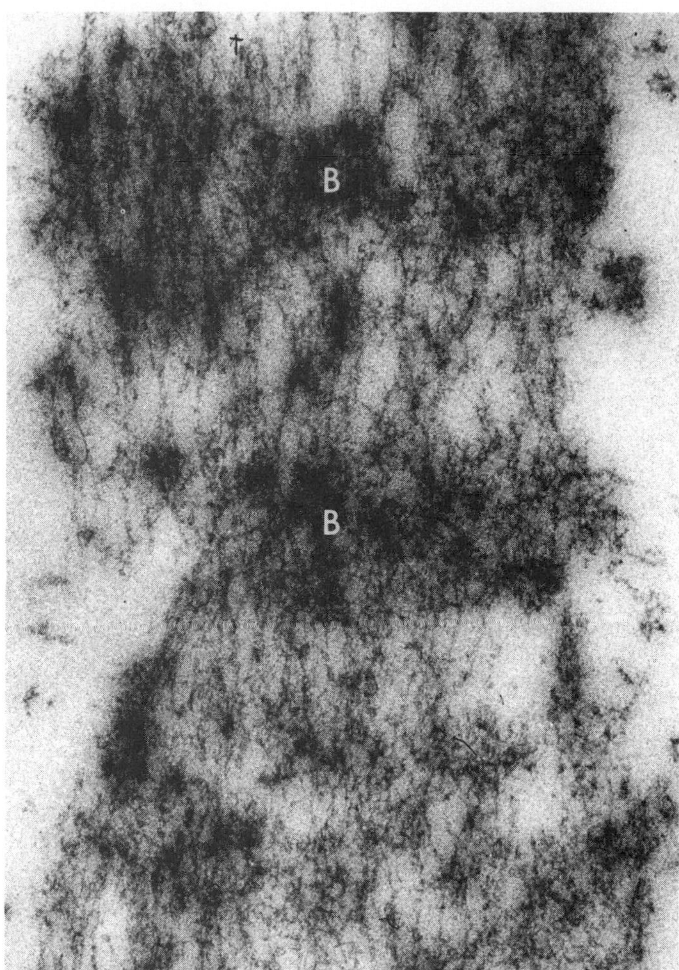

10–39. Section of a polytene chromosome from the salivary gland of Drosophila, *sectioned along the long axis of the chromosome. Chromatin fibers can be traced in both the band* (B) *and interband regions. The fibers are more closely packed or coiled in the bands. × 35,500. Courtesy of H. Swift.*

chromosome to the other and to be tightly coiled in the bands and uncoiled in the interband regions (see Fig. 10–40a).

Active sites on polytene chromosomes: Bands and puffs. Within a given species, the number and pattern of bands is precisely repeated in all of the cells containing polytene chromosomes. This is true in tissues as diverse as intestinal and salivary gland cells. Thus, differential activity of the cells in the various tissues is not reflected in differences in the number or types of bands in polytene chromosomes. However, differences can be detected in the appearance of some bands in different tissues or in the same tissue at different stages in development. In different tissues, or at different times, one or more bands are seen to swell or increase in size. At the same time, the altered bands are more diffuse, suggesting that uncoiling or extension of the fibers in the bands has taken place. In some cases, a band may be expanded into a large, brushlike structure that extends around the chromosome, well beyond the original diameter of the band. These expanded areas, called *puffs,* may be so extensive that several adjacent bands are partly dispersed and appear to contribute to the puff. However, careful investigations by Beermann have shown that even the largest puffs, called *Balbiani rings,* originate from the expansion of a single band (Fig. 10–40b and 10–41).

Through a number of experiments, puffs have been shown to be morphological evidence of the activity of a single gene. Much of this work has been carried out by Beermann and his co-workers, particularly Ulrich Clever and Claus Pelling (for detailed literature references, see Clever, 1968). Puffs are intensely stained by dyes specific for RNA. Incorporation of radioactive precursors of RNA into the puffs can be demonstrated by both light and electron microscope autoradiography. Administration of actinomycin D, which inhibits RNA synthesis, interferes with puff formation. These findings indicate that the puffs are sites of active RNA synthesis. Acidic, nonhistone proteins are also found in quantity in the puffs. These are probably complexed with RNA into ribonucleoprotein. The remainder of the chromosome, including both bands and interbands, contains histones (basic proteins). Histones are also found in the puffs, although only in small quantities compared to acidic proteins. Presumably, the histones form a regular structural part of the chromosome, while the acidic proteins are found only as a temporary part of the chromosomes in a complex with RNA synthesized in the puffs. Puffs remain unlabeled for long periods of time following administration of radioactive protein precursors, even though protein is incorporated. This suggests that the protein associated with RNA in the puffs is synthesized at some other location in the cell and that a large pool of these completed proteins probably exists in the nucleus or cytoplasm.

The RNA synthesized in a puff remains in this location for some time after synthesis, for periods

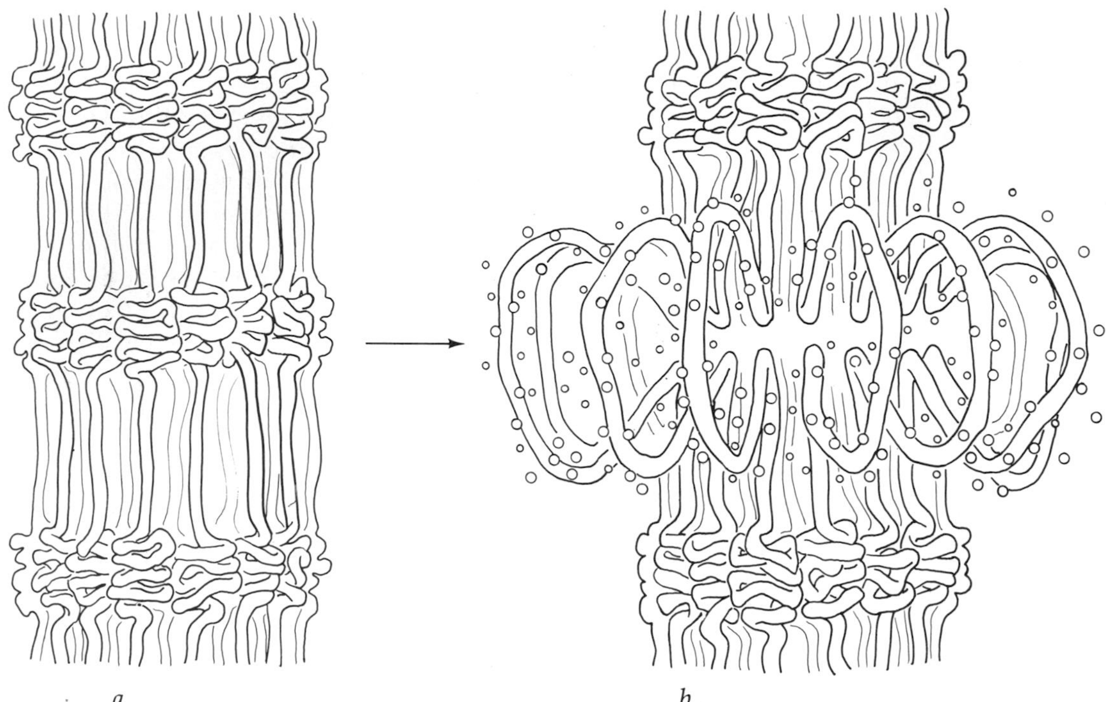

10–40. *Diagram illustrating the hypothetical relationship between bands and puffs. The individual fibers of the chromosomes, densely coiled in the bands in* a, *uncoil and extend into the surrounding nucleoplasm to form the puff in* b. *The RNA-protein product of the puff is shown diagrammatically as a granular component surrounding uncoiled fibers in the puff region. Adapted from H. Swift,* Proc. Ann. Tissue Culture Assoc. *Miami, 1965.*

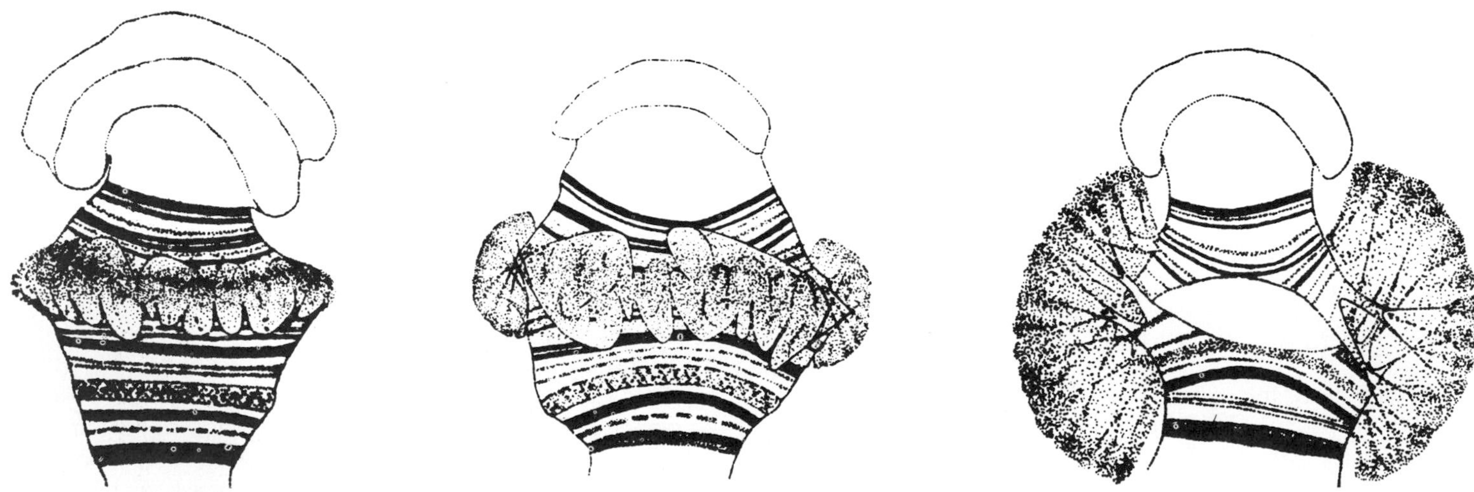

10–41. *Development of a large puff by expansion of a single band in* Chironomus tentans. *From* Protoplasmatologia 6D *by W. Beermann, 1962. Courtesy of W. Beermann and Springer-Verlag.*

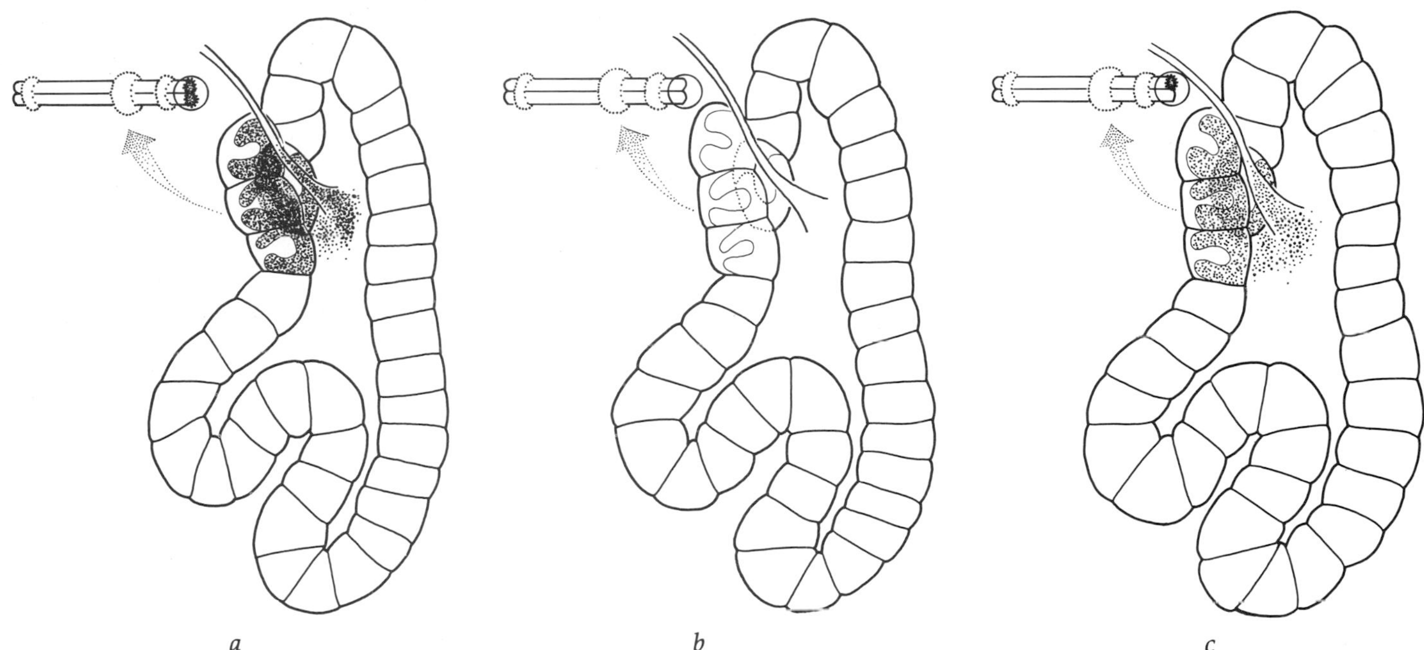

10–42. *Diagram of Beerman's experiment with* Chironomus *hybrids. In the salivary gland of* Ch. pallidivittatus (a), *four cells near the duct of the gland produce a granular secretion. Chromosome IV in these cells has a puff at one end. In the salivary gland of* Ch. tentans (b), *the corresponding cells produce a clear secretion. No puff is present at the end of chromosome IV. In hybrids between the two species* (c), *the four cells near the duct produce a secretion of reduced granularity. Chromosome IV in these hybrid cells contains a puff around only one-half of the chromosome. This half of the pair synapsed to form chromosome IV originates from the* Ch. pallidivittatus *parent. Redrawn, original courtesy of W. Beermann.*

approaching two hours. If administration of labeled precursors is followed by transfer to a medium free of label, radioactivity gradually decreases and disappears from the puff in about 4 hours. This indicates that the synthesized RNA is slowly released from the puffed site.

Puffs are thus identifiable as sites active in the synthesis of RNA. The best evidence that this RNA represents a transcribed gene has been obtained by Beermann from a study of puff formation in the larvae of two closely related species of *Chironomus.* In one species (*Chironomus pallidivittatus*), a number of cells at one end of the salivary gland secrete a granular fluid, which enters the duct passing out of the gland. In the nuclei of these cells, one of the polytene chromosomes (designated chromosome IV) has a conspicuous puff at one end. In the closely related species *Chironomus tentans,* the same group of cells secretes a clear, nongranular fluid. No puff is visible in the corresponding location on chromosome IV in *tentans.* Beermann found that in hybrids bred between the two species, the same group of cells at the end of the salivary gland produced a secretion that contained granules, but was considerably less granular than the *pallidivittatus* parent. Chromosome IV in the hybrid has a puff in the band near the tip, but in only one-half the diameter of the chromosome, presumably the half of the synapsed pair derived from the *pallidivittatus* parent (see Fig. 10–42). These experiments demonstrated that a correlation exists between the appearance of a specific puff and the presence or absence of a specific cell product and suggest that this site represents a genetic locus active in the synthesis of mRNA.

Ultrastructure of RNA synthesis and transport in polytene nuclei. The large puffs of polytene chromosomes are easily recognized under the electron microscope (Fig. 10–43), and the appearance of puffs is consistent

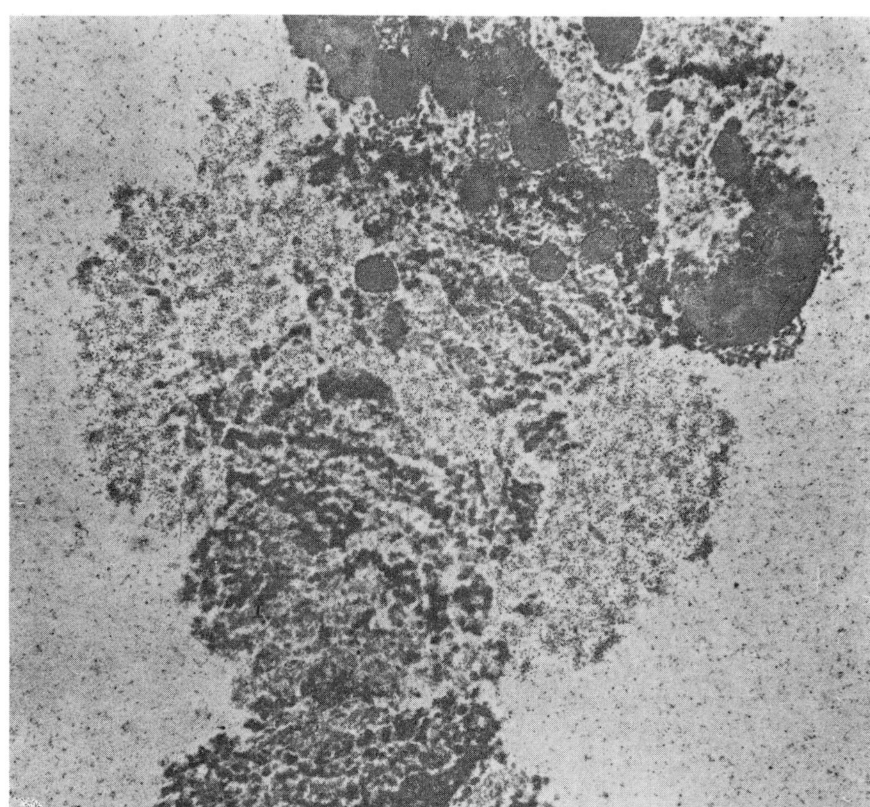

10–43. Thin section of a large puff from chromosome IV of Chironomus thummi. *Many particles are visible within the puff and in reduced numbers in the surrounding nucleoplasm.* × *4,500. Courtesy of H. Swift and B. J. Stevens.*

with the interpretations of puff structure derived from light microscopy (compare Figs. 10–43 and 10–41). The puffs appear under the electron microscope as diffuse regions in the chromosomes in which a large aggregation of small fibrils or granules is visible. The chromosome appears to have uncoiled or expanded in the puff region. The conspicuous granules or fibrils associated with the puff vary in dimensions; in the case of granular puffs, granule dimensions range from 100–1000 Å, with average dimensions falling at about 400 Å. Granules of the same dimensions and structure can also be seen in the surrounding nucleoplasm. Part of the fibrillar material visible in a puff represents chromatin fibrils that extend from the chromosome axis into the puff. Presumably these are loops of chromatin originating from the chromosome axis (see Fig. 10–40). Other fibrils, recognizably thicker than chromatin fibrils, probably represent ribonucleoprotein originating in the puff (see below).

The ultrastructure and cytochemistry of the puffs in *Chironomus* were studied by Hewson Swift and his colleagues. Swift found that the large puffs (Balbiani rings) in this organism contain great numbers of dense granules 400–500 Å in diameter and coarse 180–220 Å fibrils (Fig. 10–44). Digestion with ribonuclease removes all of the granules and the coarse fibrillar material from the puff (Fig. 10–45), indicating that these structures represent the ribonucleoprotein product of the puff. The thin 50–100 Å fibrils that remain visible after ribonuclease digestion are probably the looped chromosome fibers of the puff.

Swift noted that the interior of the large puffs contains both coarse fibrils and granules, while at the edges of the puffs only granules were present. At many points within the puffs, connections between fibrils and granules could be seen. Often these suggested that a conversion from fibrillar to granular form takes place within the puff (arrows, Fig. 10–44). Other granules of identical ultrastructure occur in abundance in the nucleoplasm and can also be seen near the nuclear envelope or in contact with the pore complexes (Fig. 10–46). Similar material, deformed into a rodlet, extends through the pores into the cytoplasm (Fig. 10–47).

On the basis of these observations, Swift proposed that the first RNA protein product of the puff is a coarse fibril 180–220 Å in diameter, which is transformed into a 400 Å granule within the puff. The particles found free in the nucleoplasm represent granules in transit

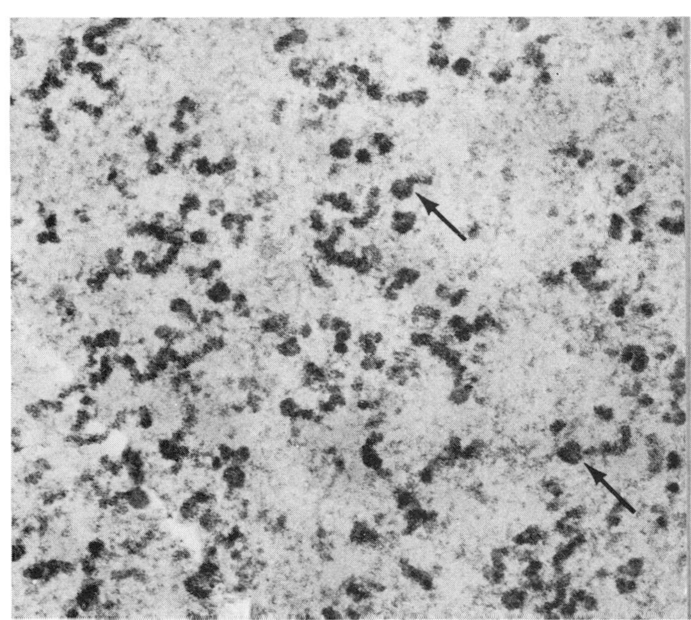

10–44. A portion of the granular material within the large puff of chromosome IV of Ch. thummi *at higher magnification. At several points (arrows), the granules are connected to short, coiled fibers (see text).* × *59,500. Courtesy of B. J. Stevens, H. Swift, and The Rockefeller University Press.* [J. Cell Biol. *31:55 (1966)*]

from the puffs to the nuclear envelope. At the pore complex, the granules extend into a rodlet that passes into the cytoplasm. Because, according to Beermann's work with puffs in *Chironomus,* the RNA synthesized at a puff probably represents mRNA, the granules seen by Swift probably represent the ultrastructural manifestation of the messenger. The 400 Å granules of the puffs resemble both in dimensions and ultrastructure the *perichromatin granules* described by Bernhard as extrachromosomal ribonucleoprotein particles of the nucleoplasm (see p. 220).

It is significant that the 400 Å granule is transformed into a rodlet of smaller dimensions as it traverses the pore complex, even though the margins of the pore formed by the membranes of the nuclear envelope are more than wide enough to accommodate a 400 Å particle. The rodlets formed from the granules apparently narrow to approximately 150 Å in the center of the pore complex. This is consistent with other work (see pp. 231–232) indicating that a channel existing in the center of the pore can accommodate particles with a maximum diameter of about 150 Å.

The ultrastructure of the nucleolus in polytene nuclei. Polytene nuclei also provide excellent material for studying the morphology of RNA metabolism in the nucleolus. Polytene nuclei often have one or more prominent nucleoli (Fig. 10–48); the position at which the nucleolus attaches to the chromosomes is clearly visible in both the light and electron microscopes. Undoubtedly, the multiple DNA cistrons coding for the 45S ribosomal precursor are concentrated into this band.

The nucleolus attached to the nucleolar organizer region completely surrounds the chromosome, forming a disc-like aggregation of material. Inside the disc, close to the chromosome, the nucleolus consists primarily of thin 30–50 Å fibrils corresponding to the fibrillar zone. This fibrillar material probably represents the 45S precursor of ribosomal RNA. Toward the periphery of the nucleolus, the fibrils are replaced by 150 Å particles of the granular zone. This zone probably represents the 28S ribosomal RNA subunit, in combination with one or more proteins. Elizabeth Hay has shown that if autoradiographs are made of thin-sectioned nucleoli in polytene chromosomes after administration of radioactive RNA precursor, the first label appears in thin fibrillar material close to the chromosome. If slightly longer periods of time are allowed to elapse, the label appears over the granular periphery of the nucleolus. Thus both the morphology and sequence of the label in the nucleoli of polytene chromosomes confirm the proposed identity of the fibrillar material with the 45S RNA precursor and the granular material with the larger ribosomal subunit.

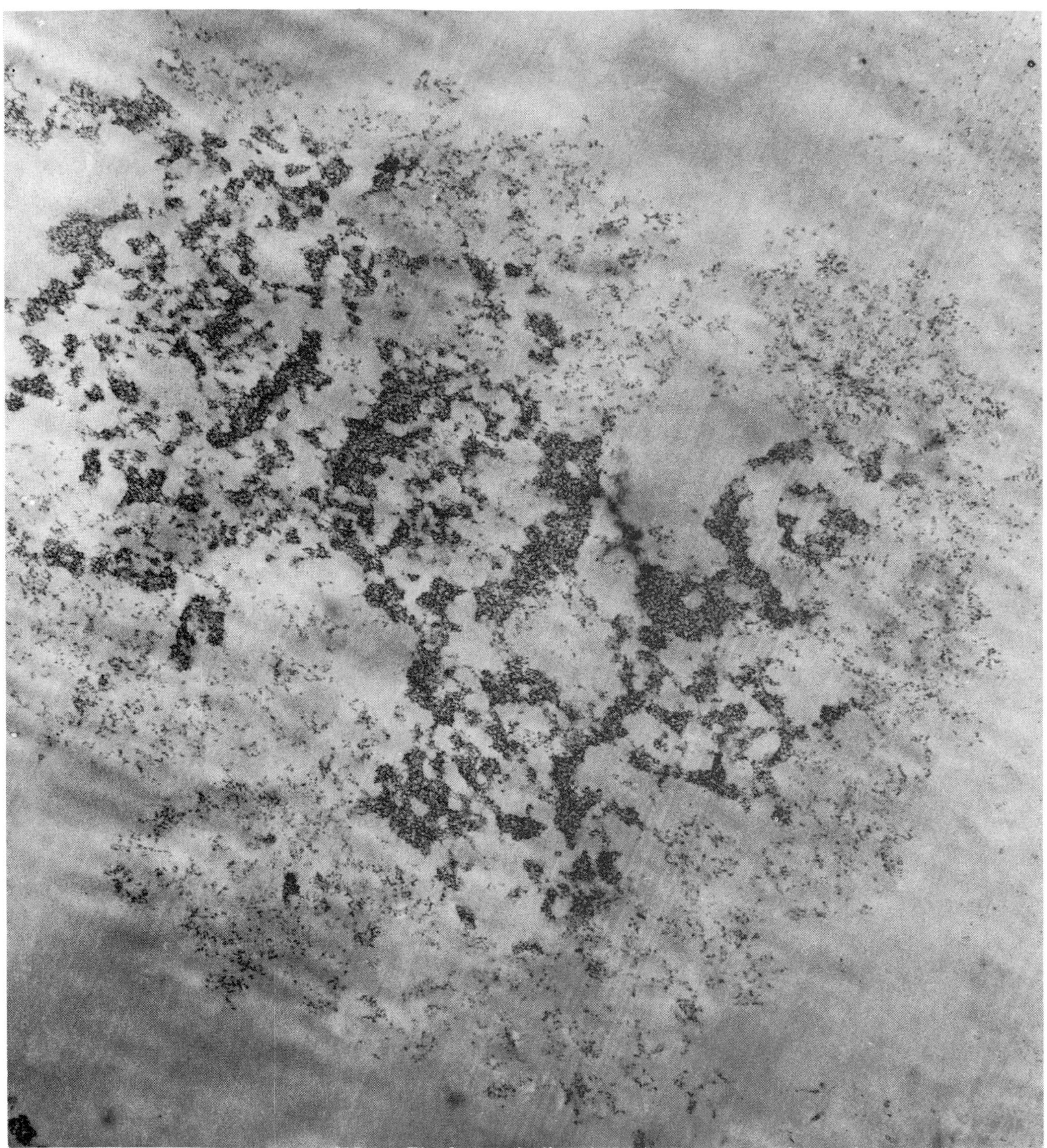

10–45. Two large puffs of Ch. thummi *after formalin fixation, digestion in ribonuclease, and staining with sodium tungstate. The residual material represents the DNA-protein of the puffs.* × *18,000. Courtesy of B. J. Stevens, H. Swift, and The Rockefeller University Press.* [J. Cell Biol. *31:55 (1966)*]

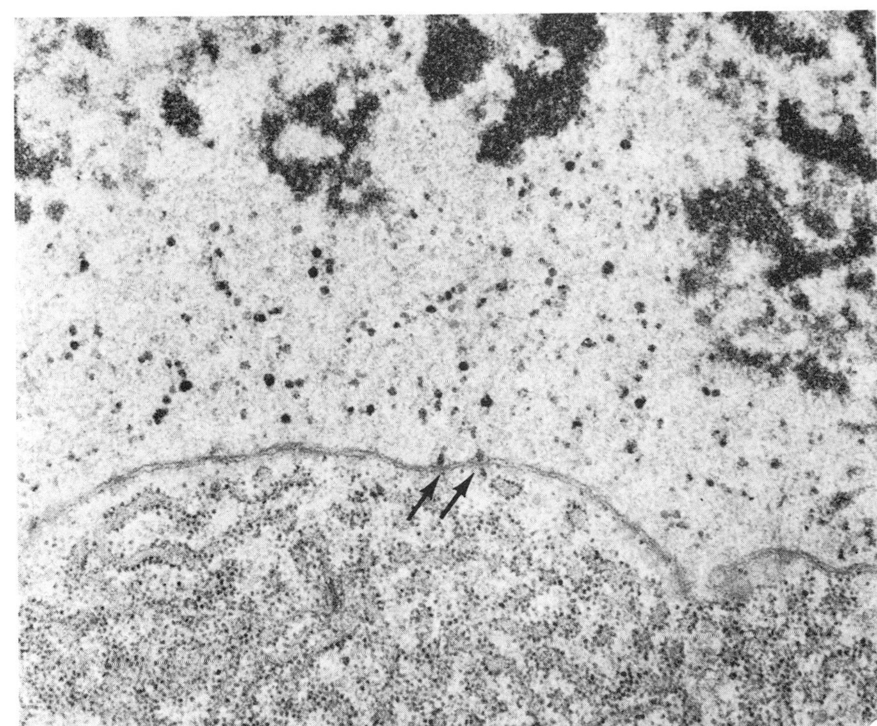

10–46. Distribution of the granules synthesized in the puffs in the nucleoplasm and in contact with the nuclear envelope (arrows). × 30,000. Courtesy of B. J. Stevens, H. Swift, and The Rockefeller University Press. [J. Cell Biol. *31:55 (1966)*]

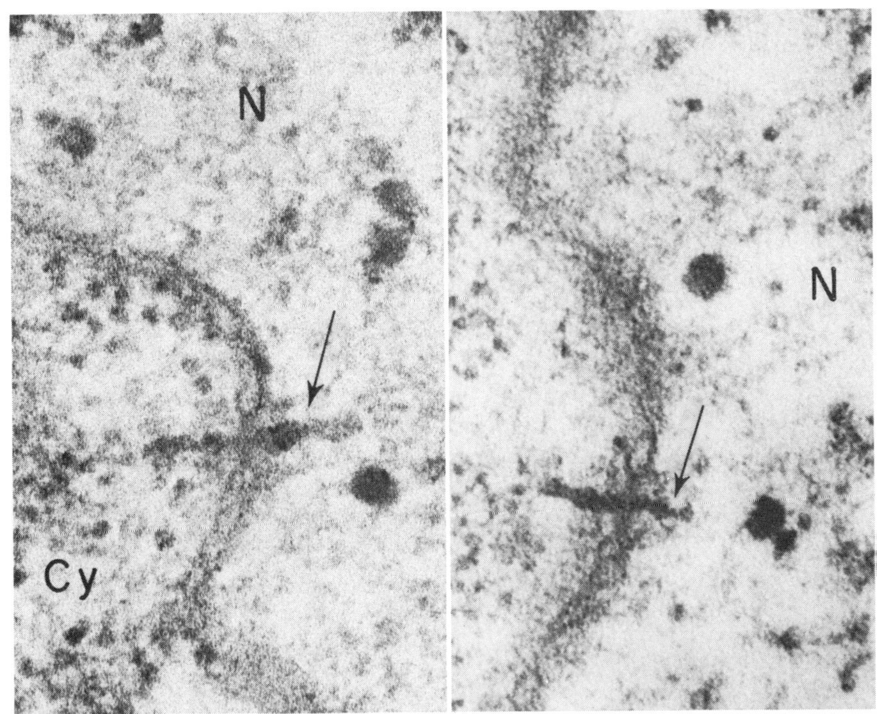

10–47. Penetration of the granules synthesized at the puffs through the nuclear envelope (arrows). × 124,000. Courtesy of H. Swift and B. J. Stevens.

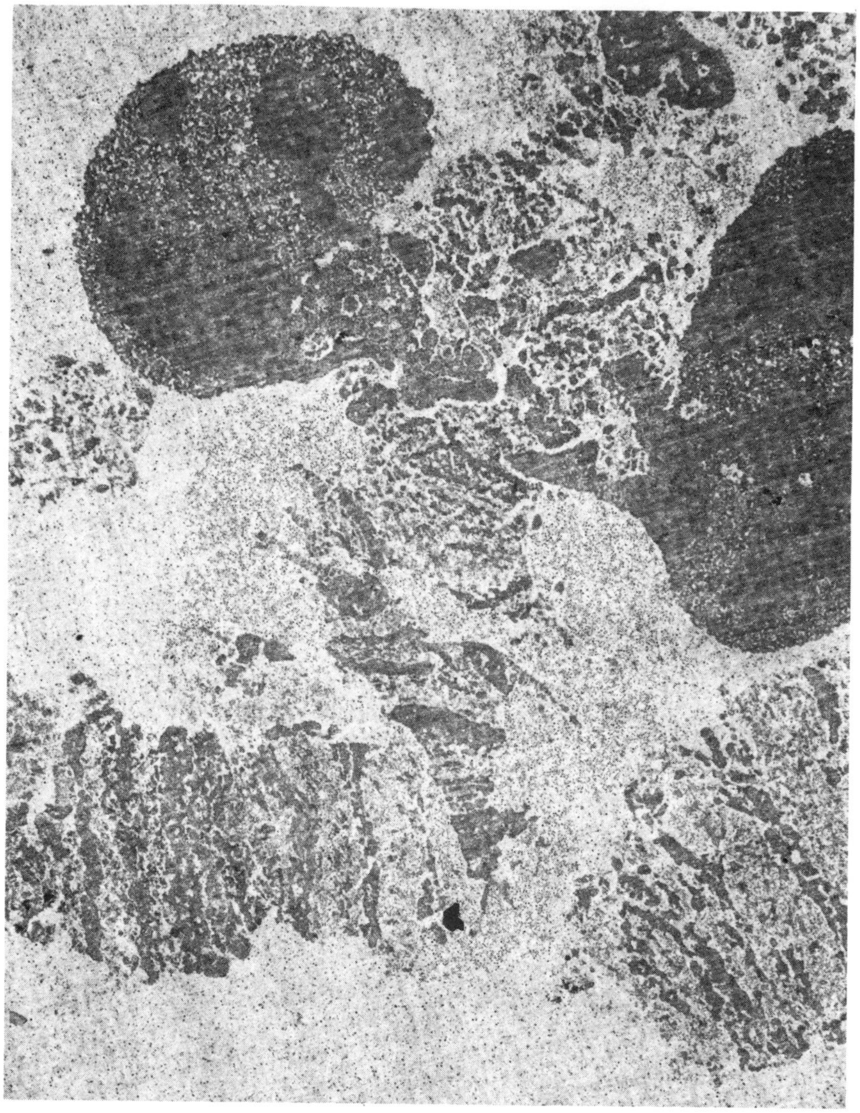

10–48. *The nucleolar organizer and nucleolus of chromosome IV in the salivary gland of* Ch. thummi. *The band associated with the nucleolus often shows a distinctive appearance because of penetration of the nucleolar material into the chromosome. Two large puffs or "Balbiani rings" are visible just below the nucleolus. × 4,900. Courtesy of B. J. Stevens, H. Swift, and The Rockefeller University Press.* [J. Cell Biol. *31:55 (1966)*]

Summary

In this chapter, the morphological appearance of the various ribonucleoprotein structures of the nucleus has been summarized, and morphological evidence has been given that these ribonucleoproteins are synthesized on the chromosomes and are transported across the nuclear envelope to the cytoplasm. Within the nucleolus, fibrils and granules consisting of ribonucleoprotein are visible. A close correspondence can be drawn between these fibrils and granules and the 45S ribosomal RNA precursor and the 28S subunit of the ribosome. Both of these rRNAs are combined with one or more proteins in the nucleolus. The other RNA species of the nucleus have not been conclusively identified with ribonucleoprotein structures visible under the electron microscope. These species include the DNA-like RNA of the nucleus (including mRNA), tRNA, and 28S and 5S rRNA. Limited evidence points to the 400 Å perichromatin granules and the closely similar ribonucleoprotein granules found in the puffs of polytene chromosomes as mRNA. These ribonucleoproteins pass through the pore complexes of the nuclear envelope to enter the cytoplasm. The pore complexes are evidently an adaptation that permits the passage of macromolecules between the nucleus and cytoplasm.

Suggestions for Further Reading

Books

Dalton, Albert J., and Francoise Haguenau, 1968. *The Nucleus.* Vol. 3, *Ultrastructure in Biological Systems.* Academic Press, New York. A collection of review articles emphasizing nuclear and nucleolar ultrastructure.

Protoplasmatologia, 1964. Vol. 5:4. A collection of articles on the structure and permeability of the nuclear envelope.

Papers and Review Articles

Alonso, P., and J. Perez-Silva, 1965. Giant chromosomes of protozoa. *Nature,* vol. 205, pp. 313–314.

Barr, H. J., and W. Plaut, 1966. Comparative morphology of nucleolar DNA in *Drosophila. Journal of Cell Biology,* vol. 31, pp. C17–C22.

Beermann, Wolfgang, 1962. Riesenchromosomen. *Protoplasmatologia,* vol. 6:D, pp. 1–161. An exhaustive review of the structure and function of polytene chromosomes; in German.

Beermann, Wolfgang, and Ulrich Clever, 1964. Chromosome puffs. *Scientific American,* vol. 210, pp. 50–58.

Bernhard, W., and Nicole Granboulan, 1968. Electron microscopy of the nucleolus in vertebrate cells. In A. J. Dalton and F. Haguenau, eds., *Ultrastructure in Biological Systems,* vol. 3, *The Nucleus.* Academic Press, New York, pp. 81–149.

Bloom, S., 1970. Structural changes in nuclear envelopes during elongation of heart muscle cells. *Journal of Cell Biology,* vol. 31, pp. 218–233.

Clever, Ulrich, 1968. Regulation of chromosome function. *Annual Review of Genetics,* vol. 3, pp. 11–30.

Fawcett, Don W., 1966. On the occurrence of a fibrous lamina on the inner aspect of the nuclear envelope in certain cells of vertebrates. *American Journal of Anatomy,* vol. 119, pp. 129–146.

Feldherr, C. M., and C. V. Harding, 1964. The permeability characteristics of the nuclear envelope at interphase. *Protoplasmatologia,* vol. 5:4, pp. 35–50.

Gall, Joseph G., 1964. Electron microscopy of the nuclear envelope. *Protoplasmatologia,* vol. 5:4, pp. 4–25.

Gall, Joseph G., 1967. Octagonal nuclear pores. *Journal of Cell Biology,* vol. 32, pp. 391–400.

Hay, Elizabeth D., 1968. Structure and function of the nucleolus in developing cells. In A. J. Dalton and F. Haguenau, eds., *Ultrastructure in Biological Systems,* vol. 3, *The Nucleus.* Academic Press, New York, pp. 2–79.

Kessel, R. G., 1968. Annulate lamellae. *Journal of Ultrastructure Research,* Supplement 10, pp. 1–82.

Loewenstein, Werner E., 1964. Permeability of the nuclear membrane as determined with electrical methods. *Protoplasmatologia,* vol. 5:4, pp. 26–34.

Monneron, A., and W. Bernhard, 1969. Fine structural organization of the interphase nucleus in some mammalian cells. *Journal of Ultrastructure Research,* vol. 27, pp. 266–268. Considers ribonucleoprotein particles of the nucleoplasm outside of the nucleolus.

Perry, Robert P. 1967. The nucleolus and the synthesis of ribosomes. *Progress in Nucleic Acid Research and Molecular Biology,* vol. 6, pp. 219–257.

Scheer, Ulrich, and W. W. Franke, 1969. Negative staining and ATPase activity of annulate lamellae of newt oocytes. *Journal of Cell Biology,* vol. 42, pp. 519–533.

Schoefl, G. I., 1964. The effect of actinomycin D on the fine structure of the nucleolus. *Journal of Ultrastructure Research,* vol. 10, pp. 224–243.

Stevens, B. J., and Hewson Swift, 1966. RNA transport from nucleus to cytoplasm in *Chironomus* salivary glands. *Journal of Cell Biology,* vol. 31, pp. 55–77.

Swift, Hewson, 1965. Molecular morphology of the chromosome. In The chromosome: structural and functional aspects. *Proceedings of the Annual Tissue Culture Association,* Miami.

11
Protein Synthesis

The site of protein synthesis in the cytoplasm is the ribosome. Each major class of RNA, messenger, transfer, and ribosomal RNA, interacts directly at the ribosome during protein synthesis. Before this interaction takes place, rRNA and tRNA undergo a series of preliminary reactions in the cytoplasm. Ribosomal RNA enters the cytoplasm in the nascent ribosomal subunits. Before protein synthesis begins, these subunits are joined into complete ribosomes. In another series of preliminary reactions, tRNAs are linked to their specific amino acids. These preliminary reactions and the subsequent events of protein synthesis are described in this chapter.

The biochemical events of protein synthesis have been localized in structures visible under the electron microscope. Ribosomes themselves are visible and can be readily identified in the cytoplasm, either freely suspended in the ground substance or attached to membranes of the endoplasmic reticulum. Newly synthesized proteins can be traced in the elements of the endoplasmic reticulum and also within elements of the Golgi apparatus and secretion vesicles of various types. These morphological aspects of protein synthesis are also discussed in this chapter. Also included is a brief survey of some hypothetical mechanisms advanced to account for the control or regulation of protein synthesis.

Preliminary Reactions of Protein Synthesis

The Assembly and Structure of Ribosomes

The large and small ribosomal subunits that enter the cytoplasm through the nuclear envelope are in essentially complete form. As noted in Chapter 10, a

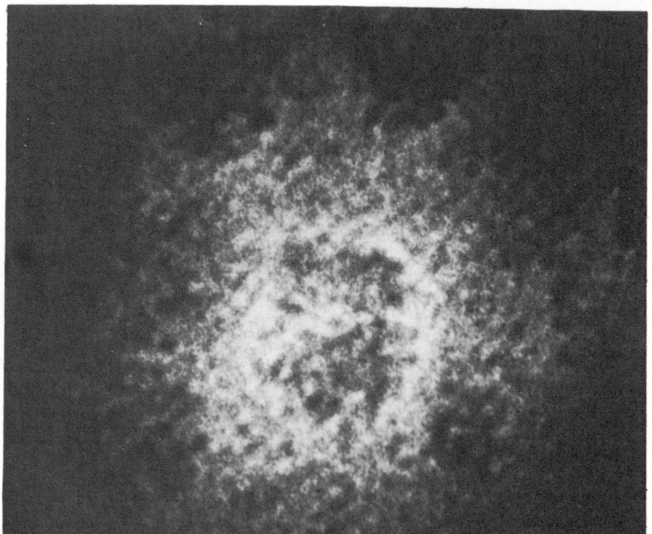

11–1. A bacterial ribosome, isolated from E. coli *and negatively stained with phosphotungstic acid.* × *980,000. From* The Ribosome *by A. S. Spirin and L. P. Gavrilova, 1969. Courtesy of A. S. Spirin and Springer-Verlag.*

very small quantity of protein is probably removed before the subunits combine into the complete ribosome. In bacteria, there are indications that the large and small subunits do not associate to form the complete ribosome until the instant of attachment to mRNA. It is not known whether this pattern of assembly extends to higher organisms. Many cell types contain a cytoplasmic pool of complete ribosomes that are unattached to mRNA. On isolation, these intact ribosomes are capable of combining with template RNA and are active in protein synthesis in cell-free systems. Thus separate, uncombined subunits do not appear to be a requirement for attachment of the messenger.

Intact ribosomes from all organisms share a number of structural features (Figs. 11–1 to 11–3). The ribosome is roughly elliptical in shape, with one end of the ellipse slightly flattened. A cleft marking the line of fusion of the large and small subunits runs at right angles to the long axis of the ribosomes. The cleft is located asymmetrically with respect to the long axis and is visible in both negatively stained (Fig. 11–1) and thin-sectioned (Fig. 11–2) ribosomes. This surface crevice can be most clearly distinguished in bacterial ribosomes. The division between the subunits of eukaryote ribosomes is less distinctly marked; eukaryote ribosomes are in general less clearly outlined against the surrounding

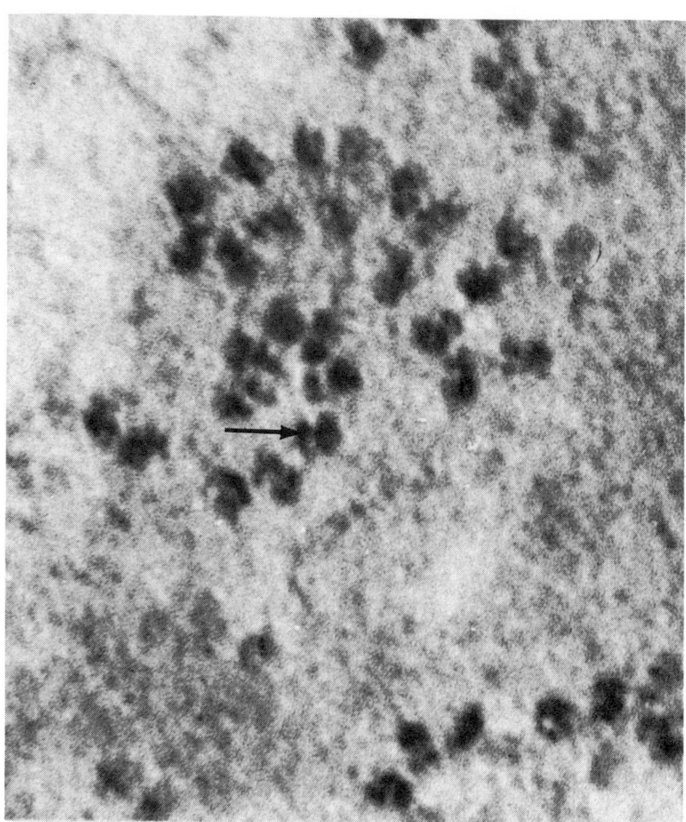

11–2. Polyribosomes in a thin-sectioned rat liver cell. A cleft marking the division between the large and small subunit is visible in many of the ribosomes. The arrow marks the smaller ribosomal subunit. × *200,000. Courtesy of N. T. Florendo.*

cytoplasmic ground substance than prokaryote ribosomes.

According to H. E. Huxley and G. Zubay, who carried out extensive investigations of bacterial ribosomes by negative staining, the smaller ribosomal subunit (arrow, Fig. 11–2) is located at the flattened end of the ribosome, fitting like a cap over the larger subunit. Subunits are sometimes reported to show surface granularity or surface striations, with a periodicity of 30 Å. The reality of this surface detail is uncertain, as it is clearly visible only in underfocused images of the ribosome and may thus result from the combined effects of diffraction and phase contrast in the electron image (see discussion in Chapter 3).

The reported dimensions of the ribosome vary according to the source and degree of hydration. Bacterial

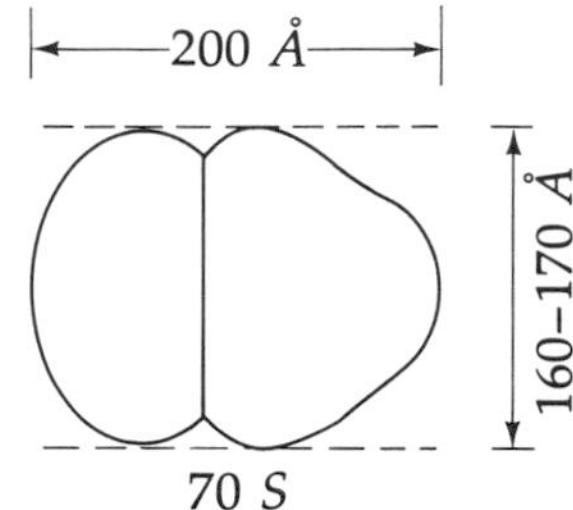

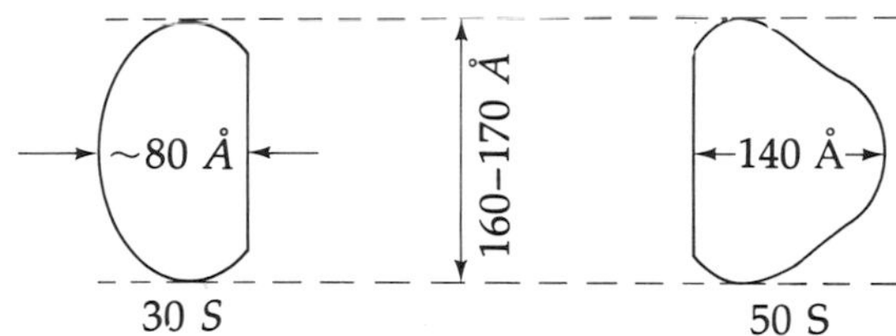

11–3. Average dimensions of the two ribosomal subunits of bacteria (dried state). Redrawn, original courtesy of A. S. Spirin and Springer-Verlag from The Ribosome.

ribosomes are approximately 200 × 170 Å in the isolated, dried state (Fig. 11–3) and 290 × 210 Å in aqueous media. The ribosomes of higher organisms are slightly larger in dimensions, 200 × 240 Å when dried, and 200–240 × 300–340 Å in a water solution. After isolation and negative staining, ribosomes from eukaryotes measure roughly 200 × 250 Å. The ribosomes of chloroplasts and mitochondria are smaller and markedly similar in dimensions to bacterial ribosomes (see Chapter 16).

The proteins of ribosomes. On analysis, the ribosomes of bacteria prove to be about 60 percent RNA and 40 percent protein by weight. No lipids or polysaccharides are found. Eukaryote ribosomes contain a smaller proportion of RNA, with the ratio of RNA to protein approaching a value of 1. The protein complement of ribosomes is exceedingly complex. Bacterial ribosomes appear to contain as many as 50–60 different proteins. Eukaryotic ribosomes show greater complexity in their protein complement. Estimates of the total number of proteins in eukaryote ribosomes are highly variable, running from 50–60 up to as many as 150 different polypeptide chains (see C. G. Kurland, 1970, and A. S. Spirin and L. P. Gavrilova, 1969). The variability in these estimates reflects the difficulties inherent in the isolation and identification of ribosomal proteins. The number of proteins carried with the ribosome on isolation depends on the *p*H of the isolating medium, the concentration of monovalent and divalent cations, and the amount of "washing" or purification of the ribosomes on extraction. Small variations in these conditions may result in the loss of ribosomal proteins or the adsorption of extraneous protein from the surrounding cytoplasm. While it is impossible to estimate the protein complement of eukaryote ribosomes with any precision, most investigators favor a range of values not much greater than the 50–60 different proteins considered the maximum for bacterial ribosomes.

Details of the binding of protein to the ribosome have been studied by T. Staehelin and M. Meselson and P. Traub and his co-workers. Ribosomal proteins fall into three classes as defined by strength of bonding to the ribosomes. About 30 percent of the protein is only very loosely bound and is easily removed. This fraction probably contains some extraneous protein, as removal does not interfere with protein synthesis in cell-free systems. Another fraction, representing about 50 percent of the total protein, can be reversibly removed by adjusting the Mg^{+2} concentration. The remaining 20 percent, called the "core" protein, is very tightly bound. If this fraction is removed, the ribosome is irreversibly destroyed. The fraction removed reversibly by adjustments of the Mg^{+2} concentration can be further subdivided into subfractions; some of these, when added to the "core" RNA-protein particles, impart the capacity to bind mRNA or tRNA or to polymerize amino acids into peptides. Apart from these limited observations, very little is known of the identity, structure, or function of the protein complement of the ribosomes.

Dissociation of the ribosome. Bacterial ribosomes can be readily dissociated into large and small subunits by adjusting the concentration of Mg^{+2} ion in the medium. At levels of 0.005M to 0.01M Mg^{+2}, bacterial ribosomes remain intact. If Mg^{+2} concentration is reduced to between 0.00025M and 0.0005M, reversible dissociation into large and small subunits occurs. At slightly lower concentrations of magnesium ion, the "splitable" protein fractions are reversibly released. As the level falls below 1000 Mg^{+2} ions per ribosome, the ribosomal subunits are irreversibly denatured; that is, the "core" proteins are removed.

The subunits of eukaryote ribosomes are more tightly bound together; simple adjustments of Mg^{+2} ion concentration are usually not sufficient to cause dissociation. It is necessary in addition to adjust the *p*H upward,

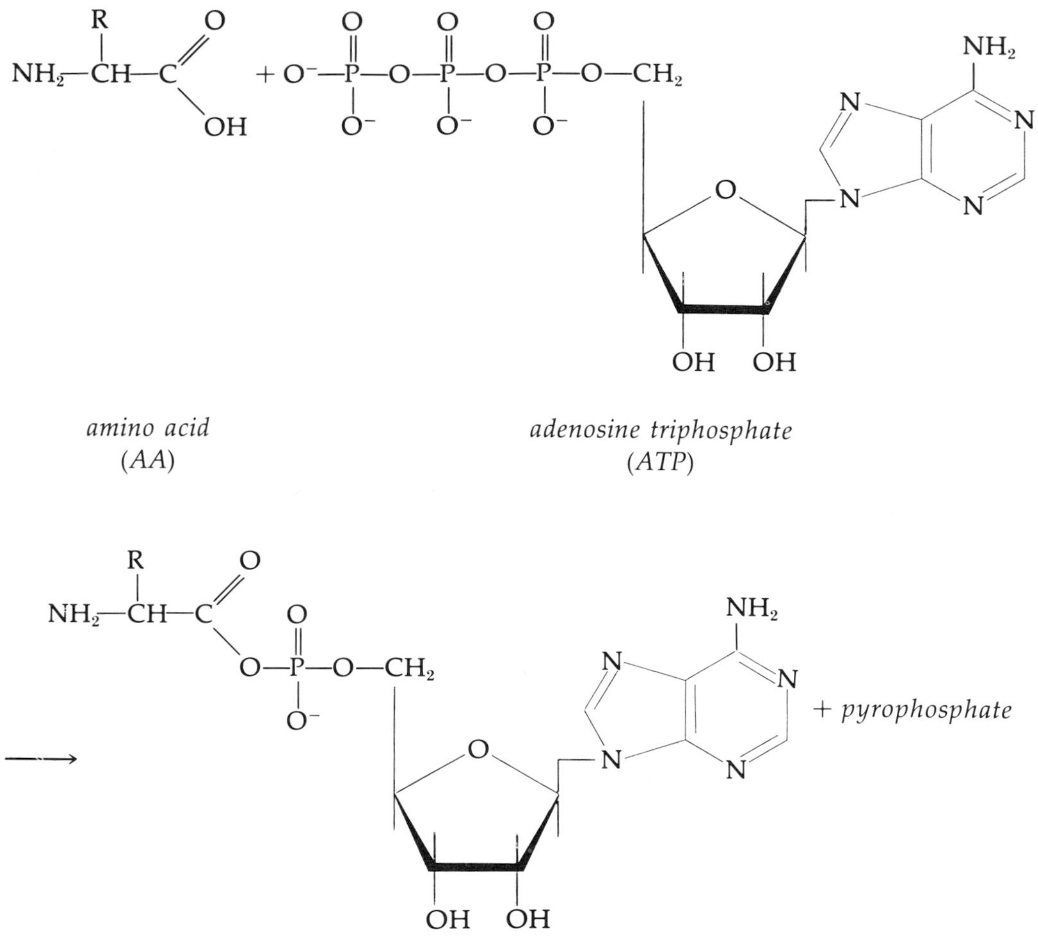

amino acid
(AA)

adenosine triphosphate
(ATP)

$\longrightarrow$ *aminoacyl adenylate* *(AA ~ AMP)* + *pyrophosphate*

11–4. Activation of an amino acid. The reaction, in which the adenosine monophosphate residue is bound to an amino acid, is catalyzed by aminoacyl-tRNA synthetase.

or increase the concentration of monovalent cations. Increased concentrations of phosphate or carbonate ions, which have a high affinity for Mg^{+2} ion, also promote the dissociation of the ribosomal subunits of higher organisms. These observations point to a possible role of Mg^{+2} ion in binding the subunits of the ribosome together.

It is not known whether the dissociation of ribosomes, evidently promoted by extraction of neutralizing Mg^{+2} ions from the phosphate groups of rRNA, represents a change in RNA-RNA interaction, RNA-protein interaction, or some combination of the two. For reversible dissociation and reassembly of subunits to occur, both the RNA and protein of the ribosome must remain intact. Enzymatic degradation of either component, for example, interferes with reassembly of the ribosome *in vitro*. This suggests that the interactions binding ribosomes together involve interactions between both RNA and protein (for detailed literature references, see Spirin and Gavrilova, 1969).

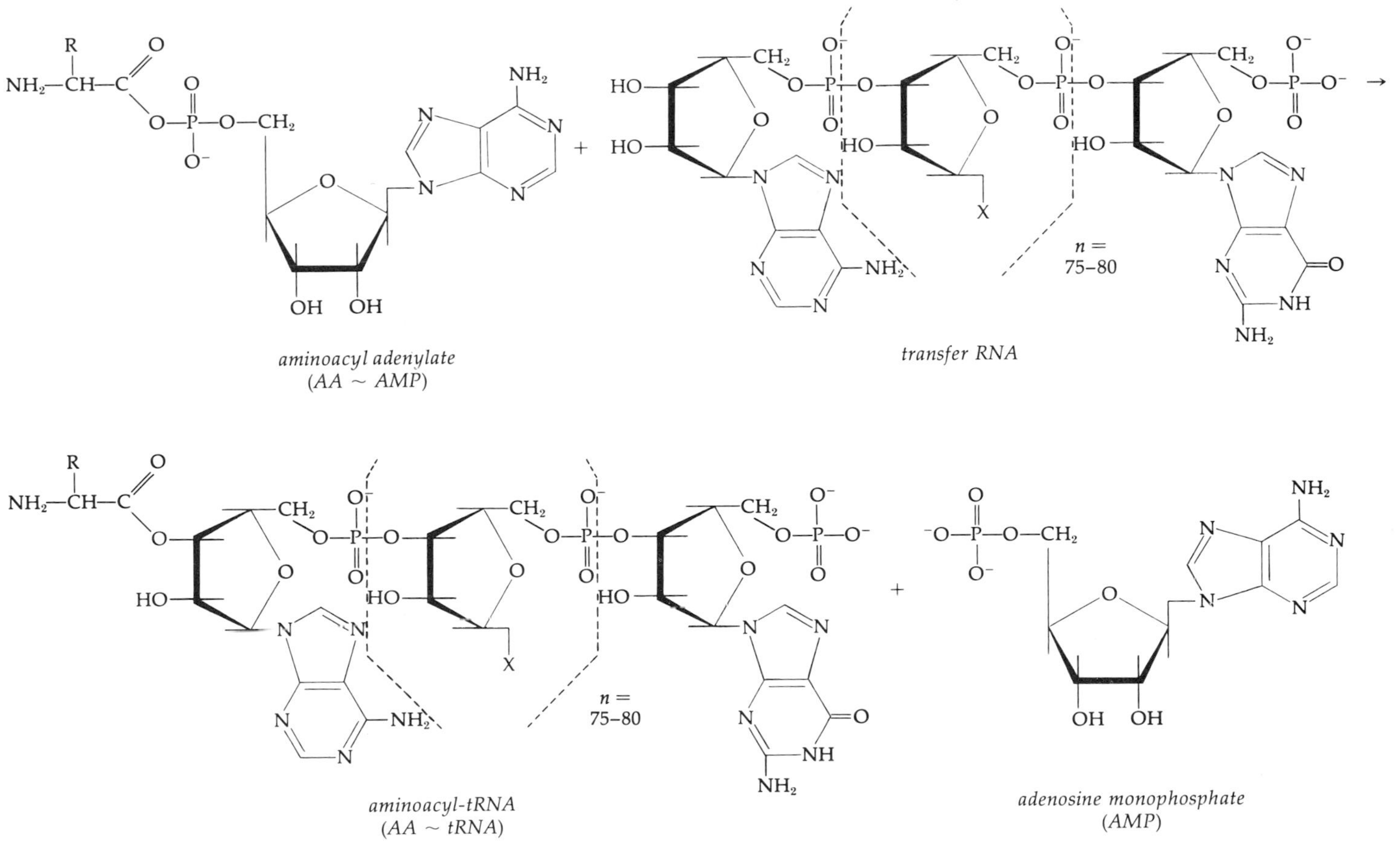

aminoacyl adenylate (AA ~ AMP)

transfer RNA

aminoacyl-tRNA (AA ~ tRNA)

adenosine monophosphate (AMP)

11–5. Binding of an activated amino acid to transfer RNA. This reaction is also catalyzed by aminoacyl-tRNA synthetase.

The Activation and Attachment of Amino Acids to Transfer RNA

The activation and attachment of amino acids to specific transfer RNA molecules occurs in two clearly defined steps in the "ground substance" of the cytoplasm. These interactions, first demonstrated by P. C. Zameçnik, M. B. Hoagland, and their co-workers, do not require the participation of ribosomes or mRNA. A major part of the energy required to form peptide bonds in protein synthesis originates in the hydrolysis of ATP, which occurs during amino acid activation; the ultimate source of this energy is in the systems synthesizing ATP in the mitochondrion. Both reactions, the activation of amino acids and the attachment of the activated amino acids to transfer RNA are catalyzed by a family of enzymes called *aminoacyl-tRNA synthetases.*

In the first step in the activation of amino acids, the carboxyl group of an amino acid interacts with the terminal phosphates of ATP, forming an aminoacyl-adenylate (AA-AMP) complex (see Fig. 11–4). This reaction is depicted in skeleton form by equation 11–1.

$$\mathrm{AA} + \mathrm{ATP} \xrightarrow[\text{synthetase}]{\text{aminoacyl-tRNA}} \mathrm{AA} \sim \mathrm{AMP} + \mathrm{PP} \qquad (11\text{–}1)$$

After binding the amino acid to AMP, the aminoacyl-adenylate complex remains attached to the surface of the enzyme. The same enzyme molecule then forms a complex with a specific tRNA molecule and catalyzes the second step in the reaction (Fig. 11–5), in which the bond with AMP at the carboxyl of the amino acid is transferred to the 3′—OH group of the terminal adenosine of the tRNA. In the second step, adenosine monophosphate (AMP) is released to the medium. In skeleton form, the second reaction proceeds as follows:

$$\text{AA} \sim \text{AMP} + \text{tRNA} \xrightarrow[\text{synthetase}]{\text{aminoacyl-tRNA}} \text{AA} \sim \text{tRNA} + \text{AMP} \quad (11\text{–}2)$$

The enzyme catalyzing the overall sequence of reactions is relatively complex and evidently contains a number of active sites. The enzyme in the free state presumably has sites for the recognition and binding of an amino acid and the ATP molecule. After binding of these substrates and catalysis of the formation of the amino acid-AMP bond, the enzyme probably undergoes a configurational transition that unmasks or activates another site specific for the binding of the correct tRNA to the aminoacyl-adenylate enzyme complex.

Each amino acid has a specific aminoacyl-tRNA synthetase. There are, therefore, at least 20 different forms of this enzyme. Possibly, there are a multiple number of enzymes for most amino acids, corresponding to the multiple tRNAs specific for these amino acids, but this does not seem to be the case. For a number of bacterial species, only one aminoacyl-tRNA synthetase has been isolated for each of the amino acids lysine, valine, leucine, isoleucine, phenylalanine, and tyrosine, even though each of these binds to more than one type of tRNA. In a very few cases, two synthetases have been isolated for a given amino acid. However, these additional synthetases, identified in the eukaryote *Neurospora* by L. Barnett and his co-workers, have proved to be of mitochondrial origin and appear to be specific for mitochondrial rather than cytoplasmic protein synthesis (for detailed literature references, see Novelli, 1967).

The reaction of amino acids with ATP and tRNA satisfies two requirements for protein synthesis. One of these is the provision of energy for the formation of the peptide bond. The second is provision of a recognition capacity between amino acids and the codons of mRNA, accomplished through the medium of the tRNA molecules attached to specific amino acids by the synthetase enzyme. These events, along with the transport of mRNA to the cytoplasm and the assembly of ribosomes from subunits, complete the preliminary interactions for the synthesis of protein in the cytoplasm.

The Mechanism of Protein Synthesis at the Ribosome

The details of protein synthesis outlined in the following paragraphs have been derived almost entirely from experiments in bacterial systems. Because this is the case, this section will describe the large and small subunits of the ribosome as the 50S and 30S subunits. Essentially the same series of events is believed to take place on the larger ribosomes of eukaryotes. The mechanism as described represents a summary of the findings of many investigators working in major laboratories in all parts of the world. The work is so extensive that it is impossible to provide literature references to all supporting experimentation. Instead a limited number of representative papers have been listed as examples. (See especially Spirin and Gavrilova, 1969, and de Man and Noorduyn, 1969, for original literature references in this area.)

Initiation

The first step in the initiation of protein synthesis is the binding of mRNA to the ribosome. If this binding is allowed to occur in cell-free systems, the messenger attaches to the smaller ribosomal subunit. This indicates that the mRNA binding site is on the 30S subunit; the messenger first binds at the 5′—OH end of the RNA chain. A loosely bound proteinaceous substance, called simply F_3, stimulates the binding of messenger to ribosome. Only one 30S subunit interacts initially with each molecule of mRNA; if this mRNA-ribosome complex is treated with ribonuclease, a chain of mRNA about 29 nucleotides in length remains resistant to the enzyme. This nuclease-resistant portion may represent the segment bound to the 30S subunit of the ribosome (see Brown and Doty, 1968, and Takanami and Zubay, 1964).

The mRNA-ribosome complex formed as the first step in protein synthesis binds tRNA molecules from the surrounding medium, each carrying a specific amino acid. The initial tRNA bound to the ribosome must contain the correct anticodon to allow base pairing with the first coding triplet at the 5′ end of the mRNA. Other factors must be present for the first binding of tRNA to the ribosome-mRNA complex, including at least two additional initiating factors (called F_1 and F_2) and guanosine triphosphate (GTP), a molecule closely similar in structure and properties to ATP. If

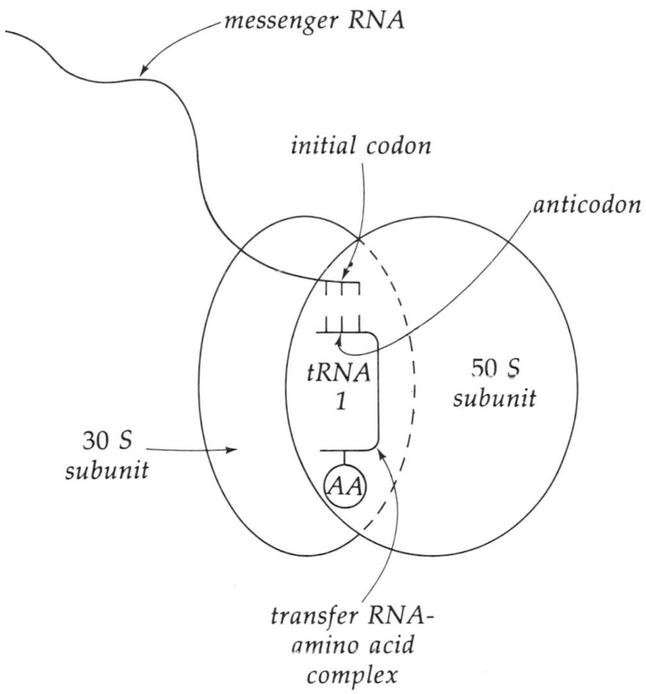

11–6. The initial complex between mRNA, the ribosome, and aminoacyl-tRNA. Binding sites for mRNA and the first aminoacyl-tRNA complex attracted from the medium are on the 30S subunit (see text).

these are present, along with the correct concentrations of ions, a mixture of ribosomes, mRNA, and aminoacyl-tRNAs will form complexes in which each ribosome is bound to one molecule of mRNA and one molecule of aminoacyl-tRNA. Separation of the ribosomal subunits by adjusting the Mg^{+2} concentration at this point reveals that both the mRNA and the aminoacyl-tRNA are bound to the smaller 30S ribosomal subunit. Therefore, the 30S subunit probably contains at least two sites, one binding mRNA and one binding the amino acid-tRNA complex. This three-way complex completes the first phase in the initiation of protein synthesis (see Fig. 11-6; Stanley et al., 1966).

The elucidation of the next step in protein synthesis presented considerable difficulties until certain details were worked out by Kjeld A. Marcker and Frederick Sanger. If ribosomes in cell-free systems are separated into subunits after the initiation phase has passed, the 30S subunit is found to be complexed with the mRNA and with a molecule of aminoacyl-tRNA as in the initiation phase described above. The growing protein chain, however, is found to be bound to the 50S subunit. If the nascent protein chain is removed from the 50S subunit, the most recently added amino acid is found to be still complexed with its tRNA. Therefore, the 50S subunit contains a site that binds specifically to tRNAs complexed with a polypeptide chain rather than a single amino acid. This 50S site, called the *peptidyl-transfer RNA site,* has little or no affinity for aminoacyl-tRNA (see Bretscher and Marcker, 1966).

Supposedly, after the first phase in initiation, the ribosome undergoes a change in configuration which allows the first aminoacyl-tRNA complex to move from its site on the 30S subunit to the binding site on the 50S subunit. As the tRNA moves to the 50S subunit, the mRNA is also shifted by one triplet, exposing a new codon at the 30S site. Movement of the aminoacyl-tRNA and messenger is known to be stimulated by the presence of GTP and at least two proteinaceous factors called "translocating factors" (see Lucas-Lenard and Lipmann, 1966).

This step brings us to the problem in initiation solved by Marcker and his colleagues. How does the first aminoacyl-tRNA complex move from the 30S to the 50S site, which is known to bind readily to peptidyl-tRNA but not to aminoacyl-tRNA? Marcker and Sanger found that in bacteria the first aminoacyl-tRNA coded for by the messenger is a modified form, which has binding affinity for both the 50S site and 30S site. This modified complex was discovered partly because many bacterial proteins (about 60 percent) contain a methionine residue as the first amino acid in the polypeptide chain. If partly synthesized proteins are separated from ribosomes extracted from bacterial cells, the initial methionine is found to contain a formyl group bound to the amino group of the amino acid. This special amino acid, *formylmethionine,* was found to be universally present as the initial amino acid of newly synthesized proteins. Subsequently, the tRNAs specific for methionine were separated into two distinct fractions, one carrying methionine and one carrying formylmethionine to the ribosome. By experiments in which synthetic trinucleotides were tested for their ability to bind to tRNAs (see the discussion of the genetic code on p. 263), Marcker found that the tRNA carrying formylmethionine was coded for by the triplets AUG or GUG in mRNA. In other experiments, it was found that these triplets code for formylmethionine only at the 5′ end of the mRNA; if located internally in the messenger, AUG codes for the usual unformylated methionine, and GUG codes for the insertion of valine into a polypeptide chain. Thus AUG and GUG, if at the 5′ end of mRNA, are *initiator codons* that indicate the placement of formylmethionine into the first position in a polypeptide chain.

The ability of the formylmethionine-tRNA complex to bind to either ribosomal subunit provides the basis for transfer of the initial tRNA-mRNA complex from the 30S to the 50S site. This movement exposes the second mRNA triplet at the 30S site. A second aminoacyl-tRNA complex from the surrounding medium then binds at this site, pairing according to the correct codon-anticodon relationship. At this instant, the total complex consists of the ribosome, the messenger, and two molecules of aminoacyl-tRNA (Fig. 11–7).

The two amino acids in this configuration are localized in a position favorable for the formation of the first peptide bond. The enzymatic activity required for the formation of the peptide linkage is associated with the 50S ribosomal subunit; supposedly an enzyme, *peptidyl synthetase,* is located within the peptidyl-tRNA binding site of the larger subunit (see Monro, 1967). The formation of the peptide bond at this site represents a substitution of a peptide bond for the ester bond between a tRNA and its attached amino acid (Fig. 11–8). Hydrolysis of the bond linking the amino acid to the first tRNA, which is broken when the peptide bond forms, releases sufficient energy for the formation of the peptide linkage. Additionally, for each peptide bond formed, one molecule of GTP is hydrolyzed to GDP, with the release of one phosphate. Hydrolysis of GTP is known to take place before the formation of the peptide bond; this suggests that the free energy released by the hydrolysis of GTP may be used in the steps just preceding formation of the peptide bond, possibly including the movement of the messenger from the 30S to the 50S site (see Conway and Lipmann, 1964).

After formation of the first peptide bond, at the instant when the first tRNA is released from the ribosome, the second tRNA, occupying the 30S site, carries a dipeptide consisting of the first two amino acids coded for by the messenger. This complex satisfies the binding requirement of the 50S site for a peptidyl-tRNA. The peptidyl-tRNA moves to the 50S site, carrying with it the second codon of the messenger. The vacant 30S site then binds another aminoacyl-tRNA complex according to the correct codon-anticodon pairing. At this instant, the total complex consists of a ribosome bound to mRNA, one molecule of aminoacyl-tRNA, and one molecule of peptidyl-tRNA (Fig. 11–9). The third amino acid, at the 30S site, is in a position

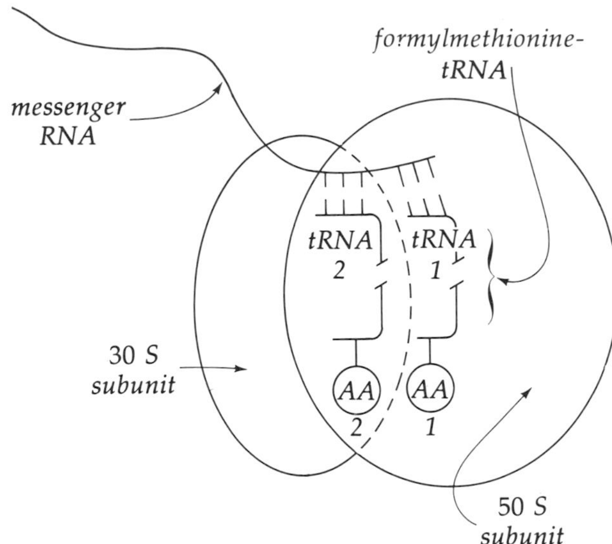

11–7. Second phase in the initiation of protein synthesis. The two aminoacyl-tRNA complexes are in a favorable position for the formation of the first peptide bond.

favorable for the formation of the second peptide bond. From this point on, the cycle repeats until all codons of the messenger have been "read" or translated.

Termination of Synthesis

The mechanism outlined above continues until the end of the messenger is reached. At this point, special conditions are required for the release of the finished protein from the ribosome. Completion of synthesis and release of the protein chain, collectively called *termination,* have been found to depend on specific codons in the messenger.

One line of evidence for the existence of terminator codons was derived from the study of protein synthesis in cell-free systems stimulated by synthetic mRNAs. If a synthetic messenger such as poly-U is added to a cell-free system,[1] the "message" is read by the ribo-

[1] Poly-U is an artificial RNA consisting of a repeating sequence of uridine nucleotides. Normally, as this sequence does not contain the initiation codons, poly-U and other synthetic polynucleotides initiate protein synthesis very slowly. However, if the concentration of Mg^{+2} ion in cell-free systems is raised from the standard 0.01M to 0.02 or 0.03M, initiation can be induced without the presence of the initiation codons AUG or GUG at the 5′ end of the template RNA (see Nakamoto and Kolakofsky, 1966).

$$\overbrace{\textit{peptide chain}\text{—NH—}\underset{}{\overset{R'}{\overset{|}{\text{CH}}}}\text{—CO—tRNA}_1'}^{\textit{peptidyl-tRNA}} + \underbrace{\text{NH}_2\text{—}\underset{R''}{\underset{|}{\text{CH}}}\text{—CO}\sim\text{tRNA}_2}_{\textit{aminoacyl-tRNA}} \rightarrow$$

$$\underbrace{\textit{peptide chain}\text{—NH—}\overset{R'}{\overset{|}{\text{C}}}\text{—CO—NH—}\overset{R''}{\overset{|}{\text{CH}}}\text{—CO}\sim\text{tRNA}_2}_{\textit{peptidyl-tRNA}} + \text{tRNA}_1$$

11–8. Formation of the peptide bond. The reaction is catalyzed by an enzymatic mechanism located in the 50S ribosomal subunit.

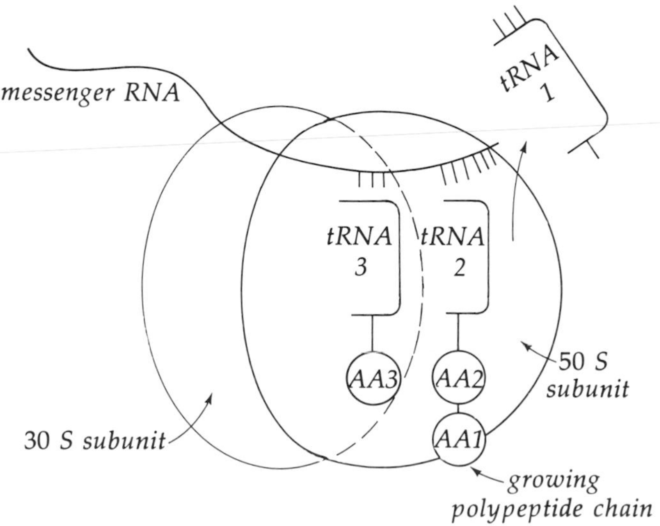

11–9. The ribosomal complex during synthesis of a protein (see text). The growing peptide chain is attached to the 50S subunit.

somes until the last codon is reached. At this point, assembly of amino acids into the polypeptide ceases without release of the chain from the ribosome. This indicates that arrival at the 3′ end of the messenger is not enough in itself to cause termination of synthesis and release of the new polypeptide chain from the ribosome and suggests that natural mRNAs probably contain a special termination sequence at the 3′ end (see Gilbert, 1963).

Further experimentation, primarily by M. S. Bretscher and his colleagues has shown that this is actually the case. Bretscher found that artificial RNA molecules containing adenine and uracil (poly-AU) in a random sequence promoted release of polypeptides of different length instead of stalling synthesis at the 3′ end of the messenger. Increasing the proportion of adenine to uracil to a ratio of 2:1 led to a maximum rate of release. This suggested that a termination codon consisting of two adenines and one uracil in an undetermined sequence exists. Similar experiments revealed that a combination of one adenine, one uracil, and one guanine also promotes the termination of synthesis with the release of the new polypeptide. These conclusions were supported by the work of M. W. Nirenberg and his associates in which synthetic trinucleotides were tested for their capacity to bind specific tRNAs. This work revealed that three codons, containing the nucleotides indicated as terminators by the artificial RNA polymers, do not bind tRNA. These triplets, called "nonsense" or *terminator* codons, are the sequences UAA, UAG, and UGA.

The nonsense codons evidently promote the termination of synthesis when inserted at any point in mRNA. The precise mechanism for termination is unknown. For a time, it was proposed that specific "terminator tRNAs" pair with the termination codons at the 30S site and cause release of polypeptide on reaching the 50S site. The search for terminator tRNAs has been unsuccessful; current hypotheses hold that as the termination codon reaches the 30S site, no tRNA is attracted from the surrounding medium. The vacant codon in some manner translocates to the 50S site and causes release of the polypeptide chain. A proteinaceous releasing factor, separable into two fractions designated R_1 and R_2, is necessary for the release of the completed peptide chain. These factors evidently catalyze the hydrolysis of the bond linking the finished polypeptide to the final tRNA. Little is known about the releasing factors except that R_1 is believed to catalyze release in the presence of the terminating codons

UAA or UAG and R_2 in the presence of the codons UAA or UGA (see Capecchi, 1967, and Scolnick et al., 1968).

After release of the polypeptide chain from the ribosome, enzymes in the cytoplasmic ground substance may remove either the formyl group from the initial methionine or the entire formylmethionine residue (see Capecchi, 1966). Through this mechanism, proteins are produced that have other amino acids at the "front" end of the polypeptide chain. The initial methionine is removed from approximately 40 percent of the proteins in the bacterium *Escherichia coli* by this mechanism.

Folding of the polypeptide chain into the finished 3-dimensional configuration may begin during synthesis on the ribosome or may await the completion of synthesis and be accomplished elsewhere. Some degree of folding probably takes place as the chain is synthesized, because polypeptide chains spontaneously assume 3-dimensional configurations in solution depending on factors such as the *p*H, temperature, and ionic character of the surrounding medium. Folding, involving the formation of covalent bonds (such as S—S linkages within the structure), probably requires the intervention of an enzymatic mechanism. The insertion of prosthetic groups, such as the polyporphyrin group of hemoglobin or cytochrome *c*, or lipid or polysaccharide groups, may occur either during or after folding of the protein. These groups may be linked electrostatically or covalently to the protein. In general, very little is known about the mechanism or locale of these interactions, except that in some cases polysaccharide or lipid groups are known to be attached within the membranous vesicles of the Golgi complex.

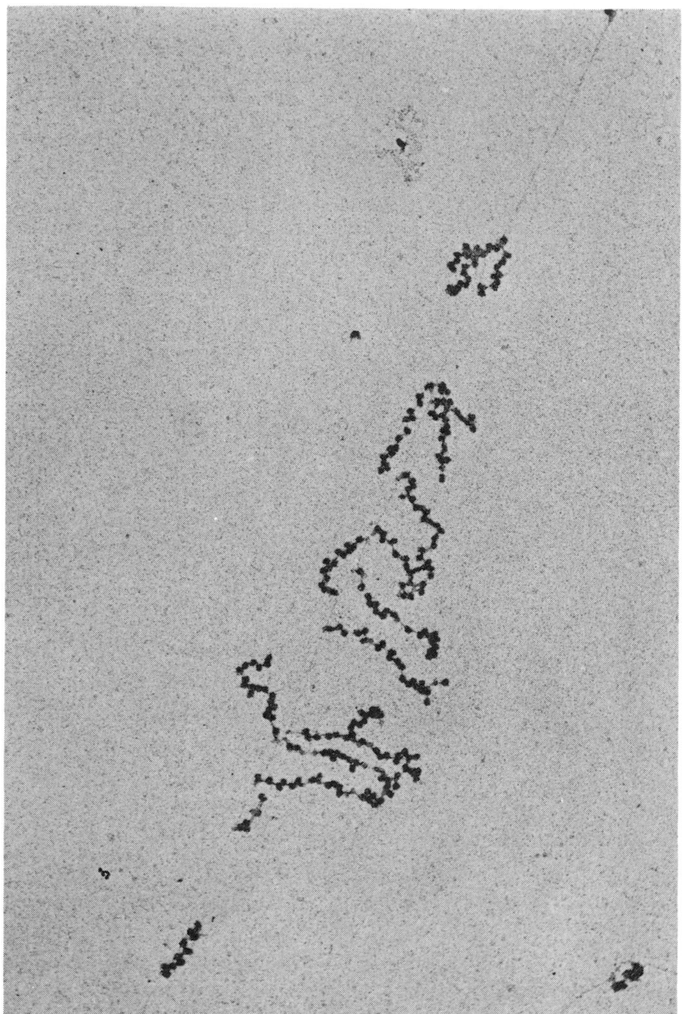

11–10. Transcription and translation in progress simultaneously in E. coli *(see text).* × *46,000. Courtesy of O. L. Miller, Jr., Oak Ridge National Laboratory, Barbara A. Hamkalo and C. A. Thomas, Jr., Harvard Medical School.*

Morphological Demonstration of the mRNA-Ribosome Complex

Oscar L. Miller has obtained remarkable electron micrographs of the total system operating in bacteria, from transcription of the messenger to protein synthesis (Fig. 11–10). To understand this figure, remember that in bacteria the chromosome is not separated from the cytoplasm by a nuclear envelope. Newly synthesized mRNA therefore immediately enters the cytoplasmic environment, and attachment to ribosomes may occur even before transcription of the messenger is complete. Miller interprets the thin strand running from top to bottom in Fig. 11–10 as the bacterial DNA. Attached to this at intervals are side branches containing rows of spherical particles. These are probably individual molecules of mRNA already attached to ribosomes. Where the side branches join to the DNA strand, small, dense particles can be seen. Miller proposes that these represent molecules of RNA polymerase engaged in transcribing the gene. Because the messenger-ribosome complexes become longer toward the bottom of the picture, the direction of transcription must be from top to bottom, and the initiation codons therefore lie at the top of the picture, just above the shortest side branches.

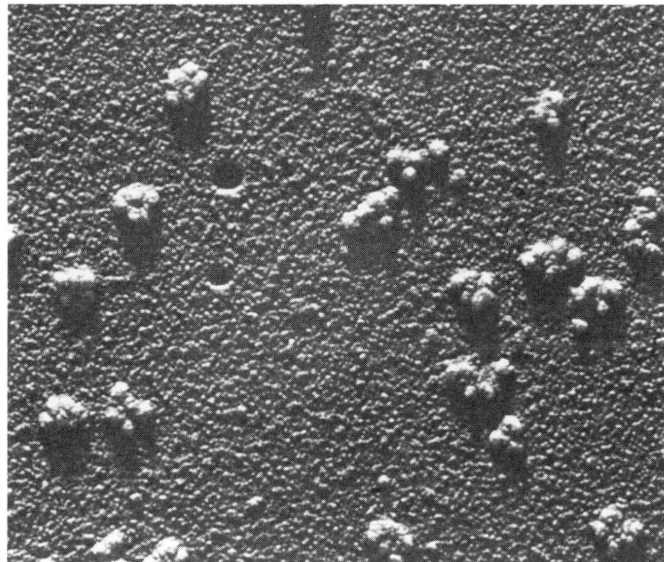

a

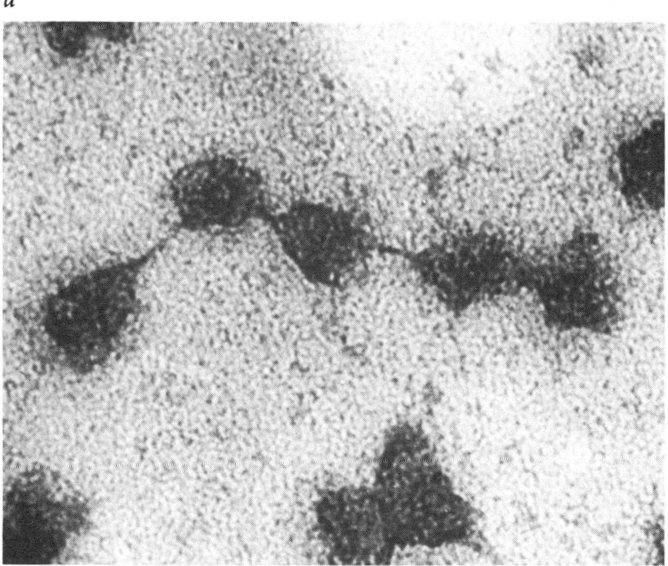

b

11–11. *Polysomes from rabbit reticulocytes.* a, *Shadowed preparation.* × *68,000.* b, *Preparation positively stained with uranyl acetate.* × *475,000. From* Science *138:1399 (1962). Copyright 1962 by the American Association for the Advancement of Science. Courtesy of J. R. Warner, A. Rich, and C. E. Hall.*

The messenger RNA molecules, with multiple numbers of attached ribosomes, are called *polysomes.* Depending on the length of the messenger, polysomes may include strings of from 5 to 7 to as many as 100 ribosomes (Fig. 11–11).

A Brief Summary of the Events in Protein Synthesis

Synthesis begins with the binding of mRNA to the ribosome, stimulated by an initiation factor, a protein called F_3. Two other proteinaceous factors, F_1 and F_2, stimulate the binding of formylmethionine-tRNA to the 30S site. GTP is required for this binding, along with an initiation codon, AUG or GUG, at the 5′ end of the messenger. The formylmethionine-tRNA then moves to the 50S site, carrying the messenger with it. These events are stimulated by the presence of GTP. A second aminoacyl-tRNA is attracted and bound to the 30S site, and a peptide bond is formed between formylmethionine and the second amino acid. This reaction is catalyzed by enzymatic activity associated with the 50S subunit and does not require the presence of GTP. However, during the movement of aminoacyl-tRNA to the 50S site and translocation of the messenger, one molecule of GTP is hydrolyzed to GDP. As the peptide bond is formed, the tRNA at the 50S site is released, and the tRNA at the 30S site, now linked to the polypeptide chain, moves to the 50S site. As this movement is accomplished, the messenger is translocated, exposing the next codon at the 30S site. The cycle then repeats until a termination codon (UAA, UAG, or UGA) is encountered. Release of the polypeptide from the ribosome requires in addition the presence of the releasing factors R_1 and R_2.

It must be emphasized that the details of protein synthesis outlined in this section have been worked out almost exclusively in bacterial systems. But from the evidence currently available, it seems very likely that protein synthesis proceeds by a very similar system in both prokaryotes and eukaryotes.

The Genetic Code

The solution of the nucleic acid code for protein synthesis has been a matter of central interest among investigators in all branches of the sciences. Basically, the problem to be solved was how codewords for twenty amino acids could be spelled out from an alphabet consisting of the four nitrogenous bases in DNA or RNA. As early as 1952, it was realized by A. Dounce and others that these bases would have to be taken in combinations of three or more to provide the capacity to code for 20 amino acids. If taken one at a time, only four different codewords could be constructed from the four nitrogenous bases in DNA or RNA ($4^1 = 4$); if taken two at a time to form two-letter codewords, only 16 amino acids could be specified ($4^2 = 16$). But if taken by threes to form three-letter codewords, 64 combinations are possible. Therefore it seemed likely that the codewords consisted of at least three nucleic acid "letters."

These considerations were taken up in greater detail in 1954 by G. Gamov, who also held that the nucleic acid code was likely to be constructed from three-letter words. Gamov pointed out additional problems that require solution if a linear code based on three-letter words exists. A code of this type could include *overlapping* or *nonoverlapping* codewords. In an overlapping code, a letter in a given position may serve as a part of two different codewords partly superimposed over each other. Thus the message ABCDEFGHI could be read ABC, BCD, CDE, DEF . . . and so forth. In a nonoverlapping code, the codewords are completely separate; a letter in any position in the message is a part of only one codeword. The same message would then be read only as ABC, DEF, GHI. But if the code is nonoverlapping, the problem of *punctuating* the code must be solved. For example, in written English, the *space* is used as an indicator of punctuation between words. One possibility for punctuation of the code is that a mechanism exists to establish the correct "reading frame" at the beginning of the coded message; once the frame is established, the message is simply read by threes until the end of the message is reached. This is called a *sequential* solution to the punctuation problem. Gamov also considered the problem of *degeneracy* or synonyms in the code. With 64 codewords available in a nucleic acid code constructed from three-letter words, 44 codewords in excess of requirements would exist. In a nondegenerate code, each amino acid would be specified by only one codeword and the remaining 44 would be "nonsense" codons. In a degenerate code, each amino acid would be specified by multiple, synonymous codewords. Gamov held that the code is probably completely degenerate and that all of the 64 codewords are assigned to amino acids. F. H. C. Crick proposed an alternate solution in 1957, in which the code was considered as nondegenerate.

These problems, including the size of the codon, degree of overlap, punctuation, and degeneracy of the code occupied the attention of investigators over the period from about 1960 to 1967. By 1967, most of the general features of the code had been solved and established by experimentation.

The size of the codeword was established experimentally by several different approaches using both genetic crosses and protein synthesis in cell-free systems. Crick and his colleagues studied a bacterial virus in which mutations can be induced by the dye *acridine orange*. This chemical causes single additions or deletions of nucleotides in the DNA of the virus. Crick reasoned that if the code is read sequentially, a single addition or deletion of a nucleotide pair in the DNA of a gene would cause a shift out of the correct reading frame. This can be explained as follows, under the assumption that three-letter codewords are used. Consider the coded message ABCABCABCABC. An *addition* at position 5 in the code would change the message to read ABCADBCABCABC. In this case the message would be read as ABC-ADB-CAB-CAB-C instead of ABC-ABC-ABC-ABC. The effect of the addition (a deletion would have a similar effect) is to disturb the reading frame of the code from the place of insertion to the end of the message. *Two* additions or deletions would also have the same effect. *Three* additions or deletions, if close enough together, might not significantly affect the message, because after the third addition or deletion the correct reading frame would be reestablished. Crick and his co-workers were able to identify single, double, and triple addition or deletion mutations in a particular gene of the bacterial virus and were able to correlate these mutations with the activity of the gene. As predicted, single and double deletions or additions usually destroyed the activity of the gene; triple mutations often resulted in restoration of activity. This established that the codewords are triplets or are at least based on multiples of three.

Another series of experiments with cell-free systems, carried out by H. G. Khorana and his co-workers, complemented Crick's conclusions precisely. Khorana was able to construct artificial RNA polymers with different repeating sequences. Repeating polymers of the form ABABABABABAB... coded only for the synthesis of a repeating *dipeptide* in cell-free systems. Assuming three-letter codewords, this is the expected result, because the code would be read $\overset{1}{\text{ABA}}$-$\overset{2}{\text{BAB}}$-$\overset{1}{\text{ABA}}$-$\overset{2}{\text{BAB}}$...; only two different "words" are contained in the message. Repeating *trinucleotides* of the form ABCABCABCABC coded for the synthesis of *homopolymers* of amino acids; that is, polypeptides containing a sequence of only one amino acid. This is also to be expected, as this message can only be read ABC-ABC-ABC-ABC.... Finally, repeating *tetranucleotide* poly-

second base of codon

first base of codon	U		C		A		G		*third base of codon*
U	UUU	Phe	UCU	Ser	UAU	Tyr	UGU	Cys	U
	UUC	Phe	UCC	Ser	UAC	Tyr	UGC	Cys	C
	UUA	Leu	UCA	Ser	UAA		UGA		A
	UUG	Leu	UCG	Ser	UAG		UGG	Try	G
C	CUU	Leu	CCU	Pro	CAU	His	CGU	Arg	U
	CUC	Leu	CCC	Pro	CAC	His	CGC	Arg	C
	CUA	Leu	CCA	Pro	CAA	Gln	CGA	Arg	A
	CUG	Leu	CCG	Pro	CAG	Gln	CGG	Arg	G
A	AUU	Ileu	ACU	Thr	AAU	Asn	AGU	Ser	U
	AUC	Ileu	ACC	Thr	AAC	Asn	AGC	Ser	C
	AUA	Ileu	ACA	Thr	AAA	Lys	AGA	Arg	A
	AUG	Met	ACG	Thr	AAG	Lys	AGG	Arg	G
G	GUU	Val	GCU	Ala	GAU	Asp	GGU	Gly	U
	GUC	Val	GCC	Ala	GAC	Asp	GGC	Gly	C
	GUA	Val	GCA	Ala	GAA	Glu	GGA	Gly	A
	GUG	Val	GCG	Ala	GAG	Glu	GGG	Gly	G

11–12. The RNA code for the amino acids. To convert to the DNA code, let U = A, G = C, C = G, and A = T. For example, the RNA codeword GAU is equivalent and complementary to the DNA codeword CTA. The codons UAG, UAA, and UGA (terminator codons) are not assigned to amino acids.

mers coded for *tetrapeptides,* chains of polypeptides containing sequences of four different amino acids. This is to be expected only if the message ABCDABCD-ABCDABCDABCDABCD is read in triplets as $\overset{1}{\text{ABC}}$-$\overset{2}{\text{DAB}}$-$\overset{3}{\text{CDA}}$-$\overset{4}{\text{BCD}}$-$\overset{1}{\text{ABC}}$-$\overset{2}{\text{DAB}}$-$\overset{3}{\text{CDA}}$-$\overset{4}{\text{BCD}}$. Khorana's results, when taken with Crick's findings, are possible only if the codewords consist of three letters. This work also establishes that the code is nonoverlapping. For example, the repeating trinucleotide ABCABCABC... would code for a polypeptide containing more than one amino acid if the codewords overlap.

One final series of experiments completed the determination of the triplet as the codeword. These experiments, carried out by Marshall W. Nirenberg and his colleagues, also allowed assignment of most of the amino acids to specific codons as well. Nirenberg synthesized all of the 64 possible trinucleotides and tested their binding one at a time to the 20 different aminoacyl-tRNAs. This work, in combination with the results of earlier experiments with artificial messengers such as poly-U and poly-AU, allowed unambiguous assignment of 61 of the 64 possible codons (Fig. 11–12)[2]. Three codons, UAG, UAA, and UGA, were found to have no coding assignments. The code is *universal* as far as has been determined; all organisms, from bacteria to man, share the same codewords and synonyms for the 20 amino acids. Nirenberg and Khorana received the Nobel Prize for their work with the genetic code in 1968.

Inspection of the code reveals a number of interesting features. The assignment of codons confirms that the code is highly degenerate; there are several synonymous codons for each amino acid. There is definite order in the degeneracy. With a few exceptions, each amino acid has either two or four synonymous codons (the exceptions are methionine and tryptophan, with one codon; isoleucine, with three; and serine and arginine, with six). Within the group of synonyms for each codeword, an additional pattern of order is noted.

[2]By convention the codons are designated in RNA rather than DNA equivalents. The RNA codon UUU for phenylalanine, for example, corresponds to the DNA codon AAA.

All amino acids with the codeword XYU (where X and Y are two letters of the code, and U is uridine) also have a synonym with the spelling XYC. That is, two synonyms exist that differ only in the third letter; this letter may only be U or C. A similar rule holds for almost all codewords of the form XYG; almost all amino acids with the codeword assignment XYG also have a synonym with the spelling XYA.

Table 11–1 Anticodon-Codon Pairing According to the Wobble Hypothesis

Anticodon	*Codon*
U (or ψ) pairs with	A or G
C pairs with	G
A pairs with	U
G pairs with	U or C
I pairs with	U, C, or A

From F. H. C. Crick, Codon-anticodon pairing: the wobble hypothesis. *J. Mol. Biol.* 19:548(1966).

Degeneracy and the Wobble Hypothesis

F. H. C. Crick has proposed that the patterns detectable in the degeneracy of the nucleic acid code are related to peculiarities in the codon-anticodon pairing between mRNA and tRNA. These pairing irregularities involve the third letter of the codeword, in which pairs other than the usual A-U and G-C combinations are formed. Crick proposed that the unusual pairing at the third position results from the freedom of this base to "wobble" in position in tRNA so that alternate hydrogen-bonding patterns are possible. These alternative positions, according to Crick, allow the unusual base pairs given in Table 11–1 to form at position three in the codon-anticodon pair. Note in the table that the base inosine in tRNA can pair with U, C, or A on the messenger. The rare base pseudouridine (ψ), which also occurs in tRNA, forms the same base pairs as U.

Crick's predictions are borne out by tests with synthetic trinucleotides, which have established that a single RNA can pair with a multiple number of codons. These codons vary in letter assignment at the third position according to the pattern predicted by Table 11–1. Thus, the degeneracy in the code may have its origin in the flexibility of codon-anticodon pairing at the third position.

The Regulation of Protein Synthesis

In almost all multicellular organisms, at least some degree of specialization of individual cells takes place. This occurs even though all somatic cells of an organism, barring mutation, contain the same genetic message. This specialization, observed as differences in the structure and function of the various cells of an organism, ultimately reflects differences in the numbers and kinds of proteins synthesized in the cytoplasm. Therefore at some point, protein synthesis is controlled or *regulated* to increase the level of some proteins and to inhibit the production of others.

Regulation may be imposed either by inhibition of synthesis of the messenger (called *transcriptional* control) or during protein synthesis in the cytoplasm (*translational* control). There is some indication that both transcriptional and translational controls operate to regulate protein synthesis in the cell. These control mechanisms and some of the evidence and speculation concerning them are discussed in turn in the following sections.

Transcriptional Control of Protein Synthesis

Transcriptional control in prokaryotes: The operon hypothesis. Particularly good evidence has been obtained for transcriptional control in bacteria. This evidence has provided the basis for the formulation of an important hypothetical mechanism for transcriptional control called the *operon hypothesis.* This hypothesis, originated by F. Jacob and J. Monod, was developed from the results of experiments carried out with the bacterium *Escherichia coli.* In this bacterium, several closely related genes control the metabolism of a group of sugars called *β-galactosides.* These genes are usually collectively called the *lactose system,* because lactose, a β-galactoside, is a major substrate for the enzymatic mechanisms coded for by the system. One enzyme, *galactose permease,* catalyzes the penetration of lactose into the bacterial cell; a second, *β-galactosidase,* hydrolyzes lactose into glucose and galactose, and a third, *galactose transacetylase,* catalyzes the attachment of an acetyl group to an intermediate in the lactose pathway. If no lactose is present in the medium surrounding wild-type *E. coli,* only a few molecules of these enzymes are found within the bacterial cytoplasm. If lactose is added to the medium, synthesis of the three enzymes begins very quickly, ultimately reaching levels of several thousand molecules per cell. Because the concentration of enzymes within the bacterium can be raised by addition of lactose to the medium outside the cell, the lactose enzymes are said to be *inducible.*

Genetic crosses revealed that the genes coding for the three enzymes of the lactose system are adjacent to one another on the *E. coli* chromosome, in the order

β-galactosidase	galactose permease	galactose transacetylase

Other genes were discovered to affect the transcription of these loci without coding for an enzyme taking part in lactose metabolism. One of these is called the i gene. A mutation destroying the activity of the i gene results in a maximum synthesis rate of the lactose enzymes whether an inducer substance is present in the medium or not. This indicates that the i gene codes for an agent that normally keeps the lactose system genes "turned off." This gene, by genetic crosses, has been located within the *E. coli* chromosomes at a point separated from the lactose system genes by several intervening, unrelated loci. Therefore the effect extended by the i gene on the lactose system must take place through an agent that diffuses through the intervening distance in the bacterial cytoplasm.

A second region of the chromosome, called the O locus, also affects the rate of synthesis of the lactose system enzymes. Some mutations in this region, which was found to be adjacent to the lactose system genes, also lead to a maximum synthesis rate of the three lactose system enzymes whether or not an inducer is present in the medium. A most significant effect of the O locus is observed in bacterial cells that are *diploids* over the part of the chromosome containing this region.[3] If the lactose system with the adjacent O gene is present in the diploid state, a mutation O^c (constitutive) causes a maximum rate of activity only in the lactose system genes *lying on the same chromosome.*

This has been determined by taking advantage of mutations of the β-galactosidase gene. Two mutations of this gene, designated z_1 and z_2, code for alternate forms of the galactosidase enzyme that can be separately identified by biochemical tests. By placing these markers on the two chromosomes of a partial diploid, it can be determined which lactose system remains inducible. Thus in the diploid

$$\begin{array}{cccc} O^+ & z_1 & y & a \\ \hline \hline O^c & z_2 & y & a \end{array} \qquad (11\text{–}3)$$

where O^+ is the normal O gene and O^c the mutant, z_1 and z_2 are the alternate forms of β-galactosidase, y is galactose permease and a is galactose transacetylase, the z_1 mutant remains inducible, and the z_2 locus is transcribed whether the inducer is present or not. This experiment shows that the O locus controls the transcription of the lactose genes lying adjacent to it on the same chromosome.

These observations of activity of the i and O loci were synthesized by Jacob and Monod into a brilliant hypothesis for the regulation of transcription in bacteria (see Fig. 11–13). They proposed that without the i gene and the O region the lactose system would be continuously active. The three lactose system genes are controlled as a unit (called the *operon*) by the gene i, termed the *regulator* gene. The regulator, continuously active, codes for the synthesis of a substance called the *repressor.* If no inducer is present, the repressor diffuses through the bacterial cell and binds to the O region, called the *operator.* As long as the repressor is bound to the operator, the genes z, y, and a of the lactose system are not transcribed.

The effect of the inducer substance is to combine with and inactivate the repressor. Jacob and Monod proposed that the repressor contains two sites, one specific for binding to the operator and one for the inducer. If no inducer substance is present the site for binding to the operator is undisturbed, and the repressor binds to the operator. If inducer is present, the repressor combines with the inducer; this combination causes a rearrangement in the repressor resulting in a loss of affinity for the operator. The genes of the operon are then transcribed continuously.

The same hypothesis also explains *inhibition* of enzyme synthesis by the presence of a substance in the

[3] Bacteria are ordinarily haploid; however, it is possible by several methods to include fragments of a second chromosome within bacterial cells. These cells are called partial diploids.

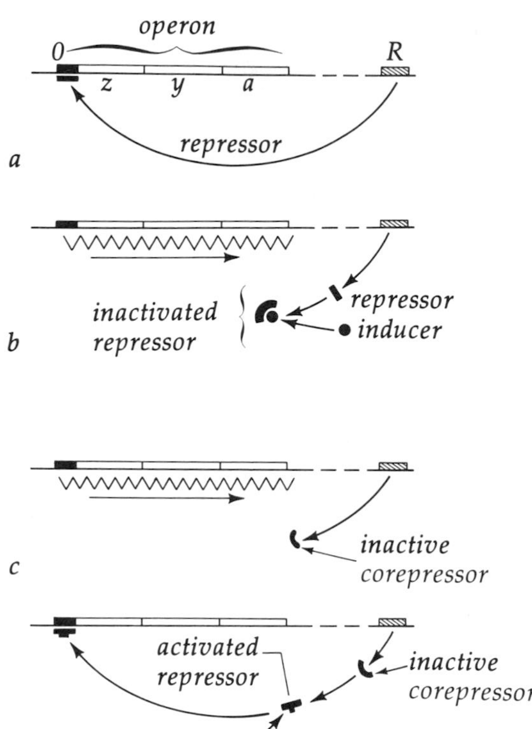

11–13. *Diagram outlining the operon hypothesis of Jacob and Monod. In induction* (a *and* b), *an* active *repressor* (a), *continually synthesized, combines with the operator and inhibits transcription of the operon. The repressor is* inactivated *by combination with the inducer* (b), *releasing inhibition of the operon. In repression* (c *and* d) *an* inactive *repressor (c), with no affinity for the operator, is continually synthesized. This repressor is activated* (d) *by combination with the end product of a biochemical pathway and binds to the operator, turning "off" the operon (see text). Redrawn, original courtesy of W. Hayes and John Wiley & Sons, Inc. from* The Genetics of Bacteria and Their Viruses.

medium, as, for example, in the repression of an enzymatic pathway for the biosynthesis of an amino acid that is available in the environment. In this case, the regulator gene is considered to produce an *inactive* rather than active repressor. The inactive repressor has two binding sites as does the active repressor, one specific for the operator gene and one for the inhibiting substance (termed the *corepressor*). The binding site for the operator is inactive or masked in the inactive repressor. The repressor does not combine with the operator, and the transcription of the genes adjacent to the operator is continuous. If the repressor combines with the inhibiting substance, a rearrangement of the molecule results that *activates* the operator binding site of the repressor. The activated repressor then combines with the operator, and transcription of the genes controlled by the operator is inhibited. These alternate forms of the basic hypothesis can be simply summarized as follows:

Induction: active repressor + inducer
→ inactive repressor: operon continuously transcribed

Repression: inactive repressor + corepressor
→ active repressor: operon not transcribed

Direct experimental support for the operon hypothesis, for which Jacob and Monod received the Nobel Prize in 1965, has been obtained through the isolation and characterization of the repressor. In 1966, W. Gilbert and B. Müller-Hill isolated and purified a protein coded for by the regulator (*i*) gene in *E. coli.* This repressor substance, a protein with a molecular weight of 150,000, combines with substrates that induce the synthesis of the enzymes of the lactose operon. Complexing of the regulator protein with the inducer destroys its affinity for the operator gene. When not bound to the inducing substance, this protein has strong affinity for and binds to DNA containing the normal operator gene for the lactose operon. Thus the activity of the protein isolated by Gilbert and Müller-Hill coincides precisely with the expected reaction of the hypothetical repressor formulated by Jacob and Monod.

Transcriptional controls in eukaryotes. The applicability of the operon concept to higher organisms is somewhat uncertain, because clustered genes of the type found in the lactose system of bacteria are rare in eukaryotes. However, this system could serve to control individual genes in the eukaryote chromosome if each individual locus is supplied with an operator.

The regulator-operator mechanism has the characteristic of sensitivity to rapid changes in the environment and allows rapid compensatory adjustments in the synthetic machinery of the cell in response to these changes. These adjustments are short term by nature and do not result in permanent alterations in the collection of genes that are transcribed. Short-term responses of this type undoubtedly occur in eukaryotes, and, as noted, these responses may be moderated by an operon mechanism. But, in addition, the cells of eukaryotes have the capacity for long-term, more or less permanent regulation of the activity of genetic loci. These long-term changes are a part of the specialization that is a characteristic feature of multicellular organisms.

Although many hypotheses have been advanced, the basis for long-term regulation of genetic activity in eukaryotic cells is unknown. This capacity for permanent differentiation is accompanied by greater complexity of the nucleus and chromosomes in eukaryotes, and most of the hypotheses proposed to account for long-term changes in the synthetic capacities of cells are based on one or another of these special features. Thus, in various proposals the histones, nonhistone proteins, or the nuclear envelope are considered as basic to transcriptional regulation in eukaryotes.

The histones and transcriptional control. The idea that histones regulate the activity of DNA is a relatively old one in molecular biology, dating back to at least 1945, when Edgar and Ellen Stedman proposed that histones inhibit and control the activity of genes. The Stedmans also incorrectly hypothesized that nuclei from different tissues contained different histones. But more recent work, particularly in the laboratory of James Bonner, has indicated that the basic presumption that the histones regulate genetic activity may be correct. Bonner and his co-workers compared the rate of RNA synthesis in cell-free systems, using DNA and nucleoprotein isolated from plants as a template. Naked DNA, in Bonner's system, was about five times as active in RNA synthesis as DNA complexed with histone. Addition of histone to the naked DNA greatly reduced its template capacity.

Bonner also attempted to study the effect of histones on the activity of a specific locus. Template chromatin was isolated from seeds (cotyledons) and from apical buds. Synthesis of a plant protein, a globulin typical of seeds, was measured when the isolated seed and apical bud chromatin were added to a cell-free system. When chromatin from the seed was used, Bonner was able to identify plant globulin by immunological methods in the protein synthesized by the cell-free system. The amount of globulin was greatly reduced if apical bud deoxyribonucleoprotein was used as template. However, removal of about half of the histones from apical bud deoxyribonucleoprotein led to a large increase in the amount of globulin synthesized. Bonner reasoned from this that the histones of apical bud deoxyribonucleoprotein repress stretches of DNA coding for globulin.

Bonner's work makes it appear that histones indeed regulate and control the activity of the genes. But some features of the molecular structure and properties of the histones make it difficult to understand how these proteins could act to regulate the transcription of genes. Precision in the control of transcription would require histones specific in binding properties for each gene. However, only a small number of histones have been identified. Even the most optimistic estimates place the total number at less than twenty. No differences are noted between the histones of different tissues of the same organism; in fact, relatively few differences show up in histones from different species. The amino acid sequence of one histone fraction (fraction IV) has recently been determined. This fraction, when isolated from two species as widely separated as the pea and calf thymus, shows only minor differences in amino acid sequence. Furthermore, no differences can be detected between the histones of "active" and "inactive" chromatin (see p. 269). In contrast to the repressors in bacterial systems, the histones are notably nonspecific in interaction and will combine readily with DNA from any source or even from totally unrelated molecular systems. For example, glycolysis or oxidative phosphorylation are also inhibited by the addition of histones.

More recently, Bonner and his co-workers have found an RNA fraction apparently covalently bound to chromosomal proteins and have proposed that the specificity required for transcriptional control is imparted to the histones by this bound RNA. In these experiments, nucleoprotein from several plant and animal species was dissociated into DNA and histone fractions by adjustments in the ionic strength of the medium. If the nucleoprotein was reconstituted immediately without further treatment, the RNA synthesized was similar to RNA coded for by freshly isolated, nondissociated deoxyribonucleoprotein. If, however, the dissociated nucleoprotein was treated with ribonuclease before reconstitution, the specificity of synthesis was found to be lost. Bonner claims from this that the RNA fraction, a low molecular weight RNA consisting of short chains of about 40 nucleotides, provides the required specificity to the histones.

These experiments have met with some criticism from workers outside the Bonner laboratory. It has been pointed out that because RNA has been observed to bind readily to many unrelated molecular species from disrupted cells, the apparent RNA-histone complexes may be an artifact. Histones bind to DNA through an electrostatic attraction between the acidic phosphate groups of DNA and the basic groups of the protein. The binding between RNA and specific DNA sequences would be expected to involve the formation of hydrogen bonds. It is not understood how any specificity imparted by RNA could counteract or modify the relatively strong and nonspecific electrostatic attraction between histones and the phosphate groups on DNA.

The various conceptual and biochemical problems surrounding the specificity of binding to DNA has left the possible functional role of the histones an open question. The histones may be purely structural, and transcriptional control may be accomplished by nonhistone proteins (see below). Whatever their function, the fact that histones from widely separated organisms are closely similar in amino acid sequence indicates that the role played by these molecules in the eukaryote chromosome is critical and that even small mutations with limited structural disturbance are lethal.

The nonhistone proteins and transcriptional control. The possible function of nonhistone proteins in the control of transcription is supported by many investigators. The bacterial repressor isolated by Gilbert and Müller-Hill is a nonbasic and therefore nonhistone protein. Conceivably, similar molecules could bind specifically to the genes of eukaryote chromosomes to turn off RNA transcription. In contrast to the histones, nonhistone proteins are satisfyingly different from tissue to tissue and between different species. But, as yet, there is little evidence that nonhistone proteins are actually part of the chromosomes; chances are good that at least a part of this protein fraction is a contaminant from the extrachromosomal nucleoplasm or cytoplasm. The addition of nonhistone proteins to DNA template in cell free systems does not appear to inhibit RNA synthesis. Thus the role of nonhistone proteins in the regulation of transcription, while a possibility, is without much direct experimental support.

Morphological Evidence of Transcriptional Control in Eukaryotes

The puffing patterns observed in the polytene chromosomes of diptera (see Chapter 10) provide morphological evidence of transcriptional control in eukaryotes. The puffs, which are sites of RNA synthesis on the chromosome, are localized and probably reflect the activity of single bands on the chromosome. Moreover, some puffs are tissue specific and therefore reflect differential activity of the chromosomes in different tissues. A good example of this type is the conspicuous puff or Balbiani ring found by Wolfgang Beermann to be present in salivary gland cells that give off a granular secretion (see the description of Beermann's experiments in Chapter 10). Other puffs have been associated with mutations. A mutation in *Drosophila* giving rise to a highly vacuolated salivary gland is reflected in the appearance of a puff at the site mapped for the mutated gene; normal, wild-type *Drosophila* do not show a puff at this site (see Slizynski, 1963).

Long-term differentiation of cells is also marked by a timed appearance and disappearance of particular puffs. The best studied of these are changes in puffing patterns associated with molting in the larva. As the molting stage approaches, some puffs characteristic of the intermolt stage disappear and new puffs appear at various points on the polytene chromosome. Molting can be artificially induced by injections of the insect hormone ecdysone. Within 30 to 60 minutes after injection of ecdysone, two new puffs appear in the polytene chromosomes. These are also found in untreated larva that are approaching a molt. A further series of new puffs appears within 5 to 72 hours if the level of ecdysone is maintained. Injection of puromycin, an inhibitor of protein synthesis, prevents appearance of this second group of puffs, even though ecdysone is maintained at high levels. This suggests that either a direct or indirect product of the first puffs induces the second set of puffs to appear. This is reminiscent of the operon mechanism in bacteria, in which a protein coded for by a gene at one locus controls the activity of other parts of the genome. Histone can be detected both in active puffs and in inactive bands of polytene chromosomes. Thus, activity in the puffs is not correlated with a loss of histones, which might be expected if histones are repressors of transcription.

Heterochromatin and Euchromatin

The differential gene activity expressed morphologically as puffs in polytene chromosomes has a possible equivalent in more typical eukaryotic cells in the differentiation between coiled and extended chromatin in interphase nuclei. These differentiated areas are seen as patches of tightly condensed aggregations of 100 Å fibers interspersed among more loosely packed chromatin (see Fig. 3–21). Autoradiography has shown that radioactive precursors of RNA are incorporated at a much higher rate in the extended, diffuse areas of the chromatin. On this basis, the extended regions have been termed "active" chromatin and the condensed regions "inactive" chromatin.

Active and inactive chromatin correspond morphologically to differences in the coiling or state of aggregation of chromosomes known classically as euchromatin and heterochromatin. In classical terms, heterochromatin is defined as chromosome material that remains tightly coiled and condensed at times, such as during interphase, when most of the chromatin is uncoiled. Heterochromatinization can frequently be correlated with inhibition of genetic activity. In the mealybug, the entire set of paternal chromosomes in the male becomes heterochromatic and remains tightly coiled during interphase; the genes in this set of chromosomes are not expressed. A similar mechanism operates in mammalian females, including humans, in which one of the two X chromosomes becomes heterochromatic. Many, but not all, genes in the heterochromatic X chromosome are inactive. Because heterochromatinization is thus correlated with the long-term inactivity of whole blocks of genes, a number of authorities have proposed that regulation of genes in differentiation is accomplished by coiling or condensation of segments of the chromosomes.

Post-Transcriptional Control in Eukaryotes

Other control mechanisms, while not transcriptional, have been proposed as acting immediately after transcription. One hypothesis holds that all genes are continuously active and that in effect there is no transcriptional control. Instead, the various messengers are selectively degraded in the nucleus after synthesis, so that the RNA products of only some loci reach the cytoplasm. Some limited support for this hypothesis may be taken from the fact that much of the DNA-like RNA transcribed on the chromosomes is degraded in the nucleus and does not reach the cytoplasm (see Chapter 10). However, studies of polytene chromosomes indicate that RNA synthesis is limited to particular regions of the chromosomes; this synthesis, expressed morphologically as puffing, can be correlated with differential genetic activity. Thus, in cases in which genetic activity can be clearly related to sites of RNA synthesis on the chromosomes, it appears that transcription is localized in particular genes rather than generally distributed over the entire length of the chromosome.

Another possibility is that the flow of mRNA to the cytoplasm is selectively controlled by the nuclear envelope. According to this hypothesis, whether particular mRNAs reach the cytoplasm for protein synthesis is "decided" in the pore complexes of the envelope. While possible, this proposal remains completely speculative; no supporting evidence exists for activity of this type in any membranous systems of the cell.

Translational Control of Protein Synthesis

Translational control of protein synthesis is known to occur in the eggs and embryos of echinoderms and amphibians. In the sea urchin, for example, synthesis of protein is arrested after oogenesis; the egg remains almost completely quiescent until fertilization. Yet the cytoplasm is packed with large quantities of RNA. After fertilization, at least some of this RNA evidently becomes active as template for much of the protein synthesized by the embryo. The mechanism for the translational control exerted before fertilization in these eggs is unknown, but indications are good that the mRNA or the ribosomes or possibly both are masked in some way by temporary combination with a protein.

Another example of translational control is provided by the single-celled alga *Acetabularia.* In this genus, the nucleus is found in the rhizoid, a root-like structure at the base of the organism. A stalk with a cap extends from the rhizoid. The cap is of distinct morphology for the species. These cells can be enucleated by simply amputating the rhizoid. If the cap is subsequently removed, the enucleated stalk is capable of regenerating a new cap. Obviously, the information required for regenerating the cap is stored in the cytoplasm of the stalk. These enucleated, capless stalks may be stored for weeks under conditions unfavorable for cap formation without destruction of the capacity for synthesizing the cap. When the stalk is transferred into favorable conditions, a new cap is formed. In this case, some form

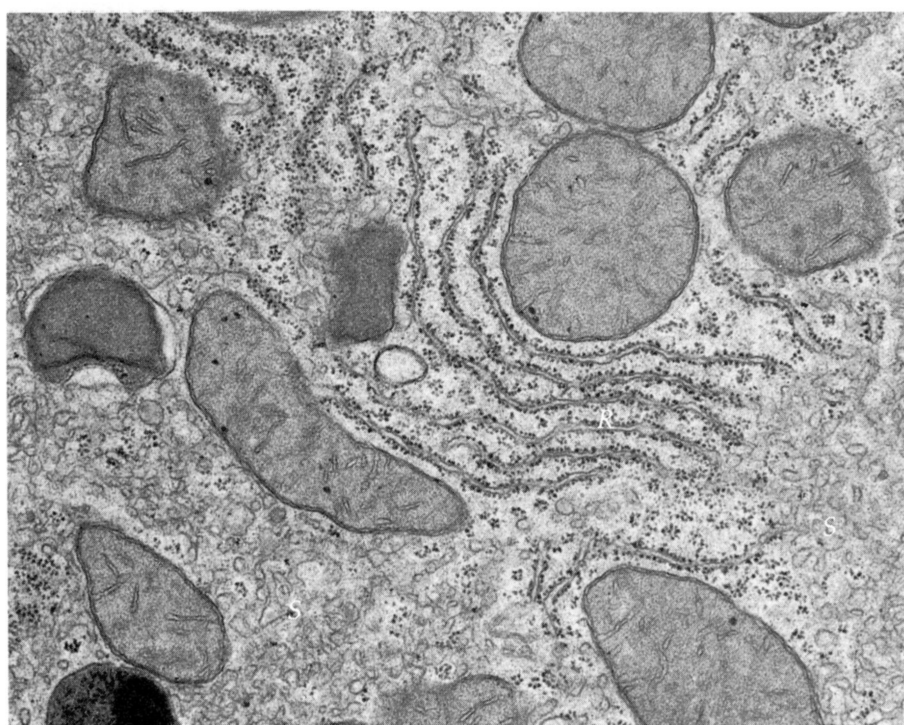

11–14. A liver cell from the hamster, showing both rough (R) *and smooth* (S) *ER.* × *26,000.*
From The Cell *by D. W. Fawcett, 1966. Courtesy of D. W. Fawcett and W. B. Saunders Company.*

of translational control that regulates the time at which the cap appears exists in the cytoplasm.

There are other possibilities for translational control, such as inhibition of synthesis by a product of the system (called "feedback inhibition"), regulation of the availability of ribosomes, and controlled destruction of the messenger by ribonuclease. Experimental support for these possible mechanisms is very limited at the present time. The molecular basis of both translational and transcriptional control and the long-term regulation of the activity of eukaryotic cells in general remains one of the most difficult and fundamental problems facing cell biologists.

Morphological Aspects of Protein Synthesis in the Cytoplasm

The Endoplasmic Reticulum

Ribosomes occur in the cytoplasm in large numbers, either freely suspended in the ground substance or tightly bound to a complex system of cytoplasmic membranes termed the *endoplasmic reticulum.* Another system of cytoplasmic membranes of similar structure but without ribosomes is given the same descriptive term. The two systems are usually distinguished by the designation *"rough" endoplasmic reticulum* (abbreviated as rough ER) for the membranes containing bound ribosomes, and *"smooth" endoplasmic reticulum* (smooth ER) for the membrane system without ribosomes. The same system of membranes was known to early light microscopists as the *ergastoplasm.*[4] In the late 1800's, the French microscopist C. Garnier noted that the ergastoplasm was intensely stained by basic dyes. Garnier also observed that the ergastoplasm of pancreatic cells disappeared in fasted animals and

[4]Originally the term "endoplasmic reticulum," coined by K. R. Porter, was synonymous with "ergastoplasm," which Garnier introduced in 1897. Subsequent electron microscope studies led to the realization that ergastoplasm or endoplasmic reticulum contained two distinct but frequently interconnected systems, the "rough" and "smooth" ER. Fritjof Sjöstrand has introduced the term *α-cytomembranes* for the rough endoplasmic reticulum, and *β-cytomembranes* for the smooth endoplasmic reticulum in order to provide a distinction between the two endoplasmic systems. However, these terms have not yet met with general acceptance.

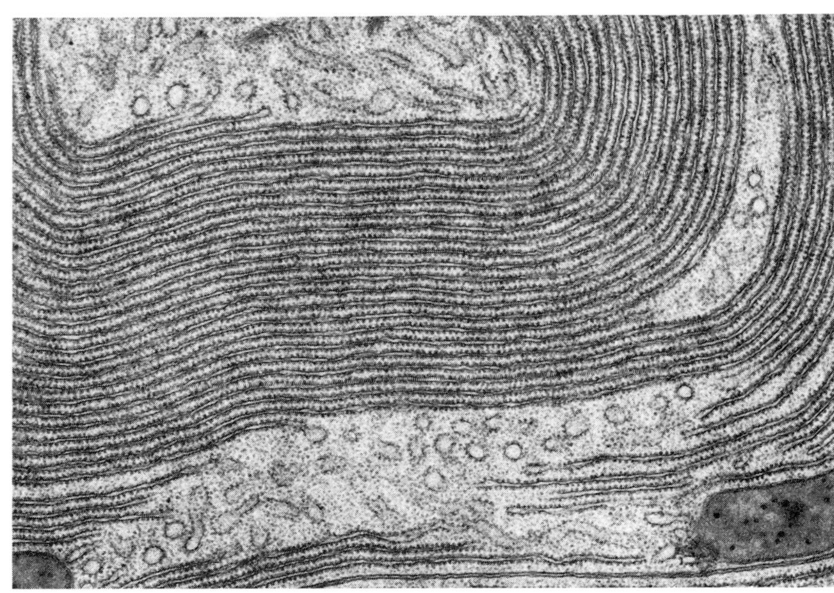

11–15. Closely spaced cisternae of the rough ER in a pancreatic cell of the bat. × *18,000. From* The Cell *by D. W. Fawcett, 1966. Courtesy of D. W. Fawcett and W. B. Saunders Company.*

reappeared after feeding. Garnier proposed from these observations that the ergastoplasm is directly involved in the secretion of digestive juice by the pancreas. Although much debate ensued, nearly a half century was to pass before additional information of any significance came to light. In 1939, T. O. Caspersson was able to show that the basophilia of the ergastoplasm was probably due to the presence of large quantities of RNA. Somewhat later, in 1950, independent studies by Caspersson and Jean Brachet showed direct connection of the ergastoplasm with protein synthesis.

Research with the electron microscope in the 1950's by Fritjof Sjöstrand, George E. Palade and Keith R. Porter revealed that the ergastoplasm consists of an extensive system of tubules, vesicles, and large, flattened vacuoles with walls made up of single membranes. Occasionally, connections could be seen between the ergastoplasm (by this time called the endoplasmic reticulum), the plasma membrane, and the nuclear envelope, giving rise to the concept that the endoplasmic reticulum formed a channel to transport running through the interior of the cell in communication with the cell exterior.

The rough endoplasmic reticulum is readily identified by the presence of large numbers of ribosomes bound to the surfaces of the component membranes (Figs. 11–14 and 11–15). The tubules, vesicles, and large flattened sacs of the rough ER may extend in a branching network through large areas of the cytoplasm. These vesicles vary in dimensions, from 250 Å to as much as 5000 Å in diameter, depending on the type, activity, or age of the cell. The largest flattened sacs (Fig. 11–15) are called *cisternae.* Sjöstrand discovered that the membranes forming the vesicles and cisternae of the rough ER are thinner than the plasma membrane, approximately 50–60 Å in diameter as opposed to 90–100 Å for the plasma membrane. These membranes are also less distinctly trilaminar than the plasma membrane and exhibit globular structure intermediate in complexity between plasma and mitochondrial membranes (see Chapter 4). The outside surfaces of the vesicles of the rough ER, on the side facing the cytoplasmic ground substance, are studded with ribosomes, often packed in regular arrays approximately 150 Å apart (Fig. 11–15). Free ribosomes are also distributed in the cytoplasm between the cisternae of the endoplasmic reticulum. In sections parallel to the surfaces of the reticulum membranes, the ribosomes are frequently seen to be lined up in rows that trace out whorls or spirals (Fig. 11–16). These patterns may represent the linkage of a group of ribosomes into polysomes by mRNA.

Cells with a highly developed rough ER often show in addition a single layer of ribosomes covering the cytoplasmic surface of the outer membrane of the nuclear envelope (Fig. 11–17). Additionally, connections can sometimes be seen between this membrane and the rough ER (arrow, Fig. 11–17). These connections have been interpreted to mean that the rough ER is derived from the outer nuclear membrane by a process of bleb-

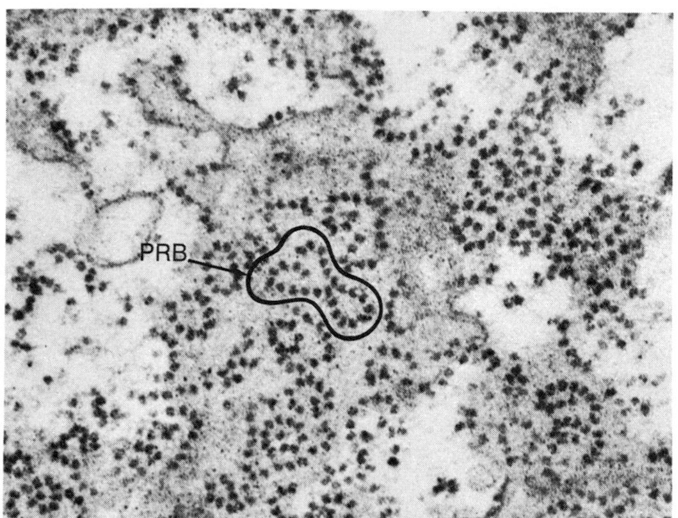

11–16. Distribution of ribosomes on the membrane surfaces of the rough ER in a plant cell. The linear arrays suggest that the ribosomes are bound into polysomes (PRB) by an attached molecule of mRNA. Subdivision of the ribosomes into large and small subunits is discernible at many points. Epidermal cell of radish root. × 59,000. From Cell Ultrastructure *by W. A. Jensen and R. B. Park, 1967. Courtesy of E. Newcomb and Wadsworth Publishing Company, Inc.*

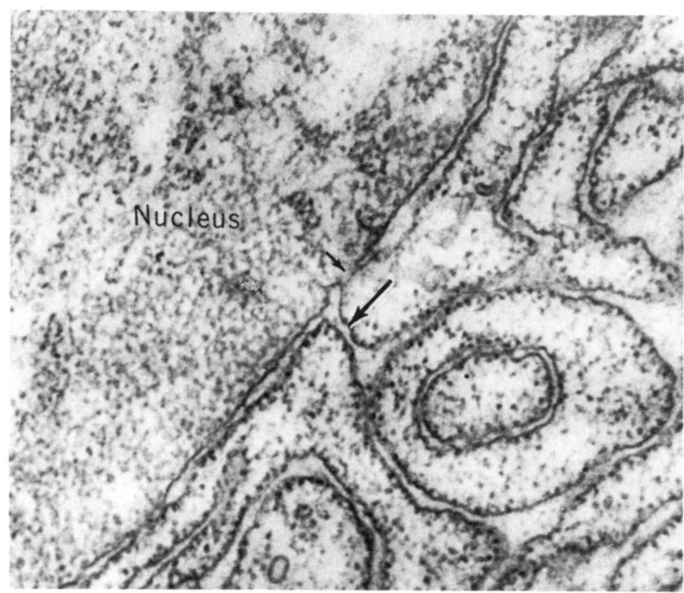

11–17. A connection between the outer membrane of the nuclear envelope and the rough ER (large arrow) in a nucleus of mouse salivary gland. The small arrow points to a pore complex. × 62,000. From The Cell *by D. W. Fawcett, 1966. Courtesy of D. W. Fawcett, H. Parks, and W. B. Saunders Company.*

bing or budding off. However, the development of rough ER has been observed to occur without any detectable involvement of the nuclear envelope; although evaginations of the outer nuclear membrane have been observed in embryonic cells in the process of developing rough ER, these are infrequent and have not been seen to appear in conjunction with the earliest appearance of the cytoplasmic membrane system. Thus, while there is an obvious structural similarity between the outer nuclear membrane and the rough ER, the precise interrelationship of the two membrane systems is not clear enough to state with any assurance that the reticulum is derived from the nuclear envelope.

The distribution of free and membrane-bound ribosomes in the cytoplasm. The relative distribution of free and membrane-bound ribosomes varies considerably depending on the cell type. Most cells contain at least some proportion of ribosomes bound to membranes of the endoplasmic reticulum. In a few types, such as muscle and kidney tubule cells, the rough ER is almost completely absent. The proportion of bound ribosomes is also relatively low in undifferentiated or embryonic cells, in which most of the ribosomes are freely suspended in clusters in the cytoplasm. In other cells, particularly in highly differentiated cells engaged in active secretion of protein, a large proportion of the ribosomes are bound to the endoplasmic reticulum. This reaches its greatest development in the tissues of mammalian organs such as the salivary gland and pancreas. In these cells, the cytoplasm becomes tightly packed with closely spaced cisternae of the rough ER (see, for example, Fig. 11–15). Plant cells also contain membranes of the rough ER. The development of the cisternae in plants almost never reaches the levels found in animal cells; in even the most complex systems the membranes are more scattered and sparsely distributed (Fig. 11–18).

The endoplasmic reticulum and protein synthesis. The role of the endoplasmic reticulum in protein synthesis was confirmed by biochemical studies in the 1950's, first carried out by George E. Palade and Philip Siekevitz. They isolated a fraction from disrupted cells

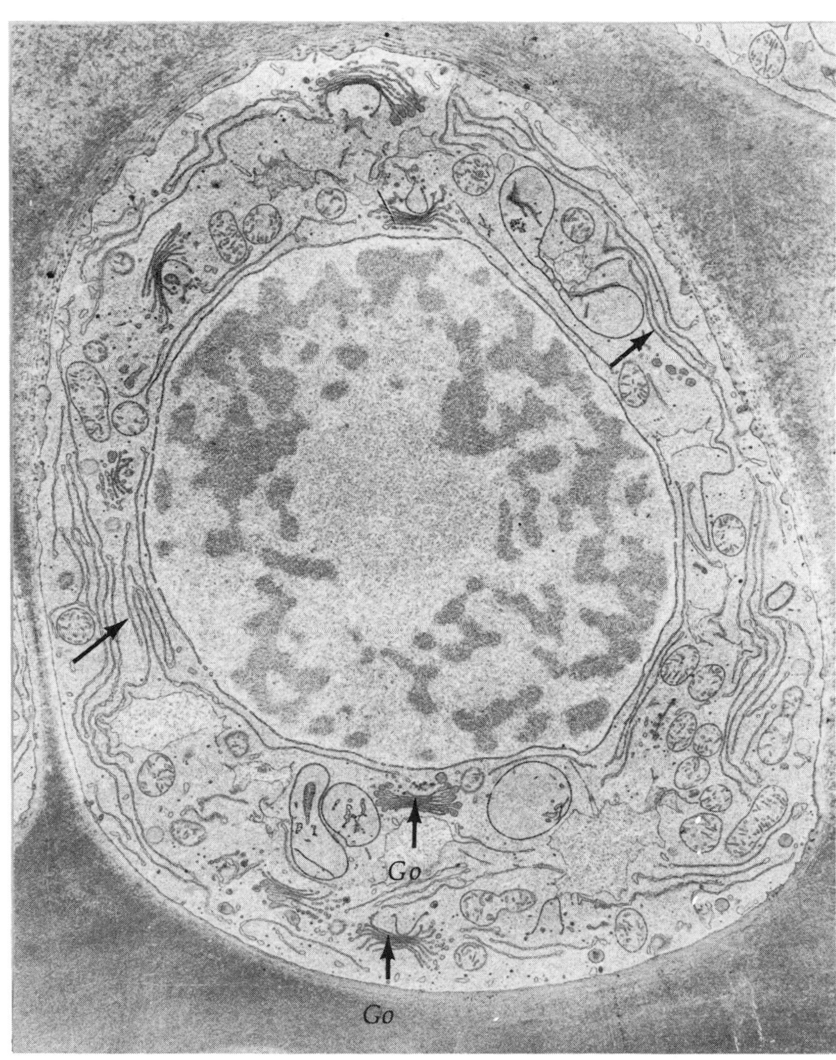

11–18. Distribution of membranes of the endoplasmic reticulum in a root cap cell of corn (arrows). Elements of the Golgi complex are also present (Go). *Potassium permanganate fixation.* × *8,200. Courtesy of W. G. Whaley, J. E. Kephart, and H. H. Mollenhauer.*

that on examination under the electron microscope proved to contain fragments of the rough ER. This fraction, called the *microsomal* fraction, contained membranes, ribosomes, and the enzymes and other factors required for protein synthesis and was capable of incorporating amino acids into polypeptides *in vitro*. Other biosynthetic pathways were also identified with the microsomal fraction, including enzymatic activity associated with carbohydrate and sterol metabolism (for a detailed summary of the biochemical properties of the microsomal fraction, see Goldblatt, 1969). It now appears that at least some of the additional biosynthetic capabilities of this fraction arise from contamination with other membrane systems, particularly smooth ER (see below).

More recently, it has been possible to verify that the endoplasmic reticulum is active in protein synthesis by autoradiographic methods. If cells actively engaged in protein synthesis are exposed for a brief period to radioactive amino acids, the first incorporation of label proves to take place over the ribosomes of the endoplasmic reticulum. Clusters of ribosomes freely suspended in the cytoplasm are also labeled at the same time. If longer periods of time are allowed to elapse between exposure to radioactive amino acids and fixation of tissues, the labeled material can be observed to have progressed through the channels of the rough ER and into vesicles or elements of the smooth ER (an experiment of this type is summarized in some detail later in this chapter). These findings confirm that the

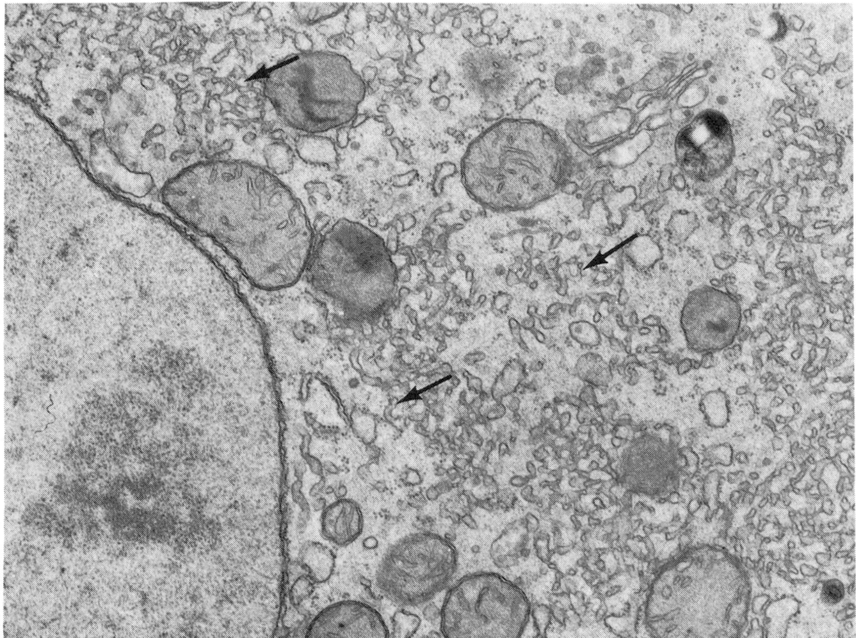

11–19. Elements of the smooth endoplasmic reticulum (arrows) in hamster liver cell. × 18,000. From The Cell *by D. W. Fawcett, 1966. Courtesy of D. W. Fawcett and W. B. Saunders Company.*

rough ER functions in the synthesis of protein and also indicate that transport of the synthesized protein through the vesicles of the ER also takes place.

Smooth Endoplasmic Reticulum

Although continuities are noted between the rough and smooth ER, the membranes of the smooth ER are distinct in both distribution and ultrastructure. The profiles of the smooth ER are tubular rather than vesicular and are of smaller dimensions than the cisternae of the rough ER, approximately 500–1000 Å in diameter. The smooth tubules are frequently packed into an interdigitated, lattice-like array in the cytoplasm (Fig. 11–19).

Sjöstrand has noted that the smooth membranes are slightly thicker than the membranes of the rough ER, about 60 Å after OsO_4 fixation and 70–80 Å after fixation in permanganate. Like the rough membranes, the smooth reticulum is less distinctly trilaminar than the plasma membrane and shows some development of globular or particulate ultrastructure (Fig. 11–20).

Both rough and smooth ER are present in most cells. The relative proportions of the two systems may vary considerably. In some types, such as cells of the pancreas, almost all reticulum membranes are rough; in others, such as epithelial cells, almost all membranes are smooth. Between these two extremes, all gradations can be found. Smooth membranes are especially abundant in cells synthesizing steroid hormones. A highly developed smooth ER is also found in association with glycogen deposits in liver cells. The enzyme glucose 6-phosphatase, important in the metabolism of glycogen, is found in quantity in extracts of smooth ER from this source. Smooth ER is also found in abundance in striated muscle; in these cells, the smooth membrane system is called the *sarcoplasmic reticulum.*

The frequent connections observed between the smooth and rough ER gave rise to the hypothesis that the primary function of the smooth membranes is transport of the raw materials and end products of synthesis. This proposal is not without experimental support. Sanford L. Palay followed the absorption of fats by cells of the intestinal villi in fasted animals. Droplets of lipid, identifiable in the electron microscope as deposits of dense material, first appeared in vesicles formed in invaginations of the plasma membrane (called *pinocytotic vesicles*). These vesicles pinched off from the plasma-membrane and appeared to merge with the smooth ER. In animals sacrificed after 30 minutes, lipid droplets were observed within the vesicles of both the smooth and rough ER. Some lipid droplets were also seen in the *perinuclear* space. These findings and the morphological continuities observed prompted the hypothesis that an intracellular "waterway" extends through the cytoplasm, from the exterior directly into the vesicles of the smooth and rough ER, and

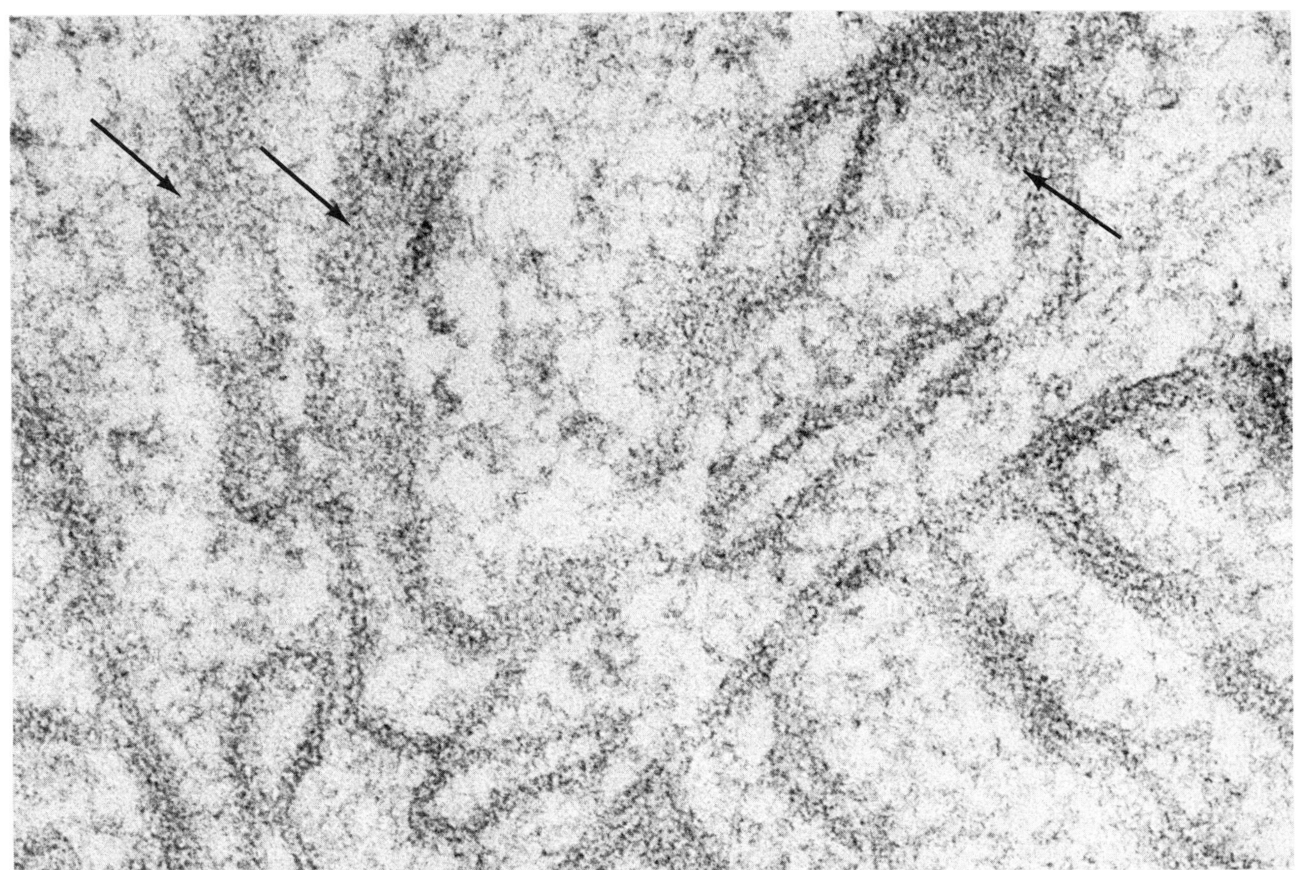

11–20. Membranes of the smooth ER (arrows) of mouse kidney at high magnification, showing the presence of particulate substructure. × *310,000. Courtesy of F. S. Sjöstrand and Academic Press, Inc.* [J. Ultrastr. Res. *9:340 (1963)*]

through these vesicles into the perinuclear space (Fig. 11–21). Through these channels materials might travel in either direction.

Palay's observations of the transport of fat droplets directly support this hypothesis. Whether this system is fundamental to the transport of metabolites in all cells is debatable, as connections between the nuclear envelope and the rough ER at one end and the rough or smooth ER and the plasma membrane at the other are not regularly seen. Many cell types do not contain extensive systems of rough or smooth ER. It is possible that this transport system is limited to specialized cells engaged in absorption or secretion, such as the cells of the intestinal villi observed by Palay.

In recent years it has become apparent that distinct biochemical functions in addition to transport can be identified with the smooth ER. The abundance of glucose 6-phosphatase in smooth membranes from liver suggests an involvement of this system in glycogen metabolism. Enzymes associated with other pathways have also been localized in the smooth ER, indicating possible involvement in the metabolism of lipids, including oxidation of fatty acids and the synthesis of phospholipids, glycolipids, and steroids. This apparent diversity of function suggests that the systems classified together as "smooth endoplasmic reticulum" may in reality represent several distinct types of membranes with characteristic structure and enzymatic machinery.

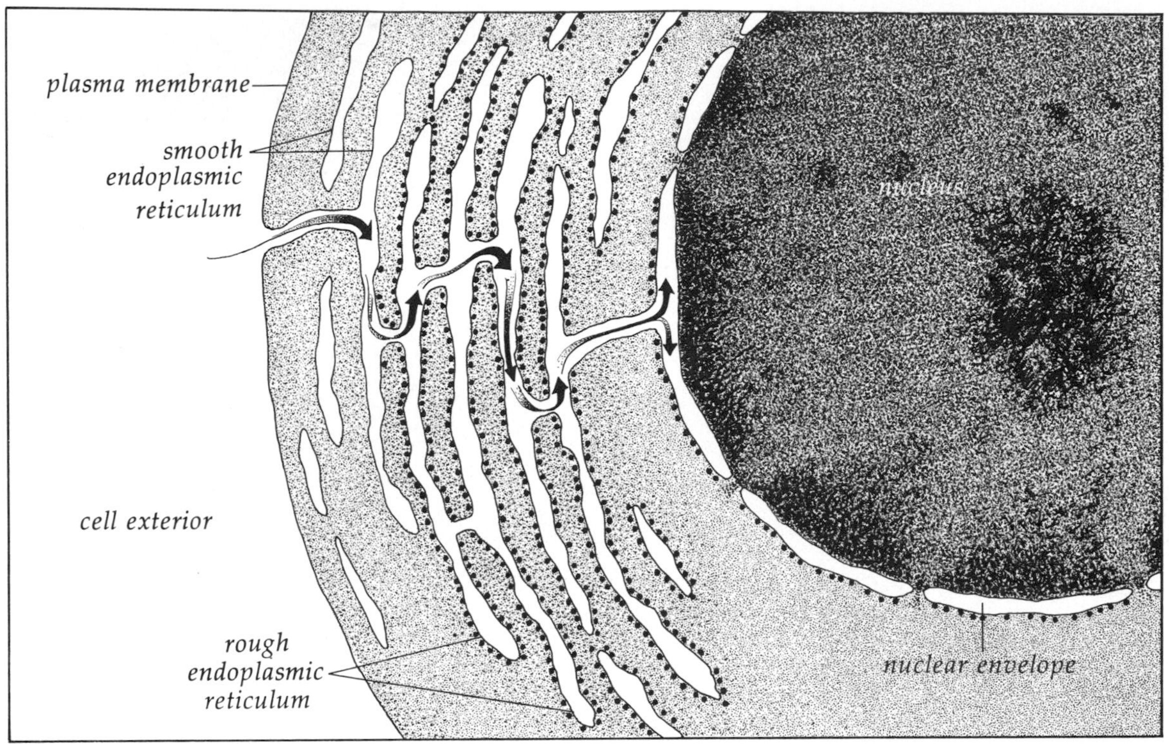

11–21. Diagram of the hypothetical "waterway" extending from the cell exterior, through elements of the smooth and rough ER to the perinuclear space (see text). A channel of this type is probably found only in highly specialized cells.

However, as will be demonstrated below, at least some of these membranes are engaged in the transport of recently synthesized protein.

The Golgi Complex

In many cells particularly secretory cells which discharge a product to the cell exterior, proteins synthesized in the endoplasmic reticulum are further modified in a distinct membranous organelle called the *Golgi complex* or *Golgi apparatus.* The subject of intensive research and debate for many years, this structure is now recognized as a constituent organelle of all eukaryotic cells. The Golgi complex was first described in 1898 by an Italian cytologist, Camillo Golgi. Golgi found that if cells were exposed to fixatives containing metals such as osmium or silver, a reticular network appeared in the cytoplasm that took up a brown or yellow stain from the metal. Other cytologists soon demonstrated that, after exposure to osmium tetroxide or silver nitrate, many cells revealed the presence of a similar network. This region, termed by Golgi the "internal reticular apparatus," gradually became known as the Golgi complex (see also footnote, p. 19). Because the complex is not visible under the bright field microscope unless special techniques are employed, many investigators were convinced that the reticular network was an artifact created by the staining method. These observational difficulties were resolved by electron microscopy in the 1950's, when A. J. Dalton and Marie D. Felix, and Fritjof Sjöstrand independently demonstrated that the Golgi complex is indeed a real and distinct cytoplasmic organelle. The complex has since been identified in unstained, living cells with phase contrast microscopy. The Golgi complex has a basically similar morphology in plant and animal cells. Under the light microscope, the complex appears as a network or collection of scale-like discs, either uniformly scattered throughout the cytoplasm or concentrated into one region of the cell. Under the electron microscope, the complex proves to consist of two membranous components (Figs. 11–22 and 11–23). The central mass of the complex is formed

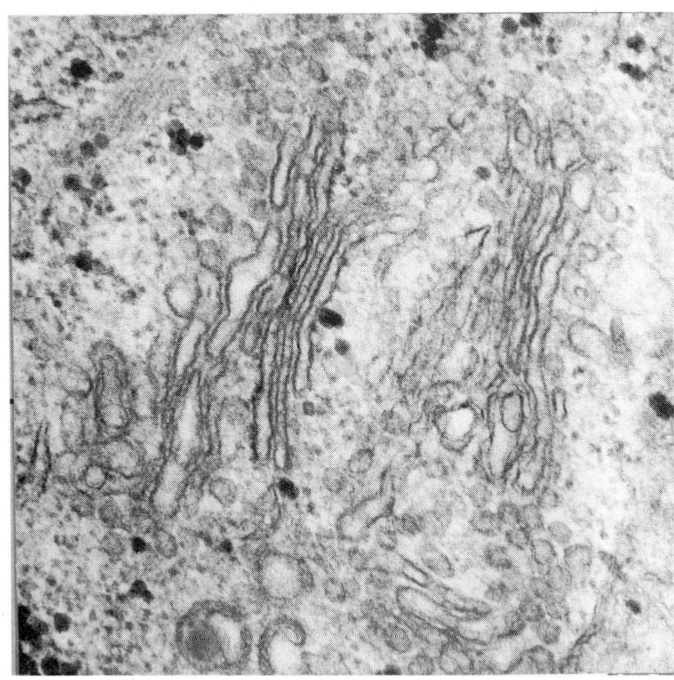

11–22. Elements of the Golgi complex in a cell from the intestinal epithelium of a chick embryo. × 53,000. Courtesy of D. R. Burgess.

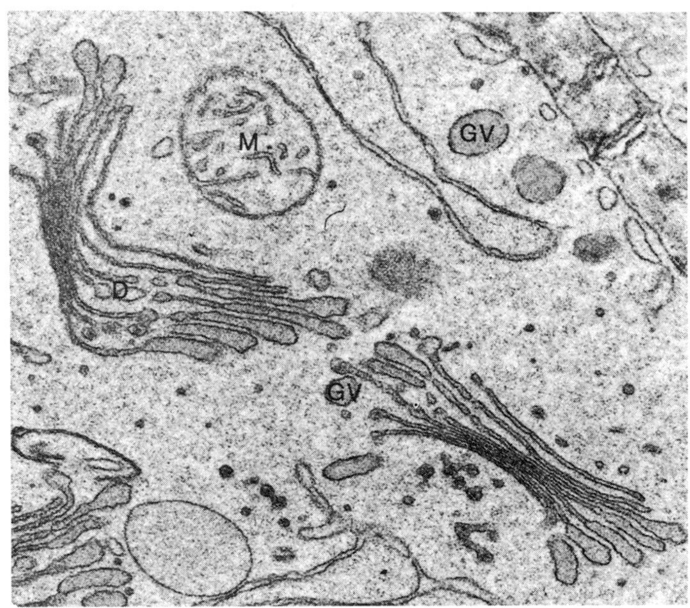

11–23. Elements of the Golgi apparatus in a root cap cell of corn fixed with potassium permanganate. Fragmentation of the flattened sacs into separate vesicles (GV) *is clearly visible.* D, *dictyosome;* M, *mitochondria. × 28,000. From* Cell Ultrastructure *by W. A. Jensen and R. B. Park, 1967. Courtesy of W. G. Whaley and H. H. Mollenhauer.*

from a stack of flattened sacs. Around this mass are dispersed a large number of spherical vesicles of different sizes.

Each flattened sac of the central mass is disc-like and consists of a single continuous membrane enclosing an interior space. The degree of flattening is such that the interior space between the opposing walls is reduced to 150–200 Å. This separation is fairly regular, except at the margins, where the interior space is increased to 600–800 Å. This gives the edges of the sacs a swollen or dilated appearance in cross section. Each Golgi complex consists of a number of these flattened sacs, usually between 4 and 8, stacked in a vertical array (see Figs. 11–22 and 11–23). Each sac is separated from its neighbor by a regular space, averaging about 200 Å in diameter. In cross section, the sacs appear to be slightly curved, giving the complex convex and concave faces. The total structure thus resembles a shallow cup in three-dimensional configuration.

Along the inner, concave face and around the periphery the complex is surrounded by small vesicles 200–1000 Å in diameter. These vesicles often occur in rows extending outward from the edges of the sacs, giving the impression that they arise by a process of budding or segmentation of the margins of the sacs. This impression is reinforced by freeze-etch preparations of the Golgi complex (Fig. 11–24) in which fractures along the surface of an individual sac of the complex show the edges to be branched or fenestrated. At times, the proliferation of vesicles by the Golgi is so extensive that much of the regular stacking of the central sacs is obscured (Fig. 11–25). Both the swollen margins of the sacs and the separated vesicles contain faintly granular material, which becomes more concentrated and electron dense in vesicles further removed from the main body of the complex.

The vesicles of the Golgi complex occur in two forms. One type, the most common, has smooth walls and is extremely variable in size, ranging in diameter from 200 to 1000 Å or more. The largest of these probably arise through fusion of smaller vesicles. A second type, less commonly found, is called a "coated" or "rough" vesicle; the walls of this form are coated with a layer of granules or short bristles on the cytoplasmic surface. Coated vesicles are more regular in size than smooth vesicles, approximately 600 Å in diameter. Both types

11–24. Replica of an element of the Golgi complex of Allium cepa, *prepared for electron microscopy by the freeze-etch technique. This micrograph gives the impression that the Golgi vesicles arise by a process of fenestration and fragmentation at the edges of the flattened sacs of the Golgi complex. × 88,000. Courtesy of D. Branton, H. Moor, and Academic Press, Inc.* [J. Ultrastr. Res. *11:401 (1964)*]

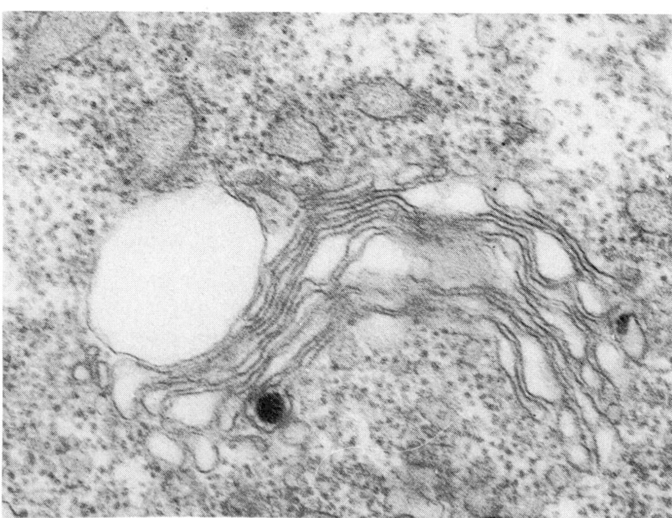

11–25. Extensively vesiculated form of the Golgi complex, from a tunicate follicle cell. × 45,000. Courtesy of R. G. Kessel.

of vesicles appear to contain membranes of slightly thicker dimensions than the sac membranes of the central mass of the Golgi complex. The functional significance of the morphological distinction between coated and smooth Golgi vesicles is unknown; both apparently arise similarly by budding of the margins of the flattened central sacs in the complex.

Although the structure of the Golgi complex is relatively uniform from cell to cell, small variations can be noted between species. Other differences are observed to occur in response to changes in physiological state of the cell. These are most often seen as an increase or decrease in the number of individual sacs of the Golgi complex, corresponding to an increase or decrease in secretory activity of the cell.

The Golgi complex is found in multiple numbers in the cell. In plants, multiple complexes are more or less evenly scattered through the cytoplasm. In animal cells, the complexes are more often localized in a mass near the nucleus. The total number of complexes is extremely variable in cells from different tissues or species. While the average number is about 20 per cell, there may be many more. For example, root tip cells from corn contain several hundred complexes; more than 25,000 per cell have been counted in the rhizoid apex of *Chara*, a green alga. Some animal tissues, such as cells of the salivary glands, may also contain complexes numbering in the thousands.

The function of the Golgi complex in protein synthesis. The involvement of the Golgi complex in the transport of newly synthesized proteins has been traced out using autoradiography. Lucien Caro and George E. Palade followed the migration of label in cells of the pancreas after administration of radioactive amino acids (see Fig. 11–26). In cells fixed immediately after a three-minute exposure to labeled amino acids, radioactivity, as expected, is localized over the rough endoplasmic reticulum (Fig. 11–26a). After longer periods of time, label was observed over the vesicles of the ER near the Golgi complex (Fig. 11–26b). Within 20 minutes after exposure to label, radioactivity was observed over the Golgi complex, including the smaller vesicles at the margins of the complex. After about one hour, only the large vesicles containing the secretion product of the pancreatic cells (called *zymogen granules*) were labeled. Finally, these vesicles could be observed to discharge their contents to the exterior. These observations indicate that newly synthesized proteins migrate

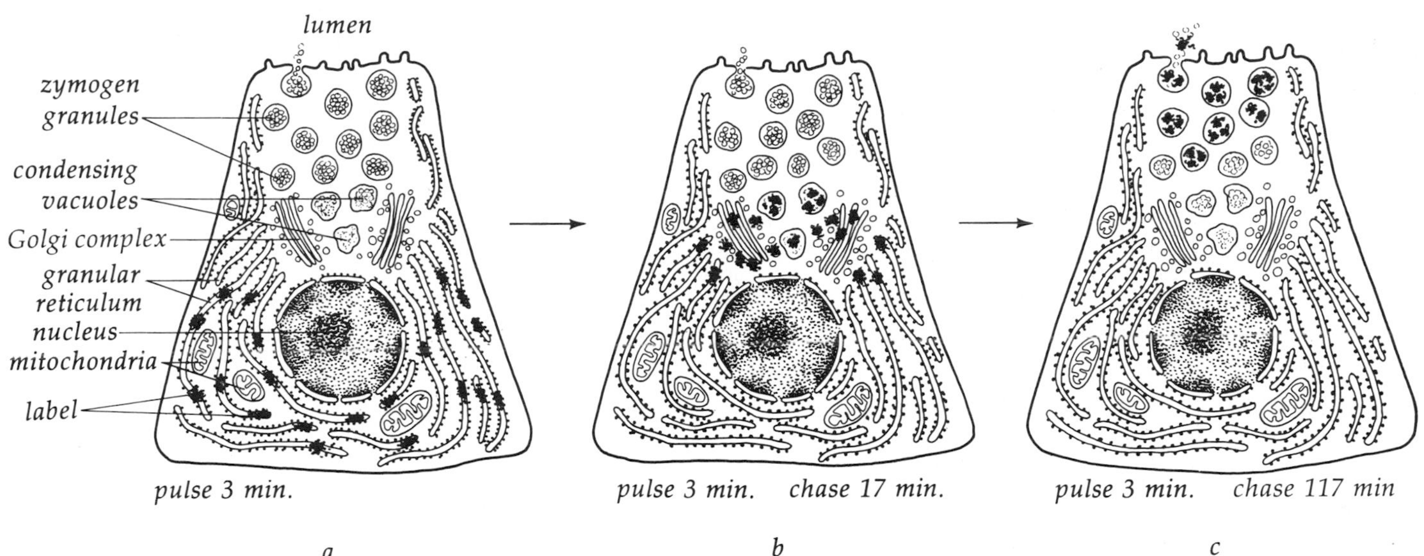

11–26. Localization of radioactivity after injection of tritiated leucine into pancreatic cells of the guinea pig. a, *3 minutes after injection; only the rough ER is labeled.* b, *3-minute exposure to label followed by a 17 minute "chase" by washing in non-labeled medium. Label is distributed over the rough ER in the vicinity of elements of the Golgi complex, and over the sacs and secretion vesicles of the Golgi complex.* c, *3-minute exposure followed by 117-minute chase. Label is centered over the zymogen granules, the secretory product of the cell. Courtesy of P. Favard. From* Handbook of Molecular Cytology, *North-Holland Publishing Company, Amsterdam, 1969.*

from the ER through the Golgi complex to the zymogen granules and eventually to the cell exterior. Other experiments using different tissues have confirmed this sequence and suggest that the following pattern is common to secretory cells: (1) incorporation of amino acids into polypeptides on the ribosomes, (2) transfer of the newly synthesized polypeptides into the vesicles of the endoplasmic reticulum, (3) migration of protein through the vesicles of the ER to the Golgi complex, and (4) concentration of the protein into secretion vesicles in the Golgi.

Many uncertainties surround the details of these steps. It is not known how polypeptides, which do not penetrate membranes readily, are transferred from the ribosomes across the cisternae membranes to the ER interior. It has been suggested that a separation, through which the newly synthesized proteins are transported to the interior of the cisternae, may exist in the reticulum membrane where it attaches to the ribosome. After migration through the cisternae to the vicinity of the Golgi complex (by yet another unknown mechanism), the synthesized proteins are transferred to the large flattened sacs of the Golgi. This is believed to occur through the medium of small vesicles, which bud off from the edges of the ER and fuse with the sacs of the Golgi complex after traversing the intervening space. No direct connections are seen between the two systems. Small vesicles of this type, called "transition elements," are visible in Fig. 11–27.

After transfer to the Golgi complex, the secretory products are believed to become concentrated into the small vesicles that bud off from the margins of the complex. It is not known to what extent the proteins secreted into the vesicles undergo biochemical modification in the Golgi. In some cells, it appears that the primary role of the complex is merely to concentrate the secreted proteins into membrane-bound vesicles. In other cells, particularly in cases in which the secretory product is a polysaccharide or a protein with a carbohydrate moiety, biochemical modification apparently takes place in the Golgi complex (see below).

The secretion product of the Golgi complex can frequently be recognized as a moderately electron-dense granular inclusion in the dilated edges of the central

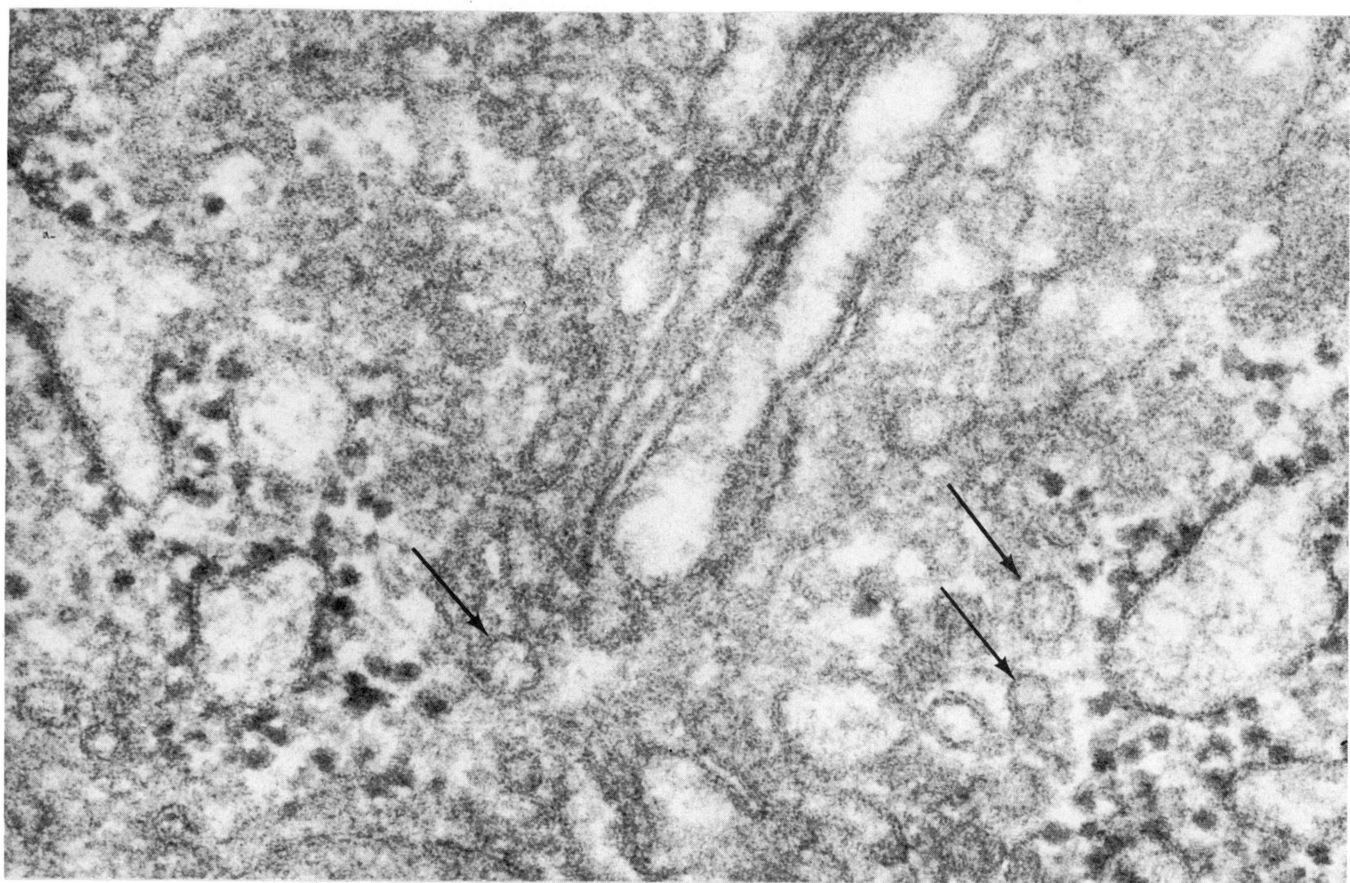

11–27. Small vesicles (in a cat pancreas) called "transition elements" (arrows), which presumably conduct materials synthesized within the rough ER to the Golgi complex. × 180,000. From Cytology and Cell Physiology, *ed. G. H. Bourne, 1964. Courtesy of F. S. Sjöstrand and Academic Press, Inc.*

sacs. This inclusion is also visible in the small peripheral vesicles and in the larger smooth-walled vesicles at a distance from the complex (Fig. 11–28). This is generally accepted as an indication that after budding off from the edges of the Golgi complex, the small peripheral vesicles progressively fuse together to form the larger smooth walled vesicles or secretion granules.

A number of experiments have directly implicated the Golgi complex in the formation of mucoproteins. Marian Neutra and C. P. Leblond followed the secretion of mucus in the goblet cells of rat intestine. Mucus contains a glycoprotein, a protein with an attached carbohydrate group. By administering radioactive glucose, Neutra and Leblond were able to locate the place where the carbohydrate subunit attached to the protein. Within 3–5 minutes after exposure of these cells to a brief pulse of label, radioactivity was located over the flattened sacs of the Golgi complex. After 20 minutes, label was observed to have migrated to small vesicles at the periphery of the complex. After 40 minutes, radioactivity was found concentrated into large, mucus-containing vesicles near the margin of the cell. These experiments show that both the synthesis of polysaccharides from glucose and the linkage of the polysaccharides to protein probably take place in the sacs of the Golgi complex. Similar experiments indicate that at least some lipids are complexed with proteins to form lipoproteins in the Golgi complex.

Other experiments have demonstrated that the synthesis of polysaccharides, or precursors of polysac-

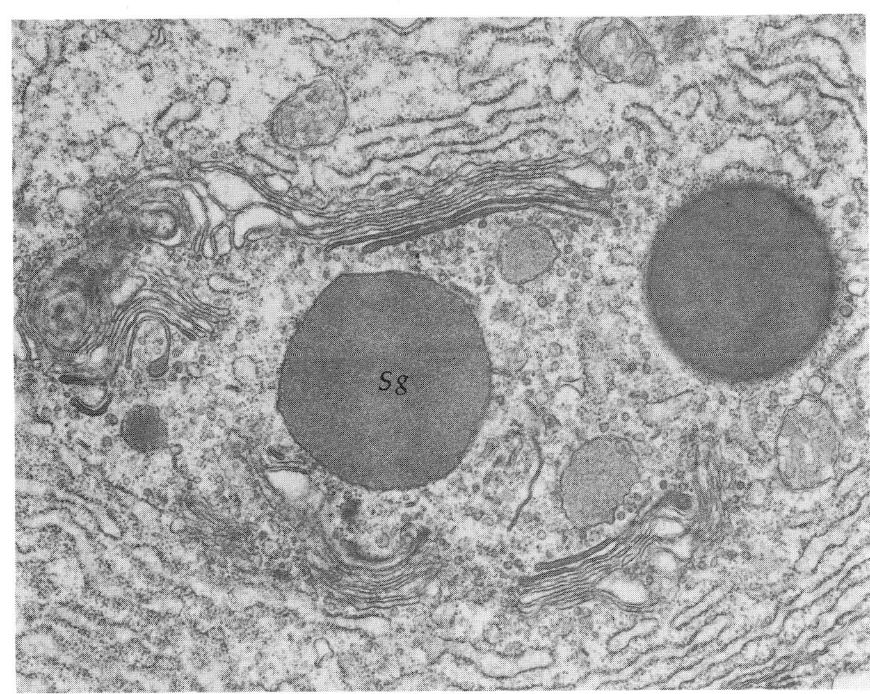

11–28. Formation of a secretion granule (Sg) *by the Golgi complex in a gastrodermal cell of a coëlenterate. A moderately electron-dense inclusion is present in peripheral elements of the Golgi complex, in small vesicles of intermediate size, and in the large secretory granule. Presumably, the smaller vesicles fuse to form the large secretion granules after budding off from the Golgi complex. × 16,000. Courtesy of H. W. Beams and R. G. Kessel.*

charide, probably also takes place in the Golgi complex. D. H. Northcote and J. D. Pickett-Heaps followed the migration of label in plant cells exposed to radioactive glucose. Incorporation of label was first detected in the Golgi complex and later in the cell wall, indicating that at least a part of the polysaccharide precursors of cell walls in plants are synthesized in the Golgi.

Many substances in addition to the products mentioned in the preceding sections are found to be secreted into vesicles in the Golgi complex. These include the granules of the pituitary secretion, granules of leucocytes, lipoprotein granules of hepatic cells, protein granules of mammary gland cells, lysosomes (see below), and the yolk granules of some oocytes. Mucoproteins and mucopolysaccharides are typically associated with vesicles secreted by the Golgi complex, particularly in the gums, slimes, and mucigens secreted by plant cells. This diversity in secretory product is probably reflected in wide differences in the enzymatic complement of the Golgi complex in different cells.

Lysosomes. One of the most studied of the highly diversified products of the endoplasmic reticulum–Golgi complex system of membranes is the *lysosome.* First identified in centrifuged extracts of the cytoplasm and later in thin sections of intact cells by Alex B. Novikoff and Christian de Duve, lysosomes consist of a large number of hydrolytic enzymes concentrated into typical cytoplasmic secretion vesicles. The enzymes of the lysosome are together capable of breaking down all major classes of biological macromolecules (see Table 11–2).

Lysosomes arise in many cells by fusion of the small vesicles that bud off from the margins of the Golgi complex by a process closely similar to the secretion of the zymogen granules of the pancreas. Synthesis of the enzymes enclosed within the vesicles can be traced to the rough ER.

Lysosomes are highly varied in shape and size in different cells. This characteristic makes identification and morphological description difficult. However, by using a cytochemical test, lysosomes may be readily identified in tissue sections. This test takes advantage of the fact that acid phosphatases are consistently found as a component of lysosomes. Cells are incubated with a suitable substrate for the phosphatases in a mixture containing lead ions. As the phosphatases hydrolyze the substrate, the lead ions are precipitated as an insoluble complex with the liberated phosphate groups. These deposits, identifiable as electron-dense clusters

Table 11–2 Acid Hydrolases Found in Rat Liver Lysosomes

Biological substrates	*Enzymes*
Phosphate esters	acid phosphatase
Proteins	cathepsin
	collagenase
	phosphoprotein phosphatase
	arylamidase
Nucleic acids	acid ribonuclease
	acid deoxyribonuclease
Polysaccharides and mucopolysaccharides	α-glucosidase
	β-N-acetylglucosaminidase
	β-glucuronidase
	β-galactosidase
	α-mannosidase
	β-xylosidase
	aryl-sulphatase
	hyaluronidase
	neuraminidase
Lipids	sphingomyelinase
	esterase

From R. Wattiaux, Biochemistry and function of lysosomes. In A. Lima-de-Faria, ed., *Handbook of Molecular Cytology*. North-Holland Publishing Company, Amsterdam.

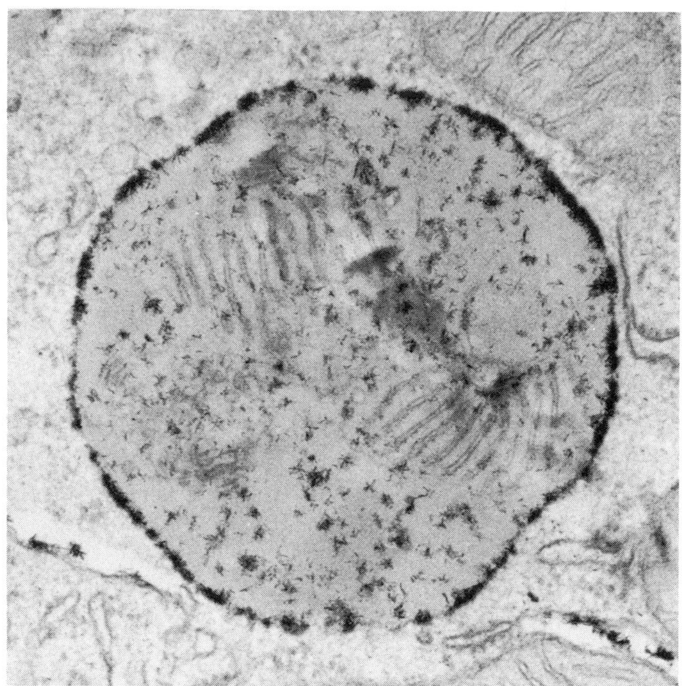

11–29. A lysosome in the cytoplasm of a mouse kidney cell identified by precipitation of lead phosphate, which is caused by the activity of a lysosomal enzyme, acid phosphatase. Remnants of several mitochondria are visible within the lysosome. × 50,000. Courtesy of F. Miller.

of lead atoms in thin sections, remain at the sites of phosphatase activity and permit identification of the lysosome (Fig. 11–29). By this means, lysosomes have been identified in all animal cells examined and in at least one plant tissue, the root cap cells of corn (Fig. 11–30). Lysosomes are especially abundant in cells that ingest large quantities of extracellular material, such as leucocytes or phagocytes. These vesicles are also prominent in the cells of tissues undergoing degenerative changes or metabolic stress, such as in starvation, aging, or hormonal stimulation.

Lysosomes enter into the hydrolysis of substrates in the cell by several different processes. One involves the digestion within pinocytotic vesicles of material brought into the cytoplasm. In pinocytosis, the plasma membrane contacts the extracellular material and invaginates, creating a pocket in which the material is trapped. This pocket then pinches off from the plasma membrane and drifts into the cytoplasm, where it fuses with one or more lysosomes. Fusion activates the enzymes contained in the lysosome, which proceed to hydrolyze the substrate contained in the pinocytotic vesicle. Fig. 11–31 shows several lysosomes in the process of fusing with a large vesicle that contains bovine hemoglobin, which has been absorbed by the cell, in this case a rat liver cell. After hydrolysis by the enzymes of the lysosome, many components of the vesicle probably enter the cytoplasm of the cell as small molecules and are used as precursors of synthesis. Any remaining undigested debris may be retained as a membrane-bound deposit in the cell (termed a *residual body*), or may be expelled to the cell exterior. In this function, the lysosomal mechanism acts as a digestive system for the cell. This mechanism is important in the defense of the cell against bacteria, virus, and toxic molecules.

In the second lysosomal process, cell organelles rather than extracellular material are digested by the lysosome. In some manner, as yet unknown, a cell organelle penetrates the lysosome membrane and is digested within the lysosome. Fig. 11–29 shows the remains of two mitochondria undergoing hydrolysis within the lysosome. The significance of autolysis or

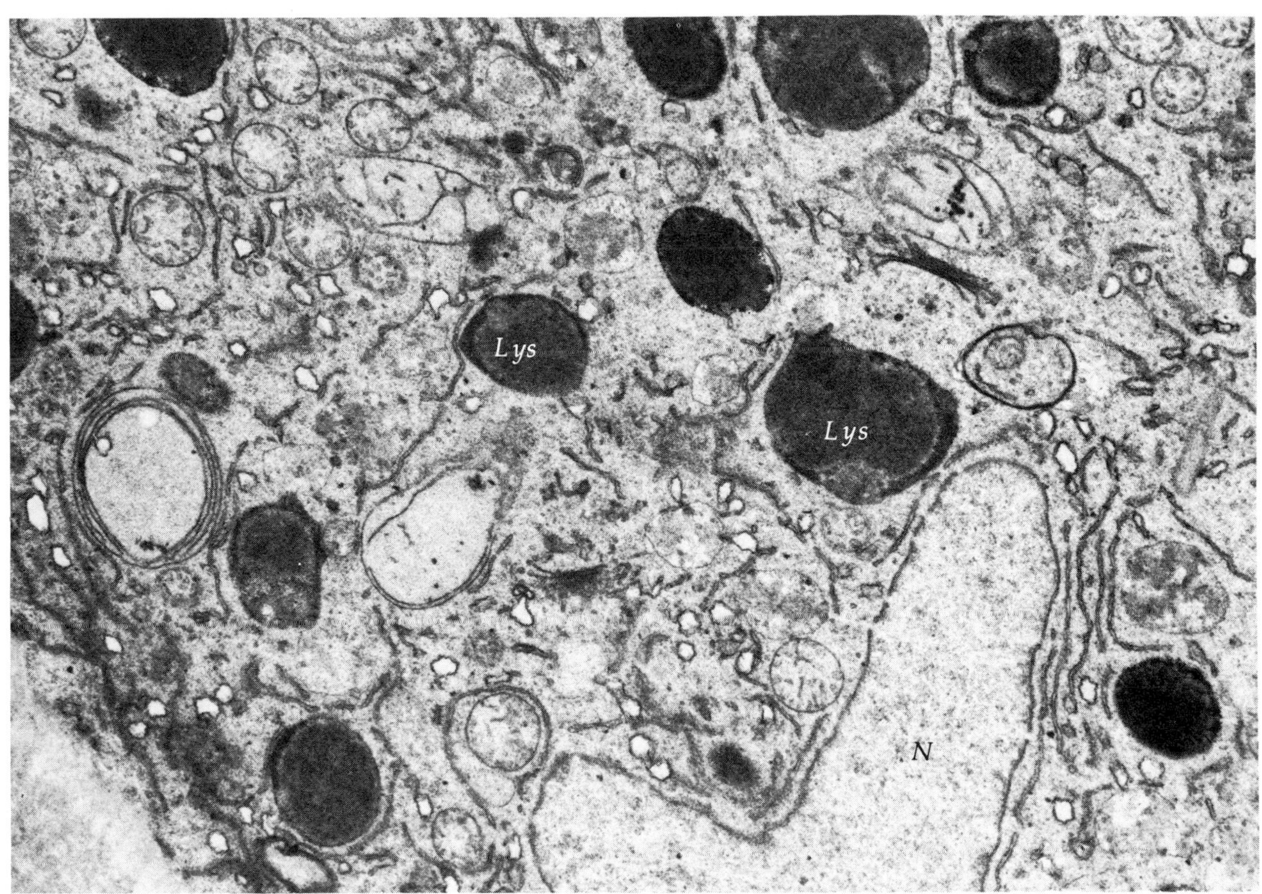

11–30. Presence of lysosome-like bodies in the cytoplasm of a root cap cell of corn. These bodies show a positive reaction for the presence of an acid phosphatase and other enzymes typical of lysosomes. Lys, *lysosome;* N, *nucleus.* × *12,500. Courtesy of Patricia Berjak and Academic Press, Inc., from* J. Ultrastr. Res. *23:233 (1968).*

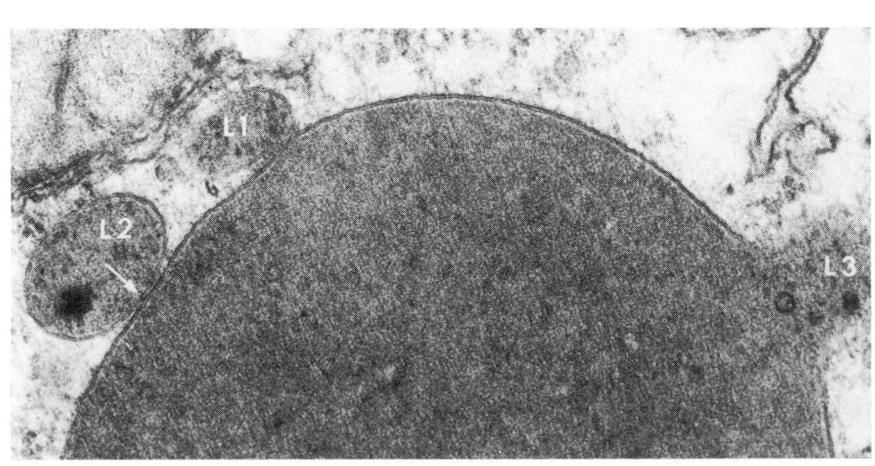

11–31. Lysosomes (L1, L2, and L3) apparently fusing with a large, membrane-bound droplet of absorbed hemoglobin. The arrow marks a point of membrane fusion between a lysosome and the hemoglobin droplet. Rat liver cell 30 minutes after injection of bovine homoglobin. × *58,000. Courtesy of S. Goldfischer, A. B. Novikoff, A. Albala, L. Biempica, and The Rockefeller University Press, from* J. Cell Biol. *44:513 (1970).*

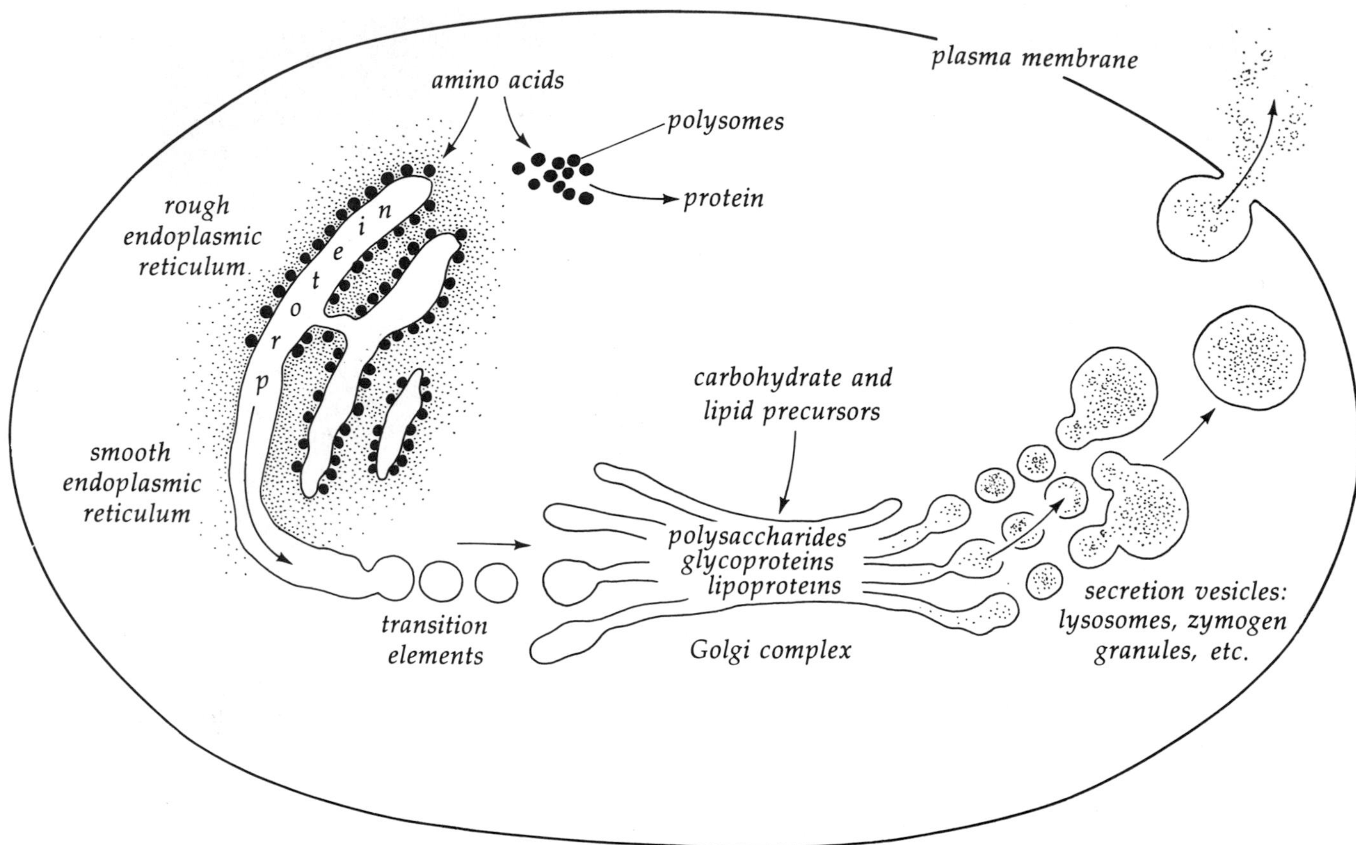

11–32. Diagram illustrating the integration of cytoplasmic elements in the synthesis of protein and the formation of secretion vesicles (see text).

self-digestion of this type is obscure, except that it probably plays an important role in the rapid breakdown of organelles in cells undergoing differentiation or physiological stress. A third lysosomal process is similar in mechanism, except that the lysosome membrane ruptures, and the hydrolytic enzymes are released into the surrounding cytoplasm. This process, if a large number of lysosomes are involved, either results in or accompanies death of the cell. In connection with this activity, Christian de Duve has termed lysosomes as "suicide bags."

Summary of the Morphological Aspects of Protein Synthesis

The morphological aspects of protein synthesis are summarized in Fig. 11–32. Proteins are synthesized on ribosomes, either freely suspended in the cytoplasm or bound to membranes of the rough endoplasmic reticulum. After synthesis, protein may enter directly into the cytoplasm or may penetrate into the cisternae of the endoplasmic reticulum. From here, the newly synthesized protein is transported through the smooth endoplasmic reticulum to the vicinity of the Golgi complex. The protein is then transferred to the complex by "transition elements," which break off from the smooth endoplasmic reticulum and fuse with the flattened sacs of the Golgi complex. After modification within the Golgi, which may include attachment of lipid or carbohydrate groups, the protein is secreted into peripheral vesicles that bud off from the margin of the complex. The peripheral vesicles gradually fuse together to form the larger secretion vesicles or granules, such as lysosomes or zymogen granules. These eventually discharge to the cell exterior by fusing with the plasma membrane or in some cases rupture to release their contents directly into the cytoplasm.

Suggestions for Further Reading

Books

Harris, Henry, 1968. *Nucleus and Cytoplasm*. Clarendon Press, Oxford.

Spirin, A. S. and L. P. Gavrilova, 1969. *The Ribosome*. Springer-Verlag, New York.

Papers and Review Articles

Beams, H. W., and R. G. Kessel, 1968. The Golgi apparatus: structure and function. *International Review of Cytology*, vol. 23, pp. 209–276.

Bonner, J., M. E. Dahmus, D. Farnbrough, R. C. Huang, K. Marushige, and D. Y. H. Tuan, 1968. The biology of isolated chromatin. *Science*, vol. 159, pp. 47–56.

Brenner, S., L. Barnett, E. R. Katz, and F. H. C. Crick, 1967. UGA: a third nonsense triplet in the genetic code. *Nature*, vol. 213, pp. 449–450.

Bretscher, M. S., H. M. Goodman, J. R. Menninger, and J. D. Smith, 1965. Polypeptide chain termination using synthetic polynucleotides. *Journal of Molecular Biology*, vol. 14, pp. 634–639.

Bretscher, M. S., and K. A. Marcker, 1966. Polypeptidyl-sRNA and aminoacyl-sRNA binding sites on ribosomes. *Nature*, vol. 211, pp. 380–384.

Brown, J. C., and P. Doty, 1968. Protein factor requirement for binding of messenger RNA to ribosomes. *Biochemical and Biophysical Research Communications*, vol. 30, pp. 284–291.

Capecchi, M. R., 1966. Initiation of *E. coli* proteins. *Proceedings of the National Academy of Sciences*, vol. 55, pp. 1517–1524.

Capecchi, M. R., 1967. Polypeptide chain termination *in vitro:* isolation of a release factor. *Proceedings of the National Academy of Sciences*, vol. 58, pp. 1144–1151.

Caro, L. G., and G. E. Palade, 1964. Protein synthesis, storage, and discharge in the pancreatic exocrine cell. *Journal of Cell Biology*, vol. 20, pp. 473–495.

Clever, Ulrich, 1968. Regulation of chromosome function. *Annual Review of Genetics*, vol. 2, pp. 11–30.

Conway, T. W., and F. Lipmann, 1964. Characterization of a ribosome-linked guanosine triphosphatase in *E. coli* extracts. *Proceedings of the National Academy of Sciences*, vol. 52, pp. 1462–1469.

Crick, F. H. C., 1966. Codon-anticodon pairing: the wobble hypothesis. *Journal of Molecular Biology*, vol. 19, pp. 548–555.

Dahmus, M. E., and J. Bonner, 1970. The role of nucleoproteins in regulation of gene function. *Federation Proceedings*, vol. 29, pp. 1255–1260.

deDuve, C., 1963. The lysosome. *Scientific American*, vol. 208, pp. 64–72.

de Man, J. C. H., and N. J. A. Noorduyn, 1969. Ribosomes: properties and function. In A. Lima-de-Faria, ed., *Handbook of Molecular Cytology*, North-Holland Publishing Co., Amsterdam, pp. 1079–1100.

Favard, P., 1969. The Golgi apparatus. In A. Lima-de-Faria, ed., *Handbook of Molecular Cytology*. North-Holland Publishing Company, Amsterdam, pp. 1130–1155.

Georgiev, G. P., 1969. Histones and the control of gene action. *Annual Review of Genetics*, vol. 3, pp. 155–180.

Gilbert, W., 1963. Polypeptide synthesis in *Escherichia coli* I: ribosomes and the active complex. *Journal of Molecular Biology*, vol. 6, pp. 374–388.

Gilbert, W., and B. Müller-Hill, 1966. The lac operator is DNA. *Proceedings of the National Academy of Sciences*, vol. 58, pp. 2415–2421.

Goldblatt, P. J., 1969. The endoplasmic reticulum. In A. Lima-de-Faria, ed., *Handbook of Molecular Cytology*. North-Holland Publishing Company, Amsterdam, pp. 1101–1129.

Hoagland, M. B., M. L. Stephenson, J. F. Scott, L. I. Hecht, and P. C. Zamecnik, 1958. A soluble ribonucleic acid intermediate in protein synthesis. *Journal of Biological Chemistry*, vol. 231, pp. 241–257.

Huxley, H. E., and G. Zubay, 1960. Electron microscope observations on the structure of microsomal particles from *E. coli*. *Journal of Molecular Biology*, vol. 2, pp. 10–18.

Kurland, C. G., 1970. Ribosome structure and function emergent. *Science*, vol. 169, pp. 1171–1177.

Lucas-Lenard, J., and F. Lipmann, 1966. Separation of three microbial amino acid polymerization factors. *Proceedings of the National Academy of Sciences*, vol. 55, pp. 1562–1566.

Marcker, K., and F. Sanger, 1964. N-Formyl-methionyl-sRNA. *Journal of Molecular Biology*, vol. 8, pp. 835–840.

Monro, R. E., 1967. Catalysis of peptide bond formation by 50S ribosomal subunits from *Escherichia coli*. *Journal of Molecular Biology*, vol. 26, pp. 147–151.

Nakamoto, T., and D. Kolakofsky, 1966. A possible mechanism for the initiation of protein synthesis. *Proceedings of the National Academy of Sciences*, vol. 55, pp. 606–613.

Neutra, M., and C. P. Leblond, 1966. Synthesis of the carbohydrate of mucus in the Golgi complex as shown by electron microscope radioautography of goblet cells from rats injected with glucose-H^3. *Journal of Cell Biology*, vol. 31, pp. 119–136.

Nirenberg, M., and P. Leder, 1964. RNA codewords and protein synthesis. *Science*, vol. 145, pp. 1399–1407.

Northcote, D. H., and J. D. Pickett-Heaps, 1966. A function of the Golgi apparatus in polysaccharide synthesis and transport in the root-cap cells of wheat. *Biochemical Journal*, vol. 98, pp. 159–167.

Novelli, G. D., 1967. Amino acid activation for protein synthesis. *Annual Review of Biochemistry*, vol. 36, pp. 449–484.

Palay, S. L., and L. J. Karlin, 1959. An electron microscope study of the intestinal villus II: the pathway of fat absorption. *Journal of Biochemical and Biophysical Cytology*, vol. 5, pp. 373–383.

Porter, K. R., 1961. The ground substance: observations from electron microscopy. In J. Brachet and A. E. Mirsky, eds., *The Cell*, vol. 2. Academic Press, New York, pp. 621–675.

Scolnick, E., R. Tompkins, T. Caskey, and M. Nirenberg, 1968. Release factors differing in specificity for terminator codons. *Proceedings of the National Academy of Sciences*, vol. 61, pp. 768–774.

Siddiqi, O., 1969. The mechanism of gene action and regulation. In E. E. Bittar and N. Bittar, eds., *Biological Basis of Medicine*, vol. IV. Academic Press, New York, pp. 43–80.

Sjöstrand, F. S., 1964. The endoplasmic reticulum. In G. H. Bourne, ed., *Cytology and Cell Physiology*. Academic Press, New York, pp. 311–375.

Slizynski, B. M., 1963. Functional changes in the polytene chromosomes of *Drosophila melanogaster*. *Proceedings of the 11th International Congress of Genetics, The Hague*, vol. 1, p. 108.

Staehelin, T., and M. Meselson, 1966. *In vitro* recovery of ribosomes and synthetic activity from synthetically inactive ribosomal subunits. *Journal of Molecular Biology*, vol. 16, pp. 245–249.

Stanley, W. M., M. Salas, A. J. Wahba, and S. Ochoa, 1966. Translation of the genetic message: factors involved in the initiation of protein synthesis. *Proceedings of the National Academy of Sciences*, vol. 56, pp. 290–295.

Takanami, M., and G. Zubay, 1964. An estimate of the size of the ribosomal site for messenger RNA binding. *Proceedings of the National Academy of Sciences*, vol. 51, pp. 834–839.

Traub, P., K. Hosokawa, G. R. Craven, and M. Nomura, 1967. Structure and function of *E. coli* ribosomes IV: isolation and characterization of functionally active ribosomal proteins. *Proceedings of the National Academy of Sciences*, vol. 58, pp. 2430–2436.

Wattiaux, R., 1969. Biochemistry and function of lysosomes. In A. Lima-de-Faria, ed., *Handbook of Molecular Cytology*. North-Holland Publishing Company, Amsterdam, pp. 1159–1178.

Woese, C. R., 1967. The present status of the genetic code. *Progress in Nucleic Acids and Molecular Biology*, vol. 7, pp. 107–172.

Woese, C. R., 1969. The biological significance of the genetic code. *Progress in Molecular and Subcellular Biology*, vol. 1, pp. 5–46.

Zamecnik, P. C., 1960. Protein synthesis. *Harvey Lectures*, vol. 54, pp. 256–281.

Unit Four *Cellular Reproduction*

12
Mitosis I: Replication of the Chromosomes

One fundamental characteristic of living matter is the ability to reproduce. In all living organisms, this capacity ultimately depends on processes taking place at the cellular level. This is true of even the simplest organisms, the viruses, which still require the mechanisms of a host cell for reproduction. The cellular processes involved in reproduction occur in three phases that are usually separate in time and are potentially independent of each other. In the first phase, during *interphase*, the cell undergoes a period of growth or synthesis in which the DNA and protein of the chromosome are replicated. Also synthesized during interphase are other materials required for the later division of the cell, such as precursors of the spindle in eukaryotes and probably a store of energy in the form of ATP. These events do not cause gross rearrangements or alterations in cell organelles and, except for nuclear enlargement and an increase in the capacity to bind the Feulgen stain, do not result in conspicuous alterations in the light or electron microscope image.

Some time after replication and stockpiling during interphase the cell enters into the second phase of reproduction. In this phase, termed *mitosis*, a complex sequence of events separates and distributes the replicated chromosomes equally into two daughter nuclei. In contrast to replication, the processes of mitosis involve relatively gross rearrangements of cell organelles. Following mitosis, in the final phase of cell reproduction, the cytoplasm is divided into two parcels, which accompany the daughter nuclei. The rearrangements taking place in this final phase, termed *cytokinesis*, are also easily visible in both the light and electron microscopes.

In most cells, the three phases of reproduction follow each other in fairly close sequence. A few cell types undergo replication without mitosis or cytokinesis;

this condition, which produces either polytene or polyploid nuclei, is comparatively infrequent but occurs regularly in insects and other organisms such as ciliate protozoa. More frequent, but still relatively uncommon, are cells that replicate and undergo mitosis without subsequent cytokinesis. This sequence, usually resulting in multinucleate cells or tissues, occurs in the fungi and some algae. In some vertebrate tissues, including the liver and other organs, these multiple nuclei fuse, producing large single nuclei that contain more than the diploid number of chromosomes (see Alfert and Geschwind, 1958).

The morphological details of mitosis and cytokinesis and the functional significance of these events have been understood since the 1880's. At this time they were correctly described and interpreted by a number of investigators, most prominent of whom were Eduard Strasburger and Walther Flemming. The years since the turn of the century have witnessed elaboration but no fundamental changes in the descriptions of mitosis and cytokinesis outlined by these early biologists. In spite of this long familiarity and the relative sophistication and power of contemporary methodology, we understand little more of the molecular interactions underlying the visible morphological processes of mitosis and cytokinesis than did Strasburger and Flemming.

The events of premitotic interphase, mitosis, and cytokinesis are described in this and the following chapter. An overview of the entire process of cell division without extensive experimental support will be given first in this chapter. The replication of chromosomes is then considered in detail; some facets of mutation and the problem of redundancy of the DNA in eukaryotic chromosomes will also be presented. In the chapter to follow, the cellular mechanisms involved in division of the chromosomes and cytoplasm will be reviewed. This will include discussion of the structure of metaphase chromosomes, the formation, structure, and function of the mitotic spindle, and the possible role of the centrioles in cell division.

An Overview of Replication, Mitosis, and Cytokinesis

The Cell Cycle

Replication of the chromosomes was originally thought to take place during the early stages of mitosis. In the early 1950's, experiments by Hewson Swift and his colleagues, in which the binding of Feulgen stain by interphase nuclei at different stages was studied quantitatively, indicated that doubling of DNA takes place during interphase. These findings were confirmed a short time later by autoradiography, first carried out by A. Howard and S. R. Pelc, and J. H. Taylor and his colleagues. Since this time, a great many investigators, using radioactive precursors of DNA and synchronized cell cultures, have established that DNA replication takes place during interphase in eukaryotic cells.

This synthesis follows a time sequence that is similar in almost all higher organisms; a nomenclature originally suggested by Pelc is widely used to designate the stages before, during, and after replication. After the close of the preceding mitosis, a period of variable length ensues in which no DNA synthesis can be detected. In this time interval, called G_1 (G = "gap," see Fig. 12–1), the nucleus therefore contains the usual somatic amount of DNA.[1] At some time late in G_1, an unknown event starts the period of DNA replication, termed the "S" (or *synthetic*) period. After replication is complete another interval, G_2, passes before mitosis begins. During G_2, the cell contains 4X the amount of DNA in a gamete. At the end of G_2, the cell undergoes mitosis and cytokinesis. Following this, the daughter cells enter G_1 of the subsequent interphase.

The length of the G_1 period is highly variable and is dependent on the physiological condition of the cell. Wide variation in G_1 is noted even within a single cell type held in tissue culture under uniform conditions. In some cells, such as actively growing yeast, there may be no G_1 period; these cells may either synthesize DNA continuously, even during nuclear division, or may enter into the S period immediately after division is complete. But in most eukaryote cells, G_1 lasts from 3–4 hours at a minimum to an indefinite period of days, months, or years. In contrast, S and G_2 are relatively unaffected by physiological conditions, and in most organisms proceed in sequence once the triggering mechanism, no doubt a part of the late G_1

[1] The amount of DNA at various stages of the cell cycle is often designated by the "C-value." The "C" is an arbitrarily chosen letter used to indicate the relative amount of DNA per nucleus at different stages. The relative quantity of DNA in the gametes of an organism is used as the base level. Gametes are therefore 1C; somatic cells before replication are 2C. After replication, during G_2, cells contain the 4C amount of DNA. In this book the values 1X, 2X, 4X are used as equivalents to the C-values.

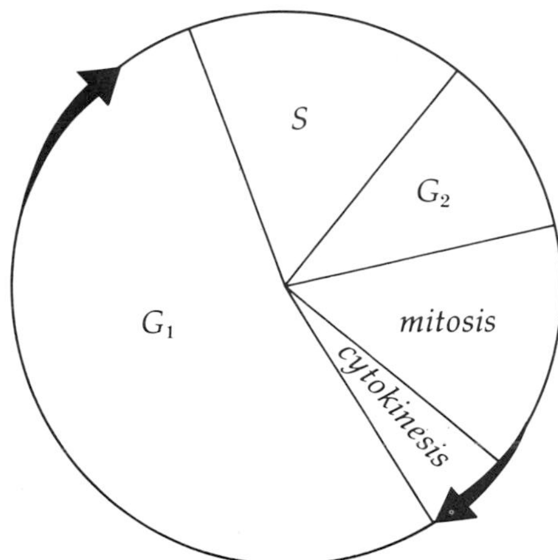

12–1. The cell cycle. G_1, S, and G_2 take place in sequence during interphase. G_1 may be of variable length, but for a given cell type S and G_2 are usually of uniform duration (see text).

period, is activated.[2] The lengths of S and G_2 are therefore fairly constant in a given cell type. In most cells DNA synthesis takes between 7 and 8 hours of the cell cycle; the subsequent G_2 period may last from 2–5 hours.

From this it can be seen, because S and G_2 are usually more or less fixed, that the triggering event for all the stages of cell division, including S, G_2, mitosis, and cytokinesis, must occur during the G_1 period. Once S begins, the subsequent stages of the cycle usually follow without delay. The identity of the initiating event has occupied the attention of scientists for many decades. It was once thought that the initiation of mitosis followed an increase in mass of the cell to a prescribed level. So many exceptions to this hypothesis have been uncovered that it is now generally accepted that although the division of cells parallels an increase in mass of the cell, this increase alone is not enough to initiate division. Whatever the triggering event is, and at this writing it remains unknown, it probably involves a sequence of interactions between the nucleus and its cytoplasmic and extracellular environment, with the activation of genes or operons that code for an initiator substance. That the initiator substance may be a protein is indicated by studies carried out in bacteria, the protozoan *Tetrahymena,* and several plants and animals. These studies show that a brief period of protein synthesis is necessary for the initiation of the S phase. For example, Y. Hotta and H. Stern followed the effects of several inhibitors of protein synthesis on microspores of the plant *Lilium.* These workers established that if protein synthesis was inhibited during G_1 cells failed to enter the S period. Cells already synthesizing DNA were unaffected by the inhibitory agents and completed the S phase.

Other investigations have shown that additional synthesis of RNA template and protein is required during S and G_2 for the cell to enter into mitosis. Inhibition of protein synthesis after the start of the S phase, while not interfering with the completion of DNA replication, prevents cells from initiating mitotic prophase. This additional protein synthesis, begun during the S phase, is quite extensive and involves both the replication of chromosomal proteins and the accumulation of components used by the cell during division. The histones are synthesized during this period, at approximately the same time the DNA replicates. Precursors of the mitotic spindle are continuously synthesized during interphase. Replication of the centrioles, if present, also usually takes place during interphase. At the end of interphase, synthesis of almost all constituents required for division of the cell is complete; synthesis of RNA, however, including both messenger and ribosomal RNA, continues until late in mitotic prophase (see, for example, Taylor, 1960).

Mitosis

Although the process is continuous, it is convenient to divide mitosis into four arbitrary stages: *prophase* (*pro* = before), *metaphase* (*meta* = between), *anaphase* (*ana* = back), and *telophase* (*telo* = end). In Fig. 12–2, these stages are shown in individual frames from a time-lapse movie of cell division. The same process in diagrammatic form is shown in Fig. 12–3. It is worthwhile keeping in mind, in the discussion that follows, that although mitosis is well-characterized morphologically, the molecular and biochemical mechanisms underlying these events remain almost completely unknown.

Prophase

During interphase, the nucleus shows little internal structural differentiation except for the nucleolus, granules, fibers identifiable as interchromosomal ma-

[2] In a very few cell types, such as lily microspores, a well-defined G_2 period may be lacking. Additionally, both G_1 and G_2 are apparently absent during the early cleavage divisions of the sea urchin embryo.

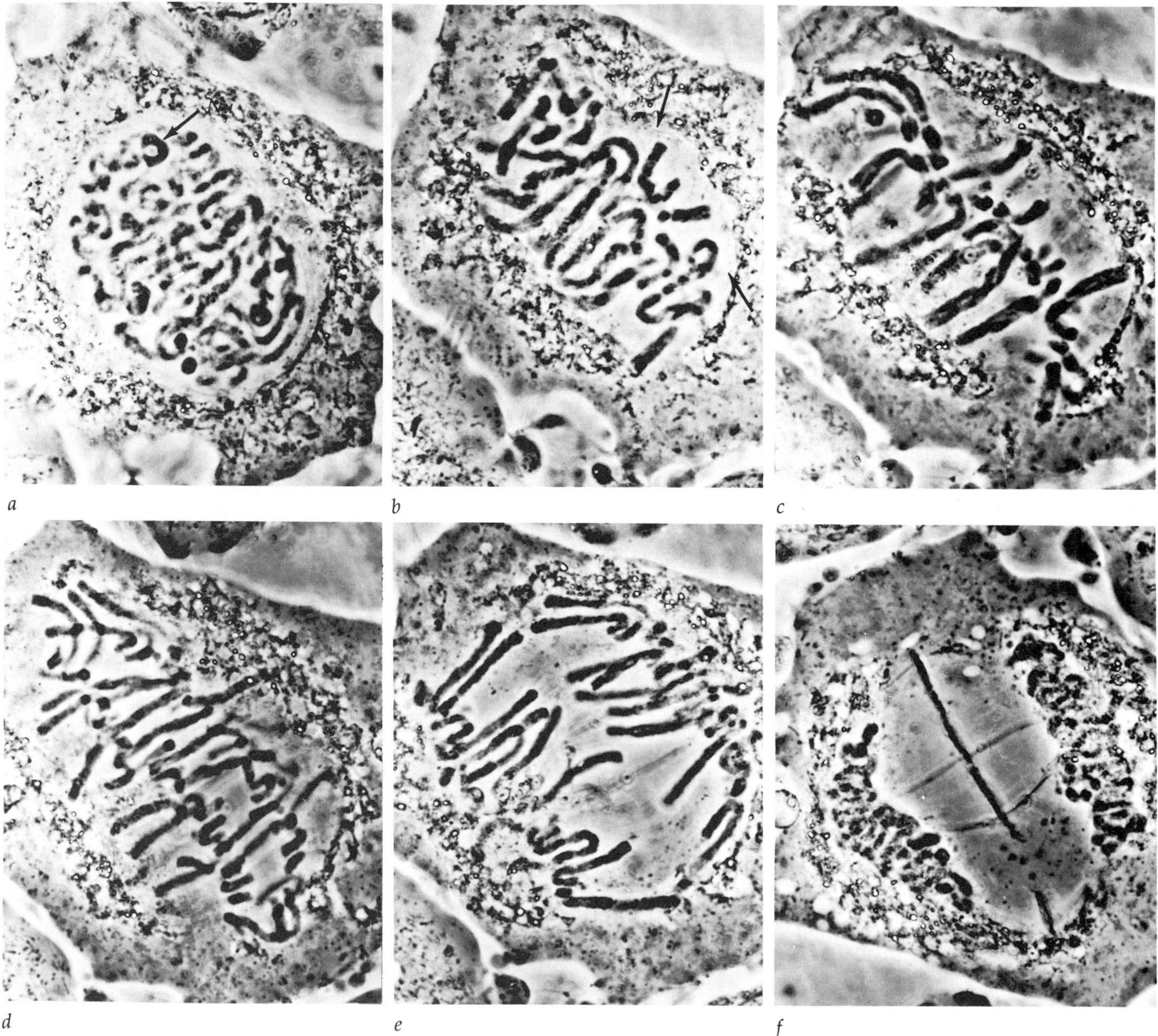

12–2. *Mitosis in living endosperm tissue of the plant* Haemanthus katherinae, *observed by phase contrast microscopy.* a, *Prophase. The chromosomes have coiled down into distinct structures. The nucleolus (Nu) is still visible at these early stages. The "clear zone" around the nucleus is the first indication of spindle formation in plants. Electron micrographs at this stage show the clear zone to be occupied by large numbers of spindle microtubules.* b, *Beginning of metaphase. The nuclear envelope has broken down but fragments (arrows) are still visible.* c, *Metaphase. The centromeres lie in a plane called the "metaphase plate" at this stage. The chromosome arms extend in random fashion to either pole.* d, e, *Anaphase.* f, *Telophase. The chromosomes at the poles have uncoiled and are no longer individually distinguishable.* ×*700. From* Cinematography in Cell Biology, *1963. Courtesy of M. Bajer and Academic Press, Inc.*

terial, and localized aggregations of the 100 Å chromatin fibers into clumps of heterochromatin. The beginning of prophase (Figs. 12–2a, 12–3a) is marked by the first appearance of chromosomes as faint threads in the nucleus. As the threads take shape, the clumps of heterochromatin gradually become indistinguishable from the rest of the chromatin. The word mitosis (*mitos* = thread) is derived from the thread-like aspect of the chromosomes as they begin to coil or fold into thicker structures at prophase. These can be seen distinctly only under the light microscope; in the ultrathin sections used in electron microscopy early prophase chromosomes appear only as scattered blocks or aggregations of chromatin (Fig. 12–4).

Under the light microscope, the chromosomes are often seen to be double as soon as they become visible in the prophase nucleus. The subunits, called *chromatids,* are the two products of replication of the chromosome. During early and mid-prophase, the body of each chromatid can frequently be seen to be made up of regular coils or gyres of chromatin. These coils are most easily seen in the light microscope (see Fig. 13–2), but are sometimes also visible in electron micrographs (Fig. 13–3b). Regular coiling of this type, involving a gyre or coil of the whole chromatid is discernible only at relatively gross levels. Within these gross coils, at the ultrastructural level, no regular pattern is ever seen in the folding of the component 100 Å chromatin fibers in either thin-sectioned or isolated chromosomes. The molecular basis for either the irregular folding of chromosome fibers at the ultrastructural level or the regular coiling of whole chromatids is unknown. Coiling and condensation continue until each chromosome is a comparatively short, rod-like body. During this process the gyres become coarser and decrease in number. As condensation becomes complete, the two chromatids of each chromosome, which may be loosely intertwined in early prophase, lie more or less parallel to each other in an almost flat plane (see Fig. 13–4). The end result is the packaging of the extended interphase chromosome into a short, compact unit.

By the end of prophase, when condensation is complete, each chromosome shows at least one constriction somewhere along its length. This constriction, called the *primary constriction,* contains the *centromeres* or *kinetochores* and marks the future point of attachment of spindle fibers at metaphase. Other constrictions, called *secondary constrictions,* may be present at other points along the chromosomes. The positions of the primary

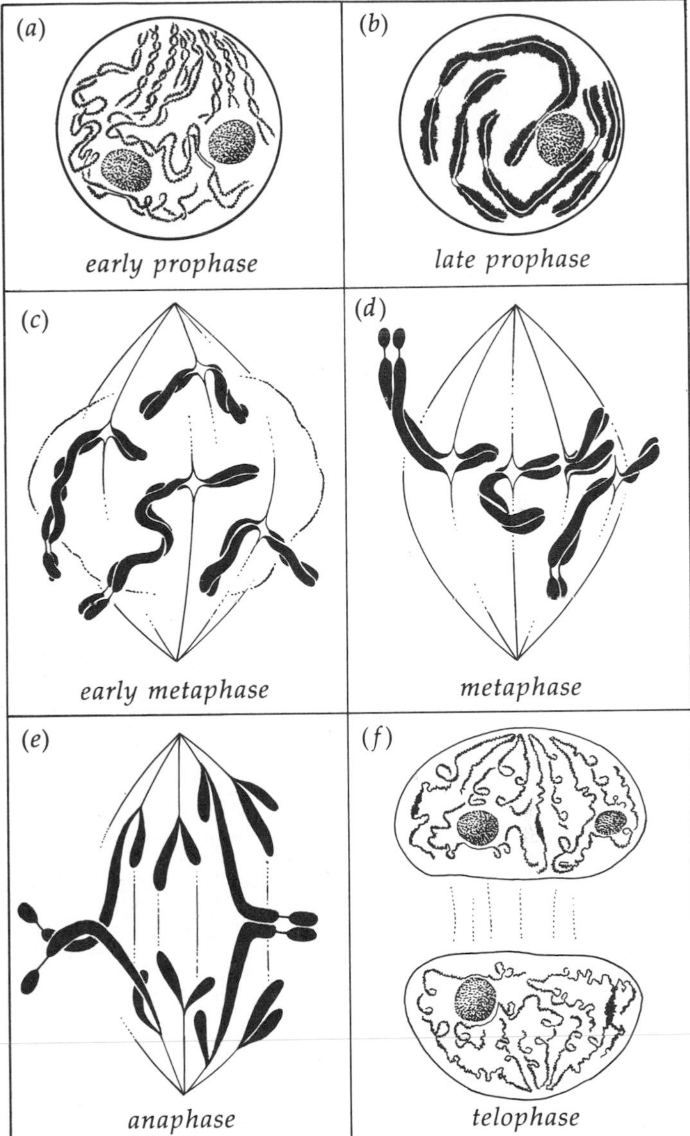

12–3. *Diagram of mitosis. Compare with Fig. 12–2. Adapted from* Chromosome Marker *by K. R. Lewis and B. John, 1963. Courtesy of K. R. Lewis, B. John, and J. & A. Churchill, London.*

and secondary constrictions, along with differences in length and shape of the chromosomes of a set, allow individual chromosomes to be identified. The collective morphology of the chromosomes at late prophase or metaphase is characteristic of a particular plant or animal species and often provides enough markers to allow identification.

As prophase continues, the nucleolus in most species becomes smaller and, by late prophase or metaphase,

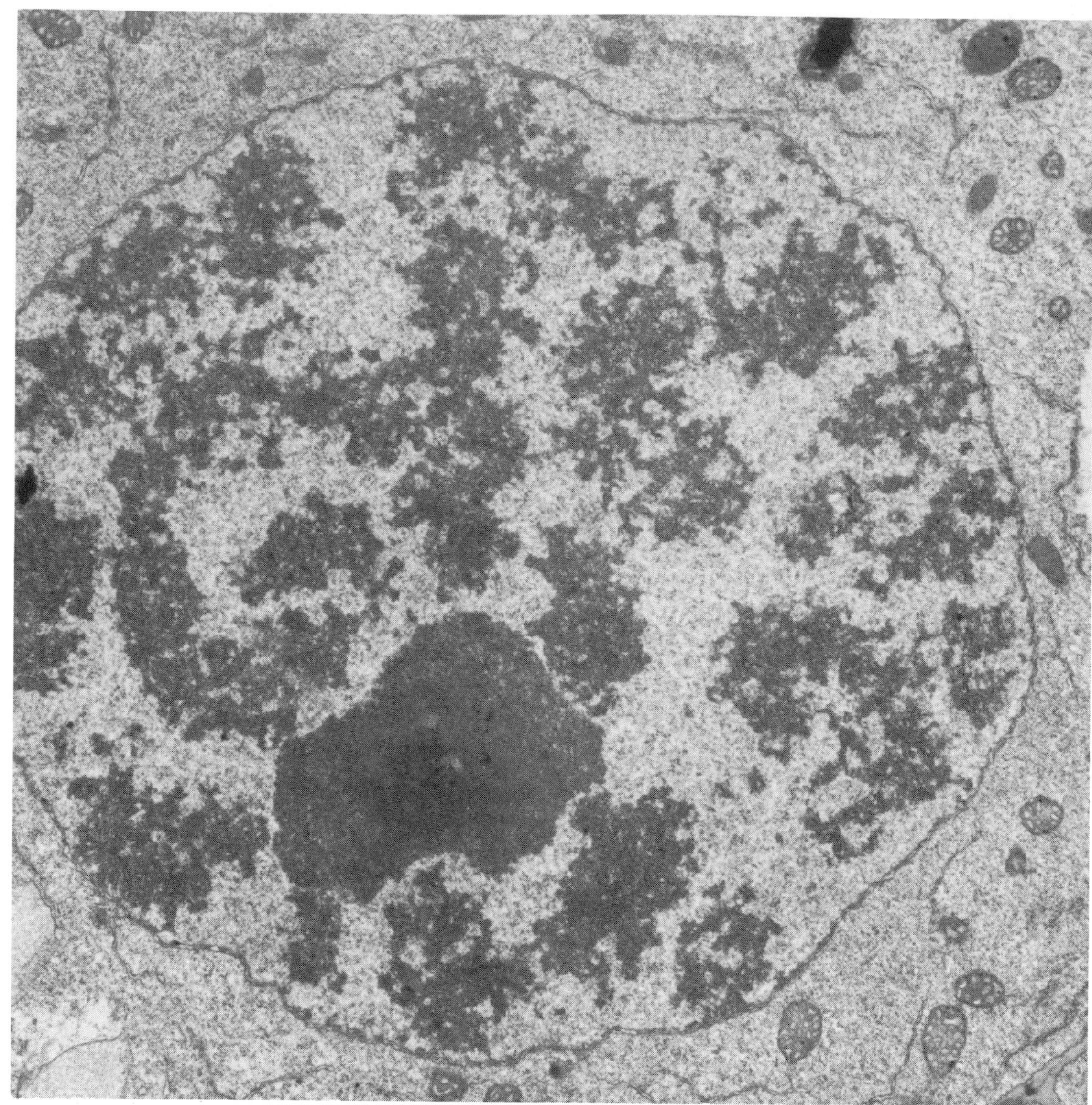

12–4. Thin sectioned prophase chromosomes in a nucleus from onion root tip meristem. Because of the thinness of the section, chromosomes are seen primarily as blocks of aggregations of chromatin. × 12,000.

From Cell Ultrastructure *by W. A. Jensen and R. B. Park, 1967. Courtesy of W. A. Jensen and Wadsworth Publishing Company, Inc.*

eventually breaks down and disappears. The nucleolar components, released as the nucleolus disintegrates, become distributed in the nucleoplasm. At the end of prophase, marked by the fragmentation of the nuclear envelope, the nucleolar material becomes indistinguishable from cytoplasmic ribonucleoprotein that invades the nuclear area. The best indications are that, except for some proteins, the nucleolar substance released during prophase is incorporated into the cytoplasm and does not enter into the reorganizing nucleolus at the completion of mitosis (for details of nucleolar reorganization after division, see Chapter 13).

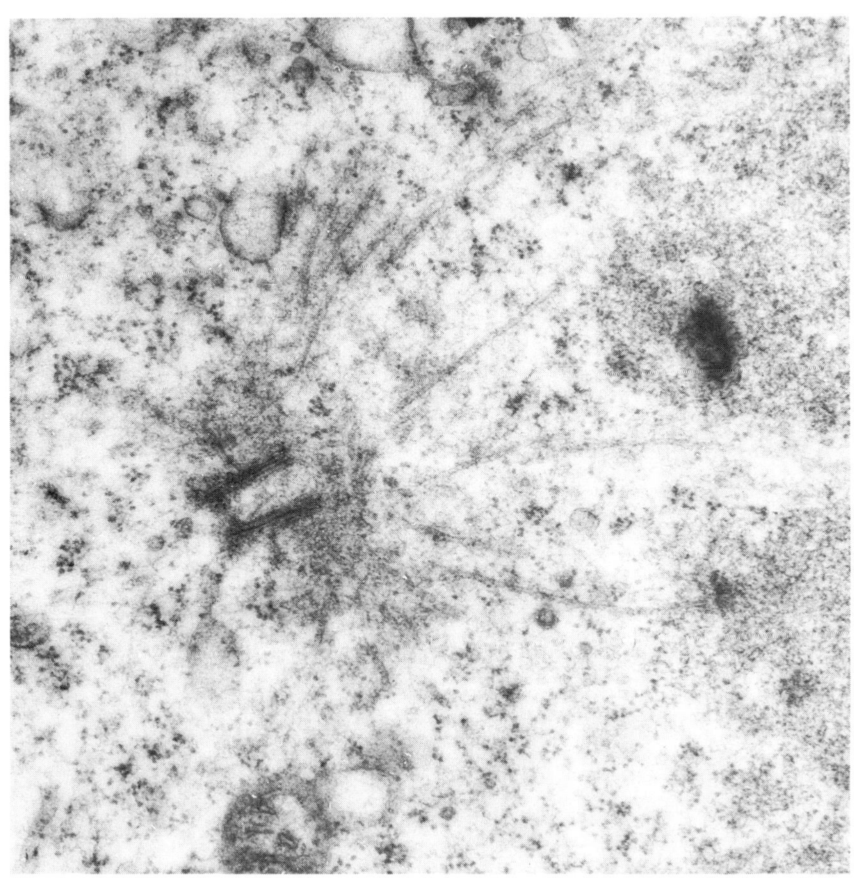

12–5. *A centriole of a cell from the spleen of a chicken. Microtubules of the aster and spindle are anchored in the cytoplasm near the centriole. Although microtubules are a primary structural unit of both the centriole and spindle, direct connections between microtubules of the two structures are not observed. × 44,500. Courtesy of J. André.*

In almost all animal cells, and in cells in plants giving rise to motile gametes, the cytoplasm just outside the nucleolus contains a pair of small, barrel-shaped bodies called *centrioles* (Fig. 12–5). These structures, which in most species are about 0.2 microns in diameter and 0.5 microns long, are just barely visible as small granules in the light microscope. In a very few organisms, such as the neuropteran insects, the centrioles may be much longer (see Friedländer and Wahrman, 1966). The centrioles are surrounded by short lengths of microtubules 150–200 Å in diameter. These tubules, which radiate outward in all directions from the centrioles, are not individually resolvable in the light microscope. However, the regular alignment of the tubules around the centrioles shapes adjacent cytoplasmic particles into a starlike array called the *aster*. This configuration is easily resolved by light optics.

The centrioles are replicated during interphase, usually during the S period. At the initiation of prophase, the aster separates into two parts; each of these contains one of the original pair and one of the newly replicated centrioles (Fig. 12–6). One pair of centrioles and the associated microtubules remain in position just outside the nuclear envelope. The other pair migrates approximately 180° around the periphery of the nucleus, coming to lie at the opposite end of the cell just outside the nuclear envelope (Fig. 12–6c). As the pairs of centrioles separate, the microtubules between them frequently lengthen accordingly, so that the two asters remain interconnected by a system of tubules at one side of the nucleus. The positions attained by the centrioles and asters at the completion of this migration locate the future division poles for the subsequent metaphase and anaphase.

Metaphase

The end of prophase and the transition into metaphase is marked in most organisms by the fragmentation and disappearance of the nuclear envelope. Three major rearrangements take place after the breakdown

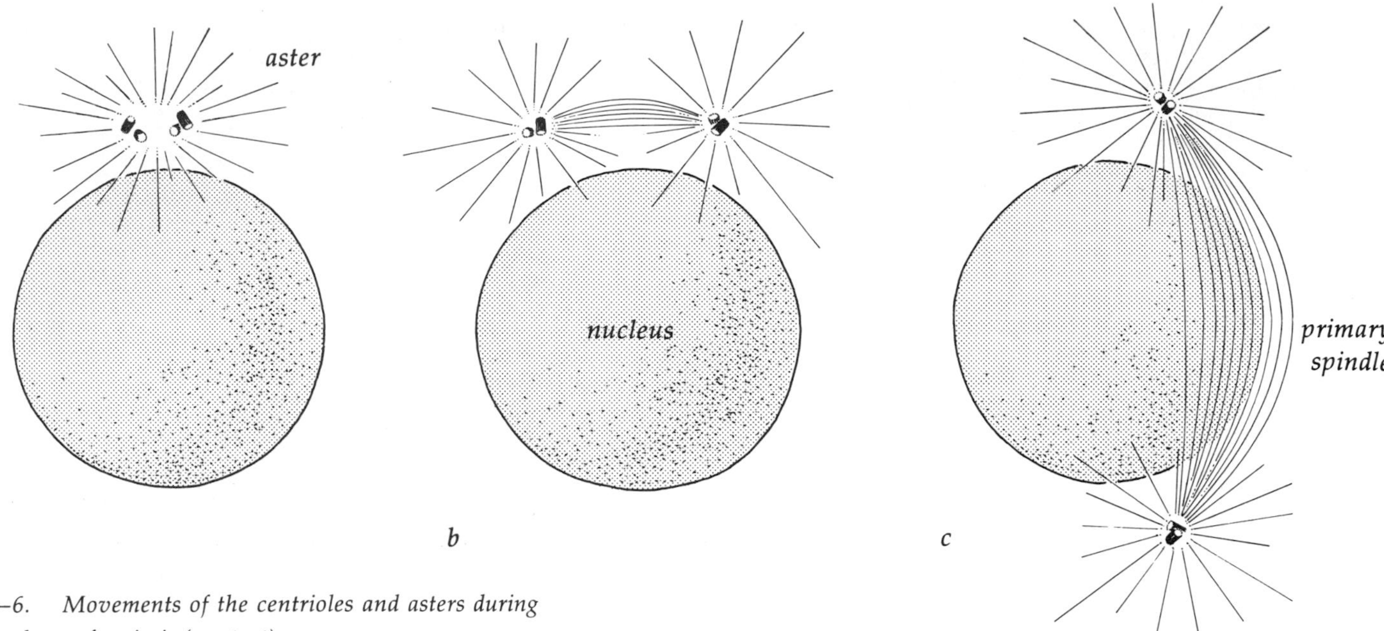

12–6. *Movements of the centrioles and asters during prophase of meiosis (see text).*

of the nuclear envelope. First, the area formerly occupied by the nucleus becomes filled with the *spindle,* a system of microtubules extending between the poles of the cell. At the same time, the chromosomes, which are randomly scattered throughout the nucleus at the time of fragmentation of the nuclear envelope, move, often with complicated oscillations, to the midpoint of the spindle. There they become arranged in a plane across the approximate center of the cell. Finally, during movement to the midpoint of the spindle, each chromosome becomes attached at the centromere to a bundle of spindle microtubules. This condition establishes metaphase: the spindle is completely formed and the chromosomes, each attached at the centromere to spindle microtubules, are aligned at midpoint of the spindle.

Formation of the spindle. Spindle formation in cells containing centrioles is initiated during prophase, at the time the asters separate. As mentioned, the microtubules between the two asters may remain connected as the asters separate and lengthen as migration continues. When migration is complete, each aster has short microtubules (called "astral rays") that radiate outward in all directions and long microtubules that extend around the nucleus and connect the two poles. The latter are called spindle fibers; the asters and interconnecting microtubules at this time form the *primary spindle* (Fig. 12–7).

In cells without asters or centrioles, the primary spindle begins to form around the outside of the nucleus in a zone a few microns thick (Fig. 12–8). This zone appears clear in the light microscope because cytoplasmic organelles and particles are excluded from the space. In the electron microscope, this zone can be seen to be occupied by large numbers of microtubules. These extend in roughly parallel array toward the ends of the cell that will become the division poles.

Whether or not the cell contains centrioles and asters, the end result of spindle formation is essentially the same. As the nuclear envelope breaks down, the developing spindle moves into the position in the cell formerly occupied by the nucleus. At this point, the spindle is broad at the midpoint and narrow at the poles and, except in a few cell types where the spindle takes up an eccentric position (for example, in oocytes), stretches almost from one end of the cell to the other. Little difference is noted in the structure of the com-

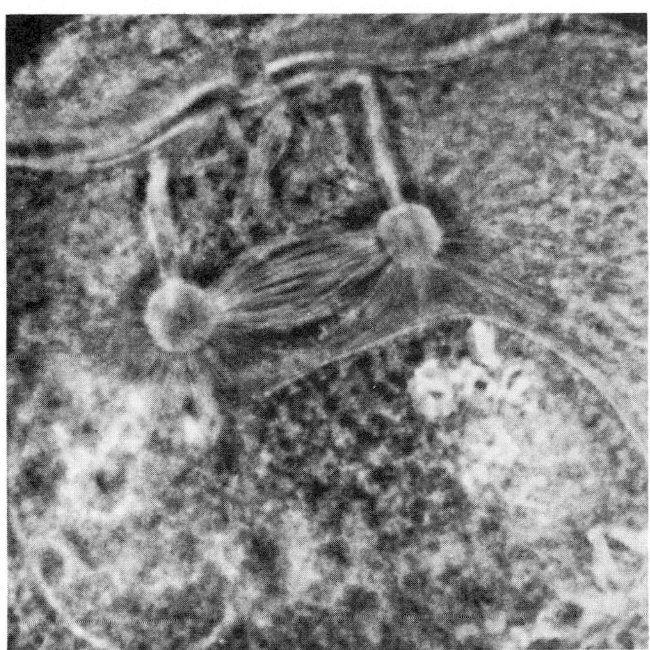

12–7. *A primary spindle in the protozoan* Barbulanympha. × *1,400. Courtesy of American Philosophical Society, from L. R. Cleveland,* Trans. Am. Phil. Soc. *43:809 (1953).*

pleted spindle whether asters are present or not, except that anastral spindles are apt to be somewhat broader at the poles.

The cells of higher plants, including all angiosperms and most gymnosperms, do not contain centrioles or asters at any part of the life cycle. Nevertheless, these cells are able to form a spindle and complete the subsequent stages of mitosis without difficulty. Therefore, the asters and the contained centrioles are not necessary for formation of the spindle. While the precise significance of the division of asters in early prophase remains unknown, it seems possible that this mechanism merely insures that the centrioles, if present in the cell line, are distributed to both daughter cells during mitosis.

Congression and attachment of the chromosomes to the spindle. After breakdown of the nuclear envelope, the chromosomes, now almost fully coiled into short double rodlets a few microns or less in length, move to the midpoint of the spindle. Although the molecular basis for this movement, called *congression* or *metakinesis,* is unknown, some investigators hold that it is moderated by early attachment to microtubules of the spindle (see p. 339). In any event, by the time the chromosomes reach the midpoint of the spindle, they have attached to spindle microtubules at the centromere.

The centromeres, already double as a result of replication of the chromosomes during interphase, attach to the spindle in such a way that the two centromeres of each chromosome face and are connected to opposite poles (Fig. 12–9). It is not known whether new microtubules are formed to make this connection or whether an interruption is made in existing pole-to-pole fibers. The connection is faintly visible in the light microscope, even though the individual spindle microtubules are too small to be resolved. This is because each chromosome connects to a bundle of from 4 to 8 or more microtubules. This composite spindle fiber is called a *chromosome fiber.*

Large numbers of spindle microtubules remain extended from pole to pole without making connections to the chromosomes. These are called *continuous fibers* to distinguish them from the chromosome fibers of the spindle. Both types of microtubules persist in the spindle throughout metaphase and anaphase; an interaction between the two types is considered by some investigators to be the basis for movement of the chromosomes by the spindle during anaphase (details of the several hypotheses of anaphase movement are given in Chapter 13).

The chromosomes do not make completely random connections to the midpoint of the spindle at metaphase. Often the smaller chromosomes of the set are concentrated toward the interior of the spindle. The larger chromosomes tend to become aligned at the periphery with the arms of the chromosomes extending into the surrounding cytoplasm. In some cases, the ends of the longest chromosome arms may bend and extend in random fashion almost to either pole. Thus it is most correct to state that at metaphase the *centromeres* of the chromosomes become aligned in a plane across the midpoint of the spindle. Arrival at this position forms what is sometimes referred to as the *metaphase plate* and completes the establishment of metaphase (See Figs. 12–2c and 12–3d).

Anaphase

Anaphase begins at the instant the two halves (chromatids) of each metaphase chromosome begin to separate and move to opposite poles of the spindle. The

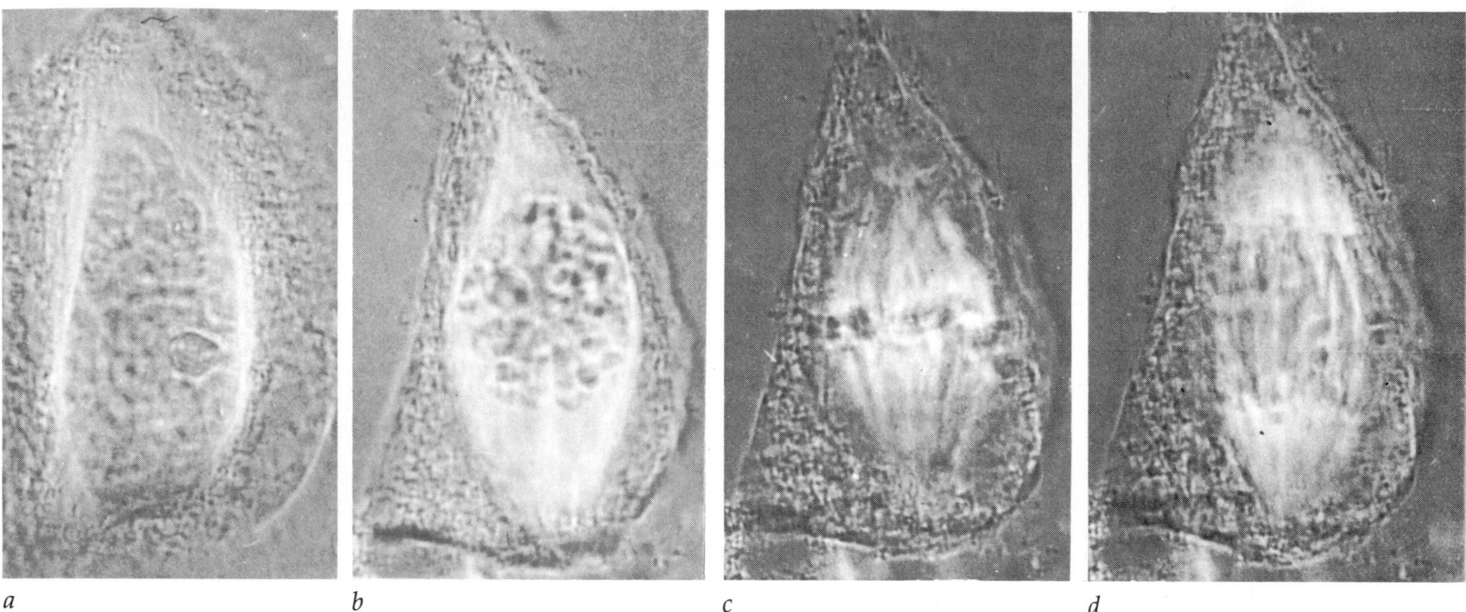

12–8. Formation of an anastral spindle in a living cell of the plant Haemanthus, *observed under polarized light. In* a, *the first appearance of the spindle is noted as a bright zone surrounding the nucleus (see also Fig. 12–2 a–c).* b, *Extension of the layer toward the poles.* c, *Metaphase.* d, *Anaphase. The spindle appears bright in polarized light because of parallel orientation of the component microtubules. × 600. Courtesy of S. Inoué and A. Bajer and the publisher.* [Chromosoma *12, 48–63 (1961)*]

physical forces underlying this movement have been a matter for speculation since the first discovery of mitosis and remain unknown to this day. Experiments with colchicine, an inhibitor of spindle formation, have shown that separation of the chromatids still takes place if the spindle is destroyed. Therefore this initial event in the onset of anaphase is separate from movement of the chromosomes to the poles and must reflect a mechanism residing in the chromosomes themselves. But subsequent progression of the chromosomes to the poles depends on the connection of spindle microtubules to the centromere; if these connections are broken chemically or by micromanipulation the movement ceases. The centromere always leads the rest of the chromosome to the poles, giving the impression that the chromosome is pulled to the pole by the spindle microtubules. If the centromere is in the approximate center of the chromosome, the chromosome is pulled into the shape of a "V," with the point of the V, the centromere, leading the way to the pole and the arms lagging behind. Other configurations, including rods or Vs with unequal arms, are produced depending on whether the centromere is positioned terminally or near one end of the chromosome (Fig. 12–10; see also Figs. 12–2e and 12–3e).

The entire set of chromosomes separates and begins anaphase movement in close synchrony. Whether the spindle fibers pull the chromosomes to the poles or the chromosomes are put in motion by some other agency, the chromosome spindle microtubules are seen to shorten to one-third to one-fifth of their original length during anaphase. The two masses of chromosomes are further separated by an increase in length of the whole spindle, which means that the continuous fibers increase in length during anaphase. The lengthening of the continuous fibers may take place at the same time or may follow the shortening of the chromosome spindle fibers; in one or another cell type, the primary source of movement may be due to either process. In either case, the chromosomes approach but do not quite reach the poles of the spindle at the completion of anaphase.

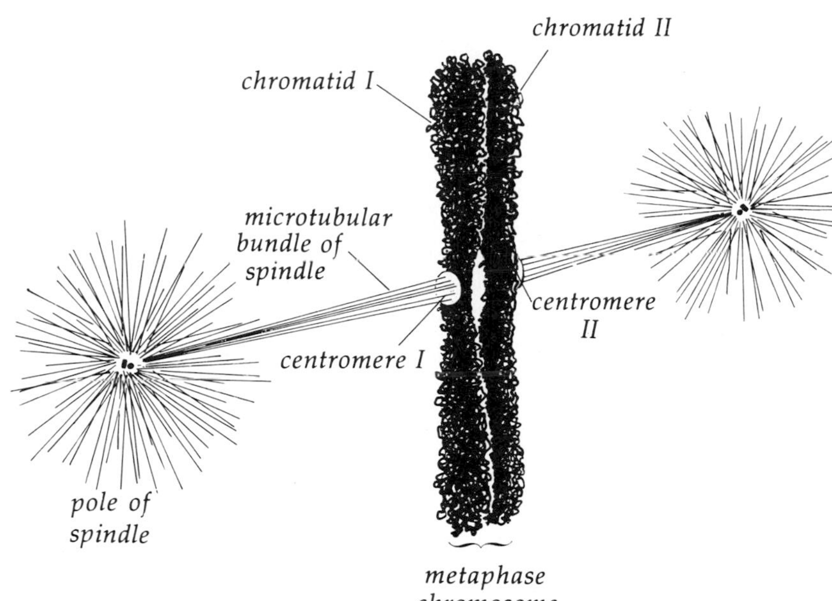

12–9. *Orientation of the centromeres at metaphase. Two centromeres are present. These are directed toward and make fiber connections to opposite poles of the spindle.*

Telophase

There is no clear-cut transition from anaphase to telophase. This final stage in mitosis is marked by the aggregation of the two groups of chromosomes into masses of chromatin in the vicinity of the poles. These masses, which subsequently form the two daughter nuclei, become surrounded by discontinuous segments of the nuclear envelope (Figs. 12–11 and 12–12). These segments progressively fuse until the chromatin mass is completely separated from the cytoplasm. As the reconstruction of the nuclear envelope takes place, the chromatin mass becomes diffuse and the nucleus begins to take on the typical interphase aspect (Figs. 12–12 and 12–13). Although the physical basis of the transition to the interphase state is unknown, it is likely that the chromosomes simply uncoil or unfold by reversing the process of condensation that occurs during prophase. As the telophase nucleus takes shape, the nucleolus is reorganized on sites on the chromosomes known as *nucleolar organizers.*

During telophase, most microtubules of the spindle disappear from the cytoplasm. In cells containing centrioles, the aster is all that remains of the spindle after the completion of telophase, except for short lengths of microtubules, which persist at the former midpoint of the spindle. In plants that lack centrioles and asters microtubules also persist at the midpoint region. In both cases, these persistent microtubules contribute to the division of the cytoplasm in the subsequent cytokinesis.

The Time Course of Mitosis

The total time for mitosis to run to completion may vary from as little as 5–10 minutes (in the early development of the *Drosophila* embryo) to several hours. Average values range between 30 minutes and 2–3 hours (see Table 12–1). Of the stages of mitosis, prophase is usually the longest. Teleophase is next longest, with metaphase and anaphase usually requiring the least time for completion (Table 12–2).

Synthesis and Respiration during Mitosis

RNA synthesis falls to undetectable levels late in mitotic prophase. These changes in the levels of RNA synthesis were first extensively studied in 1960 by J. H. Taylor, who measured the rate of RNA synthesis by following the incorporation of radioactive precursors of RNA. As synthesis terminated during late prophase, RNA was found to migrate from the nucleus to the

Table 12–1 Representative Values (in hours and minutes) for Mitotic Times in Various Types of Cells

Cell	Mitotic time
Vicia faba root meristem (19°C)	2:30 3:06
Pisum sativum (peas) root meristem (20°C)	2:57
Chick fibroblasts (38°C)	0:23
Mouse spleen (in culture)	1:30
Rat jejunum (in animal)	0:28
Jensen rat sarcoma (in animal)	0:27
Rat corneal epithelium (in animal)	1:10
Chortophaga (grasshopper) neuroblast	3:00
Drosophila egg	6:00
Psammechinus (sea urchin) embryo, 2–4 cell stage (16°C)	0:28
Psammechinus embryo 200–300 cell stage	0:32

Adapted from D. Mazia, Mitosis and the physiology of cell division. In J. Brachet and A. E. Mirsky, eds., *The Cell,* Vol. III. New York: Academic Press, 1961. Reprinted by permission. See original paper for references.

cytoplasm, so that by metaphase neither RNA nor RNA synthesis could be detected in the chromosomes. Complete cessation of RNA synthesis at late prophase was correlated with disappearance of the nucleolus. RNA synthesis was found to begin again during telophase; at the same time, the nucleolus was reformed at the organizer sites of the chromosome. These changes in the rate of RNA synthesis, followed by Taylor in cells of the Chinese hamster, have since been confirmed by a large group of experimenters working with protozoa, several higher organisms, and cells in tissue culture.

Taylor found that the synthesis of protein continues throughout mitosis, but at a greatly reduced rate. Because RNA synthesis comes to a close at the end of prophase, this protein synthesis must depend on a pool of previously synthesized messenger. One group of nuclear proteins has been found by autoradiography to migrate back and forth between the nucleus and cytoplasm during the cell cycle. L. Goldstein and later D. M. Prescott followed the migration of these proteins by growing *Amoeba* in labeled amino acids. Nuclei from these labeled cells were then transplanted to unlabeled cells. A fraction of the nuclear protein, about 40 percent, was found to migrate to the cytoplasm during mitotic prophase. During reconstitution of the nucleus in telophase, this protein fraction returned to the nucleus; this cycle was continuously repeated in subsequent divisions. The function and significance of these migrating nuclear proteins is unknown.

The oxidation of fuel substances, with synthesis of ATP, continues during mitosis, but usually at reduced

Table 12–2 Length of the Stages of Mitosis (in minutes) for Selected Cells

Cell	Prophase	Metaphase	Anaphase	Telophase
Yoshida sarcoma (35°C)	14	31	4	21
MTK sarcoma (35°C)	10	44	5	18
Mouse spleen (in culture)	20–35	6–15	8–14	9–26
Triton liver fibroblast (26°C)	18 or more	17–38	14–26	28
Chortophaga (grasshopper) neuroblast (38°C)	102	13	9	57
Pea endosperm	40	20	12	110
Iris endosperm	40–65	10–30	12–22	40–75
Micrasterias rotata (desmid)	60	21–24	6–12	3–45

From D. Mazia, Mitosis and the physiology of cell division. In J. Brachet and A. E. Mirsky, eds., *The Cell,* Vol. III. New York: Academic Press, 1961. Reprinted by permission. See this paper for original references.

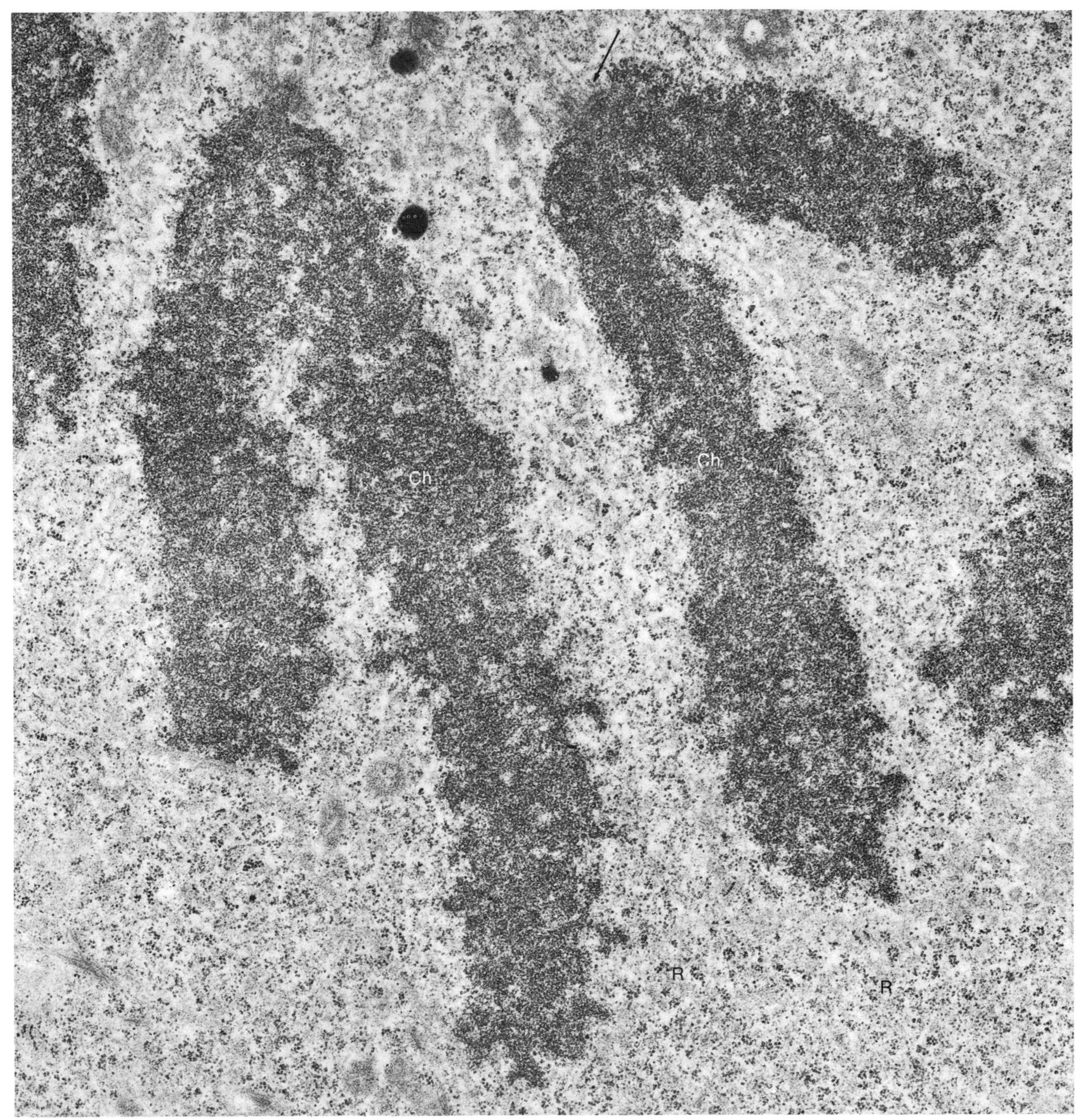

12–10. Anaphase chromosomes (ch) of the Tasmanian wallaby. The chromosome is pulled into a "V" or "J" shape depending on the location of the centromere (arrow). × 24,000. From Cell Ultrastructure *by W. A. Jensen and R. B. Park, 1967. Courtesy of B. R. Brinkley and Wadsworth Publishing Company, Inc.*

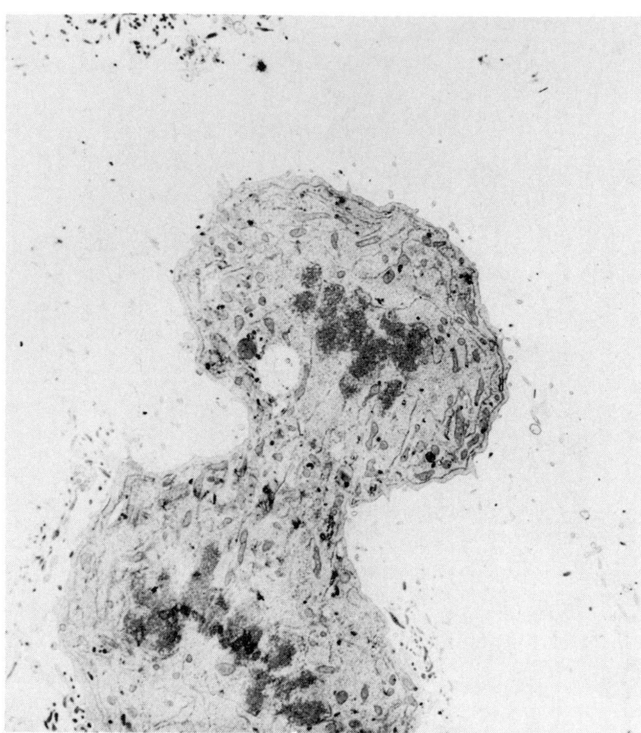

12–11. *Early telophase in a human tissue culture cell. Newly synthesized segments of the nuclear envelope are being formed over the surface of the chromosome masses at the poles of the spindle. × 2,500. Courtesy of G. G. Maul and Academic Press, Inc.* [J. Ultrastr. Res. *31:375 (1970)*]

levels. It is not known whether the lowered rate of respiration during mitosis reflects a reduction in the overall energy requirements of the cell or indicates that the mechanisms in division interfere in some way with cellular metabolism. It is also possible that the drop in respiratory rate indicates that a reservoir of ATP, accumulated during interphase, is used as an energy source during mitosis.

The Significance of Mitosis

The end result and functional significance of replication and mitosis is the production of two daughter nuclei, each with genetic capacity equivalent to the original parent nucleus. Mitosis is a distributive mechanism basic to all higher organisms. This process probably arose in evolution as an adaptation to compensate for mechanical difficulties encountered in the division and distribution of DNA codes of greater and greater length. In bacteria, in which only one chromosome of relatively short length is present, nuclear division is accomplished by a simple mechanism probably involving the plasma membrane. After replication, the two "chromatids" of the replicated bacterial chromosome make separate connections with the plasma membrane; as the cytoplasm divides, these points are pulled apart, carrying the replicated chromosomes into separate daughter cells (for further details on bacterial division, see Chapter 13). It is likely that this simple mechanism became inadequate as the coded information involved lengths of DNA too great to be restricted with mechanical convenience to a single chromosome. The appearance in evolution of the mitotic mechanism provided the capacity for the equal distribution of a replicated genome of great complexity, subdivided into a large number of subunits (chromosomes). The total process works equally well with either haploid or diploid cells, or indeed any degree of ploidy.[3]

Although many small variations are noted in mitosis in different organisms, none of these represent basic and fundamental differences in the process as outlined. In some protozoa, such as the ciliates, mitosis in the micronucleus proceeds without breakdown of the nuclear envelope. The spindle may be formed either inside or outside the envelope. If inside, spindle connections are made as in other organisms. If outside, the chromosomes become attached at the centromeres to points on the nuclear envelope, which in turn attach to spindle fibers (see Fig. 13–17). In this case, the movement of the chromosomes is accompanied by elongation and extension of the nuclear envelope to the poles. Similar processes and variations significant for the possible evolution of the spindle mechanism occur in other protozoa, in the dinoflagellates, and in yeast. These are reviewed in a more detailed discussion of the spindle in Chapter 13.

---

[3] An alternate mechanism to mitosis, termed *amitosis,* may be found under pathological conditions in some cell types, such as tumors. In this form of reproduction, the nucleus simply divides by means of a gradually deepening central furrow that eventually cuts the nucleus into two parts. To work successfully, this system requires that the total genetic information of the nucleus be present in so many duplicates that both daughter nuclei are likely to receive one or more complete copies.

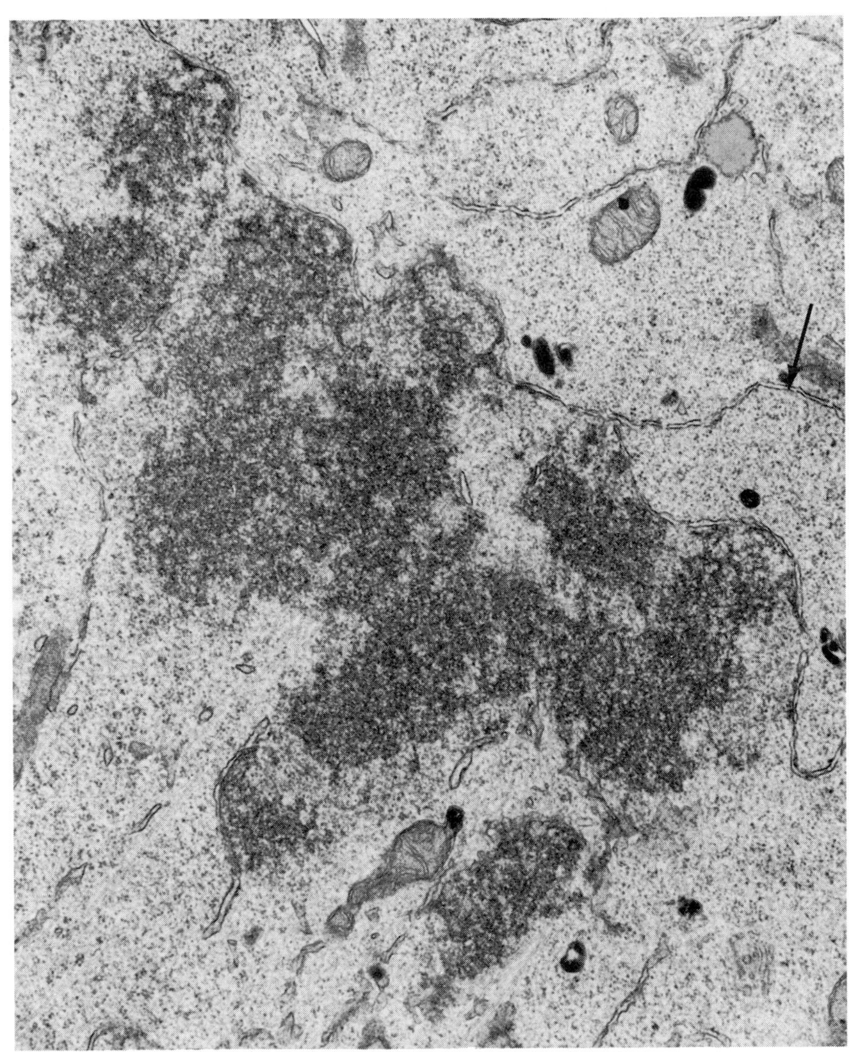

12–12. Higher magnification of a part of the chromosome mass near the pole of the spindle. Segments of the nuclear envelope are visible around much of the chromosomal material. In the early stages of nuclear envelope formation, the vesicles giving rise to the membrane system closely resemble elements of the rough endoplasmic reticulum (arrow). × 20,000. Courtesy of G. G. Maul and Academic Press, Inc. [J. Ultrastr. Res. *31:375 (1970)*]

Cytokinesis

In most cells, mitosis and cytokinesis take place in close sequence. Usually, division of the cytoplasm is initiated during mitotic anaphase. In animal cells (Fig. 12–14), a furrow appears in the plasma membrane at this time and gradually deepens until the daughter cells are completely separated. The furrow follows a plane through the former midpoint of the spindle, which is marked by short lengths of spindle microtubules that persist in this region. In plants, short lengths of microtubules also persist at telophase at the midpoint of the spindle. These persisting microtubules become invaded by a layer of small vesicles. This layer of microtubules and vesicles expands outward until the cell walls are reached. In the path of the expanding layer of vesicles, classically termed the *phragmoplast,* the plasma membrane and primary cell wall are laid down between the daughter cells. The expanding membrane and cell wall can be seen in Figs. 12–2e and 12–2f (see also Figs. 13–40 to 13–43; cytokinesis is taken up in detail in Chapter 13).

Replication of the Chromosomes

Replication during interphase entails the synthesis and duplication of all of the molecular components of the chromosome, including DNA, histones, and any

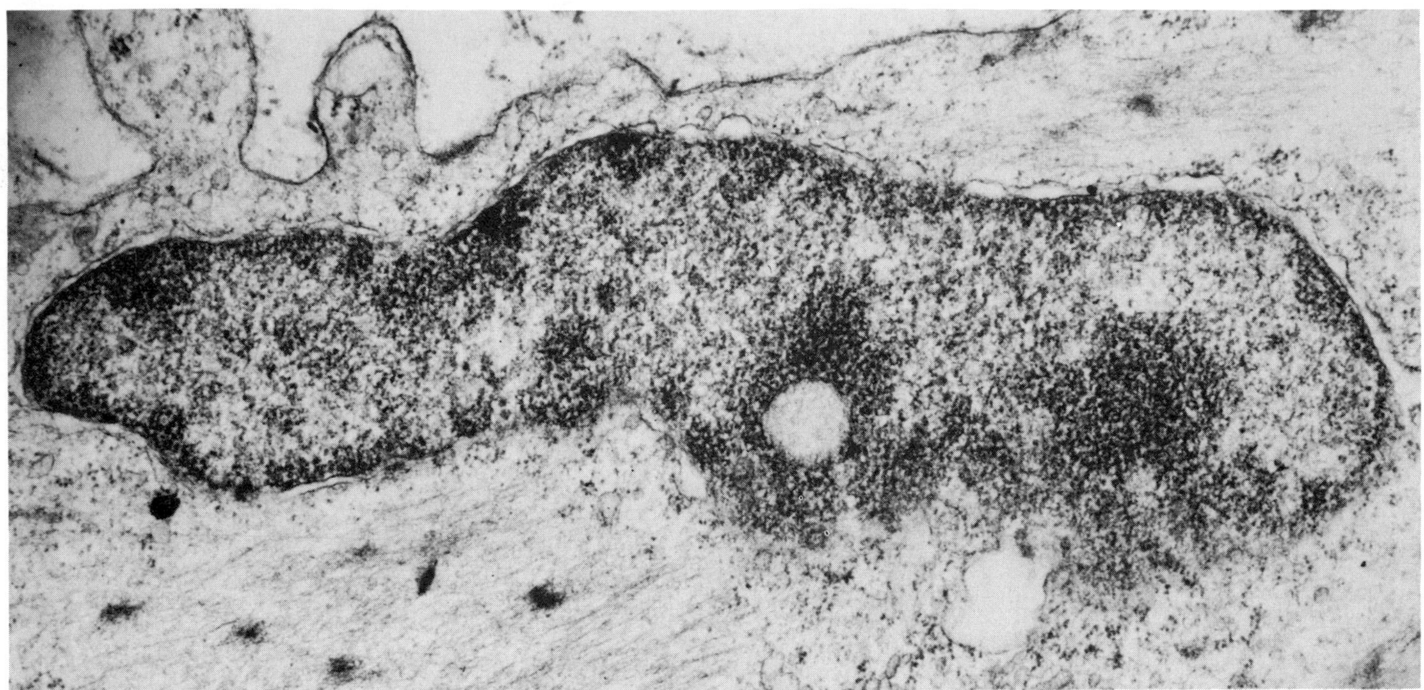

12–13. A telophase nucleus from dividing tissue of a chick embryo. Formation of the nuclear envelope is more complete than in Fig. 12–12; the chromatin has initiated transformation to the interphase state. × 37,000. Courtesy of J. L. S. Cobb.

nonhistone proteins that are present. Only the replication of DNA is understood in any detail; many uncertainties surround the synthesis mechanisms of histone and nonhistone protein.

When Watson, Crick, and Wilkins discovered the molecular structure of DNA, they pointed out that an obvious mechanism for precise replication is inherent in the structure of the molecule (Fig. 12–15; see Chapter 8 for details of the molecular structure of DNA). At any point along the DNA duplex, a nitrogenous base on one of the two stands of the molecule can pair with only one base on the complementary strand (Fig. 12–15a). If an adenine, for example, occurs on one of the two strands, only thymine will normally be found at the same level in the opposite strand. Thus, the sequence of purine and pyrimidine bases on one strand determines the sequence on the opposite strand. If the two strands are separated (Fig. 12–15b), it is obvious that each could serve as a *template* for the synthesis of the missing half (Fig. 12–15c).

This possible mode of replication, outlined by Watson, Crick, and Wilkins, set the stage for subsequent work. It was soon pointed out by Max Delbruck and Gunter S. Stent that the distribution of the newly synthesized strands could follow one of three pathways after replication (Fig. 12–16). Delbruck and Stent called the end result of these three pathways *conservative*, *semiconservative*, and *dispersive* replication. In conservative replication (Fig. 12–16a), the two strands of the original molecule, after separating and serving as templates, pair again; the newly synthesized strands likewise pair together. This produces two molecules, one composed entirely of "old" strands and one of "new." In semiconservative replication (Fig. 12–16b), the two original strands, after separating and serving as templates, remain paired with their newly synthesized complementary strands. As a result of semiconservative replication, each of the two molecules formed contains an "old" and a "new" half. The terms "conservative" and "semiconservative" used in these models imply that in both pathways the original strands, although separating and acting as templates, remain intact. In dispersive replication (Fig. 12–16c), the linear continuity of the "parent" DNA strands is broken, and the parts are distributed randomly between daughter molecules. One possible mechanism by which this might occur is shown in Fig. 12–16c. In this case, numerous breaks appear in the "parent" strands before unwinding and separation. Distribution of the broken substrands between the replicated molecules is random.

Experimental Demonstration of the Semiconservative Pathway

Two definitive experiments, one carried out in eukaryotes and one in prokaryotes, established that DNA replication at interphase follows the semiconservative pathway. One of these, reported by J. H. Taylor in 1957, involved an autoradiographic study of mitotic cells of the plant *Vicia*. Taylor grew *Vicia* roots for one generation in a medium containing radioactive thymidine and followed the distribution of label in the chromosomes in subsequent divisions. After exposure of the root tips to tritiated thymidine during the S period of interphase, Taylor noted that at the subsequent metaphase both chromatids of each metaphase chromosome were labeled (Fig. 12–17). This observation is not compatible with conservative replication and also indicates that there is more than one subunit in the chromosome at replication.

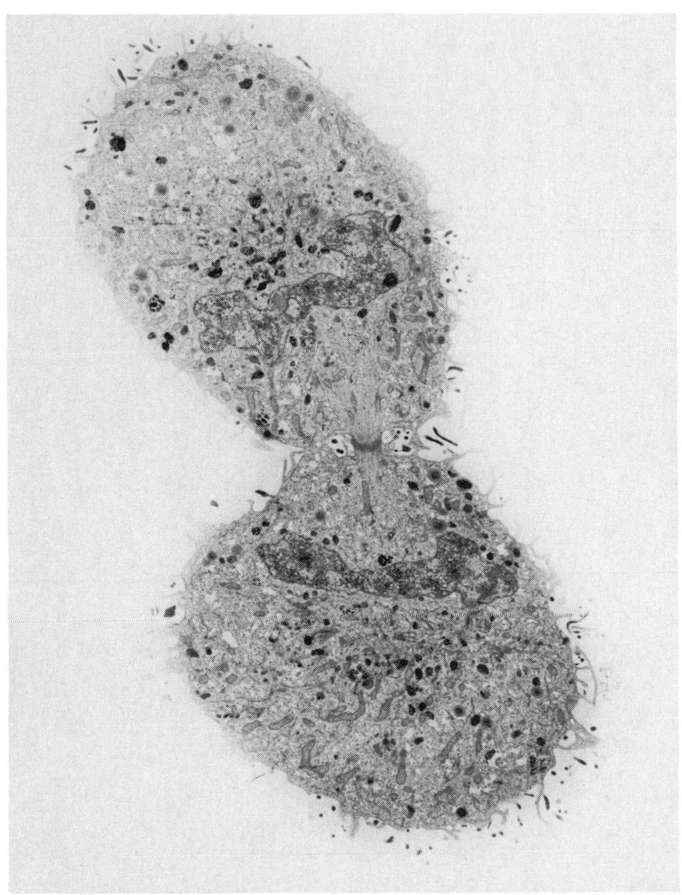

12–14. A telophase cell from human tissue culture. The narrow region marks the site of the advancing furrow, which follows a plane corresponding to the former midpoint of the spindle. Spindle microtubules and vesicles persist in this region during cytokinesis. × 4,500. Courtesy of G. G. Maul and Academic Press, Inc. [J. Ultrastr. Res. *31:375 (1970)*]

12–15. Schematic representation of the replication of DNA. a, *The "parent" duplex.* b, *Separation of strands after unwinding.* c, *Synthesis. The separated parent strands act as templates. After replication is completed, the parent DNA duplex has been precisely duplicated.*

strand I
strand II

a			b			c			
A	T		A	T		A	T	A	T
C	G		C	G		C	G	C	G
C	G		C	G		C	G	C	G
T	A	→	T	A	→	T	A	T	A
G	C		G	C		A	T	A	T
A	T		A	T		C	G	C	G
C	G		C	G		T	A	T	A
T	A		T	A					

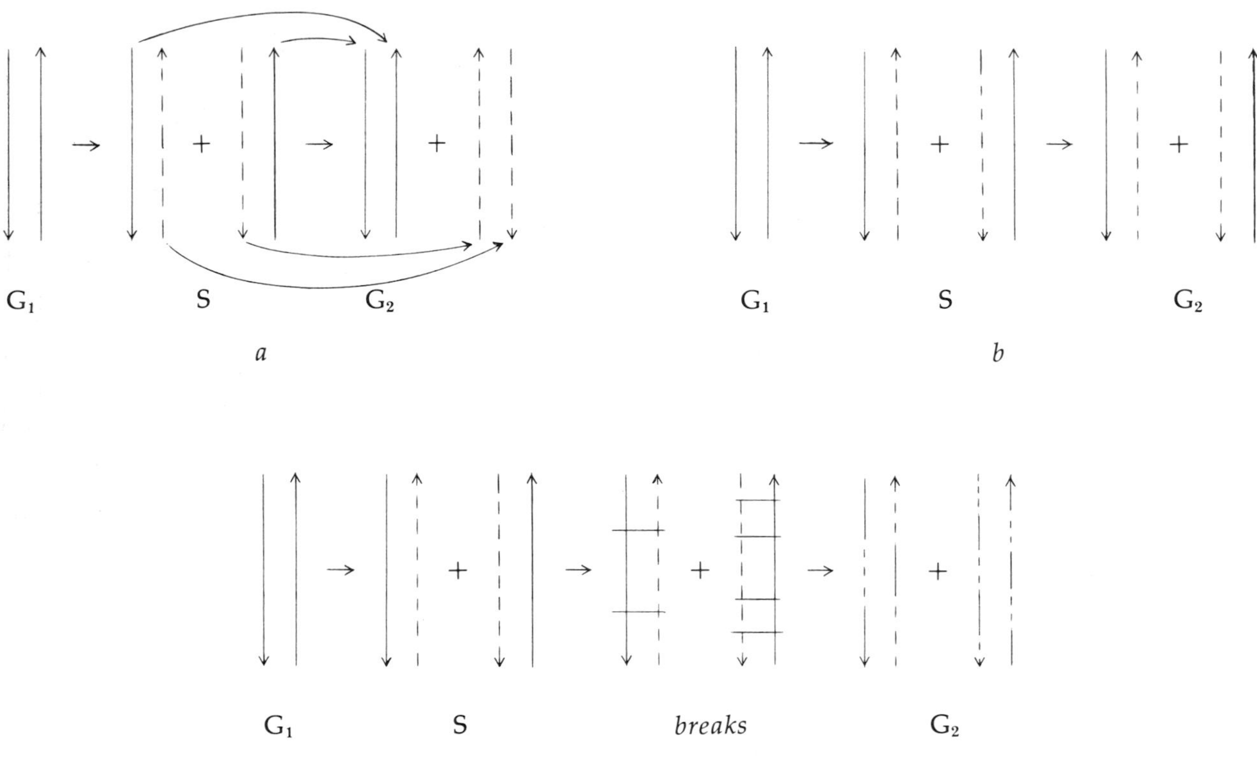

12–16. Three hypothetical pathways for DNA replication, as proposed by Delbruck and Stent (see text). a, *Conservative replication.* b, *Semiconservative replication.* c, *Dispersive replication.*

To understand these conclusions, assume that the G_1 chromosome consists of one long DNA duplex. The two nucleotide strands of this molecule separate and serve as templates for replication of complementary strands. Because this replication occurs in the presence of radioactive thymidine, the newly synthesized strands, shown as dotted lines in Fig. 12–16, will be labeled. If replication follows the conservative pathway (Fig. 12–16a), the two old (unlabeled) strands will reassociate after replication, and the two new (labeled) strands will likewise associate. During prophase, each of these DNA molecules, one completely old and one new, will coil down into the two chromatids of the metaphase chromosome. If autoradiographs are made at metaphase, only the "new" chromatid would be expected to show the presence of label. However, Taylor noted that both chromatids are labeled at the first metaphase following DNA synthesis in the presence of labeled thymidine. This observation eliminates conservative replication as a possibility.

This result also indicates that at the time of replication the chromosome behaves as if it contains at least two subunits. Assume that the chromosome contains only one subunit, which would be the case if the chromosome contained single-stranded DNA. During replication in labeled medium, this single strand would act as a template for synthesis of a single-stranded copy. During prophase, the "old" template would coil down into one of the two chromatids of the metaphase chromosome and would be unlabeled. The "new" strand, synthesized from labeled precursors, would form the other chromatid. This, as in the case of conservative replication, would be recognized as segregation of label at first metaphase, which as noted, Taylor did not observe. This result by itself does not *prove* that the chromosome contains more than one subunit, because both chromatids could be labeled in *dispersive* replication even if only one subunit is present. But further experimentation by Taylor allowed the elimination of dispersive replication as a possibility and definitely established that the chromosome replicates as if two subunits are present.

In subsequent experiments, chromosomes were examined for the distribution of label in the *second*

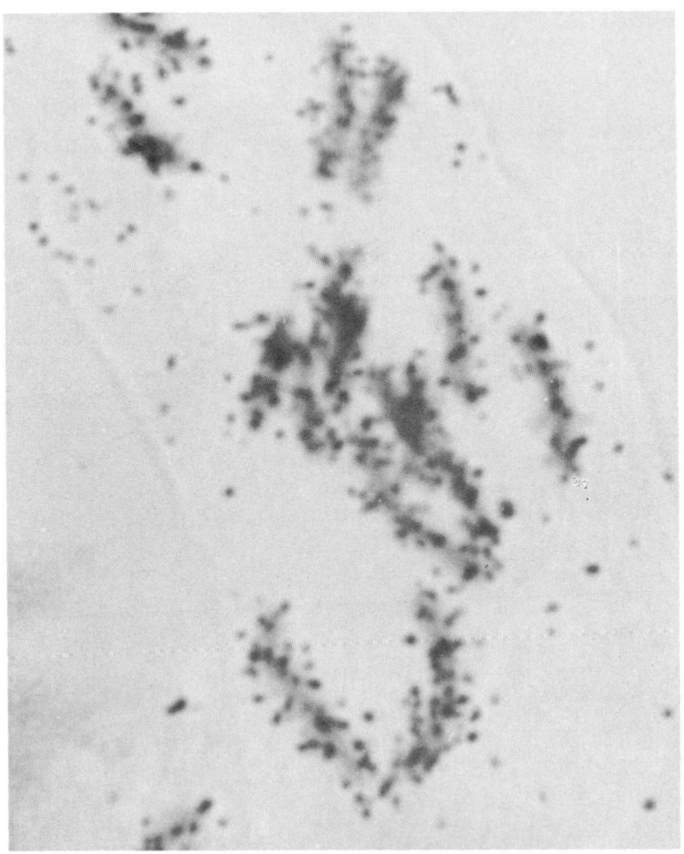

12–17. Autoradiograph of Vicia *chromosomes at the first metaphase following DNA synthesis in the presence of tritiated thymidine. Both chromatids of each chromosome are labeled.* × *2,700. From* Molecular Genetics, *ed. J. H. Taylor, 1963. Courtesy of J. H. Taylor and Academic Press, Inc.*

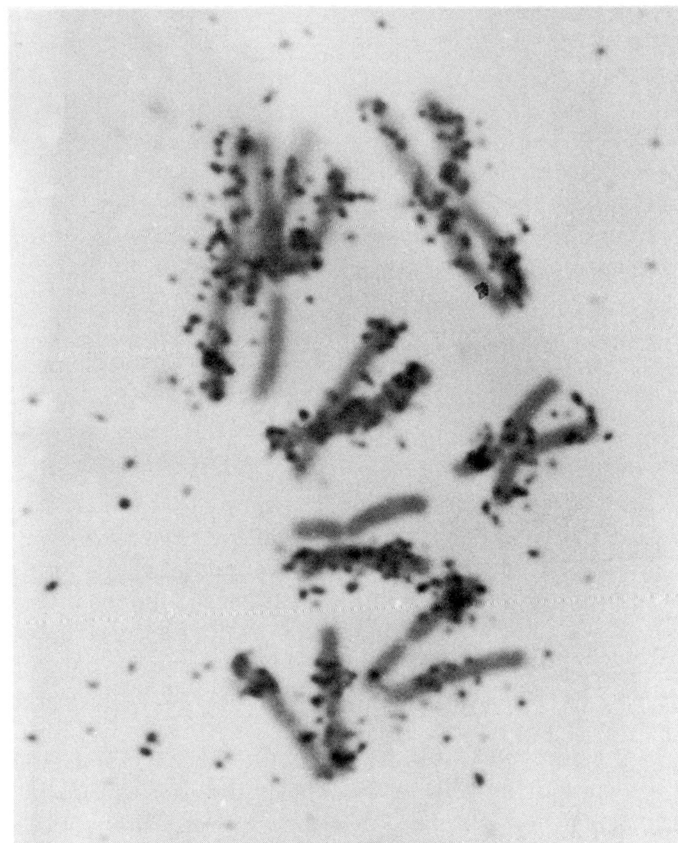

12–18. Autoradiograph of Vicia *chromosomes at the second metaphase following replication in radioactive precursor. At any level, only one of the two chromatids is labeled.* × *2,700. From* Molecular Genetics, *ed. J. H. Taylor, 1963. Courtesy of J. H. Taylor and Academic Press, Inc.*

metaphase following replication in radioactive medium. To accomplish this, cells were exposed to radioactive DNA precursor and allowed to replicate. Following replication, root tips were placed in a nonradioactive medium and allowed to proceed through mitosis and into a second cycle of replication. After this second replication in unlabeled medium, autoradiographs were made of the chromosomes at the subsequent metaphase. Taylor observed that at the second metaphase only one of the two chromatids of each metaphase chromosome was labeled (Fig. 12–18). This finding eliminated dispersive replication, because in any chromatid replicated by a dispersive pathway, chances are that a mixture of old and new subunits will always be present and that at least some label will be observed in almost all chromatids.

Taylor's observations of segregation of label at second metaphase coincide precisely with expectations if replication is semiconservative and the chromosome consists of two subunits (Fig. 12–16b). In semiconservative replication, each chromatid, at the first metaphase following replication in labeled medium, will consist of one labeled and one unlabeled subunit. During anaphase, these chromatids will be separated and pulled to opposite poles of the cell. One of these chromatids is followed diagrammatically in the subsequent interphase and mitosis in Fig. 12–19. During replication in the subsequent interphase, the two subunits of the chromatid, now a chromosome, will separate to act as templates for replication (Fig. 12–19b). Because this second replication is in unlabeled medium, the newly synthesized strands will be unlabeled.

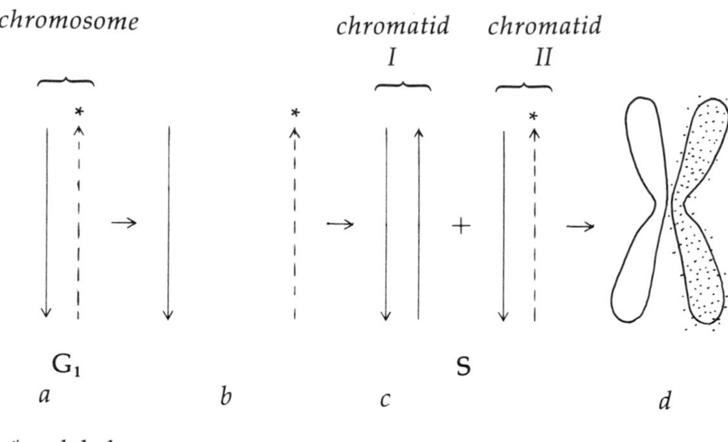

12–19. *Diagram illustrating the basis for segregation of label at second metaphase (see text).* a, *Labeled chromosome.* b, *Separation of subunits.* c, *Replication in unlabeled medium.* d, *Metaphase.*

These will remain with their template strands, only one of which was labeled (Fig. 12–19c). Therefore, only one of the two chromatids will contain labeled strands; the other chromatid will be completely unlabeled (Fig. 12–19d). This was recognized in Taylor's experiments as segregation of label between the two chromatids at the second metaphase following exposure to a radioactive precursor.

These results also show that the chromosome does not contain more than two parts that act as effective units at the time of replication. Assume that replication is semiconservative but that there are four instead of two subunits in the chromosome. This would be the case if each chromosome consists of two DNA molecules instead of one (Fig. 12–20). During replication, these four strands will separate, and each will make a labeled copy. At the first metaphase following replication, both chromatids of each metaphase chromosome will be labeled, because each consists of two labeled and two unlabeled strands. Following one of these chromatids through a subsequent replication in unlabeled medium (Fig. 12–20b) shows that each of the two chromatids produced will contain one labeled and three unlabeled strands. Even if distribution of the labeled and unlabeled strands is random between the two chromatids, a significant proportion will still show label in both chromatids (called *isolabeling*) at second metaphase. Following one chromatid through a third replication and division (Fig. 12–20c) shows that the first regular segregation of label, if four subunits are present, is expected at the *third* metaphase following a pulse of radioactive precursor.

Almost all chromosomes in Taylor's experiments showed segregation of label at second metaphase, demonstrating that the chromosome has two effective subunits at replication. A small percentage, about 5 percent, actually showed isolabeling rather than segregation of label at second metaphase; these exceptions, ascribed to various causes by Taylor, are considered significant by other investigators and form the basis of one line of argument in support of the multistranded chromosome (see Peacock, 1963, and the section on redundancy of DNA later in this chapter). Whatever the arguments concerning strandedness of the chromosome, Taylor's results show that replication is semiconservative and effectively eliminate conservative and dispersive replication in higher organisms. At the same time, they provide strong evidence that the chromosome has two effective subunits at the time of replication.

A second major series of experiments established that replication is semiconservative in prokaryotes. These experiments were performed by Matthew Meselson and Franklin W. Stahl, using the bacterium *E. coli.* Meselson and Stahl grew bacteria in a medium containing a heavy isotope of nitrogen (N^{15}) for several generations, long enough for the DNA of the bacteria to become completely labeled with N^{15} nitrogen. After growth in the N^{15} medium, the bacteria were transferred to a medium containing N^{14} nitrogen. At intervals after the transfer, DNA extracts were made and compared to N^{14} and N^{15} standards. This was done by analyzing the density of the samples in the ultracentrifuge. This technique is sensitive enough to distinguish between DNA containing the light or the heavy isotope. The results of centrifugation are shown in Fig. 12–21. DNA extracted at the time of transfer from N^{15} to N^{14} medium centrifuged down in a single

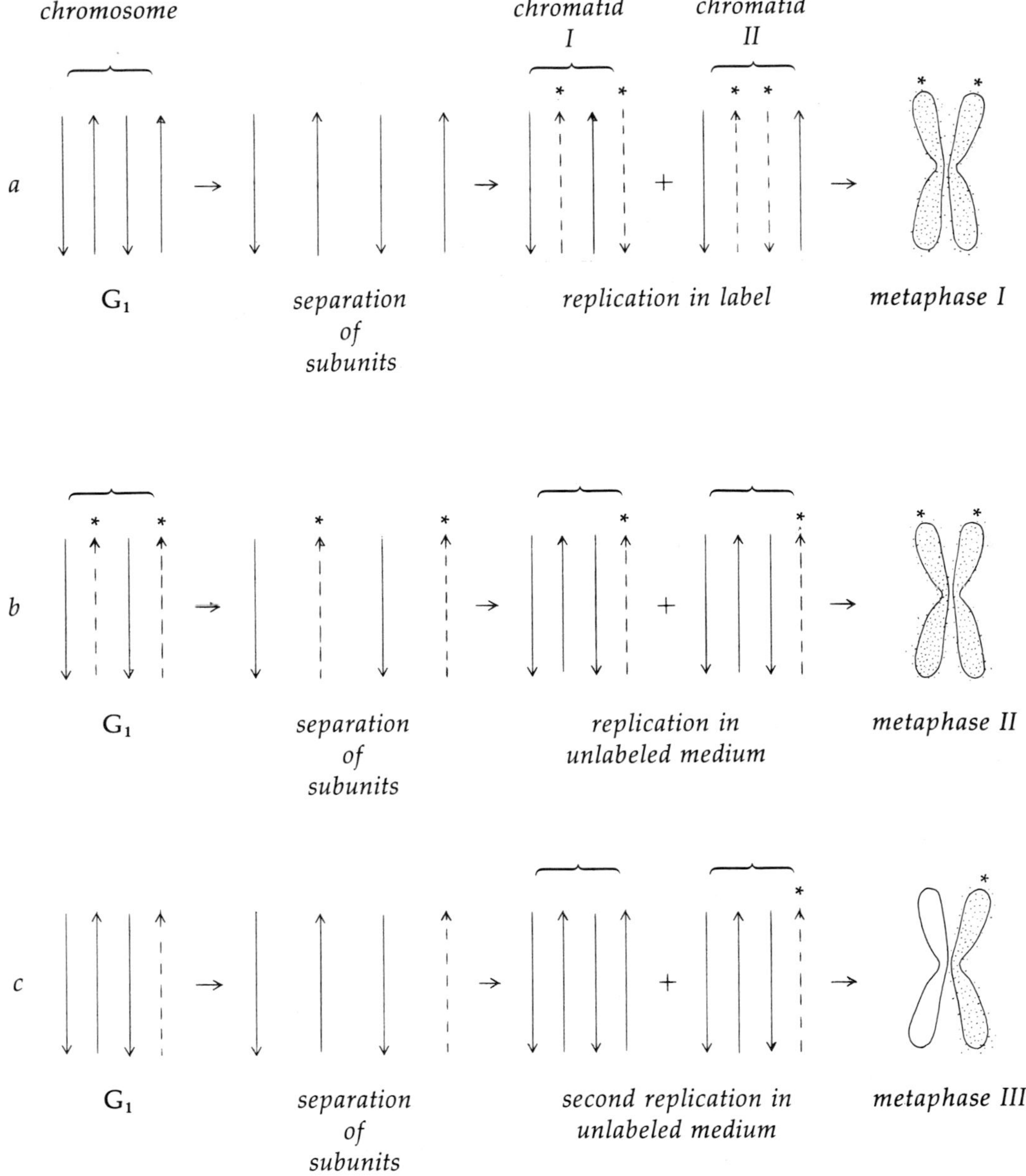

12–20. Expected segregation of label if the chromosome contains four functional subunits instead of two (see text).

band characteristic of N^{15} DNA (generation time 0, Fig. 12–21). After a period of time sufficient for most of the cells to undergo one replication (generation time 0.3 to 1.0, Fig. 12–21) the band characteristic of N^{15} DNA disappeared in DNA extracts and a new band appeared, intermediate in density between N^{15} and N^{14} DNA. After longer periods of growth (generations 1.1 to 4.1, Fig. 12–21), a new band characteristic of "pure" N^{14} DNA began to appear. The intermediate band, however, persisted and did not disappear.

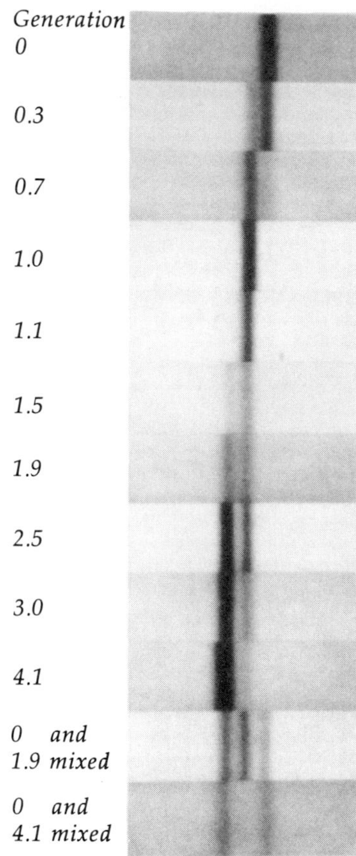

12–21. Centrifugation of DNA of E. coli *extracted at various times after transfer from N^{15} to N^{14} medium. The density increases toward the right. Bands centrifuging to a level displaced toward the left therefore represent "lighter" DNA. The three bands in the next-to-lowest frame correspond, reading from left to right, to N^{14}–N^{14} DNA, N^{14}–N^{15} hybrid DNA, and N^{15}–N^{15} DNA. Courtesy of J. Meselson and F. W. Stahl.*

These results are consistent only with semiconservative replication. If replication were conservative, two distinct bands, one characteristic of N^{14} and one of N^{15} DNA, would appear in the ultracentrifuge after one generation. This is obvious from Fig. 12–16a. In this case, the two N^{15} subunits of the bacterial DNA would separate and serve as templates for copies made with N^{14} DNA. After replication, the original "old" N^{15} DNA would reassociate, and the "new" N^{14} DNA would also associate. Thus, the two DNA products would be identifiable after one generation as separate N^{15} and N^{14} bands. The production of a single band of *intermediate* density after one generation, actually observed by Meselson and Stahl, showed that instead all of the molecules consisted of hybrids of one N^{15} and one N^{14} subunit. This result is expected if replication is semiconservative. The same outcome could result from dispersive replication; however, observation of the distribution of density after two generations rules out this possibility. If replication were dispersive, the DNA, even after many generations, would be expected to remain of intermediate density, since little pure N^{14} or N^{15} DNA would be produced. But Meselson and Stahl actually observed, after two generations, a band characteristic of "pure" N^{14} DNA in addition to the intermediate N^{15}–N^{14} "hybrid" band. This is exactly the result expected if replication is semiconservative and corresponds precisely to Taylor's observation of segregation of label at second metaphase in *Vicia*. All of the DNA molecules after the first replication, according to the semiconservative mechanism, contain one N^{15} and one N^{14} strand. At the second replication, these separate and serve as templates for the synthesis of complementary N^{14} strands. Thus, because the templates remain paired with the new strands, one set of product molecules will consist of N^{15}–N^{14} DNA and one of N^{14}–N^{14} DNA. This explains the appearance of the N^{14} band and the persistence of the N^{14}–N^{15} hybrid band after two generations of growth.

The Synthesis of DNA

Taylor's results in eukaryotes, published in 1957, and Meselson and Stahl's work with prokaryotes, published in 1958, established that replication is semiconservative. During the same period, other research efforts, most notably in the laboratory of Arthur Kornberg and his associates, were directed toward uncovering the biochemical mechanisms of DNA synthesis. Kornberg, who was awarded the Nobel prize for his work in 1959, purified an enzyme from the bacterium *E. coli* capable of polymerizing nucleotide triphosphates into DNA in cell-free systems and was able to study the mechanism of this reaction.

In cell-free systems, this enzyme, *DNA polymerase*, requires a *template* DNA, a *primer* DNA, and a supply of the four nucleotide triphosphates ATP, TTP, GTP, and CTP. The probable mechanism of synthesis is shown in Fig. 12–22. The template DNA, to be active in synthesis, must already have in place a short complementary strand a few nucleotides in length at the point at which synthesis is to begin. This short length of complementary DNA is the primer and is evidently required because the enzyme can add nucleotides only to the end of an existing chain with an exposed 3′—OH group.

At the initiation of DNA synthesis the enzyme binds to the template-primer complex near the 3′—OH group of the primer (Fig. 12–22b). The medium surrounding the enzyme-template-primer complex contains the four nucleotide triphosphates; these are equally available to the enzyme. The specific nucleotide triphosphate selected from the medium is determined by the usual base pairing rules. In Fig. 12–22, the first exposed base is an adenine; base pairing will favor the binding of a thymidine triphosphate from the medium at this point (Fig. 12–22b). The active sites on the enzyme are probably arranged in space so that the thymidine triphosphate from the medium is bound in a position adjacent to the 3′—OH end of the primer. At this point, the enzyme possibly undergoes a change in conformation that promotes the formation of a linkage between the innermost phosphate of the thymine and the oxygen attached to the 3′ carbon of the primer (Fig. 12–22c). At the same time, the two terminal phosphates of the thymidine triphosphate are hydrolyzed and released to the medium; the energy required for formation of the new phosphodiester linkage is provided by this hydrolysis. As a result of this reaction, the enzyme returns to its original conformation and moves to the next exposed base on the template. The entire cycle then repeats until the end of the template is reached.

Each nucleotide phosphate added to an exposed 3′—OH group by the enzyme provides the next 3′—OH group for the polymerizing reaction. Thus, a 3′—OH group is always present at the "new" end of the growing chain. Because the "old" end of the growing chain contains an exposed 5′ group, the chain is said to grow in a 5′ to a 3′ direction (see also Chapter 9). All of the known DNA polymerases synthesize DNA in the 5′ to 3′ direction, opening a number of questions about the possible mechanisms that operate to replicate both strands of a DNA duplex simultaneously (see below).

DNA polymerase. DNA polymerase can use as a template either double-stranded or single-stranded DNA or synthetic DNAs consisting of repeating sequences of the same or different nucleotides. There are indications that a part of the accuracy in the "copying" mechanism depends on the structure of DNA polymerase and not only on base pairing, because mutations in the gene coding for DNA polymerase in microorganisms result in an increase in "mistakes" in the bases incorporated into the newly synthesized strand (see Speyer et al., 1966).

DNA polymerase has a molecular weight slightly in excess of 100,000. Evidently only one polypeptide chain comprises the entire molecule, because denaturation releases only one chain of protein. The enzyme is complex, with at least three probable sites; one for binding to the template, one for binding of nucleotide triphosphates, and a site for attachment to the 3′—OH end of the primer.

It is not certain that the DNA polymerase isolated by Kornberg is actually the replicating enzyme in the cell. The DNAs replicated by this enzyme in cell-free systems show imperfections such as failure of the newly synthesized strands to separate or branching of the product DNA. The latter effect is observed if native double-stranded DNA is used as template. Further, the highly branched DNA synthesized cannot be separated from the template by heating, showing that template and product ("old" and "new") strands are covalently linked instead of hydrogen-bonded as expected. In spite of these difficulties, many investigators provisionally accept the Kornberg enzyme or an enzyme with similar properties as the *in vivo* DNA polymerase and regard the peculiarities noted in cell-free systems as the result of failure to precisely duplicate the physiological conditions of the intact cell.

DNA Synthesis and Semiconservative Replication

The unwinding problem. The fact that DNA synthesis is semiconservative requires that the template DNA duplex must unwind into separate strands during replication. It has often been argued that long DNA duplexes could not be unwound during replication without imposing a requirement for prohibitively high increments of energy. This is because in order to unwind, the DNA must rotate around the long axis of the molecule; this must be accomplished in the nucleoplasm, which is likely to be at least semiviscous. But calculations by C. Levinthal and H. R. Crane indicate that the energy required to unwind the template into separate strands probably represents less than 1 percent of the total required for synthesis of the phosphodiester bonds linking successive nucleotides together. This small increment in the total requirement is easily met by the energy supplied to the system by the hydrolysis of terminal phosphates from the nucleotide triphosphates added as the chain unwinds. However, it is not known whether a direct coupling exists between the polymerizing and unwinding systems.

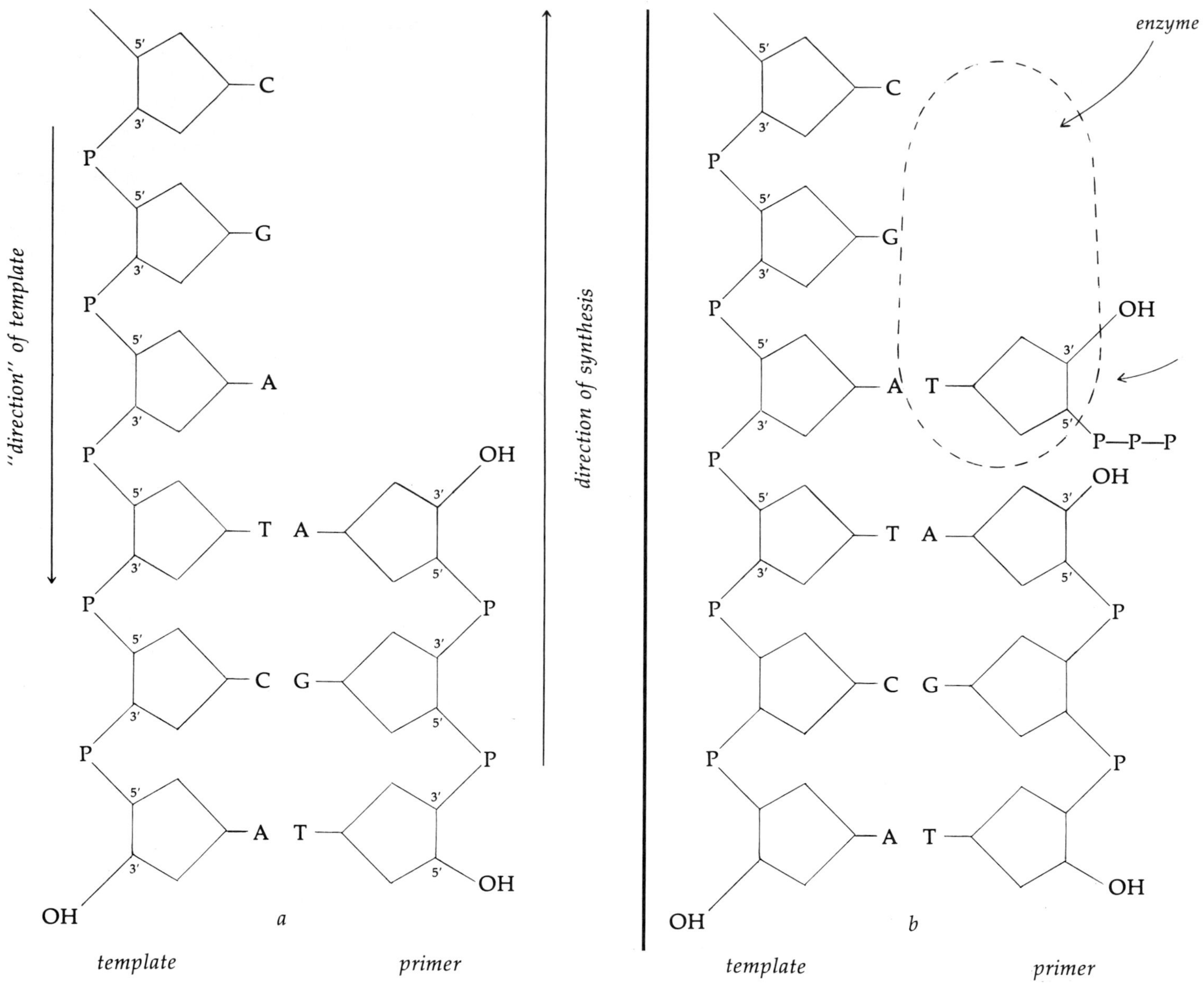

More serious problems are presented by circular DNA molecules. In bacteria and many other prokaryotes, the chromosome exists as a single closed circle of DNA. The DNA of mitochondria and possibly of chloroplasts is also circular (see Chapter 16). It is obvious that two strands of a DNA molecule covalently linked into a closed circle cannot unwind and separate. Yet, as established by the Meselson and Stahl experiments, prokaryote DNA replicates according to a semiconservative mechanism, which means that the strands of the circle do indeed unwind and come apart. This paradox has not yet been completely resolved.

John Cairns has investigated details of the replication of bacterial DNA by means of autoradiography. Cairns labeled bacterial DNA by growing cells in the presence of radioactive precursor and then isolated and purified the DNA. In these preparations, the DNA could often be isolated as an intact circle. When placed under a photographic emulsion, these circles expose a pattern of grains that can be traced under the light microscope (see Fig. 12–23).

In autoradiographs of the DNA extracted from these

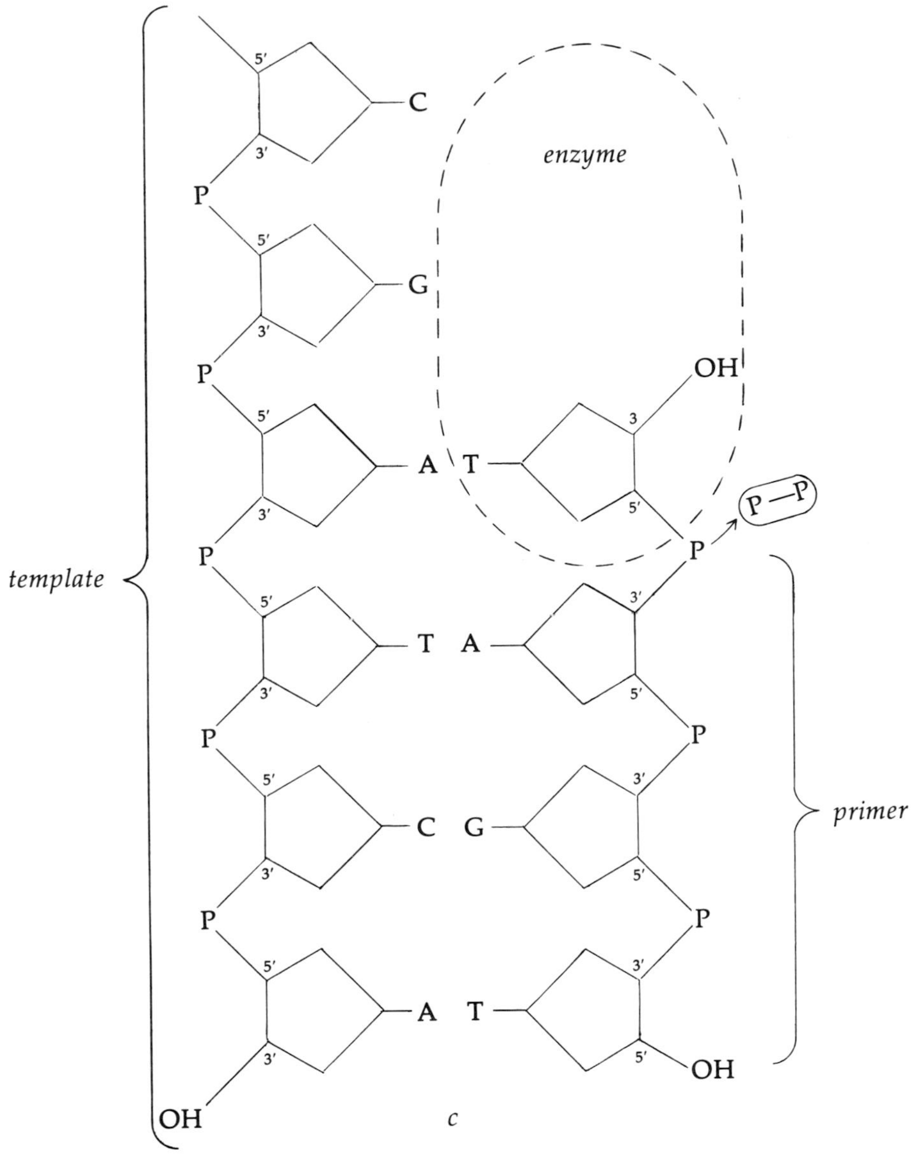

12–22. *Synthesis of DNA (see text).*

cells, a "fork" could often be seen at some point around the circle (arrow, Fig. 12–23). Cairns was able to show that the forks represent sites of replication by comparing the radioactivity on either side of the branching point. This was done by counting the number of grains exposed in the emulsion over the radioactive DNA. If labeled cells were allowed to replicate in unlabeled medium, the two branches of a fork in extracted DNA each proved to be one-half as radioactive as the part of the single molecule in front of the fork. This is the result expected if the forks represent points of synthesis and replication is semiconservative, because the newly replicated duplexes beyond the fork will contain one-half "old" and one-half "new" DNA. By measuring the total length of DNA in the circle and comparing the value obtained with the known replication time for *E. coli,* Cairns estimated that the replicating point moves along the molecule at a rate of 30 microns per minute. At this rate, the DNA would be required to unwind at approximately 10,000 RPM for replication to proceed. How can DNA in a closed circle unwind at this rate? Cairns proposes that a "swivel" exists at the point at which synthesis terminates, around which the DNA rotates to accomplish unwinding (Fig. 12–24).

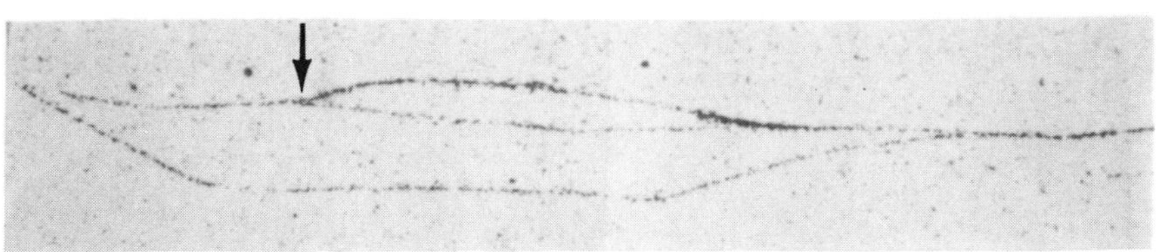

12–23. *Pattern of grains traced in a photographic emulsion by a radioactive chromosome of the bacterium* E. coli. *The chromosome has partly replicated. One of the two "forks" (arrows) represents the advancing site of replication.* × 270. *From "The Bacterial Chromosome" by J. Cairns.* Scientific American 214 *(1966). Copyright © 1966 by Scientific American, Inc. All rights reserved.*

Replication would begin at one side of the swivel and proceed around the molecule; as synthesis reaches the swivel the finished circles separate. Presumably, the swivel would also replicate at this time. As an alternative, more simple hypothesis, many investigators have proposed that unwinding of the DNA is accomplished by rotation around single-stranded breaks introduced enzymatically into the duplex (Fig. 12–24b).

Problems connected with the sequential replication of DNA. Cairns' swivel provides a hypothetical solution to the unwinding problem. But other difficulties arise from the fact that replication of the bacterial chromosome is sequential and proceeds from a single point or "fork." This can be seen directly in Cairns' autoradiographs; other less direct methods, using genetic analysis, also indicate that replication proceeds sequentially around the bacterial chromosome. The *E. coli* chromosome has been "mapped" by genetic crosses. As replication proceeds, the genes lying on the map can be seen to increase sequentially from a single to a double "dose." This can be detected by a measure of the capacity to synthesize inducible enzymes. As replication proceeds, the capacity to synthesize these enzymes doubles, each increasing in turn

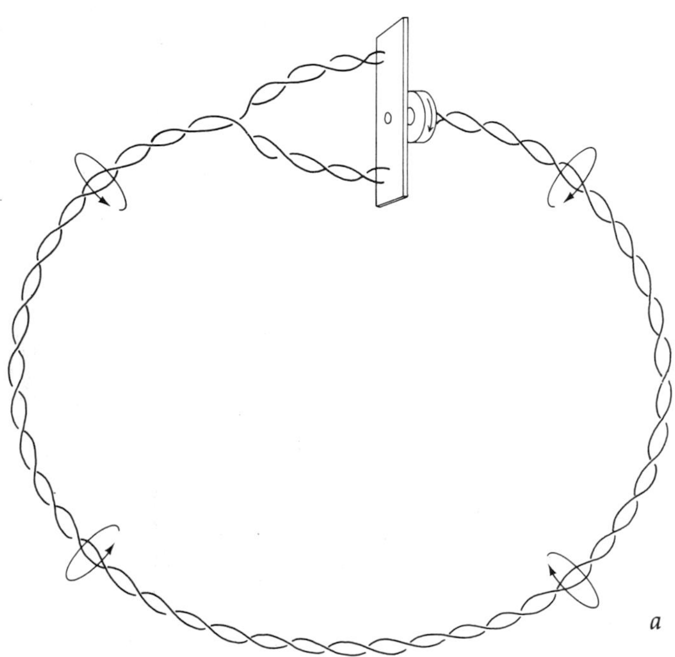

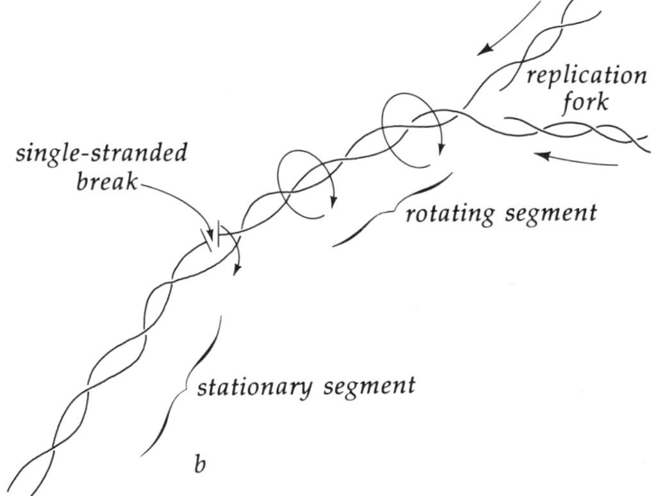

12–24. a, *Cairn's "swivel" mechanism, proposed as a model to account for the unwinding and replication of circular DNA. At the close of replication, as the fork completes the circle, the swivel is presumably replicated. Redrawn from "The Bacterial Chromosome" by J. Cairns.* Scientific American *214 (1966). Copyright © 1966 by Scientific American, Inc. All rights reserved.* b, *An alternative hypothesis, in which rotation takes place around a single-stranded break enzymatically introduced into the molecule.*

according to the sequence established for the *E. coli* map (see Helmstetter, 1968.).

Sequential replication from a single fork implies that the two strands of the template DNA unwind and are replicated simultaneously. This poses a serious problem, because the known DNA polymerases catalyze the synthesis of DNA only in the 5′ to 3′ direction. This means that the template DNA followed by the polymerase must run in the 3′ to 5′ direction, because the newly synthesized strand is opposite in direction to the template (see Fig. 12–22a). However, as the two strands of the template DNA unwind, only one is presented to the enzyme in the 3′ to 5′ direction. (This is illustrated in Fig. 12–25a.) In this figure, only the template strand on the left can be replicated by the enzyme in the 3′ to 5′ direction. Yet sequential replication from a single fork implies that synthesis must proceed along both template strands simultaneously (Fig. 12–25b).

A number of solutions have been proposed to account for this paradox. One, not too widely accepted, is that DNA polymerase in the living cell is able to catalyze synthesis in both the 5′ to 3′ and 3′ to 5′ directions. A second, even less favored than the first, is that an as yet undiscovered polymerase, which catalyzes synthesis in the 3′ to 5 direction, exists in the cell. A third, which seems most plausible and has the benefit of limited experimental support, holds that DNA synthesis is *discontinuous* on the two template strands. According to this hypothesis, the new DNA strands are replicated along the unwinding templates in short, discontinuous lengths at the fork. On the 3′ to 5′ template strand, these short segments are synthesized away from the fork; on the opposite strand, synthesis travels toward the fork (Fig. 12–25c). Thus, synthesis of the new DNA strand always proceeds in the 5′ to 3′ direction. After the short segments are synthesized in this manner, they are covalently joined by a second enzymatic mechanism.

The discontinuous synthesis model is supported by several lines of evidence. One series of experiments by Reiji Okazaki and his co-workers indicates that the DNA at the replication fork is indeed first synthesized in short chains and later linked together as proposed by the discontinuous synthesis model. Okazaki exposed replicating bacteria to radioactive DNA precursor for a very short time interval lasting only a few seconds. The bacteria were then lysed immediately and a DNA extract made. Radioactivity was found to be associated with very short pieces of DNA 1000–2000 nucleotides in length. These could be separated from the template DNA by heating, indicating linkage only by hydrogen bonds. If longer periods of time were allowed between the pulse of label and lysis of the cells, the radioactivity was found to be associated with DNA of very high molecular weight. This indicates that shortly after synthesis, the short fragments are covalently linked end to end into an uninterrupted chain as predicted by the model.

A second line of evidence comes from the existence of an enzyme, *polynucleotide ligase,* isolated in 1967 from *E. coli* by M. Gellert and also by independent investigation in a large number of other laboratories in the same year. This enzyme catalyzes the formation of a covalent linkage between the 3′ end of one polynucleotide chain and the 5′ end of an abutting chain, provided that the two chains are adjacent and in place in a double helix. Therefore, this enzyme would provide the required capacity to link together the short lengths of DNA produced by discontinuous synthesis at the replication fork. Genetic evidence indicates that polynucleotide ligase may indeed be directly involved in

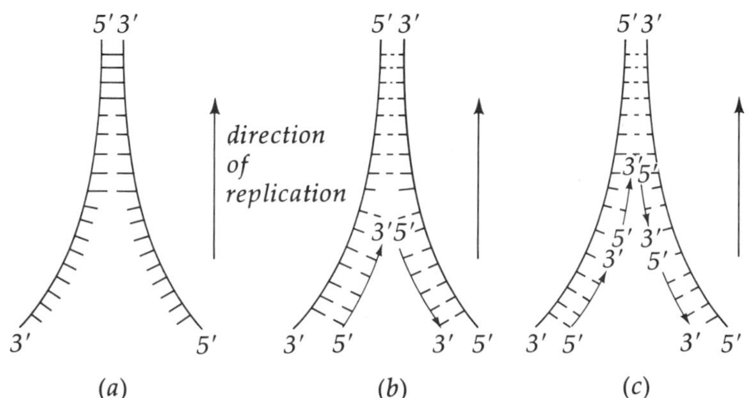

12–25. *The discontinuous synthesis model (see text).*

DNA replication. A mutated form of polynucleotide ligase has been identified by Okazaki and others in phage and in bacterial cells. In the mutant, the enzyme is inactive and DNA synthesis is blocked at elevated temperatures; if the temperature is reduced to levels at which the enzyme is active, replication proceeds normally. Thus two lines of evidence, the existence of the Okazaki fragments and the involvement of polynucleotide ligase in replication, support the discontinuous synthesis model.

Summary of the mechanism of DNA synthesis. The replication of DNA, which follows the semiconservative pathway, is catalyzed by the enzyme DNA polymerase. Synthesis proceeds sequentially along the chromosome, producing a fork, in advance of which the template DNA unwinds. Unwinding in circular DNA may be accomplished by "swivels" inserted into the chromosome or may take place around breaks introduced enzymatically into one of the two strands of the template. At the replicating point, synthesis of DNA may be discontinuous and proceed in opposite directions over short lengths of the unwound template strands. According to the discontinuous synthesis hypothesis, these short lengths are subsequently covalently linked together, probably by the enzyme polynucleotide ligase. This discontinuous synthesis at the molecular level is "seen" as unidirectional synthesis at the macroscopic levels detectable by autoradiography and genetic analysis.

Mutations

Mutations result from "mistakes" made during replication or mitosis. Mutated chromosomes contain DNA that is not identical to the parental template DNA. These mistakes, although rare, provide the ultimate source of genetic variability in organisms. Mutations arise through several mechanisms, and may involve lengths of DNA as short as one nucleotide or as long as whole chromosome arms. Alterations taking place within a single locus are called *gene mutations;* those involving longer stretches of DNA are called *chromosome mutations.*

Gene mutations occurring during replication are probably limited to insertion of an incorrect nucleotide at the point of synthesis. Substitution of a purine for a different purine or a pyrimidine for a pyrimidine is called a *transition.* A mutation of this type would take place, for example, if a thymine on the template is mistakenly paired with a cytosine. Although the mechanism generating transitions is unknown, this type of mutation could arise if a template base takes up an alternate arrangement of its atoms in space (called a *tautomer*) that allows the incorrect base pair to form. Substitution of a purine for a pyrimidine or vice versa is called a *transversion.* This will arise if the polymerase reaction mistakenly pairs a purine on the template with another purine, or forms a pyrimidine-pyrimidine pair. Although transversions evidently arise in nature, the generating mechanism is unknown. Transitions and transversions may result in amino acid substitutions in the polypeptide coded for by a gene or may cause appearance or disappearance of a codon terminating polypeptide synthesis. These mutations arise spontaneously or may be induced by chemicals or radiation.

Other types of mutations, involving *additions* or *deletions* of one or more nucleotides, may also occur during replication. Mutations of this type are more characteristic of meiosis than mitosis and are probably induced almost entirely by errors in recombination during meiotic prophase (see Chapter 14) rather than by errors in replication.

Some Notes on Replication in the Eukaryote Chromosome

Replication of DNA in higher organisms is believed to proceed by the same molecular pathways as in prokaryotes. However, several additional conditions are imposed by the structure of eukaryote chromosomes. Each chromosome contains many times as much DNA as is present in the bacterial cell. This compounds the unwinding problem, because much greater lengths of DNA are present. In addition, each eukaryote chromosome probably proceeds through replication from a multiple number of points rather than from a single site as in the bacterial chromosome. Otherwise, the large quantity of DNA present in eukaryote cells could not be replicated within the observed time limits of the S period.

John Cairns has estimated from observations made with human tissue culture cells that the replicating point travels along the eukaryote chromosome at about 0.5 microns per second. Under the assumption that the average human chromosome contains 3 centimeters of DNA (from measurements of the total quantity of DNA per cell), Cairns calculated that each chromosome must contain at least 100 sites replicating simultaneously for DNA synthesis to be completed in six hours, the length of the S period observed for human cells. Because replication from multiple sites will produce a series of discontinuous subunits in each chromosome, an additional as yet undiscovered enzymatic mechanism would be required to link the subunits together at the completion of replication.

Cairns' conclusions are supported by the results of exposure of eukaryotic cells to very brief pulses of radioactive DNA precursor during replication. At the subsequent metaphase, the chromosomes show scattered, discontinuous points of radioactivity at multiple sites. This is expected if replication proceeds simultaneously at a number of separate sites on the chromosome. Otherwise, the first label would be expected to appear uniformly at the ends, or perhaps the centromeres, of the chromosomes. The pattern observed in each chromosome is repeated from generation to generation, showing that the points at which synthesis is initiated probably reflect regular structural differentiation along the chromosome. Unlike prokaryote DNA, eukaryote DNA replication has been shown to proceed in both directions from each initiation site (see Huberman and Riggs, 1968).

The estimate made by Cairns of the number of replication sites per chromosome is conservative, because it was assumed that all chromosomes of the complement simultaneously enter into synthesis at the beginning of the six-hour S period and continue together until the close of replication. However, it has been demonstrated by autoradiography that in many eukaryotes, including humans, considerable asynchrony in the initiation time of DNA synthesis exists between the chromosomes of a set, some starting replication much later than others. Within a given chromosome, heterochromatin, if present, is also delayed in initiating replication. Some authorities, taking this asynchrony into account, have estimated that the number of replication sites per chromosome may be much greater than Cairns' estimate—perhaps as high as 1000 to 10,000—to allow completion of DNA synthesis by the end of the S period.

The unwinding problem has been approached through proposals that the long DNA helix of the eukaryote chromosome is interrupted at frequent intervals by linkers or swivel points, around which short subunits of the total length may rotate independently to unwind. Alternatively, single-stranded breaks, around which rotation and unwinding may occur, might be generated enzymatically at intervals along the chromosome. Although there is no direct evidence that linkers, swivels, or breaks actually exist or appear at the time of replication, the existence of multiple replicating points, for which there is good evidence, implies that multiple unwinding sites must exist in some form.

Replication of the eukaryote chromosome also includes synthesis of histone and probably at least some nonhistone protein. Very little is known in detail about the synthesis of these components. The synthesis of histones seems to be confined to the S period and proceeds independently of DNA synthesis. Conclusive determination of synthesis time has never been possible, primarily because the histones are quite difficult to characterize biochemically against the total complement of cellular protein. In the mid-1950's, it was shown by Max Alfert and by D. P. Bloch and G. C. Godman that parallel increases in the staining of histones and DNA could be detected during interphase. This was interpreted as indicating that DNA and histone are synthesized at the same time in the cell cycle. Later work, in which histone-like proteins were isolated from synchronized cell cultures, seemed to indicate that histones are synthesized throughout the cell cycle. However, these experiments suffer from difficulties in the elimination of extraneous protein and the lack of precision in synchrony of the cultures. More recently, Bloch has shown that the uptake of labeled lysine and arginine into chromosomal protein takes place during S and overlaps partly into the G_2 period. On extraction, the labeled protein was identified as histone. From this and the earlier experiments by Alfert and Bloch it seems most likely that histones are synthesized in close synchrony with DNA during the S period.

Results with nonhistone proteins are less definitive. Because almost all cytoplasmic proteins resemble the presumptive nonhistone proteins of the chromosome in reaction, the synthesis of chromosomal nonhistone protein cannot be detected or separated from the general background level of protein synthesis in the cell. But it seems reasonable to assume that if nonhistone proteins are indeed a part of the chromosome, these are also synthesized during the S period along with DNA and the histones. Autoradiography has shown that chromosomal proteins are distributed among daughter chromosomes dispersively rather than semiconservatively after replication (see Prescott and Bender, 1963). This suggests that much of the protein is disassembled from the chromosome at the replicating points and reassembled nonselectively along with newly synthesized protein on the replicated DNA.

Redundant or Repetitive DNA

A mutation occurring in a short length of DNA,

such as a transversion or transition, will of necessity be confined to a single gene. Mutations of this type can often be detected in the subsequent generation. For example, *Drosophila* or bacteria exposed to an agent, such as X-irradiation, that promotes mutations frequently reveal genetic alterations that can be shown to appear first in the generation immediately following the irradiated generation. This is explained most simply if the chromosomes are assumed to contain only one copy of the mutated gene. It is easy to see how a mutation could be detected if each gene is represented only once or twice, as in a diploid nucleus; if multiple copies are present, the additional unmutated copies would be expected to mask or cause a graduation in the effects of the mutated gene. Other mechanisms, such as recombination during meiosis (Chapter 14) are also most easily explained if genes are present in only one copy on each chromosome.

In recent years, at least a few genes have been discovered to occur in multiple copies in the chromosomes of many organisms. Some of these observations have already been discussed in previous chapters in connection with the description of the regions of the chromosome coding for ribosomal RNA. The cistrons for ribosomal RNA in *Drosophila*, for example, are present in about 180 copies; humans have about 2000 copies of these cistrons.

Repetitious DNA of this type, in which a sequence is repeated many times, has been termed *redundant* DNA. R. J. Britten has devised several techniques for indirectly measuring the amount of redundancy in a DNA sample. Britten's determinations of redundancy are carried out on DNA molecules that have been broken into short pieces about 500 nucleotides in length and dissociated by heating from double helices into single strands. The rate of reassociation after cooling depends on a number of factors, including salt concentration, temperature, and concentration of the DNA. If these are held constant, the rate of reassociation will depend on the percentage of collisions that results in the approach and pairing of complementary strands of a DNA molecule. If there is no redundancy, each strand is likely to have only one complementary "mate" in the solution. Therefore, only a small percentage of the collisions will cause sorting of complementary strands, and the rate of reassociation will be relatively slow. If much redundant DNA is present, the probability of collision between complementary strands is increased, and the rate of reassociation will be higher.

Britten has found that DNA samples from all organisms studied to date contain fractions that reassociate very slowly, at a base rate equivalent to nonredundant or "unique" DNA, and other DNA that reassociates very rapidly. By following the relative rate of reassociation of the "fast" fraction, Britten has been able to estimate the quantity of redundant DNA present. In calf DNA, for example, 40 percent of the DNA has been found to be redundant, with an average repeat frequency of redundant sequences of about 100,000 times. In the mouse, one stretch of DNA 300 nucleotides long has been estimated to be present in 1,000,000 copies.

There are other less direct indications that many organisms may contain a high percentage of redundant DNA. In the plant genus *Vicia* (one of the groups used by Taylor for his replication studies), different members of the genus exhibit wide differences in the quantity of DNA per nucleus. *Vicia faba* contains five times as much DNA per nucleus as *V. sativa*, even though both plants have the same number of chromosomes. The chromosomes of *V. faba* are correspondingly larger in size, but with closely similar morphology in terms of the location of primary and secondary constrictions. Other members of *Vicia* exhibit even greater differences in DNA content per nucleus, even though the chromosome number and morphology are similar throughout the genus. This situation is not uncommon in the plant and animal kingdoms. The plant genus *Aquilegia* exhibits a fortyfold difference in DNA content per nucleus between different species; in the flatworm genus *Mesostoma* differences of elevenfold are present. The extra DNA in different members of these genera probably represents redundant DNA, as it is hard to imagine that closely related organisms could differ by ratios of five to forty times in the number of different genes present.

Redundancy therefore appears to be present to some degree in many organisms and perhaps is universal in eukaryotes. What is the function of the repeated sequences? It is possible that the repeated sequences are extra copies of genes and that redundancy represents an adaptation for increased output of RNA. This is the case in the cited examples of nucleolar organizer cistrons coding for ribosomal RNA. Alternatively, some of the redundant DNA could represent a uniform "punctuation" sequence between genes or sequences signaling the initiation or termination of replication or transcription. These would be expected to be repeated many times.

Redundancy and "strandedness." The redundant DNA probably takes one of two structural configurations in the eukaryote chromosome. One possibility is that the redundancy is *linear.* Redundant stretches would then represent repeats of sequences in a single DNA molecule making up each chromosome. The second possibility is that the redundancy is *lateral,* and that the chromosome consists of a multiple number of DNA molecules arranged side by side in a sort of cable. Each DNA subunit in the chromosome "cable" would be identical in sequence.

Lateral redundancy is often termed "multistrandedness." The multistranded chromosome is actually an old concept in cytology; at one time, until the 1940's, it was taken for granted that chromosomes were multistranded. As the molecular basis for mutation became known, it was apparent that mutation is most simply explained if the chromosome is assumed to consist of a single DNA molecule. As a result, the consensus of opinion swung from the multistranded to the single-stranded (containing one DNA duplex) chromosome. The recent discovery of extensive redundancy in the DNA of eukaryotes has forced a reassessment of the single-stranded hypothesis and reopened consideration of the possibility that chromosomes might be multistranded. With the information at hand, it is impossible to completely rule out either the single-stranded or multistranded chromosome, although the weight of evidence remains in favor of single-strandedness. This evidence comes from four major sources: (1) morphological observations with the light and electron microscopes, (2) the types of aberrations caused by X-irradiation, (3) the pattern of incorporation of radioactive DNA precursors, and (4) the digestion of chromosomes with deoxyribonuclease. These will be reviewed in turn in the sections to follow.

Morphological observations of the strandedness of chromosomes. The strongest evidence in favor of the multistranded chromosome comes from light microscope observations of chromosomes at anaphase in both living and fixed material. Among the most convincing observations of this type are the phase contrast micrographs made by Andrew Bajer of living plant endosperm tissue (Fig. 12–26). In these pictures, a subdivision can be seen along the chromatids at anaphase. This longitudinal subdivision makes it appear as if the chromatid is composed of two half-chromatids (Fig. 12–26, arrows). A split of this type is also sometimes visible in electron micrographs of sectioned anaphase chromosomes (Fig. 12–27). The presence of two half-chromatids at anaphase suggests that the chromatid is already double and consists of at least two DNA duplexes at this stage. This would mean that at replication the chromosome doubles to four copies of the DNA.

Apparent multistrandedness is also visible in chromosomes isolated for electron microscopy by the surface-spreading technique (Figs. 12–28a and 12–28b; for a description of this technique, see Chapter 8). Many regions of chromosomes isolated by this method appear to contain large numbers of parallel fibers; in

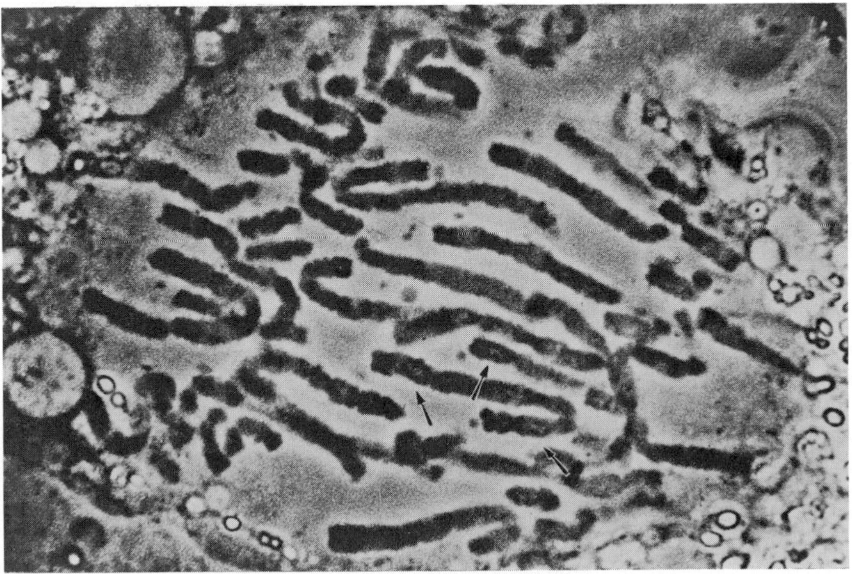

12–26. *Apparent half-chromatids (arrows) visible at anaphase in chromosomes of* Haemanthus. *Phase contrast micrograph of a living endosperm cell.* × *1,200. Courtesy of A. Bajer and the publisher.* [Chromosoma *17:291 (1965)*]

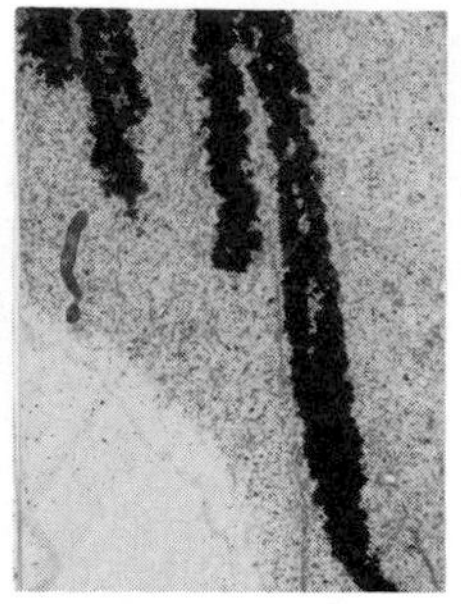

a

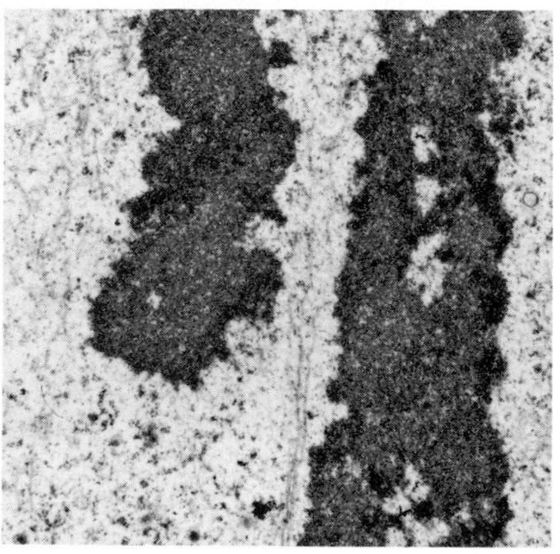

b

12–27. a, *Apparent half-chromatids in a thin-sectioned chromosome from* Haemanthus. b, *a region of* a *at higher magnification. Courtesy of C. Jensen, A. Bajer, and Academic Press, Inc., from* J. Ultrastr. Res. *26:367 (1969).*

the chromosome shown in Fig. 12–28b this apparent multistrandedness extends across the centromere region.

Proponents of the multistranded chromosome claim this morphological evidence as strong support for their view; detractors point out that the multiplicity of lateral subunits visible in these chromosomes may only be apparent and result from back-and-forth parallel folding of a single subunit. Whatever the basis for the image seen, both sides agree that chromosomes at least look multistranded under both the light and electron microscopes.

X-irradiation of chromosomes and the strandedness question. Irradiation of living cells by X-rays causes many disturbances in chromosomes, including breaks, in which a complete separation appears somewhere in an arm of the chromosome. These breaks may subsequently heal or may remain open to become visible at metaphase. If cells are irradiated in G_1, breaks include both chromatids of a chromosome at the following metaphase and extend completely across the arm at that point (Fig. 12–29a). These are called *chromosome aberrations* because they include both chromatids. If irradiation is delayed until G_2, breaks may appear in only one of the two chromatids of a chromosome at the subsequent metaphase. These are termed *chromatid aberrations* (Fig. 12–29b). In general, irradiation in G_1 regularly gives chromosome aberrations and irradiation in G_2 produces chromatid aberrations. This is interpreted by assuming that, at G_1 the chromosome consists of a single extended DNA duplex, which presents a single "target" to an X-ray. After replication, two possible targets are present, and the same amount of X-rays will be likely to produce breaks in only one of the two chromatids. This pattern of one- and two-hit breaks is considered good evidence for the single-stranded (linearly redundant) chromosome.

This evidence is in direct conflict with the observations of apparent half-chromatids seen with the light microscope at anaphase. If half-chromatids are really present at this stage, irradiation at the subsequent G_1 period should only very rarely produce breaks across whole chromosome arms. This conflict has been reconciled by the assumption that the two subunits, visible as half-chromatids at anaphase and uncoiled and extended at the subsequent G_1, become so closely associated at G_1 that the target presented to X-rays is effectively single. Support for this hypothesis has been recently obtained by Sheldon Wolff, who detected chromatid aberrations in cells irradiated in late G_1, just before the initiation of the S phase. Wolff fits this observation into the multistranded hypothesis by the assumption that the two subunits of the G_1 chromosome, so closely associated that only a single target is presented to X-rays during most of this period, dissociate and present two targets late in G_1. This separation of the half-chromatids late in G_1 is presumably connected in some way with the initiation of DNA synthesis in S.

Half-chromatids at G_1 would require that the G_2 chromosome be four-stranded, and that quarter chromosome aberrations (called *subchromatid* breaks) should sometimes be visible at metaphase. Ordinarily, as we have seen, the G_2 chromosome behaves as if two, not four, targets are presented to X-irradiation. A very few highly controversial observations have been made, in cells irradiated during prophase, of aberrations that

12–28. a, *Apparent multistrandedness in metaphase chromosomes isolated from the plant* Vicia. *The arms of the chromatids appear to consist of large numbers of parallel fibers.* × *32,500.* b, *Centromere region (arrow) of a chromosome isolated from* Vicia. *Numerous parallel fibers connect the two arms of the chromosome across the centromere. Approximately one-half of the total width of the chromosome (one chromatid) is visible in this micrograph.* × *32,500. Photographs by the author.*

have been interpreted as quarter chromosome breaks. These are seen as *anaphase bridges,* in which, as the chromatids separate, two arms remain attached at a point somewhere between the centromere and the tip. These were detected in low frequency with the light microscope in the 1950's (see, for example, Crouse, 1954); more recently, B. R. Brinkley and R. M. Humphrey have obtained excellent electron micrographs of thin-sectioned anaphase bridges (Fig. 12–30).

These bridges are interpreted according to the multistranded hypothesis as shown in Fig. 12–31. Fig. 12–31a shows the most frequently obtained result, in which chromatid breaks are produced by irradiation in G_2, even though 4 subunits may be present. This assumes that the four are so closely paired by twos that one pair may be broken as a single target by an X-ray. Fig. 12–31b shows the hypothetical origin of anaphase bridges. At the time of irradiation at late G_2 or prophase, it is assumed that in rare instances a break may include one member of each adjacent pair rather than the two members of a pair. Instead of remaining open, these breaks reheal, but incorrectly, so that the chromatids are now interconnected instead of remaining separate. At anaphase, the interconnection results in a failure of the arms to separate at the point of the fusion, producing a "bridge" of chromatin between the two poles.

This interpretation of anaphase bridges is by no means universally accepted. The bridges are extremely rare and lie at the limit of resolution of the light microscope. Even in electron micrographs the chromatids cannot be clearly traced. As a result, it cannot be determined whether the expected "crossover" of the hypothetical half-chromatids actually exists at the point where the arms attach to form the bridge. X-irradiation is known to cause a general "stickiness" of the chromosomes at metaphase. The molecular basis of this stickiness is unknown, but it is expressed as a general adherence of subparts of chromatids to the same or different chromosomes. It has been repeatedly pointed out that anaphase bridges may merely be the result of nonspecific stickiness of this type.

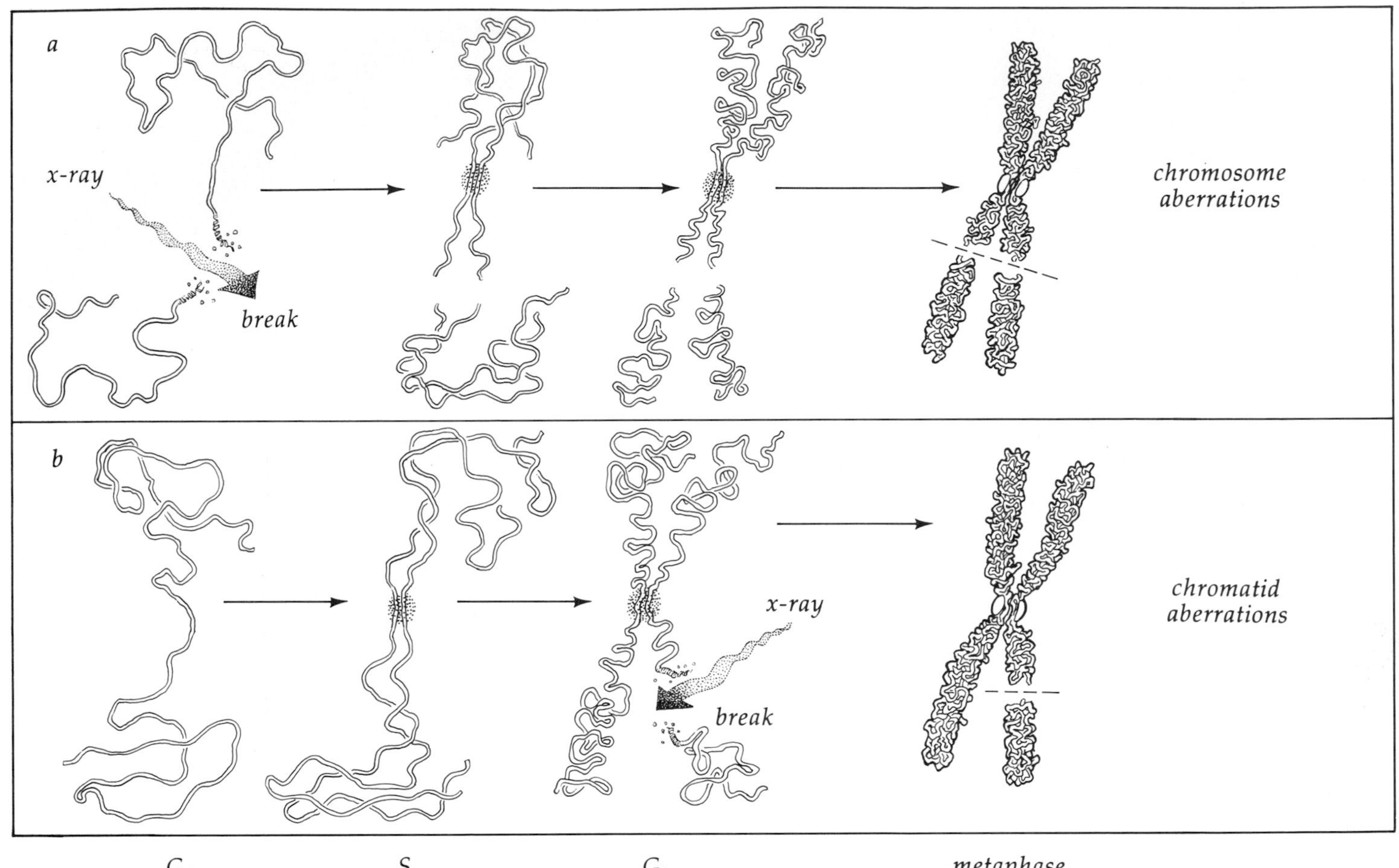

12–29. *Pattern of aberrations produced in chromosomes by X-irradiation at interphase.* a, *Irradiation at G_1.* b, *Irradiation at G_2 (see text).*

To summarize the results of irradiation studies, it may be said that the majority of observations from X-irradiation of chromosomes are most simply interpreted if the chromosome is assumed to contain a single DNA duplex at G_1 and two duplexes after replication at G_2 and that any redundancy must therefore be linear rather than lateral. A number of disturbing observations, such as the production of two-stranded breaks at late G_1 and the possible existence of half chromatid aberrations, detract somewhat from the certainty of this conclusion and suggest that lateral redundancy, at least to the extent that two parallel DNA molecules are present at G_1, may instead be present.

Results from the study of the incorporation of radioactive DNA precursors. The results of Taylor's experiments with the incorporation of radioactive precursors into chromosomes during replication have already been outlined in this chapter. This work indicates that the chromosome has two subunits at the time of replication; this is inferred from the fact that segregation of label is first seen at the second metaphase after exposure to radioactive precursor (see p. 305). Because these subunits are distributed semiconservatively between daughter chromosomes it seems most likely that they are the two strands of a single DNA duplex. Thus, the Taylor experiments are taken as strong evidence that the chromosome consists of a single long DNA molecule before replication.

As noted, the demonstration that G_1 chromosomes contain a single DNA molecule before replication depends on the pattern of segregation of label between chromatids at second metaphase (see Fig. 12–19). In the description of Taylor's work, it was mentioned that "almost all" of the chromosomes in Taylor's experiments showed segregation of label at second metaphase. The small fraction excluded by the "almost all"

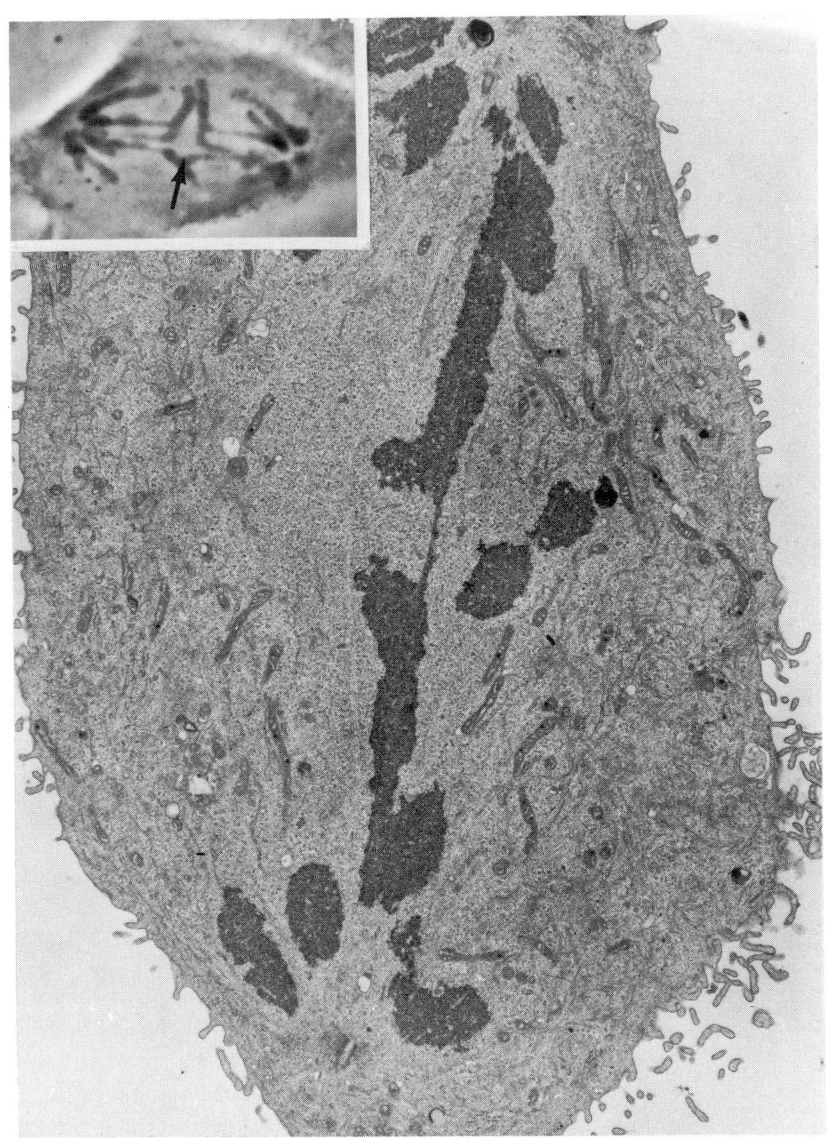

12–30. An "anaphase bridge" as seen in thin sections of kangaroo rat cells. Inset, light micrograph of a cell containing an anaphase bridge (arrow). Main figure, the same cell, sectioned for electron microscopy. Courtesy of B. R. Brinkley, R. M. Humphrey, and The Rockefeller University Press, from J. Cell Biol. *42:827 (1969).*

causes difficulties about the certainty of Taylor's results. Taylor found a few chromosomes at second metaphase in which both chromatids were labeled. This condition, called "isolabeling," if regularly observed, might indicate that a higher degree of strandedness than two subunits exists in the chromosome (see Fig. 12–20). Taylor ascribed these exceptional isolabeled chromosomes to other causes, such as multiple interchanges (breakages and fusions) between the sister chromatids of a chromosome or residual label that was incorporated at the second interphase instead of the first.

These exceptional cases were investigated in detail by W. J. Peacock, who found that most chromosomes showed segregation of label at second metaphase, as noted by Taylor. However, a significant number, about 5 percent, always showed isolabeling at second metaphase. The possibility of late incorporation of residual precursor as a cause of this isolabeling was eliminated by careful experimental controls. Peacock also showed that the possibility of multiple exchanges between the chromatids as a cause of isolabeling is remote, because these occur with an average frequency of less than one

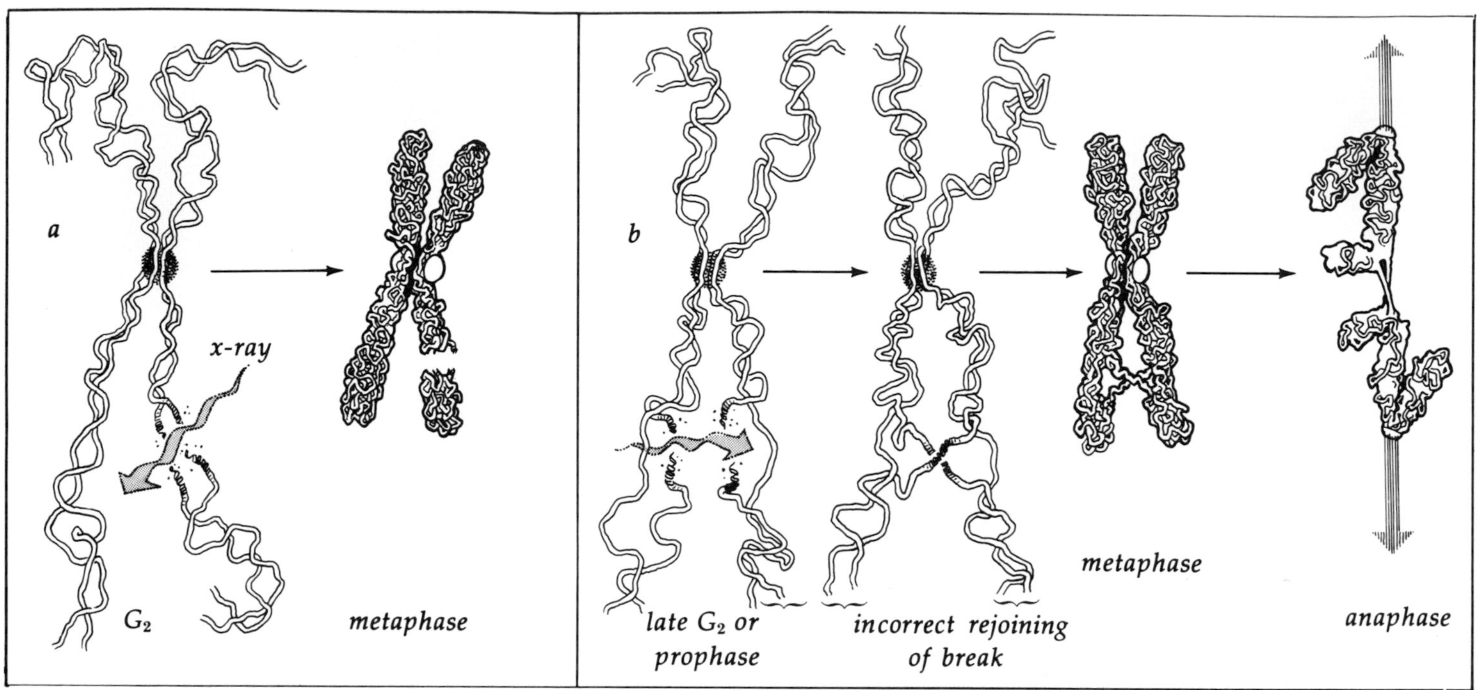

12–31. Diagram illustrating the presumed origin of anaphase bridges (see text).

exchange per chromosome arm. This would not be sufficient to account for isolabeling, which would probably require between at least 5 and 10 exchanges per chromosome arm.

Peacock maintains that multistrandedness is another possible cause of isolabeling at second metaphase. According to this hypothesis, the chromosome contains a multiple number of DNA molecules. At the time of replication, a "linker" mechanism in the chromosome normally ordains that, as the template DNA unwinds, all of the 3′ to 5′ strands of the several DNA molecules pass as a group to one daughter chromatid, and all of the 5′ to 3′ strands pass to the other. A model of this type is mechanically feasible; several have been devised and published (see for example, Fig. 12–32, a model advanced by Sheldon Wolff).

The linker mechanism is considered to operate with some degree of imperfection, so that in about 5 percent of the cases sorting out of the 3′ to 5′ and 5′ to 3′ strands is imperfect. This could result in isolabeling at second metaphase if the incorrect sorting of template occurred during second interphase, when replication takes place in unlabeled medium in the Taylor experiments. However, there is no concrete evidence of any kind for the existence of a "linker" in eukaryote chromosomes.

Because the most simple explanation of Taylor's results rests on the assumption that the chromosome consists of a single DNA duplex before replication, these experiments are generally taken as strong evidence for single-strandedness. The observations of isolabeling, as yet controversial and without satisfactory explanation, stand as a possible exception to the single-stranded argument.

Interpretation of strandedness from the results of enzymatic digestion of chromosomes. A study of the rate of the enzymatic breakdown of chromosomes undertaken by Joseph G. Gall is often advanced as the strongest evidence for single-stranded chromosome. Gall isolated prophase chromosomes from oocytes of the salamander *Triturus viridescens.* The unfixed chromosomes were placed in a chamber containing a solution of deoxyribonuclease and observed under the light microscope as digestion proceeded. Breaks were counted and scored against time as they appeared (the prophase chromosomes of amphibians are greatly extended at some stages of meiosis; see Chapter 14).

Gall reasoned that the probability of a break, if a single strand is present, is:

$$b = k_1 t \qquad (12\text{–}1)$$

where b is the number of breaks and t is time. The proportionality constant, k_1, is related to the activity of the enzyme. Gall assumed that, if a multiple number

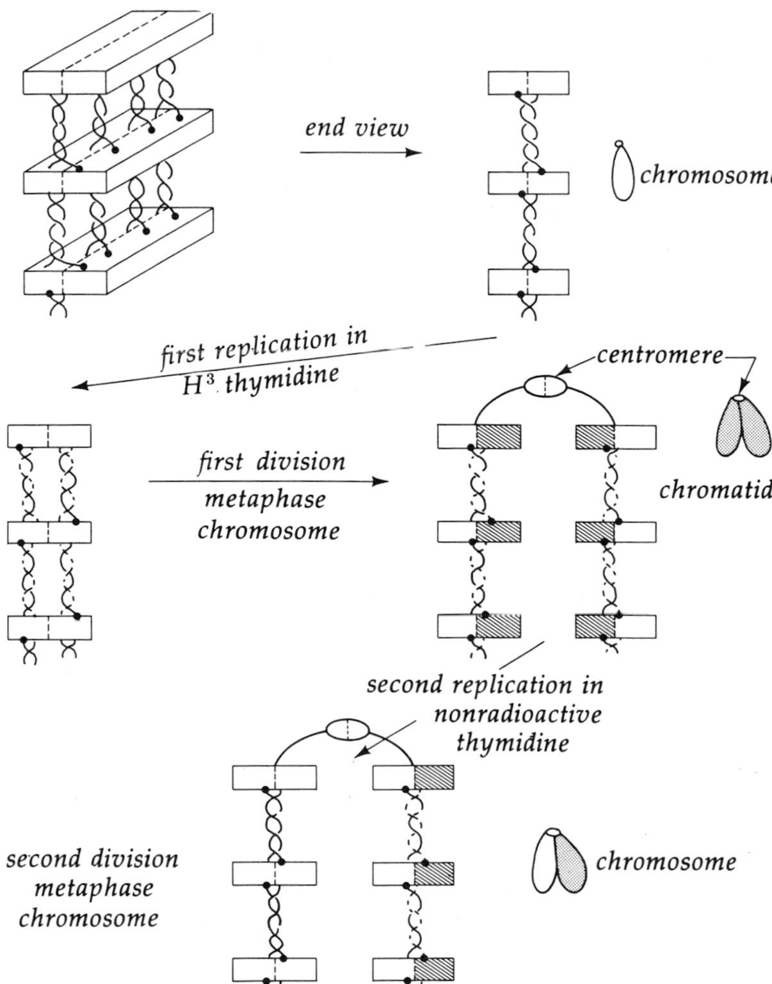

12–32. A model advanced to show that it is mechanically possible for a multistranded chromosome to show segregation of label at second metaphase. Four DNA duplexes are considered to be present, held together by "linkers" (the solid blocks). Heavy lines indicate parental DNA. Dotted lines indicate radioactive DNA synthesized in the presence of labeled precursor. During unwinding for replication, all strands of like direction (3′ to 5′ or 5′ to 3′) are held as a unit by the linkers and, after serving as a template for replication, pass to the same chromatid. Adapted, original courtesy of J. E. Trosko, S. Wolff, and The Rockefeller University Press, from J. Cell Biol. *26:125 (1965).*

of subunits are present, these are attacked independently and with equal probability by the enzyme. It was assumed further that, for a visible break to appear, all of the multiple subunits would have to be broken at or very close to the same level. Because breaks in individual strands are assumed to be independent events, the probability of breaks occurring at the same level in n strands is:

$$b = (k_1 t)^n \qquad (12\text{–}2)$$

Taking logs of both sides of equation 12–2 gives:

$$\log b = n \log k_1 + n \log t \qquad (12\text{–}3)$$

Because $n \log k_1$ is equivalent to another constant, k_2, equation 12–3 can be written:

$$\log b = n \log t + k_2 \qquad (12\text{–}4)$$

Equation 12–4 is a linear equation; therefore, a plot of $\log b$ against $\log t$ should yield a straight line of slope n.

Gall plotted the experimentally determined values for b and t, and obtained a value close to 2 for n, the number of strands in the chromatid.[4] The deoxyribonuclease used in the experiment attacks the two strands of a DNA molecule independently, giving single-stranded breaks at random points along the molecule. For the molecule to separate into two pieces, a break must occur in both strands close to the same level. Because the deoxyribonuclease used in the digestion experiments is known to have this activity, Gall postulated that the two subunits revealed in the chromatid are the two nucleotide chains of a DNA duplex. Therefore, according to Gall's results, the chromosome con-

[4]The value actually obtained by Gall was 2.6 for the chromatid axis. The departure of this value from 2 is minimized by proponents of the single-stranded chromosome and seized on as a basis for argument by the supporters of the multistranded hypothesis.

tains a single DNA molecule and is single-stranded rather than multistranded.

A general assessment of the strandedness argument. All evidence, except for Gall's work with the kinetics of digestion of chromosomes by deoxyribonuclease, leaves the question of strandedness somewhat unsettled, because in each case there are unexplained exceptions that detract from the certainty of the conclusions. When Gall's results are taken into consideration, the weight of evidence clearly falls on the side of the single-stranded chromosome; as noted, this interpretation has always provided the most simple explanation of the experimental results.

The greatest intuitive difficulties are presented to proponents of the multistranded hypothesis by the necessity to explain replication and mutation of multistranded chromosomes. Replication can be explained if linkers are assumed to operate as in the model shown in Fig. 12–32. But these models, by virtue of their complexity and the lack of evidence for the existence of linkers, have never been taken too seriously. Mutations present even greater difficulties; no reasonable models for mutation in multistranded chromosomes have been forthcoming. These conceptual difficulties alone, even if evidence did not exist in favor of the single-stranded chromosome, have been enough to eliminate multistrandedness from serious consideration in the view of most investigators. But the existence of redundancy removes some of the force from these objections to the multistranded chromosome.

Mutation, as noted, has always been the main stumbling block to the multistranded chromosome. For mutation to appear in a multistranded chromosome and be detectable in the subsequent generation, it must occur in all of the repeated copies of a gene. This would require that a mechanism operate to generate a transition or transversion in all multiple copies of a gene at precisely the same point in the sequence. This is so difficult to visualize that, up to now, mutation alone has been enough to allow a comfortable rejection of multistrandedness. But if DNA is indeed redundant, precisely such a mechanism is required, whether the redundancy is expressed as linear repeats on a single DNA molecule or multiple copies on separate DNA molecules.

Some investigators now favor a "master-slave" model, first proposed by H. G. Callan, as a working hypothesis for the generation of mutations in redundant DNA. Stated simply, the model proposes that one of the multiple copies of a repeated gene is a "master" copy. At some point during or after replication, the remaining copies of the gene, the "slave" copies, are somehow compared in sequence to the master copy. Corrections are made if the slave copies have undergone transitions or transversions not appearing in the master copy; presumably, if mutations have occurred in the master copy, these are also incorporated into the slave copies. While simple in conception, this model would require an exceedingly complex mechanism for the comparison of the master and slave copies and another for the "correction" of the slaves if differences in sequence are detected by the comparison mechanism. There is no evidence for the existence of these systems and almost no inkling of any method by which they might work. Thus, the explanations required to accommodate mutation in single-stranded chromosomes are as complex and difficult as in the multistranded chromosome if redundancy is taken into consideration. In spite of these difficulties, there are indications that a mechanism for the comparison of multiple copies of cistrons must indeed exist, whether the chromosome is single-stranded or multistranded. This is taken from the fact that the redundant sequences in DNA that code for ribosomal RNA in the nucleolar organizer region of the chromosome are preserved with few alterations in sequence in a given species and in fact have remained closely similar in the evolution of widely divergent plant and animal species.

It must be kept in mind in this line of discussion that not all of the DNA evaluated in the renaturation experiments of Britten and others is redundant. More than half of the DNA in most organisms appears by this method to contain unique, nonrepetitive sequences. Further, much of the repetitive DNA consists of short sequences that are highly simplified (see Straus, 1971). The possibility exists that many of these redundant segments are not structural genes and are not transcribed during RNA synthesis. However, at least some redundant DNA, as for example in the case of the nucleolar organizer region cited above, is functional and serves as template for RNA synthesis.

Suggestions for Further Reading

Books

John, B., and K. R. Lewis, 1969. The chromosome cycle. *Protoplasmatologia,* Volume 6b, pp. 1–125.

Kornberg, A., 1961. *Enzymatic Synthesis of DNA.* John Wiley & Sons, New York.

Lewis, K. R., and B. John, 1963. *Chromosome Marker.* Little, Brown & Co., Boston.

Swanson, C. P., 1957. *Cytology and Cytogenetics.* Prentice-Hall, Englewood Cliffs, N. J.

Wilson, E. B., 1928. *The Cell in Development and Heredity.* Macmillan, New York.

Papers and Review Articles

Alfert, M., and I. I. Geschwind, 1958. The development of polysomaty in rat liver. *Experimental Cell Research,* vol. 15, pp. 230–270.

Bajer, A. 1965. Subchromatid structure of chromosomes in the living state. *Chromosoma,* vol. 17, pp. 291–302.

Banhoeffer, F., and W. Nesser, 1969. Replication of the bacterial chromosome. *Annual Review of Genetics,* vol. 3, pp. 233–246.

Bloch, D. P., and G. Godman, 1955. A microphotometric study of the synthesis of deoxyribonucleic acid and nuclear histone. *Journal of Biophysical and Biochemical Cytology,* vol. 1, pp. 17–28.

Britten, R. J., and D. E. Kohne, 1969. Implications of repeated nucleotide sequences. In A. Lima-de-Faria, ed., *Handbook of Molecular Cytology,* North-Holland Publishing Co., pp. 38–51.

Cairns, J., 1965. The mechanics of DNA replication in bacteria. *Journal of Cell Physiology,* vol. 70, Supplement 1, pp. 65–76. Includes a skeptical analysis of the evidence concerning replication by Kornberg's DNA polymerase.

Cairns, J., 1966. The bacterial chromosome. *Scientific American,* vol. 214, pp. 37–44.

Crouse, H. V., 1954. X-ray breakage of lily chromosomes at first meiotic metaphase. *Science,* vol. 119, pp. 485–487.

Drake, J. W., 1969. Mutagenic mechanisms. *Annual Review of Genetics,* vol. 3, pp. 247–268.

Friedländer, M., and J. Wahrman, 1966. Giant centrioles in neuropteran meiosis. *Journal of Cell Science,* vol. 1, pp. 129–144.

Gall, J. G., 1963. Kinetics of deoxyribonuclease action on chromosomes. *Nature,* vol. 198, pp. 36–38.

Gellert, M., 1967. Formation of covalent circles of lambda DNA by *E. coli* extracts. *Proceedings of the National Academy of Sciences,* vol. 57, pp. 148–155.

Helmstetter, C. E., 1968. Origin and sequence of chromosome replication in *E. coli* B/r. *Journal of Bacteriology,* vol. 95, pp. 1634–1641.

Hotta, Y., and H. Stern, 1963. Molecular facets of mitotic regulation I: synthesis of thymidine kinase. *Proceedings of the National Academy of Sciences,* vol. 49, pp. 648–654.

Howard, A., and S. R. Pelc, 1951. Nuclear incorporation of P^{32} as demonstrated by autoradiographs. *Experimental Cell Research,* vol. 2, pp. 178–187.

Huberman, J. A., and A. D. Riggs, 1968. On the mechanism of DNA replication in mammalian chromosomes. *Journal of Molecular Biology,* vol. 32, pp. 327–341.

Lafontaine, J.-G., and A. Lord, 1969. Organization of nuclear structures in mitotic cells. In A. Lima-de-Faria, ed., *Handbook of Molecular Cytology.* North-Holland Publishing Co., Amsterdam, pp. 381–411.

Lark, K. G., 1969. Initiation and control of DNA synthesis. *Annual Review of Biochemistry,* vol. 38, pp. 569–604.

Levinthal, C., and H. R. Crane, 1956. On the unwinding of DNA. *Proceedings of the National Academy of Sciences,* vol. 42, pp. 436–438.

Lima-de-Faria, A. 1969. DNA replication and gene amplification in heterochromatin. In A. Lima-de-Faria, ed., *Handbook of Molecular Cytology.* North-Holland Publishing Co., Amsterdam, pp. 277–325.

Mazia, D., 1961. Mitosis and the physiology of cell division. In J. Brachet and A. E. Mirsky, eds., *The Cell,* vol. III. Academic Press, New York, pp. 77–412.

Meselson, J., and F. W. Stahl, 1958. The replication of DNA in *Escherichia coli. Proceedings of the National Academy of Sciences,* vol. 44, pp. 671–682.

Monesi, V., 1969. DNA, RNA, and protein synthesis during the mitotic cell cycle. In A. Lima-de-Faria, ed., *Handbook of Molecular Cytology,* North-Holland Publishing Co., Amsterdam, pp. 472–499.

Okazaki, R., T. Okazaki, K. Sakabe, K. Sagimoto, and A. Sugino, 1968. Mechanism of DNA chain growth I: possible discontinuity and unusual secondary structure of newly synthesized strands. *Proceedings of the National Academy of Sciences,* vol. 59, pp. 598–605.

Patau, K., and H. Swift, 1953. The DNA content (Feulgen) of nuclei during mitosis in root tip of onion. *Chromosoma,* vol. 6, pp. 149–169.

Peacock, W. J., 1963. Chromosome duplication and structure as determined by autoradiography. *Proceedings of the National Academy of Sciences,* vol. 49, pp. 793–801.

Prescott, D. M., and M. A. Bender, 1963. Synthesis and behavior of nuclear proteins during the cell life cycle. *Journal of Cellular and Comparative Physiology,* vol. 62, Supplement 1, pp. 175–194.

Richardson, Charles C., 1969. Enzymes in DNA metabolism. *Annual Review of Biochemistry,* vol. 38, pp. 795–840.

Robbins, E., and N. K. Gonatas, 1964. The ultrastructure of a mammalian cell during the mitotic cycle. *Journal of Cell Biology,* vol. 21, pp. 429–463.

Speyer, J. F., J. D. Karam, and A. B. Lenny, 1966. On the role of DNA polymerase in base selection. *Cold Spring Harbor Symposia on Quantitative Biology,* vol. 31, pp. 693–697.

Straus, N. A., 1971. Comparative renaturation kinetics in amphibians. *Proceedings of the National Academy of Sciences,* vol. 68, pp. 799–802.

Swift, H., 1965. Molecular morphology of the chromosome. In *The Chromosome, Structural, and Functional Aspects. Annual Symposium, Tissue Culture Association,* Miami, pp. 26–49. Includes a review of strandedness, with leanings toward the single-stranded chromosome.

Taylor, J. H., 1959. The organization and duplication of genetic material. *Proceedings of the 10th International Congress of Genetics,* Montreal, pp. 63–78.

Taylor, J. H., 1960. Nucleic acid synthesis in relation to the cell division cycle. *Annals of the New York Academy of Sciences,* vol. 90, pp. 409–421.

Taylor, J. H., and R. D. McMaster, 1954. Autoradiographic and microphotometric studies of desoxyribose nucleic acid during microgametogenesis in *Lilium lorgiflorium. Chromosoma,* vol. 6, pp. 489–521.

Wolff, Sheldon, 1969. The strandedness of chromosomes. *International Review of Cytology,* vol. 25, pp. 279–296. A review of the strandedness question, with leanings toward the multistranded side.

13 Mitosis II: Division of the Nucleus and Cytoplasm

In the previous chapter, an overview of replication, mitosis, and cytokinesis was presented, and details were given of the replication of the chromosomes during interphase. This chapter will provide more detailed information on subsequent events and the structures important in the division of the nucleus and cytoplasm, including the structure of the chromosome during division; organization and function of the spindle, asters, and centrioles; the fate of the nucleolus during division; and the division of the cytoplasm (cytokinesis) in animals and plants. A brief discussion of cell wall synthesis will be included as a part of the description of cytokinesis in plants.

The Structure of Chromosomes at Prophase and Metaphase

Coiling of Chromosomes to the Metaphase State

The onset of mitosis is marked by the first appearance of chromosome threads in the nucleus. These gradually become visible as the unit fibers of the chromosome coil into superstructures large enough to be resolved in the light microscope (Fig. 13–1). As mentioned in Chapter 12, no regular coils or gyres can be seen in the condensing chromosome at the level of the nucleoprotein fibers seen in the electron microscope. However, regular coiling of whole chromatid arms is often visible at prophase and metaphase (Fig. 13–2). Ordinarily, the coiling visible in the light microscope does not extend below the level of a single major coil of the chromatid arm.

It has been proposed at various times that a coil-within-a-coil pattern extends downward through

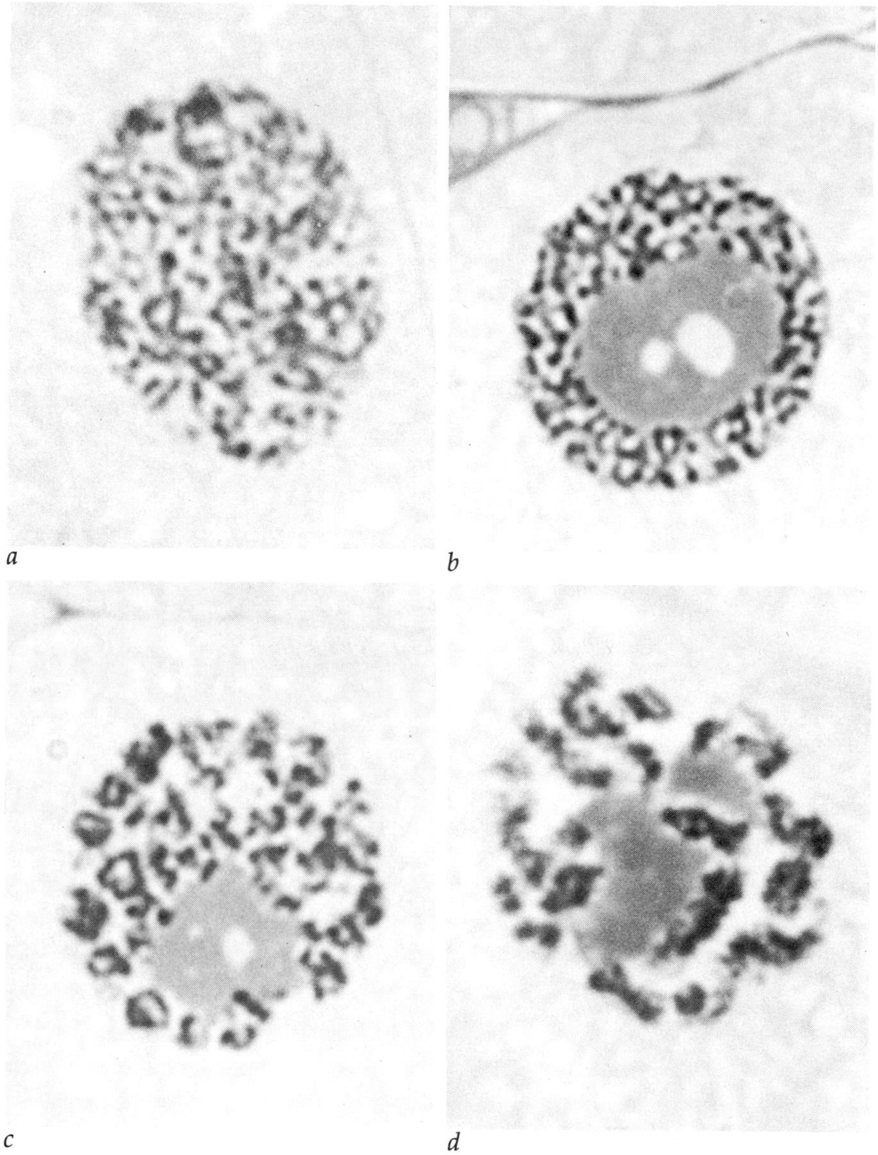

a b c d

13–1. *Progressive coiling of the chromosomes during prophase in the plant* Vicia faba. a, *Early prophase. No individual chromosomes are recognizable, but coiling and condensation have begun.* b, *Later stage in prophase. The strands of chromatin have thickened, and are more distinctly outlined against the surrounding nucleoplasm. The nucleolus occupies much of the volume of the nucleus in this section.* c, *Later stage of prophase. Coiling has progressed to the extent that individual chromosomes can be traced.* d, *Late prophase. Individual chromosomes are clearly recognizable.* × *3,500. Courtesy of J.-G. Lafontaine.*

various levels to the unit nucleoprotein fiber. According to this hypothesis, the chromosome at prophase and metaphase consists of a hierarchy of gyres from the 100 Å unit fiber up to the major coils visible in the light microscope. This does not appear likely, because regular gyres have never been seen below one or two orders of coiling of the chromatid in either the light or electron microscopes. In many species, no regular coiling of the chromatid is visible at any time during prophase or metaphase. The lack of regular gyres at the ultrastructural level does not necessarily indicate that the folding of 100 Å unit fibers is random and unordered. The pattern of folding, though irregular, may nevertheless be precisely determined by the condensation mechanism. An ordered mechanism of this type, if it exists, implies that the same pattern of folding would be observed at a given point in both members of a homologous pair. Unfortunately, the complexity of the arrangement of 100 Å fibers in metaphase chromosomes has prevented recognition of any repeated patterns of folding in either thin-sectioned or isolated chromosomes (see Figs. 13–3a and 13–3b). That ordered folding is imposed at some level is indicated by the fact that the gross morphology of metaphase

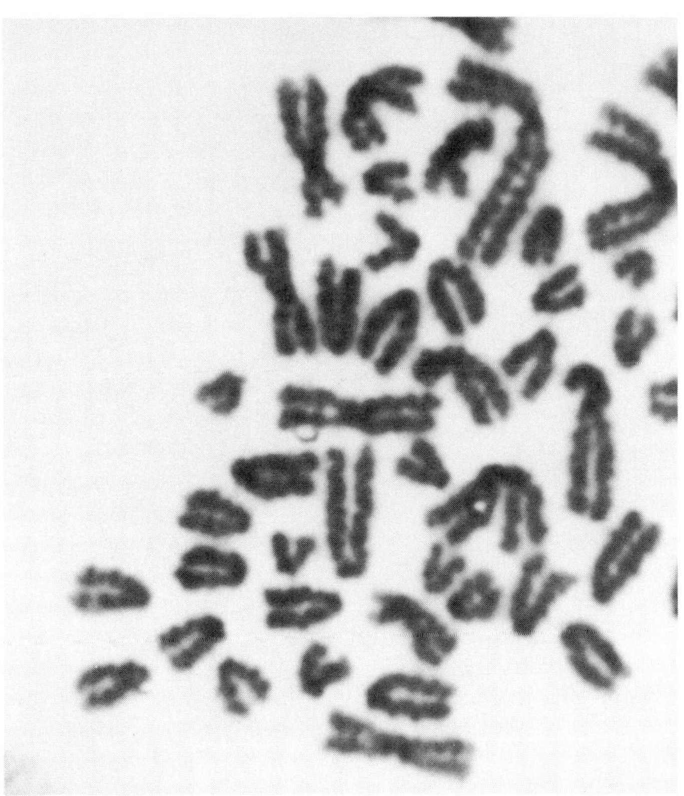

13–2. Gross coiling of chromatid arms, visible in chromosomes of the marsupial Marmosa. *Courtesy of K. M. Fisher.*

chromosomes is characteristic for a given homologous pair and is repeated from cell to cell of the same species. Thus, the regular position and morphology of features such as the primary and secondary constrictions ultimately reflect the folding pattern of the component 100 Å fibers.

The molecular basis of condensation remains unknown. The folding mechanism may be a part of the 100 Å unit fiber itself or, alternatively, may be a part of the extrachromosomal nucleoplasm that somehow acts on the fiber. One viewpoint, popular a few years ago, holds that the unit nucleoprotein fiber contains a contractile protein; this protein, by shortening, folds the fiber back on itself. Another possibility, advanced more recently, is that the fiber contains groups spaced along its length that become mutually attractive at the time coiling begins; progressive adhesion of these groups folds the chromosome into the metaphase state. These mechanisms, residing in the fiber, would be triggered by factors reaching the chromosomes through the surrounding nucleoplasm. Other hypotheses state that forces developed in the extrachromosomal nucleoplasm push or pull the fibers into the folded state. An example of this type of mechanism would be the generation of force by streaming of the extrachromosomal nucleoplasm, analogous to the cytoplasmic streaming visible in living cells under the light microscope.

All of this is merely speculation, because we have no clues to the actual mechanisms acting to fold the unit chromosome fiber. Although it is not necessarily so, it seems most likely that the "information" required for condensation of the chromosomes is built into the structure of the component fibers rather than into agents of the surrounding nucleoplasm. This would allow replication of the folding mechanism along with replication of the chromosome fiber during interphase. These "folders," spaced in a regular pattern along the unit chromosome fibers, would be progressively activated as condensation proceeds.

Coiling continues through prophase and metaphase until mid-anaphase, with progressive contraction and thickening of the chromosome arms. This progressive coiling does not depend in any way on attachment of the chromosomes to the spindle. If spindle formation is interrupted by agents such as radiation or colchicine (see Chapter 12), coiling and condensation of the chromosomes continues unhindered and may in fact produce more compact chromosomes than in untreated cells.

The Structure of the Primary Constriction: The Centromeres

The primary constriction and the centromeres. The most conspicuous linear differentiation of the metaphase chromosome is usually the primary constriction, which contains the centromeres (sometimes also called the *kinetochores*). The position of the primary constriction, which may be central, subterminal, or terminal with respect to the ends of the chromosome, is fixed for a given metaphase chromosome. Under the light microscope, the primary constriction usually appears as a localized narrowing of the chromosome (Fig. 13–4). Occasionally, one or more granules can be discerned in this constriction, spaced along the narrow region like beads on a string. The chromosome in the region of the primary constriction is faintly stained by

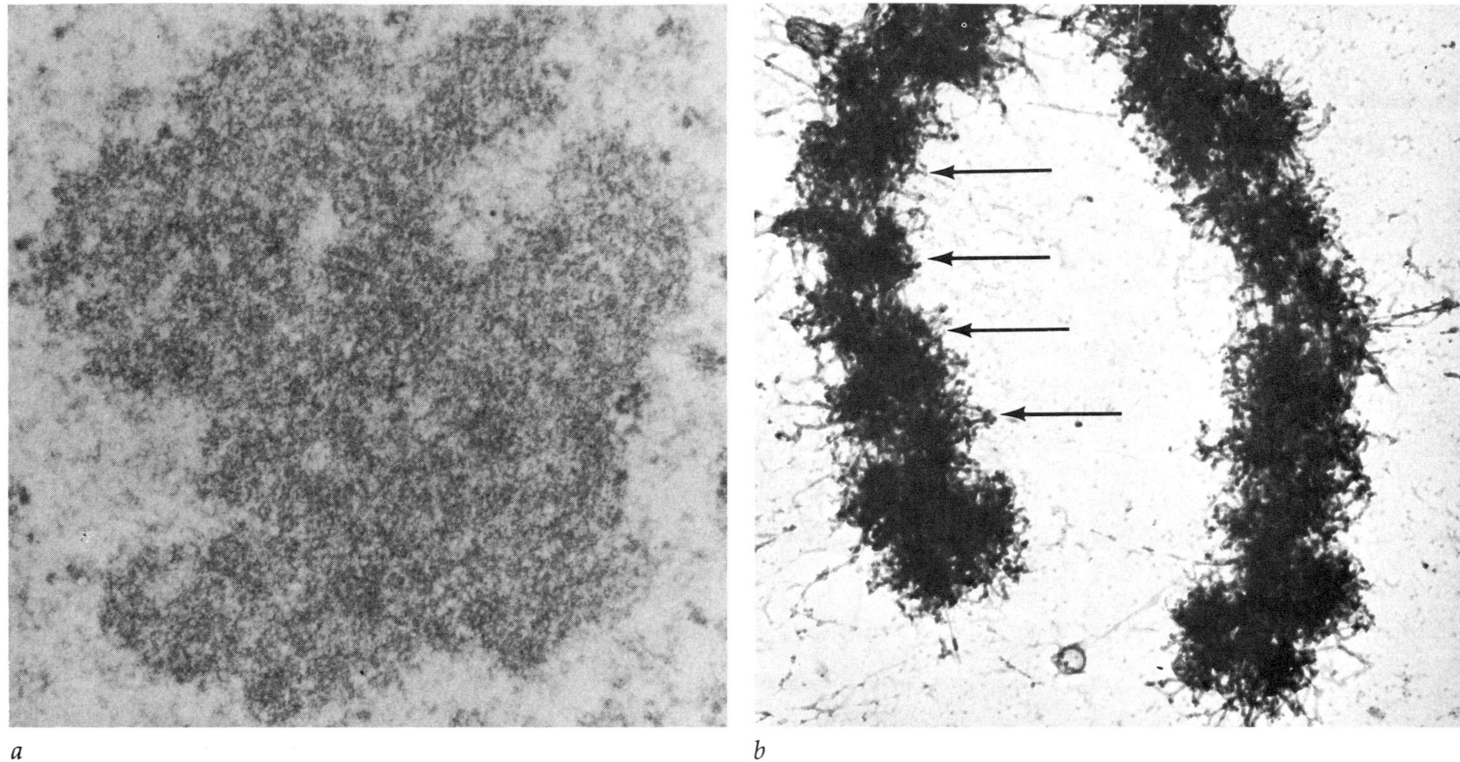

a b

13–3. a, *A thin sectioned chromosome at metaphase. No order is discernible in the pattern of folding of the component 100 Å fibers. From* Oncopeltus. × *52,000. Courtesy of B. R. Zirkin.* b, *A metaphase chromosome isolated from bovine tissue culture cells. Although the suggestion of a major coil of the chromatid is present, especially in the chromatid to the left (arrows), no order is discernible in the packing of the component fibers. × 24,500. Photograph by R. Jahn.*

the Feulgen reaction, showing that DNA and, therefore, chromatin extends through and forms a part of the structure of this region. This inference from cytochemical data is supported by the electron microscopy of both thin-sectioned and isolated chromosomes, which demonstrates that the unit fibers of the chromosome extend through and form a part of the primary constriction (see, for example, Figs. 12–28b and Figs. 13–6 to 13–9).

Until recently, no specialized components were identified in thin sections of the primary constriction. With ordinary osmium fixation, the centromere region appears to contain no structures other than the component 100 Å fibers of the chromosome. In a few cases, an aggregation of dense material could be seen, superimposed over the chromosome at the point of microtubule attachment (Fig. 13–5). In the mid-1960's, with the development of aldehyde fixatives, additional structure was seen in the primary constriction. This work has made it clear that there are two major structural subdivisions of this region: the centromeres and the underlying unit fibers of the chromosome, to which the centromeres are anchored (see below).

The centromeres can first be identified in thin sections in early prophase, before connections are made by the chromosomes to spindle microtubules. At one time, it was thought that the primary constriction contained a single centromere that doubled or split in late metaphase to initiate anaphase movement. However, electron microscopy has shown that the primary constriction contains two centromeres during prophase, one attached to each of the two chromatids of the chromosome (see Fig. 13–6a).

The structure of the centromere and primary constriction has been investigated in detail by Andrew Bajer (1968a), B. R. Brinkley and Elton Stubblefield

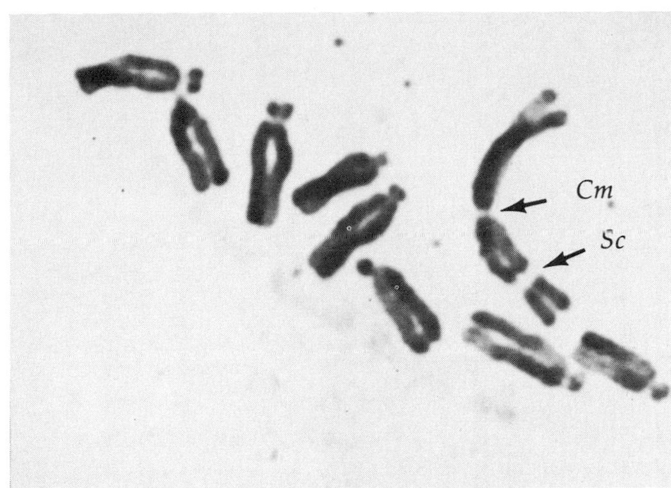

13–4. Metaphase chromosomes of the broad bean, Vicia faba. *One pair contains both a primary constriction, marking the location of the centromeres* (Cm), *and a prominent secondary constriction* (Sc). *× 1,600. Courtesy of S. Wolff.*

(1966), and Pentti T. Jokelainen (1967). Although there are minor differences on some aspects of centromere structure, these investigators agree that the centromeres are plate- or cup-like discs, 2000–2500 Å in diameter, attached to the surface of the chromatids at the primary constriction. Both centromeres of a metaphase chromosome are visible in cross section in Fig. 13–6a. In cross section (Figs. 13–6, 13–7, and 13–8), each centromere disc appears to be made up of multiple layers. Most conspicuous is a dense disc-like layer 300–400 Å in thickness and slightly convex on its outer surface. This layer is typically separated from the underlying chromatin fibers by a less dense zone 150–350 Å in diameter. The convex outer surface of the centromere disc, which faces one pole of the spindle, is covered with an electron dense "corona" of material that appears finely fibrillar or granular in thin sections. Once attachment to the spindle is made, a multiple number of spindle microtubules, usually from four to ten, become anchored in the substance of each centromere. These pass perpendicularly through the various layers of the centromere, sometimes penetrating through into the underlying chromatin fibers.

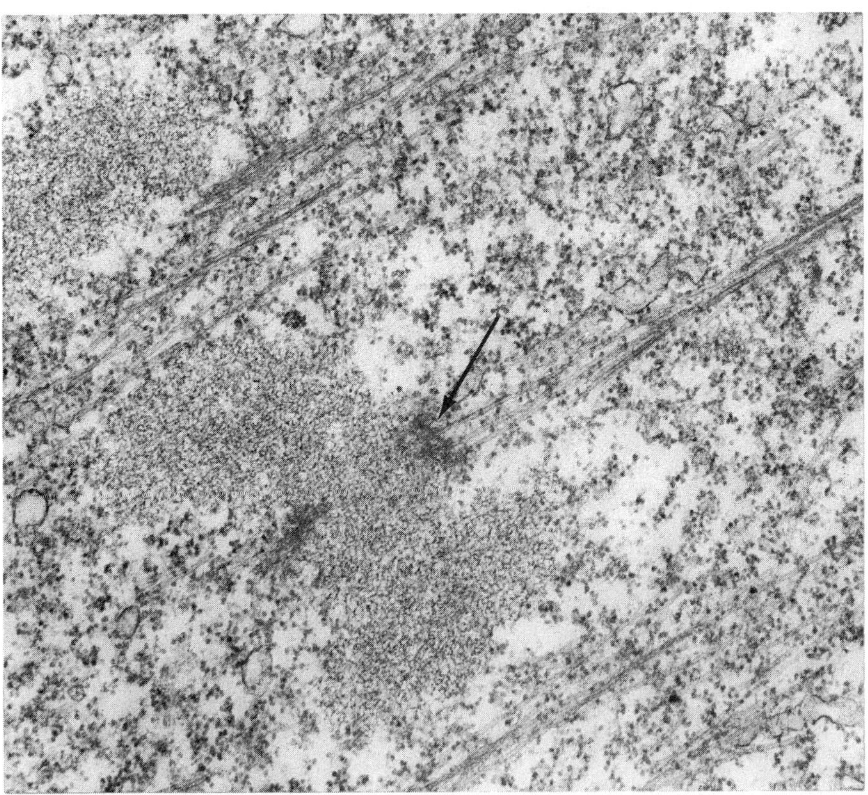

13–5. Centromere structure as seen in osmium-fixed material. Electron-dense, amorphous material is aggregated around the point of attachment of spindle microtubules to the chromosome (arrow). × 36,000. Courtesy of Patricia J. Harris and The Rockefeller University Press. [J. Cell Biol. *25:73 (1965)*]

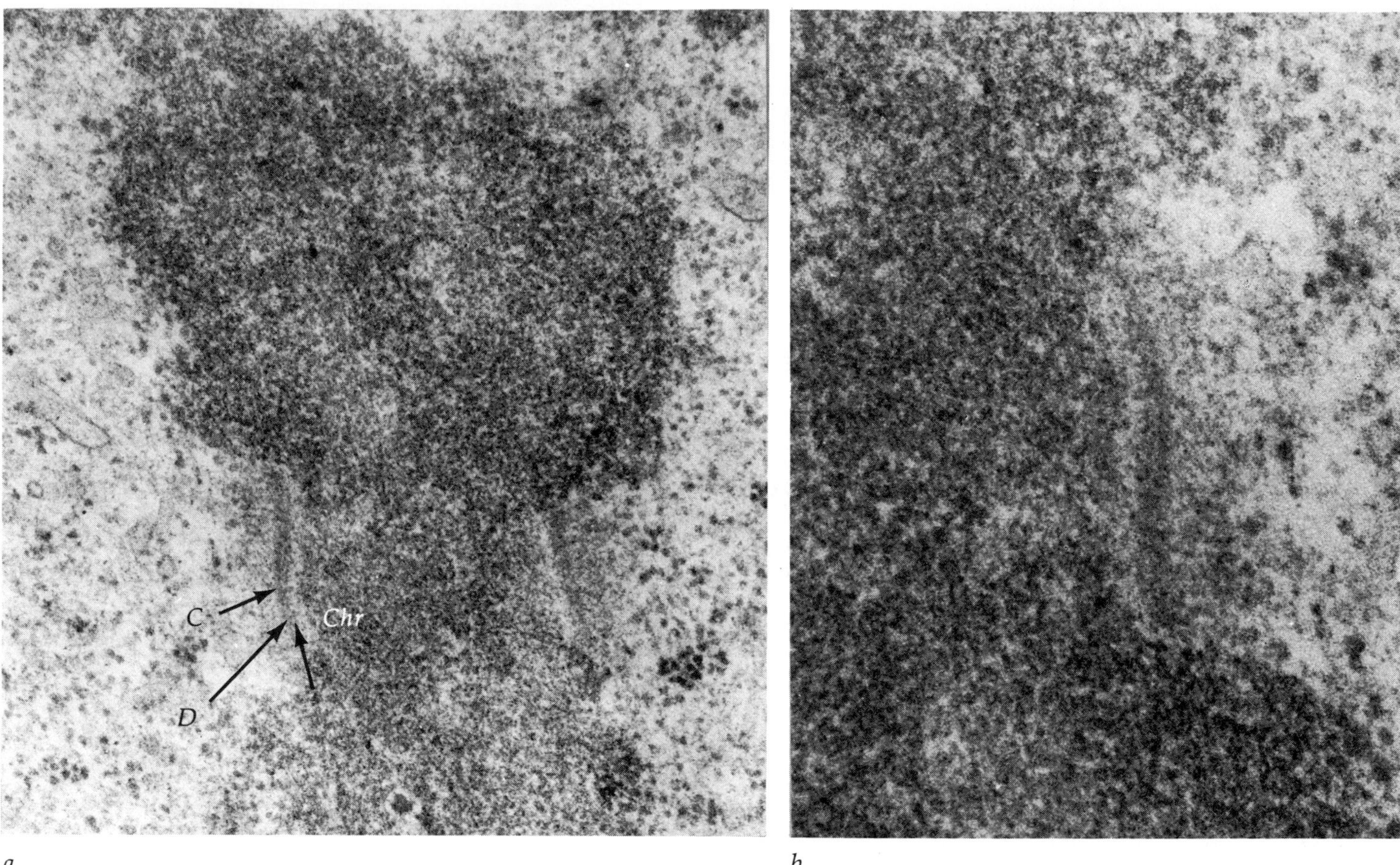

13–6. a, *Centromere structure in Chinese hamster cells after glutaraldehyde-osmium fixation.* C, *corona;* D, *centromere disc. Note the electron-lucent layer (arrow) separating the discs from the underlying chromatin* (Chr). × *47,000.* b, *One of the centromeres in* a *at higher magnification.* × *100,000. Courtesy of B. R. Brinkley, E. Stubblefield, and the publisher, from* Chromosoma *19:28 (1966).*

There is considerable evidence that the centromere may contain additional structural complexities. Jokelainen, on whose careful work much of the recent elucidation of centromere structure is based, has shown that the centromere disc may be a three-layered structure. According to Jokelainen, the centromere consists of an outer dense layer, a middle, less dense zone (corresponding to the two layers described in the section above), and an additional inner dense layer 150–300 Å in thickness, closely applied to the underlying chromatin. These three layers are clearly visible in Fig. 13–7 (one of Jokelainen's micrographs); a suggestion of the third layer (arrow) is also visible in 13–6. Additional structural differentiation is seen by Brinkley and Stubblefield in the corona of electron dense material covering the surface of the centromere (Fig. 13–6b). In this micrograph, 50–80 Å fibrils are visible in the coronal layer, extending outward at right angles to the surface of the centromere.

The layered structure of the centromere as outlined above (see also diagram in Fig. 13–11) appears in many plants and animals and may well be universal in eukaryotes. But in some organisms, even when using aldehyde fixation, little or no structural differentiation of this type is seen. It is not known whether this undifferentiated configuration (see for example Fig. 13–9) represents an alternate less structured form of the centromere or is simply material that may require more complete preservation than afforded by the aldehyde fixatives. In a few groups of organisms, such as the

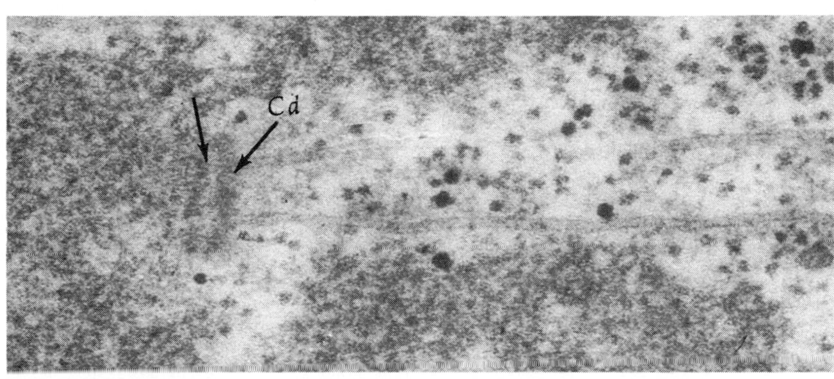

13–7. Centromere of a rat cell, fixed in glutaraldehyde-osmium. An additional dense layer (arrow) is visible between the centromere disc (Cd) *and the chromatin. × 72,000. Courtesy of P. T. Jokelainen and Academic Press Inc., from* J. Ultrastr. Res. *19:19 (1967).*

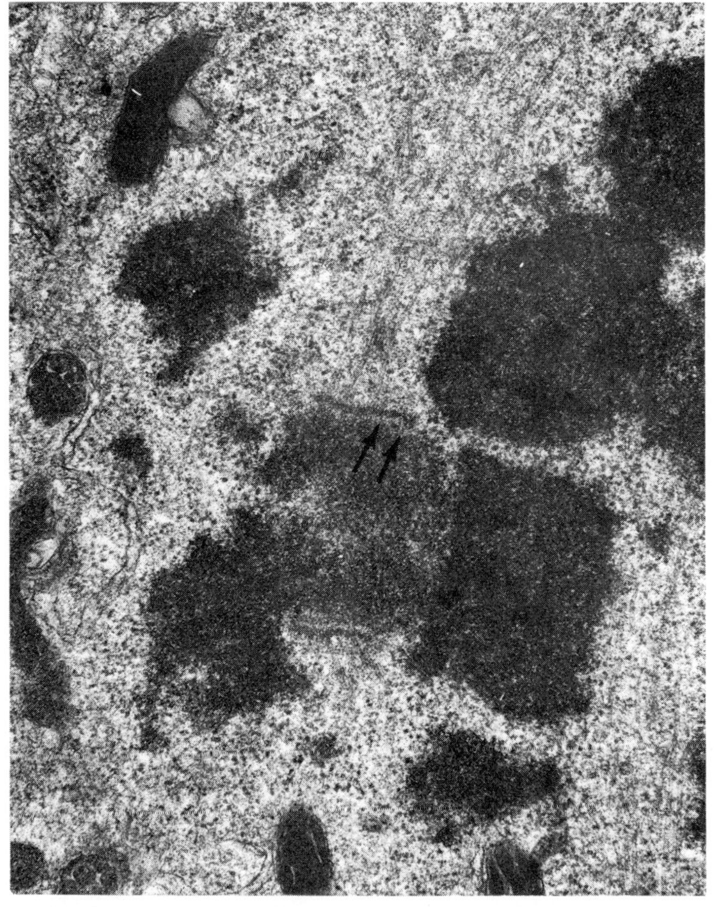

a

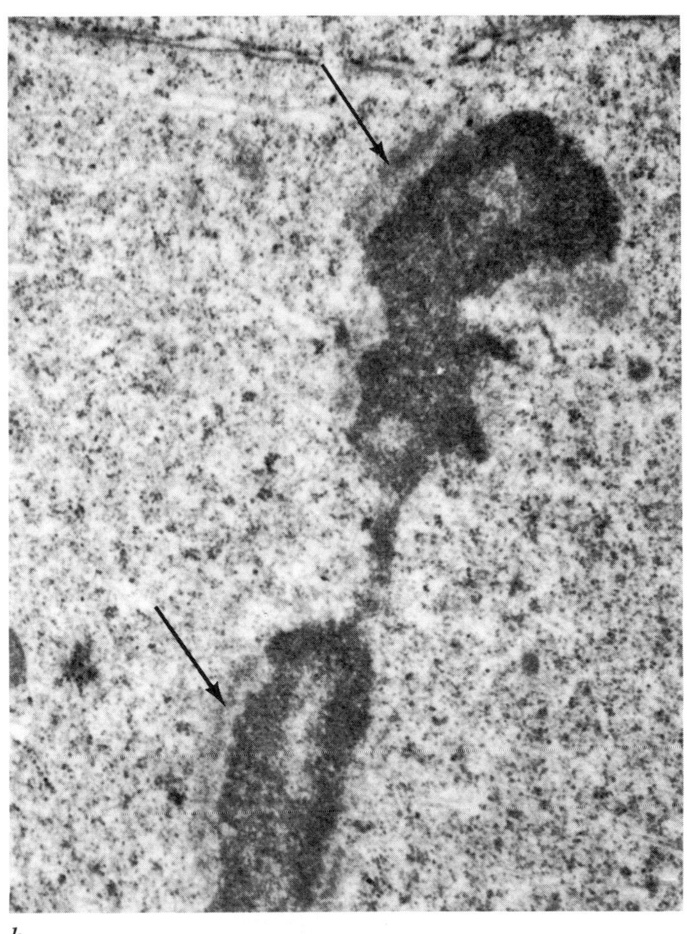

b

13–8. a, *Centromere of a Chinese hamster cell, clearly showing the relationship of a group of spindle microtubules to the centromere disc. The tubules apparently penetrate through the disc into the underlying chromatin (arrows). × 37,000. Courtesy of L. J. Journey.*

b, *Diffuse centromere of the insect* Rhodnius. *Note that a dense layer similar in structure to the centromere disc covers much of the chromosome surface (arrows). × 21,000. Courtesy of R. C. Buck and Academic Press Inc., from* J. Ultrastr. Res. *18:489 (1967).*

Hemiptera, no localized centromeres or primary constrictions occur. Instead, spindle fiber connections are made at many points along the length of the chromosome arms. Electron microscopy of this configuration (called *diffuse centromeres*) has revealed that a layer of electron-dense material evidently analogous to the centromere disc covers extensive regions of the chromosome surface. As in the centromere disc, this layer is separated from the underlying chromatin by a less dense space (Fig. 13–8b; see Buck, 1967).

The molecular identity of the material making up the centromere disc is unknown. Because the various layers can be readily distinguished from the underlying 100 Å chromatin fibers and are obviously of basically different structural composition, it seems most likely that the centromere disc does not merely represent a specialized coiling configuration of the unit fibers of the chromosome. Thus, the centromere disc probably consists of nonchromatin material that is attached to the surface of the chromosome at the primary constriction. This interpretation is supported by the fact that the components of the centromere disc are evidently more sensitive to the conditions of fixation than the 100 Å fibers of the chromatin.

Secondary Constrictions

Secondary constrictions (Figs. 13–4 and 13–11) are usually present in at least one pair of homologous chromosomes of an organism, at least in species with clearly localized centromeres. Where diffuse centromeres are present, there are no primary or secondary constrictions. Frequently, the secondary constriction marks the position of the nucleolar organizer; the nucleolus reappears at this location during reconstitution of the daughter nuclei at telophase (see below). In some cases, secondary constrictions are not associated with nucleolar function. While the latter are useful in the identification of metaphase chromosomes, their functional significance is unknown. Under the electron microscope, the secondary constriction shows no identifiable structural components other than the 100 Å fibers of the chromatin. Often, the fibers in this region appear to be less tightly coiled (Fig. 13–10). To the extent that diffuse, less densely coiled chromatin may be identified with "active" chromatin (see Chapter 11), it is possible that secondary constrictions are formed by limited regions of the chromosome that are active in RNA transcription during mitosis.

A Summary of Metaphase Chromosome Structure

The probable structure of the metaphase chromosome, including the primary constriction, centromeres, and the secondary constriction is shown diagrammatically in Fig. 13–11. This model assumes, for simplicity, that the 100 Å unit fiber of the chromosome consists of a single DNA duplex with associated histone and nonhistone protein. Spaced along the unit fiber are hypothetical "folders" that condense the chromosome from the interphase to metaphase state during prophase of mitosis. As shown in the model, the body of the chromosome is considered to be made of tightly packed unit fibers. Although it has been proposed from time to time that the metaphase chromosome contains a rod-like protein backbone to which the unit fibers become anchored during prophase condensation, no evidence of such a backbone has ever been seen in thin-sectioned or isolated chromosomes. Therefore, fibers are shown as the only structural component of the arms of the chromosome. This carries with it the implication that the primary difference between the interphase and metaphase chromosome is the degree of coiling of the unit fibers.

These fibers extend through and form the axis of the primary and secondary constriction. The centromere discs, shown as attached to the chromosome at the primary constriction, are considered to contain structures that are identifiably different from the unit fibers of the chromosome. Bundles of spindle microtubules are shown anchored in the centromeres.

Independent Movements of the Chromosomes during Prophase and Early Metaphase

The chromosomes undergo several types of independent movement before attachment to the spindle microtubules during metaphase. The first conspicuous movement occurs during early prophase, when the dividing asters migrate around the outside of the nuclear envelope. At this time, the chromosomes can sometimes be observed to follow the movement of the asters as if an attractive force extended from the asters across the nuclear envelope to the chromosomes. The basis of this attraction and movement of the chromosomes, called *polarization,* is unknown. No direct connections can be seen between the asters and the chromosomes at this time in either the light or electron microscope.

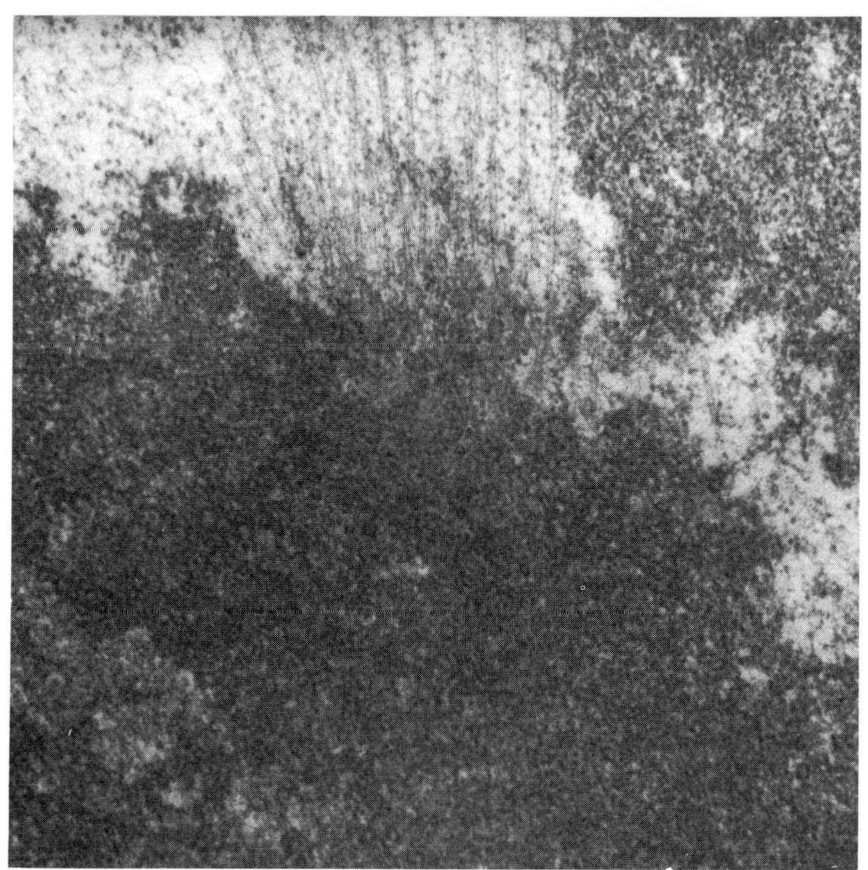

13–9. Centromere region of the plant Haemanthus. *Even though glutaraldehyde-osmium fixation has been used, no specialized structures are visible in the centromere region. Note the large number of microtubules attached at the centromere in* Haemanthus. *× 39,000. Courtesy of A. Bajer and the publisher, from* Chromosoma *25:249 (1968).*

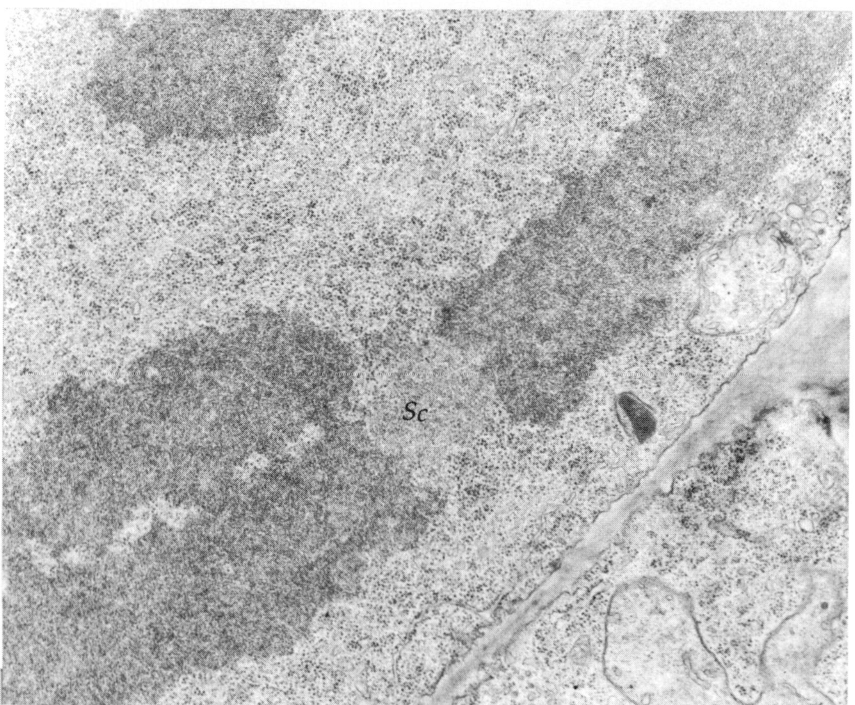

13–10. Fine structure of the secondary constriction (Sc) *in a chromosome from the plant* Vicia faba. *The constriction contains fibrillar material of lower density than the adjacent chromatin. × 18,000. Courtesy of J.-G. Lafontaine.*

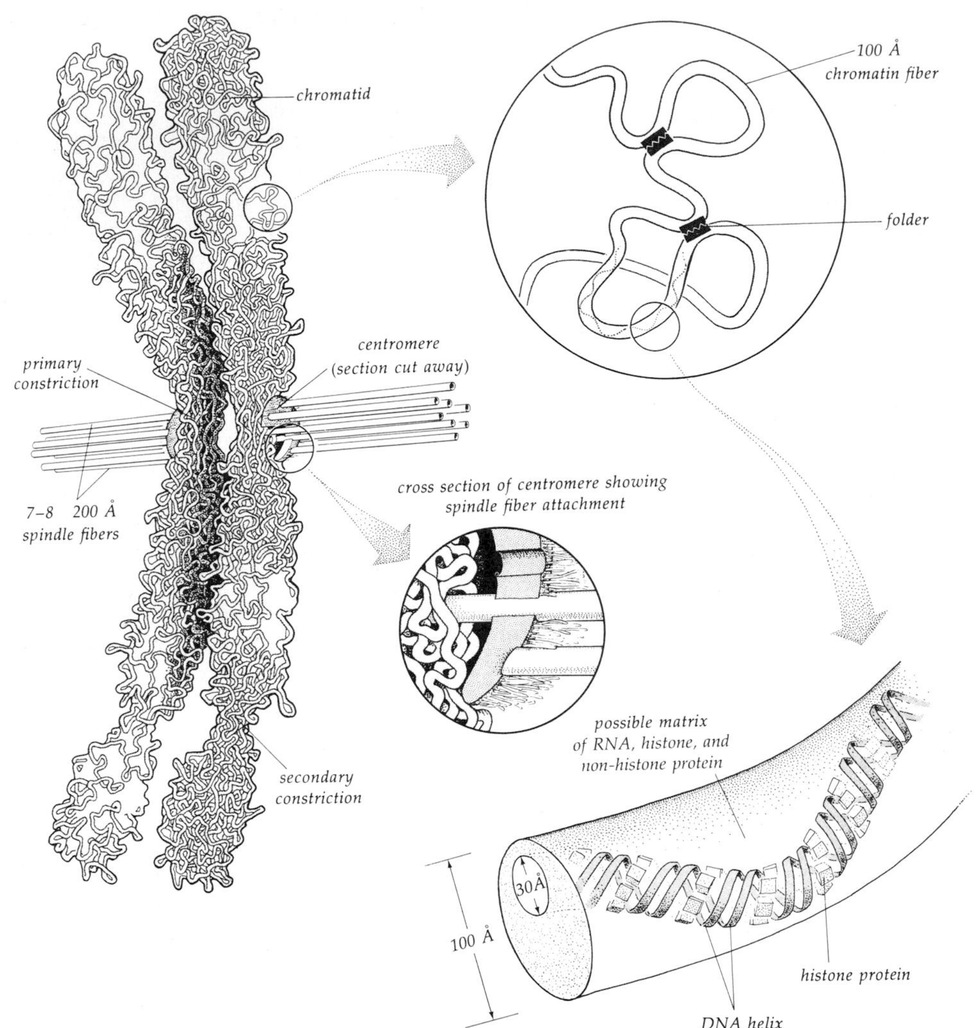

13–11. Diagram summarizing the possible structure of the metaphase chromosome. The arms of the chromatids are believed to contain only tightly packed masses of 100 Å chromatin fibers. Each chromatid probably consists of a single, folded 100 Å unit fiber and would therefore have only two free ends. Although apparently irregular, the folding of the 100 Å unit fibers is highly ordered and produces the regular morphology of the metaphase chromosome, including characteristic length and width, and placement of primary and secondary constrictions. Superimposed over the primary constriction of each chromatid is the centromere disc (lower circle). The disc is apparently not structured from chromatin fibrils and probably consists of a proteinaceous substance that becomes attached to the primary constriction during early prophase. Spindle microtubules are shown anchored in the centromere disc; some of these penetrate through the disc to contact the fibrils lying immediately below. The surface of the discs facing the poles of the spindle are covered with a fibrous "corona" of unknown composition. The basic folding pattern of the chromatin fibrils may be regulated by "folders," mutually attractive groups spaced along the unit fibers that are progressively unmasked and activated during prophase and metaphase of mitosis (upper circle). The folders and their spacing would be a part of the regular structure of the chromosome fiber and would presumably be reproduced as a part of chromosome replication during interphase. According to the model, the unit fiber contains a single nucleohistone complex consisting of DNA with associated histone (lower right). This complex is wound into a 100 Å "supercoil," with a DNA/unit length ratio about twice that of fully extended nucleohistone. A matrix, possibly containing RNA, histone, and nonhistone protein, is shown around the nucleohistone supercoil. The precise identity and manner of binding of the matrix substance is unknown (for details of fiber substructure, see Ch. 8).

Other apparently independent movements of the chromosomes are observed after the nuclear envelope breaks down at the close of prophase. When the envelope disintegrates, the chromosomes are usually randomly scattered throughout the area formerly occupied by the nucleus. As spindle formation becomes complete, the chromosomes migrate from their scattered locations to the midpoint of the spindle. This movement, called *congression,* is not always direct; the migration may involve several oscillations between the poles and the midpoint before the final position on the metaphase plate is taken up. The mechanism for the movements taking place in congression is unknown. Some authorities consider that these premetaphase migrations are carried out independently of the spindle and do not require connections between the chromosomes and spindle microtubules. Others maintain that spindle connections are made by the centromeres immediately on breakdown of the nuclear envelope and that all premetaphase movements are moderated by spindle microtubules. Attempts to study this problem using irradiation of chromosomes and kinetochores by an ultraviolet microbeam have been inconclusive. Ultraviolet irradiation destroys spindle microtubules and interferes with anaphase movement if applied during metaphase (see Bajer, 1968a). Application during congression of the chromosomes to the metaphase plate in early metaphase has produced conflicting results, with some investigators claiming interference and others detecting no effect. Thus it is not clear from these experiments what role, if any, connections of the chromosomes to spindle fibers might have in movement of the chromosomes to the metaphase plate.

An experiment by B. R. Brinkley and R. B. Nicklas, in which micromanipulation of the chromosomes was employed, gives more direct information about the possible function of microtubules in the movement of chromosomes to the metaphase plate. Using tissue from grasshoppers as experimental material, Brinkley and Nicklas detached single chromosomes from the metaphase plate with a microneedle. The detached chromosome remained motionless for a time in the cytoplasm and then returned to the metaphase plate. Electron micrographs prepared from tissue at various stages in this sequence showed that the movement returning the detached chromosome to the metaphase plate was not initiated until connections were made at the centromere to spindle microtubules. This suggests that such connections are necessary for movement of the chromosomes and possibly that the motions of the chromosomes during congression are also moderated by microtubules attached to the chromosome at the centromere.

Positioning the chromosomes at the midpoint of the spindle involves more than a simple movement to the metaphase plate. During migration, the chromosomes in many species are "sorted out" according to size so that the smaller chromosomes come to lie in the interior of the spindle. By the time the chromosomes reach the midpoint of the spindle, or soon after, both centromeres of the primary constriction have made connections to spindle microtubules. These are made in such a way that the two centromeres of a chromosome always connect to opposite poles. The basis for these movements and orientations, including the sorting out during congression and the connections made by the centromeres at metaphase remains unknown.

The Spindle

Formation of the Spindle

Several lines of evidence indicate that in both plants and animals the microtubules of the spindle are assembled from a cytoplasmic pool of building blocks in the form of ready-made protein subunits. In this sense, the spindle microtubule represents a *polymer* constructed from a pool of protein *monomers.* This was first demonstrated by H. Went and D. Mazia by means of an antibody specific for spindle protein. Went and Mazia isolated spindles in quantity from sea urchin embryos and injected a protein extracted from these spindles into a rabbit. After a few days the rabbit produced an antibody to the spindle protein. This antibody was then purified and tested against extracts made from sea urchin cells in both metaphase and interphase. Cells in both stages showed a reaction to the antibody, indicating that the protein of the spindle is present in interphase as well as in metaphase. Other more recent experiments also support the conclusion that the tubules of the spindle are constructed from a pooled protein subunit (see, for example, Robbins and Shelanski, 1969). Many agents, including exposure to cold, high hydrostatic pressure, ultraviolet radiation, and chemical agents such as colchicine, are known to disperse the spindle rapidly. Within seconds after exposure to these agents, the spindle disintegrates and disappears. Electron microscopy of these cells reveals extreme disorganization or even complete absence of spindle microtubules. Once the dispersing agent is removed, the spindle reforms rapidly, sometimes within 2–3 minutes after return of the cell to normal conditions. The speed with which the spindle is reconstructed suggests rapid polymerization of the component microtubules from a pool of monomers rather than synthesis of new spindle protein. This is supported by the fact that inhibitors of protein synthesis, such as puromycin or actinomycin D, do not slow the restoration of the spindle after disruption (for details of these experiments, see Inoué and Sato, 1967).

This information indicates that the proteins of spindle microtubules can be quickly interconverted between the polymerized tubular form and a monomer pool. This ready interconversion forms the basis of one of the more plausible hypotheses of movement of the chromosomes by the spindle (see below). The monomer pool persists throughout interphase, as shown by the experiments of Went and Mazia, Robbins and Shelanski and others; no doubt synthesis of additional monomer takes place during interphase on cytoplasmic ribosomes.

Structure of the Spindle

The arrangement of microtubules in the midpoint of the spindle is closely similar in plants and in animals (Fig. 13–12). The spindle is broadest at this point and is roughly circular in cross section at all levels. At either pole, the spindle narrows; the component microtubules converge toward these points. This narrowing is most pronounced in organisms in which the spindle fibers converge to centers formed by the asters (*astral spindles*). In the plants and the few animal cells containing no asters or centrioles (*anastral spindles*), the spindle is broader at the ends, assuming a barrel rather than spindle shape.

The spindle may contain 500–1,000 or more microtubules. In addition to tubules, various regions of the spindle may contain scattered vesicles, ribosomes, particles of smooth and rough endoplasmic reticulum, and concentrations of amorphous, electron-dense material of unknown identity. These structures are found dispersed throughout the spindle between bundles of parallel microtubules. Vesicular components are usually most concentrated at the poles of the spindle.

The structure of spindle microtubules. The microtubules of the spindle are unbranched, hollow cylinders that may reach many microns in length. While the tubules of the continuous fibers cannot usually be completely traced in thin sections, it is believed that these extend from pole to pole of the spindle. Spindle microtubules run in parallel array in relatively straight courses, giving the impression that the tubules are structurally rigid but not resistant to bending over shallow arcs.

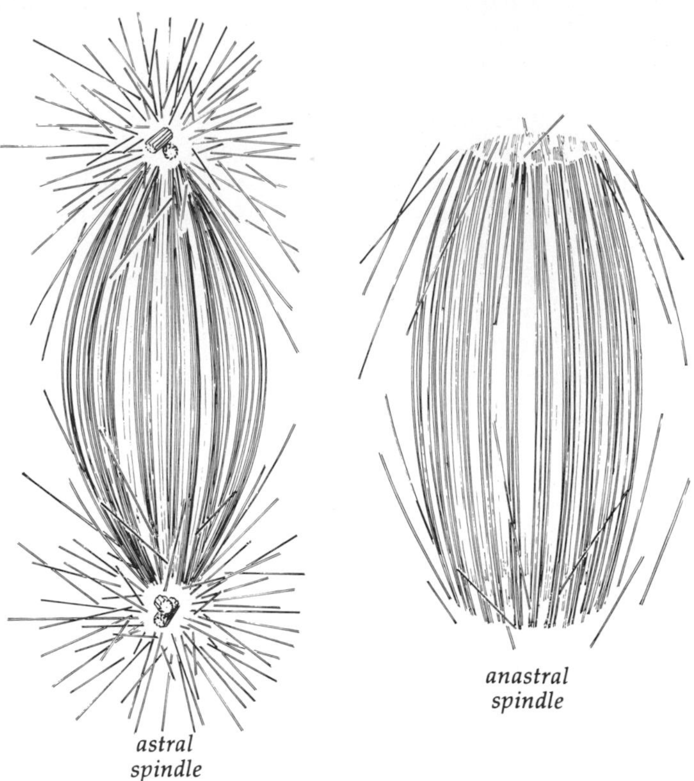

13–12. Comparison of the structure and distribution of astral and anastral spindles. Astral spindles appear in most animals, some fungi, and the gametic divisions of lower plants. Anastral spindles appear in higher plants, somatic cells in most lower plants, some protozoa, and some oocytes during meiotic divisions.

Spindle microtubules are variously reported to be 150–220 Å in diameter, slightly smaller than the tubules seen in the shafts of cilia and flagella (see Chapter 15). The diameter observed in thin sections varies depending on the preparation methods used. In methacrylate plastics, used almost exclusively as embedding media until the early 1960's, spindle microtubules appear to be 150–180 Å in diameter. In thin sections of material embedded in the more recently developed epoxy resins, the diameter appears slightly larger, approximately 180–220 Å. With the best available fixation (glutaralde-

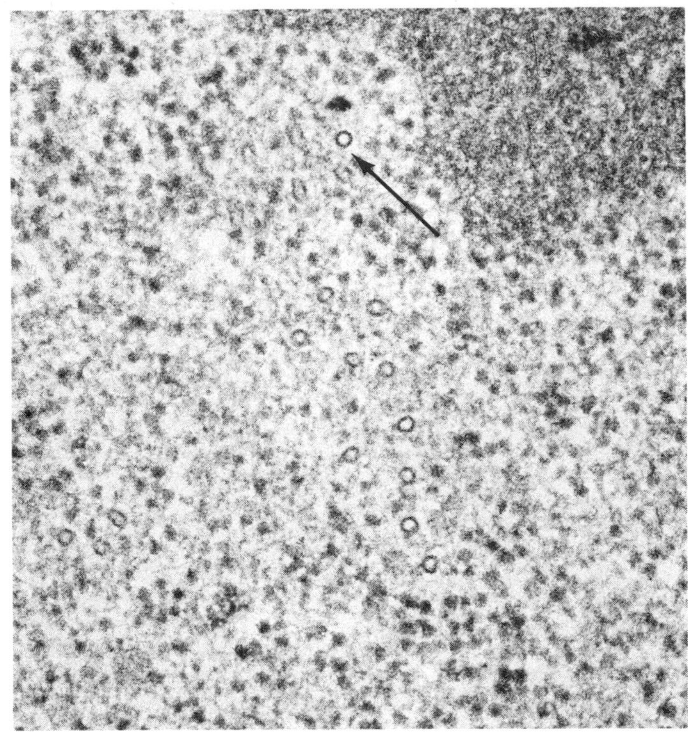

13–13. Spindle microtubules of the Swiss mouse in transverse section. The outside diameter of the individual microtubules in this micrograph is about 200 Å. Evidence of regular particulate substructure is visible in the walls of one of the microtubules (arrow). × 99,000. From The Nucleus, *ed. A. J. Dalton and F. Haguenau, 1968. Courtesy of E. de Harven and Academic Press Inc.*

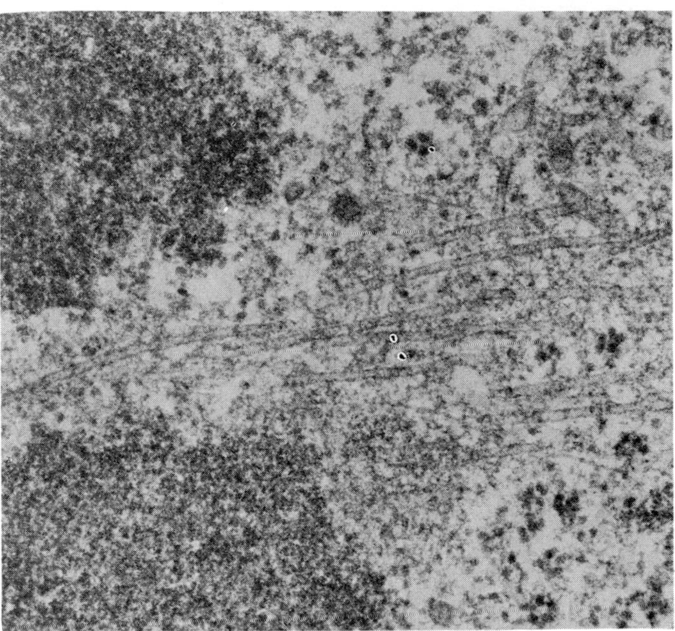

13–14. Spindle microtubules of the mouse in longitudinal section. × 63,000. From The Nucleus, *ed. A. J. Dalton and F. Haguenau, 1968. Courtesy of E. de Harven and Academic Press Inc.*

hyde-osmium) and epoxy resin embedding, the diameter of spindle microtubules is very regular, averaging 180–200 Å. The diameter is the same at all stages of mitosis, indicating that the fibers do not stretch or contract during any part of the process (see below).

Spindle microtubules are easily recognized in both longitudinal and cross sections (Figs. 13–13 and 13–14). The walls are electron-dense, and in cross section can be seen to be 40–60 Å thick. In favorable sections the tubule wall appears to be formed from a circle of subunits (arrow, Fig. 13–13) almost invariably 13 in number. The central bore of the tubule is electron-lucent. This is usually taken to mean that the contents of the tubule lumen are fluid or at least less rigid than the walls.

The tubules of the spindle, as mentioned, are intermixed with scattered collections of vesicles, ribosomes, endoplasmic reticulum, and unidentified electron-dense material. It is not known whether these components are significant in the function of the spindle or are merely passively included in the regions between the bundles of spindle microtubules. Because of the inclusion of these structures and the chromosomes, isolation and analysis of the spindle has revealed a confusing array of molecular components, including all of the major classes of macromolecules. However, some proteins are found in abundance and are believed to represent the structural subunits of spindle microtubules. This work has been aided by the binding capacities of the spindle inhibitor, colchicine.

Colchicine evidently causes disorganization of the spindle by binding to the subunits of spindle microtubules. If colchicine is added to proteins extracted from isolated spindles or if the protein from mitotic cells arrested with colchicine is analyzed, the colchicine is found to be tightly bound to a protein sedimenting at 6S. This protein has been subjected to detailed investigation by R. E. Stephens, and G. G. Borisy and E. W. Taylor.

The 6S protein, because of its affinity for colchicine, is thought to represent the basic subunit of spindle microtubules. Other protein fractions sedimenting over the range from 2S to 27S, with significant peaks at 2.3S, 3.5S, 9S, 13S, 22S, and 27S, are also found in spindle extracts. Some of these undoubtedly represent extraneous material isolated along with the spindle. But there are some good indications that at least a few of these fractions, particularly the 22S peak, may represent different aggregation states of the 6S particle. This is suggested by the inverse ratio in which the 6S and 22S fractions occur in extracts (Table 13–1). It is also possible that the 6S fraction alone is the microtubular protein, as indicated by its capacity to bind colchicine, and that the 22S fraction is a totally different protein forming a part of the matrix surrounding the spindle microtubules. The latter hypothesis is supported by observations that the 22S protein does not bind colchicine and does not react with antibodies for 6S protein. But these findings do not absolutely exclude the possibility that the 22S protein may represent an aggregate of 6S particles, because the capacity to bind colchicine and 6S antibody could be lost if extensive rearrangements occur in the conversion from 6S particle to 22S aggregate. Whether the 22S fraction is a 6S aggregate, it does not seem possible that it could directly form a microtubule subunit, because according to Stephens it is too large (150 × 200 Å, particle weight 880,000) to fit into the 40–60 Å microtubule wall (for detailed literature references in this area, see Forer, 1969).

Table 13–1 Proportions of 6S and 22S Protein in Spindle Extracts after Different Isolation Procedures

Method of isolation	*Amount of component*		*Ratio of amounts*
	6S	*22S*	*6S:22S*
Hexylene glycol	20%	80%	1:4
Dithiodipropanol	75	25	3:1
Ethanol-digitonin	83	17	5:1

Adapted from A. Forer, Chromosome movements during cell division. In A. Lima-de-Faria, ed. *Handbook of Molecular Cytology*. Amsterdam: North-Holland Publishing Co., 1969. See this paper for original references to preparation methods.

Function of the Spindle: The Anaphase Movement

Once congression of the chromosomes to the midpoint of the spindle is complete and attachments have been made between the centromeres and spindle microtubules, all further movements are made in close synchrony. At least two of the subsequent motions are distinct and are controlled by different mechanisms. These are the separation of chromatids and movement of the chromosomes to the poles.

That these are separate is indicated by experiments in which cells are arrested in metaphase by the addition of colchicine. As noted, the spindle is completely disrupted by this chemical. The metaphase chromosomes in treated cells are left dispersed in the area formerly occupied by the midpoint of the spindle. If the concentration of colchicine is maintained, these chromosomes split, more or less in synchrony, and each chromatid separates from its sister. Detachment may be incomplete, so that the chromatids remain connected in the centromere region, but frequently complete separation takes place. Subsequently, the chromosomes uncoil and form a single telophase nucleus or a group of several nuclei of varying size. A single nucleus, if produced, is functional and apparently normal except that it contains twice the normal telophase complement of chromosomes (experimental polyploids in plants are commonly derived in this manner by the use of colchicine). This experiment shows that a functionally intact spindle is not required for the separation of sister chromatids.

Poleward movement of the chromosomes, in contrast, requires that the spindle be intact. This movement, as noted, takes place in close synchrony. The synchrony is not perfect; one or more chromosomes may lag behind or reach the pole before the remainder of the group. The movement is usually at a steady or slightly decelerating rate as anaphase proceeds; all of the chromosomes, regardless of size, move at approximately the same speed toward the poles. The speed of movement, 0.2 to 5 microns per minute, is very slow in relation to other cellular movements such as muscle contraction. Even cytoplasmic streaming, a relatively slow movement, is much faster, often proceeding at as much as 250–300 microns per minute.

R. B. Nicklas, E. W. Taylor, and A. Forer, working independently, have made various calculations related to the force required to move a chromosome from the equator to the pole of the spindle. These calculations must be regarded as approximations at best, because no direct measurements can be made of the viscosity of the medium through which the chromosomes must be moved or of the effects of differences in the size and shape of the chromosomes on the force required for movement. Within the limitations imposed by the requirement to estimate the effects of these parameters, the force required to move a chromosome of average size is calculated to be approximately 10^{-8} dynes. Generation of force of this magnitude is well within the capacity of the cell; Forer estimates that application of this force over the distance from the equator to the spindle would require an expenditure of approximately 30 molecules of ATP. While this may at first seem a large quantity of ATP, it is much smaller than the energy required to synthesize the total protein of the spindle, keeping in mind that each peptide linkage requires the conversion of approximately 1 ATP to ADP and that the number of peptide bonds in each spindle microtubule runs into the millions. Taylor has estimated that the ATP required to move an entire set of chromosomes is about one billion times smaller than that required for the synthesis of the required spindle protein. That ATP is directly involved in the activity of the spindle is indicated by the fact that isolated spindles contain a specific ATPase called *dynein;* this ATPase appears to be present in highest concentrations at the poles of the spindle (see Hartmann, 1964).

Careful measurements by Hans Ris in the early 1940's showed that anaphase movement has two components, one contributed by shortening of the chromosome fibers and one by lengthening of the whole spindle. The distance between the chromosomes and the poles shortens to about one-third to one-fifth of the separation at metaphase, and the separation between the poles increases. These observations are of great significance for hypotheses of chromosome movement, because any model must take into account the fact that at anaphase one group of microtubules evidently lengthens and another group becomes shorter. The contribution of movement from the two sources to the total distance traveled by the chromosomes varies in different cell types. Most commonly, the total movement is derived from a fairly equal combination of the two (Fig. 13–15).

Hypotheses of the Anaphase Movement

With these observations of the anaphase movement in mind, we can now turn to a few of the more plausible hypotheses of anaphase movement of chromosomes by the spindle. A number of these can be disposed of on the basis of the information provided in the previous paragraphs. In a few cases, additional detracting or supporting evidence is included.

Some early hypotheses of interest. At the turn of the century and in the early years of this century, the existence of spindle fibers was considered a highly debatable point. Because of this, many hypotheses of anaphase movement involved electromagnetic fields, diffusion, cytoplasmic streaming or gel-sol transformations, all of which could conceivably move chromosomes as a group without a fiber connecting the poles to the chromosome. The electromagnetic hypotheses were abandoned because intense magnetic fields experimentally imposed on cells during mitosis are without effect on the movements of chromosomes or indeed on any phase of cell division. Diffusion or cytoplasmic streaming are equally unlikely. If currents set up by one of these processes were responsible for "washing" the chromosomes to the poles, it would seem that any other cytoplasmic particles in the vicinity would also be carried along. However, small particles and vesicles can be seen under the light microscope to move in directions opposite to the chromosomes at anaphase. Gel-sol transformations, popular in the 1930's as a source of anaphase movement, also seem very unlikely today. According to this idea, the chromosomes are pushed to the poles at anaphase by a gradually expanding gel that forms between the sister chromatids. This is made unlikely by the frequently observed exceptions to the rule that the chromosomes advance in synchrony to the poles. Often, a chromosome lags behind and advances to the pole at a distance behind the main group. This would be impossible if a rigid, expanding gel were formed in the area between the separating chromosomes. (These early hypotheses, which are of considerable historical interest, are described in detail in Schrader, 1953).

Current hypotheses of the anaphase movement. All recent hypotheses take into account the connection of chromosomes to the microtubules of the spindle. These proposed mechanisms fall into one of three categories: contraction, sliding filament, and tubule assembly-disassembly models. The contraction model, which states that the chromosomes are moved at anaphase by contraction of the spindle microtubules, is

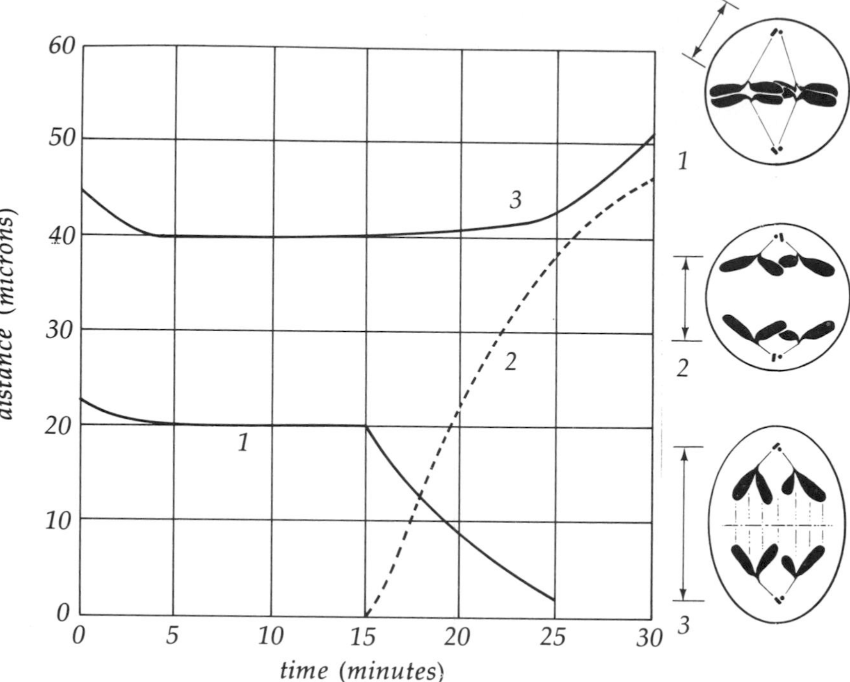

13–15. Relative contributions of chromosome and continuous spindle microtubules to anaphase movement. 1, distance separating chromosomes from the poles; 2, distance between sister chromatids; 3, total distance between the poles. At time 15, anaphase separation begins (curve 2). During this process spindle microtubules connected to the chromosomes become shorter and the continuous microtubules, which extend from pole to pole, lengthen. Redrawn from "How Cells Divide" by D. Mazia. Scientific American *205:100 (1961).*

not widely favored. Contraction implies that a molecule of the microtubule, presumably a protein, shortens by folding or coiling. It is difficult to imagine how this could happen without being accompanied by a thickening of the walls of the spindle microtubules. But electron microscopy has shown that anaphase proceeds without any perceptible thickening or change in diameter of either the microtubule or its wall. This hypothesis also fails to explain how the pole-to-pole spindle microtubules act to push the poles farther apart during anaphase.

The sliding filament model must be taken more seriously. Briefly, this hypothesis states that the two major types of spindle microtubules, the pole-to-pole and pole-to-centromere fibers, generate force at anaphase by sliding over each other. This has an analogy in the current hypothesis of muscle contraction, in which the muscle is believed to shorten during contraction by a progressive interdigitation of two classes of molecular filaments making up the muscle fiber. The force developed is considered to be produced by chemical bonds that are alternately formed and broken as the filaments slide past each other. Although analogous behavior on the part of spindle microtubules is a possibility, there is little concrete evidence to support this hypothesis. A number of investigators have detected patterns of density between microtubules that may represent cross-bridging (Fig. 13–16); these are considered by some to be evidence of interactions between the microtubules of the spindle.

This model has the advantage of accounting for both movement imparted to the chromosomes by the pole-to-chromosome fibers and lengthening of the spindle; the force required for both could be generated by the interaction between the two types of microtubules. Some difficulties are presented by the fact that for maximum generation of force, the two types of microtubules should be interspersed in something approximating a 1:1 ratio. But in many organisms, the pole-to-pole fibers form a separate bundle in the axis of the spindle, and the pole-to-centromere fibers, with the chromosomes, are arranged around the periphery (the so-called "hollow spindle"). In this case interaction between pole-to-centromere and continuous fibers would be limited to the transition zone between the two classes.

Perhaps the most difficult observations to fit into the sliding filament model occur in the protozoa. In many protozoa, the nuclear membrane remains intact during mitosis. The spindle may be either intranuclear or located entirely outside the nucleus. In one group of ciliate protozoa described by L. R. Cleveland

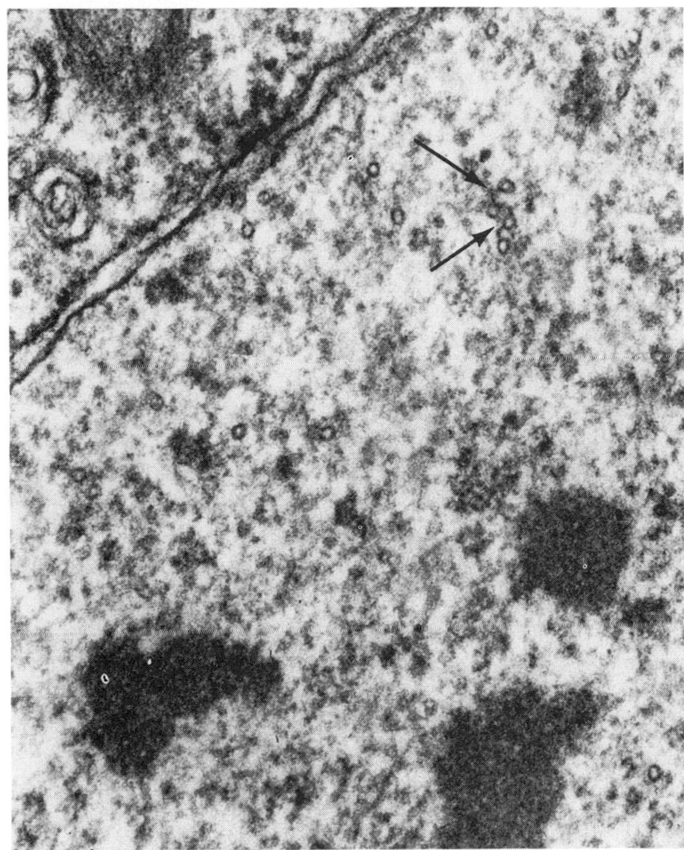

13–16. *Apparent cross-bridging (arrows) between spindle microtubules in a metaphase cell of the green alga* Blastophysa rhizopus. *× 85,500. Courtesy of H. J. Wilson and The Rockefeller University Press.* [J. Cell Biol. *40:854 (1969)*]

in the 1930's, the continuous pole-to-pole fibers form a bundle around which the nucleus is wrapped (Fig. 13–17). The chromosome-to-centromere fibers are evidently separate from the continuous fibers and make connections to the centromeres through the nuclear envelope. Because the two classes of fibers are apparently separate, it is difficult to imagine how the sliding filament model could operate in this organism if indeed no continuous fibers are located in the vicinity of the pole-to-centromere fibers. Reinvestigation of Cleveland's work with the electron microscope would be especially useful in view of the importance of these early observations for the sliding filament hypothesis.

The remaining hypothesis for the anaphase movement, the tubule assembly-disassembly model, has been advanced in its most complete form by Shinya Inoué. The hypothesis states that an equilibrium exists between the pool of spindle microtubule monomers and the polymerized filaments of the spindle. The pole-to-centromere filaments are shortened during anaphase movement by disassembly of these microtubules at the poles of the spindle. As the polar end of the tubule is disassembled, the microtubule, with its attached chromosome at the opposite end, is pulled toward the pole. The monomers removed at the polar end of the pole-to-centromere fibers either return to the monomer pool or are built into the polar ends of the continuous pole-to-pole fibers. As the monomers polymerize at the ends of the continuous fibers, the poles of the spindle are pushed farther apart. Thus the assembly-disassembly model takes both components of the anaphase movement into account. But

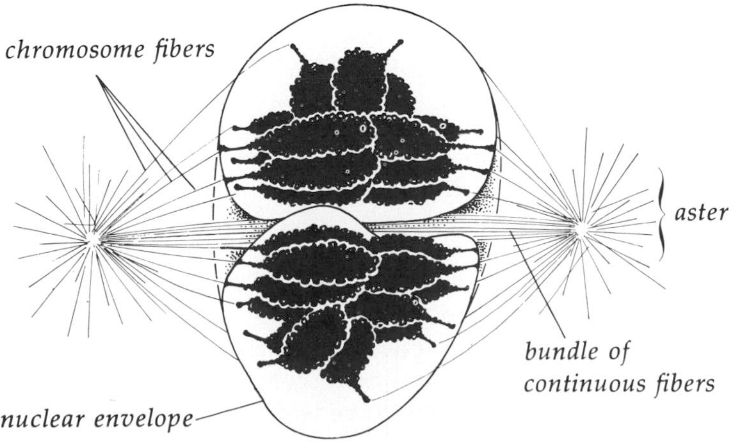

13–17. *Drawing of an anaphase nucleus in the protozoan* Barbulanympha. *The nuclear envelope remains intact during mitosis, and continuous and chromosome spindle fibers are clearly separate. Redrawn from L. R. Cleveland,* J. Morph. *85:197 (1949); original courtesy of the Wistar Institute of Anatomy and Biology.*

one thing unexplained by the model is the mechanism that generates the force for movement of the pole-to-centromere microtubules and the attached chromosomes to the poles. It is not easy to see how removal of subunits from one end of the tubule would result in movement of the whole tubule, and this is not explained in the model. On the other hand, the growth or elongation of the continuous pole-to-pole tubules by addition of subunits at the ends carries with it an explanation of the force generated. This model has many parallels in biological growth.

The tubule assembly-disassembly hypothesis has some indirect limited experimental support. The ATPase activity of the spindle is greatest at the poles; this, in the model, is the site of assembly-disassembly activity. Inoué has also noted that very dilute solutions of colchicine induce anaphase movement of the chromosomes. This fits in with the model, because colchicine in dilute solution might be expected to bind to tubule subunits, causing disassembly and pushing the equilibrium in the direction of the monomer pool without completely destroying the spindle.

Of the three current models for movement of the chromosomes at anaphase, either the sliding filament or the tubule assembly-disassembly model seem most likely to be correct. Both models have active adherents at the present time. As noted, some observations are difficult to fit into the sliding filament model, and the generation of force in the assembly-disassembly model presents some conceptual difficulties. Until concrete evidence is forthcoming, the basis of the anaphase movement of chromosomes must be regarded as undecided.

Some general observations on the significance of microtubular form for the movement of chromosomes on the spindle. Microtubules are found in many parts of the cell in addition to the spindle and occur in the cytoplasm in all stages of the life cycle. Microtubules are a conspicuous structural component of the axis of cilia and flagella from all organisms except prokaryotes. Microtubules are also found in parallel arrays in regions of amoeboid cells undergoing cytoplasmic streaming. In plants, additional collections of microtubules are found in the cytoplasm near the plasma membrane areas in which new cell wall is being laid down. In animal cells, numerous microtubules extend through the cytoplasm of the long processes of nerve cells. In the nucleated erythrocytes of many vertebrates, a band of microtubules extends around the periphery of the cell just under the plasma membrane. In at least a few cases, microtubules have been reported as an inclusion in all major cell organelles, including nuclei, mitochondria and chloroplasts, and in bacterial cells. In all cases, the structure of microtubules from these sources is closely similar, with outside diameters ranging from 150–250 Å.

The presence of microtubules in the spindle, the flagellum, and regions undergoing cytoplasmic streaming suggests that these structures are directly involved in the generation of force for movement. Their presence in other locations suggests other possible roles, including transport and skeletal support. The latter are not difficult to visualize as functions of a tubule, because this structural form is used for the same tasks by human engineers. But the connection of tubular form with the provision of motility is not obvious; as noted, no one has yet been able to derive a completely satisfactory model for the provision of motion by microtubules. (One important hypothesis is discussed in detail as a part of the description of the morphology and function of the flagellum in Chapter 15.)

The close structural similarity observed in microtubules in all parts of the cell has led to the development of a number of unifying hypotheses stating that all of the various functions of microtubules are properties of the same basic organelle. According to this idea, a spindle microtubule is the same as, for example, the microtubules of flagella. This is supported by the observation that highly diverse microtubular systems share a capacity to bind colchicine. This is true of spindle microtubules, the microtubules of nerve cells, cytoplasmic microtubules engaged in cell wall synthesis in plants, and some microtubules of the axial complex in flagella (the "central singlets" of the complex; see Chapter 15). Colchicine also has the effect of inhibiting the generation and growth of the axial complex of developing flagella.

Other evidence suggests that spindle microtubules and some other microtubular types, while possibly containing a colchicine-binding subunit closely similar to the subunits of other microtubules, may at least represent a less stably assembled configuration of the basic subunit. The microtubules of flagella, cilia, and the marginal band of erythrocytes are relatively stable during fixation and were described early in the development of electron microscopy. These microtubules are even rigid enough to be isolated from the cell with-

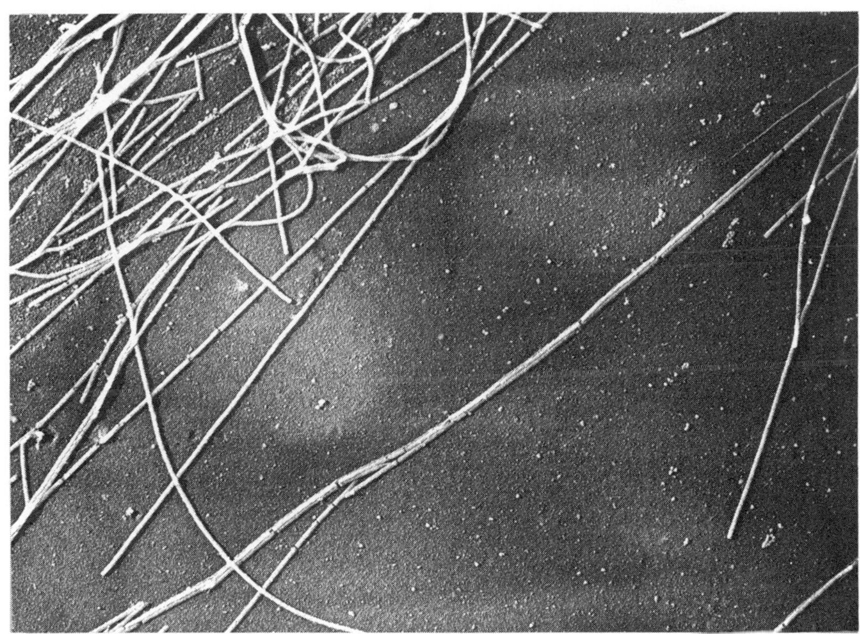

13–18. Microtubules isolated from erythrocytes of the salamander Taricha granulosa. *These tubules, which lie in a marginal band around the circumference of the erythrocyte, probably serve a supportive or "skeletal" function.* × *22,000. Photograph by the author.*

out destruction (Fig. 13–18). Spindle microtubules, in contrast, have never been isolated in this manner and are highly labile and sensitive to the conditions of preparation. These tubules are not well fixed by osmium alone and were not clearly seen and described until the development of the aldehyde fixatives. A number of other types of microtubules share this lability to the fixation method, including the scattered, single cytoplasmic microtubules and the tubules associated with cell wall deposition in plants. These observations suggest that several types of microtubules, different at least in the mode of assembly of a monomeric subunit and possibly even basically different in molecular structure, may exist in the cell. (See also the discussion of microtubular structure and function in flagella in Chapter 15.)

Some Comparative Aspects

There is no evidence that division of the nucleus in prokaryotes depends in any way on a system of microtubules. Although microtubules have been reported in bacteria, these are not frequently encountered and do not appear to be regularly present during cell division. It has been proposed instead (see Jacob et al., 1964) that the plasma membrane functions in the division of the bacterial nucleus. The mechanism is believed to work as follows. The bacterial chromosome, a single DNA molecule in the form of a circle, replicates during interphase. During, or as a part of the replicative process, the DNA attaches to the plasma membrane. Alternatively, the membrane attachment may be a regular part of chromosome structure in bacteria and as such is replicated along with the DNA during the period of synthesis. Once replication is complete, each of the two "daughter" genomes has separate attachments to the plasma membrane. During division of the bacterial cytoplasm (cytokinesis) the plasma membrane undergoes extensive growth, resulting in a gradual separation of the attachment points. As this mechanism proceeds, the daughter chromosomes are pulled apart and are finally included in separate cells at the completion of cytoplasmic division.

A number of observations have given limited support to this scheme, at least to the extent of establishing that connections between the bacterial nucleus and the plasma membrane occur in nature. If bacterial DNA is isolated from gently lysed cells, the molecule, along with DNA polymerase, is frequently attached to a fragment of the membrane, as expected from the proposed model. These attachments have also been detected in thin-sectioned bacterial cells. Antoinette Ryter, on the basis of serial sections of bacteria, has estimated that there are one to three attachments per cell between the plasma membrane and the nuclear

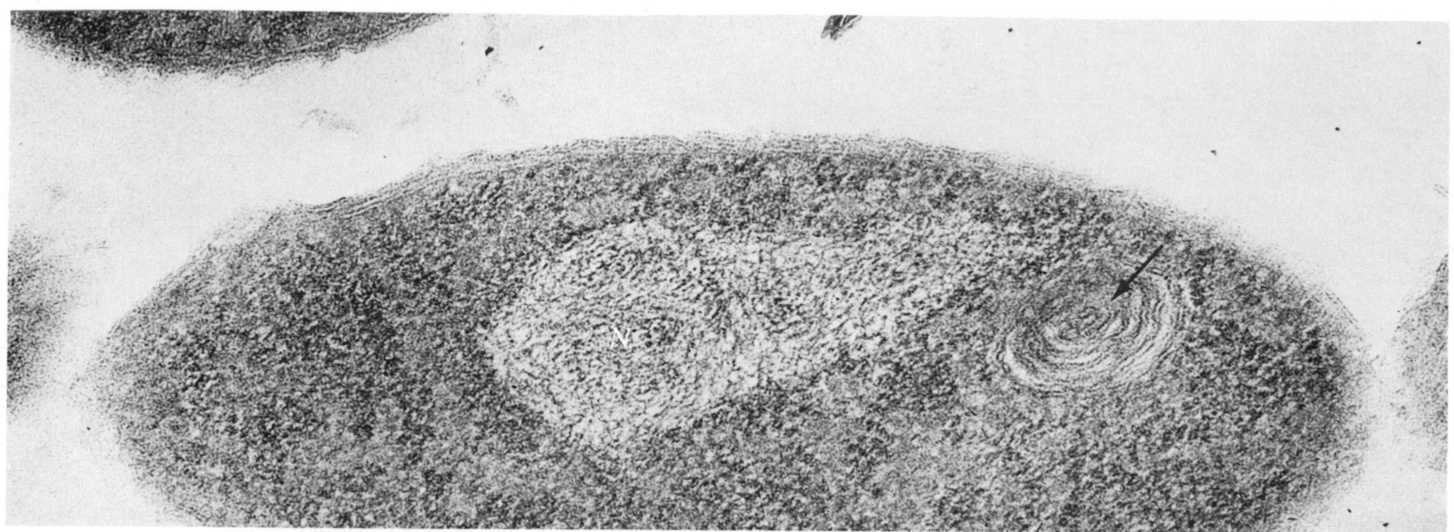

13–19. *Thin-sectioned* E. coli, *showing a nucleoid* (N) *and a closely associated mesosome (arrow).* × *100,000. Courtesy of A. Ryter.*

material. These connections either are direct or are made to a complex invagination of the plasma membrane called the *mesosome* (Fig. 13–19). The possible role of the mesosome as a fundamental part of the division mechanism is uncertain, because these structures are not always present and are common only in gram-positive bacteria. There are other indications that the mesosome may function instead as a "bacterial mitochondrion," as isolation and biochemical analysis of these structures reveals the presence of cytochromes in relatively high concentration.

Indications of the involvement of membrane systems in the division of the nucleus are also found in the dinoflagellates. This protozoan group has structural features in common with both prokaryotes and eukaryotes (see also Chapter 8 and Chapter 10). Although the dinoflagellate nucleus is separated from the cytoplasm by a nuclear envelope, the chromatin closely resembles the bacterial nucleus in ultrastructure and reaction to fixatives (the *Kellenberger reaction;* see p. 182). The dinoflagellate chromatin is scattered in the nucleoplasm in discrete, sausage-shaped deposits superficially similar to the metaphase chromosomes of more typical eukaryotes. The role of the nuclear envelope in distribution of the dinoflagellate "chromosomes" to daughter cells has been investigated in detail by Donna F. Kubai and Hans Ris, who found that, in the dinoflagellate *Gyrodinium,* the nuclear envelope remains intact during all phases of mitosis. During prophase, cytoplasmic channels are formed through the nucleus by a process of invagination of the nuclear envelope. The cytoplasm of these channels, which remains separated from the nucleoplasm by the nuclear envelope, becomes filled with parallel bundles of microtubules (Fig. 13–20). The chromosomes attach to the nuclear envelope in the interior of the nucleus along these channels (arrows, Fig. 13–20). During anaphase, the chromosomes separate and are pulled into "V" shapes, with the apex of the V at the attachment points. Kubai and Ris suggest that the membrane surrounding the channels is involved in the separation of daughter chromosomes. Although microtubules are present within the cytoplasm in the channels, no direct connections could be detected by Kubai and Ris between either the chromosomes or the nuclear envelope and the microtubules. These tubules may provide mechanical support to the nuclear channels, or they may in some manner provide the motive force for movement of the chromosomes by the nuclear envelope.

The involvement of the nuclear envelope in division of the dinoflagellate nucleus is similar to the mechanism proposed for division of the nucleus in bacteria. At the same time, the presence of microtubules in association

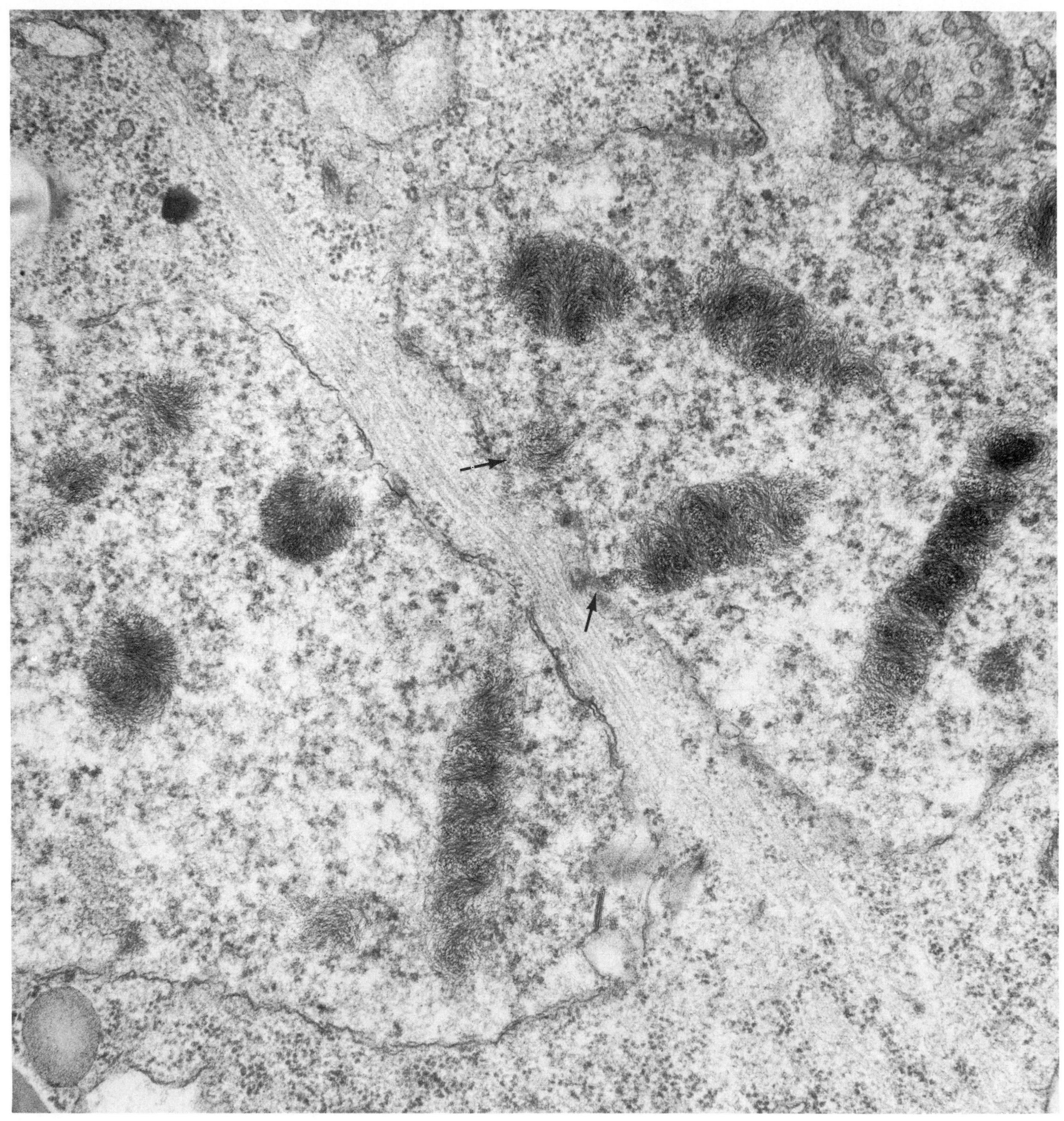

13–20. *Longitudinal section through a cytoplasmic channel in a dividing nucleus of the dinoflagellate* Gyrodinium cohnii. *Microtubules run in parallel array through the channel. Chromosomes make contact (arrows) with the nuclear envelope on the nuclear side of the channels. × 47,500. Courtesy of D. F. Kubai, H. Ris, and The Rockefeller University Press.* [J. Cell Biol. *40:508 (1969)*]

with nuclear division has obvious similarities to the more typical eukaryote mitosis. In the flagellates studied by Cleveland (see Fig. 13–17), the nuclear membrane remains intact during mitosis as it does in the dinoflagellates. A spindle forms in the cytoplasm and extends through a "channel" formed by the encirclement of the spindle by the nucleus. The chromosomes make contact with the nuclear envelope as in the dinoflagellates; however, in this case, microtubules of the spindle also make connections at the same points to the cytoplasmic side of the nuclear envelope. Here, the force for movement of the chromosomes is apparently provided by the spindle microtubules.

Comparison of these systems suggests that, in evolution, a series of adaptations may have resulted in transfer of the basis of division from membranes to the microtubules of the spindle. Assuming that bacteria represent the most primitive system, it can be proposed that in the first cells the genome was distributed by attachment of the replicated chromosomes to the membranes of the daughter cells. In these cells, as in bacteria, the entire genome was probably contained in a single chromosome. This presents fewer mechanical difficulties than division in an organism with multiple chromosomes, because as long as a daughter cell receives one chromosome the complete genome is passed on intact. As the genome of cells became more complex and was subdivided among a multiple number of chromosomes, greater precision was needed in division. In eukaryotes, this precision is supplied by two features of the mitotic system: the polarity of division, established by the parallel orientation of spindle microtubules; and the attachments made by the two centromeres of each chromosome to these microtubules, which direct the sister chromatids toward the opposite poles. Although the systems in the dinoflagellates and protozoa give no clue to the evolution of the centromere, they do suggest that an intermediate step in the evolution of the spindle was the establishment of polarity in the dividing nucleus by the first appearance of parallel bundles of microtubules. In the dinoflagellates, the motive force for division of the chromosomes is still provided by growth of a membrane, with polarity provided by microtubules; in the flagellates, both force and polarity are supplied by the spindle microtubules. In the most highly evolved system, present in the majority of eukaryotes, involvement of the membranes of the nuclear envelope is lost, and transference of the mechanics of division to a microtubular system is complete.

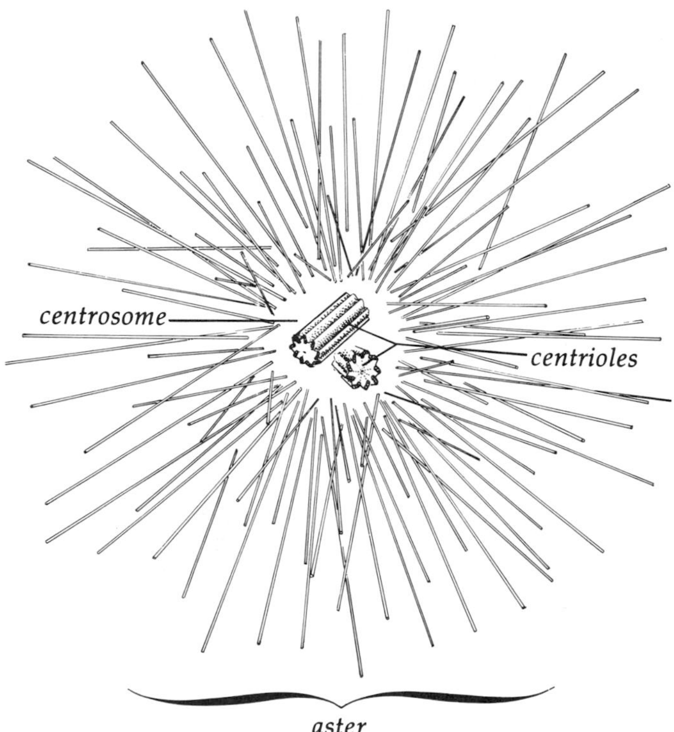

13–21. Diagram showing the relationship of the centrioles, centrosome, and aster.

Asters and Centrioles

Asters and centrioles occur at the poles of the spindle in most animals, flagellated protozoa, and at least at a few stages of the life cycle of many lower plants, including some primitive gymnosperms. Only the angiosperms are completely without centrioles and asters. The centrioles, first described in the 1870's by the Belgian cytologist Edouard van Beneden, are just visible as small granules in the light microscope. Each aster contains two centrioles during interphase; often the two members of the pair lie at right angles to each other in the center of the aster (see Fig. 13–21). Immediately surrounding the centrioles is a "clear" space, called the *centrosome* by light microscopists. This space appeared clear or empty in the light microscope because the small granules and other cytoplasmic bodies present in the periphery of the aster are excluded from this space. In the electron microscope, this space is actually occupied by the bases of the astral microtubules, which approach but do not contact the centrioles (see Fig. 13–25).

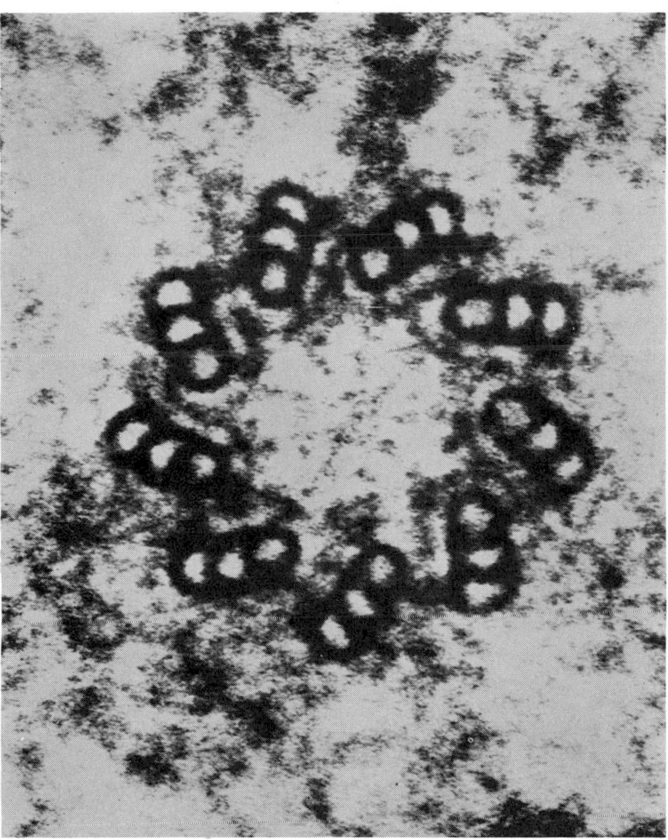

13–22. Cross section of a centriole in a mouse tissue culture cell. The "cartwheel" in the centriole lumen is faintly visible. × 270,000. From The Nucleus, *ed. A. J. Dalton and F. Haguenau, 1968. Courtesy of E. de Harven and Academic Press Inc.*

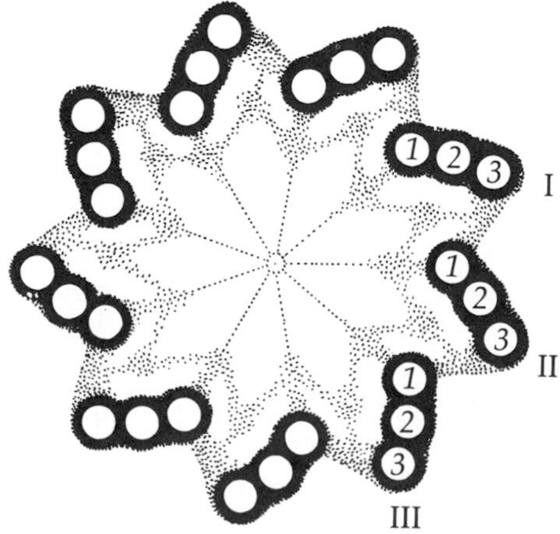

13–23. Diagram of centriole structure as determined by Markham rotation of centriole images. Redrawn from The Nucleus, *ed. A. J. Dalton and F. Haguenau, 1968. Courtesy of E. de Harven and Academic Press Inc.*

Ultrastructure of the Centriole

In the 1950's, electron microscopy revealed that centrioles exhibit a remarkably uniform morphology in all organisms in which they occur. Under the electron microscope, the centriole appears as an electron-dense, hollow cylinder about 3000–5000 Å long and 2000 Å in diameter. In cross section (Fig. 13–22), the walls of the cylinder prove to be made up of a complex system of microtubules, arranged around the periphery in nine sets, with three microtubules in each set. The three microtubules of a set are lined up in a row cocked at an angle of 30–40° to the circumference of the centriole. The microtubules, each 150–200 Å in diameter, are embedded in a electron-dense matrix of amorphous substructure.

At most levels, the centriole appears to be "hollow" or electron-lucent in the center. If sections are made near one end, the lumen contains a density pattern resembling a cartwheel, with a central hub and nine spokes. The spokes radiate from the hub and connect with the innermost tubule of each of the nine sets making up the centriole wall. The cartwheel is faintly visible in the centriole shown in Fig. 13–22. Markham rotation of the image of centrioles (see Chapter 10) reveals that the pattern of density associated with the cartwheel structure is quite complex and, with small specific variations, is closely similar in all animals and plants (Fig. 13–23). In longitudinal sections, a plug of electron-dense material is visible at one end of the centriole (Fig. 13–24). This pattern of density probably represents the cartwheel viewed from the side.

Under the electron microscope the aster proves to consist of large numbers of microtubules that radiate outward from the centrioles. The number and density of the astral microtubules is highly variable, depending on the species and the stage of the cell cycle; in many organisms the aster is diminished to only a few scattered microtubules during interphase. These tubules reach their greatest density and length during mitosis and may extend outward and contact the plasma membrane. Small vesicles and ribosomes lying below the resolution limit of the light microscope are interspersed among the microtubules of the aster and are also found in the space immediately surrounding the centrioles (Fig. 13–25). This space, the "clear" centrosome of light microscopy, sometimes also contains numerous electron-dense bodies, about 1000 Å in diameter, called *satellites.* The astral microtubules, which do not make direct connection with the microtubules of the centri-

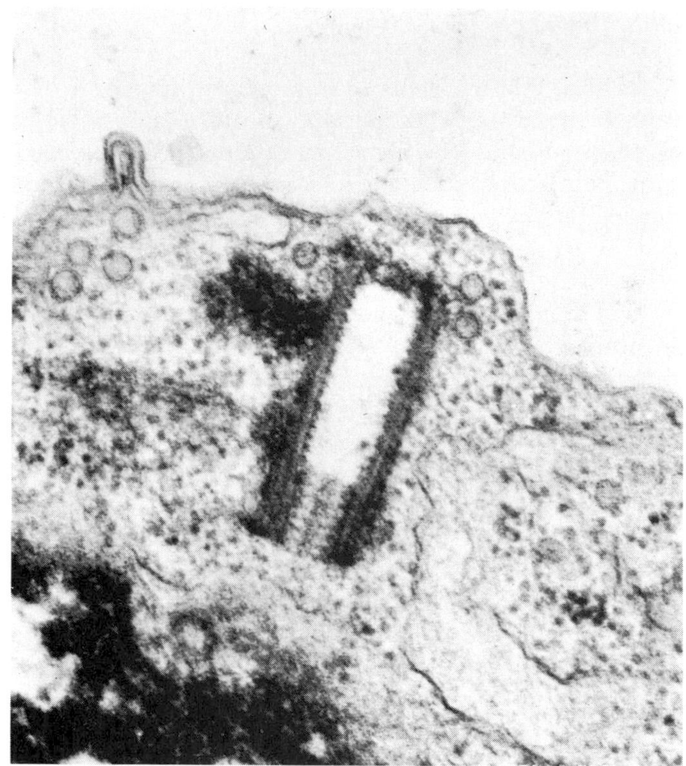

13–24. *A centriole of a spermatozoan of the plant* Nitella. *The dense material plugging one end of the lumen is probably the "cartwheel" of the centriole in longitudinal section. ×84,000. Courtesy of F. R. Turner and The Rockefeller University Press, from* J. Cell Biol. *37:370 (1968).*

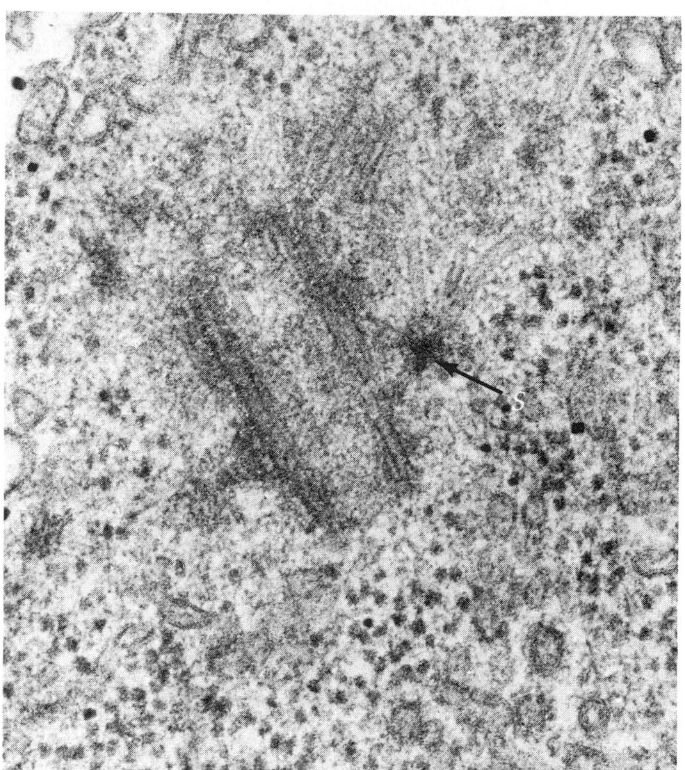

13–25. *A centriole in the mouse. A conspicuous satellite body (S, arrow) is visible near the centriole. Several microtubules of the aster are attached to the satellite. The space immediately surrounding the centriole corresponds to the "centrosphere" of classical light microscopy. ×100,000. From* The Nucleus, *ed. A. J. Dalton and F. Haguenau, 1968. Courtesy of E. de Harven and Academic Press Inc.*

oles, often terminate in the satellite bodies. These, in turn, are connected to the centrioles by thin fibrous processes (arrow, Fig. 13–25). The significance of the satellites and their microtubular connections is obscure because satellites are visible only during interphase when the proliferation of microtubules of the asters and spindle is at a minimum.

Replication of the centrioles. The aster at early interphase, during G_1, usually contains two centrioles. In most organisms, these are replicated during the subsequent S period. The ultrastructure of centriole replication was first described in 1958 by W. Bernhard and E. de Harven and later in detail by Joseph G. Gall. Gall found that during replication the centrioles move apart slightly, and each gives rise to a small, bud-like extension, the *procentriole* (Fig. 13–26). The procentriole, a short cylinder about 700 Å long, gradually increases in size and extends at right angles from the end of the parent centriole that contains the cartwheel. Although less distinctly outlined, the typical pattern of microtubules is visible in cross sections of the procentriole. During this period, a halo of amorphous, electron-dense material appears around the replicating centrioles; the satellites (always around only one of the two centrioles) are also most numerous at this time (see Fig. 13–26). The functional significance of these configurations is unknown.

From the time of its first appearance, the procentriole is separated from its parent by a 500–700 Å space. Development and lengthening of the daughter centriole

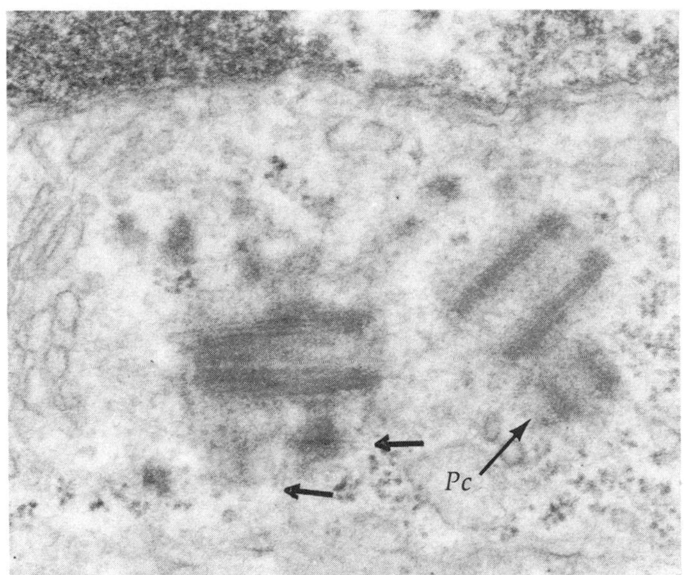

13–26. A pair of centrioles in a thymocyte cell of the rat. One of the two centrioles has a procentriole (Pc) *arising as a bud from its ventral surface. The procentriole is separated from the parent centriole by a 600 Å space in this micrograph. Several satellite bodies (arrows) are also present. × 61,000. Courtesy of R. G. Murray, A. S. Murray, A. Pizzo, and The Rockefeller University Press, from* J. Cell Biol. *26:601 (1965).*

continues through the S period and may sometimes extend through G_2 and prophase. In most organisms, centriole replication is complete by metaphase. By this time, the centrioles have been separated into two pairs, each consisting of a parent and daughter centriole, by the separation of the aster into two parts. Once the daughter centriole is mature, it may either remain oriented at right angles to the parent centriole or may take up other positions, including parallel or tandem arrays. It is interesting that colchicine has no apparent effect on the structure or replication of the centriole microtubules at any time in the life cycle of the cell.

The Role of the Centrioles and Aster during Mitosis

The centrioles and asters have been traditionally viewed as giving rise to the spindle during mitosis in spite of the fact that in many plants spindles apparently normal in function appear without the intervention of either asters or centrioles. In animals, the first appearance of the spindle follows separation of the asters. The spindle gradually forms between the separating asters; by the time these reach their position at opposite sides of the nucleus, the formation of the spindle is almost complete. The hypothesis that the asters give rise to the spindle is a natural consequence of these observations. The more recent discovery that both the spindle and the centrioles are constructed from microtubules of similar dimensions and ultrastructure also gives indirect support to this proposal.

However, it appears that even in animal cells containing asters and centrioles the spindle may form normally if the centrioles are removed. Roland Dietz has succeeded in removing the asters and centrioles from a line of dividing cells. By flattening a cell, one of the two asters at prophase was prevented from migrating to its position at the opposite side of the nucleus. The spindle in this cell formed normally, with one astral and one anastral pole. Division proceeded normally, producing one daughter cell with two asters and one cell without asters or centrioles. In the next division, the cell without asters or centrioles formed an apparently normal spindle, similar in appearance to astral spindles except for a slight broadening at the poles. The division stages proceeded normally in the anastral cell. These experiments and the absence of centrioles and asters in many plants, protozoa, and animal oocytes suggests very strongly that these structures are not directly involved in generation of the spindle.

Other less direct evidence can be taken in support of this conclusion. Although both the centriole and spindle are constructed from microtubules, no direct connections are seen at any time between spindle microtubules and the centriole. Further, the microtubules of the spindle are extremely sensitive to conditions of fixation, cold, and the presence of agents such as colchicine or ultraviolet radiation. The centriole, in contrast, is comparatively easily fixed and appears to be unaffected by agents that disrupt the spindle.

A few authorities, such as the classical cytologist Edmund B. Wilson and more recently Roland Dietz, have suggested that the mechanism involving centrioles and asters in mitosis functions primarily to insure that the centrioles are passed on to both daughter cells. Thus, instead of generating the spindle, the centrioles may simply be separated and pushed to opposite sides of the nucleus by the microtubules that appear between the dividing asters during prophase. This possible function of the astral microtubules is paralleled by the activity of the spindle during anaphase, when the poles are apparently pushed farther apart by the continuous fibers. The presence of centrioles in an organism appears to be correlated with the existence, at some time in the life cycle, of cells that possess cilia or flagella. During development of the flagellum, the centriole gives rise to the system of microtubules that forms the axis of the flagellar shaft. This centriole usually persists as the *basal body* of the flagellum (for details, see Chapter 15). Because of the importance of this function, it is not difficult to conceive that the astral mechanism appeared in evolution as a system to insure the distribution of centrioles among daughter cells. This interpretation would be strengthened if it could be established that centrioles arise only from preexisting centrioles. Unfortunately, although there are some indications that centrioles do not arise *de novo*, evidence on this point is very sketchy and somewhat contradictory. (The possibility that centrioles possess a continuous line of inheritance independent of nuclear control is discussed at length in Chapter 16.)

The Nucleolar Cycle during Mitosis

Great variations exist between different cell types in the disposition of the nucleolus during mitosis. But in almost all organisms, the nucleolus eventually breaks down and disappears at the end of prophase or during metaphase of mitosis. In a number of species, such as in many fungi and the alga *Spirogyra*, the nucleolus persists throughout mitosis. At the ultrastructural level, breakdown of the nucleolus is first detected as a loosening of nucleolar structure. The fibrillar and granular zones then become more distinctly outlined, evidently through loss of the nucleolar matrix (see Chapter 10 for details of nucleolar ultrastructure). The nucleolus becomes progressively smaller; toward the end of prophase, just before disappearance of the nucleolus, only the granular component remains (Fig. 13–27). In many species, this granular material and any persisting fibrillar material becomes dispersed into the nucleoplasm on breakdown of the nucleolus. As the spindle forms, much of this material is distributed between the microtubules of the spindle.

During telophase, the nucleolus is reorganized on one or more chromosomes, appearing first as a spherical

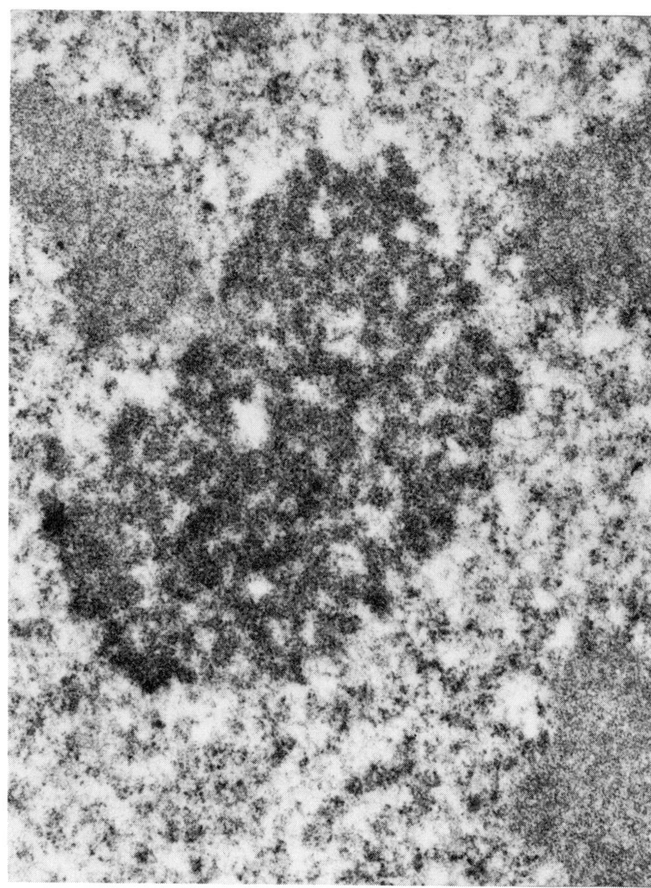

13–27. A nucleolus at late prophase in the grasshopper Chortophaga viridifasciata. *Much of the matrix material has been lost. Almost all of the mass of the nucleolus shown consists of elements of the granular zone. × 24,000. Courtesy of B. J. Stevens and The Rockefeller University Press, from* J. Cell Biol. *24:349 (1965).*

body that gradually increases in size until the typical interphase condition is reestablished. In the early 1930's, E. Heitz observed that the number of nucleoli formed at this stage (usually one or two) is a fixed characteristic of the species. A short time later, Barbara McClintock demonstrated that a specific chromosomal site is necessary for reappearance of the nucleolus at telophase. McClintock, working with maize, succeeded in obtaining individuals deficient in the nucleolar site by inducing deletions of this region of the chromosome. In organisms with the deletion, the nucleolus failed to reappear at telophase; instead, numerous small, nucleolus-like bodies were distributed among the chromosomes. McClintock concluded from this that the nucleolar site, when present, functions to "organize" the diffusely spread nucleolar material into a compact nucleolus at telophase. On this basis, McClintock termed the nucleolar site the *nucleolar organizer.*

This site is readily recognized at metaphase in most species as a prominent secondary constriction in both members of one homologous pair of chromosomes. In the polytene chromosomes of dipteran larvae (see Chapter 10) the nucleolar organizer site is clearly visible and identifiable during interphase (Fig. 13–28). In a diploid organism, the nucleolar organizers of both chromosomes of the pair containing the site are usually active. Initially, this may lead to the development of two nucleoli, but often these fuse to form a single large structure. In some organisms, including man, additional sites are distributed among the chromosomes, forming in a few species as many as ten separate nucleoli at a maximum. Even in these species, the number is often reduced through fusion.

According to J.-G. Lafontaine and A. Lord, the reorganizing nucleolus in some plant cells is at first composed mostly of 100 Å chromatin fibers. This probably represents the chromatin of the nucleolar organizer site of the chromosome. Following the first appearance of the nucleolus, the interstices between the 100 Å fibers become filled with material identifiable as the fibrillar zone of the nucleolus. Soon after this stage, the particulate component (granular zone) of the nucleolus begins to form. By late prophase, the nucleolus has begun to assume the typical interphase configuration.

Classical cytologists believed that the nucleolar substance, after breakdown of the nucleolus at prophase, became distributed over the condensing chromosomes as a "matrix." This matrix supposedly adhered to the chromosomes and was carried to the poles at anaphase. During reconstitution of the nucleus in telophase, this material, called at this stage the *prenucleolar matrix,* was thought to flow back to the nucleolar organizer site to be incorporated into the reorganizing nucleolus. This interpretation was supported by two observations. The apparent prenucleolar material, visible between the chromosomes at telophase, showed the presence of RNA when stained by light cytochemical techniques. As the nucleolus was reorganized, the prenucleolar material gradually disappeared from between the telophase chromosomes.

More recent evidence from autoradiography has eliminated the possibility that the nucleolar RNA expelled from the chromosomes during prophase returns to the reorganizing nucleolus at telophase. (Some nucleolar proteins, however, are reincorporated into the nucleolus during telophase.) In 1962, D. M. Prescott and M. A. Bender showed that if RNA is labeled during the previous interphase, autoradiography during mitosis reveals that nucleolar RNA is transferred to the cytoplasm and does not return to the nucleolus at telophase. No persistent label can be detected on the chromosomes or reforming nucleolus during the later stages of mitosis. Thus, the RNA seen during telophase in the "prenucleolar" material is newly synthesized at telophase and does not represent prophase nucleolar RNA returning to the nucleolus.

This leaves unanswered the question of the possible relationship of the so-called prenucleolar material to the reorganizing nucleolus. More recent electron microscopical evidence confirms that material analogous to the prenucleolar substance can be identified between the chromosomes at telophase and that this substance disappears as the nucleolus is reorganized. Barbara J. Stevens has carried out detailed investigation of this "prenucleolar substance." During telophase, the chromosomes at the poles of the spindle are gradually enclosed by reforming elements of the nuclear envelope. At this time, patches of granular material containing 150 Å particles can be identified at the surfaces of the uncoiling chromosomes (Fig. 13–29, arrows). This material, which stains for the presence of RNA in adjacent thick sections, is the "prenucleolar substance" of light microscopy. At higher magnifications, the prenucleolar substance proves to consist of 150 Å granules similar in morphology to the granular component of the nucleolus. As telophase progresses, this material is assembled into larger spherical masses. At the same time, the telophase nucleolus begins its reappearance at the nucleolar organizer site (Fig. 13–30). As the telophase nucleolus grows in size, the masses of prenucleolar material are diminished until, at late prophase, no prenucleolar material can be identified in the nucleus.

Two hypotheses are currently held as valid possibilities for the function of the prenucleolar substance. One considers that the prenucleolar substance may indeed form a part of the reorganizing nucleolus. Because the prenucleolar substance first appears in

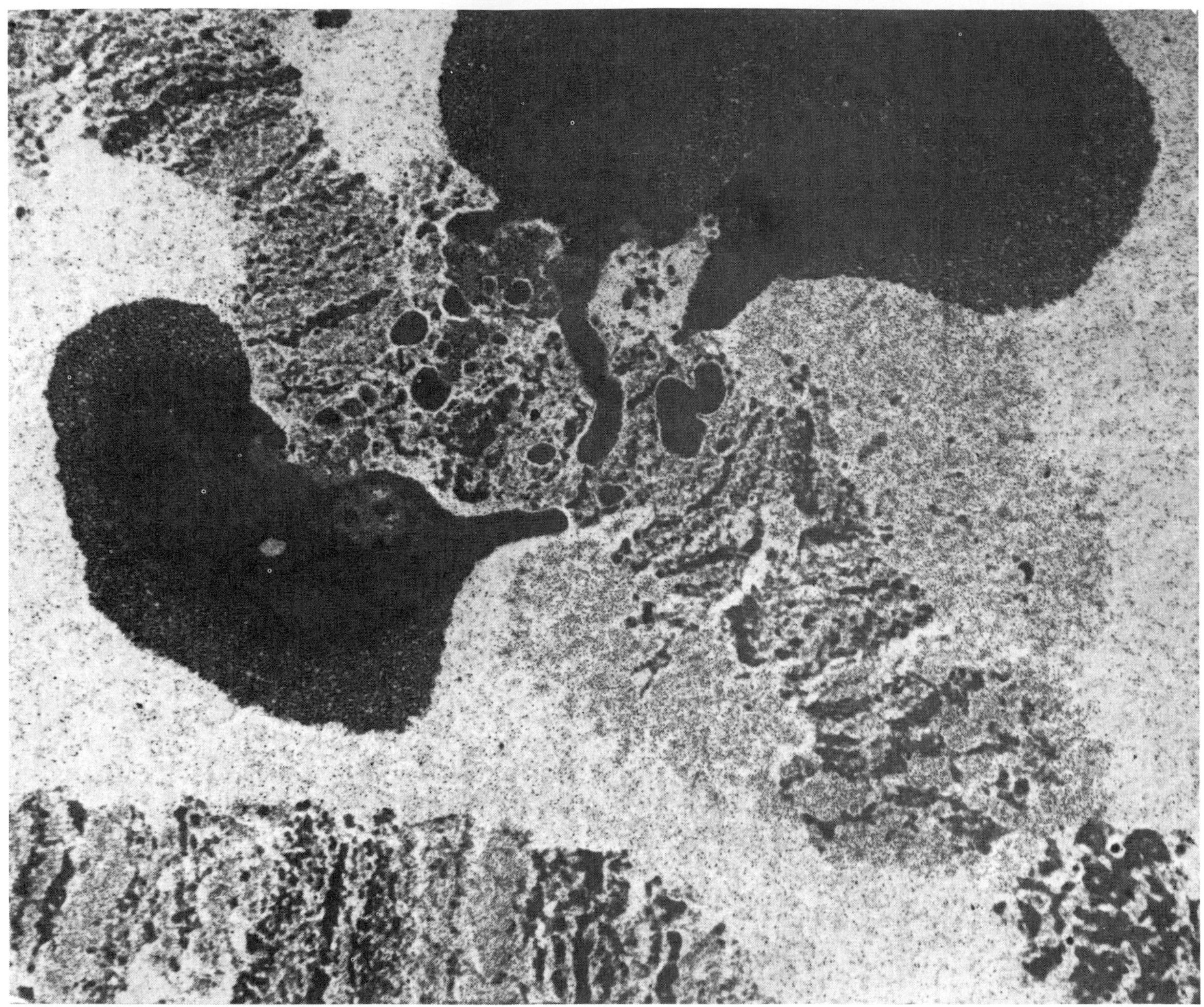

13–28. *The nucleolus and nucleolar organizer region in a salivary gland cell of* Chironomus thummi. × *7,300. From* The Nucleus, *ed. A. J. Dalton and F. Haguenau, 1968. Courtesy of J. A. Kloetzel and Academic Press Inc.*

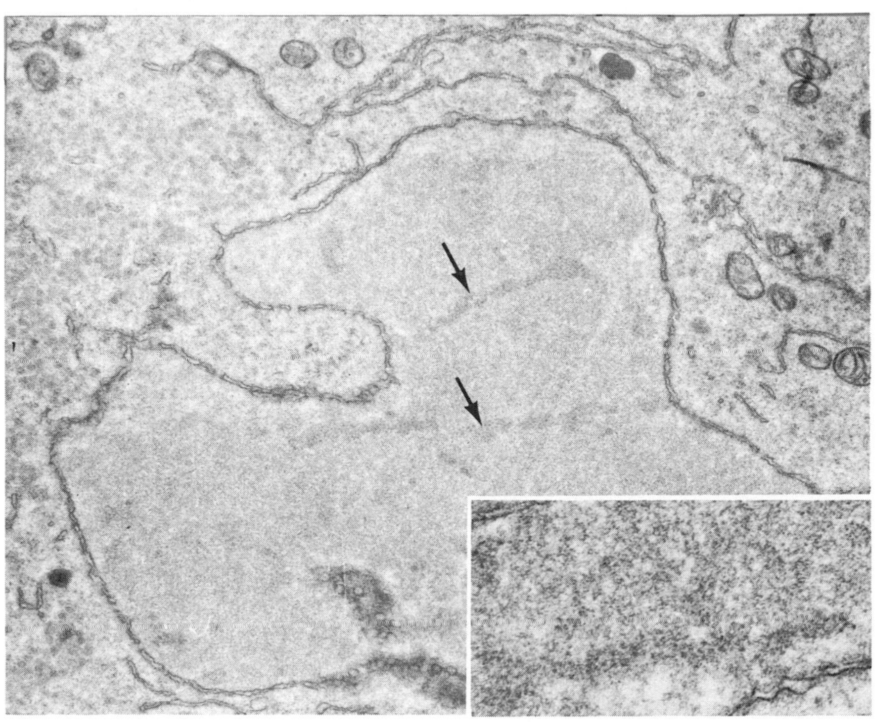

13–29. *A reorganizing nucleus at telophase in* Chortophaga viridifasciata. *At several regions corresponding to surfaces of the uncoiling chromosomes, deposits of dense granular material are present (arrows). This material is considered equivalent to the "prenucleolar substance" of classical light microscopy.* × *15,500. At higher magnification (inset), the prenucleolar substance is distinctly granular.* × *50,000. Courtesy of B. J. Stevens and The Rockefeller University Press, from* J. Cell Biol. *24:349 (1965).*

regions of the chromosome outside of the nucleolar organizer region, it does not seem likely that this material could represent the precursors of ribosomal RNA, because these substances are known to be synthesized within the nucleolar organizer site. But the prenucleolar material could possibly represent another as yet undefined RNA species that contributes to the mass of the nucleolus. The alternate hypothesis holds that the prenucleolar material is another RNA species unrelated to nucleolar function, perhaps a messenger that is synthesized during telophase and only coincidentally disappears or is dispersed during reorganization of the nucleolus.

The ultrastructural evidence is of little help in settling this problem. Although this evidence parallels the findings from light microscopy, it is impossible to determine whether the RNA-containing granular material is a precursor of the nucleolus or an unrelated species. The classical evidence is also equivocal. McClintock showed that small bodies similar to nucleoli appear in other regions of the nucleus if the nucleolar organizer site is deleted. Further, at telophase one or more chromosomes occasionally become separated from the main chromosomal mass. These, for a time, may become enclosed by a separate nuclear envelope, forming a separate nuclear fragment called a *karyomere.* In these fragments, bodies similar to nucleoli may also arise. This would appear to show that the prenucleolar material in the normal nucleus actually does contribute to the reorganizing nucleolus.

But although these bodies, called *pseudonucleoli,* are at least superficially similar to the nucleolus and contain RNA, isolation and analysis of RNA from anucleolate cells reveals that 18S and 28S rRNA are completely absent (see Brown and Gurdon, 1966). Other work (see Chapter 11) has shown that a small RNA component of the ribosome, 5S rRNA, is evidently synthesized outside the nucleolar organizer region. This RNA species, which eventually becomes complexed with the 28S rRNA subunit in the ribosome, probably enters the nucleolus after synthesis elsewhere on the chromosomes, and may represent at least a small part of the material accumulating in pseudonucleoli in anucleolate mutants. But it does not seem possible that all of the prenucleolar material could be accounted for by 5S RNA. Therefore, the identity and function of the so-called prenucleolar material must be regarded as unknown at the present time.

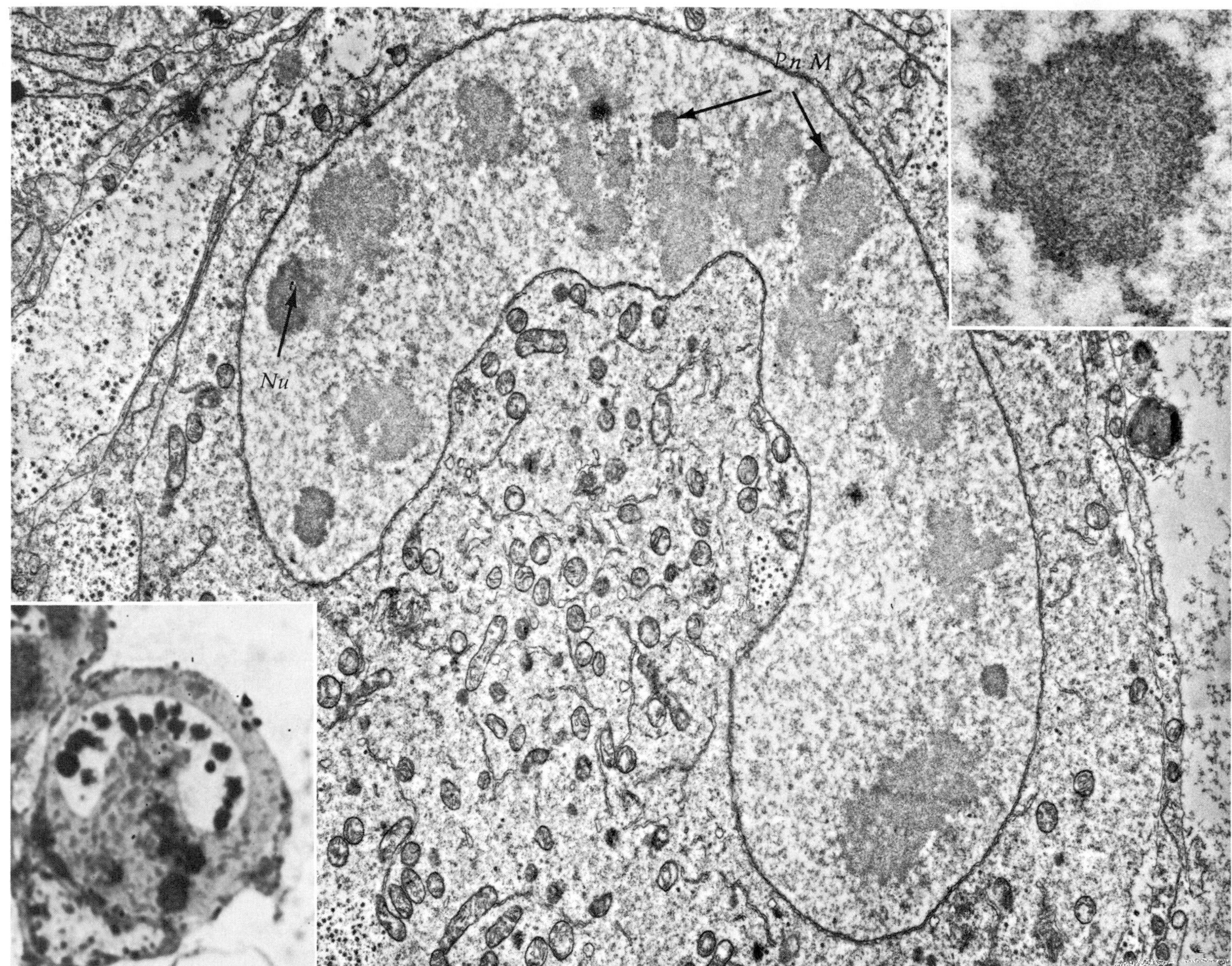

13–30. A telophase cell of the grasshopper Ch. viridifasciata. *The reorganizing nucleolus is identifiable in the inset at the lower left. The main figure is an electron micrograph of the same cell; the developing nucleolus* (Nu) *is visible at the nucleolar organizer site. Masses of prenucleolar material* (PnM) *are present at several locations in the nucleus. This material disappears as the reorganizing nucleolus increases in size. Upper right, one of the prenucleolar bodies at higher magnification, showing the conspicuously granular substructure. Lower, × 2,400. Main, × 11,000. Upper, × 51,000. Courtesy of B. J. Stevens and The Rockefeller University Press, from* J. Cell Biol. *24:439 (1965).*

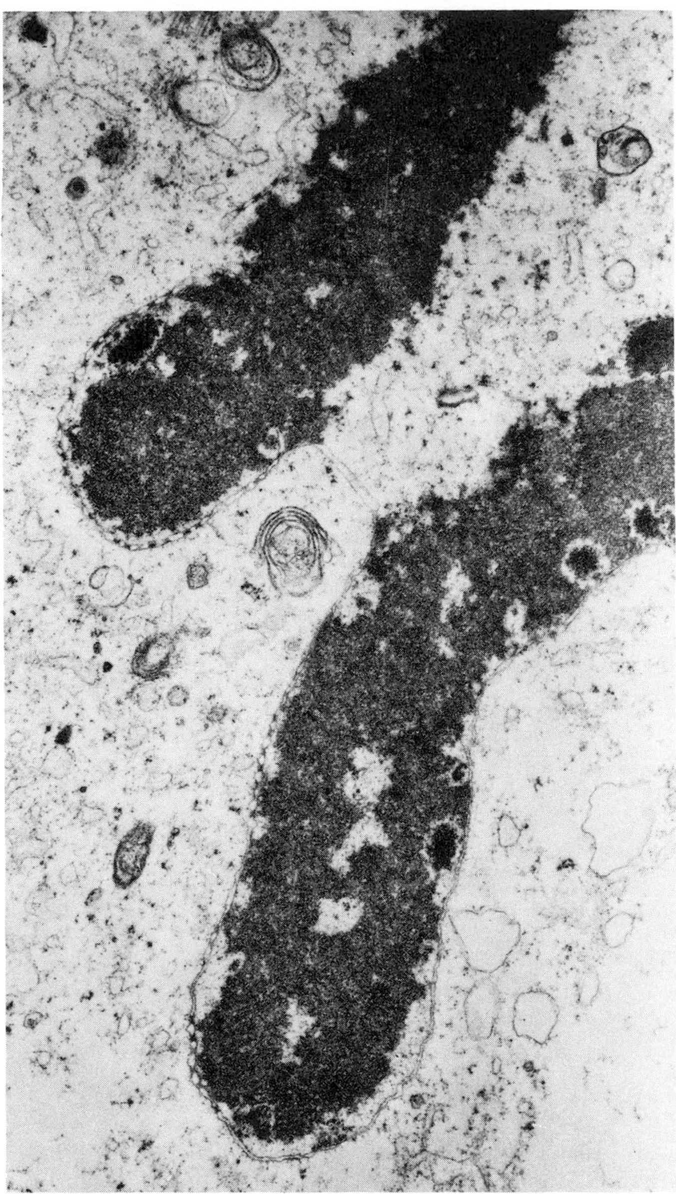

13–31. *Initial stages in the formation of the nuclear envelope around telophase chromosomes of* Haemanthus. *The developing nuclear envelope is so closely applied to the surfaces of the telophase chromosomes that little extrachromosomal material is included.* × *14,500. Courtesy of J. Molé-Bajer and the publisher, from* Chromosoma *26:427 (1969).*

The Nuclear Envelope during Mitosis

In most organisms, the nuclear envelope disintegrates during prophase and is reconstituted during telophase. Near the end of prophase, discontinuities appear in the envelope. This fragmentation continues until the envelope is separated into scattered pieces. For a time, the pore complexes are recognizable in these fragments, but soon all differentiation is lost and the segments of envelope are indistinguishable from the endoplasmic reticulum. According to Andrew Bajer, who has followed the breakdown of the nuclear envelope in living cells with phase contrast microscopy, the total disintegration of the envelope may take place in as little as two minutes during late prophase.

The envelope begins its reappearance in telophase when the chromosomes reach the poles of the spindle and begin to uncoil. At first, flattened vesicles indistinguishable from the rough endoplasmic reticulum become closely applied to the surfaces of the uncoiling chromosomes (Fig. 13–31). These increase in number and gradually coalesce until the chromosomal mass is completely enclosed by a double envelope (see Fig. 13–29). As the membranes coalesce, pore complexes develop; by the end of telophase, the uncoiling chromosomes are completely separated from the cytoplasm by a fully differentiated nuclear envelope (for a representative study see Robbins and Gonatas, 1964).

The vesicles forming the nuclear envelope, presumably derived from the endoplasmic reticulum, are initially so closely applied to the telophase chromosomes that little extrachromosomal material is included within the nucleus except for a few scattered ribosomes and an occasional spindle microtubule. By the end of telophase, when uncoiling of the chromosomes has become extensive, the nucleus contains large areas recognizable as extrachromosomal material (see Fig. 13–30). Whether this material penetrates across the nuclear envelope from the cytoplasm or is a product of the chromosomes is not known. That at least a part of this material is newly synthesized by the chromosomes is indicated by the fact that at telophase the incorporation of RNA precursors into newly synthesized RNA begins (see p. 300).

In a few organisms, such as the protozoa, the nuclear envelope remains intact during mitosis. The spindle may be extranuclear, as in the flagellates studied by Cleveland, or may be entirely intranuclear, as in the micronucleus of the ciliate *Blepharisma*, investigated

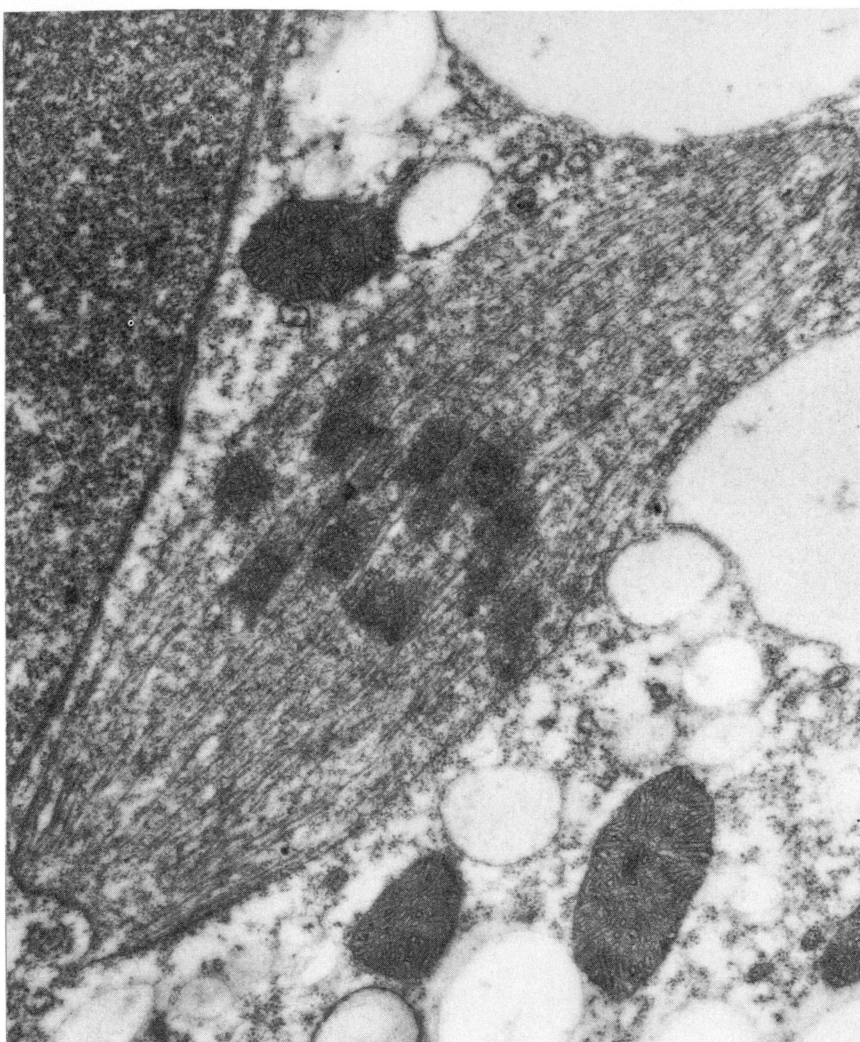

13–32. *Micronucleus from the protozoan* Blepharisma *at late metaphase or early anaphase. The spindle is located entirely within the nuclear envelope, which remains intact during mitosis in this organism.* × *32,000. Courtesy of R. A. Jenkins and The Rockefeller University Press, from* J. Cell Biol. *34:463 (1967).*

recently by R. A. Jenkins. In *Blepharisma* (Fig. 13–32) the spindle forms within the nucleus during prophase. At anaphase, the chromosomes are separated primarily by elongation of the intranuclear spindle. The new nuclear envelopes surrounding the chromosomes at the poles are reconstructed *inside* the original nuclear envelope, which remains intact until late telophase (Fig. 13–33). During this process, flattened cisternae similar to the endoplasmic reticulum are closely associated with the newly forming nucleolar envelopes (arrows, Fig. 13–33). This work indicates that the precursors of the spindle and nuclear envelope can penetrate through the membranes of the "old" envelope. This takes place in spite of the fact that pores seem to be absent from the micronuclear envelope in *Blepharisma.* This protozoan also provides an example of spindle formation without the presence of centrioles or asters. In other protozoa, some algae, and fungi with intranuclear spindles, the "old" nuclear envelope remains intact and separates into two parts during telophase by a process of furrowing or constriction at the former midpoint of the spindle.

Cytokinesis

At the end of telophase, with reconstitution of the nuclear envelope, reappearance of the nucleolus, un-

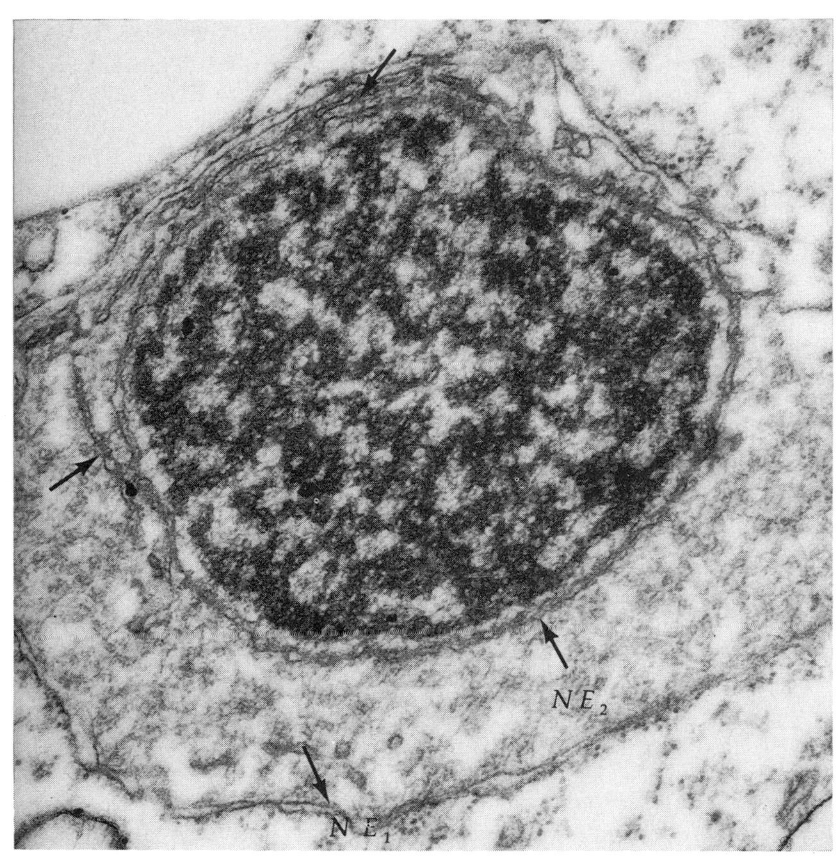

13–33. *Late telophase micronucleus in* Blepharisma. *A new nuclear envelope* (NE$_2$) *is forming around the daughter nucleus; the original nuclear envelope* (NE$_1$) *is still intact at this stage. Flattened sacs (arrows), similar to cisternae of the rough endoplasmic reticulum, are closely associated with the newly synthesized nuclear envelope.* × *67,000. Courtesy of R. A. Jenkins and The Rockefeller University Press, from* J. Cell Biol. *34:463 (1967).*

coiling of the chromosomes, and the appearance of extrachromosomal nucleoplasm, the daughter nuclei have essentially returned to the interphase state. The spindle disappears during this time, except for relatively short microtubules that remain associated with the asters (if these are present) and other microtubules that persist at the former midpoint or equator of the spindle. The latter microtubules are intimately involved in the subsequent division of the cytoplasm.

The division of the cytoplasm is usually initiated in close coordination with mitosis. In both plants and animals, the first events in cytokinesis usually take place during mid-anaphase to telophase. The precise relationship of cytokinesis to mitosis is not clear, however, because the two processes may be widely separated in time. For example, in some insect embryos and in the endosperm of plants, the nucleus may divide repeatedly without division of the cytoplasm, producing a multinucleate tissue. After a time, the cytoplasm enters rapid cytokinesis, enclosing the multiple nuclei in separate cells. In plants such as the myxomycetes and ascomycetes, the multinucleate condition persists throughout the life cycle, and no connection exists at any time between mitosis and cytokinesis. Whether cytokinesis is closely coordinated with mitosis or not, the precise events triggering the division of the cytoplasm in any case are unknown.

In the most frequently encountered situation, in which cytokinesis closely follows mitosis, the site of division of the cytoplasm is closely related to the position of the spindle. Normally, the plane of division passes through the midpoint or equator of the spindle at right angles to a line passing through the spindle axis. If the spindle is eccentrically located within the cell, the plane of division still passes through the spindle equator, resulting in unequal division of the cytoplasm. It is not clear whether the spindle determines the plane of cytokinesis or whether both the position of the spindle and the plane of division are determined by another mechanism. The position of the spindle is most clearly indicated as the determiner in animal cells. Experimental displacement of the spindle, which can be accomplished by centrifugation or micromanipulation, causes corresponding displacement of the plane of cytoplasmic division. In one such study, K. Kawamura induced a change in position of the spindle of the grasshopper *Chortophaga* by micromanipulation. He found that relocation as late as anaphase was always accompanied by a shift of the plane of division to the new location of the spindle equator.

In most cells, displacement of the spindle after late metaphase or in anaphase causes no disturbance of the site of division, which passes through a plane corresponding to the original position of the spindle equator. This makes it appear as if an event in metaphase "triggers" division of the cytoplasm; this trigger acts on the plasma membrane in a position lateral to the spindle equator. In plants, the same relationships also hold true, although the position of the plane of division appears to be fixed at earlier stages of mitosis. Curiously enough, in some plant cells a band of microtubules appears around the circumference of the cell at interphase, marking the point of the future plane of division and the midpoint of the spindle long before the spindle appears (the "preprophase band"—see below). In this case, a still earlier event marks both the position of the spindle and the future plane of cytoplasmic division.

Division of the cytoplasm in all animal cells, protozoa, and many algae occurs by a mechanism known as *furrowing*. The same process also occurs in the divisions forming pollen in dicotyledenous plants. The furrow first appears as a shallow depression in the plasma membrane around the circumference of the cell in a plane passing through the spindle equator and gradually constricts or extends toward the center of the cell (marked by the spindle axis) until the cytoplasm is cut into two parts. In almost all plant cells, cytokinesis takes place by formation of a *cell plate*, a membranous partition that also first appears on the plane marked by the spindle equator (see Fig. 12–2). This partition expands laterally until contact is made with the cell walls and the daughter nuclei are completely separated. These two major forms of division, at first glance apparently quite different, share several processes in common in the early stages of cytokinesis. In both systems, the microtubules of the spindle are intimately involved in the cytoplasmic division. These similarities and the very considerable differences will become apparent in the descriptions to follow, in which furrowing and cell plate formation will be taken up.

Furrowing

The first visible change marking the onset of furrowing is the appearance of patches of amorphous, electron-dense material at the equator of the spindle (arrows, Fig. 13–34). These appear at mid- to late anaphase, after movement of the chromosomes to the poles, and surround but do not penetrate the spindle microtubules (Fig. 13–35). As anaphase progresses into telophase, the dense material, initially in discontinuous aggregations, increases in quantity until a solid layer stretches across the spindle equator (Fig. 13–36). During mid- to late telophase spindle microtubules become disorganized and gradually disappear in the polar regions. The tubules persist and even increase in number at the equator. These remain embedded in the dense layer at the midpoint. Some tubules appear to terminate at the equator at this stage; others penetrate completely through and extend for some distance on either side of the dense layer of amorphous material. Other structures, such as a row of vesicles, may also be present in the equatorial layer. Frequently, mitochondria penetrate into spaces between the spindle microtubules on either side of the dense layer. The dense layer, with the associated microtubules and vesicles, is termed the *midbody*. This structure is large enough to be visible under the light microscope by late anaphase (for a representative study of the ultrastructure of midbody formation, see Krishnan and Buck, 1965.)

As the midbody develops, concurrent alterations take place at the cell surface. A linear depression appears in the plasma membrane, extending around the cell at the level of the midbody. In some cells, such as cleaving sea urchin embryos, this gradually deepening furrow gives the impression that a string is being tightly drawn about the midregion of the cell (Fig. 13–37). Under the electron microscope, the furrow is seen in cross section as a depression at either side of the cell opposite the midbody (see Fig. 12–14). This furrow gradually extends inward until the midbody is reached. The midbody is gradually compressed by the deepening furrow into a narrow fibrous connection between the daughter cells (Fig. 13–38). It may persist in this form for a time, but in most cells the remnants of the midbody disappear soon after completion of furrowing.

13–34. The developing midbody in human tissue culture cells. At the earliest stages, electron-dense, amorphous material (arrows) collects around remnants of microtubules at the former midpoint of the spindle. × 56,000. Courtesy of A. Krishan, R. C. Buck, and The Rockefeller University Press, from J. Cell Biol. *24:433 (1965).*

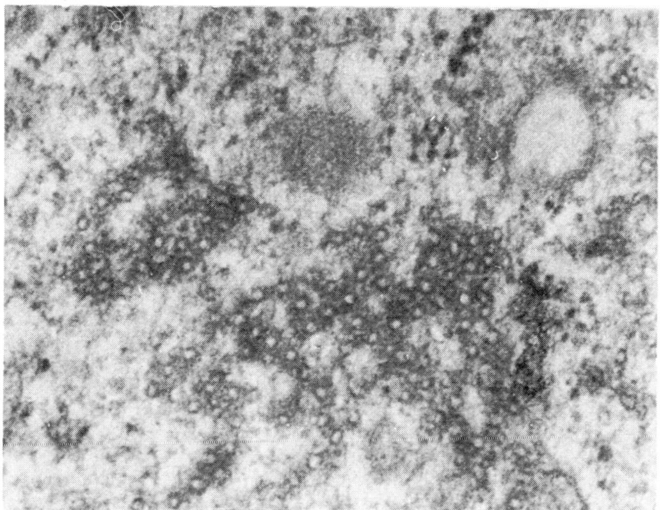

13–35. A cross section of spindle microtubules in the region of the developing midbody. Note that the dense material does not penetrate into the lumen of the microtubules. × 91,000. Courtesy of A. Krishan, R. C. Buck, and The Rockefeller University Press, from J. Cell Biol. *24:433 (1965).*

The appearance and development of the furrow, which can easily be followed under the light microscope, has intrigued and baffled biologists for a century. The molecular basis for furrowing is unknown. Hypotheses developed to explain this phenomenon have been almost as imaginative and varied as those proposed to account for the movement of the chromosomes by the spindle at anaphase. Most current ideas include the hypothesis that the furrow advances by means of a contractile mechanism. In free-living cells or in a loosely aggregated collection of cells, such as in a tissue culture, the appearance of the furrow does give the impression of a contractile mechanism acting around the perimeter of the cell. However, in closely packed cells, such as in the solid tissues of a many-celled organism, the development of the furrow may take a somewhat different form. In these cells, a row of vesicles may appear across the equator of the cell between the midbody and the plasma membrane. These simply increase in number and coalesce until the daughter cells are separated. Thus, furrowing is not always so clearly ascribable to the action of a contractile ring around the perimeter of the cell. Daniel

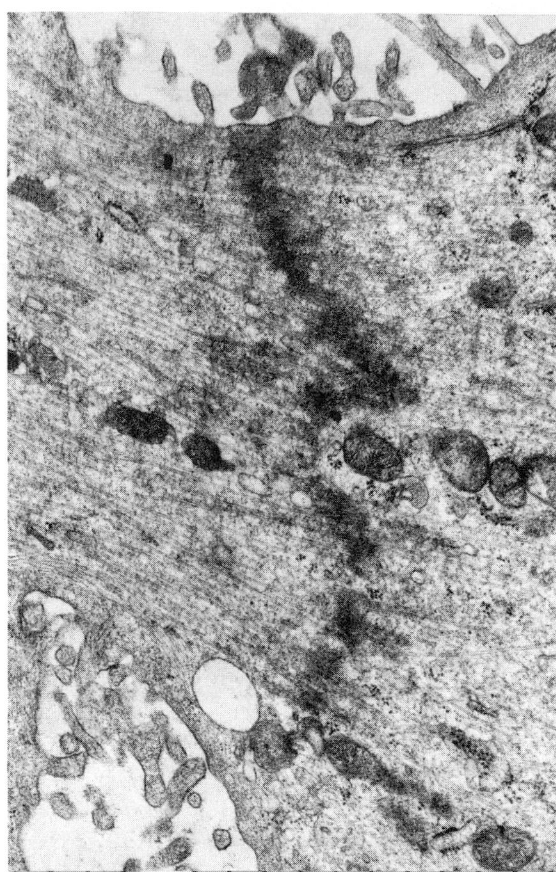

13–36. *Midbody at late telophase in a human tissue culture cell. The dense material now forms a continuous layer across the spindle midpoint.* × *26,000. Courtesy of A. Krishan, R. C. Buck, and The Rockefeller University Press, from* J. Cell Biol. *24:433 (1965).*

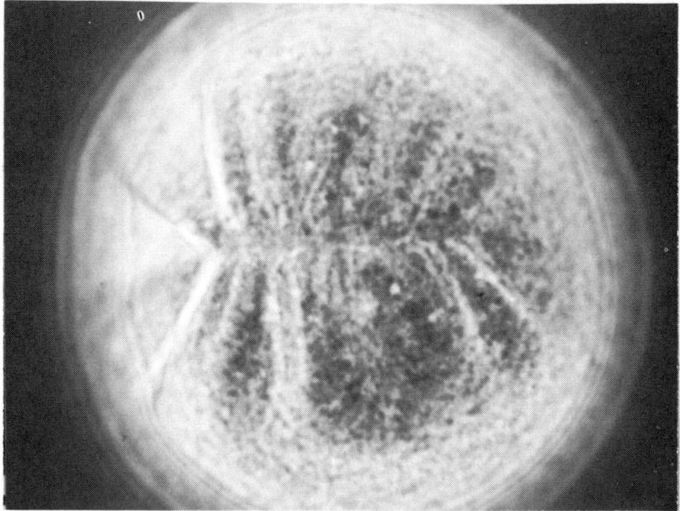

13–37. *Phase contrast light micrograph of a living* Aequorea *egg during formation of the cleavage furrow. The presence of the surface folds radiating from the furrow gives the impression that a drawstring is being tightened around the cell.* × *80. Courtesy of D. Szollosi and The Rockefeller University Press.* [J. Cell Biol. *44:192 (1970)*]

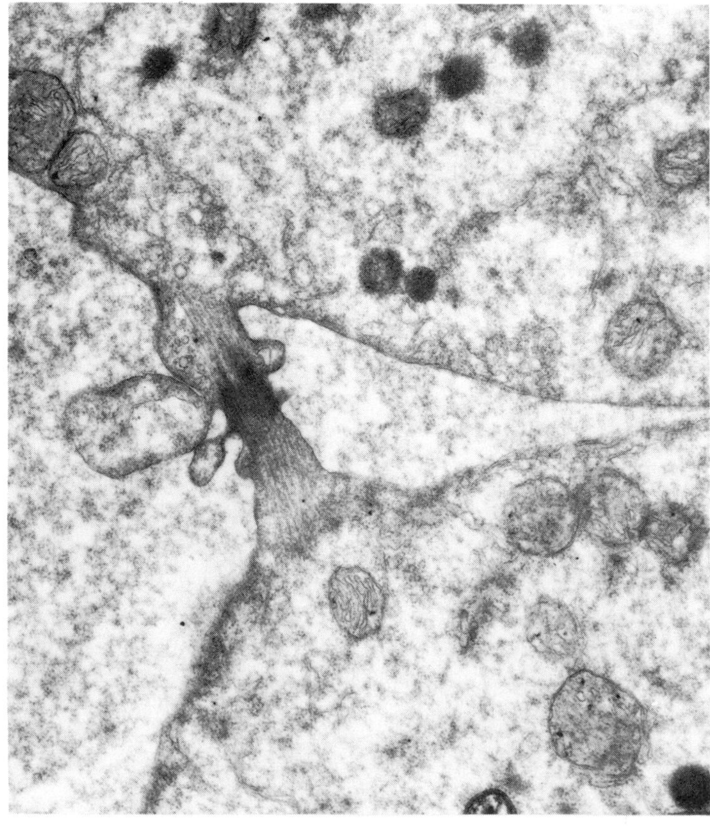

13–38. *Persistence of the midbody in dividing embryonic cells of the zebrafish* Brachydanio. × *9,000. Courtesy of R. J. Thomas and Academic Press Inc.* [J. Ultrastr. Res. *24:232 (1968)*]

Szollosi has detected a layer of 50–70 Å filaments (Fig. 13–39) at the apex of the advancing furrow in the cleaving eggs of a jellyfish, *Aequorea*, and a polychaete worm, *Armandia*. Szollosi has proposed that these filaments, by analogy with the sliding filament model for muscle contraction, provide the basis for contraction of the advancing furrow.

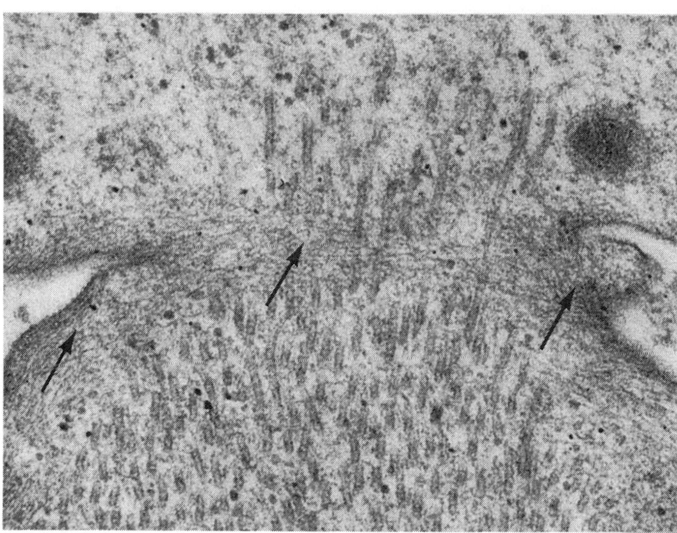

13–39. *Fine cytoplasmic filaments (arrows) at the apex of the advancing furrow in a cleaving rat egg.* × *34,000. Courtesy of D. Szollosi and The Rockefeller University Press.* [J. Cell Biol. *44:192 (1970)*]

Formation of the Cell Plate in Plants

The first events in cell plate formation are remarkably similar to formation of the midbody in cytokinesis by furrowing. Andrew Bajer has followed these developments in remarkable light microscope photographs of division in endosperm cells (Fig. 13–40 a–d). The first visible changes occur at the midpoint of the spindle, following the passage of the chromosomes to the poles at anaphase (Figs. 13–40 a and b). Vesicles or globules appear in this region, intermingled with fibers of the spindle. These gradually increase in number until a complete layer is formed across the equator of the spindle (Figs. 13–40 b and c). Under the electron microscope, the development of this layer, termed by botanists the *phragmoplast,* closely parallels the development of the midbody in furrowing cells. Scattered patches of dense material form along the midpoint of the spindle, closely applied to the surfaces of microtubules that persist in this region (Fig. 13–41). Much of this material, in contrast to the midbody, is concentrated into small membrane-bound vesicles. In many plant cells, these vesicles appear to originate from elements of the Golgi apparatus, which become numerous in regions adjacent to the developing phragmoplast. Endoplasmic reticulum is also often noted in this region. The dense material and the vesicles increase in number until a continuous layer is formed (Fig. 13–42).

At first, the phragmoplast is confined to the area of the cell formerly occupied by the midpoint of the spindle. At the earliest stages, the phragmoplast may be incomplete in the center, so that viewed from the poles it appears doughnut-shaped. The phragmoplast first grows centripetally, extending outward from the circumference of the spindle until the walls of the cell are reached. As the phragmoplast expands, additional microtubules appear at the growing edges. This process continues until the phragmoplast extends entirely across the cell. Within the phragmoplast, the vesicles increase in size, gradually fusing until the daughter cells are separated by continuous membranes in these regions (Fig. 13–43). Fusion of vesicles continues, following behind the edges of the advancing phragmoplast. When the daughter cells are completely separated by continuous plasma membranes, the configuration of membranes, dense areas, residual microtubules, and vesicles at the midregion is termed the *cell plate.* Complete formation of the cell plate may occupy from 30 minutes to two hours.

As the cell plate becomes continuous, new cell wall (see below) is laid down between the membranes separating the daughter cells (Fig. 13–44a). The new cell wall, at first very thin and highly flexible, is gradually thickened by the secretion of additional layers of wall material at the surfaces of the daughter cells (Fig. 13–44b). Protoplasmic connections between the daughter cells, called *plasmodesmata,* remain scattered throughout the new cell wall (arrows, Fig. 13–44b). These connections may consist of protoplasmic strands as narrow as 300–400 Å in diameter.

The appearance of a preprophase band of microtubules marking the future position of the cell plate was mentioned in the introduction to this section. This band (Figs. 13–45a and b), discovered by J. D. Pickett-Heaps and D. H. Northcote, disappears as the spindle forms at prophase. It may be, as Pickett-Heaps has suggested, that the microtubules of the preprophase band are incorporated into the developing spindle. This incorporation may be a factor in the positioning of the spindle and the future cell plate. But because the cell plate does not always appear in the plane marked by the preprophase band during interphase, the relationship between this microtubular structure and cytokinesis remains unclear.

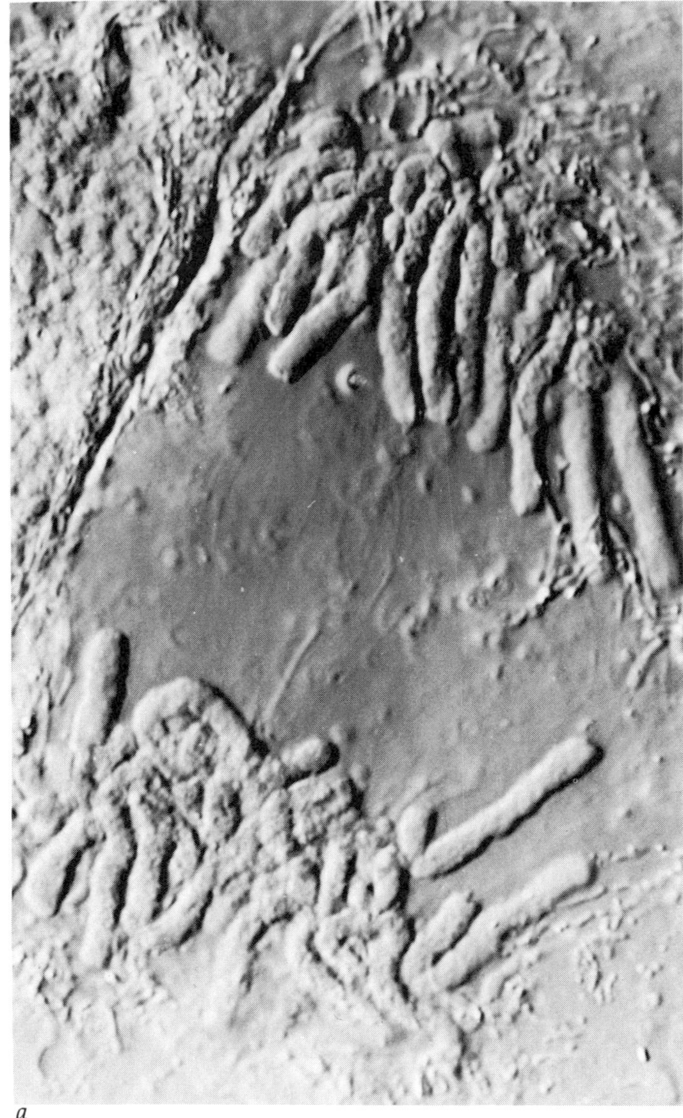

a

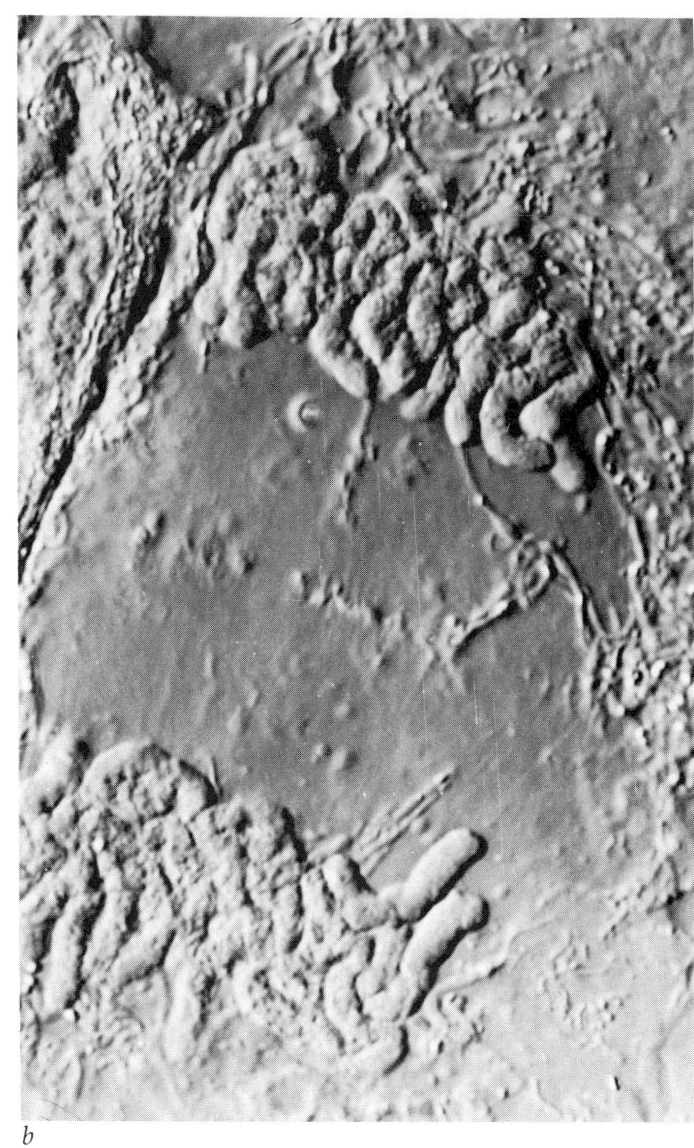

b

13–40. Phragmoplast formation, illustrated in sequential phase contrast light micrographs of a living Haemanthus *cell. × 2,000. Courtesy of A. Bajer.*

As in the case of furrowing, the molecular processes underlying formation of the cell plate are completely unknown. In both cases, microtubules, vesicles, and amorphous dense material appear in advance of the plasma membranes separating daughter cells. In plants, the Golgi apparatus and the endoplasmic reticulum are apparently involved in the production of the vesicles that form the midbody of the plant spindle (the phragmoplast). This may also be the case in cytokinesis by furrowing. However, the juxtaposition of elements of the Golgi apparatus with the developing midbody is never so obvious in animals as it is in plants.

Cytokinesis in bacteria, blue-green algae, and fungi occurs without formation of the phragmoplast and cell plate. In these organisms, the new wall first appears as an indentation in the plasma membrane around the circumference of the dividing cell. As this indentation gradually extends inward toward the center of

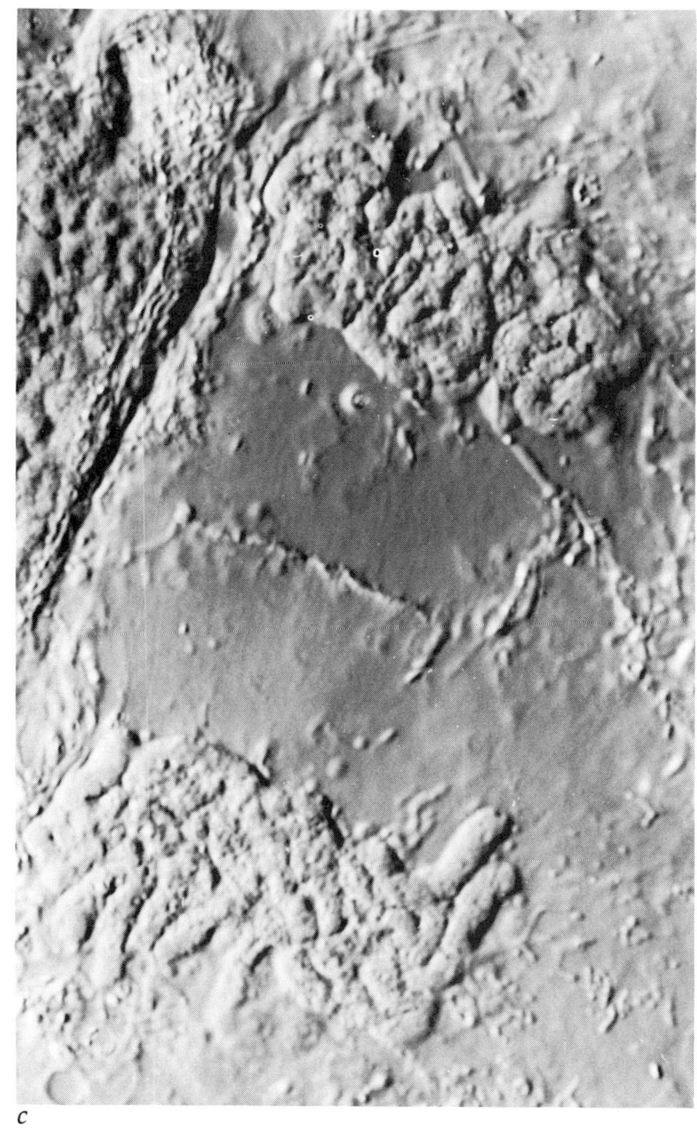
c

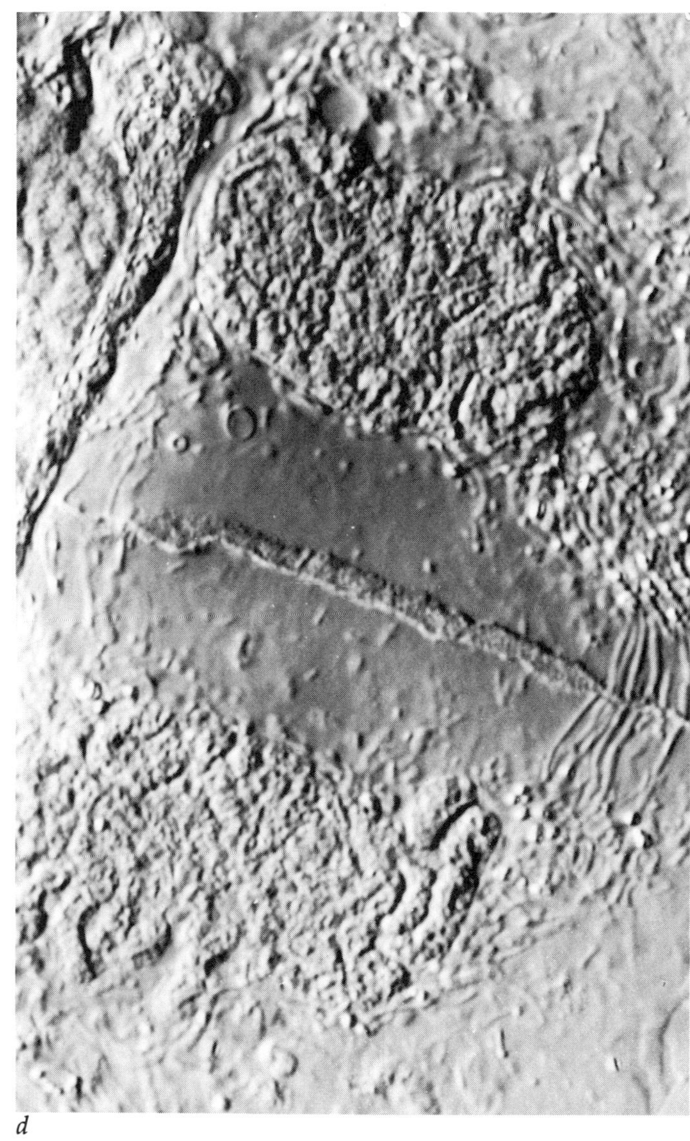
d

the cell, the new wall is laid in its path. Thus wall formation in these groups occurs in a direction opposite to the higher plants and in this sense resembles furrowing in animal tissues.

Some notes on cell wall synthesis. The first cell wall material laid down as a plate between dividing cells consists largely of *pectin* and quantities of lipid-like substances. This material is evidently secreted into vesicles of the Golgi complex; these subsequently become concentrated in the phragmoplast. As the phragmoplast develops, the vesicles fuse with the plasma membrane and discharge their contents into the intracellular space which defines the region of deposition of the cell plate. The endoplasmic reticulum also contributes to the extracellular materials laid down within the phragmoplast. The materials of the cell plate form an amorphous deposit that does not exhibit

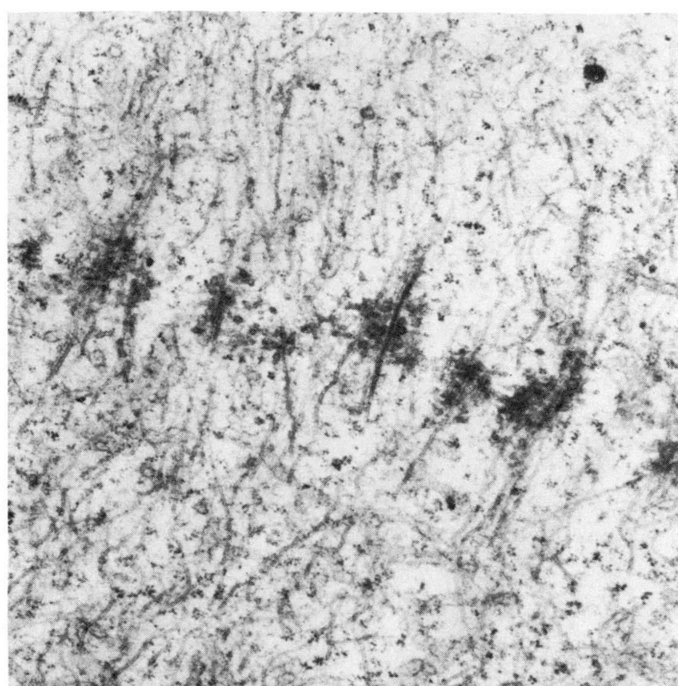

13–41. *An early stage of phragmoplast formation in* Haemanthus. *Aggregations of electron-dense, amorphous material are present around remnants of the spindle microtubules at the midpoint of the cell.* × *22,000. Courtesy of A. Bajer and the publisher.* [Chromosoma *24:383 (1968)*]

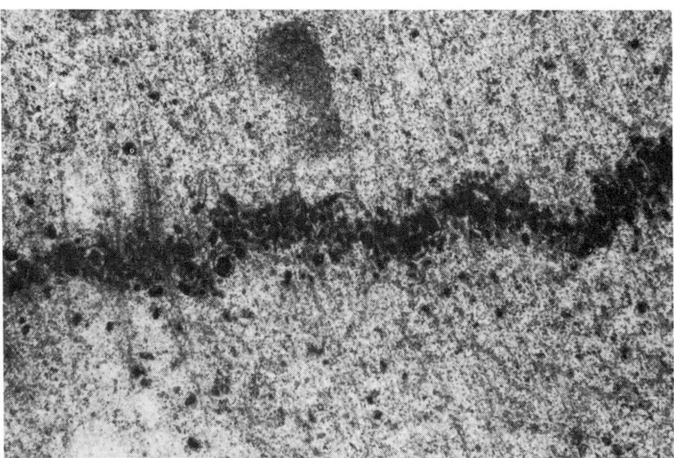

13–42. *A later stage in phragmoplast formation in* Haemanthus. *The dense material now forms a continuous layer across the midpoint of the cell.* × *17,000. Courtesy of A. Bajer and the publisher.* [Chromosoma *24:383 (1968)*]

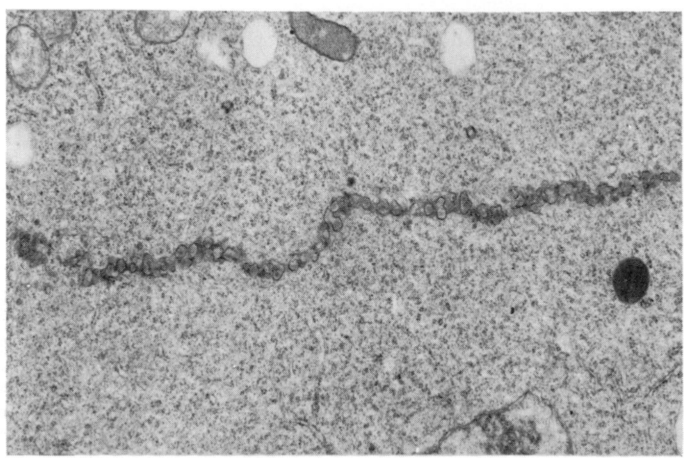

13–43. *Cell plate formation in* Haemanthus. *Fusion of vesicles in the phragmoplast has led to partial formation of plasma membranes between the daughter cells.* × *11,000. Courtesy of W. G. Whaley, M. Dauwalder, J. E. Kephart, and Academic Press Inc.* [J. Ultrastr. Res. *15:169 (1966)*]

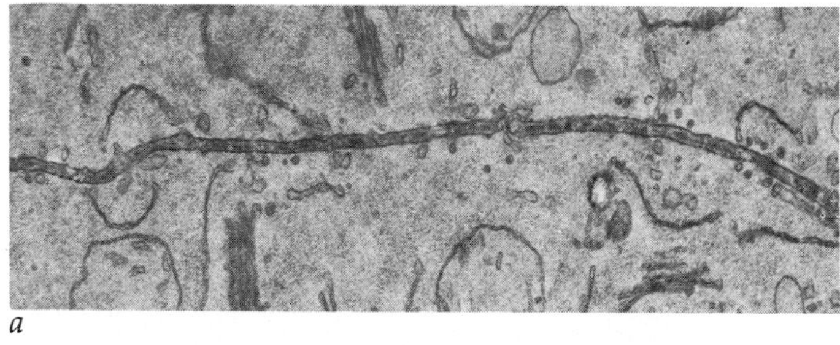

a

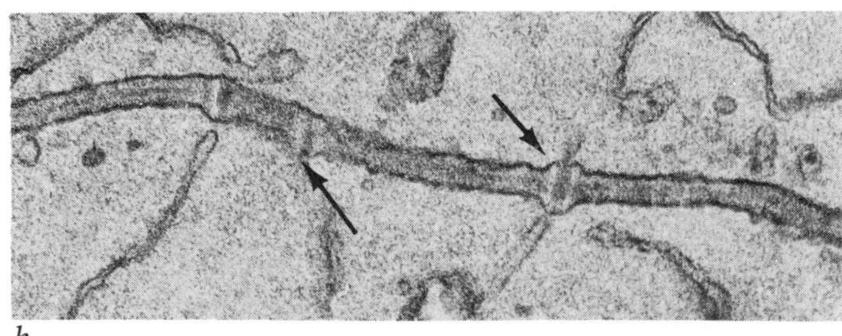

b

13–44. a *and* b, *Progressive stages in the secretion of cell wall material between the newly formed plasma membranes separating daughter cells in* Phalaris canariensis. *Protoplasmic connections called* plasmodesmata *(arrows) remain open between the daughter cells.* a, × *18,000.* b, × *45,500. Courtesy of A. Frey-Wyssling, J. F. Lopez-Saez, K. Muhlethaler, and Academic Press Inc.* [J. Ultrastr. Res. *10:422 (1964)*]

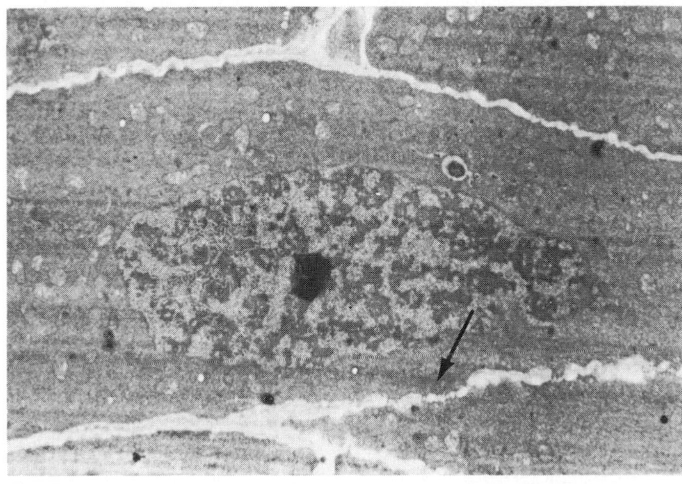

a

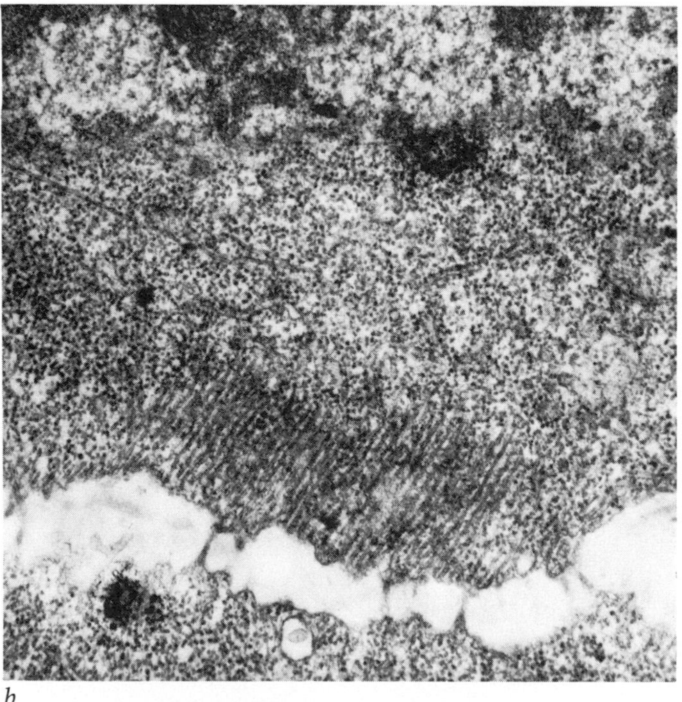

b

13–45. a, *Root meristem cell of wheat.* × *3,700. The "preprophase band" of microtubules marked by an arrow in this micrograph is shown at higher magnification in* b. × *29,000. Courtesy of J. D. Pickett-Heaps and D. H. Northcote.*

a

pectin

galacturonic acid

b

13–46. a, *Pectin, a primary constituent of the first cell wall material laid down between daughter cells in cytokinesis, is a polymerized chain of galacturonic acid subunits* (b).

evidence of a crystalline array if investigated by X-ray diffraction.

Pectin is a long polymer made up of subunits of galacturonic acid (Fig. 13–46a). These are linked together in the polymer by oxygen bridges which extend between the 1 and 4 carbons (Fig. 13–46b). The acidic COOH groups may be complexed and neutralized by various basic groups or may be esterified to methyl groups. This gives the various pectins different solubility properties. These substances are present in the primary cell walls of almost all higher plants and may also be found as a constituent of secondary walls.

Soon after the first pectin-lipid plate is formed, a layer of cellulose microfibrils is laid down on the two surfaces facing the daughter cells. These microfibrils,

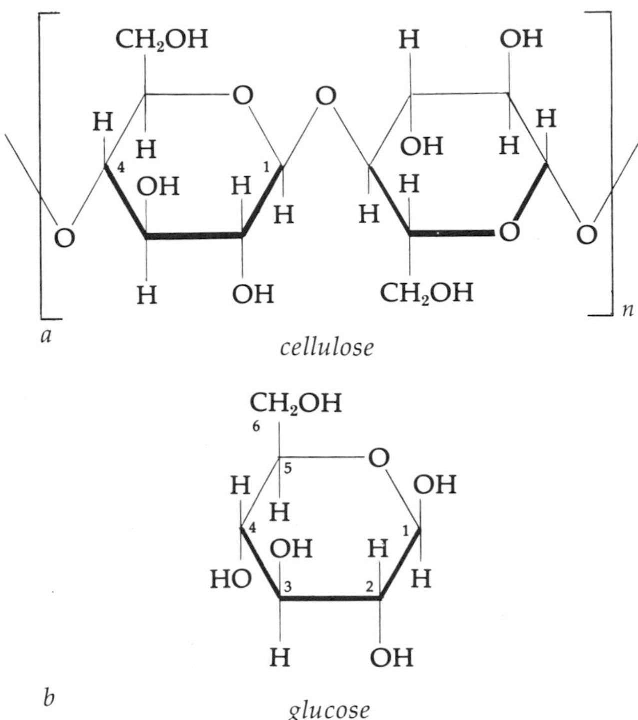

13–47. *Cellulose* (a), *a polymerized chain of glucose subunits* (b).

which are 100–300 Å in diameter and of variable length, consist of a highly ordered, semicrystalline lattice of cellulose polymers. Each individual polymer consists of a chain of glucose residues linked by 1, 4 oxygen bridges as in pectin (Fig. 13–47). Polymerization of the glucose subunits into cellulose occurs outside the plasma membrane; evidently, the enzymes and synthetic machinery for cellulose formation are contained either in the plasma membrane or in the extracellular space. Cellulose microfibrils are not observed on the cytoplasmic side of the plasma membrane. On the extracellular side, a separation of about 100 Å exists between the closest layer of cellulose microfibrils and the surface of the plasma membrane (see Fig. 13–50).

In the primary wall, the layer of cellulose is laid down in the form of a network with little or no parallel orientation of the microfibrils (Fig. 13–48). This wall remains extensible and undergoes a considerable increase in area as the plant cell increases to its final size.

13–48. *Early cell wall of a cotton cell. The network-like array of cellulose fibrils remains extensible.* × *19,000. Courtesy of A. L. Houwink.*

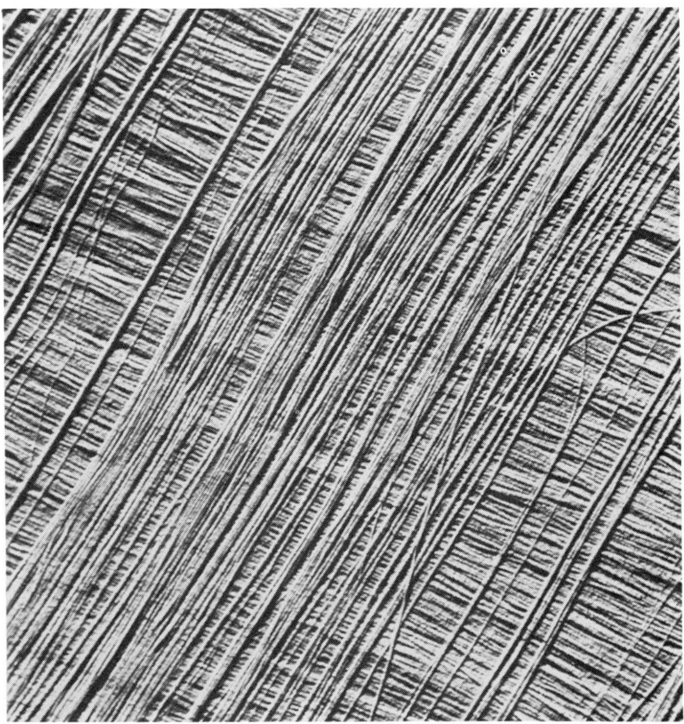

13–49. *Wall of the green alga* Chaetomorpha, *showing ordered arrangement of cellulose microfibrils.* × *21,000. Courtesy of E. Frei and R. D. Preston.*

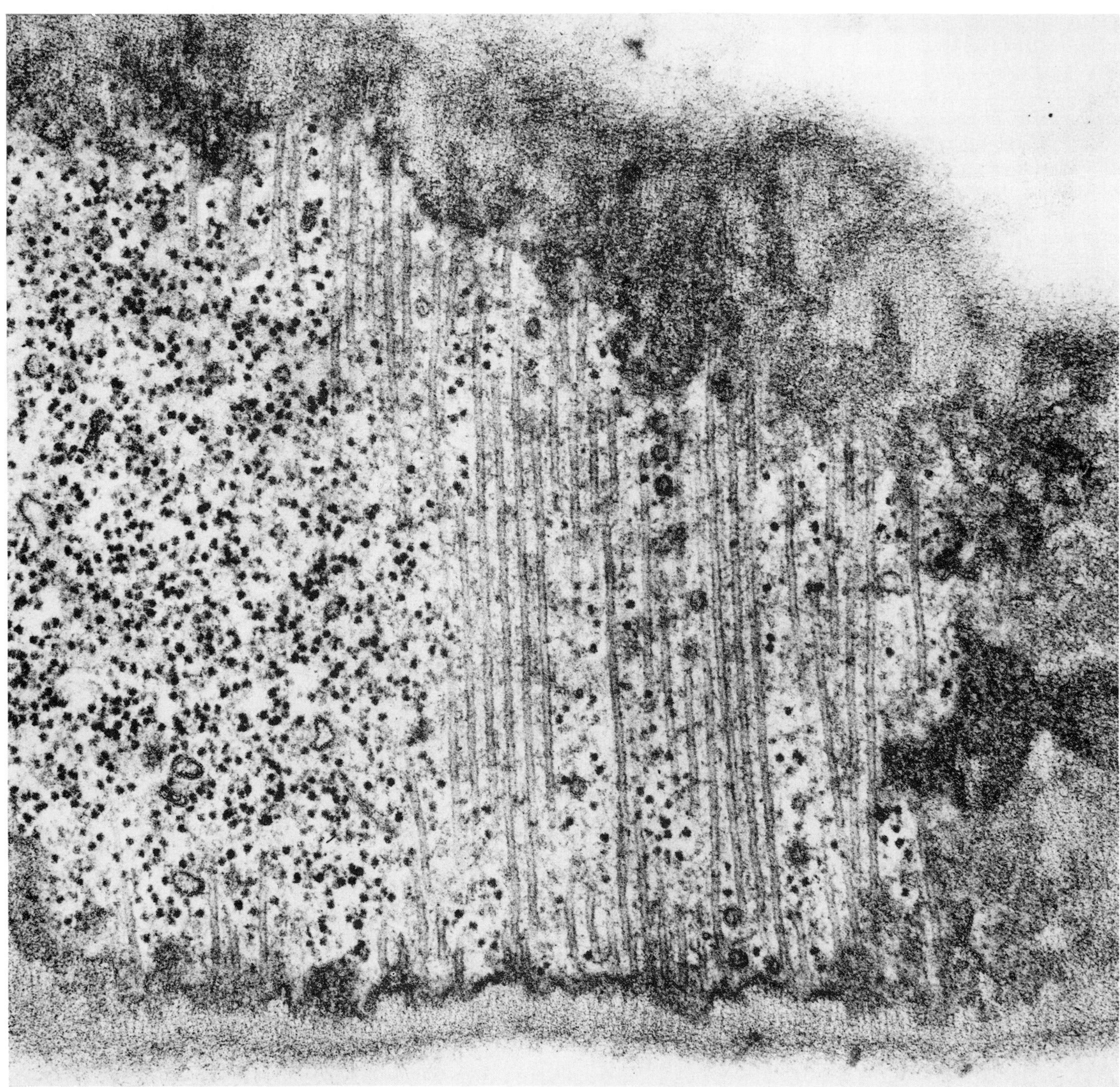

13–50. Parallel orientation of the microtubules just underneath the plasma membrane in a cell of the bean Phaseolus vulgaris. *Cellulose microfibrils of the newly synthesized wall are oriented in the same direction as the microtubules.* × *80,000. Courtesy of E. H. Newcomb.*

As full growth is reached, another layer of cellulose fibrils, embedded in a matrix of other substances, is deposited to form the secondary wall. These fibers are laid down in a more ordered pattern, often forming nearly parallel rows (Fig. 13–49). These rows, in which the individual microfibrils may approach lengths of 2–5 microns, extend around the wall at an angle to the long axis of the cell. Frequently, successive layers are deposited with the angle changed so that a pattern of crossed microfibrils is formed in alternate layers (visible in Fig. 13–49).

In 1963 M. C. Ledbetter and K. R. Porter discovered a layer of microtubules just under the plasma membrane in regions in which cellulose microfibrils are being deposited. Subsequent investigations have shown that this is a characteristic feature of cellulose synthesis in cell wall formation. Where the cellulose microfibrils are laid down in a parallel array, the microtubules just below the surface can be seen to extend in the same direction (Fig. 13–50). The function of these microtubules is unknown. Various hypotheses consider that the layer of tubules is either directly involved in the synthesis of the cellulose microfibrils or in some manner controls fibril orientation. Experimental evidence on this matter is contradictory. A number of experiments indicate that the microtubules are directly involved in orientation of cellulose microfibrils. For example, P. B. Green studied the effects of colchicine on cells of *Nitella* engaged in wall synthesis. These cells normally deposit microfibrils in a parallel array at an angle to the long axis of the cell. When colchicine was added, the parallel orientation was destroyed and the microfibrils were laid down in a random pattern. This experiment appears to demonstrate the interference with microtubule formation (colchicine destroys the microtubules associated with cell wall synthesis) affects orientation but not synthesis of microfibrils. But in a later experiment, J. D. Pickett-Heaps found that application of colchicine to wheat seedlings did not alter the ordered pattern of microtubule deposition, even though microtubules were almost completely destroyed. Gross malformation of cell wall shape, however, was noted. These experiments at least allow the conclusion that the synthesis of microfibrils is not directly dependent on the presence of microtubules. The possible influence of microtubules on cellulose microfibril orientation is still a debatable point.

The cell walls of various plants contain many substances in addition to cellulose and pectin. Many of these are polymers of sugars closely related to glucose. Mannose is found in the walls of green algae, red algae, yeast, and some higher plants; xylose polymers are found in red algae and higher plants; and galactose polymers are found in red algae. The latter polysaccharides are water soluble and readily form gels containing water in high proportion. Nearly all fungi contain chitin as a wall component; this substance also consists of a polymerized chain of glucose-like subunits. The woody tissues of higher plants contain large quantities of lignin, a complex molecule made up of polymerized subunits of sinapin or coniferyl alcohol. Lignin occurs in gymnosperms and angiosperms as a constituent of an amorphous matrix in which the cellulose microfibrils of the cell walls are embedded.

Cell wall formation in plants, and furrowing in animal cells, completes the total process of cell division. This process is initiated during interphase by unknown events, which trigger the initiation of the S period. This is followed by G_2, mitosis, and cytokinesis. Although these replication and division mechanisms are separable, in most cells the three mechanisms follow in close sequence. The division of the chromosomes and centrioles, accomplished with precision by the mitotic processes, is complemented by the more or less equal distribution of cytoplasmic organelles between daughter cells by cytokinesis. The end result is the production of daughter cells having a capacity equal to the original parent cell.

Suggestions for Further Reading

Books

Schrader, F., 1953. *Mitosis*. Columbia University Press, New York. Tabulates and discusses the many theories of anaphase movement developed up to the 1950's, including the early hypotheses of classical microscopists.

Wilson, E. B., 1925. *The Cell in Development and Heredity*. Macmillan, New York.

Papers and Review Articles

Bajer, A., 1968a. Chromosome movement and fine structure of the mitotic spindle. In *Aspects of Cell Motility, Proceedings of the 22nd Symposium of the Society for Experimental Biology,* Oxford, England, pp. 285–310.

Bajer, A., 1968b. Fine structure studies on phragmoplast and cell plate formation. *Chromosoma,* vol. 24, pp. 383–417.

Bernhard, W., and E. DeHarven, 1960. L'ultrastructure du centriole et d'autres elements de l'appareil achromatique. *Proceedings of the 4th International Conference for Electron Microscopy,* Berlin, pp. 217–227.

Birnsteil, M., 1967. The nucleolus in cell metabolism. *Annual Review of Plant Physiology,* vol. 18, pp. 25–58. Includes a good review of the nucleolus and nucleolar organizer during mitosis.

Borisy, G. G., and E. W. Taylor, 1967. The mechanism of action of colchicine: binding of colchicine-^{3}H to cellular protein. *Journal of Cell Biology,* vol. 34, pp. 525–533.

Brinkley, B. R., and E. Stubblefield, 1966. The fine structure of the kinetochore of a mammalian cell *in vitro. Chromosoma,* vol. 19, pp. 28–43.

Brinkley, B. R., and R. Bruce Nicklas, 1968. Ultrastructure of the meiotic spindle of grasshopper spermatocytes after chromosome micromanipulation. *Journal of Cell Biology,* vol. 39, p. 16a.

Brown, D. D., and J. B. Gurdon, 1966. Size distribution and stability of DNA-like RNA synthesized during development of anucleolate mutants of *Xenopus laevis. Journal of Molecular Biology,* vol. 19, pp. 399–422.

Buck, R. C., 1967. Mitosis and meiosis in *Rhodnius prolixus:* the fine structure of the spindle and diffuse kinetochore. *Journal of Ultrastructure Research,* vol. 18, pp. 489–501.

DeHarven, Etienne, 1968. The centriole and the mitotic spindle. In A. J. Dalton and F. Haguenau, eds., *Ultrastructure in Biological Systems,* vol. 3. Academic Press, New York, pp. 197–227.

Dietz, R., 1966. The dispensability of the centrioles in the spermatocyte divisions of *Pales ferruginea.* In C. D. Darlington and K. R. Lewis, eds., *Chromosomes Today,* vol. 1. Plenum Press, New York, pp. 161–166.

Forer, A., 1969. Chromosome movements during cell division. In A. Lima-de-Faria, ed., *Handbook of Molecular Cytology.* North-Holland Publishing Company, Amsterdam, pp. 553–601.

Frey-Wyssling, A., J. F. Lopez-Saez, and K. Muhlethaler, 1964. Formation and development of the cell plate. *Journal of Ultrastructure Research,* vol. 10, pp. 422–432.

Gall, J. G., 1961. Centriole replication. A study of spermatogenesis in the snail *Viviparus. Journal of Biophysical and Biochemical Cytology,* vol. 10, pp. 163–193.

Green, P. B., 1962. Mechanism for plant cellular morphogenesis. *Science,* vol. 138, pp. 1404–1405.

Hartmann, J. F., 1964. Cytochemical localization of ATP in the mitotic apparatus of *Hela* and sarcoma 180 tissue culture cells. *Journal of Cell Biology,* vol. 23, pp. 363–370.

Hepler, P. K., and W. T. Jackson, 1968. Microtubules and early stages of cell-plate formation in the endosperm of *Haemanthus katherinae. Journal of Cell Biology,* vol. 38, pp. 437–446.

Inoué, S., and H. Sato, 1967. Cell motility by labile association of molecules. *Journal of General Physiology,* vol. 50, part 2, pp. 259–292. A review of current models of anaphase movement, with details of a major hypothesis advanced by the authors to account for movement of the chromosomes on the spindle.

Jacob, F., S. Brenner, and F. Cuzin, 1964. On the regulation of DNA replication in bacteria. *Cold Spring Harbor Symposia on Quantitative Biology,* vol. 28, pp. 329–348.

Jenkins, R. A., 1967. Fine structure of division in ciliate protozoa I: micronuclear mitosis in *Blepharisma. Journal of Cell Biology,* vol. 34, pp. 463–481.

Jokelainen, Pentti T., 1967. The ultrastructure and spatial organization of the metaphase kinetochore in mitotic rat cells. *Journal of Ultrastructure Research,* vol. 19, pp. 19–44.

Kawamura, K., 1960. Studies on cytokinesis in neuroblasts of the grasshopper *Chortophaga viridifasciata. Experimental Cell Research,* vol. 21, pp. 1–18.

Kreger, D. R., 1969. Cell walls. In A. Lima-de-Faria, ed., *Handbook of Molecular Cytology.* North-Holland Publishing Co., Amsterdam, pp. 1444–1479.

Krishnan, A., and R. C. Buck, 1965. Structure of the mitotic spindle in L-strain fibroblasts. *Journal of Cell Biology*, vol. 24, pp. 433–444. Includes formation of the midbody and furrowing.

Kubai, D. F., and H. Ris, 1969. Division in the dinoflagellate *Gyrodinium Cohnii* (Schiller). *Journal of Cell Biology*, vol. 40, pp. 508–528. Mitosis in a dinoflagellate, with discussion of the possible evolution of the spindle.

Lafontaine, J.-G., 1968. Structural components of the nucleus in mitotic plant cells. In A. J. Dalton and F. Haguenau, eds., *Ultrastructure in Biological Systems*, vol. 3. Academic Press, New York, pp. 151–196.

Lafontaine, J.-G., and Lord, A., 1966. Ultrastructure and mode of formation of the telophase nucleolus in various plant species. *National Cancer Institute Monograph 23*, pp. 63–67.

Lafontaine, J.-G., and A. Lord, 1969. Organization of nuclear structures in mitotic cells. In A. Lima-de-Faria, ed., *Handbook of Molecular Cytology*. North-Holland Publishing Co., Amsterdam, pp. 381–411. An excellent survey of mitosis.

Ledbetter, M. C., and K. R. Porter, 1963. A "microtubule" in plant cell fine structure. *Journal of Cell Biology*, vol. 19, pp. 239–250.

Mazia, D., 1961. Mitosis and the physiology of cell division. In J. Brachet and A. E. Mirsky, eds., *The Cell*, vol. 3. Academic Press, New York, pp. 77–412.

Newcomb, E. H., 1969. Plant microtubules. *Annual Review of Plant Physiology*, vol. 20, pp. 253–287.

Nicklas, R. B., 1965. Chromosome velocity during mitosis as a function of size and position. *Journal of Cell Biology*, vol. 25, pp. 119–235.

Pickett-Heaps, J. D., 1967. The effects of colchicine on the ultrastructure of dividing plant cells, xylem wall differentiation, and distribution of cytoplasmic microtubules. *Developmental Biology*, vol. 15, pp. 206–236.

Pickett-Heaps, J. D., 1969. Preprophase microtubules and stomatal differentiation: some effects of centrifugation on symmetrical and asymmetrical cell division. *Journal of Ultrastructure Research*, vol. 27, pp. 24–44.

Pitelka, D. R., 1969. Centriole replication. In A. Lima-de-Faria, ed., *Handbook of Molecular Cytology*. North-Holland Publishing Co., Amsterdam, pp. 1199–1218.

Prescott, D. M., and M. A. Bender, 1962. Synthesis of RNA and protein during mitosis in mammalian tissue culture cells. *Experimental Cell Research*, vol. 26, pp. 260–268.

Robbins, E., and N. K. Gonatas, 1964. The ultrastructure of a mammalian cell during the mitotic cycle. *Journal of Cell Biology*. vol. 21, pp. 429–463.

Robbins, E., and M. Shelanski, 1969. Synthesis of a colchicine-binding protein during the *Hela* cell life cycle. *Journal of Cell Biology*, vol. 21, pp. 371–373.

Ryter, A., 1968. Association of the nucleus and the membrane of bacteria: a morphological study. *Bacteriological Reviews*, vol. 32, pp. 39–54. Division in bacteria.

Stephens, R. E., 1967. The mitotic apparatus. Physical characterization of the 22S protein component and its subunits. *Journal of Cell Biology*, vol. 32, pp. 255–275.

Stevens, B. J., 1965. The fine structure of the nucleolus during mitosis in the grasshopper neuroblast cell. *Journal of Cell Biology*, vol. 24, pp. 349–368.

Szollosi, D., 1970. Cortical cytoplasmic filaments of cleaving eggs: a structural element corresponding to the contractile ring. *Journal of Cell Biology*, vol. 44, pp. 192–209. Furrowing in an invertebrate egg.

Taylor, E. W., 1965. Brownian and saltatory movements of cytoplasmic granules and the movement of anaphase chromosomes. In E. H. Lee, ed., *Proceedings of the 4th International Congress of Rheology*. Interscience Publishers, New York.

14 Meiosis

The mechanisms for distributing chromosomes in mitosis insure that daughter cells receive identical copies of the genetic information. The chromosome number in somatic cells is kept constant by this process; genetic variability is limited to mutations arising from "mistakes" during premitotic replication of the chromosomes and mechanical difficulties in separation of sister chromatids during anaphase. The latter may result in breakage and elimination of chromosome segments. But mutations and mechanical aberrations during mitosis are comparatively rare, and mitosis may be regarded as a system for maintaining genetic constancy in all somatic cells of an organism.

During the late 1800's, it became apparent that another, modified form of cell division must operate at some time in the life cycle of sexually reproducing organisms. This appreciation developed through studies of the behavior of sperm and egg nuclei in fertilization. In 1874, L. Auerbach observed that in nematode eggs the two nuclei present just after fertilization fused to form a single nucleus. Soon after this, in 1875, Oskar Hertwig demonstrated that one of these nuclei is the egg nucleus, and the other is derived from the fertilizing spermatozoan. Two years later, Eduard Strasburger reported that an equivalent process takes place during fertilization in plants. These observations that the sperm and egg nucleus fuse during fertilization indicated that the zygote nucleus contains elements donated by both gametes. This was confirmed in 1883 by the Belgian cytologist Edouard van Beneden, who demonstrated in *Ascaris* that equal numbers of chromosomes are given to the zygote nucleus by the egg and sperm. These observations led to the first proposition that a modified form of nuclear division must take place before the fusion of gamete nuclei. It was known that the number of chromosomes in somatic cells remains constant. But unless the chromosome

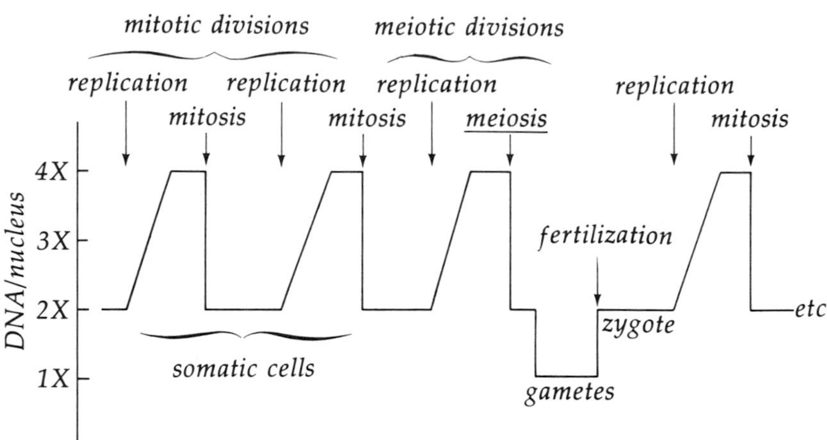

14–1. *The relative amounts of DNA per nucleus at various stages of the cell cycle. The base level, 1X, represents the amount of DNA in a gamete nucleus. Modified from C. P. Swanson, T. Merz, and W. J. Young,* Cytogenetics.

number is halved at some time during the life cycle, the fusion of gamete nuclei at fertilization would double the number of chromosomes once each generation. This was clearly stated in 1887 by August Weismann, who predicted that a reduction in chromosome number to one-half the somatic value must take place at some time during the production of gametes.

The anticipated reduction divisions were discovered very shortly after Weismann's prediction by Oskar Hertwig and T. Boveri, who found that the number of chromosomes is reduced to one-half the somatic value by a series of divisions just before maturation of the gametes. These observations were made in animal species. Within a few years, the same processes were described in plants by Strasburger and E. Overton. Final details of the specialized reduction division were filled in by J. B. Farmer and J. E. Moore, who termed the process *maiosis* from the Greek *meioun*, "to diminish." This was later changed to *meiosis* (see also the description of these events in the introductory chapter).

The overall effect of the meiotic division is simple to describe in broad outline. In mitosis, a single replication is followed by a single division of the nucleus. In meiosis, a single replication is followed by *two* divisions of the nucleus, with the end result that the gametes produced contain one-half the somatic number of chromosomes. This is paralleled by changes in the quantity of DNA per nucleus, as measured by the amount of Feulgen stain bound (Fig. 14–1). If the final gametic nucleus is considered as the base level (1X), then each cell after replication contains 4X the haploid amount of DNA. This 4X amount is found whether replication precedes a meiotic or mitotic division. If replication is followed by mitosis, the daughter nuclei each contain the 2X amount at the completion of division. If the division sequence is meiotic, the 4X value is reduced at the close of two sequential divisions to the 1X value. This is restored to the normal 2X value after fusion of the gametic nuclei during fertilization (these values are often given as 1C, 2C, and 4C instead of 1X, 2X, and 4X; see note on p. 290).

The events taking place in meiosis are the subject of this chapter. As the description of this process unfolds, it will become obvious that meiosis, in addition to providing a mechanism for precisely reducing the chromosome number to the haploid value in gametes, also represents an adaptation for generating variability among the products of the division cycle. This is in direct contrast to mitosis, in which genetic variability is kept at a minimum.

An Overview of Meiosis

Premeiotic Interphase

The cell lines destined to undergo meiosis are the product of ordinary mitotic divisions. At some point following a DNA replication, the gametic cells enter a meiotic rather than mitotic sequence. Few differences have been noted between premeiotic interphase and premitotic interphase. A large number of experiments using radioactive precursors of DNA have shown that replication of the chromosomes takes place during premeiotic interphase; this replication is closely similar to premitotic DNA replication except that it may be extended over greater periods of time. Additionally, it has been noted that replication of the X chromosome, which is delayed in premitotic S in organisms such as the mouse and grasshopper, takes place at the same time as the rest of the chromosome complement during S of premeiotic interphase.

There are indications from studies of meiosis in the lily by Herbert Stern and Yasuo Hotta that the first irreversible changes directing cells toward meiosis may take place early in G_2. Stern and Hotta found that, if placed in a culture medium, cells removed from anthers of the lily during premeiotic G_1 and S revert after a few days to ordinary mitotic divisions without entering meiosis. If allowed to enter into early G_2 before removal, an irreversible change was noted in the behavior of the cells. These G_2 cells, if placed in a culture medium, were unable to undergo a normal mitosis; although spindles were formed, the chromosomes failed to separate at anaphase. Eventually these cells reverted to an interphase-like state without further replication or division. Cells removed from the anthers later in G_2 entered an abnormal form of meiosis rather than mitosis in culture. These cells always exhibited irregularities in pairing (see below) during prophase, so that meiosis was never completed. These experiments in lily show that a change may occur early in G_2 to shift the division mechanism from the mitotic to the meiotic pattern. Although the nature of this change is unknown, the shifting mechanism probably represents the activity of one or more genes controlling the division cycle, because it is known that many facets of the meiotic process are under genetic control and can be modified by mutations (see, for example, Rees, 1961).

Cells in premeiotic interphase are similar in morphology to premitotic cells except that late in interphase the nuclei of premeiotic cells enlarge. At the same time, limited segments of the chromosomes may condense or coil and become visible under the light microscope for brief periods. These short segments are called "prochromosomes." In the organisms in which this type of late interphase condensation occurs, the prochromosomes may vary between the diploid and haploid number, suggesting that limited pairing of homologous chromosomes occurs in premeiotic interphase. The functional significance of this condensation and possible early pairing is unknown. Whatever the function, it is not likely to account for recombination (see below) because the segments that condense and pair represent only a small fraction of the total nuclear chromatin. By the end of interphase, these coiled segments usually disappear.

Prophase

Prophase of meiosis is considerably more complex than mitotic prophase. Although continuous, meiotic prophase is divided for convenience into five stages, *leptonema* (or *leptotene*), *zygonema* (or *zygotene*), *pachynema* (or *pachytene*), *diplonema* (or *diplotene*), and *diakinesis*.[1] The meaning and derivation of these terms will become apparent as the description progresses.

Leptonema. The onset of leptonema (*leptos* = fine or thin; Figs. 14–2a and 14–3) is marked by the first appearance of the chromosomes as extended, faint threads in the nucleus. This stage is initially similar to the beginning of mitotic prophase, except that the condensing leptotene chromosomes are often more extended and show a regular series of swellings and constrictions along the chromatids. This gives the chromosomes the appearance of a single row of beads. The bead-like swellings, called *chromomeres*, are probably formed simply by localized differences in the degree of coiling of the chromatin fibers (see Fig. 14–10). The significance of this coiling configuration is unknown. Chromomeres were once thought to represent the genes. However, chromomeres are too small in number (it has been estimated in the lily, for example, that there are 1500–2500 chromomeres visible in meiotic prophase) to include all of the genes, which are believed to be present in higher organisms in tens of thousands. In addition, the number and pattern of chromomeres changes as prophase proceeds.[2]

As leptonema progresses, condensation continues with further shortening and thickening of the chromosome arms. Because replication has already taken place each chromosome visible in leptonema consists of two chromatids. Typically, the division between the chromatids is not visible during leptonema and, unlike mitotic prophase, each chromosome appears to be a single unit under the light microscope. At this stage in many animals, a peculiar arrangement of the chromosomes is noted in which the ends of the chromosomes are attached to the nuclear envelope at points

---

[1]The *-nema* (*nema* = thread) is the noun form, and the *-tene* form denotes the adjective. In practice, the forms are used interchangeably to describe the stages of meiotic prophase.

[2]A frequently advanced hypothesis holds that the chromomere patterns observed in chromosomes at prophase of meiosis, at the lampbrush stage, and the bands of polytene chromosomes (see Chapter 10) are all manifestations of the same underlying arrangement of the genetic material. The coiled regions would presumably represent functional units of the chromosome and the degree of coiling the relative activity of these units. Little in the way of direct evidence exists concerning this hypothesis or the identity and significance of the proposed functional units presumably represented by the chromomeres.

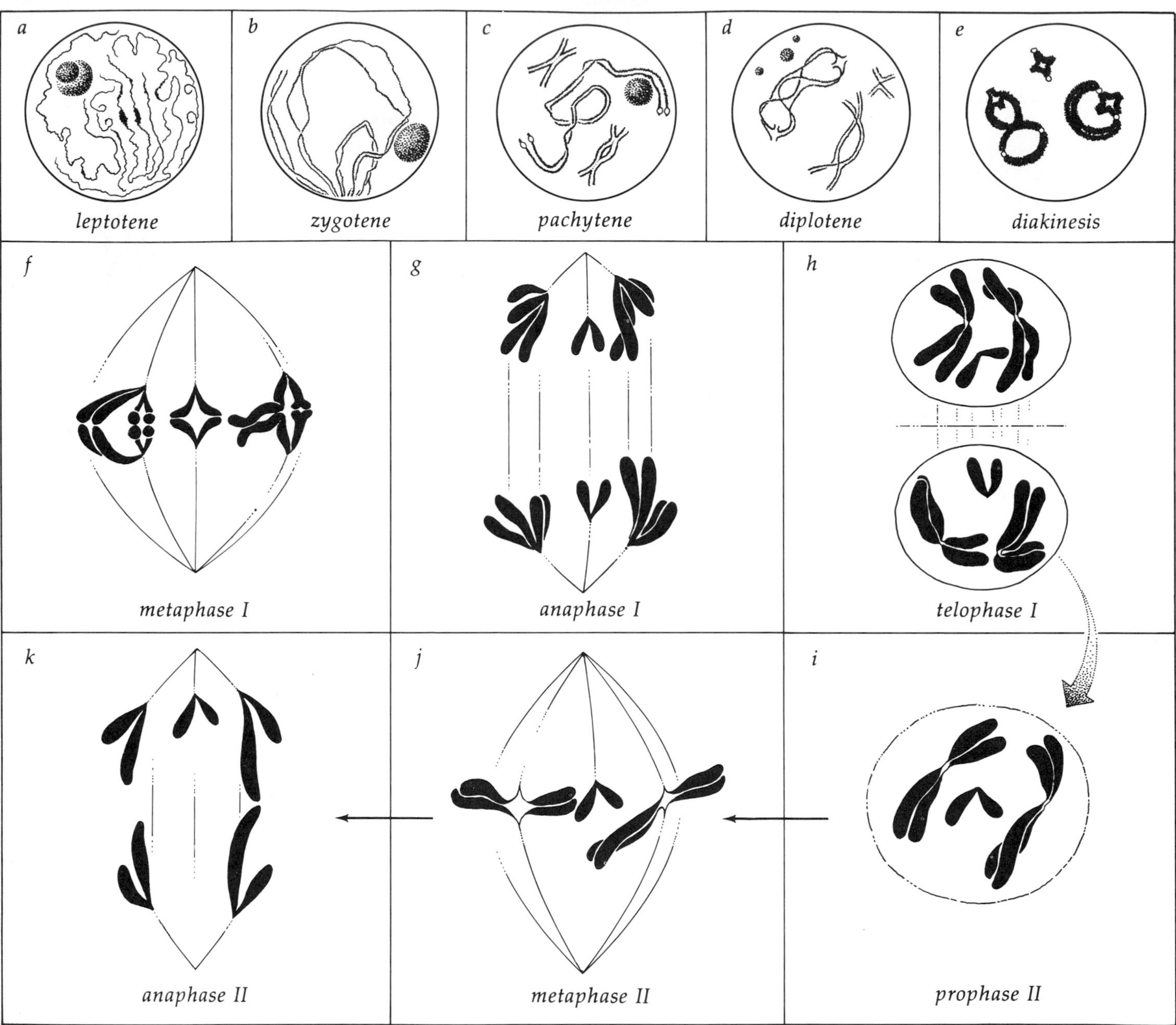

14–2. *Diagram outlining the stages of meiosis. Adapted from* Chromosome Marker *by K. R. Lewis and B. John, 1963. Courtesy of K. R. Lewis, B. John, and J. & A. Churchill, London.*

near the centrioles, which at this stage are still contained in a single aster. The remainder of the chromosome extends as a loop into the nucleoplasm.

Zygonema. Leptonema ends, and zygonema (*zygon* = yoke, adjoining; Figs. 14–2b and 14–4) begins when pairing between homologous chromosomes is initiated. Remember that in the diploid nucleus each chromosome (except for the sex chromosomes—see below) is represented twice. One member of each pair is derived from the male parent of the organism; the other is derived from the female parent. The two chromosomes of a pair, called *homologs,* have the same genes in the same order along the chromosomes in a

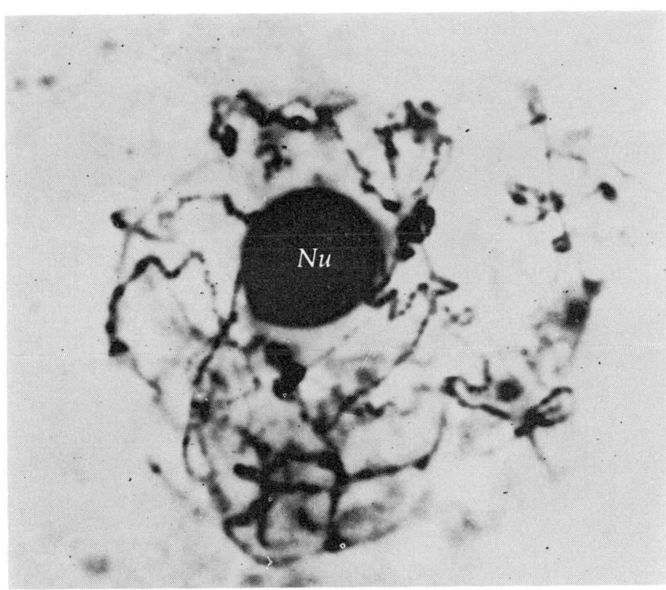

14–3. A leptotene nucleus of corn. The chromosomes have coiled into discrete, individual threads but have not yet paired. The nucleolus (Nu) *is still present. From* The Cell, *Vol. III, ed. J. Brachet and A. E. Mirsky, 1961. Courtesy of M. M. Rhoades and Academic Press Inc.*

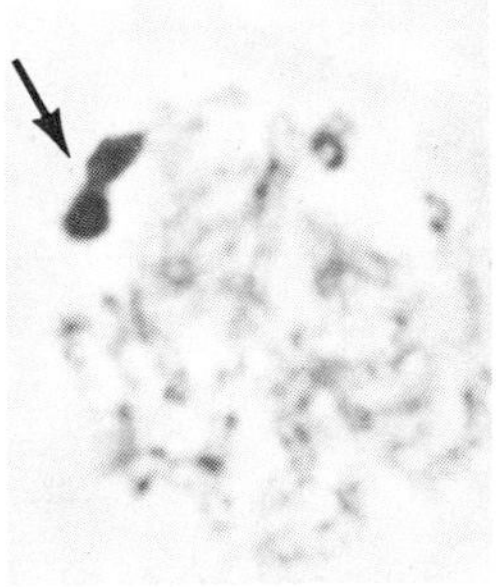

14–4. A zygotene nucleus in the milkweed bug, Oncopeltus. *The sex chromosomes (arrow) are tightly coiled at this stage.* × *2,000. Courtesy of B. John, from S. Wolfe and B. John,* Chromosoma *17:85 (1965).*

linear array, but alternate forms of the genes (alleles) may be present in either homolog. Before zygotene is initiated, the two members of each homologous pair are usually widely separated from each other in the nucleus. At the onset of zygotene, the homologs are brought together by an unknown mechanism and begin to pair closely at one or more points. The initial sites of pairing may be located anywhere along the chromosomes. In animals in which the ends of the chromosomes are attached to the nuclear envelope, pairing is frequently initiated at these terminal regions. Once contact is made, pairing extends in zipper-like fashion along the chromosomes until the homologs are closely aligned in a precise chromomere-by-chromomere association. The two homologs remain separated by a 0.15 to 0.2 micron space, barely visible in the light microscope. The synaptinemal complex is observed in this space in thin sections of zygotene-pachytene chromosomes (Fig. 14–5). This structure, visible only under the electron microscope, is in some manner related to close pairing and recombination of alleles during meiotic prophase (see p. 403).

There are two alternate arrangements of the chromosomes in the zygotene nucleus. In animals in which the terminal regions of the chromosomes are attached to the nuclear envelope at points near the aster, pairing apparently diminishes the number of loops to one-half the leptotene number. These loops, now double, extend from the nuclear envelope into the nucleoplasm in a striking pattern called the "bouquet" (see Fig. 14–33). In the alternate configuration, found in plants and in a few animals including some arthropods and other invertebrates, the paired chromosomes are distributed in the nucleoplasm in a more or less tangled mass without more than occasional attachments of chromosomes to the nuclear envelope (see Fig. 14–6).

Pachynema. Zygonema ends and pachynema (*pachus* = thick; Figs. 14–2c and 14–6) begins at the completion of pairing. Leptonema and zygonema run their course in a relatively short time, usually within a few hours. In contrast, pachynema may last for days, weeks, or even years. During all of pachynema the homologous chromosomes remain closely paired, becoming progressively shorter and thicker as condensation continues. Near the end of pachynema, this progressive coiling and condensation may result in a shortening of the overall length of the chromosomes to one-fourth to one-sixth of the leptotene dimensions. Attachment of the ends of the chromosomes to the nuclear envelope, where such attachments are present, may persist throughout pachynema.

The pachytene chromosomes are called either *bivalents* or *tetrads,* depending on whether the paired structure is considered at the level of chromosomes or chromatids. At the chromosome level, "bivalent" is used; this reflects the fact that the homologs are paired by twos to form the pachytene configuration. But each homolog is in turn double and consists of the two products of the premeiotic replication. Thus, a total of four chromatids (the "tetrad") are present in the paired homologs. This is shown diagrammatically in Fig. 14–11; actually, the subdivision of the chromosomes

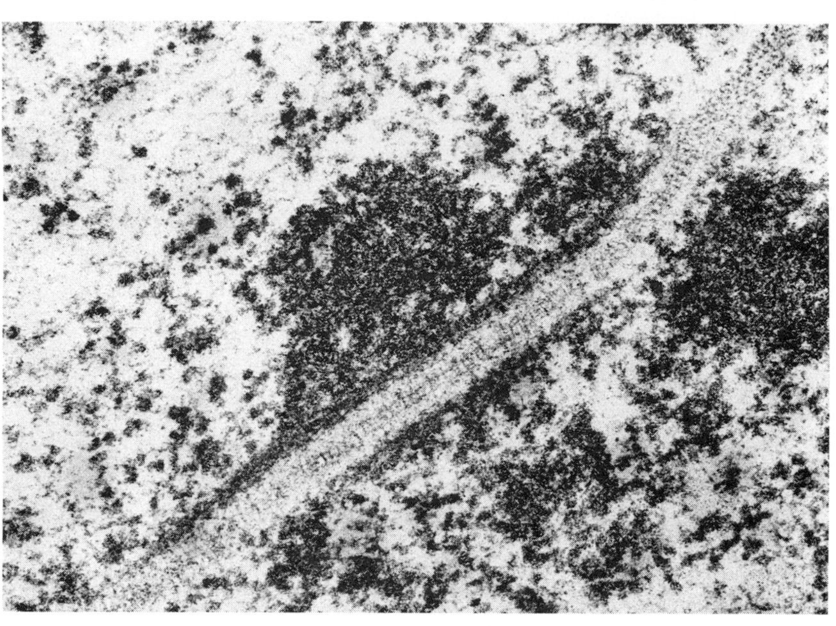

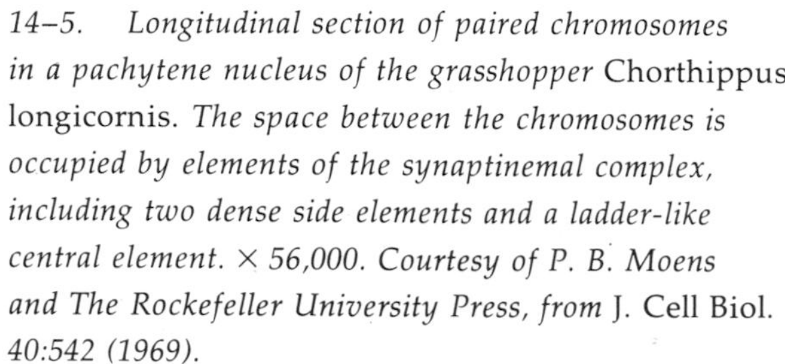

14–5. *Longitudinal section of paired chromosomes in a pachytene nucleus of the grasshopper* Chorthippus longicornis. *The space between the chromosomes is occupied by elements of the synaptinemal complex, including two dense side elements and a ladder-like central element.* × *56,000. Courtesy of P. B. Moens and The Rockefeller University Press, from* J. Cell Biol. *40:542 (1969).*

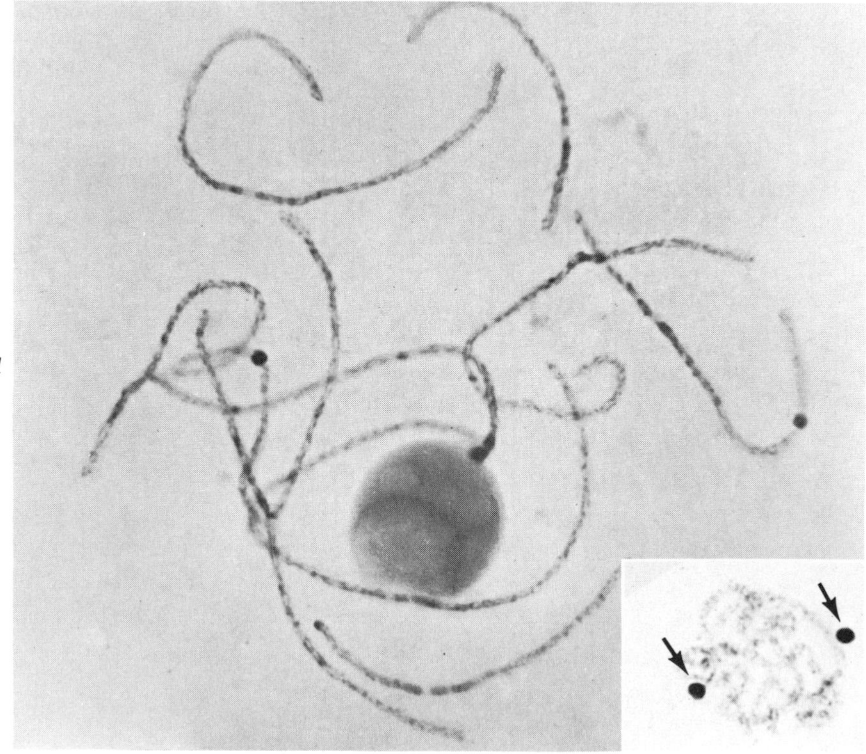

14–6. a, *Pachytene nucleus of corn. The large spherical body is the nucleolus. From* The Cell, *Vol. III, ed. J. Brachet and A. E. Mirsky, 1961. Courtesy of M. M. Rhoades and Academic Press Inc., from a preparation by D. T. Morgan, Jr. Inset, Pachynema in the milkweed bug* Oncopeltus. *The arrows point to the sex chromosomes.* × *1,300. Courtesy of B. John, from S. Wolfe and B. John,* Chromosoma *17:85 (1965).*

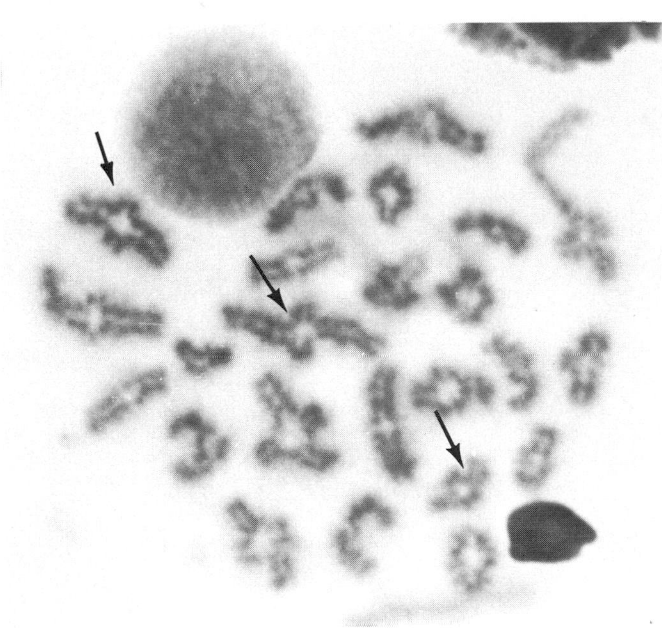

14–7. *Diplonema in the deer mouse* Peromyscus. *Chiasmata are visible at many points (arrows). Slight uncoiling has occurred, giving the chromosomes a "fuzzy" appearance.* × *2,300. Courtesy of A. M. Waterbury.*

into chromatids is not visible under the light microscope at pachynema.

Although the subdivision between chromatids cannot be discerned, the 0.15 to 0.2 micron separation between the paired homologs is maintained throughout pachynema. As mentioned, this regular space separating the homologs is occupied by the synaptinemal complex. Although it has been shown by various methods that the chromatids of homologous chromosome pairs exchange segments or *recombine* during pachynema (see p. 388), no evidence of crossovers or exchanges between the homologs can be seen at this stage in either the light or electron microscopes. This suggests that the exchanges are at the molecular level and do not at this time alter the visible arrangement of the chromosomes.

Diplonema. The end of pachynema and the onset of diplonema (Figs. 14–2d and 14–7) are marked by a widening of the separation between the paired chromosomes, sometimes to the extent that the homologs seem to repel each other. At the same time, the subdivision between chromatids becomes visible for the first time. Thus, at diplonema all four chromatids of the tetrad become visible. The relaxation of close pairing at this stage is correlated in the electron microscope image with the disappearance of the synaptinemal complex.

Instead of completely separating, the homologs at diplotene are held together at one or more points spaced along the chromosomes (Figs. 14–2d and 14–7, arrows). On close inspection, these persistent regions of contact prove to be points at which two of the four chromatids of the tetrad cross between the homologs. These crossing-places, called *chiasmata* (singular *chiasma* = crosspiece) undoubtedly represent the end result of exchanges (recombination) between the homologs that occurred previously during pachynema. The number and position of the chiasmata is variable in different individuals. However, there is normally at least one chiasma per tetrad.

Meiotic cells may remain in diplonema for long periods of time. In the human female, oocytes in the fetal ovary reach diplonema by the fifth month of prenatal life and remain arrested in this stage until shortly before ovulation, which may not take place until 12 to 50 years later.

During diplonema the chromosomes in most species uncoil at least to some extent. The degree of uncoiling, which in a few organisms may be so extensive that the appearance of an interphase nucleus is approximated (see Fig. 14–8), is correlated with intensive RNA synthesis and growth of the cell that occurs at this time. In a few species, such as the male grasshopper, uncoiling and RNA synthesis may take place as early as pachytene. This growth is most marked in the diplotene stage of the oocytes of many animal species, including amphibian, avian, and reptilian eggs. These may reach very large size during diplonema. The yolk of a hen's egg, for example, is an oocyte; growth from microscopic dimensions to this size takes place during the diplotene stage of meiosis.

In the oocytes of many species, including amphibians, birds, and sharks, the diplotene tetrad may form a peculiar structure known as the "lampbrush chromosome" (see Fig. 14–38). When in this configuration, much of the chromatin uncoils into a series of loops that extend into the nucleoplasm from the axis of the tetrad. Intensive synthesis of RNA takes place on the loops during this stage, which in amphibian oocytes may last for about one year.

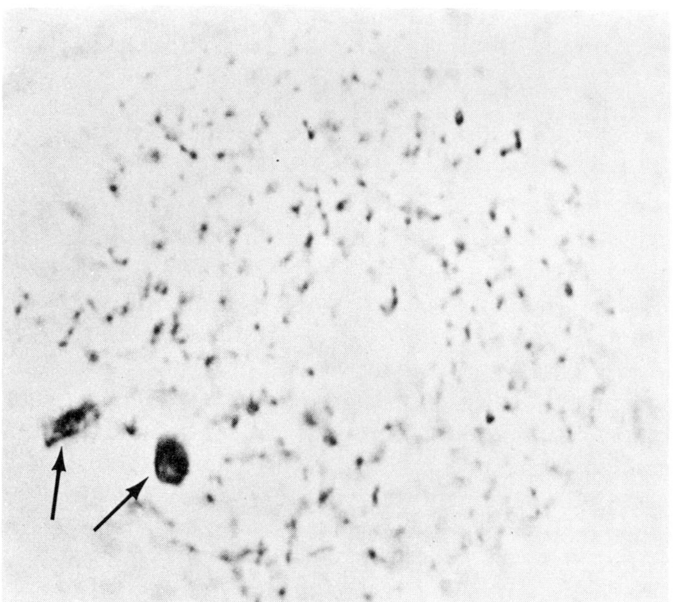

14–8. Diplotene nucleus of the milkweed bug. Uncoiling of the chromosomes has progressed to an interphase-like state. The sex chromosomes (arrows) remain tightly coiled. × 2,500. Courtesy of B. John, from S. Wolfe and B. John, Chromosoma *17:85 (1965).*

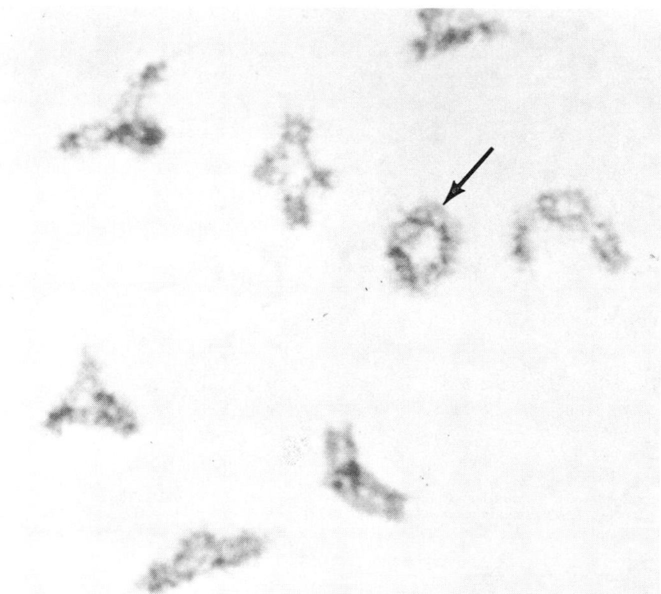

14–9. Diakinesis in the milkweed bug. The arrow points to a ring formed by a tetrad with chiasmata at the tips of the chromosomes. × 2,200. Courtesy of B. John, from S. Wolfe and B. John, Chromosoma *17:85 (1965).*

Diakinesis. Toward the end of diplonema the chromosomes, if uncoiled, return to the tightly coiled state. The transition from diplonema to diakinesis (*dia* = across; Figs. 14–2e and 14–9) is subtle, and it is not always easy to determine which of the stages is present. The chromosomes reach their greatest prophase contraction in diakinesis, in some cases becoming almost spherical. At this stage in many species the nucleolus disappears, and the tetrads become more or less evenly distributed throughout the nucleus. As diakinesis progresses, the chiasmata may move toward the ends of the chromosome arms. This process, called *terminalization,* continues until many of the tetrads are held together only at the tips of the chromosomes (arrow, Fig. 14–9).

Metaphase I of Meiosis

The beginning of metaphase I of meiosis (Fig. 14–2f and 14–10) is signaled by the breakdown of the nuclear envelope, just as in mitosis. Before this, the centrioles and asters, if present, divide and migrate to points at opposite sides of the nucleus. Formation of the spindle, whether centrioles are present or not, is the same as in mitosis (see Chapter 13). Following formation of the spindle, the tetrads, which are left scattered in the region of the cell formerly occupied by the nucleus, move to the equator of the spindle. The basis for this movement, called *congression,* is unknown.

At the equator of the spindle, the centromeres of each tetrad make attachments to microtubules of the spindle. This attachment is made in a fundamentally different way from metaphase of mitosis and provides the basis for the segregation of the homologs into different daughter cells after the first meiotic division. Recent electron micrographs taken by Peter Luykx have shown that each homolog has two centromeres corresponding to the two chromatids of the replicated chromosome (see Fig. 14–12). Thus, four centromeres are present in the tetrad. But even though two centromeres are present in each homolog, Luykx's photographs show that these make connections to only one

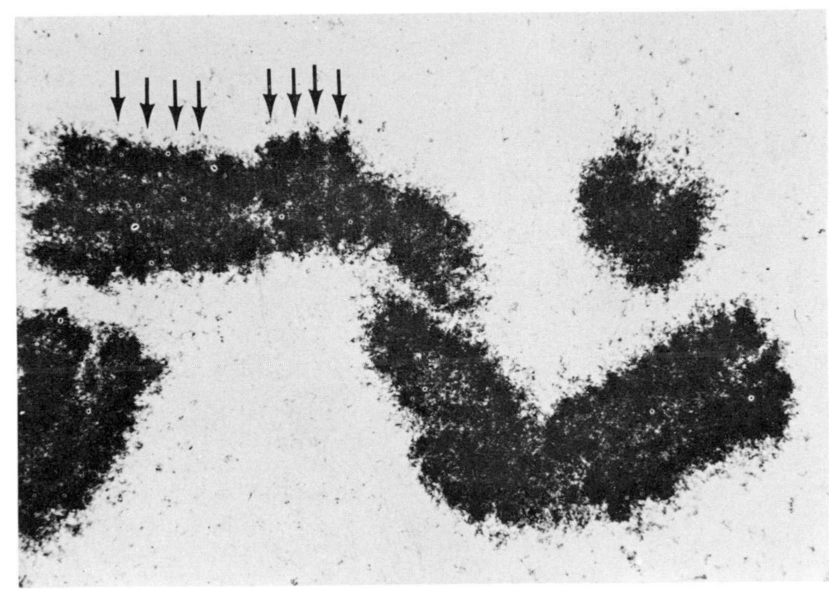

14–10. *Chromosomes of the plant* Haemanthus *at metaphase I of meiosis. Coiling of the arms into chromomeres (arrows) is obvious. Courtesy of A. Bajer.*

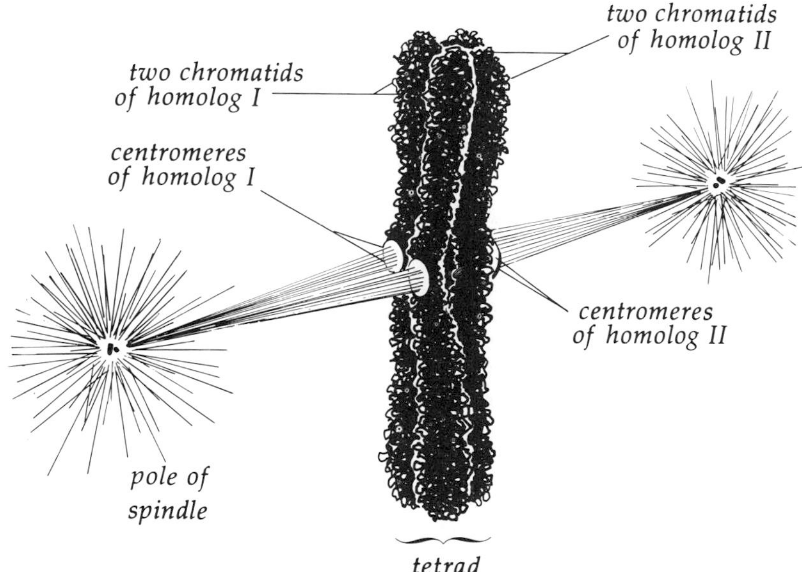

14–11. *Diagram of the tetrad at metaphase I of meiosis, showing the orientation and spindle fiber connections of the centromeres.*

pole of the spindle (see Fig. 14–11). The other homolog of a pair makes attachments to spindle microtubules leading to the opposite pole. Thus the two centromeres of each homolog act as a functional unit at metaphase I of meiosis. Luykx found that the two centromeres of a homolog may initially make connections to both poles of the spindle early in metaphase I, but the "incorrect" attachments are somehow broken by the completion of metaphase.

Anaphase and Telophase I

As a consequence of the connections made by the centromeres at metaphase I, the homologs are separated at anaphase I and pass to opposite poles of the spindle (Fig. 14–2g). As the homologs separate, any remaining chiasmata are pulled apart. At the completion of anaphase I, each pole contains only one of the two chromosomes of each homologous pair. Thus in terms of *chromosomes* the poles now contain the *haploid* number. But because each chromosome is double, both the 2X number of *chromatids* and the 2X amount of DNA are present at each pole.

At telophase I of meiosis the chromosomes may either partially revert to an interphase condition or may continue directly into the second division of meiosis without uncoiling. Wide variation between species exists in the extent and development of telophase I and interphase II. In most organisms, a transitory nuclear envelope is formed around the polar masses of chromosomes during telophase I for at least a brief period of time. For example, in grasshoppers, and the plants *Zea mays* (corn) and *Tradescantia* the chromosomes partly uncoil during telophase I, and a nuclear envelope is formed. This condition is rare in plants but is common among insects, annelids, and some vertebrates. In a few species the chromosomes uncoil so completely that they cannot be distinguished at late telophase I. This condition occurs in a number of amphibians. But in the polar divisions of most animal eggs and in spermatogenesis in the roundworm *Ascaris* and numerous insects, no nuclear envelope is formed during telophase I, and little or no perceptible uncoiling of the chromosomes occurs. Thus all gradations between complete reversion to an interphase state and no visible uncoiling are found in various species in nature.

The Second Meiotic Division

The mechanics of the second meiotic division are essentially the same as in ordinary mitosis but with two significant differences. There is no replication of chromosomes between the first and second meiotic divisions, and the chromosomes are present in the haploid number. The interphase between meiotic divisions I and II may be very short or nonexistent, depending on the degree of uncoiling of the chromosomes at telophase I. Likewise, the duration and extent of prophase II depends on whether significant uncoiling occurs in telophase I. In organisms in which little or no reversion to an interphase state takes place during telophase I, the chromosomes may pass almost directly from telophase I to metaphase II.

At the close of prophase II, the nuclear envelopes surrounding the two groups of chromosomes separated by the first meiotic division break down (if nuclear envelopes have indeed formed). Ordinarily no nucleolus appears during interphase or prophase of the second meiotic division. At metaphase II, two spindles are formed anew in the areas corresponding to the two poles of telophase I. The alignment of the chromosomes on these spindles (Fig. 14–2j) is precisely as in metaphase of mitosis. Each chromosome, consisting of two chromatids, attaches to both poles of the spindle. Luykx's micrographs have demonstrated that as these attachments are made the two centromeres of each chromosome, previously side-by-side at telophase I, separate and rotate to face the opposite poles of the spindle (Figs. 14–12a and b).

At anaphase II, the two chromatids of each chromosome are separated and move to opposite poles of the spindle. At the completion of anaphase II, each pole contains the haploid number of chromatids. The DNA quantity in these polar masses of chromosomes corresponds to one-fourth of the amount present at premeiotic G_2. At telophase II, nuclear envelopes form around the four division products, and the distinct outline of the chromosomes is lost. In animals, the four products are segregated into separate cells by furrowing and in most higher plants by the formation of cell walls. The development of these haploid cells into functional gametes (termed *gametogenesis*) is complex and varies not only between species but also between males and females of the same species (gametogenesis is discussed in detail in Chapter 16).

The Time and Place of Meiosis

Within the same species, the time course of meiosis is much longer than that of mitosis. In cells of the lily, for example, mitosis usually takes 24 hours. Meiosis in the same organism takes about two weeks. Most of the difference in the duration of the dividing stages is in meiotic prophase; in most plants and animals the stages after metaphase I pass in quick succession. Prophase I of meiosis, depending on the species, may last from a few days to a period of years. During prophase I, recombination of alleles and extensive growth of the gametic cells takes place. The extended length of meiotic prophase is probably due in large part to these two activities.

Edmund B. Wilson (1925) divided sexually reproducing organisms into three major classes in terms of the place in the life cycle occupied by meiosis (Fig. 14–13). In animals, including all multicellular forms and many protozoa, and a few lower plants, such as the rock kelp and diatoms, meiosis takes place in the last few divisions immediately before formation of the gametes

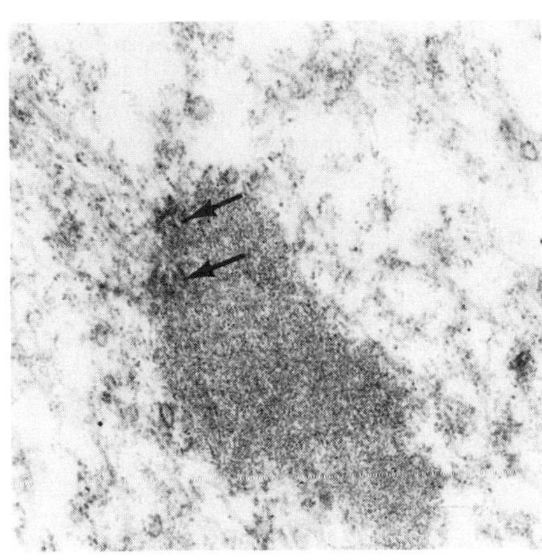

a

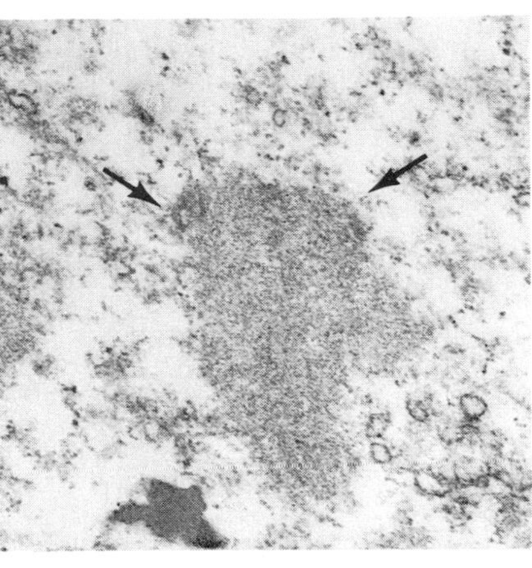

b

14–12. *Differences in the attachments of the centromeres to the spindle in the first and second meiotic divisions.* a, *During the first meiotic division in* Urechis, *both centromeres (arrows) of the homolog are directed toward the same pole.* b, *In the second division in* Urechis, *the two centromeres are directed toward opposite poles.* × *30,000. Courtesy of P. Luykx, from* Exptl. Cell Res. *39:658 (1965).*

(Fig. 14–13a). Wilson terms this sequence of events *gametic* or *terminal* meiosis; in this process, the gametes are typically formed in groups of four. In the male, all four products are functional. In the female (see Chapter 15) usually only one product of gametic meiosis is functional.

Sporic or *intermediate* meiosis (Fig. 14–13b) is found in all higher plants and some lower plants but is not known to occur in animals. In this form, typical of organisms that alternate between haploid and diploid generations, meiosis occurs at some intermediate time between fertilization and the formation of gametes. Fertilization produces a diploid generation, which grows into the sporophytic generation of the plant by mitosis. At some point meiosis occurs, producing spores rather than gametes. The haploid spores, on germination, grow into functional plants that form the alternate haploid gametophytic generation. These haploid plants are individually diversified genetically and represent the variability arising from recombination and segregation of homologous pairs during the first of the two meiotic divisions. At maturity, the haploid generation produces gametes by differentiation of cells following ordinary mitosis. All gametes from a single haploid gametophytic plant are therefore genetically identical. On fusion of gametes, the diploid, sporophytic generation is repeated. Among various plants, either the sporophytic or gametophytic phase of the life cycle may be most conspicuous. In the angiosperms and gymnosperms, the gametophyte is limited to a microscopic plant parasitic on the sporophyte, which lives for only a brief period. In a few algae, the gametophytic and sporophytic generations are closely similar in morphological appearance, but, in general, the two generations are usually conspicuously different.

A third type of meiosis occurs in fungi, some algae and diatoms, and a few protozoa. In this form, termed *zygotic* or *initial* meiosis (Fig. 14–13c) the meiotic divisions take place immediately after fertilization. Thus, the diploid phase is limited only to the fertilized egg and lasts for only a brief period of time. This form of division has been extremely important in the development of contemporary hypotheses of recombination, because in one fungal organism, it has been discovered that DNA replication takes place in the haploid, gametic form before the diploid zygote appears. Therefore, in this species a clean separation between replication and recombination has been possible. This has great significance for the possible processes by which recombination may occur.

Synthesis of Nucleic Acids and Protein during Meiosis

RNA synthesis has been detected during one or more stages of prophase I in all organisms that have been studied. This has been most clearly demonstrated by autoradiography, which has shown that radioactive precursors of RNA are incorporated throughout prophase. As the cells reach metaphase, this incorporation drops to undetectable levels (for representative studies of RNA synthesis during meiotic prophase, see Taylor, 1959, and Monesi, 1964). During zygonema and pachynema RNA synthesis can be detected by autoradiography over the entire chromosomal complement. Later, during diplonema, RNA synthesis may become concentrated in the nucleolus, indicating that a shift

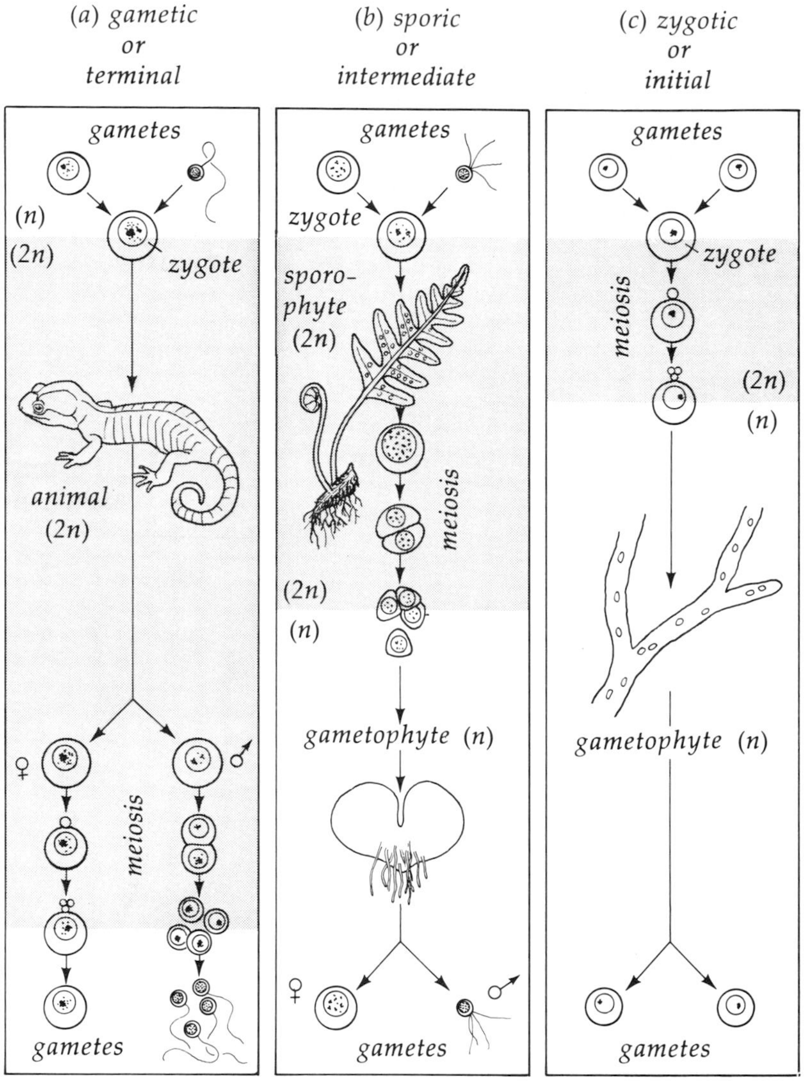

14–13. The three types of meiosis, according to Wilson (see text). The shaded portion marks the diploid phase of the life cycle. Redrawn with permission of the Macmillan Company from The Cell in Development and Heredity *by E. B. Wilson. Copyright 1925, The Macmillan Company, renewed 1953 by Anne M. K. Wilson.*

to transcription of rRNA takes place late in prophase in these species. This is particularly evident in amphibians, in which a great number of accessory nucleoli are produced at diplonema, evidently by extensive replication and release of DNA at the nucleolar organizer segment of the chromosomes (see the section on lampbrush chromosomes later in this chapter). The extra DNA associated with these nucleoli is synthesized during pachynema.

Limited DNA synthesis has also been found to occur in the lily during zygotene and pachytene. This additional DNA synthesis, recently discovered by Herbert Stern and Yasuo Hotta, appears to be clearly correlated with the phenomena of pairing and recombination during prophase I. The amount of DNA synthesized is relatively small, amounting to only 0.3–0.4 percent of the total genome. Interruption of this synthesis with inhibitors causes aberrations during meiosis, such as failure of the homologs to pair or to separate at the subsequent metaphase after pairing.

The incorporation of radioactive amino acids into protein can be detected throughout meiosis. Stern and his colleagues (see Parchman and Stern, 1969) have shown that inhibition of protein synthesis by agents such as cycloheximide causes meiotic arrest and various abnormalities depending on the stage at which cells are exposed to the inhibitor. For example, formation of chiasmata can be blocked if inhibitors are added to cells in zygotene or early pachytene. Because inhibitors of DNA synthesis also have the same effect if added at

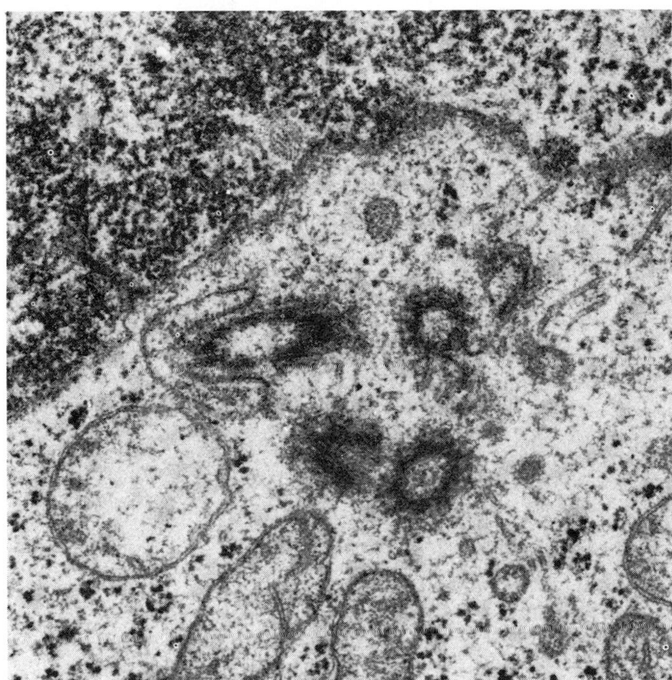

14–14. *Prophase cell of a* Locusta *spermatocyte showing the four centrioles in position just outside the nuclear envelope. × 32,000. Courtesy of P. B. Moens and the publisher, from* Chromosoma, *28:1 (1969).*

these stages, it is apparent that synthesis of both protein and DNA are necessary for the normal course of meiosis to continue. The identity of the proteins synthesized during meiosis are unknown. One new histone, identifiable by a difference in its rate of migration in an electric field, appears during meiosis in some plants (see Sheridan and Stern, 1967). The significance and function of this histone, which is found at all stages of meiosis, is unknown.

The Centrioles in Meiosis

At early prophase, in cells containing centrioles and asters, four centrioles are present in an undivided aster in the cytoplasm just outside the nuclear envelope (Fig. 14–14). Later in prophase, not long before breakdown of the nuclear envelope, the aster divides and two of the four centrioles migrate through the cytoplasm to the opposite side of the nucleus. At metaphase I, each pole of the spindle contains two centrioles. Subsequently, during prophase II, these separate to form the poles of the spindles for the second meiotic division. Each of the four cells formed as a product of the second division thus receives one centriole. However, in many animals two centrioles are present in the mature male gamete (see Chapter 15). Therefore, at some time after the second division, a replication of the centrioles must occur in these males. The precise time and place of this replication has not been detected. In the eggs of protozoa, copepods, insects, tunicates, *Amphioxus,* and many vertebrates, no asters or centrioles have been detected at any time during the meiotic divisions of the oocyte. The spindle in these forms is barrel-shaped or slightly spread rather than pointed at the poles. Thus, in structure the spindle in these anastral oocytes is similar to the spindle of higher plants.

The Sex Chromosomes during Meiosis

In many plants and animals, the males and females of a species may show differences in chromosome structure or number. The chromosomes regularly found to be different between the males and females of a species are called *sex chromosomes.* Many variations exist among different organisms, from species in which no morphological differences can be detected in the chromosomes of either sex to others in which a chromosome present in one sex is totally absent in the other. The sex chromosomes, if present, usually exhibit conspicuous differences in the degree of condensation during meiotic prophase and frequently remain tightly coiled at diplonema when the rest of the chromosomes of the complement (termed the *autosomes*) partially uncoil (see Fig. 14–8). In most species, the female contains a homologous pair of the sex chromosomes, while in the cells of the male only one of this pair is present. By convention, the chromosomes present in a pair in the female are called *X chromosomes.* In the male, the single representative of this pair, the X, may exist alone ("XO" males) or may be associated with another sex chromosome, the Y, which is not found in females. Females with a pair of X chromosomes and XY or XO males are the rule in most animals and many plants. The chromosomes of an XY male are illustrated in Fig. 14–15. In a number of groups, such as the butterflies and moths, copepods, and birds, the situation is reversed, with XX males and XY females.

In the sex with the pair of X chromosomes (called the *homogametic* sex, because all of the gametes produced will receive an X chromosome) the two X chromosomes pair normally at zygotene and pass through

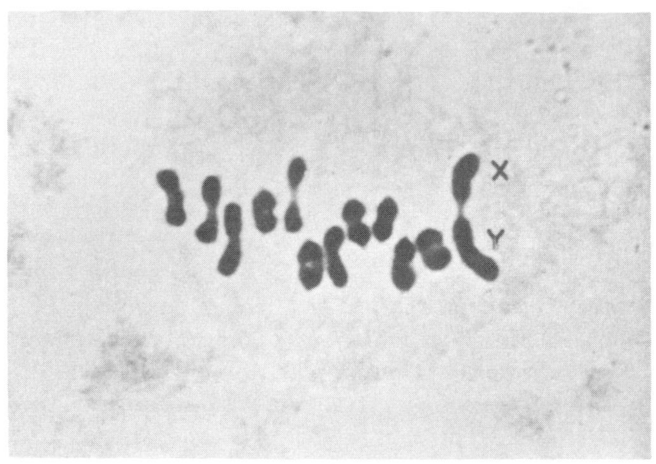

14–15. Sex chromosomes in the nucleus of a plant, Silene. *In* Silene, *male plants are the heterogametic sex.* X *and* Y *indicate the sex chromosomes. Courtesy of H. E. Warmke.* [Am. J. Bot. *33:648 (1946)*]

the meiotic divisions without exhibiting conspicuous differences from the rest of the chromosome complement, except for the noted differences in the degree of condensation. In XY individuals (the *heterogametic* sex) the X and Y chromosomes may or may not contain homologous segments capable of pairing at zygonema. If homologous segments are present, the X and Y pair at zygonema and at metaphase I may either separate as the autosomes do or, in some cases, may form centromere connections similar to mitosis. In the latter situation, both products of the first meiotic division receive an X and Y chromatid. Where no homologous segments exist between the X and the Y, the two may separate at metaphase I, going to opposite poles of the spindle, or may behave as in metaphase of mitosis, with both poles of the spindle receiving an X and a Y chromatid. In all events, the products of the second meiotic division receive either an X or a Y chromosome.

In individuals without a Y chromosome (XO), both chromatids of the single unpaired X chromosome usually pass to the same pole of the spindle at anaphase I. At anaphase II, the nucleus formed from this pole will distribute one X chromatid to each of the two products of the division; these will form X gametes. The other anaphase I nucleus, which receives no X chromosomes, divides normally to produce O gametes.

At fertilization, gametes containing an X or Y (or no Y) fuse to form the diploid zygote, with a resultant XX or XY (or XO) complement of sex chromosomes. The determination of sex following fertilization involves a complex interaction between the genes carried by sex chromosomes and the autosomes. (For a description of the mechanisms of sex determination, which is outside the scope of this book, see Mittwoch, 1967, Lewis and John, 1963, and Swanson et al., 1967.)

Recombination

During prophase I, segments of the chromosomes undergo recombination. As a result, the chromosomes at the end of meiosis are no longer identical in DNA sequence to their somatic counterparts. The change differs from random mutation in that the sequences are altered by *physical exchanges* of segments between homologous chromosomes. These exchanges take the form shown in Fig. 14–16. Consider a homologous pair with the genes *A,a,* and *B,b,* located at two points on one chromosome arm (Fig. 14–16a). One member of the pair has the alleles *A* and *B* at these loci. As a result of replication at premeiotic interphase, both chromatids of the chromosomes will have the *A* and *B* forms of the two genes at these points. The opposite member of the homologous pair has the alleles *a* and *b* at these loci. At this stage, these chromosomes are identical in sequence to chromosomes of the same homologous pair in any cell of the same organism. At zygonema, the homologs are closely paired. If recombination occurs between these loci during pachynema (Fig. 14–16b), two of the four chromatids at the close of meiosis contain the new combinations *A–b* and *a–B* (Fig. 14–16c).

The precise mechanism operating to recombine portions of chromosomes between homologs is unknown. However, the process has several characteristics that can be demonstrated by genetic crosses or biochemical tests. Recombination results from physical exchanges between homologs which, as will be demonstrated, must involve a process of breakage and reunion of chromosome segments. Each exchange event, as shown in Fig. 14–16, involves only two of the four chromatids. As a result, the end products after recombination consist of two chromatids of unchanged sequence (the *parental* chromatids) and two with altered sequence (the *recombinants*). The recombinant chromatids exchange segments reciprocally and precisely, so that no segments are lost in the exchange. In most species, the positions at which an exchange may occur

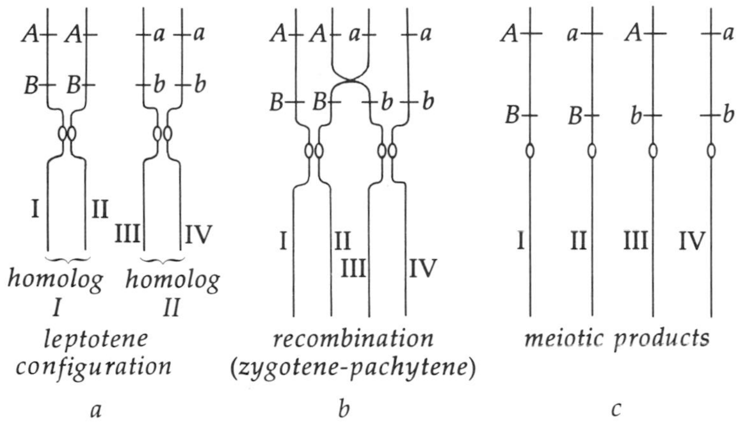

14–16. Presumed mechanism accounting for recombination during meiotic prophase (see text).

are randomly located on any chromosome arm.[3] The appearance of one exchange reduces the probability that another may occur close by on the same chromosome arm. This effect, called *positive interference,* does not extend across the centromere; the two arms of a chromosome on either side of the centromere are independent in this regard.

The precision of the exchange is one of the most astounding facts of cell biology. As one scientist, Charles A. Thomas, Jr., has aptly put it, the precision of exchange is analogous to parting the contents of two volumes and resplicing the four parts together without adding or deleting a single letter from either copy. In order for an exchange to be perfect, splicing between the two copies must take place at precisely the same letter; otherwise one volume would contain too many, and the other volume too few letters after splicing. Recombination proceeds with exactly this precision down to the level of the "letters" of the DNA sequence, the individual nucleotides. The molecular basis of this precision is unknown and remains one of the most challenging and significant problems of biology.

The characteristics of recombination described up to this point represent a level of understanding developed only very recently. Direct evidence that recombination involves breakage and exchange and that the recombination event takes place during zygonema-pachynema has been provided only in very recent years, since 1960. However, recombination has been suspected to take place in this manner by many scientists since the turn of the century, although other mechanisms have been favored in speculation and hypothesis at various times. Because tracing the development of these early hypotheses is vital to the understanding of contemporary models, some time will be spent in describing the early work in recombination.

Early Hypotheses of Recombination

The possibility that homologous chromosomes physically exchange segments at prophase of meiosis was envisaged even before linkage and recombination were first discovered by genetic crosses. In 1909, F. A. Janssens proposed that the chiasmata visible at diplonema arise through an exchange of segments between homologous chromosomes. Janssens called his model the *chiasmatype* hypothesis. Soon after this, in 1911, the American geneticist T. H. Morgan hypothesized that the exchange of chromatin between homologs resulted in recombination of genes positioned on either side of a chiasma. Morgan was also able to conclude from his experiments that the frequency of recombination between any two genes located on the same chromosome is determined by the distance separating them. The greater this distance, the greater the chance of a recombination between the two genes. Another important discovery of the period was the demonstration that each recombination produced two parental and two recombinant chromatids.

In the light of these findings, Janssens' hypothesis was restated in the 1930's by C. D. Darlington as the *partial chiasmatype* hypothesis. According to this hypothesis, recombination occurs by an exchange of segments between homologous chromatids at prophase of meiosis after replication and pairing of the chromosomes has taken place. Each exchange involves

[3] In a few species, such as the plant *Oenothera,* chiasmata are nonrandom and appear only in some localized regions of the chromosomes.

only two of the four chromatids of the synapsed homologs (the word "partial" in "partial chiasmatype" refers to this fact). The exchanges, which take place during zygonema-pachynema, first become visible at diplonema. According to the partial chiasmatype hypothesis, a 1:1 ratio is expected between visible chiasmata and recombination events.

At the time Darlington proposed the modified hypothesis it was believed that the chiasmata, on separation of the homologs at diplonema, are subjected to a strain or torsion that causes breaks to appear in the chromatids at the crossing places. On rehealing, if the broken ends rejoin reciprocally between nonsister chromatids, a recombination takes place. With the advantage of recent insight into DNA structure, it has become obvious that the torsion hypothesis of breakage and exchange is too imprecise to account for reciprocal exchange between homologous DNA molecules. It is now assumed that both breakage and exchange occur during close pairing at zygonema-pachynema before the separation culminating in the diplotene configuration is initiated. Therefore, the chiasmata visible at diplonema are the result and not the cause of breakage and exchange. In its modern form, the partial chiasmatype hypothesis is most often referred to simply as the "breakage and exchange" model. The theoretical relationship of chiasmata to recombination in the model is shown diagrammatically in Fig. 14–17.

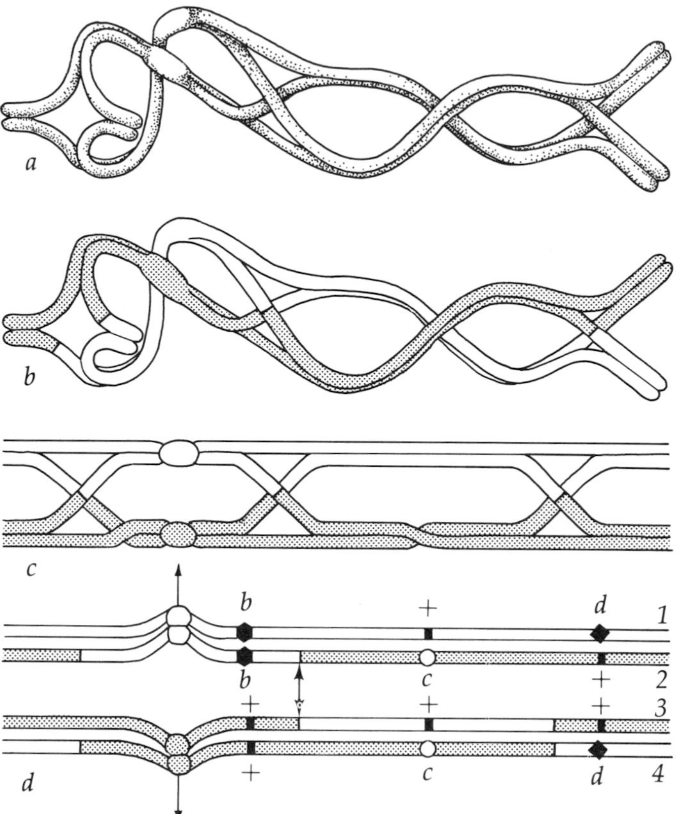

14–17. Chiasmata. a, *A pair of diplotene chromosomes with three chiasmata, as seen in cytological preparations.* b, *Interpretation of the relationship of sister chromatids to chiasmata. Sister chromatids (daughter products of a single homolog) have the same shading; homologous chromatids have different shading patterns.* c, *Similar to* b *but with the twists in the chromosomes straightened out to show the individuality of the chromatids.* d, *Expected genetic consequences of the chiasmata shown in* a *as interpreted in* c, *following terminalization of chiasmata.* b, c, *and* d, *recessive alleles at three loci in the right chromosome arm; +, the wild-type allele of each, distinguished by their positions; arrows are attached to centromeres and indicate segregation to opposite poles in anaphase of the first meiotic division; chromatids 1 and 2 separate in one second-division figure, 3 and 4 in the other, each being recovered in a separate meiotic product. Diagram and caption from* Genetic Organization, *Vol. 1, ed. E. W. Caspari and J. W. Ravin, 1969. Courtesy of S. Emerson and Academic Press Inc.*

Experimental Support for the Breakage and Exchange Hypothesis

Experimental support for the breakage and exchange hypothesis has appeared at various times since its first statement by Darlington. In 1931 two separate investigations, one carried out by Curt Stern and another by Harriett B. Creighton and Barbara McClintock, showed that recombination is correlated with rearrangement of chromatids. Both Stern, working with *Drosophila*, and Creighton and McClintock, working with maize, used mutants in which the two members of one homologous pair could be morphologically distinguished from each other. In both cases, one member of the pair had attached to one end a small portion of genetic material translocated from another chromosome (see Fig. 14–18). The effects of recombination were followed both cytologically and by means of ge-

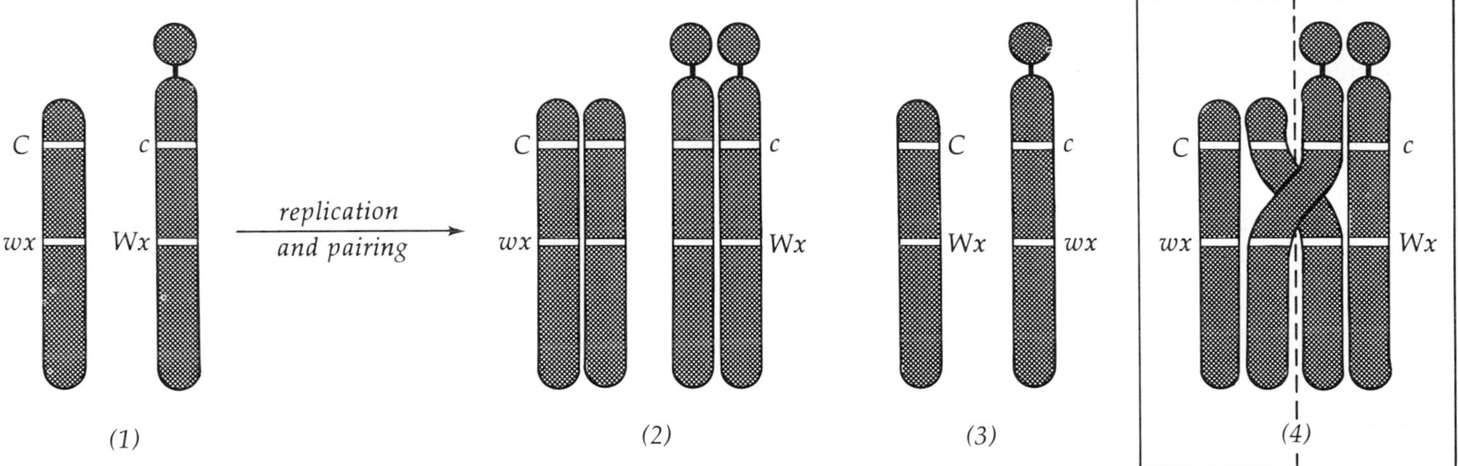

14–18. Diagram of an experiment carried out by Creighton and McClintock in corn, in which recombination was followed both genetically and cytologically. One chromosome of a homologous pair carried a small knob translocated to its tip (1); this chromosome carried the alleles c, Wx. *The other member of the homologous pair, without the knob, carried the alleles* C *and* wx. *After replication and synapsis, the homologous pair appeared as depicted in* (2). *This plant was crossed with another carrying the recessive alleles* c, w *and with completely normal chromosomes (no knob). In the progeny of the cross, knobbed chromosomes were detected with the alleles* c, wx, *and knobless chromosomes carrying the alleles* C, Wx (3). *These could arise only by the exchange event shown diagrammatically in* (4). *The movement of the knob with the* c *allele suggests that a physical exchange occurred during the recombination event. From data reported by H. B. Creighton and B. McClintock,* Proc. Natl. Acad. Sci. *U.S. 17:492 (1931).*

netic markers located on the mutated homologous pair. This work showed that, if a genetic recombination occurred between the centromere and the tip of the mutated chromosome, the recombination was accompanied by the appearance of a chiasma and a transfer of the identifiable tip material between nonsister chromatids of the homologous pair. These experiments support both the breakage and exchange hypothesis and the contention that a 1:1 correspondence exists between chiasmata and recombination. However, because only the end result and not actual act of transfer of material from one chromatid to another could be traced during prophase, the experiments did not prove that the transfer was by breakage and exchange.

Additional experimental evidence for the 1:1 correspondence of visible chiasmata with recombination came in 1932 with the work of George W. Beadle. Beadle studied a segment of a chromosome in maize in which recombination between two loci on one chromosome had been found by genetic crosses to have occurred in 12 percent of the chromatids resulting from meiosis. The same chromosome proved to contain a chiasma in the arm bearing the two loci in 20 percent of the meiotic cells examined. This is very close to the ratio expected for a 1:1 correspondence, because chiasmata in 20 out of 100 cells will produce 10 recombinants and 10 parentals for a total of 10/100 recombinants or 10 percent recombined chromatids (remember that each recombination involves only *two* of the four chromatids). The close correlation obtained by Beadle has been confirmed more recently by others. Although a perfect correspondence has never been obtained, it has been generally accepted that a visible chiasma is equivalent to a single recombination event.

The availability of radioactive or heavy isotopes has made possible the demonstration that recombination involves physical breakage and reunion of chromosomes. A number of definitive experiments have been carried out with these isotopes, both in prokaryotes and higher organisms. While meiosis does not occur in prokaryotes, recombination between parental types can be readily detected. Because no differences in the end result of recombination are detected between prokaryotes and eukaryotes, the processes involved at the molecular level are believed to be closely similar. In independent experiments in 1961, Matthew Meselson and Jean J. Weigle, and G. Kellenberger followed the recombination of DNA in bacteriophage. Bacteria were infected with an excess of phage particles of two parental types in numbers large enough to insure that many cells were likely to be infected by more than one phage particle. The two parental phage types carried different mutations, which could be separately identified by the characteristics of the colonies produced after infection of bacteria. One of the two parental types carried DNA labeled by a heavy isotope in addition to the identifying mutation. In some of the multiply infected cells, recombination took place between DNA originating from the two parental phage types. These could be isolated from colonies exhibiting new combinations of the parental mutations. In these recombinants, Meselson, Weigle, and Kellenberger found that the phage chromosome was a hybrid molecule consisting of segments of labeled and unlabeled DNA. Further, the relative proportions of labeled and unlabeled DNA in the recombinants were found to correspond to the amount of DNA expected to be contributed by each parent depending on the position of the exchange as determined by the genetic "map" of the bacteriophage. Similar experiments have been carried out with bacteria. In some experiments, the recombinant DNA was denatured by heating without destroying the hybrid nature of the DNA. This means, as the hydrogen bonds holding the two substrands of the DNA duplex together are broken by heating, that the hybrid DNA must be joined together by covalent linkages in each substrand of the DNA. These observations are compatible only with the breakage-reunion hypothesis and confirm that segments of the chromosomes are physically exchanged during recombination.

Equivalent experiments were carried out in higher organisms by J. Herbert Taylor, using meiotic cells of the grasshopper *Romalea*. Taylor injected grasshoppers with radioactive DNA precursor and followed the segregation of label in cells of the testis. These include cells entering both meiosis and premeiotic mitosis. The rationale of the Taylor experiments, which establish that breakage and exchange occur during meiosis in eukaryotes, is as follows. Consider a cell at the S period just before the last premeiotic mitosis (Fig. 14–19). Before replication begins, all DNA of the chromosomes is unlabeled (Fig. 14–19a). DNA synthesis then takes place in the presence of the injected pulse label (Fig. 14–19b); as a result of semiconservative replication (see Chapter 12), this cell enters mitotic metaphase with one substrand of each DNA duplex labeled (Fig. 14–19c). Therefore, at this stage, all chromatids will show the presence of radioactive label. After this cell enters the last interphase before meiosis and replicates in unlabeled medium (Fig. 14–19d; the pulse label can be effectively washed out of the tissue), only one of the two chromatids of each chromosome at metaphase of meiosis would be expected to show the presence of label *as a result of replication alone.* Up to this point the results are identical to Taylor's work with mitotic tissue, in which the segregation of label was first detected at the second metaphase following administration of radioactive precursor. If no recombination by breakage and exchange occurs during prophase I of meiosis, any single chromatid would be expected to be either completely labeled or unlabeled (Fig. 14–19e). If recombination occurs by exchanges of segments between labeled and unlabeled chromosomes, it should be possible to detect chromatids after the completion of prophase that contain labeled and unlabeled segments (Fig. 14–19f). Exchanges of this type were actually observed by Taylor (Fig. 14–20), in ratios approaching the expected frequencies. Because these could arise only by breakage and exchange of segments between labeled and unlabeled chromatids, this experiment established that breakage and exchange also occur during meiosis in higher organisms.

The final link in the evidence for the breakage and exchange hypothesis concerns the time at which recombination occurs. Considerable argument has hinged on this facet of recombination. Hewson Swift and Taylor independently demonstrated in the early 1950's that the replication of DNA in meiosis takes place during premeiotic interphase and not during prophase of meiosis, as some investigators had maintained (a very small fraction of the DNA, less than 1 percent, is evidently replicated during meiotic prophase; see p. 386). Is recombination associated with replication of the bulk of DNA during premeiotic interphase, or does it take place after replication during prophase of meiosis, as predicted by the breakage and exchange model? The definitive experiment in support of recombination during meiotic prophase comes from work carried out in 1966 by Jane M. Rossen and M. Westergaard, using a fungus in which zygotic or ini-

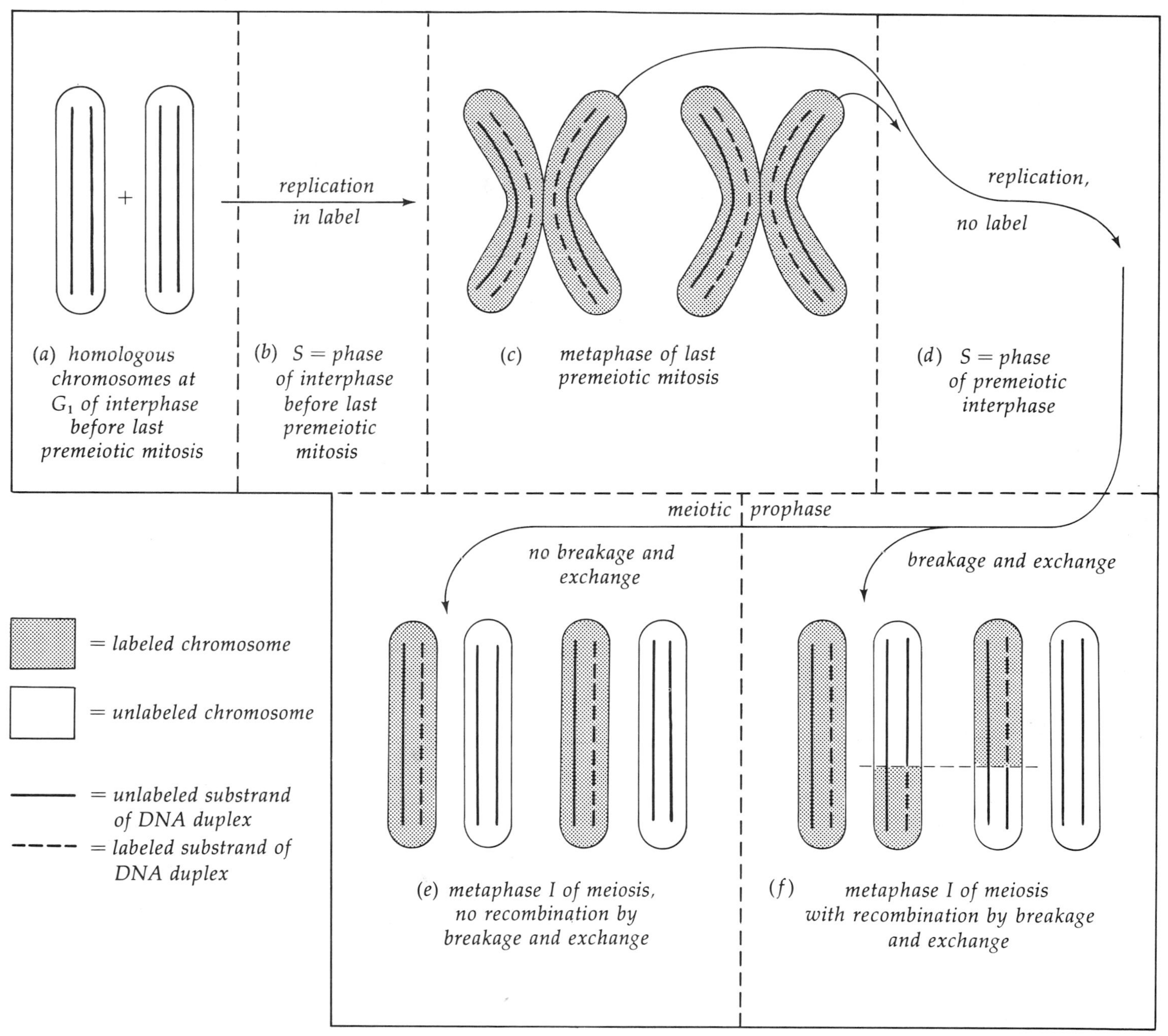

14–19. Diagram of Taylor's experimental results demonstrating recombination by the breakage and exchange mechanism in the grasshopper Romalea *(see text).*

tial meiosis occurs (see p. 385). In this fungus, an ascomycete, nuclei are haploid in the somatic tissues. The haploid gametic nuclei, derived from somatic nuclei by mitosis, fuse to form the zygote. Meiosis then takes place immediately after fusion of the gametic nuclei. The nuclei in this organism are large enough to enable measurements to be made of the relative quantities of DNA per nucleus. Rossen and Westergaard found by measuring the quantity of DNA per nucleus that replication occurred while the nuclei were

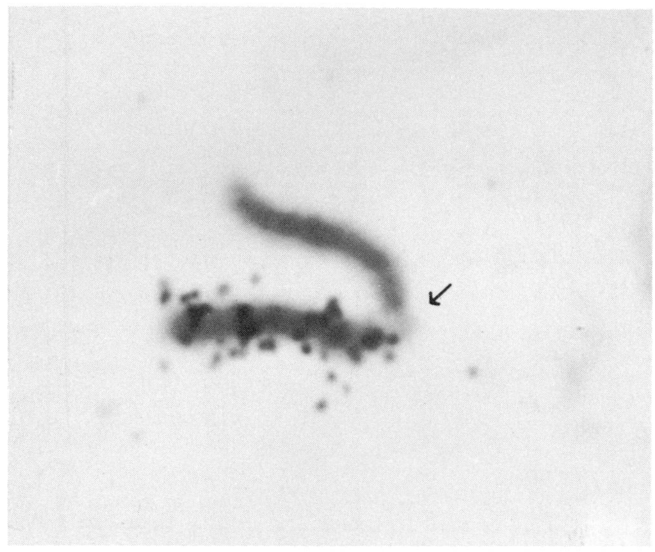

a

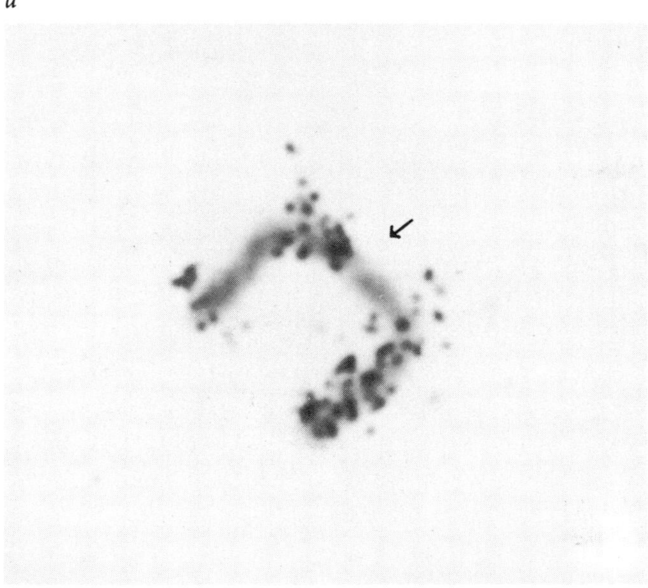

b

14–20. Chromosomes at the second metaphase of meiosis in Romalea. *The prophase I cells from which these metaphase II chromosomes were derived entered prophase with one of the two chromatids labeled. The arrow marks the position of the centromeres, which are terminal (located at the ends of the chromosomes). In* a, *no exchanges have taken place. In* b, *one reciprocal exchange has taken place. Transfer of labeled material from one chromatid to the other in this manner could only take place by breakage and exchange of chromosome segments.* × *4,000. Courtesy of J. H. Taylor and The Rockefeller University Press, from* J. Cell Biol. *25:57 (1965).*

still haploid, before fusion of the nuclei to form the zygote. Because the nuclei are haploid, no recombination could occur at this time. After fusion of the gametes, meiosis occurs immediately, with typical synapsis of homologous pairs during meiotic prophase. Appropriate genetic tests show that recombination of genetic markers occurs in this ascomycete; this could only take place during the zygotic meiosis after the replication of DNA. Thus, in this organism a clear separation is observed between replication at premeiotic interphase and recombination during prophase of meiosis. Equivalent experiments have also been carried out by others in the alga *Spirogyra,* which also exhibits zygotic meiosis.

These experiments, dating from the early 1900's to the present time, support the tenets of the breakage and exchange hypothesis: each recombination event occurs by breakage and reunion of chromatids, involves only two of the four chromatids, takes place at prophase of meiosis after replication of the chromosomes, and is reflected morphologically by a visible chiasma at diplonema. The development of the hypothesis has been presented as if experimentation and speculation in recombination has followed a straight-line path since the early 1900's. This, as the reader might suspect, has not been the case, and other hypotheses have held sway at various times, depending on the weight of contemporary evidence. One alternative hypothesis, the *copy-choice* model, must be given attention here because it was favored by a majority of biologists until quite recently. As will be seen, a considerable body of evidence now eliminates the copy choice model from serious consideration.

The Copy-Choice Model for Recombination

The copy-choice model was originally advanced by John Belling in 1931. Belling proposed that recombination occurs during replication of the chromosomes. In its original form, the Belling hypothesis assumed that both replication and recombination take place during prophase of meiosis, after pairing of homologous chromosomes (Fig. 14–21 a–c). Belling considered that the chromomeres were replicated first as separate units and are later interconnected as replication was completed. The "new" chromomeres were considered to replicate sequentially in a line outside the paired homologs (Fig. 14–21b). If the "old" homo-

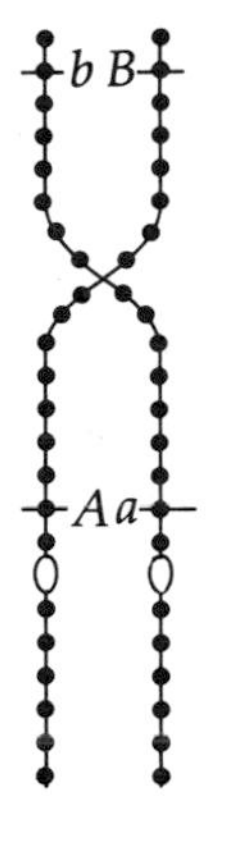

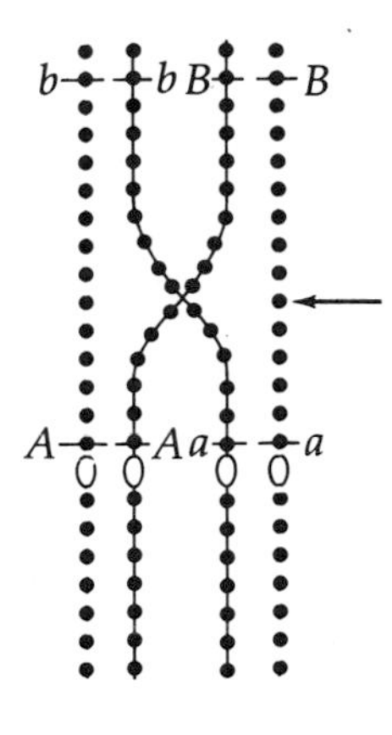

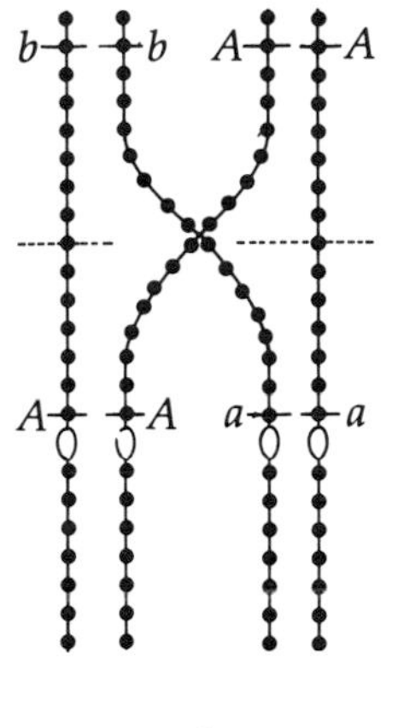

14–21. Recombination according to the copy-choice model, proposed by Belling in 1931. a, *Unreplicated, paired homologs.* b, *Replication of chromomeres.* c, *Interconnection of chromomeres. (see text.)*

logs exchanged sides by twisting at some point (arrow, Fig. 14–21b), the line of replicating chromomeres would use as a template the opposite homolog after the switch point. At the close of replication, when the "new" chromomeres were interconnected into a completed chromosome, the products would consist of two "old" parental chromatids and two "new" chromatids, showing recombined segments wherever the parental chromatids changed sides (Fig. 14–21c). Note that after replication according to this model the resulting chromatids would consist of all "new" or all "old" DNA. The model therefore requires that replication be *conservative* rather than semiconservative (see Fig. 12–16). This requirement for conservative replication, as will be seen, also applies to the modern variants of Belling's model. But in the 1930's, this could not be raised as an objection, and the model satisfied the requirement that for any single recombination event only two of the four chromatids were recombined. The model also accomplished recombination without breakage and exchange of chromatids.

The latter feature was particularly appealing to molecular geneticists in the 1950's, who felt that breakage and exchange in prophase of meiosis was too haphazard a mechanism to account for the precision of recombination. Other very rare recombinational events, such as nonreciprocal recombination (see p. 397), also seemed to be explained most simply by a copy-choice mechanism. As a result, the Belling hypothesis was disinterred and brought up to date during the 1950's and survived until the early 1960's as the favored model for recombination. In its more recent form, the hypothesis states that at the time of replication at interphase the template DNA molecules are in close proximity (Fig. 14–22a). The "new" strands in the process of synthesis may exchange templates at some point during replication (Fig. 14–22b). The end result of a switch in template would be the synthesis of recombined molecules of DNA without breakage and exchange. In Fig. 14–22, replication is indicated as conservative. But if replication is followed through by the semiconservative pathway (Fig. 14–23), recombination according to the copy-choice model cannot occur without tangling the substrands of the replicated DNA duplexes where an exchange occurs (Fig. 14–23c). These "crossover" points, in fact, could be separated *only by breakage and exchange.*

The fact that replication must be conservative for the copy-choice model to operate would also restrict all recombinants to the "new" DNA strands, even if a multiple number of recombination events takes place at successive positions along the chromosome. Multiple exchanges of this type do occur frequently in nature. Where they occur, a second exchange may or may not involve the same chromatids as the first exchange, because at any point either chromatid of a homolog may recombine with the opposite homolog of the pair. This was demonstrated in Belling's day, and, although modifications of the hypothesis could in part account for these multiple exchanges, it was responsible for the eclipse of the copy-choice model until the 1950's.

These difficulties were not enough in themselves to eliminate the model, and advocates of the mechanism felt that the incompatibility of copy-choice with semiconservative replication and multiple exchanges between alternate chromatids could somehow be ex-

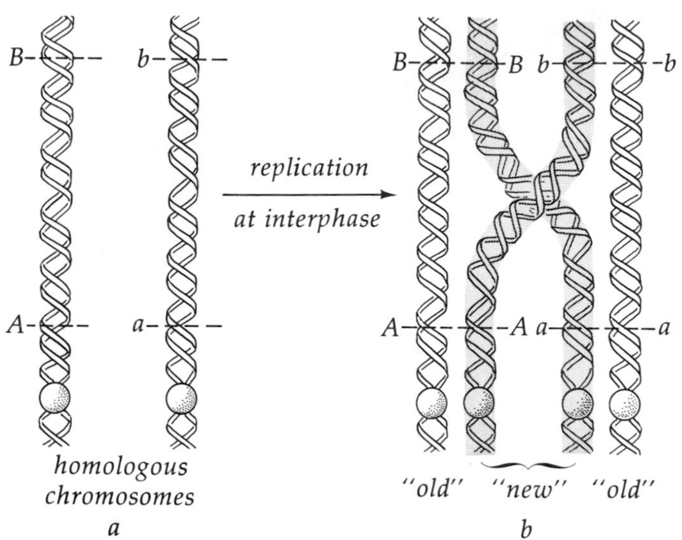

14–22. Copy-choice in terms of DNA helices. Because of the "switch" between templates, the "new" strands will be recombinants. However, the model requires that replication be conservative instead of semiconservative.

plained. But the recent experimental demonstrations that recombination does indeed occur by breakage and exchange (p. 392) and that replication and recombination are separated in time between premeiotic interphase and prophase of meiosis (p. 393) have effectively eliminated the model from serious consideration.

The Development of Molecular Models for Recombination

The detection of intragenic or allelic recombination. As mentioned earlier in this chapter, Morgan concluded in 1911 that the frequency of recombination between two genetic loci is a function of the distance separating them on the chromosome. As the distance decreases, the space available for a random "hit" by a chiasma decreases, and the probability that a recombination event will occur goes down. When very short distances are encountered between loci on a chromosome, as in the case of two genes lying in closely adjacent stretches of DNA, the possibility of detecting recombinants is very low. Chances of detecting these rare events are improved if large numbers of offspring, approaching millions, can be obtained from an organism with a short generation time and a high frequency of recombination. Unfortunately, the eukaryotes used in "classical" genetics, primarily *Drosophila* and maize, have relatively long generation times and produce relatively few offspring. As a result, classical geneticists were unable to detect recombination over very short stretches of DNA and concluded that the gene, which is many nucleotides in length, is the smallest unit of recombination.

But prokaryotes have been found to satisfy the requirements for the detection of very rare recombinational events. In addition, recombination occurs at a higher frequency per cell in prokaryotes than in eukaryotes—about 1000 times higher in phage and 100 times higher in bacteria. A few eukaryotes, such as the fungus *Neurospora,* which undergoes typical meiosis, have also provided offspring in numbers sufficient to permit extensive analysis of very rare recombinational events.

With these organisms, recombination can be detected between points separated by very short distances on the chromosome. These distances are so short that recombination can be identified *within* the boundaries of a gene. In fact, recombination has been detected within distances as short as the length of DNA occupied by a single codon. This type of recombination, termed *intragenic* or *allelic,* may appear in only one out of every 10,000 offspring or at even lower frequencies.

Special Characteristics of Intragenic Recombination

Methods sensitive enough to detect recombination within the limits of a gene have revealed a number of fine structural characteristics of the breakage and exchange mechanism that were unsuspected by classical geneticists. Three phenomena associated with recombination at this level provide clues (and present extra problems) for the possible molecular mechanism of breakage and exchange. Over very short lengths of DNA within a gene, recombination exhibits *negative interference,* is *nonreciprocal,* and is *polarized.*

Negative interference. Negative interference refers to the fact that, within very short lengths of DNA, if one recombination event occurs a second is very likely to occur close by. In eukaryotic cells in which negative interference can be detected, the multiple exchanges differ from multiple recombination events between widely separated alleles in that the same two of the four chromatids are involved in all of the closely spaced multiple exchanges.

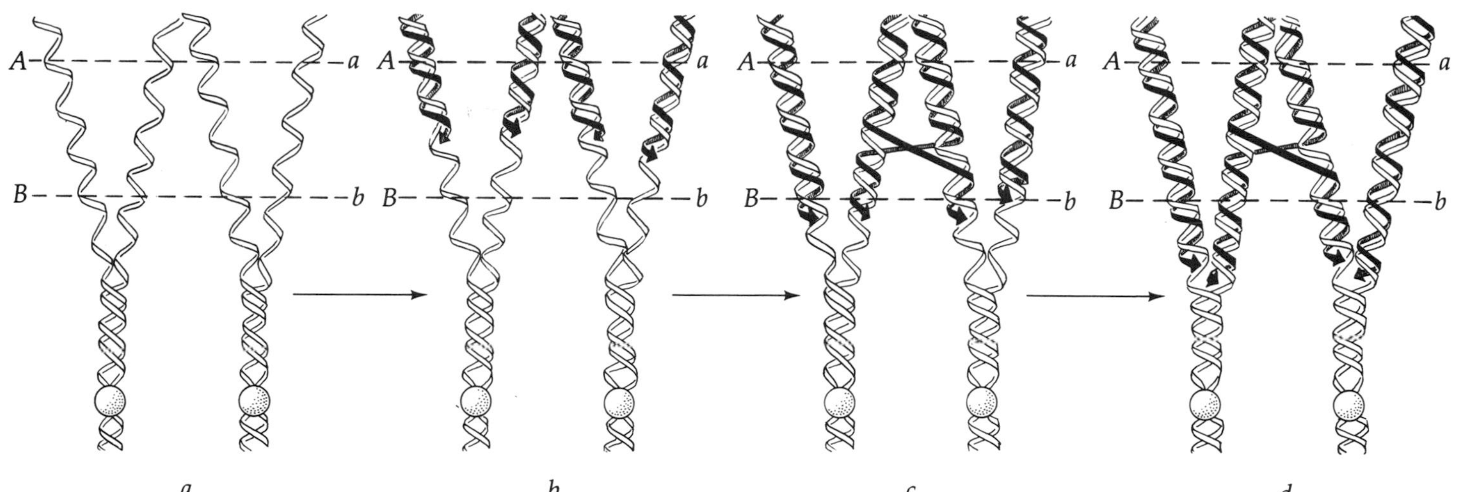

14–23. *Copy-choice in a system replicating semiconservatively.* a, *Homologous chromosomes unwind.* b, *Replication begins.* c, *"Crossover" takes place.* d, *Replication continues; the chromatids are tangled and cannot separate. As a result of semiconservative replication, any recombined chromatids resulting from a switch between templates* (c) *will be tangled. Separation is impossible unless* breakage and exchange *take place at the "crossover" point.*

The basis for negative interference is unknown. Several explanations have been advanced; most of these are based on a suggestion made in 1960 by J. R. Pritchard, who proposed that during pairing the chromosomes at most points are aligned in only approximate register. Close pairing, in the molecular sense, occurs only at random points roughly the length of a gene. Close pairing of this type would be a prerequisite for recombination. Because the probability of a close pairing event would increase as the length of DNA increases, the possibility of *two* such pairing events and the resulting recombination would also increase as the separation between alleles increases. This explains the observations of positive interference from classical genetics. If close pairing does take place over a given region, multiple exchanges between two of the four chromatids can take place with equal probability over the limits of the close pair. This would explain negative interference over short lengths of DNA.

The clustering of exchanges noted in negative interference has been observed in all forms in which methods permit the detection of allelic recombination, including phage, bacteria and fungi, and in a few cases in *Drosophila* and maize. Thus it appears to be a general phenomenon, and it is possible that wherever any recombination event occurs, there is a cluster of multiple exchanges over a short distance of the chromosome. The actual length of the chromosome involved, from crosses carried out in phage, appears to be approximately the limits of a gene. Usually a total of two to three exchanges over this length can be observed if methods of analysis permit their detection.

Nonreciprocal recombination. In exchanges detectable by classical methods, the recombinants are always reciprocal. That is, if the parents have the loci *A,a* and *B,b* located on homologous chromosomes as in Fig. 14–16 and recombination occurs between these loci, the alleles on the four chromosomes at the end of meiosis are expected to be distributed in a 2:2 proportion as follows: two parentals (*A–B* and *a–b*) and two reciprocal recombinants (*A–b* and *a–B*). However, in an exchange between very closely spaced points on the chromosome, recombination is not always reciprocal. Most commonly both of the recombinants have the same allele instead of reciprocal alleles at one locus, giving the combination *A–B*, *A*–**b**, *a*–**b**, and *a*–**b**. This is termed a 3:1 segregation instead of a 2:2 segregation of alleles.

Other unexpected combinations are also observed. Some of the most interesting of these are seen in the ascomycete fungi. In these forms the products of meiosis, the spores, are enclosed in a sac called the *ascus.* In *Neurospora,* the spores are lined up in a row in the

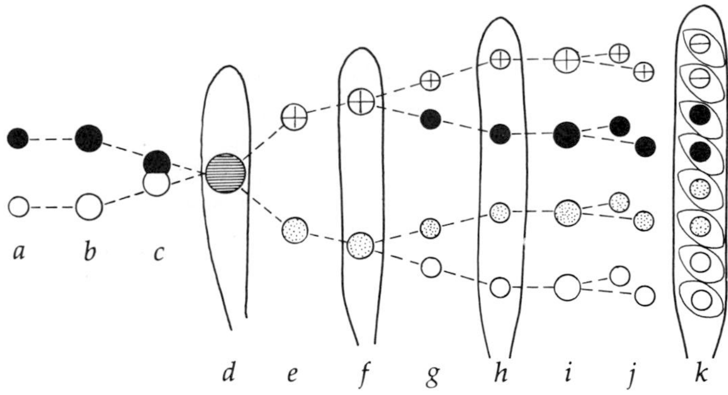

14–24. Meiosis in the ascomycete Neurospora, *an example of zygotic or* initial *meiosis (see Fig. 14–13).* a, *Haploid nuclei of opposite mating types, containing the 1X amount of DNA per nucleus.* b, *The same nuclei after DNA replication, now containing the 2X amount of DNA per nucleus. These nuclei, however, still contain the* haploid *number of chromosomes.* c, *Fusion of gamete nuclei to form the zygote nucleus, which is the only diploid nucleus in the life cycle. As a consequence of the last premeiotic DNA replication, this nucleus contains the 4X amount of DNA. The zygote nucleus is enclosed in the ascus* (d), *an elongated, tubular structure with an inside diameter not much greater than the enclosed nucleus.* e, *Anaphase I of meiosis. The nuclei remain encased in the ascus* (f). g, *Anaphase II of meiosis. The anaphase II nuclei are held in register in the ascus* (h); *because of the wall diameter they cannot slip past each other.* i, *First postmeiotic DNA replication.* j, *First postmeiotic mitosis.* k, *The eight products of the first postmeiotic mitosis enclosed in the ascus. In* Nuerospora, *one additional mitotic division takes place before maturation of the spores. Modified from* Genetic Organization, *Vol. 1, ed. E. W. Caspari and J. W. Ravin, 1969. Courtesy of S. Emerson and Academic Press Inc.*

ascus; the walls of the sac are too narrow to permit the spores to slide past each other. As a result, the products of each meiotic division remain side by side, and the products of each of the four chromatids can be traced and assigned to each spore nucleus (see Fig. 14–24). In *Neurospora,* two mitotic divisions follow meiosis, producing 16 spores in a mature ascus. However, genetic analysis is usually carried out after the first postmeiotic mitosis, using asci containing 8 spores.

At the 8-spore stage, after one postmeiotic mitosis, *reciprocal* recombination would produce 4 spores containing parental types and 4 recombinants. A *nonreciprocal* cross giving a 3:1 segregation after meiosis would produce an ascus containing a 6:2 segregation after one mitotic division. Surprisingly, asci are also found giving 5:3 and 3:5 segregations, indicating that some haploid nuclei at the end of meiosis are *hybrid* and show segregation of alleles after the first postmeiotic mitosis (no segregation is noted after the second mitotic division following meiosis). Hybrid nuclei are unexpected at the close of meiosis, because the cells are haploid and each chromatid is represented only once. This suggests that, over very short regions, the two substrands of the DNA duplex of the hybrid chromatids are not complementary and contain sequences originating from opposite homologs (Fig. 14–25). During the first mitotic interphase following meiosis these noncomplementary substrands would separate (Fig. 14–25b) and serve as templates for replication. Because they contain noncomplementary sequences over short lengths, the resulting chromatids would be slightly different in sequence (Fig. 14–25c). If the hybrid regions contained a genetic marker included in a cross, these differences in sequence would be recognized as nonreciprocal segregation, giving a 5:3 or 3:5 distribution of alleles at the 8-spore stage.

Nonreciprocal recombination occurs without deletions or additions in any recombinant or parental chromatids, so that each chromosome is fully functional and the precision of breakage and exchange is preserved. This phenomenon, at first so baffling and apparently contradictory, has been successfully explained by several models based on the breakage and exchange hypothesis (see below).

Polarization. Polarization means simply that exchanges are more likely to take place at some points along a DNA molecule than at others. These points of favorable exchange are so closely spaced that detection is impossible unless methods permit comparison of the relative rates of recombination between very closely spaced locations within a gene. The points at which nonreciprocal recombination occur are also found to be polarized; that is, nonreciprocal recombination is more likely to take place in some lengths of the DNA of a gene than in others.

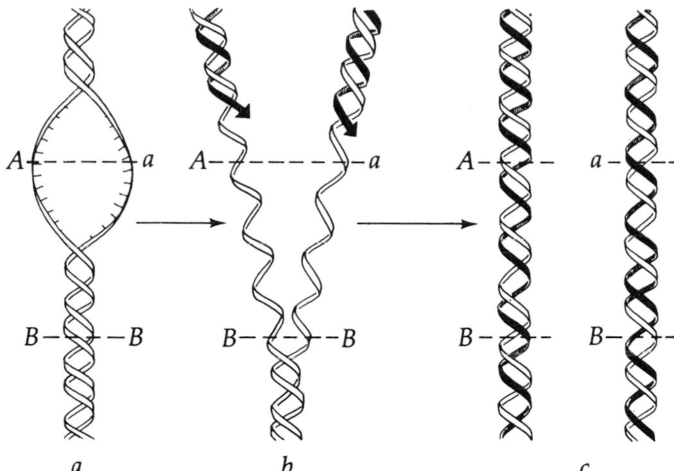

14–25. *Hypothesis accounting for segregation following the first postmeiotic mitosis in* Neurospora. a, *A segment of the DNA. Over a short region including the* A *locus, the two strands of the duplex are noncomplementary and have originated from opposite homologs. One strand contains either the "sense" or "missense" sequence coding for the* A *allele; the opposite strand has the sequence for the* a *allele. Because these are noncomplementary sequences, the structure of the double helix is disturbed at this region.* b, *Unwinding of the strands during the first postmeiotic replication.* c, *Completion of replication. The noncomplementary "sense" and "missense" strands of the* A *locus have served as templates. The resulting daughter duplexes carry different alleles at the* A *locus.*

Molecular Models for Recombination

Negative interference, nonreciprocal recombination, and polarization have been successfully explained by a number of models. Although these models differ in detail, all have in common some basic assumptions about the breakage and exchange mechanism. Out of these models, proposed independently in the early 1960's by Matthew Meselson, J. H. Taylor, Robin Holliday, and H. L. K. Whitehouse, a composite scheme has been synthesized that at present enjoys wide favor among cell biologists.

The composite model (Fig. 14–26) assumes that two of the four chromatids (or two DNA molecules of opposite parentage in the case of prokaryote recombination) approach and pair in such a way that regions carrying different alleles are in close register (Fig. 14–26a). The pairing at this stage is either a precise base-by-base alignment or closely approximates this condition. Next, openings are made by deoxyribonucleases in each of the four substrands of the two chromatids (Fig. 14–26b). Note that the single-strand openings, called "nicks," do not have to be in exactly corresponding locations in the homologous DNA substrands. Exact correspondence is not necessary for operation of the model and indeed permits explanation of nonreciprocity. Following the opening of single-stranded nicks, the two DNA molecules unwind to expose four lengths of single-stranded DNA. Each strand has a complementary mate in one of the two single-stranded regions liberated from the DNA molecule of the opposite homolog. While the molecules are in the broken state, one or more of the exposed single-stranded ends may be "nibbled" or shortened by other deoxyribonucleases. The complementary strands from opposite chromatids then pair by precise alignment of complementary nucleotides and wind into a double helix (Fig. 14–26d). When this happens, a length of *hybrid* DNA is produced (lengths between the brackets in Fig. 14–26d). These hybrid (also called *heteroduplex*) regions therefore consist of a "sense" and "missense" strand originating from opposite parents and may not be completely complementary. At these regions of jointure between the newly paired strands, gaps or overlaps are produced because of the inexact location of the single-strand nicks produced in step 14-26b. The overlaps are then excised and the gaps filled by repair synthesis. Finally, the substrands of the recombined molecules are covalently linked by ligases (arrows, Fig. 14–26e).

The lack of precise register in "nicking" will account for nonreciprocal recombination over short lengths of the rejoined DNA. As noted, the regions of jointure produced by rewinding strands originating from opposite homologs (the hybrid regions between the brackets in 14–26d) will consist of the sense or missense strands of different alleles. These regions may contribute to nonreciprocal exchange. Synthesis of new DNA to fill in gaps left by the imprecision in nicking and any nibbling that takes place will also contribute to nonreciprocal recombination. Fig. 14–27 shows one possible result of hybrid overlap, excision, and repair synthesis, and illustrates how the nonreciprocal ratios can be successfully explained by the model. In order to understand this figure, simply regard the solid lines

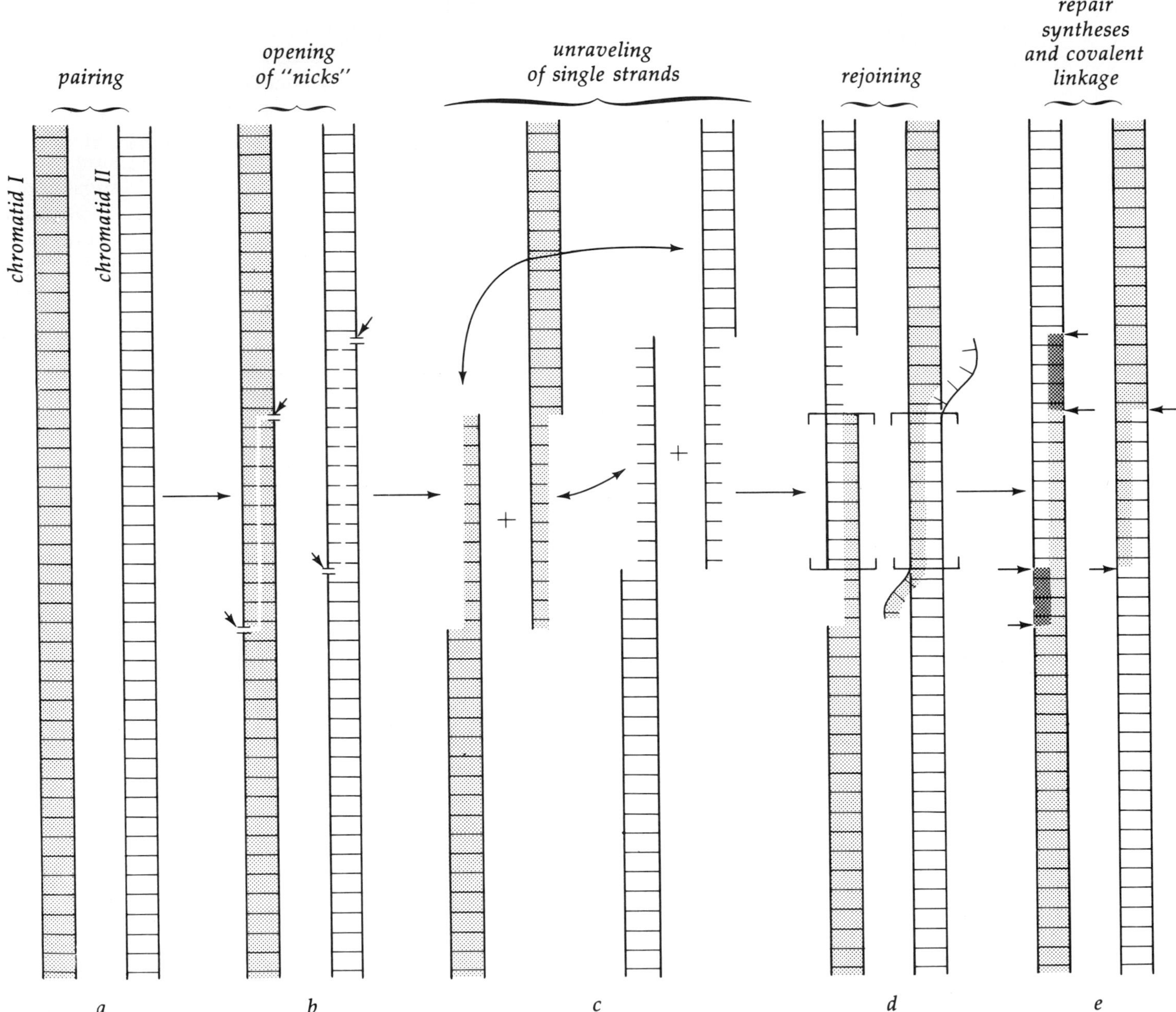

14–26. A composite model successfully explaining intragenic recombination (see text). Modified from C. A. Thomas, Jr., Prog. Nucleic Acid Res. Mol. Biol. *5:315 (1966).*

as DNA substrands carrying a dominant allele + and the dashed lines as carrying the alternate allele –. Recombination has occurred between the two middle DNA duplexes designated as "recombinant" types; the outside duplexes are nonrecombinant parental molecules. Breakage and exchange has taken place according to the outlined molecular model. The darkly shaded areas indicate regions where parent DNA, originally carrying a sequence for the – allele, was "nibbled" away; repair synthesis has taken place in these

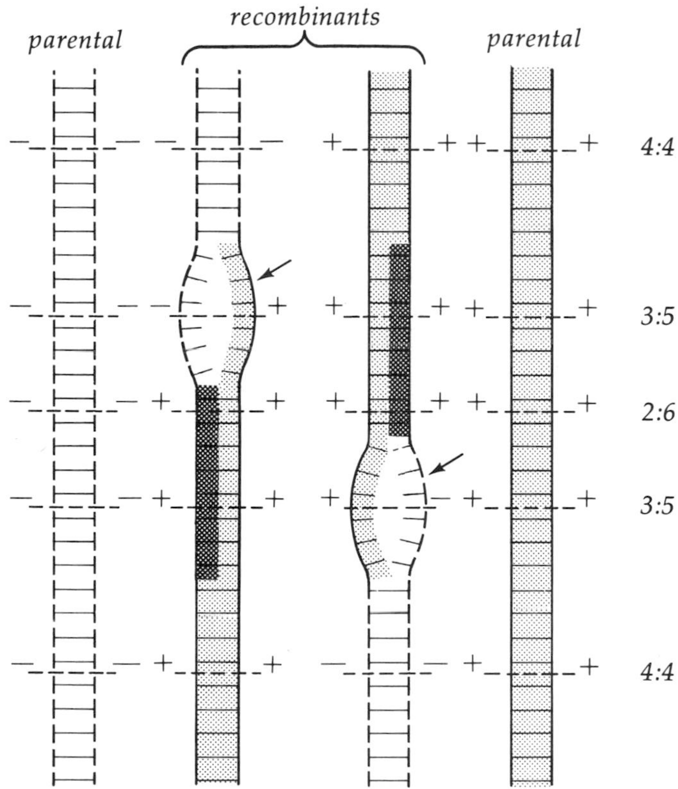

14–27. *Explanation of the origin of nonreciprocal recombination (see text). Adapted from M. Meselson, in R. A. Brink, ed.,* Heritage from Mendel. *Madison: University of Wisconsin Press, 1967, p. 81, by permission of the author.*

regions, using the + sequence as template. At two points, indicated by arrows, a short "hybrid" overlap region, containing noncomplementary sequences originating from opposite homologs, survived without excision or repair. By counting across the strands of the duplexes and scoring the totals of the dashed and solid lines, the relative proportions of the + and − alleles can be obtained; these give the 4:4, 5:3, 6:2, 5:3, and 4:4 ratios indicated. At the first mitosis following this recombination event, each substrand will serve as a template for replication. The end result will be the production of eight *DNA duplexes*, or chromatids, with the ratio of + and − alleles indicated by counting across *substrands* at the immediate postmeiotic stage illustrated in Fig. 14–27.

Although the model explains the observations of allelic recombination, many problems surround the details. The basis for the initial pairing remains unknown. Although the attraction between complementary bases can account for the required precision of pairing in breakage and exchange, such pairing, as far as is known, is possible only between *single-stranded* segments of DNA. How is the initial pairing between intact double helices accomplished (equivalent to the situation in Fig. 14–26a)? No known molecular basis exists for a specific base-by-base attraction between intact double helices that have not undergone unwinding. It is possible that short segments of DNA undergo localized separation of the two substrands and that initial pairing occurs by alignment of these regions. Thus alignment may precede instead of follow breakage as shown in the model. However, mechanical problems exist in this event, because for base-by-base pairing to occur DNA substrands must wind about each other into a double helix. Unless breaks are first introduced into the parent strands, unwinding from one helical arrangement in homologs and rewinding into a hybrid helix in a recombinant is impossible. This paradox has not yet been answered by a satisfactory model.

After pairing, either precise or approximate, is established, a second related problem is presented by the location of the initial "nicks." One attractive feature of the model is that nicks do not have to take place in exactly corresponding positions in the two homologs. However, they do have to appear in *approximately* the right position to produce lengths of single-stranded DNA capable of complementary pairing. It is difficult to imagine how any such system could operate unless the molecules are first brought into close register by at least approximate base-by-base alignment of homologous regions. As we have seen, however, such alignment seems unlikely unless breaks are introduced *before* pairing to allow unwinding.

In spite of these difficulties, experimental evidence exists to show that many features of the model are at least possible. All enzymes required for the breakage and rejoining are known and well characterized. Deoxyribonucleases called *endonucleases* exist that can cause single-stranded "nicks" in a DNA duplex. However, there is no good evidence that any of these are actually involved in recombination; mutations affecting endonucleases do not appear to inhibit or modify recombination. *Exonucleases* that "nibble" the ends of single-stranded DNA are also well known, and in this case there is evidence from phage genetics that mutations in this enzyme inhibit recombination. Exonucleases are also expected to figure in the excision of surplus single-stranded DNA from regions of overlap left after rejoining. Regions of DNA that are left single-stranded by "gaps" after recombinants have rejoined can be repaired by DNA polymerase. In fact, there are good indications that the Kornberg DNA polymerase (implicated in DNA synthesis—see p. 310) actually functions as a repair rather than replication enzyme. Finally, ligases (see p. 315) can accomplish the covalent linkage of recombinants once repair is complete. Once again, studies with ligase-deficient mutants have failed to show any correlation with recombination. But it has been demonstrated that mutants lacking *both* DNA polymerase and ligase produce recombinant DNA linked together only by hydrogen bonds rather than covalent linkages. This suggests that both enzymes are actually involved in recombination; however, the exact relationship of these observations to the proposed model is not clear.

The repair of segments of DNA molecules during replication would require that a relatively small amount of DNA synthesis should be detectable during prophase of meiosis. While the bulk of the DNA is replicated in premeiotic interphase, as we have seen, evidence does exist that a very small quantity of DNA is synthesized during zygonema-pachynema, at precisely the stage recombination is believed to take place. This evidence, obtained by Stern and Hotta from experiments with synchronously dividing cells of lily, indicates a close relationship between the observed DNA synthesis and recombination. Stern and Hotta showed that DNA synthesis amounting to 0.3 percent of the total DNA occurs during meiotic prophase. The DNA synthesized was shown by hybridization experiments to be scattered throughout the chromosomes, as would be expected if the synthesis is associated with randomly placed recombination events. Exposure of the cells during zygonema to an inhibitor of DNA synthesis resulted in fragmentation of the chromosomes at later stages of meiosis. This fits in with the model, suggesting that "nicks" opened during the early stages of recombination were left unrepaired as a result of the inhibition of DNA synthesis. Other effects were noted, such as failure of chromosomes to separate properly at anaphase I and II. Experiments with bacteria have indicated that limited DNA synthesis compatible in amount with the repair expected in bacteria can also be detected during recombination in these organisms. Therefore, there is evidence that the repair synthesis expected according to the model actually takes place.

A Summary of Recombination

Whether approached on the level of classical or molecular genetics, recombination has proved to involve close pairing of homologous chromosomes, with breakage, exchange, and rejoining of segments of the DNA. Classical cytology and genetics carried out in eukaryotes established that exchanges are reciprocal and take place at random points along the chromosomes. In eukaryotes, each exchange involves only two of the four chromatids. These exchanges become visible as chiasmata at diplonema. Taken with the observation that recombination takes place by exchange of chromosome segments, these observations provide the basis for the breakage and exchange hypothesis.

From molecular genetics come the observations that recombination can occur within the limits of a gene. Methods sensitive enough to detect exchanges within so short a segment of DNA also reveal that although no deletions or additions appear in the total DNA of the recombinant molecules, exchanges are often nonreciprocal with respect to frequency of mutants in the recombined chromosomes. These methods also reveal that exchanges are polarized and exhibit negative interference. Negative interference is explained by assuming that close molecular pairing occurs only at limited points scattered throughout the chromosomes. Where molecular pairing takes place, multiple exchanges, usually two or three, may take place. Polarization can be accounted for by the assumption that some segments are more likely to break and be involved in exchanges after close pairing occurs.

Nonreciprocal exchange forms the main basis for derivation of the molecular model for recombination. The model proposes that single-stranded breaks are introduced in the substrands of each of the two recombinant molecules. These breaks are not required to appear in precisely analogous locations in the two molecules, but must be in approximately the right places to produce four single-stranded "ends" with complementary sequences. The two molecules then unravel between the nicks, exposing the four single-stranded ends, which reassociate by twos into recombinant molecules. Where gaps or overlaps occur, depending on the imprecision in location of the original nicks, repair synthesis or excision takes place, which may result in either the elimination or duplication of mutant sequences. Winding of DNA from opposite homologs into a hybrid duplex and repair synthesis may produce aberrant segregation ratios in the recombinant chromosomes and thus account for nonreciprocal exchanges.

The molecular and classical models for recombination are entirely compatible and can be interrelated by assuming that where an exchange occurs in higher organisms, detectable at this level as recombination *between* genes and reflecting the various characteristics of classical breakage and exchange, the molecular model operates at the actual point of physical breakage and reunion of the chromosomal DNA. This implies that in all recombination, whether in prokaryotes or in higher organisms, the phenomena of nonreciprocal segregation, polarization, and negative interference may appear over very short lengths of the chromosome.

The Morphology of Recombination in Eukaryotes

The Synaptinemal Complex

In all eukaryotic cells with normal meiosis and recombination, the close pairing of homologous chromosomes in zygonema-pachynema is correlated with the appearance of the *synaptinemal complex* (also termed *synaptonemal complex, axial complex,* or *core*). Besides being of special interest in recombination, this structure also represents one of the few configurations of the chromosome in which the 100 Å unit fibers are arranged into a superstructure with order visible in the electron microscope.

The synaptinemal complex was first described in 1956 in independent reports by Montrose J. Moses and Don W. Fawcett. At first, the complex was thought to be confined to spermatocytes, but subsequent work proved that the structure is present in both oocytes and spermatocytes at prophase of meiosis. Moses and a number of other investigators established that the complex is closely associated with the synapsed homologs by comparison of thick sections for light microscopy with adjacent thin sections for electron microscopy (Fig. 14–28 a–c). These comparisons demonstrate that the synaptinemal complex occupies the space between the paired homologs (Fig. 14–28c). In the light microscope, this space can be just resolved as a longitudinal separation between the homologs that extends along the entire length of the paired chromosomes.

An extensive series of investigations has established that the complex occurs universally in all cells undergoing meiosis. The basic morphology of the complex in longitudinal section (see Figs. 14–29 and 14–30) is remarkably uniform in all plants and animals and consists of three parallel, electron-dense elements separated by less dense spaces, all of extremely regular dimensions within a species. The two *side elements* (refer to Fig. 14–30a), which vary in width in different species from 200–800 Å, appear to be made up of electron-dense coarse granules or fibers of slightly larger dimensions than 100 Å unit fibers of the chromatin. The side elements are regular in dimensions and do not vary greatly in width in a given species. The chromatin surrounding the side elements may be either aggregated into dense, irregular masses (see Fig. 14–28c) or so diffusely extended that no distinct outline is discernible (see Fig. 14–30).

The side elements are separated by a regular central space that may vary in different organisms from 200–300 Å to as much as 1000–1250 Å. In species with synaptinemal complexes of the maximum width, the two side elements plus the central space are therefore about the same total diameter as a centriole. Along the middle of the central space lies the electron-dense *central element,* separated by an equal distance from either side element. The central element takes two basic forms. In all plants and animals except the insects, the central element is indistinct and appears to be merely a region of overlap of thin fibers originating from the side elements (arrow, Fig. 14–29). In many insects the central element takes on a more ordered appearance (Figs. 14–30a and 14–30b). In these forms, the central element appears ladder-like, with two longitudinal, parallel side components 250–300 Å apart connected by cross striations regularly spaced at 200–300 Å intervals. In some species, three longitudinal components are present in the central element. The third is spaced along the center line of the central element, equidistant from the sides of the "ladder" (Fig. 14–30b).

In longitudinal sections cut at 90° to the plane of section shown in Fig. 14–30, the central element in the ordered forms appears as five to six layers of electron-dense dots in longitudinal rows. Because these dots probably represent cross sections of the "rungs" of the ladder-like array, five to six ladders are probably stacked one on top of the other to make up the ordered form of the central element. Where this form of the central element appears, both the longitudinal components and the rungs of the ladders are of smaller dimensions than 100 Å chromatin fibers and approximate 50–75 Å.

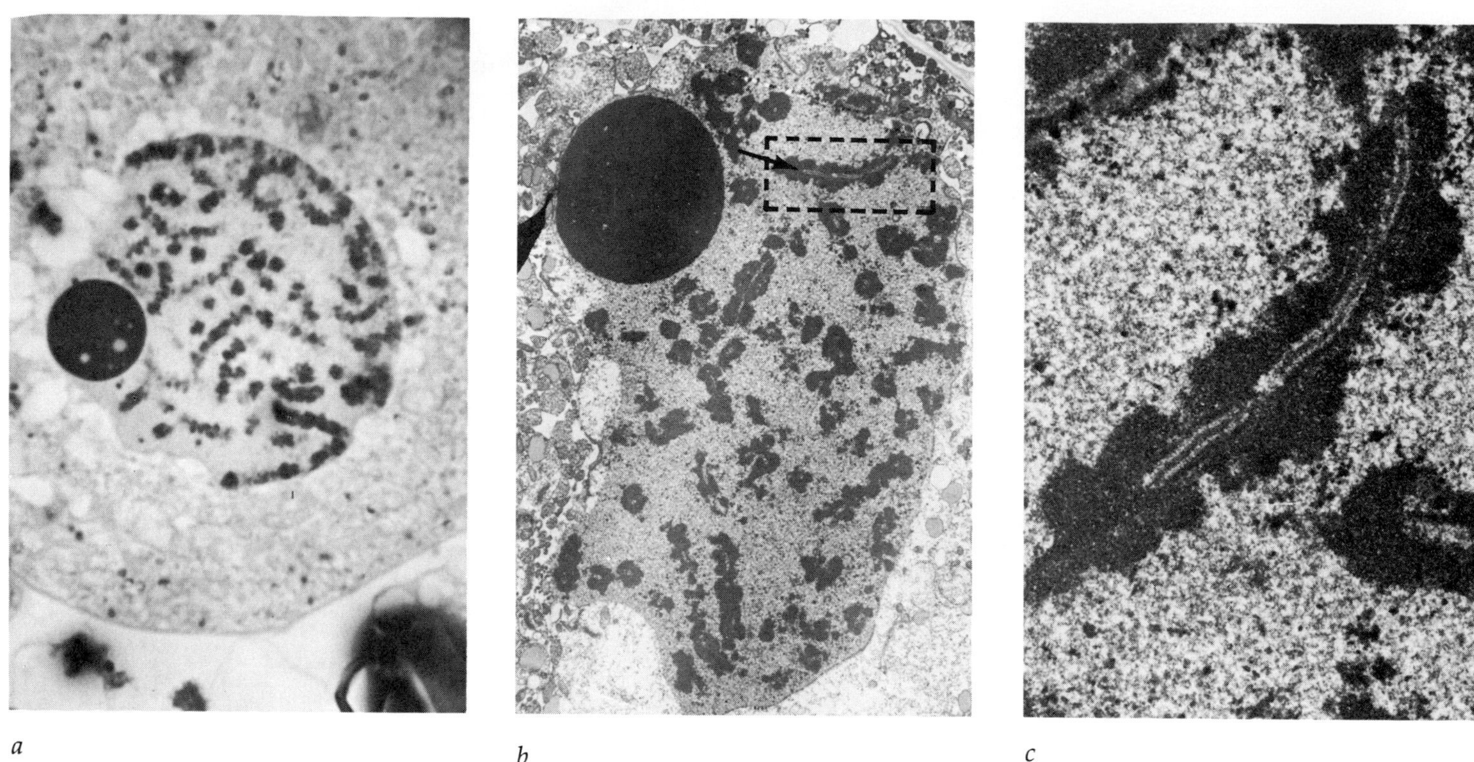

14–28. *Localization of the synaptinemal complex in the space between the synapsed chromosome in the plant* Lilium longiflorum. a, *A light micrograph of a pachytene nucleus. × 1,600.* b, *An adjacent section of the same nucleus at low magnification under the electron microscope. The synaptinemal complex (arrow) occupies the space between the paired homologs. × 4,000.* c, *The region enclosed by dotted lines in* b *at higher magnification. × 21,500. Courtesy of P. B. Moens and the publisher, from* Chromosoma *23:418 (1968).*

In all forms of the synaptinemal complex very thin fibers cross the central space and connect the central element with the side elements. These *transverse fibers*, which range in diameter from 60–70 Å down to invisibility, lie more or less at right angles to the side and central elements. In insects with ordered central elements, the transverse fibers are also ordered and are present in numbers approximately equal to the rungs of the "ladder."

In cross section (Fig. 14–31) the total complex is 350–400 Å in thickness. Frequently, the central element is less distinct and of smaller dimensions than the side elements. The side elements are spaced to either side of the central element, giving the impression that the whole complex is a flat, ribbon-like structure lying between the synapsed homologs.

In many plants and animals, the synaptinemal complex terminates at the inner membrane of the nuclear envelope (Fig. 14–32). This connection finds its counterpart in light microscopy, in which the synapsed homologs can be seen to make terminal attachments to the nuclear envelope. At the point of attachment of the complex to the envelope, the side elements are often greatly thickened and more electron dense.

Peter B. Moens has made a careful study of the arrangement of the synaptinemal complexes and the synapsed homologs in the bouquet stage in locusts. This study clearly demonstrates the relationship of the complex to the synapsed homologs. Fig. 14–33 shows a composite diagram made from serial sections of a single nucleus. The connections to the nuclear envelope are obvious. Note also that the complexes are twisted on themselves in the loop, which extends from the envelope. This is equivalent to the relational spiral of the homologs often observed at pachytene in light microscope preparations.

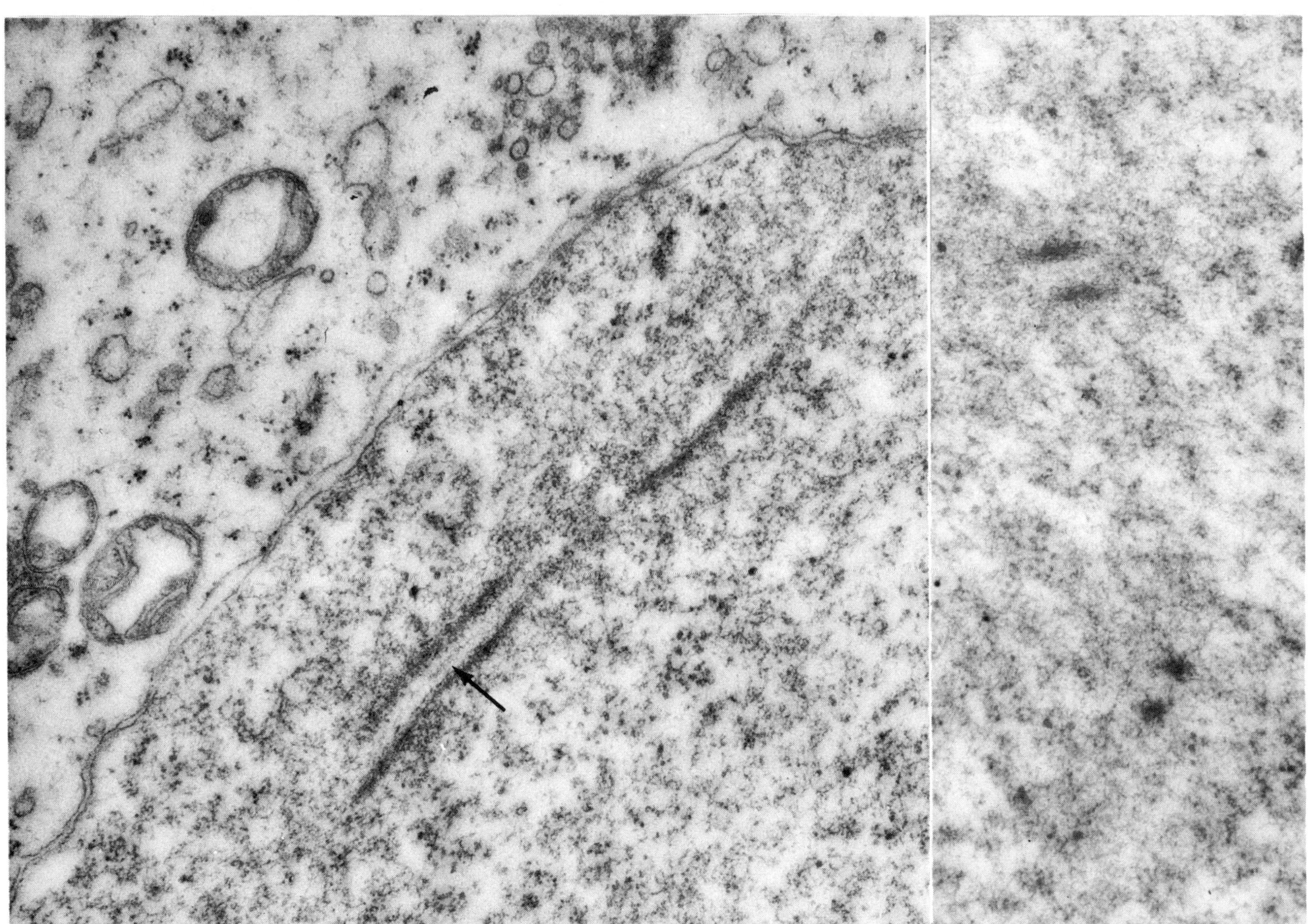

14–29. A synaptinemal complex in a rat spermatocyte nucleus, showing the more commonly observed indistinct form of the central element (arrow). Two complexes are shown in cross section at the right. × 50,000. Photograph by the author.

Cytochemistry of the synaptinemal complex. The synaptinemal complex has been investigated by several techniques for electron microscopy that give limited indications of the chemical composition of this structure. In 1964, James R. Coleman and Montrose J. Moses analyzed the DNA and protein components of the complex and the surrounding chromatin, using material fixed in glutaraldehyde without post-fixation by osmium. Because tissue has little or no electron density after aldehyde fixation, the attachment of metal stains could be clearly traced. Coleman and Moses stained the tissue with a salt of the heavy metal *indium,* which under some conditions attaches selectively to nucleic acids (see Chapter 3). After staining with indium, the chromatin fibers surrounding the complex and the inner border of the side elements facing the central element became electron dense. Much of the substance of the side elements failed to take up stain. However, stained fibrils apparently originating from the surrounding chromatin were observed to penetrate into

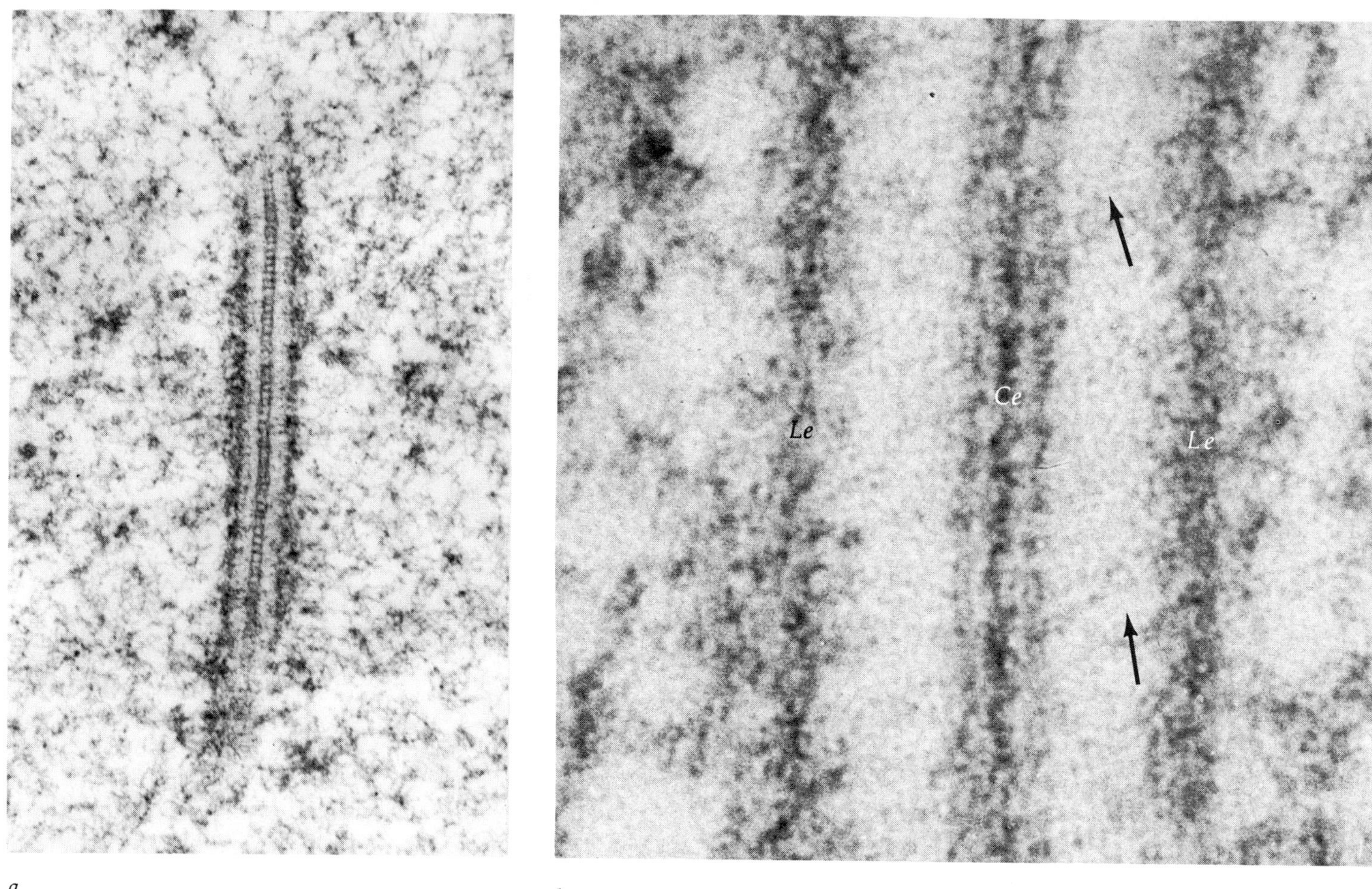

14–30. a, *A synaptinemal complex from* Philaenus spumaris, *in which a highly ordered, ladder-like central element occurs. × 60,000. Courtesy of P. L. Maillet and R. Folliot.* b, *Synaptinemal complex of* Gryllus argentinus, *showing a central element with three longitudinal components. Transverse fibers (arrows) can be faintly seen in the region separating the side and central elements.* Le, *lateral elements;* Ce, *central element. × 440,000. Courtesy of R. Wettstein and J. R. Sotelo.*

and cross the side elements. Very faint density could also be seen in the thin transverse fibers connecting the side and central elements.

If the material was digested with deoxyribonuclease before staining, no density could be detected in either the synaptinemal complex or the surrounding chromatin. If the same digested tissue was subsequently stained with a less selective electron stain such as lead hydroxide, the side and central elements of the complex and the surrounding chromatin became visible. Coleman and Moses interpreted this pattern as the result of binding of stain by protein, considered a component common to all of these structures. Another cytochemical study, using a solution of phosphotungstic acid in alcohol as a selective stain for protein, also indicates that protein, possibly a basic type similar to histone, is a constituent of the synaptinemal complex (see Sheridan and Barnett, 1967). These experiments suggest that the complex consists of a framework of protein, across which fibers originating from the surrounding

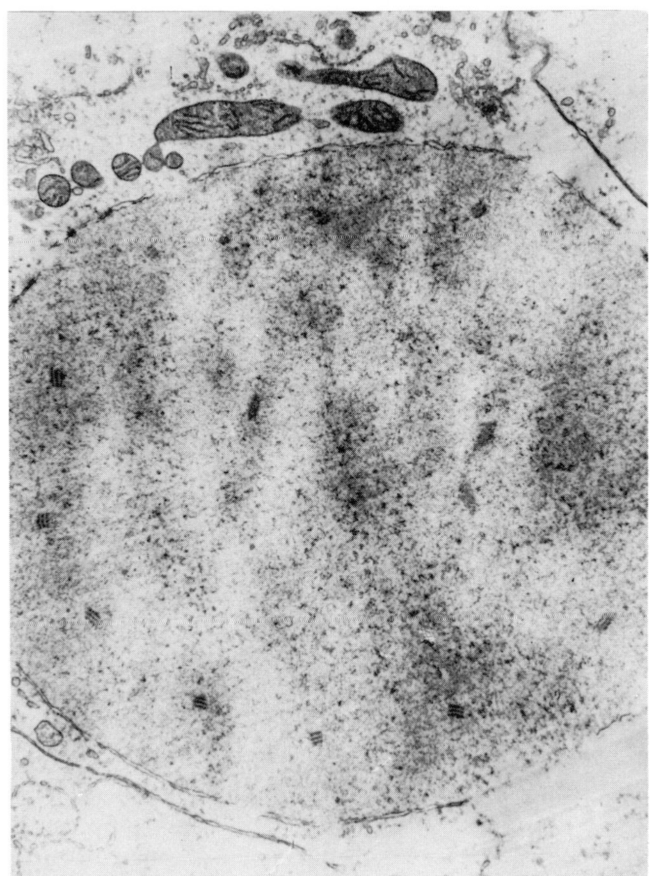

14–31. *The synaptinemal complex in cross section. In this species, the insect* Philaenus spumaris, *the synapsed homologs are dispersed in a regular circle. Twenty-three chromosomes are present in the male. Twenty-two of these pair to form eleven tetrads during meiosis. These are visible in the micrograph. The remaining chromosome, the* X (Philaenus *is an* XO *male), does not enter into pairing and is not seen in this micrograph. × 8,500. Courtesy of P. L. Maillet and R. Folliot.*

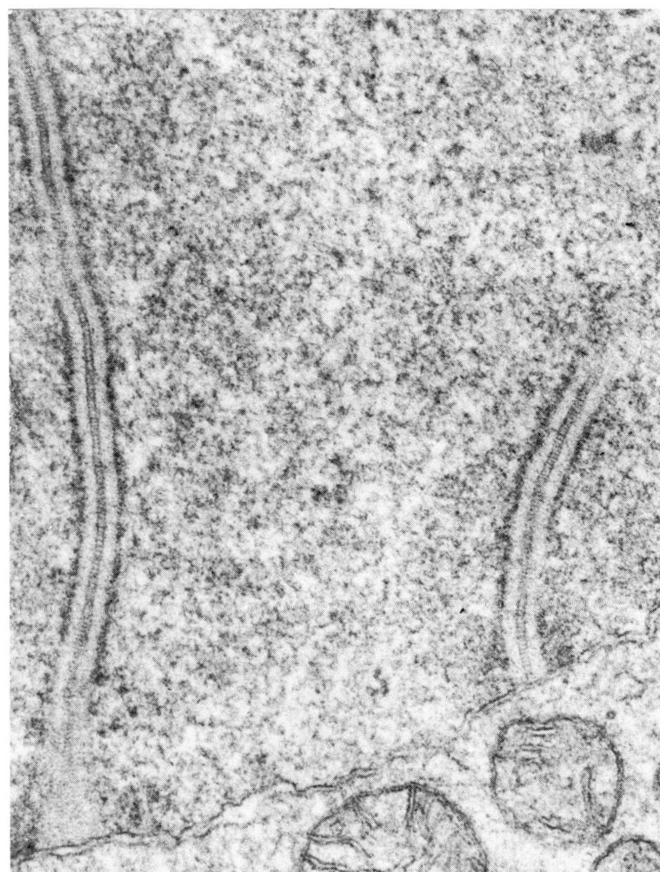

14–32. *A nucleus of* Philaenus, *showing attachment of the synaptinemal complex to the nuclear envelope. × 35,000. Courtesy of P. L. Maillet and R. Folliot.*

chromatin extend. The pattern of development of the complex (see below) also suggests that this is the case.

The Relationship of the Synaptinemal Complex to Pairing and Recombination

Much circumstantial evidence connects the synaptinemal complex with pairing and recombination of the homologs. First, there is the correspondence in position of the complex with the synapsed homologs at pachynema. Further, when light and electron microscope images of the same cell are compared, the complex is shown to lie precisely between the synapsed homologs in the region in which close pairing and recombination are expected to take place. The complex is first seen to appear at zygonema, is present throughout pachynema, and disappears from the chromosomes at the onset of diplonema when close pairing ceases. In leptonema, before synapsis, single side elements are visible in many regions of the nucleus. These appear to join together, with formation of the central element, at zygotene. Thus both the location of the complex, and the synchronization of its appearance and disappearance with the onset and conclusion of

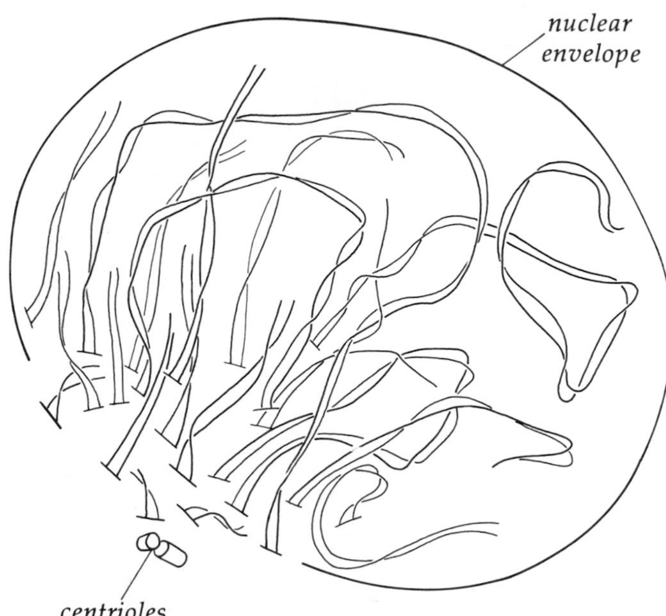

14–33. *The spatial arrangement of synaptinemal complexes in the nucleus of* Locusta migratoria *at pachytene. This diagram was constructed by combining the images of 100 serial sections of the same nucleus. Redrawn, original courtesy of P. B. Moens and the publisher, from* Chromosoma *28:1 (1969).*

close pairing indicate an intimate involvement with pairing and recombination.[4]

Other evidence (summarized in Table 14–1) links the complex more directly with recombination. Gunter F. Meyer found that in male *Drosophila,* in which pairing but no recombination occurs, no synaptinemal complexes are present. In female *Drosophila,* in which normal pairing and recombination take place, the complex is present. However, in females with a mutation suppressing recombination the complex was found to be absent. In addition, in the insects *Tipula caesia* and *Phryne fenestralis,* no chiasmata are seen at diplonema in the light microscope; in the electron microscope, the synaptinemal complex is found to be absent in both species. But in another tipulid fly, *Tipula oleracea,* the regular presence of chiasmata at diplonema is correlated with the presence of the synaptinemal complex at zygonema-pachynema. These findings clearly implicate the complex in recombination.

A few observations have been made of species lacking chiasmata at diplonema that nevertheless contain synaptinemal complexes at zygonema-pachynema. This is true, for example in a hybrid form of the tomato. Rather than eliminating the possibility that the complex is involved in recombination, these exceptional observations probably indicate that although the complex appears to be necessary for recombination to occur, presence of the complex by itself does not insure that recombination will take place.

A final and very important piece of evidence linking the synaptinemal complex with recombination is provided by work with inhibitors of DNA synthesis. The investigations of Hotta and Stern (see p. 386), mentioned in the previous section on recombination, demonstrated that a small amount of DNA synthesis takes place at prophase. Inhibition of this synthesis interferes with the normal function of chromosomes during meiosis. Thomas F. Roth and Mishio Ito followed the effects of inhibition of DNA synthesis at prophase on the synaptinemal complex in the same organism (the lily) used by Hotta and Stern. Roth and Ito found that inhibition of DNA synthesis at leptonema also inhibited the formation of the synaptinemal complex. If inhibition was delayed until late leptonema or early zygonema, a reduced number of complexes appeared in the lily nuclei. Thus, inhibition of the DNA synthesis at prophase, which is likely to be a part of the recombination mechanism, also inhibits the formation of the synaptinemal complex.

The Problem of Polycomplexes

The various lines of evidence listed in the foregoing paragraphs have led to the general acceptance of the hypothesis that the synaptinemal complex is involved in some manner in the mechanisms of recombination. This point of view received a temporary setback with the discovery in insects of structures apparently identical to the synaptinemal complex at stages following prophase I (see Table 14–2). These complex-like structures, called *polycomplexes,* were sometimes found to persist through the second meiotic division and even into the early stages of maturation of the haploid gametes, where pairing of homologs could not possibly occur. These observations at first seemed incompatible with the hypothesis that the synaptinemal complex is morphological evidence of pairing and recombination.

[4]For literature references to the many papers linking the synaptinemal complex to recombination, see Moses, 1968. For a representative study, in which all points outlined in this paragraph are included in a single paper, see Moens, 1968b.

Table 14–1 Evidence for Associating Synaptinemal Complex Formation with Meiotic Recombination in the Diptera

Species and condition	*Visible pairing*	*Chiasma formation or genetic crossing over*	*Synaptinemal complexes*
Drosophila melanogaster male	+	−	−
Drosophila melanogaster female	+	+	+
Drosophila melanogaster female (mutant C3G)	−	−	−
Phryne fenestralis	+	−	−
Tipula caesia	+	−	−
Tipula oleracea	+	+	+

From M. J. Moses, Synaptinemal complex. *Ann. Rev. Genet.* 2:363 (1968) with data presented by G. F. Meyer at the 3rd European Regional Conference on Electron Microscopy, Prague, 1964.

But in recent years, careful studies of the morphology, time of appearance, and location of polycomplexes has allowed their existence to be reconciled with the probable involvement of the synaptinemal complex in recombination.

The structure of polycomplexes. Polycomplexes have so far been found only in insects and a few other species such as crab *Cancer* and possibly corn microspores.[5] In the most common configuration, polycomplexes appear to be made up of a large number of alternating side and central elements (Fig. 14–34). These are usually found in groups numbering from 3 or 4 to 20 or more but have been counted in masses numbering up to 60 repeated side and central elements. The collection of polycomplexes may be relatively flat, with no greater thickness than the ordinary synaptinemal complex (Fig. 14–35a). In other cases, each element may be much thicker than usual, so that the polycomplex mass takes on the appearance of a multilayered sandwich (Fig. 14–35b). Even more bizarre configurations occur, including tubular complexes (Fig. 14–35c) or alternating side and central elements in a semicrystalline array (Fig. 14–35d).

The structure and dimensions of polycomplexes in longitudinal section are usually identical to the synaptinemal complexes occurring in the same insect. However, chromatin fibers may be either present or absent from areas immediately surrounding the polycomplexes. If present, chromatin fibers are found only in close association with the outermost side elements of a mass of polycomplexes. According to most investigators, the largest masses of polycomplexes rarely show any direct connection to chromosome fibers, even at the periphery.

The relationship of polycomplexes to the synaptinemal complex. Careful observation of the time of appearance of polycomplexes shows that these structures are not observed until the close of pachynema. Thus, the disappearance of synaptinemal complexes from the synapsed homologs is correlated with the first appearance of polycomplexes. This has led to the suggestion, made by many investigators, that polycomplexes are residual synaptinemal complexes that separate from the chromosomes at the close of pachynema and persist for a time in masses in the nucleus. According to this hypothesis, the sequence of appearance of the various elements of the synaptinemal complexes and polycomplexes is as follows. First, single side elements are seen at leptonema. At zygonema, these are aligned, with formation of the central element to form the synaptinemal complex. The complex persists throughout pachynema and is then stripped from the chromo-

[5]The observation of polycomplexes in the crab were made in 1968 by Susan G. Langreth (Ph.D. thesis, University of Chicago, unpublished). Polycomplex-like structures were seen in corn microspores by Hewson Swift and Barbara J. Stevens (Nucleo-chromosomal interaction in microspores of maize. *National Cancer Institute Monograph* 23:145, 1966).

Table 14–2 Stages at which Polycomplexes and Multiple Synaptinemal Complexes Have Been Observed

	Primary spermatocyte	*Division I*	*Secondary spermatocyte*	*Division II*	*Spermatid*
Gryllidae (crickets)					
Gryllus argentinus	+	+	+	+− (telophase)	+
Myogryllus verticalis	(observations presumably consistent with the above)				
Gryllus sp.	(observations presumably consistent with the above)				
Gryllus campestris	+				+
Gryllus bimaculatus	+				+
Gryllus domesticus	+				+
Acrididae (grasshoppers)					
Laplatacris dispar	−				−
Aleua lincata	−				−
Other					
Blaptica dubia (cockroach)					+
Periplaneta americana (cockroach)	−				
Acanthopachylus aculeatus (arachnid)	+	+			−
Philaenus spumaris (homopteran)	+				
Oncopeltus fasciatus (hemipteran)	+				

From M. J. Moses, Synaptinemal complex. *Ann. Rev. Genet.* 2:363(1968). See this paper for original references.

somes and dismantled at the onset of diplonema. At this stage in a very few organisms the disassembled subunits of the complexes are reorganized into polycomplexes and persist in this form through the subsequent stages of meiosis and early gametogenesis. In all cases, any residual complexes disappear by the time gametes are mature.

In connection with this hypothesis it has been pointed out by Thomas F. Roth that the ratio of side to central elements in polycomplexes and in the synaptinemal complex is different. In the polycomplex, the side and central elements are found in a 1:1 ratio instead of the 2:1 ratio seen in the synaptinemal complex. To account for this discrepancy Roth has suggested that approximately one-half of the side elements in the synaptinemal complex consists of chromatin; the other half is part of the protein of the lateral element. As the complex is stripped from the chromosome, the residual side element is reduced in thickness by one-half. These join side-by-side with the side elements of other residual complexes during formation of the polycomplex to form a composite lateral element with dimensions equivalent to the intact side elements of the synaptinemal complex. Thus, according to Roth's suggestion, the polycomplex actually represents side and central elements in the combination [½S½S]—C—[½S½S]—C—[½S½S]... instead of the apparent S—C—S—C—S—C—S configuration.

If the proposed sequence of events is correct, the presence of the polycomplex after pachynema does not invalidate the hypothesis that the synaptinemal complex is involved in recombination, because the polycomplex would represent a storage place for residual complexes disassembled from homologs at the close of synapsis. The polycomplex may thus represent another manifestation of the monomer-polymer assembly-disassembly system observed in other structures of the cell, as in the case of spindle microtubules. Very possibly, a subunit of the synaptinemal complex is synthesized in large quantities some time before zygotene (Fig. 14–36a); these are "polymerized" into the lateral elements of the synaptinemal complex

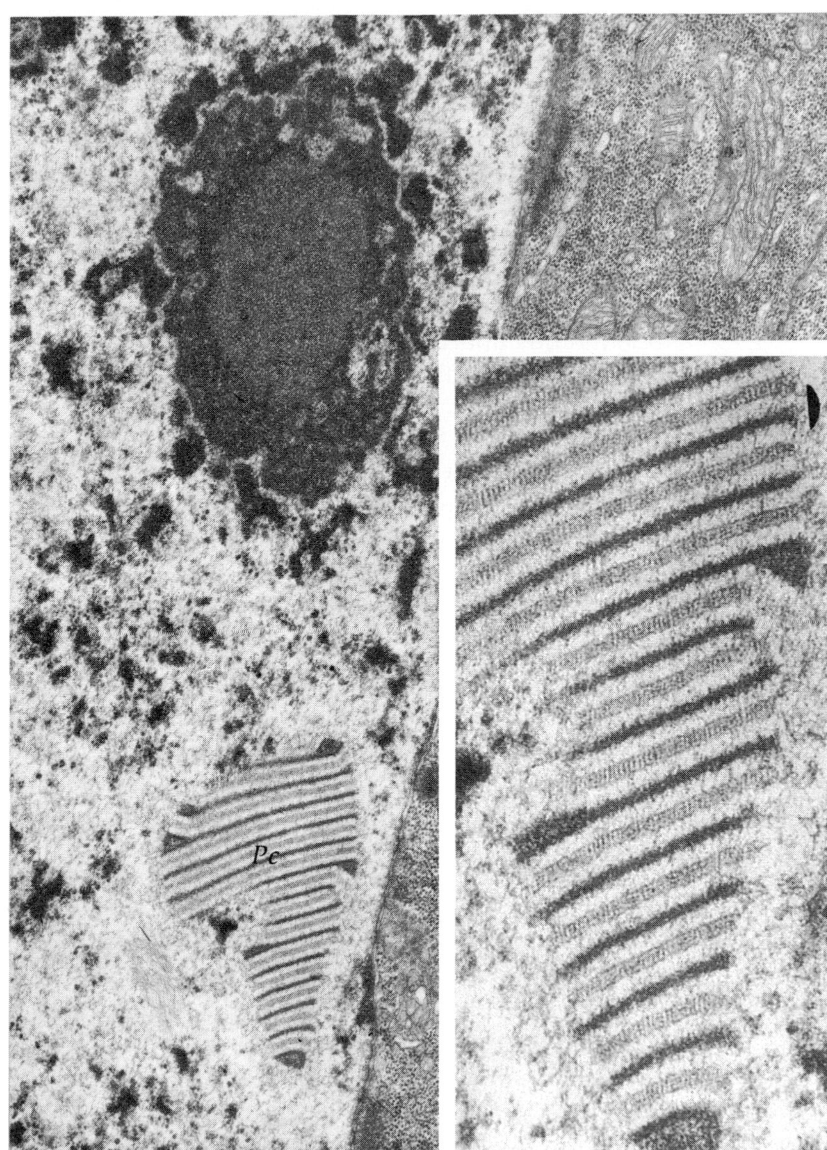

14–34. Polycomplexes (Pc) in the nucleus of the mosquito, Aedes aegypti. *× 20,000. Inset, the polycomplex mass in a at higher magnification. × 59,000. Courtesy of T. F. Roth.*

(14–36b) which pair to complete the complex at zygonema (14–36c); at diplonema these are disassembled into "monomers" (14–36d). In most organisms, the monomer is broken down at this time; however, in a few insects the monomers are reassembled for a time into the polymer structure, possibly as a means of storage, until breakdown of the monomers is completed (14–36e).

Before leaving the polycomplexes and their probable relationship to the synaptinemal complex, one additional observation must be mentioned. In one case, polycomplexes were observed by Thomas F. Roth in the nuclei of nurse cells in the ovaries of the mosquito. Because the nurse cells are nonmeiotic tissue and remain diploid throughout the development of the mosquito ovary, the appearance of polycomplexes in these nuclei seemed at first to invalidate the hypothesis that the synaptinemal complex is a part of the recombination mechanism. An explanation for this observation has been provided by Montrose J. Moses (1968), who points out that in *Drosophila*, oocytes are formed as a part of a cluster of 16 cells. At first, two of the 16 cells are pro-oocytes and contain synaptinemal complexes. These two form oocytes, while the remainder of the

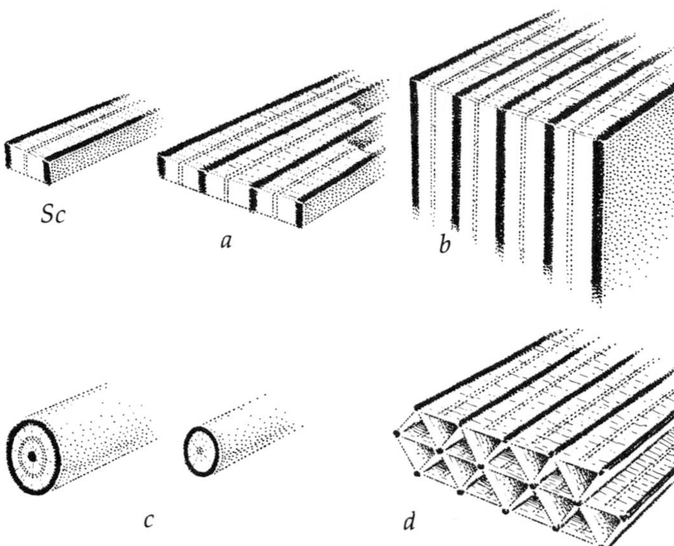

14–35. *Various forms taken by polycomplexes in insects. Sc, the pachytene synaptinemal complex.* a, *Polycomplexes in the form of a flat sheet.* b, *A multilayered sandwich.* c, *Tubular.* d, *A crystal-like lattice. Diagram modified from M. J. Moses,* Ann. Rev. Genet. *2:363 (1968).*

cells become nurse cells. Near the end of oogenesis in the cluster, one of the two oocytes reverts to a nurse cell, leaving 15 nurse cells and one oocyte. The oocyte reverting to nurse cell could therefore be expected to contain residual elements of the synaptinemal complex. If, as Moses points out, the development of the oocyte is at all similar in the mosquito, the appearance of occasional nurse cells containing residual polycomplexes is not surprising. (Other reports of partial entry of nurse cells into meiosis are described in Chapter 15.)

Thus all reported observations of polycomplexes are either consistent with the interpretation that these structures are probably derived from residual elements of the synaptinemal complex or are at least explainable in terms of this hypothesis. With this in mind, we will now return to the synaptinemal complex and give some attention to the possible role of this structure in the mechanism of recombination.

The Synaptinemal Complex and Recombination: Some Possibilities

The synaptinemal complex is located between the homologs in zygonema-pachynema in the region where close pairing and recombination are expected to take place. The homologs, although closely paired, remain separated by the total width of the synaptinemal complex. "Crossovers" of the lateral elements are never seen to take place; the separation and dimensions of the lateral elements and central elements are extremely regular in any species. This distance, 1500–2000 Å, is too great for any known intermolecular interactions; the DNA molecules involved in a recombination must approach more closely than the limits imposed by the width of the complex. The formation of hydrogen bonds, expected to be important in the pairing of the substrands of DNA duplexes during breakage and exchange, would require the approach of homologous chromosomes to distances at most less than 10 Å. Therefore it seems likely that the unit chromosome fibers must spin out from the masses of chromatin at the sides of the synaptinemal complex, traverse the lateral elements, and extend into the central element of the complex. In this 250–300 Å central element, close pairing in the molecular sense is most likely to occur. The faintly staining transverse fibers may represent unit fibers of the chromosomes extending into the region of close pairing, breakage, and exchange in the central element.

The completed synaptinemal complex first appears when homologs become closely paired. No doubt this initial pairing is only an approximation in the molecular sense, because, as mentioned, no intermolecular forces such as the attraction between complementary base pairs are known to operate over such distances. In fact, the forces that bring the homologous chromosomes together from their initial wide separation in the leptotene nucleus to establish even the initial "rough" pairing at zygotene remain unknown. Once the chromosomes are aligned on the framework provided by the synaptinemal complex, the process of close molecular pairing is expected to begin. This might take place in the following manner. The initial pairing positions homologous chromomeres at approximately the same level on opposite sides of the synaptinemal complex (Fig. 14–37a). At this level, fibers spin out from the homologous chromatids, threading through the side elements of the synaptinemal complex (Fig. 14–37b). The chromosome fibers originating from opposite sides of the complex approach to within molecular distances in the central element. Within the central element, comparison of the sequences of nucleotides in the paired chromosome fibers takes place. The chromosome fibers from opposite sides of the complex are considered to be in movement, so that a continuous comparison of nucleotide sequences takes place. This movement of fibers through the central element, with comparison

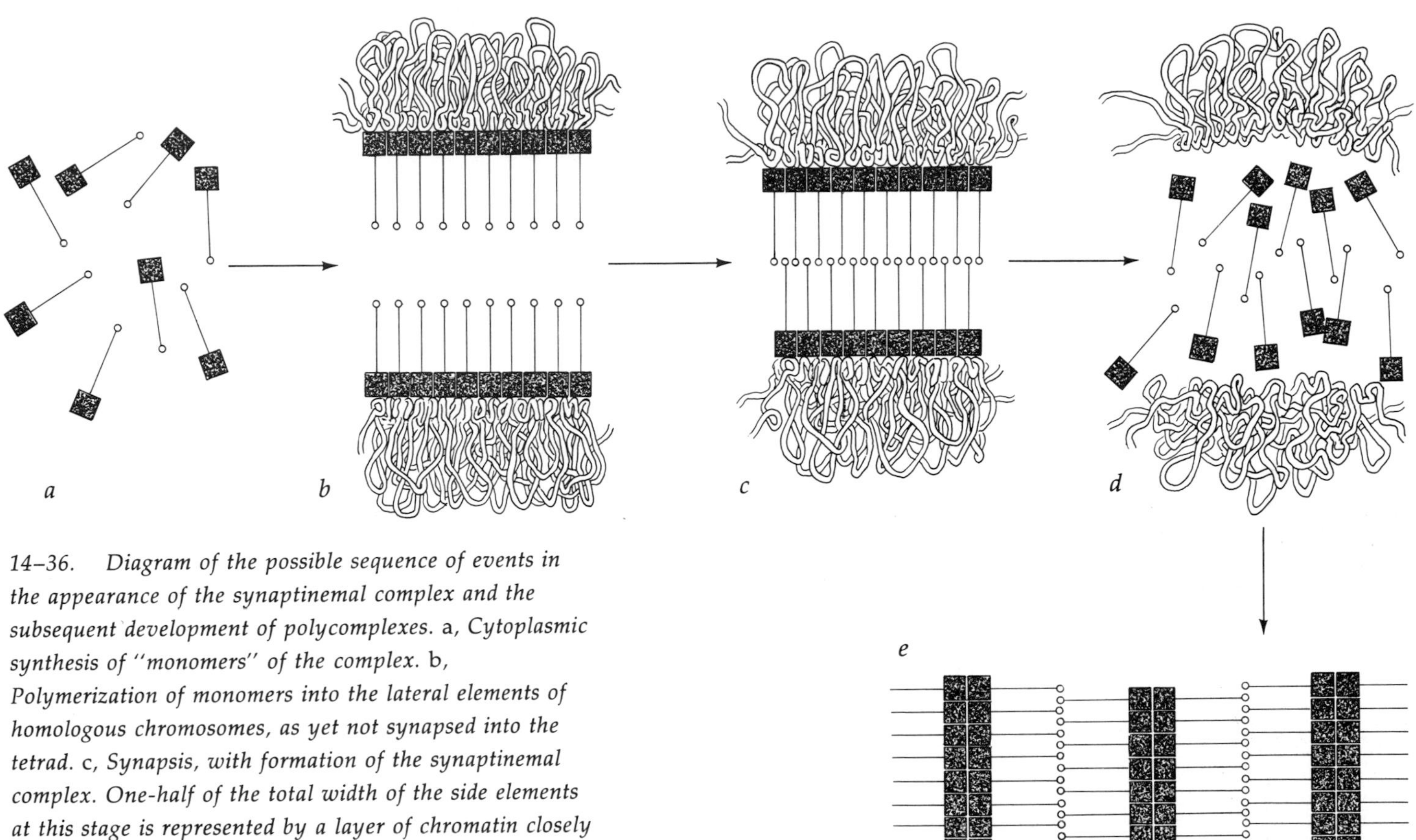

14–36. Diagram of the possible sequence of events in the appearance of the synaptinemal complex and the subsequent development of polycomplexes. a, *Cytoplasmic synthesis of "monomers" of the complex.* b, *Polymerization of monomers into the lateral elements of homologous chromosomes, as yet not synapsed into the tetrad.* c, *Synapsis, with formation of the synaptinemal complex. One-half of the total width of the side elements at this stage is represented by a layer of chromatin closely applied to the polymerized subunits of the synaptinemal complex.* d, *Disassembly of monomers from the homologs at diplonema. Close pairing ceases at this time.* e, *Temporary assembly of the detached monomers into polycomplexes.*

of sequences from homologous chromatids, is considered to be a "searching" mechanism that seeks for homologous sequences in the two chromosome fibers (Fig. 14–37c). Eventually, homologous sequences may approach and become paired. At this point, movement ceases and the fibers are locked in position by close-pairing forces. The probability of a successful matching of homologous sequences of this type is expected to be low and random at any specific site.

Once molecular pairing is established between homologous sequences at any site in the central element, recombination of DNA segments could take place at this location. Within this closely paired segment, exchanges could be polarized, nonreciprocal, and multiple, satisfying the requirements of intragenic recombination. The structure of the central element even provides a framework for alignment of DNA with the required enzymes. The alignment of enzymes on the central element would allow a mechanism for the introduction of "nicks" into the homologous DNA molecules in approximately the right positions for breakage and exchange. Thus, all of the features of the molecular model for recombination could be localized and would operate within the framework of the central element of the synaptinemal complex.

Toward the end of pachynema, as breakage and exchange is completed, the disassembly of elements of the synaptinemal complex begins. At this time, the recombined chromosome fibers extending across the space occupied by the complex begin to fold or coil. This process continues, with gradual thickening at the chiasma, until, at diplonema, the exchange point becomes visible as a crossover of chromatid fibers in the light microscope. Because extensive coiling in this region would involve considerable movement of chromosome fibers, the actual point at which the chiasma

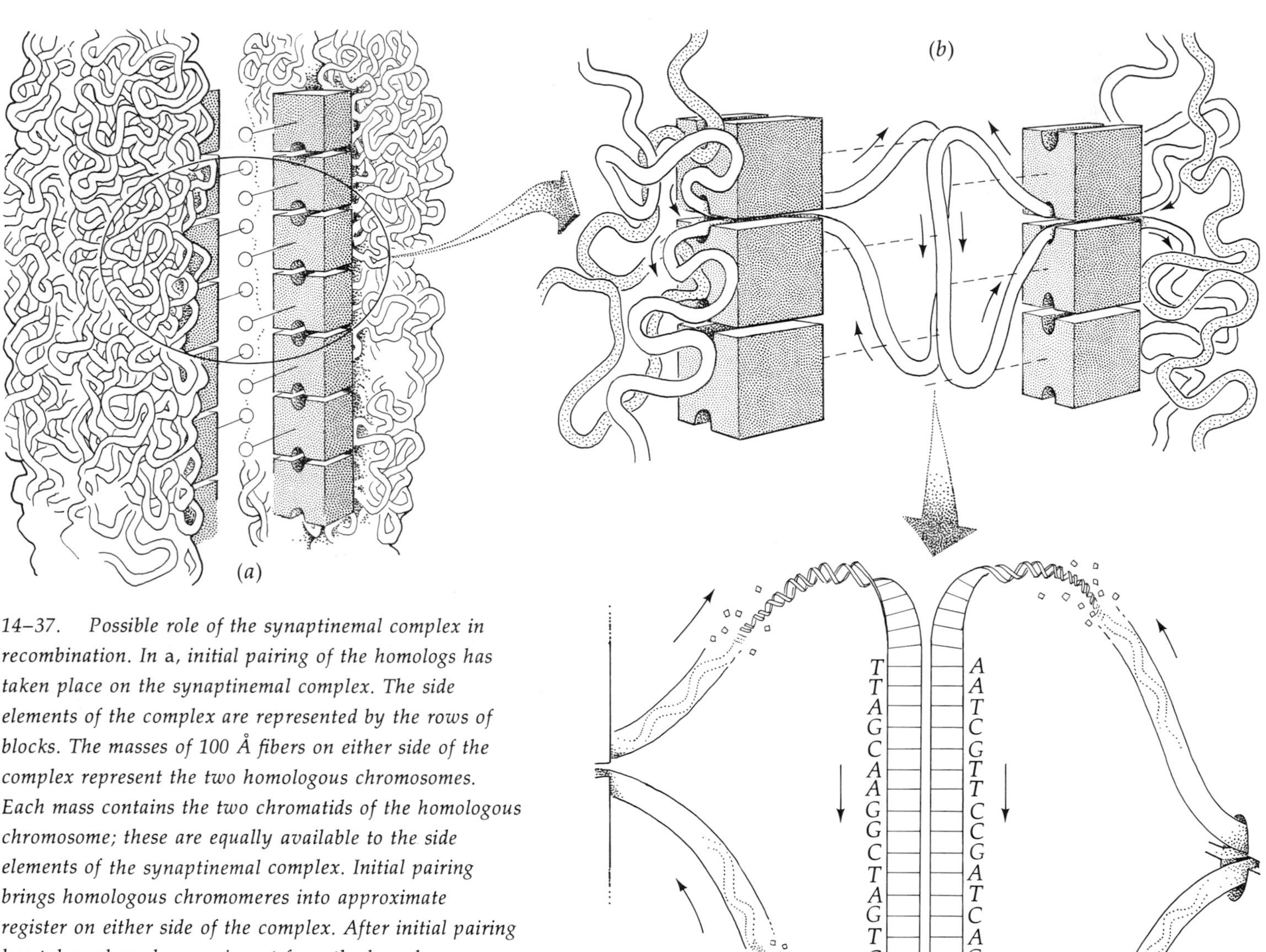

14–37. Possible role of the synaptinemal complex in recombination. In a, *initial pairing of the homologs has taken place on the synaptinemal complex. The side elements of the complex are represented by the rows of blocks. The masses of 100 Å fibers on either side of the complex represent the two homologous chromosomes. Each mass contains the two chromatids of the homologous chromosome; these are equally available to the side elements of the synaptinemal complex. Initial pairing brings homologous chromomeres into approximate register on either side of the complex. After initial pairing has taken place, loops spin out from the homologous chromomeres and extend into the central element (*b *in the diagram). At any level a loop originates from only one of the two chromatids on either side of the complex. The loops are in continuous movement. They approach to within molecular distances in the central element of the synaptinemal complex. A continual comparison of sequences takes place in the central element until homologous DNA sequences become aligned (*c *in the diagram). Once alignment has taken place, subsequent events follow the pattern for molecular recombination (see Fig. 14–26). Note from Fig. 14–26 that the model requires only that the base sequences of homologous DNA be brought into approximate register. Loop comparison of this type takes place at all levels of the synaptinemal complex. Successful alignment of homologous DNA sequences at any level is a random event and is low in probability, producing on the average about one successful alignment per chromosome arm. The duration of pachynema reflects the total time required for the fiber length contained in the chromomeres to pass through the comparison mechanism in the central element.*

first becomes visible would only approximate the location of the original molecular exchange.

Thus the synaptinemal complex can be conveniently fitted into the phenomena of initial pairing, close pairing, recombination, and chiasmata formation and can even be related to the features of both classical and intragenic recombination. That the complex is involved in at least some of these processes seems certain; whether all of the listed mechanisms take place with the direct intervention of the complex remains to be seen.

Lampbrush Chromosomes: A Configuration of the Diplotene Nucleus

At the close of pachynema, when close pairing ceases and recombination is completed, the meiotic cells in many species enter a period of growth. This is especially marked in oocytes, particularly in the developing eggs of birds and amphibians. During this period, which takes up much of the diplotene stage, the chromosomes partially or completely uncoil into a state reminiscent of the interphase nucleus. This stage is known as "diffuse diplotene."

Some authorities have maintained that all meiotic cells, in both plants and animals, undergo at least a partial reversion to the interphase state at diplonema, with at least some uncoiling of the chromosomes. This may vary from only limited uncoiling, reflected as a slight loss of distinct outline in the chromosomes, to complete reversion to an interphase-like state. In oocytes from a number of species, including many arthropods, sharks, amphibians, reptiles, birds, and other animal groups, a peculiar configuration of the diplotene chromosomes appears at this time. This coiling arrangement was first described in 1892 by J. Rückert, who called the chromosomes "lampbrush chromosomes" because they resembled the large brushes used to clean the chimneys of kerosene lamps. For better or worse, the name has survived; from the time of their first discovery, lampbrush chromosomes have been the subject of intensive research. In spite of this, many questions remain as to their function during diplonema.

The Structure of Lampbrush Chromosomes

It was soon found by examination and micromanipulation of lampbrush chromosomes under the light microscope that the "bristles" of the brushlike configuration are actually loops that extend out from the axes of the paired chromosomes (Figs. 14–38 and 14–41). The double axis, which represents the two homologous chromosomes, is held together by chiasmata at widely spaced points (arrows, Fig. 14–38). At early diplonema, the four strands of the tetrad are usually visible in the chromosomes. But in the lampbrush configuration, no subdivision of the double axis into chromatids is visible except at very limited regions. However, the loops are paired on each of the axes, so that at any level four loops of closely identical morphology (two on each axis) are present.

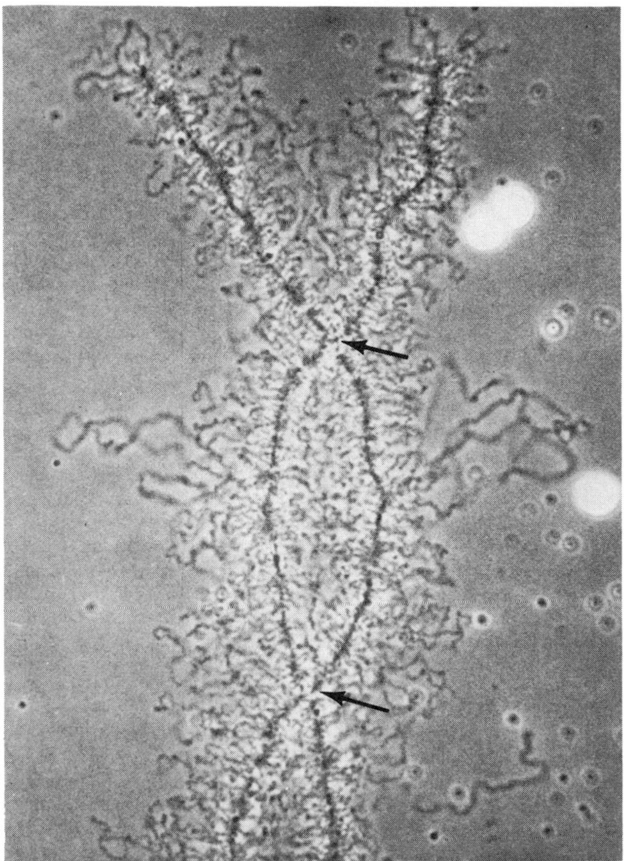

14–38. *Phase contrast light micrograph of lampbrush chromosomes isolated from the newt* Triturus viridescens. *The two main axes are the two homologous chromosomes at diplonema, still loosely held in a tetrad by chiasmata (arrows). Multiple, paired loops extend outward from the axes of the homologs.* $\times$ *470. Courtesy of J. G. Gall.*

If the lampbrush chromosomes are stretched by micromanipulation, it can be seen that each pair of

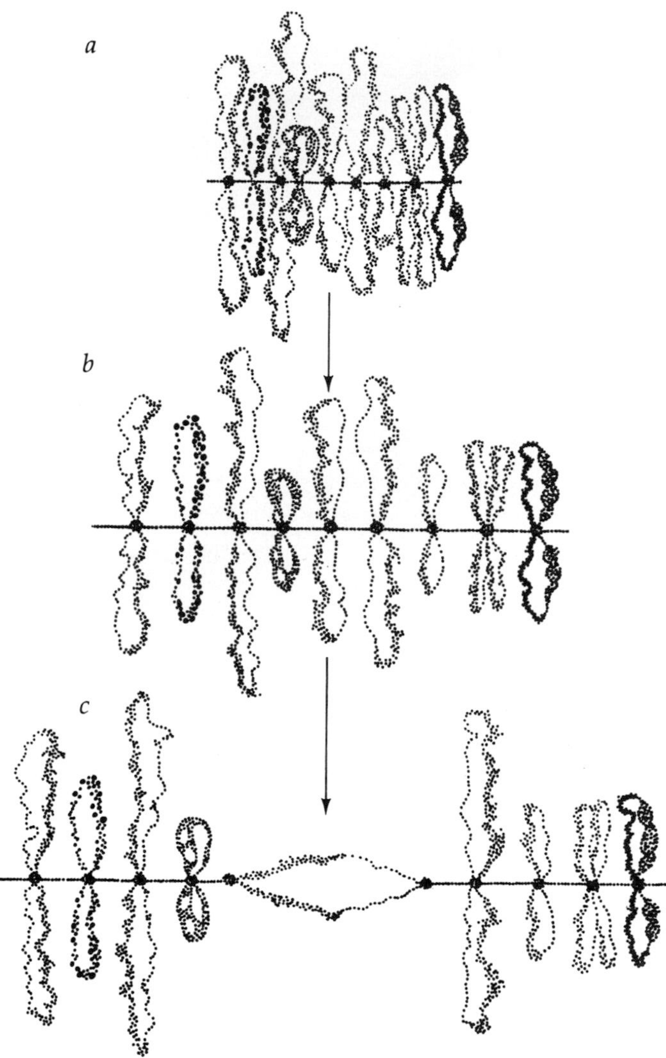

14–39. Formation of a double bridge in a lampbrush chromosome stretched by micromanipulation. The appearance of the bridge suggests that the two side loops are a part of the main axis of the chromosome and represent portions of the two chromatids of the homolog that have "spun out" from the main axis. Redrawn, original courtesy of H. G. Callan, Internat. Rev. Cytol. *15:1(1963).*

loops is inserted into the axis at a chromomere (Fig. 14–39a). If the chromosomes are stretched further (Fig. 14–39b and 14–39c), a chromomere may separate into two parts, forming a gap bridged by the two arms of the loop pair (Fig. 14–39c). This has been interpreted to mean that the loops are part of the continuous axis of the chromosome and represent a portion of the chromomere that has "spun out" from the main axis.

The chromomeres and pairs of loops number in the thousands in a complete set of chromosomes. The pairs of loops can be distinguished morphologically by differences in size, thickness, and the appearance of a coating or matrix that appears to cover the surface of each loop. This matrix may appear granular, fibrillar, or amorphous under the light microscope. While the loops of lampbrush chromosomes cannot be traced in thin sections for electron microscopy, isolated preparations (Fig. 14–40) often show that the loops are distinctly fibrillar. The lampbrush chromosomes are many times larger than either pachytene or late diplotene chromosomes from the same organism and may extend up to one millimeter in length.

Cytochemical experiments, undertaken by Joseph G. Gall and H. G. Callan have indicated that the matrix of the loops represents a coating of ribonucleoprotein synthesized during diplonema. The loops react intensely to stains for RNA under the light microscope; treatment with ribonuclease greatly reduces the thickness of the loop matrix. Proteases, such as trypsin or pepsin, also cause a significant reduction in the diameter of the loops. Deoxyribonuclease breaks the loops and the chromosome axis into short lengths. These experiments have been interpreted by Gall and Callan to mean that the loop consists of an axis of DNA surrounded by a matrix of ribonucleoprotein. The visibility of the loops depends on the matrix of ribonucleoprotein, which increases the diameter to levels within the limits of resolution by the light microscope.

Experiments with radioactive precursors bear out these interpretations. Gall and Callan, in both cooperative research and independent investigation, have shown that the loops become labeled very quickly if cells are exposed to a radioactive precursor of RNA. Radioactive precursors of protein cause label to appear somewhat more slowly in the loops. It is thought likely that the RNA of the loops is synthesized in this location, because RNA is known to be synthesized on DNA. But no intact ribosomes are present inside the nucleus; unless the protein of the matrix is synthesized by an as yet unknown mechanism, without the intervention of ribosomes, the protein found in this location must originate in the cytoplasm.

The matrix on many of the loops is asymmetrical, showing a difference in thickness at the two ends of the loop near the point of insertion into the basal

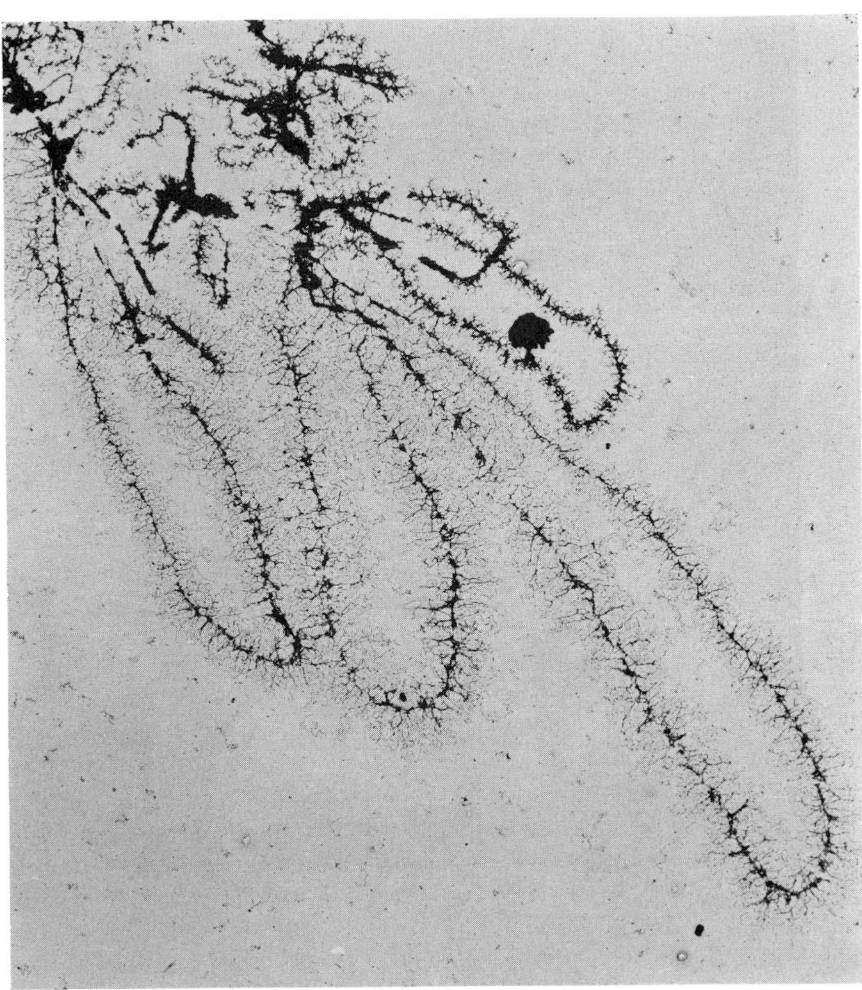

14–40. Electron micrograph of several loops of a lampbrush chromosome isolated from the Triturus viridescens. *The loops are distinctly fibrillar.* × *2,900. Courtesy of O. L. Miller, Oak Ridge National Laboratory.*

chromomere (Fig. 14–41). The pattern of label, if radioactive precursors of protein are used, has no particular relationship to the thick or thin ends of the loops; label for protein appears over all regions of the loop simultaneously. In contrast, exposure of the cells to radioactive precursors for RNA may show a very curious pattern in the sequence of appearance of the label. This pattern is seen most clearly in a pair of giant, granular loops that appear in several amphibians, including *Triturus cristatus*, the crested newt (Fig. 14–42a and 14–42b). Gall and Callan have shown that within a few hours after exposure of oocytes to tritiated uridine the first label is detected in the giant loops at the point of insertion of the thin end into the chromomere (arrows, Fig. 14–42a). After about two days, if the supply of tritiated uridine is maintained (the precursor is injected into the body cavity of the animals), a short zone extending from the narrow tip is labeled. Within 4–5 days, the label has progressed halfway around the loop; after 7 days about two-thirds of the giant loops are labeled. The loops become totally labeled after exposure to precursor for 10–14 days. As the point of insertion of the thick end of the loop into the axis is reached, the newly synthesized nucleoprotein is released from the chromosome. Gall and Callan have proposed that this pattern of label must result from one of two mechanisms; either the RNA matrix is synthesized at the point of insertion and subsequently moves around the loop by sliding along the DNA axis, or both axis and matrix move simultaneously, passing from the point of attachment of the thin end around the circumference of the loop.

It has not yet been possible to determine experimentally which of these alternatives actually accounts

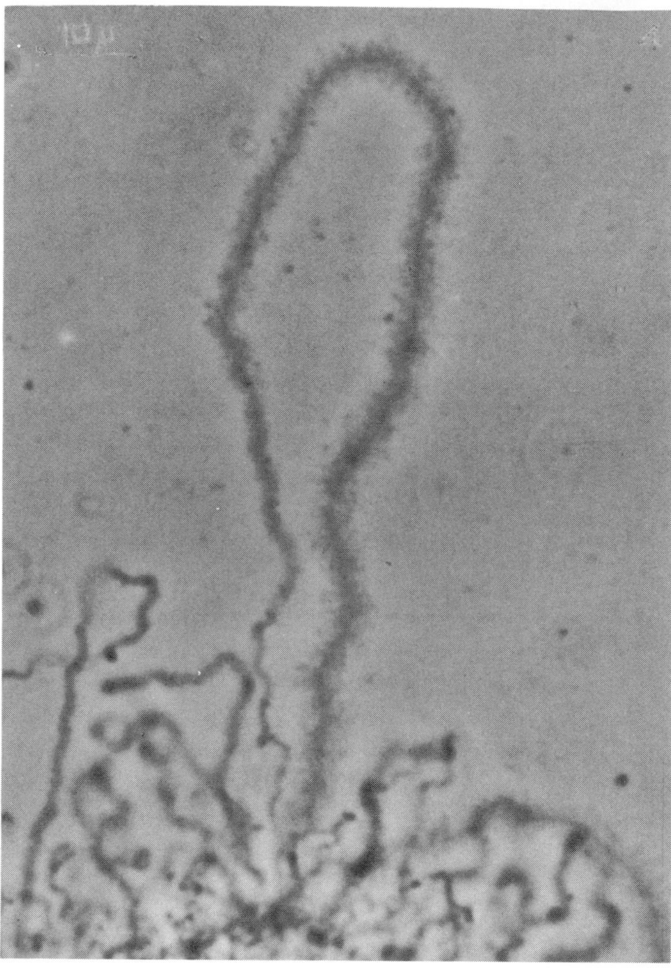

14–41. *A loop of the lampbrush chromosome of* Triturus cristatus. *The loop becomes progressively thicker in the clockwise direction.* × *1,300. Courtesy of H. G. Callan.*

for movement of the label around the giant loops. On purely intuitive grounds, Gall and Callan have favored the interpretation that both the axis and the ribonucleoprotein product are in movement. According to this hypothesis, the loop is continuously spinning out from the chromomere at the thin end and recoiling back at the thick end. The chromomere is thus considered to be bipartite, with one segment uncoiling and the other segment coiling, much like supply and takeup reels. The separation of chromomeres into two parts in stretched lampbrush chromosomes (see Fig. 14–39), with the gap bridged by the loop arms, supports this interpretation.

This hypothesis has led to a highly speculative but most interesting set of calculations. Gall and Callan point out that the total duration of the lampbrush stage in *T. cristatus* is about 200 days. If the giant loops are continuously moving and a complete circuit of a point on the axis from the origin around the loop to the point of insertion takes about 10 days (as indicated by the pattern of label), then at any given time about 10/200 = 1/20 of the total length of DNA wound into the chromomere is visible. Assuming further that all loops in the entire set of lampbrush chromosomes follow a similar pattern of movement,[6] then at any given time about one-twentieth of the total DNA in the cell is unwound into the loops. Gall estimated the total length of DNA visible in the loops to be 50 centimeters; if this represents one twentieth of the total, the oocyte nucleus would be expected to contain a total of 10 meters of DNA. These calculations are in surprisingly close agreement with the length expected for the known quantity of DNA per cell in the same amphibian.

The Functional Significance of Lampbrush Chromosomes

The function of the diplotene chromosomes during the lampbrush stage remains unknown. Undoubtedly, the synthesis of ribonucleoprotein at this stage is related to the growth of the oocyte; much of the RNA synthesized during diplotene is stored in the egg cytoplasm until fertilization, to be used during the early growth of the embryo. It is not at all clear, however, why the entire genome should be active in RNA synthesis at this stage. No part of the chromosome complement is without loops, and all loops seem to function in the synthesis of RNA. If Gall and Callan are correct in their assumption that the loops are in continuous movement, uncoiling and recoiling into the chromomeres, almost all of the total gene complement would be transcribed at some time during the diplotene stage. But it does not seem possible that the RNA

[6]The giant granular loops are atypical in that sequential label is not detectable in the smaller loops. Label in these loops appears over the whole structure almost simultaneously. However, Callan has detected sequential label of the small loops in oocytes recovering from actinomycin D blockage of RNA transcription.

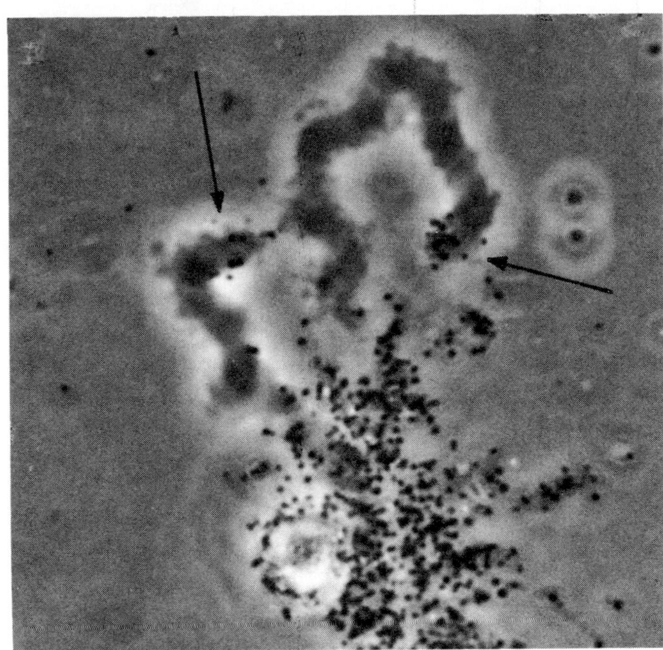

a

14–42. a, *Autoradiograph of the giant loops of the lampbrush chromosomes of* Triturus cristatus *after 4 hours' exposure to radioactive uridine. Exposed grains of the photographic emulsion are deposited over a short region at the insertion of one end of the loops (arrows).*

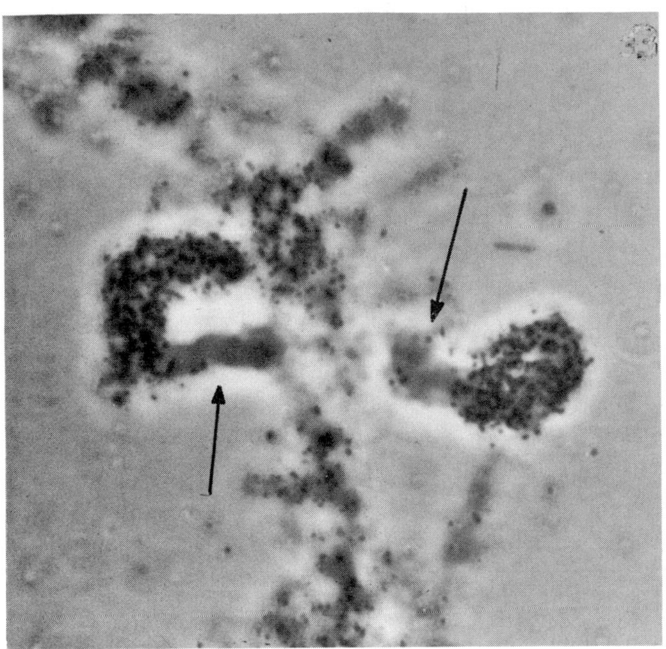

b

b, *The same pair of loops in a chromosome removed from the animal after 6 days exposure to tritiated uridine. The label has progressed about halfway around the loops. The thick ends of the loops (arrows) are not yet labeled. Courtesy of H. G. Callan.*

transcribed from the entire genome is required for the early development of the embryo. Further, most of the RNA stored in the amphibian oocyte appears to be ribosomal rather than messenger RNA (see below).

Callan has proposed that the activity of the chromosomes at this stage is related to the redundancy of DNA in the genome. According to Callan, the continual movement of the loops is part of a mechanism in which duplicate copies of a gene (the "slave" copies) are compared to a "master" copy. Corrections are made if any differences in sequence due to mutation are revealed by the comparison method. However, it is not obvious why intensive synthesis of ribonucleoprotein should be a part of this mechanism. There is no evidence supporting the Callan scheme at this time; the functional significance of the total activity of the genome during the lampbrush stage of diplonema remains obscure.

Nucleolar Synthesis during the Lampbrush Stage

In many amphibians, the development of the lampbrush configuration is accompanied by the production of large numbers of accessory nucleoli. If the RNA of the oocyte is analyzed by DNA-RNA hybridization or centrifugation, about 97–98 percent of the total proves to be rRNA. Therefore, even though the whole genome appears to be active during diplonema, most of the product of the synthetic machinery reaching the cytoplasm is rRNA. This is reflected in the morphology of the egg cytoplasm, which becomes packed with ribosomes toward the close of diplonema.

The nuclei of amphibians may contain as many as 600–1000 nucleoli during diplonema. By various means, including isolation and centrifugational analysis, each of the accessory nucleoli has been shown to contain a core of DNA that is a duplicate copy of the nucleolar site of the chromosome.

The synthesis of the extra DNA associated with the accessory nucleoli can be traced to pachytene. For example, during pachytene in the toads *Xenopus* or *Bufo*, a large quantity of DNA, synthesized on the chromosomes, becomes concentrated at one side of the nucleolus into a structure known classically as the "nuclear cap" (Fig. 14–43). In favorable preparations, this cap

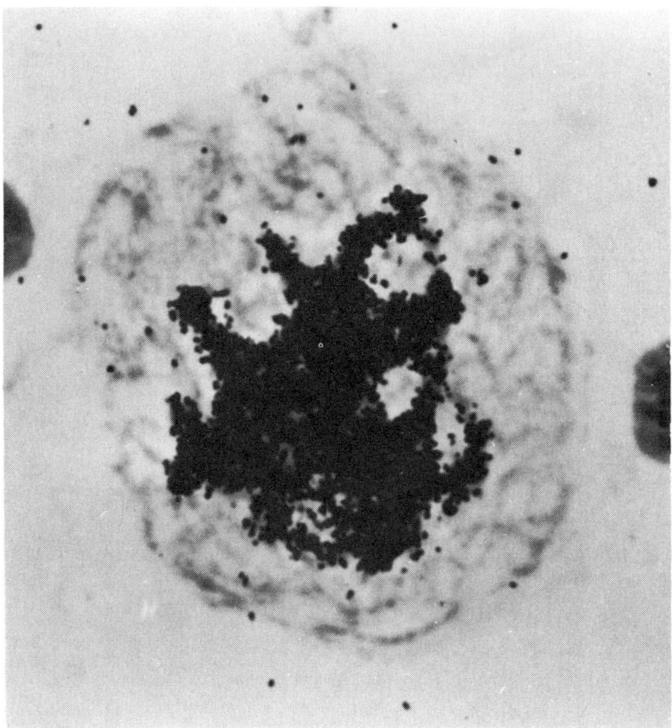

14–43. *Autoradiographs of the "nuclear cap" of* Xenopus. *A large quantity of DNA coding for ribosomal RNA becomes concentrated at one side of the nucleus during pachynema. This DNA was detected autoradiographically by hybridization* in situ *with radioactive ribosomal RNA.* × *1,800. Courtesy of J. G. Gall.*

can be seen to be completely separate from the chromosomes. Centrifugational analysis, carried out by Joseph G. Gall, has shown that the extra DNA has a density typical for the nucleolar organizer. At the onset of diplonema, the nuclear cap disintegrates, and the fragments develop into accessory nucleoli.

The accessory nucleoli are active in the synthesis of rRNA. Oscar Miller has succeeded in isolating the accessory nucleoli of the amphibian *Triturus* for electron microscopy. In Miller's preparations, these nucleoli, when dispersed by treatment with dilute salt solutions or distilled water, appear as long fibers less than 100 Å in diameter. Each fiber, which may form a closed circle ranging from 35 microns to 5 millimeters in diameter, is periodically coated with matrix material (Fig. 14–44). Miller found that the axis of the circle can be broken by deoxyribonuclease; ribonuclease removes the matrix material. Exposure of oocytes to tritiated uridine before isolation of the accessory nucleoli produces label only over the fibers of the matrix material. On the basis of these results, Miller proposes that the axis of the circle is a DNA molecule that duplicates the nucleolar organizer segment of the chromosome; the matrix represents recently synthesized RNA-protein. Thus the segments containing matrix probably represent nucleolar genes active in RNA synthesis.

The appearance of these segments at higher magnification (Fig. 14–45) is consistent with this interpretation. According to Miller, the axial filament underlying the matrix is probably one copy of the nucleolar organizer site and codes for 45S RNA. Each ribonucleoprotein fiber of the matrix, extending at right angles to the axial fiber, is presumably a partially transcribed molecule of 45S RNA. At the base of each matrix fibril is a dense granule that may represent a molecule of RNA polymerase engaged in the synthesis of RNA. The shortest ribonucleoprotein fibrils, at one end of the matrix, are probably the product of polymerase molecules recently initiating transcription. At the opposite end, the longest matrix fibrils represent nearly complete molecules of 45S RNA. If Miller's interpretations are correct, this remarkable micrograph is one of the very few examples of a gene caught in the act of transcription (another is Miller's photograph of an active gene from a bacterium; see Fig. 11–10).

The extensive synthesis of rRNA and the development of the lampbrush configuration are adaptations of the cellular machinery directed toward the synthesis of materials necessary for the maturation of functional gametes. Although this discussion of diplotene synthesis has concentrated on oocytes, it is certain that equivalent processes are also a part of the development of spermatocytes. In males of some species of *Drosophila*, for example, a limited development of loops closely similar to lampbrush chromosomes originates from segments of the Y chromosome at diplonema. Mutation of the Y chromosome causing absence of one or more of the loops results in varying morphogenetic effects on the spermatozoa, all of which produce nonfunctional sperm (see Hess, 1968).

The synthesis of RNA in oocytes and spermatocytes ends at the close of diplonema. At this time, the loops on lampbrush chromosomes, if present, retract and the chromosomes coil down into the diakinesis configuration. The accessory nucleoli and the nucleoli regularly located at the organizer segment of the chromosome disintegrate and disappear from the cell by the time prophase draws to a close. Many oocytes remain ar-

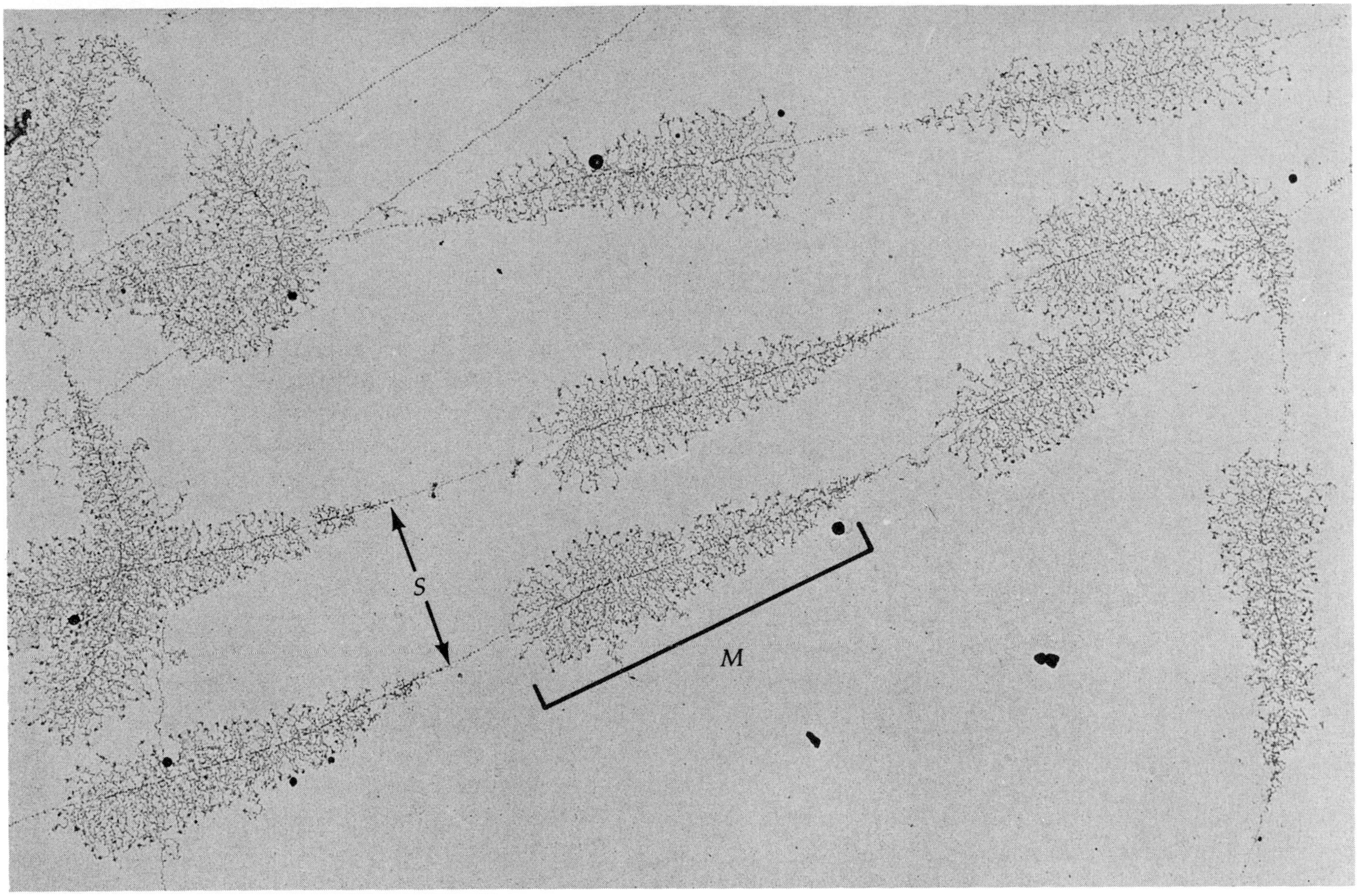

14–44. *Dispersed nucleoli isolated from* Triturus viridescens. *The axial strands are believed to represent the DNA of the cistrons coding for ribosomal RNA. The fibrillar matrix* (M) *coating the axial strands* (S) *represents ribosomal RNA-protein (see text).* × *22,500. Courtesy of O. L. Miller, Jr., and Barbara R. Beatty, Oak Ridge National Laboratory.*

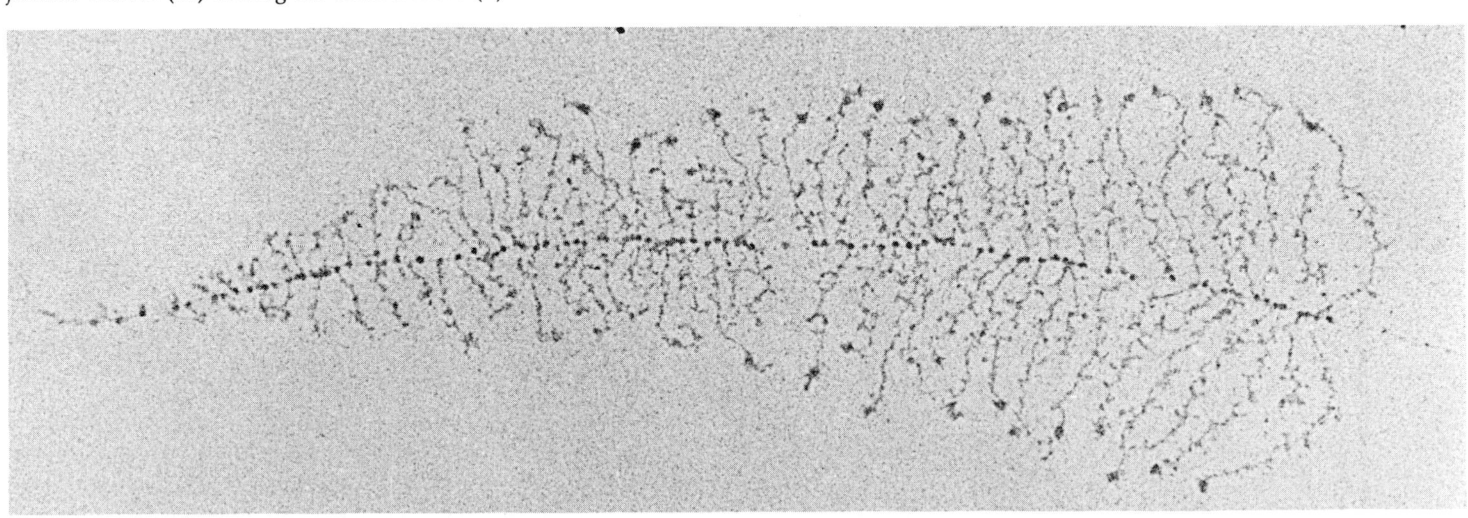

14–45. *A segment of the nucleolar material in Fig. 14–44 at higher magnification (see text).* × *60,000. Courtesy of O. L. Miller, Jr., and Barbara R. Beatty, Oak Ridge National Laboratory.*

rested in development at late prophase; spermatocytes continue through and complete the remaining stages of meiosis. In any case, both oocytes and spermatocytes subsequently undergo some morphological changes, which complete the conversion into functional gametes. These changes and representative events in fertilization are the subject of the following chapter.

Suggestions for Further Reading

Books

Lewis, K. R., and B. John, 1963. *Chromosome Marker.* Little, Brown & Company, Boston.

Mittwoch, Ursula, 1967. *Sex Chromosomes.* Academic Press, New York.

Swanson, C. P., 1967. *Cytology and Cytogenetics.* Prentice-Hall Inc., Englewood Cliffs, N.J.

Swanson, C. P., T. Merz, and W. J. Young, 1967. *Cytogenetics.* Prentice-Hall, Inc., Englewood Cliffs, N.J.

Wilson, E. B., *The Cell in Development and Heredity.* Macmillan, New York.

Papers and Review Articles

Bodmer, W. F., and A. J. Darlington, 1969. Linkage and recombination at the molecular level. In E. W. Caspari and A. W. Ravin, eds., *Genetic Organization,* vol. 1. Academic Press, New York, pp. 223–265.

Brown, D. D., and I. B. Dawid, 1968. Specific gene amplification in oocytes. *Science,* vol. 160, pp. 272–279. Description of the DNA associated with ribosomal RNA synthesis in oocytes.

Callan, H. G., 1963. The nature of lampbrush chromosomes. *International Review of Cytology,* vol. 15, pp. 1–34.

Callan, H. G., 1969. Biochemical activities of chromosomes during the prophase of meiosis. In A. Lima-de-Faria, ed., *Handbook of Molecular Cytology,* North-Holland Publishing Company, Amsterdam, pp. 540–552.

Coleman, J. R., and M. J. Moses, 1964. DNA and the fine structure of synaptic chromosomes in the domestic rooster (*Gallus domesticus*). *Journal of Cell Biology,* vol. 23, pp. 63–78.

Emerson, S., 1969. Linkage and recombination at the chromosome level. In E. W. Caspari and A. W. Ravin, eds., *Genetic Organization,* vol. 1. Academic Press, New York, pp. 267–360.

Grell, Rhoda F., 1969. Meiotic and somatic pairing. In E. W. Caspari and A. W. Ravin, eds., *Genetic Organization,* vol. 1. Academic Press, New York, pp. 361–492.

Hess, O., 1968. The function of the lampbrush loops formed by the Y chromosome of *Drosophila hydei* in spermatocyte nuclei. *Molecular and General Genetics,* vol. 103, pp. 58–71.

Kellenberger, G., M. L. Zichichi, and J. J. Weigle, 1961. Exchange of DNA in the recombination of bacteriophage λ. *Proceedings of the National Academy of Sciences,* vol. 47, pp. 869–878.

Luykx, P., 1965. Kinetochore-to-pole connections during prometaphase of the meiotic divisions in *Urechis* eggs. *Experimental Cell Research,* vol. 39, pp. 658–668. One of the very few descriptions of centromere structure during meiosis.

Meselson, Matthew, 1967. The molecular basis of genetic recombination. In R. A. Brink, ed., *Heritage from Mendel.* University of Wisconsin Press, Madison, pp. 81–104.

Meselson, M., and J. J. Weigle, 1961. Chromosome breakage accompanying genetic recombination in bacteriophage. *Proceedings of the National Academy of Sciences,* vol. 47, pp. 857–868.

Meyer, F., 1964. A possible correlation between the submicroscopic structure of meiotic chromosomes and crossing over. *Proceedings of the 3rd European Regional Conference on Electron Microscopy,* Prague, pp. 461–462.

Miller, O. L., and B. R. Beatty, 1969. Portrait of a gene. *Journal of Cellular Physiology,* vol. 74, Supplement 1, pp. 225–232. Remarkable photographs of isolated nucleolar genes.

Moens, P. B., 1968a. The fine structure of meiotic chromosome polarization and pairing in *Locusta migratoria* spermatocytes. *Chromosoma,* vol. 28, pp. 1–25. The synaptinemal complex and its attachments to the nuclear envelope, studied by serial sections of a single nucleus.

Moens, P. B., 1968b. The structure and function of the synaptinemal complex in *Lilium longiflorum* sporocytes. *Chromosoma*, vol. 23, pp. 418–451.

Monesi, V., 1964. Ribonucleic acid synthesis during mitosis and meiosis in the mouse testis. *Journal of Cell Biology*, vol. 22, pp. 521–532.

Moses, M. J., 1968. Synaptinemal complex. *Annual Review of Genetics*, vol. 2, pp. 363–412. An exhaustive and excellent review of the complex.

Parchman, L. G., and H. Stern, 1969. The inhibition of protein synthesis in meiotic cells and its effect on chromosome behavior. *Chromosoma*, vol. 26, pp. 298–311.

Rees, H., 1961. Genotypic control of chromosome form and behavior. *Botanical Review*, vol. 27, pp. 288–318.

Rhodes, M. M., 1961. Meiosis. In J. Brachet and A. E. Mirsky, eds., *The Cell*, vol. 3. Academic Press, New York, pp. 1–75.

Richardson, C. C., 1969. Enzymes in DNA metabolism. *Annual Review of Biochemistry*, vol. 38, pp. 795–840. Pages 829–830 of this article contain a valuable criticism of current hypotheses of the role of enzymes in recombination.

Rossen, J. M., and M. Westergaard, 1966. Studies on the mechanism of crossing over II: meiosis and the time of chromosome replication in the Ascomycete *Neottiella rutilans*. *Comptes Rendus Laboratoire Carlsberg*, vol. 35, pp. 233–243.

Roth, T. F., 1966. Changes in the synaptinemal complex during meiotic prophase in mosquito oocytes. *Protoplasma*, vol. 61, pp. 346–386.

Roth, T. F., and M. Ito, 1967. DNA-dependent formation of the synaptinemal complex at meiotic prophase. *Journal of Cell Biology*, vol. 35, pp. 247–255.

Sheridan, W. F., and R. J. Barnett, 1967. Cytochemical studies of chromosomal ultrastructure. *Journal of Cell Biology*, vol. 35, p. 125a.

Sheridan, W. F., and H. Stern, 1967. Histones of meiosis. *Experimental Cell Research*, vol. 45, pp. 323–335.

Stern, H., and Y. Hotta, 1967. Chromosome development during development of meiotic tissue. In L. Goldstein, ed., *The Control of Nuclear Activity*, Prentice-Hall, Inc., Englewood Cliffs, N.J., pp. 47–76. Describes the authors' experiments demonstrating limited DNA synthesis during zygonema-pachynema.

Stern, H., and Y. Hotta, 1969. Biochemistry of meiosis. In A. Lima-de-Faria, ed., *Handbook of Molecular Cytology*, North-Holland Publishing Company, Amsterdam, pp. 520–539.

Taylor, J. H., 1959. Autoradiographic studies of nucleic acids and proteins during meiosis in *Lilium longiflorum*. *American Journal of Botany*, vol. 46, pp. 477–484.

Taylor, J. H., 1965. Distribution of tritium-labeled DNA among chromosomes during meiosis I: spermatogenesis in the grasshopper. *Journal of Cell Biology*, vol. 25, pp. 57–67.

Taylor, J. H., 1967. Patterns and mechanisms of genetic recombination. In J. H. Taylor, ed., *Molecular Genetics*, part 2. Academic Press, New York, pp. 95–135.

Taylor, J. H., 1969. The structure and duplication of chromosomes. In E. W. Caspari and A. W. Ravin, eds., *Genetic* Organization, vol. 1. Academic Press, New York, pp. 163–221.

Thomas, C. A., 1966. Recombination of DNA molecules. *Progress in Nucleic Acid Research and Molecular Biology*, vol. 5, pp. 315–341.

Westergaard, M., 1964. Studies on the mechanism of crossing over I: theoretical considerations. *Comptes Rendus Laboratoire Carlsberg*, vol. 34, pp. 359–405.

Wolfe, S. L., 1969. Molecular organization of chromosomes. In E. E. Bittar and N. Bittar, eds., *The Biological Basis of Medicine*, vol. 4. Academic Press, London, pp. 3–42. Contains a review of the synaptinemal complex and polycomplexes.

15
Gametogenesis and Fertilization

Meiosis reduces the chromosome number and quantity of DNA per nucleus to precisely one-half of the value at G_1 of premeiotic diploid cells. This also results in the reduction of chromosomes from the diploid to the haploid number; the two members of each chromosome pair are separated and appear singly in the products of meiosis.[1] In animals, the haploid "generation," represented by the gametes, is short lived. If fertilization is successfully accomplished, fusion of the gametes returns the chromosomes to the diploid level. In plants, the haploid generation may exist for greater lengths of time and become elaborated by mitotic divisions into a many-celled, haploid organism. But at some time, gametes are differentiated within the haploid plant, following ordinary mitotic divisions. Fusion of these gametes returns the chromosome complement to the diploid value and reestablishes the alternate diploid generation.

This alternation between haploidy and diploidy, a characteristic of all sexually reproducing organisms, is maintained by alternate meiosis and fertilization. For the process to run full cycle, the gametic nuclei produced after meiosis must be transported and brought together so that fertilization may occur. The difficulties presented by transport and union of gametes have different solutions in various plants and animals.

This chapter is concerned with the cellular mechanisms evolved to accomplish gametic transport and

---

[1]In many plants, more than one pair of each chromosome type may be present. These plants, called *polyploids,* follow the same pattern of alternation of generations; the multiple pairs of chromosomes in cells of the premeiotic generation are each separated during meiosis and appear as single chromosomes in the alternate postmeiotic generation.

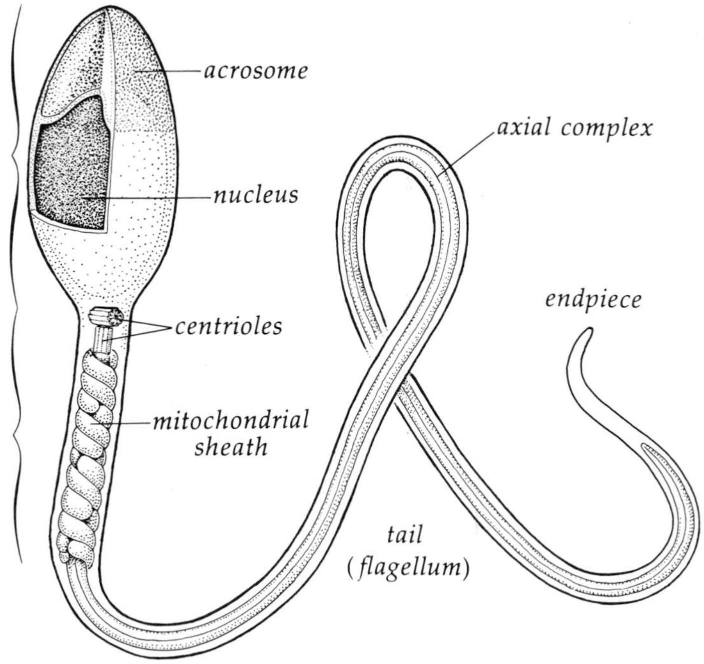

15–1. Diagram illustrating the basic morphology of an animal spermatozoan.

union. Because so much variation exists, particularly in plants, it is impossible to provide more than a few examples of the major systems. With this in mind, gametogenesis and fertilization are first discussed in plants and animals in which the male gamete (the sperm) is motile and the female (the egg) is nonmotile. The chapter closes with a description of the highly specialized systems of gamete formation and fertilization in higher plants.

Spermiogenesis

Morphological Changes in Development of Spermatozoa in Animals

In almost all male animals the completion of anaphase II is immediately followed by metamorphosis of the four products of meiosis into mature sperm. The metamorphosis, called *spermiogenesis,* includes extensive changes in both nucleus and cytoplasm. Before spermiogenesis is initiated the four products of meiosis (called *spermatids*) are not conspicuously different in morphology from ordinary somatic cells. Each contains all of the major cytoplasmic organelles, including mitochondria, the Golgi complex, ribosomes, centrioles, and complete membrane systems around the nucleus and at the cell surface. Within the nucleus, fibrils and granules closely similar to the interphase configuration are visible. Nucleolar material, however, may not be recognizable. During metamorphosis into mature sperm, each organelle undergoes extensive changes; eventually, the nucleolus and much of the cytoplasmic mass is eliminated from the cell.

The mature spermatozoon in animals is usually a greatly elongated cell, consisting typically of a head, middle piece, and tail (Fig. 15–1). Almost the entire mass of the sperm head is formed by the nucleus, which may be flattened, greatly elongated, hooked, or twisted into a spiral. The morphology of the head is so diverse and species-specific that many animals can be identified by the morphology of the sperm head alone (see Figs. 15–2 and 15–3). Under the electron microscope, the nucleus of the mature sperm usually appears to be uniformly electron-dense and shows no evidence of the fibrils or granules present in somatic nuclei. A nuclear envelope consisting of two concentric membranes enclosing a greatly attenuated perinuclear space is often visible, but pore complexes rarely occur.

Only a thin layer of cytoplasm surrounds the nucleus in the sperm head. The usual cytoplasmic structures and organelles, such as mitochondria, endoplasmic reticulum, Golgi complex, and ribosomes are absent from this layer. At the anterior end of the sperm

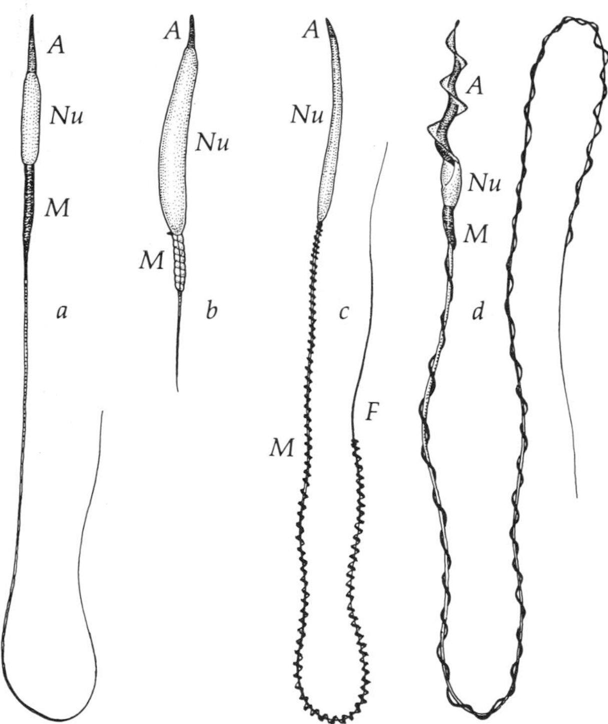

15–2. *Variations in sperm morphology. These spermatozoa are from representative species of birds and reptiles.* A, *acrosome,* F, *flagellum,* M, *middle piece,* Nu, *nucleus.* a, Chameleon, b, Testudo *(tortoise),* c, Columba *(pigeon),* d, Chloris *(finch).*

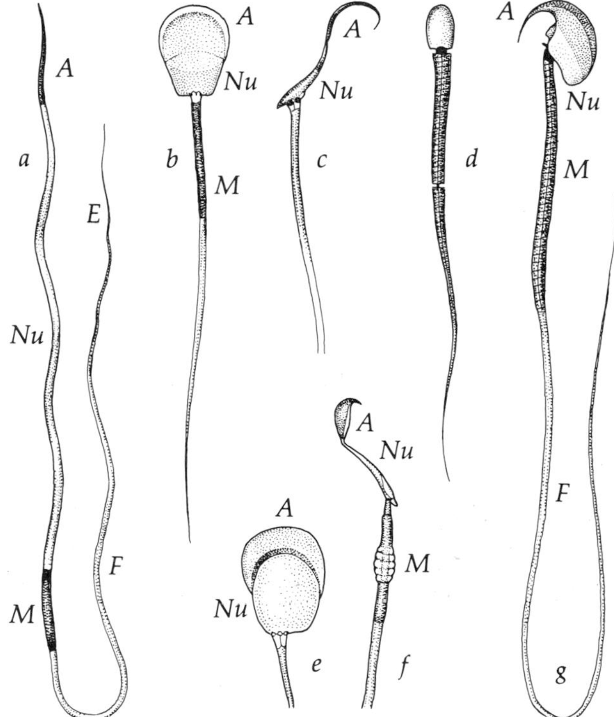

15–3. *Mammalian sperm.* A, *acrosome,* E, *endpiece of flagellum,* M, *middle piece,* Nu, *nucleus.* a, Echidna *(spiny anteater);* b, Cervus *(deer);* c, Sciurus *(squirrel);* d, Verspugo *(bat);* e, f, *Guinea pig;* g, Mus *(mouse).*

head, just underneath the plasma membrane, is the *acrosome* (Fig. 15–1), a small cap-like structure that arises from the Golgi complex during spermiogenesis. The acrosome, which functions in the penetration of the egg by the sperm during fertilization, shows wide variations in size and morphology between different species.

The plasma membrane surrounds the nucleus and acrosome, extending toward the posterior as a sheath around the middle piece and tail of the spermatozoan. The middle piece (see Fig. 15–1) consists primarily of a sheath of mitochondria surrounding the tail axis just behind the head. These mitochondria are frequently highly modified, and in some organisms, such as butterflies and moths, can be identified as mitochondria only by tracing their developmental origins during spermiogenesis (see p. 436). Other conspicuous organelles of the middle piece are the centrioles and a variety of structures associated with the attachment of the tail axis to the base of the nucleus. Two centrioles are normally present. One lies close to the base of the nucleus just outside the nuclear envelope (the *proximal centriole*). The second, the *distal centriole,* is located behind the first and gives rise to the axial complex of the tail (see Fig. 15–23).

The tail, extending from the distal end of the middle piece, contains little more than an *axial complex* with a sparse sheath of cytoplasm enclosed by the plasma membrane. The axial complex arises from the distal centriole and extends from the base of the nucleus through the middle piece to the tip of the tail. This complex, visible under the light microscope as a central fiber running the length of the tail, consists of a regular arrangement of microtubules (see Fig. 15–21) that in cross section appears as an outer circle of nine double tubules (called *peripheral doublets*) surrounding a central pair of single microtubules (*central singlets*); the entire configuration is called the "9 + 2 pattern." This arrangement of the microtubules of the sperm tail is remarkably uniform throughout the animal and plant kingdoms, and with few exceptions is found in the axis of all flagella from any source except the prokaryotes. Other structures, called *accessory fibers* (see Fig. 15–25) may also be present in the tail cytoplasm surrounding the axial complex.

The basic arrangement of the sperm cell into head, middle piece, and tail is typical of almost all animals. Nonflagellate sperm (those lacking a tail) occur in a few species of nematodes, some crustacea, a few arachnids (mites), and some myriapods. Usually, the absence of the tail is correlated with the absence of the organelles of the middle piece, including mitochondria and centrioles. Multiple sperm tails are found very rarely and appear in only a few species, such as flatworms (Rhabdocoeles) and as abnormal types in the snail *Viviparus* and other animals normally bearing more typical sperm. Other less extensive modifications of the basic arrangement into head, middle piece, and tail are also found in a number of species among the barnacles, copepods, myriapods, and a few insects and annelids.

The organelles of the mature sperm are formed and take up their positions during a series of regular rearrangements initiated as early as anaphase II of meiosis. The centrioles move to the side of the nucleus that will form the posterior of the sperm head; the mitochondria of the spermatid move to the same position. The Golgi complex forms a series of vesicles that fuse to form the *proacrosomal granule*, a structure that later in spermiogenesis develops into the acrosome. After secretion of the proacrosomal granule, the Golgi complex moves to the posterior end of the developing sperm cell and is later eliminated near the completion of spermiogenesis along with most of the residual cytoplasm of the spermatid. During this time, the nucleus undergoes a progressive elongation and reduction in volume. Within the nucleus, the chromatin fibrils coalesce into granule-like aggregations or sheets, which eventually fuse into a uniformly dense mass. At the same time, the extrachromosomal fibers, granules, and other structures of the nuclear sap are eliminated, often being expelled into the cytoplasm by blebbing of the nuclear envelope. At the close of spermiogenesis, the condensed sperm nucleus, although greatly elongated or flattened, may be reduced to as little as 0.5 percent of the volume of the original spermatid nucleus. These morphological changes of the nucleus are almost always accompanied by a replacement of the histones typical of somatic cells by more basic histones or by the basic protein *protamine*.

The distal centriole, as noted, gives rise to the 9 + 2 axial complex of flagellar microtubules. As this complex elongates, it carries with it an extension of the cytoplasm and plasma membrane, eventually forming the tail. As these developments take place, the mitochondria take up a position near the base of the nucleus. In some animals, particularly the insects, the mitochondria fuse at this time into a composite structure known as the *nebenkern* (*neben* = near, *kern* = nucleus). Whether or not the nebenkern is formed, the mitochondria gradually extend outward in a sheath surrounding the developing sperm tail to form the middle piece. In many vertebrates and some invertebrates, the mitochondria of the middle piece become arranged end to end in a tightly packed spiral of very shallow pitch. Within the sheath, the cristae of the mitochondria may disappear or become highly modified, or the matrix may fill with a dense, paracrystalline substance. The end result in these cases, termed the *mitochondrial derivative*, bears little resemblance to typical somatic mitochondria.

These morphological changes in development of the sperm will be taken up in more detail in the sections to follow. Although the functional significance of many of the details of sperm morphology is unknown, the developmental changes obviously provide motility to the mature sperm and package the nuclear contents in a form that is transportable with a minimum expenditure of energy.

The development of the sperm nucleus. Early in spermiogenesis, after completion of telophase II, spermatid nuclei still contain chromatin fibers and interchromosomal material closely similar to somatic nuclei (Fig. 15–4). At this early stage, the sex chromosome can still be distinguished in many spermatids as a densely coiled mass of chromatin. Remnants of the nucleolus may also persist, often in association with the sex chromosome and residual elements of the synaptinemal complex. These persistent associations of sex chromosome, nucleolus, and synaptinemal complex in the early spermatid are sometimes collectively termed the "sex vesicle."

Later in spermiogenesis, before elongation of the nucleus begins, condensation of the nuclear material is initiated. Often this condensation is correlated with

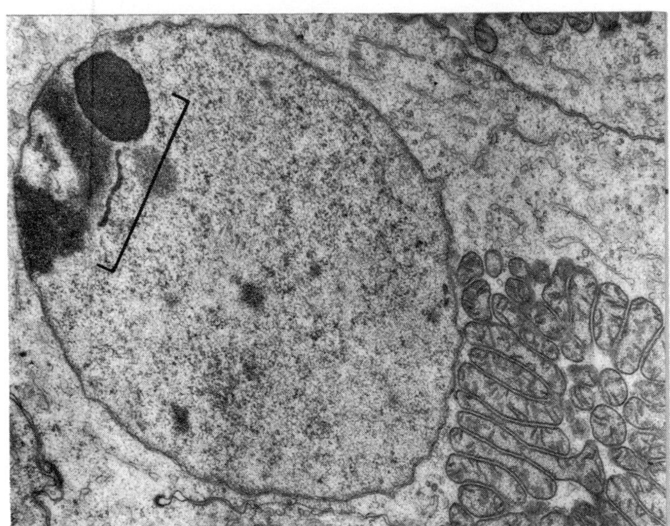

15–4. *An early stage in spermiogenesis in the house cricket* Acheta domestica. *Interchromosomal material is still identifiable, interspersed between the chromatin. The tightly coiled sex chromosome is associated with the nucleolus and an "inclusion body" into a structural complex classically termed the "sex vesicle" (bracket).* × *11,000. Courtesy of J. S. Kaye, R. McMaster-Kaye, and The Rockefeller University Press,* [J. Cell Biol. *31:159 (1966)*]

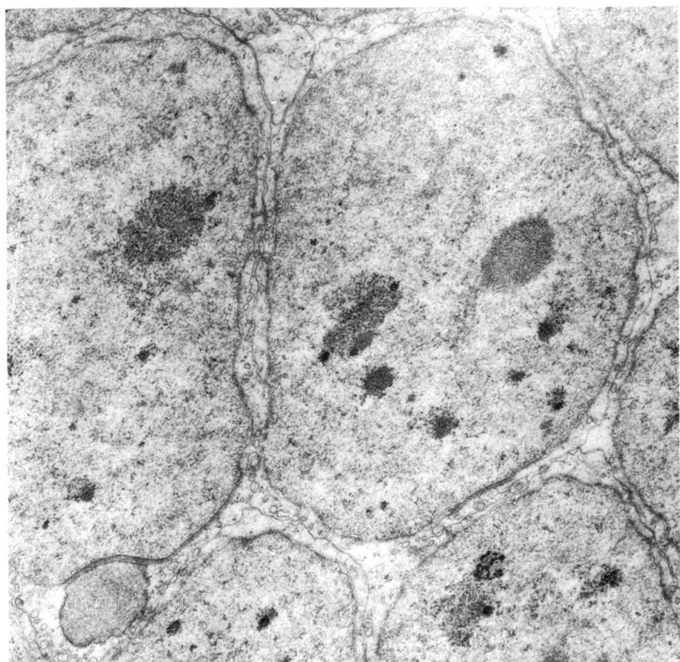

15–5. *A later spermatid in* Acheta. *Interchromosomal material has disappeared from the nucleus by this stage.* × *12,000. Courtesy of J. S. Kaye, R. McMaster-Kaye, and The Rockefeller University Press,* [J. Cell Biol., *31:159 (1966)*]

disappearance of the interchromosomal granules and fibrils. Jerome S. Kaye and Rachel McMaster-Kaye have shown that in the house cricket the disappearance of interchromosomal fibrils and granules coincides with the loss of nonhistone protein from the spermatid nucleus. In Fig. 15–4, a nucleus of the house cricket, extrachromosomal material is readily identified; at a later stage (Fig. 15–5), when nonhistone protein can no longer be detected in the nucleus by cytochemical tests, this material has disappeared. This coincidence suggests that some of the granular material may be nonhistone protein.

Later, in many animals and in some plants (see Fig. 15–29), the chromatin of the spermatid nucleus becomes concentrated in a dense layer just inside the nuclear envelope (Fig. 15–6). The significance of this stage, called the "shell stage," is unknown. At this time, the interior of the nucleus is almost free of chromatin fibers, and the central space is occupied by unidentified proteinaceous material. Some thickening or coarsening of the chromatin fibrils may also be observed at this time. Kaye estimates that the diameter of the chromatin fibrils in the shell stage shown in Fig. 15–6 has increased from the 80 Å typical of the early spermatid to 130–170 Å. The physical basis for this increased diameter is obscure; side-by-side aggregation, coiling, and folding have all been proposed as possible mechanisms underlying the change.

Soon after these modifications take place the spermatid nucleus begins to elongate. As elongation proceeds, the chromatin material aggregates into thick fibers, coarse granules, or an interconnected system of sheets or lamellae (Figs. 15–7, 15–8, and 15–9). As these structures form, the last traces of the sex chromosome and nucleolus disappear. The lamellae may be highly ordered and often produce striking patterns in transversely sectioned nuclei. In the snail, *Viviparus*, for example, the lamellae are regularly separated by spaces 100 Å in diameter; in other organisms lamellae form a highly regular honeycomb configuration. The thickness of the sheets or lamellae, which are believed to be formed by aggregation of chromatin fibrils, has been reported to vary in different species from as little

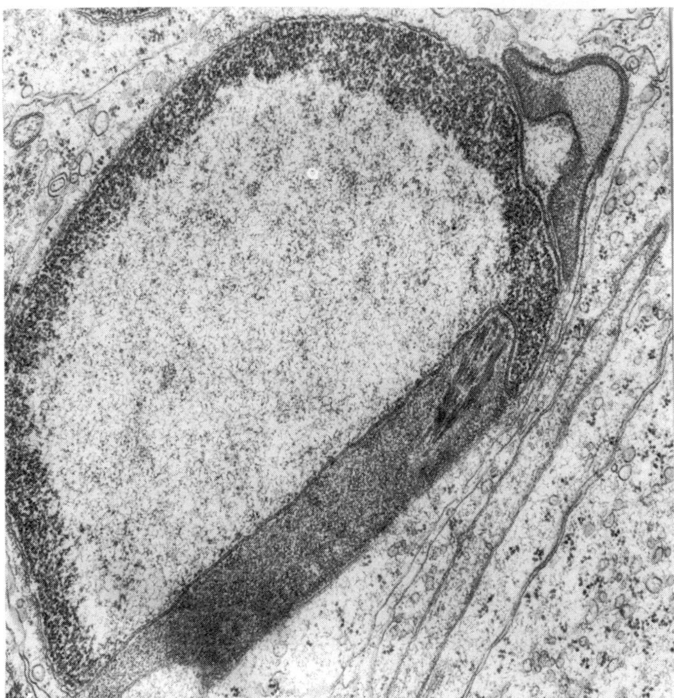

15–6. *Shell stage in the house cricket* Acheta. *The chromatin is concentrated into a dense layer just under the nuclear envelope.* × *12,000. Courtesy of J. S. Kaye, R. McMaster-Kaye, and The Rockefeller University Press,* [J. Cell Biol., *31:159 (1966)*]

as 70 Å to 250–300 Å. The formation of these structures in the nucleus of the spermatid is usually correlated with a change in the nucleoprotein component of the nucleus from somatic histones to arginine-rich histones or protamines (see below).

The granules or sheets become more densely packed as elongation of the spermatid nucleus proceeds until, at the completion of spermiogenesis, the nucleus appears as a uniformly dense mass (Fig. 15–10). Although lamellae or other periodic structures cannot be discerned in this mass in the mature spermatid nucleus, other lines of evidence indicate that a high degree of order persists. Freeze-etch preparations of sperm nuclei show the presence of lamellae even though these cannot be seen in thin sections (see Koehler, 1966). X-ray diffraction and analysis of the birefringence of sperm nuclei also indicate that the nucleoprotein is ordered into a semicrystalline array (see discussion of the X-ray diffraction of sperm nuclei in Chapter 8).

Condensation of the chromatin fibrils by aggregation into thick fibers or granules occurs in most vertebrates, including amphibians, birds, and mammals. In a few species, such as the frog, clumping is not pronounced, and the chromatin fibrils simply become more closely packed until the intervening spaces disappear. This process is also typical of a few invertebrates, including the earthworm and some molluscs. Highly ordered lamellae as an intermediate stage in condensation commonly appear only in snails and insects; this pattern is absent from vertebrates except for the lizards and snakes. (For a complete tabulation of the known variations in the conversion of nuclear structure during spermiogenesis, see Kaye, 1969.)

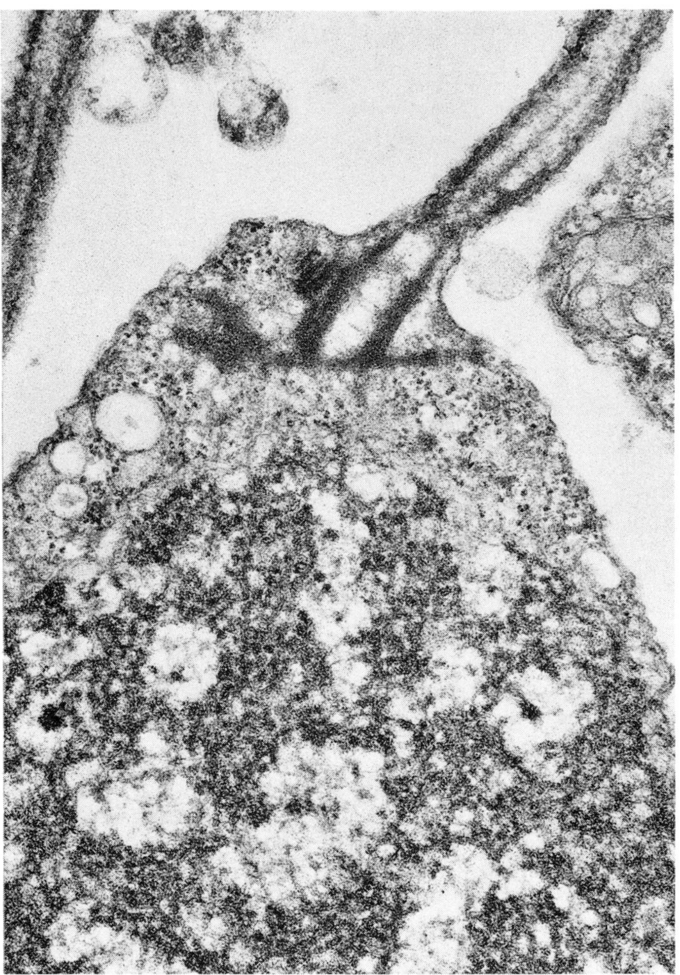

15–7. *Aggregation of nuclear material into coarse granules in spermiogenesis in a nucleus of the sea urchin* Arbacia. × *46,000. Courtesy of F. J. Longo, E. Anderson, and Academic Press Inc.* [J. Ultrastr. Research, *27:486 (1969)*].

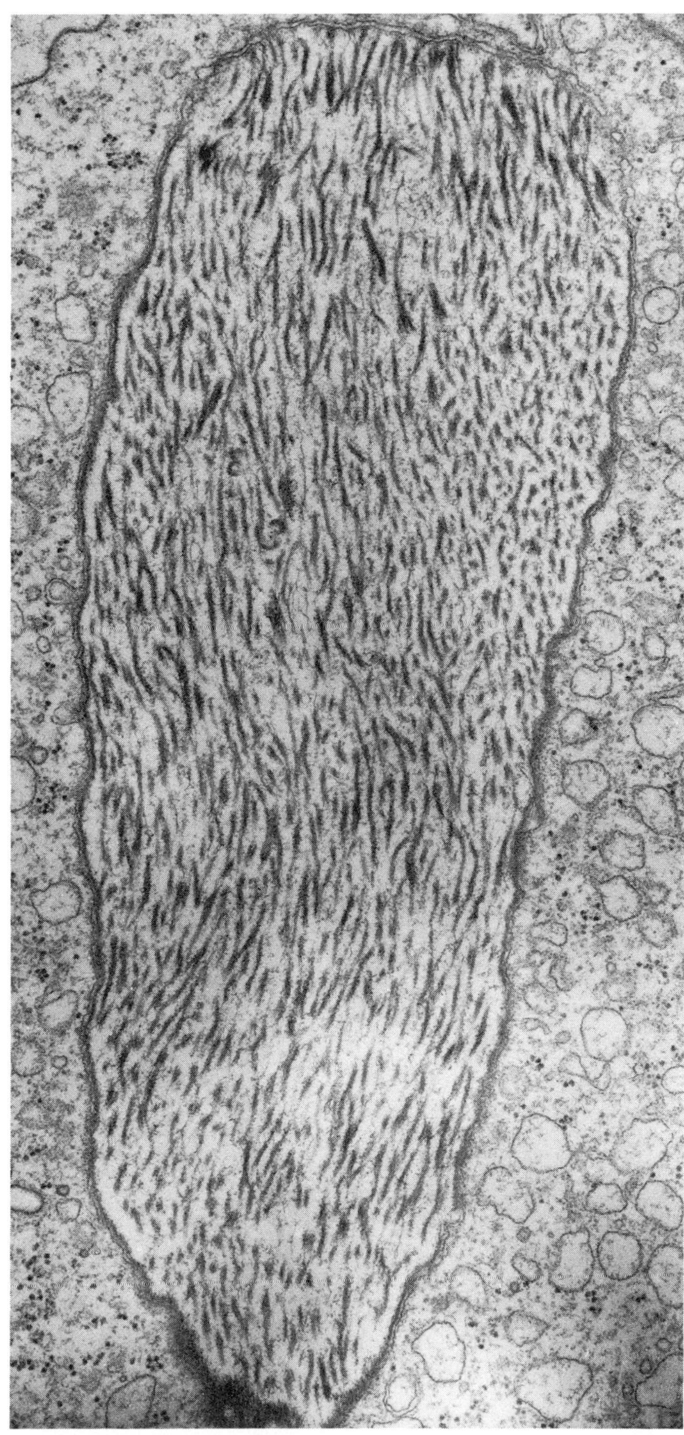

15–8. A nearly longitudinal section of a spermatid nucleus of the snail Viviparus contectoides, *showing aggregation of chromatin into coarse fibers during spermiogenesis. × 31,000. From* Handbook of Molecular Cytology, *ed. A. Lima-de-Faria, 1969. Courtesy of J. S. Kaye and North-Holland Publishing Company, Amsterdam.*

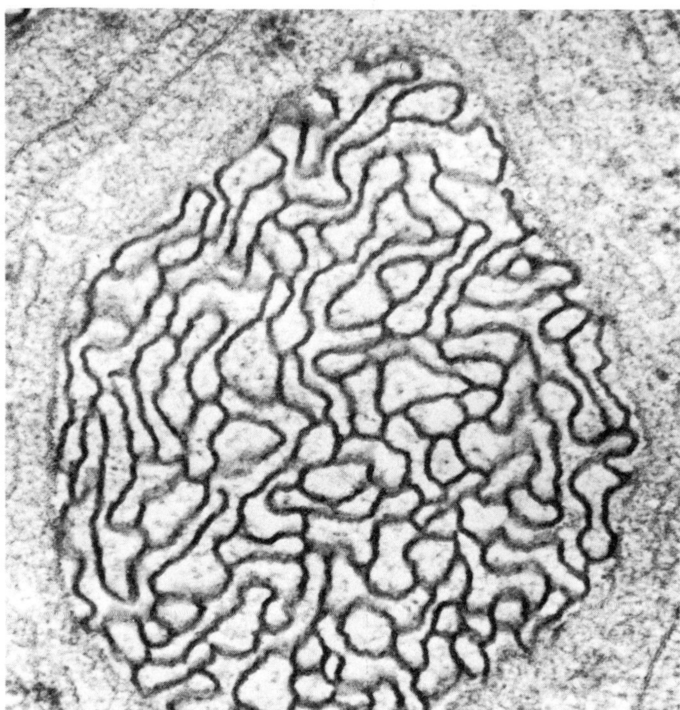

15–9. Aggregation of chromatin into interconnected sheets in a spermatid of a grasshopper. × 60,000. From Handbook of Molecular Cytology, *ed. A. Lima-de-Faria, 1969. Courtesy of J. S. Kaye and North-Holland Publishing Company, Amsterdam.*

The histone-protamine shift in maturing sperm. Friedrich Miescher, who first identified the nucleic acids in the 1800's (see Chapter 8), found that the nucleic acid in salmon sperm nuclei is complexed with a strongly basic protein. This protein, later called protamine, has different properties from the histone proteins associated with DNA in somatic cells. In the 1950's, Max Alfert discovered that somatic histones are replaced by more basic proteins late in spermiogenesis. This change is first detected as an increase in the quantity of arginine in relation to other amino acids. In different organisms, the increase of arginine proceeds to various degrees of completeness, from slight increases up to the high levels found in some protamines, in which two-thirds of the amino acids are arginine residues. In some species, little or no histone of the somatic type remains in the mature sperm nucleus.

These changes in nuclear proteins appear at the same time that morphological rearrangements of the chroma-

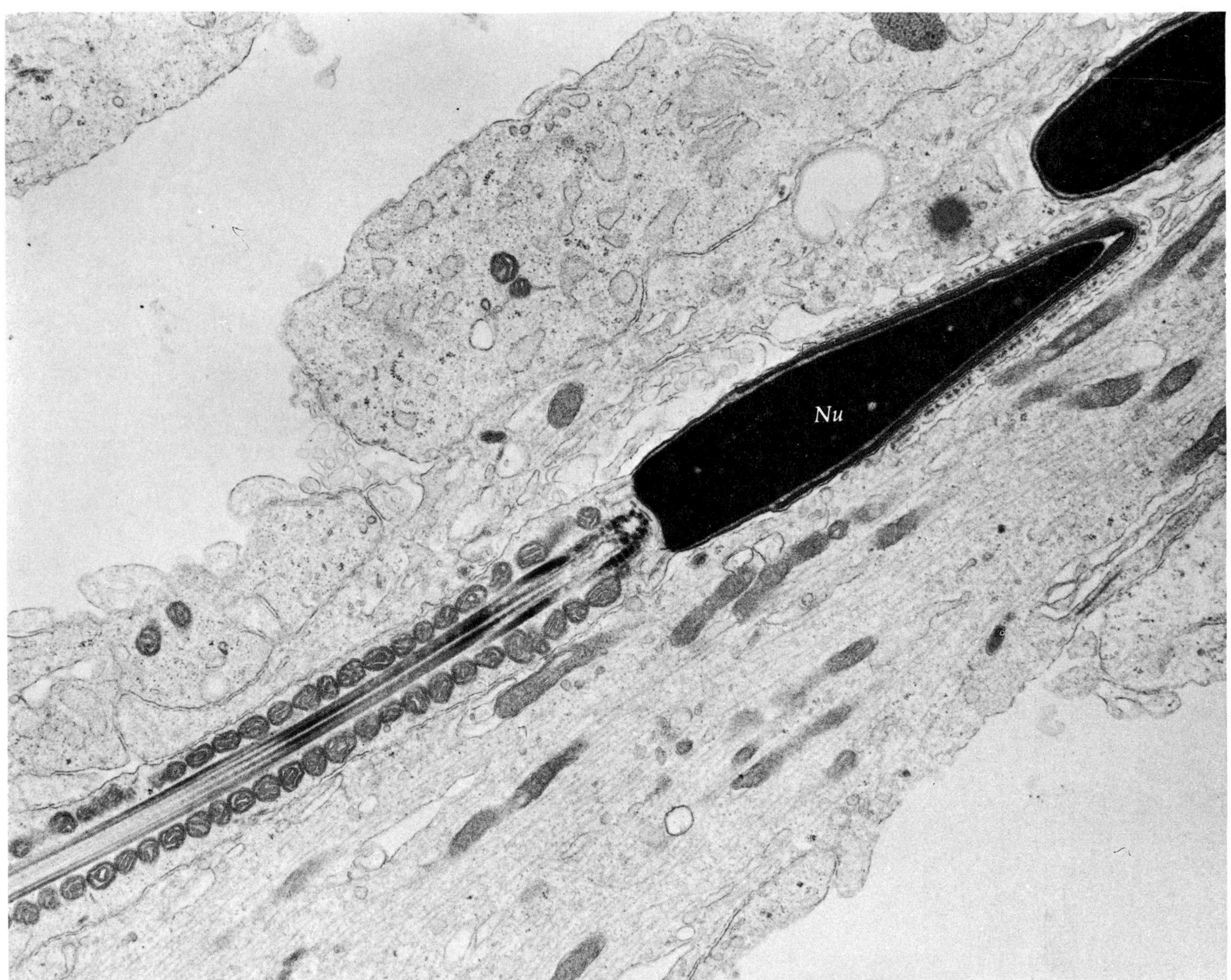

15–10. Nearly mature spermatozoan of a marmoset. The nucleus (Nu) *is uniformly dense and no longer shows internal differentiation into fibrils.* × *18,000. Courtesy of J. B. Rattner, B. R. Brinkley, and Academic Press Inc.* [J. Ultrastr. Research *32:316 (1970)*]

tin become visible in sperm nuclei under the electron microscope. David P. Bloch, working with the snail *Helix,* found that the appearance of lamellae in the nucleus was correlated with an increase of arginine in the nuclear histones. Later in development, as metamorphosis into the compact, uniformly dense nucleus of the mature sperm became complete, he noted that the arginine-rich histones were replaced by protamine.

Bloch has classified the proteins of the mature sperm into categories based on the degree of replacement of somatic type histone by arginine-rich types. These categories have been designated by representative genus names because separation and chemical identification of the various arginine-rich proteins has not yet been possible. At one end of the spectrum is the *Salmon* type, with a protamine as the major protein; arginine,

Table 15–1 Distribution of Nucleoprotein Types in the Sperm of Animals and Plants

	Salmon	*Mouse*	*Mytilis*	*Rana*	*Crab*
Roundworms (*Ascaris*)		X			
Flatworms (*Planaria*)				X	
Echinoderms			X	X	
Annelids	X		X	X	
Arthropods					
Crustacea				X	X
Insects		X		X	
Molluscs	X	X	X		
Vertebrates					
Fish	X		X	X	
Amphibians	X		X	X	
Birds	X				
Mammals		X			
Plants*				X	

*One report exists of *Salmon* type protein in the sperm of *Sphaerocarpus*, a liverwort.

Constructed from information in D. P. Bloch, A catalog of sperm histones. *Genetics*, Suppl. 61:93 (1969).

making up two-thirds of the total residues, is the sole basic amino acid present. The *mouse* type is rich in arginine, but less so than in protamine; here the proteins are larger and more complex. The *Mytilus* (surf clam) type is intermediate in arginine content, and in complexity falls between somatic histones and protamines. The *Rana* (leopard frog) type is very similar to somatic histones, and a fifth type, the *crab*, contains little or no basic protein. These categories are arbitrary, and undoubtedly more subtle gradations than these five exist in nature. The distribution of these protein types among a number of animals and a few plants is given in Table 15–1.

Several hypotheses have been advanced regarding the significance of the histone-protamine shift, including proposals that an increase in arginine-rich protein is a factor in the condensation and streamlining of the sperm head, protects the chromatin from osmotic or other damage during fertilization, or simply inhibits template activity of the sperm chromatin. But according to Bloch, who has carefully analyzed a considerable body of evidence, there seems to be little relationship between the type of protein present and the degree of condensation or shape of the sperm head. Similarly, on the molecular level, the amount of arginine does not appear to be related to the degree of crystalline packing in the DNA-protein of the mature sperm. Also apparently unrelated are the relative longevity of the sperm, or whether fertilization is internal or external. The latter point might be important, because the solubility of nucleoprotein in fresh or salt water varies with the relative percentage of arginine. Bloch notes that although closely related groups of animals tend to have similar kinds of sperm nucleoproteins, there is no evidence of any evolutionary trend in the distribution pattern of sperm protein types. The fact that histones in some animals are apparently not replaced by more basic proteins during spermiogenesis demonstrates that genetic inactivation of sperm DNA does not require the histone-protamine transition. Thus, at present, the significance of replacing histone by arginine-rich proteins or protamine is unknown. It is interesting that in evolution the somatic histones have been conserved and show relatively few differences between species as compared to the variety of proteins found in sperm nuclei.

Development of the acrosome. The acrosome, a complex organelle peculiar to the sperm cell, is consistently found in animal spermatozoa from all sources except a few insects (caddis flies) and the higher teleost fishes. An analysis of the role of the acrosome during fertilization (see p. 464) has shown that its function is the same in all cases: enzymes and other substances concentrated into the acrosome dissolve a pathway for the sperm through the surface coats of the egg during fertilization. Among the acrosomal enzymes important in this function are *hyaluronidase* and *zona lysin*. Hyaluronidase specifically attacks *hyaluronic acid*, a constituent of the "intercellular cement" found on the cell exterior. This substance binds together the follicle cells surrounding some eggs. Zona lysin is specific in activity for the substance of the zona pellucida, another exterior coat of the egg (see below). Also present in the acrosome as major components are several glycolipids and polysaccharides of unknown function.

The acrosome is bounded by a continuous single membrane and lies at the anterior end of the sperm in a space between the plasma membrane and the nuclear envelope (Fig. 15–11). Frequently, the plasma, acrosomal, and nuclear membranes are so closely spaced that they appear fused. In other sperm, as in the example shown in Fig. 15–11, a layer of "periacrosomal material" surrounds the acrosome and clearly separates the acrosomal membrane from the adjacent plasma and nuclear membranes. In many species, the space between the acrosome and the anterior tip of the nucleus is occupied by a fibrous or rod-like structure. This structure, classically termed the *perforatorium*, may be so extensive that it extends into a deep invagination in either the surface of the nucleus (as in Fig. 15–11) or the base of the acrosome. During fertilization, the perforatorium in some species, such as the sea urchin, extends into a fiber that contacts the egg surface and is considered to aid in penetration or attachment of the sperm to the egg.

The interior of the acrosome is filled with an electron-dense granular or fibrillar substance. This material is often differentiated into complex substructures (see Fig. 15–16). The relationship of this morphological differentiation to the function of the acrosome is unknown. During fertilization the acrosome is ruptured or lysed, and the acrosomal substance is expelled to the exterior of the cell.

The acrosome first appears during spermiogenesis as a collection of vesicles arising from the Golgi complex (Fig. 15–12), usually at the anterior end of the developing sperm cell near the nucleus. These vesicles gradually fuse into a single large vesicle, which forms the proacrosomal granule (Fig. 15–13). When synthesis is complete, the granule moves from the Golgi complex and attaches to the nuclear envelope at the anterior end of the cell (Fig. 15–14). Subsequently, the developing acrosome undergoes internal differentiation, often developing extensive substructure as in Figs. 15–15 and 15–16 (the sequence of micrographs in Figs. 15–12 to 15–16 outline acrosome development in the house cricket).

At maturity, the acrosome remains separated from the surrounding cytoplasm by a continuous single membrane. This structural arrangement, the manner of derivation from the Golgi complex, and the presence of the contained lytic enzymes suggests that the acrosome may be a highly specialized lysosome, with its digestive activity directed toward the egg surface.

After synthesis of the elements of the acrosome, the Golgi complex moves to the posterior end of the spermatid, eventually to be sloughed off late in spermiogenesis along with other remnants of the cytoplasm. It is interesting that in spermiogenesis residual elements such as the Golgi and large volumes of cytoplasm are simply eliminated by blebbing of the cell surface; apparently lysosomal destruction or other mechanisms of internal disassembly do not take place within the developing sperm.

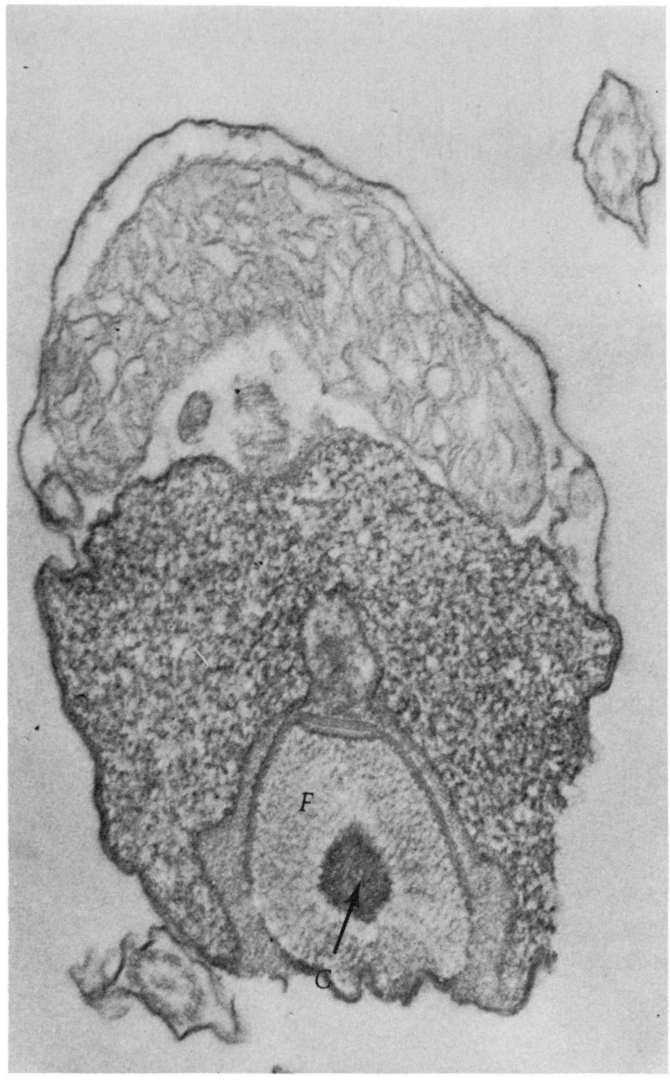

15–11. The acrosome of the starfish Asterias amurensis. F, *fibrillar material:* G, *acrosomal granule.* × *65,000. Courtesy of Y. Hagiwara and Academic Press Inc.* [J. Ultrastr. Research *18:551 (1967)*]

Formation of the nebenkern and middle piece. Soon after initiation of spermiogenesis, the mitochondria aggregate into a mass at the posterior of the nucleus. The subsequent metamorphosis of these mitochondria may vary from simple aggregation to development of highly modified mitochondrial derivatives. In some

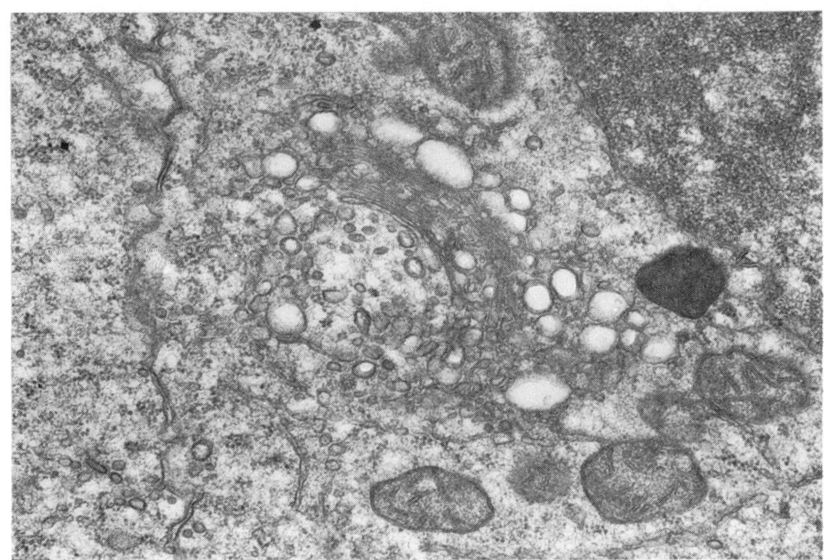

15–12. Acrosome formation in the house cricket, at an early stage of spermiogenesis. The Golgi complex, located near the nucleus, is active in secretion of numerous vesicles. These contribute to formation of the proacrosomal granule. × 23,000. Courtesy of J. S. Kaye and The Rockefeller University Press, from J. Cell Biol. *12:411 (1962).*

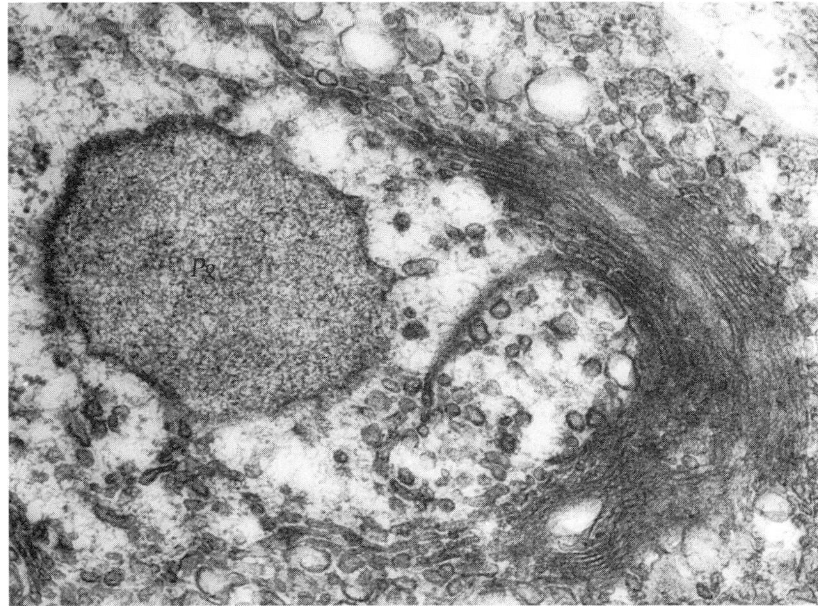

15–13. Acrosome formation in the house cricket. Golgi vesicles have fused to form the proacrosomal granule (Pg). *These initial developments take place at the posterior end of the nucleus. × 45,000. Courtesy of J. S. Kaye and The Rockefeller University Press, from* J. Cell Biol. *12:411 (1962).*

pulmonate snails, the mitochondria are simply distributed unchanged in the region of juncture between the head and tail of the sperm. In mammals, mitochondria are formed into a spiral wrapped around the axis of the sperm tail in the middle piece. Individual mitochondria retain identity in the spiral, and abut or slightly overlap in an end-to-end alignment. The cristae of the mitochondria of the mammalian spermatozoan become modified into concentric layers parallel to the outer mitochondrial membrane. Otherwise, their morphology remains similar to somatic mitochondria.

In other species, mitochondria are fused into a composite structure, the nebenkern, before distribution into the sheath of the middle piece. The fine structure of the development of the nebenkern and its subsequent metamorphosis into the mitochondrial derivative in insects was extensively studied by J. André and later by David M. Phillips (the micrographs shown in

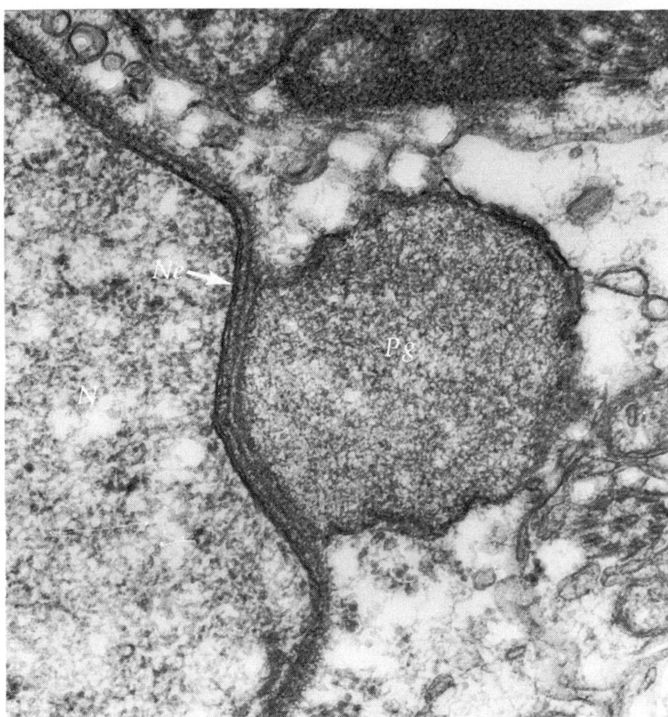

15–14. Attachment of the proacrosomal granule to the surface of the nucleus. N, *nucleus;* Ne, *nuclear envelope;* Pg, *proacrosomal granule.* × *52,000. Courtesy of J. S. Kaye and The Rockefeller University Press, from* J. Cell Biol. *12:411 (1962).*

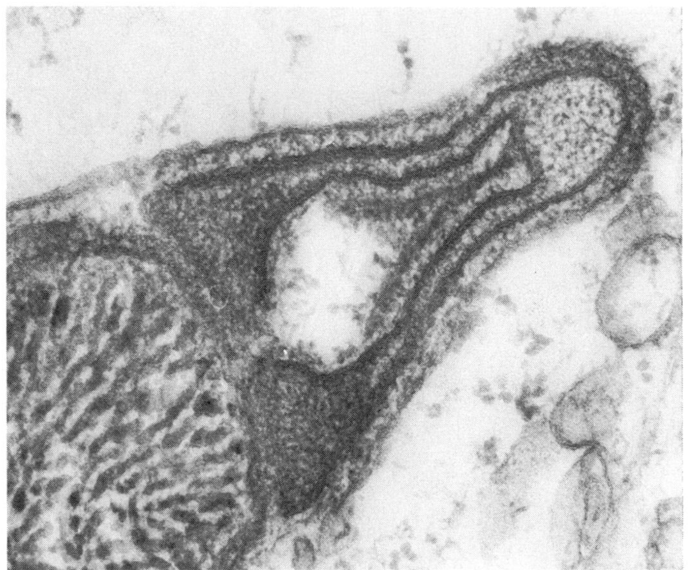

15–15. Differentiation of the acrosome of the house cricket at an intermediate stage. A deep invagination has appeared in the surface of the developing acrosome facing the nucleus. The acrosome has the shape of a hollow cone. Within, the "inner cone" (a structure analogous to the perforatorium) has initiated development. By this time, the acrosome has migrated to the anterior end of the developing spermatid. Note the aggregation of chromatin into coarse, longitudinal fibers. × *65,000. Courtesy of J. S. Kaye and The Rockefeller University Press, from* J. Cell Biol. *12:411 (1962).*

Figs. 15–17 to 15–20 are from a study of insect spermiogenesis; Phillips, 1970). In insects, the nebenkern develops as a fused mass of interlocking mitochondria at the base of the nucleus (Fig. 15–17). By carefully tracing the contortions of the fused outer membranes of the mitochondrial mass, it can be shown that at the completion of fusion, the nebenkern consists of only two large interdigitated mitochondria. At this stage, the internal structure of the nebenkern is still closely similar to somatic mitochondria. After development to this point, the two fused mitochondrial masses separate, become distributed around the axis of the flagellum (Fig. 15–18), and progressively elongate into the sheath of the middle piece. The two halves of the nebenkern usually take up positions on opposite sides of the flagellum. By the time development is complete, the middle piece in insects extends nearly to the tip of the sperm tail. In most insects, the internal structure of the mitochondrial sheaths then undergoes some degree of reorganization. In the mitochondrial derivatives shown in Fig. 15–19, the cristae have become arranged into regularly spaced, short extensions that extend perpendicularly into the matrix. The matrix is continuous and filled with dense matter. In the derivative toward the top of the micrograph, periodicity is apparent in the matrix; this is characteristic of mitochondrial derivatives in insects. This periodicity may vary from 90 to 300–400 Å and is occasionally seen in a "herringbone" pattern. In cross section, the mitochondrial derivatives in different species frequently take on highly individual arrangements (Fig. 15–20). As Phillips points out, the functional significance of derivative formation, which is without parallel in somatic cells, remains unknown.

The middle piece in all animal sperm consists primarily of the mitochondrial sheath. At the region of juncture of the base of the sperm tail with the nucleus, various dense structures, termed *basal plates, nuclear caps,* or *centriolar adjuncts* are found (see Figs. 15–17

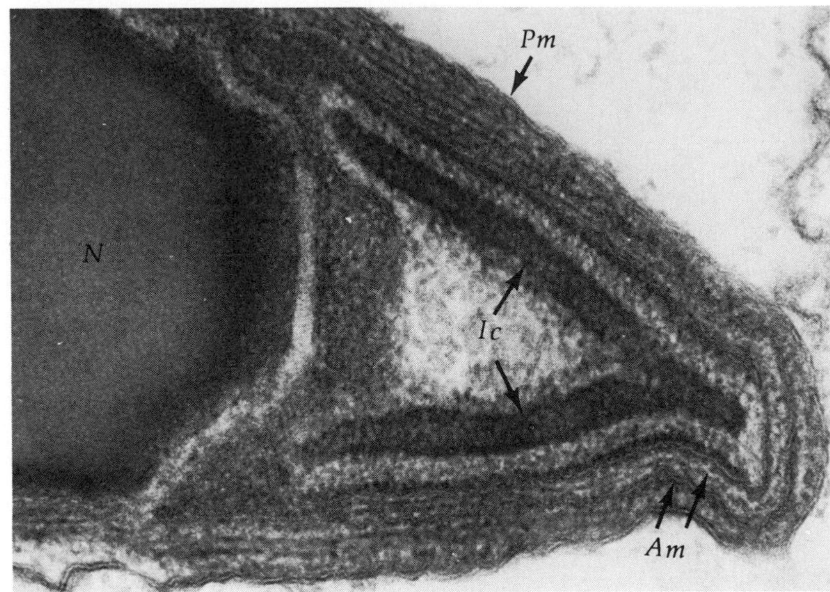

15–16. *Mature acrosome of the house cricket.* Pm, *plasma membrane;* Am, *acrosomal membrane;* Ic, *inner cone (perforatorium);* N, *nucleus. By this stage, the nucleus has become uniformly dense.* × *106,000. Courtesy of J. S. Kaye and The Rockefeller University Press, from* J. Cell Biol. *12:411 (1962).*

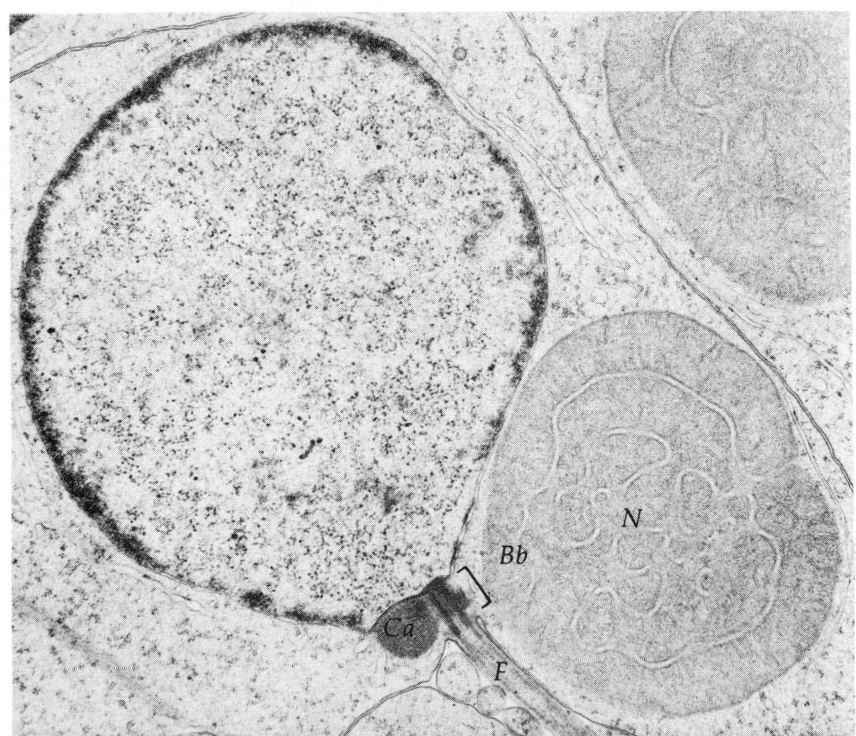

15–17. *An early stage in the development of the nebenkern* (N) *in the insect* Euschistus. *At this stage, progressive fusion of mitochondria leads to a composite structure consisting of two large, interdigitated mitochondria.* Ca, *centriolar adjunct;* Bb, *basal body;* F, *flagellum.* × *14,000. Courtesy of D. M. Phillips and The Rockefeller University Press.* [J. Cell Biol. *44:243 (1970)*]

and 15–18). These structures are believed to serve as an anchor or place of attachment of the sperm tail to the head. Nuclear caps are commonly found in vertebrate sperm, except for the human. The distal end of the middle piece in some groups, including mammals, is marked by a ring-like, dense structure of unknown function termed the *annulus* or *ring centriole.*

Sperm tails, centrioles, and sperm motility: Structure of the axial complex of the sperm tail. The axial complex (or *axoneme*) of the sperm tail is constructed from a system of microtubules that extends from the base of the head to the tip of the tail. The arrangement of the microtubules of the axoneme is remarkably uniform and is perhaps the most regularly repeated morphologi-

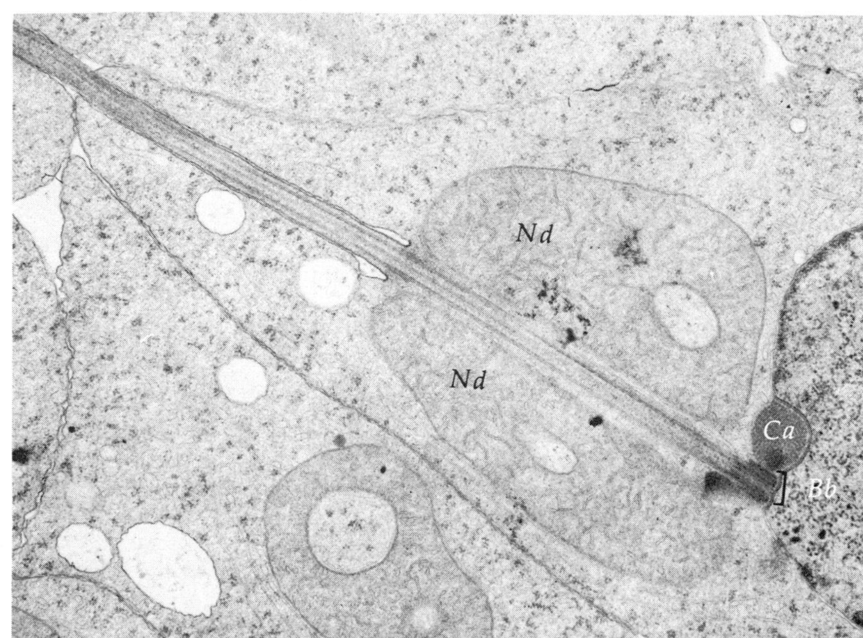

15–18. A later stage in the development of mitochondrial derivatives in Euschistus. *The two nebenkern derivatives* (Nd) *have taken up positions on opposite sides of the flagellum.* Ca, *centriolar adjunct;* Bb, *basal body.* × *10,000. Courtesy of D. M. Phillips and The Rockefeller University Press.* [J. Cell Biol. *44:243 (1970)*]

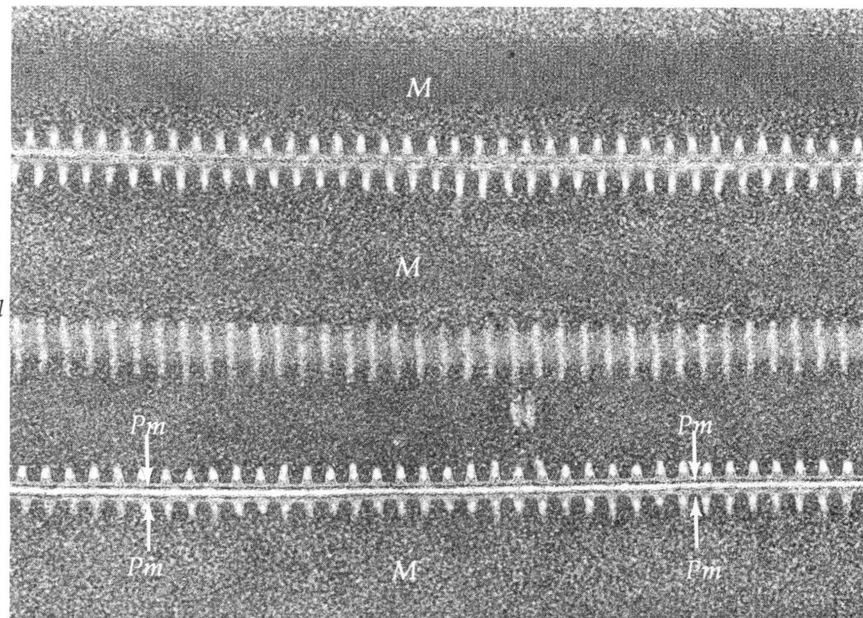

15–19. A longitudinal section through the mitochondrial derivatives of three closely aligned sperm tails of a beetle. Pm, *plasma membrane;* M, *matrix. The matrix of the mitochondrial derivative at the top of the picture shows a regular transverse periodicity.* × *92,000. Courtesy of D. M. Phillips and The Rockefeller University Press.* [J. Cell Biol. *44:243 (1970)*]

cal pattern in living organisms. This complex is found in the axis of sperm tails, flagella, and cilia from any source except bacteria—including all protozoan, plant, and animal species. The microtubules of the axial complex are arranged into a cylinder, consisting of nine peripheral double tubules (*doublets*) arranged in a circle around two single central tubules (Figs. 15–21 and 15–22). The two central singlets are separated by a space approximately 50–100 Å wide. The complete system of microtubules in the axoneme ranges from 1400 to 2500 Å in diameter and averages about 2000 Å in most species. At this diameter, it is just visible under the light microscope as a fiber in the axis of the sperm tail.

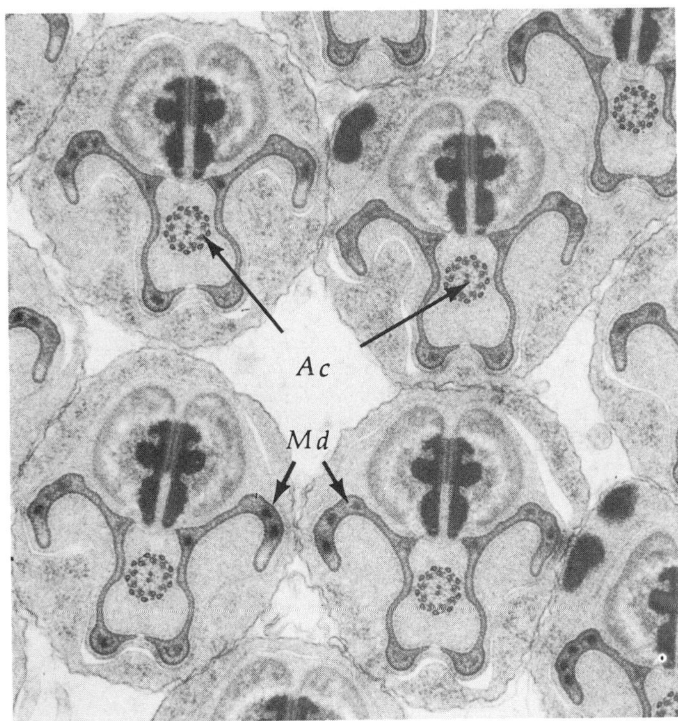

15–20. *A group of spermatids of the plant hopper* Acanalonia, *sectioned at a level in the middle piece. The mitochondrial derivatives* (Md) *exhibit a highly complex morphology, which is characteristic of the species.* Ac, *axial complex.* × *21,000. Courtesy of D. M. Phillips and The Rockefeller University Press.* [J. Cell Biol. *44:243 (1970)*]

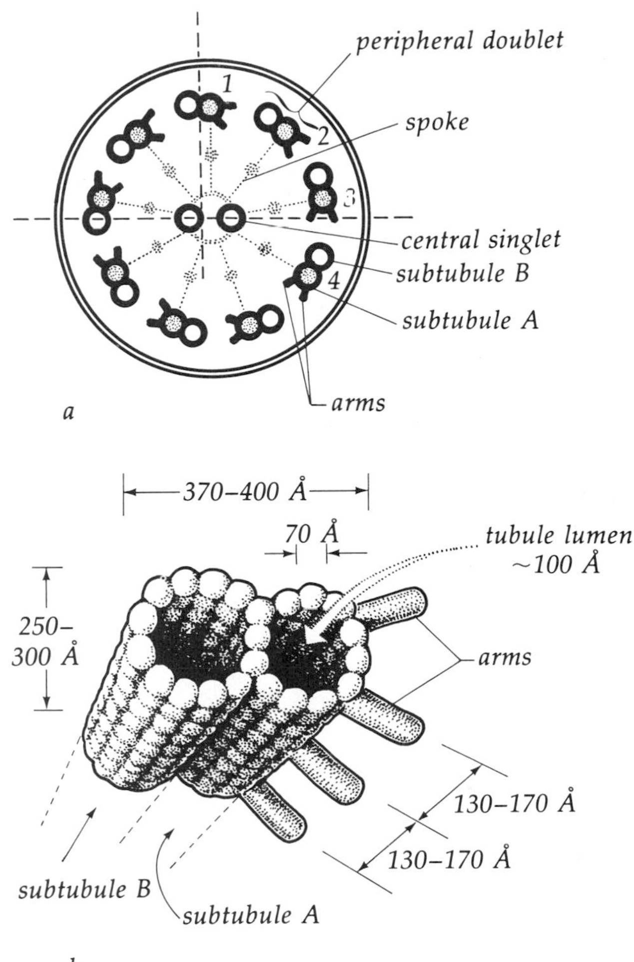

15–21. a, *The "9 + 2" arrangement of microtubules and accessory structures in the axial complex.* b, *Details of tubule ultrastructure in the peripheral doublet.*

The outer circle of doublets is deployed in such a way that if a line is drawn through the centers of the two central singlets, only one doublet of the circle is located on a perpendicular to this line (Fig. 15–21a). This pair is designated as "doublet 1." The rest of the doublets are numbered consecutively to 9, in the direction pointed by a pair of short, dense arms attached to one side of each doublet. The tubule of the doublet bearing the arms, often slightly larger and with a denser lumen, is designated *subtubule A.* This subtubule is connected to the central pair of singlets by a faintly outlined *spoke.* The second tubule of the doublet, *subtuble B,* bears no appendages and has an apparently structureless lumen.

The arms attached to one side of subtubule A project outward at a slight angle. In longitudinal section, the arms can be seen to be periodic structures, spaced at intervals of about 130–170 Å. If the transversely sectioned axial complex is viewed from the base toward the tip of the tail, the arms extend from each subtubule A in a clockwise direction.

Higher magnification (Figs. 15–21b and 15–22) reveals that the wall of the microtubules of the axoneme is constructed from a single layer of spherical subunits approximately 70 Å in diameter. Each tubule in cross section consists of a circle of 12–13 of these subunits. The fused tubules of the peripheral doublets share 3–4 subunits in a common wall, giving a total comple-

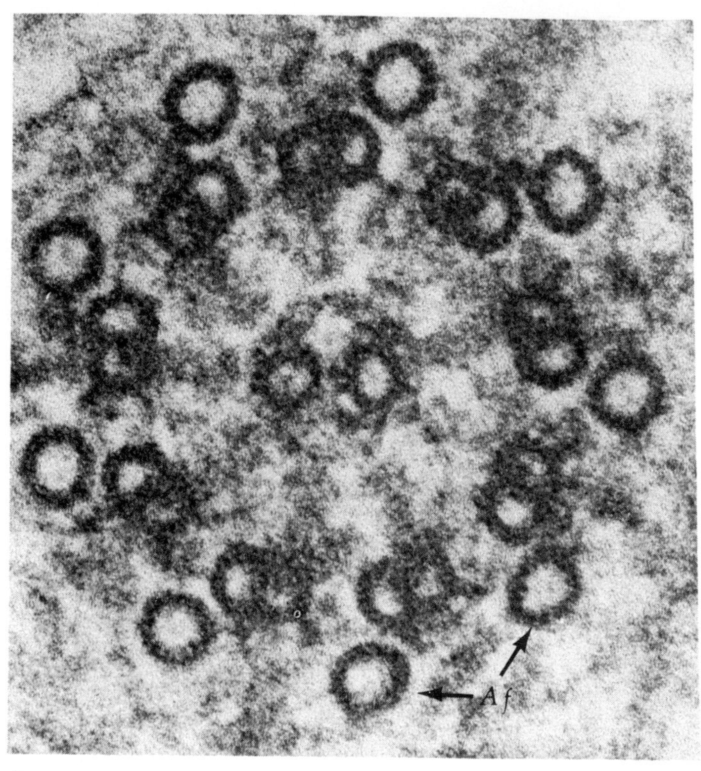

a

b

15–22. a, *The axial complex of the field cricket* Acheta assimilis. *An accessory fiber* (Af) *is located near the A subtubule of each peripheral doublet.* × *310,000. Courtesy of J. S. Kaye and The Rockefeller University Press.* [J. Cell Biol. *45:416 (1970)*] b, *Isolated microtubules of a flagellum, prepared for electron microscopy by negative staining with phosphotungstic acid. The arrangement of a particulate subunit in longitudinal rows is clearly visible.* × *140,000. Courtesy of R. Barton and Academic Press Inc.* [J. Ultrastr. Res. *20:6 (1967)*]

ment of 22–23 subunits for each doublet. In negatively stained preparations of extracted tubules, the subunits appear to lie in straight longitudinal rows in the walls (Fig. 15–22b). In these preparations, each subunit appears as a dark sphere with a central 35 Å core that does not take up stain. For this reason, the diameter of the tubule subunits is often given as 35–45 Å. Because the walls of the whole tubule and the subunits in thin sections appear 70 Å wide, the larger dimension is given here.

At the base of the flagellum, the A and B subtubules of the doublets are continuous with the A and B microtubules of the centriole (Fig. 15–23). The outermost microtubule of each centriole triplet ends at the point of juncture of the axial complex to the centriole. The central singlets of the axial complex arise from a dense granule or plate at this point of juncture. At the opposite end of the axial complex, the A and B subtubules of the doublets dissociate into separate microtubules and end at different levels near the tip of the tail. The total length of the axial complex may vary from as little as 1 micron to several millimeters. In *Drosophila,* the sperm tail is several millimeters long and may in fact be longer than the whole adult animal. Spermatozoa 10 millimeters long have been reported in another insect, the back swimmer *Notonecta.*

• *Some exceptions to the 9 + 2 pattern of the axial complex.* Exceptions to the 9 + 2 pattern of microtubules in the axial complex are exceedingly rare. Among the few, a number are significant in discussions of sperm motility. Some of these are sperm in which the central singlets are missing, giving a 9 + 0 pattern. These are found in the sperm tails of the mayfly *Chloeon,* the fish

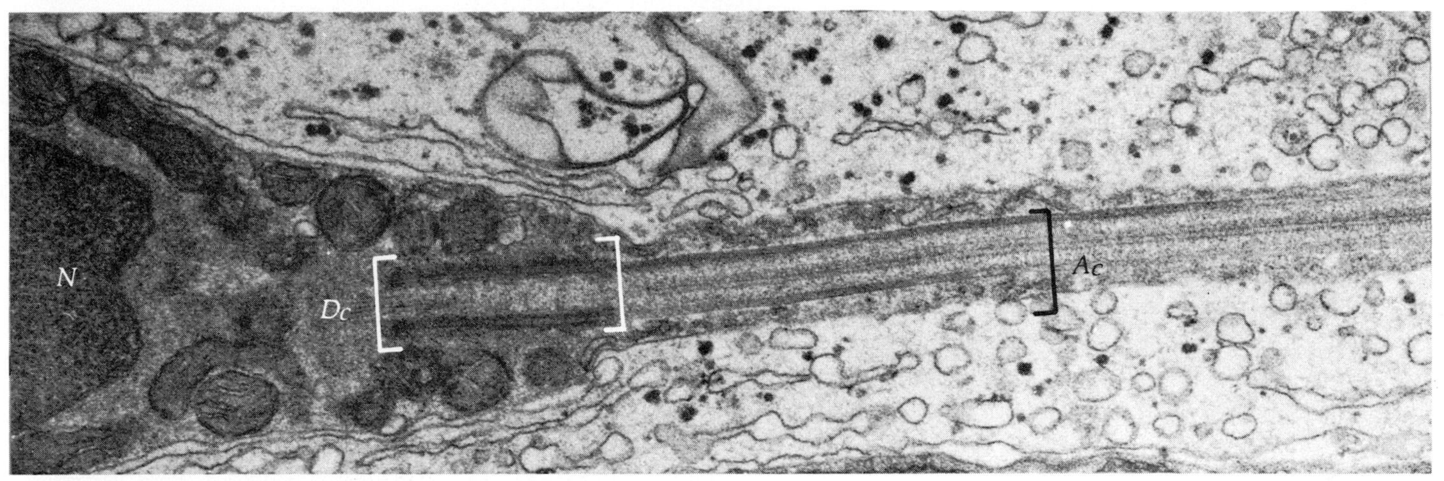

a

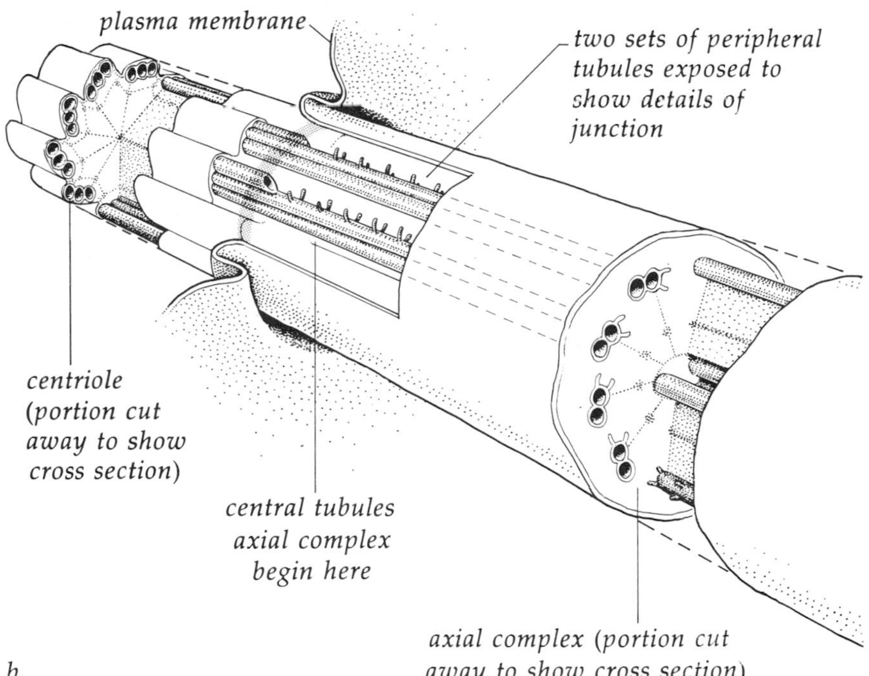

b

15–23. a, *The distal centriole (basal body) and axial complex of a spermatozoan from the frog* Rana. Dc, *distal centriole;* Ac, *axial complex;* N, *nucleus.* × *32,000. Courtesy of B. R. Zirkin.* b, *Diagram showing the relationship of the microtubules of the basal centriole and the axial complex.*

Lycondontis, and the annelid worm *Myzostomum.* Because these sperm are motile and appear to progress by normal flagellate movement, it is unlikely that the central singlet tubules are a critical part of the motile apparatus. Other forms have a 9 + 1 pattern, with the single central tubule highly modified into a complex cylinder with multiple concentric walls. In the flatworms, in which the latter pattern is common, only spermatozoa show a 9 + 1 axial complex. Other flagellated cells of the organism, such as the ciliated cells of the epithelium, possess the usual 9 + 2 pattern. A few other variations are found in insect species, such as the 9 + 7 axial complex of the caddis fly. In the treehopper, with perhaps the most bizarre modification of the basic pattern yet discovered, the microtubules of the flagellum are subdivided into four parts, each enclosed by a separate membrane. Three of these contain two doubles and two singles, and one has three doubles and three singles, for a total of nine doublets and nine singlets.

The sperm of coccids retain no vestige of the 9 + 2 complex, although systems of microtubules are a pre-

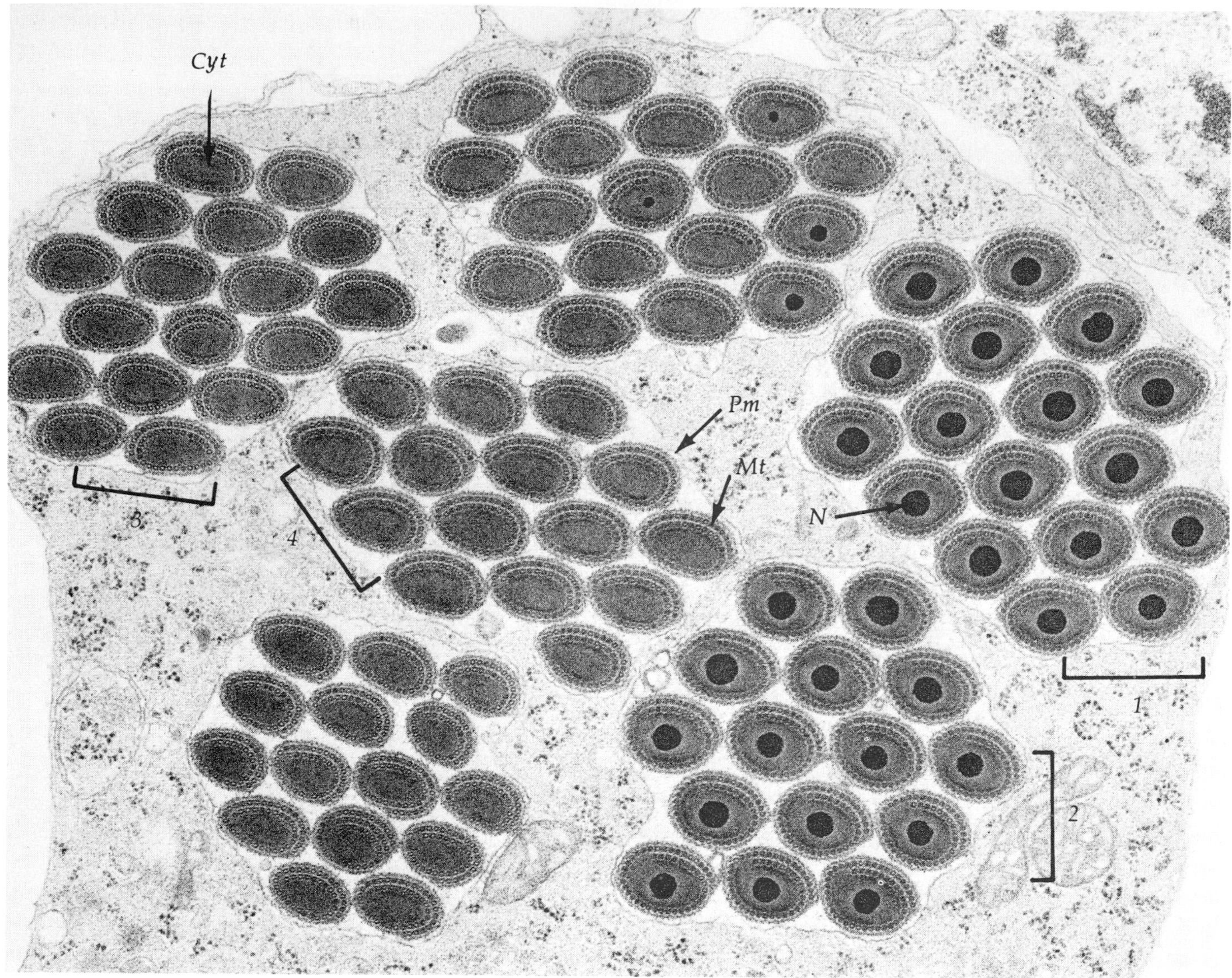

15–24. Cross section of part of the testis of the scale insect Parlatoria oleae. *A number of sperm bundles are present, each consisting of from 15 to 17 individual spermatozoa. Each spermatozoon is delineated by a plasma membrane* (Pm) *and a layer of microtubules* (Mt). *The microtubules, which lie just below the plasma membrane, overlap to form a double layer around approximately one-half of each cell. In the bundles sectioned at more anterior regions (bundles 1 and 2) the nucleus* (N) *is visible as a dense core without separation from the surrounding cytoplasm by a nuclear envelope. At more posterior levels (bundles 3 and 4) only dense, apparently undifferentiated cytoplasm* (Cyt) *is present. × 43,000. Courtesy of W. G. Robison Jr. and The Rockefeller University Press.* [J. Cell Biol. *29:251 (1966)*]

dominant part of the cell structure. These sperm are actively motile. The sperm (Fig. 15–24) occur in bundles of 10–20 closely packed individuals. The position of individual spermatozoa is marked by a sheath of 45–50 microtubules that extends the length of the bundle. The anterior half of each microtubular sheath encloses a core of nucleoprotein. No nuclear envelope is present. Similarly, there are no recognizable mitochondria, acrosomal membranes, or centrioles. These organisms have been extensively investigated by W. Gerald Robison, who has observed that on release from the bundle, each spermatozoan has the capacity for movement throughout its entire length. Because the only internal structure extending the length of the sperm is the microtubular sheath, Robison considers it likely that the sheath is the source of motility.

Accessory fibers. The sperm tails of many animals contain 9 additional longitudinal components spaced immediately outside each doublet of the axial complex. These additional components, called *accessory fibers,* are most prominent in mammalian sperm but are also found in insects and some birds, molluscs, and snakes. In mammals (Fig. 15–25) the accessory fibers are dense structures of considerable mass. The nine accessories are of unequal size and shape, giving the mammalian sperm tail an asymmetrical aspect in cross section. Characteristically, fibers 1, 5, 6, and 9 (numbered according to the adjacent doublet of the axial complex) are larger and extend farther toward the tip of the tail. Near the tip, the accessory fibers taper to smaller diameters and phase out in an order related to the basal diameter, the smaller fibers disappearing first (see Fig. 15–25).

Little is known of the origin of the accessory fibers in development of the sperm tail. In insects, Phillips has traced the first appearance of the accessories to an arm extending from subtubule B of the peripheral doublets. This arm subsequently detaches and develops into an accessory fiber.

The function of the accessory fibers is unknown. The nine accessories occur almost exclusively in animals with internal fertilization. Because this is the case, it has been proposed that these fibers are motile organelles providing additional power for movement of the sperm through the more viscous medium of the reproductive tract. This is supported by selective isolation of components of the sperm tail (see below), which has revealed that the accessory fibers contain an ATPase in combination with protein similar to the proteins of striated muscle.

The generation of the axial complex during spermiogenesis. The initiation of the growth of the axial complex from the microtubules of the distal centriole may take place as early as anaphase II of meiosis. One of two patterns is followed in the development of the tail and axial complex. If the distal centriole is located deep within the cytoplasm of the spermatid, a vesicle appears over the advancing margin of the developing axial complex. As the microtubules of the axial complex

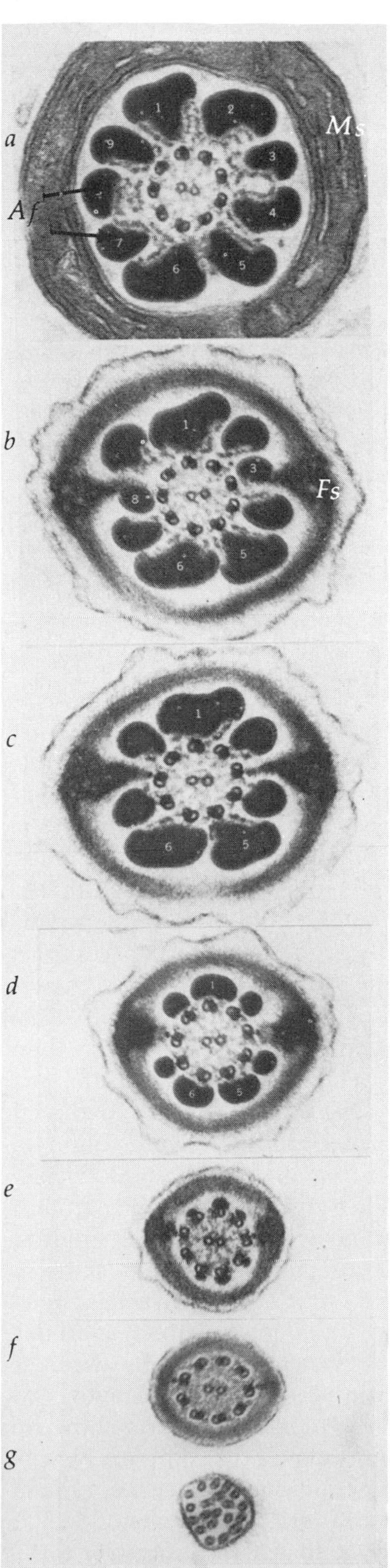

15–25. Sperm tail of a guinea pig, transversely sectioned at different levels from the anterior end (a) *to the tip* (g). *The accessory fibers* (Af) *are identified by the number of the adjacent doublet of the axial complex. The accessory fibers extend nearly to the tip of the tail.* Ms, *mitochondrial sheath;* Fs, *fibrous sheath.* × *50,000. Courtesy of D. W. Fawcett.*

elongate, the vesicle invaginates and spreads over the tip of the complex (Fig. 15–26), eventually reaching and fusing with the plasma membrane. Alternatively, at the initiation of flagellar development, the distal centriole may be located just below the surface of the plasma membrane. In this case, the growing fibers of the axial complex advance into a gradually lengthening extension of the plasma membrane.

When development is complete, the distal centriole usually remains attached to the microtubules of the axial complex at the base of the tail (see Fig. 15–23). When in this position, the centriole is frequently termed the *basal body, basal granule,* or *kinetosome.* Rarely, the centrioles may not persist after spermiogenesis; Phillips has found that in many insects, although the sperm are completely motile, no trace of either the proximal or distal centriole can be found in the mature spermatozoan.

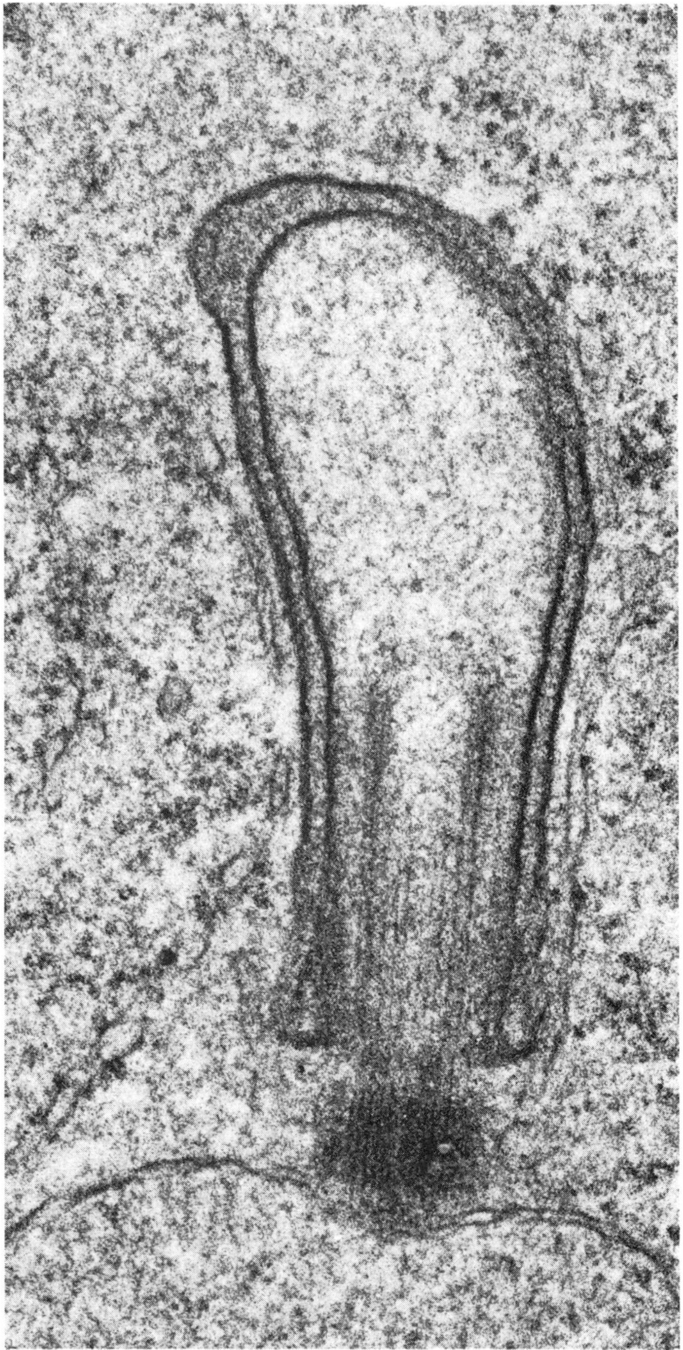

15–26. *An initial stage in the development of the flagellum of the mosquito* Culex. *The developing axial complex, in longitudinal section in this micrograph, is encased in an elongated, cup-like vesicle.* × *82,000. Courtesy of D. M. Phillips and The Rockefeller University Press.* [J. Cell Biol. *44:243 (1970)*]

Some Hypotheses Concerning the Molecular Basis of Flagellar Motion

Molecular structure of the flagellum. Sperm tails and other flagella can be removed and concentrated into quantities great enough for chemical analysis. I. R. Gibbons, who has pioneered this work; has been able to dissect the components of the sperm tail into fairly pure samples of matrix, doublets, and central singlets. This was done by separating the axial complex from the membrane and matrix components by extraction with detergents (to break down membranes) followed by centrifugation. The doublets were then separated and purified by taking advantage of the fact that central singlets are slightly more easily dissolved in some media. The purified doublets were found to contain a protein sedimenting at 4S to 6S, with a molecular weight of approximately 100,000. A similar protein subunit has been identified with the central singlets. These proteins resemble the muscle protein actin in amino acid composition. Although much has been made of this similarity, there is no direct evidence that the two systems are closely related. The parallels in amino acid composition are only approximate, and antibodies prepared against muscle proteins show no reaction to purified protein of the axial complex. Nevertheless, most authorities still assume at least a general relationship in structure and function between flagellar proteins and the actin-myosin system of muscle.

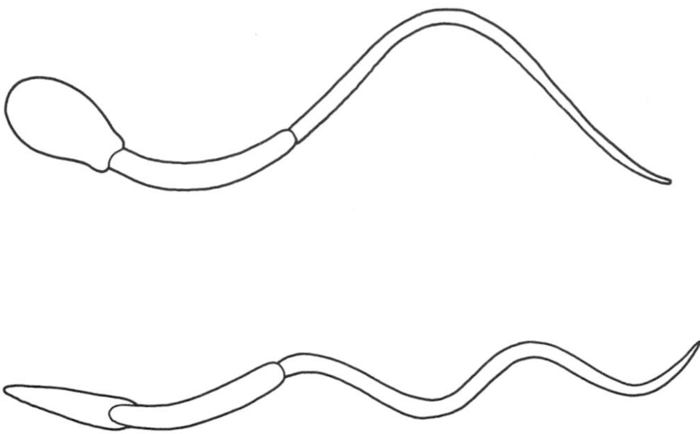

15–27. The components of the complex flagellar beat in bull sperm. The top wave, in a plane parallel to the dorsoventrally flattened head, is of larger amplitude and lower frequency. The wave illustrated at the bottom is in a plane perpendicular to the head and is of smaller amplitude and higher frequency. Redrawn from Fertilization, *Vol. I, ed. C. B. Metz and A. Monroy, 1969. Courtesy of L. Nelson and Academic Press, Inc.*

Some differences are noted between singlet and doublet protein. The greater susceptibility of the central singlets to solution has been noted. Colchicine (see p. 341), which interferes with the generation of the axial complex, is bound by the subunits of the central singlets but not the peripheral doublets. Each 4-6S subunit of the singlets evidently contains one binding site for colchicine (see Shelanski and Taylor, 1968).

An enzymatic protein capable of catalyzing the hydrolysis of ATP can be extracted from the peripheral doublets. This protein, called *dynein,* exists either as a 14S monomer or a 30S dymer. Addition of dynein to extracted doublets results in reappearance of the arms on the A subtubules of the doublets. From this it has been concluded that the arms consist primarily of an ATPase. By comparing the total quantity of dynein per flagellum with the ATP turnover and frequency of flagellar beat, it has been estimated that each dynein molecule hydrolyzes one molecule of ATP during each flagellar beat.

The molecular basis of sperm motility. In bull sperm, most frequently used for quantitative investigations of flagellar motion, the flagellar beat is the sum of two vibrations (Fig. 15–27). One wave, perpendicular to the flattened head, is faster in frequency and smaller in amplitude. The second, parallel to the head, is of slower frequency and larger amplitude. The two waves combine to produce a complex three-dimensional undulation. The motion is propagated in waves that normally pass from base to tip. However, reversal of the propagation from tip to base can occur. The normal pattern gives the impression that the movements originate in flexures or bending of the base of the sperm tail.

Peter Satir has carefully followed the movement of flagella in relation to the microtubules of the axial complex. He determined that flexure of the shaft of the flagellum occurs without contraction or shortening of the microtubules of the axial complex. This was done by analyzing sections made near the tips of straight and bent flagella. At the end of an unbent flagellum, the doublets at both sides of the flagellum extend within equal distances of the tip. In a flagellum bent to one side, the doublets on the side toward the direction of bending are found to extend farther into the tip (Fig. 15–28). This is consistent with the interpretation that the doublets do not contract on bending of the flagellum, because the radius of curvature of the doublets on the outside of the bend (*a* in Figure 15–28) is greater than the radius of the doublets on the inside of the bend (*b* in Fig. 15–28). Therefore, the "outside" doublets traverse a greater length along the flagellum than the "inside" doublets and will appear shorter at the tip if the doublets do not contract.

Satir has interpolated these findings into a sliding filament model for flagellar motion. In the model, the motion is produced by the doublets sliding past each other, in analogy with the sliding filament model for muscle contraction. The energy for the sliding motion, which results in forceful bending of the flagellum, is provided in the model by hydrolysis of ATP, accomplished by enzymes contained in the arms of the doublets.

Satir's model has the advantage of applicability to any known flagellar system, whether the standard 9 + 2 pattern is present or not. The sliding filament mechanism could work as well in the relatively unorganized microtubular sheath of coccid sperm, in the 9 + 0 or 9 + 1 systems, or in the highly aberrant divided axial complex of the treehopper. Whatever the basis of flagellar movement, it obviously does not specifically depend on the presence of peripheral doublets, central singlets, or even the basal centriole, because motile sperm that lack these structures are known.

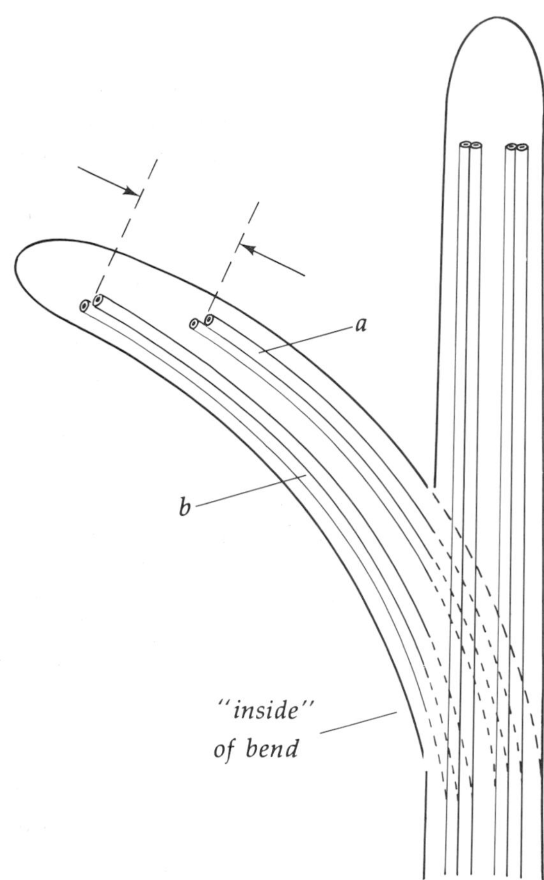

15–28. *Diagram illustrating the extension of the peripheral doublets in the axial complex during bending of the flagellum. In an unbent flagellum, the peripheral doublets on all sides of the axial complex extend an equal distance into the tip. In a flagellum bent to one side, the doublets on the inside of the bend* (b) *extend farther into the tip than those on the outside of the bend* (a). *This indicates that the doublets do not shorten in length or contract during bending of the flagellum (see text). Modified, courtesy of P. Satir and The Rockefeller University Press, from* J. Cell Biol. *39:77 (1968).*

Biochemistry of flagellar movement. By analyzing the respiration of separated sperm heads and tails, it has been possible to determine that the flagellum accounts for about 70 percent of the total energy dissipated by active spermatozoa. If glycolysis and respiration are blocked by inhibitors, movement of the tail ceases when stored quantities of ATP are exhausted. If the blocks to either glycolysis or respiration are removed, movement begins again. All sperm contain the necessary biochemical apparatus for mitochondrial respiration. Although the mitochondrial substructure may be highly modified, the biochemical mechanisms in oxidation, electron transport, and phosphorylation closely resemble the systems of somatic cells.

As far as is known, the metabolites in sperm are primarily stored in the form of lipid. Many additional substances from the surrounding medium can also be used as fuels. The reproductive tracts of mammals contain oxidizable materials; for example, in man the seminal fluid ejaculated with the sperm contains fructose and citric acid in high concentration. The enzymatic apparatus for absorption and metabolism of these substrates is present in quantity and in active form in mammalian spermatozoa. Under artificial conditions, several other sugars and various fatty acids and amino acids can also be metabolized by sperm.

Spermiogenesis in Plants

The plant spermatozoon. Many lower plant species elaborate motile gametes that are remarkably similar to animal spermatozoa in both structure and development. In plant sperm, often called *antherozoids* or *spermatozoids,* the nucleus elongates during development but usually extends into a spiral form. During nuclear morphogenesis, a shell stage and other nuclear rearrangements similar to animal sperm may appear. No structures equivalent to the acrosome have been observed.

Plant sperm are regularly multiflagellate. The flagella, usually two in number, are located at the anterior end of the cell and trail back along the side of the nucleus. In some plants the flagella are more numerous, reaching levels of a hundred or more in the fern *Marsilea* (see below). During plant spermiogenesis the flagella arise from centrioles that are closely similar to their animal counterparts. In most plants these centrioles apparently arise *de novo* in cells destined to give rise to gametes, first appearing in the cytoplasm one or two divisions before differentiation into sperm.

Spermatozoids resembling animal spermatozoa occur in some algae and in mosses and liverworts. Motile sperm with multiple flagella are also found in ferns and primitive gymnosperms such as *Zamia* and *Ginkgo.* The sperm cells of *Zamia* are probably the largest motile

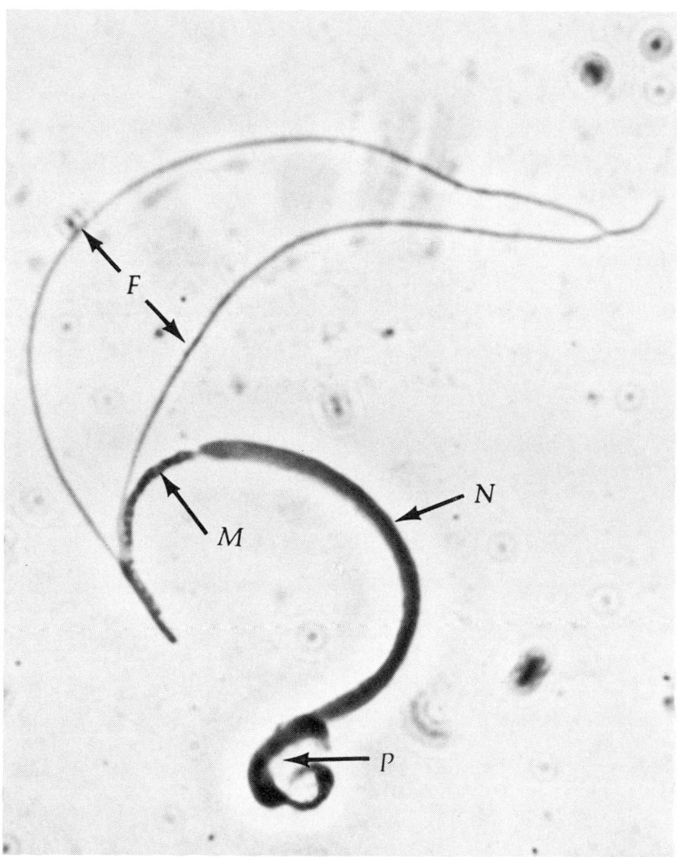

15–29. A light micrograph of the mature spermatozoan of the plant Nitella. *F, flagella; M, mitochondria; N, nucleus, P, plastids.* × *27,000. Courtesy of F. R. Turner and The Rockefeller University Press, from* J. Cell Biol. *37:370 (1968).*

gametes of any animal or plant and, with a diameter of more than 300 microns, are readily visible to the naked eye. Higher gymnosperms produce male gametes similar in overall structure to the motile gametes of *Zamia* and *Ginkgo*, except that centrioles and flagella do not occur and the gamete is nonmotile. The evolutionary transition from motile gamete to pollen is complete in the most advanced gymnosperms such as *Pinus*, in which the male gamete is reduced to a nucleus carried to the egg by the growth of a pollen tube.

The male and female gametes of lower plants do not differentiate from immediate postmeiotic cells as in animals. Instead, postmeiotic cells differentiate into spores that, on germination, produce the gametophyte generation of the plant. The haploid gametophyte increases in size by mitotic divisions, and at some point groups of cells in the sex organs of the mature plant differentiate into male and female gametes. Because this differentiation follows mitosis instead of meiosis, all sperm and eggs of a gametophyte are genetically identical except for differences arising from mutation.

Spermiogenesis in Nitella, *a green alga.* The development of sperm in the alga *Nitella* is representative of the processes occurring in the lower plants. In *Nitella*, studied by F. R. Turner, the mature sperm is biflagellate (Fig. 15–29). The two flagella are attached to the anterior end of the cell and extend posteriorly. The spermatozoid is divided into three regions. About a fourth of the cell at the anterior end is occupied by a mass of mitochondria. The nucleus takes up the middle half of the cell, and a group of plastids is located at the posterior tip.

The early anaphase cell destined to become a spermatozoid is undifferentiated and contains two centrioles at the pole of the spindle. Soon after the last cell division, the chromatin of the nucleus aggregates into a dense layer just inside the nuclear envelope (Fig. 15–30). At this stage, the nucleus appears closely similar to the shell stage in animal spermiogenesis (see Fig. 15–6). During this stage in *Nitella* elements of the Golgi complex are conspicuous but do not form secretory vesicles analogous to the proacrosomal granule. As spermiogenesis continues, the nucleus elongates and the chromatin aggregates into dense filaments oriented roughly parallel to the longitudinal axis of the nucleus (Fig. 15–31). At the same time, the mitochondria and plastids separate into two groups which move to opposite ends of the elongating nucleus. As these developments take place, the Golgi elements gradually disappear. At about this time, a group of microtubules appears near the centrioles at the anterior end of the spermatid and gradually lengthens and extends as a sheath over the elongating nucleus. By the time the sperm is mature, the microtubular sheath covers one-half the circumference of the cell and extends the full length of the sperm. The individual microtubules lie in a single layer, parallel to the long axis of the sperm (Fig. 15–32). Turner concludes that this sheath provides a cytoskeletal framework on which the nucleus elongates and the mitochondria and plastids become arranged. A closely similar sheath of microtubules called the *manchette* appears in animal sperm, as for example in the developing spermatozoa of the earthworm

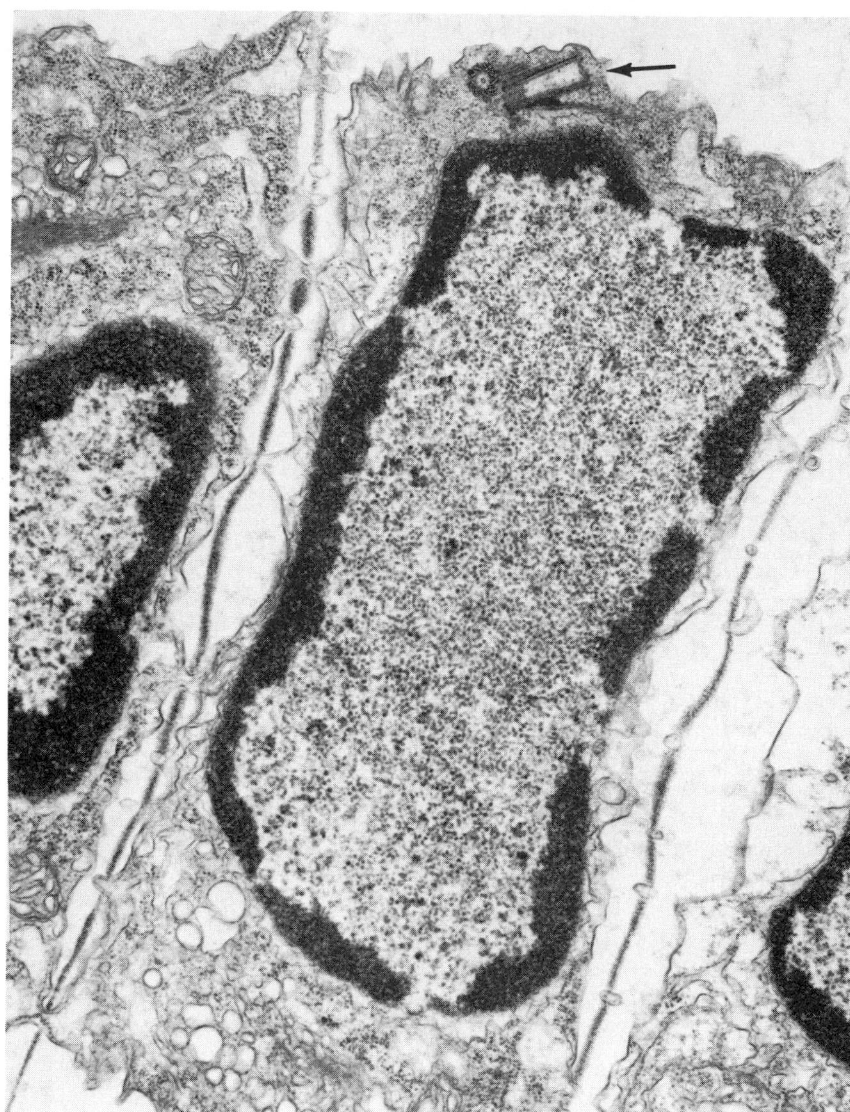

15–30. An early "shell" stage in Nitella. *Centrioles (arrow) are visible at the end of the nucleus in one cell. Compare with Fig. 15–6. × 21,000. Courtesy of F. R. Turner and The Rockefeller University Press, from* J. Cell Biol. *37:370 (1968).*

Lumbricus. In animal sperm the manchette is also thought to provide force for elongation of the nucleus during spermiogenesis.

Both centrioles lengthen slightly and give rise to flagella. The flagellar axial complex (Fig. 15–32) is typical in morphology except that in *Nitella* the arms of subtubule A are not clearly defined. The centrioles at the base of each flagellum are closely similar in morphology to their animal counterparts (Fig. 15–33). At the juncture between the centriole and the axial complex of the flagellum, cross sections reveal a star-like pattern of spokes (Fig. 15–34). The presence of this stellate pattern in the transition region between centriole and axial complex is apparently characteristic of plant flagella.

As spermiogenesis continues, the nuclear filaments coalesce, finally producing a uniformly opaque nucleus in the mature sperm. The mitochondria remain collected at the anterior end of the sperm at the base of the flagella but do not form a flagellar sheath as in the middle piece of animal sperm. These mitochondria retain somatic morphology. The plastids remain concentrated at the posterior end of the sperm cell. During maturation of the spermatozoid, the plastids become filled with dense granules. At maturity (Fig. 15–35), the sperm becomes motile and is released through a

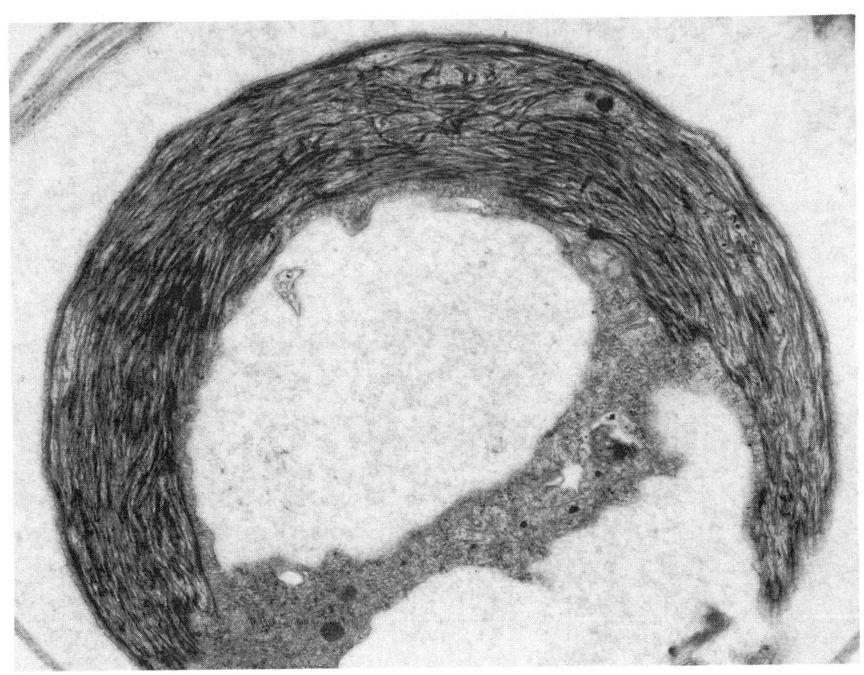

15–31. A longitudinal section through the nucleus of Nitella *at an intermediate stage in spermiogenesis. The chromatin has aggregated into coarse fibrils. Compare with Fig. 15–8. × 18,000. Courtesy of F. R. Turner and The Rockefeller University Press, from* J. Cell Biol. *37:370 (1968).*

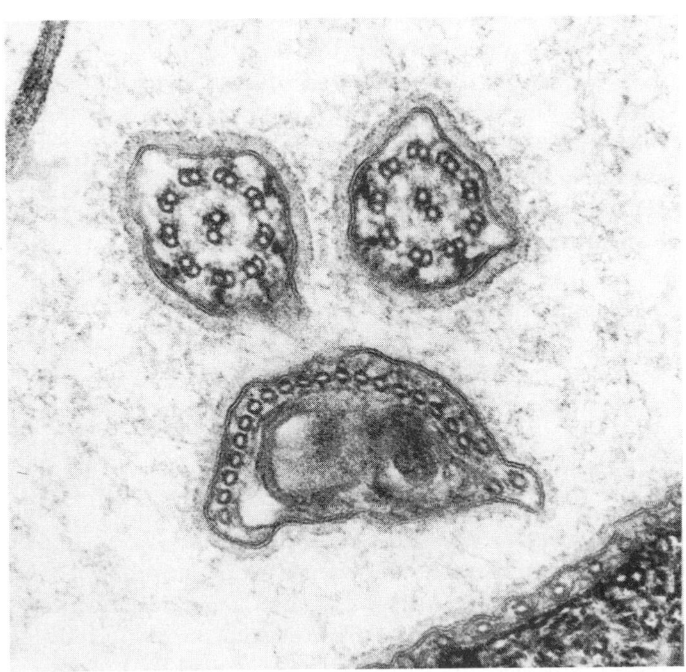

15–32. A cross section of the anterior end of a Nitella *spermatozoan, near the origin of the two flagella. × 86,000. Courtesy of F. R. Turner and The Rockefeller University Press, from* J. Cell Biol. *37:370 (1968).*

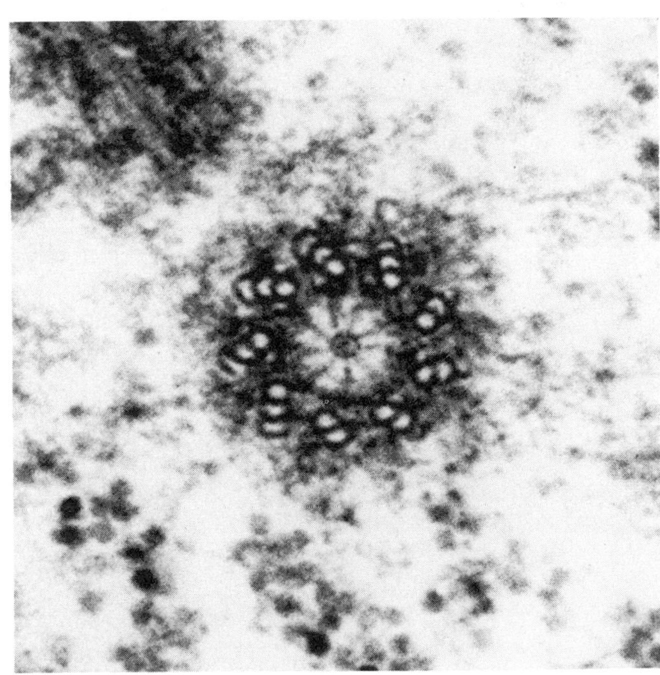

15–33. A centriole of Nitella. *The "cartwheel" structure is visible in the lumen of the centriole. × 150,000. Courtesy of F. R. Turner and The Rockefeller University Press, from* J. Cell Biol. *37:370 (1968).*

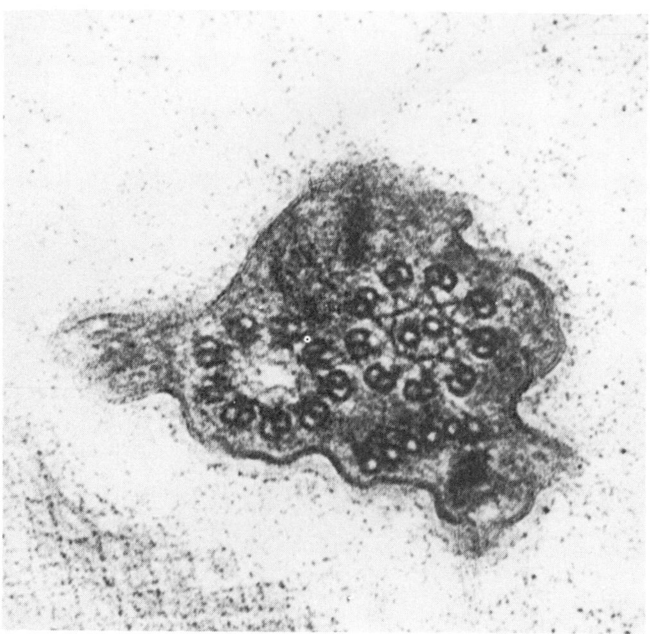

15–34. *A section of the two centrioles (basal bodies) of* Nitella *at the anterior end of the spermatozoan. The centriole on the right has been sectioned in the region of transition between the basal centriole and the flagellum. The star-like pattern is typically found in this transition region in plant spermatozoa. × 104,000. Courtesy of F. R. Turner and The Rockefeller University Press, from* J. Cell Biol. *37:370 (1968).*

pore in the cell wall which originally enclosed the developing spermatid.

Centriole replication during spermiogenesis in Marsilea. Spermatogenesis in the fern *Marsilea* has been of interest to cytologists since the late 1800's because the "centrioles" in this species, called *blepharoplasts,* are large and apparently arise *de novo* during the mitotic divisions just preceding spermiogenesis. Recently, the blepharoplasts of *Marsilea* were reinvestigated by Ikuko Mizukami and Joseph G. Gall. Mizukami and Gall found that these giant structures consist of a mass of procentrioles. The blepharoplasts are visible at either side of the nucleus (arrows) in Fig. 15–36. At higher magnification, the compound nature of the blepharoplasts is obvious (Fig. 15–37). The individual units closely resemble animal procentrioles in size and structure; as is typical of procentrioles, the triplet microtubules of the walls and the central hub structure are indistinct (arrow, Fig. 15–37). No evidence of a mature centriole in or near the procentriolar mass could be found, either before or during development of the blepharoplast. After the last division before spermiogenesis, each procentriole elongates and develops clearly defined microtubular structure. At the same time, the mass of centrioles becomes less tightly packed. When reorganization of the nucleus begins later in spermiogenesis, the centriolar mass gradually breaks down and the individual centrioles, numbering from 110 to 150, take up positions around the nucleus just outside the nuclear membrane. As the nucleus elongates the layer of centrioles comes to lie just under the plasma membrane. Each centriole at this stage gives rise to a flagellum, producing at maturity a multiflagellate sperm (Fig. 15–38).

The development of the masses of procentrioles in *Marsilea* is of particular interest because mature centrioles are not present during any part of the process. Thus, in *Marsilea,* procentrioles do not arise from previously existing mature centrioles. Mizukami and Gall were unable to determine when the procentrioles first appeared, or whether the presence of preexisting procentrioles is necessary for the replication of the masses first visible in the last few divisions before spermiogenesis. The development of the centrioles in *Marsilea* has a parallel in animal spermiogenesis in the multiflagellate sperm of the snail *Viviparus,* but in the snail the two masses of procentrioles grow around a central pair of mature centrioles (see Gall, 1961).

A Summary of Spermiogenesis

The completion of spermiogenesis in both plants and animals produces a single cell with the chromosomal material concentrated into a package occupying only a fraction of the volume of the somatic nucleus. In most sperm, the nucleus appears structureless and opaque under the electron microscope, and internal differentiation into the fibrils and granules typical of the interphase nucleus is lost. The metamorphosis of the sperm nucleus is usually accompanied by replacement of some or all of the histones by more basic proteins containing a higher proportion of arginine residues. As these changes are completed, the differentiation of the nuclear envelope into pore complexes disappears.

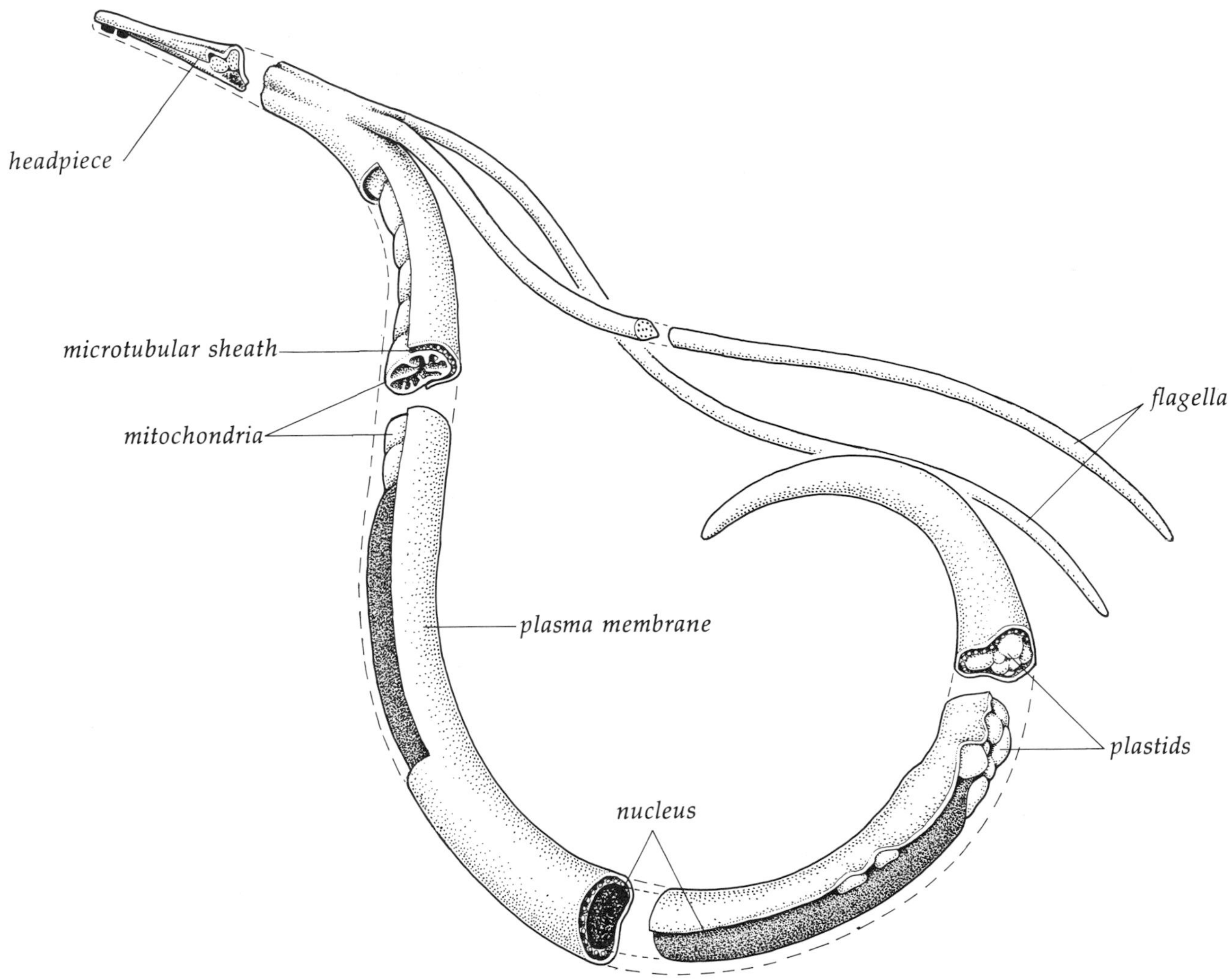

15–35. A diagram of the ultrastructural features of a Nitella *sperm. Redrawn, original courtesy of F. R. Turner and The Rockefeller University Press, from* J. Cell Biol. *37:370 (1968).*

The cytoplasm of the sperm cell also becomes condensed in volume. The structures and organelles retained are specialized primarily for providing motility to the sperm cell, or in animal sperm, for penetration of the surface coats of the egg. Motility is provided by the flagellum, a complex structure containing microtubules usually arranged into the characteristic axial complex of nine peripheral doublets in a circle surrounding a central pair of singlets. Also present in some sperm are nine accessory fibers or microtubules arranged in a circle just outside the peripheral doublets. Although the role of the axial complex in flagellar motion is unknown, an important hypothesis considers that the peripheral doublets slide past each other to provide the force for motion, in a mechanism analogous to the sliding filament model for muscle contraction. In keeping with this hypothesis, ATPase activity is associated with the axial complex and the accessory fibers if these are present. During development of the sperm the axial complex arises from the microtubules of a centriole, which usually persists as the basal body of the flagellum in the mature spermatozoan.

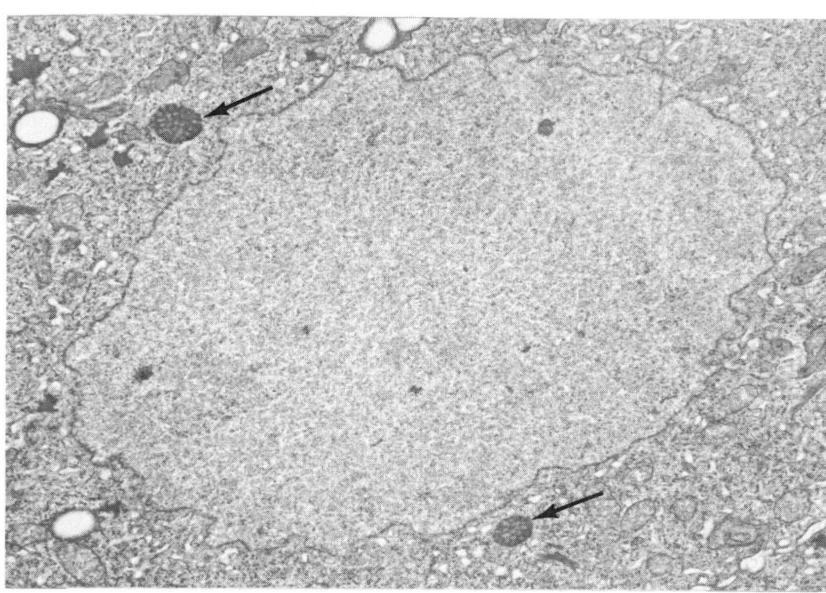

15–36. Section of a cell that will later give rise to spermatozoa in the plant Marsilea. *Two large* blepharoplasts, *each consisting of a mass of procentrioles, are located to either side of the nucleus (arrows). × 12,000. Courtesy of I. Mizukami, J. G. Gall, and The Rockefeller University Press, from* J. Cell Biol. *29:97 (1966).*

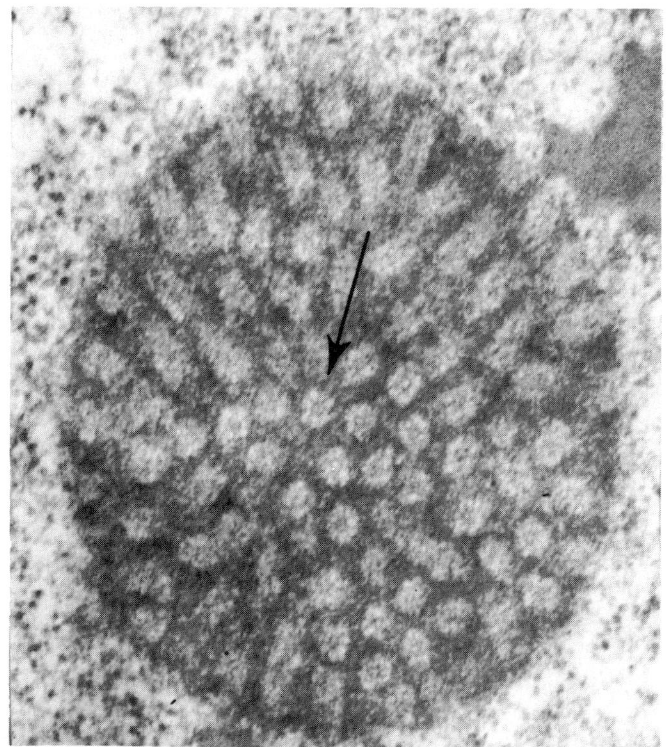

15–37. A blepharoplast of Marsilea *at higher magnification. Where the centrioles are cut transversely (arrow), the hub and spoke arrangement is discernible. × 85,000. Courtesy of I. Mizukami, J. G. Gall, and The Rockefeller University Press, from* J. Cell Biol. *29:97 (1966).*

Mitochondria are closely associated with the sperm tail, either in a form closely similar to somatic mitochondria or modified into specialized structures such as the nebenkern or mitochondrial derivatives. The mitochondrion may be concentrated at the base of the flagellum or may extend for some distance into the tail as a sheath around the axial complex.

In animal sperm, enzymes with lytic activity for the surface coats of the egg are concentrated into the acrosome. In the mature spermatozoan, the acrosome is positioned at the anterior end of the cell, frequently as a cap-like structure covering the tip of the nucleus. During spermiogenesis the acrosome arises from secretion vesicles of the Golgi complex. At maturity, the acrosome consists of a single membrane enclosing the acrosomal enzymes, primarily hyaluronidase and zona lysin.

All other structures are eliminated from the cytoplasm during spermiogenesis, so that at maturity the sperm cell consists of little more than a greatly condensed nucleus, the acrosome, and the flagellar apparatus.

Oogenesis

The egg in animals and most lower plants is, in contrast to the sperm, a nonmotile cell. The conspicuous adaptations of the egg are concerned with the storage

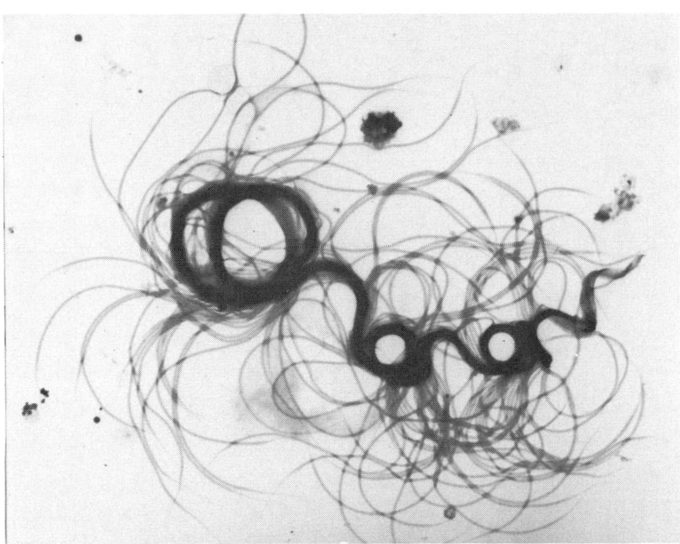

15–38. *A mature spermatozoan of* Marsilea. *Multiple flagella occur over the entire length of the spermatozoan.* × *2,600. Courtesy of I. Mizukami, J. G. Gall, and The Rockefeller University Press, from* J. Cell Biol. *29:97 (1966).*

of information and nutrient substances required for the early growth of the embryo after fertilization.

In animals, differentiation of the oocyte into the egg begins much earlier in the meiotic process than spermiogenesis and differs fundamentally from gametogenesis in the male in that only one of the four haploid products of meiosis becomes a functional gamete. Because morphogenesis of the egg cannot be separated from the events of meiosis, both processes in the female are generally included under the term *oogenesis.*

Growth of the egg begins during prophase I, after meiosis has advanced to the late pachytene and early diplotene stages. At this time, the nucleus and cytoplasm synthesize a variety of substances to be stored in the egg cytoplasm. This stage is marked in some animals by the appearance of the lampbrush configuration of the chromosomes, which at this stage are still loosely paired into tetrads. Great quantities of RNA are transcribed on the chromosomes, including a large proportion of rRNA (for a complete description of the morphology and function of the chromosomes during diplonema in the oocyte, see Chapter 14).

In plants, the development of the egg is less complex because oogenesis, which takes place after mitosis, does not involve the synthesis of nutrients to be stored in the egg cytoplasm. Usually, in lower plants, the egg differentiates simply as a somewhat enlarged cell, which either remains enclosed for a time in the tissues of the gametophyte or is released into the surrounding aqueous medium. In either case, the egg contains chloroplasts that become active in synthesis soon after fertilization and thus does not require large quantities of stored nutrients. The complex events in gametogenesis and fertilization in the higher plants will be discussed in a separate section later in this chapter.

Oogenesis in Animals

In the developing oocytes of animals, the synthetic activity of the nucleus during prophase is reflected by a great increase in the activity and growth of the cytoplasm. Large numbers of ribosomes are synthesized. These may either become attached to membranes of the endoplasmic reticulum or remain freely suspended in the ground substance. Both the free and attached ribosomes are for a time intensively engaged in the synthesis of proteins. Lipids and polysaccharides are also synthesized in large quantities in the oocyte cytoplasm. Most of the protein and lipid is stored in membrane-bound structures called *yolk bodies* or *yolk platelets;* the cytoplasm of the eggs of all animals except mammals becomes closely packed with these bodies. The activity of the synthetic machinery of the cytoplasm is marked by extensive proliferation of the various cytoplasmic organelles, including mitochondria, the Golgi complex, and annulate lamellae.

In addition to the internal synthesis of yolk substances, enlargement of the egg cytoplasm also results from the entry of proteins, lipids, and polysaccharides from outside the oocyte. Some of these materials are absorbed through the plasma membrane of the oocyte from the surrounding extracellular fluids. Other finished materials, including even intact yolk bodies, Golgi, and mitochondria, are transported into the egg cytoplasm from surrounding *follicle* or *nurse* cells through direct cytoplasmic bridges. The final outcome of this prophase synthesis in some animals may be growth of the oocyte to relatively immense proportions.

The growth period of the egg during diplonema may extend from a few days to years, depending on the species. During this time, the nucleus, which also greatly increases in size, is sometimes termed the *germinal vesicle.* As the oocyte attains its final dimensions, synthesis ceases and the rate of metabolism falls

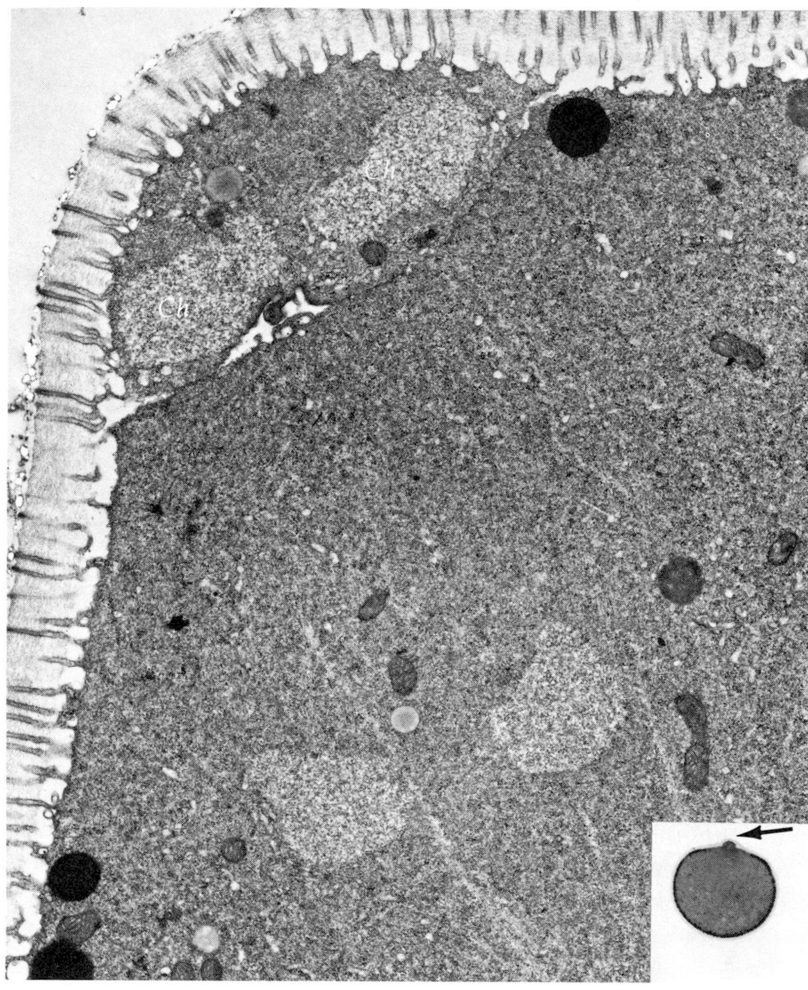

15–39. *An egg with first polar body from the surf clam* Spisula. *A light micrograph of the egg and polar body (arrow) is shown in the inset. The main figure shows the polar body and the underlying egg cytoplasm at higher magnification under the electron microscope. The chromosomal material* (Ch) *in the polar body is the result of the first meiotic division. Just below the polar body, in the egg cytoplasm, the second meiotic division can be seen in progress.* × *12,000; inset* × *250. Courtesy of F. J. Longo, E. Anderson, and Academic Press Inc. from* J. Ultrastr. Res. *33:495 (1970).*

to relatively low levels. The eggs of some animals are ovulated at this stage and remain in prophase I until fertilization. In others, the eggs are shed at metaphase I and remain arrested at this stage until fertilization. Other eggs, including those of most mammals, are blocked at metaphase II. Only in a very few species, such as the sea urchin, is meiosis completed before shedding of the eggs. In all cases, the mature unfertilized egg is a quiescent cell, relatively inert in terms of respiration and synthesis. The molecular nature of the blocks to metabolism and completion of meiosis in the unfertilized egg is unknown.

Formation of polar bodies. No matter what the stage of meiotic arrest, the events subsequent to metaphase I are essentially the same in all animals after the resumption of meiosis. The germinal vesicle breaks down, and the chromosomes and spindle take up a position near the plasma membrane. At anaphase and telophase I the accompanying division of the cytoplasm is unequal, with the result that one of the two prophase II nuclei is separated into a small, abortive cell known as a *polar body* (Fig. 15–39). The second division of cytoplasm at anaphase and telophase II is also unequal, producing a second polar body. During the second division, the first polar body in different species either disintegrates, remains quiescent, or also divides. The result, in various animals, is a single large cell containing the majority of the cytoplasm, with one, two, or three polar bodies attached at one side. After the final division, the remaining single egg nucleus, now called the *pronucleus,* sinks to the interior of the egg.

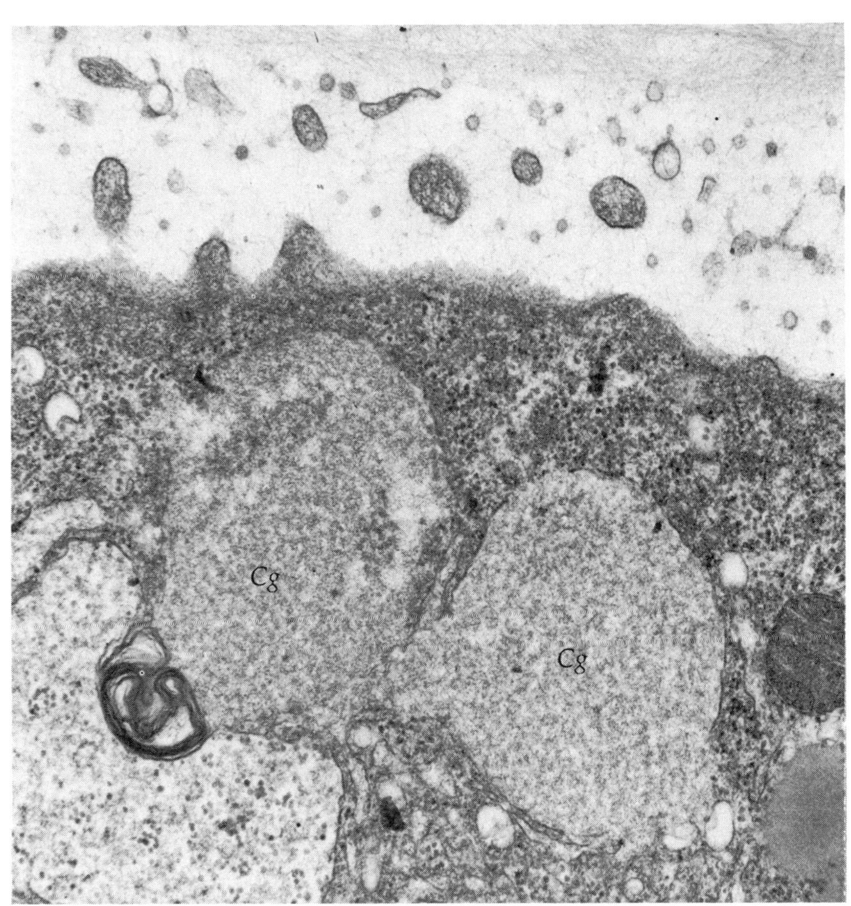

15–40. Cortical granule (CG) *from an unfertilized* Xenopus *egg. The granules often appear to consist of several layers of alternating dense and lucent material, or are distinctly granular.* × *27,000. Courtesy of R. D. Grey and J. L. Hedrick.*

The polar bodies, which must be regarded as nonfunctional, abortive eggs, vary considerably in degree of development in various animals. In mammals, the first polar body fails to divide during the second meiotic division of the oocyte and soon disintegrates, leaving the egg with one or two anaphase II polar bodies at maturity. The polar bodies of mammalian eggs are relatively large and may represent up to one-twentieth of the volume of the mature oocyte. In the rat, well-defined nuclei, with chromosomes approaching the interphase state, appear in the polar bodies. In a few groups, such as insect and crustacean eggs, distinct polar bodies are not formed. Instead, the polar chromosomes are enclosed by membranes, forming polar nuclei that remain nonfunctional in the peripheral egg cytoplasm. In one rather curious instance, the polar nuclei of coccids regularly become infected by a fungus and remain functional in a symbiotic role.

The peripheral envelopes of the egg. The mature eggs of different animals exhibit various differentiations of the periphery and a diverse array of surface coats. Adding to the difficulties presented by this complexity is a confusion in terminology resulting from the use of different names for analogous structures in frequently studied subgroups, such as mammalian, insect, or amphibian eggs. Only a few of the most commonly used terms will be given here.

As a rule, a layer of cytoplasm at the periphery of the egg just inside the plasma membrane becomes free of cytoplasmic inclusions except for membrane-bound structures called *cortical granules* (Fig. 15–40). The cortical granules remain in the clear layer just beneath the plasma membrane until fertilization. In many animals, this layer of the cytoplasm shows the first visible reaction to the penetration of the sperm. At this time the cortical granules rupture and contribute their contents to the formation of the surface coats of the fertilized egg (see also p. 469). Little is known of the origin of cortical granules except that they are synthesized in deeper layers of the cytoplasm, probably with the participation of the endoplasmic reticulum

15–41. a, *A section through the vitelline coat of an unfertilized* Xenopus *oocyte. 1, perivitelline space containing microvilli* (Mv); *2, vitelline coat; 3, jelly layer.* × *60,000. Courtesy of R. D. Grey and J. L. Hedrick.*

and the Golgi complex. These granules occur in the eggs of most animals, including the mammals, but are not found in a few, such as the brine shrimp *Artemia* and the aschelminth *Priapulus.*

Surface envelopes of the egg. All eggs are surrounded by a continuous plasma membrane. In many animals, the egg surface is thrown into many folds and microvilli, giving the plasma membrane a highly irregular outline in thin sections (see Fig. 15–39). In mammals, the plasma membrane of the egg is often termed the *vitelline membrane.* In most animals, a transparent, gelatinous coat, closely applied to the surface of the plasma membrane, immediately surrounds the egg. In invertebrates, this coat is unfortunately also given the term "vitelline membrane." Recently, the suggestion has been made that the terms *vitelline coat, capsule,* or *envelope* be used for this layer, because no cellular membranes form a part of its structure. This exterior coat differs in structure from species to species and sometimes, although very rarely, does not occur at all. In some organisms, the vitelline coat is highly complex and is constructed from several layers of distinct ultrastructure (Fig. 15–41). In mammals, the analogous egg coat is called the *zona pellucida* (see Fig. 15–42). As far as can be determined, the zona pellucida is entirely comparable to the vitelline coat of other vertebrate and invertebrate eggs.

15–42. a, *Diagram of a mammalian oocyte, illustrating typical ultrastructural features.* N, *nucleus;* NCL, *nucleolus;* MVB, *multivesicular body;* M, *mitochondria;* GC, *Golgi complex;* V, *vesicles;* ER, *endoplasmic reticulum;* MV, *microvilli;* ZP, *zona pellucida;* FE, *follicle cells;* PO *and* PFE, *extensions of the oocyte and follicle cells into the zona pellucida;* FGC, *follicle cell Golgi complex;* IB, *inclusion body;* D, *junction between oocyte and follicle cell membranes. Courtesy of E. Anderson, H. W. Beams, and Academic Press Inc.* [J. Ultrastr. Res. 3:432 *(1960)*]. b, *A mouse oocyte.* N, *nucleus;* Nu, *nucleolus;* Pm, *plasma membrane;* Zp, *zona pellucida. Immediately surrounding the egg is a layer of follicle cells.* × *2,800. From* An Atlas of Ultrastructure *by J. A. G. Rhodin, 1963. Courtesy of J. A. G. Rhodin and W. B. Saunders Company.*

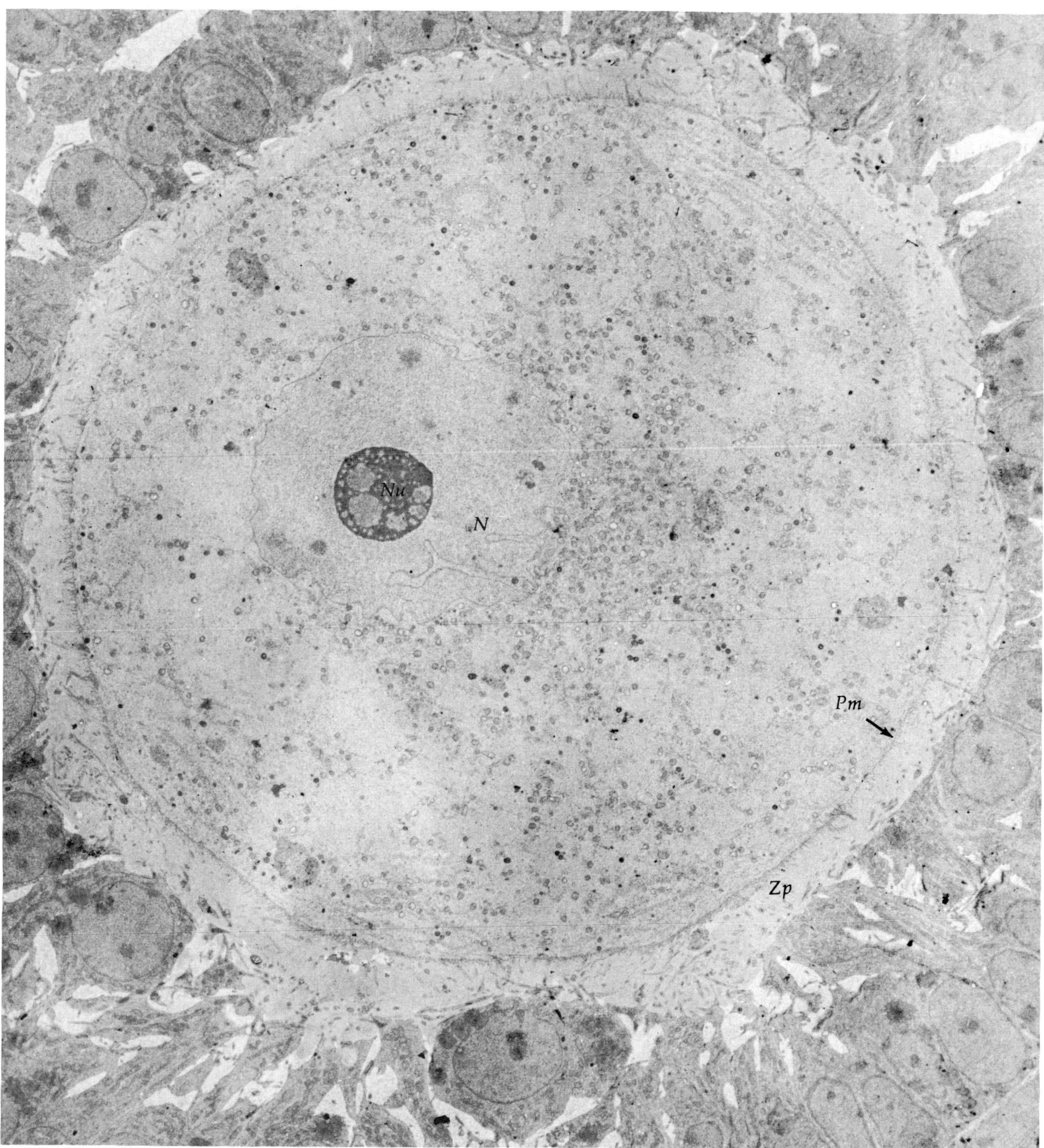

b

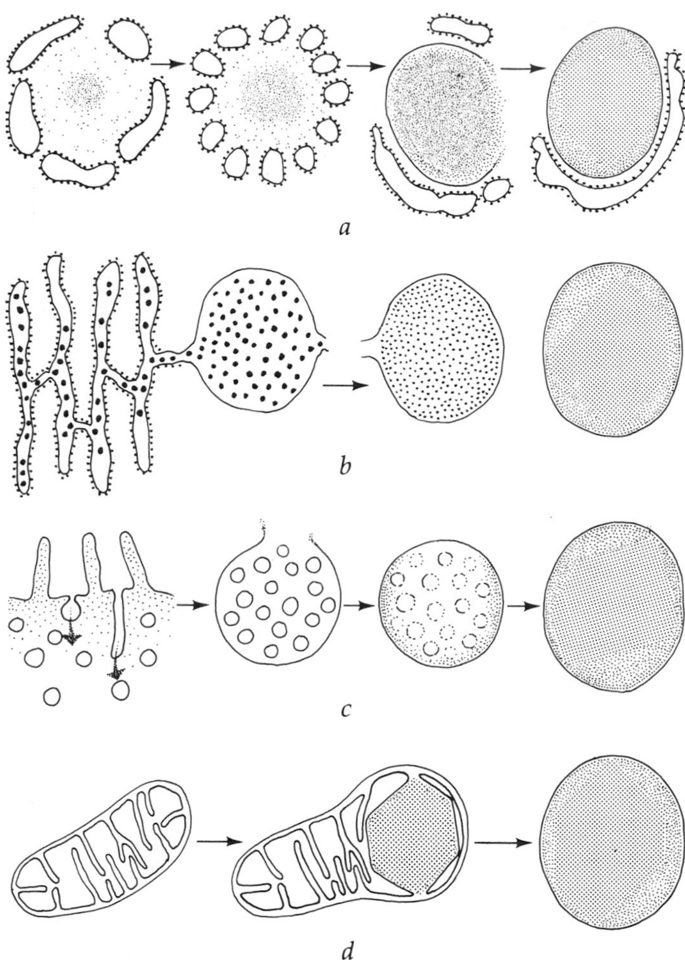

15–43. *Diagram illustrating four patterns of yolk formation, as classified by Norrevang (see text).* a, *From vesicles arising near elements of the endoplasmic reticulum.* b, *From vesicles in direct contact with the* ER. c, *From pinocytotic vesicles.* d, *Within mitochondria. Redrawn, original courtesy of A. Norrevang.* [Intern. Rev. Cytology *23:118 (1968)*]

It is not known whether the egg, the surrounding follicle cells, or the nurse cells give rise to the vitelline coat of the egg. During oogenesis, processes from both the developing oocyte and the follicle cells extend into and across the vitelline coat (giving rise to still another name for this layer, the "zona radiata"). Under the electron microscope, the vitelline coat appears finely fibrous or granular and contains polysaccharides and glycoproteins as major constituents. This coat, which probably functions to protect the egg from mechanical and chemical injury, may persist after fertilization and remain until considerable development of the embryo has taken place.

Other envelopes or coats may be found outside the vitelline coat. These layers, secreted by follicle or nurse cells or by organs of the oviduct, take on a great variety of forms in different species. In insects and some fish eggs an impervious layer called the *chorion* is secreted around the egg by follicle cells. This layer, which forms the outside coat of the mature egg, contains a minute passage, the *micropyle,* for penetration of the sperm during fertilization. The albumen (the "white" of a hen's egg) and shell of avian and reptilian eggs are analogous structures, secreted around the egg during passage through the oviduct. In amphibians, molluscs, and echinoderms, this layer takes the form of a gelatinous mass that surrounds the freshly laid eggs.

Follicle and nurse cells. In most animals, the oocyte is surrounded during development by one or more cells called *follicle* or *nurse* cells. No clear distinction exists between follicle and nurse cells, although nurse cells are fewer in number and may not be as closely applied to the egg surface as follicle cells. At one end of the distribution of types are the nurse cells of insects, which may number from one to ten or more and may be either closely associated with the egg or merely connected by a narrow protoplasmic bridge. Nurse cells of this general type are common in invertebrates. In some groups, particularly in the arthropods, the nurse cells have the same mitotic origin as the eggs and sometimes show structures imitative of the developing oocyte. This may even extend to synapsis of the chromosomes, with development of fragments of the synaptinemal complex (see the discussion on the presence of polycomplexes in the nuclei of nurse cells in Chapter 14). In all cases, the development of the nurse cells as eggs is abortive, and meiosis is nonfunctional and incomplete. Nurse cells in some insects, such as *Drosophila,* may become highly polyploid during development of the egg. Frequently a direct continuity is established between the cytoplasm of the egg and the nurse cells. Through these connections materials, including intact cell organelles and yolk bodies, may be seen to pass. In a few groups, including some tunicates, coelenterates, and molluscs, the developing oocyte may be seen to engulf whole nurse cells during the extensive growth period of the egg cytoplasm.

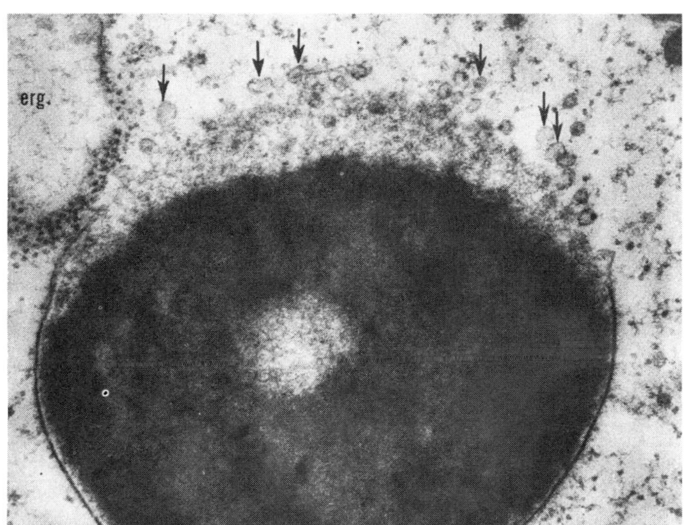

15–44. *Yolk formation in an oocyte of* Priapulus. *The developing yolk body, which is partially enclosed by a membrane, is surrounded by elements of the endoplasmic reticulum* (erg). *The arrows point to small vesicles that are probably incorporated into the yolk body.* × *19,000. Courtesy of A. Norrevang.* [Intern. Rev. Cytology *23:118 (1968)*]

If the cells adjacent to the egg occur in greater numbers and completely surround the egg, they are termed follicle cells. This more common condition, is encountered in vertebrates, including the mammals (Fig. 15–42a). These cells, in derivation, are not usually of common mitotic lineage with the oocyte and do not normally show evidence of abortive meiosis. In vertebrates, for example, the follicle cells are of mesodermal origin and are derived from the somatic tissues of the ovary, while the germ cells are of endodermal origin. Both the follicle cells and the enclosed oocyte throw out processes or microvilli that interdigitate across the vitelline coat of the egg. These are diagramed in Fig. 15–42a and can be seen in section in Fig. 15–42b. In some cases, the processes of the oocyte and follicle cells are interconnected by direct cytoplasmic bridges. More commonly, the follicle cells remain separated from the cytoplasm of the egg by intact plasma membranes. By the use of both radioactive tracers and antigen-antibody reactions, substances can be shown to pass from the cytoplasm of the follicle cells to the oocyte. Presumably, these substances are primarily nutritive. However, it is possible that a quantity of inactive RNA, stored in the egg cytoplasm, is also contributed by the follicle cells.

The follicle cells are often aggregated into a mass around the egg, stabilized by an intercellular cement containing the polysaccharide hyaluronic acid. This follicular coat, termed in mammals the *cumulus oophorus*, is readily dispersed by an enzyme released during fertilization from spermatozoa. This enzyme, hyaluronidase, is found in quantity as one constituent of the acrosome. The follicle cells may or may not persist around the egg after oogenesis; in mammals, the coating of follicle cells ruptures as ovulation takes place. If these cells are thrown off, the protective envelope formed by the vitelline layer remains around the egg. Frequently, as in the chorion of insects, an additional surface layer is secreted by the follicle cells before dispersal.

The Secretion and Deposition of Yolk during Oogenesis

Yolk bodies, often called yolk platelets, arise by several different mechanisms during oogenesis. In an excellent review of oogenesis, Arne Norrevang (1968) has classified yolk deposition as occurring by one or more of four different processes (Fig. 15–43). In one process (Figs. 15–43a and 15–44), yolk bodies arise more or less freely in the cytoplasm without direct connection with the endoplasmic reticulum or Golgi complexes. Although direct connections do not exist, elements of the ER and Golgi are in close proximity, and it is assumed that these organelles take part in this type of yolk deposition. This pattern is seen in the frog *Rana* and the aschelminth *Priapulus*, among other species. In a second type (Fig. 15–43b), yolk bodies arise in direct connection with rough and smooth endoplasmic reticulum. In this process the cisternae of the rough endoplasmic reticulum become filled with a granular deposit that can be followed into continuous channels of the smooth ER. Frequently, the Golgi complex is also directly involved in the secretion process. This material accumulates in membrane bound vesicles that eventually pinch off from the smooth ER and fuse to form the yolk bodies. In a few species such as the crayfish and snail, this route of deposition can be followed by the migration of conspicuous disc-shaped 600–1000 Å granules that first appear in the cisternae of the rough ER (Fig. 15–45). These granules pass into elements of the smooth ER, finally to be incorporated into yolk bodies. As the granules are packed into yolk bodies, their distinctive ultrastructure is lost.

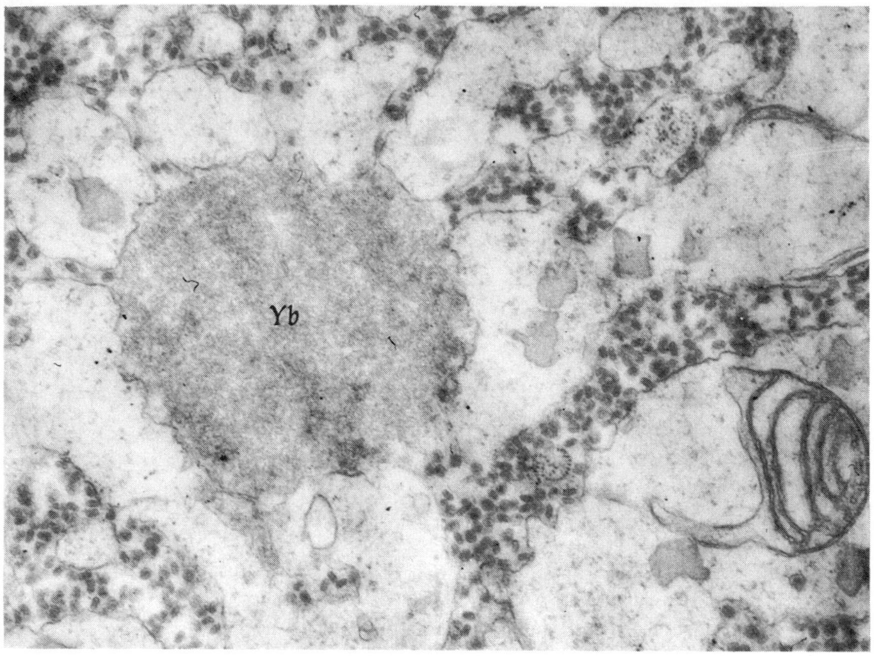

15–45. Yolk body formation through direct connection with the endoplasmic reticulum. A nearly mature yolk body (Yb) *is directly connected with membranes of the smooth ER; the granular inclusions originate in the rough ER.* × *24,000. Courtesy of H. W. Beams, R. G. Kessel, and The Rockefeller University Press.* [J. Cell Biol. *18:621 (1963)*]

In both these patterns of synthesis, the assembly sites of yolk granules are attended by large numbers of mitochondria. In some oocytes, the number of mitochondria may become very large. In *Priapulus,* Norrevang has estimated that the number of mitochondria may reach 40,000 at the height of yolk synthesis. Also present in most oocytes during yolk deposition are elements of annulate lamellae (see Chapter 10). Although these are present before and during yolk synthesis, the functional role of these elements in oogenesis remains obscure.

Fundamentally different from the synthesis mechanism depicted above is the formation of yolk bodies by a process known as *pinocytosis* (Fig. 15–43c and Fig. 15–46). In this process, yolk materials are taken into the oocyte directly from the extracellular medium in minute pits or invaginations that appear in the plasma membrane. Once formed, these inpocketings with their enclosed contents pinch off from the plasma membrane and sink into the cytoplasm, where they fuse into yolk bodies with other, similarly formed vesicles. The substances taken in by pinocytosis are secreted by follicle or nurse cells or are carried from more distantly located secretory cells to the site of oogenesis by the circulatory system.

An additional route for the deposition of yolk is the unexpected and little understood formation of yolk bodies within mitochondria (Figs. 15–43d and 15–47). In amphibians, in which this process has been most clearly demonstrated, the yolk is deposited in a crystalline array between the inner and outer membranes of the mitochondrion but apparently not in the matrix. As deposition proceeds, the cristae gradually disappear and the mitochondrion is transformed into a yolk body. Nothing is known of the role of the mitochondrion in this process, the origin of the material deposited in the intermembrane space, or whether the yolk is synthesized within the mitochondrion or merely transported across the outer mitochondrial membrane to be accumulated inside into the yolk body (see, for example, Balinsky and Devis, 1963).

The nutrient material stored in the yolk body may consist of protein, phospholipid, sometimes polysaccharides, or mixtures of these substances. In spite of considerable variation in the chemical makeup of the enclosed material, the substructure of yolk bodies is remarkably uniform in different animal species (Fig. 15–48). The yolk body usually consists of a dense crystalline core that at higher magnification shows a periodicity at 75–85 Å. The crystalline core, often regularly angular in outline, fills up most of the yolk body. A small amount of less structured *matrix* is usually observed between the core and the surrounding membrane.

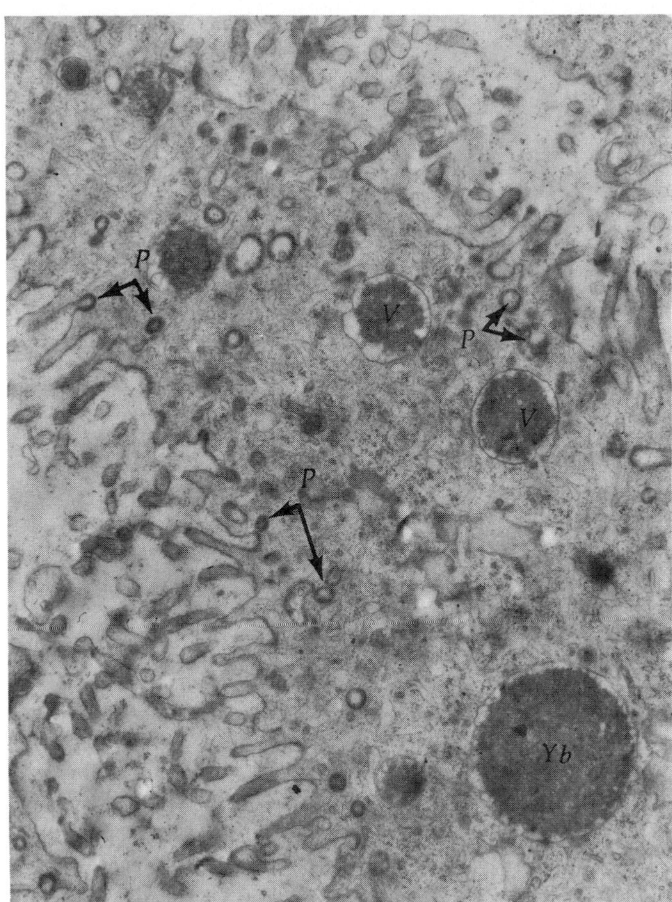

15–46. Yolk body formation by incorporation of pinocytotic vesicles in an oocyte of the milkweed bug Oncopeltus. P, *small vesicles forming by evaginations of the plasma membrane;* V, *larger vesicles at deeper levels of the cytoplasm presumably formed by fusion of the small pinocytotic vesicles;* Yb, *yolk bodies formed by progression fusion of the smaller vesicles.* × *16,000. From* Cellular Membranes in Development, *ed. M. Locke, 1964. Courtesy of H. W. Beams, R. G. Kessel, and Academic Press Inc.*

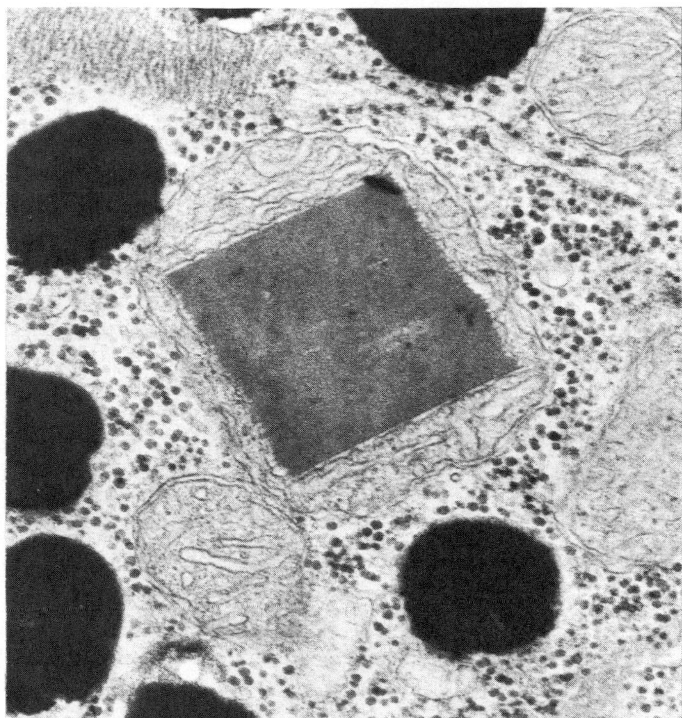

15–47. Formation of a yolk body within a mitochondrion of the frog Rana. *The yolk body is enclosed within the intercristae space and remains separated from the matrix by the membranes of the cristae.* × *49,000. Courtesy of P. B. Armstrong.*

Yolk bodies of this type are observed in most animals, except mammals, teleost fishes, and birds. No proteinaceous yolk is secreted in mammals. Instead, large numbers of membrane-bound aggregations of small vesicles appear in the cytoplasm as the oocytes mature. These structures, termed *multivesicular bodies* (Fig. 15–49), are roughly 1000–6000 Å in diameter and contain smaller membranous vesicles 100–500 Å in diameter. Some authorities have maintained that multivesicular bodies are analogous to the yolk bodies of other animals and that nutrients for the developing embryo are stored inside. However, no experimental demonstration of the relationship of multivesicular bodies to yolk bodies has been forthcoming, and the significance of these structures remains unknown. Similar bodies are sometimes observed in ordinary somatic cells; their function in this instance is also obscure.

In addition to yolk bodies, the cytoplasm of many eggs also becomes packed with lipid droplets. These droplets, which are not separated from the surrounding cytoplasm by a membrane, usually consist of concentrations of triglycerides and phospholipids. These droplets appear in close proximity to mitochondria and other elements of the cytoplasm, including the endoplasmic reticulum and the Golgi apparatus. However, the role, if any, of these organelles in the synthesis and assembly of lipid droplets is unknown.

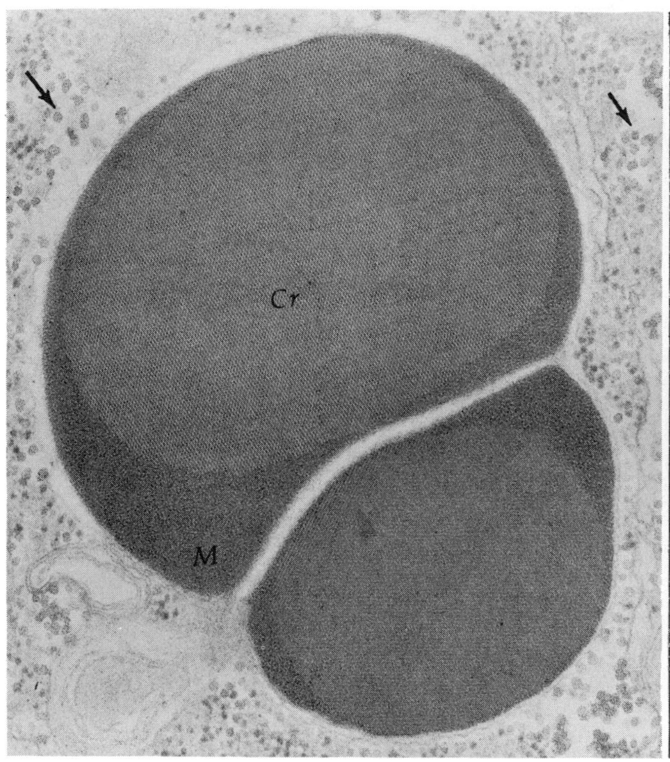

a

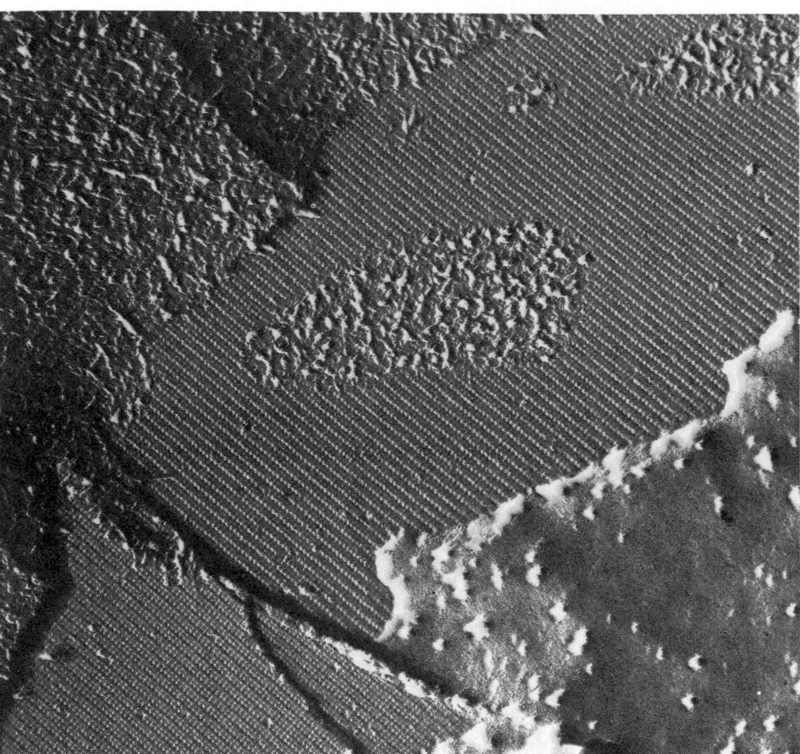

b

15–48. a, *A yolk body of the amphibian* Xenopus. *Typically, the yolk body contains an angular crystalloid* (Cr) *embedded in a matrix substance* (M). *A regular periodicity is visible in the crystalloid. Numerous glycogen granules (arrows) are visible in the cytoplasm surrounding the yolk body.* × *49,000. Courtesy of P. B. Armstrong.* b, *Fractured surface of a yolk body of* Xenopus *prepared by the freeze etch method. Periodicity of the crystalloid produces striking patterns in these preparations.* × *84,000. Courtesy of R. L. Leonard.*

The mature egg, with the cytoplasm packed with yolk, lipid droplets, ribosomes, and inactive RNA and the nucleus arrested in various stages of meiosis, is shed from the ovary and awaits penetration of the sperm. The meeting of sperm and egg triggers a sequence of events beginning with penetration of the coats of the egg by the sperm and culminating in the union of the male and female pronuclei to produce the zygote nucleus and initiate the development of the embryo.

Fertilization

Fertilization in Animals

The process of fertilization in animals can be conveniently subdivided into a number of more or less separate steps controlled by different mechanisms in the sperm and egg. Initial approach, if the egg is mature and the sperm is from the same species, is followed by tight attachment of the sperm to the surface envelopes of the egg. Attachment, a species-specific interaction, appears to be moderated by surface components of both the egg and sperm, similar in action to an antigen-antibody reaction. After attachment, the sperm penetrates through the egg envelopes, finally reaching the plasma membrane of the egg. Passage of the spermatozoan through the outer extracellular envelopes of the egg is accomplished in almost all animals by the activity of enzymes liberated from the acrosome, which expels its contents to the exterior during or just before attachment of the sperm to the egg. Once the fertilizing sperm reaches the plasma membrane of the egg, continuities between sperm and egg membranes are formed, and the egg and sperm cytoplasm become

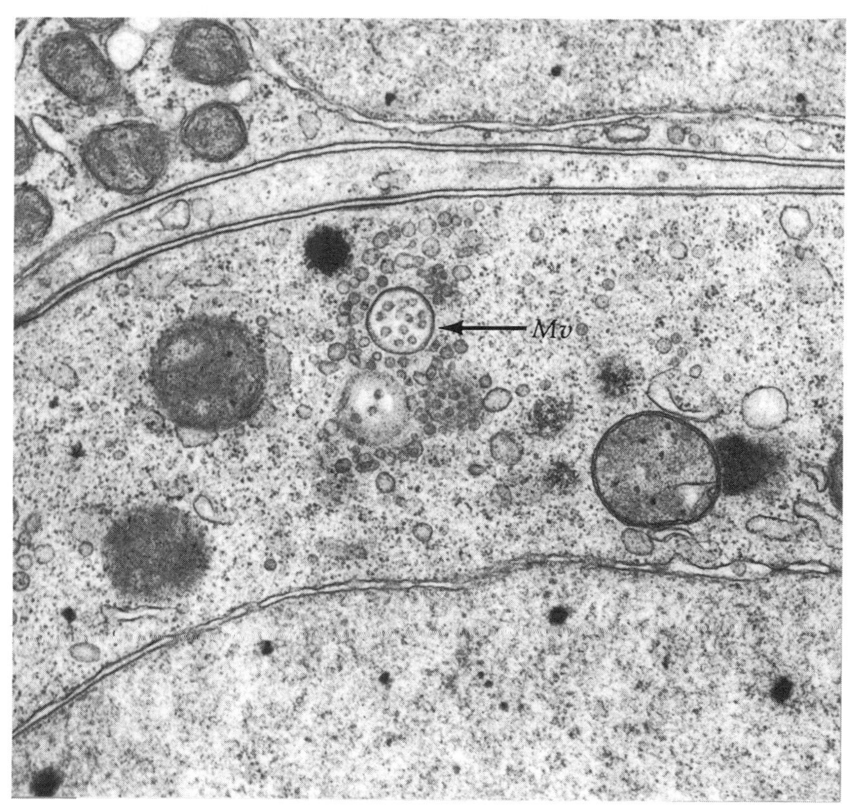

15–49. *A multivesicular body* (Mv) *in the cytoplasm of the mouse oocyte. The multivesicular body is surrounded by small Golgi vesicles.* × *24,000. From* An Atlas of Ultrastructure *by J. A. G. Rhodin, 1963. Courtesy of J. A. G. Rhodin and W. B. Saunders Company.*

continuous. As a result of this union, the sperm nucleus is gradually engulfed along with other organelles of the sperm including the centrioles, mitochondria, and axial complex of the tail. As the sperm nucleus sinks into the egg cytoplasm, its chromatin becomes transformed morphologically from the uniformly dense aspect typical of spermatozoa to the somatic condition. At this stage, it is known as a *pronucleus*. Finally, the male pronucleus fuses with the egg pronucleus. These steps—approach, attachment, penetration, absorption of the sperm, and fusion of pronuclei—will be considered in the sections to follow, along with the reaction of the egg to fertilization and what little is known of the biochemistry of these events.

Approach of the sperm to the egg. The physiological basis for the approach of sperm to the vicinity of the egg has been a subject of extensive debate and much experimentation since the turn of the century. Many investigators consider it likely that the sperm reaches the egg simply by random, unguided movement. According to this viewpoint, no sperm "attractants" are secreted by the egg or adjacent cells such as the follicle cells. This attitude has developed because efforts to identify sperm attractants have met with little success in animals. Perhaps the only unequivocal demonstration of such a substance was carried out by R. L. Miller, who identified a substance secreted by the female of the coelenterate *Campanularis* capable of attracting sperm both *in vivo* and *in vitro*. The attractant was found to be species-specific and extractable. Identification of sperm attractants of this type is the exception rather than the rule among animals, and the initial reactions of fertilization seem to await the random collision of sperm and egg.

Attachment of the sperm to the egg surface. Contact between sperm and egg results in an immediate and firm attachment of the sperm head to the outer surface of the egg investments. This attachment, which resembles an antigen-antibody reaction, follows only if egg and sperm are from the same species and also depends on maturity of the egg. The immediate binding of sperm to egg is probably moderated by the presence of active groups on both the sperm plasma membrane and the outer envelopes of the egg. The identity of the active egg-sperm attachment factors is unknown. Experiments dating back many years have identified a substance present in the jelly coats of eggs, such as the sea urchin and amphibians, that is capable of coagulating a suspension of active sperm. This substance, known classically as "fertilizin," has since been identified as a protein. Supposedly, another protein, or perhaps another smaller molecule, an "antifertilizin," capable of specific interaction with the fertilizin, exists on the surface of the sperm. Fertilizins have been identified in many species, and it is considered very likely that "fertilizin-antifertilizin" systems are widespread in nature as the basis of attachment of sperm to the egg. In mammals, in which much work has been done on antigens of the egg and sperm, it has become apparent that both the sperm and the egg investments contain a spectrum of specific antigens that function in recognition and attachment during fertilization (see Metz, 1967).

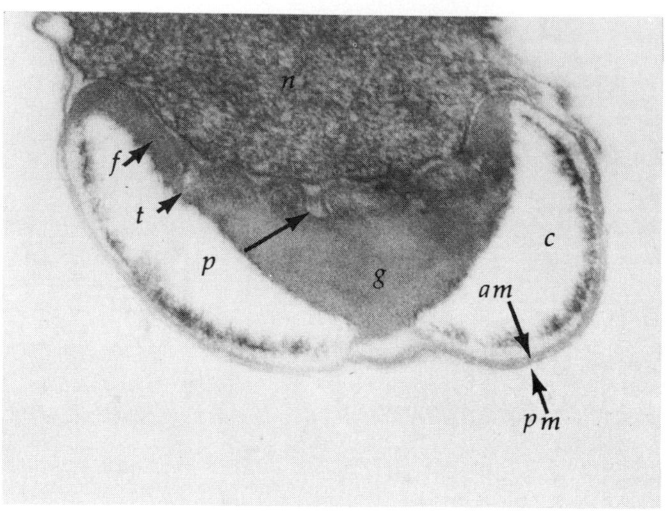

15–50. The unreacted acrosome of a Hydroides *sperm.* n, *nucleus;* p, *perforatorium or periacrosomal material;* t, *invagination of surface of the acrosome facing the nucleus;* g, *acrosomal granule;* c, *cavity of acrosomal vesicle;* f, *fine granular or fibrillar material of the acrosome;* am, *acrosomal membrane;* pm, *plasma membrane. × 64,000. Courtesy of L. H. Colwin and A. L. Colwin, from* J. Biophys. Biochem. Cytol. *10:211 (1961).*

Penetration of the outer envelopes of the egg: The acrosome reaction. Very quickly following attachment to the outer surface of the egg, the sperm penetrates through the various surface coats to reach the plasma membrane of the egg. In all animal eggs without an impermeable chorion, as for example in the annelids, molluscs, crustaceans, echinoderms, and vertebrates, the penetration mechanism works through the activity of enzymes released by rupture of the acrosome. In contrast to the specificity of initial attachment of sperm to egg, the enzymes released by the acrosome are relatively unspecific in activity. The two major enzymes present, hyaluronidase and lysin, are capable of dissolving the outer envelopes of all animal eggs with either a jelly coat or an investment of follicle cells. The acrosome reaction is the most immediate of the series of sperm-egg interactions following attachment of the sperm to the egg surface. In some groups, particularly the mammals, rupture of the acrosome may be initiated even before attachment of the sperm. In this case, the initiating substances are believed to be present in the female reproductive tract close to the site of the egg.

The acrosome reaction, one of the more spectacular interactions during fertilization, has been extensively studied by both light and electron microscopy. The reaction in invertebrates has been thoroughly investigated by Arthur L. Colwin and Laura Hunter Colwin. In the annelid *Hydroides*, the intact, unreacted acrosome (Figs. 15–50 and 15–51) is a complex structure exhibiting internal differentiation into a central, dense granule and a peripheral, finely fibrillar cortex. The Colwins have shown that, immediately on contact with the outer envelope of the egg, the acrosome membrane fuses with the sperm plasma membrane at several points, opening continuous channels from the acrosome to the exterior of the cell (Fig. 15–52). Through these channels, which open to the surface of the sperm, the enzymes of the acrosome flow and attack the substances of the egg coats. As the sperm advances through the egg coats, with the anterior end leading, the granule and cortex material of the acrosome gradually diminish and disappear (Fig. 15–53). At the edges of the opening into the acrosome, the acrosomal membrane and sperm plasma membrane are continuous (arrow, Fig. 15–53), so that the nucleus and cytoplasm of the sperm remain separated from the surrounding material of the egg coat by unbroken membranes. As the sperm head approaches the plasma membrane of the egg, fine villi or evaginations arise from the inner acrosomal membrane, which remains intact over the anterior surface of the nucleus. These villi, just beginning to form at the stage

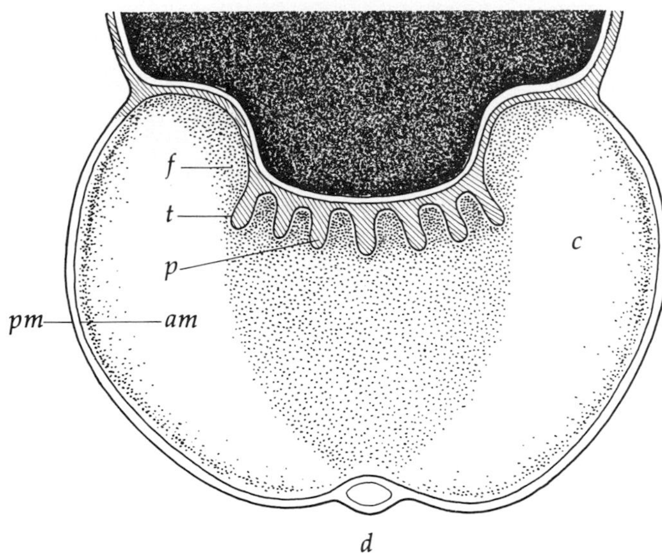

15–51. Diagram of the structures visible in Fig. 15–50. (See Fig. 15–50 for key.) The structure labeled d *is the apical vesicle, a structure of unknown function. Redrawn, original courtesy of L. H. Colwin and A. L. Colwin, from* J. Biophys. Biochem. Cytol. *10:211 (1961).*

shown in Fig. 15–53, extend and make contact with the plasma membrane of the egg (Fig. 15–54). The egg plasma membrane responds to this contact by the formation of small pits or depressions that interdigitate with the villi from the sperm. Fusions appear at these points between the two plasma membranes, and eventually the membranes of egg and sperm become continuous (Fig. 15–55). At least initially, the cytoplasmic border of the egg at this point, therefore, consists of a mosaic of patches of sperm and egg plasma membranes. As continuity is established, the cytoplasm of the egg flows outward around the sperm cell and gradually engulfs the nucleus, forming a "fertilization cone" visible in the light microscope. The principal stages of sperm penetration are diagramed in Fig. 15–56. The Colwins have demonstrated that the *Hydroides* pattern is typical of most invertebrate and many vertebrate species.

In *Hydroides*, according to the Colwins, nine seconds elapse between the first attachment of the sperm to the outer egg coats and fusion of sperm and egg membranes. In the sea urchin, the acrosome reaction occurs in about one second; a further two minutes is required for traversal of the jelly coats of the egg.

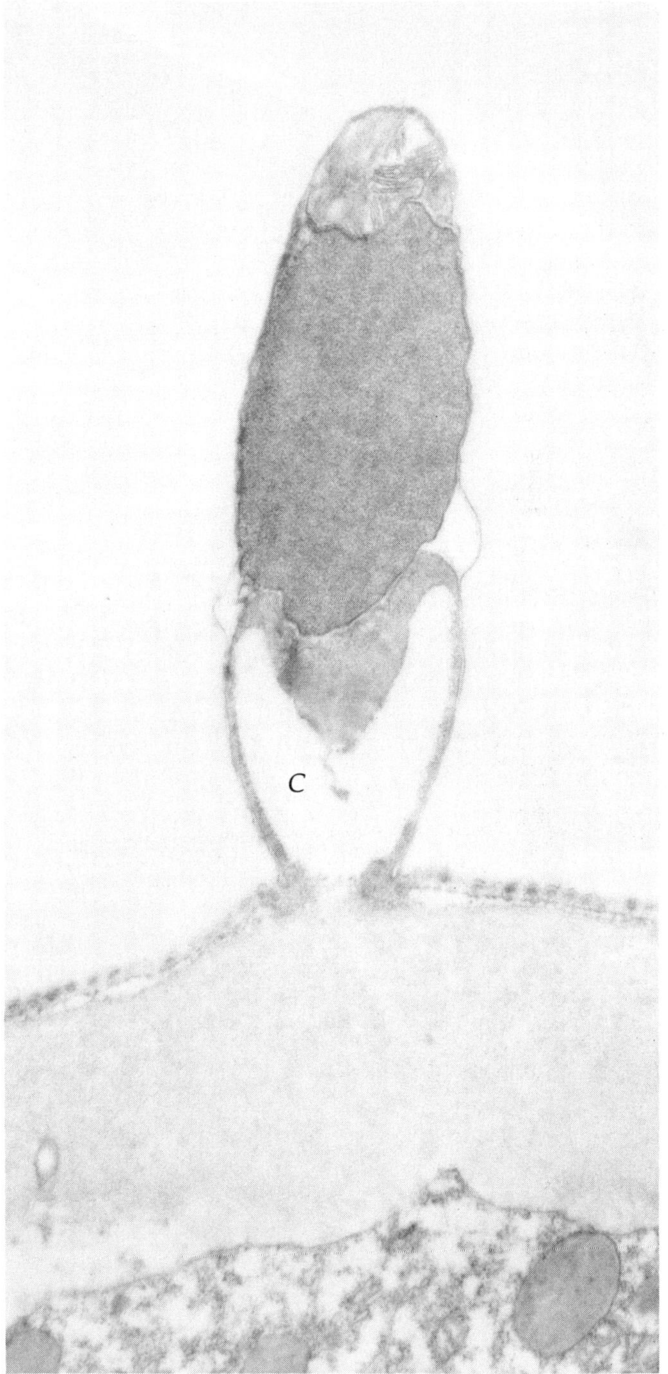

15–52. Initial contact of a Hydroides *spermatozoan with the egg surface. Continuities between the acrosome and plasma membranes open the acrosomal interior* (C) *to the outside of the sperm cell.* × *47,000. Courtesy of L. H. Colwin and A. L. Colwin, from* J. Biophys. Biochem. Cytol. *10:211 (1961).*

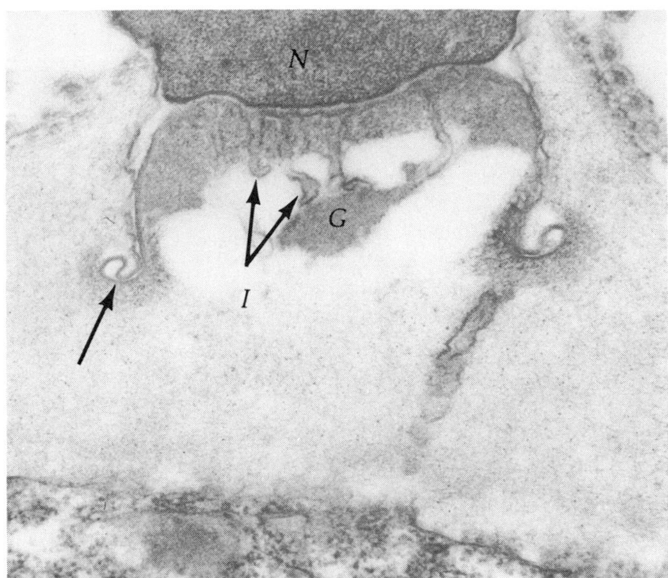

15–53. Modifications of the acrosome in Hydroides *during movement of the sperm through the egg envelope. The acrosomal granule* (G) *has decreased in size. At the edges of the opening into the acrosomal interior, the acrosome membrane is continuous with the plasma membrane of the spermatozoan (arrow).* N, *nucleus;* I, *tubular invaginations of the inner acrosomal membrane.* × *53,000. Courtesy of L. H. Colwin and A. L. Colwin, from* J. Biophys. Biochem. Cytol. *10:211 (1961).*

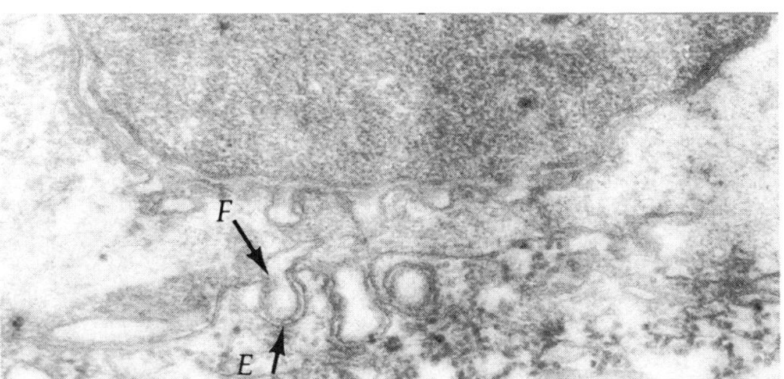

15–54. A later stage during fertilization in Hydroides. *The tubular invaginations of inner acrosomal membrane* (F) *interdigitate with indentions of the egg plasma membrane* (E). × *73,500. Courtesy of L. H. Colwin and A. L. Colwin, from* J. Biophys. Biochem. Cytol. *10:211 (1961).*

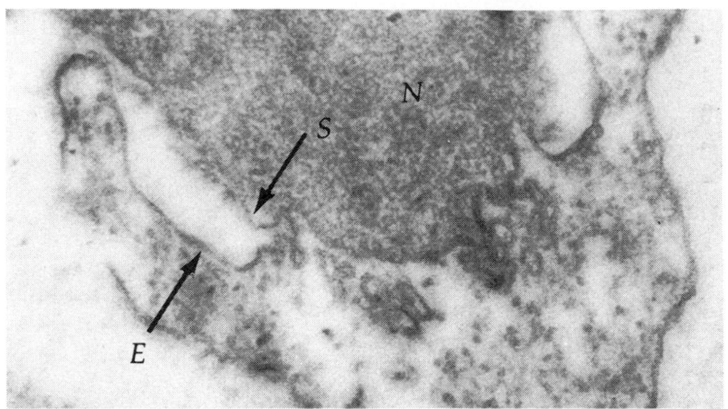

15–55. Fertilization in Hydroides, *at a stage after fusion of the sperm and egg membranes.* E, *egg plasma membrane;* S, *sperm plasma membrane;* N, *sperm nucleus.* × *72,000. Courtesy of L. H. Colwin and A. L. Colwin, from* J. Biophys. Biochem. Cytol. *10:211 (1961).*

In invertebrates and some vertebrates the acrosome reaction is accompanied by extension of a filament from the base of the acrosome (Fig. 15–57). This filament, which arises from the space between the acrosome and the nucleus, extends rapidly through the egg coats and contacts the membrane of the egg. The filament as it extends remains surrounded by the plasma membrane of the sperm cell. The acrosomal filament, although of unknown structure and function, is believed to be a keratin-like protein that establishes a firm connection between the sperm and the plasma membrane of the egg. In at least a few instances, the initial fusion of egg and sperm membranes has been seen to take place at the tip of the acrosomal filament.

The acrosome reaction in mammals is similar in many ways to the pattern observed in *Hydroides* and other invertebrates. R. Yanagimachi and Y. D. Noda have studied fertilization in the golden hamster. In this species, the unreacted acrosome extends as a cap around the anterior half of the nucleus (see Figs. 15–58 and 15–59). As the sperm contacts the layer of follicle cells around the egg, multiple openings appear in the membranes separating the acrosome from the cell exterior (Figs. 15–58 and 15–60). These channels are formed by fusions between the outer acrosomal and plasma membranes as in *Hydroides*. As the channels form, the acrosomal contents disperse into the surrounding medium. In the golden hamster, as in many mammals, the acrosome is differentiated into an anterior cap and a sleeve-

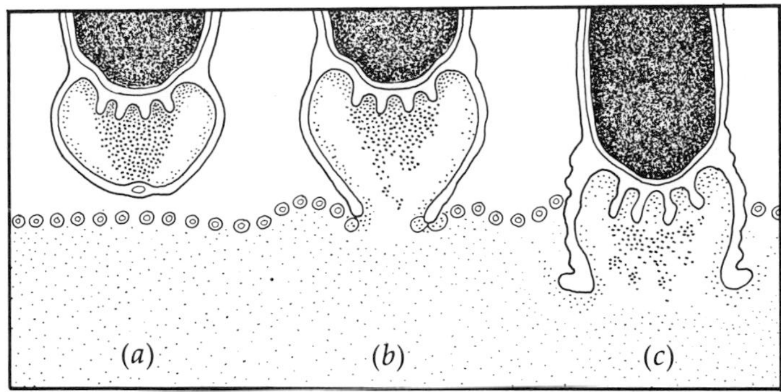

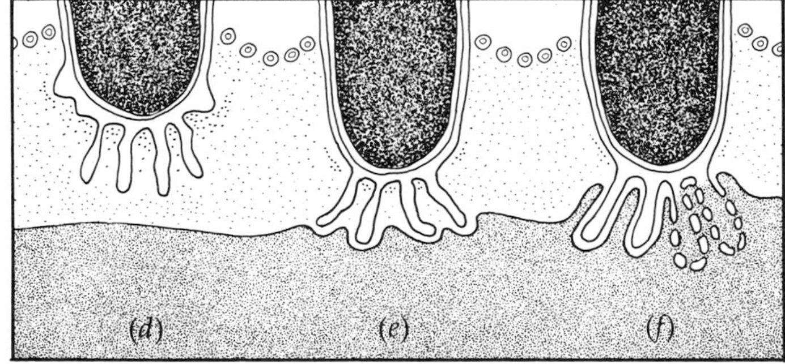

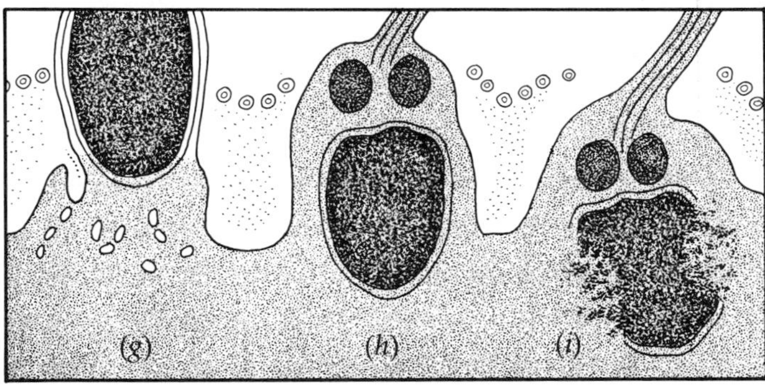

15–56. *A summary of the various stages in penetration of the egg by the sperm in* Hydroides *(see text).* a, *Sperm before contact with the egg surface.* b, *Sperm after initial contact with the egg. The acrosome has opened to the exterior.* c *and* d, *Successive stages in the extension of the inner acrosomal membrane into tubules as the sperm penetrates the outer envelope of the egg.* e, *Interdigitation of tubules of the sperm with invaginations in the egg plasma membrane.* f *and* g, *Progressive fusion of sperm and egg plasma membranes.* h, *Extension of egg cytoplasm around the sperm to form the fertilization cone.* i, *Absorption of the sperm nucleus into the egg cytoplasm. Transformation of the sperm nucleus into the male pronucleus has begun. Redrawn from* Fertilization, *Vol. I, ed. C. B. Metz and A. Monroy, 1967. Courtesy of L. H. Colwin, A. L. Colwin, and Academic Press Inc.*

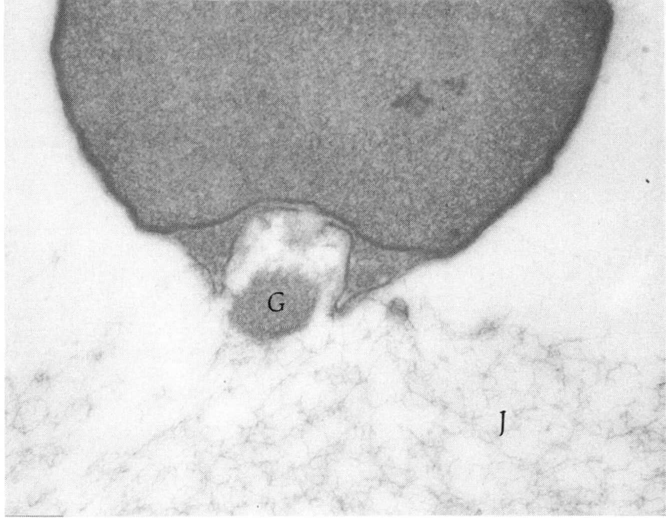

a

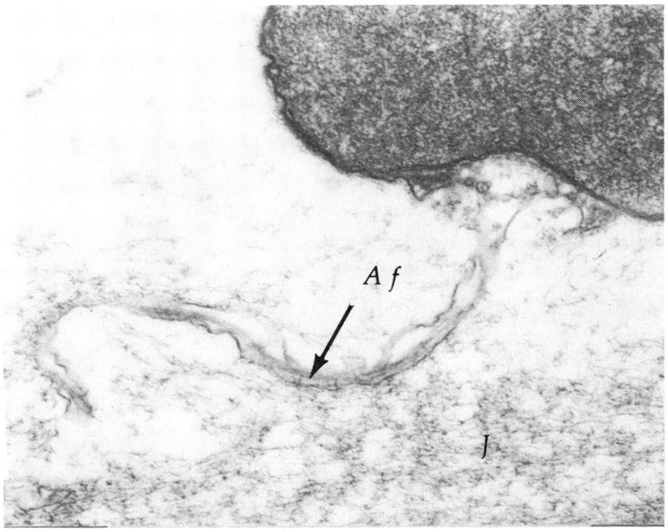

A f
J

b

15–57. *The acrosome reaction in* Saccoglossus, *a hemichordate, in which an acrosomal filament extends from the periacrosomal space to contact the egg surface. The extended filament remains covered with an extension of the inner acrosomal membrane. In* a *the acrosomal reaction has just begun; in* b *the acrosomal filament has extended into the jelly coat of the egg.* Af, *acrosomal filament;* G, *acrosomal granule;* J, *jelly coat of the egg.* a, *× 40,000.* b, *× 51,000. Courtesy of A. L. Colwin and L. H. Colwin.*

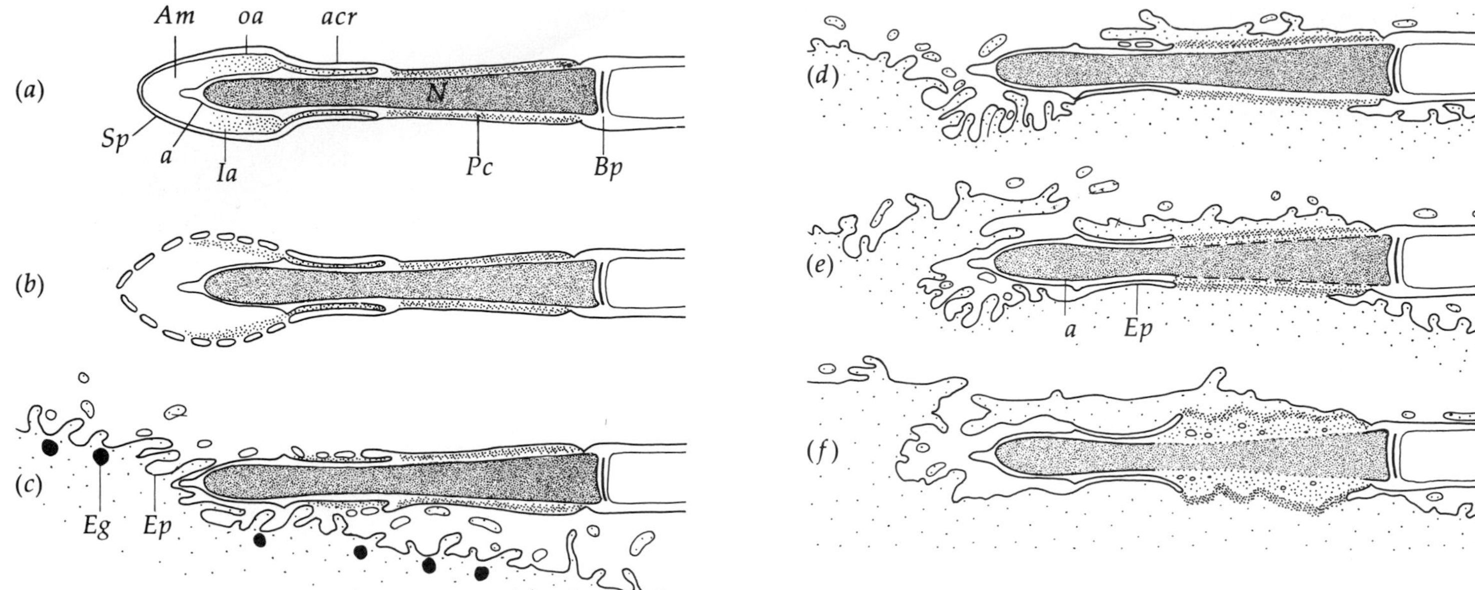

15–58. *Diagram of the stages in sperm penetration in the hamster.* a, *An unreacted sperm.* b, *Fusion of acrosomal and plasma membranes with resulting perforation of the acrosome.* c, *Contact of the spermatozoan with the egg plasma membrane.* d, *Fusion of egg and sperm membranes (shaded region).* e *and* f, *Stages in adsorption and conversion of the sperm nucleus.* Sp, *sperm plasma membrane;* oa, *outer acrosomal membrane;* a, *inner acrosomal membrane;* Am, *acrosomal material;* N, *nucleus,* Pc, *perinuclear cytoplasm,* Bp, *basal plate;* Ep, *egg plasma membrane;* Eg, *cortical granules. Diagram courtesy of R. Yanagimachi, Y. D. Noda, and Academic Press Inc., from* J. Ultrastr. Res. *31:465 (1970).*

like collar that extends posteriorly over the nucleus. While no direct evidence exists, it is generally held that the cap contains hyaluronidase. The collar or more posterior portions of the acrosome are believed to contain lysin. In general, the cap region of the acrosome disperses first. This is followed by liberation of the collar material as the sperm penetrates more deeply into the follicle cells and approaches the egg. The sperm then lies flat on the surface of the zona pellucida and in about 30 minutes passes through the zona to the plasma membrane of the egg (Fig. 15–61). During these stages of penetration, the tail of the sperm maintains the flagellar beat. Soon after contact, microvilli extend from the egg plasma membrane and gradually surround the sperm nucleus (Figs. 15–58 d–f and 15–62). At this point, the beat of the sperm tail ceases, continuities are established between egg and sperm membranes, and the cytoplasm of the egg gradually flows around and engulfs the sperm nucleus.

In animal eggs possessing a hardened, relatively impermeable exterior chorion, such as in insects, cephalopod molluscs, and fishes, the sperm passes to the immediate egg surface through a narrow channel in the chorion known as the micropyle. In teleost fishes, the micropyle is wide enough to admit only a single sperm cell to the egg surface (Fig. 15–63). The absence of the acrosome in fishes is possibly related to the existence of the specialized channel for sperm penetration to the egg surface.

Reactions of the egg to sperm penetration. Penetration of the sperm to the egg is followed immediately by a series of practically instantaneous reactions of the egg. The most conspicuous of these is the appearance of a fluid-filled separation between the plasma membrane of the egg and the egg envelopes. Often, as in many invertebrates, the envelopes of the egg can be seen to lift a few microns from the plasma membrane to produce

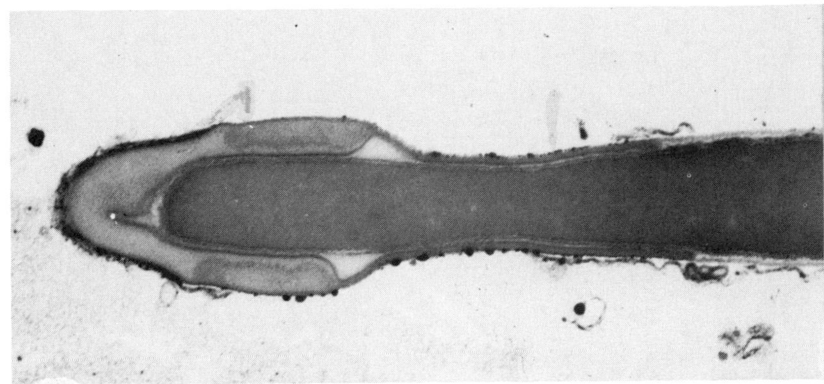

15–59. *Unreacted sperm of the hamster, equivalent to Fig. 15–58* a. × *29,000. Courtesy of R. Yanagimachi, Y. D. Noda, and Academic Press Inc., from* J. Ultrastr. Res. *31:465 (1970).*

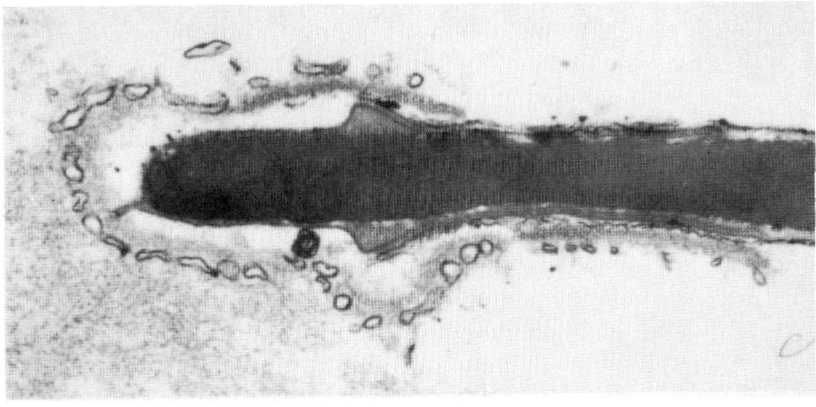

15–60. *Stage in sperm penetration in the hamster, equivalent to Fig. 15–58* b. *Continuities have appeared between the acrosome and plasma membrane of the spermatozoan, exposing the acrosomal contents to the exterior of the cell.* × *31,000. Courtesy of R. Yanagimachi, Y. D. Noda, and Academic Press Inc., from* J. Ultrastr. Res. *31:465 (1970).*

this space. Although this raised surface coat does not consist of cellular membranes, it is frequently termed the "fertilization membrane." The elevation of the surface coat of the egg is frequently accompanied by disappearance of the granules from the cortical layer of the egg cytoplasm. The cortical granules, which lie just below the plasma membrane of the unfertilized egg (Fig. 15–64a), become irregular in outline at this time and open to the egg surface, releasing their contents to the exterior (Fig. 15–64b). The discharge of cortical granules passes quickly in a wave over the egg surface, beginning at the penetration point of the fertilizing spermatozoan and extending around the egg. In a number of species, such as the marine worm *Nereis*, the material expelled from the cortical granules to the exterior forms a coat of jelly around the fertilized egg. In most species, the site at which the sperm enters remains visible for a time after fertilization as a cone of cytoplasm that extends into the fluid-filled space surrounding the egg.

The egg becomes impermeable to additional sperm almost immediately after attachment of the fertilizing sperm to the egg surface. This "block to polyspermy," as it is called, is often ascribed to the material discharged by the cortical granules because the development of the block and cortical granule discharge are closely coincident in time. However, completion of cortical granule discharge has been observed to progress more slowly than the block to polyspermy, which is practically instantaneous over the whole surface of the egg. Therefore, the surface changes, which also involve a loss of the sperm attachment mechanism, are believed by many investigators to progress by a separate process possibly similar to the conduction of impulses over the surface of nerve fibers. This is supported by investigations that have revealed very rapid changes, occurring at the instant of fertilization, in the permeability to ions and transient electrical changes over the egg surface, similar to the events in propagation of the nerve impulse (see Tyler, Monroy, and Metz, 1956).

Besides the changes observable at the surface of the egg, internal rearrangements of the egg cytoplasm and nucleus, including movements of organelles and changes of viscosity of the cytoplasm, can also be seen

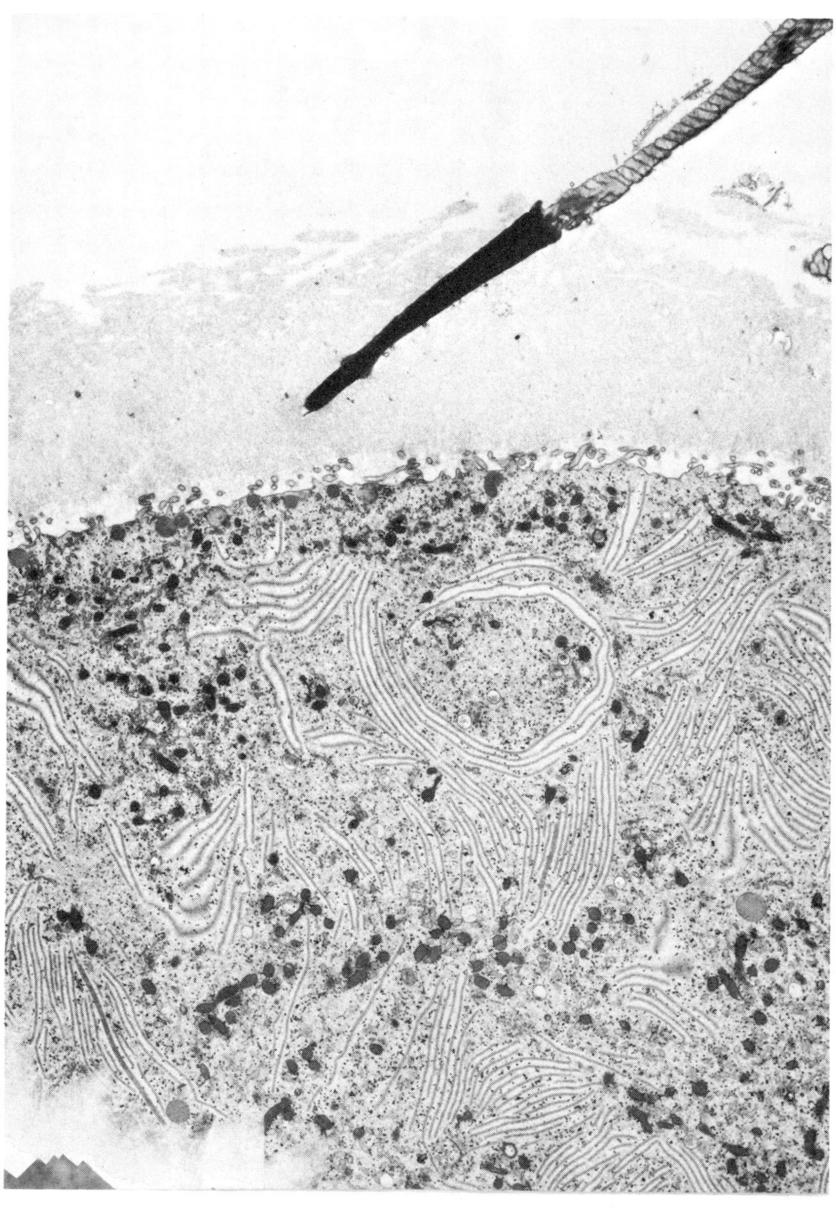

15–61. Penetration of the hamster sperm through the zona pellucida to the plasma membrane of the egg. × 6,800. Courtesy of R. Yanagimachi, Y. D. Noda, and Academic Press Inc., from J. Ultrastr. Res. *31:465 (1970).*

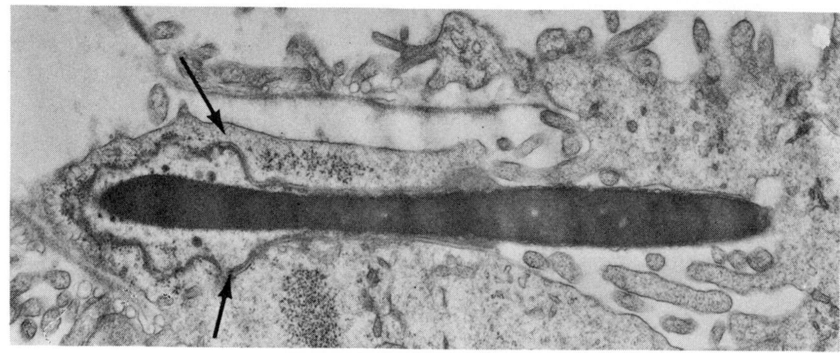

15–62. Initial stage in fusion of egg and sperm plasma membranes (arrows). × 20,000. Courtesy of R. Yanagimachi, Y. D. Noda, and Academic Press Inc., from J. Ultrastr. Res. *31:465 (1970).*

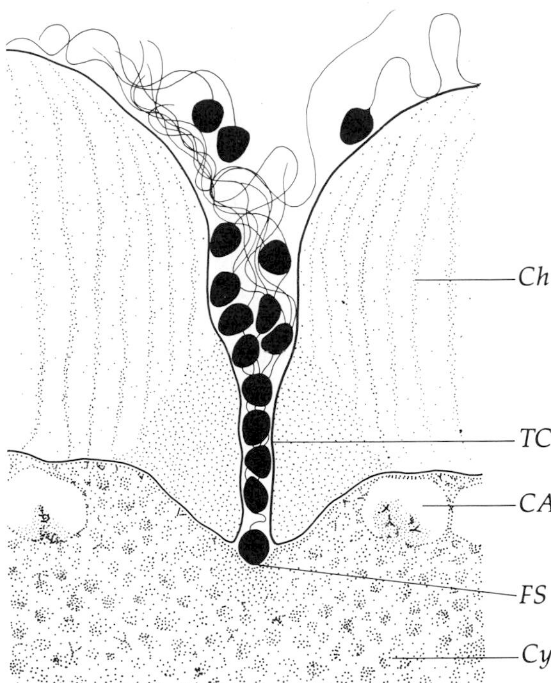

15–63. *Diagram of a trout egg during fertilization. The narrow channel* (TC) *through the chorion* (Ch) *is just wide enough to admit a single fertilizing spermatozoan* (FS) *to the egg plasma membrane.* CA, *cortical vesicle;* Cy, *egg cytoplasm. Redrawn, original courtesy of A. S. Ginsburg.*

to occur in many species. In eggs in which meiosis is arrested at prophase I or subsequent stages, attachment and penetration of the fertilizing sperm cell is followed by resumption of nuclear division and formation of the mature female pronucleus.

Fertilization is also followed by biochemical changes, immediately reflected as an abrupt increase in respiration of the egg. This initiation of active metabolism, first noted by Otto Warburg in 1910, is measurable as an increase in oxygen consumption and takes place concurrently with the morphological changes of the egg surface following penetration of the sperm. Protein synthesis, which proceeds in the unfertilized egg at an exceedingly low level, also responds to fertilization by a sudden increase in rate (see, for example, Hultin, 1952). In some unknown manner, mRNA molecules, present in the egg cytoplasm in an inactive state before fertilization, are "unmasked" and, immediately after penetration of the sperm, become active as templates for the synthesis of protein (for detailed literature references to the many experiments in this area, see Monroy and Tyler, 1967). Other biochemical changes, such as an increase in ATP synthesis and appearance of the enzymes and intermediates of glycolysis and respiration, take place almost immediately on fertilization and continue throughout embryonic development. Both the nature of the arrest of metabolic activity of the egg before fertilization and the release from arrest after penetration of the sperm remain obscure.

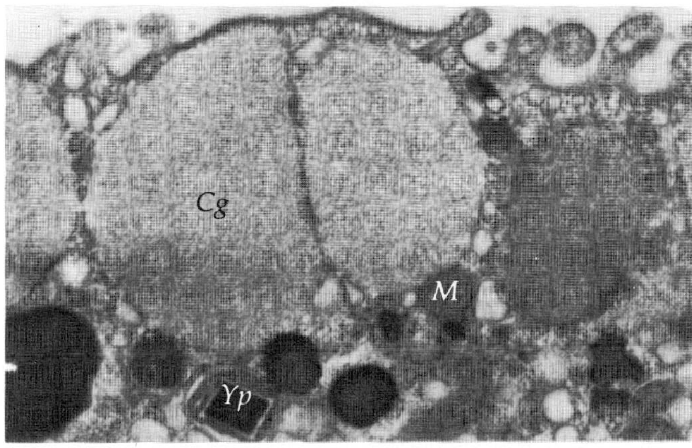

a

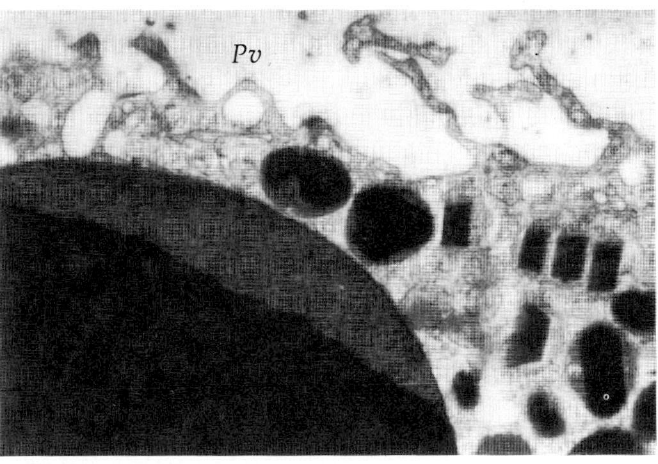

b

15–64. *Stages in the cortical granule reaction in the frog* Rana pipiens. *In* a, *the cortical granules* (Cg) *are still intact.* b, Rana *oocyte 13 minutes after fertilization. A cortical granule has just fused with the plasma membrane, releasing its contents into the perivitelline space* (Pv). *Other granules have become irregular in outline.* M, *mitochondria;* YP, *yolk bodies or platelets.* a, *× 18,500.* b, *× 13,000. Courtesy of N. E. Kemp. N. L. Istock, and The Rockefeller University Press.* [J. Cell Biol. *34:111 (1967)*]

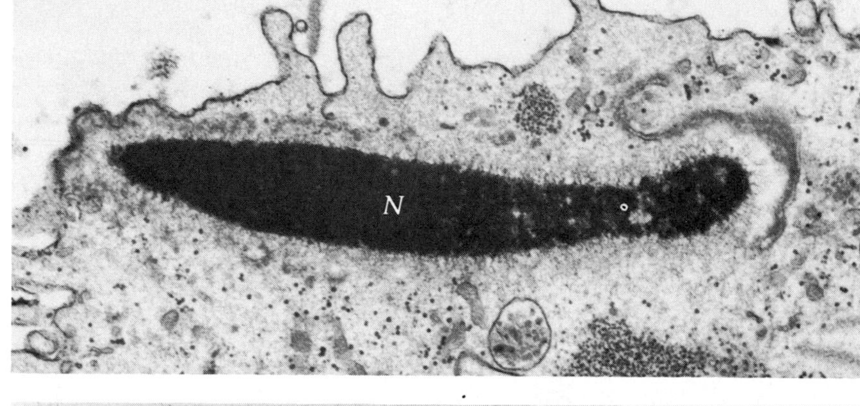

15–65. A sperm nucleus (N) *of the hamster after absorption into the egg cytoplasm. Transformation of the nucleus from the apparently homogeneous aspect of the spermatozoan to an interphase-like state is just beginning. Note that the sperm nucleus is "naked"; no nuclear envelope is present at this stage. × 15,000. Courtesy of R. Yanagimachi, Y. D. Noda, and Academic Press Inc., from* J. Ultrastr. Res. *31:465 (1970).*

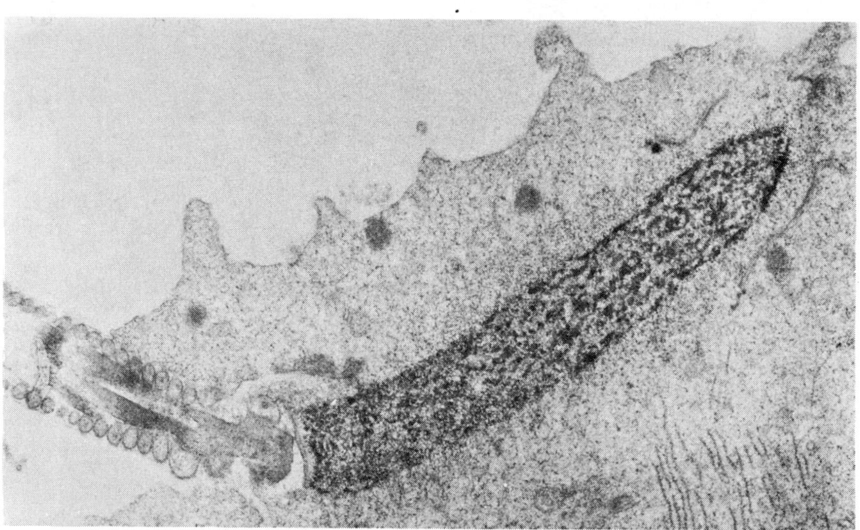

15–66. A later stage in transformation of the sperm nucleus in the rat. × 20,000. From Fertilization, *Vol. II, ed. C. B. Metz and A. Monroy, 1969. Courtesy of L. Piko and Academic Press Inc.*

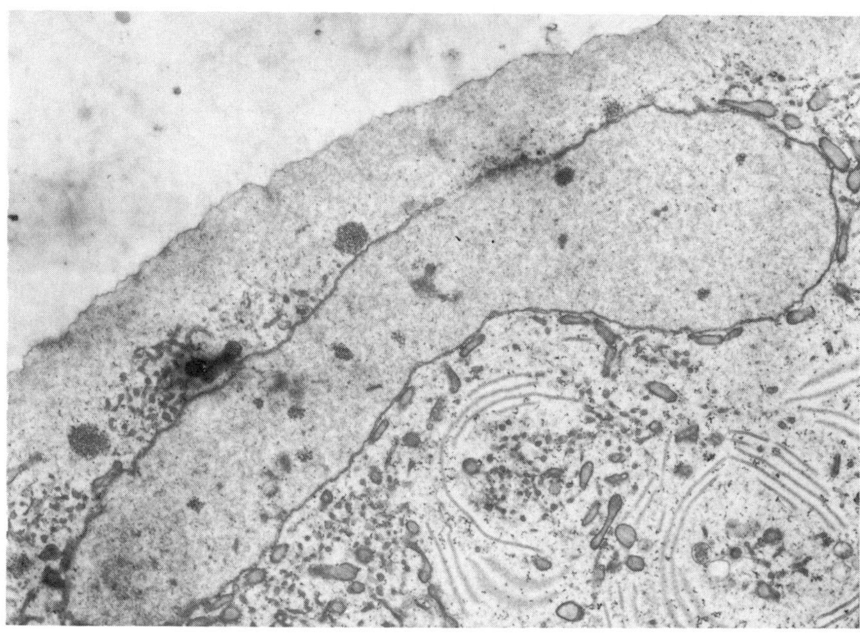

15–67. Sperm nucleus in the hamster, at a stage when transformation into the interphase-like state is nearly complete. A nuclear envelope has formed, and is nearly continuous. × 10,000. Courtesy of R. Yanagimachi, Y. D. Noda, and Academic Press Inc., from J. Ultrastr. Res. *31:465 (1970).*

Absorption of the nucleus and other organelles of the sperm cell. Following the establishment of continuities between the plasma membranes of sperm and egg, the egg cytoplasm gradually surrounds the sperm and invests the nucleus and other sperm organelles. The incorporation of the sperm proceeds from anterior to posterior, starting with the nucleus and progressing to the midpiece and the tail. In general, the entire sperm eventually enters the egg, but many exceptions are noted in which at least the tail is left outside the plasma membrane of the egg, such as in the annelid *Nereis*. Variation within a single species is noted here; in the sea urchin *Arbacia*, for example, the tail may either be incorporated into the egg cytoplasm or may remain outside the plasma membrane of the egg.

It is impossible in most cases to trace the fate of the cytoplasmic organelles of the sperm except for the distinctive axial complex of the tail, which usually disintegrates soon after fertilization, but may in some cases persist in the fertilized egg into the early cleavage stages. In one study (see Szollosi, 1965), the sperm mitochondria remained associated with a persistent axial complex of this type. Because of this location and their conspicuously smaller size, the sperm mitochondria could be traced in the egg cytoplasm after fertilization. In this case, the sperm mitochondria disintegrated after the first egg cleavage and did not appear to contribute to the mitochondrial population of the embryo. The same conclusions were reached by W. A. Anderson, who was able to follow the sperm mitochondria of *Paracentrotus* after fertilization because of their larger size, modified cristae, and greater electron density. These sperm mitochondria, after absorption into the egg cytoplasm, were observed to undergo degradation and eventually disappeared.

As the sperm nucleus sinks into the egg cytoplasm, the chromatin reverts from the dense, apparently structureless mass typical of the mature sperm to a condition approximating the interphase state (Figs. 15–65, 15–66, and 15–67). Transformation of the nucleus takes from 10 minutes to about a half hour in most species. Curiously enough, the sperm nucleus, immediately after entry into the egg cytoplasm, does not seem to possess a nuclear envelope (Figs. 15–65 and 15–66). As the chromatin undergoes transformation, vesicles similar to elements of the endoplasmic reticulum collect around the borders of the naked nucleus and gradually coalesce into a continuous double membrane. As the vesicles join together, pore complexes differentiate; by the time transformation of the nucleus is complete, the nuclear envelope is also fully formed. During these developments, the sperm nucleus gradually increases in size, partly through accumulation of interchromosomal material.

Very little is known about the molecular basis of the morphological changes in the sperm nucleus after absorption into the egg. David P. Bloch and H. Y. C. Hew found that soon after fertilization protamine is no longer detectable in the sperm nucleus of the snail *Helix*. The protamine is replaced by basic histones, which differ from both protamine and somatic histone in composition. These nuclear proteins, termed *cleavage histones* by Bloch and Hew, were found in the egg nucleus as well as the sperm nucleus. The cleavage histones persisted through early embryonic development up to the gastrula stage, at which time conversion to the adult somatic histones was noted. Presumably, the histones acquired by the transforming sperm nucleus are synthesized in the egg cytoplasm.

Fusion of egg and sperm nuclei. Once acquisition of a nuclear envelope and development of an interphase-like state is complete, the enlarging sperm nucleus migrates to the vicinity of the egg nucleus and fuses with it. All conceivable degrees of fusion are observed in various animal species. In the sea urchin, fusion is complete, with no trace remaining of the separate structures of egg and sperm pronuclei. Complete fusion of this type, although often observed in plants, is rare in animals. More commonly, a single spindle is formed in the egg cytoplasm and the egg and sperm nuclei differentiate separately into the metaphase state. The chromosomes from the two nuclei attach to the spindle and frequently remain in separate groups until telophase of the first cleavage mitosis is completed. In a few species, such as the fresh water crustacean *Cyclops*, the two groups of chromosomes retain their identity, forming bilobed nuclei until well after gastrulation of the embryo. Eventually, in all species, the distinction between the maternal and paternal chromosomes is lost, and fusion becomes complete.

Because the nuclei enter a mitotic division during fusion, a replication of the chromosomes of both egg and sperm nuclei must take place just before this time, between the first entry of the sperm nucleus and prophase of the first cleavage division. Very little is known of the time and place of this first "S" phase of the pronuclei in most species. This first replication was detected in one species, the mouse, by M. Alfert (1950), who demonstrated that DNA synthesis takes place in the haploid nuclei before fusion. Among species in which the sperm and egg nuclei differentiate into the metaphase state before fusion, this replication must also take place while the nuclei are still haploid; evidently, in the sperm nucleus, conversion of the nucleoprotein from protamine- or arginine-rich protein to "cleavage histones" is followed very closely or accompanied by DNA synthesis.

Traditionally, the centrioles found in the asters of the first cleavage division were thought to be derived from the sperm. As we have seen, the insect sperm described by Phillips (see p. 444) do not contain identifiable centrioles at maturity. Therefore it seems unlikely that the asters or centrioles of the fertilized egg are derived from the sperm in these insects. In other instances, such as in protozoa, copepods, and many vertebrates, no asters or centrioles are visible in oocytes during meiosis and subsequent maturation of the egg; here, the centrioles of the embryo must originate *de movo* or from the fertilizing spermatozoan. Evidently, the centrioles and asters of the first cleavage division may be derived from either the egg or the sperm, depending on the species in question.

Gametogenesis and Fertilization in the Higher Plants

In many plants, the egg remains embedded within the tissues of the parent before and for a time after fertilization. In this condition, the egg surface is "naked"; that is, the egg does not possess the various surface layers typical of animals. Internally, the eggs of plants are also less complex than animal eggs, owing to the fact that no large quantities of yolk or other nutrient reserves are stored within the egg cytoplasm. This is related partly to the dependence of the egg on the gametophyte for nourishment and partly to the ability of the egg to form sugars and other substances by photosynthesis after fertilization. The history of the egg during development is also more simple than in animals because the egg differentiates directly from a haploid cell following a mitotic rather than meiotic division.

The development of motile sperm in *Nitella*, described earlier in this chapter, is more or less typical of most lower plant groups, including some algae, liverworts, mosses, and ferns. Here it is seen that, in contrast to the egg, the sperm is closely similar in function and morphology to animal sperm. However, in higher plants the male gamete diverges from this pattern and is reduced to a relatively undifferentiated cell without a specialized motile apparatus.

E. B. Wilson (1928) pointed out that the evolution of the reduced sperm of the higher plants can be traced back through intermediate types from the motile, free-swimming sperm cells of lower plants. Motile sperm, complete with flagella, are present in the primitive gymnosperms such as the cycads and *Ginkgo*. However, these sperm cells, although motile, are not released into the external environment of the plant. Instead, they are carried to the female gametophyte in *pollen* (the male gametophyte) and, after germination of the pollen grain, are carried within a pollen tube that penetrates into the ovary of the female. The sperm cell is then released to swim only in the fluids surrounding the egg in the ovary. In the higher gymnosperms and angiosperms, a complete sperm cell, with both cytoplasm and nucleus, enters the vicinity of the egg by a similar route, but motility through flagella and the associated organelles is lost. Additionally, in the higher gymnosperms and angiosperms *two* sperm cells are released from the pollen tube into the female gametophyte. In the angiosperms, one of these fertilizes the egg nucleus, forming the zygote. The second fuses with one or more nuclei near the egg to form the *endosperm*, which after extensive growth serves as a nutritive tissue for the developing embryo. Thus, in its ultimate development, fertilization in plants is double.

Gametogenesis in the higher plants. While the egg nucleus is relatively simple in comparison to the female gamete in animals, the events leading up to differentiation of the egg in the gametophyte are complex and highly varied in different gymnosperms and angiosperms. Meiosis in plants is *intermediate* (see Chapter 14) and gives rise in the sporophyte to *spores*, which germinate and after mitotic division produce the gametophyte. In higher plants, the conspicuous form of the plant is the sporophyte; this includes all of the familiar flowering trees, shrubs, and plants. Meiosis in these diploid sporophytes gives rise to spores, just as in lower plants. However, the spores are not released from the sporophyte but instead germinate and grow into the gametophyte in the tissues of the sporophyte. After a series of mitotic divisions, either egg or sperm nuclei differentiate within the gametophyte.

In the most common sequence in higher plants, the female gametophyte develops in the *ovary* of the sporo-

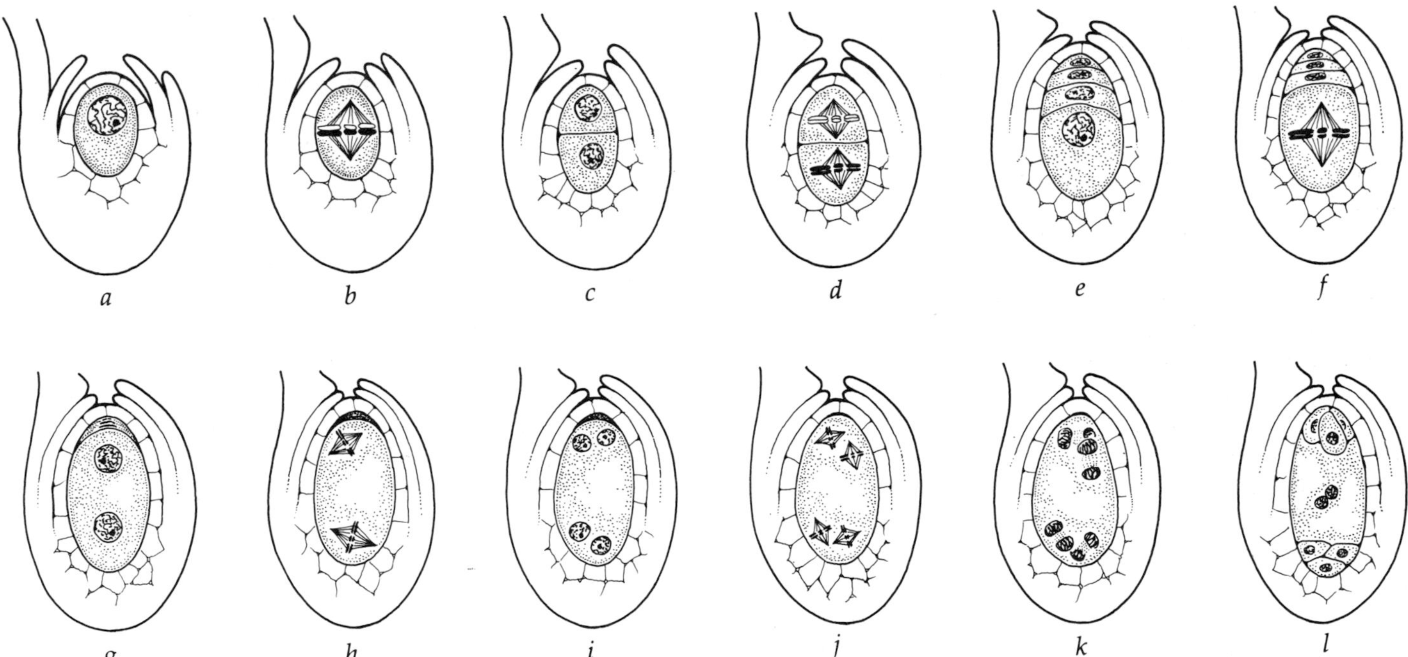

15–68. Development of the female gametophyte in a typical angiosperm (see explanation in text). Redrawn, original courtesy of C. P. Swanson.

phyte from a large spore called the *megaspore.* The megaspore originates from meiosis and is the sole survivor of the four haploid products of the meiosis (Fig. 15–68 a–e). The haploid megaspore immediately germinates within the ovary to form the female gametophyte and undergoes three sequential mitotic divisions, producing a total of eight cells (Fig. 15–68 f–k). One of these differentiates into the egg. The remaining seven differentiate into tissues of the gametophyte. Two of the seven take up a position at the entrance to the ovary. These cells, when fully differentiated, are called *synergids.* During fertilization, the two synergids are the first cells of the gametophyte to receive the pollen tube, and they also function in the transport of nutrients to the egg and developing embryo. Three more cells of the developing gametophyte move to the end opposite the synergids. These cells, called *antipodals,* degenerate in most plants without taking an active role in fertilization or embryonic development. The remaining two of the seven cells take up a position near the egg. The nuclei of this pair, which may fuse together or remain separate but closely aligned, eventually fuse with one of the two sperm nuclei to form the triploid endosperm. At the completion of its growth, the female gametophyte is a small plant, often with only eight cells, which remains parasitic on the tissues of the parent sporophyte (Fig. 15–68l). The ovary bearing the female gametophyte is a part of the central structure of the flower of the sporophyte (Fig. 15–69). Connected to the ovary by the *style* is the *stigma,* the structure that first receives the male gametophyte.

The egg contained within the female gametophyte is a complete cell. The egg cells of higher plants vary considerably in size and structure. These may be highly vacuolated, as in cotton, or contain almost unvacuolated cytoplasm, as in corn. All eggs are apparently completely surrounded by a plasma membrane and possess a wall over at least the part of the cell facing the synergids.

The male gametophyte develops from a small spore, called the *microspore,* in another part of the flower

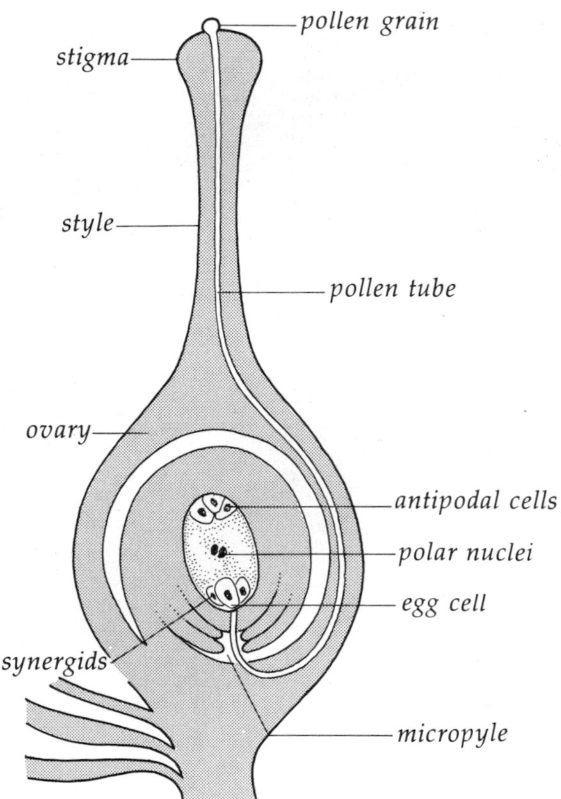

15–69. Diagram illustrating fertilization in an angiosperm. The pollen tube, arising from the pollen grain, penetrates the tissues of the stigma, style, and ovary, eventually reaching the micropyle of the ovary.

called the *anther*. As in megasporogenesis, the microspore is a product of meiosis in the sporophyte; in this case, all four of the resultant haploid spores germinate into male gametophytes (the pollen). Only part of the development of the male gametophyte takes place in the anther. In this initial development, the haploid nucleus divides once to form two nuclei. One of these, the *generative* nucleus, divides again after germination of the pollen grain to produce two sperm cells. The other nucleus, called the *vegetative* nucleus, regulates the growth of the pollen tube after germination. The mature pollen grain develops a hard, impermeable coat and on release from the anther, represents a microscopic plant with only two cells (Fig. 15–70). These cells remain in a state of metabolic arrest until the pollen grain is passively brought in contact with the stigma of the flower by wind, water, or insects.

Growth of the male gametophyte resumes after germination of the pollen grain on the stigma of a flower of the correct species. Almost immediately after contact with the stigma, the quiescent state of the pollen is broken. This is detected as an abrupt increase in respiration; enzymes are activated within the pollen and a small amount of DNA-like RNA, probably an mRNA, is synthesized (see, for example, Stanley and Linskens, 1964; Tano and Takahashi, 1964). The coat of the pollen grain splits, and from one side the pollen tube extends outward and penetrates into the tissues of the stigma (Fig. 15–71). The growing pollen tube is similar in ultrastructure to the typical plant cell. Mitochondria, plastids, endoplasmic reticulum, and vesicles of various sizes are present in the cytoplasm of the tube. During growth of the pollen tube only the advancing end contains cytoplasm; at regular intervals, the older portions of the tube become empty and are sealed off by partitions. At all times, the growing tube is surrounded by a primary cell wall approximately 1000 Å thick.

A part of the growth of the pollen tube is accomplished by the absorption of water from the stigma and style. Although other nutrients are probably also absorbed from the tissues of the host sporophyte, the tube contains quantities of stored materials sufficient for most of its growth. This has been demonstrated by experiments in which pollen is stimulated to germinate under isolation *in vitro*. Under these conditions, with only a supply of water present, the pollen tubes reach lengths approximating the growth under natural conditions in the style. Addition of actinomycin D does not significantly hinder early growth of the tube; evidently, the messenger and other RNAs necessary for initial tube growth are synthesized during development of the pollen in the anther. Continued growth of the tube, however, is actinomycin D-sensitive and evidently depends on mRNA synthesis by the vegetative nucleus. Growth of the pollen tube continues until the micropyle of the ovary and the synergid cells of the female gametophyte are reached (for references to the physiology of growth of the pollen, see Linskens, 1967).

15–70. A pollen grain, the mature male gametophyte of the angiosperms. Gn, *generative nucleus;* Vn, *vegetative nucleus.* × *6,500. Courtesy of W. A. Jensen.*

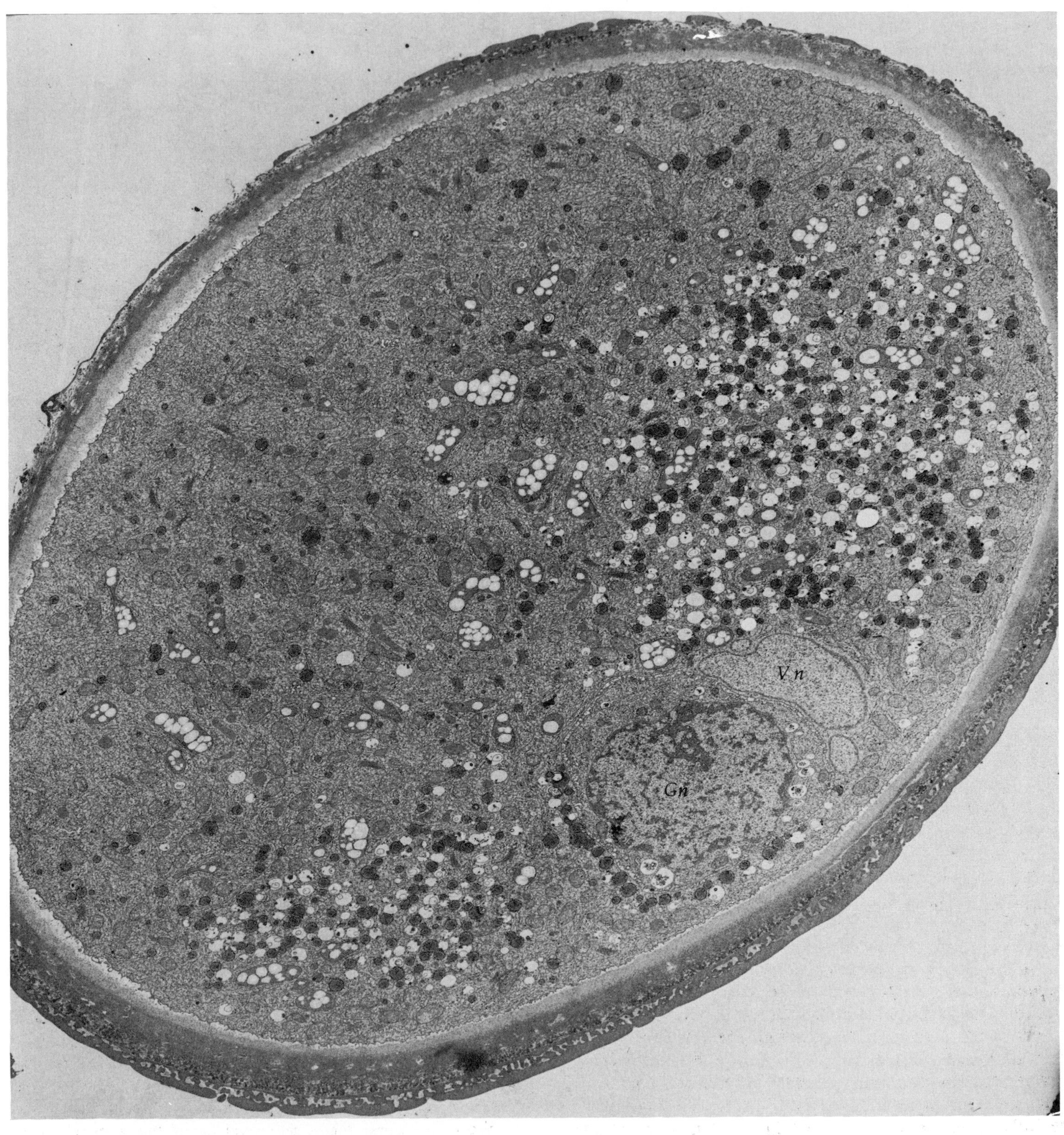
Vn
Gn

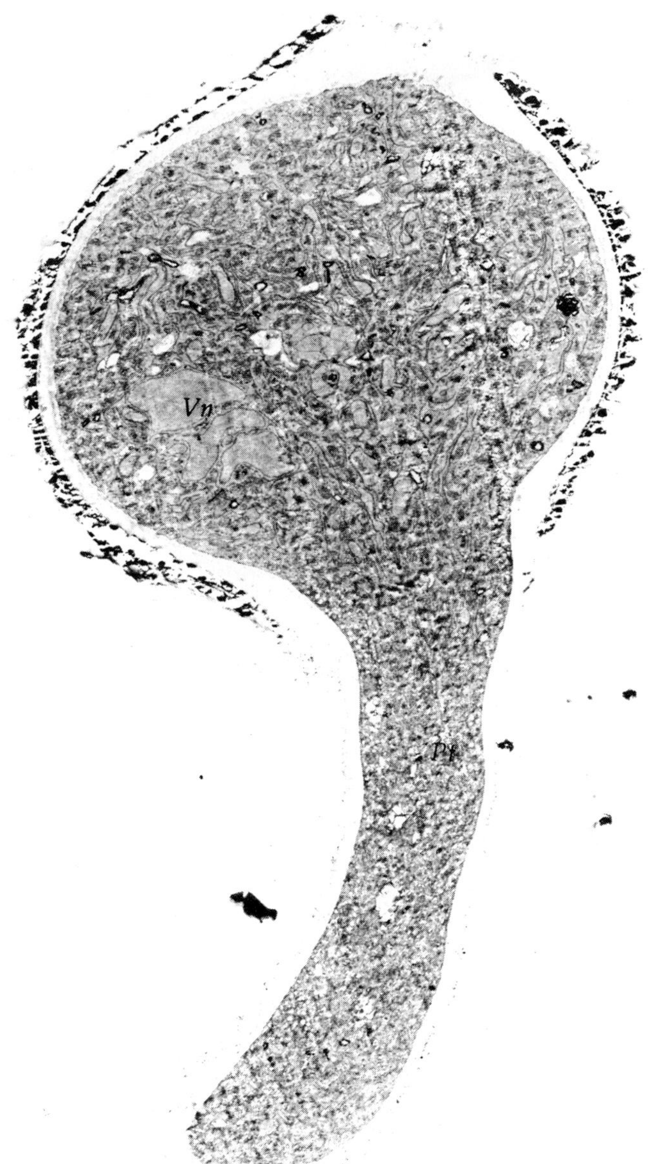

15–71. *A germinating pollen grain.* Vn, *vegetative nucleus;* Pt, *pollen tube.* × *2,400. Courtesy of M. M. A. Sassen.*

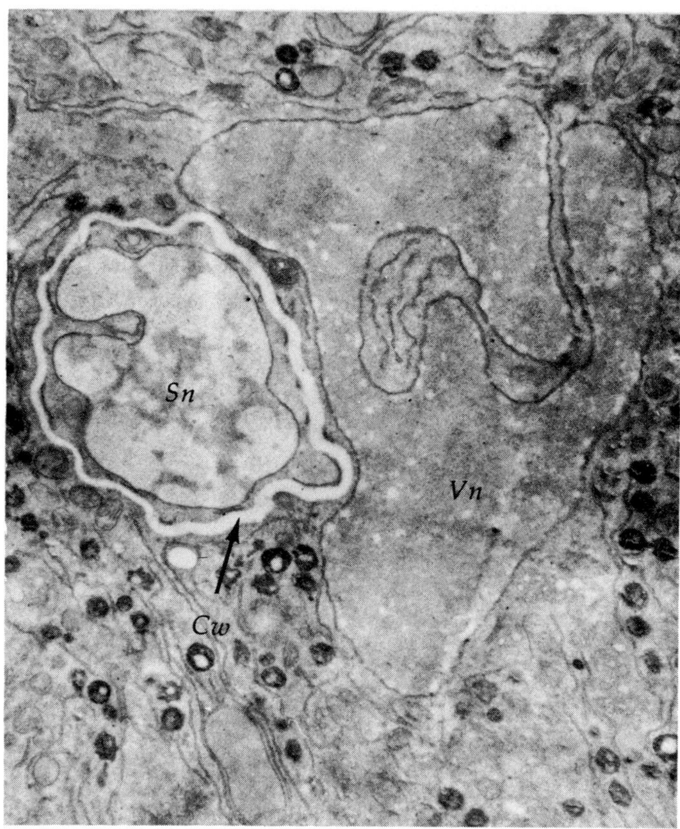

15–72. *The sperm nucleus* (Sn) *within the cytoplasm of the pollen tube. In this example, the cytoplasm associated with the sperm nucleus is clearly separated from the cytoplasm of the pollen tube by a cell wall* (Cw). Vn, *vegetative nucleus.* × *10,000. Courtesy of M. M. A. Sassen.*

As the pollen tube grows down into the style and approaches the ovary, the nucleus of the generative cell (Fig. 15–72) divides mitotically to produce the two sperm of the mature male gametophyte. The sperm are true cells surrounded by a continuous plasma membrane, but they contain a greatly reduced cytoplasm with only a few mitochondria, a limited endoplasmic reticulum, and a few vesicles (Fig. 15–73). No plastids have been observed in the sperm of flowering plants (for a representative study of sperm morphology, see Jensen and Fisher, 1968). Although a cell wall may surround the generative cell before division to produce the two sperm nuclei, this wall is lost when the sperm are formed.

Fertilization of the female gametophyte. The pollen tube normally penetrates into the female gametophyte at the region of junction of the two synergids. After penetration, the tube enters one of the two synergids. Then, the two sperm nuclei and some of the surrounding cytoplasm are released into the cytoplasm of the synergid through a pore in the end of the tube. At about the same time, the nucleus of the synergid en-

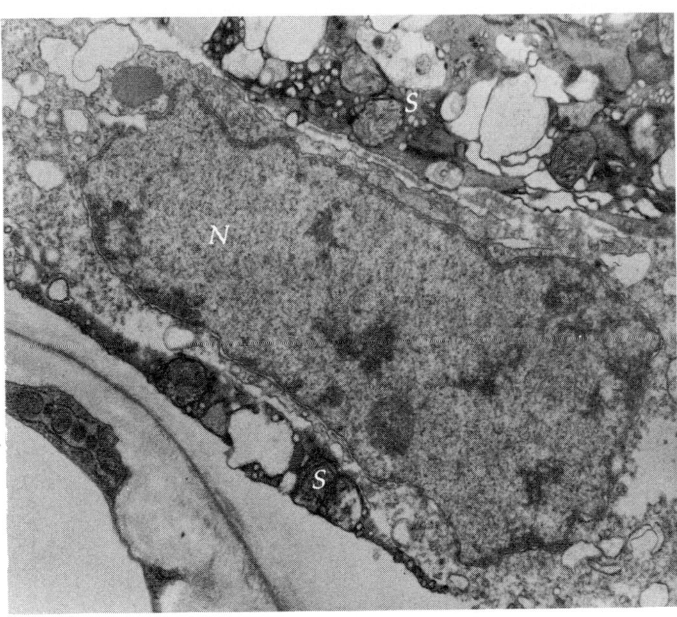

15–73. *A portion of a pollen tube of cotton containing the sperm nucleus* (N). *Tissue of the style* (S) *surrounds the pollen tube.* × *11,500. Courtesy of W. A. Jensen.*

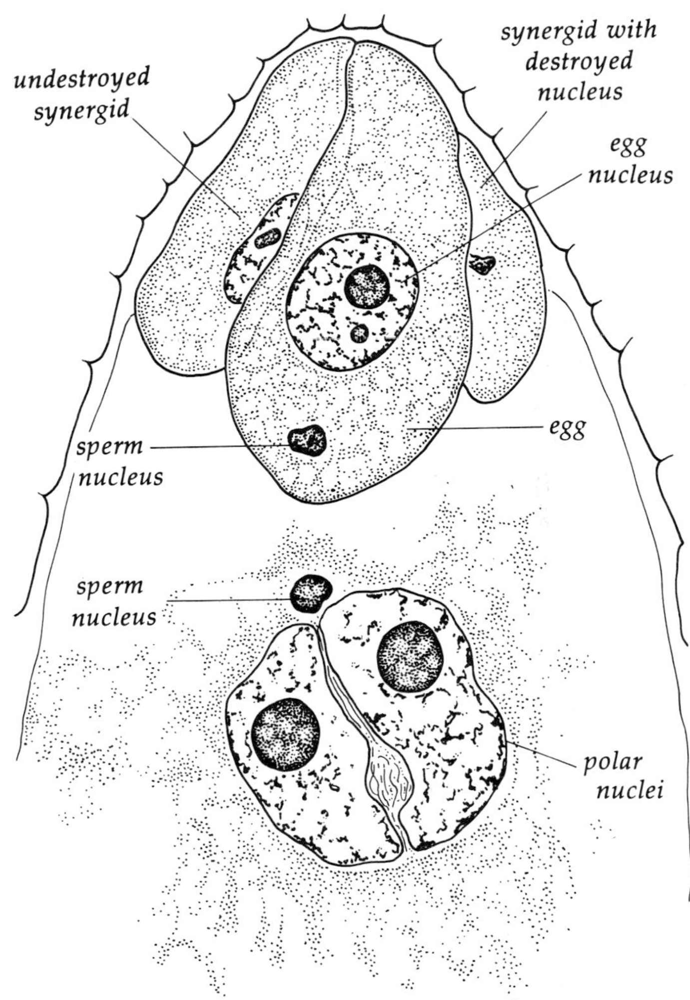

15–74. *A diagram of the female gametophyte of* Triticum vulgare *30 minutes after pollination. One of the two sperm nuclei is suspended within the egg cytoplasm; the second has approached the two polar nuclei. The latter will fuse with the second sperm nucleus to form the triploid endosperm. From* Fertilization, *Vol. II, ed. C. B. Metz and A. Monroy, 1969. Courtesy of M. M. A. Sassen and Academic Press Inc.*

tered by the tube degenerates and disappears. Almost immediately after this the two sperm nuclei, in many plants elongated or spiraled in shape, pass from the synergid and penetrate more deeply into the female gametophyte. Whether the sperm nuclei are capable of autonomous movement or are passively transported by cytoplasmic streaming of the female gametophyte has been a subject of debate for many years and is still not resolved. Fig. 15–74 shows the gametophyte of *Triticum* at this stage of fertilization. In this species, the egg cell is located to one side of the synergids. One sperm nucleus has approached and entered the egg cell; the second lies near the secondary pair of nuclei from which the endosperm will be derived.

Fusion of the egg and sperm nucleus follows a pattern similar to animal species. In various plants, the two nuclei may fuse completely before the first embryonic mitosis, as in sea urchins, or during or after alignment of the chromosomes on the spindle of the first division. Fusion of the other nucleus with the secondary nuclei to form the triploid endosperm proceeds by a similar mechanism. The fertilized egg nucleus subsequently divides mitotically to produce the embryo, which becomes embedded in masses of tissue derived from the endosperm. The embryonic sporophyte, with the endosperm and surface coats derived from the ovary of the parent sporophyte, eventually forms the *seed.* Much diversity exists in nature in the relative contributions of endosperm and parent sporophyte tissue to the mass of the seed.

The development of the endosperm in higher plants has a curious parallel in a few animal species. In the parasitic chalcid wasp *Litomastix,* for example, the polar bodies fuse to form diploid or triploid tissue that divides mitotically, forming a mass of cells that surrounds the developing embryo. While this tissue thus resembles the endosperm of plants in developmental history, its precise functional role is obscure.

In both plants and animals, the first division of the fertilized egg nucleus is followed by a regular sequence of divisions and rearrangements of the developing tissues of the embryo. "Directions" for the early phases of this complex development in animals are partly derived from mRNAs laid down in the cytoplasm much earlier, during prophase of the meiotic division giving rise to the egg. Subsequent development in animals (usually following the gastrula stage) and probably the entire developmental sequence in plants is moderated by mRNAs synthesized by the chromosomes of the embryo. Because the total development of the embryo is exceedingly complex and highly variable among the different species of plants and animals, space does not permit even an outline of these events here.

Suggestions for Further Reading

Books

Austin, C. R., 1968. *Ultrastructure of Fertilization.* Holt, Rinehart and Winston, New York.

Metz, C. B., and A. Monroy, 1967. *Fertilization,* vols. I and II. Academic Press, New York. Several articles from these volumes are referenced below.

Swanson, C. P., 1957. *Cytology and Cytogenetics.* Prentice-Hall, Englewood Cliffs, N.J.

Wilson, E. B., 1928. *The Cell in Development and Heredity.* The Macmillan Company, New York.

Papers and Review Articles

Afzelius, B. A., 1969. Ultrastructure of cilia and flagella. In A. Lima-de-Faria, ed., *Handbook of Molecular Cytology.* North-Holland Publishing Company, Amsterdam, pp. 1219–1242.

Alfert, M., 1950. A cytochemical study of oogenesis and cleavage in the mouse. *Journal of Cellular and Comparative Physiology,* vol. 36, pp. 381–410.

Alfert, M., 1956. Chemical differentiation of nuclear proteins during spermatogenesis in the salmon. *Journal of Biophysical and Biochemical Cytology,* vol. 2, pp. 109–114.

Anderson, W. A., 1968. Structure and fate of the paternal mitochondrion during early embryogenesis of *Paracentrotus lividus. Journal of Ultrastructure Research,* vol. 24, pp. 311–321.

André, J., 1962. Contribution à la connaissance du chondriome. Étude de ses modifications ultrastructurales pendant la spermatogénèse. *Journal of Ultrastructure Research,* supplement 3, pp. 1–185.

Balinsky, B. I., and R. J. Devis, 1963. Origin and differentiation of cytoplasmic structures in the oocytes of *Xenopus laevis. Acta Embryologica Morphologica Experientia,* 6:55–108.

Beams, H. W., 1964. Cellular membranes in oogenesis. In M. Locke, ed., *Cellular Membranes in Development.* Academic Press, New York, pp. 175–219.

Bloch, D., 1969. A catalog of sperm histones. *Genetics,* supplement 61, pp. 93–111.

Colwin, L. H., and A. L. Colwin, 1967. Membrane fusion in relation to sperm-egg association. In C. B. Metz and A. Monroy, eds., *Fertilization,* vol. 1. Academic Press, New York, pp. 295–367.

Dan, Jean C., 1967. Acrosome reaction and lysins. In C. B. Metz and A. Monroy, eds., *Fertilization,* vol. 1. Academic Press, New York, pp. 237–293.

Fawcett, D. W., 1961. Cilia and flagella. In J. Brachet and A. E. Mirsky, eds., *The Cell,* vol. II. Academic Press, New York, pp. 217–297.

Gall, J. G., 1961. Centriole replication. A study of spermatogenesis in the snail *Viviparus*. *Journal of Biophysical and Biochemical Cytology*, 10:163–193.

Hultin, T., 1952. Incorporation of N^{15} labeled glycine and alanine into the proteins of developing sea urchin eggs. *Experimental Cell Research*, vol. 3, pp. 494–496.

Jensen, W. A., and D. B. Fisher, 1967. Cotton embryogenesis: double fertilization. *Phytomorphology*, vol. 17, pp. 261–269.

Jensen, W. A., and D. B. Fisher, 1968. Cotton embryogenesis: the sperm. *Protoplasma*, vol. 65, pp. 277–286.

Kaye, Jerome S., 1969. The ultrastructure of chromatin in nuclei of interphase cells and in spermatids. In A. Lima-de-Faria, ed., *Handbook of Molecular Cytology*. North-Holland Publishing Company, Amsterdam, pp. 361–380.

Kaye, J. S., and R. McMaster-Kaye, 1966. The fine structure and chemical composition of nuclei during spermiogenesis in the house cricket. *Journal of Cell Biology*, vol. 31, pp. 159–179.

Koehler, J. K., 1966. Fine structure observations in frozen-etched bovine spermatozoa. *Journal of Ultrastructure Research*, vol. 16, pp. 359–375.

Linskens, H. F., 1967. Fertilization mechanisms in higher plants. In C. B. Metz and A. Monroy, eds., *Fertilization*, vol. II. Academic Press, New York, pp. 189–253.

Mann, T., 1967. Sperm metabolism. In C. B. Metz and A. Monroy, eds., *Fertilization*, vol. I, Academic Press, New York, pp. 99–116.

Metz, C. B., 1967. Gamete surface components and their role in fertilization. In C. B. Metz and A. Monroy, eds., *Fertilization*, vol. I. Academic Press, New York, pp. 163–236.

Mizukami, I., and J. G. Gall, 1966. Centriole replication II: sperm formation in the fern *Marsilea* and the cycad *Zamia*. *Journal of Cell Biology*, vol. 29, pp. 97–111.

Monroy, A., and A. Tyler, 1967. The activation of the egg. In C. B. Metz and A. Monroy, eds., *Fertilization*, vol. I. Academic Press, New York, pp. 369–412.

Nelson, L., 1967. Sperm motility. In C. B. Metz and A. Monroy, eds., *Fertilization*, vol. I. Academic Press, New York, pp. 27–97.

Norrevang, A., 1968. Electron microscopic morphology of oogenesis. *International Review of Cytology*, vol. 23, pp. 113–186.

Phillips, D. M., 1970. Insect sperm: their structure and morphogenesis. *Journal of Cell Biology*, vol. 44, pp. 243–277.

Piko, L., 1967. Gamete structure and sperm entry in mammals. In C. B. Metz and A. Monroy, eds., *Fertilization*, vol. II. Academic Press, New York, pp. 325–403.

Robison, W. G., 1966. Microtubules in relation to the motility of a sperm syncytium in an armored scale insect. *Journal of Cell Biology*, vol. 29, pp. 251–265.

Rosen, W. G., 1968. Ultrastructure and physiology of pollen. *Annual Review of Plant Physiology*, vol. 19, pp. 435–462.

Satir, P., 1968. Studies on cilia III: further studies on the cilium tip and a "sliding filament" model of ciliary motility. *Journal of Cell Biology*, vol. 39, pp. 77–94.

Schultz, P., and W. A. Jensen, 1968. *Capsella* embryogenesis: the egg, zygote, and young embryo. *American Journal of Botany*, vol. 55, pp. 807–819.

Shelanski, M. L., and E. W. Taylor, 1968. Properties of the protein subunit of central-pair and outer-doublet microtubules of sea urchin flagella. *Journal of Cell Biology*, vol. 38, pp. 304–315.

Sleigh, M. A., 1969. Physiology and biochemistry of cilia and flagella. In A. Lima-de-Faria, ed., *Handbook of Molecular Cytology*. North-Holland Publishing Company, Amsterdam, pp. 1243–1258.

Stanley, R. G., and H. F. Linskens, 1964. Enzyme activation in germinating *Petunia* pollen. *Nature*, vol. 203, pp. 542–544.

Szollosi, D. G., 1965. The fate of sperm middle-piece mitochondria in the rat egg. *Journal of Experimental Zoology*, vol. 159, pp. 367–378.

Tano, S., and H. Takahashi, 1964. Nucleic acid synthesis in growing pollen tubes. *Journal of Biochemistry*, vol. 56, pp. 578–580.

Turner, F. R., 1968. An ultrastructural study of plant spermatogenesis. Spermatogenesis in *Nitella. Journal of Cell Biology,* vol. 37, pp. 370–393.

Tyler, A., A. Monroy, and C. B. Metz, 1956. Fertilization of sea urchin eggs. *Biological Bulletin,* vol. 110, pp. 184–195.

Yanagimachi, R., and Y. D. Noda, 1970. Ultrastructural changes in the hamster sperm head during fertilization. *Journal of Ultrastructure Research,* vol. 31, pp. 465–485.

16
Cytoplasmic Inheritance

The preceding chapters in this unit have outlined the mechanisms whereby the total genetic complement of an organism is maintained during reproduction. In vegetative reproduction, both the chromosome number and the combination of genetic alleles peculiar to an individual are preserved intact by mitosis. In sexual reproduction, the chromosome number is maintained by the combined effects of meiosis and fertilization. Variability results from recombination in prophase I and the random assortment of chromosomes during anaphase I and II and is complemented by the chance union of gametes during fertilization. Because of the basic order of the chromosomal mechanisms that recombine and distribute genes in meiosis, inheritance of nuclear genes follows patterns that can be described in mathematical terms. These patterns are the subject matter of classical or Mendelian genetics.

Very shortly after the rediscovery of Mendel's results at the turn of the century, a few observations were made of a pattern of inheritance that appeared to be completely independent of the nucleus. In 1909, a German botanist, C. Correns, reported the results of his studies of color inheritance in the "four-o'clock," *Mirabilis*. In some individuals of this species, segments of the plant fail to turn green. Some leaves are completely white or contain green and white segments. Within the cells of the bleached segments, the chloroplasts are colorless. By choosing flowers from different regions of these plants, Correns was able to obtain eggs or pollen originating from white, green, or mixed segments. Crosses between these flowers revealed the curious fact that inheritance of color is determined entirely by the maternal parent and does not follow Mendelian patterns of segregation. If flowers on green segments were fertilized, the offspring were always green, whether pollen was obtained from flowers on green, white, or mixed segments. If the female was white, all seedlings were white; flowers or mixed segments produced green, white, or mixed offspring. The male parent had no apparent effect on the inheritance of green color. Once established, the trait for color was maintained. Correns noted that the inheritance of greening followed the pattern of inheritance of cytoplasm in plants. The pollen donates little or no cytoplasm; the plastids and other cytoplasmic components of the zygote are derived almost exclusively from the egg. On this basis, Correns proposed that the inheritance of color in this case was due to a *cytoplasmic* rather than nuclear factor.

Other examples of cytoplasmic inheritance were unearthed during the first half of this century. These occur in a great variety of organisms, principally in the higher plants, algae, and protozoa. Kirk and Tilney-Bassett (1967) list more than 30 examples of the cytoplasmic control of plastid color similar to *Mirabilis*. In these purely genetic studies of cytoplasmic inheritance, it was not possible to localize the inheritable factors in any specific organelle. The extranuclear genes were simply regarded as located somewhere in the cytoplasm.

Cytoplasmic control of inheritance has also been proposed on the basis of cytological evidence. Much of the work by early cytologists in this area is intuitive; in fact, every major organelle of the cytoplasm has at some time since the 1800's been claimed to possess an independent lineage. More recent cytological findings and the genetic evidence support the independent inheritance of only two organelles, the mitochondrion and the chloroplast. Even in these organelles, independence from nuclear genes is evidently only partial. In addition, some limited evidence suggests that centrioles may also contain independent genes.

The genetic, structural, and functional evidence for the cytoplasmic inheritance of factors located in the mitochondrion and chloroplast will be considered in this chapter, along with an evaluation of the much weaker case for the centriole. Although the earliest significant work was done with the inheritance of the chloroplast, the mitochondrion will be taken up first because characterization of the mitochondrial system in recent years has been most complete.

The Cytoplasmic Control of Inheritance in the Mitochondrion

The Genetic Evidence

The genetic evidence for cytoplasmic control of inheritance in mitochondria comes from experiments with yeast and *Neurospora*. If yeast is grown on agar plates, colonies that grow more slowly than normal strains frequently appear. These mutants, called *petites*, were studied in detail by Boris Ephrussi in the early 1950's. Crosses between petites revealed two separate patterns of inheritance. One group, called *segregational petites*, behaved in typical Mendelian fashion; genes controlling the appearance of these aberrant colonies could be located on the chromosomes and showed linkage to other nuclear genes. A second group proved to be independent of nuclear control. The most important of these, called *suppressive* petites, follow a strictly cytoplasmic pattern of inheritance.

The inheritance of suppressive petites has been studied in a type of sexual reproduction in yeast that produces *heterokaryons*. In yeast, sexual reproduction takes place by fusion of two haploid cells. Ordinarily, on conjugation of these cells, the nuclei fuse to form a diploid zygote. However, in some strains, only the cytoplasm fuses. The nuclei remain separate in the mixed cytoplasm, forming the heterokaryon. The heterokaryon subsequently divides, and the haploid nuclei are distributed unchanged between the daughter cells. If a yeast carrying the suppressive petite trait along with a nuclear mutation is crossed with a normal to form a heterokaryon, the colonies derived after division of the heterokaryon are sometimes normal and sometimes petite. By using the nuclear marker, it can be determined that the formation of the suppressive petite colonies is independent of the type of nucleus present, whether derived from the original normal or petite parent. Suppressive petites are therefore controlled by a cytoplasmic factor (see Wright and Lederberg, 1957).

Biochemical analysis of the petite yeast, carried out by Ephrussi and his colleagues, revealed that metabolism of substrates proceeds only by the glycolytic pathway. Respiration and oxidative phosphorylation are inactive; isolation of the components of the electron transport system revealed that the petite cells lack cytochromes *a* and *b*. Mitochondria in these cells were also altered in morphology. Electron microscopy of the petite cells showed that the inner mitochondrial membranes were aberrant, transformed from the usual pattern of folds into irregular whorls.

The petite mutation has other interesting characteristics. Normal yeast can be converted to the petite form by exposure of colonies to ultraviolet light. The effective wavelengths are those absorbed strongly by nucleic acids. Acridine dyes, which bind strongly to DNA, also induce the petite mutation. Taken together, genetic and biochemical experiments with yeast suggest that a cytoplasmic factor, probably a nucleic acid, causes the suppressive petite mutation. The primary site of the deficiencies caused by the mutation is in the mitochondrion. However, these findings do not by themselves prove that the mutation is in a genetic system carried in the mitochondrion; the responsible factor could be either mitochondrial or in the extramitochondrial cytoplasm.

M. B. Mitchell and H. K. Mitchell have identified a similar group of slow-growing mutants in the fungus *Neurospora*. The strains, called *poky* mutants, are characterized by slow growth, absence of cytochromes *a* and *b*, and alterations in the morphology of inner mitochondrial membranes. Sexual reproduction in *Neurospora* takes place by the fusion of "male" nuclei carried in *conidia* with "female" nuclei in a larger structure termed the *protoperithecium*. Offspring of crosses between poky and normal strains are stable and follow a strictly maternal line of inheritance, with descendants of the cross exhibiting only the mitochondrial condition of the cytoplasm of the protoperithecium. This pattern of inheritance follows the maternal cytoplasm, no matter what rearrangements of chromosomal genes take place in the meiosis following union of "male" and "female" nuclei. Thus no linkage of the poky trait to any chromosomal genes can be established. As with petite mutations in yeast, a very high incidence of the poky trait can be induced by ultraviolet light and acridine dyes.

In both the petite yeast and the poky mutants of *Neurospora*, the cytoplasmic trait affecting mitochondrial structure and function could be located either inside or outside the mitochondrion. Some indication that the trait is actually carried within the mitochondrion was provided in 1965 by E. G. Diacumakos. Diacumakos isolated mitochondria from a poky mutant and purified the mitochondrial suspension by centrifugation. These mitochondria, when injected by micropipette into normal strains, frequently induced the appearance of a stable mutation to the poky trait. Injection of nuclei or purified DNA from poky strains had no effect.

While providing stronger evidence, these results do not conclusively establish that the cytoplasmic factors controlling the petite and poky traits are inside the mitochondrion, because it is impossible to eliminate the chance that a cytoplasmic contaminant was carried along with the isolated mitochondria or absorbed on the mitochondrial surface during isolation. This prompted the use of more direct methods, including attempts to localize DNA within the mitochondrion.

The Search for DNA in the Mitochondrion

The genetics of the poky and petite strains in fungi pointed to the likelihood that the responsible cytoplasmic factors are located within the mitochondrion. The capacity of mutagens, such as ultraviolet light and the acridine dyes, to induce the poky and petite traits implicated nucleic acids as the molecular basis for the mitochondrial factor. This evidence, taken with the establishment of DNA as the molecular basis of heredity in the nucleus, made it seem most probable that the cytoplasmic genes under investigation are represented by molecules of DNA located somewhere in the mitochondrion.

Early cytological work, largely discounted by later cell biologists as exceptional observations, had already indicated that DNA is contained in at least some mitochondria. One of these observations was made in trypanosomes (Fig. 16–1), a group of single-celled, parasitic protozoans responsible for several important diseases, including African sleeping sickness and Chagas' disease. At one stage of its life cycle, the trypanosome contains three major organelles typically collected together in one region of the organism (Fig. 16–2). In the trypanosome shown, a centriole forms the basal body of the single flagellum, which extends from the anterior end of the cell. Just beyond the basal body is a conspicuous spherical or lobed organelle termed the *kinetoplast*. Behind this structure is the single nucleus. During division of the trypanosome, each of these structures divides in sequence, beginning with the basal body. Soon after discovery of the Feulgen reaction, E. Bresslau and L. Scremin in 1924 showed that the kinetoplast of trypanosomes contains a compact

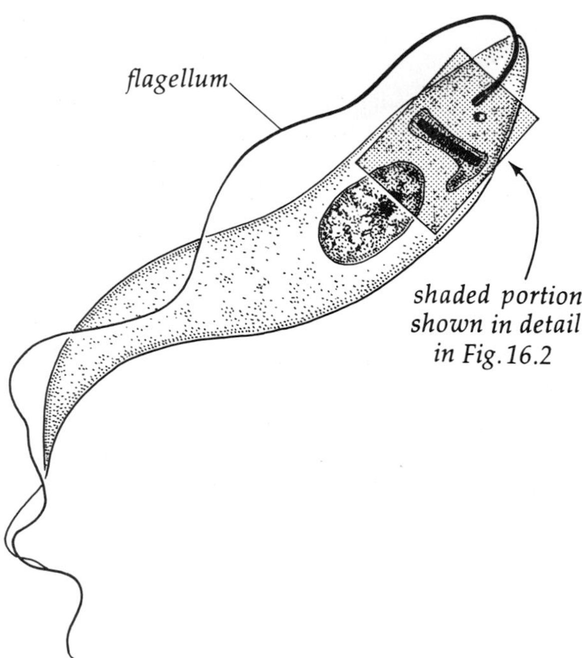

16–1. *A typical trypanosome.*

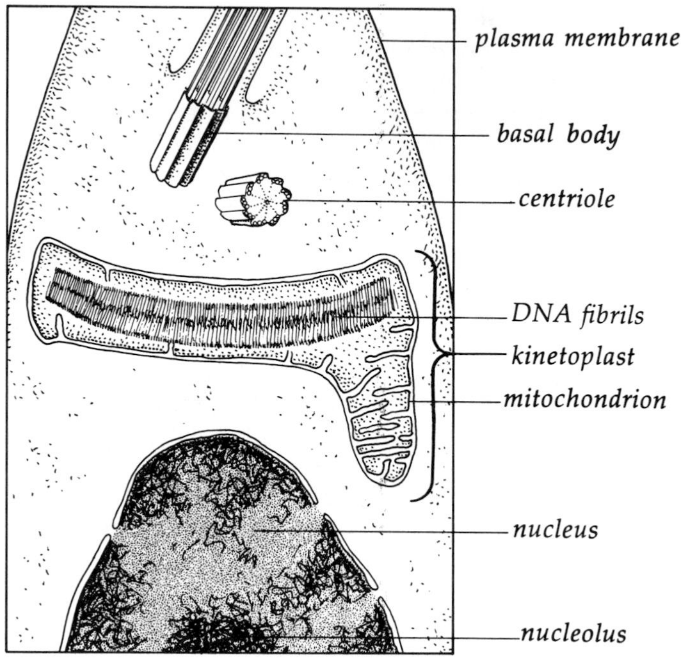

16–2. *Relationship of the nucleus, kinetoplast, and centriole or basal body in a trypanosome. Redrawn, original courtesy of P. R. Burton, D. G. Dusanic, and The Rockefeller University Press.* [J. Cell Biol. *39:318 (1968)*]

mass that stains intensely for DNA. Although the derivation of the kinetoplast at this time was uncertain, later ultrastructural investigations by M. Steinert demonstrated conclusively that this organelle exhibits more or less typical mitochondrial structure and contains identifiable cristae, matrix, and inner and outer membranes. Experiments using tritiated thymidine show that the DNA mass of the kinetoplast incorporates radioactive label just before division of the cell, at the same time that replication of the chromosomes occurs in the nucleus. Interestingly enough, the kinetoplast is sensitive to acridine dyes; exposure of trypanosomes to these substances frequently results in destruction and elimination of the DNA mass (see Simpson, 1968). The inner membranes of kinetoplasts lacking DNA are distorted into whorls as in the petite and poky mutants of yeast and *Neurospora;* respiratory activity of the cells is greatly reduced or absent.

In *Trypanosoma lewesii,* investigated recently by Paul R. Burton and Donald G. Dusanic, the DNA of the kinetoplast consists of a mass of fibers oriented parallel to the long axis of the cell (Fig. 16–3). With osmium fixation, the fibers appear to be 30–80 Å in diameter. Glutaraldehyde fixation reveals an organization of the fibers into 100–160 Å tubules with walls 40–60 Å thick. Cells fixed after growth in media containing tritiated thymidine show a distribution of radioactivity throughout the fibrous mass.

A second early discovery of DNA in mitochondria was made by Igor B. Dawid, who worked with the eggs of the amphibians *Xenopus* and *Rana*. These eggs were found to contain 300–500 times as much DNA as somatic cells of the same organism. Although indirect, Dawid's experimental findings indicated that the extra DNA of these amphibian eggs is mitochondrial. Purified extracts containing mitochondria as the only structures recognizable by electron microscopy retained the extra DNA of the oocyte. Two mitochondrial enzymes were used to measure the amount of mitochondrial substance present; the concentration of DNA in the extracts paralleled the distribution of these enzymes. An additional finding indicated that the DNA of these preparations was actually situated *inside* the mitochondria: the extra DNA was stable and protected from attack by deoxyribonuclease only as long as the mitochondria remained intact.

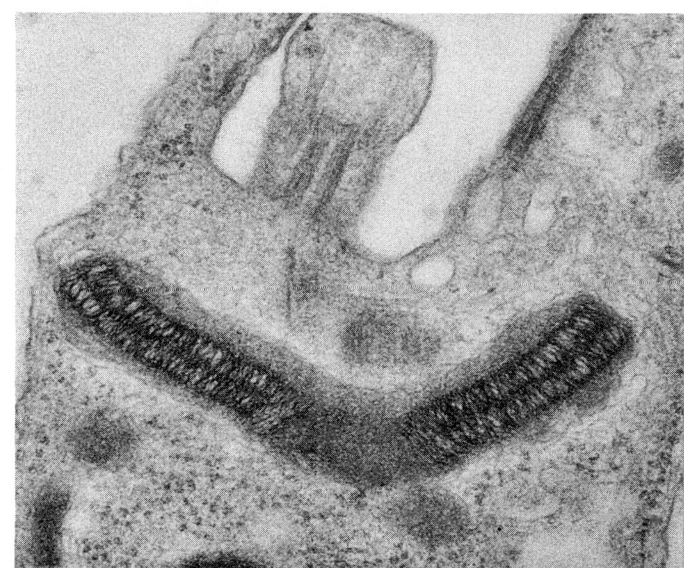

16–3. A dividing kinetoplast of Trypanosoma lewesii. *The dense fibrous material in the kinetoplast can be identified as DNA by autoradiography and the Fuelgen stain. ×53,500. Courtesy of P. R. Burton, D. G. Dusanic, and The Rockefeller University Press.* [J. Cell Biol. *39:318 (1968)*]

The discovery of DNA in the kinetoplast of trypanosomes and mitochondria of amphibian oocytes was not thought to be of general significance until other work began to reveal the presence of DNA as a regular constituent of both mitochondria and chloroplasts. This work was precipitated by the first clear demonstration (by Hans Ris and W. Plaut in 1962) of DNA deposits in the single chloroplast of *Chlamydomonas,* a unicellular alga. Ris and Plaut found 50 Å filaments in scattered locations in the stroma of the *Chlamydomonas* chloroplasts. Adjacent thick sections of this material, when stained for light microscopy by the Feulgen reaction, showed that these regions contain DNA; previous digestion with deoxyribonuclease eliminated the thin fibers and the DNA stain from the chloroplasts. Shortly after this discovery, Ris was able to demonstrate fibrils of the same morphology in the matrix of mitochondria from *Micromonas* (Ris, 1962). Ris' discovery and the chloroplast work sparked renewed investigation into the substructure of mitochondria. Fibers of the type described by Ris were subsequently found in mitochondria from many different plants and animals.

The fibers are more evident in mitochondria from embryonic or rapidly growing cells (Fig. 16–4); however, it has become generally accepted that DNA fibers of this type are probably a constituent of all mitochondria. Shadowed preparations of mitochondria spread on a water surface usually show the presence of from 1 to 5 separate pieces of DNA per mitochondrion (Fig. 16–5).

The DNA fibrils of the mitochondrion typically lie in the matrix. Apparent connections between the DNA fibers and the inner mitochondrial or cristae membranes are often seen. While the reported dimensions vary, the diameter of the fibrils within the DNA masses averages about 30–40 Å, considerably below the 100 Å characteristic of chromosomal DNA-protein fibers. This reduced diameter and other evidence indicates that the DNA of the mitochondrion is probably "naked," that is, not associated with a protein, in a condition similar to the bacterial chromosome. Hewson Swift and his colleagues have shown that the reaction of mitochondrial DNA to fixatives closely resembles bacterial DNA. Fixation by osmium tetroxide buffered by the standard solutions causes pronounced clumping of the mitochondrial DNA masses (Fig. 16–6); Kellenberger fixation (see Chapter 8) produces uniformly thin filaments 30–40 Å in diameter (Fig. 16–7). These reactions closely follow the patterns of fixation of the bacterial nucleus. The findings point to a close similarity between bacterial and mitochondrial DNA and suggest that mitochondrial DNA is probably not complexed with histone or other proteins.

The Physical Properties of Mitochondrial DNA

The amount of DNA associated with the mitochondrial fraction varies in quantity from as little as 0.2 percent of total cellular DNA in cells cultured from mouse liver and 1 percent in chick liver to 15 percent in diploid yeast, 20 percent in haploid yeast, and more than 99 percent in amphibian oocytes. In most animals, the mitochondrial DNA fraction can be readily identified by its behavior in the ultracentrifuge because it usually spins down at a slightly different rate from nuclear DNA. In cases where nuclear and mitochondrial DNA approach the same density, the mitochondrial fraction can be characterized by first isolating and purifying the mitochondria. These studies have revealed a number of physical characteristics common to the mitochondrial DNA of animals, plants, and protozoa.

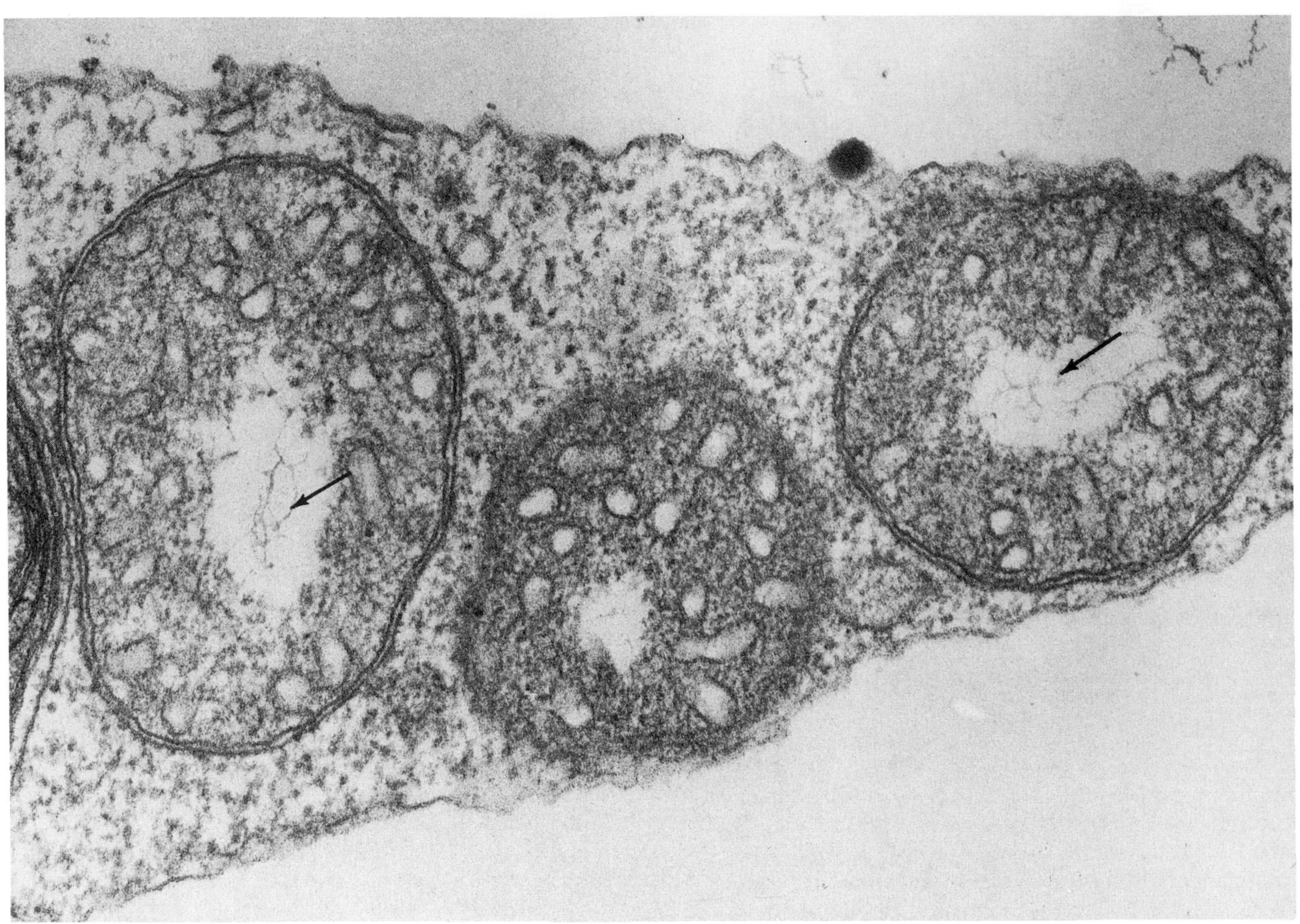

16–4. *DNA fibrils (arrows) within the matrix of mitochondria of the brown alga* Egregia menzieskii. *×80,000. Courtesy of T. Bisalputra, A. A. Bisalputra, and The Rockefeller University Press.* [J. Cell Biol. *33:511 (1967)*]

Table 16–1 shows the distribution of density values for mitochondrial and nuclear DNA from a wide variety of organisms. The differences in density noted in the table reflect variations in the relative proportions of guanine-cytosine and adenine-thymine base pairs in the DNA molecules. As the ratio of G-C pairs decreases, the DNA becomes lighter. It can be seen from this table that, although the relative densities of nuclear and mitochondrial DNA vary independently, the densities of mitochondrial fractions from closely related organisms are roughly the same. Mitochondrial DNA from the fungi is slightly lighter than nuclear DNA from the same group; the two fractions in amphibia and mammals are approximately the same. Birds have mitochondrial DNA of slightly greater density than nuclear DNA. In the higher plants, the values for mitochondrial DNA are considerably higher.

Mutations such as the poky and petite strains in fungi result in extensive changes in the density of mitochondrial DNA. Although the contour length of the

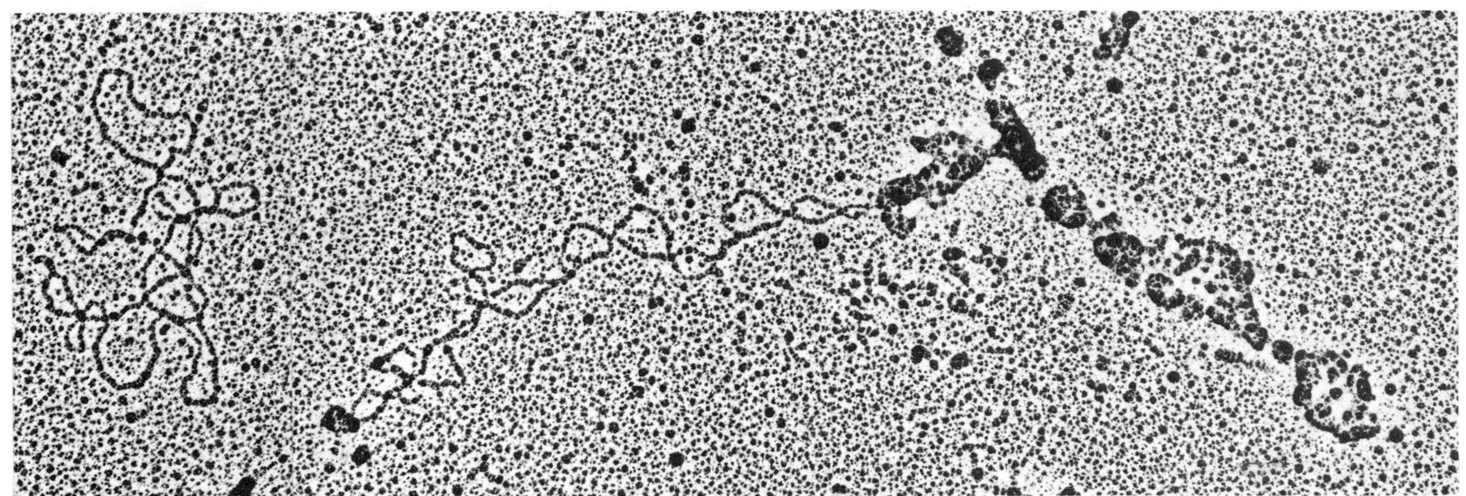

16–5. Two circular DNA molecules (from mouse fibroblasts) isolated from mitochondria disrupted by osmotic shock. The molecules have been shadowed by a technique in which the specimen is rotated through 360° during evaporation of the metal. This technique greatly increases the visibility of the DNA duplex. $\times$ *49,000. From* Science *165:25 (1969). Copyright 1969 by the American Association for the Advancement of Science. Courtesy of M. M. K. Nass.*

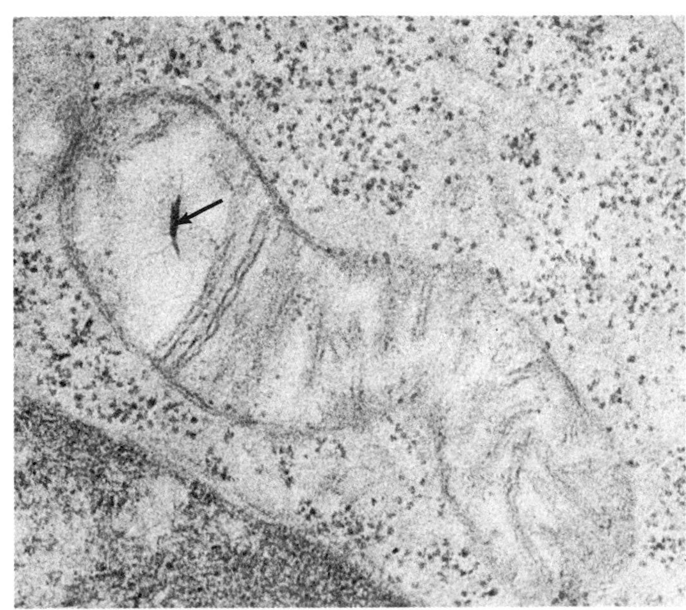

16–6. Mitochondrion in an erythroblast of the liver of a mouse embryo after fixation in osmium tetroxide by the standard techniques. The DNA deposit in the matrix (arrow) is coagulated into a dense clump. $\times$ *43,000. Courtesy of H. Swift and D. M. Phillips.*

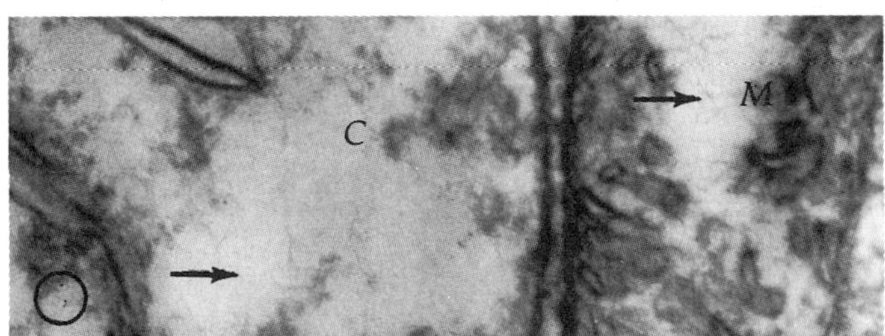

16–7. Mitochondrion (M) *from a leaf of Swiss chard after fixation in Kellenberger's fixative. Note the fine filaments in the matrix (arrow). An adjacent chloroplast* (C) *also shows fine filaments in the stroma.* $\times$ *56,500. Courtesy of N. Kislev, H. Swift, L. Bogorad, and The Rockefeller University Press.* [J. Cell Biol. *25:327 (1965)*]

DNA circles (see below) is unchanged, these mutations shift the density as much as 10 percent, indicating relatively drastic rearrangements of the DNA sequence, of much greater magnitude than a point mutation. In one mutant, practically all G-C pairs are eliminated, producing an almost pure A-T polymer. The mechanism underlying mutational changes of this extent in DNA remains unexplained.

Table 16–1 *Buoyant Densities of DNA*

Organism	Nuclear DNA	Mitochondrial DNA
Fungi		
Yeast	1.700	1.685
Neurospora crassa	1.712	1.701
Amphibians		
Rana pipiens	1.702	1.702
Xenopus laevis	1.700	1.702
Siredon mexicanum	1.704	1.695
Necturus maculosus	1.707	1.695
Birds		
Duck	1.700	1.711
Pigeon	1.700	1.707
Chick	1.701	1.708
Mammals		
Rat	1.703	1.701
Mouse	1.701	1.701
Guinea pig	1.700	1.702
Rabbit	1.701	1.703
Beef	1.704	1.703
Lamb	1.703	1.703
Higher plants		
Onion	1.688	1.706
Sweet potato	1.692	1.706
Cabbage	1.692	1.706
Mung bean	1.691	1.706
Red bean	1.693	1.707
Beet	1.695	1.705
Swiss chard	1.689	—
Tobacco	1.690	—

Adapted from H. Swift, and D. R. Wolstenholme, Mitochondria and Chloroplasts. In A. Lima-de-Faria, ed., *Handbook of Molecular Cytology*. Amsterdam: North-Holland Publishing Company, 1969. Courtesy of H. Swift, P. R. Wolstenholme, and the publisher. See this article for original references.

The DNA isolated from mitochondria, when heated, is denatured and separates at 100°C into single nucleotide chains, showing that in the native form the mitochondrial DNA is a Watson-Crick double helix. Electron microscopy of mitochondrial DNA from animals reveals that the molecules are circles of two types (Fig. 16–8), either "open," without twists in the circle at any point, or highly twisted. Measurement of the outside contours of these circles, either twisted or open, discloses a constant included DNA length of from 4.5 to 5.5 microns (see Table 16–2; for detailed literature references to the many papers in this area, see Swift, 1965).

Table 16–2 *Contour Length of Circular DNA from Animal Mitochondria*

Species	Average circumference according to various authors*
Chordata	
Mammalia	
Man	4.81
Ox	5.3
	5.1
Sheep	5.4
Rat	5.1
	4.9
	5.4
Mouse	5.1
	4.96
	4.74
	5.24
Guinea pig	5.6
Aves	
Chick	5.35
	5.55
	5.1
	5.26
Duck	5.1
Amphibia	
Frog (*R. pipiens*)	5.56
Toad (*X. laevis*)	5.40
Osteichthyes	
Carp	5.4
	5.4
Echinodermata	
Echinoidea	
Sea urchin (*L. pictus*)	4.45
Arthropoda	
Insecta	
Fly (*M. domestica*)	5.2

*From P. Borst and A. M. Kroon, Mitochondrial DNA. *Int. Rev. Cytol.* 26:107 (1969). For original references see this article.

The "open" circles, which sediment at 27S, and the twisted circles, sedimenting at 39S, are considered to be alternate forms of the same molecule, because twisted circles can be converted to open circles by brief treatment with an endonuclease or heat denaturation. Although it is uncertain which form of the molecule is present in the native state in the intact cell, the twisted form is favored because denaturation by enzymatic

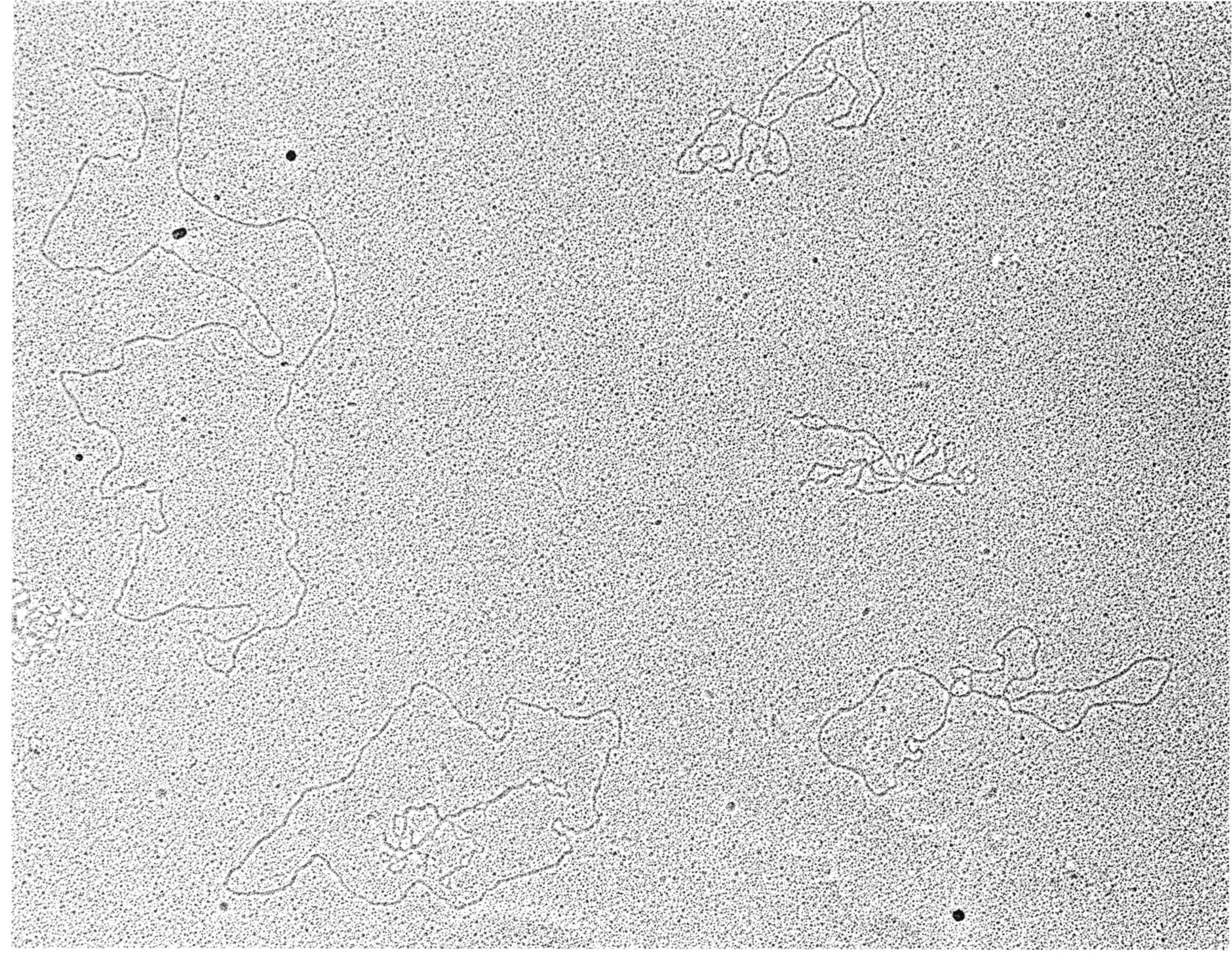

16–8. DNA circles isolated from mouse fibroblast mitochondria. Upper left, an untwisted circle. Upper, middle and lower right, three circles with various numbers of "supertwists" (see explanation in text). The molecules have been spread on a water surface in a film of protein and shadowed from all sides by rotation through 360° during evaporation of the shadowing metal. ×39,000. From Science *165:25 (1969). Copyright 1969 by the American Association for the Advancement of Science. Courtesy of M. M. K. Nass.*

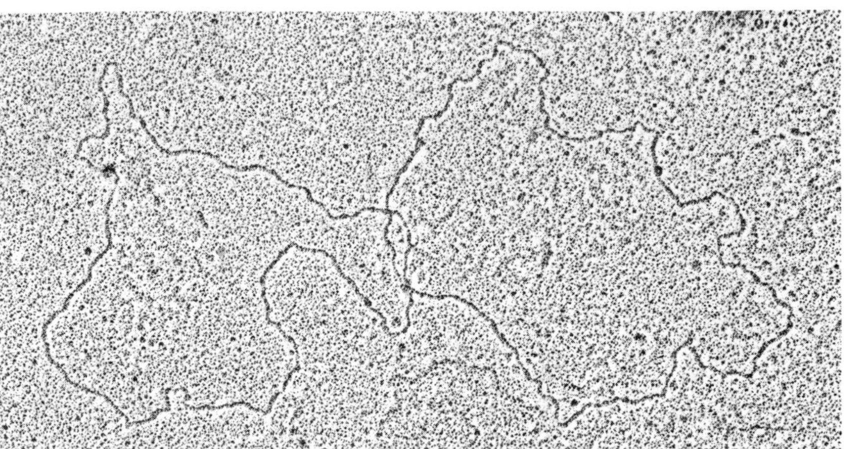

16–9. Interlocked DNA circles isolated from mitochondria of mouse fibroblast cells. Circles of this type are encountered at higher frequencies in mitochondria from rapidly growing and malignant cells. ×46,000. From Science *165:25 (1969). Copyright 1969 by the American Association for the Advancement of Science. Courtesy of M. M. K. Nass.*

digestion or heat induces conversion to the open form. The number of twists in the undenatured form is a matter of debate, because variations in degree of twisting result from changes in the ionic concentration of the isolation medium. (The number of twists in the circle is directly related to the number of turns per unit length of the Watson-Crick double helix; this in turn is partly a function of the ionic constituents of the suspending medium.) A. M. Kroon, in his excellent review of mitochondrial DNA structure (1969) gives what he terms a "ballpark estimate" of 35 twists per DNA circle in the native state.

A small percentage of the DNA molecules isolated from mitochondria are linked into chains of two or more units (Fig. 16–9). About 1 percent of the circles are linked in this way in mitochondrial DNA isolated from normal tissue. In cells of some tumors, the proportion of linked circles is much higher, approaching 10–20 percent. The significance of these multimers, as they are called, is obscure. In the circular chromosomes of prokaryotes, it is known that two circular DNA molecules can become linked together during breakage and exchange in recombination. Because recombination between mitochondrial DNA molecules has recently been detected, it seems possible that the multimers of both normal and malignant tissue may arise from breakage and exchange of segments between molecules of mitochondrial DNA.

The structural configuration of mitochondrial DNA in protozoa, fungi, and plants is much less certain. Only *linear* DNA molecules of varying sizes have been isolated from the mitochondria of protozoa and plants. Mitochondrial DNA from the protozoan *Tetrahymena,* for example, forms a homogeneous population of linear molecules between 17 and 18 microns in length. Mitochondrial DNA isolated from higher plants is extremely heterogeneous (see, for example, Wolstenholme and Gross, 1968). In a recent study of mitochondrial DNA from the lily, lengths from less than 1 to more than 35 microns were described. Although the bulk of the DNA from this source fell at lengths greater than 10 microns, no suggestion of a modal or unit length was obvious. Even wider distributions have been noted in the red bean, in which lengths up to 62 microns were obtained. Part of the variability in plants no doubt arises from the fact that chloroplast DNA (see below) is difficult to distinguish from mitochondrial DNA and probably appears in the mitochondrial preparations as a contaminant. This is significant, because chloroplasts regularly contain larger quantities of long DNA than do mitochondria. Other likely sources of the variability in the mitochondrial DNA of plants are breakage of longer molecules during isolation and end-to-end aggregation of the broken segments.

Because the linear molecules could result from breakage of DNA circles during isolation, it cannot be stated with certainty whether the native mitochondrial DNA of plants and protozoa is linear or circular. Thus, both the possible circularity and the length of mitochondrial DNA of plants is very much in doubt, except for the probability that the native length probably exceeds the 5 microns typical of animals.

An intermediate condition appears to be present in yeast. Mitochondrial DNA in this fungus is uniformly 4.5–5 microns long, and can be isolated either as circles or linear molecules. Circular yeast mitochondrial DNA can be converted to the linear form by heating to 70°C; cooling and storage at 50°C for two hours causes reappearance of the circles. This has been interpreted to mean that the ends of the yeast mitochondrial DNA molecules consist of short single-stranded segments, with complementary sequences capable of forming base pairs if the opposite ends become associated. At favorable temperatures, hydrogen-bonding between these segments may join the ends to form circles. This is in contrast to the situation in animals, in which the mitochondrial DNA is linked into circles by covalent bonds (see Shapiro et al., 1968). Mitochondrial DNA isolated from *Neurospora* is in the linear form, with uniform lengths of 6–7 microns. As in the case of plants and protozoa, the native condition of mitochondrial DNA in the fungi remains uncertain.[1]

Replication of Mitochondrial DNA

Isolated mitochondria, if suspended in a medium containing a supply of the four nucleotide triphosphates, are capable of incorporation of these precursors into stable DNA of high molecular weight (see, for example, Wintersberger, 1966). This can also be demonstrated by autoradiography *in vivo;* exposure of rapidly growing cells to tritiated thymidine before fixation reveals a cytoplasmic label over mitochondria (Fig. 16–10). Active replication of the mitochondrial DNA in rapidly growing tissue would be expected to reveal circles with replication "forks" as described by Cairns for bacteria (see Fig. 12–23). Circles of this type, showing forks (see Fig. 16–11), were recently isolated from mitochondria of rapidly growing rat liver tissue by R. H. Kirshener, D. R. Wolstenholme, and N. J. Gross. These findings imply that because DNA is evidently synthesized in the mitochondrial interior, DNA polymerase, ligase, and other factors associated with DNA replication are also present.

With some difficulty, it has been shown that the replication of mitochondrial DNA is probably semiconservative. Edward Reich and David J. L. Luck followed the pattern of DNA replication in *Neurospora,* using organisms grown for a period in a medium containing heavy nitrogen (N^{15}). After transfer to N^{14} medium, according to the semiconservative mechanism, DNA molecules after the first replication would be expected to consist of $N^{15}N^{14}$ hybrids (for a fuller

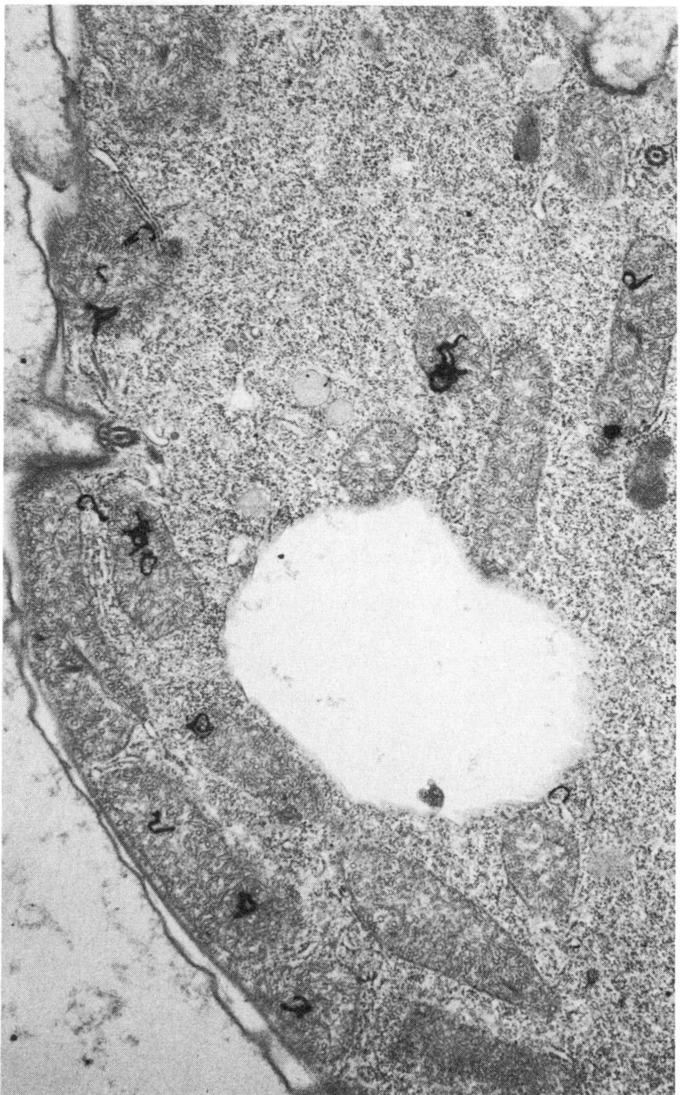

16–10. *Autoradiograph of the protozoan* Tetrahymena pyriformis *after exposure to tritiated thymidine. Exposed grains are centered over mitochondria, indicating incorporation of this DNA precursor within the mitochondrion.* ×*13,500. Courtesy of R. Charret, J. André, and The Rockefeller University Press.* [J. Cell Biol. *39:369 (1968)*]

explanation of this experimental method, an adaptation of the Meselson-Stahl technique used to study DNA replication in bacteria, see Chapter 12). Quite unexpectedly, mitochondria isolated after one cycle of replication still contained N^{15} DNA without significant quantities of the lighter isotope. Hybrids, made up of

[1]Recent work (C. P. Hollenberg, P. Borst, R. W. J. Thuring and E. F. J. Van Bruggen, Size, structure, and genetic complexity of yeast mitochondrial DNA. *Biochim. Biophys. Acta 186:*417, 1969) indicates that yeast mitochondrial DNA may exist in the form of 27 micron circles. In this case the observations of Shapiro et al. would represent the result of interaction of broken segments of the mitochondrial DNA.

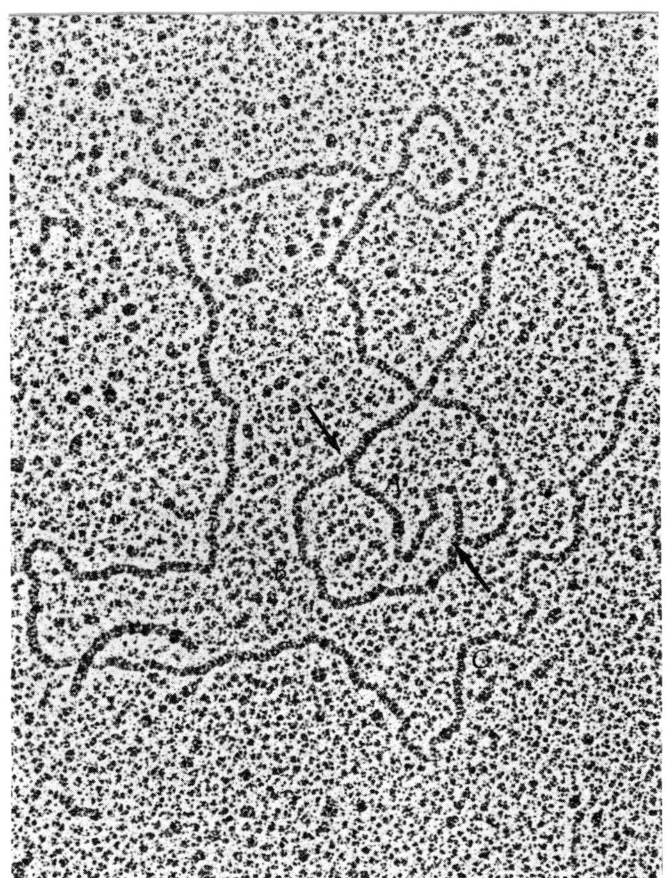

16–11. Partially replicated DNA circles from rat liver mitochondria. The two forks visible in the molecule (arrows) represent the origin and advancing site of replication. The two segments A *and* B *are of equal length, and* C *is of a different length. In all of the replicating mitochondrial DNA circles isolated, the sum of either* A *or* B *and* C *is equivalent to the total included length of an unreplicated circle. ×96,000. Courtesy of R. H. Kirschner, D. R. Wolstenholme, and N. J. Gross, from* Proc. Natl. Acad. Sci. U.S. *60:1466 (1968).*

one N^{15} and one N^{14} strand, did not appear in quantity until after three divisions. Although the eventual identification of $N^{15}N^{14}$ hybrids indicates that replication is semiconservative, the delay in appearance of the hybrids is unexplained. Luck has proposed, as a working hypothesis, that mitochondria contain a storage pool of DNA precursors that requires several generations to dilute out after transfer to N^{14} medium.

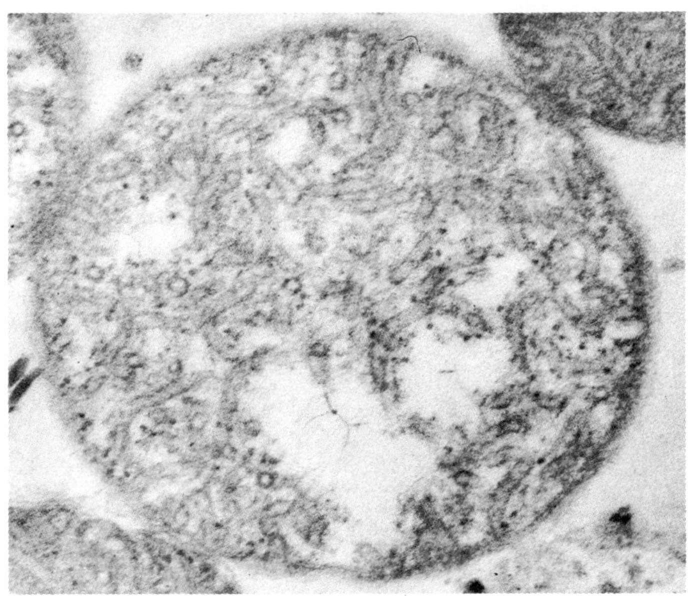

16–12. A mitochondrion isolated from Tetrahymena. *Numerous ribosome-like particles are visible in the matrix. ×47,000. Courtesy of H. Swift.*

RNA in the Mitochondrion

For many years, cytologists reported the presence of scattered particles slightly smaller than cytoplasmic ribosomes in the matrix of the mitochondrion (Fig. 16–12). The spherical particles, about 150–200 Å in diameter, take up stains that are selective for nucleic acids, such as uranyl acetate, and are removed from the matrix by digestion with ribonuclease. Thus, the cytological evidence suggested that a ribosome-like particle is present in the mitochondrial interior.

These results have been supported by biochemical analysis of mitochondrial preparations, which reveal that significant quantities of RNA can be extracted from mitochondria. This RNA is protected from digestion by added ribonuclease as long as the membranes of the isolated mitochondria are intact, indicating that the RNA is carried inside the mitochondrion.

The bulk of the RNA extracted from mitochondria sediments out into peaks resembling cytoplasmic rRNA. But the sedimentation constants for mitochondrial RNA are often determined to be significantly less than the 18 and 28S values characteristic for eukaryotic ribosomal subunits. In a few instances,

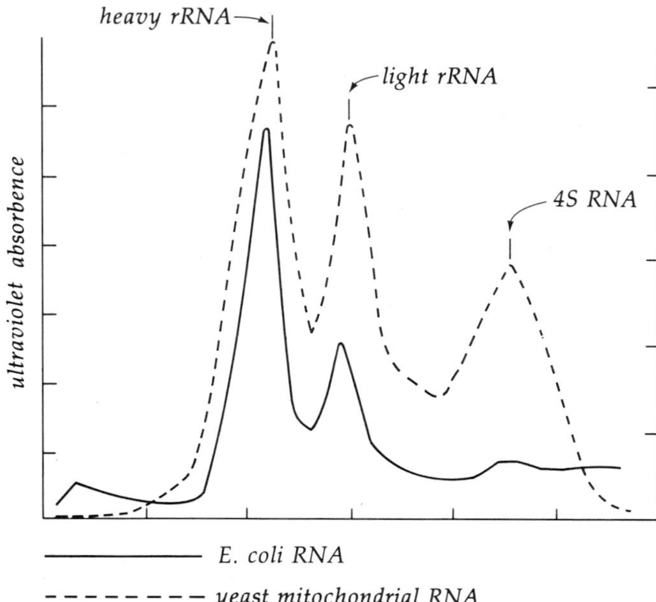

16–13. Comparison of sedimentation pattern of yeast mitochondrial RNA (dotted line) and RNA from the bacterium E. coli *(solid line). Heavier species are located to the left on the graph. The two peaks for rRNA correspond very closely for the bacterial and mitochondrial preparations, giving values of 23S and 16S for* E. coli *and 22S and 15S for the mitochondria. The lightest peak to the right is 4S RNA. Courtesy of M. Fauman, M. Rabinowitz, and G. Getz, from* Biophys. Biochim. Acta *182:355 (1969).*

such as in yeast, the rate of sedimentation closely approximates the lighter subunits of bacterial ribosomes (Fig. 16–13). Similar values have been obtained for human tissue (see Attardi et al.). In other cases, the sedimentation constant for mitochondrial RNA falls somewhere between bacterial and eukaryote organisms and can be clearly distinguished from the peaks for cytoplasmic ribosomes. This is true of rat liver and hamster, two favorite sources for mitochondrial RNA in animals (see, for example, Dubin and Brown, 1967). In general, the mitochondrial RNA of animals, protozoa, and fungi gives peaks at 17S and 23S, slightly heavier than bacterial ribosomes but considerably lighter than cytoplasmic subunits. Data from plants are so limited at this writing that no estimations of subunit size can be given for mitochondrial RNA in these organisms.

In addition to the peaks characteristic of the large and small ribosomal subunits, a significant quantity of RNA isolated from mitochondria sediments down at 4S. This RNA fraction, intensively investigated by W. E. Barnett and his colleagues, proves to be capable of selectively binding amino acids, indicating that it represents a mitochondrial tRNA. Barnett has been able to identify different "soluble" RNAs in the 4S fraction, specific in binding affinity for 18 of the 20 amino acids. In at least three cases, these mitochondrial tRNAs are apparently different in structure and specificity from their cytoplasmic counterparts. For example, a leucyl-tRNA associated with the mitochondrial fraction of *Neurospora* showed a base-pairing affinity only for artificial RNA polymers containing uridine-cytosine sequences (poly UC). Cytoplasmic (extramitochondrial) leucyl-tRNA, in addition to the poly UC affinity, also binds uridine-guanine and uridine (poly UG and poly U) polymers. The altered specificities of the mitochondrial tRNAs also extend to the corresponding synthetase enzymes (aminoacyl-tRNA synthetase; see the section on activation of amino acids in protein synthesis in Chapter 11). These findings make it seem likely that mitochondria contain a unique set of tRNAs.

Isolated mitochondria suspended in a medium containing a supply of nucleoside triphosphates have been found capable of incorporating these substances into mitochondrial RNA. E. Wintersberger has shown that if radioactive precursors are added to isolated yeast mitochondria, RNAs sedimenting at 4S, 17S, and 23S are specifically labeled, showing that these fractions are probably synthesized within the mitochondrion. Other experiments using hybridization techniques also support this conclusion. These experiments have shown that mitochondrial RNA, but not cytoplasmic RNA, hybridizes with mitochondrial DNA.

Thus the mitochondrion probably contains two of the three major RNA classes, rRNA and tRNA; these are likely to be synthesized *in situ* on mitochondrial DNA. Messenger RNA has proved just as elusive in the mitochondrion as in other parts of the cell. Wintersberger has shown that with very brief pulses of radioactive precursor, an RNA fraction sedimenting at about 16S takes up label in isolated yeast mitochondria. This quickly labeled fraction is unstable; if extraction of RNA after a pulse label is delayed for only a very short time it can no longer be detected. It has been proposed that this unstable fraction represents a mitochondrial mRNA, because bacterial mRNAs are characterized by instability and are broken down soon after synthesis. This coincidence of instability cannot be considered as definitive evidence for the synthesis of an mRNA in the mitochondrion. Additional evidence of an indirect nature comes from the inhibitory effect of actinomycin D on protein synthesis in isolated mitochondria (see below). Because this antibiotic acts by interfering with the polymerization of RNA on a DNA template, it is considered likely that the inhibitory effect in mitochondria is obstructing the synthesis of mitochondrial mRNA coding for the manufacture of mitochondrial proteins. All of this evidence is indirect; there is no direct evidence for a mitochondrial mRNA. No mitochondrial RNA has been recovered with the capacity to stimulate the synthesis of specific mitochondrial proteins in cell-free systems.

Possible Functions of the Nucleic Acids of Mitochondria

The function of the mitochondrial DNA and RNA system remains something of a mystery. Without a doubt, the mitochondrial DNA codes for its own replication and for the transcription of at least rRNA and tRNA. Given this capability, the question remains as to whether there are any additional products coded for by the mitochondrial system. In spite of a great quantity of experimental work directed toward this question, only partial answers have been obtained.

The synthesis of protein in isolated mitochondria. Amino acids added to a freshly isolated suspension of mitochondria are linked into proteins that become incorporated inside the mitochondrion. In addition to a supply of amino acids, this system requires the presence of a substrate capable of oxidation by the mitochondria. Other experiments, using inhibitors of phosphorylation, also indicate that the protein synthesis mechanism is closely coupled to the synthesis of ATP by the electron transport system. D. B. Roodyn and his colleagues were able to show that, if radioactive amino acids are added to a suspension of mitochondria, the protein synthesized is incorporated into the insoluble portion of the inner mitochondrial membranes. The outer membrane fraction remains unlabeled. The soluble matrix also remains without label. Because of the insolubility of the labeled fraction, it has not been possible to identify the proteins synthesized by the system except that the radioactive material is associated with the structural, rather than enzymatic proteins bound to the inner membranes. To date, no label has been found in any enzyme of either the citric acid cycle, electron transport system, or phosphorylation systems.

Protein synthesis has been reported in mitochondria isolated from fungi, animals, and plants. By the use of several inhibitors, the synthesis of protein in mitochondria has been shown to proceed by a mechanism distinguishable from cytoplasmic synthesis. The antibiotic *chloramphenicol,* which interferes with the association of ribosomal subunits in bacteria but not in eukaryotes, completely inhibits mitochondrial protein synthesis but has no effect on synthesis by extramitochondrial ribosomes. The disruption of mitochondrial protein synthesis by actinomycin D has already been mentioned; this antibiotic does not interfere with the synthesis of protein by extramitochondrial ribosomes of the cytoplasm. The opposite effect is noted for the antibiotic *cycloheximide,* which interrupts the attachment of ribosomes to mRNA in eukaryote cytoplasm but has little effect on bacterial or mitochondrial systems. Thus the interesting structural parallels between the DNA and ribosomal systems of mitochondria and bacteria are also reflected in the mechanisms for protein synthesis.

The coding capacity of mitochondrial DNA. No proteins other than the structural proteins of the inner membranes have been implicated in mitochondrial synthesis. Even though the petite and poky mutants of fungi lack cytochromes *a* and *b,* these proteins fail to take up label in normal mitochondria supplied with radioactive amino acids. This suggests that these proteins are synthesized outside the mitochondrion. Many other mitochondrial proteins, such as cytochrome *c* and the enzymes malic dehydrogenase, aconitase, and other enzymes of the citric acid cycle have definitely been established by their Mendelian patterns of inheritance as controlled by nuclear genes. What then is the likelihood that mitochondrial DNA codes for proteins and other substances of the mitochondrion besides the structural proteins of the inner membranes, rRNA, and tRNA?

An indirect answer to this question lies in the quantity of DNA contained in the mitochondrion. On the average, this quantity is about 1 percent of the DNA of a bacterial cell. The uniform circles of the mitochondria of animals contain DNA with a molecular weight of 10–11 million, about enough for 15,000 base pairs.

Assuming that all the circles in a mitochondrion represent copies of the same genome, transcription of rRNA, a likely function of mitochondrial DNA, would take up about one-third of the circles, because a ribosome contains RNA chains totaling about 5000 nucleotides in length. At this point, insufficient DNA is left to code for the proteins of the ribosome. Although estimates vary, synthesis of both the RNA and all of the protein of the ribosome would require DNA approximately twice the length of the mitochondrial DNA circle. Coding for 20 tRNAs, in addition to the rRNA, would require another 10 percent of the mitochondrial DNA circle. However, insufficient space would then be left for 20 molecules of aminoacyl-tRNA synthetase.

Even if the mitochondrial DNA codes only for protein, the capacity of the molecule does not begin to meet total mitochondrial requirements. The 15,000 base pairs are equivalent to a coding capacity for about 25 proteins of molecular weight 25,000. The total mitochondrial requirement, if all the enzymes and structural proteins are considered, far exceeds this capacity. Just one functional unit of the enzymes of the electron transport system, for example, is estimated to lie near a molecular weight of 2,000,000. Thus, coding capacity in excess of the amount required for rRNAs and tRNAs of the mitochondrion seems very unlikely.

The Origin and Continuity of Mitochondria

The presence of DNA, whether limited in coding capacity or not, brings up the question of the continuity of the mitochondrion. Does the DNA of the mitochondrion exist and replicate with complete independence from the nucleus? If so, this suggests that mitochondria probably arise only from preexisting mitochondria. On the other hand, if mitochondrial DNA can be replicated at will from a master copy located somewhere in the chromosomes, it is possible that mitochondria can be constructed *de novo* by the cell from unorganized precursors.[2]

This question, whether mitochondria arise *de novo* or only from preexisting mitochondria by division is a very old one in cytology. Arguments on one side or the other extend back at least 80 years. In contemporary terms, the question boils down to whether the DNA of the mitochondrion is an independent entity that must be replicated and passed on to daughter mitochondria for the line to survive in cells or whether a master copy exists in the nucleus from which the mitochondrial segment can be copied at will to form new mitochondria.

Cytological evidence is generally in favor of the interpretation that mitochondria arise only from preexisting mitochondria. In living tissue observed by phase contrast microscopy, mitochondria can be seen to divide, and incidentally, to fuse together again. The single-celled alga, *Micromonas,* contains a single mitochondrion, chloroplast, and nucleus. Division of the nucleus is accompanied by synchronous division of the mitochondrion and chloroplast. In this organism, therefore, the single mitochondrion can be clearly seen to arise through division of a preexisting mitochondrion (see Manton and Parke, 1960).

In another experiment, carried out by Bernard Tandler and his co-workers, mice were placed on a diet low in riboflavin. Within cells of the liver of these mice, mitochondria became greatly reduced in number. After return to a normal diet, the mitochondrial complement was quickly restored. Frequently, in these cells, mitochondria apparently in the process of division could be identified (Fig. 16–14). At the apparent point of division, a membranous partition extends completely across the mitochondrion. This evidence and the observation of mitochondrial division in *Micromonas,* suggests that mitochondria arise from preexisting mitochondria by division.

Attempts to eliminate *de novo* origin as a possibility have been inconclusive. In 1963, D. J. L. Luck followed the distribution of radioactive choline in a choline-requiring strain of rapidly growing *Neurospora.* Growth of these organisms in choline, a precursor of membrane lipids, produced cells containing completely labeled mitochondria. Following transfer to nonradioactive medium, a gradually increasing population of completely unlabeled mitochondria would be expected to appear if mitochondria arise *de novo.* If new mitochondria are derived only by division of preexisting mitochondria, all mitochondria would be labeled; the amount of label, if present in stable form in the membranes, would simply be diluted roughly equally among daughter mitochondria as the total mass increases. Luck

[2]A similar mechanism operates in the nuclei of amphibian oocytes at prophase of meiosis, during which the nucleolar organizer site synthesizes hundreds or even thousands of copies of the DNA coding for the ribosomal subunits (see Chapter 14). These are released from the chromosome and enter the nucleoplasm, where they form the "cores" of the multiple nucleoli of the oocyte.

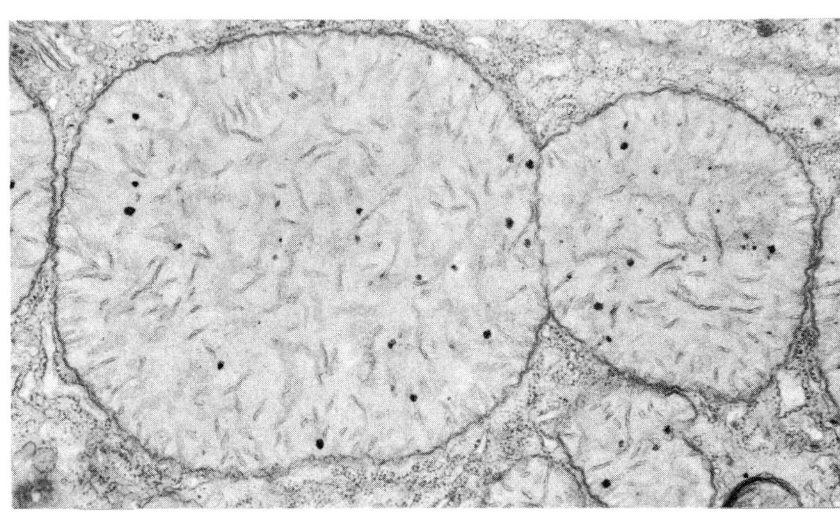

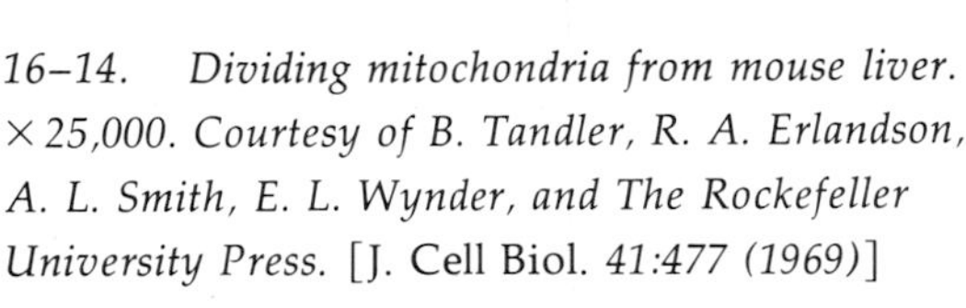

16–14. Dividing mitochondria from mouse liver. ×25,000. Courtesy of B. Tandler, R. A. Erlandson, A. L. Smith, E. L. Wynder, and The Rockefeller University Press. [J. Cell Biol. *41:477 (1969)*]

observed that after transfer to unlabeled medium, radioactivity was distributed throughout all mitochondria in growing cells; no population of unlabeled mitochondria appeared. This experiment, which would rule out *de novo* formation of mitochondria, is somewhat weakened by the fact that membrane lipids are unstable and turn over, at least to some extent, between different membranes (see Akiyama and Sakagami, 1969). Thus a totally "new" mitochondrion, even if constructed entirely from unlabeled membrane precursors, might nevertheless be expected to pick up some label from lipids by exchange with previously existing membranes. Fusion of mitochondria, observed under phase contrast of living cells, would also lead to the same result. Therefore, Luck's results are not conclusive evidence against the *de novo* production of mitochondria.

For a time, it was thought that yeast grown anaerobically contained no mitochondria. When grown without an oxygen supply, yeast metabolizes substrates by the glycolytic pathway. No typical mitochondria are present in the cytoplasm, and the enzymes of the citric acid cycle or the electron transport system cannot be detected. If transferred to an aerobic environment, mitochondria and the associated respiratory enzymes appear, apparently *de novo*. However, Hewson Swift and his co-workers demonstrated that with aldehyde fixation small vesicles consisting of a double membrane but no cristae were visible in the cytoplasm of aerobically grown yeast (Fig. 16–15). These particles, if isolated, proved to contain DNA of density characteristic of mitochondria. The vesicles could incorporate radioactive amino acids; this incorporation was insensitive to cycloheximide and was inhibited by chloramphenicol. These observations led Swift to propose that the small vesicles are *promitochondria* that contain DNA and a limited inner membrane system but no enzymes of citric acid cycle or oxidative phosphorylation. On transfer of yeast to an aerobic medium, these promitochondria presumably differentiate into typical mitochondria by proliferation of cristae and synthesis or assembly of the respiratory enzymes.

Thus, all available evidence—the observations of dividing mitochondria in living tissue, the synchronized division of nucleus and mitochondrion in *Micromonas*, the observations of apparently dividing mitochondria in the electron microscope, the promitochondria observed by Swift in anaerobic yeast, and, with some reservations, the choline incorporation experiments of Luck—points to the conclusion that mitochondria do not arise *de novo*. While none of this evidence absolutely rules out *de novo* origin, the weight is definitely in favor of derivation by division of previously existing mitochondria.

Summary

The genetic experiments with the poky and petite strains of yeast have established that at least a part of the genetic control of the mitochondrion resides in the cytoplasm. Cytochemistry of mitochondria prepared for both light and electron microscopy demonstrates the presence of DNA in the mitochondrion; in the electron microscope, this DNA takes the form of 30–40 Å filaments that are closely similar to bacterial nuclei in

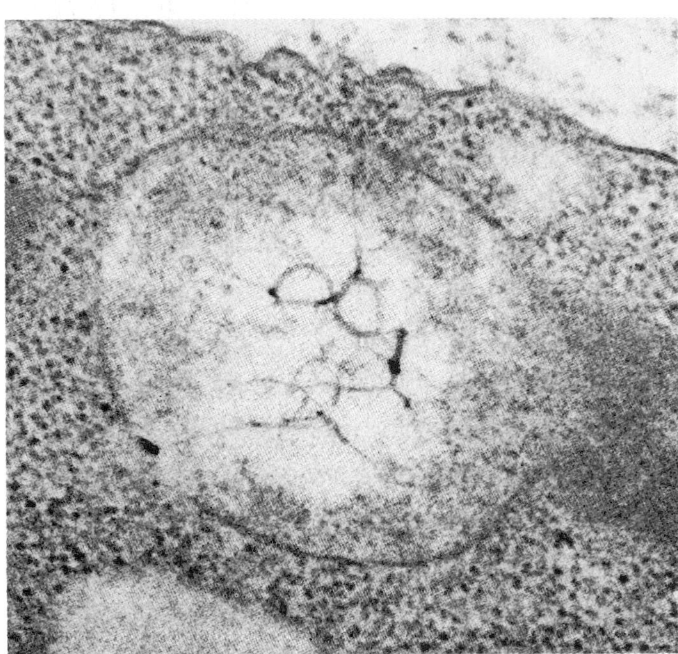

16–15. *Yeast promitochondrion after growth under anaerobic conditions. No cristae are present, but the matrix contains filaments of DNA. ×110,500. From* Handbook of Molecular Cytology *edited by A. Lima-de-Faria. Courtesy of H. Swift and North-Holland Publishing Company, Amsterdam.*

reaction to fixatives and stains. On isolation, the DNA of animal mitochondria consists of circular Watson-Crick double helices with contour length of 4.5–5.5 microns. From 1 to 5 of these circles may be present in a single mitochondrion. In addition to the DNA, the mitochondrion possesses ribosomes and tRNA. The enzymes for DNA replication, RNA transcription, and protein synthesis are probably also present. Although information is incomplete, it has been established that this system is likely to contain at least the information required for the synthesis of rRNA, tRNA, and one or more structural proteins of the inner mitochondrial membranes. Because the coding capacity of the DNA is very limited in comparison to the total protein complement of the mitochondrion, it is highly unlikely that more than a very small fraction of mitochondrial structure is coded for by mitochondrial DNA. This is supported by the fact that many enzymes of the mitochondrion, such as cytochrome *c* and the enzymes of the citric acid cycle, can be shown to be coded for by nuclear genes and to follow Mendelian patterns of inheritance. Thus, the control of mitochondrial synthesis by mitochondrial DNA is only partial. In spite of this, the best evidence is that mitochondria arise only by division of preexisting mitochondria.

The probable origin of mitochondria by division of preexisting mitochondria, along with the many similarities in size, shape, type of DNA and ribosomes, and sensitivity to inhibitors of protein synthesis between mitochondria and bacteria, has engendered wide discussion of the hypothesis that mitochondria arose in evolution by adaptation of an ancient prokaryotic symbiont to permanent residence in the cytoplasm. As the next part of this chapter will show, a similar case can be made for the evolution of chloroplasts from an original prokaryotic symbiont. Because it is more profitable to consider the possible evolutionary origins of the two organelles together in the same section, this most interesting topic will be deferred until after a description of the genetic and structural bases of cytoplasmic inheritance in the chloroplast.

The Cytoplasmic Control of Inheritance in the Chloroplast

Genetic Evidence

Correns' work with *Mirabilis*, already described in the introduction to this chapter, was one of the first demonstrations of the extranuclear control of inheritance in chloroplasts. These experiments established that a cytoplasmic factor, inherited through the maternal line, controlled the greening of chloroplasts. A large number of cytoplasmic mutations of this type have been identified in plants. In general, these mutations are passed on through the maternal cytoplasm, segregate in non-Mendelian fashion, and show no linkage to any nuclear gene.

A few additional examples of cytoplasmic inheritance of factors controlling the chloroplast are worth considering. One of the most well-studied is the inheritance of the "IOJAP" mutation in corn. This mutant gene, when homozygous, causes abnormalities in both color and structure of chloroplast. Typical Mendelian segregation and linkage to known chromosomal genes establish that the IOJAP mutation is nuclear. However, once mutated, the chloroplasts breed true, even if they are carried by crosses into the cytoplasm of progeny with the normal genotype. In this

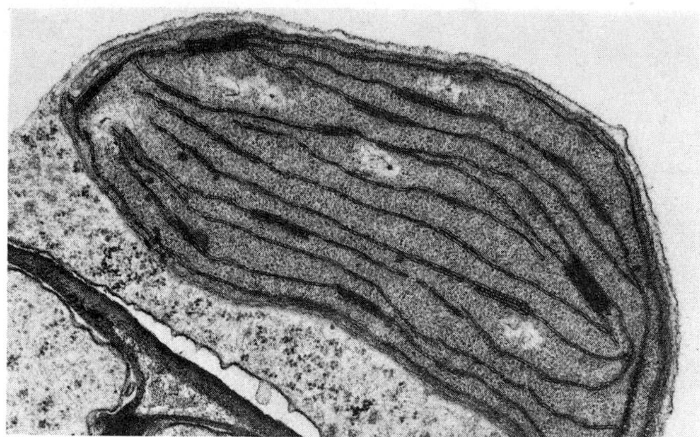

16–16. A normal chloroplast from a potentially green region of IOJAP maize. ×19,500. Courtesy of L. K. Shumway and T. E. Weier.

example, a nuclear gene in some way causes permanent mutation of a chloroplast gene. The mutant chloroplasts, in addition to exhibiting abnormalities in color, also show structural irregularities.

In an electron microscope study of the IOJAP mutation, L. K. Shumway and T. E. Weier demonstrated that the aberrant chloroplasts lack the normal system of internal membranes (Figs. 16–16 and 16–17). In a few cases, both mutant and normal plastids were contained in the same cell (Fig. 16–18). These could arise in crosses between wild type "males" and IOJAP "females" by the occasional introduction of normal chloroplasts through the pollen. The maintenance of both types of chloroplasts in the same cell provides strong evidence that the control of some parts of chloroplast structure and function is autonomous and independent of the genetic environment provided by the nucleus.

Related experiments, with a somewhat different approach, have also indicated cytoplasmic control of chloroplast inheritance. Aharon Gibor and Sam Granick irradiated the flagellate protozoan *Euglena* with an ultraviolet microbeam. If wavelengths at 2600 and 2800 Å were used (the wavelengths mainly absorbed by nucleic acids), irradiation induced loss of green pigment and other aberrations in *Euglena* chloroplasts. Gibor and Granick found that irradiation of the nucleus alone had no effect on the chloroplasts, but irradiated cells in which the nucleus was shielded from the ultraviolet microbeam developed a high incidence of chloroplast abnormalities. Isolation and biochemical analysis of the bleached chloroplasts from the mutated *Euglena* revealed that at least three enzymes of the Calvin cycle and two components of the photosynthetic electron transport system were absent. Additional disruptions of the synthetic pathways for the chlorophyll and carotenoid pigments were also noted. Once established, the mutant chloroplasts breed true, and no reversion to the normal type is noted in subsequent generations. The effectiveness of ultraviolet light at 2600 and 2800 Å in inducing these chloroplast mutations suggests involvement of nucleic acids; the limitation of this effect to extranuclear regions provides strong evidence that the affected genetic material is located somewhere in the cytoplasm.

DNA in the Chloroplast

The foregoing evidence alone does not allow localization of the cytoplasmic factors controlling chloroplast structure and function. But as early as the 1930's, it was occasionally reported that Feulgen staining indicated the presence of DNA in chloroplasts. These results were not taken seriously at the time, because Feulgen staining generally yields negative results in the cytoplasm. This is not surprising, even in view of contemporary results confirming the presence of DNA in both mitochondria and chloroplasts, because the concentrations found are too low to be consistently detected by the Feulgen stain.

By the late 1950's, it was still held in most quarters that the chloroplast does not contain DNA as a regular constituent. The first important evidence to the contrary came with the discovery by C. Ralph Stocking and Ernest M. Gifford that chloroplasts incorporate radioactive DNA precursors that can be detected *in situ* by electron microscope autoradiography. In spite of these findings, the presence of DNA remained at issue until 1962, when Hans Ris and W. Plaut obtained positive results with a battery of techniques employing both light and electron microscopy. Using the unicellular alga *Chlamydomonas* as experimental material, Ris and Plaut showed by electron microscopy that the chloroplasts in this organism contained collections of 25–30 Å fibrils that could be removed by prior digestion in deoxyribonuclease. Positive results for light microscopy were also obtained by Feulgen staining (Fig. 16–19) and the use of the dye acridine orange, which fluoresces a bright yellow-green color when complexed with DNA. These results indicated that the DNA molecules are located in several regions of the stroma.

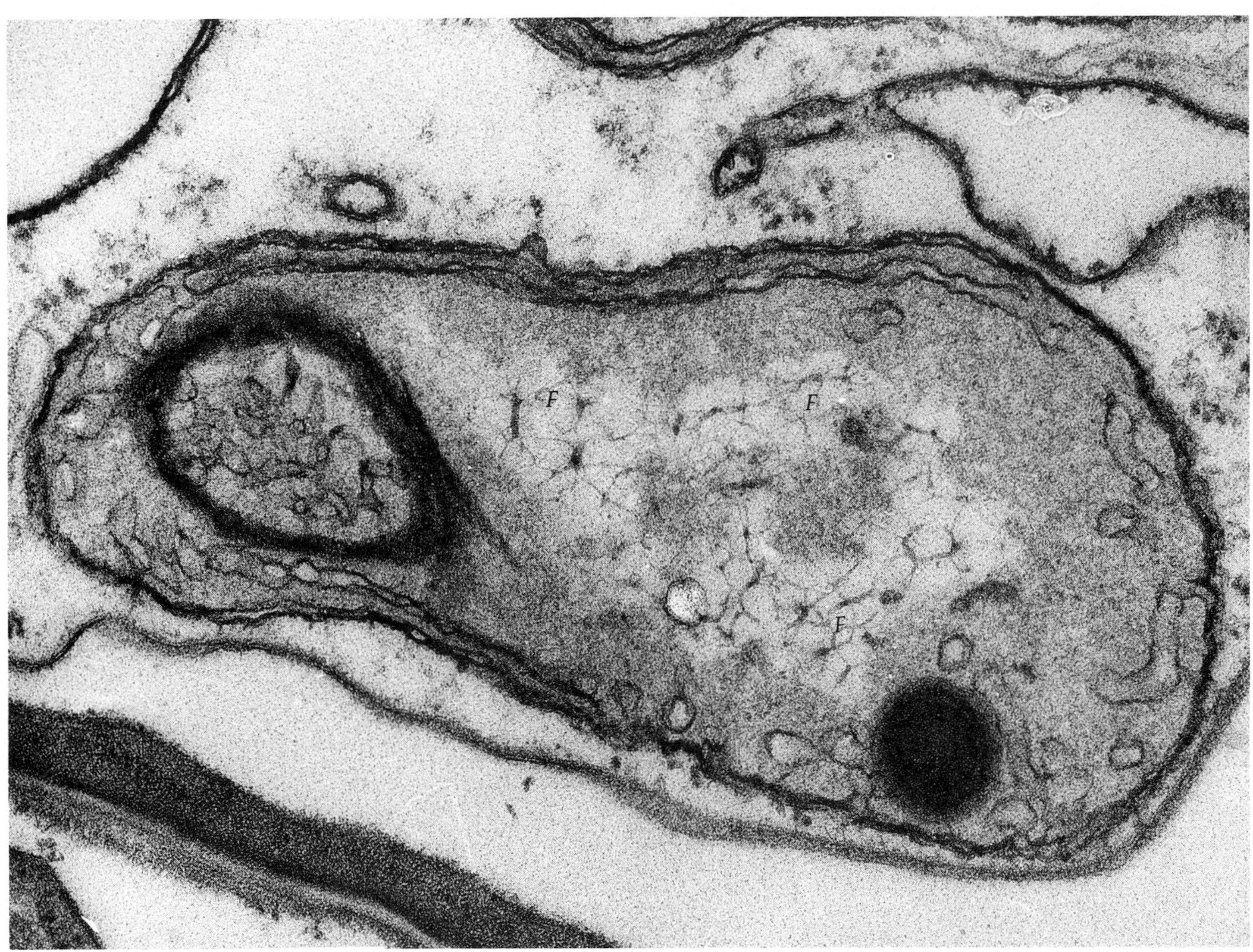

16–17. An aberrant chloroplast from a white portion of an IOJAP mutant. Internal thylakoid membranes are completely absent. F, *presumed DNA fibrils.* ×*63,500. Courtesy of L. K. Shumway and T. E. Weier.*

As mentioned earlier in this chapter, Ris was also successful in demonstrating the presence of DNA in mitochondria at the same time.

These important discoveries were confirmed and extended in the chloroplasts of higher plants by Naomi Kislev, Hewson Swift, and Laurence Bogorad. With osmium fixation, the collections of DNA in the chloroplast appeared as thickened, dense clumps (Fig. 16–20). Fixation by the Kellenberger technique for bacterial DNA yielded much thinner fibers in these regions, running from about 50 Å down to invisibility (Fig. 16–21). This suggests that the chloroplast DNA is of the bacterial type, uncomplexed with protein. These investigators also followed the binding of uranyl ion in formalin-fixed chloroplasts treated with nucleases. Formalin fixed tissue, if no postfixation with osmium

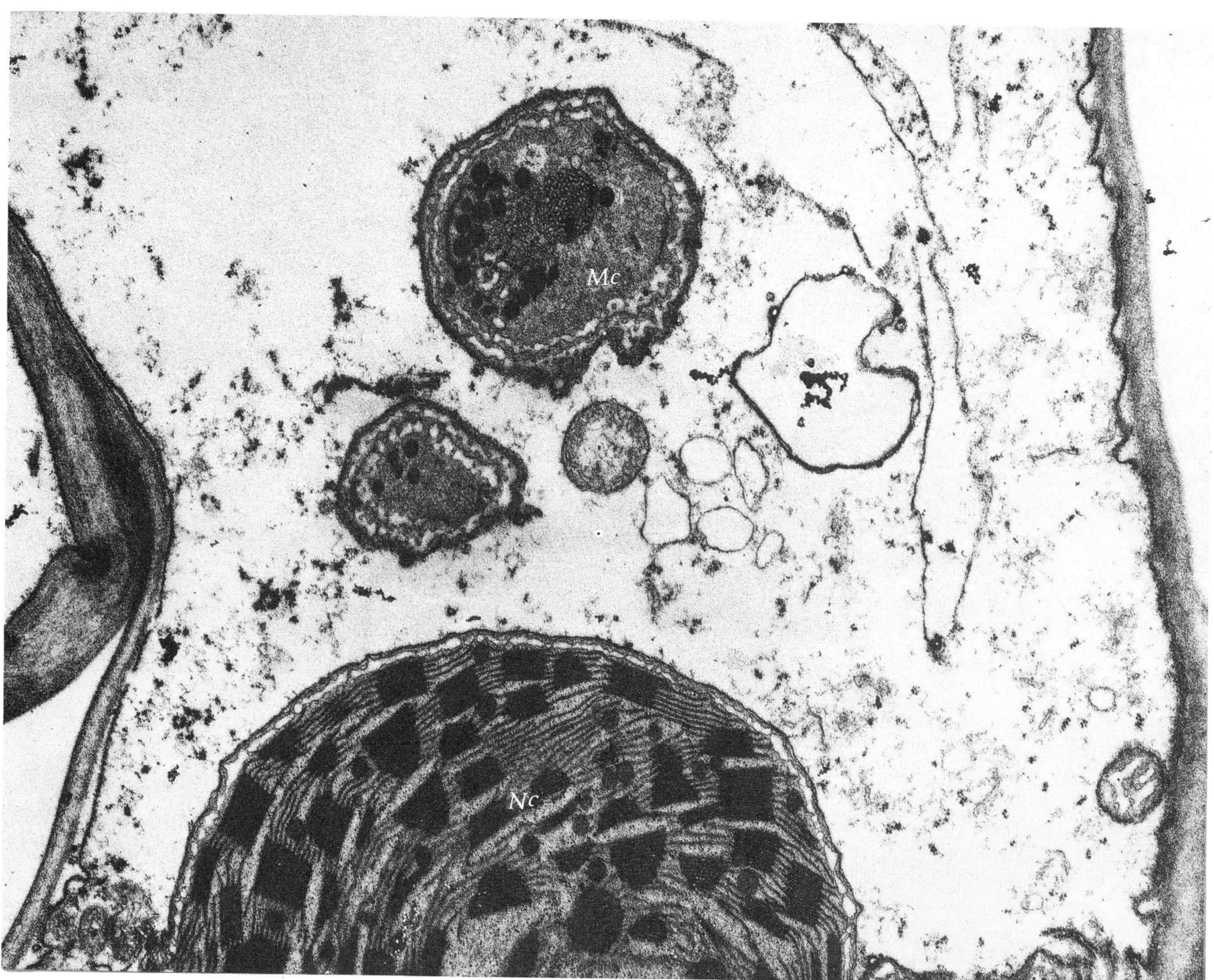

16–18. Region of a single cell of an IOJAP leaf containing both a normal (Nc) *and a mutant* (Mc) *chloroplast.* ×*16,500. Courtesy of L. K. Shumway and T. E. Weier.*

is employed, has almost no density in the electron microscope. If these preparations were digested with ribonuclease and stained with uranyl ion, which has affinity for the nucleic acids, a filamentous component, presumably DNA, took up density in scattered regions of the stroma (Fig. 16–22). Digestion with deoxyribonuclease removed these filaments; chloroplasts treated with deoxyribonuclease alone showed the presence of numerous ribosome-like particles after staining (Fig. 16–23). Thus, the cytological evidence indicates that DNA and RNA are regular components of the chloroplasts.

Chloroplast DNA is typically found to be in the form of thin fibers 25–50 Å in diameter, with morphology and distribution closely similar to the DNA of mitochondria. As in mitochondria, the DNA fibrils are more

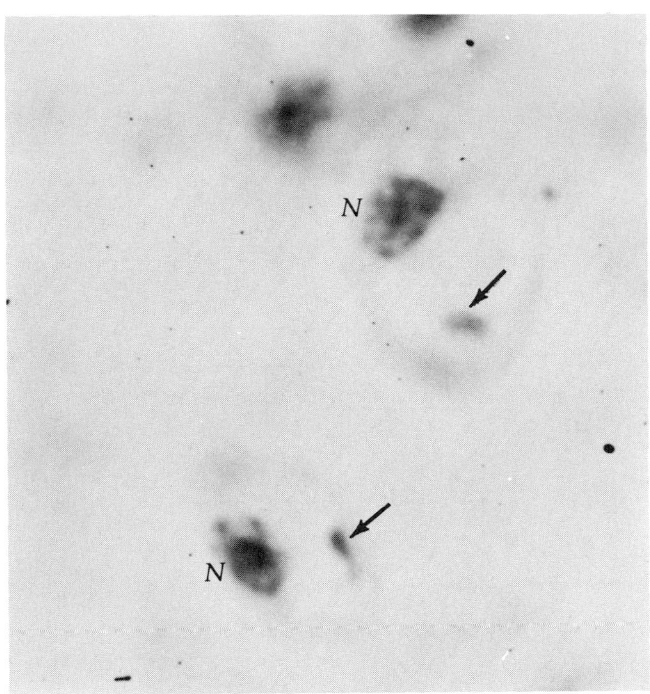

16–19. *Two cells of the alga* Chlamydomonas *stained with the Feulgen reaction.* N, *nucleus. Staining density is also associated with the chloroplast (arrows). Courtesy of H. Ris, W. Plaut, and The Rockefeller University Press.* [J. Cell Biol. *13 (1962)*]

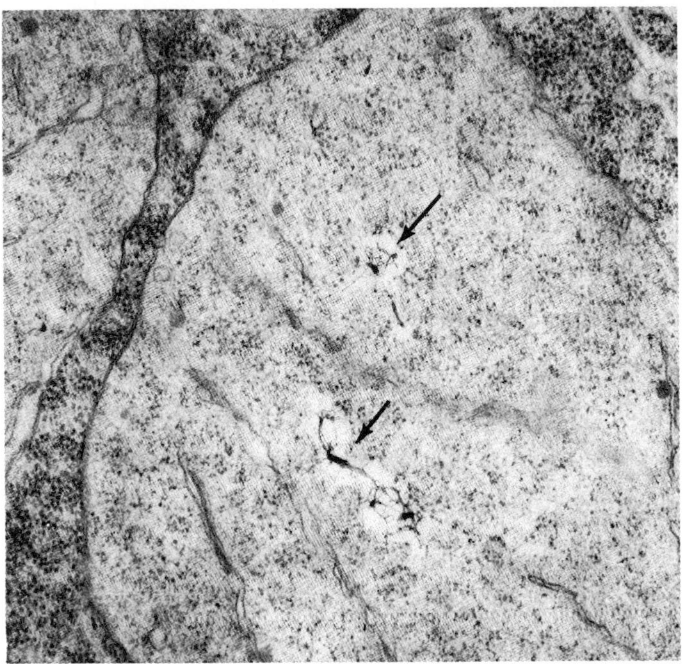

16–20. *An etiolated bean leaf after osmium fixation. The DNA of the chloroplasts has coagulated into dense clumps (arrows).* ×*32,500. Courtesy of H. Swift.*

conspicuous in chloroplasts from rapidly growing or embryonic tissue. Although a few connections between these fibrils and the inner membranes of the chloroplast are seen, it is not known whether these are a regular feature of the arrangement of DNA in the chloroplast.

Centrifugation of DNA isolated from plant cells also reveals a distinct DNA fraction specifically associated with the chloroplast. In most plants, this approach has not been as successful as with mitochondria because contamination from the nucleus is difficult to eliminate. The DNA in intact mitochondria is insensitive to attack by deoxyribonuclease, probably because of the barrier provided to the enzyme by the inner and outer membranes. Therefore, it is easy to eliminate contaminating nuclear DNA from mitochondrial preparations by simply treating the extracts with deoxyribonuclease while mitochondrial membranes are still intact. Unfortunately, this technique destroys *both* nuclear and chloroplast DNA at any stage of the

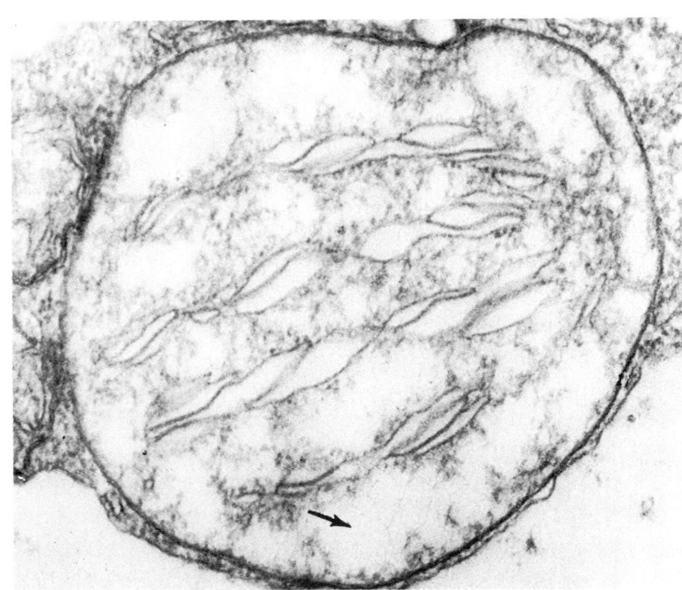

16–21. *Chloroplast from Swiss chard after Kellenberger fixation. Fine fibrils are present in the stroma (arrow).* ×*41,500. Courtesy of N. Kislev, H. Swift, L. Bogorad, and The Rockefeller University Press.* [J. Cell Biol. *25:327 (1965)*]

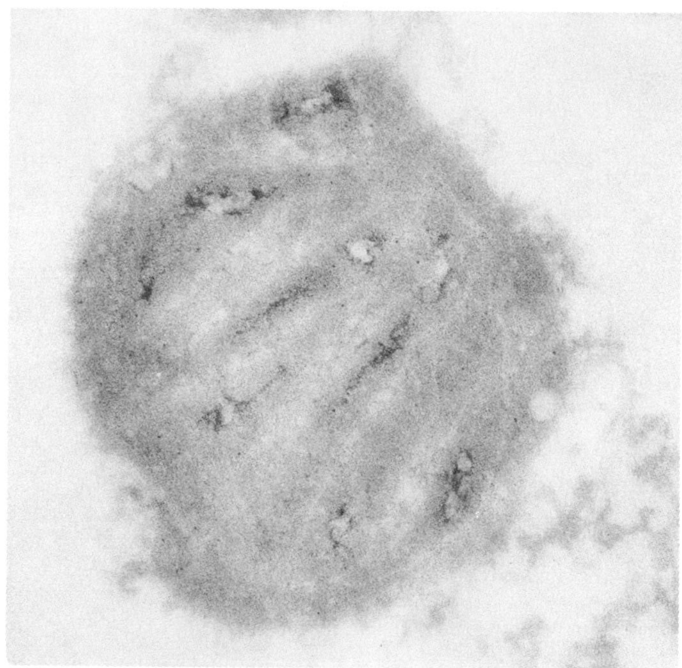

16–22. Chloroplast from Swiss chard after extraction with ribonuclease. Fine fibrils, presumably DNA, are present in the stroma. ×51,000. Courtesy of H. Swift.

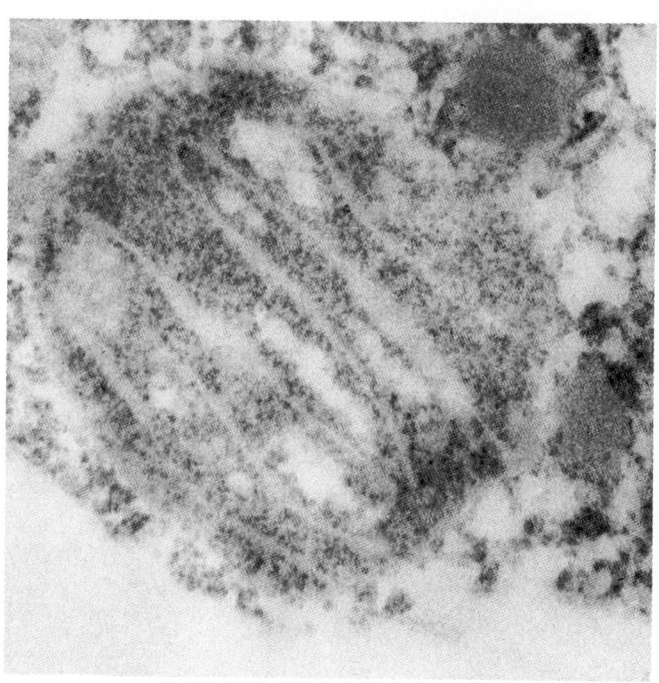

16–23. Chloroplast from Swiss chard after extraction with deoxyribonuclease. The filamentous material has been removed; density is concentrated instead in ribosome-like particles. ×40,000. Courtesy of H. Swift.

isolation procedure; evidently, chloroplast membranes are permeable to the enzyme. Therefore, it has not been possible in most plants to obtain preparations of chloroplast DNA that approach the purity of equivalent mitochondrial extracts. In spite of these limitations, a DNA fraction has been identified that is increased in concentration with greater degrees of purification of isolated chloroplasts (Fig. 16–24); partial physical characterization of this fraction has been possible (see below). Even with the difficulties encountered in isolation it is obvious that chloroplasts contain a much greater quantity of DNA than mitochondria. This quantity approximates the amount found in a bacterial cell and exceeds the concentration of DNA in mitochondria by about 100 times.

The difficulties in the purification of isolated chloroplast DNA are greatly reduced in the unicellular alga *Acetabularia*. In *Acetabularia*, the single nucleus is located in a root-like rhizoid at one end of the cell. The nucleus can be eliminated simply by cutting off the rhizoid. The remainder of the plant, which can be maintained in the enucleated state for as long as three months, contains numerous chloroplasts. Aharon Gibor and his colleagues and Jean Brachet extracted DNA from the enucleated stalks of these plants. In these preparations, freed of contamination from chromosomal DNA by the enucleation technique, it was possible to confirm that a specific DNA fraction is associated with the chloroplast. The concentration of this DNA, about 1×10^{-16} to 10×10^{-16} grams per chloroplast, is somewhat lower than the quantities determined for other species, in which the elimination of nuclear DNA is less certain. In these other species, estimates run from 1×10^{-15} to 1×10^{-14} grams; undoubtedly nuclear contamination is partly responsible for the higher values.

The properties of chloroplast DNA. The DNA isolated from chloroplasts of *Acetabularia* and other plants is usually linear, varying in size from small pieces to

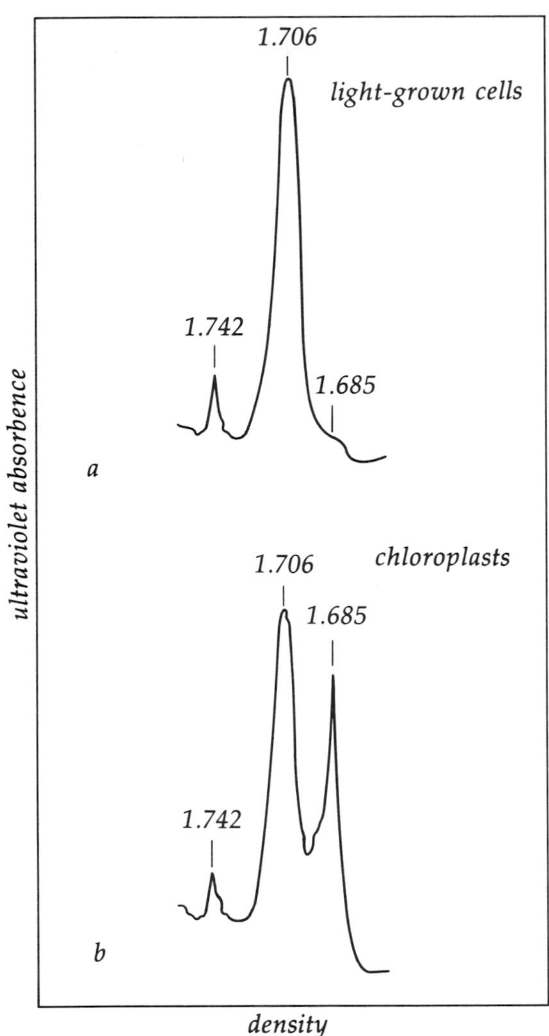

16–24. a, Whole cell DNA from Euglena. *b, chloroplast fraction. DNA at a density of 1.685 is enriched in the chloroplast fraction. The peak at 1.742 represents DNA of known density added to calibrate the gradient. Redrawn, original courtesy of M. Edelman, C. A. Cowan, H. T. Epstein, and J. A. Schiff.*

lengths of up to 150 microns or more. Because of the diversity of sizes found, it has not been possible to arrive at an estimation of a unit length for chloroplast DNA or to determine whether the DNA in a chloroplast exists as a single molecule or in several pieces. DNA preparations from tobacco and spinach chloroplasts yielded linear molecules up to 150 microns long when spread on a water surface; in *Acetabularia*, molecules up to 50 microns in length were found. While estimates vary, the lower limit of unbroken chloroplast DNA molecules may be as high as 50 microns. Therefore, although the exact lengths have not been determined, indications are that chloroplast DNA is probably much longer than the 4.5 to 5.5 microns typical of mitochondrial DNA in animals. One observation of DNA circles from *Euglena* chloroplasts exists (see Manning *et al*, 1971); in this case the total length of DNA included in the circles was 40 microns.

In preparations isolated on a water surface, the DNA of chloroplasts broken by osmotic shock is released as a tangled mass containing relatively long pieces of DNA (Fig. 16–25). Few broken ends are visible in these masses. What appear as broken ends at lower magnification prove to be folded loops of DNA (Fig. 16–26). These loops show "supertwists" resembling the supercoils found in mitochondrial DNA circles isolated by the same methods. In the study, carried out by Beverly R. Green and Hugh Burton, from which the micrographs shown in Figs. 16–25 and 16–26 were taken, DNA masses were found in 20 to 40 percent of the ruptured chloroplasts.

Other physical properties of chloroplast DNA are closely similar to mitochondrial DNA. No histones or other proteins are complexed with the chloroplast DNA. Denaturation by increased temperature follows the pattern expected for a Watson-Crick double helix. Chloroplast DNA can be distinguished from chromosomal DNA by differences in density, absence of methylated bases, and rapid renaturation after denaturation by heating to 100°C. These properties again are also closely similar to mitochondrial DNA. The base composition of DNA from the chloroplasts of higher plants is remarkably uniform from species to species (Table 16–3). All extracts analyzed from this source have a greater density than chromosomal DNA from the same species. In the alga, the reverse is true, with chloroplast DNA exhibiting lighter densities than the corresponding nuclear DNA. This distribution of densities closely resembles the situation in mitochondria, in which the DNA from closely related groups is approximately the same but may either be heavier or lighter than chromosomal DNA from the same species.

Replication of chloroplast DNA. If supplied with the four nucleotide triphosphates, isolated chloroplasts are capable of synthesizing stable DNA of high molec-

16–25. Chloroplast DNA isolated from Acetabularia mediterranea. *Few free ends are visible in the mass, indicating that the DNA is of considerable length.* ×*29,000. From* Science *168:981 (1970). Copyright 1970 by the American Association for the Advancement of Science. Courtesy of B. R. Green and H. Burton.*

ular weight. If a radioactive precursor such as tritiated thymidine is added to the suspending medium, the DNA synthesized is radioactive and, on extraction, exhibits the characteristics of chloroplast DNA, including typical density, typical guanine-cytosine ratio, and rapid renaturation after heating (see, for example, Spencer and Whitfeld, 1967a; N. S. Scott et al., 1968). In intact cells, radioactive precursor is also incorporated into a stable, insoluble fraction in chloroplasts. This replication is continuous and can be shown to be independent of the cyclic synthesis of chromosomal DNA. The capacity for internal replication implies that DNA polymerase, ligase, and other enzymatic mechanisms required for DNA synthesis are present in the chloroplast interior. These findings exactly parallel the results of equivalent studies with mitochondria.

RNA in the Chloroplast

The experiments designed by Ris and Swift and their colleagues to detect the presence of DNA in the chloroplast also revealed that the stroma contains large quantities of RNA in the form of ribosome-like particles (see Fig. 16–23). Particles of this type are readily seen in the stroma, especially in chloroplasts maturing from the proplastid to mature stages. These particles, which are sensitive to ribonuclease digestion, are more clearly outlined than the equivalent structures in mitochondria, and like mitochondrial ribosomes are of slightly smaller dimensions than cytoplasmic ribosomes (Fig. 16–27). The particle diameters, 150–200 Å, fall within the range typical of prokaryotic ribosomes.

Extraction and analysis reveals that most RNA of the chloroplast is of the ribosomal type, with base ratios similar to cytoplasmic RNA. This RNA spins down in the ultracentrifuge in more discrete bands than mitochondrial RNA, giving uniform peaks in most plants at about 68–70S for whole ribosomes and 16 and 23S for the RNA subunits (Table 16–4). While the 70S value is typical of whole prokaryote ribosomes, the 16 and 23S obtained for the two RNA subunits is close to the values expected for the equivalent subunits of the eukaryotic ribosomes of plants. Thus, the chloroplast ribosomes appear to share some features with

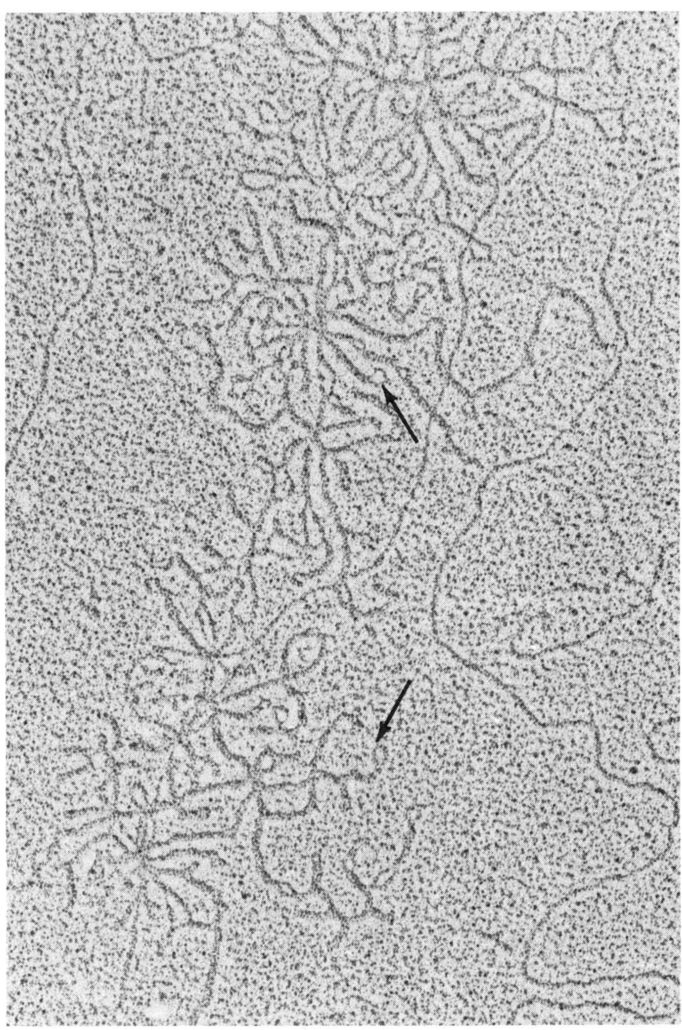

16–26. *Chloroplast DNA from* Acetabularia mediterranea. *A large number of "supertwists" is visible in this preparation. The arrows point to loops at the ends of twisted regions.* ×*36,500. From* Science *168:981 (1970). Copyright 1970 by the American Association for the Advancement of Science. Courtesy of B. R. Green and H. Burton.*

both eukaryotic and prokaryotic ribosomes. The 16 and 23S constants represent the sedimentation of the purified RNA of the subunits. With attached protein, the chloroplast subunits sediment at velocities approximating the 30 and 50S peaks of bacterial ribosomes (see Table 16–4). Therefore, although some discrepancies are found, the similarities in dimensions between chloroplast and bacterial ribosomes in electron micro-

Table 16–3 Density of Nuclear and Chloroplast DNA in CsCl, in grams/cc

Organism	*Nuclear DNA*	*Chloroplast DNA*
Higher plants		
Marigold (*Tagetes patula*)	1.692	1.702
Tobacco (*Nicotiana tabacum*)	1.696	1.706
Sweet potato (*Ipomoea batalas*)	1.692	1.706
Mung bean (*Phaseolus aureus*)	1.691	1.706
Onion (*Allium cepa*)	1.688	1.706
Spinach (*Spinacia oleracea*)	1.692	1.695
Beet (*Beta vulgaris*)	1.695	1.705
Swiss chard (*Beta vulgaris* var. *cicla*)	1.690	1.700
Lettuce (*Latuca sativa*)	1.692	1.695
Broad bean (*Vicia faba*)	1.692	1.695
Sweet pea (*Lathyrus odoratus*)	1.692	1.695
Algae and Protista		
Euglena gracilis	1.707	1.686
Chlamydomonas reinhardii	1.723	1.695
Chlorella ellipsoida	1.716	1.695

Adapted from H. Swift and D. R. Wolstenholme, Mitochondria and chloroplasts. In A. Lima-de-Faria, ed., *Handbook of Molecular Cytology*. Amsterdam: North-Holland Publishing Company, 1969, and P. Bòorst & A. M. Kroon, Mitochondrial DNA. *Int. Rev. Cytol.* 26:107 (1969). See these papers for original references.

graphs are also reflected in the sedimentation velocities. Other similarities are found in the concentrations of Mg^{+2} ion required to prevent breakdown into subunits. Eukaryotic ribosomes are stable at relatively low concentrations of Mg^{+2} ion, at levels of from 0.00025 to 0.0005M (see Chapter 11). At these levels, both bacterial and chloroplast ribosomes dissociate into subunits. Higher Mg^{+2} ion concentrations, at 0.005M to 0.01M, are required to prevent dissociation (for detailed literature references in this area, see Smillie and Scott, 1969.)

Additional peaks at 4S and 5S are found in RNA isolated from chloroplasts. Very recent work has identified a number of tRNA species in association with this fraction. The possibility that the 4S fraction represents a unique set of chloroplast tRNAs, although likely, has not yet been determined. As in mitochondria, an unstable RNA sedimenting at about 11S is identifiable in preparations isolated quickly after exposure of chloroplasts to a pulse label with tritiated uridine. It is possible that this fraction represents mRNA (see Spencer and Whitfeld, 1967b).

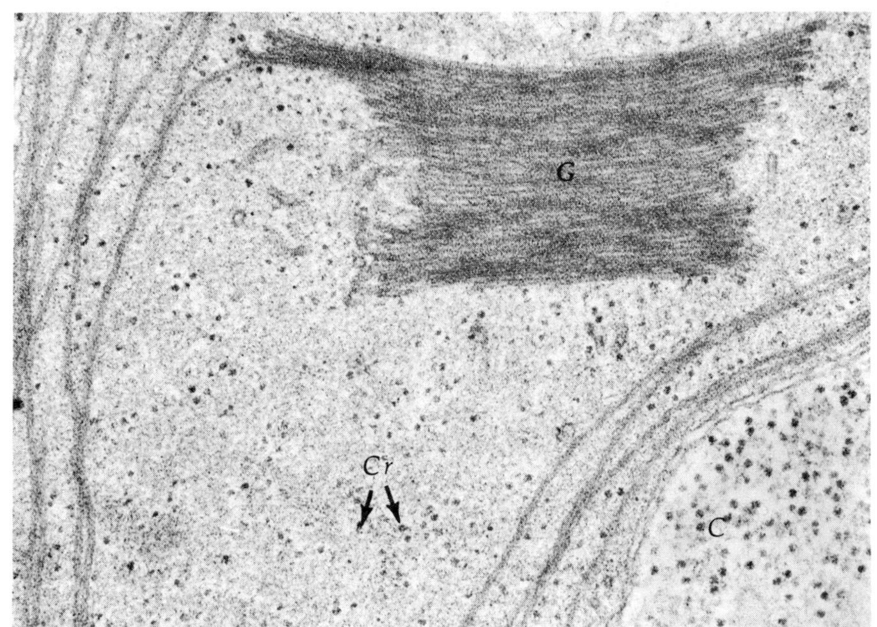

16–27. Interior of a chloroplast from a mutant of Chlamydomonas reinhardi. *The chloroplast ribosomes* (Cr), *distinctly outlined in the stroma, are clearly smaller in dimensions than the ribosomes of the cytoplasm* (C) *outside the chloroplast. G, granum. ×62,500. Courtesy of U. W. Goodenough, R. P. Levine, and The Rockefeller University Press.* [J. Cell Biol. *44:547 (1970)*]

Incorporation of precursors into a stable RNA has been demonstrated in chloroplasts isolated from many species, including broad bean, tobacco, cabbage, *Acetabularia,* and *Euglena.* This incorporation depends on a supply of the four nucleoside triphosphates and on DNA, as can be shown by the sensitivity of the transcription process to deoxyribonuclease and actinomycin D. These findings and the ability of isolated chloroplasts to replicate DNA indicate that the various polymerases, ligases, and other factors required for the synthesis of nucleic acids are probably a part of chloroplast structure.

Possible Functions of the Nucleic Acids of the Chloroplast

Chloroplasts are able to incorporate radioactive amino acids into proteins in both the intact cell and after isolation. In contrast to the negative results obtained to date with mitochondrial systems, ribosomes extracted from chloroplasts are capable of synthesizing protein if supplied with activated amino acids, mRNA, and other factors required for cell-free systems. Spiral whorls of ribosomes, visible in thin sections of chloroplasts (Figs. 16–28 and 16–29), provide additional evidence of endogenous protein synthesis. As in mitochondrial systems, protein synthesis in chloroplasts follows the pattern of sensitivity to antibiotics exhibited by bacteria, with inhibition by actinomycin D and chloramphenicol but not cycloheximide.

Very few clues to the identity of the proteins synthesized in the chloroplast are available. Work with radioactive amino acids has revealed that instead of insoluble proteins of the inner membranes as in mitochondria, the chloroplast system incorporates label into *soluble* proteins, including enzymes of the Calvin cycle (see Spencer, 1965). Inhibition of the chloroplast system with chloramphenicol produces deficiencies in the proteins of the Calvin cycle and the photosynthetic mechanisms, including enzymes of the light-dependent electron transport systems. However, synthesis of most of the same proteins is also blocked by inhibition of *cytoplasmic* ribosomes with cycloheximide. The enzymes of the Calvin cycle also occur in the cytoplasm outside the chloroplast. There is no evidence that the Calvin cycle enzymes synthesized within the chloroplast are different in structure or function from their cytoplasmic counterparts.

This situation is made even more baffling by the fact that most of the proteins affected by inhibition of the chloroplast synthetic machinery are controlled by genes localizable on the chromosomes. In fact, most or all substances of the chloroplast, except for the RNA and possibly the ribosomal proteins of the chloroplast, are evidently controlled by nuclear genes.

The apparent control of most chloroplast proteins by nuclear genes has prompted a number of investigators

Table 16–4 "S" Values of Chloroplast and Cytoplasmic Ribosomes

Species	Chloroplast ribosomes Whole	Chloroplast ribosomes Subunits	Chloroplast ribosomes RNA	Cytoplasmic ribosomes Whole	Cytoplasmic ribosomes Subunits	Cytoplasmic ribosomes RNA
Euglena gracilis	43	30,23		49	36,27	
	60	36,30	19,14	70		
	43	30,17	23,16	65	47,38	
Chlamydomonas reinhardii	70			80		
Acetabularia mediterranea		28,18				
Spinacia oleracea (spinach)	66	47,33				
	70	50,30		80		
Nicotiana tabacum (tobacco)	70					
Pisum sativum	70	45,32		80	56,40	
Brassica pekinensis (Chinese cabbage)	68			83	55,50	24,16
Phaseolus vulgaris (pinto bean)	70		23,16	80		25,16
	67					
Triticum vulgare	70					
Clivia miniata	78					
Chenopodium album (lamb's quarters)	67	48.7,30.4				

Adapted from R. M. Smillie and N. S. Scott, Organelle biosynthesis. *Prog. Molec. Subcell. Biol.* I: 136 (1969). See this paper for original references.

to conclude that chloroplast DNA codes only for the synthesis of rRNA and protein and tRNA. Other segments of the chloroplast DNA may possibly represent regulatory genes. With the evidence at hand, little can be said about these conclusions except that the chloroplast DNA probably has at least sufficient coding capacity to carry out these functions. The smallest values reported for the total quantity of DNA per chloroplast, 1×10^{-16} grams (in *Acetabularia*) would allow coding for RNA and about 150 proteins of molecular weight 27,000. The larger amounts reported for the chloroplasts of other plants, including bean, spinach, beet, and *Chlamydomonas,* vary between 2×10^{-15} and 10×10^{-15} grams. At these higher levels, if accurate, the coding capacity is increased to 3000–16,000 proteins of molecular weight 27,000. Although it is hazardous to draw conclusions from calculations of this type, there seems to be more than enough DNA in the chloroplast to code at least for rRNA, and tRNA, and ribosomal proteins. At the higher levels, capacity exists to code for additional substances, including mRNAs for a large number of different proteins. In view of the genetic evidence, which indicates nuclear control of most chloroplast protein, and the probability that a large part of the additional DNA originates from nuclear contamination, these possible additional functions must be regarded at the moment as speculative.

The Continuity of Chloroplasts in Dividing Cells

Many observations have been made of chloroplasts arising by division of preexisting chloroplasts. This process is best seen in algal cells containing only a few chloroplasts. The example provided by *Micromonas,* in which the single mitochondrion and chloroplast divide in synchrony with the nucleus and the remainder of the cytoplasm, was mentioned in the section on the origin and continuity of the mitochondrion. Other observations date back into the 1800's. At about the turn of the century, C. van Wisselingh described an abnormal strain of *Spirogyra* with one normal and one mutant chloroplast in each cell. The mutant chloroplast, which lacked a pyrenoid body, was seen to divide during cell division along with the normal chloroplast; all daughter cells received one chloroplast of each type. In this example, the mutant chloroplast appeared to possess an independent lineage and was clearly derived by division of a preexisting chloroplast. In *Euglena,* the chloroplast population is divided approximately equally between daughter cells. Soon after division of the cytoplasm, the chloroplasts divide, reestablishing the normal complement. The numerous light microscope observations of chloroplast division are complemented by electron micrographs of chloroplasts in the process of division (Fig. 16–30).

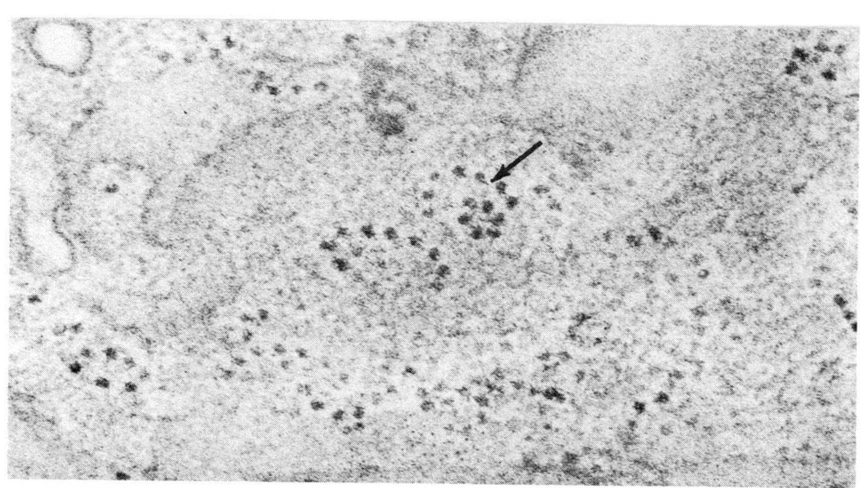

16–28. Chloroplast polysomes in the bean Phaseolus vulgaris. *In several of the polysomes a thin electron-dense strand (arrow), presumably an mRNA, connects the individual ribosomes of a whorl. ×117,000. Courtesy of H. Falk and The Rockefeller University Press.* [J. Cell Biol. *42:582 (1969)*]

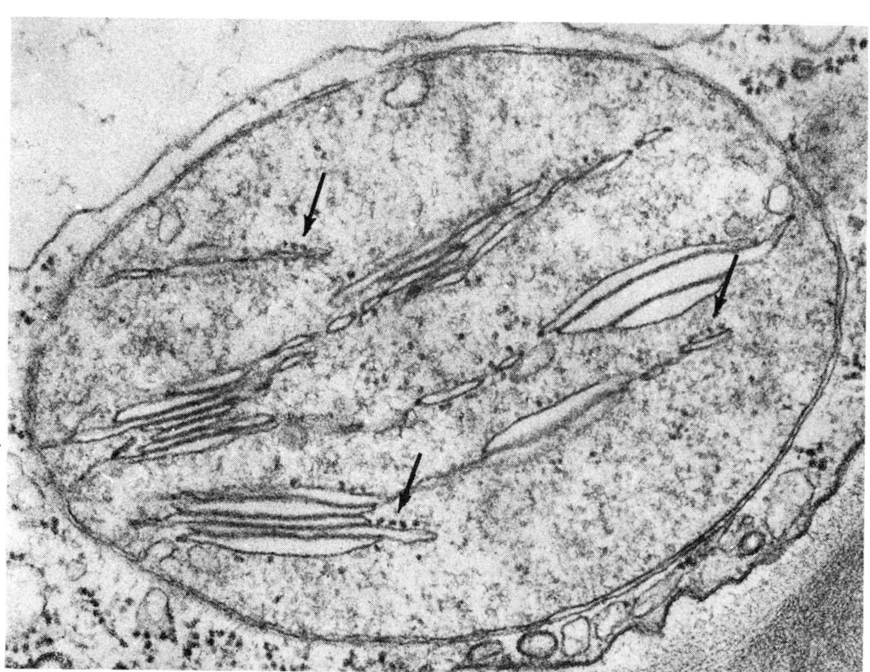

16–29. Thylakoids of the bean in transverse section. Polysomes (arrows) are apparently attached to the stromal side of the thylakoid membranes. ×66,000. Courtesy of H. Falk and The Rockefeller University Press. [J. Cell Biol. *42:582 (1969)*]

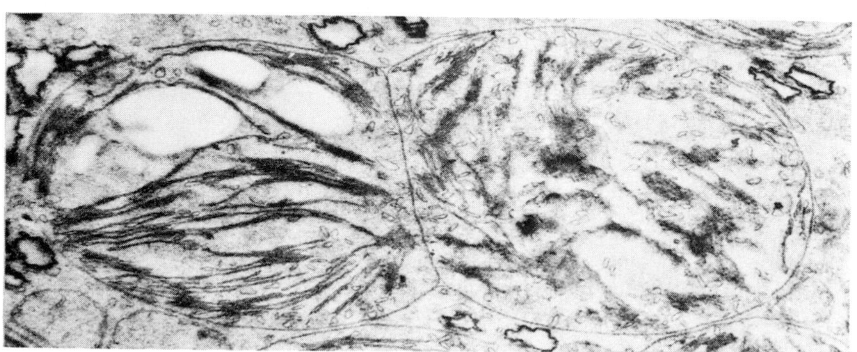

16–30. A dividing chloroplast of a fern. ×13,000. Courtesy of E. Gantt, H. J. Arnott, and The Rockefeller University Press. [J. Cell Biol. *19:446 (1963)*]

The appearance of the chloroplast in this figure, in which a membranous septum extends along the line of division, is remarkably similar to the structure of dividing mitochondria (Fig. 16–14).

Although these descriptive observations do not prove that chloroplasts arise only by division of preexisting chloroplasts, the absence of any evidence to the contrary makes it likely that this is actually the case. This state of affairs, it will be remembered, is analogous to that of the mitochondrion.

Summary

The extensive genetic experiments in cytoplasmic inheritance in plants have established that chloroplasts are partially independent of control by nuclear genes. DNA can be demonstrated in chloroplasts by cytochemical techniques for both light and electron microscopy. In the electron microscope, the DNA of the chloroplast appears as tangled masses of 30–40 Å fibrils embedded in clear areas of the stroma. When extracted, chloroplast DNA appears as linear molecules of variable size, from very short pieces up to at least 150 microns. The total quantity of DNA per chloroplast has been estimated to range from 1×10^{-16} to 1×10^{-14} grams. The extent to which nuclear contamination contributes to the spread of these values for length and total quantity cannot be evaluated at the present time. Chloroplast DNA, purified by centrifugation, can be distinguished from nuclear DNA by its distinctive density, rapid renaturation, and absence of methylated bases. No histones or other proteins are found complexed with chloroplast DNA. The chloroplast also appears to contain rRNA and possibly tRNA; both these RNAs and the DNA can be synthesized by the chloroplast *in vivo* and *in vitro*. Protein synthesis can also be detected in chloroplasts both in living cells and in the isolated state. By the use of radioactive precursors, the proteins synthesized have been identified with the enzymes of the soluble fraction. Because many proteins of the chloroplast are coded for by nuclear genes, the control of synthesis by chloroplast DNA is considered only partial and is possibly limited to the RNA and protein of ribosomes. Cytological observations suggest that chloroplasts are derived only from preexisting chloroplasts by division. Some of these observations and comparisons between nuclei, mitochondria, and chloroplasts are summarized in Table 16–5.

Cytoplasmic Inheritance of the Centriole

The centriole is considered by many cell biologists to possess an independent line of inheritance. In contrast to the genetic basis for cytoplasmic inheritance

Table 16–5 Some Comparisons between Nuclei, Mitochondria, and Chloroplasts

	Nuclei	*Mitochondria*	*Chloroplasts*
Presence of DNA	+	+	+
Histones and nonhistone proteins	+	−	−
Methylated bases	+	−	−
Rapid renaturation	±	+	+
Circular DNA		+ animals; ± plants	−?
Length of DNA		4 or 5 to 5.5μ	? to 150μ
Proteins synthesized		"insoluble"	"soluble"
Inhibition by chloramphenicol	cytoplasmic ribosomes not inhibited	+	+
Inhibition by cycloheximide	cytoplasmic ribosomes inhibited	−	−

of chloroplasts and mitochondrion, this conclusion rests solely on cytological observations of the replication pattern of centrioles during cell division. At this time the "daughter" centriole is seen to arise as a small bud, the procentriole, at one side of the mature "parent" centriole. In animals, daughter centrioles arise almost exclusively by this route. This pattern suggests that replication is partially or completely under the control of the parent centriole. This inference is reinforced by observations of centriole replication in ciliate protozoans. These protozoans have multiple flagella that cover most of the surface of the cell. Each individual flagellum carries at its base a centriole (the "basal body"). At the time of cell division, the system of centrioles basal to the flagellar apparatus replicates autonomously, without apparent connection with the independent division cycle of other centrioles associated with the nucleus. Electron micrographs made before division of the cell show a small procentriole replicating near each mature basal body. By the time the cell divides, these procentrioles have become fully mature and the total number of centrioles has doubled.

An experiment performed by T. M. Sonneborn and his colleagues demonstrated that the ciliary system is at least partially independent of nuclear control. A small patch of the pellicle of *Paramecium* was removed by microsurgical techniques and grafted back onto the surface in a reversed position. In the region of the graft, the ciliary beat was reversed. The reversed pattern was replicated and maintained in daughter cells. Because no nuclear changes could have resulted from the graft, this experiment indicates that the information required for assembly and function of the ciliary system is located in the pellicle and not the nucleus (see Beisson and Sonneborn, 1965).

On the basis of observations of this type and an exhaustive investigation of the pattern of centriole replication in animals, Andre Lwoff proposed that centrioles arise only from preexisting mature centrioles. This hypothesis implies that centrioles possess an independent information system coding for this replication and are therefore genetically autonomous.

Subsequent discoveries have made it necessary to modify at least a part of Lwoff's hypothesis. These more recent studies have revealed that procentrioles arise in some organisms without the presence or involvement of mature centrioles or possibly even procentrioles. The best examples come from work with plants and protozoans known as ameboflagellates.

In the lower plants, somatic cells do not usually contain centrioles. However, many of these forms have motile gametes with two or more flagella (see Chapter 15). At the base of the flagella are typical centrioles, indistinguishable from their counterparts in animal cells. These centrioles have been shown to arise, apparently *de novo*, during the last few cell divisions before spermiogenesis from cell lines without recognizable mature centrioles or even procentrioles. Ikuko Mizukami and Joseph Gall found in the plants *Zamia* and *Marsilea* that, a few divisions before spermiogenesis, procentrioles first appear in a mass near the nucleus. These gradually develop into mature centrioles without any evidence of serial replication from a "master" procentriole; no mature centriole is present at any time during this development. In an equivalent study, F. R. Turner showed that centrioles are present only in the last few divisions before spermiogenesis in the plant *Nitella;* these first appear as single procentrioles near the nucleus (for a more complete summary of these reports, see Chapter 15).

The nature of this work makes it difficult to establish whether the procentrioles in these plants arise *de novo* or are derived from preexisting procentrioles maintained somewhere in the cytoplasm in the parent line of cells. Procentrioles are difficult to recognize and can easily escape detection in the cytoplasm, even if serial sections of the whole cell are made. One study, carried out by F. L. Renaud and Hewson Swift in the water mold *Allomyces,* indicates that procentrioles may indeed be maintained in somatic cells. In *Allomyces,* flagellated reproductive cells appear at one stage in the life cycle. The vegetative cells of this organism do not contain mature centrioles. However, Renaud and Swift noted that procentrioles were present in vegetative cells, carried in a depression of the nuclear envelope. These are apparently nonfunctional except in cells differentiating into motile gametes. Whether or not the system in *Allomyces* exists in other fungi and plants remains to be seen.

At the very least, these observations in plants require a revision of Lwoff's hypothesis that centrioles arise only by replication of previously existing mature centrioles. Other observations in protozoa indicate that not even procentrioles are required for the appearance of centrioles. The most significant of these has been made in the ameboflagellates, a group in which alternation between ameboid and flagellated stages takes place. In one genus, *Tetramitus,* studied by Darryll E. Outka and Byron C. Kluss, the two stages are very different in morphology. The ameboid form has no flagella and is a relatively undifferentiated cell (Fig. 16–31a). The flagellated stage (Figs. 16–31b; 16–32) is

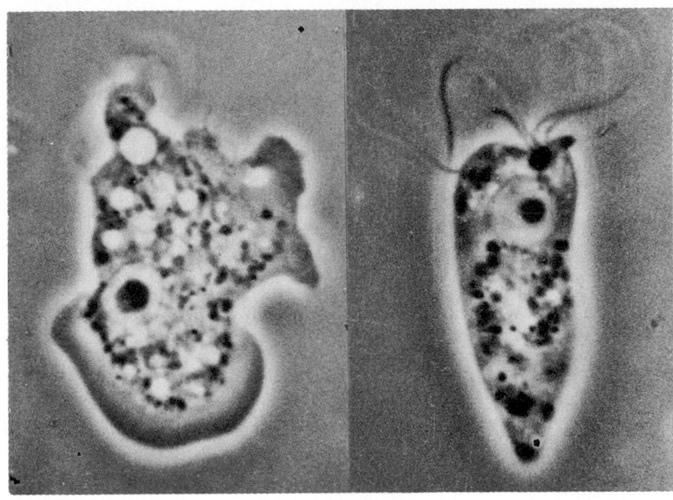

16–31. *The amoeboflagellate protozoan* Tetramitus. a, *ameboid form.* b, *flagellate form.* ×1900. *Courtesy of D. E. Outka, B. C. Kluss, and The Rockefeller University Press, from* J. Cell Biol. *35:323 (1967).*

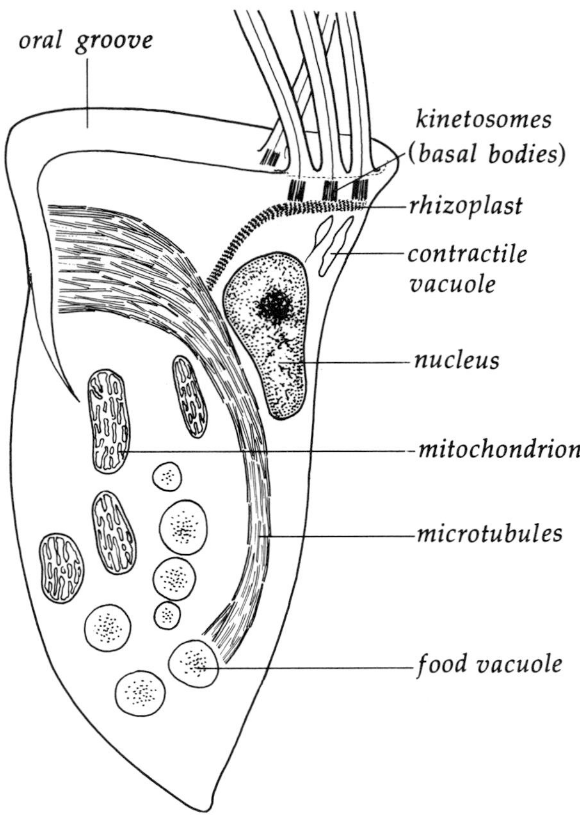

16–32. *Diagram of the flagellated form of* Tetramitus. *Courtesy of D. E. Outka, B. C. Kluss, and The Rockefeller University Press, from* J. Cell Biol. *35:323 (1967).*

highly complex, including four basal bodies and flagella.

Outka and Kluss were unable to detect centrioles or procentrioles in serial sections of the ameboid form. By adjusting the culture medium, amebae can be induced to differentiate into the flagellate form. During the transformation, which takes about two hours, centrioles appear at the anterior end of the cell near the nucleus and form the basal bodies for developing flagella. In electron micrographs of *Tetramitus* at this stage a process of the nucleus can be seen to extend into the region occupied by the developing centrioles (Fig. 16–33). The centrioles are apparently assembled directly from microtubules, with no previous development of a procentriole. When the assembly is completed, the centriole is identical in structure to the centrioles and basal bodies of other cells (Fig. 16–34). Once transformed into the flagellar form, *Tetramitus* may continue to divide without reversion to the ameboid form. During division the new basal bodies continue to be assembled *de novo*, without the usual procentriolar bud extending from the base of the old centriole. A similar system has been described by others in the ameboflagellate *Naegleria*. While the existence of a procentriole or possibly less organized precursor cannot be completely ruled out, this study makes it seem likely that centrioles can very possibly be assembled entirely *de novo*, without the involvement of either mature centrioles or procentrioles.

DNA in the Centriole

The possibility that centrioles may be autonomous has encouraged a search for DNA in this organelle. To date, these efforts have been inconclusive. Part of this work has been done by isolation of large quantities of basal bodies from protozoans such as the ciliate *Tetrahymena*. Usually, the cells are broken up and the membranes dissolved by exposure to a detergent. Purification of centrioles, released by dissolution of the membranes, is then accomplished by centrifugation. Analysis of these preparations reveals the presence of 0 to 3 percent or more DNA. On the basis of experiments of this type, the quantity of DNA per centriole

16–33. A longitudinal section through the flagellated stage of Tetramitus, *showing the developing basal bodies* (B). *A process of the nucleus (arrow) extends toward the basal bodies.* ×*34,500. Courtesy of D. E. Outka, B. C. Kluss, and The Rockefeller University Press, from* J. Cell Biol. *35:323 (1967).*

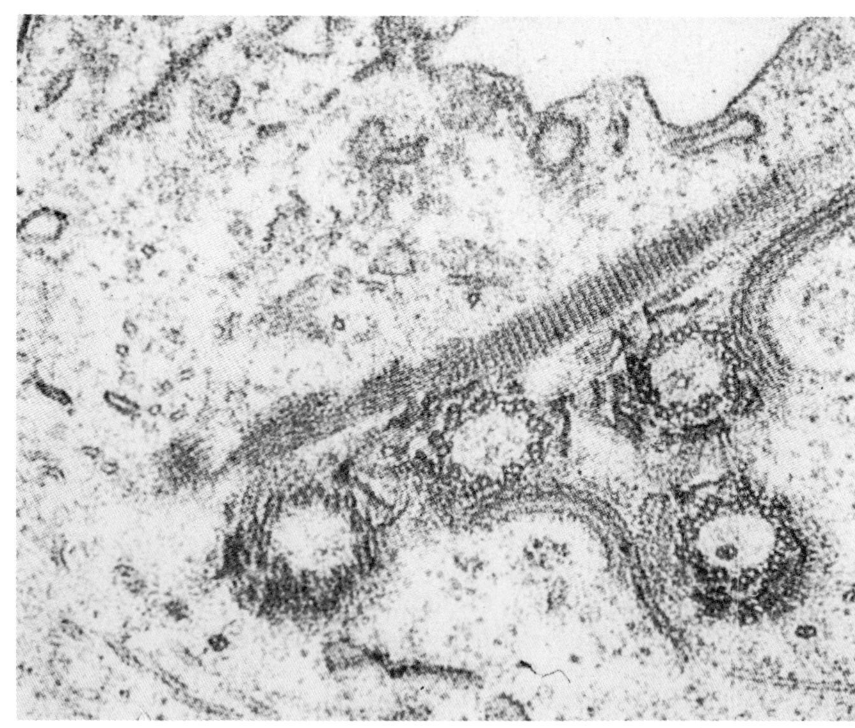

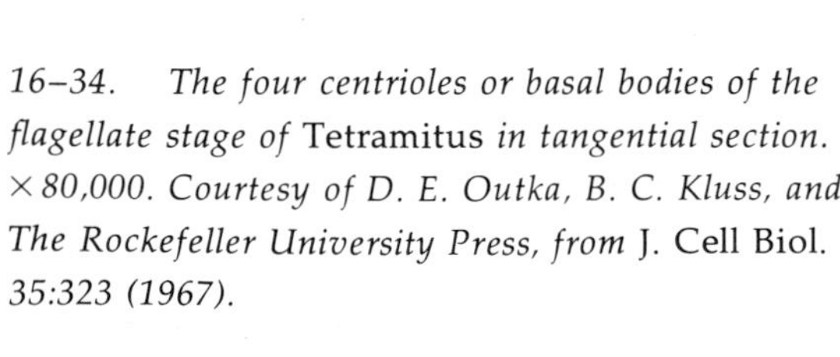

16–34. The four centrioles or basal bodies of the flagellate stage of Tetramitus *in tangential section.* ×*80,000. Courtesy of D. E. Outka, B. C. Kluss, and The Rockefeller University Press, from* J. Cell Biol. *35:323 (1967).*

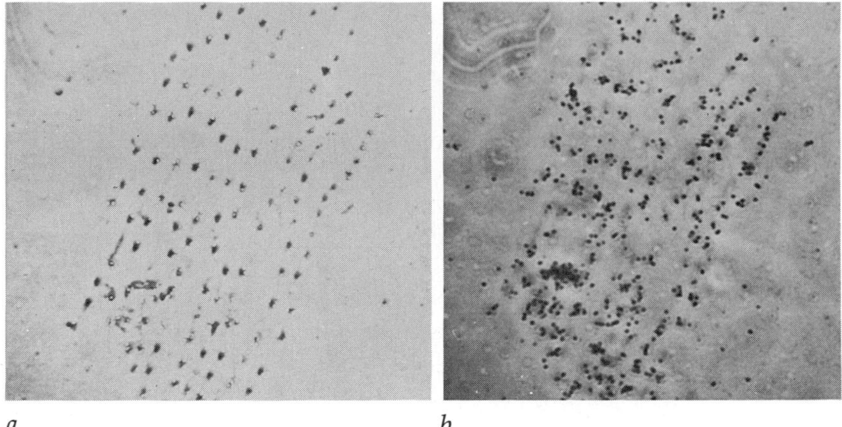

a b

16–35. *Fragment of a pellicle from* Paramecium *after growth of the organism in tritiated thymidine.* a, *Focused on the level of the basal bodies of the fragment.* b, *Focused on the grains exposed in the photographic emulsion by incorporated radioactive precursor. The location of the grains corresponds to the position occupied by the basal bodies (centrioles). ×1000. Courtesy of J. Smith-Sonneborn and W. Plaut.*

has been estimated to approach 1×10^{-16} to 1×10^{-17} grams. At this level, the DNA concentration in the centriole approaches the quantities found in mitochondria and chloroplasts. In spite of this, no fibrils suggestive of DNA have been observed in sections of centrioles. In experiments following the uptake of radioactive DNA precursor by autoradiography, evidence of label is rarely seen over the centrioles except in isolated pellicles of protozoa.

If the centrioles isolated from ciliates are examined under the electron microscope, these preparations prove to contain large quantities of contaminating cellular debris, among which DNA originating from broken mitochondria or the nucleus is likely to be found. Therefore, the possibility of contamination of these preparations cannot be ruled out. One series of experiments, by Linda H. Hufnagel, supports the possible nuclear origin of DNA in the isolated pellicles. Hufnagel isolated pellicles from *Paramecium* and measured the density of DNA extracted from this preparation. The pellicular DNA was found to coincide with the density of nuclear DNA from the same organism. Hufnagel also found that the isolated pellicles have a strong affinity for DNA added to the isolating medium, indicating that the DNA in the centriole preparations is possibly nuclear in origin.

Very different conclusions were drawn in independent experiments carried out by J. Randall and Claire Disbrey, and Joan Smith-Sonneborn and W. Plaut. In both experiments, pellicles were isolated from ciliate protozoans and stained with acridine orange. A bright yellow-green fluorescence, characteristic of the acridine orange-DNA complex, was noted at points corresponding to the rows of basal bodies. Pellicles from cells grown in medium containing tritiated thymidine showed a pattern of label at sites corresponding to the rows of basal bodies below the surface (Fig. 16–35). In contrast to Hufnagel's findings, Smith-Sonneborn and Plaut could not detect a significant binding affinity in the isolated pellicles for extraneous DNA.

From these contradictory experiments it is impossible to be certain that the centriole contains DNA. Although cytological observations suggest that, in the majority of animals, centrioles arise only from preexisting centrioles, exceptions, in which centrioles appear to arise *de novo*, can be found in both plants and animals.

No genetic evidence exists to support cytoplasmic inheritance in the centriole. Although mutations in flagellar structure have been described (14 of these are known in *Chlamydomonas*), these have proved to follow ordinary Mendelian patterns of inheritance and thus are controlled by nuclear genes. Therefore, in contrast to chloroplasts and mitochondria, the evidence at the present time does not clearly support an independent cytoplasmic line of inheritance in centrioles.

The Possible Evolutionary Origins of Mitochondria and Chloroplasts

Many parallels exist between the structure and biochemistry of chloroplasts and mitochondria and of prokaryotes such as the bacteria and blue-green algae. The DNA of chloroplasts, mitochondria, and prokaryotes is not complexed with detectable quantities of proteins such as the histones and is collected into aggregations without being separated from the immediate environment by membranes. The DNA of bacteria and mitochondria and at least the chloroplasts of Euglena is circular. Other parallels are found in the structure of ribosomes, which are of equivalent size in prokaryotes, mitochondria, and chloroplasts, and in the function of these ribosomes in protein synthesis. Particularly striking in the latter instance are the close similarities in response to inhibitors of protein synthesis such as chloramphenicol and cycloheximide. In addition to these biochemical relationships, chloroplasts and blue-green algae show striking analogies in structure. In both cases, the photosynthetic apparatus is bound to thylakoids; the suspension of these structures in the chloroplast stroma is much like the arrangement of the photosynthetic membranes found in the blue-green algae.

These similarities have led to the proposition that mitochondria and chloroplasts have their evolutionary origin in ancient prokaryotes that became established as symbionts in the cytoplasm of ancestral eukaryotes. This is by no means a new hypothesis in cell biology. The biologist C. Mereschkowski proposed in 1905 that chloroplasts represent an adapted symbiont; other investigators advanced the same idea during these early years. With recent discoveries of the molecular parallels between mitochondria, chloroplasts, and prokaryotes, the adapted symbiont idea has been resurrected and has gained wide acceptance.

The symbionts giving rise to chloroplasts and mitochondria may have found their way into the cytoplasm simply by being devoured by ancestral eukaryotes. Although this may seem to be an unlikely route of entry, many parallels can be found in modern organisms for the origin of symbiosis through this route (see below). Because mitochondria but not chloroplasts are found in all eukaryotic cells, it seems most likely that the ingestion of the symbiont giving rise to mitochondria occurred first. Later, a branch of the primitive eukaryotes already containing mitochondria ingested a photosynthesizing prokaryote, giving rise to the subgroup ancestral to the plants.

The possible steps in the origin of these organelles have been outlined by Lynn Margulis as a part of an elegant general hypothesis of eukaryote evolution. Along with many other scientists, Margulis proposes that the primitive atmosphere was probably composed of hydrogen, water vapor, ammonia, and methane or CO_2 as a carbon source. Oxygen was presumably absent from this mixture. In the primitive environment, about 3000–5000 million years ago, the first cellular organisms arose. The rigors of the primitive environment are considered by Margulis to have favored the evolution of two major types of single-celled organisms: primitive photosynthesizers, using the energy of light to synthesize ATP and fix carbon dioxide, and primitive nonphotosynthetic anaerobes, existing by the breakdown of reduced substances. The latter probably used final electron acceptors of intermediate oxidation-reduction potential, such as sulfate or nitrate. Both primitive groups of cells probably resembled contemporary prokaryotes in structure.

As the photosynthesizers increased in number, oxygen began to appear in the atmosphere, given off as a byproduct of the photosynthetic reactions. Eventually, as oxygen became more available in the environment, a group of primitive anaerobes adapted to the use of this substance as a final electron acceptor. This conversion to aerobic respiration in some prokaryotes and the development of heterotrophs feeding on other cells as a method of survival set the stage for the appearance of the mitochondrion. This is considered by Margulis to have happened when an anaerobic heterotroph, limited in the efficiency of its respiration to the use of an intermediate electron acceptor, engulfed a primitive aerobe. Because of the greater efficiency in respiration provided to the heterotroph, the ingested aerobe survived in the cytoplasm as a symbiont. As the relationship became established, many duplicate functions of the symbiont were taken over by the host cell and the symbiosis became obligate. At the same time, the host anaerobe became dependent for survival on the respiratory capacity of the ingested aerobe. This gradual process of mutual adaptation culminated in the transformation of the cytoplasmic aerobe into the mitochondrion.

This new cell type, a heterotroph containing a symbiotic aerobe in its cytoplasm, gave rise to the eukaryotic line. Other features of the eukaryotic cell, including separation of the nucleus from the cytoplasm by the nuclear envelope, evolution of more complex structure in the chromosomes, and appearance of the mitotic apparatus for distributing the chromosomes to daughter cells during division, arose at about the same time. Some time after the progress of the eukaryote ancestors to this level, a second major symbiotic relationship became established in a eukaryote subgroup as a result of the ingestion of a photosynthetic prokaryote. As this symbiosis became obligate, the photosynthetic prokaryote gradually evolved into the chloroplast, founding the eukaryote line that led to modern plant cells.

Contemporary organisms offer some striking examples of symbiotic relationships closely equivalent

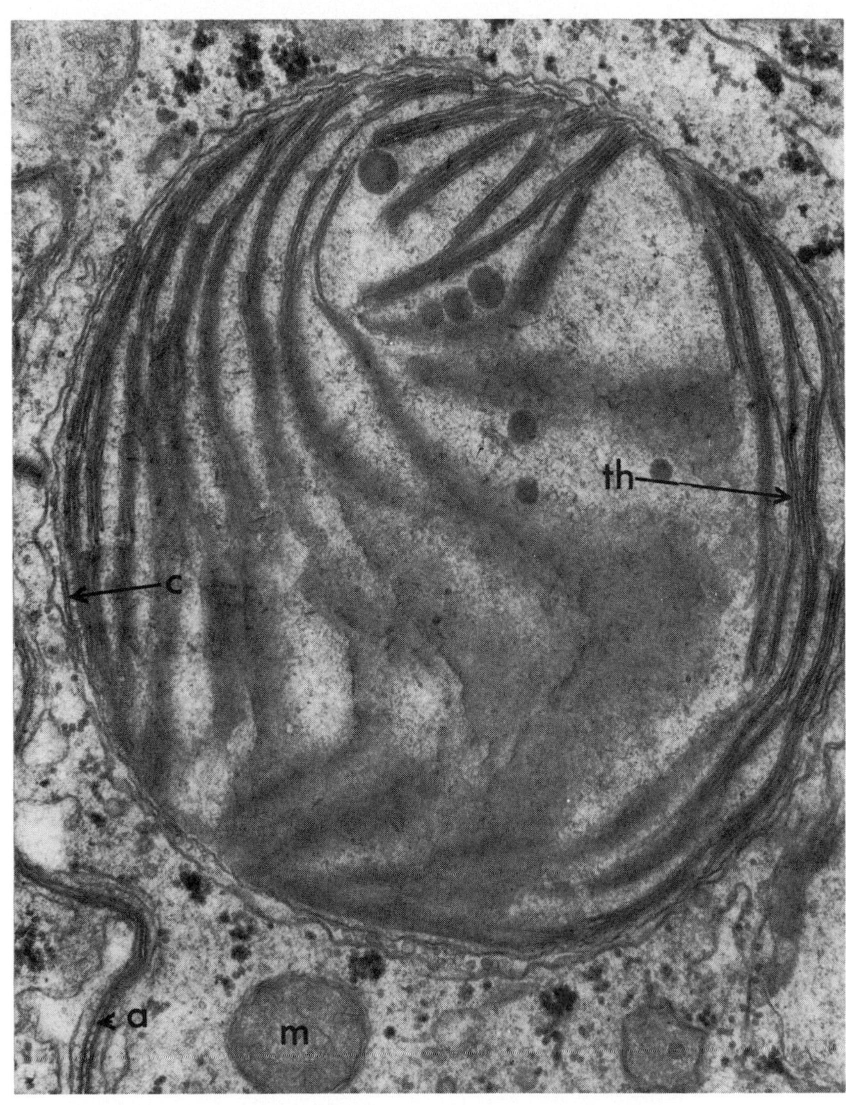

16–36. A chloroplast in the cytoplasm of a cell in the digestive gland of Elysia viridis. c, *chloroplast boundary membranes;* th, *thylakoids;* m, *mitochondrion;* a, *cell junction.* ×*40,000. Courtesy of R. K. Trench, R. W. Greene, B. G. Bystrom, and The Rockefeller University Press.* [J. Cell Biol. *42:404 (1969)*]

to the ancestral mitochondrial and chloroplast associations proposed by Margulis. Peter H. Raven, another scientist who has given consideration to the possible symbiotic origin of mitochondria and chloroplasts, has listed more than 150 genera of invertebrates (comprising eight phyla) containing different algae as photosynthetic symbionts. Blue-green algae, which occur as symbionts in amoebae, flagellated protozoa, and diatoms, can in some cases hardly be distinguished from chloroplasts. In fact, some argument exists among taxonomists as to whether some of these organisms are animals harboring a photosynthetic symbiont or true plant cells containing chloroplasts. For example, the cryptomonad *Cyanophora paradoxa* is a flagellated, protozoan-like unicellular organism with a proteinaceous cell coat and a contractile vacuole. Within the cytoplasm of this species are chloroplast-like bodies that are considered by some authorities to be symbiotic blue-green algae and by others typical eukaryote chloroplasts.

These examples can be used to show that symbiotic relationships of the type proposed by Sagan actually occur in nature. A further example demonstrating that symbiosis can originate through ingestion and cytoplasmic adaptation of an organism or organelle is provided by certain marine gastropods. In the molluscs *Tridachia, Elysia,* and others, chloroplasts are present in the cytoplasm of cells lining the cavities of the digestive system (Fig. 16–36). These chloroplasts can be shown by the use of labeled carbon dioxide to be active in photosynthesis. The labeled carbohydrates and other compounds are found to diffuse into the other tissues of the mollusc. These chloroplasts appear to originate from algae eaten by the gastropods. The larvae of these animals are free of chloroplasts until an intermediate feeding stage (the *veliger*) develops. Presumably, the chloroplasts of algae ingested at this time are absorbed into the cytoplasm of the cells lining the gut and become established as organelle symbionts. Robert K. Trench and his associates, who described the symbiotic relationship in *Tridachia* and followed the fixation and distribution of radioactive carbon, found that the chloroplasts in the gut cells were maintained in functional state in animals kept away from a fresh supply for six weeks. Therefore, the chloroplasts survive in the cytoplasm for at least this length of time. This example of the development of symbiosis between a heterotroph and an ingested photosynthetic organelle is obviously equivalent to the mechanisms proposed by Margulis for the evolutionary origin of mitochondria and chloroplasts.

Other parts of the scheme find contemporary parallels in parasitism. Parasites are often greatly simplified organisms, and lack many systems required for a free existence. The gradual reduction of complexity in parasites is considered to occur through mutations that eliminate many of the functions duplicated and carried out by the host organism until the parasitic existence becomes obligatory. In the same way, the mutually advantageous association between the ancient heterotrophs and the engulfed aerobes and photosynthesizers became obligatory through mutations eliminating functions duplicated in either member of the symbiotic relationship.

Suggestions for Further Reading

Books

Ephrussi, B., 1953. *Nucleo-Cytoplasmic Relations in Microorganisms.* Clarendon Press, Oxford.

Jinks, J., 1964. *Extrachromosomal Inheritance.* Prentice-Hall, Englewood Cliffs, New Jersey.

Kirk, J. T. O., and R. A. E. Tilney-Bassett, 1967. *The Plastids.* W. H. Freeman, San Francisco.

Margulis, Lynn, 1970. *Origin of Eukaryotic Cells.* Yale University Press, New Haven, Connecticut.

Roodyn, D. B., and D. Wilkie, 1968. *The Biogenesis of Mitochondria.* Methuen and Company, London.

Papers and Review Articles

Akiyama, M., and T. Sakagami, 1969. Exchange of mitochondrial lecithin and cephalin with those in rat liver microsomes. *Biochimica et Biophysica Acta,* vol. 187, pp. 105–112.

Attardi, G., Y. Aloni, B. Attardi, D. Ojala, L. Pica-Mattoccia, D. L. Robberson, and B. Storrie, 1970. Transcription of mitochondrial DNA in HeLa cells. *Cold Spring Harbor Symposia on Quantitative Biology,* vol. 35, pp. 599–619.

Beale, G. H., 1969. The role of the cytoplasm in heredity. In E. E. Bittar and N. Bittar, eds., *Biological Basis of Medicine,* vol. IV. Academic Press, London, pp. 81–107.

Beisson, J., and T. M. Sonneborn, 1965. Cytoplasmic inheritance of the organization of the cell cortex in *Paramecium aurelia. Proceedings of the National Academy of Sciences,* vol. 53, pp. 275–282.

Borst, P., and A. M. Kroon, 1969. Mitochondrial DNA: physiochemical properties, replication, and genetic function. *International Review of Cytology,* vol. 26, pp. 107–190.

Burton, P. R., and D. G. Dusanic, 1968. Fine structure and replication of the kinetoplast of *Trypanosoma lewesii. Journal of Cell Biology,* vol. 39, pp. 318–331.

Dawid, I. B., 1966. Evidence for the mitochondrial origin of frog egg cytoplasmic DNA. *Proceedings of the National Academy of Sciences,* vol. 56, pp. 269–276.

DeHarven, E., 1968. The centriole and the mitotic spindle. In A. J. Dalton and F. Haguenau, eds., *Ultrastructure in Biological Systems,* vol. 3, *The Nucleus.* Academic Press, New York, pp. 197–227.

Dubin, D. T., and R. E. Brown, 1967. A novel ribosomal RNA in hamster cell mitochondria. *Biochimica et Biophysica Acta,* vol. 145, pp. 538–540.

Granick, S., and A. Gibor, 1967. The DNA of chloroplasts, mitochondria and centrioles. *Progress in Nucleic Acid Research and Molecular Biology*, vol. 6, pp. 143–186.

Green, B. R., 1970. *Acetabularia* chloroplast DNA: electron microscopic visualization. *Science*, vol. 168, pp. 981–982.

Hufnagel, L. A., 1966. Fine structure and DNA of pellicles isolated from *Paramecium aurelia*. In *Proceedings of the 6th International Congress for Electron Microscopy*, Kyoto, Japan, pp. 239–240.

Kislev, N., H. Swift, and L. Bogorad, 1965. Nucleic acids of chloroplasts and mitochondria in swiss chard. *Journal of Cell Biology*, vol. 25, pp. 327–344.

Kroon, A. M., 1969. DNA and RNA from mitochondria and chloroplasts (biochemistry). In A. Lima-de-Faria, ed., *Handbook of Molecular Cytology*. North-Holland Publishing Company, Amsterdam, pp. 943–971.

Lefort-Tran, M., 1969. Cytochemistry and ultrastructure of chloroplasts. In A. Lima-de-Faria, ed., *Handbook of Molecular Cytology*. North-Holland Publishing Company, Amsterdam, pp. 1047–1076.

Manning, J. E., D. R. Wolstenholme, R. S. Ryan, J. A. Hunter, and O. C. Richards, 1971. Circular chloroplast DNA from *Euglena gracilis*. *Proceedings of the National Academy of Sciences*, vol. 68, pp. 1169–1173.

Manton, I., and M. Parke, 1960. Further observations on small green flagellates, with special reference to possible relatives of *Chromulina pusilla*. *Journal of the Marine Biology Association*, vol. 39, pp. 275–298.

Mitchell, M. B., and H. K. Mitchell, 1956. A nuclear gene suppressor of a cytoplasmically inherited character in *Neurospora crassa*. *Journal of General Microbiology*, vol. 14, pp. 84–89.

Nass, M. M. K., 1969. Mitochondrial DNA: advances, problems, and goals. *Science*, vol. 165, pp. 25–35.

Outka, D. E., and B. C. Kluss, 1967. The ameba-to-flagellate transformation in *Tetramitus rostratus*. *Journal of Cell Biology*, vol. 35, pp. 323–346.

Pitelka, D. R., 1969. Centriole replication. In A. Lima-de-Faria, ed., *Handbook of Molecular Cytology*. North-Holland Publishing Company, Amsterdam, pp. 1199–1218.

Raven, P. H., 1970. A multiple origin for plastids and mitochondria. *Science*, vol. 169, pp. 641–646.

Reich, E., and D. J. L. Luck, 1966. Replication and inheritance of mitochondrial DNA. *Proceedings of the National Academy of Sciences*, vol. 55, pp. 1600–1608.

Renaud, F. L., and H. Swift, 1964. The development of basal bodies and flagella in *Allomyces arbusculus*. *Journal of Cell Biology*, vol. 23, pp. 339–354.

Ris, H., 1962. Ultrastructure of certain self-dependent cytoplasmic organelles. *Proceedings of the 5th International Conference for Electron Microscopy*, Philadelphia, p. xx-1.

Ris, H., and W. Plaut, 1962. Ultrastructure of DNA-containing areas in the chloroplast of *Chlamydomonas*. *Journal of Cell Biology*, vol. 13, pp. 383–391.

Schatz, G., 1970. Biogenesis of mitochondria. In E. Racker, ed., *Membranes of Mitochondria and Chloroplasts*. American Chemical Society Monograph 165, pp. 251–314.

Schiff, J. A., and H. T. Epstein, 1965. The continuity of the chloroplast in *Euglena*. In M. Locke, ed., *Proceedings of the 24th Symposium of the Society for Developmental Biology*. Academic Press, New York, pp. 131–189.

Scott, N. S., V. C. Shah, and R. M. Smillie, 1968. Synthesis of chloroplast DNA in isolated chloroplasts. *Journal of Cell Biology*, vol. 38, pp. 151–157.

Shapiro, L., L. I. Grossman, J. Marmur, and A. K. Kleinschmidt, 1968. Physical studies on the structure of yeast mitochondrial DNA. *Journal of Molecular Biology*, vol. 33, pp. 907–922.

Shumway, L. K., and T. E. Weier, 1967. The chloroplast structure of IOJAP maize. *American Journal of Botany*, vol. 54, pp. 773–780.

Simpson, L., 1968. Effect of acriflavin on the kinetoplast of *Leishmania tarentolae:* mode of action and physiological correlates in the loss of kinetoplast DNA. *Journal of Cell Biology*, vol. 37, pp. 660–682.

Smillie, R. M., and N. S. Scott, 1969. Organelle biosynthesis: the chloroplast. *Progress in Molecular and Subcellular Biology*, vol. 1, pp. 136–202.

Smith-Sonneborn, J., and W. Plaut, 1967. Evidence for the presence of DNA in the pellicle of *Paramecium*. *Journal of Cell Science*, vol. 2, pp. 225–234.

Spencer, D., 1965. Protein synthesis by isolated spinach chloroplasts. *Archives of Biochemistry and Biophysics*, vol. 111, pp. 381–390.

Spencer, D., and P. R. Whitfeld, 1967a. DNA synthesis in isolated chloroplasts. *Biochemical and Biophysical Research Communications*, vol. 28, pp. 538–542.

Spencer, D., and P. R. Whitfeld, 1967b. Ribonucleic acid synthesizing ability of spinach chloroplasts and nuclei. *Archives of Biochemistry and Biophysics*, vol. 121, pp. 336–345.

Steinert, M., 1960. Mitochondria associated with the kinetonucleus of *Trypanosoma mega*. *Journal of Biophysical and Biochemical Cytology*, vol. 8, pp. 542–546.

Swift, H., 1965. Nucleic acids of mitochondria and chloroplasts. *American Naturalist*, vol. 99, pp. 201–227.

Swift, Hewson, and D. R. Wolstenholme, 1969. Mitochondria and chloroplasts: nucleic acids and the problem of biogenesis (genetics and biology). In A. Lima-de-Faria, ed., *Handbook of Molecular Cytology*. North-Holland Publishing Company, Amsterdam, pp. 972–1046.

Trench, R. K., R. W. Greene, and B. G. Bystrom, 1969. Chloroplasts as functional organelles in animal tissue. *Journal of Cell Biology*, vol. 42, pp. 404–417.

Wintersberger, E., 1966. Occurrence of a DNA-polymerase in isolated yeast mitochondria. *Biochemical and Biophysical Research Communications*, vol. 25, pp. 1–7.

Wolstenholme, D. R., and N. J. Gross, 1968. The form and size of mitochondrial DNA of the red bean *Phaseolus vulgaris*. *Proceedings of the National Academy of Sciences*, vol. 61, pp. 245–252.

Wright, R. E., and J. Lederberg, 1957. Extranuclear transmission in yeast heterokaryons. *Proceedings of the National Academy of Sciences*, vol. 43, pp. 919–923.

17 The Origin of Cellular Life

No recognizable intermediates exist today between inanimate matter and cellular life. It is therefore impossible to trace out the steps in the evolution of living from nonliving matter in terms of contemporary organisms or suborganisms. Although virus particles at first seem to stand as an exception to these statements, these structures clearly represent a degenerate form of once-cellular life that has become parasitic and dependent on more highly organized cells for existence.

Thus anyone interested in the origin of life is provided only with cells or their products as living forms for investigation. But even the simplest cells are highly organized and contain complex molecular systems that are almost completely interdependent. This complexity and interdependence makes it difficult to conceive of mechanisms that could account for their spontaneous appearance from nonliving matter or to imagine how one system could have appeared in advance of another in the course of evolution. In spite of these difficulties, a number of working hypotheses for the origin of life have been advanced and even subjected to limited experimental tests.

These hypotheses fall into several groups according to the sequence of stages by which cellular life is presumed to have first appeared. The first stage in this sequence was the formation of the earth and the primitive atmosphere. In the second stage, quantities of "organic" molecules were abiogenically synthesized through the action of inanimate forces on the primordial earth and atmosphere. The group of hypotheses dealing with this stage of evolution has been successfully tested by experimentation. In the third stage, these organic molecules became concentrated into aggregates or clusters capable of chemical interaction with the environment. These aggregates presumably used as an energy source either light or the supply of organic molecules produced in the environment at earlier stages. As these clusters of molecules formed, some were more successful than others in using the available energy sources and became predominant types. During this process, the quality of life appeared in these aggregates, including the capacity for growth and eventually self-replication. In this fourth stage in the origin of life, evolution was gradually transformed from chemical to organic. In the final stage, natural selection led to the appearance of adaptations producing the cellular level of organization. These adaptations would include, as a minimum, a limiting membrane, a chromosome, in which information required for synthesis and self-replication is stored, and a translation system capable of using the information of the chromosome to produce required molecules not available in the environment. Also present would be molecular systems that convert the chemical energy of absorbed metabolites or light energy into a form useful in the cell.

These stages and some of the hypotheses advanced to account for them will be discussed in this chapter. Before entering into this discussion, however, it is useful to consider what is meant by "life" in terms of the origin of cells. It is not difficult to define the quality of life if the characteristics of present-day cells are used as a guide: a collection of molecules is alive if it has the capacity for growth, reproduction, and movement, and can respond to outside stimuli by compensatory changes in internal processes. But if the intermediate, primordial living systems are considered, a definition of life is much more difficult. Presumably, the transformation from nonliving to living was gradual and no sudden event took place that resulted in an instantaneous conversion into the living state, with simultaneous appearance of all of the characteristics of cellular life. Therefore, any attempt to define a point of transformation from nonliving to living in this gradual transformation must of necessity be arbitrary. However, several definitions of primordial life have been attempted, and these are useful as starting points for the discussion of the origins of life.

Life has been defined in terms of systems ranging from simple to relatively complex. K. M. Madison has proposed a simple definition in which "life [is] a group of chemical systems in which free energy . . . released as a part of the reactions of one or more of the systems . . . is used in the reactions of one or more of the remaining systems." According to this definition the quality of life requires only the coupling of exergonic to endergonic reactions. Life at this level of complexity could exist without the involvement of either organic molecules or organic catalysts. A more complex set of conditions is imposed by J. Perret, who holds that "life is a potentially self-perpetuating open system of linked organic reactions, catalyzed stepwise and almost isothermically by complex and specific organic catalysts, which are themselves produced by the system." This definition presupposes a relatively advanced state, because "organic catalysts," or enzymes, are included. Because these are internally produced by the system, a mechanism for the storage and translation of information for their synthesis would also be required.

The definition to be used in this chapter leans toward the level of complexity assumed by the latter interpretation and uses as a basis the transition from chemical to organic evolution, which is presumed to have taken place during an intermediate stage in the appearance of cellular life. For the purpose of this chapter, a system is considered to have the quality of life if it is (1) capable of synthesis; (2) contains an information system that directs this synthesis toward an increase in mass of the system; and (3) is mutable. The inclusion of mutability as a minimum requirement is related to the transition from chemical to organic evolution. The presence of an information system capable of directing synthesis implies that an increase in mass would occur through the addition of *like* molecules to the system. But for organic, as opposed to chemical, evolution to occur, stable variations that are recorded in the information system (mutations) must appear and be presented to the environment. Without this mechanism natural selection and adaptation cannot occur.

Therefore, according to this admittedly arbitrary definition, a collection of molecules would be considered alive if it is capable of using an energy source to undergo controlled growth and is mutable. Living organization of this complexity represents an intermediate level between the first spontaneous aggregations of complex, interacting organic molecules and the cell. With this definition, we will now consider the hypotheses of the origin of life at this level of complexity.

The Primordial Earth

Opinions differ as to the probable constituents of the atmosphere and surface of the earth at the time of its origin. Hypotheses dealing with this question fall into two general groups, depending on whether the earth is considered to have a fiery origin or to have solidified through condensation of clouds of cosmic dust. The fiery origin theory, lately fallen from favor, states that the matter of the planets was originally pulled from the sun by a closely passing star. These globules of hot gases then gradually cooled and condensed into the planets. This hypothesis, first proposed by Sir James Jeans in England, has been replaced in recent years by the theory that the solar system and planets condensed from interplanetary dust. Originally proposed by Emmanuel Kant, this "agglomerative" theory has been disinterred and brought up to date by O. J. Schmidt, C. F. Z. von Weizsacker, and H. C. Urey. In its contemporary form, the theory states that stars and planetary systems are continuously forming from and disintegrating into clouds of cosmic gas and dust particles. Most of the rarified substance of the cosmic clouds consists of hydrogen gas; helium and neon are also found in small quantities in the clouds. Because of the temperatures in interstellar space, which approach absolute zero, all other substances are present as solid particles. These include water, methane, and ammonia in the frozen state, and particles of other elements and compounds including metallic iron and nickel and silicates, oxides, sulfides, and carbides of these and other metals. Gas and dust clouds of this constitution are still present in space and can be detected and analyzed by such methods as rocket probes and spectroscopy.

The agglomerative theory considers that our solar system condensed from a large cloud of this type. Rapid condensation of most of the cloud led to internal pressures and heat that initiated the conversion of matter to energy; this thermonuclear release created the sun. The rest of the condensed cloud became suspended as discs of material around the sun, extended outward into space. These discs, analogous to the rings presently visible around the planet Saturn, gradually condensed into the planets. As these bodies formed, internal temperatures rose, due to solar heating, gravitation, and the effects of increasing internal pressures. This heating led to the melting and stratification of the solid substances of the condensing dust rings. Heating due to gravitational energy is considered by some authorities to have generated internal temperatures as high as 1000–3000°C at initial stages of formation of the earth, with resulting melting and stratification of metallic components in the core. The lighter compounds, such as water, silicates, and oxides, would tend to rise to the surface of the condensing sphere.

According to the agglomerative theory, the earth cooled soon after its formation to temperatures approaching modern conditions. At the surface of the cooled earth were particles and rocks of various sizes, including silicates and oxides of the various metals. Water, probably initially present as steam, eventually condensed into the primordial lakes and seas. Some idea of the variety of compounds likely to be present at the cooled surface has been gained by analysis of meteorites. The results of one analysis of this type are shown in Table 17–1.

The Primordial Atmosphere

Water vapor and hydrogen are expected to have been present as components of the primary atmosphere surrounding the earth. At the surface, reaction of water with the metallic carbides, nitrides, and sulfides of the crust would produce large quantities of the gases methane and ammonia and smaller quantities of hydrogen sulfide.

Whether the primitive atmosphere developed any significant concentrations of carbon dioxide is a matter of some debate. One group of theorists holds that any carbon dioxide present would quickly be converted to carbonates and form a part of the solid crust of the earth. Others maintain that significant quantities would always be present, arising as a part of the gases expelled from erupting volcanoes.

Whether carbon dioxide was present does not affect the possibility of spontaneous generation of organic compounds because it appears that these can be formed as long as the primitive atmosphere contained a carbon source in the form of methane or a closely related gas. The presence of large quantities of water, methane, and ammonia gave the primordial atmosphere a strongly "reducing" character; any oxygen present is believed to have been quickly bound into oxides on surface rocks and particles. Confirmation of this hypothesis of the primordial atmosphere is obtained by analyzing the atmospheres of other planets of the solar system by rocket probes and spectroscopy. These analyses indicate the lack of oxygen and the presence of methane, ethylene, or acetylene on Venus, Mars, and Jupiter, and ammonia in quantity on at least Venus.

Thus the primitive earth, with a crust containing large quantities of the oxides, silicates, sulfates, and so forth of various metals is believed to have been surrounded by a reducing atmosphere containing water vapor, ammonia, and a carbon source such as methane or carbon dioxide. Liquid surface water was also present in rivers, streams, lakes, and oceans, all at temperatures approaching present-day levels. The effects on this primordial earth of natural sources of energy, including heat, visible and ultraviolet radiation from the sun, and electrical discharge during storms are believed to have induced the spontaneous, abiogenic synthesis of complex organic molecules from these simple precursors.

Table 17–1 Composition of the Orgueil Meteorite

Component	Percent composition	
	Not dried	Dried at 110°C
SiO_2	24.475	26.0310
Fe_2O_3	13.324	14.2360
FeO	17.924	19.0630
MgO	8.163	8.6711
S	4.369	4.6466
SO_4	2.195	2.3345
NiO	2.450	2.6057
CaO	2.183	2.3220
Al_2O_3	1.175	1.2498
Na_2O	1.244	1.3230
K_2O	0.307	0.3265
MnO	1.815	1.9302
Cr_2O_3	0.225	0.2392
CoO	0.085	0.0904
Cl	0.073	0.0776
P	trace	trace
NH_3	0.098	0.1042
H_2O	5.975	—
H_2O^{+}*	7.345	7.8120
Carbon compounds	6.027	6.4100
Total	96.442	99.4728

*H_2O^+ designates water obtained above 110°C.

From B. Nagy, W. G. Meinschein, and D. J. Hennessy, Mass spectroscopic analysis of the Orgueil meteorite: evidence for biogenic hydrocarbons. *Ann. N. Y. Acad. Sci.* 93: 1961, pp. 27–35.

The Spontaneous Appearance of Organic Molecules on the Primordial Earth

In 1924, the Russian biochemist A. I. Oparin published a short booklet outlining his hypotheses of the origins of life on earth. Essentially the same ideas were independently published a few years later by the English scientist J. B. S. Haldane. Although Oparin and Haldane reasoned from the premise that the primordial earth had a fiery origin, both nevertheless arrived at the conclusion that the primitive atmosphere was reducing and contained quantities of water, ammonia, and a hydrocarbon gas such as methane. These gases were exposed to frequent electrical discharges during the violent storms that are presumed to have swept over the primitive earth at this time. Through the input of energy arising from the bolts of lightning, interaction between the various gases produced a variety of organic molecules that were carried down with the rain and gradually accumulated in the waters of the ancient lakes and seas. As the concentrations of these substances increased, interactions between them and with the elements and compounds exposed at the surface of the earth led to the production of still more complex molecules and also to hydrolysis and reduction. The end result was the spontaneous production of a great variety of complex organic molecules.

Both Oparin and Haldane assumed that gaseous oxygen was absent from the primitive atmosphere. This, along with the absence of microorganisms, is of great significance, for oxidation and fermentation are the chief routes by which accumulated organic molecules are broken down in the contemporary environment. Supposedly, the abiogenically synthesized organic substances persisted for greater lengths of time in the primitive environment and became more and more concentrated in lakes and seas. According to Haldane, the ancient seas finally reached a consistency equivalent to "hot, dilute soup."

It is doubtful that these substances could actually reach high concentrations in the open seas, because most of the presumed organic reactions are reversible, and the usual equilibria between reactants and products would be expected to have operated in primitive times as well as the present. But this is not of great significance, because other processes, such as evaporation or adsorption (see below) could lead to localized regions of higher concentration. What is significant is the proposition that organic molecules could arise by

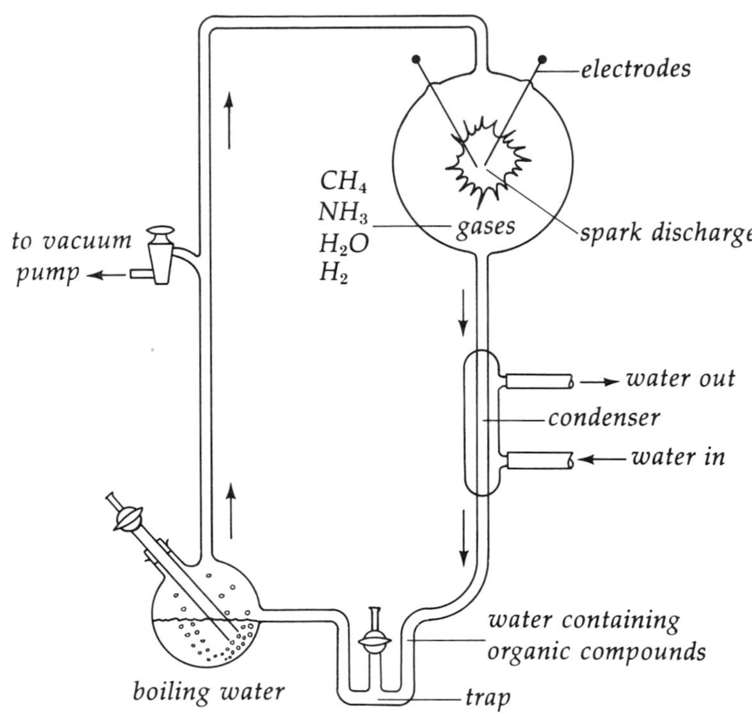

17–1. The Miller apparatus (see text). Courtesy of S. L. Miller.

the processes envisioned by Oparin and Haldane. These hypotheses have met with direct experimental support.

The Miller experiments. In 1953, S. L. Miller, working in the laboratory of H. C. Urey, devised a simple apparatus to test the effect of electrical discharges on a simulated primitive atmosphere containing a mixture of water vapor, methane, ammonia, and hydrogen. The apparatus (Fig. 17–1) was set up so that water vapor and the three gases flowed through a large chamber containing continuously sparking electrodes. Just below the sparking chamber, the connecting tubing was cooled by a water jacket to induce the condensation of any substances produced by the effects of the electrical discharge. Operation of the apparatus for one week yielded a surprising variety of organic compounds, including several amino acids and even organic polymers. Table 17–2 lists the compounds formed in a typical series of experiments.

Other experiments based on systems closely similar to the Miller apparatus but using other gases and different sources of energy have revealed that many other substances are readily formed under conditions approximating the primordial environment. For example, if hydrogen sulfide is added to the mixture of gases included by Miller, sulfur-containing compounds including thiourea, thioacetamide, and thiocyanate have been obtained. Other methods, including the effects of sparking on solutions and heating substances such as hydrogen cyanide with ammonia, have produced still other organic compounds including almost all of the naturally occurring amino acids. Significantly, irradiation of solutions of formaldehyde by ultraviolet light was found by C. Ponnamperuma and R. Mariner to yield deoxyribose and ribose sugars. A basic requirement for all of these experiments is the presence of a reducing "atmosphere"; if oxygen is included in the mixture of gases, yields of organic molecules are exceedingly low or nonexistent.

In the first series of experiments, purines and pyrimidines were not found among the spontaneously synthesized organic molecules. But in later work, the presence of these compounds was detected. Ponnamperuma and his co-workers found that subjecting hydrogen cyanide (considered by many scientists to

Table 17–2 Organic Compounds Formed in Sample Runs in the Miller Apparatus, Yield in Moles × 10^5

	Run 1	*Run 3*	*Run 6*
Glycine	63 (2.1)*	80 (0.46)*	14.2 (0.48)*
Alanine	34	9	1.0
Sarcosine	5	86	1.5
β-alanine	15	4	7.0
α-aminobutyric acid	5	1	–
Methylalanine	1	12.5	–
Aspartic acid	0.4	0.2	0.3
Glutamic acid	0.6	0.5	0.5
Iminodiacetic acid	5.5	0.3	3.9
Iminoacetic-propionic acid	0.5	–	–
Formic acid	233	149	135
Acetic acid	15.2	135	41
Propionic acid	12.6	19	22
Glycolic acid	56	28	32
Lactic acid	31	4.3	1.5
α-hydroxybutyric acid	5	1	–
Succinic acid	3.8	–	2
Urea	2	–	2
Methylurea	1.5	–	0.5
Sum of yields of compounds listed	15%	3%	8%

*Percent yield of glycine, based on carbon placed in the apparatus as methane.
Reprinted with permission from S. L. Miller, in A. I. Oparin, ed., *The Origin of Life on the Earth* (I.U.B.: Sympos. Series). New York: Pergamon Press, 1959.

be a probable component of the primordial atmosphere) at 10^{-4} M to ultraviolet radiation yielded both adenine and guanine. Other methods produced quantities of uracil, the nucleoside adenosine, and even the nucleotides AMP, ADP, and ATP (see, for example, Ponnamperuma et al., 1963; Lowe et al., 1963).

It has also been possible to induce the spontaneous production of protein-like polymers of amino acids and end-to-end linkage of nucleotides into nucleic acids. In one series of experiments, S. W. Fox and his colleagues produced protein-like polypeptide chains by heating mixtures of amino acids to 160–210°C for several hours. Other workers, including Ponnamperuma, have detected the synthesis of polypeptide chains in experiments in which the starting substances were simple ingredients expected to be present on the primitive earth. Nucleic acids have also been induced to form spontaneously in mixtures of nucleotides if high concentrations were held at 50–60°C and mixed with a "polyphosphate ester" made by dissolving phosphorous pentoxide in ether and other organic solvents (see Schramm et al, 1962; for additional literature references to the abiogenic synthesis of organic molecules see Keosian, 1964).

Thus it has been repeatedly demonstrated that the spontaneous synthesis of organic molecules, as proposed by Oparin and Haldane, is likely to have occurred under the conditions believed to be extant in the atmosphere and on the surface of the primordial earth. With the variety of compounds detected to date in experiments designed to test the Oparin-Haldane hypothesis, it is held likely that in the millions of years over which the period of abiogenic synthesis extended, all major classes of molecules of living systems would have appeared, including lipids, sugars, polysaccharides, amino acids, polypeptides, nucleotides, and nucleic acids. As noted, the absence of microorganisms and the presumed lack of oxygen in the primordial atmosphere would greatly reduce the rate of breakdown of these substances as compared to the contemporary environment. The accumulation of these compounds set the stage for the next phase in the evolution of life, the spontaneous aggregation of the abiogenically synthesized molecules into collections that took on one or more characteristics of life.

The Concentration of the Organic Molecules of the Primitive Environment into Functional Aggregates

As a preliminary to the spontaneous assembly of organic molecules into discrete aggregates, it can be assumed that some degree of initial concentration may have occurred by evaporation of water from lakes, inland seas, tidal pools, and marshlands. From this point, aggregation may have proceeded by one of several routes. At least three major hypotheses have been proposed as tentative mechanisms by which this aggregation might have occurred: Oparin's *coacervate* theory, the *proteinoid microsphere* theory of S. W. Fox, and the *adsorption hypothesis* advanced by J. D. Bernal.

Coacervates and microspheres. Oparin assumes that the abiogenic synthesis of molecules proceeded as far as the polymerization of long polypeptides of essentially random sequence. These polypeptides, with varying patterns of positive and negative charges depending on the amino acid residues included, would be expected to come together into larger aggregates if mutually attractive surface groups were exposed. This aggregation might extend to the formation of particles of colloidal size. Under the proper conditions, these colloidal particles can themselves become mutually attractive, producing still larger aggregations. A large aggregate of this type, including a combination of several smaller colloidal particles, is called a *coacervate*. If large numbers of charged groups are located at the surface of a coacervate, adjacent water molecules become arranged into a discrete film or boundary layer around the colloidal particles. The combined effects of the charged groups and the boundary layer can lead to the selective diffusion of substances from the surrounding medium into a coacervate.

A somewhat similar process of aggregation has been envisioned by Fox, who has shown that if spontaneously synthesized "proteinoids" are heated in water, small spherical particles separate from the solution on cooling. These particles, termed by Fox "microspheres", are similar to coacervates in reactivity and can selectively concentrate substances from the surrounding medium.

Bernal's adsorption hypothesis. A major difficulty with both Oparin's coacervate theory and Fox's microspheres is that abiogenic synthesis must proceed to the level of a protein-like polymer. These polypeptides must occur in relatively high concentrations for aggregation to occur by either method. But it seems doubtful that protein-like substances, even if produced in any quantity by spontaneous synthesis, could have reached the required levels of concentration except at limited regions in the primitive environment.

According to Bernal, the first spontaneous aggregation occurred instead through the adsorption of organic substances on particles of clay in the mud around tidal flats and estuaries. Once adsorbed to the surface of the mineral clays, the properties and activity of many organic molecules are significantly altered. Further, some clays consisting of metallic compounds have catalytic properties and enhance the breakdown of organic molecules adsorbed from the medium. The aggregations of organic molecules would not be suspended in the open sea, but concentrated into mud flats at the edge of seas, rivers, and lakes. These beds could reach extended size, so that the areas available for spontaneous generation of the initial properties of life would be widely distributed.

Aggregation by the Spontaneous Formation of Primitive Membranes

An additional possibility, not too frequently advanced, is that the first aggregation occurred by the spontaneous formation of films or micelles of lipid-like organic molecules. These substances, if present in any degree of concentration, will spontaneously segregate in aqueous media to form particles consisting of folded bilayers (see Chapter 4). These bilayers exhibit many properties of living membranes and selectively adsorb various substances, including proteins or polypeptides, from the surrounding medium. Substances adsorbed to the surface of lipid bilayers are significantly altered in properties; the spontaneous arrangement of organic molecules on bilayers could have catalytic effects and serve to alter the rates of breakdown or polymerization of the adsorbed molecules.

All of the listed mechanisms of aggregation have been demonstrated by experiment to be feasible, and one or more of these processes may have actually taken place in the initial stages of aggregation of molecules from the primitive environment. Supposedly, within this mixture of aggregated molecules, the first life processes were sorted out. This stage, the next step in the origin of life, presents great conceptual difficulties and to date has not been explained satisfactorily by any hypothesis.

The Origin of Life in the Primitive Aggregates

Whether the initial aggregation took place through the formation of coacervates, microspheres, or adsorption on clays or lipid bilayers, it is presumed that a group of organic molecules became arranged so that a reaction mechanism was set up. Some of these mechanisms might remove hydrogens and electrons from reduced substances with the release of free energy. These reactions in turn may have spontaneously become coupled with reaction systems requiring energy, such as the polymerization of amino acids into a random polypeptide. By this route, the aggregate could increase in mass. Ultimately, continued increase in size would lead to fragmentation. In this way, a primitive pattern of growth and reproduction could have appeared.

The appearance of organic catalysts. Adsorption of these aggregates to clays or bilayers might be expected to increase the rate of the internal reaction mechanisms.

Clays containing metals or metallic ions would probably be effective as catalysts. As Melvin Calvin has pointed out, the metallic iron atom has properties analogous to some enzymes. Electron transfer as carried out by the cytochromes (see p. 113) can be accomplished by the reversible transformation between ferrous and ferric iron. Iron in these alternate states is also capable of increasing the rate of decomposition of hydrogen peroxide and thus has properties similar to the enzyme peroxidase. Other metallic ions, such as magnesium, zinc, copper, cobalt, and molybdenum, also have catalytic properties. The capabilities of these ions as catalysts are greatly altered depending on the state of binding to other substances. Calvin uses the example of the peroxidase activity of iron. Iron in the pure state has relatively weak peroxidase activity, but some ferrous hydroxides may show an increase of activity over the metallic form of about 100 times. If bound into the protoporphyrin ring (see Fig. 17–2), the activity is increased about a thousand times; when attached to a specific protein, the peroxidase activity is increased a millionfold.

pyrrole + formaldehyde

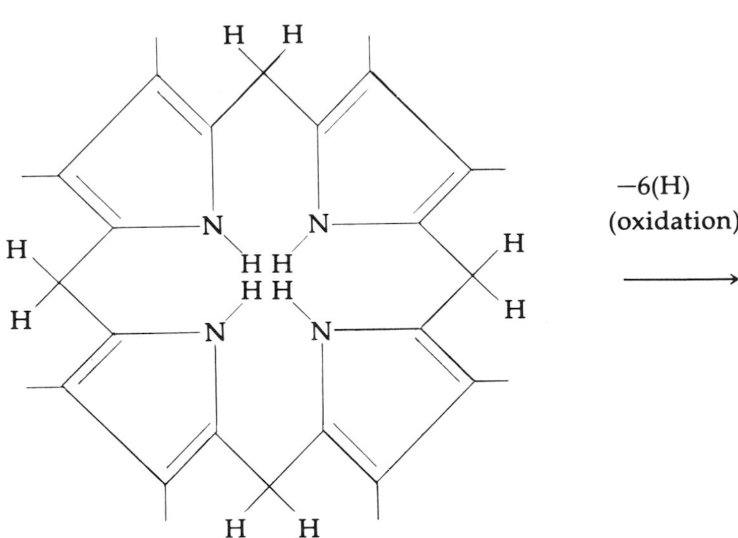

−6(H)
(oxidation)

porphyrin ring system

17–2. Formation of the porphyrin ring by condensation of pyrrole with formaldehyde. From Chemical Evolution *by M. Calvin, 1969. Courtesy of M. Calvin and Clarendon Press, Oxford.*

In addition to possessing catalase or peroxidase activity, metallic ions may also catalyze other types of reactions. By their ability to form coordinate bonds with two or more atoms or groups of atoms, metallic ions can bring potentially interacting molecules closer together in space. As noted, a metallic ion may act as an oxidation-reduction catalyst. Finally, metallic ions attract electrons strongly and can distort the spatial arrangement of the atoms of bound molecules, possibly reducing the energy of activation required for reactions such as hydrolysis.

Using these potential interactions as a guide, it is possible to envision a route of evolution of organic catalysts from inorganic ions. A metallic ion, such as ferrous or ferric iron, may have been adsorbed into a primitive organic aggregate. Catalysis of a reaction yielding free energy may have taken place, followed by spontaneous coupling with reaction systems requiring energy, such as the polymerization of amino acids into a random sequence. At some point, combination between some of these polymers and the inorganic catalysts may have taken place. This combination, although nonspecific, could possibly lead to stabilized binding of the metal to the aggregate and increased catalytic activity. Further increases in catalytic rate would follow the partial or complete transference of catalytic properties to the attached protein. With this transfer, the developing organic catalyst could take on the additional property of specificity. But for evolution to proceed to this point, an information system directing the controlled synthesis of protein would also be required in the aggregate.

Origins of the information system. The progenitors of the information system may have first arisen as a part of the initial reaction mechanisms yielding free energy to the aggregate. If nucleotides were available in the solution surrounding the primitive aggregate, these substances may first have entered the system as molecules coupling exergonic to endergonic reactions. Thus the first protein-nucleic acid association may have taken the form of a primitive enzyme-coenzyme interaction in the aggregate. The necessary ATPase activity could have been present in weak form in the aggregate. Fox, for example, has been able to detect the spontaneous appearance of ATP-splitting activity in his microspheres.

Under the proper conditions, occasional polymerization of the nucleotides into nucleic acids would occur spontaneously in the aggregates. Some combinations of nucleotides in the randomly sequenced nucleic acid chains may have favored the adsorption and retention of particular amino acids and polypeptides, leading to further growth and increased ability to compete for substrates in the environment. At some point, specificity developed in the interaction between short sequences of nucleic acids and specific amino acids or polypeptides, leading eventually to a primitive coding system.

The possible means by which this coding relationship between the sequence of nucleotides in nucleic acids and the order of amino acids in polypeptides first appeared is the most difficult conceptual problem for scientists concerned with the evolution of life. This problem is central to understanding the origins of life, because once the information appears and is linked to the synthesis of proteins of ordered sequence, the subsequent development of greater complexity and specificity through mutation and natural selection is relatively easy to visualize. Unfortunately, at the present time, a reasonable hypothesis to account for the development of the coding interrelationship between nucleic acids and protein has yet to appear.

The major difficulty presented by the coding system as it is known from contemporary organisms is the close interdependence between the precision of the code and the specificity of enzymatic proteins. The precise translation of the nucleic acid sequence into a sequence of amino acids in a protein depends ultimately on the attachment of the correct amino acid to a tRNA by the synthetase enzyme during protein synthesis (see Chapter 11). The accuracy of this attachment is a property of the synthetase enzyme. The specificity of the synthetase group of enzymes in turn resides in the sequence of amino acids in the component polypeptide chains. But this specific combination of amino acids is ultimately determined by the sequence of nucleotides in the gene coding for the protein. It is this circular interdependence between the nucleic acid code and the specificity of enzymatic proteins that makes it so difficult to see how one system could arise before the other. Yet the complexity of the total mechanism is such that the spontaneous appearance of both systems simultaneously is almost impossible to imagine. This unsolved paradox remains the most critical for our understanding of the origins of life.

Once the mutable information directing the synthesis of proteins of specific sequence appeared, further developments would follow as a consequence of natural selection. At some time, either initially if the original aggregation of organic molecules first took place by the spontaneous formation of lipid bilayers or at a later point if by adsorption on clays or other processes, the organic aggregate would become surrounded by a semipermeable lipid film, eventually reinforced by protein. The early, if not initial appearance of membranes seems a likely possibility on purely intuitive grounds, because these structures are as fundamental to contemporary living systems as the nucleic acid code and proteinaceous catalysts.

At this level of development, the primitive living system would possess the beginnings of cellular organization: a mutable coding system, consisting of a nucleic acid polymer; a translation system, directing the synthesis of protein of specific sequence according to directions supplied by the coding system; a catalyzed reaction system, coupling the exergonic breakdown of adsorbed organic substances to the endergonic reaction systems synthesizing nucleic acids and protein and probably using nucleotide phosphates as coupling factors; and finally, separation of the systems carrying out these functions from the surrounding environment by a continuous membrane. No doubt some or all of these processes were initially organized on the primitive membrane. Further adaptations would lead to the development of enzymes of greater specificity, with or without attached metallic ions, and assembly of the apparatus for synthesis of protein into ribosomes.

Provision of energy for the maintenance of these primitive systems, as noted, would rest on breakdown of organic molecules absorbed from the environment. Most of the exergonic reactions would involve removal of hydrogen and electrons from reduced substances, with molecules of intermediate oxidation-reduction potential, such as sulfate or nitrate serving as final electron acceptors. Thus respiration would be anaerobic. If the primitive organisms became widely distributed, continuing respiration would gradually exhaust the supply of reduced organic molecules in the environment. Because the spontaneous, abiogenic synthesis of these materials would probably lag behind use by the primitive anaerobes, exhaustion would have eventually led to the disappearance of life except for a new mechanism that must have emerged soon after the first development of life: photosynthesis. With the appearance of photosynthesis and the consequent release of large quantities of oxygen as a byproduct into the environment, the atmosphere and the primitive earth experienced a fundamental change from a "reducing" to "oxidizing" nature.

The Origins of Photosynthesis and the Change from a Reducing to Oxidizing Atmosphere

Regarded in the most basic terms, photosynthesis has two essential features. One is the splitting of water, to yield oxygen, hydrogen ions, and electrons:

$$2H_2O \rightarrow O_2 + 4H^+ + 4e^- \qquad (17\text{-}1)$$

In a closely related mechanism, the electrons removed from water are increased in potential through the adsorption of quanta of visible light:

$$4e^- + \text{light} \rightarrow 4e^{-*} \qquad (17\text{-}2)$$

where e^{-*} indicates electrons of relatively high potential. The electrons, raised in chemical potential in this manner, then enter into the "work" of the cell, either as a direct source of free energy through the electron transport system coupled with ATP synthesis (photophosphorylation) or in the reduction of CO_2:

$$CO_2 + 2H^+ + 2e^- \rightarrow (CH_2O) + H_2O \qquad (17\text{-}3)$$

Thus the electrons raised in potential through the initial reaction of photosynthesis served as a continuously replenishable substitute for the dwindling supply of reduced substances in the primordial environment.

The mechanism by which contemporary plants carry out reaction 17–1 is still not understood. Presumably, an as yet undiscovered enzyme catalyzes the splitting of water to release hydrogen ions, electrons, and oxygen. In most contemporary photosynthesizing organisms, reaction 17–2 is mediated by chlorophyll, a porphyrin containing a metallic ion, Mg^{+2}, as a part of its active group. Presumably, in the first primitive photosynthesizers, both the splitting of water and the increase in potential of the removed electrons was accomplished by inorganic catalysts or primitive conjugates of inorganic ions with a polypeptide. This was possibly aided by the fact that in the primitive atmosphere, the radiation falling on the surface of the earth contained more energy concentrated in the ultraviolet wavelengths; these wavelengths today are largely absorbed by the ozone layer of the outer atmosphere. A photon of ultraviolet light has sufficient energy to "split" a molecule of water directly, without the aid of a catalyst. Therefore, the catalytic requirements of the first photosynthesizers were probably initially less stringent until oxygen and ozone began to reach higher levels in the atmosphere.

At the same time, the primitive photocatalysts eventually leading to chlorophyll must have made their appearance. It has been shown in the abiogenic synthesis experiments that the building blocks of the porphyrin ring system can appear spontaneously. Melvin Calvin has provided a scheme by which the porphyrin ring could be spontaneously generated by condensation of pyrrole and formaldehyde (Fig. 17–2). Possibly, abiogenic synthesis of this type led to the first biological molecules capable of absorbing light and releasing an electron excited to a higher energy level.

The appearance of photosynthesis led to the establishment of large numbers of primitive organisms living as autotrophs rather than heterotrophs. As these organisms increased in number, the continual release of oxygen into the atmosphere caused a fundamental change from a reducing to an oxidizing environment that persists to the present day. With the appearance of oxygen in the atmosphere and the incessant activities of heterotrophs, it is doubtful that abiogenically synthesized organic substances could ever again reach significant levels. Therefore, the spontaneous appearance of life is probably impossible under the conditions present today.

The presence of oxygen also led to adaptations enabling a group of primitive cells to use this substance as a final electron acceptor. With this development, the appearance of aerobic respiration, the organization of cellular life probably soon reached levels approximating present-day prokaryotes in complexity.

The Appearance of Eukaryotes

Subsequent developments producing eukaryotic cells have already been outlined as a part of Chapter 16, in which the inheritance and possible origins of mitochondria and chloroplasts are discussed. Briefly, the hypotheses outlined in Chapter 16 state that a primitive aerobe, using oxygen as a final electron acceptor, was ingested by a primitive anaerobic heterotroph. The aerobe was tolerated and persisted in the cytoplasm of the anaerobe as a symbiont. Eventually, as most of the nonrespiratory functions of the aerobe were taken over by the host cell and as the host became dependent on the greater respiratory efficiency of the aerobe, the symbiosis became obligate. During the same period, other adaptations of the eukaryotic cell appeared, including the nuclear envelope, greater complexity of the chromosome, and the mitotic mechanism for division of the nucleus. Shortly after these developments took place, ingestion of a primitive, prokaryote-like photosynthesizer by an early eukaryote founded the subgroup leading to plants.

The Time Course of the Evolution of Life

The course of organic evolution from the beginnings of the Cambrian period extends over 700 million years. But fossils of relatively highly organized, multicellular forms are found in the earliest Cambrian strata, indicating that the origins of life date from a much earlier period. Although somewhat controversial, remains of primitive algae-like cells have been found embedded in rocks between 2000 and 2500 million years old (see Barghoorn and Schofp, 1965). If these fossils indeed represent the earliest cells, about 2000 million years would be allowed for the precellular stages in the evolution of life, because the oldest rocks of the crust of the earth were formed about 4000 million years ago.

In this span of time, 2000 million years, extending from soon after solidification of the earth's crust up to the formation of the first known cells, evolution proceeded through the phases of abiogenic synthesis, aggregation, and appearance of the fundamental processes of life. Although the possibility of spontaneous development of these mechanisms is difficult to conceive, this time span, 2000 million years, is such that

> given so much time, the "impossible" becomes possible, the possible probable, and the probable virtually certain. One has only to wait: time itself performs the miracles.[1]

Suggestions for Further Reading

Books

Bernal, J. D., 1967. *The Origin of Life*. World, New York.

Calvin, M., 1969. *Chemical Evolution*. Oxford University Press, New York.

Kenyon, D. H., 1969. *Biochemical Predestination*. McGraw-Hill, New York.

Keosian, J., 1964. *The Origin of Life*. Reinhold, New York.

---

[1]From G. Wald, The origin of life. *Scientific American* 191:45 (1954).

Papers and Review Articles

Barghoorn, E. S., and F. P. Schofp, 1965. Microorganisms from the Gunflint Chert. *Science*, vol. 147, pp. 563–577.

Fox, S. W., and K. Harada, 1960. Thermal copolymerization of amino acids common to proteins. *Journal of the American Chemical Society*, vol. 82, pp. 3745–3751.

Granick, S., 1957. Speculations on the origins and evolution of photosynthesis. *Annals of the New York Academy of Sciences*, vol. 69, pp. 292–308.

Haldane, J. B. S., 1929. The origin of life. *New Biology*, vol. 16, pp. 12–27. A reprint of the essentials of this paper can be found in the book by J. D. Bernal listed above.

Lowe, C. V., M. W. Rees, and R. Markham, 1963. Synthesis of complex organic compounds from simple precursors: Formation of amino acids, amino-acid polymers, fatty acids, and purines from ammonium cyanide. *Nature*, vol. 199, pp. 219–222.

Madison, K. M., 1953. The organism and its origin. *Evolution*, vol. 7, pp. 211–227.

Miller, S. L., 1953. A production of amino acids under possible primitive earth conditions. *Science*, vol. 117, pp. 528–529.

Oparin, A. I., 1938. The origin of life. Macmillan, London; republished by Dover Publications, New York, 1953. A reprint of this pamphlet can be found in the volume by J. D. Bernal listed above.

Perret, J., 1952. Biochemistry and bacteria. *New Biology*, vol. 12, pp. 68–96.

Ponnamperuma, C., and R. Mariner, 1963. The formation of ribose and deoxyribose by ultraviolet irradiation of formaldehyde in water. *Radiation Research*, vol. 19, p. 183.

Ponnamperuma, C., R. M. Lemmon, R. Mariner, and M. Calvin, 1963. Formation of adenine by electron irradiation of methane, ammonia, and water. *Proceedings of the National Academy of Sciences*, vol. 49, pp. 737–740.

Schmidt, O. J., 1944. A meteoric theory of the origin of the earth and planets. *Doklady Akademiia Nauk SSSR*, vol. 45, pp. 229–233.

Schramm, G., H. Groetsch, and W. Pollmann, 1962. Nonenzymatic synthesis of polysaccharides, nucleosides, and nucleic acids and the origin of self-reproducing systems. *Angewanate Chemie, International Edition*, vol. 1, pp. 1–7.

Urey, H. C., 1951. The origin and development of the earth and other terrestrial planets. *Geochimica et Cosmochimica Acta*, vol. 1, pp. 209–277.

von Weizsacker, C. F. Z., 1944. Über die Entstehung des Planetsystems. *Zeitschift für Astrophysik*, vol. 22, pp. 319–355.

Wald, G., 1954. The origin of life. *Scientific American*, vol. 191, pp. 45–53.

Index

aberration
  chromatic, 4
  spherical, 4
  spherical, effect on electron lenses, 50–52
accessory fibers of flagellum, 428, 443
accessory pigments, chloroplast, 137–139, 147
Acetabularia, 269–270, 504, 505, 508, 509
acetyl-Co A
  formation, 110
  in glyoxylate cycle, 125
acidic proteins, 171
aconitase, 110
acridine orange, 262
acrosome, 427–434, 459, 462, 464–468
  acrosome reaction, 464–468
  development, 433, 434
  in fertilization, 462, 464–468
  filament, 466
  function, 433–434
  proacrosomal granule, 428, 434
  structure, 433–434
actinomycin D, 209, 233
  effect on nucleolar ultrastructure, 219
  effect on perichromatin granules, 220
  and puff formation, 241
active site of enzymes, 98, 99
activity coefficient, 92n.
additions, in mutation, 316
adenine, 110, 166, 195ff.
adenosine triphosphatase (ATPase)
  in accessory fibers of sperm tail, 443
  in annulate lamellae, 237
  dynein (flagellar ATPase), 343, 445
  localization in mitochondria, 121
  in oxidative phosphorylation, 116–118
  in pore complex, 234
  in spindle, 343, 346
adenosine triphosphate (ATP)
  in amino acid activation, 255–256
  ATP-ADP cycle, 94–95
    in cellular metabolism, 108ff.
    in chloroplasts, 143ff.
  in cell cycle, 289, 300, 302
  in cellular oxidation, 102, 106ff.
  chemical hypothesis of synthesis, 116
adenosine triphosphate (ATP) (continued)
  chemiosmotic hypothesis of synthesis, 116–118, 148–149
  in flagellum, 445
  localization in metochondrion, 121, 122–123
  Mitchell hypothesis of synthesis, 116–118
  in movement of chromosomes by spindle, 343
  photophosphorylation, 146, 147, 148–149
  structure, 109
  synthesis in fertilized egg, 471
  synthesis in mitochondria, 115–118
Aequorea, 364
Afzelius, B. A., 234, *480*
Akiyama, M., *518*
aldehyde-osmium fixation for electron microscopy, 58–59
aldolase, 108
Aldridge, W. G., 61, *65*
Alfert, M., 164, 317, *327*, 431, 474, *480*
allelic recombination, 396–403
Allomyces, 512
Aloni, Y., *518*
Alonso, P., *249*
alpha-cytomembranes, 270n.
alpha helix, 34–36
  in histones, 172, 173
  in membranes, 74
alpha-ketoglutaric acid, 112, 119
Altmann, R., 8, 17, 101, 102
ameboflagellates, 512
amino acids
  activation and attachment to tRNA, 255–256
  codon assignments, 263
  formation in photosynthesis, 151–152
  and the genetic code, 261–264
  oxidation, 101, 118–119
  properties, 31–32
  in protein synthesis, 192–194
  structure, 31–33
aminoacyl-tRNA in protein synthesis, 256–258
aminoacyl-tRNA synthetase, 203–204, 255, 256
amino group of amino acids, 31
amitosis, 302n.
Amoeba, 299, 300
Amphioxus, 387
amyloplast, 21
anaphase bridges, 321
anaphase movement of chromosomes, 342–347
anastral spindle, 340
Anderson, W. A., 473, *480*
André, J., 435, *480*
Angstrom (Å), 9
Ångstrom, A. J., 9
annulate lamellae
  cytochemistry, 237
  development, 236
  function, 237–238
  granules, 237
  in oogenesis, 453, 460
  pore complexes, 234–238
  structure, 234–237
annulus
  of pore complex, 16, 225–227, 229, 231, 234, 235
  of spermatozoan, 437
antherozoid, 446
anticodon, 192, 193, 205
antifertilizin, 464
antipodal cell, 475
Aquilegia, 318
Arbacia, 473
arginine in histones, 172
Armandia, 364
Arnold, A., 145
Arnon, D. I., 143
Artemia, 456
Ascaris, 6, 375, 384
aspartic acid, 119, 151
aster, 295, 296, 299, 350
  in meiosis, 387
  in mitosis, 353–354
  and polarization of chromosomes, 336
  structure, 351
astral rays, 296

astral spindle, 340
asymmetry of RNA transcription, 198–199
atmungsferment, 102
Attardi, G., *518*
Auerbach, L., 375
Austin, C. R., *480*
autoradiography, 175, 181, 197, 217, 273, 290
  of bacterial replication, 313
  of lampbrush chromosomes, 416–418
autosomes, 387
Avery, O., 164, 165, *188*
Avron, M., *159*
axial complex, flagellar, 427, 428, 433–446
  colchicine binding, 445
  development, 428, 443–444
  exceptional patterns, 440–443
  and mobility, 437–446
  in plant spermatozoa, 446ff.
  protein components, 444–445
  structure, 437–443
axial complex of meiotic chromosomes. *See* synaptinemal complex
axoneme, flagellar. *See* axial complex

Bacillus subtilis, 209
bacteria, 10, 11
  chromosome replication, 313
  chromosome structure, 181–185
  nucleoid division, 302, 347–348, 350
  nucleoid ultrastructure, 180–185
  photosynthesis, 133–134, 153–154
  ribosomes, 252–254
bacteriochlorophyll, 137, 153
  P890, 153
bacteriophage, 165
Baer, K. E. von, 5
Bahr, G. F., *188*
Bajer, A., 319, *327*, 332, 359, 365, 373
Balbiani, E. G., 239
Balbiani rings, 241
Balinsky, B. I., *480*
Banhoeffer, F., *327*
Barclay, R. K., *188*
Barghoorn, E. S., *532*
Barnett, L., 256, *285*
Barnett, R. J., *423*
Barnett, W. E., 495
Barr, H. J., *249*
basal body. *See* centriole
basal granule. *See* centriole
basal plate of spermatozoan, 436
base pairs in DNA, 169
Bassham, J. A., 149, *159*
Baudhuin, P., 124, *126*
Bauer, H., 239
Beadle, G. N., 391
Beale, G. H., *518*
Beams, H. W., *285*, *480*
Bear, R. S., *82*
Beatty, B. R., *422*
Beer, M., *65*
Beermann, W., 240, 241, 243, 245, *249*, 268
Beevers, H., 124, *126*
Beisson, J., *518*
Belling, J., 394
Belling hypothesis of recombination, 394–396
Benda, C., 102
Bender, M. A., *328*, 355, *374*
Beneden, E. van, 6, 350, 375
Benson, A. A., 142, 149, *159*
Bernal, J. D., 526, 527, *531*
Bernhard, W., 216, 217, 220, 221, 222, 232, 245, *249*, 352, *373*
Berzelius, Jons Jacob, 7
beta-cytomembranes, 270n.
beta-galactosidase, 264, 265
beta-galactoside, 264, 265
bilayer, lipid, 29, 30. *See also* phospholipid, bilayers
  interaction with OsO_4,$KMnO_4$, 71–72, 74
  in membranes, 70ff.
  single, physical properties, 72
  ultrastructure, 71–72
  X-ray diffraction, 70–71
biliproteins, 139
birefringence, 75
Birnstiel, M., *211*, *373*
bivalents, 379ff.
Blackmann, F. F., 143
Blepharisma, 359, 360
blepharoplast, 450
Bloch, D. P., 317, *327*, 432, 433, 473, *480*
Bloom, S., 228
blue-green algae, 10, 11, 13
  nucleoid ultrastructure, 180–185
  photosynthesis in, 133–134, 153–154
Bodmer, W. F., *422*
Bogorad, L., 501, *519*
Bolle, A., *211*
Bonner, J., *188*, 203, 207, *285*
Borisy, G. G., 341, *373*
Borries, B. von, 49
Borst, P., *126*, *518*
bouquet stage of meiosis, 379, 404
Boveri, T., 6, 376
Brachet, J., 208, 213, 271, 504
Branton, D., 79, *82*, 144, *159*
breakage and exchange in recombination, 390–394, 395
Breidenbach, R. W., 124, *126*
Brenner, S., *285*, *373*
Bresslau, E., 485
Bretscher, M. S., 259, *285*
Bridges, C. B., 7, 163
Brinkley, B. R., 321, 332, 334, 339, *373*
Britten, R. J., 318, *327*
Brown, D. D., 208, *211*, *373*, *422*
Brown, J. C., *285*
Brown, R. E., *518*
Brown, Robert, 5
Buchner, Eduard, 8
Buchner, Hans, 8
Buck, R. C., *373*, *374*
Bufo, 419
Burton, H., 505
Burton, P. R., 486, *518*
Busch, H., 209
Bystrom, B. G., *520*

C-value, 290n.
Cairns, J., 181, 313, 314, 316, 317, *327*, 493
Callan, H. G., 326, 416, 417, 418, 419, *422*
Calvin, M., 149, *159*, 528, *531*, *532*
Calvin cycle, 149–150, 151, 152, 153
Campanularis, 463
Cancer, 409
Cannon, W. A., 7
Capecchi, M. R., *285*
capsule, 11
carbohydrate
  attachment to proteins in Golgi complex, 280
  oxidation, 101, 102, 105–108
  photosynthesis, 149–150
  synthesis in Golgi complex, 280–281
carboxydismutase, 141, 143, 149
carboxyl group of amino acids, 31
cardiolipin, 68
Caro, L. G., 278, *285*
carotenes, 137–139, 147
carotenoid pigments, 21, 137–139
  structure, 138
Caskey, T., *286*
Caspersson, T. O., 271
catalase, 124, 125
cell cycle, 290, 292ff., 299
cell division, early description, 5
cell membrane. *See* plasma membrane
cell surface, 19
cell plate in cytokinesis, 362, 364–369
cell theory, 5–6, 163
cellulose, 11, 21, 370, 372
cell wall
  bacterial, 11
  in blue-green algae, 12
  in cytokinesis, 365–372
  defined, 11, 21
  early hypotheses, 9ff.
  effects of colchicine on synthesis, 372
  primary, 21, 370
  secondary, 21, 372
  synthesis, 367–372
centrifugation, analytical, 201
centriolar adjunct, of spermatozoa, 437
centriole
  and the axial complex, 427, 440ff.
  in the blepharoplast, 450
  DNA in, 513–515
  in fertilization, 474
  inheritance, 511–515
  in meiosis, 295, 387
  in mitosis, 353–354
  in plants, 446ff.
  replication, 291, 295, 352–353, 450
  in spermatozoa, 427
  and sperm motility, 437, 445
  and spindle formation, 296–297
  ultrastructure, 351
centromere
  attachment to spindle, 297, 333

centromere (continued)
and chromosome movement, 298
corona, 333, 334
defined, 293
diffuse, 336
in meiosis, 382, 384
structure, 331–336
centrosome, 350, 351
cephalin, 28
cerebroside, 69
Chagas' disease, 485
Chance, B., 98, 114, 115, 116
Chapman, D., *82*, *83*
Chara, 278
Chargaff, E., 166
Chase, M. C., 165, *188*
chemical hypothesis of ATP synthesis, 116
chemical potential, 91–93
chemiosmotic hypothesis of ATP synthesis, 116–118, 148–149
chiasma, 381, 386ff.
in recombination, 389ff.
terminalization, 382
chiasmatype hypothesis of recombination, 389–390
Chironomus, 230, 240, 243, 244, 245
chitin, 372
Chlamydomonas, 131, 133, 500, 509, 515
Chloeon, 440
chloramphenicol, 496, 508
Chlorella, 153
chlorophyll, 21, 136, 137–139, 142, 147
light absorption, 137
P700, 145, 146, 147, 148
structure, 138
chloroplast
defined, 9, 10, 21
development, 154–157
DNA in, 130, 155, 184, 487, 500–509
early description, 128
fret, 128
granum, 128ff.
inheritance, 483, 484, 499–511
intermembrane space, 131
lamellae, 128
lipids, 68, 69, 70, 79, 134, 136–137
membrane structure, 134–142
origin, 509–511, 515–518
osmiophilic granules, 130, 133, 136
photosynthesis, 142–154
pigments, 137–139, 147
proteins, 134–136
protein synthesis in, 508–509
pyrenoid body, 130
ribosomes, 130–131
RNA in, 501–502, 506–508
starch grains, 130
stroma, 128ff.
stromal lamellae, 128
structural protein, 134–136
structure, 127–142
thylakoid, 128ff.
cholesterol, 29, 31, 68, 69, 70, 136
choline, 28, 68
chondriosome, 13
chondroids, 13
chorion, 458, 468
Chortophaga, 362
chromatid
aberrations, 320
coiling, 293
in meiosis, 381ff.
pattern of labeling in replication, 305–308
separation at mitotic anaphase, 297–298
subchromatid organization, 319–320
chromatin
active, 267, 269
changes during spermiogenesis, 428–433, 447ff.
dinoflagellate, 184–185
early description, 7, 8, 14
effects of fixation on structure, 58
euchromatin, 269
in fertilization, 473
heterochromatin, 269
inactive, 267, 269
isolated, 177–180
nucleolar, 215, 216, 217
ultrastructure, 174–180
chromatin body, prokaryotic, 11
chromatophore, 13, 133–134
chromocenter, 240n.
chromomeres, 377, 379, 394, 412, 415, 416, 418
chromoplast, 21
chromosome
aberrations, 320
attachment to spindle, 297
bacterial, 180–184, 302, 313
bivalent, 379ff.
centromere, 293, 297, 298, 331–336, 382, 384
chiasma, 381, 382, 386ff., 389ff.
chromomeres, 377, 379, 394, 412, 415, 416, 418
condensation, coiling, 178, 293, 329–331, 377–381
congression, 297, 339
dinoflagellate, 184–185
early description, 6
fiber, of spindle, 297
homologs, 378ff.
interaction with OsO_4, 59
isolation, 177–180
lampbrush, 381, 415–419
linkers in, 324
metaphase, 293, 329–338
mitotic, 291ff.
movement by spindle, 298, 342–347
movements, independent, 336–339
multistranded, 308, 319–326
mutations, 316
pairing, 378–379
polarization, 336
proteins, 317. *See also* histones, nonhistone protein
recombination, 388–403
replication, 290–291, 303–317
sex chromosomes, 387
strandedness, 308, 319–326
synapsis, 378–379
synaptinemal complex, 379, 381, 403–415
chromosome (continued)
tetrad, 379ff.
ultrastructure, 174ff.
unit fibers of, 174–180
X chromosome, 269
chromosomes, polytene
activity, 241–245
degree of polyteny, 240
development, 240
Feulgen staining, 239
nucleolus, 245–248
puffs, 241–245
RNA synthesis, 241–245
structure, 238–245
ultrastructure, 243–245
chromosome theory of heredity, 7
Chromulina, 103
cis-aconitate, 110
cisternae
of endoplasmic reticulum, 19
of rough ER, 271
cistrons
in RNA transcription, 192ff.
RNA, 208–209
citrate synthetase, 110
citric acid cycle, 102, 104, 105, 110–112, 121
Claude, A., 102
Clayton, R. K., 146, *159*
cleavage histones, 473
Cleveland, L. R., 344, 345, 350, 359
Clever, U., 241, *249*, *285*
coacervate, 526, 527
code, DNA. *See* genetic code
codon, 192, 261–264
initiator, 257–258
nonsense, 259
terminator, 204, 258–260
coenzyme A, 109, 110ff.
in fatty acid oxidation, 118
in oxidation of proteins, 118–119
in photosynthesis, 153
structure, 109
coenzyme Q, 113ff., 136, 154
coiled bodies, of nucleoplasm, 221–223
coiling, 329–331
and heterochromatin, 269
during meiosis, 377–381
and regulation, 269
colchicine, 298, 331, 339
and anaphase movement, 346
binding to flagellar axial complex, 445
binding to microtubules, 346
binding to spindle protein, 341–342
and derivation of polyploids, 342
effect on cell wall synthesis, 372
Coleman, J. R., 405, *422*
Colwin, A. L., 464, 465, *480*
Colwin, L. H., 464, 465, *480*
complementary base pairs, DNA, 169
condensation of chromosomes. *See* coiling
congression, 297, 339
conservative replication, 304–310
continuous fibers of spindle, 297
Conway, T. W., *285*

copy-choice hypothesis of recombination, 394–396
core of meiotic chromosomes. *See* synaptinemal complex
corepressor, 266
core protein, ribosomal, 253
Corey, R. B., 35
corona of centromere, 333, 334
Correns, C., 7, 483, 484, 499
cortical granules, 455–456, 469
Crane, H. R., 313, *327*
Craven, G. R., *286*
Creighton, H., 390
Crick, F. H. C., 165, 166, 169, *189*, 262, 263, 264, *285*, 304
Criddle, R. S., 121, 134, 136, *159*
cristae, mitochondrial, 17, 103ff., 121
critical point method, 177
Crouse, H. V., *327*
cumulus oophorus, 459
Cuzin, F., *373*
Cyanophora, 517
cyclic photosynthesis, 148
cycloheximide, 496, 508
Cyclops, 473
cytochrome oxidase, 147
cytochromes, 73, 102, 112ff., 147, 484, 485, 496
cytochromoid *c*, 154
cytokinesis, 289, 290, 299, 303
  by cell plate formation, 362, 364–369
  by furrowing, 362–364
  in prokaryotes, 366–367
  relationship to mitosis, 361
  relationship to position of spindle, 361–362
cytoplasm
  defined, 16ff.
  inheritance, 483ff.
cytosine, 166, 195ff.

Daems, W. T., *126*
Dahmus, M. E., *285*
Dalton, A. J., 249, 276
Dan, J. C., *480*
Danielli, J. F., 74
dark reactions of photosynthesis, 143, 149–153
Darlington, A. J., *422*
Darlington, C. D., 389, 390
Davson, H., 74
Dawid, I. B., *422*, 486, *518*
DCMU [3-(3,4-dichlorophenyl) 1,1-dimethylurea], 145
de Duve, C., 124, *126*, 281, 284, *285*
de Harven, E., 352, *373*, *518*
dehydrogenase. *See* specific dehydrogenases
Delbruck, M., 304
deletions, in mutation, 316
de Man, J. C. H., *285*
deoxyribonuclease (DNAase), 165, 175, 324, 325
  endonuclease, 401–402
  exonuclease, 401–402
  and recombination, 399–402
deoxyribonucleic acid (DNA)
  bacterial replication, 347–348
  bacterial ultrastructure, 180–183
  C-value, 290n.
  chloroplast, 130, 155, 184, 487, 500–506, 508–509
  discontinuous synthesis hypothesis, 315–316
  discovery, 8
  establishment as hereditary molecule, 164–165
  heteroduplex, 399
  hybridization, 199, 206–207
  linkers in, 324
  mechanism of synthesis, 310–316
  mitochondrial, 103, 104, 184, 485–494, 496–498
  molecular structure, discovery, 165–168
  mutations, 316
  nucleolar, 208–211, 217
  polarity, 199
  primer, 310–311
  proteins associated with, 169–173
  recombination, 388–403, 408
  redundancy, 290, 317–326
  replication, 169, 290–291, 303–316, 347–348
  semiconservative replication, 304–310
  structure, 165–169
  synthesis during fertilization, 473–474
  synthesis during meiotic prophase, 386, 402, 419–420
  template, 310–312
deoxyribose, 165ff.
desmosome, 19, 20
Deuel, H. J., *39*
Devis, R. J., *480*
de Vries, H., 7
Diacumakos, E. G., 485
3-(3,4-dichlorophenyl) 1,1-dimethylurea (DCMU), 145
Dickerson, R. E., 31, *39*
Dietz, R., 353, *373*
differentiation, 173
diffraction
  in electron microscopy, 54, 56–57
  in phase contrast light microscopy, 46
  X-ray, 70
diffuse centromere, 336
diffuse diplotene, 415
diglyceride, 26
dihydroxyacetone phosphate, 108, 118
dinoflagellates, 184–186
  division, 348
  nuclear envelope, 234
dipeptide, 32
diphosphatidyl glycerol, 68
diphosphoglycerate kinase, 108
diploids, partial, 265
Disbrey, C., 515
dispersive replication, 304, 306
disulfide linkage, 35, 37
  interaction with OsO_4, 59
DNA polymerase, 310–313
  attachment to membranes, 347
DNA polymerase (continued)
  in recombination, 402
  and sequential replication, 315
Doty, P., 36, *39*, *285*
Dounce, A., 261
Drake, J. W., *327*
Drosophila, 7, 163, 206, 208, 209, 230, 239, 240, 268, 299, 318, 390, 407, 408, 411, 420, 458
Dubin, D. T., *518*
DuPraw, E. J., 179
Dusanic, D. G., 486, *518*
Duysens, L. N. M., *159*
dynein (flagellar ATPase), 343, 445

ecdysone, 268
Edstrom, J. E., 208
EDTA-uranyl technique for RNA staining, 220, 232
egg
  development in animals, 453–462
  fertilization, 462–480
  polar bodies, 454, 455
  surface envelopes, 455–458
electron microscope, 49–57
  construction, 49, 52–53
electron scattering in electron image formation, 54–56
electron stains, 60–62
electron transport
  coupling to ATP synthesis, 116–118
  energetics, 114
  localization in mitochondrial structure, 121
  mitochondrial, 104, 112–118
  in photosynthesis, 147–149
  sequence of carriers in mitochondrion, 115
elementary particles in inner mitochondrial membranes, 122–123
Elodea, 487
Elysia, 517
Emerson, R., 143, 145, 146, 147
Emerson, S., *422*
endonuclease (DNAase), 401–402
endoplasmic reticulum (ER)
  cisternae, 271, 274
  and cytokinesis, 365, 366, 367
  description, 14
  development, 271–272
  and lysosomes, 281–284
  protein synthesis in, 272–273
  rough, 19, 270–274, 275, 278
  smooth, 19, 270, 274–276
  structure, 79
  transport through, 176, 274–275, 278–280
  in yolk formation, 459
endosperm, 474, 475, 479, 480
Engelman, D. M., 83
enthalpy, 92
entropy, 90–91, 92
enzymatic catalysis, 95–99
  mechanism, 98–99
enzymes. *See also* specific enzymes
  active site, 98, 99
  discovery, 7–8

enzymes (continued)
enzyme-substrate complex, 98, 99
inducible, 265
of the lysosome, 281–282
mechanism of catalysis, 98–99
turnover number, 97
Ephrussi, B., 484, *518*
Epstein, H. T., *519*
equilibrium constant, 92
ergastoplasm, 270
Escherichia coli, 260, 264, 265, 266, 308, 310, 313, 314
ester linkage in lipids, 26–27
ethanolamine, 28, 29, 68
etioplast, 21, 154, 155–157
euchromatin, 269
Euglena, 131, 500, 505, 508, 509
eukaryote, 19, 23, 24
evolution of nucleus, 184–185
Euplotes, 176
Everett, G. A., 205, *211*
excited state of electron in light absorption, 144
exciton, 146
exonuclease (DNAase), 401–402
extrachromosomal material, 14
in prokaryotes and dinoflagellates, 185

F_1, F_2, F_3, in protein synthesis, 256
FAD (flavine adenine dinucleotide), 112
Farmer, J. B., 6, 376
Farnbrough, D., *285*
fast green staining of basic proteins, 174
fats
oxidation, 118
saturation, 27, 28
synthesis in chloroplasts, 153
fatty acids, 26, 27
distribution, 27
in membranes, 68
oxidation, 101, 105, 108, 118
oxidation in smooth ER, 275
synthesis in chloroplasts, 153
Favard, P., *285*
Fawcett, D. W., 229, *249*, 403, *481*
Feldherr, C. M., 231, *249*
Felix, M. D., 276
Fernández-Morán, H., 75, 122
ferredoxin, 147, 148, 154
fertilization, 6, 462–480
acrosome reaction in, 464–468
attachment of sperm to egg during, 463
biochemical changes during, 471
in plants, 474–480
fertilization cone, 465
fertilization membrane, 469
fertilizin, 464
Feughelman, M., 172, *188*
Feulgen, R., 8
Feulgen reaction, 8, 164, 174, 175, 181, 217, 239, 289, 290, 485, 500
fibrillar part (zone) of nucleolus, 215–216
fibrous lamina of nucleus, 229
Finean, J. B., *83*
Fischer, E., 7, 98
Fisher, D. B., *481*
flagellin, 13
flagellum
accessory fibers, 428, 443
bacterial, 13
colchicine binding, 445
development, 428, 443–444
exceptional patterns, 440–443
in plant spermatozoa, 446ff.
protein components, 444–445
and sperm mobility, 437–446
structure, 437–443
flavine adenine dinucleotide (FAD), 112
flavoproteins, 102, 112ff., 147, 154
Fleischer, S., 73, *83*
Flemming, W., 6, 8, 290
Fluck, D. J., *82*
fluorescence, 144
focal length, 42
focal point, 42
follicle cell, 453, 458–459, 464, 466, 468
Forer, A., 342, *373*
formylmethionine, 257, 260
Fox, S. W., 526, 527, 529, *532*
Franke, W. W., *249*
free cell formation, hypothesis of, 5, 6
free energy, 91–93, 94ff.
changes in mitochondrial electron transport, 114
standard free energy change, 93
freeze etching, 63–65
of chloroplasts, 141
of Golgi complex, 277
of membranes, 79
of sperm heads, 430
freeze substitution, 103
effect on mitochondrial membranes, 103
fret, chloroplast, 128
Frey-Wyssling, A., *373*
Friedlander, M., *327*
fructose 1,6-diphosphate, 108
fructose 6-phosphate, 107
Fuhs, G. W., *188*
fumaric acid, 112
fumerase, 112
furrowing in cytokinesis, 362–364

G_1, G_2
of cell cycle, 290–291
of premeiotic interphase, 377
galactolipids, synthesis in chloroplast, 153
galactose permease, 264, 265
galactose transacetylase, 265
galacturonic acid, 369
Gall, J. G., 177, 209, 213, 226, *249*, 324, 325, *327*, 352, *373*, 416, 417, 418, 420, 450, *481*, 512
Gall technique for isolation of chromosomes for electron microscopy, 177–180, 181
Gamov, G., 262
ganglioside, 69
Garnier, C., 270, 271
Gavrilova, L. P., *211*, *285*
Geis, I., *39*
Gellert, M., 315, *327*
gene in RNA transcription, 192ff.
gene mutations, 316
generative nucleus, 476
genetic code, 192, 257
codon assignments, 263
degeneracy in, 262, 263–264
solution, 261–264
wobble hypothesis, 264
genophore, 11
Georgiev, G. P., *211*, *285*
Geren, B., 454
germinal vesicle, 454
Geschwind, I. I., *327*
Gestelland, R. F., *211*
Gibbons, I. R., 444
Gibbs, W., 91
Gibbs free energy, 92
Gibor, A., 500, 504, *518*
Gifford, E. M., 500
Gilbert, W., 266, 268, *285*
Ginkgo, 446, 474
glucose
and cell wall synthesis, 370
efficiency of cellular oxidation, 118
incorporation in Golgi complex, 281
oxidation, 105, 106–110, 118
photosynthesis, 149–150
glucose 6-phosphatase in smooth ER, 274, 275
glucose 6-phosphate, 107
glutamic acid, 119
glyceraldehyde phosphate, 102
glycerol, 26, 27, 28, 118
in photosynthesis, 153
glycerophosphoric acid, 28
glycogen, 12, 274, 275
glycolipids, 68–69, 136, 275, 433
glycolysis, 105, 106–110, 112, 118
reversal of pathway in photosynthesis, 150
in spermatozoa, 446
glycoproteins, 39, 280, 458
glyoxisomes, 19, 124–126
glyoxylate cycle, 124–125
Godman, G. C., 317, *327*
Goldblatt, P. J., *285*
Goldstein, L., 300
Golgi, C., 19, 276
Golgi complex (apparatus), 14, 19, 260
and cell wall synthesis, 367–372
and cytokinesis, 365, 366
and development of the acrosome, 433–434
function, 278–281
and lysosomes, 281–284
number and distribution, 278
in oogenesis, 453, 456
polysaccharide synthesis in, 280–281
in spermiogenesis, 427, 428, 433–434
structure, 276–278
transition elements, 279
and yolk formation, 459
Gonatas, N. K., *328*, *374*
Goodman, H. M., *285*
Gorter, E., 74

Granboulan, N., 249
Granick, S., 128, 500, *518*, *532*
granular part (zone) of nucleolus, 214ff.
granum, chloroplast, 21, 128ff.
Green, B. R., 505, *518*
Green, D. E., 73, *83*, 114, 121, 123
Green, P. B., 372, *373*
Greene, R. W., *520*
Grell, R. F., *422*
Grendel, F., 74
Grew, N., 4, 128
Griffith, F., 165
Groetsch, H., *532*
Gross, N. J., 493, *520*
Grossman, L. I., *519*
ground state of electron in light absorption, 144
guanosine triphosphate (GTP) in protein synthesis, 256–258
Gunning, B. E. S., 156, *159*
Gurd, F. R. N., *39*
Gurdon, J. B., 208, *211*, *373*
Gyrodinium, 348

Haguenau, F., *249*
Haldane, J. B. S., 524, 525, 526, *532*
Hamilton, L. D., *188*
Hanahan, D. J., *39*
Harada, K., *532*
Harding, C. V., *249*
Harris, H., *211*, *285*
Hartman, P. E., *211*
Hartmann, J. F., *373*
Hatch, M. D., 151
Hatch-Slack cycle, 151–152
Haurowitz, F., *39*
Hay, E. D., 175, 217, 245, *249*
Hayashi, M. N., *211*
Hecht, L. I., *286*
Heitz, E., 128, 239, 354
Helix, 432, 473
helix, DNA, 166–169
Helmstetter, C. E., *327*
Henn, F. A., *83*
Hensen, W. A., 481
Hepler, P. K., *373*
Hershey, A. D., 165, 188
Hertwig, O., 6, 375, 376
Hess, O., *422*
heterochromatin, 269, 293
heterokaryon, 484
Hew, H. Y. C., 473
hexokinase, 106
Heywood, S. M., 202, *211*
Hill, R., 143
Hill reaction, 143, 149
Hillier, J., 49
Hirudo medicinalis, 229
histone-protamine shift in spermiogenesis, 431–433
histones, 164, 167, 170, 172
  cleavage histones, 473, 474
  in fertilization, 473, 474
  fractions, 173, 267
  meiotic, 387
histones (continued)
  in nucleoprotein complex, 172–173
  in polytene chromosomes, 241
  and regulation of protein synthesis, 203, 267–268
  replacement in spermiogenesis, 428, 430, 431–433
  synthesis in cell cycle, 291, 317
Hoagland, M. B., 255, *286*
Hogeboom, G. H., 18, 102
Holley, R. W., 205
Holliday, R., 399
homologs, 378ff.
homopolymer, DNA, 199–200
Hooke, Robert, 3
Hooper, C. W., *188*
Hosokawa, K., *286*
Hotta, Y., 291, *327*, 377, 386, 402, 408, *423*
Howard, A., 290, *327*
Huang, R. C., 285
Huberman, J. A., *327*
Hufnagel, L. A., 515, *519*
Hughes, A., *24*
Hultin, T., 481
Humphrey, R. M., 321
Hunter, J. A., *519*
Hurwitz, J., 197
Huxley, H. E., 252, *286*
hyaluronic acid, 433, 459
hyaluronidase, 433, 459, 464, 468
hybridization, DNA-RNA, 199, 206–207
Hydroides, 464, 465, 466

image formation in electron microscopy, 54–57
  diffraction in, 54, 56–57
  effects of physical aperture on, 55
  focus, effects of, 56
  interference in, 56
  phase contrast in, 54, 56–57
  scattering in, 54–56
  in shadowed preparations, 62–63
image formation in light microscopy, 42–44
image plane, 43
image, real, 43
image, virtual, 43–44
indium chloride stain, 61, 405
inducer, 265, 266
inducible enzymes, 265
induction, 265–266
inductive resonance in photosynthesis, 145–146
Ingram, V. M., 211
initiation of protein synthesis, 256–258
initiator codons, 257–258, 260
inosine, 204
inositol, 28, 29, 68
Inoué, S., 345, 346, *373*
interchromatin fibrils, 14
interchromatin granules, 14, 220–223
interference, 46
  in electron microscopy, 56
  in phase contrast microscopy, 46
intermembrane space
  chloroplast, 131
  mitochondrial, 103ff.
intragenic recombination, 396–403
intramitochondrial granules, 17, 103
IOJAP mutant of corn, 499–500
isocitric acid, 110
isolabeling, 308, 323–324
Ito, M., 408, *423*

Jackson, W. T., *373*
Jacob, F., 264, 265, 266, *373*
Jagendorf, A. T., 118, 149, *159*
Janssens, F. A., 389
Jeans, J., 523
Jenkins, R. A., 360, *373*
Jensen, W. A., 481
Jinks, J., 518
John, B., *327*
Jokelainen, P. T., 333, 334, *373*

Kalckar, H., 102
Kamen, M. D., *159*
Kant, E., 523
Karan, J. D., *328*
Karlin, L. J., *286*
karyomere, 357
Katz, E. R., *285*
Kawamura, K., 362, *373*
Kaye, J. S., *188*, 429, *481*
Kellenberger, E., 182, *188*
Kellenberger, G., 392, *422*
Kellenberger fixation
  of bacterial DNA, 182–184
  of chloroplast DNA, 501
  of mitochondrial DNA, 487
Kenyon, D. H., *531*
Keosian, J., 531
Kessel, R. G., 236, *249*, *285*
Khorana, H. G., 262, 263
kinetochore. *See* centromere
kinetoplast, 485–487
kinetosome. *See* centriole
Kirk, J. T. O., *159*, 484, *518*
Kirshner, R. H., 493
Kirtley, M. E., *99*
Kislev, N., 501, *519*
Kleinschmidt, A. K., 181, *519*
Klotz, I., *99*
Kluss, B. C., 512, 513, *519*
Koehler, J. K., *481*
Kohne, D. E., *327*
Kok, B., *159*
Kolakofsky, D., *286*
Kölliker, R. A. von, 5
Korn, E. D., *83*
Kornberg, A., 310, 313, *327*
Kornberg enzyme in recombination, 402
Koshland, D. E., 98, 99
Krebs, H., 102, 110
Krebs cycle. *See* citric acid cycle
Kreger, D. R., *373*
Krishnan, A., *374*
Kroon, A. M., 492, *518*, *519*

Kubai, D. F., *188*, 348, *374*
Kung, H. K., 205, *211*
Kurland, C. G., *286*

lactose system, 264–266
Lafontaine, J.-G., *327*, 355, *374*
lamella, chloroplast, 128
lampbrush chromosome, 381, 415–421
  function, 416–419
  structure, 415–416
Langridge, R., *188*
Lark, K. G., *327*
Lavoisier, A., 7
lead citrate as stain for electron microscopy, 60–61
Leblond, C. P., 280, *286*
Ledbetter, M. C., 372, *374*
Leder, P., *211*, *286*
Lederberg, J., *520*
Leeuwenhoek, Anton van, 4, 128
Lefort-Tran, M., *519*
Lehninger, A. L., *99*, 102, 115, 116, *126*, *159*
Lemmon, R. M., *532*
Lenny, A. B., *328*
leptonema, 377–378ff.
lecithin, 28, 68
leucoplast, 21, 154
Levinthal, C., 313, *327*
Lewis, K. R., *327*, *422*
life
  defined, 522
  origin, 521ff.
ligase (DNA ligase) in recombination, 399–402
light
  absorption, 144–145
  conversion to chemical energy in photosynthesis, 145–146
light microscope, 41–49
light reactions, photosynthesis, 143–149
lignin, 21, 372
Lilium, 291
Lima-de-Faria, A., *24*, *327*
linkers, in eukaryote chromosome, 324
Linskens, H. F., *481*
lipid, 26–29
  chloroplast lipids, 134, 136–137
  droplets in eggs, 461
  interaction with OsO_4, 59, 76
  in membranes, 68ff.
  metabolism in smooth ER, 275
  in mitochondria, 121
  neutral, 26–28
  synthesis in the chloroplast, 153
lipid bilayer. *See* bilayer, lipid
Lipmann, F., *285*, *286*
lipoproteins, 39. *See also* proteins
  synthesis in Golgi complex, 280
Litomastix, 480
Loening, U. E., *211*
Loewenstein, W. E., 230, *249*
Lopez-Saez, J. F., *373*
Lord, A., *327*, 355, *374*
Lovern, J. A., *39*
Lowe, C. V., *532*
Lucas-Lenard, J., *286*
Luck, D. J. L., 493, 494, 497, 498, *519*
Lumbricus, 448
Luykx, P., 382, 384, *422*
Lwoff, A., 512
Lycondontis, 441
lysin, 464, 468
lysine in histone, 172
lysosome, 19, 98
  function, 282–284
  identification, 281–282
  relationship to acrosome, 434
  residual bodies, 282

McCarty, M., *188*
McClintock, B., 354, 355, 357, 390
MacLennan, D. H., 121
Macleod, C. M., *188*
McMaster-Kaye, R., *188*, *328*, 429, *481*
Madison, J. T., 205, *211*
Madison, K. M., 522, *532*
Madjilov, A. A., *211*
magnification, 42, 43
  and resolution, 52
malate dehydrogenase, 112
malic acid, 112, 151
Malkin, S., *159*
manchette of spermatozoa, 442, 443, 447–448
Mann, T., *481*
Manning, J. E., *519*
mannose, 372
Manton, I., *519*
Marcker, K. A., 257, *285*, *286*
Margulis, L., 516, 517, 518
Mariner, R., 525, *532*
Markham, R., *532*
Markham rotation, 226–227
  of the centriole image, 351
Marmur, J., *519*
Marsilea, 446, 512
Marushige, K., *285*
master-slave hypothesis, 419
matrix
  mitochondrial, 17, 103
  of the nucleolus, 215, 216–217
Mazia, D., *328*, 339, 340, *374*
megaspore, 475
meiosis
  anaphase and telophase I, 383
  DNA synthesis during, 386, 402
  early description, 6, 375–376
  gametic, 385
  in higher plants, 475
  intermediate, 385
  metaphase I, 382–383
  morphological aspects, 403–406
  in *Neurospora*, 398
  in oogenesis, 453–454
  overview, 376–384
  premeiotic interphase, 376–377
  prophase I, 377–382
  recombination in, 388–403
  second meiotic division, 384
  sporic, 385
  synthesis of RNA and protein during, 387
meiosis (continued)
  terminal, 385
  time course, 384
  zygotic, 385, 393
membrane
  artificial, 73, 74
  bacterial, 73
  chloroplast, 134ff.
  Danielli-Davson model, 74, 75, 76
  evolution, 80
  lamellar model, 68, 74–77
  lipids, 26–29, 68–72, 73, 74–77
  mitochondrial, 103, 121–123
  OsO_4 binding, 71ff.
  particulate model, 68, 74–77
  phospholipid cholesterol complex in, 69–70
  polysaccharides, 68
  protein-lipid ratio, 73
  proteins, 72ff.
  Robertson model, 76–79
  unit membrane model, 76–79
Mendel, G., 7, 163
Menke, W., 128
Menninger, J. R., *285*
Mereschkowski, C., 516
Mertz, T., *422*
Meselson, M., 253, *286*, 308, 310, 313, *328*, 392, 399, *422*, *286*
mesosome, 13, 348
Mesostoma, 318
messenger RNA (mRNA)
  attachment to ribosome, 252, 256
  identification, 201–202
  in initiation of protein synthesis, 256–258
  in mitochondria, 495–496
  overview of function, 192–194
  properties, 200
  sedimentation, 201, 202–203
  size, 202–203
  structure, 200
  in termination of protein synthesis, 258–260
metakinesis, 297
metaphase
  chromosomes, structure, 329–338
  of mitosis, 295–297
  plate, 297, 339
  spindle formation, 296–297
methylation of tRNA, 205
Metz, C. B., *480*, *481*
Meuller, P., *83*
Meves, F., 102
Meyer, G. F., 407, *422*
microbodies, 19, 124–126
microfibrils, cellulose, 370, 372
Micromonas, 103, 128, 131, 487, 497, 509
micron, 9
micropyle, 458, 468
microscope
  electron, 49–57
  light, 41–49
  phase contrast, 44–47
microsome, 273
microspore, 475

microtubules, 62
in cell wall synthesis, 372
centriole, 351
and cytokinesis, 362–366
fixation, 346–347
flagellar, 427, 437ff.
manchette of spermatozoa, 442, 443, 447–448
and motility, 346
and movement of chromosomes, 298, 343–347
occurrence, 346
preprophase band, 362
spindle, 296, 297, 298, 299, 333, 339–347
microvillus, 19
midbody in cytokinesis, 362, 365, 366
Miescher, F., 8, 164, 170, 431
Miller, O. L., 260, 420, *422*
Miller, R. L., 463
Miller, S. L., 525, *532*
millimicron, 9
Mirabilis, 483, 484, 499
Mirsky, A. E., 164
missense strand of DNA, 192n.
Mitchell, H. K., 485, *519*
Mitchell, M. B., 485, *519*
Mitchell, P., 116, 117
Mitchell hypothesis, 148–149
mitochondrion
cristae, properties and function, 121
defined, 9, 10, 17–18
derivative, mitochondrial, of spermatozoa, 428, 434–436
DNA in, 103–104, 184, 485–494, 496–498
early description, 101–102
electron transport in, 112–118
in fertilization, 473
inclusions, 103–104
inheritance, 484–498
lipids, 68, 70, 79
nebenkern, 428, 434–436
number, 103
occurrence, 103
origin, 497–498, 515–518
oxidation in, 105–118
oxidative phosphorylation in, 115–118
protein synthesis in, 496
regulation of function, 123
RNA in, 494–497
in spermatozoa, 427, 428, 446ff.
structure, 102–105, 121
in yolk formation, 459–460
mitosis, 6, 289, 290, 291
anaphase, 297–298
asters and centrioles in, 350–354
condensation and coiling of chromosomes in, 329–331
and cytokinesis, 360ff.
metaphase, 295–297
metaphase chromosomes, 331–339
nuclear envelope cycle in, 359–360
nucleolar cycle in, 354–357
overview, 291–303
prophase, 291–295
respiration during, 300–302
mitosis (continued)
significance, 302
spindle in, 339–350
telophase, 299
time course, 299, 300
Mittwoch, U., *422*
Mizukami, I., 450, *481*, 512
Moens, P. B., 404, *422*, *423*
Mohl, H. von, 5, 128
monamine oxidase, 121
Monesi, V., *328*, *423*
Monneron, A., *249*
Monod, J., 264, 265, 266
Monro, R. E., *286*
Monroy, A., *480*, *481*
Moore, J. E., 6, 376
Morgan, T. H., 389, 396
Morowitz, H. J., *99*
Moses, M. J., 175, *188*, 403, 405, 411, 412, *422*, *423*
mucoprotein, 39
Muhlethaler, K., *373*
Müller-Hill, B., 266, 268, *285*
multistrandedness of chromosomes, 308, 319–326
multivesicular body, 461
Munn, E. A., *126*
mutations, 316, 317–318
chromosome, 316
gene, 316
myelin, 69, 73, 75, 76, 79
Myzostomum, 441

NAD. *See* nicotinamide adenine dinucleotide
NADP. *See* nicotinamide adenine dinucleotide phosphate
Naegleria, 513
Nägeli, K. von, 5
Nakemoto, T., *286*
nanometer, 9
Nass, M. M. K., *519*
nebenkern, 428, 434–436
negative interference, 396, 399ff.
negative staining for electron microscopy, 62
of chloroplast membranes, 141
of mitochondrial membranes, 122–123
of the nuclear envelope, 226
Nelson, G. J., *83*
Nelson, L., *481*
Nereis, 469, 473
Nesser, W., *327*
Neumann, J., *159*
Neurospora, 256, 396, 397, 398, 484, 485, 486, 493, 495, 497
Neutra, M., 280, *286*
neutral proteins, 171
Newcomb, E. H., *374*
Nicklas, R. B., 339, 342, *373*, *374*
nicotinamide, 110
nicotinamide adenine dinucleotide (NAD), 108ff.
in regulation of the mitochondrion, 123
structure, 109
nicotinamide adenine dinuncleotide phosphate (NADP)
in oxidation, 112
in reduction and synthesis, 120
reduction in photosynthesis, 146ff.
structure, 120
Nirenberg, M. W., 259, 263, *286*
Nitella, 372, 447–450, 512
Noctiluca, 234
Noda, Y. D., 466, *481*
Nomura, M., *286*
noncyclic photosynthesis, 148
nonhistone protein, 171
in puffs, 241
and regulation of protein synthesis, 267–268
replacement during spermiogenesis, 429
synthesis during cell cycle, 317
nonreciprocal recombination, 396, 397–398, 399ff.
nonsense codons, 259
Noorduyn, N. J. A., *285*
Norrevang, A., 459, 460, *481*
Northcote, D. H., 281, *286*, 365
Notonecta, 440
Novelli, G. D., *286*
Novikoff, A. B., *126*, 281
nuclear cap, 419–420
of spermatozoa, 436
nuclear envelope, 185, 271
breakdown in mitotic interphase, 293, 295–296
cycle during mitosis, 359–360
defined, 14, 16
dinoflagellate, 234
and division of chromosomes, 349–350
in fertilization, 473
fibrous lamina, 229
fixation, 59
in meiosis, 378, 379, 384, 404
passage of nucleoprotein through, 231–234, 245
perinuclear space, 224
permeability, 225, 229–234
pore complexes, 225–228
potential, 230
reconstruction in anaphase, 299
and regulation, 269
resistance, electrical, 230
in spermatozoa, 426
structure, 224–228
and the synaptinemal complex, 404
nuclear equivalent, prokaryotic, 11
nucleic acids
discovery, 8
indium chloride stain for, 61
interaction with OsO_4, 59
in mitochondria, 485ff.
nuclein, 8, 164
nucleohistone, 169, 180
electron microscopy, 174–180
structure, 171–173, 175–176
nucleoid, prokaryotic, 11
bacterial, division, 347–348, 350

nucleolar organizer, 14, 208, 220, 336, 355ff.
  in meiosis, 419–420
  in mitosis, 299
  of polytene chromosomes, 245
nucleolus, 14, 174
  accessory, 386, 419–420
  chromatin, 215, 216, 217
  cycle during mitosis, 291–294, 299, 354–357, 384ff.
  cytochemistry, 216–217
  fibrillar part (zone), 215–216
  granular part (zone), 214–216
  inclusions, 216
  interaction with OsO_4, 59
  and lampbrush chromosomes, 419–420
  matrix, 215
  nucleonema, 214, 215–216
  organizer, 14, 208, 220, 245, 299, 336, 355ff., 419–420
  pars amorpha, 214, 215–216
  in polytene nuclei, 245–248
  pseudonucleolus, 357
  role in synthesis of ribosomes, rRNA, 208–211, 217–219, 245
  segregation by actinomycin D, 219
  structure, 213–220
  vacuoles, 216
nucleonema of nucleolus, 214
nucleoprotamine, 169–172
  structure, 171–172
  X-ray diffraction, 171–172
nucleoproteins, 39, 167, 169ff.
  effects of salt concentration on, 176
  electron microscopy, 174–180
  ribosomal, 210–211
  structure, 171–173, 175–176
  X-ray diffraction, 171–172, 173, 176, 177
nucleotide, 165, 169
nucleotide triphosphates, 197
  in DNA synthesis, 310ff.
  in RNA synthesis, 197–198
nucleus, 14–16
  chromocenter, 240n.
  coiled bodies, 221–223
  dinoflagellate, 184–185
  early description, 6–7, 8
  euchromatin, 269
  extrachromosomal ribonucleoprotein, 220–223
  in fertilization, 473, 479
  fibrous lamina, 229
  fixation, 58
  generative, 476
  heterochromatin, 269
  interchromatin granules, 220–223
  isolation, 177–180
  perichromatin fibers, 221–223
  perichromatin granules, 220–223
  polytene, 238–248
  prokaryote, 180–185
  pronucleus, 454, 463
  protein synthesis in, 203–211
  spermatozoon, 426, 428, 433, 446ff.
  structure, 173–180, 223
  vegetative, 476

numerical aperture, 48
nurse cell, 453, 458
Nwagwu, M., 202, *211*

Ochoa, S., 197, *286*
Ohba, Y., *188*
Ojala, D., *518*
Okazaki, R., 315, 316, *328*
Okazaki, T., *328*
Okazaki fragments, 315–316
Oodegonium, 133
oogenesis, 452–462, 474–475
  in animals, 453–462
  formation of polar bodies, 454, 455
  in higher plants, 474–475
  secretion of yolk, 459–461
Oparin, A. I., 524, 525, 526, *532*
operator, 265, 266
operon hypothesis, 264–266, 268
osmiophilic granule
  of chloroplast, 21, 130, 133, 136, 155
  of mitochondrion, 103
osmium tetroxide fixation, 58, 59
  of bacterial DNA, 182–184
  of chloroplast, DNA, 501
  of chromosomes, 174, 176
  of endoplasmic reticulum, 270
  of lipid bilayers, 71–72, 74, 76
  of lipids, 76
  of membranes, 71ff.
  of mitochondrial DNA, 487
  of mitochondrial membranes, 103
  of myelin, 75ff.
  of perinuclear space, 224
  of pore complexes, 225, 226
  of proteins, 76
Outka, D. E., 512, 513, *519*
Overton, E., 6
oxaloacetic acid, 110, 112, 151, 152
oxidation, cellular, 105ff.
oxidative phosphorylation, 102, 105, 112–118
  location in mitochondrion, 121

Painter, T. S., 239
pairing of chromosomes, 378–379, 399ff.
Palade, G. E., 102, 103, 271, 272, 278, *285*
Palay, S. L., 274, 275, *286*
Palmer, K. J., *82*
palmitic acid, 118
pantothenic acid, 110
Paracentrotus, 473
Paramecium, 515
Parchman, L. G., *423*
Pardon, J. F., *188*
Park, L., *159*
Park, R. B., *82*, 139, 140, 141, 143, *159*
Parke, M., *519*
pars amorpha of nucleolus, 214
partial chiasmatype hypothesis, 389–390
partition, chloroplast, 128, 142
Pasteur, L., 8
Patau, K., *328*
Pauling, L., 35

Peacock, W. J., 323, *328*
Pease, D. C., *65*
pectin, 21, 367
Pelc, S. R., 290, *327*
Pelling, C., 241
Pelomyxa, 103
Penman, S., 209, 213
pentose cycle, 149
pentose shunt, 119
peptide bond, 32–34
  discovery, 7
  formation, 193, 255, 258
peptidyl synthetase, 258
Perez-Silva, J., *249*
perforatorium, 434
periacrosomal material, 433
perichromatin fibers of nucleoplasm, 221–223
perichromatin granules of nucleoplasm, 220–223, 232, 245
perinuclear space, 16, 19, 224, 274
peroxidase, 98
peroxisomes, 19, 124
Perret, J., 522, *532*
Perry, R. P., 208, 209, 210, *211*, 213, *249*
petite, mutant of yeast, 484, 486
phase contrast
  in electron microscope image formation, 54, 56–57
  in light microscopy, 44–47
phenylalanine, 192, 193, 194
Phillips, D. C., 98
Phillips, D. M., *188*, 435, 436, 443, 444, 474, *481*
phosphatidic acid, 28, 29
phosphatidyl
  choline, 29
  ester, 28
  glycerol, 28, 136
  group, in membranes, 68
  inositol, 28
phosphodiesterase, 205
phosphoenolpyruvate, 108
phosphofructokinase, 108
phosphoglucomutase, 108
phosphogluconic acid, 119, 120
3-phosphoglyceraldehyde, 108, 110, 123, 149
phosphoglyceric acid, 149, 151
phospholipid, 26, 28–29, 136
  bilayers, 70–72, 73
  interaction with OsO_4, 71–72, 74
  in membranes, 68ff.
  metabolism in smooth ER, 275
  in mitochondrial membranes, 121
  phospholipid-cholesterol complex, 69–71
  properties of suspensions in water, 70–72, 73
  X-ray diffraction, 70
phosphomonoesterase, 205
phosphopyruvic acid, 151
phosphoryl 1, 3-phosphoglyceric acid, 149
phosphorylation, oxidative, 115–118
phosphotungstic acid, (PTA)
  as stain for electron microscopy, 61

phosphotungstic acid (continued)
  interaction with inner mitochondrial membranes, 122–123
photophosphorylation, 146, 147, 148–149
photosynthesis
  in bacteria and blue-green algae, 154
  conversion of light to chemical energy, 145–146
  dark reactions, 143, 149–153
  early description, 142–143
  light absorption in, 144–145
  light reactions, 143, 144–149
  localization in chloroplast fine structure, 153
  overall reactions, 143–144
  photosynthetic unit in, 134ff., 145–146, 147
  primary chemical products, 146
  reaction center, 145–146
  systems I and II, 147–149
photosynthetic unit, 134ff., 145–146, 147
phragmoplast, 303, 365–366
Phryne, 408
phycobilins, 147
Pica-Mattoccia, L., *518*
Pickett-Heaps, J. D., 281, *286*, 365, 372, *374*
pigments, chloroplast, 137–139
pinocytosis, 282, 460
pinocytotic vesicles, 274
Pinus, 447
Pitelka, D. R., *374*, *519*
plasma membrane, 10
  in cytokinesis, 303
  and division of bacterial nucleoid, 302, 347–348, 350
  lipid components, 68, 69, 76, 79
plasmodesmata, 21, 365
plastoquinones, 136, 147
Plaut, W., *249*, 487, 500, 515, *519*
poky, mutant of *Neurospora*, 485, 486
polar bodies, 454, 455
polarity of DNA strands, 199
polarization of chromosomes, 336
polarized light analysis of chloroplasts, 142
polarized recombination, 396, 398, 399ff.
pole pieces in electron lenses, 49
pollen, 474, 476
pollen tube, 475, 476, 478
Pollmann, W., *532*
polycomplexes of synaptinemal complex
  in nonmeiotic cells, 411–412
  relation to synaptinemal complex, 409–411
  structure, 408–409
polynemy. *See* strandedness of chromosomes
polynucleotide ligase, 315
polypeptide, 32
  synthesis, 192–194
polysome, 202, 261, 271
polytene chromosomes
  activity, 241–245
  degree of polyteny, 240
  development, 240
  Feulgen staining, 239
  nucleolus, 245–248
  puffs, 241–245
  and regulation, 268–269
polytene chromosomes (continued)
  RNA synthesis, 241–245
  structure, 238–245
Pon, N. G., 139, 140, 143, *159*
Ponnamperuma, C., 525, 526, *532*
pore complex
  of annulate lamellae, 234–238
  annulus, 225–227
  diameter, 226–227
  diaphragm, 225
  and fibrous lamina, 229
  granules, 233
  negative staining, 226
  of nuclear envelope, 16, 225–229
  numbers, 228
  passage of ribonucleoprotein through, 231–234, 245
  permeability, 229–234, 245
  symmetry, 226–227
porphyrin ring, 113, 137
Porter, K. R., 102, 128, 270n., 271, *286*, 372, *374*
positive interference, 389
postfixation for electron microscopy, 58
potassium permanganate
  effect on mitochondria, 103
  fixation of the endoplasmic reticulum, 274
  fixation of the nuclear envelope, 224–225, 230
  as fixative for electron microscopy, 59
  reaction with myelin, 75ff.
potential, in cellular oxidations, 105–106
prenucleolar matrix, 355, 357
preprophase band, 362, 365ff.
Prescott, D. M., *211*, 300, *328*, 355, *374*
Priapulus, 456, 459, 460
Priestley, J., 7, 142
primary constriction. *See* centromere
primary spindle, 296
Pritchard, J. R., 397
proacrosomal granule of spermatozoa, 428, 434
procentriole, 352, 450
prochromosome, 377
prokaryote
  defined, 10, 23, 24
  DNA replication, 308–316
  nucleoid, division, 347–348, 350
  nucleoid, structure, 180–185
  photosynthesis, 133–134, 153–154
prolamellar body, 155
promitochondrion, 498
pronucleus, 454, 463, 471
proplastid, 154–157
prosthetic group, 113
  in mitochondrial electron transport, 113, 114
protamine, 8, 164, 167, 428, 430
  binding to DNA, 169–172
  in fertilization, 473, 474
  in histone-protamine shift in spermiogenesis, 431–433
proteinoids, 526
proteins
  acidic, 171
proteins (continued)
  in association with DNA, 169–172
  chloroplast, 134–136
  enzymatic, 97
  interaction with OsO_4, 59, 76
  in membranes, 68, 72ff.
  in mitochondrial membranes, 121
  neutral, 171
  nucleoproteins, 169–172
  oxidation, 118–119
  residual, 171
  ribosomal, 191, 192, 210–211, 253–254
  of spindle, 341–342
  structural, 72, 73, 121, 134–136, 137
  structure, 31–37
  in yolk bodies, 460
protein synthesis
  amino acid activation, 255–256
  bacterial, 256–257
  during cell cycle, 300
  in chloroplasts, 152, 153
  in fertilized egg, 471
  initiation, 256–258
  mechanism, 256–270
  in mitochondria, 496
  during mitosis, 386, 387
  morphological aspects, 270–285
  nuclear, 203, 211
  overview, 192–194
  regulation, 264–270
  termination, 258–260
protochlorophyll, 155–156
pseudonucleolus, 357
pseudouracil, 204, 205
puffs in polytene chromosomes, 268
Pullman, M. E., *126*
Purkinje, J. E., 5
puromycin, 210
pyrenoid body, 130
pyruvate phosphokinase, 108
pyruvic acid, 108
  oxidation, 110
pyruvic acid dehydrogenase, 110

quantasomes, 139–140, 153
quantum efficiency of photosynthesis, 151
quantum, light, 144
  efficiency of use in photosynthesis, 145
quinones, 136, 147

R_1, R_2, in protein synthesis, 259
Rabl, C., 6
Racker, E., *82*, 123, *126*
Rana, 459, 486
Randall, J., 515
Rasch, E., 240
Raven, P. H., 517, *519*
reaction center in photosynthesis, 145–146, 147
real image, 43
recombination, 381, 388–403
  allelic, 396–403
  breakage and exchange hypothesis, 390–394, 395

recombination (continued)
copy-choice hypothesis, 394–396
early hypotheses, 389–390
intragenic, 396–403
molecular models, 396–403
partial chiasmatype hypothesis, 389–390
semiconservative replication and, 395–396
synaptinemal complex in, 407–408, 412–415
redox potentials. *See* reduction-oxidation potentials
reduction
of carbon dioxide in photosynthesis, 127, 142, 146, 149–150
cellular, 105ff.
reduction-oxidation potentials
in electron transport, 113–115
and free energy, 113–115
in photosynthetic electron transport, 147–149
redundancy of DNA, 317–326
lateral, 319
linear, 319
and strandedness, 319–326
Rees, H., *423*
Rees, M. W., *532*
refraction of light, 41, 42
refractive index, 41, 42
in phase contrast microscopy, 45
regulation, 264–270
operon hypothesis, 264–266
transcriptional controls, 199–200, 203, 264–269
translational controls, 269–270
regulator, 265, 266
regulatory RNA, 267
Reich, E., 493, *519*
Remak, R., 5
Renaud, F. L., 512, *519*
replication of DNA, 191, 290–291, 303–317
repression, 265–266
repressor, 265, 266
residual body of lysosome, 282
residual proteins, 171
resolution
in electron microscopy, 48, 49, 50–52
in light microscopy, 47–48
and magnification, 52
respiration, cellular, 7, 8, 105
aerobic, 106
anaerobic, 106
early discoveries, 101–102
during mitosis, 300–302
in spermatozoa, 446
Revel, J. P., 175
Rhodes, M. M., *423*
riboflavin, 113
ribonuclease
pancreatic, 205
structure, 37, 38
T_1, 205
ribonucleic acid (RNA)
in annulate lamellae, 237, 238
centrifugational analysis, 201
discovery, 8
in endoplasmic reticulum, 271
ribonucleic acid (RNA) (continued)
in lampbrush chromosomes, 416ff.
messenger (mRNA), 192ff., 201–203
in mitochondria, 494–496
regulatory, 203, 267–268
ribosomal (rRNA), 192ff., 206–211
staining for electron microscopy, 220
structure, 194–196
synthesis, 196–200
synthesis in cell cycle, 199, 291
synthesis during meiosis, 381, 385–386, 416
synthesis in nucleolus, 208–211
synthesis on polytene chromosomes, 241–245
transfer (tRNA), 192ff., 203–206
ribonucleoprotein, 191, 201, 210–211
in lampbrush chromosomes, 416–420
in polytene chromosomes, 244–245
ribosomal, 253
structures of nucleoplasm, 220–223
transport from nucleus to cytoplasm, 231–234
ribose, 110, 194–195
ribosomal RNA (rRNA)
bacterial, 207
chloroplast, 506–508
eukaryotic, 206, 207
maturation, 209–211
maturation and nucleolar morphology, 217–219, 245
methylation, 206, 210
mitochondrial, 494–497
overview of function, 192–194
structure and properties, 200, 201, 206–208
synthesis during meiotic prophase, 419–420
ribosome, 13, 14, 18, 19, 192
assembly of subunits, 251, 254
bacterial, 252ff.
chloroplast, 130–131, 155, 506
dissociation, 253–254
in endoplasmic reticulum, 270–274
interaction with OsO_4, 59
mitochondrial, 103, 494–497
on nuclear envelope, 271
in nucleus, 203, 211
polysomes, 261, 271
proteins, 253
in protein synthesis, 256–260
structure, 252–253
subunits, 207, 210–211
synthesis, 208–211
ribothymine, 204
ribulose, 1,5-diphosphate, 149, 150, 151, 152
ribulose 1,5-diphosphate carboxylase, 143
ribulose monophosphate, 150
ribulose 5-phosphate, 120, 121
in pentose shunt, 120
Richards, B. M., *188*
Richards, O. C., *519*
Richardson, C. C., *328, 423*
Richardson, J. P., *211*
Riggs, A. D., *327*
ring centriole of spermatozoa, 437
Ris, H., 10, 164, 179, 184, *188*, 343, 348, *374*, 487, 500, 501, 506, *519*
Ritossa, F. M., 208, *211*
RNA methylase, 199
RNA polymerase, 192, 194, 197, 198, 199ff., 260, 420
Robberson, D. L., *518*
Robbins, E., *328*, 340, *374*
Robertson, J. D., 76, 78, 79
Robinson, G. W., *159*
Robison, W. G., 442, 443, *481*
Romalea, 392
Roodyn, D. B., 496, *518*
Rosen, W. G., *481*
Rossen, J. M., 392, 393, *423*
Roth, T. F., 408, 410, 411, *423*
rough ER. *See* endoplasmic reticulum
Rouser, G., *83*
Rückert, J., 415
Rudin, O., *83*
Rusgis, C., *211*
Ruska, E., 49
Ryan, R. S., *519*
Ryber, A., 347
Ryter, A., *188, 374*

S of cell cycle, 290–291, 295, 376
S unit, 201
Sachs, J. von, 128
Sagimoto, K., *328*
Sakabe, K., *328*
Sakagami, T., *518*
Salas, M., *286*
salivary gland chromosomes. *See* polytene chromosomes
Salser, W., *211*
Sanger, F., 257, 286
sarcoplasmic reticulum, 274
satellites of centriole, 351–352
Satir, P., 445, *481*
Sato, H., *373*
scattering, electron, in electron microscope image formation, 54–56
Schatz, G., *126, 519*
Scheer, U., *249*
Schiff, J. A., *519*
schizomycetes, 184
Schleiden, M. J., 5, 6, 163
Schmidt, O. J., 523, *532*
Schmitt, F. O., 75, 76, *82*
Schneider, W. C., 102
Schoefl, G. I., 219, *249*
Schofp, F. P., *532*
Schrader, F., *372*
Schramm, G., *532*
Schultz, P., *481*
Schwann, T., 4, 5, 163
Scolnick, E., *286*
Scott, J. F., *286*
Scott, N. S., *519*
Scremin, L., 485
Sechaud, J., *188*
secondary constriction of chromosome, 293, 336

sedimentation constant, 201
sedimentation velocity, 201
Seeds, W. E., *188*
semiconservative replication, 304–310
  experimental demonstration, 305–310
  and recombination, 395–396
  the unwinding problem, 313
Senior, A. E., 121
sense strand of DNA, 192n., 199
sequencing of tRNA, 205–206
sex chromosomes, 387
  coiling, 387
  in meiosis, 387–388
sex vesicle, 428
shadowing in electron microscope specimen preparation, 62–63
Shah, V. C., *519*
Shapiro, L., *519*
Shelanski, M., 340, *374, 481*
shell stage of spermiogenesis, 429, 447
Sheridan, W. F., *423*
Shumway, L. K., 500, *519*
Siddiqui, O., *286*
Siekevitz, P., 272
Simon, G., *83*
Simpson, L., *519*
Singer, M. F., *211*
Sjöstrand, F. S., *65*, 75, 79, *83*, 102, 103, 121, *126*, 270n., 271, 274, 276, *286*
Slack, C. R., 151, *159*
Slater, E. C., 116
Sleigh, M. A., *481*
Slizynski, B. M., *286*
Smillie, R. M., *519*
Smith, J. D., *285*
Smith-Sonneborn, J., 515, *519*
smooth ER. *See* endoplasmic reticulum
soluble RNA. *See* transfer RNA
Sonneborn, T. M., 512, *518*
specimen preparation for electron microscopy, 57–65
  autoradiography, 175
  critical point method, 177
  EDTA method for RNA staining, 220
  freeze-etch, 63–65
  isolation methods, 61–62, 177–179
  Kellenberger fixation, 182–184, 501, 487
  stains, 60–62
specimen thickness, effect on resolution, 57
Spencer, D., *519, 520*
spermatids, 426
spermatozoa
  development, 426–437, 446–450
  nonflagellate, 428
  plant, 446–450
  structure, 426–428, 446–450
  X-ray diffraction of sperm nuclei, 171–172
spermatozoid, 446
spermiogenesis, 426–452
  in animals, 426–437
  cytoplasmic changes, 434–437
  nuclear changes, 428–433
  in plants, 446–450
  shell stage, 429, 447
Speyer, J. F., *328*
spherical aberration, 4, 50–51, 52
  effect on resolution in electron microscopy, 50–52
sphigomyelin, 68
Spiegelman, S., 208, *211*
spindle
  anastral, 340
  astral, 340
  comparative aspects, 347
  in cytokinesis, 362–366
  dinoflagellate, 348
  energetics, 342–343
  evolution, 350
  extranuclear, 302
  formation, 296–297, 339–340, 353–354
  hollow, 344
  inclusions, 340, 341
  movement of chromosomes, 298, 342–347
  primary, 296
  protein components, 341–342
  role of asters and centrioles in formation, 353–354
  structure, 340–342
Spirin, A. S., *211, 285*
Spirogyra, 128, 131, 354, 394, 509
Staehelin, T., 253, *286*
Stahl, F. W., 308, 310, 313, *328*
Stahl, R. C., *211*
stains for electron microscopy, 60–62
  EDTA method for RNA stain, 220
Stanley, R. G., *481*
Stanley, W. M., *286*
starch grains, chloroplast, 128, 130
Stedman, Edgar, 267
Stedman, Ellen, 267
Steinert, M., 486, *520*
Stent, G. S., 304
Stephens, R. E., 341, 342, *374*
Stephenson, M. L., *286*
Stepka, W. A., *159*
Stern, C., 390
Stern, H., 291, *327*, 377, 386, 402, 408, *423*
steroids, 29, 31, 275
sterols, 26, 29, 31, 68, 136
Stevens, A., 197
Stevens, B. J., *249*, 355, *374*
stickiness of metaphase chromosomes, 321
Stocking, C. R., 500
Stoeckenius, W., 71, 72, 74, *83*
Stokes, A. R., *158*
Storrie, B., *518*
strandedness of chromosomes, 318–325
  evidence from enzymatic digestion, 324–325
  evidence from isotope incorporation, 322–324
  evidence from X-irradiation, 320–322
  morphological observations, 319–320
  and redundancy, 319–325
Strasburger, E., 6, 290, 375
Straus, N. A., *328*
stroma, chloroplast, 21, 128ff., 153
stromal lamellae, 128
structural protein, 72, 73, 121, 134–136, 137
Stubblefield, E., 332, 334, *373*
subchromatid aberrations, 320, 321
succinic acid, 112
Sugino, A., *328*
sulfatide, 69
Sutton, W. S., 7
Svedberg (S) unit, 201
Swanson, C. P., *327, 422, 480*
Sweers, H. E., *159*
Swift, H., 164, 184, *189*, 240, 244, 245, *249*, 290, *328*, 392, 487, 498, 501, 506, 512, *519, 520*
synapsis, of chromosomes, 378–379, 399ff.
synaptinemal complex, 379, 381, 403–405
  cytochemistry, 405–406
  in nonmeiotic cells, 411–412
  occurrence, 403
  polycomplexes, 411–412
  in recombination, 407–408, 412–415
  structure, 403–404
synergid, 475, 478, 479
synthetase (aminoacyl-tRNA synthetase), 203–204, 255, 256, 495, 529
system
  defined, 88
  reversible, 90
system I, system II, in photosynthesis, 147, 148ff.
  location in chloroplast membranes, 153
  prokaryote, 153–154
Szollosi, D., 364, *374, 481*

Takahashi, H., *481*
Takanami, M., *286*
Tandler, B., 497
Tano, S., *481*
Taylor, E. W., 341, 342, *373, 374, 481*
Taylor, J. H., 290, 299, 300, 305, 306, 307, 308, 310, 322, 323, 324, *328*, 392, 399, *423*
terminalization of chiasmata, 382
termination of protein synthesis, 258–260
terminator codons, 258–260
tetrad, 379ff.
Tetrahymena, 291, 492, 513
Tetramitus, 512–513
Theorell, H., 102
Thomas, C. A., 389, *423*
Thompson, T. E., *83*
thylakoid, chloroplast, 128ff., 134
  prokaryote, 133–134
  ultrastructure, 139–142
thymidine, tritiated, 175, 306
thymine, 166, 195, 196ff.
Tilney-Bassett, R. A. E., *159*, 484, *518*
Tipula, 408
Ti Tien, H., *83*
Tompkins, R., *286*
tonofibrils, desmosome, 19, 20
Tradescantia, 384
transcription, 191
  asymmetry, 198–199
  controls, 264–269
  direction, 198
  initiation, 199–200
  mechanism, 197–198

transcription (continued)
  overview, 192–194
  termination, 199–200
transfer RNA (tRNA)
  anticodon, 192, 193, 205
  attachment to amino acids, 255–256
  chloroplast, 507
  identification and purification, 203–204
  methylation, 204
  mitochondrial, 495
  number, 204
  overview of function, 192–194
  properties, 200, 203–206
  rare bases, 204
  role in protein synthesis, 256–259
  structure, 196, 200, 203–206, 207
transforming agent, 164–165
transition elements of Golgi complex, 279
transition in mutation, 316, 317
translation. *See* protein synthesis
translocating factors in protein synthesis, 257
transversion in mutation, 316, 317
Traub, P., 253, *286*
Trench, R. K., 518, *520*
Trichonympha, 344
Tridachia, 517, 518
triglyceride, 26, 27
triosephosphate isomerase, 108
Triticum, 479
Triturus, 324, 417, 418, 420
trypanosome, 485–486
Tschermak, E. von, 7
Ts'O, P., *188*
Tuan, D. Y. H., *285*
Turner, F. R., 447, *481*, 512
Tyler, A., *481*

ubiquinone, 113, 114
uninemy. *See* strandedness of chromosomes
unit fiber of chromosome, 174–180
  coiling or folding, 293, 329–331
unit membrane, 76–79
uracil, 195, 196ff.
uranyl ion (UO_2^{+2}) in Kellenberger fixation, 182, 183
uranyl (UO_2^{+2}) staining for electron microscopy, 60–61, 62
  of chloroplasts, 501–502
  of phospholipid bilayers, 71–72
  of RNA, by EDTA technique, 220, 232
Urey, H. C., 523, 525, *532*
uric acid dehydrogenase, 124
uridine, tritiated, 113, 114

vacuole, 19
  nucleolar, 216
  plant, 21, 23
Vance, W. W., 49
Vandenheuvel, F. A., 69, *83*
Van Niel, C. B., 143
vegetative nucleus, 476
Vicia, 305, 310, 318
Virchow, R., 5
virtual image, 43–44
vitamin K, 136, 147
vitelline coat or membrane, 456, 458, 459, 464
Viviparus, 429, 450

Wahba, A. J., *286*
Wahrman, J., *327*
Wald, G., *532*
Wallach, D. F. H., *83*
Warburg, O., 8, 102, 108, 119, 143, 471
Watson, J. D., 165, 166, 169, *188*, *189*, 304
Watson, M. L., 61, *65*
Wattiaux, R., *286*
Weber, C. S., 208, *211*
Weier, T. E., 128, 139, 142, *159*, 500, *519*
Weigle, J. J., 392, *422*
Weismann, A., 6, 7, 376
Weiss, S. B., 197
Weizsacker, C. F. Z. von, 523, *532*
Went, H., 339, 340
Wescott, W. C., *83*
Westergaard, M., 392, 393, *423*
Wettstein, D. von, 156, *159*
Whitehouse, H. L. K., 399
Whitfeld, P. R., *519*, *520*
Wieland, H., 8, 102
Wilkie, D., *518*
Wilkins, M. H. F., 165, 166, 169, 173, *188*, *189*
Williams, G. R., 115, 116
Wilson, E. B., 7, *327*, 372, 384, 385, *422*, 474, *480*
Wilson, H. R., *188*
Wintersberger, E., 495, *520*
Wischnitzer, S., *65*
Wisse, E., *126*
Wisselingh, C. van, 509
wobble hypothesis, 264
Woese, C. R., *286*
Wöhler, F., 7
Wolfe, S. L., *189*, *423*
Wolff, S., 320, *328*
Wolstenholme, D. R., 493, *519*, *520*
Wright, R. E., *520*

X chromosome, 269, 376, 387
Xenopus, 206, 208, 209, 419, 486
X-ray diffraction
  of chloroplasts, 142
  of nucleoprotein, 172
  of sperm nucleoproteins, 171–172

Yanagimachi, R., 466, *481*
yolk body, 453, 459–461
Young, W. J., *422*

Zabin, I., *39*
Zamecnik, P. C., 255, *286*
Zamia, 446, 512
Zea mays, 384
Zichichi, M. L., *422*
Zobel, C. R., *65*
zona lysin, 433
zona pellucida, 433, 456, 468
zona radiata, 458
Zubay, G., 252, *286*
Zygnema, 131
zymogen granule, 278